AA001004

2010 IEEE Radio Frequency Integrated Circuits Symposium

(RFIC 2010)

Anaheim, California, USA
23-25 May 2010

IEEE Catalog Number: CFP10MMW -PRT
ISBN: 978-1-4244-6240-7

Copyright © 2010 by the Institute of Electrical and Electronic Engineers, Inc
All Rights Reserved

Copyright and Reprint Permissions: Abstracting is permitted with credit to the source. Libraries are permitted to photocopy beyond the limit of U.S. copyright law for private use of patrons those articles in this volume that carry a code at the bottom of the first page, provided the per-copy fee indicated in the code is paid through Copyright Clearance Center, 222 Rosewood Drive, Danvers, MA 01923.

For other copying, reprint or republication permission, write to IEEE Copyrights Manager, IEEE Service Center, 445 Hoes Lane, Piscataway, NJ 08854. All rights reserved.

This publication is a representation of what appears in the IEEE Digital Libraries. Some format issues inherent in the e-media version may also appear in this print version.

IEEE Catalog Number:	CFP10MMW-PRT
ISBN 13:	978-1-4244-6240-7
ISSN:	1529-2517

Additional Copies of This Publication Are Available From:

Curran Associates, Inc
57 Morehouse Lane
Red Hook, NY 12571 USA
Phone: (845) 758-0400
Fax: (845) 758-2633
E-mail: curran@proceedings.com
Web: www.proceedings.com

TABLE OF CONTENTS

RSU5A RFIC PLENARY

RSU5A-1: RF Power Amplification: Can CMOS Deliver? 1
David J. Allstot
RSU5A-2: The Universal Connector: RF Application Trends Over the Next Decade 2
Gregory L. Waters

RMO1A CELLULAR TRANSCEIVERS

RMO1A-1: A 28mW WCDMA/GSM/GPRS/EDGE Transformer-Based RMO1A-2: Receiver in 45nm CMOS 3
Danielle Griffith, Venkatesh Srinivasan, Salvatore Pennisi, Vijay Rentala, Yu Su, Swaminathan Sankaran, Imtinan Elahi, Sreekiran Samala, Halil Kiper, Bijit Patel, Siraj Akhtar, Dan Edmondson
RMO1A-2: An EDGE Transmitter with Mitigation of Oscillator Pulling 7
Imran Bashir, Robert Bogdan Staszewski, Oren Eliezer, Khurram Waheed, Vasile Zoicas, Nir Tal, Jaimin Mehta, Meng-Chang Lee, Poras T. Balsara, Bhaskar Banerjee
RMO1A-3: An All-Digital Offset PLL Architecture 11
Robert Bogdan Staszewski, Sudheer Vemulapalli, Khurram Waheed
RMO1A-4: An Interstage Filter-Free Mobile Radio Receiver with Integrated TX Leakage Filtering 15
Rastislav Vazny, Werner Schelmbauer, Harald Pretl, Stefan Herzinger, Robert Weigel
RMO1A-5: A SAW-Less CMOS TX for EGPRS and WCDMA 19
Kurt Hausmann, Jeff Ganger, Mark Kirschenmann, George B. Norris, Wayne Shepherd, Vivek Bhan, Daniel B. Schwartz

RMO1B RF CMOS MODULATORS & RECEIVERS

RMO1B-1: A Quadrature Charge-Domain Filter with an Extra In-Band Filtering for RF Receivers 23
Ming-Feng Huang
RMO1B-2: A Low-Power Receiver Down-Converter with High Dynamic Range Performance 27
Diptendu Ghosh, Ranjit Gharpurey
RMO1B-3: A Multiphase PWM RF Modulator Using a VCO-Based Opamp in 45nm CMOS 31
Min Park, Michael H. Perrott
RMO1B-4: Spurious Noise Reduction by Modulating Switching Frequency in DC-to-DC Converter for RF Power Amplifier 35
Eung Jung Kim, Chang-Hyuk Cho, Woonyun Kim, Chang-Ho Lee, Joy Laskar
RMO1B 5: A Rail to Rail Input Receiver Employing Successive Regeneration and Adaptive Cancellation of Intermodulation Products 39
Edward A. Keehr, Ali Hajimiri

RMO1C FREQUENCY GENERATION AND SYNTHESIS

RMO1C-1: A D-Band PLL Covering the 81--82GHz, 86--92GHz and 162--164GHz Bands 43
Shahriar Shahramian, Adam Hart, Anthony Chan Carusone, Patrice Garcia, Pascal Chevalier, Sorin P. Voinigescu
RMO1C-2: An Integrated Frequency Synthesizer for 81--86GHz Satellite Communications in 65nm CMOS 47
Zhiwei Xu, Qun Jane Gu, Yi-Cheng Wu, Heng-Yu Jian, Frank Wang, Mau-Chung Frank Chang
RMO1C-3: Low-Noise Fractional-N PLL Design with Mixed-Mode Triple-Input LC VCO in 65nm CMOS 51
Yuanfeng Sun, Xueyi Yu, Woogeun Rhee, Sangsoo Ko, Wooseung Choo, Byeong-ha Park, Zhihua Wang
RMO1C-4: A Wideband Millimeter-Wave Frequency Doubler-Tripler in 0.13µm CMOS 55
Shadi Saberi Ghouchani, Jeyanandh Paramesh

RMO1C-5: 200GHz CMOS Prescalers with Extended Dividing Range via Time-Interleaved Dual Injection Locking 59
Qun Jane Gu, Heng-Yu Jian, Zhiwei Xu, Yi-Cheng Wu, Mau-Chung Frank Chang, Yves Baeyens, Young-Kai Chen

RMO1D W-BAND AND ABOVE

RMO1D-1: Transmitter Chipset for 24/77GHz Automotive Radar Sensors 63
Vittorio Giammello, Egidio Ragonese, Giuseppe Palmisano

RMO1D-2: A 94GHz Passive Imaging Receiver using a Balanced LNA with Embedded Dicke Switch 67
Leland Gilreath, Vipul Jain, Hsin-Cheng Yao, Le Zheng, Payam Heydari

RMO1D-3: 94GHz Silicon Co-Integrated LNA and Antenna in a mm-Wave Dedicated BiCMOS Technology 71
R. Pilard, D. Gloria, F. Gianesello, F. Le Pennec, C. Person

RMO1D-4: A 3G-Bit/s W-Band SiGe ASK Receiver with a High-Efficiency On-Chip Electromagnetically-Coupled Antenna 75
Jason W. May, Ramadan A. Alhalabi, Gabriel M. Rebeiz

RMO1D-5: A 325GHz Frequency Multiplier Chain in a SiGe HBT Technology 79
Erik Öjefors, Bernd Heinemann, Ullrich R. Pfeiffer

RMO2A RFID CIRCUITS

RMO2A-1: Semi-Active High-Efficient CMOS Rectifier for Wireless Power Transmission 83
Stephen T. Kim, Taejoong Song, Jaehyouk Choi, Franklin Bien, Kyutae Lim, Joy Laskar

RMO2A-2: A Single-Chip CMOS UHF RFID Reader Transceiver 87
Runxi Zhang, Chunqi Shi, Yihao Chen, Wei He, Ping Xu, Shuai Xu, Zongsheng Lai

RMO2A-3: Near Zero Turn-On Voltage High-Efficiency UHF RFID Rectifier in Silicon-on-Sapphire CMOS 91
Paul T. Theilmann, Calogero D. Presti, Dylan Kelly, Peter M. Asbeck

RMO2A-4: A Low Power Low Cost Fully Integrated UHF RFID Reader with 17.6dBm Output P1dB in 0.18μm CMOS Process 95
Jingchao Wang, Chun Zhang, Zhihua Wang

RMO2A-5: Far-Field RF Powering System for RFID and Implantable Devices with Monolithically Integrated On-Chip Antenna 99
Soheil Radiom, Majid Baghaei-Nejad, Guy Vandenbosch, Li-Rong Zheng, Georges Gielen

RMO2B WIDEBAND LNAS

RMO2B-1: A 2--1100MHz Wideband Low Noise Amplifier with 1.43dB Minimum Noise Figure 103
Mohamed El-Nozahi, Ahmed A. Helmy, Edgar Sánchez-Sinencio, Kamran Entesari

RMO2B-2: A 0.045mm^2 0.1--6GHz Reconfigurable Multi-Band, Multi-Gain LNA for SDR 107
Arnd Geis, Yves Rolain, Gerd Vandersteen, Jan Craninckx

RMO2B-3: A Linearity-Enhanced Wideband Low-Noise Amplifier 111
Kihwa Choi, Tamal Mukherjee, Jeyanandh Paramesh

RMO2B-4: Power Efficient Distributed Low-Noise Amplifier in 90nm CMOS 115
Brecht Machiels, Patrick Reynaert, Michiel Steyaert

RMO2B-5: A Wide-Band RF Front-End with Linear Active Notch Filter for Mobile TV Applications 119
Seung Hwan Jung, Kang Hyuk Lee, Young Jae Lee, Hyun Kyu Yu, Yun Seong Eo

RMO2C MILLIMETER-WAVE VCOS

RMO2C-1: A mm-Wave Arbitrary 2^N Band Oscillator Based on Even-Odd Mode Technique 123
Alvin Hsing-Ting Yu, Sai-Wang Tam, David Murphy, Tatsuo Itoh, Mau-Chung Frank Chang

RMO2C-2: A 24GHz and 60GHz Dual-Band Standing-Wave VCO in 0.13μm CMOS Process 127
Liang Wu, Alan W.L. Ng, Lincoln L.K. Leung, Howard C. Luong

RMO2C-3: Multi-Band Multi-Standard Local Oscillator Generation for Direct Up/Down Conversion Transceiver Architectures Supporting WiFi and WiMax Bands in Standard 45nm CMOS Process 131
Ram Sadhwani, Assaf Ben Bassat, Adil A. Kidwai, Shahar Rivel

RMO2C-4: A 0.13μm CMOS Local Oscillator for 60GHz Applications Based on Push-Push Characteristic of Capacitive Degeneration 135
Tino Copani, Hyungseok Kim, Bertan Bakkaloglu, Sayfe Kiaei

RMO2C-5: A Switched-Capacitor mm-Wave VCO in 65nm Digital CMOS 139
Mohammad Nariman, Reza Rofougaran, Franco De Flaviis

RMO2D RECONFIGURABLE PA CONCEPTS

RMO2D-1: Multi-Mode WCDMA Power Amplifier Module with Improved Low-Power Efficiency Using Stage-Bypass 143
Gary Hau, Mahendra Singh

RMO2D-2: A 3.4GHz to 4.3GHz Frequency-Reconfigurable Class E Power Amplifier with an Integrated CMOS-MEMS LC Balun 147
Leon Wang, Tamal Mukherjee

RMO2D-3: A Q-Band 6W MMIC Power Amplifier with 3-Way Power Combination Circuit 151
Hiroshi Otsuka, Kazuhisa Yamauchi, Koji Yamanaka, Shin Chaki, Kazuhiko Nakahara, Kunihiro Endo, Akira Inoue, Yoshihito Hirano

RMO2D-4: A Multi-Band Reconfigurable Power Amplifier for UMTS Handset Applications 155
Unha Kim, Kyutae Kim, Junghyun Kim, Youngwoo Kwon

RMO2D-5: High-Efficiency Reconfigurable RF Transmitter for Wireless Sensor Network Applications 159
Francesco Carrara, Giuseppe Palmisano

RMO3A CMOS WIDEBAND TRANSCEIVER ICS

RMO3A-1: A Triband 65nm CMOS Tuner for ATSC Mobile DTV SoC 163
Sanghoon Kang, Huijung Kim, Jeong-Hyun Choi, Jae-Hong Chang, Jong-Dae Bae, Wooseung Choo, Byeong-ha Park

RMO3A-2: A Multi-Standard Multi-Band Tuner for Mobile TV SoC with GSM Interoperability 167
Huijung Kim, Sanghoon Kang, Jae-Hong Chang, Jeong-Hyun Choi, Hangun Chung, Jungwook Heo, Jong-Dae Bae, Wooseung Choo, Byeong-ha Park

RMO3A-3: A 2.2mW Regenerative FM-UWB Receiver in 65nm CMOS 171
Nitz Saputra, John R. Long, John J. Pekarik

RMO3A-4: A 75 *pJ/Bit* All-Digital Quadrature Coherent IR-UWB Transceiver in 0.18μm CMOS 175
Enrique Barajas, Didac Gómez, Diego Mateo, José Luis Gonzalez

RMO3A-5: A 90nm-CMOS, 500Mbps, Fully-Integrated IR-UWB Transceiver using Pulse Injection-Locking for Receiver Phase Synchronization 179
Changhui Hu, Patrick Y. Chiang, Kangmin Hu, Huaping Liu, Rahul Khanna, Jay Nejedlo

RMO3B ADVANCED CHARACTERIZATION OF MM-WAVE COMPONENTS

RMO3B-1: The ``Load-Thru'' (LT) De-Embedding Technique for the Measurements of mm-Wave Balanced 4-Port Devices 183
Zhiming Deng, Ali M. Niknejad

RMO3B-2: RF-Pad, Transmission Lines and Balun Optimization for 60GHz 65nm CMOS Power Amplifier 187
Sofiane Aloui, Eric Kerherve, Robert Plana, Didier Belot

RMO3B-3: 200GHz f_T SiGe HBT Load Pull Characterization at mm-Wave Frequencies 191
Luciano Boglione, Richard T. Webster

RMO3B-4: A Miniature 26/77GHz Dual-Band Branch-Line Coupler Using Standard 0.18μm CMOS Technology 195
Yu-Sheng Lin, Cheng-Ying Hsu, Huey-Ru Chuang, Chu-Yu Chen

RMO3B-5: Compact Transformer Power Combiners for Millimeter-Wave Wireless Applications 199
Yi Zhao, John R. Long, Marco Spirito

RMO3C ADVANCED DEVICE TECHNOLOGIES & DESIGN TECHNIQUES

RMO3C-1: Integration of Multi-Standard Front End Modules SOCs on High Resistivity SOI RF CMOS Technology 203
F. Gianesello, S. Boret, B. Martineau, C. Durand, R. Pilard, D. Gloria, B. Rauber, C. Raynaud

RMO3C-2: Vast-Fast Low-Triggering LTdSCR ESD Protection Structure for RF ICs in CMOS 207
Jian Liu, Lin Lin, Xin Wang, Hui Zhao, He Tang, Qiang Fang, Albert Wang, Hongyi Chen, Haolu Xie, Siqiang Fan, Bin Zhao, Gary Zhang

RMO3C-3: A Cost-Competitive High Performance Junction-FET (JFET) in CMOS Process for RF & Analog Applications 211
Yun Shi, Robert M. Rassel, Richard A. Phelps, Panglijen Candra, Douglas B. Hershberger, Xiaowei Tian, Susan L. Sweeney, Jay Rascoe, BethAnn Rainey, Jim Dunn, David Harame

RMO3C-4: Tunability of Bulk Acoustic Wave Filters Using CMOS Transistors: Concept, Design and Implementation 215
M. El Hassan, Eric Kerherve, Yann Deval, J. B. David, Didier Belot

RMO3C-5: A Layout Efficient, Vertically Stacked, Resonator-Coupled Bandpass Filter in LTCC for 60GHz SOP Transceivers 219
Rony E. Amaya

RMO3C-6: Co-Design Considerations for Frequency Drift Compensation in BAW-Based Time Reference Application 223
S. Razafimandimby, D. Petit, P. Bar, S. Joblot, J.-F. Carpentier, J. Morelle, C. Arnaud, G. Parat, Patrice Garcia, C. Garnier

RMO3D SWITCH & SWITCH-MODE TECHNOLOGIES

RMO3D-1: High Efficiency and Wideband Envelope Tracking Power Amplifier with Sweet Spot Tracking 227
Dongsu Kim, Jinsung Choi, Daehyun Kang, Bumman Kim

RMO3D-2: A 150MHz, 84% Efficiency, Two Phase Interleaved DC-DC Converter in AlGaAs/GaAs P-HEMT Technology for Integrated Power Amplifier Modules 231
Han Peng, V. Pala, T. P. Chow, Mona Hella

RMO3D-3: A 0dBm 10Mbps 2.4GHz Ultra-Low Power ASK/OOK Transmitter with Digital Pulse-Shaping 235
Xiongchuan Huang, Pieter Harpe, Xiaoyan Wang, Guido Dolmans, Harmke de Groot

RMO3D-4: A Linear-in-dB SiGe HBT Wideband High Dynamic Range RF Envelope Detector 239
Hsuan-yu Marcus Pan, Lawrence E. Larson

RMO3D-5: Cellular Antenna Switches for Multimode Applications Based on a Silicon-on-Insulator Technology 243
Ali Tombak, Christian Iversen, Jean-Blaise Pierres, Dan Kerr, Mike Carroll, Phil Mason, Eddie Spears, Todd Gillenwater

RMO4A ULTRA LOW POWER RECEIVERS AND TRANSMITTERS

RMO4A-1: A 120µW Fully-Integrated BPSK Receiver in 90nm CMOS 247
Han Yan, Jose Gabriel Macias-Montero, Atef Akhnoukh, Leo C. N. de Vreede, John R. Long, John J. Pekarik, Joachim N. Burghartz

RMO4A-2: A Fully Integrated 2.4GHz CMOS Diversity Receiver with a Novel Antenna Selection 251
Yong-Il Kwon, Sang-Ku Park, T. J. Park, Hai-Young Lee

RMO4A-3: A 90µW MICS/ISM Band Transmitter with 22% Global Efficiency 255
Jagdish Pandey, Brian Otis

RMO4A-4: A 2mW CMOS MICS-Band BFSK Transceiver with Reconfigurable Antenna Interface 259
Seungkee Min, Sridhar Shashidharan, Mark Stevens, Tino Copani, Sayfe Kiaei, Bertan Bakkaloglu, Sudipto Chakraborty

RMO4A-5: A 1.8 to 2.4GHz 20mW Digital-Intensive RF Sampling Receiver with a Noise-Canceling Bandpass Low-Noise Amplifier in 90nm CMOS 263
Joonhee Lee, Jaewook Kim, SeongHwan Cho

RMO4B OPTIMIZED DESIGN TECHNIQUES FOR RF FRONT-END BUILDING BLOCKS

RMO4B-1: A Differential 4-Path Highly Linear Widely Tunable On-Chip Band-Pass Filter...................................267
Amir Ghaffari, Eric A. M. Klumperink, Bram Nauta

RMO4B-2: A CMOS Wide-Bandwidth High-Power Linear-in-dB Variable Attenuator Using Body Voltage Distribution Method 271
Yan-Yu Huang, Wangmyong Woo, Chang-Ho Lee, Joy Laskar

RMO4B-3: A 17GHz Transformer-Neutralized Current Re-Use LNA and Its Application to a Low-Power RF Front-End..275
Sandipan Kundu, Jeyanandh Paramesh

RMO4B-4: A Self-Healing 2.4GHz LNA with On-Chip S_{11}/S_{21} Measurement/Calibration for in-situ PVT Compensation...279
Karthik Jayaraman, Qadeer Khan, Baoyong Chi, William Beattie, Zhihua Wang, Patrick Y. Chiang

RMO4B-5: A Low Power LNA Using Miniature 3D Inductor Without Area Penalty of Passive Components...283
Akira Tanabe, Ken'ichiro Hijioka, Hirokazu Nagase, Yoshihiro Hayashi

RMO4C TEMPERATURE COMPENSATED OSCILLATORS

RMO4C-1: A 65nm CMOS DCXO System for Generating 38.4MHz and a Real Time Clock from a Single Crystal in 0.09mm²...287
Danielle Griffith, Fikret Dülger, Gennady Feygin, Ahmed Nader Mohieldin, Prasanth Vallur

RMO4C-2: A 50ppm 600MHz Frequency Reference Utilizing the Series Resonance of an FBAR 291
Julie Hu, Lori Callaghan, Richard Ruby, Brian Otis

RMO4C-3: An Electronically Temperature-Compensated 427MHz Low Phase-Noise AlN-on-Si Micromechanical Reference Oscillator 295
Hossein Miri Lavasani, Wanling Pan, Farrokh Ayazi

RMO4C-4: A Wide Tuning 1.3GHz LC VCO with Fast Settling Noise Filtering Voltage Regulator in 0.18µm CMOS Process...299
Hiroshi Akima, Aleksander Dec, Ken Suyama

RMO4C-5: A Wide-Range VCO with Optimum Temperature Adaptive Tuning 303
Behzad Saeidi, Joshua Cho, Georgi Taskov, Aaron Paff

RMO4D SILICON MILLIMETER-WAVE AMPLIFIERS

RMO4D-1: A 60GHz Transformer Coupled Amplifier in 65nm Digital CMOS...........................307
Michael Boers

RMO4D-2: A Stage-Scaled Distributed Power Amplifier Achieving 110GHz Bandwidth and 17.5dBm Peak Output Power...311
Jiashu Chen, Ali M. Niknejad

RMO4D-3: DC Hot Carrier Stress Effect on CMOS 65nm 60GHz Power Amplifiers 315
T. Quémerais, L. Moquillon, V. Huard, J.-M. Fournier, P. Benech, N. Corrao

RMO4D-4: A Layout-Based Optimal Neutralization Technique for mm-Wave Differential Amplifiers 319
Zhiming Deng, Ali M. Niknejad

RMO4D-5: A 100GHz Transformer-Coupled Fully Differential Amplifier in 90nm CMOS 323
Noël Deferm, Patrick Reynaert

RTU1B CMOS MILLIMETER-WAVE 60/24GHZ RADIO

RTU1B-1: A 68--82GHz Integrated Wideband Linear Receiver Using 0.18µm SiGe BiCMOS 327
Austin Ying-Kuang Chen, Yves Baeyens, Young-Kai Chen, Jenshan Lin

RTU1B-2: A 24GHz Low-Power Fully Integrated Receiver with Image-Rejection using Rich-Transformer Direct-Stacked/Coupled Technique...........................331
Nobuhiro Shiramizu, Takahiro Nakamura, Toru Masuda, Katsuyoshi Washio

RTU1B-3: A 60GHz CMOS Receiver Front-End with Integrated 180° Out-of-Phase Wilkinson Power Divider 335
Chi-Chen Chen, Yo-Sheng Lin, Jen-How Lee, Jin-Fa Chang

RTU1B-4: Coherent Parametric RF Downconversion in CMOS 339
Zhixing Zhao, Jean-François Bousquet, Sebastian Magierowski

RTU1B-5: 60GHz Broadband Image Rejection Receiver Using Varactor Tuning 343
Jihoon Kim, Wooyeol Choi, Youngrak Park, Youngwoo Kwon

RTU1C CMOS PAS

RTU1C-1: A Discrete Resizing and Concurrent Power Combining Structure for Linear CMOS Power Amplifier 347
Jihwan Kim, Hyungwook Kim, Youngchang Yoon, Kyu Hwan An, Woonyun Kim, Chang-Ho Lee, Kevin T. Kornegay, Joy Laskar

RTU1C-2: A Single-Chip 2.4GHz Double Cascode Power Amplifier with Switched Programmable Feedback Biasing under Multiple Supply Voltages in 65nm CMOS for WLAN Application 351
Mingyuan Li, Ali Afsahi, Arya Behzad

RTU1C-3: A 31dBm, High Ruggedness Power Amplifier in 65nm Standard CMOS with High-Efficiency Stacked-Cascode Stages 355
Stephan Leuschner, Sandro Pinarello, Uwe Hodel, Jan-Erik Mueller, Heinrich Klar

RTU1C-4: Analysis and Design of a Wideband High Efficiency CMOS Outphasing Amplifier 359
M. C. A. van Schie, M. P. van der Heijden, M. Acar, A. J. M. de Graauw, Leo C. N. de Vreede

RTU1C-5: A Highly Efficient 5.8GHz CMOS Transmitter IC with Robustness Over PVT Variations 363
Eun-Hee Kim, Jeong-Ki Choi, Seok-Oh Yun, Jinho Ko, Kwyro Lee

RTU1D EMERGING ARCHITECTURES IN DIGITAL FREQUENCY SYNTHESIS

RTU1D-1: A 700µA, 405MHz Fractional-N All Digital Frequency-Locked Loop for MICS Band Applications 367
S. Shashidharan, W. Khalil, Sudipto Chakraborty, Sayfe Kiaei, Tino Copani, Bertan Bakkaloglu

RTU1D-2: A 2MHz Bandwidth Δ-Σ Fractional-N Synthesizer Based on a Fractional Frequency Divider with Digital Spur Suppression 371
Pin-En Su, Sudhakar Pamarti

RTU1D-3: A 6fJ/step, 5.5ps Time-to-Digital Converter for a Digital PLL in 40nm Digital LP CMOS 375
J. Borremans, K. Vengattarmane, Jan Craninckx

RTU1D-4: A 6GHz Direct Digital Synthesizer MMIC with Nonlinear DAC and Wave Correction ROM 379
Danyu Wu, Gaopeng Chen, Jianwu Chen, Xinyu Liu, Lixin Zhao, Zhi Jin

RTU1D-5: A 10GHz 8-Bit Direct Digital Synthesizer Implemented in GaAs HBT Technology 383
Gaopeng Chen, Danyu Wu, Zhi Jin, Jin Wu, Xinyu Liu

RTU2A WLAN TRANSCEIVERS AND COMPONENTS

RTU2A-1: Dual-Band CMOS Transceiver with Highly Integrated Front-End for 450Mb/s 802.11n Systems 387
S. Gross, T. Maimon, F. Cossoy, M. Ruberto, G. Normatov, A. Rivkind, N. Telzhensky, R. Banin, O. Ashckenazi, A. B. Bassat, S. Zaguri, G. Hara, M. Zajac, N. Shahar, S. Shahaf, H. Yousef, E. Mor, Y. Eilat, A. Nazimov, Z. Beer, A. Fridman, O. Degani

RTU2A-2: A CMOS Transceiver with Internal PA and Digital Pre-Distortion for WLAN 802.11a/b/g/n Applications 391
Chia-Jun Chang, Po-Chih Wang, Chih-Yu Tsai, Chin-Lung Li, Chiao-Ling Chang, Han-Jung Shih, Meng-Hsun Tsai, Wen-Shan Wang, Ka-Un Chan, Ying-Hsi Lin

RTU2A-3: Highly Linear SOI Single-Pole, 4-Throw Switch with an Integrated Dual-Band LNA and Bypass Attenuators 395
Chun-Wen Paul Huang, Lui Lam, Mark Doherty, William Vaillancourt

RTU2A-4: A 6.1GS/s 52.8mW 43dB DR 80MHz Bandwidth 2.4GHz RF Bandpass ΔΣ ADC in 40nm CMOS 399
Julien Ryckaert, Arnd Geis, Lynn Bos, Geert Van der Plas, Jan Craninckx

RTU2A-5: Single-Chip WiFi b/g/n 1x2 SoC with Fully Integrated Front-End & PMU in 90nm Digital CMOS Technology 403
J.C. Jensen, Ram Sadhwani, Adil A. Kidwai, B. Jann, A. Oster, M. Sharkansky, I. Ben-bassat, Ofir Degani, S. Porat, Amir Fridman, H. Shang, C. Chu, A. Ly, M. Smith

RTU2C MILLIMETER-WAVE ARRAYS

RTU2C-1: A 44GHz 8-Element Phased-Array SiGe HBT Transmitter RFIC with an Injection-Locked Quadrature Frequency Multiplier 407
Sunghwan Kim, Prasad S. Gudem, Lawrence E. Larson

RTU2C-2: A Thirty Two Element Phased-Array Transceiver at 60GHz with RF-IF Conversion Block in 90nm Flip Chip CMOS Process 411
Emanuel Cohen, Claudio Jakobson, Shmuel Ravid, Dan Ritter

RTU2C-3: A 16-Element Phased-Array Receiver IC for 60GHz Communications in SiGe BiCMOS 415
Scott K. Reynolds, Arun S. Natarajan, Ming-Da Tsai, Sean Nicolson, Jing-Hong Conan Zhan, Duixian Liu, Dong G. Kam, Oscar Huang, Alberto Valdes-Garcia, Brian A. Floyd

RTU2C-4: A 24GHz Phased-Array Receiver in 0.13µm CMOS using an 8GHz LO 419
Satwik Patnaik, Ramesh Harjani

RTU2C-5: Wafer-Scale W-Band Power Amplifiers Using On-Chip Antennas 423
Yusuf A. Atesal, Berke Cetinoneri, Ramadan A. Alhalabi, Gabriel M. Rebeiz

RTU2D RF MODELING FOR SWITCH AND PA APPLICATIONS

RTU2D-1: Application of BSIMSOI MOSFET Model to SOS Technology 427
James Roach, Lee-Wen Chen, Peter Clarke, Francis M. Rotella

RTU2D-2: Modeling of SOI FET for RF Switch Applications 431
Tzung-Yin Lee, Sunyoung Lee

RTU2D-3: A High Power CMOS Differential T/R Switch Using Multi-Section Impedance Transformation Technique 435
Hyun-Woong Kim, Minsik Ahn, Ockgoo Lee, Chang-Ho Lee, Joy Laskar

RTU2D-4: Exploitation of Active Load-Pull and DLUT Models in MMIC Design 439
D. M. FitzPatrick, T. Williams, J. Lees, J. Benedikt, S. C. Cripps, P. J. Tasker

RTU2D-5: A Mixed-Signal Load-Pull System for Base-Station Applications 443
Mauro Marchetti, Rob Heeres, Michele Squillante, Marco Pelk, Marco Spirito, Leo C. N. de Vreede

RTUIF RFIC INTERACTIVE FORUM

RTUIF-01: A 228µW Injection Locked Ring Oscillator based BPSK Demodulator in 65nm CMOS 447
Qiang Zhu, Yang Xu

RTUIF-02: A 0.13µm CMOS Wireless Reflector for Phase Sweep Cooperative Diversity 451
Jean-François Bousquet, Sebastian Magierowski, Geoffrey Messier, Zhixing Zhao

RTUIF-03: Design Methodology and Comparison of Rectifiers for UHF-Band RFIDs 455
Francesco Mazzilli, Prakash E. Thoppay, Norbert Jöhl, Catherine Dehollain

RTUIF-04: A CMOS Ultra-Wideband Radar Transmitter with Pulsed Oscillator 459
Sungeun Lee, Sanghoon Sim, Songcheol Hong

RTUIF-05: 900MHz/1800MHz GSM Base Station LNA with Sub-1dB Noise Figure and +36dBm OIP3 463
Domine Leenaerts, Jos Bergervoet, Jan-Willem Lobeek, Marek Schmidt-Szalowski

RTUIF-06: A 4.35mW +22dBm IIP3 Continuously Tunable Channel Select Filter for WLAN/WiMax Receivers in 90nm CMOS 467
Mostafa Savadi Oskooei, Nasser Masoumi, Mahmud Kamarei, Henrik Sjöland

RTUIF-07: Wideband Trans-Impedance Filter Low Noise Amplifier 471
Mikko Kaltiokallio, Aarno Pärssinen, Jussi Ryynänen

RTUIF-08: A Wideband High-Linearity Mixer in 0.5µm InP DHBT Technology 475
Mark Stuenkel, Milton Feng

RTUIF-09: A High Gain Wideband 77GHz SiGe Power Amplifier 479
Roee Ben Yishay, Roi Carmon, Oded Katz, Danny Elad

RTUIF-10: A Broadband Differential Cascode Power Amplifier in 45nm CMOS for High-Speed 60GHz System-on-Chip 483
Morteza Abbasi, Torgil Kjellberg, Anton J. M. de Graauw, Edwin van der Heijden, Raf Roovers, Herbert Zirath

RTUIF-11: A CMOS LC VCO with Novel Negative Impedance Design for Wide-Band Operation 487
Chang-Hsi Wu, Guan-Xiu Jian

RTUIF-12: An 80GHz Range Synchronized Push-Push Oscillator for Automotive Radar Application 491
Chama Ameziane, Thierry Taris, Yann Deval, Didier Belot, Robert Plana, Jean-Baptiste Bégueret

RTUIF-13: Millimeter Wave CMOS VCO with a High Impedance LC Tank 495
Seung Wan Chai, Jaemo Yang, Bon-Hyun Ku, Songcheol Hong

RTUIF-14: Controlled Dither in 90nm Digital to Time Conversion based Direct Digital Synthesizer for Spur Mitigation 499
S. Talwalkar, T. Gradishar, B. Stengel, G. Cafaro, G. Nagaraj

RTUIF-15: 2--4 and 9--12Gb/s CMOS Fully Integrated ILO-based CDR 503
O. Mazouffre, R. Toupe, M. Pignol, Yann Deval, Jean-Baptiste Bégueret

RTUIF-16: A 22.5dB Gain, 20.1dBm Output Power K-Band Power Amplifier in 0.18μm CMOS 507
Chi-Cheng Hung, Jing-Lin Kuo, Kun-You Lin, Huei Wang

RTUIF-17: A 40% PAE Linear CMOS Power Amplifier with Feedback Bias Technique for WCDMA Applications 511
Hamhee Jeon, Kun-Seok Lee, Ockgoo Lee, Kyu Hwan An, Youngchang Yoon, Hyungwook Kim, Dong Ho Lee, Jongsoo Lee, Chang-Ho Lee, Joy Laskar

RTUIF-18: A Switching-Mode Amplifier for Class-S Transmitters for Clock Frequencies up to 7.5GHz in 0.25μm SiGe-BiCMOS 515
Stefan Heck, Martin Schmidt, Alexander Bräckle, Frieder Schuller, Markus Grözing, Manfred Berroth, Hans Gustat, Christoph Scheytt

RTUIF-19: SiGe Power Amplifier ICs for 4G (WIMAX and LTE) Mobile and Nomadic Applications 519
V. Krishnamurthy, K. Hershberger, B. Eplett, J. Dekosky, H. Zhao, D. Poulin, R. Rood, E. Prince

RTUIF-20: Self-Matched ESD Cell in CMOS Technology for 60GHz Broadband RF Applications 523
Chun-Yu Lin, Li-Wei Chu, Ming-Dou Ker, Tse-Hua Lu, Ping-Fang Hung, Hsiao-Chun Li

RTUIF-21: The Impact of MOSFET Layout Dependent Stress on High Frequency Characteristics and Flicker Noise 527
Kuo-Liang Yeh, Chih-You Ku, Jyh-Chyurn Guo

RTUIF-22: A Novel Low-Profile Low-Parasitic RF Package using High-Density Build-Up Technology 531
Chien-Cheng Wei, Ming-Chien Lin, Chin-Ta Fan, Ta-Hsiang Chiang, Ming-Kuen Chiu, Shao-Pin Ru, Nan Ni, Albert Cardona

RTUIF-23: A High Quality Factor Varactor Technology Evaluation 535
Romain Debroucke, Sebastien Jan, Jean-François Larchanché, Christophe Gaquière

RTUIF-24: Power Improvement for 65nm nMOSFET with High-Tensile CESL and Fast Nonlinear Behavior Modeling 539
Chia-Sung Chiu, Kun-Ming Chen, Guo-Wei Huang, Shu-Yu Lin, Bo-Yuan Chen, Cheng-Chou Hung, Sheng-Yi Huang, Cheng-Wen Fan, Chih-Yuh Tzeng, Sam Chou

RTUIF-25: RF Benchmark Tests for Compact MOS Models 543
G. D. J. Smit, A. J. Scholten, D. B. M. Klaassen

RTUIF-26: A 1.8V 74mW UHF RFID Reader Receiver with 18.5dBm IIP3 and -77dBm Sensitivity in 0.18μm CMOS 547
Xuguang Sun, Baoyong Chi, Chun Zhang, Ziqiang Wang, Zhihua Wang

Author Index

SESSION LIST

RSU5A   RFIC Plenary

RMO1A   Cellular Transceivers
RMO1B   RF CMOS Modulators & Receivers
RMO1C   Frequency Generation and Synthesis
RMO1D   W-band and Above
RMO2A   RFID Circuits
RMO2B   Wideband LNAs
RMO2C   Millimeter-Wave VCOs
RMO2D   Reconfigurable PA Concepts
RMO3A   CMOS Wideband Transceiver ICs
RMO3B   Advanced Characterization of mm-Wave Components
RMO3C   Advanced Device Technologies & Design Techniques
RMO3D   Switch & Switch-mode Technologies
RMO4A   Ultra Low Power Receivers and Transmitters
RMO4B   Optimized Design Techniques for RF Front-end Building Blocks
RMO4C   Temperature Compensated Oscillators
RMO4D   Silicon Millimeter-Wave Amplifiers

RTU1B   CMOS Millimeter-Wave 60/24GHz Radio
RTU1C   CMOS PAs
RTU1D   Emerging Architectures in Digital Frequency Synthesis
RTU2A   WLAN Transceivers and Components
RTU2C   Millimeter-Wave Arrays
RTU2D   RF Modeling for Switch and PA Applications
RTUIF   RFIC Interactive Forum

RFIC Schedule 2010

The RFIC Symposium will be held in Anaheim, CA in the Anaheim Convention Center (ACC). The headquarters hotel is the Anaheim Hilton Hotel which is adjacent to the ACC. The RFIC Plenary and Reception will be held on Sunday May 23, starting at 5:30pm.

The RFIC Symposium is held as part of Microwave week. It is followed by the IMS Symposium and Exhibition and by ARFTG. Attendees of RFIC are invited to attend the IMS Plenary Session which will be held on Tuesday Morning, May 25, 2010.

Saturday May 22, 2010

02:00pm - 06:00pm	Registration – ACC

Sunday May 23, 2010

07:00am - 06:00pm	Registration – ACC
06:30am - 08:30am	Speakers Breakfast – ACC, Room 304D
07:00am - 08:00am	Workshop Breakfast – ACC, Meeting Foyers
08:00am - 05:00pm	Workshops and Tutorials – ACC
12:00pm - 01:00pm	Workshop Lunch – ACC, Meeting Foyers
05:30pm - 06:40pm	RFIC Plenary – ACC, Room 210ABCD
07:00pm - 09:00pm	RFIC Reception – ACC, Room 213BCD

Monday May 24, 2010

07:00am - 05:00pm	Registration – ACC
06:30am - 08:30am	Speakers Breakfast – ACC, Room 304D
07:00am - 08:00am	Attendee Breakfast – ACC, Meeting Foyers
07:00am - 05:00pm	Speaker's Prep Room – ACC, Room 203A
08:00am - 05:10pm	RFIC Technical Sessions (see pages 10 to 41)
08:00am - 09:40am	RMO1A, RMO1B, RMO1C, RMO1D
09:40am - 10:10am	Break
10:10am - 11:50am	RMO2A, RMO2B, RMO2C, RMO2D
11:50am - 01:20pm	RFIC Panel and Lunch – ACC, Room 210CD
11:50am - 01:20pm	RFIC-TPC Lunch – ACC, Room TBD
01:20pm - 03:00pm	RMO3A, RMO3B, RMO3C, RMO3D
03:00pm - 03:30pm	Break
03:30pm - 05:10pm	RMO4A, RMO4B, RMO4C, RMO4D

Tuesday May 25, 2010

07:00am - 05:00pm	Registration – ACC
07:00am - 08:00am	Speaker's Breakfast – ACC, Room 304D
07:00am - 08:00am	Attendee Breakfast – ACC, Meeting Foyers
07:00am - 05:00pm	Speaker's Prep Room – ACC, Room 203A
08:00am - 04:00pm	RFIC Oral and IF Sessions (see pages 42 to 61)
08:00am - 09:40am	RTU1B, RTU1C, RTU1D
11:50am - 01:20pm	RFIC Panel and Lunch – ACC, Room 210CD
12:00pm - 01:00pm	RFIC Steering Committee Meeting – ACC, Huntington ABC
01:20pm - 03:00pm	RTU2A, RTU2C, RTU2D
02:00pm - 04:00pm	RTUIF – ACC, Rooms 208AB, 209AB

Message from the General Chair

Yann Deval

On behalf of the Steering Committee, I would like to welcome you to the RFIC Symposium!

The 2010 RFIC Symposium maintains its reputation as one of the foremost IEEE technical conferences dedicated to the latest innovations in RFIC development for wireless and wireline communication IC's. Running in conjunction with the International Microwave Symposium and Exhibition, the RFIC Symposium adds to the excitement of Microwave Week with three days focused exclusively on RFIC technology and innovation. The RFIC symposium will be held at the Anaheim Convention Center, May 23-25, 2010.

The RFIC Symposium will start on Sunday with half-day and full-day workshops, covering a large breadth of topics. Some of the topics include: SiGe HBTs towards THz operation, power management for integrated RF circuits, challenges and techniques for 3G/4G multi-mode front end designs and silicon-based design techniques for millimeter-wave applications. Don't miss out on this great opportunity to expand your horizons!

Sunday evening activities continue at 5:30pm with RFIC Plenary Session. Two renowned speakers will share their views on the direction and challenges that the RF IC industry will be facing. The first speaker is Professor David Allstot from the University of Washington, and the second speaker is Gregory Waters, Executive Vice President of Skyworks Inc. In addition to the keynote addresses, the best student paper awards are presented in the Plenary Session. The highly anticipated RFIC Reception will follow immediately after the Plenary Session, providing a relaxing time for all to mingle with old friends and catch up on the latest news.

The technical program includes oral sessions, an Interactive Forum (poster session), and two exciting lunch panel sessions. The oral presentation sessions start on Monday, May 24th with four parallel sessions throughout the morning and the afternoon. The oral sessions continue on Tuesday, May 25th synchronized with the IMS technical Program. The Interactive Forum will be held on Tuesday afternoon. This forum is the perfect place to have an opportunity to have more detailed technical discussions with the authors. Panel Sessions are also planned at lunch time on Monday and Tuesday, the topic being respectively "The Challenges, Competitions and Future Prospect of 60 GHz" and "Future of High-Speed I/O: Electrical, Optical, or Wireless?".

The RFIC Symposium concludes on Tuesday allowing participants to attend the IMS and ARFTG as well as plenty of time to visit the exhibit hall.

The RFIC organization is thankful to the IMS2010 team, without whom we could not make this conference successful. Most of all, we are particularly thankful to all the technical contributors to the RFIC Symposium. We look forward to your participation. Please continue to make this conference so vibrant within the wireless industry!

I look forward to seeing you in Anaheim!

Yann Deval
General Chair
2010 RFIC Symposium

Message from the Technical Program Committee Chairs

David Ngo   **Chris Rudell**

On behalf of the Technical Program Committee, welcome to the 2010 IEEE Radio Frequency Integrated Circuits (RFIC) Symposium. This is a leading-edge IEEE technical conference dedicated to the advancement of integrated circuits and sub-systems for RF, wireless, broadband communications, and many other emerging applications. The RFIC Technical Program Committee has worked diligently to select the best papers and assemble an excellent technical program this year. The symposium's main features include several tutorial workshops, a Plenary Session, an Interactive Forum and numerous technical paper sessions.

This year the RFIC Symposium begins on Sunday, May 23rd with workshops at the advanced and tutorial level addressing topics which challenge present day RF IC designers with respect to design techniques in advanced silicon technologies, design and integration of ICs for new emerging wireless application and the latest advances in circuit and system simulation. The Plenary Session will be held on Sunday evening, following the workshops, at which time the General Chair will present best paper awards to the top three student manuscripts of this year's conference.

Two leading experts from the RFIC community will share their views during the plenary session. The first speaker is Professor David James Allstot from the University of Washington, who will present his vision of one of the last great challenges of RF CMOS integration in his talk, "Power Amplification: Can CMOS Deliver?" The second speaker, Gregory L. Waters, Executive Vice President and General Manger, Skyworks Solutions, Inc., will discuss "The Universal Connector: RF Application Trends Over The Next Decade"

The RFIC Reception will follow immediately after the plenary session, providing a relaxing time for all to mingle with old friends and catch up on the latest news. In addition to the technical sessions on Monday and Tuesday, the RFIC Symposium also features panel sessions and many workshops. Monday's lunch panel session entitled "The Challenges, Competitions and Future Prospect of 60GHz" has panelists from both industry and academia debating the future of high data rate wireless networks. Tuesday's lunch panel session, "Future of High-Speed I/O: Electrical, Optical or Wireless?", is posed to stimulate interactive discussions with the audience.

The interest in RFIC technology, and the venue offered by the Symposium to showcase the latest advancements, continues to make the RFIC Symposium the technical forum of choice for both industry and academia, to meet, discuss results and exchange ideas. The 2010 RFIC Technical Program Committee will continue to work tirelessly toward the goal of strengthening the technical quality and scope of the program, while maintaining and improving the legacy left by previous Symposia. Of course the success of our conference would simply not be possible without the many contributions of all the authors who put enormous effort each year, to contribute both outstanding presentations and excellent manuscripts. On behalf of the entire Steering and Technical Program Committees, we thank everyone for attending the conference.

We hope you enjoy the 2010 RFIC Symposium!

David Ngo and Chris Rudell
Technical Program Chairs
2010 IEEE RFIC Symposium

Steering Committee

Yann Deval, IMS Lab, *General Chair*

David Ngo. RFMD, *TPC Chair*

Jacques C. Rudell, Univ. of Washington, *TPC Co-Chair*

Albert Jerng, Ralink, *Finance Chair*

Kevin McCarthy, Univ. College Cork, *Workshops Chair*

Georg Boeck, Berlin Institute of Technology, *Digest & CD ROM Chair*

Kevin Kobayashi, RFMD, *Publicity Chair*

Bertan Bakkaloglu, Arizona State Univ., *Panel Sessions Chair*

Yuhua Cheng, SHRIME Peking Univ., *Transactions/Guest Editor*

Brian Floyd, North Carolina State University, *Student Papers Chair*

Albert Wang, UC Riverside, *Secretary*

Noriharu Suematsu, Tohoku University, *Asia Pacific Liaisons*

Takao Inoue, Univ. of Texas, Austin, *Webmaster*

Larry Whicker, LRW Associates, *Conference Coordinator*

Advisory Board

Natalino Camilleri • Fazal Ali • Reynold Kagiwada • Sayfe Kiaei
David Lovelace • Joseph Staudinger

Executive Committee

Tina Quach • Stefan Heinen • Luciano Boglione
Jenshan Lin • Yann Deval

Technical Program Committee

Ali Afsahi, *Broadcom Corp.*

Fazal Ali, *Qualcomm*

Walid Ali-Ahmad, *MediaTek Inc.*

Bertan Bakkaloglu, *Arizona State University*

Jean-Baptiste Begueret, *University of Bordeaux, IMS Lab*

Didier Belot, *ST Microelectronics*

Paul Blount, *Custom MMIC Design*

Georg Boeck, *Berlin Institute of Technology*

Luciano Boglione, *University of Massachusetts Lowell*

Pierre Busson, *ST Microelectronics*

Natalino Camilleri, *Alien Technology*

Sudipto Chakraborty, *Texas Instruments*

Glenn Chang, *MaxLinear*

Jing-Hong Chen, *Analog Devices*

Nick Cheng, *Skyworks Solutions*

Yuhua Cheng, *Shrime Peking University*

Brian Floyd, *North Carolina State University*

Ranjit Gharpurey, *University of Texas, Austin*

Aditya Gupta, *Northrop Grumman*

Timothy Hancock, *MIT Lincoln Lab*

Andre Hanke, *Infineon Technologies AG*

Stefan Heinen, *RWTH Aachen University*

Frank Henkel, *IMST GmbH*

Tian-Wei Huang, *National Taiwan University*

Lars Jansson, *Tumbledown Technical Inc.*

Albert Jerng, *Ralink*

Waleed Khalil, *Ohio State University*

Jaber Khoja, *Microtune, Inc.*

Sayfe Kiaei, Arizona State University, *Connection One*

Bumman Kim, *Pohang University of Science and Technology*

Kevin Kobayashi, *RFMD, Inc.*

Larry Kushner, *Intersil Corp.*

Youngwoo Kwon, *Seoul National University*

Chang-Ho Lee, *Samsung*

Domine Leenaerts, *NXP Semiconductor*

Donald Y. C. Lie, *Texas Tech. University*

Fujiang Lin, *IME Singapore*

Jenshan Lin, *University of Florida*

Louis Liu, *CT Communication Technologies*

Ting-Ping Liu, *Nuvoton Technology*

David Lovelace, *ON Semiconductor*

Danilo Manstretta, *University of Pavia*

Kevin McCarthy, *University College Cork*

Srenik Mehta, *Atheros Communications*

Jyoti Mondal, *Northrop Grumman*

Sule Ozev, *Arizona State University*

Stefano Pellerano, *Intel Corporation*

Tina Quach, *Consultant*

Sanjay Raman, *Virginia Tech*

Madhukar Reddy, *MaxLinear*

Bill Redman-White, *NXP Semiconductor*

Eli Reese, *TriQuint Semiconductor*

Mark Ruberto, *Intel Corporation*

Francis Rotella, *Peregrine Semiconductor*

Carlos Saavedra, *Queen's University, Canada*

Derek Shaeffer, *InvenSense, Inc.*

Osama Shana'a, *MediaTek Corporation*

Eddie Spears, *RFMD, Inc.*

Robert Staszewski, *Delft University of Technology*

Joseph Staudinger, *Freescale Semiconductor Inc.*

Bob Stengel, *Motorola*

Freek van Straten, *NXP Semiconductor*

Noriharu Suematsu, *Tohoku University*

Julian Tham, *Arda Technologies*

Bruce Thompson, *Motorola Labs*

Albert Wang, *University of California, Riverside*

Haolu Xie, *Fujitsu Microelectronic, Tempe/Fudan University, Shanghai*

Li-Wu Yang, *SMIC*

Patrick Yue, *University of California, Santa Barbara*

Gary Zhang, *Skyworks Solutions*

Herbert Zirath, *Chalmers University*

RFIC 2010 Steering Committee

Albert Jerng
Finance

Bertan Bakkaloglu
Panel Sessions

Georg Boeck
Digest & CD-ROM

Kevin McCarthy
Workshops

Yuhua Cheng
Transactions

Kevin Kobayashi
Publicity

Brian Floyd
Student Papers

Albert Wang
Secretary

Noriharu Suematsu
Asian Liaisons

RFIC 2010 Executive Committee

Tina Quach   Jenshan Lin   Luciano Boglione   Stefan Heinen

RFIC 2010 Advisory Committee

Fazal Ali   Natalino Camilleri   Reynold Kagiwada

Sayfe Kiaei   David Lovelace   Joe Staudinger

Technical Program Committee Members

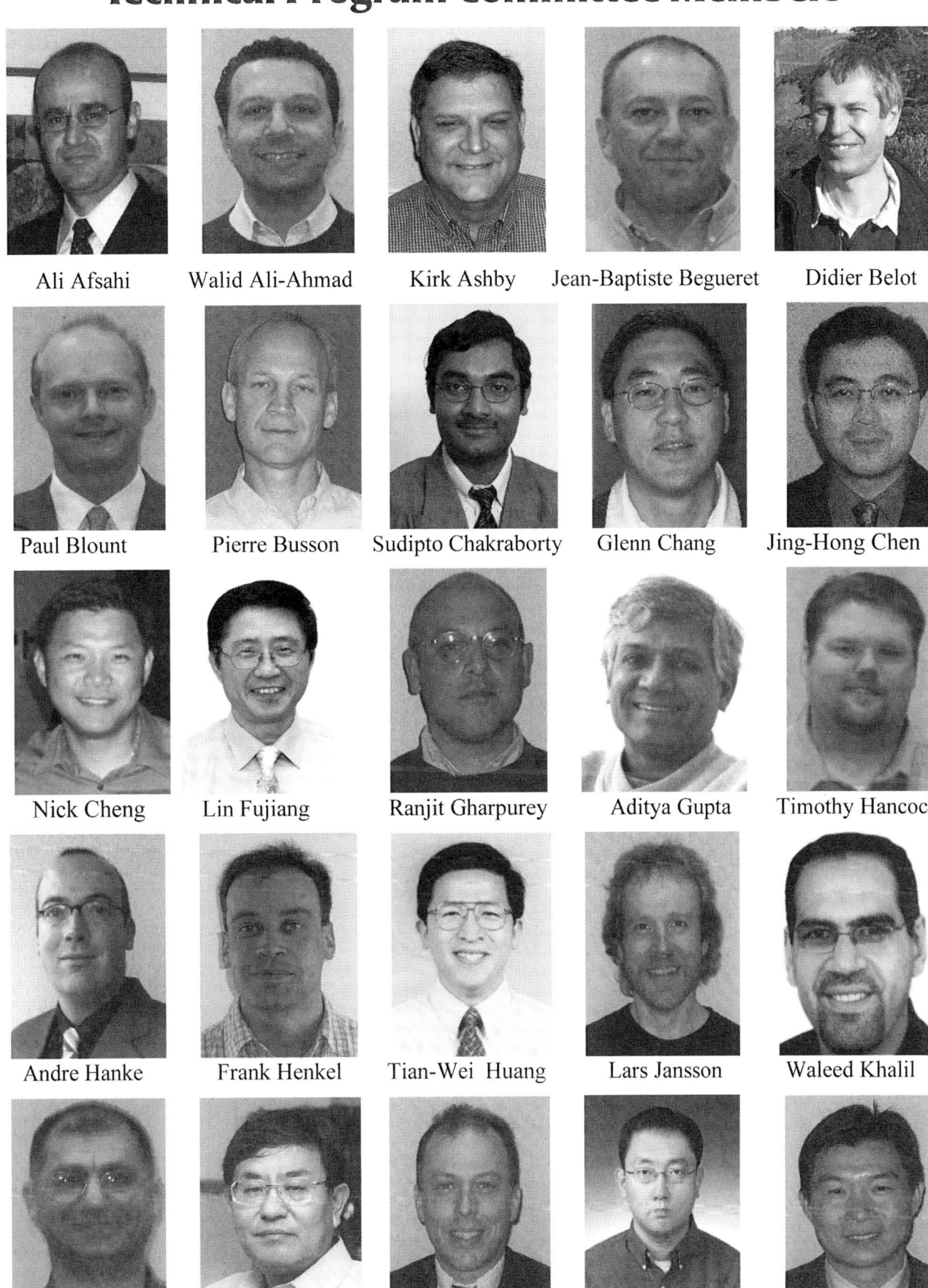

Ali Afsahi   Walid Ali-Ahmad   Kirk Ashby   Jean-Baptiste Begueret   Didier Belot

Paul Blount   Pierre Busson   Sudipto Chakraborty   Glenn Chang   Jing-Hong Chen

Nick Cheng   Lin Fujiang   Ranjit Gharpurey   Aditya Gupta   Timothy Hancock

Andre Hanke   Frank Henkel   Tian-Wei Huang   Lars Jansson   Waleed Khalil

Jaber Khoja   Bumman Kim   Larry Kushner   Youngwoo Kwon   Chang-Ho Lee

Technical Program Committee Members

 Domine Leenaerts

 Donald Lie

 Ting-Ping Liu

 Danilo Manstretta

 Srenik Mehta

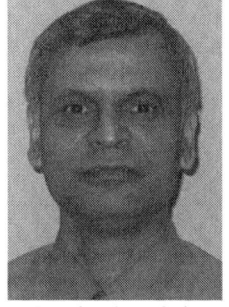 Jyoti Mondal

 Sule Ozev

 Stefano Pellerano

 Sanjay Raman

 Madhukar Reddy

 Bill Redman-White

 Eli Reese

 Mark Ruberto

 Francis Rotella

 Carlos Saavedra

 Derek Shaeffer

 Osama Shana'a

 Eddie Spears

 Robert Staszewski

 Bob Stengel

 Freek van Straten

 Julian Tham

 Bruce Thompson

 Haolu Xie

 Li-Wu Yang

Technical Program Committee Members

Patrick Yue   Gary Zhang   Herbert Zirath   Ken O (SSCS Liaison)

Conference Organization Staff

Takao Inoue   Rob Shaver   Larry Whicker
Website Chair   Paper Submission   Publications

RFIC PANEL SESSIONS

Monday, May 24, 2010
12:00 PM - 01:10 PM • Room 210CD

The Challenges, Competitions and Future Prospect of 60 GHz

Chair/Moderator:
> **SK Yong,** Marvell Semiconductor, Inc.
> **Myron Hattig,** Intel Corporation

Panelist and Affiliation:
> **C. Patrick Yue,** UC Santa Barbara

Panelists: **1. Jason Trachewsky,** Senior Technical Director and Broadcom
> Fellow, Broadcom Corporation
> **2. Scott Reynolds,** Manager, IBM TJ Watson
> **3. Myron Hattig,** Director of WLAN Standards, Intel Corporation
> **4. Raja Banerjea,** Principal Architect, Marvell Semiconductor, Inc.
> **5. Michiaki Matsuo,** Senior Manager/Chief Engineer, Panasonic Corporation
> **6. Jisung Oh,** Principle Engineer/Director, Samsung Electronics

Sponsor: **RFIC**

Panel Session Abstract: The ever growing demand for multi-gigabit data rates to support variety of new applications has pushed to the emergence of 60 GHz radio technology. Significant R&D work in the past decade have demonstrated the viability of wideband 60 GHz CMOS RFIC circuit and transceiver, which were difficult if not impossible to realize in the past, have now become a reality for commercialization. The momentum is further intensified by the heavily harmonized regulations and frequency allocation globally that allow higher EIRP limit and operation of huge unlicensed (i.e. 7 GHz) bandwidth in the 60 GHz band.

As a result, various standards (IEEE 802.11ad and IEEE 802.15.3c) and industry alliances (WirelessHD™ and WiGig Alliance) have emerged to deliver the promise of gigabit wireless solution. Multiple standard solutions could lead to two contradictory effects: On one hand, competition could lead to better 60 GHz products and drives the cost down towards commoditization. On the other hand, competition could create market confusion and co-existence issues among different products if not handled correctly. To date, among the various different standards, only 60 GHz products based on WirelessHD™ solution that supports wireless transmission of full HD contents has reported to hit the high end TV market in Jan 2009. Other 60 GHz products are under rigorous development and in the pipeline for productization. However, the question remains on their timeline in delivering the promise of gigabit experience to the customers.

In addition, Wi-Fi based IEEE 802.11n solution has started to enter the market for audio/video distribution on top of the widespread used of wireless Ethernet. Built strongly upon a broad ecosystem and interoperability among billions of Wi-Fi devices, Wi-Fi centric solution is set to evolve into gigabit data rate range with the recent development in IEEE 802.11.ac. This could potentially yet another solution that serves the similar applications and thus creates competition in market place with 60 GHz.

However, the distinct characteristics of Wi-Fi (2.4/5 GHz) and 60 GHz provide a different deployment perspective in which both technologies could be complementary rather than competing to each other. Such complementary technology requires multi-band radios that allow fast and seamless session transfer between them whenever the performance of the current radio deteriorates or an enhanced performance could be achieved.

In this panel, industry leaders and experts will discuss the challenges ahead of full scale commercialization of 60 GHz technology including implementation, tug-of-war among competitive standards, co-existence issues and future direction of 60 GHz.

Tuesday, May 25, 2010
12:00 PM - 01:10 PM • Room 210CD

Future of High-Speed I/O: Electrical, Optical, or Wireless?

Chair/Moderator:
Jacques C. Rudell, University of Washington

Co-Organizer:
Sam Palermo, Texas A&M

Panelists and Affiliations:
1. **Ali Hajimiri,** Caltech
2. **Byunghoo Jung,** Purdue University
3. **Jared Zerbe,** Rambus
4. **Sam Palermo,** Texas A&M
5. **Ronald Ho,** Sun Microsystems
6. **Daniel Kucharski,** Luxtera

Sponsor: **RFIC**

Panel Session Abstract: The rising power consumption associated with microprocessors realized in nanometer length silicon processes has placed a fundamental limit on core clock rates. This has lead to new advanced microprocessor architectures which seek to increase computational power by replicating the number of cores on a single die. Processors currently under development are estimated to use as many as 128 cores integrated on the same IC, leaving the routing of data via high-speed signaling from core-to-core, core-to-cache, or core-to-off-chip memory as a critical aspect of modern microprocessor performance. What is the future of high-speed I/O? Will the future demand for higher data rate I/Os come through incremental advances of all-electrical integrated transceivers, or will a new breed of high-speed I/O come to life in the form of either integrated optical (nanophotonics) transceivers, or perhaps mmWave wireless transceivers. Come hear a panel of experts debate what the future holds for high-speed signaling.

IMS LUNCH PANEL SESSIONS

Monday, May 24, 2010
12:00 PM - 01:10 PM • Room 210AB

Hubbert's Peak, The Coal Question, and Climate Change

Chair/Moderator:

David Rutledge, Tomiyasu Professor of Electrical Engineering,
California Institute of Technology

Tuesday, May 25, 2010
12:00 PM - 01:10 PM • Room 201AB

Silicon at THz Frequencies: A Reality or a Dream?

Chair/Moderator:

Prof. Gabriel M. Rebeiz, University of California, San Diego

Wednesday, May 26, 2010
12:00 PM - 01:10 PM • Room 210AB

Semiconductor Technology Impact on Microwave and Millimeter Wave Markets

Chair/Moderator:

Doug Lockie, Gigabeam

12:00 PM - 01:10 PM • Room 210CD

Standardizing Attributes for RF and Microwave Components and Assemblies – The Time Is Now?

Chairs/Moderators:

Dr. Chandra Gupta, Aeroflex-KDI, **Dr. Paul Khanna,** Phase Matrix

Thursday, May 27, 2010
12:00 PM - 01:10 PM • Room 210AB

On-Die Synthesized Inductors: Boon or Bane?

Chair/Moderator:

Jim Wight, Carleton University

12:00 PM - 01:10 PM • Room 210CD

RF GaN Reliability: Where Does the Technology Stand?

Chairs/Moderators:

Frank Sullivan, Raytheon, **Bernie Geller,** Vadum, Inc.

WORKSHOPS AND SHORT COURSES

Workshops and Tutorials are offered on Sunday, Monday and Friday of Microwave Week. They are distinguished by the following features:

- Advanced Level Workshops (designated as WSA, WSB, etc.) present the state of the art to specialists who are already experienced in the topic area.
- Tutorial Level Workshops (designated as TSB, RSC, etc.) are targeted towards educating attendees in new areas of microwave technology, reviewing material that is primarily a revision of previously published information.

All Workshops and Tutorials will be held at the Anaheim Convention Center. Specific room assignments will be announced at check-in.

RFIC AND IMS SPONSORED SUNDAY WORKSHOPS AND SHORT COURSES

08:00 AM-12:00 PM

WSA: Software Defined Radio for Microwave Applications.

Reviewed by: **MTT-9, MTT-20**
Organizers: **Jeffrey Pawlan,** Pawlan Communications;
Hermann Boss, Rohde & Schwarz

Abstract: Software Defined Radio (SDR) is the most significant innovation and change to radio communications since 1990. From HF to microwave frequencies, it allows the use of a fixed hardware platform to change bands, frequencies, and modulation types without any change in the hardware at all. This synergistic combination of analog and digital microwave hardware combined with software has significantly improved performance, allowed for great flexibility to ever-changing modulations and standards, shortened development time, and reduced cost. This workshop will make SDR understandable and applicable to microwave engineers.

It will begin with a clear explanation of how SDR works and its evolution through several generations of refinements. You will see and hear SDR in action with several live demonstrations of operating hardware and software along with test equipment. Your future work as a microwave engineer will be put in perspective with the current and future radio requirements.

Actual space communications SDR hardware and software will be demonstrated by the second speaker who is from JPL / NASA.

Cognitive Radio, a new related field, will be presented by a third speaker. SDR makes it possible to dynamically assess spectrum activity and change the modulation format to allow multiple signals to co-exist on the same frequency without interference or jamming.

This workshop will be practical and emphasize weak signal communications and commercial applications.

Speakers:
1. **Jeffrey Pawlan,** Pawlan Communications, "Software Defined Radio for Weaksignal and Commercial Applications"
2. **James Lux,** JPL, "The Adoption of SDR by NASA for Space Communications"
3. **Vasu Chakravarthy**, Air Force Research Laboratory/Sensors Directorate, "An Introduction to Cognitive Radio and its Implementation"

01:00 PM-05:00 PM

WSB: Advances in Filtering and Sampling for Integrated Transceivers

Reviewed by: **RFIC, MTT-9, MTT-20**
Organizers: **Tom Riley,** Kaben Wireless Silicon Inc.

Abstract: Blocker and interference filtering is a key issue in highly integrated Software Defined Radio (SDR) Receivers. If blockers can be removed prior to the ADC and conversion to the digital domain, power and area in the ADC can be greatly reduced. This workshop will show how Analog, sampled-time signal processing can be used to implement highly selective FIR, IIR and spatial FIR filters. N-path filtering can be used to design high bandwidth filters using low bandwidth analog components. Component mismatch, timing jitter and other sources of error that can affect receiver performance will be discussed. Linearity enhancement techniques for filters will be presented, as well as wideband RF front-end circuit techniques. For added blocker rejection, notched Delta-Sigma data converters are presented.

Following each speaker's presentation, the floor will be opened for interactive discussion with the audience.

Speakers:
1. **Tom Riley,** Kaben Wireless Silicon Inc., "Advances in Discrete-Time Analog Filtering"
2. **Bogdan Staszewski,** Technical University of Delft, "Discrete-Time Receiver"
3. **Asad Abidi,** University of California, Los Angeles, "A Discrete-Time Wideband Receiver for Software-Defined Radio"
4. **Dr. Martin Snelgrove,** Kapik Integration Inc., "Interference Mitigation in Receivers"

08:00 AM-05:00 PM

WSC: Interference, Noise and Coupling Effects in Modern SoC and SiP Products: Issues, Problems and Solutions

Reviewed by: **RFIC, MTT-6, MTT-12**
Organizers: **Jan Niehof,** NXP Semiconductors
Matthias Locher, ST-Ericsson
Oren Eliezer, Xtendwave

Abstract: The focus of this interactive workshop will be on resolving noise and self-interference problems: on-chip coupling effects, chip-package co-design, substrate issues, noise (inherent and external), coupling-aware RFIC floor planning, digitally assisted solutions for interference problems, EMC (chip and board level), design practices, and CAD/EDA modeling capabilities to effectively analyze and address these effects. Recognized companies and partnerships active in the semiconductor industry will present actual issues encountered in their designs and the solutions/design-practices used to address them, including key lessons learned. Interactive discussions will be facilitated to exchange valuable ideas for the benefit of participants and the industry at large.

Speakers:

1. **Nikos Haralabidis,** Broadcom, "Self-Interference in Multi-Standard RF SoC Transceivers"
2. **Dietolf Seippel,** Infineon, "Floor Planning of Complex Baseband-Radio SoCs in Consideration of Cross Talk Prevention"
3. **Matthias Locher,** ST-Ericsson, "A Bottom-Up Design and Verification Approach for Coexistence in Multi-System SoCs"
4. **Ayman Fayed,** Iowa State University, Oren Eliezer, Xtendwave, "System-level Methodology for the Power Management System Design in Complex SoCs: Minimize the Impact of Interference Through the Supply"
5. **Jonathan Jensen,** Intel, "Isolation and Coexistence Challenges – A Single-chip Bluetooth/WiFi Combo Example"
6. **Jan Niehof,** NXP Semiconductors, "Interference Issues and Coupling Effects in RF Products"
7. **Ravi Subramanian,** Berkeley Design Automation, "Advances in CAD: Simulation & Analysis of RF SoCs"

08:00 AM–05:00 PM

WSD: Ultra-Wideband (UWB) Technology: The State-of-the-Art and Applications

Reviewed by: **MTT-15, MTT-16**
Organizers: **Zhizhang (David) Chen,** Dalhousie University, Canada;
Hong (Jeffery) Nie, University of Northern Iowa, USA

Abstract: Since the FCC issued a Report and Order allowing license-free use of 0-960 MHz and 3.1-10.6GHz frequency bands in 2002, extensive research and development efforts have been made worldwide in utilizing these ultra-wide-band (UWB) frequency allocations for applications such as microwave imaging, high-speed short-range wireless communications, and wireless sensor networks. Despite a couple of setbacks in its commercialization, UWB technology and application continue to advance. More UWB algorithms and hardware design approaches are emerging. This workshop presents the latest developments of various UWB technologies, paving the way for ultimate realization of practical UWB systems. The workshop provide insight into (a) the operating principles and limitations of UWB systems, (b) design and test of UWB antennas, components, RF front-ends and transceivers, (c) the state-of-the-art in UWB signal processing algorithms, and (d) future trends in the UWB systems and their applications. This workshop will be beneficial to students, engineers and researchers who want to learn about the current status of UWB and related designs, tests and applications, and who want to follow and understand the recent developments and advanced applications of UWB.

Speakers:

1. **Dave Michelson,** University of British Columbia, Canada, "Deployment of UWB Wireless Systems in Industrial Environments"
2. **Zhining Chen,** Institute for Infocomm Research, Singapore, "Miniaturization of Ultra-Wideband Antennas"

3. **Ke Wu, Serge O. Tatu** and **Renato G. Bosisio,** École Polytechnique de Montréal, Canada, "Multi-Port Interferometers for UWB Transceiver Systems and Applications"
4. **Natalia Nikolova,** McMaster University, Canada, "Direct Methods for Detection and Imaging with Microwave Measurements in the Ultra-wide Band"
5. **Aly Fathy** and **Mohamed R. Mahfouz,** et. al., University of Tennessee, Knoxville, USA, "Recent Trends and Advances in UWB Positioning"
6. **Hong Nie,** University of Northern Iowa, USA, "Code Shifted Reference UWB transceiver and Its Applications for Intra-Vehicle Control and Communication"
7. **Zhizhang (David) Chen,** Dalhousie University, Canada"UWB Reference-based Impulse Radio Systems and Hardware Design Issues"

08:00 AM-05:00 PM
WSE: High Speed Signal Integrity Workshop

Reviewed by: **MTT-11, MTT-12**
Organizers: **Brett Grossman,** Intel Corporation
              Mike Resso, Agilent Technologies

Abstract: The triple play of voice, video, and data continues to demand ever greater bandwidths from devices and interconnects. This requirement is driving the challenges faced by the signal integrity engineer into a realm which may seem somewhat familiar to the microwave engineer. However, the challenges associated with frequency content, coupled with the density of signals, and the need to fit into relatively low cost consumer products, are a unique set of constraints which drive these solutions.

This signal integrity workshop will feature presentations which discuss practical case studies, as well as more fundamental and theoretical signal integrity research. You are welcome to attend and listen to industry and academic experts describe several of the latest developments in the field of high speed signal and power integrity.

Speakers:
1. **Paul Huray,** University of South Carolina, "Bridging the Gap"
2. **Michael Hill,** Intel Corporation, "Microprocessor Power Integrity – Metrologies and Future Challenges"
3. **Heidi Barnes,** Verigy, "The Art of VNA Calibrations for Measuring Low Loss PCB Components"
4. **Matthew Claudius,** Intel Corporation, "End Use Model Correlation"
5. **Bob Schaefer,** Agilent Technologies, "Comparison of Fixture Removal Techniques for Connector and Cable Measurements"
6. **Jim Rautio,** Sonnet Software "Measurement and Analysis of Substrate Dielectric Constant Anisotropy"
7. **Evan Fledell,** Intel Corporation, "Passive Interconnect Frequency Domain Characterization for Mixed-Medium and Vertical Interconnect Systems"
8. **Leung Tsang,** University of Washington, "Electromagnetic Modeling of High Speed Vertical Interconnect on Chip-Package-Board"

08:00 AM–05:00 PM

WSF: GaN for High Power, High Bandwidth Applications, Finally Fulfilling the Promise

Reviewed by: **MTT-5, MTT-6, MTT-7**
Organizers: **Bill Vassilakis,** Empower RF Systems
David W. Runton, RF Micro Devices

Abstract: GaN circuits have, for a long time, promised to enable amplifier applications that have not been possible with GaAs or LDMOS such as higher temperatures of operation, large operating bandwidths, and higher operating power. Material quality problems have slowed the progress on the delivery of such applications leaving many wondering when GaN would displace incumbent technologies. GaN has also defied declining cost trends of semiconductors, due to higher processing costs, smaller wafers, and lower yields. In the commercial market, dollars per watt delivered has long dominated in the selection of technology. Other factors such as efficiency, the ability to pre-distort, and linearity have been secondary. GaN is at last emerging as a serious contender for both commercial and military applications, as we see more demand for power, efficiency and larger bandwidths of operation. As other technologies are reaching inherent limits, GaN is finally ready for prime time.

Speakers:
1. **Norihiko Ui,** Sumitomo Electric Device Innovations, "Power and High Efficiency GaN-HEMTs for Cellular Base Station Applications"
2. **Oualid Hammi** and **Fadhel M. Ghannouchi,** iRadio Lab, Department of Electrical and Computer Engineering, University of Calgary, Canada, "Power Amplifiers for Wireless Communication Infrastructure"
3. **Dr. James J. Komiak**, BAE Systems Electronic Solutions, "Progress in High Power GaN HEMT Power Amplifiers for Wideband Applications"
4. **Simon Wood,** Cree Inc., "Trends in high power GaN transistors and MMICs"
5. **Bumman Kim,** Pohang University of Science and Technology, "Highly Efficient Saturated Power Amplifier based on GaN – A class P amplifier"
6. **David W. Runton**, RF Micro Devices, "Defining Application Spaces for High Power GaN"
7. **Rik Jos,** NXP Semiconductors, "GaN HEMT and their Commercial RF Power Applications"

08:00 AM-12:00 PM

WSG: MOSFET Modeling for RFIC Design Based On the Industry-Standard PSP Model

Reviewed by: **RFIC, MTT-6**
Organizers: **Kevin McCarthy,** University College Cork;
Weimin Wu, Arizona State University

Abstract: This workshop will present an overview of the state-of-the art in MOSFET modeling for the design of CMOS Radio Frequency ICs using modern nanometer-scale CMOS. It focuses on the industry-standard PSP (MOSFET) and MOSVAR (varactor) models. The workshop will review the fundamentals of both models and demonstrate the highly-accurate RF simulation capabilities they provide for RFIC designs. The workshop will also show how the PSP model can be extended to SOI and Multi-Gate devices, which will become of increasing importance to RFIC design.

Speakers:
1. **Gert-Jan Smit,** NXP Semiconductors, "The PSP Compact MOSFET Model: Physical Background and Benefits for RFIC Design"
2. **Brandt Braswell,** Freescale Semiconductor, "Deployment of an Advanced MOSFET Model in an Industrial Context"
3. **James Victory,** Sentinel IC Technologies, "MOSVAR – A PSP-Derived MOS Varactor Model"
4. **Weimin Wu,** Arizona State University "PSP-Based Modeling of SOI and Multi-Gate MOS Devices"

08:00 AM-05:00 PM

WSH: Power Management for Integrated RF Circuits: Challenges and Solutions

Reviewed by: **RFIC, MTT-6**
Organizers: **Ayman Fayed,** Iowa State University
Waleed Khalil, Ohio State University
Oren Eliezer: Xtendwave

Abstract: The recent expansion in the use of mobile communications and multi-media devices has fueled the demand for various wireless/RF transceivers to be integrated in a single SoC with the digital processing circuitry and power management functions. As battery life in mobile devices is critical, and with these transceivers typically not operating directly from the battery, regulating and delivering power to them in an efficient manner is becoming a bottleneck. Since power delivery efficiency and implementation cost on one hand, and noise and regulation quality on the other hand are two contradictory factors in traditional power management circuits, RF loads present a great challenge due to their high sensitivity to their power supply quality. This workshop will discuss the challenges and tradeoffs that power management designers have to make when designing for RF loads while maintaining high efficiency and cost-effectiveness.

Speakers:
1. **Ayman Fayed,** Iowa State University, "Challenges in Integrated Power Management for Analog, RF, and mixed-signal SoCs"
2. **Keith Kunz,** Texas Instruments, "Integrated DC-DC Converters in Nanometer CMOS RF SOCs"
3. **Bertan Bakkaloglu,** Arizona State University, "Low-noise Switched-mode and Low-dropout Linear Regulators for RF Applications"
4. **Siamak Abedinpour,** Freescale, "An Overview of Integrated Power Management Circuits for Portable RF applications"
5. **Sam Palermo,** Texas A&M University, "Supply Regulation Techniques for Frequency Synthesizers"
6. **David Allstot, Jeffery Walling,** University of Washington, "Supply Regulators in Class-E/G/H CMOS Power Amplifiers"
7. **Ram Sadhwani,** Intel, "Direct Powering of RF and Analog Circuits from DC-DC Converters"
8. **Ahmed Emira,** Newport Media, "DC-DC Converters Noise Considerations in RF SoCs"

08:00 PM-05:00 PM
WSI: Substrate Integrated Circuits (SICs)

Reviewed by: **MTT-8, MTT-12**
Organizers: **Maurizio Bozzi,** University of Pavia, Italy
Ke Wu, Ecole Polytechnique (Université de Montréal), QC, Canada

Abstract: Substrate integrated circuits (SICs) are probably the most promising candidate for the design and implementation of low-cost and high-density millimeter-wave integrated circuits and systems in the next decades. SICs, which integrate planar and non-planar structures together, are able to offer a compact, low-loss, flexible, high integration density, and cost-effective solution for integrating active circuits, passive components and radiating elements on the same substrates including multilayered geometries regardless of technological platforms such as PCB, LTCC, MHMICs, MMICs and even CMOS processes. In this way, the concept of System-in-Package (SiP), widely adopted in the design of RF/microwave circuits, can be extended to System-on-Substrate (SoS) for up-higher frequency ranges. This technological concept can be extended to terahertz and optoelectronic domains.

The aim of this workshop is to provide an overview of the current trends of research and development in the field of SICs, including modeling methods, innovative structures, design techniques and technological issues.

Speakers:
1. **Ke Wu,** Ecole Polytechnique (Université de Montréal), Canada, "State-of-the-art and Future Perspective of Substrate Integrated Circuits"
2. **Tatsuo Itoh,** University of California, Los Angeles, USA , "Progress in Composite Right/Left Handed Structures based on Substrate Integrated Waveguide"
3. **Vicente E. Boria-Esbert,** Polytechnic University of Valencia, Spain, "Computer-Aided Design Tools of Passive Circuits in Substrate Integrated Waveguide Technology"
4. **Jens Bornemann,** University of Victoria, Canada, "Multilayered Substrate-Integrated Waveguide Couplers"
5. **Maurizio Bozzi,** University of Pavia, Italy, "Full-Wave Analysis and Equivalent-Circuit Modeling of SIW Components"

6. **Ruey-Beei Wu,** National Taiwan University, Taiwan, "Development of LTCC mm-wave Passive Components for SoP Wireless Applications"
7. **Apostolos Georgiadis,** Centre Tecnològic de Telecomunicacions de Catalunya, Spain, "Oscillator and active antenna design in SIW technology"
8. **Roberto Vincenti Gatti,** University of Perugia, Italy, "SIW Components and Solutions for Large Electronic Beam Steering Arrays"
9. **Stepan Lucyszyn,** Imperial College, United Kingdom, "Substrate Integrated Metal-Pipe Rectangular Waveguides"

08:00 AM-05:00 PM

WSJ: Re-configurable Multi-Radios at the Nanoscale

Reviewed by: **RFIC, MTT-6, MTT-20**
Organizers: **Gernot Hueber,** DICE, Linz, Austria
Robert Bogdan Staszewski, Delft University of Technology, The Netherlands
Stefan Heinen, RWTH Aachen University, Aachen, Germany

Abstract: Advances in CMOS fabrication technology have enabled the use of CMOS in today's RF transceivers for wireless communications. Multi-band and multi-mode radios covering the diversity of communication standards from 2G GSM, 3G UMTS, to 4G LTE impose unique challenges on the RF-transceiver design due to limitations of reconfigurable RF components that meet the demanding cellular performance criteria at costs that are attractive for mass market applications. Nanoscale CMOS on one hand features the possibility for implementing a significant computational power and complex functionality directly on a single IC, on the other hand it shows poor raw performance or RF circuits compared to other technologies. The focus of this workshop is on the challenges the cellular standards pose on future multi-radio integration in nanoscale CMOS, along with a thorough discussion of advanced techniques for receivers and transmitters towards integration in a multi-radio SoC or SiP. Approaches include novel architectures, highly configurable analog circuitry, digitally assisted and enhanced analog/RF modules and the integration of digital signal processing into the traditionally purely analog front-ends.

Speakers:
1. **Gernot Hueber,** DICE, Austria, "Flexible RF Transceivers for 4G Systems"
2. **Ali M. Niknejad,** UC Berkeley, CA, " High Dynamic Range Wide Bandwidth Building Blocks for Multi-Mode CMOS"
3. **Vito Giannini,** IMEC, Leuven, Belgium, "The Green-Scalable Revolution of Nanoscale Software-Defined Radios"
4. **Jaques C. Rudell,** University of Washington, WA, "Nanometer CMOS Transceiver Design Enters the Era of "Co-Existence" and the SDR"
5. **Hooman Darabi,** Broadcom, Irvine, CA, "Radio Architectures for 2/3/4G Highly Integrated Cellular Applications"
6. **Francois Rivet,** IMS Lab, University of Bordeaux & Atlantic Innovation ES, France, "Towards Software Radio Receiver"
7. **Ali Hajimiri,** Caltech, CA, "Electromagnetically Reconfigurable Radios: Antenna Meets Digital"

8. **Frank Op 't Eynde,** Audax-Technologies Ltd., Wilsele, Belgium, "Unsolved Issues in SDR RF Frontends"

9. **Larry Larson,** University of California, San Diego, CA, "Low-Power Transmitters in Nanoscale CMOS"

10. **Robert Bogdan Staszewski,** TU Delft, The Netherlands, "Advances in Digital RF Architectures"

08:00 AM–05:00 PM

WSK: Multi-Mode Front End Design Challenges and Techniques

Reviewed by: **RFIC, MTT-6, MTT-20**
Organizers: **Edward Spears,** RFMD
                  Nick Chang, Skyworks Solutions

Abstract: With the proliferation of data services, mobile device original equipment manufacturers (OEMs) are presented with new, unprecedented challenges and demands from both mobile operators and consumers. Mobile operators require customized handsets and mobile devices to meet various consumer roaming needs, and the issue of rapid customization has fallen to OEMs who must configure these complex 3G devices to function in multiple frequency bands and operating modes (GSM, EDGE, WCDMA, HSPA+, with LTE on the horizon). As the number of bands and band combinations grow, frequency flexibility and signal routing at the platform level have increased in importance as critical parameters for 3G mobile device development. This sets up an unprecedented challenge for front-end suppliers who are challenged to design a broad portfolio of high-performance, multi-band, multimode front ends and components that offer frequency flexibility, ease of implementation, size reduction, and low current consumption. Presentations in this workshop will focus on the design challenges to meet these multi-mode front end requirements along with the required advancements in device technology and design techniques to meet the overall bandwidth and efficiency requirements. Design techniques of linearization, efficiency enhancement, power detection and controls will be covered in design examples utilizing various technologies such as GaAs HBT, CMOS, Silicon-on-Insulator and Silicon Germanium.

Speakers:
1. **Ville Vintola,** Nokia, "OEM Prospective for Multi-mode Solutions"
2. **Ray Arkiszewski,** RFMD, "GaAs HBT Multi Mode Amplifiers"
3. **David Ripley,** Skyworks Solutions, "Multi-mode, Multiband Power Amplifiers and Serial Bus Interface Standards"
4. **Larry Larson,** University of California at San Diego, "Design Techniques for Broadband Efficient Linear Power Amplifiers for Multi-Mode Wireless Applications"
5. **Dan Nobbe,** Peregrine, "Multimode Antenna Switch Modules"
6. **Nadim Khlat,** RFMD, "Tunable Front Ends Performance Benefits"
7. **Pasi Tikka,** Epcos, "Multimode Filter and Switch Modules"

08:00 AM-05:00 PM

WSL: Silicon-Based Technologies for Millimeter-Wave Applications

Reviewed by: **MTT-6, MTT-16, RFIC**
Organizers: **Jitendra Goel,** Raytheon Company
Lance Wei-Min Kuo, Raytheon Company
Didier Belot, STMicroelectronics
Eric Kerhervé, IMS Lab
Georg Boeck, Berlin Institute of Technology

Abstract: Traditionally, millimeter-wave (MMW) circuits utilizing only III-V technologies have been employed in low-volume, high-performance products. With the recent progress of highly scaled Si-based (SiGe and CMOS) technologies achieving f_t and f_{max} beyond 200 GHz, the application space of Si-based technologies has broadened from digital, analog, RF, and microwave domains to include MMW applications. The workshop will focus on MMW applications such as imaging (94 GHz and 140 GHz), automotive radar (LRR at 77 GHz and SRR at 79 GHz), and wireless high data rate communications (W-HDMI at 60 GHz). It gives an overview of recently developed architectures, circuit design techniques, and antenna configurations to meet the demanding performance specifications of MMW applications.

Speakers:
1. **Ali Hajimiri,** California Institute of Technology, "Si Millimeter-Wave Systems"
2. **Gabriel M. Rebeiz,** University of California at San Diego, "Ultra-Low Power Millimeter-Wave Phased Arrays and Gbps Communications Systems Using On-Chip Antennas"
3. **M. C. Frank Chang,** University of California, Los Angeles, "60-130 GHz Circuit/System Developments Based on Super-Scaled CMOS"
4. **Tian-Wei Huang** and **Huei Wang,** National Taiwan University, "Millimeter Wave Broadband Multi-Gigabit CMOS Transceiver Design"
5. **Scott K. Reynolds,** IBM T. J. Watson Research Center, "Millimeter-Wave Circuits and Systems Work at IBM Research"
6. **Piet Wambacq,** IMEC "CMOS Radio Integration for High-Data rate 60 GHz Applications"
7. **Ali Niknejad,** UC Berkeley, "mm-Wave Medical Imaging Using a 94 GHz Time-Domain Ultrawideband Synthetic Imager (TUSI)"
8. **Ullrich Pfeiffer,** University of Wuppertal, "Silicon Process Technologies for Emerging Terahertz Applications"
9. **Pierre Busson,** STMicroelectronics, "60 GHz W-HDMI Transceiver"
10. **Joy Laskar,** Georgia Tech, "mmW Digital CMOS Radio Solutions for Ultra-Low Power, High Resolution Sensing and High Bandwidth Connectivity"
11. **Cathia Laskin,** University of Toronto, "140 GHz Imaging"

08:00 AM-05:00 PM

WSM: RF Packaging Solutions for Wireless Communication Platforms

Reviewed by: **MTT-12, MTT-20, RFIC**
Organizers: **Telesphor Kamgaing,** Intel Corporation
Vijay Nair, Intel Corporation
Clemens Ruppel, TDK-EPC

Abstract: In order to satisfy the decreasing form factor and increasing functionality demand from novel devices such as netbooks and smartphones, it is imperative to create a platform, where different radios and digital logic have to co-exist. This ultimate goal can only be achieved by overcoming various significant challenges at the silicon, packaging and testing levels. This full day workshop will focus on recent research and development work that will enable future ultra-small form factor computing and communication devices that incorporate one or multiple radios on the same platform. Various technology ingredients and packaging solutions for 60GHz, WiFi, WiMAX, Bluetooth, GPS and 3G/4G radios among others will be addressed by leading industrial and academic experts in the field.

Speakers:
1. **Vijay Nair,** Intel Corporation, "Multi-protocol Multi-radio Wireless Platform Integration Challenges"
2. **Joy Laskar,** Georgia Institute of Technology, "Development of Millimeter-Wave QFN: CMOS, PCB and Phased Array"
3. **Anh-Vu Pham,** University of California, Davis, "Development of Ultra-small Wireless Passive Modules Using 3-D Organic Metamaterials"
4. **Telesphor Kamgaing,** Intel Corporation, "Package Level Realization of Passives for Multiradio Wireless Modules"
5. **Clemens Ruppel,** TDK-EPC, "Front-End Integration for Multi-Band, Multi-Standard Mobile Phones Based on LTCC"
6. **William Chappell,** Purdue University, "Silicon on Silicon Packaging Using Self-aligned Interconnects"
7. **Walter De Raedt,** IMEC, "3D Heterogeneous Integration Techniques for Wireless Devices"
8. **Kevin Slattery,** Intel Corporation, "RF Interference in Small Form Factor Devices"

08:00 AM–05:00 PM
WSN: The State of Art of Microwave Filter Synthesis, Optimization and Realization

Reviewed by: **MTT-8, MTT-16**
Organizers: **Ming Yu,** COM DEV, Canada
John Bandler, McMaster University

Abstract: Today systems require increasingly sophisticated microwave filters and multiplexers. The designer often faces the challenges of compromising between several contrasting requirements. This workshop will present a comprehensive overview of the state of the art of microwave filter synthesis, optimization and realization. Recent advances in some of the most promising application areas of microwave filters; innovative solutions concerning both design approaches and technological achievements will also be presented.

Speakers:
1. **Dick Snyder,** RS Microwave, USA, "Phase Shift, Delay, Anomalous Dispersion, and Meta-Materials: Implications for Future Filter Designs"
2. **John W. Bandler,** McMaster University, Canada, "Advanced Optimization Techniques for Modern Filter Design – From Newton to Space Mapping"
3. **Smain Amari** and **F. Seyfert,** RMC and INRIA, France, "New Development in the Synthesis and Design of Microwave Filters of Arbitrary Bandwidth"
4. **K Zaki, C. Wang,** University of Maryland, USA, "Dielectric Resonator and LTCC Filters"
5. **Jen-Tsai Kuo,** National Chiao Tung University, Taiwan, "Microwave Planar Filter Technologies"
6. **G. Macchiarella** and **S. Tamiazzo,** Politecnico di Milano, Milano, Italy, "Advanced Filter Technologies for Wireless Base Stations"
7. **Ming Yu,** COM DEV, Canada, "Advanced Filter/Multiplexer Technologies for Satellite Transponders"
8. **Ian Hunter,** University of Leeds, UK, "Advanced Tunable and Reconfigurable Filters"
9. **Vicente E. Boria-Esbert, Carlos P. Vicente-Quiles,** University of Valencia, Spain "Prediction Models of RF Breakdown Effects in Passive Components for Satellite Payloads"

IMS SPONSORED MONDAY WORKSHOPS

08:00 AM-05:00 PM
WMA: SiGe HBTs Towards THz Operation

Reviewed by: **MTT-4, MTT-7, MTT-11**
Organizers: **Paulius Sakalas,** TU Dresden, Germany, and FPL Semiconductor
Physics Institute, Lithuania; **Michael Schroter,** RFnano Corporation and
UC San Diego, USA

08:00 AM-05:00 PM
WMB: Advances in Photovoltaic Solar Cell Technology and its Possible Applications in Microwave Communications Systems as an Energy Source

Reviewed by: **MTT-4, MTT-10, MTT-16**
Organizers: **Aly E. Fathy,** University of Tennessee; **Samir El-Ghazaly,** National Science Foundation;
Fuad Abulfotuh, University of Alexandria

08:00 AM-05:00 PM
WMC: Recent Advancements and Challenges in mm-Wave Applications and Systems

Reviewed by: **MTT-6, MTT-12, MTT-16**
Organizers: **Amin Rida,** Georgia Institute of Technology; **Manos Tentzeris,**
Georgia Institute of Technology; **ae-Seung Lee,** Toyota Research Institute North America

08:00 AM-05:00 PM
WMD: New Microwave Devices and Materials Based on Nanotechnology

Reviewed by: **MTT-15, IMS2010**
Organizers: **Luca Pierantoni,** Università Politecnica delle Marche, Ancona, Italy; **Fabio Coccetti,**
LAAS-CNRS Toulouse, France; **Christophe Caloz,** École Polytechnique de Montréal,
Montréal, Canada; **George W. Hanson,** University of Wisconsin-Milwaukee, WI, USA

08:00 AM-12:00 PM

WME: High-Power-Density Packaging of Gallium Nitride

Reviewed by: **MTT-5, MTT-6, MTT-12**
Organizers: **Rüdiger Quay,** Fraunhofer Institute Applied Solid-State Physics, Freiburg; **Bernie Geller,** Vadum Inc, North Carolina; **Frank Sullivan,** Raytheon Company

01:00 PM-5:00 PM

WMF: High Efficiency High Power Microwave Amplifiers for High Data Rate Space Communications

Reviewed by: **MTT-5, MTT-7, MTT-16**
Organizers: **Dr. Kavita Goverdhanam,** U.S. Army – CERDEC, Fort Monmouth, NJ; **Dr. Rainee N. Simons,** NASA Glenn Research Center, Cleveland, OH

08:00 AM-12:00 PM

WMG: Ultra-high Speed Microwave and Photonic Devices and Systems: How Will They be Tested?

Reviewed by: **MTT-3, MTT-11**
Organizers: **Stavros Iezekiel,** University of Cyprus; **Ron Reano,** Ohio State University

08:00 AM-05:00 PM

WMH: 3D Microwave and Millimeter-Wave Packaging

Reviewed by: **MTT-6, MTT-12**
Organizers: **John A. Pierro,** Telephonics Corporation, USA ; **Debabani Choudhury,** Intel Corporation, USA

08:00 AM-05:00 PM

WMI: Making Reliable Measurements at Millimeter and Submillimeter Wavelengths

Reviewed by: **MTT-4, MTT-11**
Organizers: **Nick Ridler,** National Physical Laboratory (NPL), UK; **Andrej Rumiantsev,** SUSS MicroTec Test Systems GmbH, Germany

08:00 AM-05:00 PM

WMJ: Recent Advances in Reconfigurable Filters

Reviewed by: **MTT-8, MTT-21**
Organizers: **Dimitrios Peroulis,** Purdue University; **Raafat Mansour,** University of Waterloo

08:00 AM-05:00 PM

WMK: RF MEMS for Antennas and Integrated RF Front End

Reviewed by: **MTT-14, MTT-16, MTT-21**
Organizers: **John Papapolymerou,** Georgia Institute of Technology; **Art Morris,** WiSpry; **Hector De Los Santos,** NanoMEMS Research; **James C. Hwang,** Lehigh University

IMS SPONSORED FRIDAY WORKSHOPS

08:00 AM-12:00 PM
WFA: The Expanding Role of GaN in RF Systems

Reviewed by: **MTT-5, MTT-6, MTT-16**
Organizers: **Jim Sowers,** Space Systems/Loral; **Jay Banwait,** Northrop Grumman

08:00 AM-12:00 PM
WFB: Wireless Power Transmission

Reviewed by: **MTT-5, MTT-16, IMS2010**
Organizers: **Debabani Choudhury,** Intel Corporation; **John A. Pierro,**
Telephonics Corporation

08:00 AM-05:00 PM
WFC: Millimeter-Wave SiGe/CMOS and III-V Chips for Imaging Systems

Reviewed by: **MTT-4, MTT-6, MTT-16**
Organizers: **Gabriel M. Rebeiz,** University of California, San Diego;
Sorin Voinigescu, University of Toronto; **Vipul Jain,** Sabertek

08:00 AM-05:00 PM
WFF: New Theories, Applications and Practices of Electromagnetic Field Simulators

Reviewed by: **MTT-1, MTT-15**
Organizers: **Zhizhang (David) Chen,** Dalhousie University, Canada;
Poman So, University of Victoria, Canada

08:00 AM-05:00 PM

WFG: Emerging Optical Modulator Technologies for RF Photonics

Reviewed by: **MTT-3, MTT-16, IMS2010**

Organizers: **Ronald M. Reano,** Ohio State University; **Dieter Jäger,** Universität Duisburg-Essen

08:00 AM-12:00 PM

WFH: How to Start a Microwave Business

Reviewed by: **MTT-19, IMS2010**

Organizers: **Fred Schindler,** RF Micro Devices; **Mike Golio,** Golio Pubs

08:00 AM-12:00 PM

WFI: Practical Metamaterial RF and Antennas for Commercial Application

Reviewed by: **MTT-15, MTT-16, MTT-20**

Organizers: **Maha Achour,** RAYSPAN Corporation

IMS/ATRFG SPONSORED WORKSHOPS AND SHORT COURSES
(all workshops held in ACC)

Full Day   08:00am-05:00pm

Coding	Room	Day	Description
SC-1	TBD	Sun	**Theory and Design of Phase Locked Loops** *Instructors:* **L.Dayaratna,** Lockheed Martin; **Dean Banerje**e, National Semiconductor; **Cicero S. Vaucher, NXP** Semiconductors; **P. White,** Applied Radio Labs; **Ron Reedy,** Peregrine Semiconductor
SC-2	TBD	Sun	**Low Phase Noise Oscillators: Lecture (Theory and Design) and Laboratory** *Instructor:* **Jeremy K.A. Everard,** BAE Systems/Royal Academy of Engineering Research Professor in Low Phase Noise Signal Generation, Department of Electronics, University of York, UK.

Half Day   08:00am-12:00pm

Coding	Room	Day	Description
SC-2A	TBD	Sun	**Low Phase Noise Ocillators: Lecture Only** *Instructor:* **Jeremy K.A. Everard,** BAE Systems/Royal Academy of Engineering Research Professor in Low Phase Noise Signal Generation, Department of Electronics, University of York, UK.

Half Day   01:00pm-05:00pm

Coding	Room	Day	Description
SC-3	TBD	Sun	**Mictowave Packaging and Manufacturing 101** Organizer: **lan Lindner,** L-3 Communications, Narda Microwave West MTT Affiliation: MTT-12

SOCIAL EVENTS

SUNDAY, MAY 23, 2010

RFIC Reception • 07:00pm-09:00pm
Anaheim Convention Center, Room 213BCD

Immediately following the RFIC Plenary Session is the RFIC Reception to be held in adjacent ROOM 213BCD at the Anaheim Convention Center. This social event is a key component of the RFIC Symposium, providing an opportunity to connect with old friends, make new acquaintances, and catch up on the wireless industry. Admittance is included with RFIC Symposium registration. Additional tickets can also be purchased separately at registration.

MONDAY, MAY 24, 2010

IMS 2010 Welcome Reception • 06:00pm-08:00pm
Hilton Hotel, Sunset Deck

All Microwave Week attendees and exhibitors are invited to attend a reception hosted by IMS 2010.

TUESDAY, MAY 25, 2010

Special Luncheon for Chuck Swift • 12:00pm-02:00pm
Anaheim Convention Center, Room AR1 & 2

The IMS 2010 will hold a Special Luncheon on May 25, 2010, to celebrate Chuck Swift's 52 years of service in support of the Los Angeles Chapter of the Microwave Theory and Techniques Society. The chapter has sponsored monthly technical meetings since 1952 and periodic national/international 3-5 day meetings since 1970, such as the IMS meeting being held in Anaheim on May 23-28, 2010. Chuck formed his business, C. W. Swift & Associates, in July 1958. Since 1958, Chuck, his business and his family have supported 450 meetings and 7 IMS symposia. IMS 1989 stands out as the best performance, where Chuck put on a show of shows. The Luncheon will be held on Tuesday May 25 from 12:00 (noon) to 2:00 pm, in the Convention Center, in Rooms AR1 & 2 (near the Arena). The Luncheon is a full sit-down lunch.

Admission is $35 per person; sign-up is through the IMS 2010 Registration.

Women in Microwaves Engineering (WIM) Reception • 05:30pm-07:30pm
Uva Bar, 1580 Disneyland Drive, Downtown Disney

Meet with old friends as well as make new connections to the growing community of women who make a career in the field of high-technology. Enjoy good food, cool beverages and warm conversation at the WIM Social Event. Join us at the outside patio area of the Uva Bar in the center of the Downtown Disney entertainment district.

Student Reception • 07:00pm-09:00pm
Hilton Hotel, Room California B

Mix and mingle with fellow students from across the globe!

SOCIAL EVENTS (continued)

TUESDAY, MAY 25, 2010 (continued)

Ham Radio Social – 06:00pm-09:00pm
Hilton Hotel, California A

While enjoying a buffet and open bar, the attendees will have the opportunity to see the accomplishments of amateur radio operators who have skillfully designed and built transceivers for use from VHF to high millimeter wave bands. Some of these transceivers were made from surplus and commercially available components and some are state-of-the-art new designs including SDR. Several will be on display and their builders will be there to answer questions.

All conference attendees are welcome. You will find that amateur radio operators are utilizing their allocated frequency spectrum for very important uses and you may be interested in obtaining your license so you too can test your new designs and microwave propagation.

WEDNESDAY, MAY 26, 2010

Industry Hosted Cocktail Reception • 05:00pm-6:00pm
Anaheim Convention Center Exhibition Floor

Symposium Exhibitors will host a cocktail reception. Complimentary beverage tickets will be included in the registration packages.

MTT-S Awards Banquet • 07:00pm-10:00pm
Hilton Hotel, California Room

The MTT-S Awards Banquet includes a fine dinner, major society awards presentation, and entertainment. This years entertainment will be provided by String Theory. String Theory is an exceptional music performance drawing on the very space of the performance by transforming architecture into musical instruments and then playing the building. The result is a visually stunning landscape in which the performance unfolds. Tickets can be purchased at the time of registration.

THURSDAY, MAY 27, 2010

MTT-S STUDENT AWARDS LUNCHEON • 12:00pm-02:00pm
Hilton Hotel, California B

All students are invited to attend this luncheon which recognizes recipients of the MTT-S Undergraduate Scholarships, MTT-S Graduate Fellowships, IMS2010 Student Volunteers, IMS2010 Student Paper Awards, and the winners and participants of the IMS2010 Student Design Competitions.

MTT-S Graduates of the Last Decade (GOLD) Reception • 05:30pm-07:00pm
300 Anaheim

IEEE Graduates of the Last Decade (GOLD) was created in 1996 as a membership program to help students transition to young professionals within the larger IEEE community. MTT-S GOLD activities began at the IMS2007 meeting in Honolulu, HI. GOLD makes up approximately 10% of the MTT-S population and are a valuable part of the community. Join us for food, beverages and bowling at 300 Anaheim! Directions and additional information can be found at http://www.threehundred.com/anaheim.html.

GUEST PROGRAM

HOSPITALITY SUITE: Enjoy Southern California hospitality by joining us in the Hospitality Suite in the Garden Room of the Sheraton Hotel. Grab your guest badge and come have breakfast in the morning or snacks later in the day. Meet friends, make friends, kick back and relax. There will be a special area for children with toys and games. Guest Tours will depart from the Hospitality Suite.Open Sunday, May 23 through Thursday, May 27 7:30am to 3:30pm.

GUEST TOURS / RECREATIONAL ACTIVITIES

Guests can register for the tours both online and at the Hospitality Suite located in the Garden Room at the Sheraton Park Hotel. To register for tours online please visit: http://www.pra-tours.com/IEEE.

LOS ANGELES COUNTY MUSEUM OF ART, PETERSEN AUTOMOTIVE MUSEUM & LA BREA TAR PITS
Sunday, May 23, 2010

Suggested Itinerary

10:00 AM	Depart Sheraton Park Hotel
11:30 AM	Arrive LACMA and Petersen – Free time to explore both museums
1:30 PM	Lunch at LACMA
2:30 PM	Arrive Page Museum and La Brea Tar Pits- guided tour
4:00 PM	Depart Museum
5:30 PM	Return to Sheraton Park Hotel

Time: **10:00 AM-5:30 PM**　　　　　　　　　　　　　　Price: **$125 per person**

NEWPORT HARBOR CRUISE
Sunday, May 23, 2010

Suggested Itinerary

1:30 PM	Depart Sheraton Park Hotel
2:00 PM	Newport Harbor Cruise
2:45 PM	Free time to explore Balboa Island
4:00 PM	Board coach and depart for hotel
4:30 PM	Return to Sheraton Park Hotel

Time: **1:30 PM-4:30 PM**　　　　　　　　　　　　　　Price: **$64 per person**

QUIET ON THE SET!
Monday, May 24, 2010

Suggested Itinerary:

9:00 AM	Depart Sheraton Park Hotel
9:45 AM	Arrive Warner Bros. Studio
10:00 AM	Watch film on the Studio's history
10:30 AM	Docent guided tram tour
12:45 PM	Lunch at Warner Brothers cafeteria Commissary
1:45 PM	Depart Warner Bros. Studio
2:30 PM	Return to Sheraton Park Hotel

Time: **9:00 AM-2:30 PM**　　　　　　　　　　　　　　Price: **$138 per person**

RECREATIONAL ACTIVITIES (continued)

A PRESIDENTIAL PEEK
Monday, May 24, 2010

Suggested Itinerary:

1:00 PM ....Depart Sheraton Park Hotel
1:30 PM...Docent guided tour of the Nixon Library, based on (2) hours
3:30 PM...Free time on own at the Nixon Library
4:30 PM ....Depart the Nixon Library
5:00 PM ....Return to Sheraton Park Hotel

Time: **1:00 PM-5:00 PM**　　　　　　　　　　　　　　　　Price: **$60 per person**

IN VINO VERITAS - WINE COUNTRY OF TEMECULA
Tuesday, May 25, 2010

Suggested Itinerary:

8:15 AM ....Depart Sheraton Park Hotel
9:45 AM...Ponte Winery
11:45 AM ....Free time in the Visitor's Center
12:15 PM..Depart Ponte Winery
12:30 PM.....................................Lunch and tasting at Wilson Creek Winery & Vineyards
2:30 PM ....Free time in Visitor's Center
3:00 PM ....Depart Temecula
4:30 PM ....Return to Sheraton Park Hotel

Time: **8:15 AM-4:30 PM**　　　　　　　　　　　　　　　　Price: **$140 per person**

BRUSHSTROKES OF LAGUNA
Tuesday, May 25, 2010

Suggested Itinerary:

10:30 AM ....Depart Sheraton Park Hotel
11:00 AM ....Art walk with local artist, based on (1½) hours
12:30 PM ....Free time for browsing and shopping in Laguna Beach
2:00 PM..Depart Laguna Beach
2:30 PM ....Return to Sheraton Park Hotel

Time: **10:30 AM-2:30 PM**　　　　　　　　　　　　　　　　Price: **$96 per person**

THE GLITZ AND GLAMOUR ...
AN INSIDE LOOK
Wednesday, May 26, 2010

Suggested Itinerary:

10:00 AM ....Depart Sheraton Park Hotel
11:00 AM ....Arrive Hollywood, begin Historic Hollywood tour
12:15 PM ....Tour ends – free time in Hollywood
1:00 PM ....Depart Hollywood
1:15 PM...Lunch at uWink
2:15 PM..Free time on Rodeo Drive
3:30 PM ....Depart for Hotel
4:30 PM ....Return to Sheraton Park Hotel

Time: **10:00 AM- 4:30 PM**　　　　　　　　　　　　　　　　Price: **$108 per person**

RECREATIONAL ACTIVITIES (continued)

WINDOWS OF DISCOVERY AT BOWERS MUSEUM

Wednesday, May 26, 2010

Suggested Itinerary:

11:15 AM ....Depart Sheraton Park Hotel
11:30 AM ....Free time at Bowers Museum
1:30 PM ....Depart the Bowers Museum
1:45 PM ....Return to Sheraton Park Hotel

Time: **11:15 AM-1:45 PM**   Price: **$60 per person**

LANDMARKS OF LONG BEACH

Thursday, May 27, 2010

Suggested Itinerary:

8:45 AM ....Depart from Sheraton Park Hotel
9:15 AM...Arrive Aquarium of the Pacific Self Guided Tour
10:45 AM ....Depart for Queen Mary
11:00 AM..Arrive Queen Mary Self Guided Tour
12:00 PM ....Lunch at Promenade Cafe
1:30 PM ....Ghost and Legends Show
2:15 PM ....Board coach and depart for hotel
2:45 PM ....Return to Sheraton Park Hotel

Time: **8:45 AM-2:45 PM**   Price: **$135 per person**

SECRETS OF THE SEA

Thursday, May 27, 2010

Suggested Itinerary:

9:00 AM ....Depart Sheraton Park Hotel
9:30 AM ....Arrive at Crystal Cove
Beach Walk
11:30 AM...Board coach and depart for hotel
12:00 PM ....Return to Sheraton Park Hotel

Time: **9:00 AM-12:00 PM**   Price: **$97 per person**

THE GETTY CENTER

Friday, May 28, 2010

Suggested Itinerary:

9:30 AM ....Depart Sheraton Park Hotel
10:30 AM ....Self-guided tour at the Getty Center,
based on (3) hours Lunch on Getty Center lawn
1:30 PM ....Depart the Getty Center
2:30 PM ....Return to Sheraton Park Hotel

Time: **9:30 AM-2:30 PM**   Price: **$81 per person**

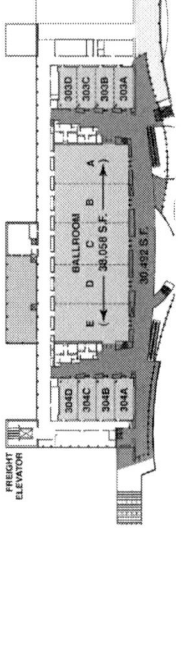

ANAHEIM
ORANGE COUNTY
VISITOR & CONVENTION BUREAU

800 W. Katella Avenue | P.O. Box 4270
Anaheim, CA 92803 | 714.765.8888
meeting.inquiry@anaheimoc.org
anaheimoc.org

ANAHEIM
CONVENTION
CENTER

Specification Summary:
- Exhibit space: more than 800,000 sf
- Meeting & Ballroom space: 152,821 sf
- Prefunction areas: more than 200,000 sf

Exhibit Halls: 813,607 total sf

	Hall A	Hall B	Hall C	Hall D	Hall E
	718 10'x10' Booths 736 Total Booths	718 10'x10' Booths 736 Total Booths	759 10'x10' Booths 791 Total Booths	1140 10'x10' Booths	650 10'x10' Booths
Ceiling Heights	18' 6" – 24'	20' 6" – 25'	20' 6" – 25'	25'	13' 8"
Theater (Capacity)	12,000	12,750	12,900	15,000	N/A
Banquet (Capacity)	9,000	9,200	9,700	13,800	8,000

Specifications are subject to change. A copy of the Anaheim Convention Center "Policies, Rules and Regulations" is available upon request.

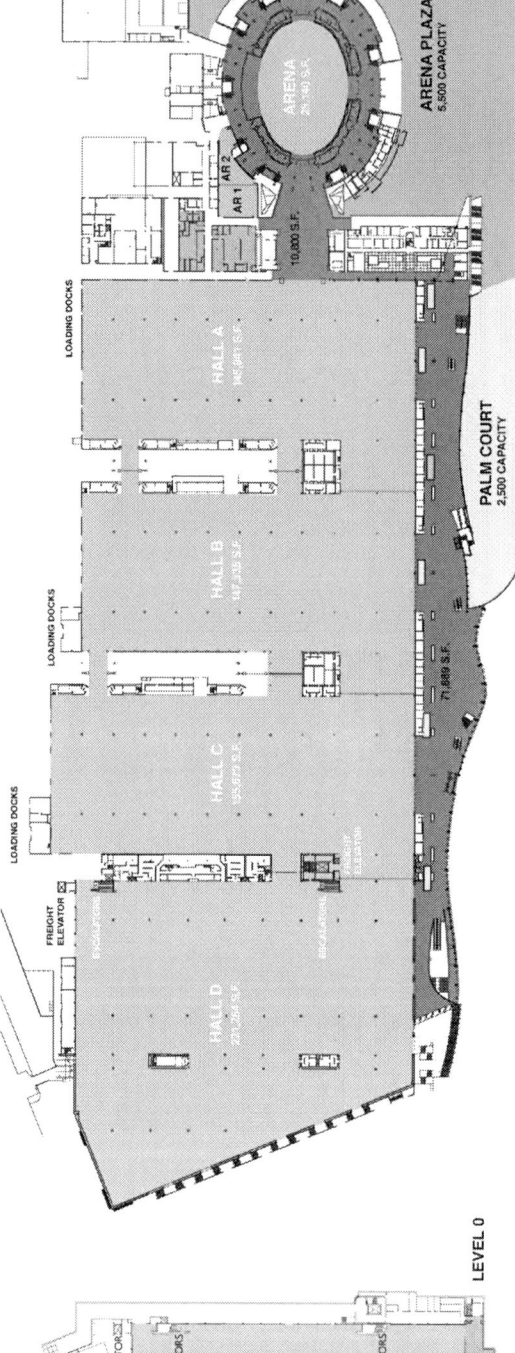

©2009 Anaheim/Orange County Visitor & Convention Bureau | anaheimoc.org

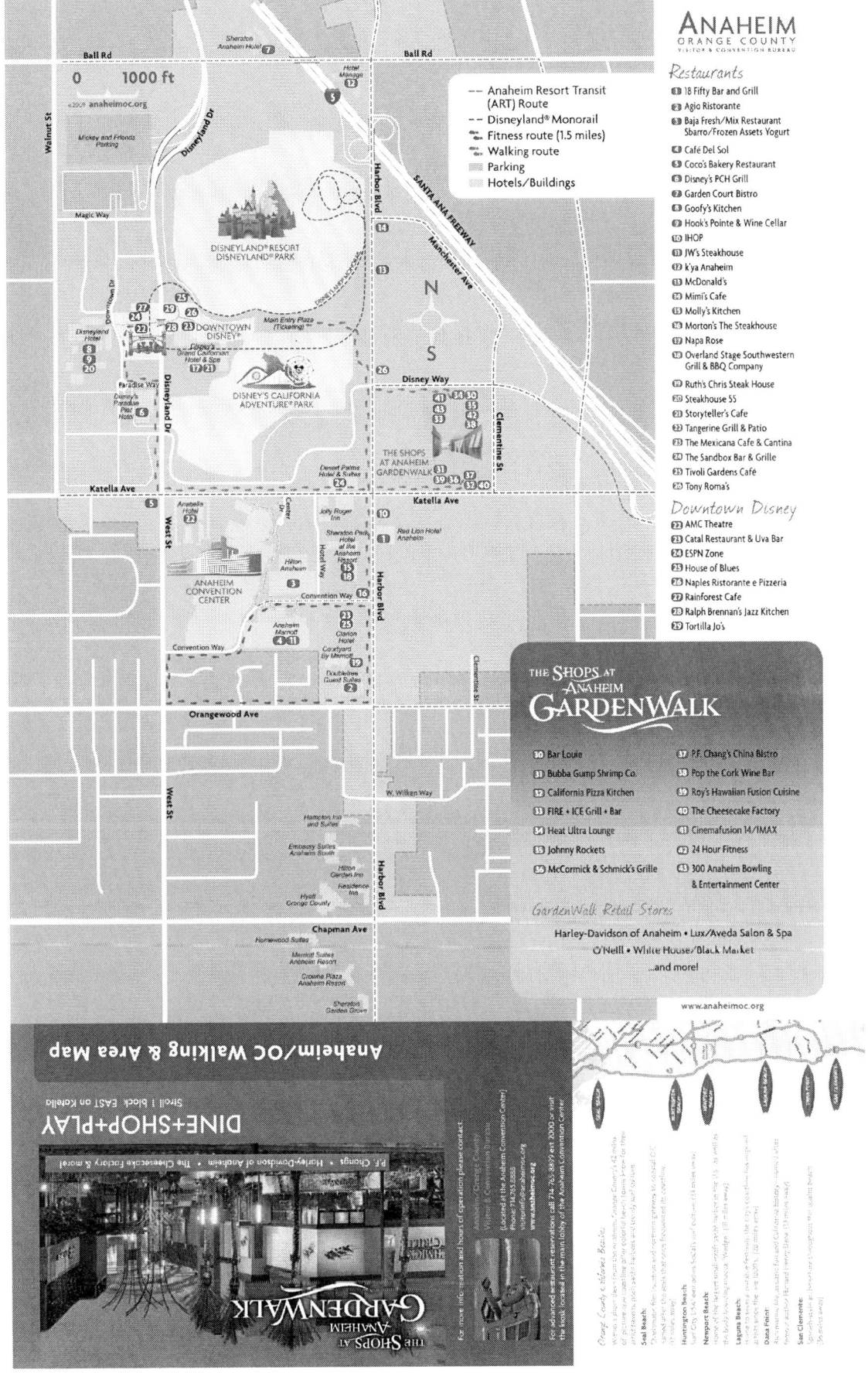

ANAHEIM
ORANGE COUNTY
VISITOR & CONVENTION BUREAU

Restaurants

1. 18 Fifty Bar and Grill
2. Agio Ristorante
3. Baja Fresh/Mix Restaurant Sbarro/Frozen Assets Yogurt
4. Café Del Sol
5. Coco's Bakery Restaurant
6. Disney's PCH Grill
7. Garden Court Bistro
8. Goofy's Kitchen
9. Hook's Pointe & Wine Cellar
10. IHOP
11. JW's Steakhouse
12. k'ya Anaheim
13. McDonald's
14. Mimi's Cafe
15. Molly's Kitchen
16. Morton's The Steakhouse
17. Napa Rose
18. Overland Stage Southwestern Grill & BBQ Company
19. Ruth's Chris Steak House
20. Steakhouse 55
21. Storyteller's Cafe
22. Tangerine Grill & Patio
23. The Mexicana Cafe & Cantina
24. The Sandbox Bar & Grille
25. Tivoli Gardens Café
26. Tony Roma's

Downtown Disney

22. AMC Theatre
23. Catal Restaurant & Uva Bar
24. ESPN Zone
25. House of Blues
26. Naples Ristorante e Pizzeria
27. Rainforest Cafe
28. Ralph Brennan's Jazz Kitchen
29. Tortilla Jo's

THE SHOPS AT ANAHEIM GARDENWALK

30. Bar Louie
31. Bubba Gump Shrimp Co.
32. California Pizza Kitchen
33. FIRE + ICE Grill + Bar
34. Heat Ultra Lounge
35. Johnny Rockets
36. McCormick & Schmick's Grille
37. P.F. Chang's China Bistro
38. Pop the Cork Wine Bar
39. Roy's Hawaiian Fusion Cuisine
40. The Cheesecake Factory
41. Cinemafusion 14/IMAX
42. 24 Hour Fitness
43. 300 Anaheim Bowling & Entertainment Center

GardenWalk Retail Stores

Harley-Davidson of Anaheim • Lux/Aveda Salon & Spa
O'Neill • White House/Black Market
...and more!

www.anaheimoc.org

Plenary Speaker 1:

David J. Allstot –
Boeing-Egtvedt Chair Professor
Department of Electrical Engineering
University of Washington

RF Power Amplification: Can CMOS Deliver?

Abstract: The total energy consumed by cellular telephones in the United States is currently estimated at about 750,000 times the energy used by an average home in one year. Moreover, about 7,500 tons of CO_2 are emitted into the atmosphere.

The RF power amplifier dissipates a large fraction of the total power because of its low efficiency. Despite more than two decades of intensive research, the challenge of on-chip RF PAs with high efficiency in digital-friendly CMOS technologies has not been met.

Switching PA topologies with relatively high efficiency have gained momentum for use in CMOS RF transceivers, and relatively high output power is being delivered using power combining techniques with several PA cells. Supply regulation techniques have enabled higher efficiency when amplifying non-constant envelope modulated signals.

This talk will cite leading-edge designs and on-going research to assess the remaining challenges for CMOS RF power amplifiers.

About David J. Allstot:

David J. Allstot received the B.S. from the Univ. of Portland in 1969, the M.S. from Oregon State Univ. in 1974 and the Ph.D. from the Univ. of California, Berkeley in 1979.

He has held several industrial and academic positions and has been the Boeing-Egtvedt Chair Professor of Engineering at the Univ. of Washington since 1999. He was Chair of the Dept. of Electrical Engineering from 2004 to 2007.

Dr. Allstot has advised approximately 100 M.S. and Ph.D. graduates, published about 275 papers, and received several awards for outstanding teaching and graduate advising. Awards include the 1980 IEEE W.R.G. Baker Award, 1995 IEEE Circuits and Systems Society (CASS) Darlington Award, 1998 IEEE International Solid-State Circuits Conference (ISSCC) Beatrice Winner Award, 1999 IEEE CASS Golden Jubilee Medal, 2004 IEEE CASS Technical Achievement Award, 2005 Semiconductor Research Corp. Aristotle Award, and 2008 Semiconductor Industries Assoc. University Research Award. His service includes: 1990-93 Assoc. Editor and 1993-95 Editor of IEEE TCAS II, 1990-93 Member of Technical Program Committee of the IEEE CICC Conference, 1992-95 Member, Board of Governors of IEEE CASS, 1994-2004, Member, Technical Program Committee, IEEE ISSCC, 1995-97, 2001, 2003-04, Member, Executive Committee of IEEE ISSCC, 1996-2000 Short Course Chair of IEEE ISSCC, 2000-2001 Distinguished Lecturer, IEEE CASS, 2001 and 2008 Co-General Chair of IEEE ISCAS, 2006-2007 Distinguished Lecturer, IEEE Solid-State Circuits Society and 2009 President of IEEE CASS.

978-1-4244-6240-7/10 $26.00 © 2010 IEEE

Plenary Speaker 2:

Gregory L. Waters –
Executive Vice President and General Manager, Skyworks Solutions, Inc.

The Universal Connector: RF Application Trends Over the Next Decade

Abstract: RF technology has enjoyed a significant expansion in consumer electronics and everyday appliances over the past two decades. This presentation will outline key new opportunities and requirements for the RF industry to assume a much greater application reach. This talk will outline why RF growth will accelerate in non-traditional markets, and the key technical and commercial problems that must be solved to enable this. We will conclude with examples of how this growth will affect industry R&D practices, and result in a different business model for leading RF firms.

About Gregory L. Waters:

Gregory L. Waters, 49, is executive vice president and general manager, front-end solutions for Skyworks Solutions, Inc. He joined the company in April 2003. Prior to joining Skyworks, he served as senior vice president of Strategy and Business development at Agere Systems, and previously held positions there as Vice president of the Wireless Communications business, and Vice president of the Broadband Communications business. Prior to this, he held a variety of senior management positions within Texas Instruments, including director of Network Access Products and Director of North American sales.

Waters received a bachelor's of science in engineering from the University of Vermont, and a master's in computer science from Northeastern University.

A 28mW WCDMA/GSM/GPRS/EDGE Transformer-Based Receiver in 45nm CMOS

Danielle Griffith, Venkatesh Srinivasan, Salvatore Pennisi, Vijay Rentala, Yu Su,
Swaminathan Sankaran, Imtinan Elahi, Sreekiran Samala, Halil Kiper, Bijit Patel, Siraj Akhtar, and
Dan Edmondson

Texas Instruments, Dallas, TX, 75243, USA

Abstract — **A transformer-based receiver designed in 45nm CMOS that meets WCDMA, GSM, GPRS, and EDGE system requirements is presented. The receiver requires no interstage SAW filters and consumes 20mA from 1.4V. The use of a transformer at the LNA output helps achieve high linearity by lowering the voltage swing while simultaneously providing current gain. The analog back end is implemented with two cascaded gain stages and a 2nd order $\Sigma\Delta$ ADC. The receiver has a gain of 60dB, noise figure of 3.0dB and an IIP2 of >+50dBm on both I+Q channels. The die area is 1.35mm^2 for 4 bands.**

Index Terms — **45nm, CMOS, EDGE, GSM, radio receiver, transformer, WCDMA**

I. INTRODUCTION

Modern cellular handsets must support multiple bands and standards. High linearity receivers are needed to support all standards with the same circuitry and no interstage SAW filters [3]. Much work has been published recently on multi-band, multi-standard receivers [1]-[4], but the power consumption is typically higher to achieve sufficient linearity to remove the SAW filters. This paper presents a transformer-based receiver in 45nm CMOS that achieves high linearity simultaneously with low power consumption and small die area. It meets all WCDMA, GSM, GPRS, and EDGE (GGE) system requirements with no interstage SAW filters.

II. RECEIVER ARCHITECTURE

The receiver architecture is as shown in Fig.1. The receiver consists of multi-band low noise amplifiers (LNA) with independent inputs to allow for band specific matching. The output of the LNA's are combined at the transformer input whose secondary output is directly coupled to the passive mixers. Traditional receiver architectures use an active stage between the LNA and mixing quad to provide current gain. The use of a transformer over such approach is advantageous primarily from a linearity perspective [6]. Given large input and output voltage swings due to blockers, the active stage introduces significant non-linearities that need to be reduced at the expense of power consumption. The transformer being passive does not introduce non-linearities and consumes no power. This is a significant advantage given the tight linearity requirements for multi-standard receivers. With regards to the overall system gain, the transformer can be designed to provide the required current gain.

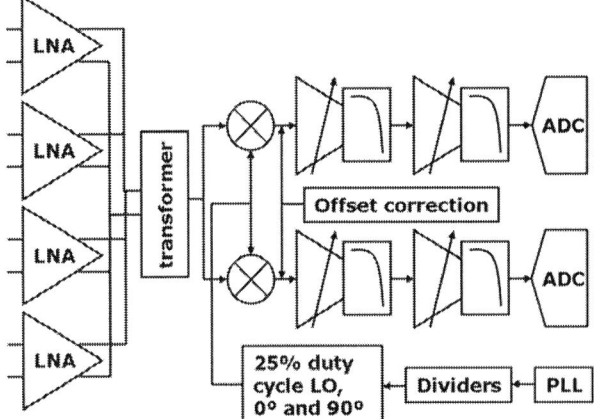

Fig. 1.   Multi-band, multi-standard receiver architecture

Receiver architectures such as in [8] that couple the LNA's output directly to the passive mixer driven by a 50% duty cycle LO require the use of AC coupling capacitors. This is to avoid any DC current leakage/low frequency inter-modulation components from propagating to the mixer inputs. This also helps isolate the base band amplifiers. In this approach, using a transformer along with a 25% duty cycle LO driving the passive mixer alleviates these concerns. With the transformer providing no DC coupling, there is no need for explicit AC coupling capacitors. This results in area savings as the capacitors are typically implemented as metal capacitors due to high linearity requirement. A 25% duty cycle LO provides I/Q isolation. This also increases the impedance looking back from the input of the base-band amplifier. This relaxes the noise requirement of the base-band amplifier when compared to a 50% duty cycle LO approach [8]. However, this is partially offset by the transformer as the transformer's current gain is accompanied by an impedance transformation that lowers the impedance.

Virtual ground is maintained at the output of the mixers with the help of a trans-impedance amplifier with tunable gain and filter corner frequency. Another gain stage provides the required system gain to bring the signal to the ADC reference level. The ADC is a 2nd order continuous time sigma delta running at 500MHz clock frequency. The following sections provide the design of the circuits in more detail.

A. Low Noise Amplifier Design & Transformer

The LNA is a fully differential, low noise, common-source amplifier that is inductively degenerated. Four LNA transconductance inputs are implemented, one each for the DCS, PCS, IMT2K, and WCDMA bands IV and X. These bands extend from 1805MHz to 2170MHz. Each of the 8 degeneration inductances are implemented using the package trace, and therefore require no extra die area. Each trace is ~1nH. EM package simulation was done to ensure the positive and negative trace inductances for each band match for best linearity performance. All four transconductance stages connect directly to a single shared cascode stage.

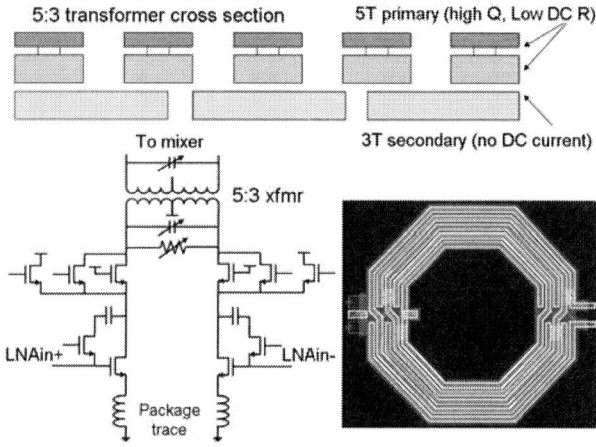

Fig. 2. Simplified circuit schematic for the LNA with transformer load. Cross section of the 5:3 transformer.

The output of the LNA is loaded with a fully differential center-tapped transformer as shown in Fig. 2. The transformer is implemented horizontally with stacked layers of metal [5], so the area is similar to a load inductor. The transformer has a turns ratio of 5:3, which provides current gain. The primary side of the inductor, which carries dc current and has the most impact on LNA noise figure (NF), is implemented in two stacked higher layers of metal for best Q. The primary side center tap is connected to supply to provide DC bias to the LNA. The secondary side of the transformer carries no DC current and is therefore implemented in a lower level of metal. No center tap is implemented on the secondary side. The 5:3 turns ratio was chosen so that parasitic metal and via

resistance matches precisely due to an equal number of crossovers. The coupling factor between primary and secondary is 0.95. The transformer has a binary weighted capacitor array across both the primary and secondary terminals for tuning the resonant frequency.

Voltage gain through the LNA can be adjusted through cascode current steering, load resistor adjustment, and a bypass switch. These three options allow the gain to be varied from 0dB to 26dB in 2dB steps.

B. Passive Mixer

The receiver uses a passive mixer as shown in Fig. 3. Because the mixer carries no dc current, the flicker noise contribution is extremely low. High mixer linearity is achieved by ensuring that the mixer is terminated in a low impedance by the base-band amplifier. The mixer is driven with a local oscillator (LO) where each waveform has a 25% duty cycle. Compared to a traditional LO signal with a 50% duty cycle, this provides a true isolation between I & Q channels, increases the mixer conversion gain by 3dB, lowers the 1/f noise contribution, and improves IP2 [7].

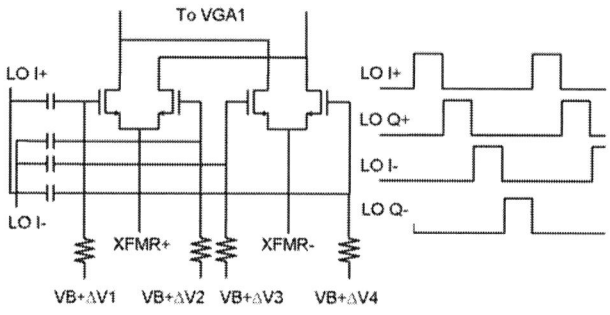

Fig. 3. Simplified circuit schematic for the mixer (I only shown), 25% duty cycle LO, and offset correction.

The mixer switch transistor size is a trade-off between the LO buffer power consumption and transistor mismatch (consequently IP2). To improve IP2 while keeping LO buffer power consumption at a minimum, the LO waveforms are ac coupled to the gates of the mixer transistors with DC bias provided by a programmable resistor string. The DC value can be changed by 0.5mV in 100 steps over 825-875mV. To address interaction between the I and Q channels, each mixer gate is independently programmable to allow for flexibility in setting a differential gate bias on I/Q [1].

C. Analog Back End

The analog back-end that interfaces to the passive mixer is as shown in Fig. 4. It consists of two fully differential

978-1-4244-6240-7/10 $26.00 © 2010 IEEE

variable gain amplifiers (VGA) that also provide first order filtering followed by a 2nd order sigma-delta modulator. Also present is a digital-to-analog converter (DAC) for offset cancellation. Two VGAs are necessary in order to meet both the gain and filtering requirements of a multistandard receiver.

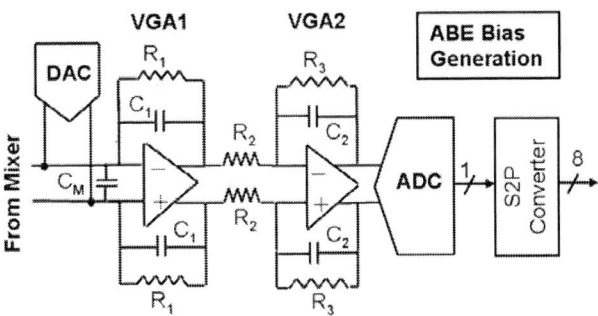

Fig. 4. Analog back end for the receiver

The first variable gain amplifier (VGA1) is a current-to-voltage converter that provides a first order filtering. The nominal cut-off frequency of VGA1 is set to 250kHz for the GGE mode and 2.5MHz for the WCDMA mode. The amplifier is designed to be a 2-stage amplifier with a Class AB output stage and is stabilized using Miller compensation. A continuous-time resistively sensed common-mode feed-back loop sets the output common-mode of VGA1. The output common-mode voltage also sets the DC level at the output of the mixer. The DC gain of the amplifier in VGA1 is designed such that it presents low input impedance at the mixer output to improve linearity. Capacitor C_M is added at the input of VGA1 to preserve the low impedance at frequencies beyond the band-width of the VGA. The value of C_M is chosen at an optimum value that balances between the stability, band-width & power consumption of the VGA and system linearity for high frequency blockers. As stated in Section II, the use of a transformer tightens the noise requirement of VGA1. This is addressed by employing techniques such as source degenerating the input transistors that improve noise without power penalty. The amplifier is optimized such that it provides a very low input referred noise of ~2nV/rt-Hz.

An offset cancellation DAC is added at the input of VGA1 to cancel out DC offsets due to various sources including LO leakage, VGA1 offsets & VGA2 offsets. This ensures that the system does not saturate due to the offsets as well as improving receiver IP2. An 8-bit current steering DAC is used for the offset cancellation.

The 2nd variable gain amplifier (VGA2) is a traditional voltage-in/voltage-out gain stage. The variable gain is

achieved by increasing the feed-back resistor in relation to the input resistor. The maximum value of R_2 is set by noise considerations and is chosen such that it optimizes the drive capability of VGA1 as well. The amplifier is similar to VGA1 but considerably more relaxed in the design for noise but more stringent in terms of linearity, setting the amplifier band-width and therefore power consumption. The filter pole frequency is set at 350kHz for GGE mode and 3.5MHz for WCDMA mode. VGA2 has a maximum programmable gain range of 24dB.

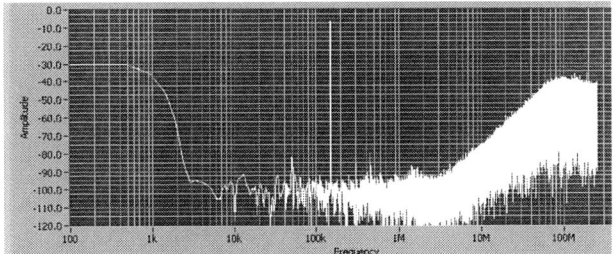

Fig. 5. FFT of the ADC's output for a 150kHz input signal of 64uApk-pk differential at the input of VGA1.

Sigma-Delta modulators are advantageous from an area and power stand-point for multi-mode receiver applications. For this reason, the ADC used in the receiver chain is a 2nd order sigma-delta modulator. The ADC is clocked at 500MHz and is designed to meet an 80dB dynamic range for the GSM mode and a 73dB dynamic range for the WCDMA mode.

III. MEASUREMENT RESULTS

Fig. 5 shows the experimental results for the analog back-end tested independently. The second-order noise shaping of 40dB/dec from the sigma-delta modulator is clearly evident from Fig. 5. The equivalent spot noise in the GSM mode is ~2.4nV/rt-Hz and ~3nV/rt-Hz in the WCDMA mode. Fig. 6 shows the IP2 of both I and Q channels as a function of bias point on the mixer gates. An IP2 of >+50dBm can be achieved on I and Q channels. Measurements for the entire receiver chain are shown in Table 1 along with published data. The measured return loss was -17dB. The die area, including four LNA inputs, all bond pads, I/Q mixers, VGA1, VGA2, feedback DAC, and ADC is 1.35mm^2.

V. CONCLUSION

In conclusion, a complete multi-band receiver in 45nm CMOS has been presented that supports both WCDMA and GSM/GPRS/EDGE standards. The receiver achieves low power consumption and high linearity by using a transformer based architecture. The analog back end is

978-1-4244-6240-7/10 $26.00 © 2010 IEEE

TABLE I: MEASURED RESULTS AND PUBLISHED DATA

Parameter	[3][1]	[1][2]	[4][1]	[2][4]	[8]	This Work
Voltage Gain	32dB	61.5dB	44dB	30dB	84dB	60dB
NF	3.4dB	2.2dB	3.4dB	3.1dB	2.6dB	3.0dB
NF, -30 dBm blocker at 80 MHz offset	3.7dB	2.39dB	N/A	N/A	2.6dB	3.2dB
NF, -30dBm blocker at 3MHz offset	N/A	3.8dB[2a]	N/A	N/A	N/A	8.5dB
IIP3 (10 MHz, 20 MHz offset blockers)	1.3dBm	1.55dB[2b]	N/A	N/A	2dBm[5]	4.3dBm
IIP3 (3.5MHz, 5.9MHz offset blockers)	-1.9dBm	N/A	N/A	-12dBm	-10.5dBm[5]	-5.5dBm
IIP3 (800 kHz, 1.5 MHz offset blockers)	-9.5dBm	4dB[2b]	N/A	N/A	N/A	-6.5dBm
IIP2 (79 MHz, 80 MHz offset blockers)	51dBm	90dBm[2c]	65dBm[3]	>39dBm[2c]	>50dBm[2c]	>50dBm
Power Consumption	27.3mW	22.6mW	37.8mW	12mW	38.4mW	28mW
Supply voltage	1.4V	1.5V	2.1V	1.5V	1.2V	1.4V

[1]Up to the first VGA. [2]Does not include ADC. [2a] ±6MHz offset GMSK blockers. [2b]Additional NF impairment due to IM3. [2c]190MHz offset blockers. [3]45MHz offset blockers. [4]LNA + Mixer only. [5]Blocker offset frequencies not available.

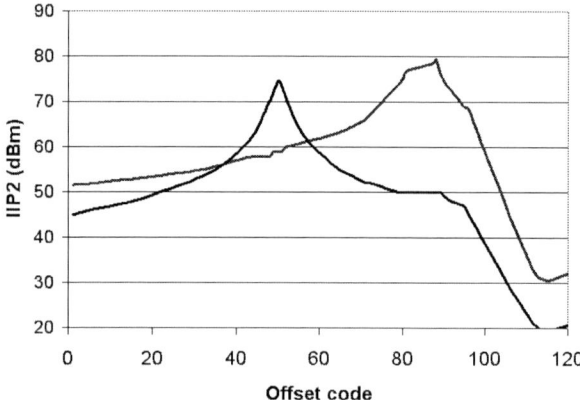

Fig. 6.   IP2 vs. offset correction code for I and Q channels. Code 64 is zero point (no offset added).

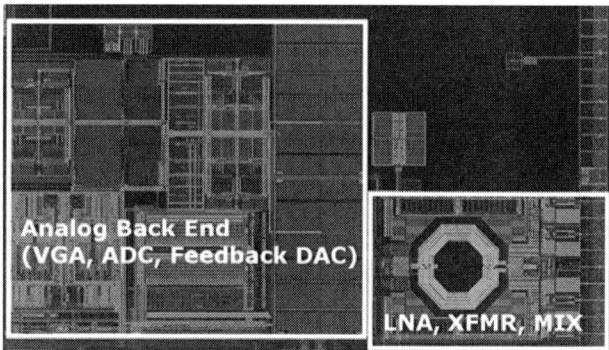

Fig 7. Layout snapshot. Die area is 1.35mm² in 45nm CMOS.

programmable to meet both GGE and WCDMA requirements. The total area for 4 bands is only 1.35mm² while the power consumption is 28mW, which is very competitive compared to other published results as shown in Table 1.

ACKNOWLEDGEMENT

The authors would like to acknowledge the support of their management and colleagues at Texas Instruments.

REFERENCES

[1]   D. Kaczman, M. Shah, M. Alam, M. Rachedine, D. Cashen, L. Han, A. Raghavan, "A single-chip 10-band WCDMA/HSDPA 4-band GSM/EDGE SAW-less CMOS receiver with DigRF 3G interface and +90dBm IIP2," *IEEE J. Solid-State Circuits*, vol. 44, no. 3, March 2009, pp. 718-738.

[2]   Y. Feng,, G. Takemura, S. Kawaguchi, P. Kinget, "Design of a high performance 2-GHz direct-conversion front-end with a single-ended RF input in 0.13μm CMOS" *IEEE J. Solid-State Circuits*, vol. 44, no. 5, May 2009, pp. 1380-1390.

[3]   N. Yanduru, D. Griffith, K. Low, P. Balsara, "RF receiver front-end with +3dBm out-of-band IIP3 and 3.4dB NF in 45nm CMOS for 3G and beyond," *IEEE RFIC Conf*, June 2009, pp. 9-12.

[4]   N. Kim, L.. Larsona, V. Aparin, "A Highly Linear SAW-Less CMOS Receiver Using a Mixer With Embedded Tx Filtering for CDMA", *IEEE J. Solid-State Circuits*, vol. 44, no. 8, August 2009, pp. 2126-2137.

[5]   S. Akhtar, R. Taylor, P. Litmanen, "A high magnetic coupling, low loss, stacked balun in digital 65nm CMOS," *IEEE RFIC Symposium*, June 2009, pp. 513-516.

[6]   C. Hermann, M. Tiebout, H. Klar, "A 0.6-V 1.6-mW transformer-based 2.5-GHz downconversion mixer with +5.4-dB Gain and -2.8-dBm IIP3 in 0.13μm CMOS," *IEEE Trans. on Microwave Theory and Techniques.*, vol. 53, no. 2, Feb. 2005, pp. 488-495.

[7]   R. Pullela, T. Sowlati, D. Rozenblit, "Low flicker-noise quadrature mixer topology," *IEEE ISSCC*, 2006, pp. 1870-1879.

[8]   A. Mirzaei, X. Chen, A. Yazdi, J. Chiu, J. Leete, H. Darabi, "A Frequency Translation Technique for SAW-Less 3G Receivers", *2009 Symp. on VLSI Circuits*, June 2009, pp. 280-281.

An EDGE Transmitter with Mitigation of Oscillator Pulling

Imran Bashir[1], R. Bogdan Staszewski[2], Oren Eliezer[3], Khurram Waheed[3], Vasile Zoicas[3], Nir Tal[3],
Jaimin Mehta[3], Meng-Chang Lee[3], Poras T. Balsara[1], Bhaskar Banerjee[1]

[1]University of Texas at Dallas, Richardson, TX 75083, USA
[2]Delft University of Technology, Delft, Netherlands
[3]Texas Instruments, Dallas, TX 75243, USA

Abstract—We propose a polar transmitter architecture that is robust to modulation-induced injection pulling of its RF oscillator by means of a built-in self compensation. A mathematical model is presented for the injection pulling mechanism, which incorporates a digitally-controlled delay circuit that minimizes injection pulling by adjusting the overall phase shift in the parasitic path between the final amplitude modulation stage (aggressor) and the RF oscillator (victim). The technique is verified in a 65-nm CMOS GSM/GPRS/EDGE SoC demonstrating compliant error vector magnitude (EVM) and modulation spectral-mask performance over process and temperature.

I. INTRODUCTION

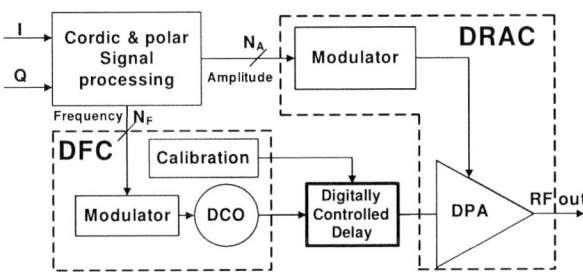

Fig. 1. Digital polar transmitter incorporating a digitally-controlled delay for mitigation of DCO frequency pulling.

The transmitter presented in this paper is based on a polar architecture in which both the phase and amplitude paths are fully digital [1] [2]. The amplitude information is applied to a digital-to-RF amplitude converter (DRAC) block while the frequency information is applied to the digital-to-frequency converter (DFC) block. The DFC block consists of the all-digital PLL coupled with the digitally controlled oscillator (DCO). The ADPLL uses a two-point modulation scheme in which the phase of the information signal is applied both at the phase detector input and directly at the DCO. The DRAC block contains the digital power amplifier (DPA), which is an array of NMOS transistors tied together at their drains. This circuit is preceded by the proposed digitally-controlled delay (DCD) block, which serves to set the optimal phase relationship between the DPA output signal and the DCO oscillations to minimize the impact of the AM signal level in the former on the phase lock of the latter.

The injection pulling experienced in the DCO, the victim circuit, caused by interference originating in the DPA, the aggressor circuit, can degrade the transmitter's performance to the extent that it fails its targeted specifications. The DPA is highly non-linear and generates a relatively strong second harmonic component, which is equal to the DCO oscillation frequency. Note that the oscillator is followed by a divide-by-2 to generate the high-band carriers or a divide-by-4 to generate the low-band carriers. When operating in the high-bands (1.7–1.9 GHz), this second harmonic is not only centered around the operating frequency of the DCO (3.4–3.8 GHz) but also shares its frequency modulation, due to the inherent polar architecture of the transmitter. This is in contrast to conventional quadrature modulators, where the LO is not phase/frequency modulated and can therefore only coincide with the center frequency of a modulated harmonic of the output signal rather than being frequency synchronous with it. This inter-relationship between the aggressor and the victim signals is what allows the phase-adjustment in the DCD to be effective in mitigating this self-interference.

II. BACKGROUND AND PRIOR RESEARCH

All previous work related to mitigation of injection pulling of an oscillator has been focused on the root cause. The conventional approach is to increase the isolation between the amplifier and the oscillator or to reduce/eliminate the interfering harmonic (e.g., an offset LO architecture). The work done in [3] focuses on three different mechanisms of coupling between the amplifier and the oscillator: resistive (substrate), magnetic (between inductors and bond wires), and capacitive (between interconnects). Various experiments were done in this work to establish the contribution of each coupling path and identify the dominant one.

Adler's early work [4] provides a mathematical expression of pulling experienced by an oscillator from an independent source. The work in [5] applies Adler's principles to a phase-locked oscillator under injection. The nature of the problem is slightly different in the presented work because the interfering AM signal generated by the DPA is not an independent source but is frequency-synchronous with the victim signal, as previously noted.

A similar analysis to that in [5] is applied here for the characteristics of the observed injection pulling. Of particular interest here is the dependency of the extent of oscillator pulling upon the phase shift applied to the aggressor signal (by means of the DCD), which is studied using the developed model. This dependency is exploited in the proposed transmitter architecture to achieve substantial reduction in injection pulling by conditioning of the aggressor's phase to mitigate its impact on the victim in a calibration step. The optimal phase shift to be applied varies with temperature and carrier frequency, thus necessitating active compensation. The calibration and compensation schemes do not require any additional hardware overhead as the digitally-intensive Digital RF Processor (DRP^{TM}) architecture has the required resources built-in, such as the capability to monitor and analyze the ADPLL's phase-error signal, which serves to establish the nature of phase perturbations experienced by the DCO as a result of the pulling.

III. THE AM-TO-FM CONVERSION IN THE DCO

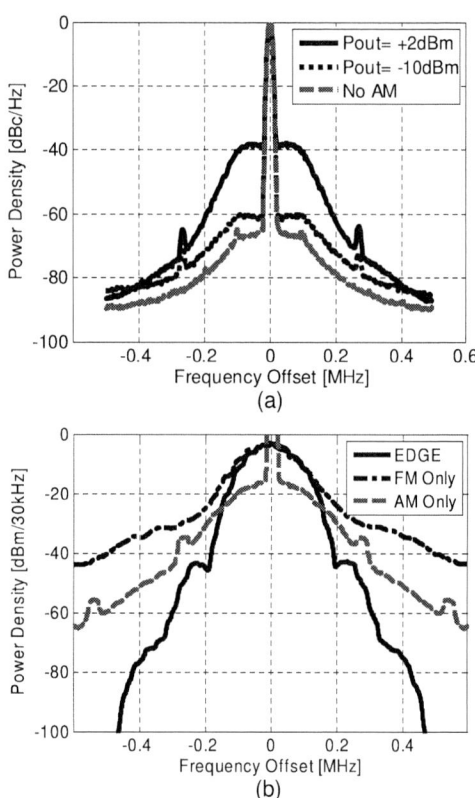

Fig. 2. (a) Measured spectrum at DPA input when the FM component is turned off; (b) Simulated EDGE spectrum.

Fig. 2 (a) shows the spectrum of the ADPLL output (DPA input) when the EDGE FM component is turned off, i.e. only the EDGE AM component is applied. When the EDGE AM component also is turned off (in the lowest trace), the ADPLL spectrum only exhibits noise from its internal sources, namely the reference, time-to-digital converter (TDC), and the DCO, shaped by the loop filter [1]. When the transmitter undergoes AM, a parasitic modulation is induced on the DCO that is visible at its spectrum. The extent of this parasitic modulation is shown to worsen with the increase in output power from

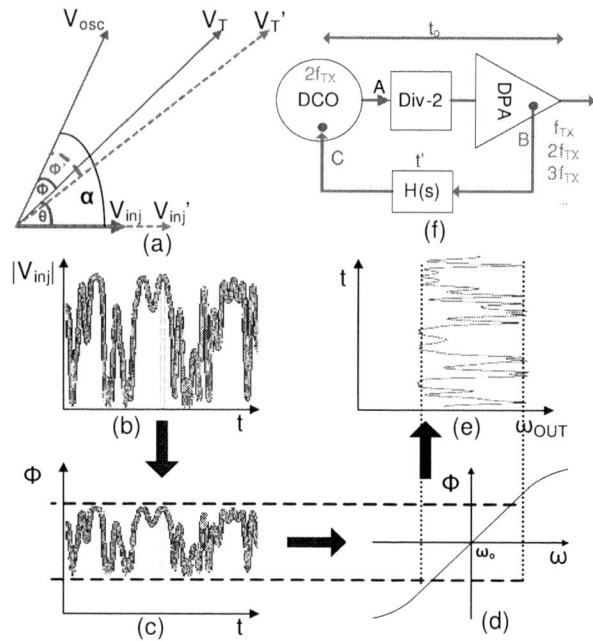

Fig. 3. AM-FM conversion mechanism: (a) Phasor representation of signals in DCO; (b) Amplitude of injected signal (EDGE-AM) over time; (c) Angular displacement over time; (d) Phase response of resonant network; (e) Output DCO frequency over time; (f) Block diagram showing interference mechanism.

-10dBm to +2dBm, confirming that the transmitter final stage is the aggressor. Fig. 2 (b) shows the simulated spectra of the modulated carrier with EDGE composite (amplitude and phase), phase/frequency modulation only (AM turned off) and AM only (FM turned off). Note the spur near the 280-kHz offset (the EDGE symbol rate) that is present on the EDGE AM spectrum is also observed on the FM spectrum of the interfered DCO, indicating an AM-to-FM conversion mechanism.

This conversion may be attributed to the Barkhausen criterion for oscillation, according to which the total phase shift that a feedback signal must undergo in the feedforward amplifying and feedback elements must be an integer multiple of 2π. This is because the amplitude-modulated aggressor signal, denoted V_{inj} in Fig. 3, effectively creates a phase shift ϕ in the oscillator, for which this criterion is satisfied at a slight frequency shift as per the almost-linear phase-frequency relationship of Fig. 3(d).

The block diagram of Fig. 3(f) illustrates the interference mechanism, where the aggressor signal, generated by the DPA couples to the DCO through a coupling path represented by a transfer function H(s). Since there is no frequency translation stage between the DCO and the DPA, the aggressor signal from the DPA and the victim signal V_{osc} in the DCO are at the same frequency, allowing them to be represented as phasors on the same vector diagram. The aggressor signal, after passing through H(s), is attenuated and has a certain phase offset with the DCO signal denoted by α in Fig. 3(a). As a result of this injection, the vector sum V_T suffers an angular displacement

978-1-4244-6240-7/10 $26.00 © 2010 IEEE   8

ϕ from V_{osc}. This additional phase that is introduced in the feedback oscillator will violate Barkhausen criteria at ω_o causing the oscillation to shift to another frequency along the curve shown in Fig. 3(d) where this criterion would be satisfied. The amplitude modulation on the aggressor from the DPA, shown in Fig. 3(b), will result in a correspondingly time varying phase displacement ϕ as shown in Fig. 3(c) and consequently a similarly time varying ω_{OUT}, as shown in Fig. 3(e). In this way, AM-FM conversion is experienced by the transmitter's oscillator.

IV. MATHEMATICAL MODEL

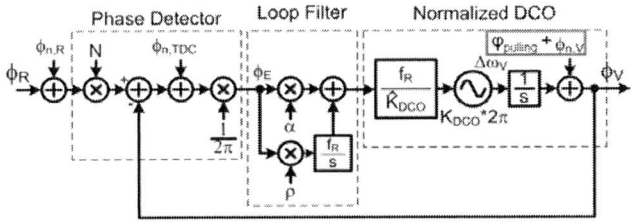

Fig. 4. ADPLL transfer function incorporating the DCO injection pulling model.

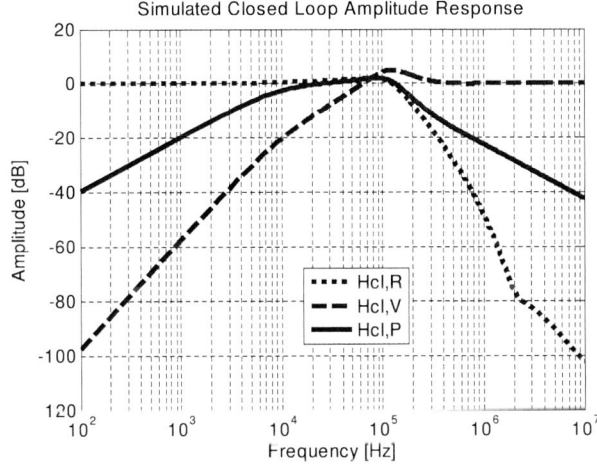

Fig. 5. ADPLL closed-loop magnitude response (H_{cl}) for the reference (R), DCO (*variable*, V), and AM-FM distortion (*pulling*, P).

The purpose of the following analysis is to establish the mathematical equation for the parasitic modulation or phase-noise due to injection pulling in frequency domain. In the given scenario, the instantaneous output frequency ω_{OUT} is given by Adler's analysis [4].

$$\omega_{out}(t) = -\frac{V_{inj}(t)}{V_{osc}} \cdot \frac{\omega_{osc}(t)}{2 \cdot Q} \cdot sin(\alpha) + \omega_{osc}(t) \quad (1)$$

where Q is the quality factor of the DCO LC tank. In [4], the oscillator was impressed by an independent source of a slightly different frequency and hence the angle between the aggressor and the victim α shown in Fig. 3(a) would vary over time due to their frequency difference, contrary to the case here, where

α is constant. However, the envelope of the injected signal is varying over time due to the AM modulation of the aggressor signal. Since the oscillator frequency is controlled by a PLL, we apply a similar analysis as presented in [5].

$$\omega_{osc}(t) = \omega_c + K_{DCO} \cdot OTW(t) \approx \omega_c \quad (2)$$

where K_{DCO} is the DCO gain of the varactor bank used during the modulation and OTW is the discrete-time digital oscillator tuning word. The approximation is based on the fact that the frequency modulation generated by the ADPLL in response to the error generated by injection pulling is significantly smaller than the carrier frequency ω_c. Substituting in Eq. 1 and taking the Laplace transform of the phase component due to injection pulling, the magnitude of the phase perturbations caused by the pulling are shown to be linearly proportional to the magnitude of the interfering signal, inversely proportional to the magnitude of the oscillations and the quality factor Q, and dependent upon the phase α through the sine function:

$$\varphi_{pulling}(s) = -\frac{\omega_c \cdot sin(\alpha)}{2 \cdot Q \cdot V_{osc}} \frac{V_{inj}(s)}{s} \quad (3)$$

where $V_{inj}(s)$ or $V_{inj}(j\omega)$ is the spectrum of the AM signal shown in Fig. 2(b). This phase-noise source is applied to the ADPLL transfer function to determine its impact on the EVM and spectrum at the transmitter output. Fig. 4 shows the ADPLL transfer function for noise sources $\phi_{n,R}$ (reference), $\phi_{n,TDC}$ (TDC), $\phi_{n,V}$ (DCO), and $\varphi_{n,pulling}$ (AM-FM). The closed-loop transfer function from the DCO to the output is high-pass [1] but the expression for phase noise due to pulling has an integration term which results in a band-pass transfer function for this phase-noise source [5]. Fig. 5 shows the simulation of closed-loop magnitude response of $H_{cl,R}$ (low pass), $H_{ol,V}$ (high pass) and $H_{cl,P}$ (band pass) with parameters that result in a loop bandwidth of 120kHz and a phase margin of $44°$. Based on Eq. 3, the expression for the phase at the ADPLL output in frequency domain is given by:

$$\varphi_{out,pulling}(s) = -\frac{\omega_c \cdot sin(\alpha)}{2 \cdot Q \cdot V_{osc}} (V_{inj}(s) \cdot H_{cl,P}) \quad (4)$$

V. MITIGATION OF DCO PULLING THROUGH DELAY/PHASE ADJUSTMENT

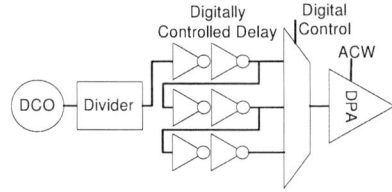

Fig. 6. Digitally-controlled delay to mitigate the DCO injection pulling.

With the expression of phase distortion $\varphi_{out,pulling}$ due to the DCO injection pulling derived in Eq. 4, an important observation can now be made. The extent of the parasitic

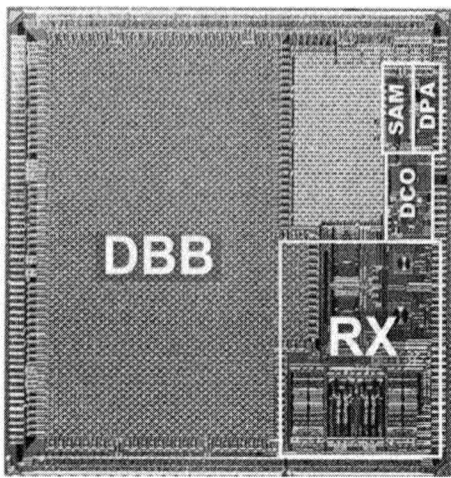

Fig. 7. Chip micrograph of the 65 nm GSM/GPRS/EDGE SoC.

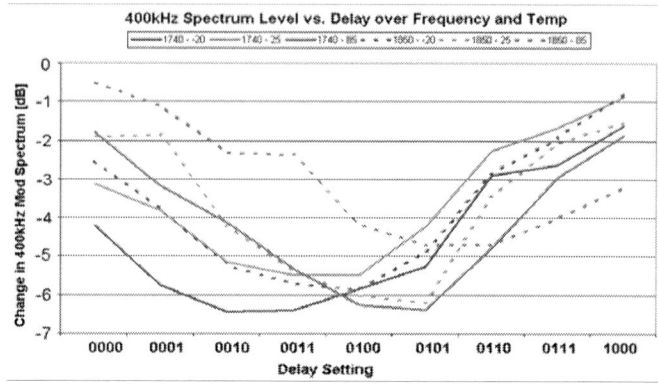

Fig. 9. Modulated spectrum at 400-kHz offset measured at transmitter output vs. delay.

modulation is a function of the angle α between the victim (DCO) V_{osc} and the amplitude-modulated aggressor (DPA) V_{inj}. As α approaches zero or π, $\varphi_{out,pulling}$ approaches zero. Consequently, we propose a digitally-controlled delay (DCD) stage between the DCO and the DPA, as shown in Fig. 6, that can be used to adjust the phase of the second harmonic of the DPA such that α is close enough to zero or π. This will require a calibration procedure, which could be done autonomously using built-in self-test (BIST) techniques. The extent of parasitic modulation, perceived by the loop as a phase error, will be visible at the phase detector output in a closed-loop operation. Thus, the optimal delay is determined during calibration through analysis of the digital phase error (PHE). As shown in Fig. 6, the DCD comprises of cascaded delay units and a multiplexer, which is digitally controlled to select the output where the total phase shift is closest to the optimal value. Since the DCD may only need to shift the signal up to $180°$ to reach the optimum, this may be, at most, a delay of about 150ps (half the period of a 3.4GHz oscillation). For a buffer delay of about 30ps, this range can be covered by about 5 stages, offering a phase resolution finer than $40°$.

VI. MEASURED RESULTS

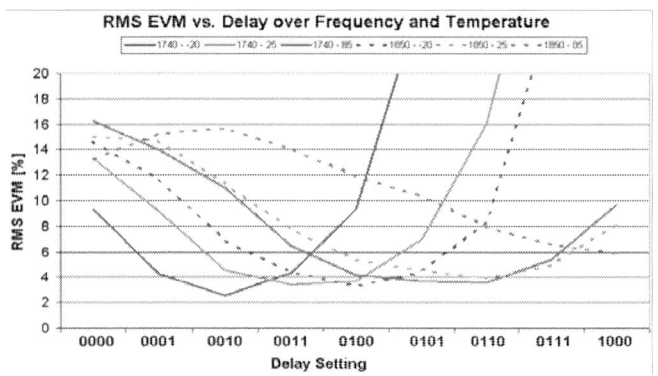

Fig. 8. EVM measured at transmitter output vs. delay.

The proposed techniques, together with the DCD circuit in the direct RF path, were incorporated into a GSM/GPRS/EDGE SoC fabricated in 65 nm CMOS and shown in Fig. 7. The EDGE EVM and modulated spectrum performance at 400kHz offset as a function of delay are shown in Fig. 8 and Fig. 9, respectively, for various frequencies and temperatures. An equation-based compensation is applied over temperature using a precisely measured inverter delay as a temperature sensor [1]. After applying the calibration and compensation to a previously incompliant transmitter, the EVM of the transmitter was limited to 2-3% and the spectral density at 400-kHz offset to below -62 dBc, thus demonstrating the effectiveness of the proposed solution.

VII. CONCLUSION

A fully-integrated and low-cost GSM/GPRS/EDGE transmitter with built-in compensation for oscillator pulling is presented. The mechanism behind its AM-FM parasitic modulation is described and is supported by a mathematical model showing that it depends on the constant phase-shift between the aggressor and victim signals. A novel digitally-controlled delay (DCD) stage is proposed to adjust this phase in order to mitigate the injection pulling experienced in the DCO. The solution is fabricated on a SoC with results demonstrating a transmitter that meets its targeted specification.

REFERENCES

[1] R. B. Staszewski, J. L. Wallberg, S. Rezeq, et al., "All-digital PLL and transmitter for mobile phones," *IEEE Journal of Solid-State Circuits*, vol. 40, no. 12, pp. 2469–2482, Dec 2005.

[2] J. Mehta, R. B. Staszewski, O. Eliezer, et al., "A 0.8mm^2 all-digital SAW-less polar transmitter in 65nm EDGE SoC," *ISSCC Dig. Tech. Papers*, sec. 3.2, Feb. 2010.

[3] S. Bronckers, G. Vandersteen, L. De. Locht, G. Van der Plas, Y. Rolain, "Study of the different coupling mechanisms between a 4GHz PPA and a 5-7 GHz LC-VCO," *IEEE Radio Frequency Integrated Circuits Symposium*, pp. 475–478, June 2008.

[4] R. Adler, "A study of locking phenomena in oscillators," *Proceedings of the IEEE*, vol. 61, pp. 1380–1385, Oct. 1973.

[5] C. J. Li, C. H. Hsiao, F. K. Wang, T. S. Horng, K. C. Peng, "A rigorous analysis of local oscillator pulling in frequency and discrete-time domain," *IEEE Radio Frequency Integrated Circuits Symposium*, pp. 409–412, June 2009.

RMO1A-3

An All-Digital Offset PLL Architecture

Robert Bogdan Staszewski[1], Sudheer Vemulapalli, Khurram Waheed

Texas Instruments Inc, Dallas, TX 75243, USA

[1]Technische Universiteit Delft, The Netherlands

Abstract— **We propose an all-digital *offset* PLL architecture in which the RF oscillator output is frequency translated through rotation of its quadrature phases before being fed back for the phase comparison with the frequency reference. This eliminates spurious tones caused by the finite resolution of the phase detection process when the synthesized frequency is very close to the integer-N multiple of the reference frequency. The phase detection in the ADPLL is performed by a time-to-digital converter (TDC), whose typical resolution of 10–30 ps is sufficient for the GSM-quality RF operation. While the TDC quantization noise does not normally produce significant phase noise degradation, the near-integer-N condition makes the loop ill-behaved such that the total quantization energy falls close to dc and will not get filtered by the loop filter. In addition, due to the frequency relationship change between aggressors and victims, an important class of spurs due to parasitic coupling is also eliminated. The hardware overhead is very small and the digital implementation does not degrade other RF parameters. The technique is validated in a 65-nm CMOS transceiver.**

I. INTRODUCTION

In the past several years, there has been a great deal of research on applying digitally-intensive and digital signal processing (DSP) approaches to efficiently implement RF wireless circuits in deep-submicron or nanometer-scale digital CMOS technologies. In particular, the RF frequency synthesis has seen a successful adoption of an all-digital PLL (ADPLL) [1], [2], in which the phase/frequency detector with the charge pump and the voltage-controlled oscillator of the traditional PLL have been replaced with a time-to-digital converter (TDC) and a digitally-controlled oscillator (DCO), respectively.

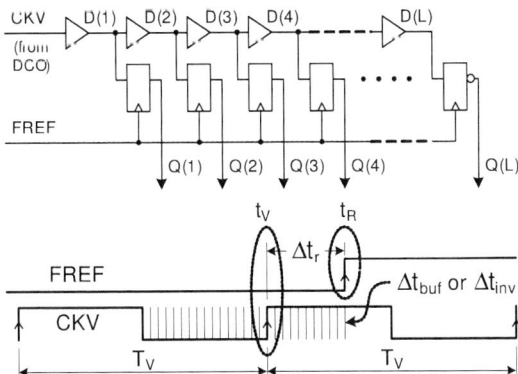

Fig. 1.   TDC: simplified schematic view (top); signal timing (bottom). The raw Q output is converted into binary word represented as Δt_r. The delay elements are inverters but shown as non inverting buffers for simplicity sake.

An undesirable consequence of *finite*-resolution converting functions is the introduction of quantization noise. As such, as shown in Fig. 1, the TDC produces a decoded digital integer Q proportional to the timing difference between the significant edges of the reference clock (FREF) of frequency f_R and variable clock (CKV) of frequency f_V, but with a certain quantization:

$$Q = round(\frac{t_R - t_V}{\Delta t_{res}}) = round(\frac{\Delta t_r}{\Delta t_{res}}) \qquad (1)$$

where, t_R and t_V are significant edge timestamps of the reference and variable clocks, respectively, Δt_{res} is the TDC timing resolution, and $round$ is the quantizing operation of rounding to the next (or, alternatively, previous) integer. Typical resolution ranges as achieved in [1], [2] are $\Delta t_{res} = 15$–30 ps and are obtained merely through the fine lithography and excellent device matching of the advanced CMOS fabrication process with no particularly special design or layout techniques. Fig. 1 illustrates an example of such simplicity.

The above quantization values are fine enough to generally guarantee mobile phone quality of RF transmit and receive operations. Only a very small fraction of the RF channels would exhibit an ill-conditioned frequency relationship in which the quantization energy does not satisfy the white-noise assumption and can produce performance degradation larger than expected from the well-behaved distribution. For example, in exact or near integer-N channels ($N = FCW = f_V/f_R$), the quantization energy of TDC is mostly concentrated in tones close to dc. The direct ADPLL frequency modulation further causes a continuous shift of these tones in correspondence to the modulation commands. These low-frequency tones, if they fall within the pass-band of the PLL, cannot be attenuated by the loop filter and, consequently, they modulate the DCO output, thus distorting the intended modulation. If the spurious tones fall between 300-700 kHz, then they affect the narrow-band spectrum mask, potentially exceeding the limits defined for it for GSM/EDGE transmitters. Similar parasitic DCO modulation for an ADPLL-based GMSK transmitter operating at integer-N channels is explained in [3]. The detailed nature of the ill-behaved spurious tones is analyzed in [4].

Recent research activities attempt to lower the TDC quantization noise by improving the TDC resolution. They include: TDC with fractional resolution [5], noise-shaped TDC [6], TDC with precise calibration and mismatch correction as well as clock doubler [7], TDC with time amplification [8], and TDC with doubling the resolution through additional row of flip-flop registers operating on a delayed clock [9].

In this paper, we propose a *simple* fully-digital method to mitigate the abovementioned low-frequency quantization noise

978-1-4244-6240-7/10 $26.00 © 2010 IEEE

of the TDC. This is done through avoidance by fundamentally changing the frequency relationship of the two signals seen by the TDC. An additional effect of avoiding the undesired frequency relationship is lowering of spurious tones due to parasitic coupling.

II. TDC-BASED ADPLL

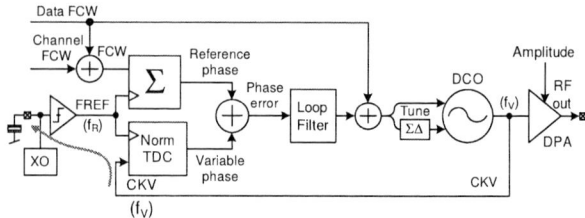

Fig. 2. All-digital PLL frequency synthesizer with a wideband frequency modulation capability.

Fig. 2 shows an RF transmitter based on the all-digital PLL (ADPLL) frequency synthesizer with a digital direct frequency modulation capability. The channel and data frequency control words are in the frequency command word (FCW) format defined as the fractional frequency division ratio N with a fine frequency resolution limited only by the FCW wordlength. The exact integer-N channels would be largely free from the abovementioned phase error quantization effects, but here they will result in a time-variant non-zero value of fractional FCW due to the data frequency modulation [3]. In Fig. 2 the simple TDC of Fig. 1 is augmented with the normalizing multiplier [10] that produces a fixed-point variable phase, which is then subtracted from the reference phase in order to produce the digital phase error. The phase error is then filtered by a digital loop filter and then normalized by the DCO gain in order to correct the DCO phase/frequency in a negative feedback manner with loop behavior that is independent from process, voltage and temperature. The coarse basic DCO resolution K_{DCO} is significantly improved through $\Sigma\Delta$ dithering of its varactors.

III. TDC RESOLUTION QUALITY PROBLEMS

The TDC quantization of phase error estimation affects the phase noise at the ADPLL output. Under the large signal assumption (spanning multiple quantization levels) the variance of the timing uncertainty is: $\sigma_t^2 = \Delta t_{res}^2/12$. The phase noise is obtained by normalizing the standard deviation of the timing error to the unit interval and multiplying by 2π radians: $\sigma_\phi = 2\pi\sigma_t/T_V$. The total phase noise power is uniformly spread over the span from dc to the Nyquist frequency. The single-sided spectral density is, therefore, expressed as $\mathcal{L} = \sigma_\phi^2/f_R$. Since the closed loop transfer function from the TDC to the ADPLL RF output is unity within the loop bandwidth, the phase noise spectrum at the output due to the TDC timing quantization is

$$\mathcal{L} = \frac{(2\pi)^2}{12}\left(\frac{\Delta t_{res}}{T_V}\right)^2 \cdot \frac{1}{f_R} \qquad (2)$$

Substituting Δt_{res} = 30 ps, f_V = 1.8 GHz, T_V = 556 ps, f_R = 26 MHz, we obtain $\mathcal{L}$ = -94.3 dBc/Hz, which is adequate even for GSM applications. Eq. (2) reveals that the TDC phase noise contribution could be minimized by improving the TDC timing resolution and increasing the sampling rate. Next generations of nanoscale CMOS processes can only bring reductions in Δt_{res} at a scaling rate of 0.7x with each CMOS node.

Eq. 2, describing the effect of the TDC resolution on the ADPLL phase noise, is only valid under the "large signal" assumption in which the timing difference between the FREF and CKV clock edges continuously exercises different quantization levels. This assumption does not hold very well at or near integer-N channels at which the fractional part of the FCW is close to zero (but not *exactly* zero in case of CW or no modulation). The more obvious solution would be to improve the basic TDC resolution [5], [6], [7], [8], [9] as established by the most stable regenerative delay in CMOS, i.e., the inverter delay, typically involving design challenges and increased power consumption. The proposed solution, realized at greater simplicity and without modifications to the core TDC circuitry, is to avoid the problem altogether by selectively moving away from the integer-N relationship.

IV. PARASITIC SPURS

A certain class of coupling spurs, which were found particularly troublesome, were described in detail in [3] for another ADPLL-based transmitter. The mechanism is as follows (see Fig. 2): The RF clock on the feedback path to the TDC (aggressor) gets parasitically coupled into the slicer input of the FREF crystal oscillator (XO). By means of subharmonic modulation of the sinusoidal FREF oscillator waveform, a jitter will be created on the FREF digital clock. The coupling mechanism creates significant distortion of the modulated RF waveform at the integer-N channels when the jitter energy passes through the low-pass loop filter and modulates the DCO. Due to the relentless push towards integration and continued reduction of device feature size, this class of spurs is likely to become even more prevalent in the near future. Fortunately, the above mentioned frequency shift of the feedback clock would completely avoid this issue.

V. OFFSET PLL ARCHITECTURE

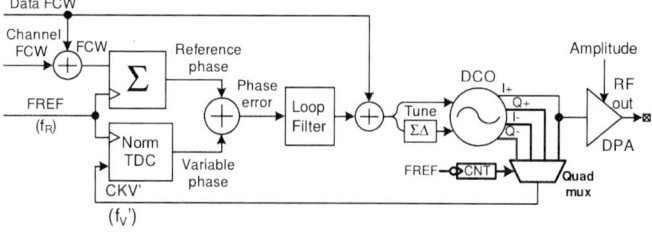

Fig. 3. All-digital offset PLL architecture that realizes frequency translation via continual rotation of the selected DCO quadrature phase. The additional circuitry augmenting the original ADPLL block diagram of Fig. 2 is drawn in bold lines.

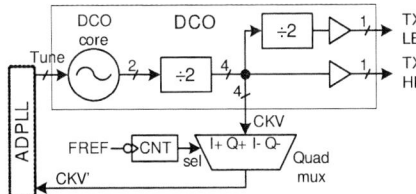

Fig. 4. Details of the DCO system comprising the DCO core operating at 2x of the high-band frequency for generating the quadrature phase clocks. The quadrature clock rotation via the digital mux performs the frequency translation.

The proposed solution to the ill-conditioned integer-N TDC behavior is to frequency translate ($f_V \rightarrow f'_V$) the feedback clock locally at the TDC while, of course, preserving the global f_V / f_R frequency relationship. Since the four clock phases of RF oscillator are always naturally available or required (2x frequency operation of the oscillator core, quadrature receiver mixer, etc.) it would appear most straightforward to change the selected quadrature phase at regular intervals. If the next quadrature phase is selected with each FREF clock via a 2-bit modulo-4 up counter, then the frequency translation will be $-f_R/4$. The implementation is illustrated in Fig. 3 with the additional DCO details shown in Fig. 4. The counter runs on the negative edge of FREF. The 4:1 quad mux is realized using pass-gates and fully-static gates and requires no special RF considerations. The area overhead is very small. The potential timing violations during the mux selection change are of no consequence since the TDC is mostly inactive except at the vicinity of the rising FREF edges. Consequently, the system-level operation of the DCO phase change is hitless.

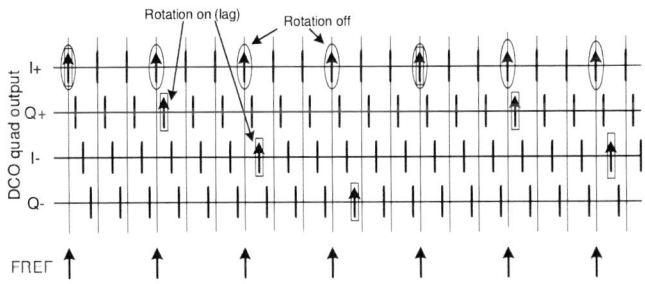

Fig. 5. Timing diagram of quadrature phase (i.e., I+, Q+, I-, Q-) rotation vs. no rotation. FCW = 3.0, FCW' = 3.25.

Fig. 5 illustrates a timing diagram of the quadrature clock rotation that realizes the negative frequency translation by $f_R/4$. The phase error comparison of the DCO clock is performed on every rising edge of the FREF clock. The FREF and DCO edges being active in the comparison process are marked with vertical arrows.[1] The selection of the DCO clock phase changes by 90° with each new FREF edge. The selected DCO phases are marked by the enclosing thin rectangles. A reference case of no rotation is included for comparison in which the arrows are enclosed in ovals. In this example, the

[1]Note that this specific implementation of the TDC-based phase detection process uses the closest *preceding* DCO clock edge, rather than the closest edge, however the two cases are equivalend with one clock-cycle delay.

frequency division ratio N or FCW = f_V / f_R = 3.0 and the feedback clock will keep on selecting the same phase, which in this case is I+. The TDC-based phase detector will see a constant input, with all the artifacts. During the quadrature rotation, however, the comparison edges will experience constant rotation of one-quarter of the DCO cycle with each FREF edge. This is equivalent to a frequency translation of 1/4 of the reference frequency, and the corresponding FCW' (seen by the phase detector) will be FCW' = FCW - 1/4. The 4-level counter mode (increment/decrement) that controls the mux selection determines the frequency translation direction (down/up). For a great majority of the channels that do not experience the near-integer-N condition, the counter is reset to I+. Since the phase detection process uses only one DCO edge at every comparison cycle, the quadrature phase switchover can be done hitlessly if the counter output changes around the falling FREF edge.

VI. MEASURED RESULTS

Fig. 6. Chip micrograph of 65-nm CMOS.

The proposed solution has been implemented in a 65-nm CMOS process technology. Fig. 6 shows a micrograph area of the frequency synthesizer and transmitter. The clock rotation mux and the counter-based mux selector are part of the digital logic section of ADPLL. The technique has been successfully verified by comparing the RF performance before and after the engagement of the frequency translation of the feedback clock.

Fig. 7 shows the spectrum of the un-modulated carrier at 467.7 kHz above the closest FREF harmonic. The center of the plot lies two GSM channels (i.e., 400 kHz) above the FREF harmonic, but the all-ones GSM data makes an additional 67.7 kHz shift. Due to the close separation of the DCO frequency and FREF harmonic, and thus resulting beating tones in the TDC, the spectrum (blue curve) normally experiences excessive distortions. The frequency separation was deliberately chosen to be not too small, as the distortions would be masked by the high power of the close-in phase noise, and not too large, as the generated tones would be well outside of the PLL filtering range. The engagement of the frequency translation of the feedback clock (red curve) largely eliminates the distortions.

Even though the frequency modulation tends to spread around the spurious tones and further mask them with the modulating energy, there are certain frequency offsets that are still vulnerable to these distortions. The modulating energy keeps on decreasing away from the carrier until about 350 kHz

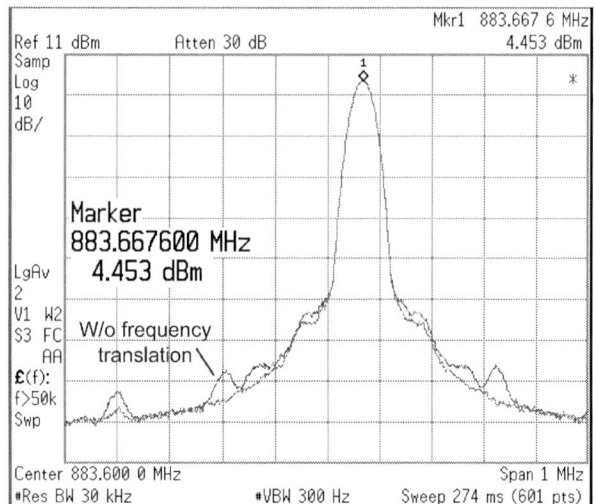 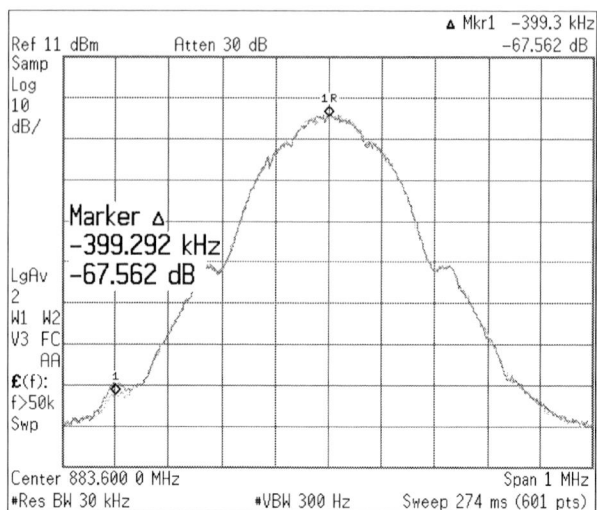

Fig. 7. Measured un-modulated spectrum at 883.6 MHz carrier (with an additional 67.7 kHz frequency offset due to all-one modulation) without (blue curve) and with (red curve) the frequency translation of the feedback clock. The 23rd harmonic of the 38.4 MHz reference frequency is 400 kHz below at 883.2 MHz. The spurious tones at the reference harmonic due to the parasitic coupling and other spectral distortions due to insufficient randomization of the TDC output are largely eliminated with the frequency translation.

Fig. 8. Measured GMSK modulated spectrum at 883.6 MHz carrier without (violet curve) and with (blue curve) the frequency translation of the feedback clock. The 23rd harmonic of the 38.4 MHz reference frequency is 400 kHz below at 883.2 MHz. The spectral mask degradation at the reference harmonic shows 2–3 dB of improvement.

away where it is so low that the phase noise and spurious contributors can start to dominate. At the same time, the 400 kHz frequency offset is one of the most vulnerable in the modulating spectrum specification. Larger frequency offsets present less of a problem due to the quickly decaying power of the parasitic coupling and stronger filtering. Consequently, the above makes it imperative to test the system behavior at the 400-kHz offset. Thus, Fig. 8 shows the GSM modulated spectrum at the 400 kHz above the reference frequency harmonic. The resulting energy at -400 kHz offset from the carrier exhibits a strong tone, which gets reduced by 2–3 dB (to -67.5 dB with respect to the 0 kHz offset) when the proposed frequency translation is engaged. The modulation distortion was measured to reduce from 2.2° to 1.8° rms. The transmitter is fully compliant with the GSM/EDGE standard and no adverse degradation in other RF specifications was observed.

VII. CONCLUSION

In this paper, we propose a new all-digital PLL architecture, in which the RF clock has an option of a simple frequency translation in order to avoid various issues pertaining to the near-integer-N frequency relationship between the synthesizer RF clock and the frequency reference clock. The main issues associated with near-integer-N operation are: (1) insufficient randomization of the TDC quantization noise, whose energy falls in-band and thus cannot be filtered by the loop filter, thus degrading the modulation distortion and possibly violating the modulation mask; (2) spurious tones due to a certain type parasitic coupling. The elective frequency translation by one-fourth of the reference frequency in either direction simply breaks the near-integer-N frequency relationship thus

fundamentally avoiding the problem. The effectiveness of the proposed technique is successfully demonstrated in a 65-nm CMOS transceiver.

REFERENCES

[1] R. B. Staszewski, J. Wallberg, S. Rezeq, et al., "All-digital PLL and transmitter for mobile phones," *IEEE J. Solid-State Circuits*, vol. 40, iss. 12, pp. 2469–2482, Dec. 2005.

[2] R. B. Staszewski, D. Leipold, O. Eliezer, et al., "A 24mm² quad-band single-chip GSM radio with transmitter calibration in 90nm digital CMOS," *Proc. of IEEE Solid-State Circuits Conf.*, sec. 10.5, pp. 208–209, 607, Feb. 2008.

[3] O. Eliezer, B. Staszewski, I. Bashir, S. Bhatara, and P. T. Balsara, "A phase domain approach for mitigation of self-interference in wireless transceivers," *IEEE Journal of Solid-State Circuits*, vol. 44, iss. 5, pp. 1436–1453, May. 2009.

[4] S. D. Vamvakos, R. B. Staszewski, M. Sheba, and K. Waheed, "Noise analysis of time-to-digital converter in all-digital PLLs," *Proc. of Fifth IEEE Dallas Circuits and Systems Workshop: Design, Application, Integration and Software (DCAS-06)*, pp. 87–90, Oct. 2006, Dallas, TX.

[5] R. Tonietto, E. Zuffetti, and R. Castello, "A 2MHz bandwidth low noise RF all digital PLL with 12ps resolution time-to-digital converter," *Proc. of European Solid-State Circuits Conf.*, pp. 150–153, Sept. 2006.

[6] C.-M. Hsu, M. Z. Strayer and M. H. Perrott, "A low-noise, wide-BW 3.6GHz digital $\Sigma\Delta$ fractional-N synthesizer with a noise-shaping time-to-digital converter and quantization noise cancellation," *Proc. of IEEE Solid-State Circuits Conf.*, pp. 340–341, Feb. 2008.

[7] C. W. Wu, E. Temporiti, D. Baldi and F. Svelto, "A 3GHz fractional-N all-digital PLL with precise time-to-digital converter calibration and mismatch correction," *Proc. of IEEE Solid-State Circuits Conf.*, pp. 344–345, Feb. 2008.

[8] M. Lee, M. E. Heidari, A. A. Abidi, "A low-noise, wideband digital phase-locked loop based on a new time-to-digital converter with subpicosecond resolution," *VLSI Symposium on Circuits*, pp. 112–113, June 2008.

[9] J. Tangudu, S. Gunturi, S. Jalan, J. Janardhanan, R. Ganesan, D. Sahu, K. Waheed, J. Wallberg, R. B. Staszewski, "Quantization noise improvement of time to digital converter (TDC) for ADPLL," *Proc. of 2009 IEEE Intl. Symp. on Circuits and Systems*, pp. 1020–1023, May 2009.

[10] R. B. Staszewski, S. Vemulapalli, P. Vallur, J. Wallberg, and P. T. Balsara, "1.3 V 20 ps time-to-digital converter for frequency synthesis in 90-nm CMOS," *IEEE Trans. on Circuits and Systems II*, vol. 53, no. 3, pp. 220–224, Mar. 2006.

An Interstage Filter-Free Mobile Radio Receiver with Integrated TX Leakage Filtering

Rastislav Vazny[1], Werner Schelmbauer[1], Harald Pretl[1], Stefan Herzinger[2], and Robert Weigel[3]

[1] Danube Integrated Circuit Engineering GmbH, A-4040 Linz, Austria
[2] Infineon Technologies AG, 81677 Munich, Germany
[3] Institute for Electronics Engineering Friedrich-Alexander-University Erlangen-Nuremberg, Cauerstr. 9, D-91058 Erlangen, Germany; 81677 Munich, Germany

Abstract — A multi-standard, multi-band fully-integrated interstage filter-free receiver with integrated auto-centered TX leakage filtering is fabricated in a 65 nm CMOS technology. The measured TX selectivity in UMTS band II is 9.1 dB, the receiver gain at 1.96 GHz is 54.1 dB, and NF is 3.68 dB. The 0.5 dB reference sensitivity degradation caused by TX leakage is reached at a TX power level of -20 dBm/3.84MHz.

Index Terms — UMTS receiver, On-chip filtering

I. INTRODUCTION

In recent years, the removal of external interstage SAW (Surface Acoustic Wave) or BAW (Bulk Acoustic Wave) filters together with the external LNA's (Low Noise Amplifier) has become a mandatory feature for a state-of-the-art full-duplex capable wireless receiver. Since the duplex filters which are currently available on the market can provide only certain isolation of around 55 dB between the Receiver (RX) and Transmitter (TX) the linearity and noise requirements of the RX has to be increased. The design for such tough requirements results in increased current consumption. An alternative way of implementation is the usage of on-chip filtering which relaxes the RX block requirements.

In the past, several attempts have been made to implement on-chip TX leakage filtering with Q-enhanced filters [1], [2]. The technical feasibility of such high-Q integrated filters is questionable since the centering of the filter and also the control of the Q value are very problematic, especially over PVT conditions.

In this paper we present an interstage filter-free receiver with automatically centered and fully integrated TX leakage filtering which can be turned off in case of reduced TX output power.

II. ARCHITECTURE AND IMPLEMENTATION

In Fig.1 the proposed multi standard multi-band direct conversion receiver (DCR) architecture with integrated TX leakage filtering is shown. The receiver architecture

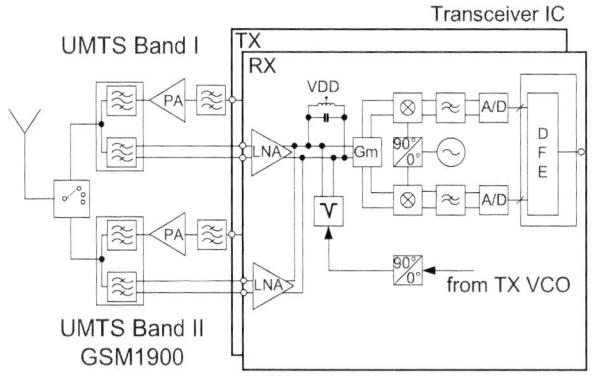

Fig. 1: Receiver Architecture.

allows co-banded operation, i.e. UMTS Band II and GSM1900 operation is possible using one physical input port. The inductively degenerated LNA topology has been chosen due to its good linearity properties with reasonable power consumption [3]. All LNA's are connected in the current domain, and are sharing one common LC tank. The center frequency of the LC tank is tuned by switching capacitors in parallel to the inductor. Therefore the LC tank can cover a wide tuning range of 800 MHz. The TX leakage signal coming from the power amplifier (PA) is filtered out before reaching the mixer stage by the integrated notch filter, which is centered by the internal TX local oscillator (LO) signal.

A. Requirements

Analysis shows, that the TX leakage signal at the RX input port, compared with other blocking signals from the 3GPP specification [4], represents the strongest blocking signal for the RX. Assuming a duplex filter with TX to RX selectivity performance of -55 dB and 4.5 dB TX front-end insertion loss, a TX leakage signal P_{TX} of -26.5dBm/3.84 MHz is present at the RX input at TX output power of 24 dBm at the antenna.

One of the major drawbacks of the DCR receiver architecture are the tough second order linearity

---

978-1-4244-6240-7/10 $26.00 © 2010 IEEE

requirements [5]. For a worse-case sensitivity degradation of 0.5 dB due to TX IM2 noise, the TX IM2 noise should be at least 9 dB below the RX noise level of -104.1 dBm (Thermal noise + RX noise figure). The required RX IIP2 can be calculated as:

$$
\begin{aligned}
IIP2_{TX} &= 2(P_{TX} - 3) - N_{IM2} + CF_{IM2} \\
&= 2(-26.5\text{dBm} - 3) + 113.1\text{dBm} - 7.7\text{dB} \qquad (1) \\
&= 46.4\text{dBm}
\end{aligned}
$$

where CF_{IM2} of -7.7 dB is the correction factor between two tone IIP2 characterization and real modulated UMTS UL channel [5].

B. On-chip TX leakage filtering

Basically the TX leakage filtering, as depicted in Fig. 2, relies on the impedance translation properties of a passive mixer. The notch filter (or band-stop) topology has been chosen since the effect on the wanted signal gain of the receiver is less pronounced than in a band-pass architecture. From the equations presented in [6] and after some mathematical operations, the input impedance Z_{RF}, of a passive mixer terminated with a baseband impedance Z_{BB} can be calculated (for the fundamental harmonic) as:

$$
Z_{RF} = \frac{4}{\pi^2} \left[Z_{BB}(\omega_{LO} + \omega_m) + Z_{BB}^*(\omega_{LO} - \omega_m) \right] \qquad (2)
$$

Note that R_{SW}, the resistance of the mixer switching transistors was neglected. Eqn. (2) shows the translation of the baseband impedance Z_{BB} to the RF domain around the frequency f_{LO}. In case the notch filter is centered at the TX frequency, the LO frequency of the translation mixer should be equal to $f_{LO} = f_{TX}$. For minimum affect on the RX gain, Z_{BB} at the duplex distance[1] should be ideally infinite ($Z_{BB(\omega = 2\pi f dpx)} = \infty$). To generate such a baseband impedance characteristic a trans-impedance amplifier (TIA) has been used.

The input impedance of a TIA can be written as:

$$
Z_{in} = \frac{R_f}{1 + A_{OL(\omega)}} \qquad (3)
$$

where R_f is the feedback resistor and $A_{OL}(\omega)$ is the open loop gain of the operational amplifier. For a typical $A_{OL}(\omega)$ value of 80 dB at $\omega = 0$ and an R_f of 1 kΩ Z_{in} is approximately 0 Ω. At $\omega = 2\pi f_{dpx}$ and $A_{OL}(\omega)$ is around 0 dB the resulting Z_{in} is ~ R_f. Therefore with the TIA the desired base-band impedance can be realized. Furthermore with proper control of $A_{OL}(\omega)$ of the TIA, the bandwidth of the notch filter can be tuned. For the TIA

---

[1] In UMTS Band II the duplex distance f_{dpx} = 80MHz.

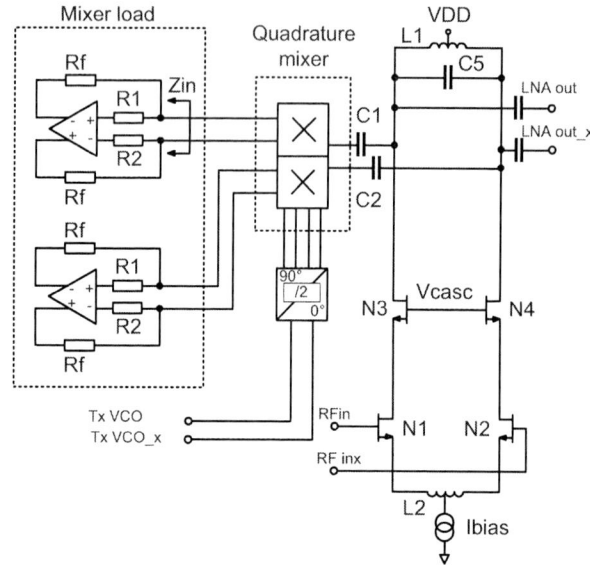

Fig. 2: LNA with integrated TX leakage filtering (biasing not shown).

implementation a class-AB operational amplifier has been chosen.

Unfortunately, the mixer doesn't translate only the impedance from $f_{LO} - f_{TX}$, but also from the $f_{LO} + f_{TX}$ frequency. To avoid the impedance degradation of the TIA at high frequencies, resistors R_1 and R_2 have been used. The degradation is caused by the capacitive parasitics of the input transistors of the operational amplifier. By inserting R_1 and R_2, the open loop gain $A_{OL}(\omega)$ is not affected in the low frequency range.

In case the phase difference between the TX leakage signal and the LO signal reaches 90°, the notch filter input impedance gets very high and the notch filter doesn't work properly. This issue can be overcome with the use of a quadrature mixer together with two TIA loads as can be seen from Fig. 2. The drawback of such a solution is the doubled power consumption compared to single mixer solution, and the generated high impedance characteristic by the TIA has been halved.

C. Mixer and BB blocks

The implemented mixer topology of the receiver consists of a Gm input stage and a passive current mixer [7] as shown in Fig. 3. The passive current mixer has been chosen due to its very good flicker noise performance to fulfill the tough noise requirements in GSM mode, because the signal is directly down-converted to the baseband domain. A TIA is used to convert the current signal to the voltage baseband (BB) signal. The TIA is simultaneously acting as a first channel-select filter. All BB operational amplifiers are implemented as class-AB

978-1-4244-6240-7/10 $26.00 © 2010 IEEE   16

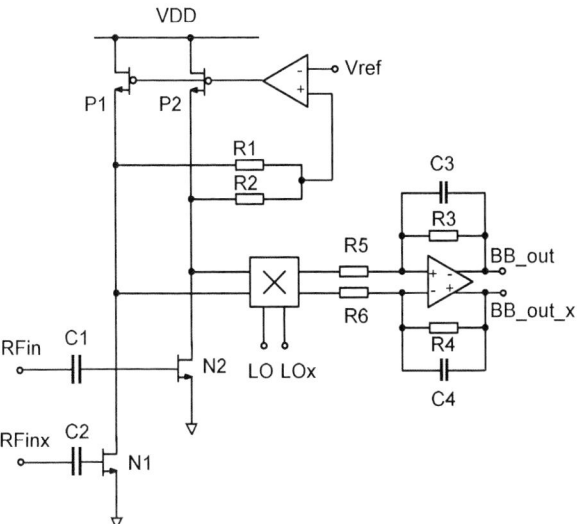

Fig. 3 Mixer topology.

amplifiers. The receiver includes also an Analog to Digital Converter (ADC) together with digital front end as presented in [7].

III. MEASUREMENT RESULTS

To verify the proposed concepts and theory, an IC was fabricated in a 65 nm standard CMOS technology. The photography of the fabricated IC is shown in Fig. 4. The integrated transmitter is not shown, since it is not a part of this work. The receiver is housed in a wafer-level package.

In Fig. 5 the measured receiver gain in UMTS Band II is shown. The measured selectivity at an offset of 80 MHz below the receive frequency is 9.1 dB. The notch filter is automatically centered at the TX frequency due to the impedance translation concept. The measured RX gain

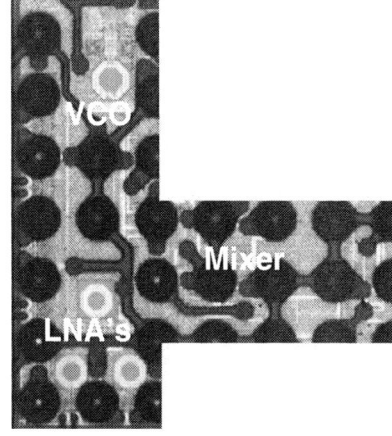

Fig. 4 Chip photograph (RX section).

drops about 2.3 dB when the notch filter is switched on. The gain drop in this case is caused by the layout parasitics. Due to the fact, that this gain drop happens in the LNA, the receiver NF increases from 2.9 dB when the filtering is off to 3.68 dB when the TX filtering is powered on.

The second and third-order linearity of the RX has been measured by a two tone measurement. The measured half-duplex IIP3 (the first tone is located at 1880 MHz and the second at 1920 MHz) with TX filtering of the receiver is -5 dBm, and is dominated only by the IIP3 performance of the LNA, since the filtering circuit is located at the LNA output.

The duplex IIP2 has been measured with tones located at 1879.5 MHz and 1880.5 MHz. The measurement result is shown in Fig. 6. The IIP2 reaches the value of +63.4 dBm, which is a 13.9 dB improvement compared to +49.5 dBm without the TX notch filter.

The measurement of reference sensitivity is shown in

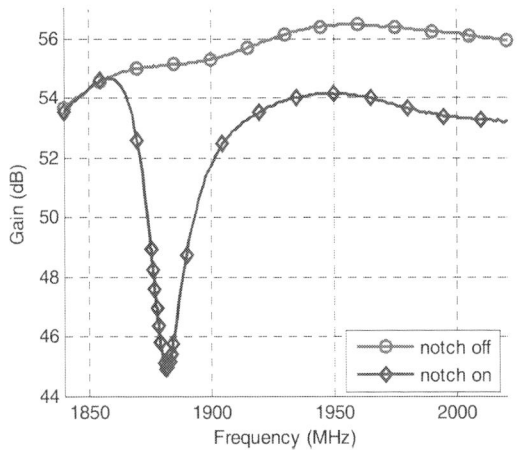

Fig. 5.   Measured RX gain. Notch filter auto centered at the mid TX frequency of UMTS Band II.

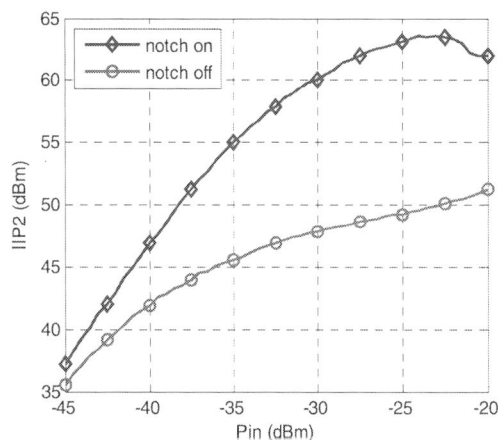

Fig. 6.   Measured duplex input referred second-order intercept point.

978-1-4244-6240-7/10 $26.00 © 2010 IEEE

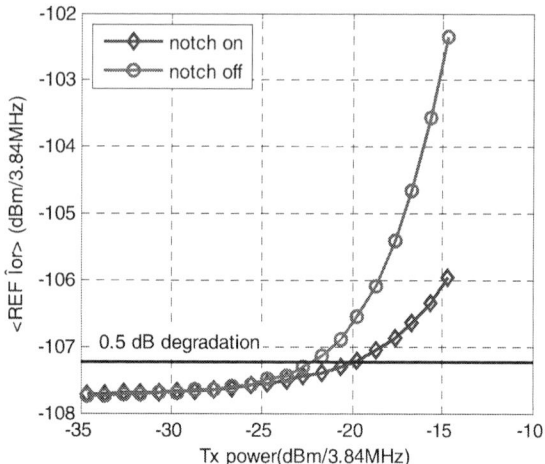

Fig. 7. UMTS Band II reference sensitivity measurement (BER<10^{-4}).

Fig. 7. To do a fair comparison, the gain in the case with disabled notch filter has been reduced to the gain level when the notch filter is powered on. The 0.5 dB reference sensitivity degradation caused by TX leakage signal is reached at a TX power level of - 20dBm/3.84 MHz with integrated filtering.

The power consumption of the whole RX has been

TABLE I
MEASUREMENT RESULTS SUMMARY

PERFORMANCE	This work		[1]*	[2]*	Units
	w/o filtering	w/ filtering			
Frequency	1960	1960	1960	1900	MHz
Gain (analog)	56.4	54.1	21.8	24.7	dB
Selectivity	1.3	9.1	9	15	dB
NF	2.9	3.68	4	8.8	dB
IIP2	+49.5	+63.4	-	-	dBm
IIP3	-10.5	-5	-1.52		dBm
S11	<20	<20	<20	<20	dB
Power consumption	40	75.6	21.58	375	mW
Required Calibration	no	no	yes	yes	-
CMOS Technology	65	65	65	65	nm

*) Only LNA measurement with integrated filtering

measured under nominal conditions as 75.6 mW including LNA, mixer, BB. The power consumption of the TX leakage filtering is dominated by the power consumption of the TX LO distribution, since the LO path has length of 3.1 mm. Generally, the additional power consumption of

the TX leakage filtering can be neglected, since the filtering is powered up only for the highest TX output power. All measurement results, together with comparison of other work are summarized in Table I.

VII. CONCLUSION

A multi-standard, fully-integrated receiver with integrated TX leakage filtering fabricated in 65 nm CMOS technology is presented. The integrated TX leakage filter is automatically centered by design and needs no further tuning or calibration. The integrated filtering allows the removal of the SAW interstage filters, and relaxes the requirements of the duplex filters. Measurements showed, that the present receiver meets the 3GPP specifications with enough margin.

ACKNOWLEDGEMENT

The authors gratefully acknowledge the assistance and support of Mr. Hans Pletzer, Dr. Chistoph Dufrene, Peter Noest, Karl Floimayr, Klaus Gasser and Martin Tegeler.

REFERENCES

[1] D. Bormann, T.D. Werth, C. Schmits, S. Heinen, "A 1.3 V, 65nm CMOS, Coilless Combined Feedback LNA with Integrated Single Coil Notch Filter," *IEEE RFIC Symp. Dig.*, pp. 311-314, June 2009.

[2] T.D. Werth, C. Schmits, S. Heinen, "Active Feedback Interference Cancellation in RF Receiver Front-Ends," *IEEE RFIC Symp. Dig.*, pp. 379-382, June 2009.

[3] P. Sivonen, A. Parssinen, "Analysis and optimization of packaged inductively degenerated common-source low-noise amplifiers with ESD protection," *IEEE Trans. Microwave Theory & Tech.*, vol. 53, Issue 4, April 2005 Page(s): 1304 – 1313.

[4] 3rd Generation Partnership Project TS 25.101 User Equipment radio transmission and reception (FDD) Technical Specification Group Radio Access Network December 2006, v7.6.0. [Online]. Available: *www.3gpp.org/ftp/Specs/html-info/25101.htm*

[5] Ali-Ahmad Y. Walid, "Effective IM2 estimation for two-tone and WCDMA modulated blockers in zero-IF," [Online]. Available: *http://rfdesign.com*, April 2004.

[6] A. Mirzaei, H. Darabi, J.C. Leete, Chen Xinyu, K. Juan, A. Yazdi, "Analysis and Optimization of Current-Driven Passive Mixers in Narrowband Direct-Conversion Receivers," *IEEE J. Solid-State Circuits*, vol. 44, pp. 2678-2688, October 2009.

[7] J. Zipper, G. Hueber, A. Holm, "A single-chip UMTS receiver with integrated digital frontend in 0.13 μm CMOS," *EEE ISCAS Symp. Dig.*, pp. 972 - 975, May 2008.

A SAW-less CMOS TX for EGPRS and WCDMA

Kurt Hausmann, Jeff Ganger, Mark Kirschenmann, George B. Norris, Wayne Shepherd, Vivek Bhan, and Daniel B. Schwartz.

Fujitsu Microelectronics of America, 2100 E. Elliot Rd, Tempe, AZ, 85284.

Abstract — A 90 nm CMOS TX path architected for operation without inter-stage SAW filters is shown. The SAW elimination strategy is purely low noise design but the architecture still achieves DG.09 weighted TX current drain of 50 mA from the battery. The combination of a passive interleaved switching mixer plus digital gain control allows 2% EVM at 2 dbm and 4.2% at -78 dbm.

Index Terms — CMOS Integrated Circuit, Mobile Communication, Transceiver, Transmitters, Code division multi-access, Surface acoustic wave filters

I. REQUIREMENTS

The elimination of inter-stage SAW filters for WCDMA RF lineups has been driven by the proliferation of multi-banded phones. Triple band WCDMA is now a requirement for several domestic markets and the frequent addition of international roaming only drives this count higher. A typical WCDMA RF subsystem has 3 db of loss between the output of the PA and the antenna port, 26 db of PA gain, and a minimum isolation between TX and RX in the RX band of 48 db. The modulator must thus achieve < -158 dbm/Hz at the duplex offset or equivalently -159 dbc/Hz to support 24 dbm at the antenna with less than -180 dbm/Hz at the input to the RX without an inter-stage SAW [2][3]. In addition, EVM <= 4% over the entire power range is often required to maximize HSUPA throughput.

II. ARCHITECTURE

The architecture of the TX chain is detailed in Fig.1. The TX analog BB is driven by a 14 bit current mode /R-2R hybrid DAC with 12 bit effective resolution at 62.4 MHz. The performance of the DAC both relaxes the requirements on the TX BB filter and allows 22 db of digital gain control by the preceding multiplier without significant EVM degradation. The TX BB filter is a 3'rd order Butterworth whose source follower output stage drives the input of the Interleaved Switching Mixer. Carrier feed through is controlled by

zeroing the dc offset (dcoc) at the output of the source followers with the mixers enabled. Since dc offsets are the primary contributor to carrier feed through in this architecture it can be corrected to better than 60 dbc at full power by the dc offset correction subsystem. The RF voltage at the output of the ISM is amplified into RF power by a differential Segmented Variable Gain Amplifier. The first generation design described here has six separate SVGA/balun pairs, two for EGPRS and four for WCDMA. EDGE is supported through an equivalent RF path to WCDMA with changes only to the bandwidth of the BB filter and the DAC clock rate. GMSK is supported by FM modulation of the TX through a separate divider and buffer in parallel with the EDGE ISM/SVGA. The provisioning of a separate path for GMSK/EDGE is due only to the need for flexibility in the choice of RF PA architecture and not any limitation of the TX topology.

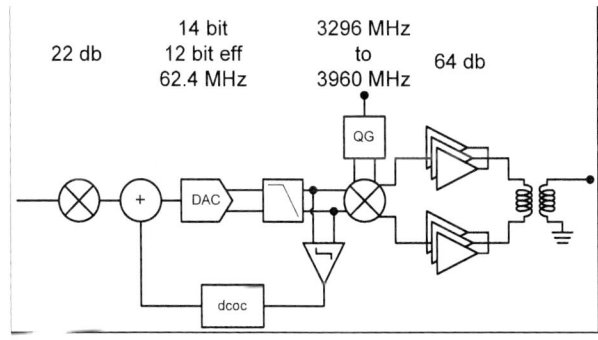

Fig. 1. The complete TX architecture from the digital mixer to the balun. The text along the top of the figure provides the dynamic range for the two gain control points plus some ancillary performance information.

III. CIRCUIT DETAILS

A. ISM

The topology of the ISM and is shown in Figure 2. Unlike the more common current mode designs where I and Q are up-converted by parallel paths [?][3][4][5] the ISM directly generates the RF sequence of voltages I, Q, -I, -Q, ... The virtues of this approach are threefold.

First and foremost, it is a passive voltage mode mixer and as a result adds only LO phase noise to the BB signal. The LO generation is accomplished by using the output of the quad gen to gate the VCO or VCO/2 signal, thereby generating four ¼ cycle non-overlapping pulses that sequentially select the desired input. This method reduces noise contribution over using combinatorial logic to generate the interleaved pulses [1]. The ISM mixer switch gates are AC coupled to allow an offset DC bias to be applied and thus a tradeoff between linearity and insertion loss. Secondly, it can support large voltage swings without compromising linearity, 1.4V peak differential for the design discussed here, thus maximizing SNR. Third, separate RF paths for each channel are not required because the I and Q channels are sampled and summed in time. It is advantageous to reduce to a single RF path quickly to minimize opportunities for channel mismatch and to save RF current. In addition, mixer transistor clock feed through produces a tone at 4X the LO frequency so the RF contribution to carrier leakage is intrinsically very low.

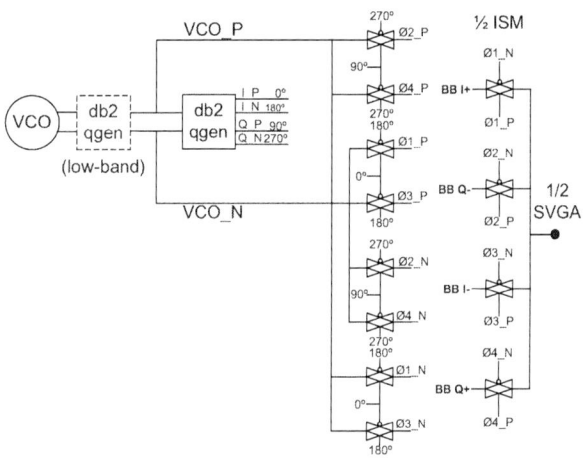

Fig. 2. Topology of one half of the Interleaved Switching Mixer. The pulses enabling the mixer segments are obtained by gating VCO output with the quad gen output via pass gates. The result is a periodic sequence of quarter cycle LO pulses that select I, Q, -I, -Q in sequence.

B. SVGA

RF power gain is provided by segmented variable gain amplifiers (SVGA), each comprised of switched CMOS inverters with capacitive feedback. In other SAW-less architectures equivalent functionality is achieved by segmenting the modulator itself [4]. In order to achieve the desired output power and linearity without dropping the impedance presented to the balun by the SVGA, the 3 highest gain stages are operated directly from the 2.7V supply. In order to support the targeted output power into a 2:1 mismatch presented at the output, the 2.7V SVGA stages use 21Å RF transistors for the gain elements, but are connected to the balun via 50Å cascode transistors. The lower gain stages are simple CMOS inverters without the cascode transistors and use a 1.6V supply that is internally derived from an external 1.8V dc to dc converter. Each stage is biased in Class A operation to maximize linearity and minimize RF signal injection into the grounds and supplies. Reduced supply voltage and stage periphery are used to minimize the current of the Class A stages as power is backed off.

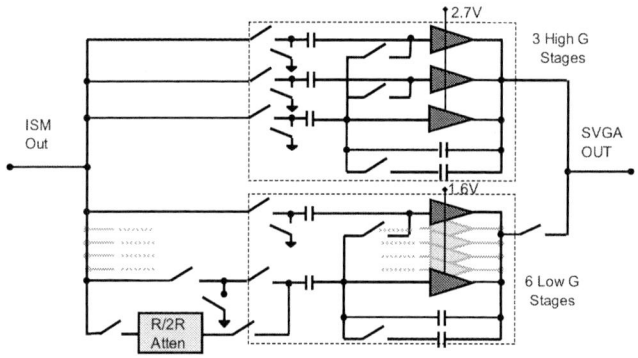

Fig. 3 The architecture of the Segmented Variable Gain Amplifier. The actual design is fully differential with a balun for differential to single ended conversion.

Each added SVGA stage covers a ~2.8dB gain step, with additional gain adjustment via 4 switchable capacitive feedback states for both the 2.7 and 1.6V circuits, providing a total of 32 programmable gain states. At the lowest power levels the smallest 1.6V CMOS stage is driven by an R/2R attenuator for an additional 32 gain steps. 16 of the available 64 gain states are typically selected for power control and are optimized for 3G TPC accuracy. As stages are disabled and isolated for gain cutbacks, the net input capacitance goes down with decreasing gain, thus lowering the drive requirements from the ISM and BB filter with a concomitant reduction in current drain.

C. TX BB

The TX BB filter is a 3'rd order Butterworth with a passive pole at its input and a source follower output stage. The 3 db corner is nominally at 5.6 MHz MHz for 3G and 1.5 MHz for 2G. The topology relies heavily on an RC filter between the output of the op amp and the gate of the source followers to achieve the far out noise target of -178 dbV/Hz as required for SAW-free operation. The combination of routing and the input capacitance of the SVGA at maximum gain present a load of 0.8 pF at the output of the ISM. Thus at maximum gain the source follower in each leg of the modulator must deliver a minimum of 2 X 2 GHz X 0.8 pF X 1.4 V = 4.5 mA to the input of the ISM as derived from the 2.7V supply. To minimize current drain the source follower tail current is steered from leg to leg by the source follower input level. To further minimize the baseband current drain the tail current magnitude is scaled with RF frequency, worse case peak-average ratio, baseband drive level, and SVGA gain.

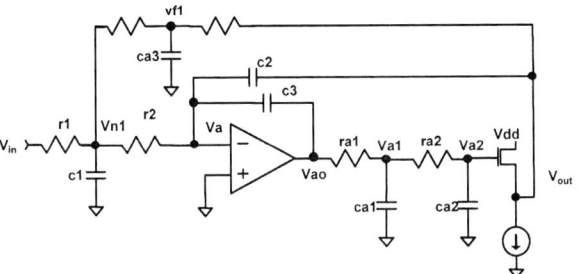

Fig. 4   Single ended half of the TX baseband filter. In order to maximize headroom the amplifier operates directly from the 2.7V supply. Not shown is the additional circuitry required to support steering of the tail currents.

IV. RESULTS

The circuit has been implemented as a sub-system in a single chip 90 nm CMOS HEDGE transceiver. The device has 6 layers of metal and is packaged as a FC-LGA. In order to optimize LO phase noise the VCO inductors are integrated into the HDI substrate.

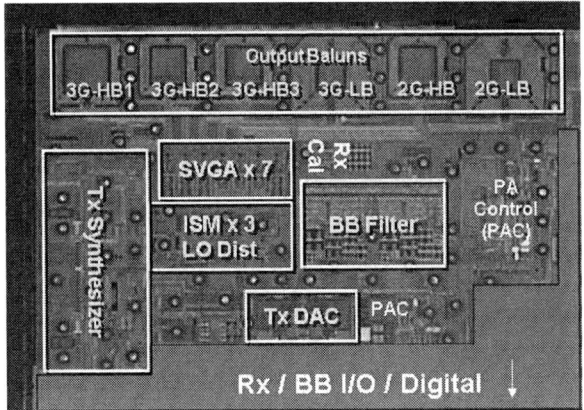

Fig 5   Die photograph of the TX section of the transceiver

An important feature of the architecture is amount and resolution of the RF gain control. The preferred gain control strategy is to set the baseband gain to the highest level consistent with the linearity goals at maximum power and cut back the SVGA gain until the R-2R attenuator is deployed. Fine resolution is achieved by using gain substitution with the digital baseband gain offsetting coarse RF gain steps. The measured band I duplex offset noise with this gain control strategy is shown in Fig. 6 below. At maximum power, the relative noise density is -163.6 dbc/Hz, a value which drops 0.74 db per db of output power in good agreement with simulation. This performance is readily understood by observing that the input noise figure of the SVGA segments is nearly constant; the result is the SVGA output noise closely tracks the SVGA gain.

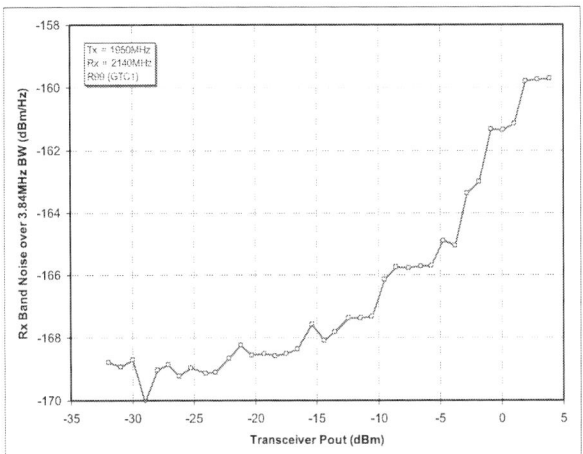

Fig. 6   RX noise for band 1 as a function of output power.

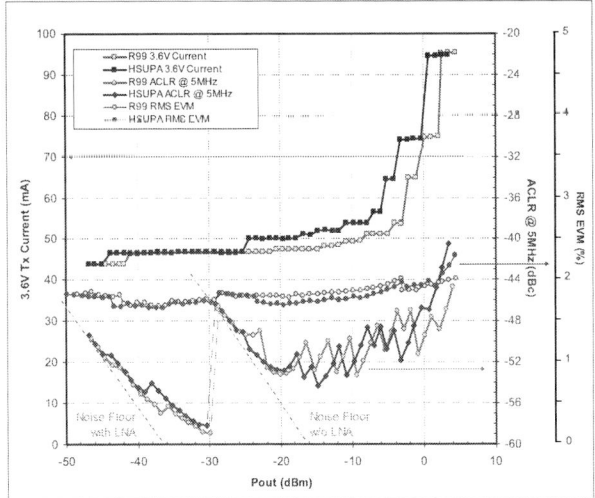

Fig. 7   Measured performance vs Pout for band I. Current is from a 3.6V battery with 90% switcher efficiency.

Parameter	Min	Typ	Max	Unit
RX band noise	-161.7	-159.8	-158.7	dbm
EVM @ -78 dbm	2.0	4.2	8.8	%
EVM @ +2 dbm	1.8	2.0	2.4	%
Gain variation : V & T	-1.0	0.0	0.7	db
ACLR R99 @ 2 dbm	-52.3	-50.2	-47.8	db
ACLR HSUPA @ 2 dbm	-45.7	-43.9	-42.3	db
DG.09 weighted current		47		mA

Fig. 8   Additional performance characteristics for band I over temperature and voltage on three parts. The temperatures are -30C, 25C, and 85C and voltages +/- 5%.

In addition to the excellent noise performance under back-off, other critical RF parameters also perform well, especially the EVM as shown in Fig. 7 and Fig. 8. The mechanism which enables this performance is the combination of the inherently low carrier feed-through of the ISM combined with modest reliance on baseband gain control. The compromise is the ACLR flattens under back-off but since it flattens to roughly -52 dbc against the 3GPP target of -33 dbc radio level performance is not compromised.

A final factor in determining the robustness of a SAW-less TX is the additional margin the lineup requires to absorb variations with PVT. As seen in Fig. 8, the RX band noise maintains 0.8 db of margin at -85 C and the gain variation is extremely tight with only 1.7 db variation the full range.

V. ACKNOWLEGEMENTS

Amongst the many people who contributed to the design and architecture special note must be given to Larry Connell, Andrew Raclaw, Patrick Rakers, Kurt Sakamoto, Kent Jaeger, and Charles Dozier. Additional thanks for measurement support from Junguang Dai and Rick Sherman.

VI. REFERENCES

[1] Xin He and Jan van Sinderen, "A 45nm Low-Power SAW-less WCDMA Transmit Modulator Using Direct Quadrature Voltage Modulation", ISSCC Dig. Tech. Papers, pp. 120-121, Feb. 2009.
[2] A. Hadjichristos, M. Cassia, et al, "Single-Chip RF CMOS UMTS/EGSM Transceiver with Integrated Receiver Diversity and GPS", ISSCC Dig. Tech. Papers, pp. 118-119, Feb. 2009.
[3] T. Sowlati, B. Agarwal, et al, "Single-Chip Multiband WCDMA/HSDPA/HSUPA/ EGPRS Transceiver with Diversity Receiver and 3G DigRF Interface Without SAW Filters in Transmitter / 3G Receiver Paths", ISSCC Dig. Tech. Papers, pp. 116-117, Feb. 2009.
[4] C. Jones, B. Tenbroek, et al, " Direct Conversion WCDMA Transmitter with -163 dbc/Hz noise at 190 MHz", ISSCC Dig. Tech. Papers, pp. 336-337, Feb. 2009.
[5] M. Farazian, B. Asuri, Yu Zhao, L. E. Larson, "A Dual-Band CMOS CDMA Transmitter Without SAW and Driver Amplifier", 2009 IEEE Radio Frequency Integrated Circuits Symposium, pp. 523-526, May 2009.

A Quadrature Charge-Domain Filter with an Extra In-Band Filtering for RF Receivers

Ming-Feng Huang

Information and Communications Research Laboratories, Industrial Technology Research Institute, HsinChu, Taiwan, R.O.C.

Abstract— **A quadrature charge-domain filter (QCDF) is proposed. This QCDF, based on the input phases, could provide a stop-band filtering and support an in-band filtering of noise. Through the use of FIR coefficient, the alias-band rejection (ABR) and in-band suppression (IBS) could be well controlled. Measurement shows the 27.1dB ABR and 31.3dB IBS from out-of-phase signals and in-phase signals, respectively. This QCDF with 36MHz bandwidth and 7.5dBm IIP$_3$ is fabricated on the 90nm CMOS process.**

Index Terms— **Charge-domain filter (CDF), Quadrature Charge-domain filter (QCDF), Time-interleaving CDF (TICDF), Anti-alias filer (AAF)**

I. INTRODUCTION

The evolution of CMOS process limits headroom due to dropping in supply voltage. This causes either poor linearity or bad conversion gain at reasonable power consumption [1]. The switched-capacitor (SC) sampler is good for discrete-time process to integrate with digital logic. Moreover, SC sampler requires low headroom for operation. Hence, the SC sampler is one of solution to relax the low-voltage design in the advance process.

In RF receivers, the charge-domain filter (CDF) is employed in signal filtering, frequency down-conversion, quadrature process, or decimation [2]-[4]. For instance, based on low-pass filter (LPF), the property of IIR filter is adopted to attenuate out-of-band signal for Bluetooth applications [2]. Based on band-pass filter (BPF), the property of FIR filter with quadrature process is proposed for WLAN applications [3]-[4]. Nevertheless, they are hard to suppress in-band noise, such as non-linear tones, pre-stage in-band noise, and leakages. Besides, since the multi-stage of CDFs or pre-stage LNA is usually required to relax the concerns of attenuation and noise floor, the issue of in-band noise is serious especially in low-voltage applications. For those reasons, a prototype of QCDF with an extra in-band filtering of noise is proposed. This QCDF could provide quadrature process, go a BPF to suppress the stop-band signal, and perform a band-reject filter (BRF) to suppress the in-band noise, simultaneously. The advantages of the proposed QCDF are verified through the chip implementation.

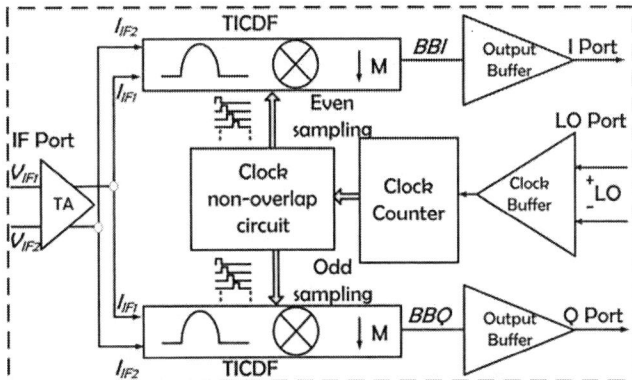

Fig. 1 The proposed QCDF with in-band filtering of noise

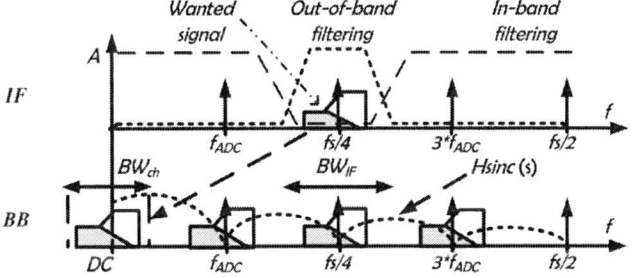

Fig. 2 The frequency diagram of QCDF

II. QUADRATURE CHARGE-DOMAIN FILTER

The proposed quadrature charge-domain filter (QCDF) is shown in Fig. 1. The transconductance amplifier (TA) transfers the input-voltage signals (V_{IF1} and V_{IF2}) into currents (I_{IF1} and I_{IF2}) for time-interleaving charge-domain filters (TICDFs). Through the even and odd sampling, TICDFs provide a quadrature process, filtering, decimation, and frequency down-conversion [3]-[4]. Here, TICDFs go a BPF if V_{IF1} and V_{IF2} are out-of-phase signals. On the contrary, TICDFs run a BRF if V_{IF1} and V_{IF2} are in-phase signals. The output buffers for measurement are also included. In order to produce the corresponding phases for an accurate operation of TICDFs, the digital part contained clock buffer, clock counter, and clock non-overlap circuits is adopted.

The frequency diagram of QCDF is shown in Fig. 2. The QCDF could down-convert the IF signal to baseband (BB). For quadrature process, both the intermediate

978-1-4244-6240-7/10 $26.00 © 2010 IEEE

frequency (f_{IF}) and sampling rate (f_S) are constricted as (1). According to FIR coefficients, QCDF with even and odd sampling could achieve the quadrature outputs at I and Q paths, respectively [3]-[4].

$$f_{IF} = \left| (\frac{1}{4} \pm \frac{M}{2}) f_S \right| \qquad M = 0,1,2,\dots \qquad (1)$$

Owing to the phase of input signals (V_{IF1} and V_{IF2}), the FIR coefficients could be organized. For example, QCDF could validate the differential inputs to be wanted signals and trap the in-phase inputs to be noise. This assumption is easy to implement because the front-end circuit or general CDF in RF receivers provides differential outputs and requests multi-stage structure to low the noise floor or to attenuate stop-band noise. Hence, in Fig. 2, QCDF go a BPF if the inputs are out-of-phase signals, and QCDF run a BRF if inputs are in-phase signals. After down-conversion from f_{IF} to DC, the SC sampler, referred to rectangular clocks and output-sample rate (f_{ADC}), could provide an *SINC* filtering ($H_{sinc}(s)$) to reduce the stop-band noise as shown in Fig. 2. According to (2), a lower f_{ADC} increased the number of the folding frequency (f_{fold}) and thus added the folding noise in band. However, a higher f_{ADC} degraded the flexibility in designing QCDF due to the limited-coefficients. Therefore, to design QCDF, the trade-off between f_{ADC} and flexibility should be concerned.

$$f_{fold} = f_{IF} \pm h f_{ADC} \qquad h = 1,2,3\dots \qquad (2)$$

In this design, we employ the 268MHz f_{IF}, 1072MS/s f_S, and 134MS/s f_{ADC} for an RF sampling receiver on 802.11g applications [3]. The detailed operations of mainly sub-circuit for QCDF would be described below.

A. Charge-Domain Filter (CDF)

In Fig. 3, CDF selected FIR coefficients and referred to signal phases of inputs from V_{in1} and V_{in2} could decide the bandwidth and attenuation. The FIR filter is organized by charging input currents (I_{In1} and I_{In2}) on capacitors (C_{n1-nN} and C_{p1-pN}) and integrating charges into C_o. Note that C_{n1-nN} and C_{p1-pN} according to out-of-phase inputs are for negative and positive coefficients, respectively. On the contrary, C_{n1-nN} and C_{p1-pN} according to in-phase signals are for all positive coefficients. A tunable IIR filter embedded in CDF is shown in Fig. 3 (a). Through V_{IIR} controlled with a DC level, M_{IIR} decides the integrated-history charges on C_{IIR}, achieving an effective capacitor connected to C_o in parallel to adjust the IIR filter. For bandwidth concern, this CDF could release the integrated-history charges by altering both of M_{RR} and M_{EN} to go the FIR filtering only.

In this design, according to input-signal phases, we employed the FIR coefficients as follows:

$$H_{FIR,COM}(z) = (2z^{-1} + 8z^{-7} + 22z^{-11} + 8z^{-15} + 2z^{-21})$$
$$+ (3z^{-3} + 18z^{-9} + 18z^{-13} + 3z^{-19}) \qquad (3)$$

$$H_{FIR,DIFF}(z) = (2z^{-1} + 8z^{-7} + 22z^{-11} + 8z^{-15} + 2z^{-21})$$
$$- (3z^{-3} + 18z^{-9} + 18z^{-13} + 3z^{-19}) \qquad (4)$$

Where, $H_{FIR,COM}(z)$ and $H_{FIR,DIFF}(z)$ are for in-phase signals and out-of-phase signals, respectively. The $H_{FIR,COM}(z)$ performs a BRF, which could provide in-band suppression (IBS) of 25dB at least. The $H_{FIR,DIFF}(z)$ performs a BPF, which could achieve a bandwidth, alias-band rejection (ABR), and stop-band attenuation (SBA) around 152MHz, 33dB, and 48dB, respectively. Note that the ABR means the ratio of down-converted signals between wanted signal and noise below Nyquist frequency. The SBA means the ratio of down-converted signals between wanted signal and stop-band noise. Actually, the SBA is better than ABR according to *SINC* filtering and f_{ADC}. Under the definition of tap-length with ratio of periodic-pulse width to $1/f_S$ as shown in Fig. 3 (b), we adopted tap-length of 24 in CDF design to achieve f_S decimated to 44.67MS/s. Therefore, the time-interleaving technique should be employed for a 134MS/s f_{ADC} to reduce the folding noise.

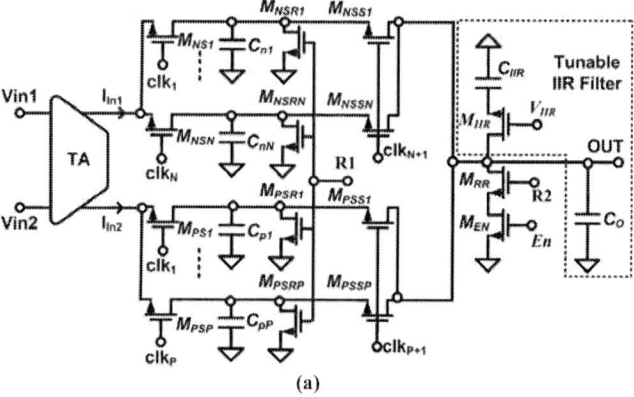

(a)

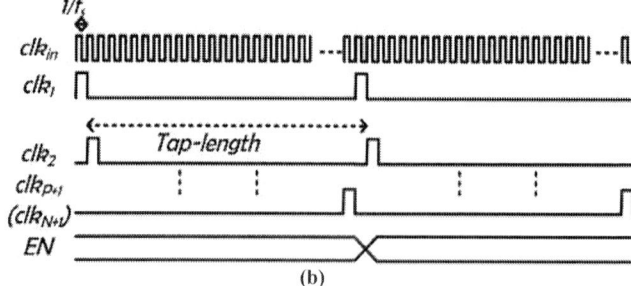

(b)

Fig. 3 The charge-domain filter (CDF): (a) schematic and (b) clocks

B. Time-Interleaving Charge-Domain Filter (TICDF)

In Fig. 4, TICDF combined with TA is presented for QCDF of I or Q path. Three CDFs construct a TICDF and merge the tunable IIR filters and output capacitor (C_O).

The reset clocks (R_{a1}-R_{a3}) and enable clock (*EN*) are options of FIR/IIR filters for wide-band or narrow-band applications. Table I shows the organization of timing control versus FIR coefficients, where a_n, *S*, and *Ru* mean the timings of FIR coefficients, integration, and charge reset, respectively. Moreover, the coefficient items of *COM_In* and *Diff_In* mean the inputs with in-phase signals and out-of-phase signals, respectively. For I-phase or Q-phase TICDF, there are three branches for interleaving the CDFs. Each branch provides 24 slots for FIR coefficients, integration, and charge reset. Owing to tap-length equal to 24, each CDF decimates f_S to 44.67MS/s. Through The time-interleaving with 8 slots among CDFs as Table I, we could provide a periodic output with ($8/f_S$), increasing f_{ADC} to 134MS/s. Therefore, the noise-folding issue as (2) could be relaxed, and QCDF could get a useful f_{ADC}.

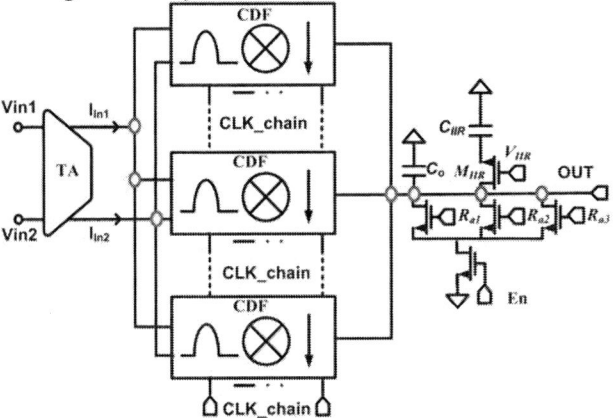

Fig. 4 The time-interleaving charge-domain filter (TICDF)

TABLE I.   FIR COEFFICIENTS AND TIMING CONTROL OF QCDF

1	2	3	4	5	6	7	8	9	10	11	12	13	14	15	16	17	18	19	20	21	22	23	24	CLK
0	2	0	3	0	0	0	8	0	14	0	22	0	18	0	8	0	8	0	0	0	3	0	2	0	0	COM_In Coel. (a_n)
0	2	0	-3	0	0	0	8	0	-14	0	22	0	-18	0	8	0	-8	0	0	0	-3	0	2	0	0	Diff_In Coel. (a_n)
a1/Ru	a2	a3	a4	a5	a6	a7	a8	a9	a10	a11	a12	a13	a14	a15	a16	a17	a18	a19	a20	a21	a22	a23	S	I1
a17	a18	a19	a20	a21	a22	a23	S	a1/Ru	a2	a3	a4	a5	a6	a7	a8	a9	a10	a11	a12	a13	a14	a15	a16	I2
a9	a10	a11	a12	a13	a14	a15	a16	a17	a18	a19	a20	a21	a22	a23	S	a1/Ru	a2	a3	a4	a5	a6	a7	a8	I3
S	a1/Ru	a2	a3	a4	a5	a6	a7	a8	a9	a10	a11	a12	a13	a14	a15	a16	a17	a18	a19	a20	a21	a22	a23	Q1
a16	a17	a18	a19	a20	a21	a22	a23	S	a1/Ru	a2	a3	a4	a5	a6	a7	a8	a9	a10	a11	a12	a13	a14	a15	Q2
a8	a9	a10	a11	a12	a13	a14	a15	a16	a17	a18	a19	a20	a21	a22	a23	S	a1/Ru	a2	a3	a4	a5	a6	a7	Q3

C. Quadrature Process for QCDF

TICDFs, sampled charges with even and odd clocks for I and Q paths, could provide the same $H_{FIR,COM}(z)$ and $H_{FIR,DIFF}(z)$ but obtain the orthogonal outputs [3]-[4]. This is because TICDFs could derive a phase delay with 90 degrees from two samplers by a time delay of $1/f_S$ upon an input frequency as (1). In Table I, we organized CDFs for quadrature process. Two TICDFs contain CDFs of *I1-3*

and *Q1-3* for I and Q paths, respectively. Pay attention to CDFs of *I1* and *Q1* in pair; thus they provided quadrature outputs with 44.67MS/s $f_{S,OUT}$. Furthermore, three pairs of CDFs using time-interleaving as above achieved not only 134MS/s $f_{S,OUT}$ but also quadrature outputs. As a result, two TICDFs constructed an analog part for QCDF could possess the talents for quadrature mixer and filters.

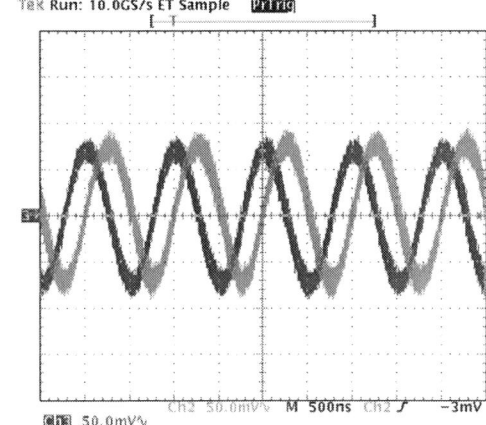

Fig. 5 The time-domain waves with output frequency of 1MHz

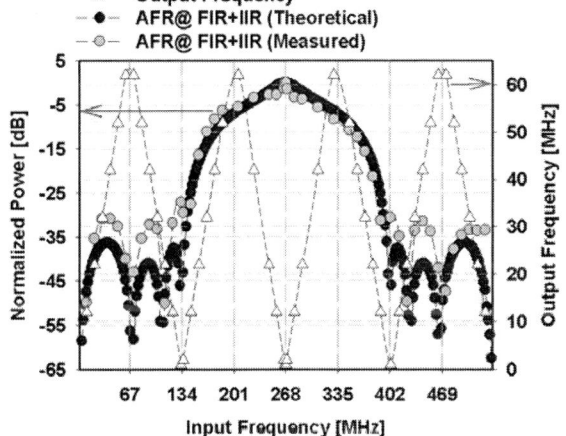

Fig. 6 The measured AFR for out-of-phase signals

III.   MEASUREMENT RESULTS

The proposed QCDF was fabricated on 90nm CMOS. By using 269MHz f_{IF} and 1072MS/s f_S, we obtained a 134MS/s f_{ADC} and 1MHz down-converted frequency as shown in Fig. 5. This demonstrates the quadrature process of QCDF. Through modulating the input frequency from DC to $f_S/2$, we monitored the power upon the corresponding output frequency for ABR. The alias-frequency responses (AFRs) among down-converted frequencies from DC to $f_{ADC}/2$ are measured for out-of-phase signals (in Fig. 6) and in-phase signals (in Fig. 7), respectively. For out-of-phase signals through IIR-filtering compensation, QCDF obtained the ABR larger than 27.1dB and got the bandwidth of 36MHz. For in-

978-1-4244-6240-7/10 $26.00 © 2010 IEEE   25

phase signals, QCDF obtained IBS better than 31.3dB at input frequency of 269MHz. This means that the wanted signals with differential phases could be well down-converted to baseband. Moreover, the common-mode noise could be attenuated once its frequency is identical to that of wanted signals.

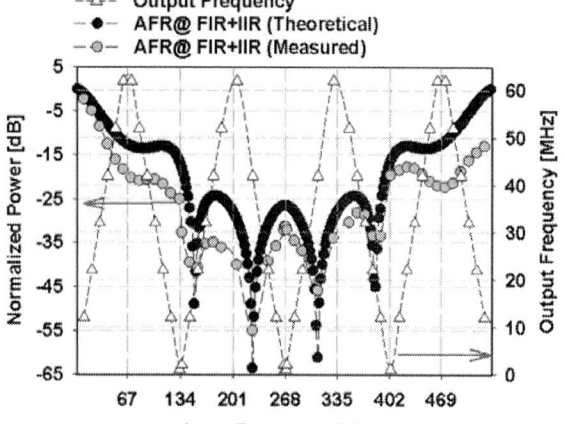

Fig. 7 The measured AFR for in-phase signals

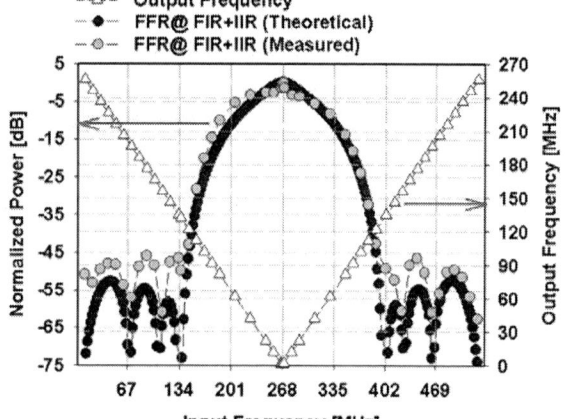

Fig. 8 The measured FFR for out-of-phase signals

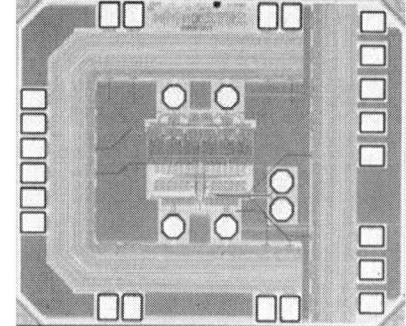

Fig. 9 Chip photomicrograph of QCDF

In Fig. 8, we measured a fundamental-frequency response (FFR) among down-converted frequencies from DC to $f_s/4$. The QCDF obtained the SBA better than 45dB and 4.7dB gain. Moreover, IIP$_3$ was measured around 7.5dBm. The whole chip of QCDF, including the clock

circuit, TICDFs, and output buffer as Fig. 1, consumes only 8.7mA power current from a 1.2V supply. In Fig. 9, the chip photograph of QCDF is presented, where the chip area is about 0.98mm^2. The above performance of QCDF is summarized in Table II for this work in detail.

TABLE II.   SUMMARY OF THE PROPOSED QCDF

	This work	[3]	[4]
Input Sampling Rate [MS/s]	1072	1072	1072
Signal Bandwidth [MHz]	~36	~22	~44
Gain [dB]	4.7	-1	17.2
IIP$_3$ [dBm]	+7.5	+5.5	0
Alias-Band Rejection (ABR)[dB]	>27.1	>18	>32.51
In-Band Suppression (IBS)[dB]	>31.3	0	0
Stop-Band Attenuation (SBA) [dB]	>45	N.D.	>53
Output Noise PSD [dBm/Hz]	-126	-131	-135
Power Supply [V]	1.2	1.8	1.2
Current [mA] (w/i Output Buffer)	8.7	48.33	7.7
Technology (CMOS)	90nm	180nm	90nm

IV.   CONCLUSIONS

In this work, we presented a quadrature charge-domain filter (QCDF), which well combined the talents for quadrature mixer and AAF. Besides, QCDF by controlling both of input phases and FIR coefficients could reduce the in-band in-phase noise, such as common-mode signals and even-harmonic signals. Such ability could relax the low-voltage designs in the advance process, when multi-stage structure is required. The demonstration proved that QCDF could provide IBS, ABR, SBA, decimation, down-conversion, and quadrature outputs. Therefore, this QCDF with respect to tap-length of FIR coefficient is suitable for RF receivers.

V.   ACKNOWLEDGMENTS

The author wants to thank Dr. Szu-Hsien Wu for assistance in chip measurement, and Dr. Tzu-Yi Yang for valuable suggestions to this work.

REFERENCES

[1] A. J. Annerma, et al., "Analog Circuits in Ultra-Deep-submicron CMOS," IEEE J. Solid-State Circuits, vol. 40, pp. 132-143, Jan. 2005.

[2] K. Muhammad, et al.," A Discrete-Time Bluetooth Receiver in a 0.13μm Digital CMOS Process,"IEEE Int'l. Solid-State Circuits Conf. (ISSCC), Feb. 2004, pp. 268-269.

[3] D. Jakonis, K. Folkesson, J. Dbrowski, P. Eriksson, and C. Svensson, "A 2.4-GHz RF sampling receiver front-end in 0.18-um CMOS," IEEE J. Solid-State Circuits, vol 40, pp. 1265-1277, June 2005.

[4] Ming-Feng Huang, Lai-Fu Chen, and Tzu-Lun Chiu, "A Quadrature Charge-Domain Filter with Frequency Down-Conversion and Filtering for RF Receivers," IEEE Radio Frequency Integrated Circuits Symposium, June 2009, pp. 547-550.

RMO1B-2

A Low-Power Receiver Down-Converter with High Dynamic Range Performance

Diptendu Ghosh and Ranjit Gharpurey
University of Texas at Austin

Abstract — **A low-power down-converter that uses a passive current-commutating mixer for frequency translation, while sharing the bias current between the RF and baseband stages is presented. An active noise shaping network is implemented to reduce low-frequency noise at the output. Linearity is enhanced through the use of non-linear feedback. The design, implemented in a 0.18 μm CMOS technology, achieves conversion gain of 35 dB, NF of 9.8 dB, in-channel OIP3 of 15.8 dBV while consuming 2.1 mA from a 1.8 V supply.**

Index Terms — **Active noise reduction, bias current sharing, down-converter, linearization, non-linear feedback.**

I. INTRODUCTION

Receiver front-ends that enhance the achievable dynamic range per unit power are highly attractive for a wide range of wireless applications, including PAN, LAN and cellular systems, as well as emerging applications related to medical telemetry and monitoring. In this work we describe a low-power, dynamic-range optimized down-converter topology for such applications. The design utilizes a current-commutating passive mixer for frequency translation. This topology was employed due to its high dynamic range ([1][2]). Typical implementations of current-commutating passive mixers use a distinct RF input transconductor stage and baseband transimpedance amplifier with independent bias currents. The reported design, however, shares the bias current between the RF and baseband stages, thereby reducing the power dissipation.

Current-reuse techniques can significantly improve the power efficiency of down converters, e.g., [3], [4]. A key differentiating aspect of the bias sharing arrangement in this design is that in principle the full voltage headroom allowed by the supply voltage is available for both the baseband and RF stages, even though the bias current is shared. This is in contrast to techniques that use a stacked bias current sharing, where the available supply voltage is split across multiple stages.

II. PRINCIPLE OF OPERATION

A. Overview

Passive CMOS current-mode mixers (Fig. 1) operate by commutating the RF current provided by an input transconductor using an AC-coupled switching mixer core. The RF transconductor is typically implemented as a common-source or common-gate stage. A differential input is assumed in Fig. 1. A low-impedance baseband termination is employed to

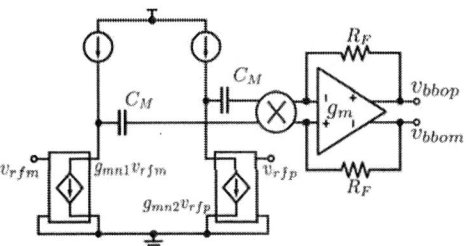

Figure 1: Conventional Passive Current-mode Mixer

minimize the voltage swing at the source and drain terminals of the on-state switches which ensures higher linearity [2] than voltage-mode passive mixers. Since the switches operate with no DC, their flicker noise contribution is minimal. The flicker noise at the output is dominated by that of the baseband transimpedance stage which typically consists of an OPAMP, or transconductor, with resistive feedback. The RF transconductors and the baseband transimpedance amplifier typically use distinct bias paths in these designs.

B. The Basic Bias-current-shared Topology

The power dissipation in the conventional current-mode passive mixer can in principle be reduced by sharing the bias current between the RF transconductor and the baseband transimpedance amplifier. One approach for implementing such a design is to vertically stack the baseband and RF stages, with an explicitly defined intermediate ground. This approach however reduces the headroom available for each of the stages. A different approach is demonstrated in this work. The design shares the bias current between the RF and baseband stages without increasing the voltage headroom requirement. We describe the operating principle of a basic implementation of this design below (Fig. 2). Techniques for enhancing performance of the basic topology and eliminating sources of degradation that arise from bias current sharing are described in the following section.

A differential common-gate RF transconductor stage consisting of NMOS transistors M_{N1} and M_{N2} is employed. The RF input signal is converted into a current by the transconductance of these devices. Capacitors C_M couple the RF current into a passive double-balanced switching mixer, implemented using NMOS devices $M_{NS1} - M_{NS4}$. The baseband stage consists of PMOS devices M_{P1} and M_{P2} that also serve as the loads for the RF input devices. A resistance R_{FB} is connected from the gate to the drain of each PMOS device. A capacitor C_{RF} is connected across the gates of M_{P1} and M_{P2}.

The capacitor C_{RF} plays a critical role in this design.

978-1-4244-6240-7/10 $26.00 © 2010 IEEE

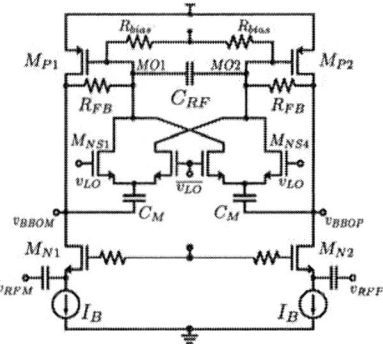

Figure 2: Basic bias-current-shared topology using a common-gate input

We first examine the effect of C_{RF} on differential-mode operation. At baseband, this capacitor is nearly an open. Therefore, looking into the gates of M_{P1} and M_{P2}, a low input impedance is observed due to the shunt-shunt feedback through R_{FB}. Observed differentially, this impedance is of the order of $2\{R_{bias}||(1/g_{mp(1,2)})\}$. The impedance observed looking into the drains of M_{P1} and M_{P2} is approximately $2(R_{FB} + R_{bias})/R_{bias}(1/g_{mp(1,2)})$, ignoring the impedance looking back into the passive mixer from nodes $MO1$ and $MO2$ at baseband. At RF, on the other hand, this capacitor presents a low impedance, and therefore the gates of M_{P1} and M_{P2} ($MO1$ and $MO2$) are shorted together. Consequently the feedback through R_{FB} is disabled at high frequencies. As such looking into the drains of M_{P1} and M_{P2}, a high impedance of value $2\{R_{FB} \| r_{op}\}$ is observed to the first order, where r_{op} is the small-signal output resistance of the PMOS devices.

For common-mode operation, the capacitor C_{RF} is floating both at RF and at baseband. Thus the feedback through R_{FB} is effective at both bands and the devices M_{P1} and M_{P2} present a low common-mode impedance at both bands.

We now trace the signal path at RF. As mentioned above, the RF drain currents of M_{N1} and M_{N2} see a high differential impedance of $2\{R_{FB} \| r_{op}\}$ looking into the drains of M_{P1} and M_{P2}. On the other hand, the passive mixer along with the capacitors C_M, presents an impedance of $(8/\pi^2)(R_{bias}||(1/g_{mp(1,2)})) + 2/j\omega_{RF}C_M$. The first term ($\propto g_{mp(1,2)}^{-1}$) arises due to the frequency translation of the baseband impedance at its output to RF. Since the impedance presented by the passive mixer is significantly smaller than that seen at the drains of M_{P1} and M_{P2}, the RF current preferentially flows into the passive mixer.

The passive mixer downconverts the RF signal to a dif-

ferential baseband current which sees a low impedance of $2\{R_{bias}||(1/g_{mp(1,2)})\}$ looking into the gates of M_{P1} and M_{P2}. The low load impedance ensures current-mode operation of the passive mixer. The shunt-shunt feedback converts this baseband current to a baseband voltage at the drains of M_{P1} and M_{P2} (v_{BBOM} and v_{BBOP} respectively), scaled by the transimpedance of R_{FB}.

The circuit operation at RF and baseband is illustrated in Fig. 3a and Fig. 3b respectively. The RF and baseband stages each perform dual tasks. The devices M_{P1} and M_{P2} are used as baseband transimpedance amplifiers and also operate as high-impedance loads for the RF stage. The input devices M_{N1} and M_{N2} operate as RF transconductors while simultaneously providing a high-impedance load at baseband. As a consequence of this dual functionality, the RF and baseband stages share the available voltage headroom, while also sharing the bias current. Unlike a stacked bias-sharing approach, the supply voltage is not split across stages.

III. DESIGN IMPLEMENTATION

Three key design techniques are utilized in the the basic topology of Fig. 2 to enhance performance. The complete design is shown in Fig. 4. These techniques are implemented such that there is no headroom penalty relative to the basic topology.

A. Network for Suppression of Baseband Noise

The RF devices $M_{N(1,2)}$ are seen to be directly coupled to the baseband in Fig. 2. Thus low-frequency noise and offsets in the RF transconductor appear at the output, unlike in the topology of Fig. 1.

The input RF devices M_{N1} and M_{N2} need to be sized using a fine channel length ($L = 0.18\mu m$ in this design) in order to ensure that the the required input transconductance can be achieved while maximizing f_T. The small device size and high f_T imply that the parasitic device capacitance is minimized, which helps to improve broadband noise performance as well. On the other hand, the short channel length of the RF devices implies that their low-frequency flicker noise contribution is large.

To minimize this source of noise, the impedance of the bias current network needs to be maximized, since this impedance provides source degeneration to the input devices. Although NMOS current mirror devices can be employed to achieve this, the noise of the bias devices would then appear at the output. Large devices can be used to reduce bias noise, however the associated capacitive parasitics would severely load the input.

Instead of NMOS current mirrors, resistors R_{in} are thus used to bias the input devices, since these do not exhibit significant flicker noise. Headroom constraints, however, impose an upper limit on the value of R_{in} which limits the attenuation of the flicker noise of the input devices.

The low-frequency differential-mode resistive degeneration is further increased without headroom penalty or significant noise penalty, through the use of an active negative resistance

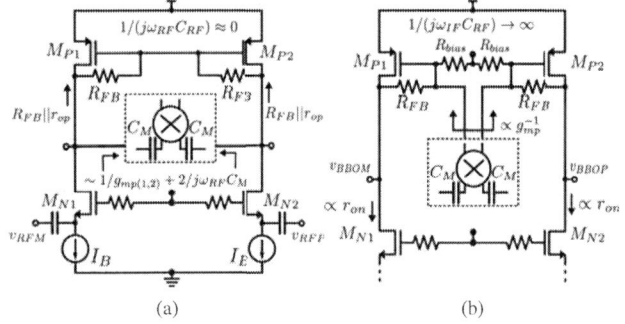

Figure 3: Differential signal flow in the basic bias-current-shared topology at (a) RF (b) Baseband

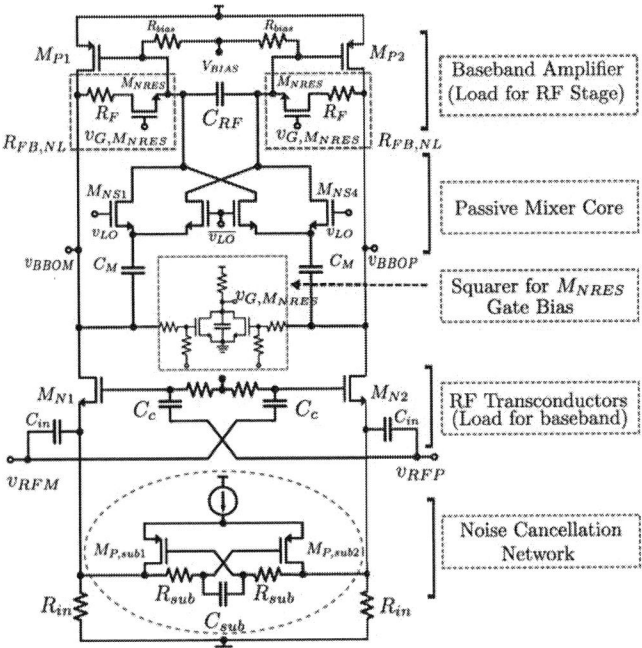

Figure 4: Schematic of Common-gate Down-conversion Mixer

("Noise Cancellation Network" in Fig. 4) in shunt with the bias resistors R_{in}. This is implemented using a cross-coupled PMOS pair $M_{P,sub(1,2)}$ employing devices biased in weak inversion. To ensure that the cross-coupled pair does not impact the RF input impedance, a low-pass network composed of resistors R_{sub} and C_{sub} is employed. This LPF renders the negative resistance ineffective at RF by providing a high-frequency short at the gates of $M_{P,sub1}$ and $M_{P,sub2}$. Additionally, since the cross-coupled devices are biased in weak inversion, they have a low f_T and therefore cannot provide significant gain at RF.

At baseband, the capacitor C_{sub} is effectively an open, and the cross-coupled devices present an effective negative resistance of $-2/g_{mp,sub(1,2)}$ looking into their drains. The parallel combination of R_{in} and the cross-coupled pair results in a differential degeneration of $2R_{in}/(1 - g_{mp,sub(1,2)}R_{in})$ at baseband. This can be made large by an appropriate choice of $g_{mp,sub(1,2)}$, thereby reducing noise contribution of M_{N1} and M_{N2}. The cross-coupled pair itself sees a small low-frequency load of $2\{(1/g_{mn(1,2)})\|R_{in}\}$, set by the source impedance of M_{N1} and M_{N2} and the resistors R_{in}, which ensures its stability.

The active cross-coupled network reduces only the differential-mode noise at the output. The common-mode low-frequency noise term is attenuated by the degeneration provided by R_{in} itself. It is further attenuated since the outputs are observed differentially.

It should be noted that this network does not impact the linearity performance of the circuit. The signal swing across the gates of the cross-coupled devices is negligibly small at RF since C_{sub} is almost a short. Additionally, the baseband signal swing at the source nodes of M_{N1} and M_{N2} is negligible.

B. Linearity Enhancement

The linearity limitation in a passive current-mode mixer design typically arises from the RF transconductor and the baseband sections, since the switching core itself is very linear. Typically the linearity performance becomes increasingly challenging at baseband prior to the channel select filter, as the signal is progressively amplified along with the interferers.

Signal dependent feedback is utilized to enhance the linearity in the baseband transimpedance stage and is implemented by using Voltage Controlled Resistors (VCR) in the feedback path. The VCRs, denoted by $R_{FB,NL}$ (Fig. 4), each consist of a series combination of a linear resistor of value R_F and an NMOS device M_{NRES} biased in the triode-region. The non-linearity of the feedback is controlled by modulating the gate-bias of M_{NRES} ($v_{G,M_{NRES}}$) using the output of a squarer circuit, which is implemented using two drain connected common-source devices. This introduces a square law voltage dependent term in the feedback resistor $R_{FB,NL}$ such that its resistance is increased as the amplitude increases. Modulation of the resistor provides gain expansion, which compensates for gain compression in the amplifier for large inputs. A significant aspect of this technique for linearization is that it causes no degradation of noise, since the noise of the non-linear resistor M_{NRES} is identical to that of a linear resistor of the same nominal value. Further the magnitude of the non-linear resistor is typically significantly smaller than the linear resistor. A voltage V_{BIAS} fed through the resistors R_{bias} sets a stable DC voltage at the drains of M_{P1} and M_{P2}.

C. Gain Enhancement

The common-gate input stage provides inherent broadband impedance matching, but the resistive nature of the match degrades the voltage gain and noise figure since the source resistance is matched to the inverse of the device transconductance. To mitigate this, cross-coupling capacitors C_C are connected across the gates of M_{N1} and M_{N2} which help to increase the conversion gain and improve the noise figure [5]. The cross-coupling also ensures that the RF inputs are applied differentially across the gate-to-source of each of the input devices. The net differential input impedance seen by the source is $(1/g_{mn(1,2)})\|\{2(R_{in}\|R_{sub})\}$, where C_{in} is assumed to be a short at RF.

The cross-coupled devices M_{Psub1} and M_{Psub2} do not impact circuit operation at RF. However, they help to increase the gain at baseband, since differential degeneration provided by them at the source nodes of the RF devices significantly increases the low-frequency impedance looking into the drains of M_{N1} and M_{N2}.

IV. MEASUREMENTS

The design was implemented in a commercial 0.18-μm CMOS process. The ICs were housed in 48 pin TQFP packages and measured on FR4 printed circuit boards. The design used an external balun for feeding RF and LO. The total loss (balun and connectors) was measured using a dedicated test fixture and was found to be 3.3 dB at the desired frequencies. The measured performance was corrected for this loss.

The measured conversion gain, was 35 dB at 1 GHz and varied by approximately ±1.5 dB from 800 MHz – 1.1 GHz. The 3-dB bandwidth at baseband was approximately 2 MHz.

978-1-4244-6240-7/10 $26.00 © 2010 IEEE

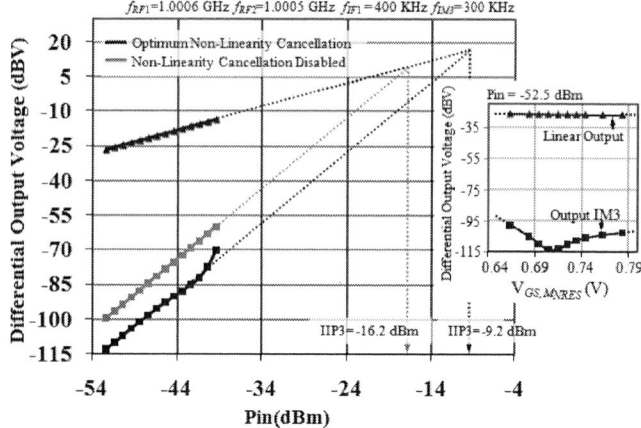

Figure 5: Measured IIP3 with and without non-linearity cancellation

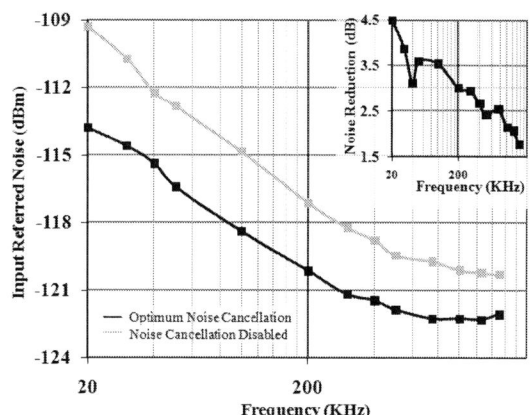

Figure 6: Input-referred noise variation for cross-coupled degeneration ON and OFF

This bandwidth limitation arose from the loading by the capacitance of the cables and the PCB trace. The in-channel 1-dB compression point is fundamentally set by the available voltage swing at the output nodes. The compression point was found to be nearly 0 dBVp at the output, which implies an input compression of approximately -25 dBm for 35 dB gain. The input P1dB can be increased by reducing R_F to lower the gain. The in-band IIP3 (using tones at offset 400 kHz and 500 kHz from the LO), was -9.2 dBm (Fig. 5). The output IM3 demonstrated a 14 dB improvement with non-linearity cancellation. The inset of Fig. 5 shows the IM3 as a function of bias applied to the MOS resistor M_{NRES}. The DSBNF at 1 MHz was 9.8 dB. The low-frequency noise decreased by about 3-4 dB with the cross-coupled degeneration enabled (Fig. 6). The flicker noise corner frequency was approximately 160 kHz, which can be reduced if longer channel length PMOS devices are used in the baseband transimpedance stage. The bias current requirement was 2.1 mA from a 1.8 V supply. The calculated FOM [6] for the mixer is 19.1 dB. The die microphotograph is shown in Fig. 7. In the practical implementation, high-linearity unity-gain PMOS buffers were utilized in cascade with the above design to drive the external signals on board. These buffers are expected to be significantly smaller in an integrated receiver implementation, where the down-converter may be followed by an integrated filter or analog-to-digital converter. The differential LO is

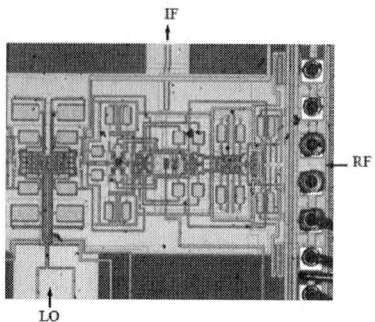

Figure 7: Die Photograph

buffered on-chip using a cascade of two inverting buffers. The bias for the buffered LO signal was chosen to optimize the noise and conversion gain performance. The mixer including the LO buffers had an area requirement of 0.2 mm². ESD protection was included in the design.

V. CONCLUSION

A passive current-mode mixer which reuses the current between the RF transconductor and the baseband transimpedance amplifier to minimize power is described. The topology allows for current sharing, while merging the operating voltage domains of the RF and baseband stages, that is, without the requiring an explicit intermediate AC ground between the stacked RF and baseband stages. This feature is expected to make the design suitable for low-voltage applications. Circuit techniques are introduced for reducing the low-frequency noise of the input transconductor and for improving the IM3 performance with a low power overhead. It is anticipated that the down-converter design will find application in systems where minimizing the power requirement is critical, such as those for sensor networks, or for ISM band portable systems. Given the high FOM of the design, with suitable optimization, it can be anticipated that the topology can be used in systems with even greater dynamic range requirement such as cellular front-ends.

REFERENCES

[1] E. Sacchi, I. Bietti, S. Erba, L. Tee, P. Vilmercati, and R. Castello, "A 15 mW, 70 kHz 1/f corner direct conversion CMOS receiver," in *Proc. CICC*, Sept. 2003, pp. 459-462.

[2] Y. Feng, G. Takemura, S. Kawaguchi, and P. Kinget, "Design of a High Performance 2-GHz Direct-Conversion Front-End With a Single-Ended RF Input in 0.13 μm CMOS," *IEEE J. Solid-State Circuits*, vol. 44, no. 5, pp. 1380–1390, May 2009.

[3] A. Liscidini, A. Mazzanti, R. Tonietto, L. Vandi, P. Andreani, and R. Castello, "Single-Stage Low-Power Quadrature RF Receiver Front-End: The LMV Cell," *IEEE J. Solid-State Circuits*, vol. 41, no. 12, pp. 2832–2841, Dec. 2006.

[4] V. Vidojkovic, J. van der Tang, A. Leeuwenburgh, and A. van Roermund, "A low-voltage folded-switching mixer in 0.18-μm CMOS," *IEEE J. Solid-State Circuits*, vol. 40, no. 6, pp. 1259–1264, June 2005.

[5] X. Li, S. Shekhar, and D. Allstot, "Gm-boosted common-gate LNA and differential colpitts VCO/QVCO in 0.18-μm CMOS," *IEEE J. Solid-State Circuits*, vol. 40, no. 12, pp. 2609–2619, Dec. 2005.

[6] J. Deguchi, D. Miyashita, and M. Hamada, "A 0.6V 380μW -14dBm LO-input 2.4GHz double-balanced current-reusing single-gate CMOS mixer with cyclic passive combiner," in *IEEE ISSCC Dig.*, Feb. 2009, pp. 224-225.

RMO1B-3

A Multiphase PWM RF Modulator Using a VCO-Based Opamp in 45nm CMOS

Min Park*, Michael H. Perrott[†]

*Massachusetts Institute of Technology, Cambridge, MA, USA
(now at Maxim Integrated Products, Sunnyvale, CA, USA)

[†]SiTime Corporation, Sunnyvale, CA, USA

Abstract—A VCO-based RF modulator employing multiphase *Pulse Width Modulation* (PWM) is presented. The proposed RF modulator encodes the baseband signal into a set of multiphase PWM signals which are generated by a VCO-based opamp. The use of PWM avoids broadband quantization noise which is produced by $\Sigma\Delta$ modulation used in other RFDAC-based modulators. The prototype IC is fabricated in a 45nm CMOS process, consumes 54.3 mW, and has an active area of 0.126 mm^2. The measured results satisfy the 802.11g WLAN spectral mask, and EVM for 10-MHz, 64 QAM OFDM at 2.4 GHz is –30 dB.

Index Terms—Pulse width modulation, PWM, Multiphase PWM, RFDAC, RF modulator.

I. INTRODUCTION

Modern CMOS processes have brought many advantages for digital circuits such as high density and high speed. In contrast, analog circuits must deal with reduced supply voltages, higher levels of mismatch, and reduced intrinsic gain. More importantly, these analog circuits are often placed within chips that are largely digital in nature, which creates challenges in achieving a seamless design flow for the overall system. For these reasons, there has been much interest in achieving analog functionality with reduced analog design effort.

In this paper, we examine this goal in the context of achieving an RF modulator structure that avoids the need for CMOS transistors with highly linear transconductance and high intrinsic gain. To do so, full swing signals are utilized whose information is encoded in the form of pulse widths. By using time as the primary signal domain, we take advantage of the steadily improving transition times offered by modern CMOS processes.

Fig. 1 provides an overview of several different methods to achieve a CMOS RF modulator. In each case, we assume that the LO signal is composed of a simple set of switches that are driven with a large-swing signal corresponding to the carrier frequency. However, the data devices can be driven in a variety of ways. In part (a), the data signal consists of a small-signal analog waveform representing the baseband signal, and linear transconductors are required for the data devices. In part (b), the data signal is composed of a full-swing signal that is produced by a $\Sigma\Delta$ modulator with the baseband signal as its input [1], [2]. The impact of $\Sigma\Delta$ modulation is to produce broadband quantization noise which must be kept adequately small to satisfy transmitter spectral mask requirements. The quantization noise can be reduced by the combination of having a high clock rate for the $\Sigma\Delta$ modulator, high frequency

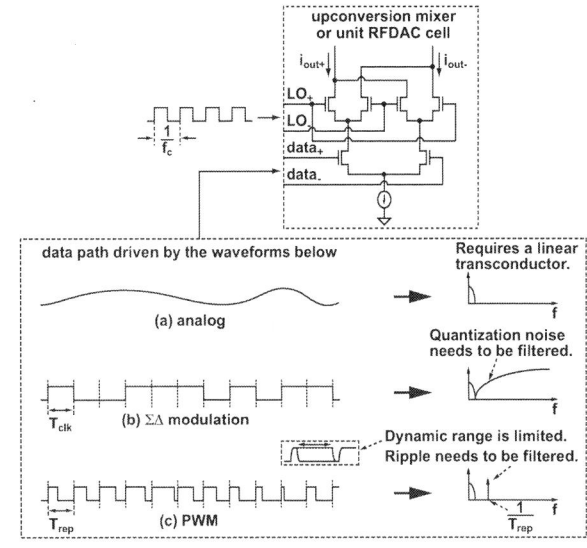

Fig. 1. Various modulation methods for RF upconversion. (a) analog, (b) $\Sigma\Delta$ modulation, (c) PWM.

bandpass filtering of the modulator output [1], [2], or multi-level quantization through the use of several parallel modulator stages with excellent matching [3]. Unfortunately, utilization of bandpass filtering or multi-level modulation brings non-trivial analog design challenges, and simply using a high clock rate for the $\Sigma\Delta$ modulator will often prove inadequate to achieve spectral mask requirements of the modulator

Part (c) of Fig. 1 shows the use of PWM to achieve a full-swing signal for the data path [4]. Unlike $\Sigma\Delta$ modulation which maintains a constant time period for data samples, PWM allows continuous adjustment of its pulse width in analog fashion. As such, the broadband quantization noise of $\Sigma\Delta$ modulation is avoided and replaced with ripple whose primary frequency content occurs at the pulse repetition frequency. A key goal is to place the pulse repetition frequency high enough in value to be easily filtered at the modulator output. However, due to finite transition times of the PWM edges, a high pulse repetition frequency also leads to reduced dynamic range [5].

As shown in Fig. 2, multiphase PWM improves the tradeoff between achieving high frequency ripple and wide dynamic range. By adding PWM signals that are offset in time with each other, the resulting *effective* ripple frequency is dramatically increased compared to the base pulse repetition rate of the individual signals [6], [7]. The addition of such multiphase

978-1-4244-6240-7/10 $26.00 © 2010 IEEE

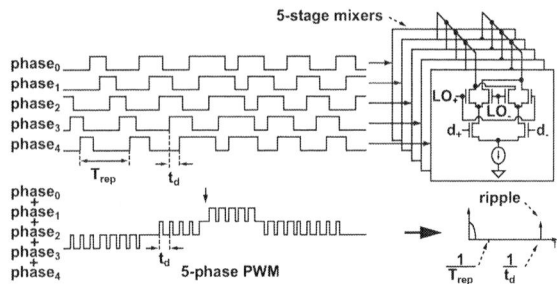

Fig. 2. The waveforms and spectrum of multiphase PWM signals.

PWM signals is achieved by using several mixer stages in parallel. While a $\Sigma\Delta$-based mixer could also utilize parallel stages to achieve a multi-level implementation, any mismatch between stages can lead to noise folding and spur generation in the modulator output. In contrast, when using the multiphase PWM approach, mismatch between stages simply leads to increased leakage of the base pulse repetition frequency (which is relatively high in frequency) to the modulator output.

The use of multiphase PWM is not a new concept, but previous approaches have involved fairly analog intensive circuit structures. As an example, the approach presented in [8] employs triangular waveform generators and multilevel comparators. In this paper, we propose the use of a relatively simple VCO-based opamp to achieve multiphase PWM which offers several implementation advantages in modern CMOS technology. Using this VCO-based opamp, we demonstrate a multiphase PWM modulator that satisfies the 802.11g WLAN spectral mask.

II. VCO-BASED OPAMP

To explain the concept of a VCO-based opamp, we begin by illustrating a unity gain amplifier that is achieved with two voltage-controlled ring oscillators, a set of XOR phase detectors and current DACs, and a simple RC filter as shown in Fig. 3. The two VCOs act as integrators from input voltage to output phase, and the XOR phase detectors and current DACs convert the phase difference between the two VCO outputs into an overall current signal which is then filtered to form the output voltage. One VCO tuning input is controlled by the input voltage, V_{in}, and the other VCO tuning input is controlled through feedback from the output voltage. While the phase detection operation of the VCO opamp has ripple that must be filtered out, the frequency of the ripple is quite high due to the multiphase principle discussed earlier in this text. As such, the ripple can be substantially attenuated with a low area RC filter.

In effect, the VCO-based opamp is a phase-locked loop (PLL) that maintains the same frequency for each VCO. As the input tuning voltage of one of the VCOs is altered according to the input, the other VCO tuning voltage tracks the input by virtue of feedback. Assuming perfect matching between the VCOs, a perfect match in input and output voltages is achieved since phase-locking forces equal VCO frequencies. Also, since the feedback signal is formed as the average of the

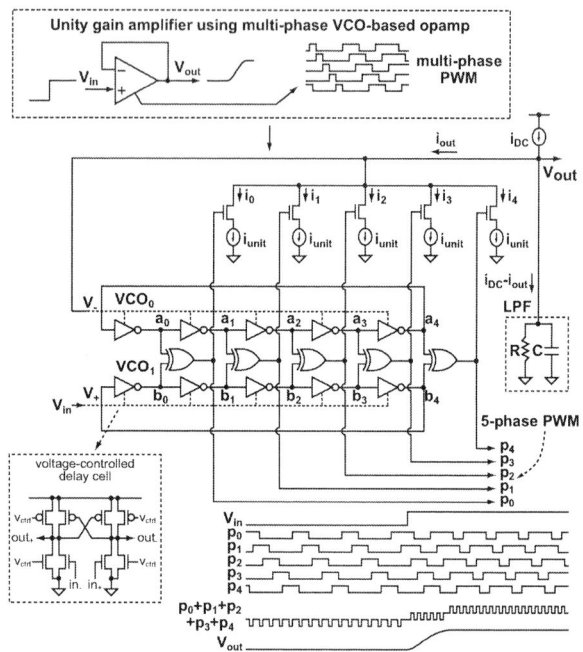

Fig. 3. A unity gain amplifier using a multiphase VCO-based opamp.

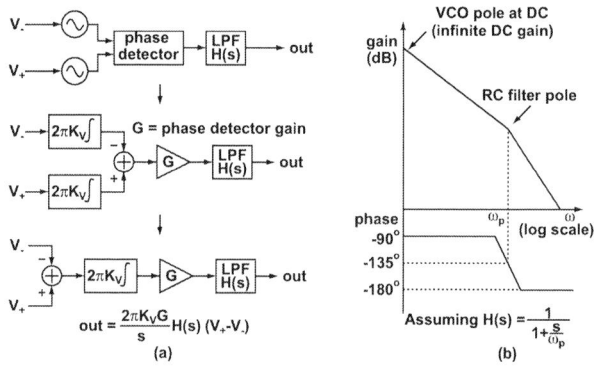

Fig. 4. (a) VCO-based opamp and its model. (b) The open loop Bode plot of a VCO-based opamp when the LPF is a simple RC filter.

multiphase PWM signals, the duty cycle of those multiphase PWM signals also track the input. As such, the VCO-based opamp is seen as an elegant structure for generating multiphase PWM signals. The pulse repetition rate of the individual PWM signals is set by the VCO frequency as controlled by the input. The effective ripple of the addition of the multiphase PWM signals is $\frac{1}{t_d}$, where t_d is set by the delay element used in the ring VCOs. Since a delay of a few tens of picoseconds can be easily achieved in modern CMOS processes, the effective ripple frequency of the VCO-based opamp can be set on the order of tens of GHz. As mentioned earlier, a simple RC filter is sufficient to suppress such ripple.

To better understand the VCO-based opamp, a simplified model is shown in Fig. 4 along with its associated open loop Bode plot. As shown by the model, the VCO-based opamp effectively integrates the difference in input voltages. The Bode plot in Fig. 4 reveals a fairly simple open loop response

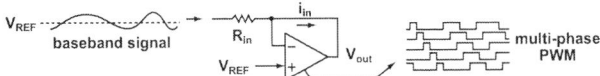

Fig. 5. Multiphase PWM generation using a VCO-based opamp.

consisting of the integrator and RC filter pole. To achieve robust, stable operation, the unity gain crossover frequency should be set less than the value of the RC filter pole.

A very interesting advantage of the VCO-based opamp is that it has infinite DC gain in its open loop transfer function due to the voltage-to-phase integration operation of the VCO circuits. In contrast, traditional opamp structures must resort to fairly complicated techniques such as nested Miller topologies [9] to achieve both high open loop DC gain and proper compensation since the intrinsic gain of devices is steadily degrading in modern CMOS processes.

However, there are, of course, disadvantages to the proposed VCO-based opamp structure. One key issue is that there is a limited locking range offered by the phase detectors within the VCO-based opamp, and one needs to worry about avoiding cycle slips during steady-state operation as encountered with other PLL circuits. The limited locking range imposes constraints on the allowable voltage and frequency variation at the input [10]. Another issue is that a practical ring VCO has a nonlinear tuning characteristic for its gain, K_V. This nonlinearity can lead to distortion at the output of the unity-gain VCO-based opamp when the input is varied at high frequencies. More importantly, large changes in K_V can lead to stability problems since the unity gain crossing frequency of the VCO-based opamp is a direct function of K_V.

III. PROPOSED RF MODULATOR EMPLOYING A VCO-BASED OPAMP

When utilizing the VCO-based opamp for a multiphase PWM modulator, it is advantageous to use the unity-gain amplifier configuration shown in Fig. 5. In this case, the input of the amplifier is set by a reference voltage V_{REF} that can be chosen for optimal performance of the modulator (such as setting an appropriate PWM repetition rate or achieving a desired nominal K_V value). A key advantage offered by this constant input voltage is that variation in the nominal K_V value is greatly reduced so that stable operation is easily achieved, and the PWM repetition frequency is fairly constrained.

Due to the feedback action of the opamp, the output voltage is also held at V_{REF}. However, if we inject current into the output, the pulse widths of the PWM signals generated by the VCO-based opamp must appropriately adjust their value in order to maintain the output voltage at V_{REF}. Therefore, we can easily create an input voltage to multiphase PWM generator by simply connecting a resistor between the baseband voltage signal and VCO-based opamp output.

While the proposed VCO-based multiphase PWM generator shown in Fig. 5 alleviates large variations in nominal K_V, the nonlinearity of the VCO K_V characteristic limits distortion

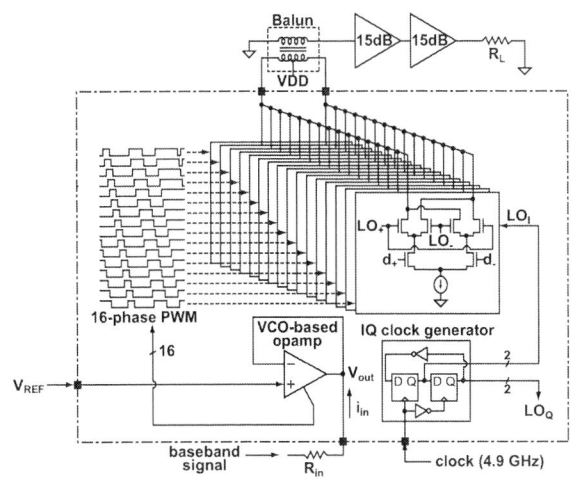

Fig. 6. Simplified block diagrams of the proposed VCO-based RF modulator chip. Only the in-phase signal path is shown for simplicity.

performance for large amplitude, high frequency input signals. To explain, the negative feedback loop of the multiphase PWM generator suppresses the VCO nonlinearity by keeping the VCO tuning voltage constant when the input signal is slowly varying. However, the feedback loop cannot instantaneously track a large amplitude, fast-varying input signal variation. The frequency dependency of nonlinearity suppression is a key design consideration for wideband modulation as pursued in this paper. As will be shown by measured results later in this paper, its impact can be reasonably mitigated for moderate performance applications such as WLAN.

The overall architecture of the proposed RF modulator is shown in Fig. 6. Both I and Q channels are included in the actual chip, but Fig. 6 shows only the I channel signal path for simplicity. Most of the signal paths are differential except for the baseband signal path. The digital blocks employ full-swing pseudo differential logic.

16-stage mixer cells are employed in the prototype chip. The 16-phase PWM signals generated from the VCO based opamps drive the 16 mixer cells to create the desired RF signal. One advantage of this architecture is that the ring VCOs and phase detectors sequentially drive the current DACs and mixer cells, such that *Dynamic Element Matching* (DEM) is achieved for free. A similar DEM effect in a VCO-based A/D converter is reported in [11]. Accordingly, the matching requirement for the mixer cells and current DACs is reasonably relaxed.

An on-chip frequency divider generates the IQ LO clocks. Note that IQ mismatch occurs due to non-idealities of the frequency divider such as duty cycle and delay mismatch [12], and due to gain/phase mismatch of the VCO-based opamps for the IQ channels. Compensation of this IQ mismatch is achieved in the prototype by manual gain and phase predistortion in an off-chip baseband signal generator.

IV. MEASURED RESULTS

The proposed VCO-based RF modulator is implemented in a 45nm CMOS process with an active area of 0.126 mm^2. The

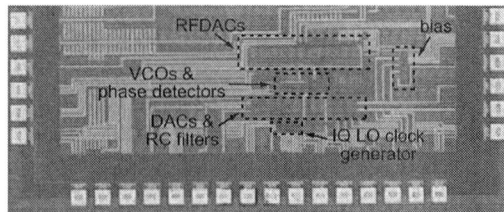

Fig. 7. Die photograph of the implemented VCO-based RF modulator.

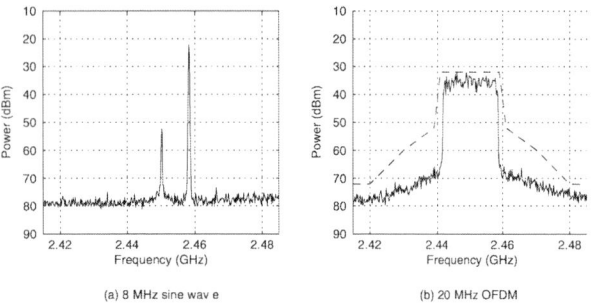

(a) 8 MHz sine wave

(b) 20 MHz OFDM

Fig. 8. (a) Measured output spectrum with a 8-MHz sine wave, (b) Measured 20-MHz OFDM output spectrum compared to 802.11g spectral mask.

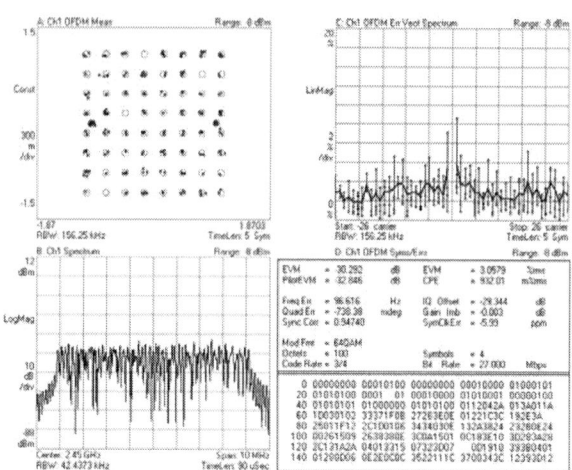

Fig. 9. Measured EVM with a 10-MHz OFDM signal for 802.11a WLAN.

die photograph is shown in Fig. 7. Total power consumption of the chip is 54.3 mW including LO clock buffer, mixers, bias, VCOs and other digital circuits. The IQ baseband signals are generated by an arbitrary function generator. The output signal from the chip is amplified by two 15-dB amplifiers before the signal is measured by a spectrum analyzer.

Fig. 8 (a) shows the modulator's in-band output spectrum at 2.45 GHz when the input signal is an 8-MHz sine wave. IQ mismatch compensation is done manually on the input baseband IQ signals in order to suppress image signals. Fig. 8 (b) shows the 20-MHz OFDM output spectrum centered at 2.45 GHz. The spectrum is compared with an 802.11g spectral mask as a reference.

For the EVM test, a 10-MHz, 64 QAM OFDM signal for 802.11a WLAN is used as a baseband signal. 10-MHz of bandwidth is chosen since the test equipment is limited to this value. The generated source signals are predistorted for IQ mismatch compensation in a PC. The predistorted baseband signals are loaded into a vector signal generator, which sends the baseband IQ signals to the chip. The measured EVM with IQ mismatch compensation is about –30 dB, as is shown in the fourth quadrant of Fig. 9. The plot in the first quadrant of Fig. 9 shows the EVM at each frequency bin and reveals that the highest EVM values occur close to the carrier frequency, which is likely due to upconverted flicker noise of the VCOs. The more detailed measurement results can be found in [10].

V. CONCLUSION

A multiphase PWM RF modulator in 45nm CMOS technology was presented which achieves upconversion of a baseband signal by using multiphase PWM signals generated by a VCO-based opamp. The prototype chip demonstrates that the proposed RF modulator is capable of upconverting a 20-MHz,

64 QAM OFDM signal at 2.4 GHz while meeting 802.11g output spectral mask requirements.

ACKNOWLEDGMENT

Texas Instruments Inc. provided chip fabrication. The authors would like to thank Dennis Buss, Alice Wang, and Chih-Ming Hung, Texas Instruments Inc. for their support.

REFERENCES

[1] A. Jerng and C. G. Sodini, "A wideband $\Delta\Sigma$ digital-RF modulator for high data rate transmitters," *IEEE J. Solid-State Circuits*, vol. 42, no. 8, pp. 1710–1722, Aug. 2007.

[2] S. M. Taleie, Y. Han, T. Copani, B. Bakkaloglu, and S. Kiaei, "A 0.18µm CMOS fully integrated RFDAC and VGA for WCDMA transmitters," *Proc. IEEE RFIC Symposium*, pp. 157–160, June 2008.

[3] P. Eloranta, P. Seppinen, S. Kallioinen, T. Saarela, and A. Pärssinen, "A multimode transmitter in 0.13µm CMOS using direct-digital RF modulator," *IEEE J. Solid-State Circuits*, vol. 42, no. 12, pp. 2774–2784, Dec. 2007.

[4] J. S. Walling, H. Lakdawala, Y. Palaskas, A. Ravi, O. Degani, K. Soumyanath, and D. J. Allstot, "A class-E PA with pulse-width and pulse-position modulation in 65nm CMOS," *IEEE J. Solid-State Circuits*, vol. 44, no. 6, pp. 1668–1678, June 2009.

[5] S. Kuo, "Linearization of a pulse width modulated power amplifier," *S.M. Thesis*, Massachusetts Institute of Technology, 2004.

[6] S. Abedinpour, B. Bakkaloglu, and S. Kiaei, "A multistage interleaved synchronous buck converter with integrated output filter in 0.18µm SiGe Process," *IEEE Trans. Power Electron.*, vol. 22, no. 6, pp. 2164–2175, Nov. 2007.

[7] T. Carosa, R. Zane, and D. Maksimović, "Scalable digital multiphase modulator," *IEEE Trans. Power Electron.*, vol. 23, no. 4, pp. 2201–2205, July 2008.

[8] B. A. Weaver, "A new, high efficiency, digital, modulation technique for AM or SSB sound broadcasting application," *IEEE Trans. Broadcasting*, vol. 38, issue 1, pp. 38–42, March 1992.

[9] J. H. Huijsing and D. Linebarger, "Low-voltage operational amplifier with rail-to-rail input and output ranges," *IEEE J. Solid-State Circuits*, vol. SC-20, no. 6, pp. 1144–1150, Dec. 1985.

[10] M. Park, "Time-based circuits for communication systems in advanced CMOS technology," *Ph.D. Thesis*, Massachusetts Institute of Technology, 2009.

[11] M. Z. Straayer and M. H. Perrott, "A 10-bit 20MHz 38mW 950MHz CT $\Sigma\Delta$ ADC with a 5-bit noise-shaping VCO-based quantizer and DEM circuit in 0.13µm CMOS," *Proc. IEEE Symposium on VLSI Circuits*, pp. 246–247, June 2007.

[12] C. D. Renter and M. Steyaert, *High data rate transmitter circuits: RF CMOS design and techniques for design automation*, Kluwer Academic Publishers, 2003.

Spurious Noise Reduction by Modulating Switching Frequency in DC-to-DC Converter for RF Power Amplifier

Eung Jung Kim[1], Chang-Hyuk Cho[2], Woonyun Kim[2], Chang-Ho Lee[2], Joy Laskar[1]

[1]Georgia Electronic Design Center, Georgia Institute of Technology, Atlanta, GA 30308, U.S.A
[2]Samsung Design Center, Atlanta, GA 30308, U.S.A

Abstract — **The presence of a spectrum spreading effect in a DC-to-DC converter operating in Pulse-Width Modulation (PWM) with a modulated switching frequency is analyzed. A step-down DC-to-DC converter prototype with a digital PWM ramp signal modulator is implemented in a standard CMOS 0.18-μm process. The step-down DC-to-DC converter has an 6-bit up/down binary counter to vary its switching frequency between 2.2 MHz and 4.4 MHz in 64 steps as means of decreasing spurious noise peaks at the output of the switch converter. The measurement results show that the spurious switch noise peak is reduced by 12 dB when a monotonic frequency stepping with the up/down counter is used, and it is shown that the additional switching frequency modulation functionality does not degrade the overall efficiency and the closed loop operation significantly.**

Index Terms — **DC-to-DC converter, modulated PWM, spread spectrum, switch power supply.**

I. INTRODUCTION

As a consequence of the increasing data rate and the limitations on wireless frequency bands, wireless transmitters utilize various linear modulation schemes. A linear transmitter often requires a linear radio frequency power amplifier (PA) because both phase and amplitude modulation are used to transmit a signal. In a wireless transmitter, the PA consumes the largest portion of overall power, so maintaining a high efficiency for the PA throughout all operating points is very critical. To overcome the low operating efficiency of a linear PA at backed-off powers, a switching power supply is widely used because of its higher efficiency over linear regulators [1]. The Pulse-Width Modulation (PWM) with a constant switching frequency makes it easy to predict where the switching noise would appear in the frequency domain, therefore the PWM control is widely used as a control scheme for a DC-to-DC converter supplying power to an RF PA.

However, the conventional PWM control scheme with a constant switching frequency inherently generates switching noise of which the spectral components are mostly confined in the fundamental switching frequency and its harmonics[2]. This confined spectral energy of switching noise creates high-level spurs in the output of

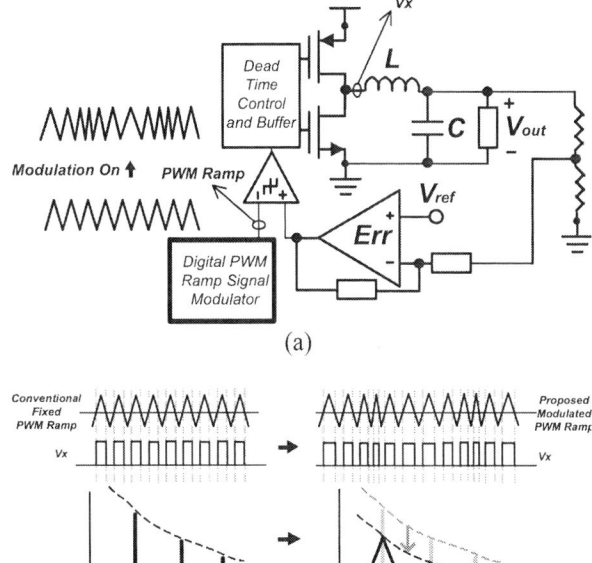

(a)

(b)

Fig. 1. (a) Block diagram of the DC-to-DC converter with digital PWM ramp signal modulator (b) spectrum spreading effect from modulation of PWM ramp signal.

a PWM controlled DC-to-DC converter. These large spurs are mixed with the modulated carrier in PA. From this frequency mixing, a large spurious noise is generated, and it can cause a failure for spectral emission specification [3].

Therefore, to use a DC-to-DC converter with an radio frequency PA, its switching noise must be minimized so as not to deteriorate the performance of the RF transmitter.

In some case, a large LC-filter does not filter switching noise enough so that even a higher order LC-filter with more external components or surface mount devices inside of multilayer laminates are used to enhance the filtering, but the number and the size of external components should be minimized for less overall system size and cost [1],[4], [5].

This paper presents a technique for reducing and broadening the frequency components of the switching

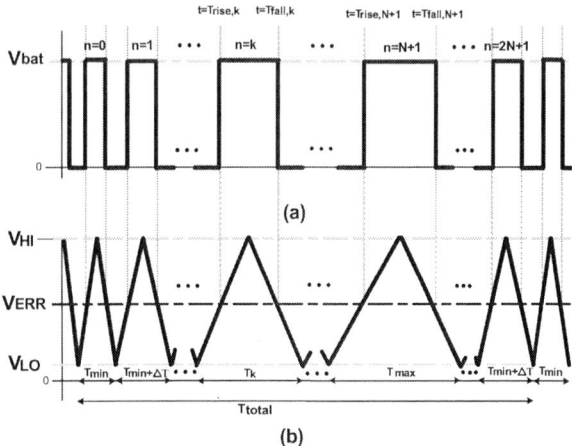

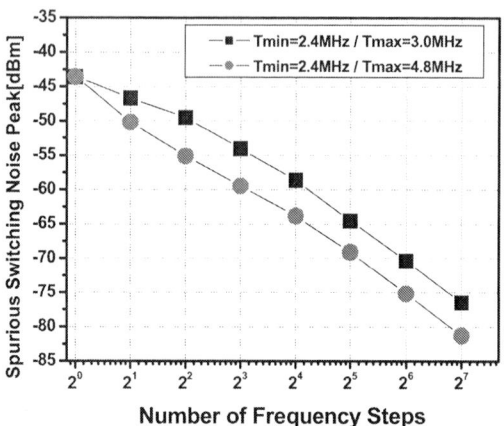

Fig. 2. (a) Pulse-width modulated square wave output with a constant duty cycle (b) triangular PWM ramp signal with varying frequency.

Fig. 3. Calculated spurious noise peak for different number of frequency steps and for different frequency ranges.

noise in a step-down DC-to-DC converter. By modulating the frequency of the triangular PWM ramp signal of which the frequency is controlled by the digital PWM ramp signal modulator, the spurious noise can be effectively spread over a wider frequency range, and the peak magnitude of the spurs can be reduced as in Fig. 1. (b). To minimize the in-balance of the inductor current at the output LC-filter due to time-varying switching frequency, small frequency steps are used, and the oscillator is designed to generate a triangular PWM ramp signal that has an equal rise and fall time throughout the switch frequency modulation.

In Section II, the spectral energy of switching noise with monotonic switching frequency stepping is mathematically analyzed. Circuit implementation of DC-to-DC converter with the digital PWM ramp signal modulator is described in Section III. Experimental results are shown in Section IV, and the conclusion is drawn in Section V.

II. ANALYSIS

A modern wireless communication system is highly affected by determinable interferences such as periodic switching noise. Therefore, if the peak energy-density spectrum of switching noise could be spread over a wider frequency range with a smaller peak value, then the noise would consist of less determinable interferences [5].

From Parseval's theorem, the energy-density spectrum of a discrete signal that determines how the energy is distributed in the frequency domain can be found. The total energy contained in a signal, summed across a period, is equal to the total energy of the signals' Fourier transform, summed for all of its frequency components [6].

Common PWM schemes use a periodic triangular or saw-shaped signal with a fixed frequency to control the output voltage of a DC-to-DC converter. Repetitive switching at a fixed frequency generates a large switching noise of which all the energy–density spectrum is confined within the fundamental switching frequency and its harmonics.

With monotonic switching frequency stepping, the switching period is increased or decreased between T_{min} and T_{max} with a time step of ΔT. If there are N steps between T_{min} and T_{max}, the integration over T_{total} (T_{total} = T_{min} + (T_{min} + ΔT) + (T_{min} + $2 \cdot \Delta T$) + $\cdots$ + (T_{min} + $N \cdot \Delta T$) + T_{max} + (T_{min} + $N \cdot \Delta T$) + (T_{min} + $(N-1) \cdot \Delta T$) + $\cdots$ + (T_{min} + ΔT)) can be divided into 2N+2 parts. Any switching period between T_{min} and T_{max} can be generalized with an index, K, by using $T_{rise,k}$ and $T_{fall,k}$, and the fact that the periodic square wave signal has only two states, zero or V_{bat}, as shown in Fig. 2. The Fourier series coefficient in the square wave can be simplified and calculated as

$$ X_n = \frac{V_{bat}}{T_{total}} \sum_{k=0}^{2N+1} \int_{T_{fall,k}}^{T_{rise,k}} e^{-2j\pi (\frac{n}{T_{total}})t} dt \quad (1) $$

which presents the spectral energy of the switching noise in the frequency domain.

Using (1), the spectral switching noise power is calculated with a different number of frequency steps and a different set of T_{min} and T_{max} assuming V_{bat}=3.6V, duty cycle=0.5, L=2.2μH, C=1.0μF, and Load=5Ω.

Fig. 3 shows that, as the number of switching step, which is N in (1), is doubled, the spurious switching noise peak is decreased by $20 \cdot \log(2) \approx 6$ dB. Increasing the range of switching frequency modulation also reduces the peak of noise spur, but these results are based on the assumption that all the LC components at the output filter and the switches are ideal, and the in-balance of the inductor current due to the time-varying switching frequency is not considered.

If the switching period is increased by ΔT in every one cycle until it gets to T_{max}, assuming the duty cycle is constant at d_1, turn-on time is increased by $d_1 \cdot \Delta T$ in the next cycle, but turn-off time is increased by $(1 - d_1) \cdot \Delta T$ in the next cycle. When $d_1 = 0.25$, the mismatch between the turn-on time and the turn-off time in the next cycle is $0.5 \cdot \Delta T$. Even though the timing mismatch is resolved as the switching period is decreased from T_{max} to T_{min}, the cumulated in-balance of the inductor current can go over the tolerance limit when ΔT is too large. Therefore, ΔT and the range of frequency stepping should be carefully chosen to avoid a device failure and undesired transient and frequency components [7].

III. ACHIEVING SPREAD SPECTRUM

To modulate the frequency of the triangular PWM ramp signal, an up/down counter and a current steering Digital-to-Analog Converter (DAC) are implemented. The up/down counter is designed to monotonically increase or decrease its output bits between the minimum and the maximum. The output of the up/down counter will steer the current output of the DAC. Each bit of the DAC is weighted differently to increase the actual resolution of the current output.

The output current of DAC, I_{DAC}, is combined with a DC bias current and generates a modulated current, I_{mod}, in Fig. 4. The level of I_{mod} determines the charging time and the discharging time at the load capacitor in the digital PWM ramp signal modulator as shown in Fig. 5.

For PWM control, a triangular signal or a saw-shaped signal are commonly used, but to keep the inductor current under control and preserve the duty-cycle constant throughout the switching frequency modulation, triangular PWM ramp signal should be always shaped to have the same rise and fall time. If the PWM ramp signal is saw-shaped, then the mismatch between turn-on time and turn-off time cannot be resolved as described in Section II. The digital PWM ramp signal modulator changes the charging and discharging time by the same amount because the top and bottom current mirrors are referencing the same current, I_{mod}.

By utilizing the digital PWM ramp signal modulator, the switching frequency is modulated without severely disturbing the in-balance of the inductor current at the LC-filter, but this scheme still has a low frequency pattern. These low frequency components could be appeared at the switching noise spectrum. Randomizing the modulation of the switching frequency can eliminate this low frequency

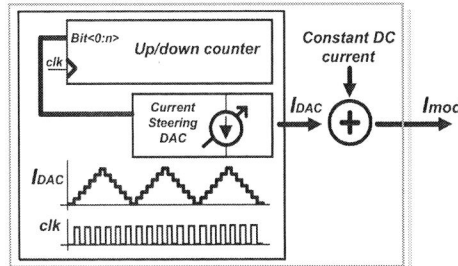

Fig. 4. Block diagram of digital current modulator.

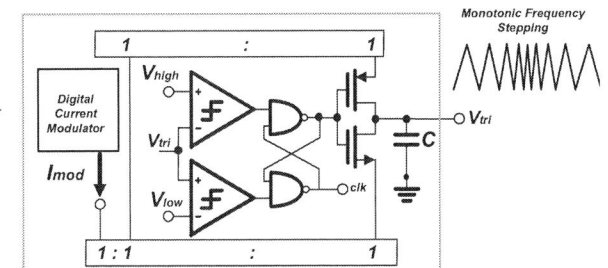

Fig. 5. Block diagram of digital PWM ramp signal modulator.

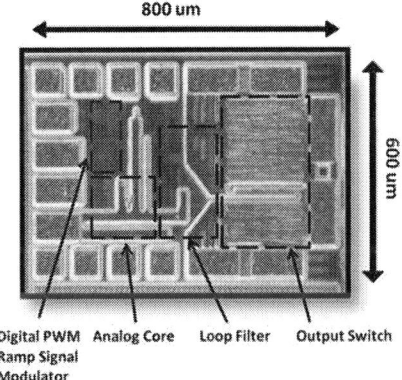

Fig. 6. Micro-photograph of the step-down DC-to-DC converter with a digital PWM ramp signal modulator.

noise [8]. However, when the frequency of the PWM ramp signal is changed in a true random fashion, the switching frequency can be abruptly changed from the minimum to the maximum in the next cycle. This can cause an adverse side effect as analyzed in Section II, and, as a result, the actual noise-power spectrum may not decrease but increase.

IV. EXPERIMENTAL RESULTS

The implemented DC-to-DC converter with the digital PWM ramp signal modulator is shown in Fig. 6. The DC-to-DC converter was implemented using a standard CMOS 0.18-µm process, and the total area of the

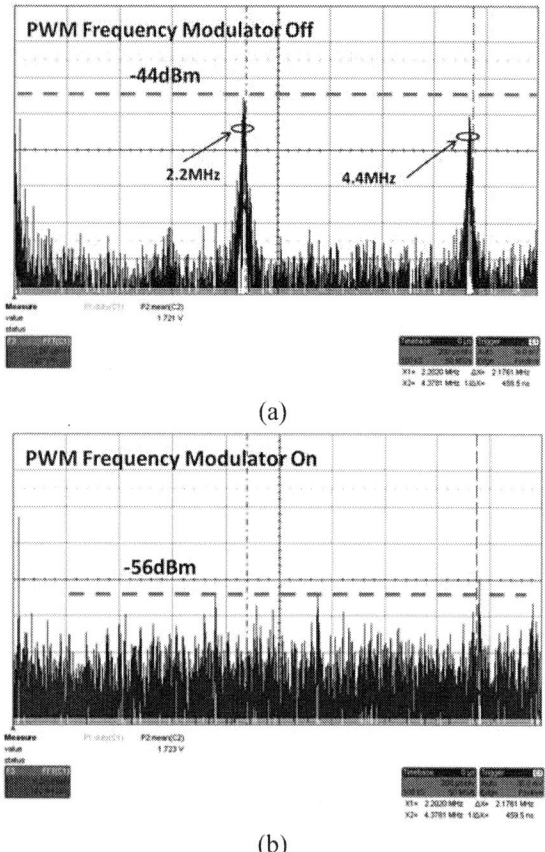

(a)

(b)

Fig. 7. (a) The output spectrum of the DC-to-DC converter using a fixed 2.2 MHz switching frequency (b) The output spectrum of the DC-to-DC converter with monotonic frequency modulation between 2.2 MHz and 4.4 MHz with 64 frequency steps.

prototype DC-to-DC converter is 800×600 μm², and the area of digital PWM ramp signal modulator is 250×100 μm². The digital PWM ramp signal modulator consumes about 5 % of the total area including wire-bonding pads.

A 3GHz digital oscilloscope was used to sample the output signal of the DC-to-DC converter in the time domain, and the internal FFT function of the digital oscilloscope was used to calculate and plot the frequency spectrum of the switching noise power. Measurements were performed at $V_{bat} = 3.4$ V, duty cycle = 0.5, and V_{out} = 1.72 V. When the switching frequency of the DC-to-DC converter is fixed at 2.2 MHz, the peak switching noise at the fundamental switching frequency is about -44 dBm as shown in Fig. 7. When the digital PWM ramp signal modulator with a 6-bit up/down converter is turned on and the switching frequency steps 64 times between 2.2 MHz and 4.4 MHz with a frequency step size of about 35 kHz, the peak noise is reduced by 12 dB from -44 dBm to -56 dBm. A 4.7μH inductor and a 1.0μF capacitor are used for the output LC-filter, and a 5 Ω load was used to measure

the noise spectrum. The results show that the proposed PWM ramp signal modulator can significantly reduce the peak of spectral noise power. The maximum efficiency of the DC-to-DC converter was 92%.

V. CONCLUSION

The spurious noise spreading mechanism by modulating the switching frequency of a step-down DC-to-DC converter is fully analyzed, and the design of the digital PWM ramp signal modulator is implemented. From the measurement results, the proposed digital PWM ramp signal modulator with low-power digital circuits in a DC-to-DC converter can reduce the peak of the spurious noise-power by 12 dB without significantly consuming extra power and area. The digital PWM ramp signal modulator consumes only 5 % of the overall system area and does not use much of static power. The overall efficiency of the DC-to-DC converter was not sacrificed, and by preserving the nature of the switching operation of the PWM control, the modulated PWM control does not affect the stability of the closed-loop operation of the DC-to-DC converter.

REFERENCES

[1] J. Lee, J. Potts, and E. Spear, "DC/DC converter powered power amplifier module for WCDMA application," IEEE Radio Frequency Integrated Circuits Sump. Dig. 2006.

[2] J. Kitchen, I. Deligoz, S. Kiaei, and B. Bakkaloglu, "Polar SiGe Class E and F Amplifier Using Switch-mode Supply Modulation," IEEE Trans. Microw. Theory and Tech., vol.55, no.5, pp.845-856, May 2007.

[3] J. C. Pedro, J. A. Garcia, P. M. Cabral, "Nonlinear Distortion Analysis of Polar Transmitters," IEEE Trans. Microwave Theory & Tech., vol. 55, no. 12, December 2007.

[4] Tae-woo Kwak, Min-chul Lee, Bae-kun Choi, Hanh-Phuc Le, Gyu-Hyeong Cho, "A 2W CMOS Hybrid Switching Amplitude Modulator for EDGE Polar Transmitters," ISSCC Dig. Tech. Papers, pp. 518-519, 2007

[5] Feng lin, Che. D. Y, "Reduction of power supply EMI emission by switching frequency modulation," IEEE Trans. Power Electronics, vol. 9, Issue 1, pp. 132 – 137, January 1994.

[6] A. V. Oppenheim, R. W. Schafer, "Discrete-Time Signal Processing," Prentice Hall, 1999

[7] Jau-Horng Chen, Pang-Jung Liu, Chen, Y.-J. E, " A spurious emission reduction technique for power amplifier using frequency hopping DC-DC converters," IEEE RFIC Symp. pp. 145 – 148, June 2009.

[8] Drissi. K. E. K, Luk. P. C. K, Bin Wang, Fontaine. J, "Effects of symmetric distribution laws on spectral power density in randomized PWM," IEEE Power Electronics Letters., pp. 41-44, June 2003.

A Rail-To-Rail Input Receiver Employing Successive Regeneration and Adaptive Cancellation of Intermodulation Products

Edward A. Keehr and Ali Hajimiri

California Institute of Technology, Pasadena, CA 91125

Abstract — **A direct conversion receiver is demonstrated which operates in the presence of a rail-to-rail (+12.4dBm) out-of-band blocker and a -16.3dBm blocker, where the ICP1 is +12.5dBm and the uncorrected extrapolated IIP3 is +33.5dBm. IM distortion is adaptively cancelled via feedforward loops which are digitally expanded to reproduce higher order nonlinear reference terms. Cancellation improves input-referred total IM distortion by over 24dB, resulting in an extrapolated IIP3 of +45.3dBm.**

Index Terms — **Adaptive equalization, feedforward cancellation, mixed-signal linearization, nonlinear circuits, RF receivers, wireless communications.**

I. INTRODUCTION

Throughout the history of radio, receivers and their blocks have been considered to exist within the realm of small-signal circuit design. Metrics such as ICP1 were meant to denote boundaries on the regions of operation and typically limited the maximum signal handling of the receiver input to less than 1V. In this work, we propose a method to overcome these limitations and implement a radio receiver that operates in the large-signal regime.

II. SYSTEM CONCEPTS

The concept by which the proposed receiver accomplishes large signal operation is an extension of that presented in [1]. Such a receiver would need to handle the presence of many higher-order intermodulation (IM) products generated from a large input signal. The dominant IM products are regenerated at RF in an alternate nonlinear receiver path, downconverted, digitized, and finally used to cancel IM products in the nominally linear (main) receiver path via adaptive filtering. Shown in Fig. 1, the proposed receiver contains two nonlinear receiver paths, corresponding to even and odd order IM products. Once downconverted and digitized, the even and odd order IM products can be successively multiplied in the digital domain to regenerate higher order IM products.

For example, IM2 products can be squared to generate an approximation to IM4 products. IM2 and IM3 products can be multiplied to generate an approximation to IM5 products.

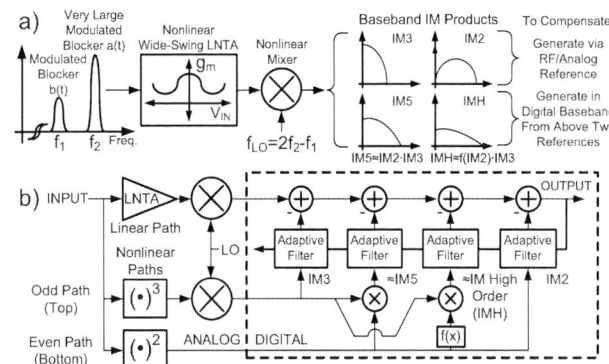

Fig. 1. Successive regeneration and adaptive feedforward cancellation of IM products at baseband implemented in this work. a) Concept. b) Simplified system block diagram.

This approximation approaches an equality when one of the blocker signals is much larger than all of the rest. As illustrated in Fig. 1, two blocker signals with complex envelopes $a(t)=a_I(t)+ja_Q(t)$ and $b(t)=b_I(t)+jb_Q(t)$ generate a number of different baseband IM products. Dropping the *(t)* for brevity, the baseband nonlinear terms can be derived:

$$IM2 \propto a_I^2 + a_Q^2 + b_I^2 + b_Q^2 \tag{1}$$

$$IM3_I \propto a_I^2 b_I + 2a_I a_Q b_Q \quad a_Q^2 b_I \tag{2}$$

$$IM3_Q \propto -a_I^2 b_Q + 2a_I a_Q b_I + a_Q^2 b_Q \tag{3}$$

The IM4 and IM5 terms in the linear path are:

$$IM4 \propto IM2^2 + 2(a_I^2 + a_Q^2)(b_I^2 + b_Q^2) \tag{4}$$

$$IM5_I \propto IM3_I((a_I^2 + a_Q^2) + \tfrac{3}{2}(b_I^2 + b_Q^2)) \tag{5}$$

$$IM5_Q \propto IM3_Q((a_I^2 + a_Q^2) + \tfrac{3}{2}(b_I^2 + b_Q^2)) \tag{6}$$

While the IM4 and IM5 reference terms are:

$$IM4_{REF} \propto IM2^2 \tag{7}$$

$$IM5_{I,REF} \propto IM3_I \, IM2 \tag{8}$$

$$IM5_{Q,REF} \propto IM3_Q \, IM2 \tag{9}$$

It can be seen that as the ratio $|a(t)|/|b(t)| \to \infty$, the reference terms approach the linear path corruptive terms and at this point large cancellation ratios can be achieved.

Some strong nonlinearities may not be well-represented by a truncated polynomial basis of intermodulation products. In this case, the IM2 terms may be modified by a fixed look-up table (LUT) representing a static function $f(x)$ prior to multiplication with IM3 terms.

Although the proposed receiver only achieves large cancellation ratios for large ratios of $|a(t)|/|b(t)|$, it is important to note that this condition is precisely the same as that of many important nonlinear blocking problems. For example, in FDD communications systems with relaxed PA/LNA isolation, the TX leakage appears as the dominant blocker to the receiver. Implantable medical sensors that receive power wirelessly may also need to demodulate a small data signal in the presence of a dominant power transfer signal [2]. In addition, radar systems can benefit from being able to handle a single very large intentional jamming signal.

III. RF/ANALOG RECEIVER ARCHITECTURE AND CIRCUITS

A. Linear Path Receiver Architecture

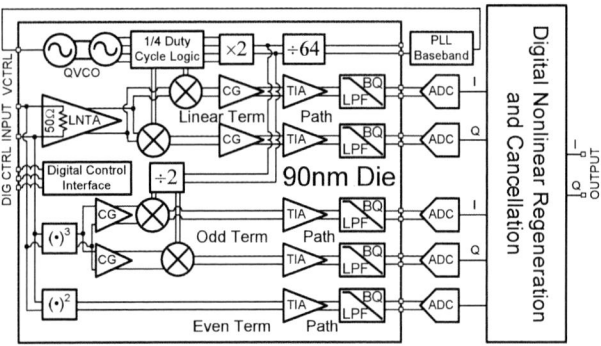

Fig. 2.　Proposed receiver architecture.

Figure 2 shows the architecture of the complete receiver and block diagram of the RF front end. In order to maximize the large-signal handling capability of the receiver, the input signal is immediately converted into current by a low-noise transconductance amplifier (LNTA). This current is then directly converted by a set of quadrature passive mixers to baseband, where it is filtered by large capacitors. In order to isolate the I and Q downconversion chains with minimal voltage swing at the LNTA output, a ¼-phase passive mixer scheme was used.

The noise generated by the transimpedance amplifier (TIA) in a passive mixer system is a well-known problem in cases such as this, where the impedance looking back up into the passive mixer is low. In order to provide a high

input impedance to the TIA, it is preceded by a common-gate (CG) buffer, thereby lowering its effective noise contribution. A 2nd-order active RC biquad (BQ) was utilized to both buffer the TIA and to complete a 3rd-order Chebychev low-pass anti-aliasing filter.

The VCO oscillates at the LO frequency in order to minimize the out-of-band phase noise floor for a given power dissipation. This is an atypical choice due to the fact that it promotes LO-RF and RF-LO coupling, increasing DC offset and decreasing IIP2, respectively. However, in this architecture DC offset is compensated by adding a differential static current to the first OTA virtual ground in the BQ. IM2 products are ultimately cancelled using the scheme described in Section II.

B. Linear Path Receiver Circuits

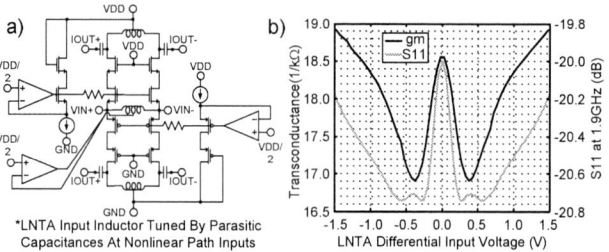

*LNTA Input Inductor Tuned By Parasitic Capacitances At Nonlinear Path Inputs

Fig. 3a.　Simplified LNTA schematic showing biasing OTAs.
Fig. 3b.　Static LNTA simulations: differential g_m and S_{11} as a function of differential input voltage, for Vdd=1.5V.

A differential rail-to-rail input receiver accommodates a signal at its input whose amplitude is nearly equal to the supply voltage before reaching ICP1. To accomplish this task, a push-pull CG-LNTA is introduced in Fig. 3a. The transconductance (and hence input impedance) remains relatively constant over a rail-to-rail input, as shown in Fig. 3b, guaranteeing that the absolute magnitude of odd-order IM products generated for large signals also remains roughly constant. However, because this transconductance function is not well-approximated by a truncated polynomial basis, the digital back-end must use LUTs representing exponential functions of the form $f(x)$, $g(x) = e^{-|k|x}$ to compensate for IM products produced in the LNA.

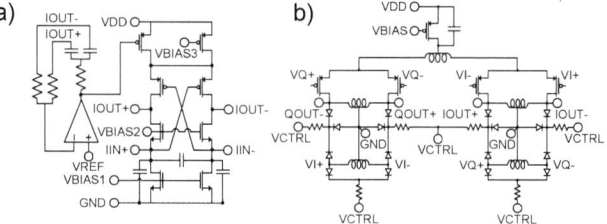

Fig. 4a.　TIA CMOS common gate buffer schematic.
Fig. 4b.　VCO schematic.

978-1-4244-6240-7/10 $26.00 © 2010 IEEE

The current gain of the CG buffer preceding the TIA is nearly doubled by using a cross-coupled CMOS architecture, shown in Fig. 4a, in which the CG-device current is re-used in a pair of PMOS common-source amplifiers. The buffer's large center capacitor of 335pF filters out large downconverted blocker signals. Reciprocal mixing noise due to the very large blocker and VCO phase noise is addressed via the use of the Q-doubling dual-LC tank 90° phase-shift QVCO shown in Fig. 4b. As described in [3], the dual tanks provide additional filtering of out-of-band phase noise.

C. Nonlinear Path Receiver Architecture and Circuits

In contrast to the architecture presented in [1], the nonlinear path inputs are derived directly from the receiver input, as this is the only point in the receiver at which some version of the RF input is commonly available as a voltage of appreciable magnitude. The odd path IM term generator utilizes a multistage architecture as in [4] but with a CMOS input stage in order to increase the dynamic range for a given current. The even path IM term generator consists of a buffered canonical squaring transconductor that feeds directly into a TIA.

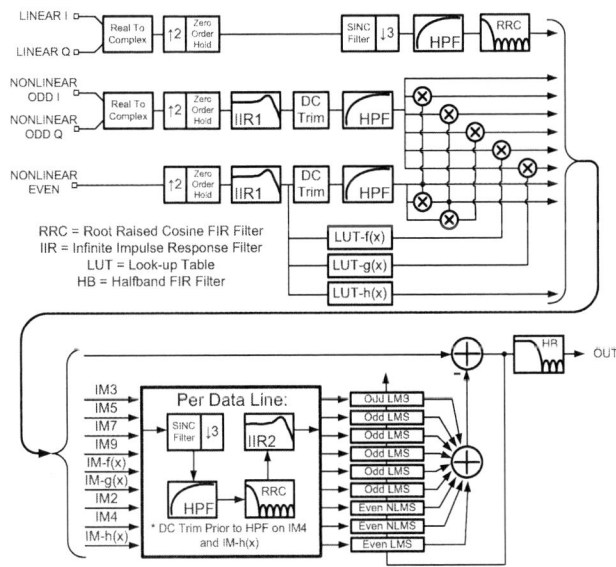

Fig. 5.   Digital back end architecture.

IV. DIGITAL BACK END

The analog outputs of the receiver are captured by 12 bit discrete ADCs running at 25MHz. For this proof-of-concept demonstration, the digital back end (DBE) is implemented in a fixed-point software model, with its architecture shown in Fig. 5. The nonlinear path inputs are

upsampled and filtered prior to successive nonlinear reference generation to ensure that unwanted higher-order nonlinear terms do not alias into the signal band. This filtering also compensates for the small amount of group delay distortion present in the nonlinear path baseband filtering. After this process is complete, an approximate digital model of the analog linear path baseband filter removes undesired residue from these operations and helps to better match the linear and nonlinear paths.

The remaining difference between the linear and nonlinear path transfer functions is fine-tuned via LMS adaptive equalizers modified to compensate for I/Q mismatch [1]. Quantized-NLMS adaptive equalizers [1] modified to divide by the square root of the norm were placed on the IM2 and IM4 lines to reduce gradient noise amplification for large signal levels. A function $h(x)=tanh(k \cdot x)$ in the even order path was found to improve cancellation performance. The exponential functions $f(x)$ and $g(x)$, along with $h(x)$, are implemented with 256-element LUTs. The complete nonlinear path circuitry utilizes 39 and 342 16-bit multipliers running at 50MHz and 16.66MHz, respectively. Based on the results of [5] and assuming that the multipliers dominate the power consumption, the extra digital circuitry and adaptive filters would consume about 12mA and 41.5mA under a 1.3V supply for the even and odd nonlinear paths, respectively. In practice, these quantities would be dramatically reduced by time-averaging, as correction is only required under infrequent blocking conditions.

V. MEASUREMENT RESULTS

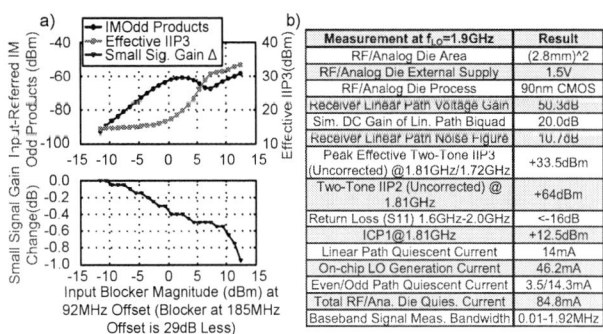

Measurement at f_{LO}=1.9GHz	Result
RF/Analog Die Area	(2.8mm)^2
RF/Analog Die External Supply	1.5V
RF/Analog Die Process	90nm CMOS
Receiver Linear Path Voltage Gain	50.3dB
Sim. DC Gain of Lin. Path Biquad	20.0dB
Receiver Linear Path Noise Figure	10.7dB
Peak Effective Two-Tone IIP3 (Uncorrected) @1.81GHz/1.72GHz	+33.5dBm
Two-Tone IIP2 (Uncorrected) @ 1.81GHz	+64dBm
Return Loss (S11) 1.6GHz-2.0GHz	<-16dB
ICP1@1.81GHz	+12.5dBm
Linear Path Quiescent Current	14mA
On-chip LO Generation Current	46.2mA
Even/Odd Path Quiescent Current	3.5/14.3mA
Total RF/Ana. Die Quies. Current	84.8mA
Baseband Signal Meas. Bandwidth	0.01-1.92MHz

Fig. 6a.   Two-CW tone IIP3 test: measured input referred error, effective IIP3, and ICP1 of standalone RF/Analog die.
Fig. 6b.   Baseline receiver performance metrics.

The non-monotonic nature of the LNTA nonlinearity is apparent in the results of the two-tone measurement shown in Fig. 6a. In this case, a large CW blocker at 92MHz LO frequency offset is swept with a smaller blocker at 185MHz offset while the LO is set to run at

1.9GHz. As predicted in Section IIIb, although the small signal gain of the receiver is negligibly reduced, the magnitude of the IM products at the output remains roughly constant. By extrapolating at each point of the sweep, an effective IIP3 metric can be obtained, reaching a peak of +33.5dBm for a rail-to-rail blocker at the input.

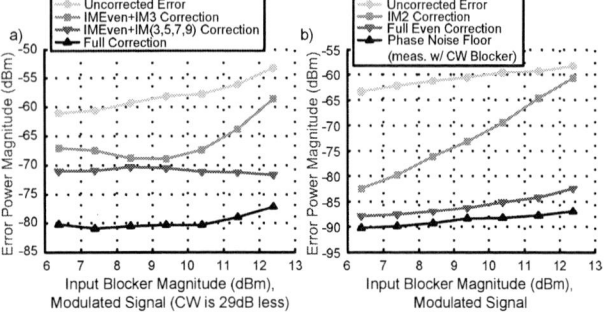

Fig. 7. Measured input-referred error with various degrees of cancellation. a) Even and odd order cancellation for two-signal blocking. b) Even order cancellation for large QPSK-modulated blocker and phase noise floor (measured with a CW blocker).

The linearity performance of the receiver is also tested under modulated blocking conditions by applying a large QPSK signal at 2MSPS along with a smaller CW blocker at LO offsets of 92 MHz and 184 MHz, respectively. The measured cancellation performance for different levels of applied correction is shown in Fig. 7. At the worst-case full-correction value in Fig. 7a, an extrapolated IIP3 metric of +45.3dBm is obtained after de-embedding residual even-order products. The correction performance for peak blocking as a function of QPSK modulation bandwidth is shown in Fig. 8a, while the convergence behavior of the adaptive algorithm is shown in Fig. 8b.

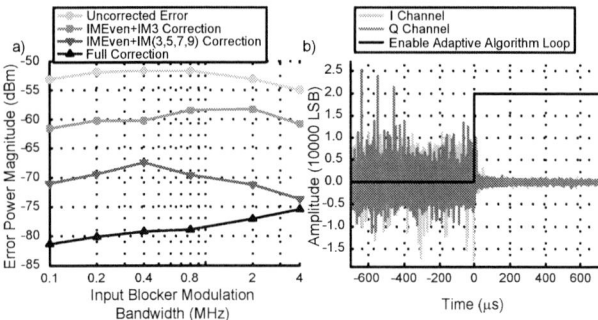

Fig. 8. a) Measured input-referred error as function of modulation bandwidth for +12.4dBm QPSK blocker / -16.3dBm CW blocker. b) Convergence behavior of full adaptive algorithm for +11.4dBm 2MSPS QPSK blocker / -16.3dBm CW blocker.

The RF/Analog die is shown in Fig. 9. The chip is fully ESD protected and is controlled by a digital interface.

VI. CONCLUSION

A large-signal handling direct conversion receiver has been demonstrated in a 90nm RF CMOS technology. It achieves an out-of-band ICP1 of +12.5dBm and a peak uncorrected IIP3 of +33.5dBm. Adaptive feedforward cancellation of modulated IM products yields over an order of magnitude improvement in input-referred error.

ACKNOWLEDGEMENT

The authors thank UMC for chip fabrication and H. Mani for assembly assistance. This work was supported by the Lee Center for Advanced Networking.

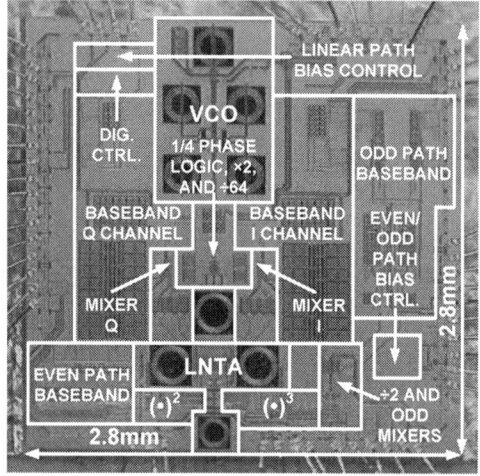

Fig. 9.   RF/Analog chip die photo.

REFERENCES

[1] E. A. Keehr and A. Hajimiri, "Equalization of Third-Order Intermodulation Products in Wideband Direct Conversion Receivers," *IEEE J. Solid-State Circuits*, vol. 43, no. 12, pp. 2853-2867, Dec. 2008.

[2] W. Liu, M. Sivaprakasam, G. Wang, M. Zhou, J. Granacki, J. Lacoss, and J. Wills, "Implantable Biomimetic Microeletronic Systems Design," *IEEE Engineering in Medicine and Biology Magazine*, pp. 66-74, Sept. 2005

[3] A. M. ElSayed and M. I. Elmasry, "Low-Phase-Noise *LC* Quadrature VCO Using Coupled Tank Resonators in a Ring Structure," *IEEE J. Solid-State Circuits*, vol. 36, no. 4, pp. 701-705, Apr. 2001.

[4] E. A. Keehr and A. Hajimiri, "Analysis of Internally Bandlimited Multistage Cubic-Term Generators for RF Receivers," *IEEE Trans. Circuits and Systems - I*, vol. 56, no. 8, pp. 1758-1771, Aug. 2009.

[5] S.K. Hsu, S.K. Mathew, M. A. Anders, B. R. Zeydel, V. G. Oklobdzija, R. K. Krishnamurthy, S. Y. Borkar, "A 110 GOPS/W 16-bit multiplier and reconfigurable PLA loop in 90-nm CMOS," *IEEE J. Solid-State Circuits*, vol. 41, no. 1, pp. 256-264, Jan. 2006.

A D-Band PLL Covering the 81-82 GHz, 86-92 GHz and 162-164 GHz Bands

Shahriar Shahramian[1], Adam Hart[1], Anthony Chan Carusone[1], Patrice Garcia[2],
Pascal Chevalier[2], Sorin P. Voinigescu[1]

[1] Edward S. Rogers Sr. Dept. of ECE, University of Toronto, Toronto, ON M5S 3G4, Canada,
[2] STMicroelectronics, 850 Rue Jean Monnet, F-38926 Crolles, France

Abstract — **This paper describes the highest frequency PLL reported to date. It achieves the widest locking range and the lowest phase noise of -93.8 dBc/Hz at 90 GHz and 78.9 dBc/Hz at 163 GHz, both measured at a 100-kHz offset. The PLL was fabricated in a 0.13-μm SiGe BiCMOS process and covers the 81-82 GHz, 86-92 GHz, and 162-164 GHz bands. It integrates on a single die a fundamental-frequency 86-92 GHz Colpitts VCO, a differential push-push 160-GHz Colpitts VCO with quadrature outputs at 80 GHz, a programmable divider chain, charge-pump, and all loop filter components. The single-ended PLL output power is -3 dBm at 90 GHz and -25 dBm at 164 GHz and consumes 1.25 W from 1.8-V, 2.5-V and 3.3-V supplies. The chip occupies 1.1mm x 1.7mm including pads.**

Index Terms — **Charge pump, loop filter, mm-wave, phase-locked loop, VCO, 160-GHz ICs, SiGe BiCMOS.**

I. INTRODUCTION

Several W-band and D-Band transceivers have been recently demonstrated in SiGe BiCMOS and CMOS technologies for a variety of radar sensor [1, 2], imaging [3, 4], radio [2] and serial 107-Gb/s Ethernet [5] applications. In all these applications, a low-noise phase-locked loop (PLL) is either mandatory or desirable.

Although mm-wave silicon PLLs have recently been reported [6, 7, 8, 9], none of them operate above 100 GHz and none of the fully integrated PLLs meet the automotive radar requirements of < -80 dBc/Hz phase noise at 100-kHz offset. This work demonstrates the first D-band PLL in silicon and investigates its phase noise performance for different reference frequencies ranging from 600 MHz to 6 GHz and for PLL bandwidths between several hundred kilo-Hertz and several mega-Hertz.

II. PLL ARCHITECTURE AND CIRCUIT DESIGN

As illustrated in Fig. 1, the PLL is equipped with two VCOs, a fundamental 90-GHz Colpitts VCO and a push-push Colpitts VCO with simultaneous quadrature outputs at 80 GHz and differential outputs at 160 GHz [4]. A mm-wave selector chooses the desired VCO signal and is proceeded by the LO distribution. The PLL prescaler comprises a Miller stage [10] followed by seven static frequency divider stages which enable a selectable divide ratio of 16, 32, 64, or 128. The divider output and the reference frequency signal are fed into a phase-frequency detector (PFD) followed by a second-order loop filter (LF) whose output is directly connected to the varactor control nodes of both VCOs. All signals are differential.

Fig. 1: PLL block diagram.

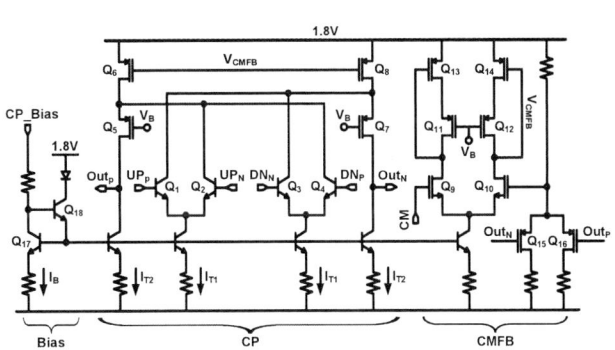

Fig. 2: Charge pump schematic.

978-1-4244-6240-7/10 $26.00 © 2010 IEEE

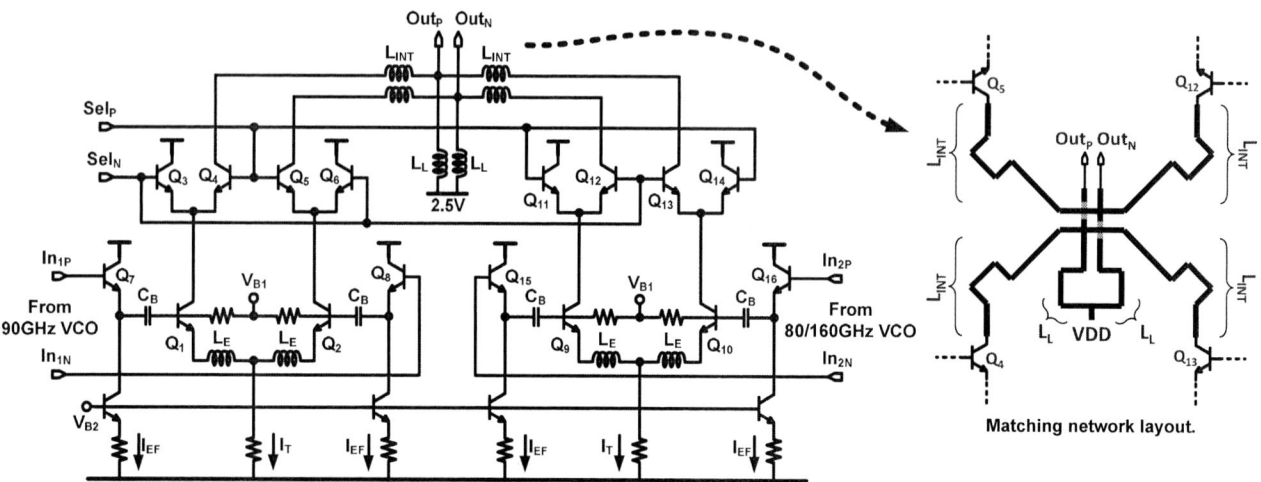

Fig. 3: Schematic and critical layout path of the mm-wave VCO selector.

A. Charge Pump

The charge pump (CP) is implemented using a BiCMOS differential folded-cascode opamp (Fig. 2) with SiGe HBT inputs and common gate 0.13-μm PMOSFETs. A common-mode feedback (CMFB) circuit is used to match the common-mode level of the CP to the VCO common-mode tank voltage. The charge pump tail current can be adjusted externally from 70 μA to 700 μA to vary the PLL bandwidth. The integrated loop filter consists of a pair of 180-pF capacitors, 500-Ω resistors, and 40-pF dampening capacitors.

B. mm-Wave VCO Selector

Fig. 3 shows the schematic of the mm-wave selector. Each half-circuit consists of a pair of emitter-followers

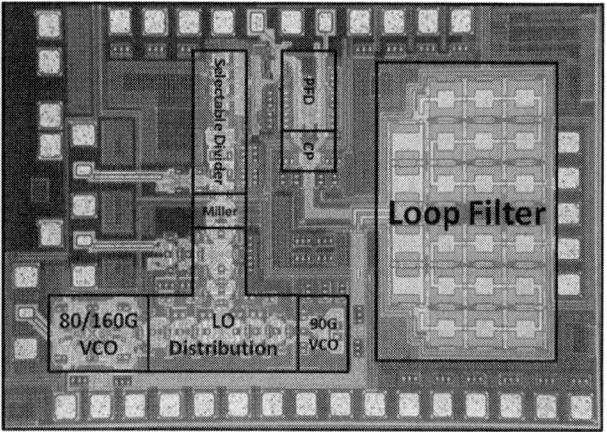

Fig. 4: Die photo of the PLL.

(Q_{7-8}) and a variable gain amplifier (Q_{1-6}). Sel_P and Sel_N select the desired VCO signal. The layout arrangement of the matching network (L_{INT} and L_L) at the output of the selector is sketched in Fig. 3.

The LO distribution amplifiers employ the same half-circuit as the selector, but without current steering transistors Q_3 and Q_6.

III. FABRICATION

The circuit was fabricated in STMicroelectronics' 0.13-μm SiGe BiCMOS process with HBT f_T and f_{MAX} of 230 GHz and 280 GHz, respectively. All critical circuit blocks (i.e. CP, LF, VCOs and LO distribution) use separate power supplies and are surrounded by deep n-well isolation and stacked metal shields (M1 to M6). The photo of the 1.1mm x 1.7mm die is shown in Fig. 4.

IV. MEASUREMENT RESULTS

The PLL operates from a 2.5-V supply except for the charge pump and the 80/160 GHz quadrature VCO which use 1.8 V, and 3.3 V, respectively. The total power consumption is 1.15 W or 1.25 W depending if the 90-GHz or the 160-GHz VCO is activated. The LO distribution, 90-GHz and 80/160-GHz VCOs consume 600 mW, 125 mW, and 225 mW respectively. The remaining power is dissipated in the divider chain, charge pump and output drivers.

The PLL locking range is from 86 GHz to 92 GHz for the first VCO, and 81 GHz to 82 GHz, as well as 162 GHz to 164 GHz, for the push-push VCO. An Agilent E8257D signal source is used as a reference signal and an Agilent

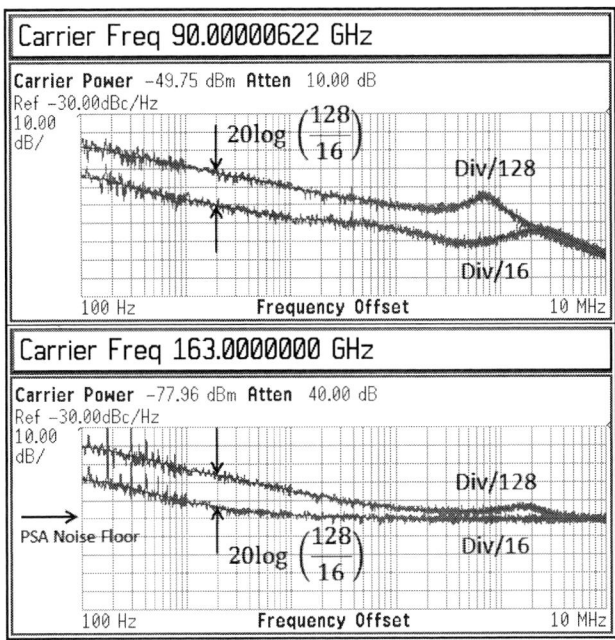

Fig. 5: *(Top)* Measured PLL phase noise at 90 GHz for divider ratios of /16 and /128. *(Bottom)* Measured PLL phase noise at 163 GHz for divider ratios of /16 and /128.

E4448A PSA, equipped with phase noise personality software, is employed to conduct phase noise measurements. The phase noise of the reference signal is approximately constant (-120 dBc/Hz at 100-kHz offset) between 600 MHz and 6 GHz. Therefore, by selecting different divider ratios in the prescaler, the phase noise variation can be observed to track the reference frequency.

Fig. 5 shows the measured phase noise of the PLL at 90 GHz and at 163 GHz.

The single-ended PLL output power is -3 dBm at 90 GHz and -25 dBm at 163 GHz. At 90 GHz, the PLL achieves a phase noise better than -81 dBc/Hz at 100-kHz offset for all divider ratios. At 163 GHz, due to the large losses of the external mixer and the low VCO signal power, the phase noise of the PLL corresponding to the /16 divide ratio is below the noise floor of the PSA for offsets greater than 3-kHz. However, with divide-by-32 ratio the measured phase noise rises above the setup noise floor and is -78.9 dBc/Hz at 100-kHz offset. Furthermore, Fig. 6 shows the measured phase noise of the PLL at various frequencies, offsets and divider ratios. It can be observed that the phase noise is approximately constant for each VCO throughout its entire locking range and that doubling the divider ratio degrades the PLL phase noise by the theoretical 6 dBc/Hz value. The in-band phase noise of the PLL is dominated by the reference signal. This suggests that the PLL phase noise may be improved by using a reference oscillator with better phase noise characteristics.

To investigate the effect of the PLL divider ratio on phase noise, the charge pump current is kept constant for measurements shown in Fig. 5. This causes the PLL loop bandwidth to decrease for higher divider ratios. Furthermore, by adjusting the charge pump current, the loop bandwidth of the PLL at 90 GHz can be varied from 0.54 MHz to 1.72 MHz, and from 1.72 MHz to 5.4 MHz for /16 and /128 divider ratios, respectively. Similar results were also obtained using the 80/160-GHz VCO. Fig. 7 shows the measured PLL phase noise at 81.5 GHz

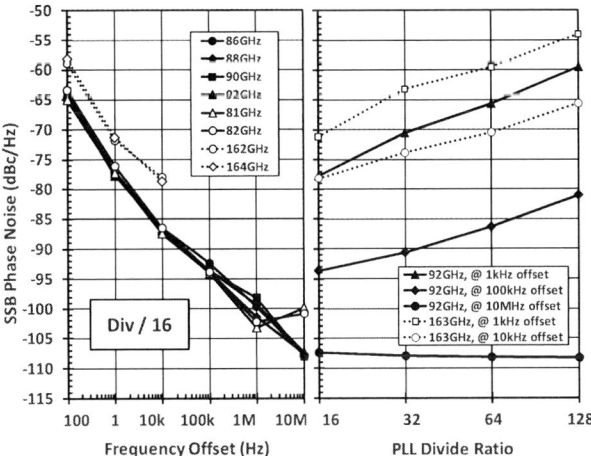

Fig. 6: *(Left)* Measured phase noise of the PLL at various frequencies versus frequency offset. *(Right)* Measured phase noise of the PLL at 92 GHz and 163 GHz and at various offsets versus the PLL divide ratio.

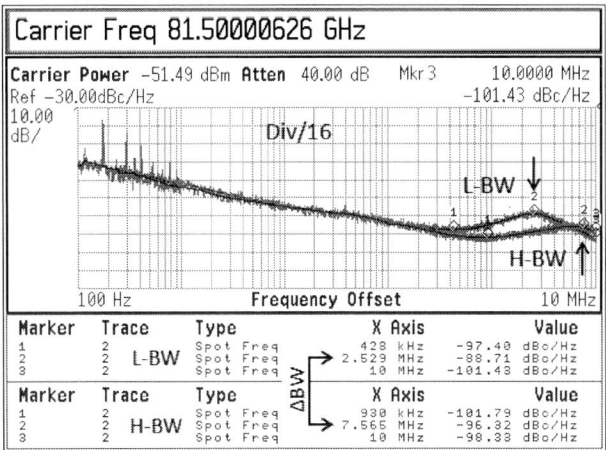

Fig. 7: Measured phase noise of the PLL at 81.5 GHz and with /16 divide ratio for two different PLL bandwidths.

Table 1: Comparison table of the state-of-the-art PLLs.

	[6]	[7]	[8]	[9]	This Work
Frequency	45.9-50.5GHz 91.8-101GHz	73.4-73.7GHz	95.1-96.5GHz	79.4GHz	81-82GHz 86-92GHz 162-164GHz
Divide Ratio	/512	/32	/256	/64	/16, /32, /64, /128
Phase Noise (dBc/Hz) @100kHz	-63.5 @ 50GHz****	-88 @ 73.5GHz	N/A	-81 @ 79.4GHz	-93.8 @ 90GHz* -78.9 @ 163GHz**
Phase Noise (dBc/Hz) @1MHz	-72 @ 50GHz	N/A	-75.2 @ 96GHz	-86 @ 79.4GHz	-98.2 @ 90GHz*
Output Power (dBm)	-10 @ 50GHz -22 @ 100GHz	N/A	-26.8 @ 96GHz	N/A	-3 @ 90GHz -25 @ 163GHz
Supply	1.5V	1.45V	1.2V, 1.3V	5.5V	1.8V, 2.5V, 3.3V
Power Consumption	57mW	88mW***	43.7mW	N/A	1.15W to 1.25W
Technology	0.13µm CMOS	90nm CMOS	65nm CMOS	SiGe	0.13µm SiGe BiCMOS
Chip Area	1.1 mm x 0.75mm	1mm x 0.8mm	1mm x 0.7mm	N/A	1.1mm x 1.7mm

* Using /16 divide ratio. ** Using /32 divide ratio. *** Not including output buffers. **** Measured at 50kHz offset.

and with /16 divide ratio for two different PLL bandwidths.

IV. CONCLUSIONS

To the best of the authors' knowledge, the PLL reported here achieves the highest frequency of operation, the largest locking range and the lowest phase noise of all PLLs operating above 60 GHz to date. A comparison of its performance to the state-of-the-art is provided in Table 1.

ACKNOWLEDGEMENT

This work was funded by an NSERC Strategic Grant Project. Equipment was provided through OIT, NSERC and CITO grants. We'd like to thank Katya Laskin for 160-GHz VCO design, Bernard Sautreuil for technology access, and Jaro Pristupa and CMC for CAD tools.

REFERENCES

[1] S.T. Nicolson, *et al.,* "Single-Chip W-band SiGe HBT Transceivers and Receivers for Doppler Radar and Millimeter-Wave Imaging," *JSSC*, vol. 34, pp. 2206 – 2217, Oct., 2008.

[2] E. Laskin, *et al.,* "A 140-GHz Double-Sideband Transceiver with Amplitude and Frequency Modulation Operating over a few Meters," *BCTM*, pp. 178 – 181, Oct., 2009.

[3] E. Laskin, *et al.,* "Nanoscale CMOS Transceiver Design in the 90–170-GHz Range," *MTT*, pp. 3477-3490, Dec., 2009.

[4] E. Laskin, *et al.,* "165-GHz Transceiver in SiGe Technology," *JSSC*, vol. 43, pp. 1087 – 1100, May, 2008.

[5] M. Moller, "Challenges in Cell-Based Design of Very-High-Speed Si-Bipolar IC's at 100 Gb/s," *BCTM*, pp. 106-114, Oct., 2007.

[6] C. Cao, *et al.,* "A 50-GHz Phase-Locked Loop in 0.13-µm CMOS," *JSSC*, vol. 42, pp. 1649-1656, Aug., 2007.

[7] J. Lee, *et al.,* "A 75-GHz Phase-Locked Loop Generator in 90-nm CMOS Technology," *JSSC*, pp. 1414-1426, Jun., 2008.

[8] K.-H. Tsai, *et al.,* "A 43.7mW 96 GHz PLL in 65nm CMOS," *ISSCC*, pp. 276 – 277, Feb., 2009.

[9] C. Wagner, *et al.,* "PLL architecture for 77-GHz FMCW radar systems with highly-linear ultra-wideband frequency sweeps," *MTT*, pp. 399–402, Jun., 2006.

[10] S.T. Nicolson, et al., "A Low-Voltage SiGe BiCMOS 77-GHz Automotive Radar Chipset," *MTT*, pp. 1092 – 1104, May, 2008.

RMO1C-2

An Integrated Frequency Synthesizer for 81-86GHz Satellite Communications in 65nm CMOS

Zhiwei Xu, Qun Jane Gu, Yi-Cheng Wu, Heng-Yu Jian, Frank Wang, Mau-Chung Frank Chang

University of California, Los Angeles, CA 90025, USA

Abstract — **We present an integrated frequency synthesizer in 65nm CMOS to enable the 81-86GHz satellite communication transceiver. The frequency synthesizer is inserted in a two-step zero-IF millimeter-wave transceiver with LO_{RF} at 70-78GHz and LO_{IF} at 1/8 of LO_{RF} to cover the desired entire frequency bands. It also features coarse phase rotation to endow beam forming capabilities for the intended communication system. The phase noise is < -83dBc/Hz at 1MHz offset as extrapolated from measured value at 1/8 of the VCO frequency (~9.4GHz). The measured reference spur is <-49dBc. Total synthesizer power consumption including LO buffers and phase rotators is 65mW at 1V power supply and the compact layout has rendered small synthesizer core area of 0.16mm^2.**

Index Terms — **Frequency Synthesizer, injection locking divider, multi-modulus programmable divider, millimeter-wave transceiver, phase rotation, VCO .**

I. INTRODUCTION

The 81-86GHz frequency segment of W band (75–110 GHz) is allocated by the International Telecommunications Union (ITU) to satellite services to alleviate the crowded spectrum for microwave frequency satellite communications. Because of the abundant bandwidth to enable the wireless communication at a very high data rate, it is drawing increasing interest from the commercial satellite operators.

Traditional W band applications are based upon discrete components or exotic technologies, such as GaAs and InP HBTs, which are bulky and expensive. Fortunately, recent advances in deep-scaled CMOS technology make it feasible to realize ~100GHz millimeter-wave integrated circuits operating with >180GHz f_T devices [1], which has been recently exemplified by a fixed division ratio PLL in [2]. The unsurpassed CMOS integration capability allows the incorporation of various functions, such as RF modulation and demodulation, digital processing, encryption and decryption, etc., into a single chip, thus fulfilling the quest for low-cost, low-power and light weight communication systems for satellite services.

Within RF/millimeter-wave communication systems, the frequency synthesizer is one of the key building blocks, which is required to generate a low-phase noise and stable local oscillation (LO) signal for signal modulation and demodulation. Numerous research efforts

[3-6] have been made in the past to design frequency synthesizers for millimeter-wave communications. Ref. [3] implemented a 15-18GHz synthesizer and a frequency tripler to avoid large division ratio and cumbersome high frequency LO routing to both receiver and transmitter in SiGe technology. Ref. [4] developed a 77-81GHz synthesizer in SiGe as well to support short-range pulsed-radar transceiver, with a fixed division ratio and relaxed quadrature LO matching requirements. Ref. [5] and [6] reported two stand-alone frequency synthesizers running at 40GHz in 130nm CMOS and 57-66GHz in 45nm CMOS, respectively, to support V-band applications.

In this paper, we present an integrated frequency synthesizer in 65nm CMOS with fully-programmable division ratios and phase rotators to accomplish 81-86GHz satellite communication that simultaneously boasts wide frequency range, low power consumption and small chip area. It employs a chain of complementary injection locking dividers to realize the demanded high-frequency prescaler, and an injection locking buffer with high drivability to facilitate a compact and high performance transceiver.

II. TRANSCEIVER ARCHITECTURE AND FREQUENCY PLAN

Direct conversion architecture can realize a compact RF transceiver with power efficient RF modulation/demodulation in lower GHz RF operations. However, the higher the operation frequency, the more sensitive the circuits are to process variations in deep-scaled CMOS. For example, the vast LO quadrature mismatches, especially the phase mismatch due to component mismatches, asymmetrical capacitive/magnetic coupling effects can seriously impede direct conversion transceiver performance in millimeter-wave frequency bands. Polyphase filter may be exploited to alleviate the problem at the price of high power consumption but lead to low LO drivability that may dramatically deteriorate the transceiver linearity and noise figure performance. Additional digital baseband calibration may be applied but often with limited range and consuming extra power.

Instead of chasing challenging millimeter-wave frequency quadrature LO operation in the direct conversion, a two-step zero-IF architecture is adopted in

978-1-4244-6240-7/10 $26.00 © 2010 IEEE

this work to lower the quadrature operation to about 10GHz frequency range (i.e. LO_{IF}), which permits LO signals to be generated through a resistive load CML divider-by-2 circuit with small mismatches. This architecture not only enables more accurate quadrature LO signal generation, but also renders more compact realization without extra area-consuming inductors in deep-scaled CMOS. It is also easier to harness various LO phases more accurately to implement a more sophisticated beam-forming front end. Fig. 1 depicts the intended transceiver architecture for 81-86GHz satellite communications.

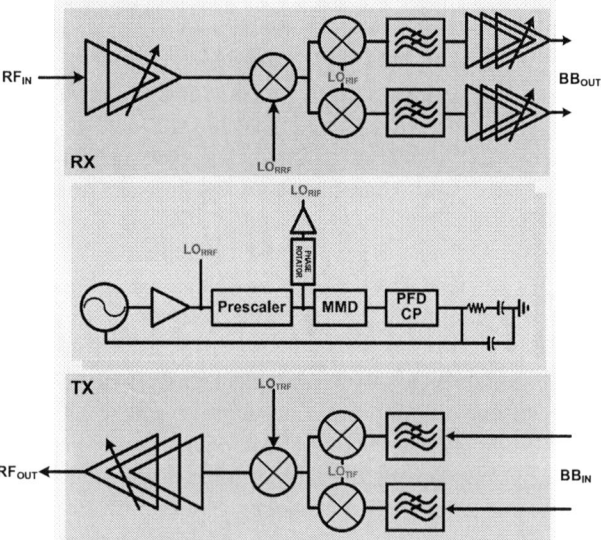

Fig. 1. Integrated synthesizer with phase rotators and T/R LO buffers to support 81-86GHz Transceiver for satellite communications.

A frequency synthesizer contains VCO at 70-78GHz as the LO_{RF} and 1/8 of its frequency as the LO_{IF} to cover the intended 81-86GHz frequency bands. The following phase rotator at LO_{IF} supports 0^0, 90^0, 180^0 and 270^0 coarse phase-tuning capability that facilitates baseband processing with various combinations of I+Q, I-Q, -I+Q and –I-Q. Integer-N frequency synthesizer architecture is chosen due to its superior phase noise performance and small form factor, which can sufficiently meet the channel requirement of required wide band applications. LO drivability to TRX up/down conversion mixers in millimeter-wave frequency is fairly challenging so we use two dedicated synthesizers for receiver and transmitter, respectively to alleviate this issue.

III. CIRCUIT IMPLEMENTATION

The proposed type II, 3rd order frequency synthesizer consists of a reference buffer, a pseudo differential PFD with charge pump, a second order loop filter, a voltage controlled oscillator (VCO) with injection locked buffer, an asynchronous multi-modulus divider (MMD), an additional phase rotator and a LO_{IF} buffer, as shown in Fig. 1. We place the injection locking buffer, RF mixers and the prescaler in close proximity to minimize the routing load, which not only eliminates extra buffers but also alleviates undesirable inter-stage matching at the millimeter-wave frequency. Consequently, we can use a single buffer to drive both prescaler and RF mixers with smaller insertion area and an estimated 30% power-saving in comparison with that of using two separate buffers.

The phase rotator and LO_{IF} buffers are also located in close proximity with IF mixers to allow well-matched resistive load buffer with sufficient drive and good matching. To ensure low power operation and circuit reliability, a single 1V supply is used for the entire synthesizer. The supply is also separated from the receiver and transmitter supplies for better isolation. Fully and/or pseudo differential implementations are exploited in the entire synthesizer design to further attenuate invading spurs from both the supply and ground and the spur emission from the synthesizer itself.

A. VCO and Injection Locking Buffer Amplifier

The VCO, as shown in Fig. 2(a), adopts LC cross-coupled pair by using NMOS as the switching transistors due to its higher f_T. A resistive bias is used at the top instead of the PMOS current source to avoid the device flicker noise from coupling into the tank to degrade the phase noise through AM-to-PM conversion. Accumulation mode NMOS varactor is used to fine-tune the oscillation frequency while a three bit switch controlled MIM capacitors are used to achieve a wide tuning range to cover desired frequency bands from 70GHz to 78GHz. The same switch capacitors are used in subsequent injection locking buffers and dividers to line up VCO, injection locking buffer and divider operating frequencies. In millimeter-wave VCO design, the tank quality (Q) is often determined by that of switch capacitors and varactors rather than that of inductors, which typically have a high Q of >20. To attain a good quality factor in switch capacitor for both on/off states, a hybrid π-switch and moderate $C_{max}/C_{min}(\sim2)$ are adopted for optimized phase noise performance.

According to our simulations, LO_{RF} buffer must support larger than 600mV$_{p-p}$ signal output to drive both RF mixer and prescaler. An inductive loading cascode structure is used to reduce the capacitive load to the preceding VCO by mitigating the Miller effect. One positive feedback cross coupled pair is inserted in Fig. 2(b) as an injection locking buffer to further boost the signal strength. Post-layout simulations show combined VCO and the injection locking buffer can achieve 70~81GHz tuning range with 1.4GHz/V K_{VCO} and larger than 650mV$_{p-p}$ LO$_{RF}$ outputs.

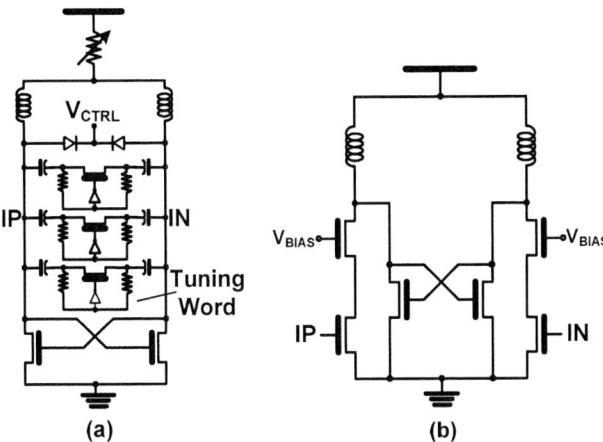

Fig. 2. Schematic of (a) VCO and (b) Injection Locking LO$_{RF}$ Buffer Amplifier.

B. Prescaler Divider and Phase Rotator

One challenge in millimeter-wave frequency synthesizer is the high speed prescaler design, which mandates an extensive tuning effort due to lack of accurate models for passive/active components and interconnects. To obtain good performance with design success on the first try, several practices have been employed besides a robust divider design: 1) Careful floor-planning of the prescaler chain and optimization of connections between them; 2) Simulation of passive components and routing paths by using EM tools, such as ADS momentum, SONNET and HFFS; 3) Extraction of device parasitics and their inclusion in simulations and verifications. As a result, simulations have achieved a deviation of less than 5% in comparison with the actual measured results.

Fig. 3(a) sketches the complementary injection locking divider. Differential circuit realization not only achieves robust design without the imbalance problem typically in a single-ended counterpart, but also utilizes both polarities to increase the locking range by injecting signals into the tank via both NMOS and PMOS devices. The injection signal threshold and effectiveness can be adjusted through the device body voltage. Such injection locking structure is adopted in the first two stages of prescalers. Their currents are controlled by a two-bit DAC, and the locking range is further enlarged by another three-bit switch-controlled MIM capacitors.

Beam forming has proven to be an effective method of boosting the transmission output power and improving the receiver sensitivity. A coarse phase rotator is designed, as shown in Fig. 3(b), to support four different phase options followed by amplitude controlled resistive load LO$_{IF}$

buffer. This phase rotator can further obtain finer phase step by adding phase interpolators.

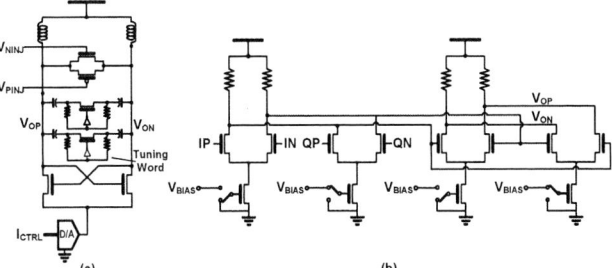

Fig. 3. Schematic of (a) prescaler divider and (b) phase rotator.

C. Programmable Divider Chain

A multi-modulus divider in fully differential structure is implemented to provide programmable division N_{div} from 128 to 248. The total division ratio of the loop equals $8 \times N_{div}$. Fig. 4 manifests the asynchronous multi-modulus divider and depicts the detailed CML logic circuit. Compared with the synchronous implementation, it consumes less power and smaller real estate. However, it may introduce excessive noise due to noise accumulation through cascade stages. A re-synchronization circuit is thus used to counter the accumulated noise along the chain by realigning the signals.

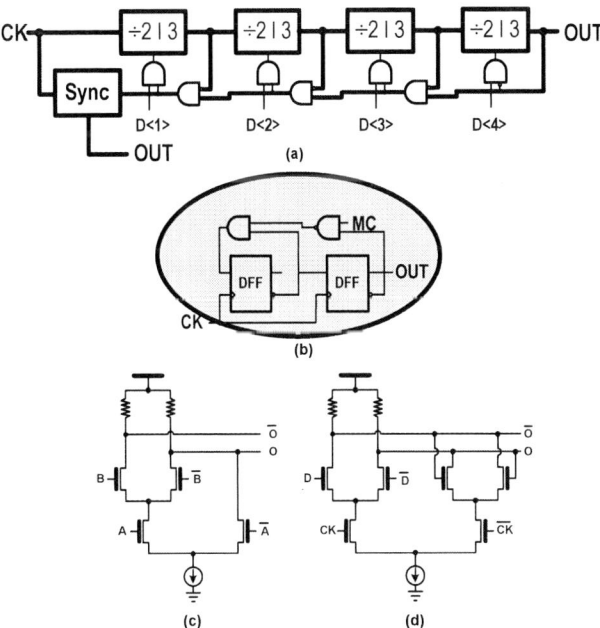

Fig. 4. (a) Multi-modulus programmable counter and its building block circuitry, including (b) ÷2/3 module, (c) "And2" logic and (d) latch

Afterwards, a pseudo-differential reference buffer, a PFD with the charge pump driven by programmable current from 50μA to 400μA, as well as a second order

978-1-4244-6240-7/10 $26.00 © 2010 IEEE

on-chip loop filter, are employed to complete the frequency synthesizer with 300 KHz loop bandwidth.

IV. EXPERIMENTAL RESULTS

Fig. 5 shows the die photo of an 81-86GHz transceiver. Within it, the integrated frequency synthesizer occupies a core area of 0.16mm^2 and burns 65mW power from a 1V supply. Fig. 6(a) demonstrates the 86GHz output from the transmitter and Fig. 6(b) demonstrates a baseband output tone from the receiver. Both of them confirm the proper function of the integrated frequency synthesizer. Fig. 6(c) draws the transmitter output frequency curve versus various control bits, which covers the desired communication frequency bands with safety margins (79GHz~88GHz). Additionally, Fig. 6(d) reveals the synthesizer phase noise measured at 1/8 of VCO frequency f_{VCO} (i.e. 9.4GHz, equals to 1/9 of TX output frequency f_{OUT}) from a separate test chip, which can be extrapolated into -83dBc/Hz @1MHz in 75GHz.

Fig. 5.   Die photo of the transceiver and frequency synthesizer.

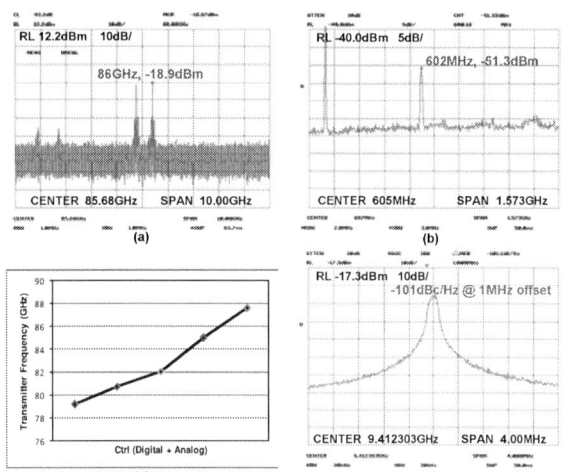

Fig. 6.   Measured (a) signal spectrum at 86GHz from transmitter and (b) baseband output signal from receiver when injecting a 85GHz CW tone; (c) frequency tuning curve at TX output and (d) the phase noise performance @ f_{VCO}/8 (9.4GHz)

V. CONCLUSION

An integrated frequency synthesizer which enables the transceiver operation at 81-86GHz for satellite communication is realized in 65nm general purpose (GP) CMOS. Table I summarizes measured performance versus those of prior developments in similar frequency range [2, 5 and 6] based on deep-scaled CMOS technologies. This frequency synthesizer clearly demonstrates its performance advantages in wider frequency range, better phase noise, and more compact design to facilitate the transceiver SoC for intended satellite communication. Its slightly higher power consumption than that of Ref. [2] is primarily due to its much wider frequency coverage and extra integration of phase rotators and T/R LO buffers.

	[2]	[5]	[6]	This Work
Type	Fundamental, Independent	Fundamental, Independent	Fundamental, Independent	Offset, Integrated
Center Freq. (GHz)	95.1~96.5	40	57~66	70-78
Divisions	256	256/258/260/262	512~8184	1024~1984
Phase Noise (dBc/Hz)	-75.9dBc/Hz @1MHz	-75.3dBc/Hz @1MHz	-75dBc/Hz @1MHz	-83dBc/Hz @1MHz
Reference Spurs (dBc)	-51.8	-46	-42	-49
Supply (V)	1.2	1.5	1.1	1.0
Power (mW)	43.7	105	76	65
Area (mm^2)	0.7 (WiPad)	0.56 (core), 1.69(WiPad)	0.82 (WiPad)	0.16 (core), 0.31 (WiPad)
Technology	65nm CMOS	0.13um CMOS	45nm CMOS	65nm CMOS

TABLE I.   PERFORMANCE SUMMARY AND COMPARISON

ACKNOWLEDGEMENT

The authors wish to acknowledge the chip fabrication and support of TSMC Inc. and the test setup help from Adrian Tang.

REFERENCES

[1] Y. Tagro, D. Gloria, S. Boret, G. Dambrine, "MMW Lab In-Situ to Extract noise Parameters of 65nm CMOS Aiming 70~90GHz Applications" *IEEE RFIC Symp. Dig.*, pp. 397-400, June 2009.

[2] K.H. Tsai, S.I. Liu, "A 43.7mW 96GHz PLL In 65nm CMOS" *Proceedings of ISSCC*, pp. 276-277, February 2009.

[3] B.A. Floyd, "A 15 to 18-GHz Programmable Sub-Integer Frequency Synthesizer for a 60-GHz transceiver," *IEEE RFIC Symp. Dig.*, pp. 529-532, June 2007.

[4] V. Jain, F. Tzeng, L. Zhou, P. Heydari, "A Single-Chip Dual-Band 22-to-29GHz/77-to-81GHz BiCMOS Transceiver for Automotive Radars," *Proceedings of ISSCC*, pp. 308-309, February 2009.

[5] C.C. Hung, D.S. Shen, S.I. Liu, "A 40GHz Fractional-N Frequency Synthesizer in 0. 13μm CMOS," *IEEE RFIC Symp. Dig.*, pp. 295-298, June 2008.

[6] K. Scheir, G. Vandersteen, Y. Rolain, P. Wambacq, "A 57-to-66GHz Quadrature PLL in 45nm Digital CMOS," *Proceedings of ISSCC*, pp. 494-495, February 2009.

RMO1C-3

Low-Noise Fractional-*N* PLL Design with Mixed-Mode Triple-Input LC VCO in 65nm CMOS

Yuanfeng Sun[1], Xueyi Yu[2,4], Woogeun Rhee[1], Sangsoo Ko[3], Wooseung Choo[3], Byeong-Ha Park[3], and Zhihua Wang[1]

[1]Institute of Microelectronics, [2]Electronic Engineering, Tsinghua University, Beijing, China
[3]MSC Development Team, Samsung Electronics, Yongin-City, Gyeonggi-Do, Korea
[4]Now with Marvell Technology, Shanghai, China

Abstract — **This paper presents a low-noise $\Delta\Sigma$ fractional-*N* PLL utilizing a mixed-mode triple-input LC VCO. An analog dual-path VCO control relaxes the nonlinearity problem of the $\Delta\Sigma$ fractional-*N* PLL, while a combination of discrete and continuous tuning methods for coarse-tuning control significantly alleviates the noise coupling problem caused by the high gain coarse-tuning path. A 3.6GHz $\Delta\Sigma$ fractional-*N* PLL implemented in 65nm CMOS exhibits nearly –100dBc/Hz in-band noise contribution and –53dBc in-band fractional spur performances from a 1.8GHz carrier.**

Index Terms — **CMOS integrated circuits, LC-VCO, phase-locked loops, phase noise, voltage-controlled oscillators**

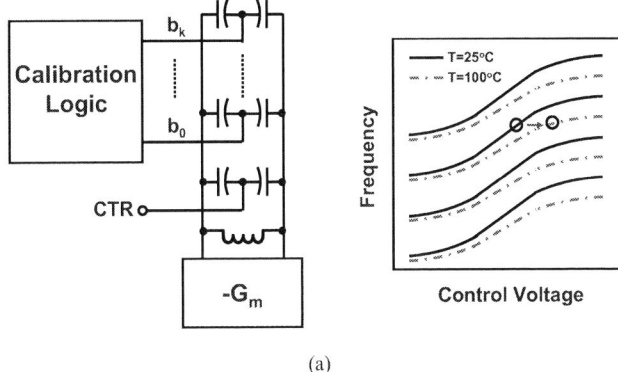

(a)

I. INTRODUCTION

A frequency synthesizer is one of the key building blocks in the transceiver to determine the overall noise performance. Designing a low-noise frequency synthesizer with broad tuning range is critical for multi-standard transceivers. The $\Delta\Sigma$ fractional-*N* PLL not only provides superior performance over conventional integer-*N* PLLs but also enables digital phase modulation. For those reasons, the $\Delta\Sigma$ fractional-*N* synthesizer has been widely used in modern wireless transceivers. However, the $\Delta\Sigma$ fractional-*N* PLL suffers from charge pump nonlinearity which is usually not an issue for the integer-*N* PLL. Also, the digital phase modulation based on the $\Delta\Sigma$ fractional-*N* PLL requires stringent bandwidth control with digital compensation [1] or very tight VCO gain control with two-point modulation [2].

Providing high open-loop dc gain and continuous frequency tuning without digital calibration logic, a dual-path control architecture has been well adopted in recent PLL design [3]-[6]. The dual-path PLL exhibits a small control voltage range in the loop filter, resulting in improved charge pump linearity. Also, the VCO gain variation can be little since the tuning voltage of the fine-tuning varactor is near the middle of the tuning curve. Fig. 1(a) shows a typical LC VCO with a digitally programmable varactor array, overcoming the tuning range problem of the LC VCO. However, the drawback of this approach is VCO gain variability over temperature, which cannot be calibrated by the digital logic as illustrated in Fig. 1(a). In the dual-path PLL design, the control voltage range can be controlled by the dc gain of the linear amplifier (LA). Since the tuning voltage of the fine-tuning varactor is near the middle of the tuning curve with a very small control

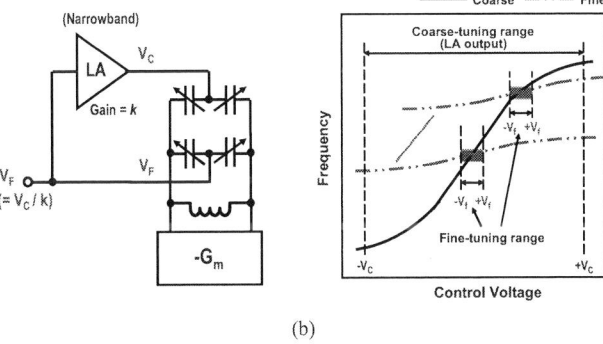

(b)

Fig. 1. LC VCO architecture comparison: (a) with calibration logic, (b) with continuous dual-path control.

voltage range V_f as shown in Fig. 1(b), the dual-path PLL can achieve good CP linearity as well as small VCO gain variation. Hence, the dual-path LC VCO can offer stable VCO gain and enhanced gain linearity over process and temperature, which is different from the LC VCO having multiple varactors.

II. DESIGN CONSIDERATIONS

In theory, the dual-path VCO can achieve much better noise performance than the conventional multi-band VCO since the fine-tuning gain can be set to a smaller value and noise contribution from the high-gain coarse-tuning path can be suppressed by narrowband filtering. However, the dual-path PLL, in practice, has difficulty in achieving low noise performance since the high gain analog coarse-tuning path is

978-1-4244-6240-7/10 $26.00 © 2010 IEEE

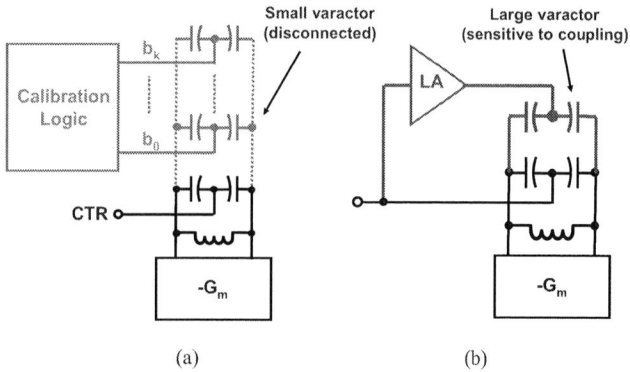

Fig. 2. Coupling path comparison: (a) single-path and (b) dual-path.

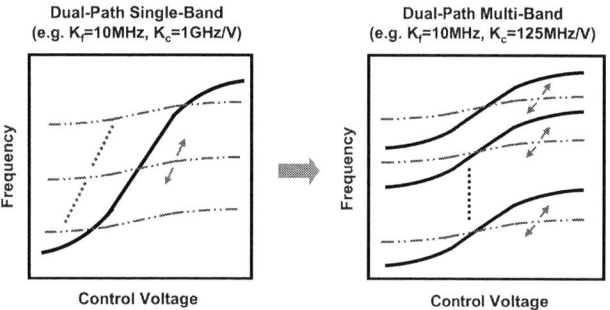

Fig. 3. Partitioned coarse-tuning control.

more vulnerable to noise coupling and self-induced noise than the digital band-switching path of the conventional single-path VCO as illustrated in Fig. 2.

To overcome the inherent problem of having a very high gain ratio between the coarse-tuning path and the fine-tuning path, a partitioned coarse-tuning control is considered for low-noise dual-path VCO design [7]. As shown in Fig. 3, instead of having a single-band coarse tuning curve to cover most of the tuning range, a combination of discrete as well as continuous tuning methods is used for the dual-path control. For example, to achieve a 500MHz tuning range with the input control voltage range of 100mV, we need the coarse-tuning gain of 1GHz/V even with the LA gain of 5. When eight coarse-tuning curves are used with a 3-bit input control, the coarse-tuning gain can be reduced to 125MHz/V. The small fine-tuning gain and the narrow input control voltage range are maintained since they can be independently controlled in the dual-path design.

Table I shows PLL architecture comparison among a PLL using a conventional LC VCO with a single varactor (single-path PLL/single-input VCO), a PLL using a conventional LC VCO with multiple varactors and calibration logic (single-path PLL/two-input VCO), a PLL using a dual-control LC VCO with coarse and fine varactors (dual-path PLL/two-input VCO), and the proposed PLL (dual-path PLL/three-input VCO). With mixed-mode coarse-tuning control, the proposed dual-path PLL achieves reduced coupling sensitivity in the coarse-tuning path while maintaining enhanced linearity in the fine-tuning path.

Table I
ARCHITECTURE COMPARISON

	Single-Path PLL Single-Input VCO	Single-Path PLL Two-Input VCO	Dual-Path PLL Two-Input VCO	Dual-Path PLL Three-Input VCO
Control Mode	Analog	Analog & Digital	Analog	Analog & Digital
Tuning Method	Continuous	Discrete	Continuous	Hybrid
Tuning Range	Poor	Good	Good	Good
Tuning over Temp.	Poor	Poor	Good	Fair
CP Linearity	Poor	Fair	Good	Good
Coupling Immunity	Poor	Good	Poor	Fair

III.   CIRCUIT DESIGN

Fig. 4 shows a block diagram of the proposed fractional-N PLL. The reference frequency of 26MHz is used. A fixed divide-by-2 circuit is used to generate the 1.8GHz quadrature outputs from a 3.6GHz VCO. The PLL employs a single high frequency divide-by-4 CML divider and multiple single-ended phase shifters to realize a low-power multi-modulus divider (MMD). The CML divider consisting of cascaded toggle flip-flops alleviates the output loading to the VCO buffer. The quadrature output phases generated by the CML divider are fed into the phase shifter followed by 5-stage single-ended 2/3 prescalers. A singled-ended charge pump is used with an external 3rd-order loop filter and the lowpass filter after the LA is integrated.

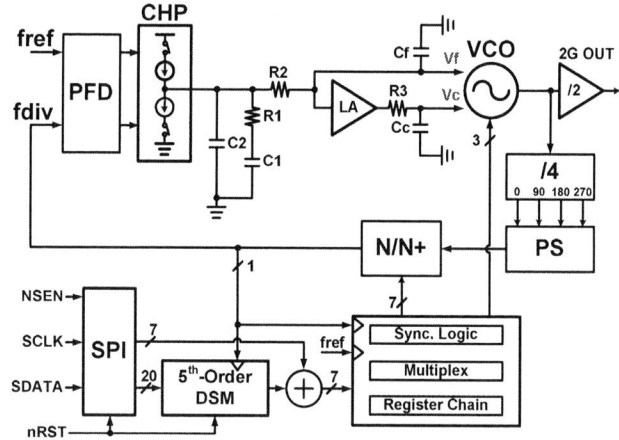

Fig. 4. PLL block diagram.

The LA in the coarse tuning path, as shown in Fig. 4, is based on an op amp with resistor feedback to provide the accurate dc gain of 5. An RC filter at the output of the LA is used to make the coarse-tuning bandwidth much narrow, minimizing the noise contribution from the coarse-tuning path. To reduce the area of the RC filter, a high resistor value of 5MΩ is used. With a 40pF capacitor, a 5kHz corner

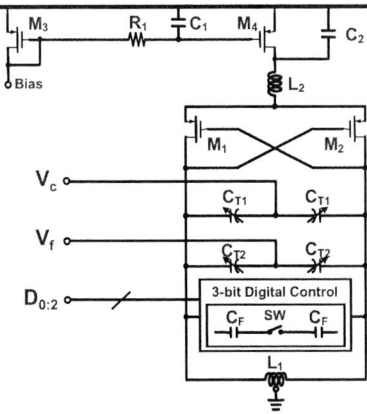

Fig.5. Schematic of mixed-mode three-input LC VCO.

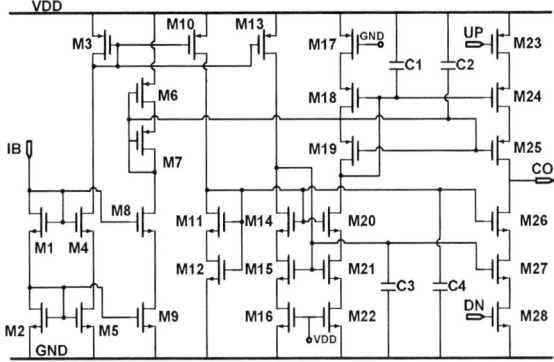

Fig. 6. Schematic of charge pump.

frequency can be designed. The thermal noise contribution of the resistor at low frequencies will be suppressed by the PLL open loop gain, while high-frequency noise will be filtered by the *RC* filter itself [6].

In this work, a single-loop 5th order $\Delta\Sigma$ modulator is used with 20-bit input to improve the fractional spur performance [8]. The noise transfer function is given by $(1 - z^{-1})^5/(1 - z^{-1} + 0.5z^{-2})$. Butterworth based noise transfer function makes the 5th-order modulator possible in the 4th-order PLL with moderate bandwidth. Compared to a 5th-order MASH modulator, the output bit spread of the single-loop modulator is reduced by more than 60%, which is good to mitigate the charge pump nonlinearity effect.

Fig. 5 shows the schematic of the mixed-mode three-input LC VCO [7]. A PMOS-based topology is used for low $1/f$ noise characteristics, *N*-well isolation, and ground-referenced output. The *LC* tank is composed of a symmetric inductor L_1 and a capacitive network of *N*-channel MOS varactors C_T and MIM capacitors C_F. The MIM capacitor instead of the accumulation-mode varactor used for C_1 helps to reduce substrate noise coupling. The parallel *LC* circuit consisting of a symmetric inductor L_2 and an MIM capacitor C_2 are tuned to two times the operating frequency, suppressing the up-conversion mechanism of the base-band noise from transistor M_4. The ratio of the coarse-tuning gain to the fine-tuning gain is about 14.

Fig. 6 shows the schematic of the charge pump. It is well known that charge pump nonlinearity causes noise folding and degraded fractional spur performance in the $\Delta\Sigma$ PLL design. With the dual-path VCO control, the output voltage range of the charge pump is significantly reduced. Accordingly, the stacked current mirrors design can obtain better dc headroom and linearity with longer channel lengths. Large values of the decoupling capacitors are used at the gate of the current mirrors for low noise operation. It also improves the switching speed of the source-switching charge pump, which results in better charge pump linearity.

IV. EXPERIMENTAL RESULTS

A 3.6GHz prototype fractional-*N* PLL is implemented in 65nm CMOS. A chip micrograph is shown in Fig. 7. The active core area is about 0.90mm^2. The LA with the lowpass RC filter occupies less than 0.02mm^2.

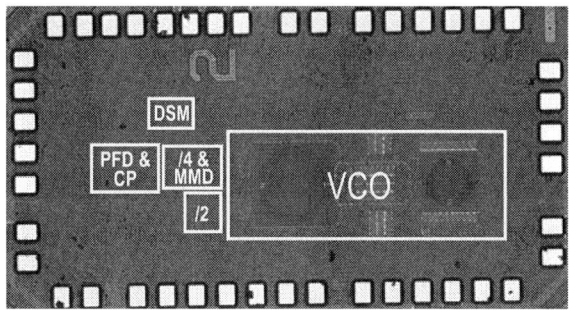

Fig. 7. Chip micrograph.

Fig. 8 shows the measured output spectrum of the divide-by-2 output exhibiting the PLL bandwidth of about 120kHz. To evaluate the fractional spur performance, the VCO frequency is set to 3562.1MHz, which are 100kHz away from the 3562MHz (= 137 x 26MHz). As a result, an integer-boundary spur at 100kHz offset from the carrier frequency is observed. The fractional spur levels of –52.7dBc and –74.8dBc are achieved at 100kHz and 500kHz offset frequencies, respectively, at the output frequency of 1781.05MHz.

Fig. 9 shows the reference spur performance. With the loop bandwidth of 120kHz, the spur level of –66dBc at 26MHz offset frequency is measured. This shows that the high-gain coarse tuning path does not contribute much to the reference spur and that the voltage ripple in the loop filter is well suppressed by the lowpass filter at the output of the LA as expected.

Since the in-band noise performance is limited by the VCO phase noise performance, the PLL bandwidth is further increased to observe the charge pump noise contribution. Fig. 10 shows the measured phase noise performance with the widened PLL bandwidth. The in-band phase noise of nearly – 100dBc/Hz is achieved when the PLL bandwidth is set to 400kHz. The noise plateau at 2–4MHz is due to the $\Delta\Sigma$ quantization noise. With the nominal bandwidth of 120kHz, the out-of-band phase noise of –127dBc/Hz is observed at 3MHz offset frequency, which can be also approximately evaluated from Fig. 9 showing the spectrum with the resolution bandwidth (RBW) of 100kHz.

The prototype synthesizer consumes 27.5mW at 3.6GHz output. The tuning range of the PLL is from 3.03 GHz to 3.67GHz with the coarse-tuning VCO gain of 140MHz/V where the small-signal VCO gain is only 10MHz/V. A summary of the measured results is given in Table II.

V. CONCLUSION

A 3.6GHz low-noise $\Delta\Sigma$ fractional-N PLL implemented in 65nm CMOS is presented. By employing a 3-bit partitioned coarse-tuning control, a tuning range of 3.03–3.67GHz is achieved with only 10MHz/V fine-tuning VCO gain and 140MHz/V coarse-tuning VCO gain. With reduced coupling sensitivity in the coarse-tuning path and enhanced linearity in the fine-tuning path, the $\Delta\Sigma$ fractional-N PLL exhibits nearly –100dBc/Hz in-band noise contribution and –53dBc in-band fractional spur performances from a 1.8GHz carrier.

REFERENCES

[1] S. Pamarti, L. Jansson, and I. Galton, "A wide-band 2.4GHz delta-sigma fractional-N PLL with 1-Mb/s in-loop modulation," *IEEE J. Solid-State Circuits*, vol. 39, pp. 49-62, Jan. 2004.

[2] S. E. Meninger and M. H. Perrott, "A 1MHz bandwidth 3.6GHz 0.18μm CMOS fractional-N synthesizer," *IEEE J. Solid-State Circuits*, vol. 41, pp. 966-980, Apr. 2006.

[3] R. Nonis, N. Da Dalt, P. Palestri, and L. Selmi, "Modeling, design, and characterization of a new low jitter analog dual-tuning LC-VCO PLL architecture," *IEEE JSSC*, vol. 40, pp. 1303-1309, June 2005.

[4] A. Loke, *et al.*, "A versatile 90-nm CMOS charge-pump PLL for SerDes transmitter clocking," *IEEE JSSC*, vol. 41, pp. 1894-1907, Aug. 2006.

[5] T. Wu, P. Hanumolu, K. Mayaram, and U. Moon, "A 4.2 GHz PLL frequency synthesizer with an adaptively tuned coarse loop," in *Proc. IEEE CICC*, Sept. 2007, pp. 547-550.

[6] W. Rhee, *et al.*, "A uniform bandwidth PLL using a continuously tunable single-input dual-path LC VCO for 5 Gb/s PCI Express Gen2 application," in *Proc. IEEE A-SSCC*, Nov. 2007, pp. 63-66.

[7] Y. Sun, *et al.*, "Dual-path LC VCO design with partitioned coarse-tuning control in 65nm CMOS," to appear in *IEEE MWCL*, Mar. 2010.

[8] X. Yu *et al.*, "A $\Delta\Sigma$ fractional-N frequency synthesizer with customized noise shaping for WCDMA/HSDPA applications," *IEEE J. Solid-State Circuits*, vol. 44, Aug. 2009, pp. 2193-2201.

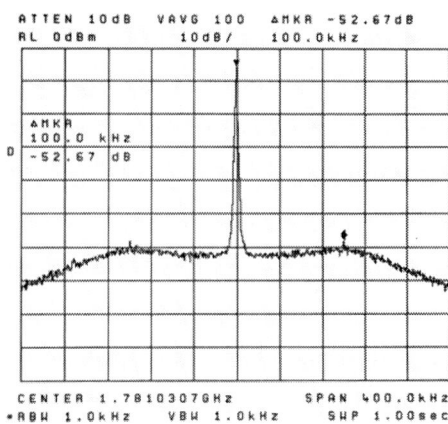

Fig. 8. Measured output spectrum with in-band fractional spur.

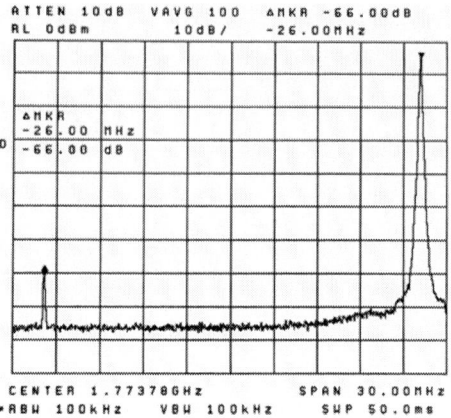

Fig. 9. Measured reference spur.

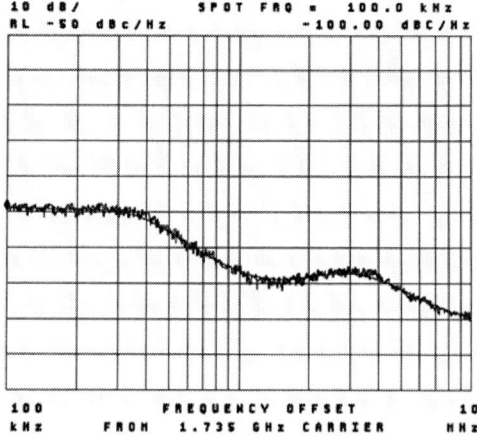

Fig. 10. Phase noise performances with widened PLL bandwidth.

Table II
MEASURED PERFORMANCE SUMMARY

Process	65 nm CMOS
Supply Voltage	1.2 V (PD/VCO with 1.8 V)
Power Dissipation	27.5 mW
Active Area	0.90 mm^2
Tuning Range	3.03 ~ 3.67 GHz
Frequency Resolution	< 100 Hz
Reference Clock	26 MHz
Bandwidth	~ 120 kHz
In-band Noise Contrib.[1,2]	< -100 dBc/Hz
Out-band Phase Noise[1]	-127 dBc/Hz @ 3 MHz offset
Reference Spur[1]	< -66 dBc
Fractional Spur[1]	< -52.7dBc @100kHz < -74.8dBc @500kHz

[1]Measured at 1.8 GHz (/2 output)
[2]Measured with widened bandwidth (~400kHz)

RMO1C-4

A Wideband Millimeter-Wave Frequency Doubler-Tripler in 0.13-µm CMOS

Shadi Saberi Ghouchani, Jeyanandh Paramesh

Department of Electrical and Computer Engineering, Carnegie Mellon University, Pittsburgh, PA

Abstract — **A combined frequency doubler and tripler is proposed for wideband millimeter wave frequency generation in CMOS. The circuit consists of a push-push FET frequency doubler along with a single-balanced mixer based frequency tripler. The frequency doubler-tripler can generate frequencies in the range of 23-48 GHz with more than -20dBm output power into 50Ω. The conversion gains of the doubler and tripler are measured to be -2.6dB and -12.3 dB, respectively, with a 0dBm input at 14.4GHz. Fabricated in 0.13-µm CMOS, the circuit has an active area of 600x440 µm². The frequency multipliers consume 12.6mW dc power from 1.2V supply, while the output buffers consume 11.9mW.**

Index Terms — **Baluns, frequency conversion, millimeter wave, mixers, negative resistance.**

I. INTRODUCTION

In mm-wave transceivers operating at 60 GHz and above in CMOS, there exists a great need for wideband, low noise local oscillators. Due to the low gain of CMOS transistors that operate close to their f_T, and the poor quality of passive elements, including varactors, fundamental frequency mm-wave VCOs are typically power hungry to provide sufficient voltage swing and phase noise. The tuning range is often very limited due to significant parasitic capacitors. Also, the center frequency is more sensitive to process variations and parasitics, which can be substantial at mm-wave frequencies.

To overcome the above shortcomings of mm-wave VCOs, an alternative approach is considered. Frequency multiplication of a lower frequency VCO to generate the final output frequency can relax most VCO requirements. Since the VCO now oscillates at a fraction of the output frequency, power consumption is greatly reduced. The overall phase noise is most likely improved despite the $20log_{10}N$ (dB) degradation caused by frequency multiplication [1]. The tuning range is extended by the multiplication factor, while sensitivity to parasitics and process variations is greatly reduced.

Frequency multiplication can be performed using nonlinear devices such as FETs, diodes, varactors to generate higher harmonics, and then using filters to suppress undesired ones at the output [2]. Sub-harmonic injection locking of an oscillator can also achieve

frequency multiplication [1], [3]. Another approach is to mix the fundamental signal with itself or its harmonics and up-convert to a higher order harmonic [4], [5].

In this paper the first approach is used to generate the second harmonic with a push-push FET frequency doubler, while a single-balanced mixer has been employed to generate the third harmonic $3f_0$ through mixing of the doubler output $2f_0$ with the fundamental input f_0. Since both doubler and tripler outputs are available, the frequency multiplier can cover a very wide band at mm-wave frequencies. For instance, with a fundamental input ranging from 12-18 GHz, the frequency doubler-tripler can generate frequencies over a continuous 24-54 GHz range.

II. CIRCUIT DESIGN

A. Frequency Doubler and Tripler

The schematic of a push-push FET frequency doubler is shown in Fig. 1(a). When the gates of M1 and M2 transistors are driven by a differential LO signal, the doubler output current I_{out2}, which is the sum of drain currents of M1 and M2, will be at twice the input frequency. Since the fundamental frequency term and other odd drain current harmonics are 180° out of phase, they cancel at the output, while the 2nd and higher order even harmonics, generated due to square law nonlinearity of the FETs, will add in phase. The transistors are biased near their threshold voltage to maximize the conversion gain of the doubler [2]. The doubler output load is tuned at the 2nd harmonic to suppress the 4th harmonic and the fundamental tone arising from LO leakage and other non-idealities such as mismatch.

The 3rd harmonic is generated by mixing the 2nd harmonic from doubler output with the fundamental input. Fig. 1(b) shows a single-balanced mixer where the push-push doubler is used as the transconductance stage, and the gates of the differential LO switches M3 and M4 are driven by same input signal as the doubler. The differential output of the tripler is tuned to the 3rd harmonic, and improves the fundamental frequency rejection at the drains of M3 and M4. A transformer balun

978-1-4244-6240-7/10 $26.00 © 2010 IEEE

is used at the output of frequency tripler to convert the differential output to single-ended for measurement purposes. The LO switches *M3* and *M4* are biased near threshold to increase conversion gain. Since the doubler does not consume a lot of current, there is no need for large switching devices, which helps reduce the parasitic capacitance. The center tap of inductor *L2* (V_{b2}) is biased at 0.7V to allow sufficient headroom for the doubler.

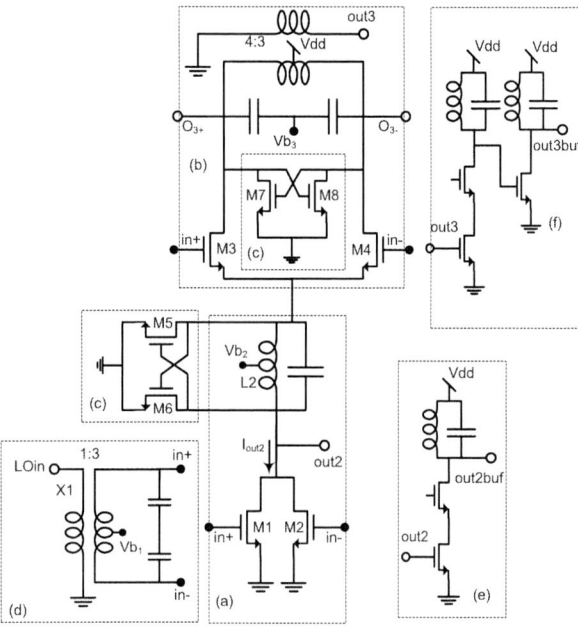

Fig. 1.   Schematic of a) push-push FET frequency doubler b) single-balanced mixer as frequency tripler c) negative resistance cross-coupled pair for gain enhancement d) input transformer balun e) doubler buffer f) tripler buffer.

B. Conversion Gain Enhancement

LC tanks have poor quality factor at mm-wave frequencies, which results in reduced voltage swing when used as tuned loads. To increase the output impedance of the doubler and the tripler, two cross-coupled pairs (*M5-M6*, *M7-M8*) are inserted across the LC tanks as shown in Fig. 1(c). The negative resistance of the cross-coupled pairs improves the effective quality factor of the tanks and thus increases the gain. It is important to size the cross-coupled transistors to avoid instability due to negative resistance.

C. Undesired Harmonics

The doubler output (denoted *out2* in Fig. 1(a)) will not have odd harmonics since it is a common-mode node. Also, any LO leakage is further rejected by the LC tank of the doubler. The dominant undesired harmonic at the

doubler output is the 4th harmonic which is filtered out by subsequent stages. The tripler output, tuned to the 3rd harmonic, will significantly reject the fundamental component generated from the lower sideband of mixing products, LO leakage from switches. However, 2nd harmonic is generated at the tripler output due to: (1) common-mode to differential conversion caused by differential load imbalance and device mismatch; (2) LO self mixing. This is more significant at the higher end of the frequency band where the tripler has more gain for 2nd harmonic than 3rd harmonic. Phase and amplitude imbalance in the balun at higher frequencies will also contribute to the common-mode to differential conversion at the single-ended output.

D. Impedance Matching

The differential LO signal (*in+*, *in-*) is generated using a transformer balun at the input (Fig. 1(d)). The 1:3 transformer also serves as the input matching network to match the input impedance of the frequency doubler and tripler to the 50Ω source. The output of the doubler is buffered with a single cascode stage to drive a 50Ω load. The buffer, tuned to the 2nd harmonic provides less than 3dB gain, and is more narrowband than the output of the doubler. This helps to further reject undesired harmonics, but also limits the output bandwidth. A 4:3 transformer balun is used at the output of the tripler to convert and match the differential output to single-ended input of the buffer. Both transformers use vertical coupling with stacked conductors; the primary uses the topmost metal while the secondary uses the next two metal layers to reduce loss. The tripler buffer consists of two stages to reduce the capacitive loading on the 4:3 transformer. The buffer gain is less than 0.6 dB when driving a 50Ω load.

III. MEASUREMENT RESULTS

The proposed frequency doubler-tripler was designed and fabricated in 0.13-μm 8-metal RF CMOS process (Fig. 2). The active die area is 600x440 μm². The measurements were performed with on-wafer probing. The input LO is provided from a signal generator, and the output is measured using a spectrum analyzer whose input range is extended to 60 GHz by waveguide harmonic mixers. The frequency doubler-tripler consumes 12.6mW from a 1.2V supply. The doubler buffer consumes 2.7mW from 1.2V, while the tripler buffer consumes 9.2mW from 1V.

The conversion gain of the frequency doubler and tripler with 0dBm input power are shown in Fig. 3. The measured frequency doubler gain is consistent with simulation, while the measurement result of frequency

978-1-4244-6240-7/10 $26.00 © 2010 IEEE

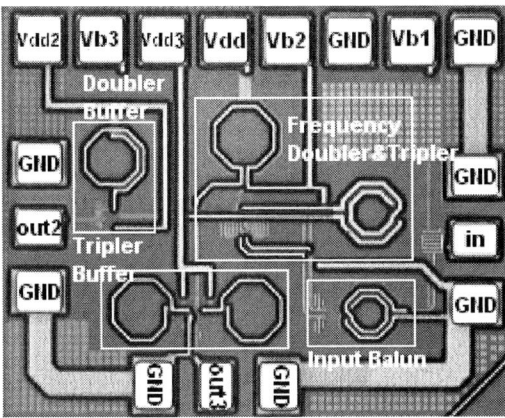

Fig. 2.   Chip microphotograph of frequency doubler-tripler.

tripler gain is better than simulation – this is mainly due to increase in tripler buffer gain, as shown later in Fig. 7(b). The maximum gain of doubler is -2.6dB at 14.4GHz input, and the tripler's maximum gain is -11.4dB at 13.9GHz.

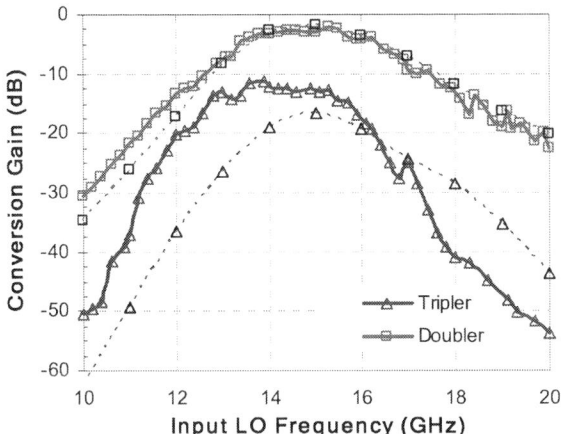

Fig. 3.   Conversion gain of the frequency doubler and tripler with a 0dBm LO input (Dashed lines show simulation).

The output spectra of the frequency doubler and tripler at input frequency of 14.4 GHz are shown in Fig. 4. The markers show -6.01 dBm at 28.8 GHz at the doubler output, and -14.95 dBm at 43.2 GHz at the tripler output. Note that the power spectrum analyzer reading does not include cable losses of around 3dB.

The harmonic content of the outputs of the frequency doubler and tripler up to 60 GHz are shown in Fig. 5. The doubler has more than 24 dB rejection for all harmonics in the band of interest 12-18 GHz. The tripler has more than 40dB rejection for the fundamental tone in the frequency range 12-16 GHz. However, as predicted earlier the 2nd harmonic at the output of frequency tripler becomes significant at the end of the band. In Fig. 5 it can be seen

that, beyond 16.5 GHz, the 2nd harmonic has more power than the 3rd harmonic which limits the operating range of the tripler. If the output of the tripler can be measured differentially, this result will probably improve.

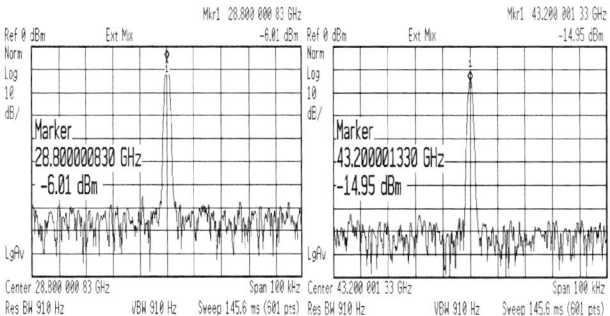

Fig. 4.   Screen shot of output spectrum of the frequency doubler and tripler at input frequency of 14.4 GHz.

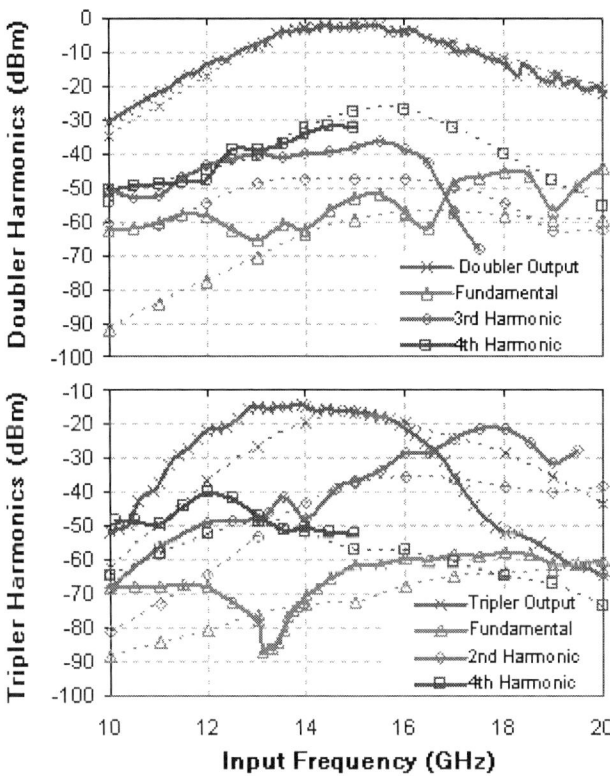

Fig. 5.   Power of undesired harmonics at the output of frequency doubler and tripler with a 0dBm LO input.

The phase noise measurement was limited to the spectrum analyzer's maximum frequency of 26.5GHz, since harmonic mixing will degrade the phase noise of the signal. As a result, only the phase noise of the frequency doubler was measured. In Fig. 6, phase noise of the doubler output is compared with the input LO frequency at

978-1-4244-6240-7/10 $26.00 © 2010 IEEE

13GHz. The phase noise degradation in the frequency doubler is consistent with the theoretical value of 6 dB.

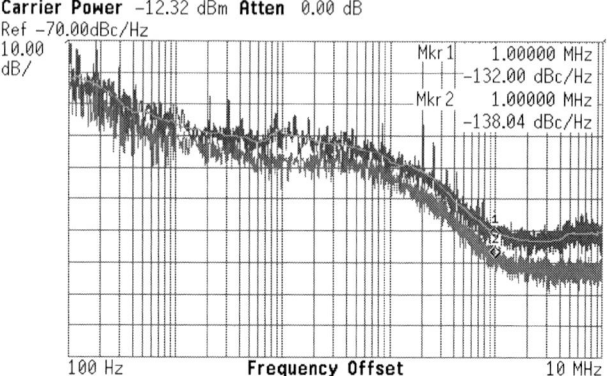

Fig. 6.   Phase noise comparison of frequency doubler output with the LO input as reference.

Input impedance matching with the transformer balun, and the gains of the doubler and tripler buffers were measured with a vector network analyzer and are plotted in Fig. 7(a)-(b). While measurement shows doubler buffer gain is lower than simulation by 1.2dB, the tripler buffer has increased by 3.3dB from simulation.

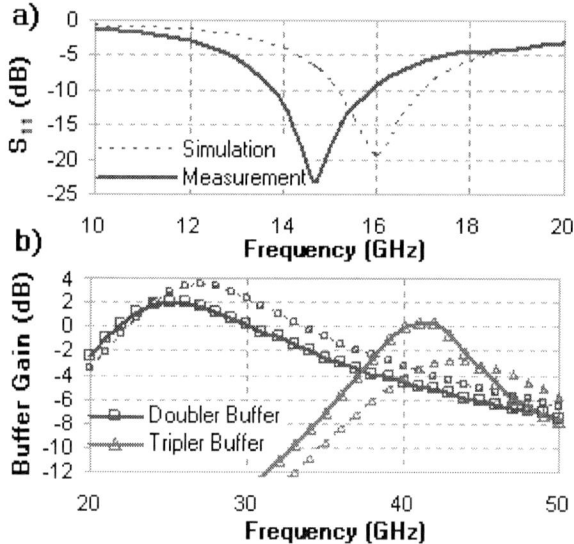

Fig. 7.   Measurement vs. simulation result of a) input impedance matching b) doubler and tripler buffer gain.

IV. CONCLUSION

A wideband mm-wave frequency doubler-tripler is proposed and demonstrated in a 0.13-μm CMOS process. A push-push FET frequency doubler combined with a

TABLE I
MEASURED PERFORMANCE SUMMARY OF FREQUENCY DOUBLER-TRIPLER

Technology	0.13-μm CMOS (2 thick layers)
Power consumption	12.6mW (core) 11.9mW (buffers)
Output frequency range	Doubler: 23-37 GHz
	Tripler: 36-48 GHz
Conversion gain	$-20dB < G_{c2} < -2.6dB$
	$-20dB < G_{c3} < -11.4dB$
Fundamental suppression	Doubler > 30dB
	Tripler > 40dB
2nd or higher harmonic suppression	Doubler > 24dB
	Tripler > 10 dB
Active chip area	600x440 μm²

single-balanced mixer generates both 2nd and 3rd harmonic simultaneously. Negative resistance cross-coupled pairs across output LC tanks enhance the conversion gain. The total operating frequency range of the doubler-tripler is 23-48 GHz with more than -20dBm output power delivered to 50Ω, when driven with a 0dBm input. The proposed topology for frequency multiplication covers 25 GHz bandwidth, not achievable by any fundamental frequency VCO, while it consumes 12.6mW of power and phase noise degradation is only due to frequency multiplication. The designed frequency doubler-tripler is an excellent alternative for wideband frequency synthesis in mm-wave transceivers or source generators.

ACKNOWLEDGMENT

The authors thank R. Bishop , A. Ravi, S. Pellerano and M. Mansuri of Intel for technical discussions. Ms. Saberi is supported by SRCEA/Intel Doctoral Fellowship.

REFERENCES

[1] W. L. Chan, and J. R. Long, "A 56-to-65 GHz injection-locked frequency tripler with quadrature outputs in 90 nm CMOS," *IEEE J. Solid-State Circuits*, vol. 43, no. 11 Dec. 2008, pp. 2739–2746.

[2] S. A. Maas, *Nonlinear Microwave and RF Circuits*, 2nd ed. Norwood, MA: Artech House, 2003.

[3] L. Zhang, D. Karasiewicz, B. Cifctioglu, and H. Wu "A 1.6-to-3.2/4.8 GHz dual-modulus injection-locked frequency multiplier in 0.18μm digital CMOS," *IEEE RFIC Symp. Dig.*, pp. 427-430, June 2008.

[4] B. R. Jackson, F. Mazzilli, and C. E. Saavedra, "A frequency tripler using a subharmonic mixer and fundamental cancellation," *IEEE Trans. Microwave Theory & Tech.*, vol. 57, no. 5, pp. 1083-1090, May 2009.

[5] C. Leifso, and J. Nisbet, "A Monolithic 6 GHz Quadrature Frequency Doubler With Adjustable Phase Offset," *IEEE J. Solid-State Circuits*, vol. 41, no. 2, pp. 405-412, Feb 2006.

RMO1C-5

200GHz CMOS Prescalers with Extended Dividing Range via Time-Interleaved Dual Injection Locking

Qun Jane Gu[*], Heng-Yu Jian[*], Zhiwei Xu[*], Yi-Cheng Wu[*], Mau-Chung Frank Chang[*],
Yves Baeyens[**] and Young-Kai Chen[**]

[*]University of California, Los Angeles, CA, USA
[**]Alcatel-Lucent/Bell-Labs, Murray Hills, NJ, USA

Abstract — **An unique time-interleaved dual injection locking scheme has been devised to enable ultra high-speed and low-power frequency division with extended frequency locking range. To prove the concept, two frequency dividers (or prescalers) have been realized in 65nm digital CMOS: one divides continuously from 158GHz to 195GHz (or 21% locking range) with input signal < 0dBm and the other divides from 181GHz to 208GHz (or 14% locking range) with input signal < -1dBm. Both prescalers consume < 2.5mW at 1V supply and contribute negligible phase noise. These test results set the highest F.O.M. (2721 and 2188 GHz²/mW, respectively) for prescalers implemented in any semiconductor technology up to this date, which in both cases is almost 10 times higher than that of prior arts.**

Index Terms — **high speed prescaler/frequency divider, time-interleaved dual injection locking, figure-of-merit.**

I. INTRODUCTION

Electromagnetic radiation windows located at W- and G-band (i.e. 94GHz, 148GHz and 220GHz) has attracted increasing interest for the implementation of multi-gigabit/sec wireless communication and through-fabric/fog imaging systems. Continuous scaling has increased cut-off frequencies (f_t and f_{max}) of CMOS beyond 200GHz in 65nm CMOS and opened possibilities for various applications in these emerging areas. The prescaler after VCO is one of the most challenging building blocks in wireless communication and imaging systems due to very stringent requirements for high dividing frequency, wide locking range, high input sensitivity, and low power consumption.

In traditional prescaler designs, high dividing frequency and wide locking range often impose mutually conflicting requirements on resonant tank characteristics. Recently reported CMOS high-frequency dividers have clearly revealed such tradeoffs: Ref. [1] achieved a 14% locking range with lower than 100GHz dividing frequency; while Ref. [2, 3] achieved higher dividing frequencies (120GHz and 130GHz, respectively) but with limited locking range of about 7%.

To pave the way for realizing portable mm-Wave or Terahertz radio, radar and/or imaging systems, we have devised a time-interleaved dual injection scheme to ease prescaler design tradeoffs and successfully demonstrated G-band prescalers (up to 208GHz) with extended dividing ranges (up to 21%).

II. CONVENTIONAL SCHEMES

From the frequency divider or prescaler's architecture point of view, high frequency injection locking schemes can be divided into two categories: Ring Oscillation based Injection locking frequency divider (RO-ILFD) and LC based Injection locking frequency divider (LC-ILFD). Ref [4] has carefully analyzed both types of injection locking schemes in terms of their operating mechanisms and design concerns to conclude that LC-ILFD is more suitable for high frequency division.

However, it is very challenging for conventional LC-ILFD to simultaneously achieve both high dividing frequency and wide locking range. This is because an injection locking divider's free-running oscillation frequency is determined by the tank self-resonant frequency of $\omega_o = 1/\left(\sqrt{L_{tank}C_{tank}}\right)$. To achieve higher dividing frequency, smaller L_{tank} and C_{tank} are required. However, Barkhausen gain criteria for oscillation demands a high resonant tank parallel impedance of $H_N = \omega_o L_{tank} Q$. With the constraint of choosing smaller L_{tank} for higher frequency, the tank impedance can only be boosted by maximizing the quality factor Q, which narrows the locking range of the divider [4].

In order to develop a more effective and efficient approach in ultra-high frequency prescaler design, a novel time-interleaved dual injection scheme is presented in this paper to pave the road for future mm-Wave and Terahertz communication and imaging systems to achieve concurrently high dividing frequency, wide injection range, low power consumption and small silicon real estate.

III. PROPOSED PRESCALER ARCHITECTURE

A. Important Observations

Before discussing the newly devised time-interleaved

978-1-4244-6240-7/10 $26.00 © 2010 IEEE

dual injection scheme and its insertion to a prescaler circuit, let us first review two traditional types of injection schemes which have never been clearly differentiated by past research: the voltage injection scheme and the current injection scheme. These two injection schemes operate based on completely distinct mechanisms that in reality have prevented them from working with each other in prior prescaler designs.

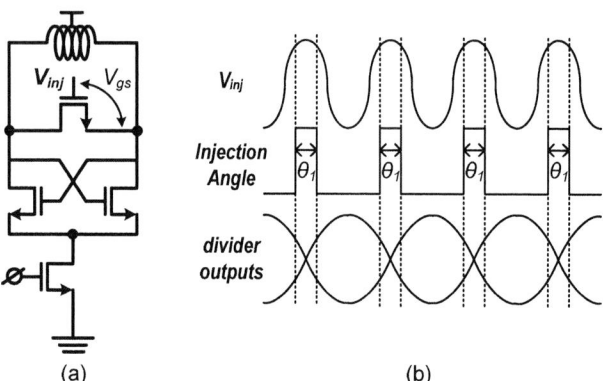

Fig. 1. (a) Voltage injection prescaler circuit insertion and (b) its effective injection angle θ_1 indicated around prescaler output's crossing period.

One circuit insertion example according to the voltage injection scheme is shown in Fig. 1(a). A voltage signal V_{inj} is injected through a NMOS mixer that shunts outputs of the crossing couple. As the injection voltage increases and the V_{gs} starts to exceed the device threshold, the mixer turns on and introduces a low impedance path to pull its source and drain (or the cross-coupled outputs) voltages closer. As a result, when voltage injection occurs at an instance outside of the output crossing time period of the prescaler, the voltage injection tends to pull outputs toward it. In case the prescaler's natural oscillation frequency is close to half of the injection frequency, such an effect will ultimately align the prescaler's output frequency and its phase with the voltage injection signal, as shown in Fig. 1(b). Consequently, the prescaler's output zero-crossings will naturally be synchronized (or locked) with voltage injection time zones, represented by the voltage injection angle θ_1.

On the other hand, the circuit insertion example for the current injection scheme is shown in Fig. 2(a). A current signal I_{inj} is injected via the current source of the crossing couple pair. During the positive (or negative) current injection cycle, the increased (or decreased) source current would split unequally to the resonant tank and increase (or decrease) the voltage difference between prescaler outputs. Provided the prescaler's natural oscillation frequency is close to half of the current injection frequency, the prescaler's output maximum (or minimum)

points will be synchronized with effective current injection time zones, represented by the current injection angle θ_2, as shown in Fig. 2(b). Nonetheless, during the prescaler's output zero-crossing periods, I_{inj} flows as a common mode current which is distributed evenly to the prescaler's outputs without any push-or-pull locking effect.

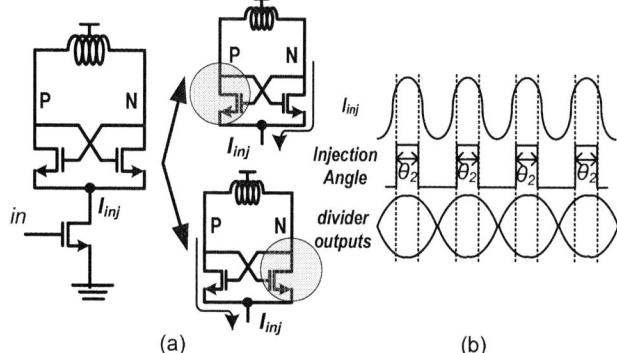

Fig. 2. (a) Current injection prescaler circuit insertion and (b) its effective injection angle θ_2 indicated around prescaler output's positive (or negative) peaking period.

B. Proposed Circuits

The aforementioned analyses have indicated that neither voltage nor current injection can synchronize or lock the prescaler's outputs at all time periods with its limited injection angle. It is therefore beneficial to interleave voltage and current injections at the prescaler's different output periods to extend its effective injection angle and consequently widen its locking range.

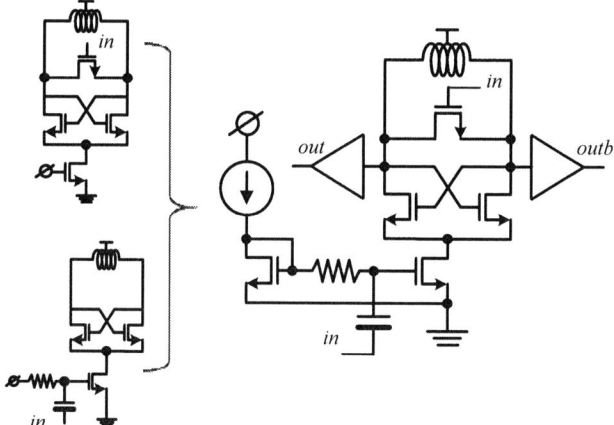

Fig. 3. A time-interleaved injection locking prescaler with enhanced injection angle.

The proposed dual-injection prescaler circuit topology is shown in Fig.3. The first injection is accomplished by injecting the signal voltage at the gate of the mixer that

978-1-4244-6240-7/10 $26.00 © 2010 IEEE

shunts the cross-coupling pair's outputs. The second is accomplished by injecting signal current through the cross-coupling pair's common source node. The voltage and current injections will interleave in time to cover two different injection locking periods: one for the zero-crossing period and the other for the positive/negative peaks. By combining both of them, which are time-interleaved to each other, the prescaler's overall injection angle is substantially greater than either one individually.

Physical circuit simulations also confirm the effectiveness of such a combination: separate voltage and current injections achieve a locking range of 11.5% and 4%, respectively. By interleaving both types of injections, the prescaler obtains up to 16% locking range at 180GHz input frequency and -4dBm input power. To validate this time-interleaved dual injection locking scheme, two prescalers have been implemented in 65nm CMOS technology by varying the value of the loading inductors.

IV. MEASUREMENT RESULTS

It is quite challenging to generate prescaler input signals at the G-band with sufficient power. Fig.4 shows our measurement setup. The signal from an external frequency synthesizer drives a multiply-by-3 and a subsequent power amplifier to generate signal at the W-band, which is then fed into a multiply-by-2 to produce the needed G-band input signal for prescaler wafer-probing. After dividing by 2, the prescaler output is down-converted by a mixer to feed into the spectrum analyzer to complete the test.

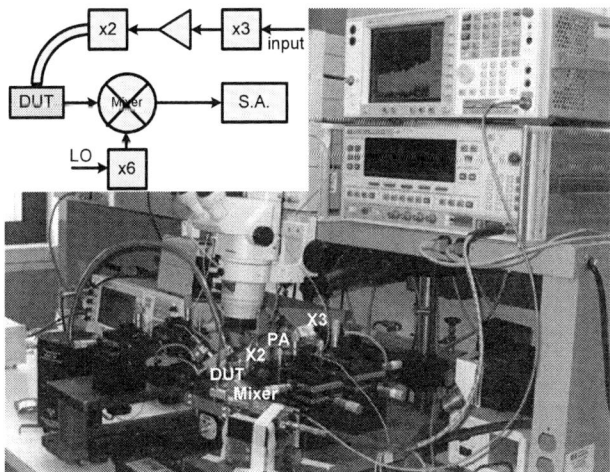

Fig. 4.   Measurement setup.

Both prescalers' measurement results are shown in Figs. 5 and 6, respectively. The input sensitivity of the first prescaler is plotted by its minimum input power versus the input frequency. The measured locking range in Fig. 5(a) is over 37GHz (158GHz~195GHz, or 21%) with < 0dBm

input power. Its lowest and highest frequency output spectra are also shown in Fig. 5(b). Correspondingly, the second prescaler's input sensitivity is plotted in Fig. 6(a) with 27GHz (181GHz~208GHz, or 14%) locking range with < -1dBm input power. Its lowest and highest frequency output spectra are shown in Fig. 6(b).

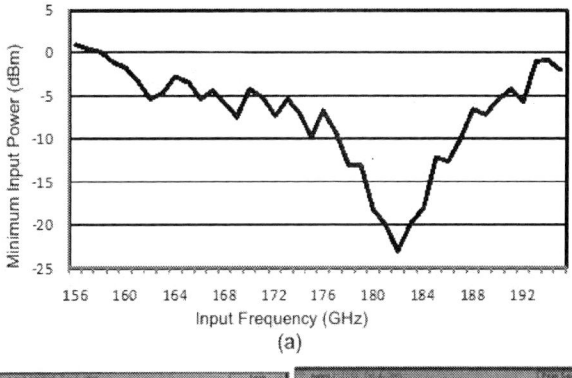

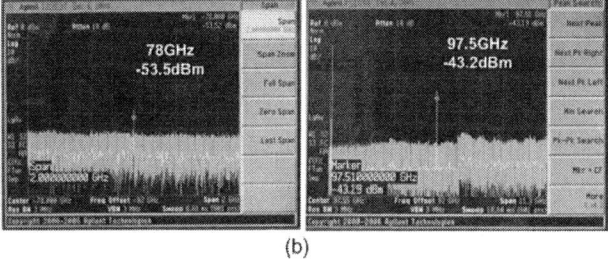

Fig. 5.   FD-V1 measured results (a) input sensitivity (b) output spectrums at lowest/highest input frequencies of 156/195GHz.

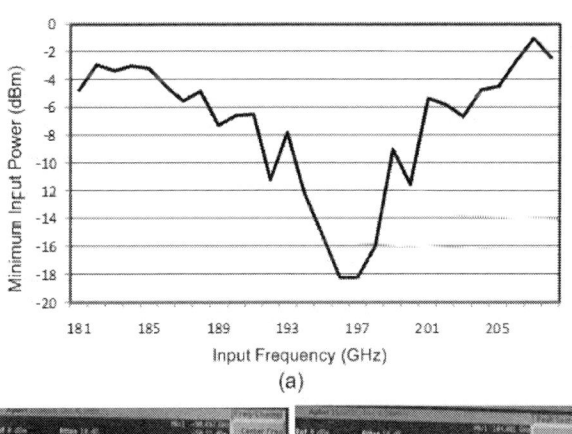

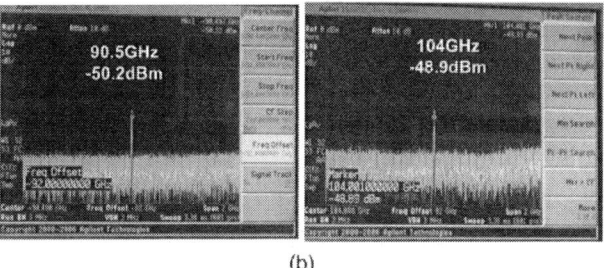

Fig. 6.   FD-V2 measured results (a) input sensitivity (b) output spectrums at lowest/highest input frequencies of 181/208GHz

Phase noise measurement results are shown in Fig.7, which indicate -91.7dBc/Hz and -91.6dBc/Hz@100KHz offset for both prescalers, respectively. They are limited by the input source phase noise of -107dBc/Hz @100KHz at 1/6 of the prescaler output frequency. The added 15.4dB phase noise is mainly due to the frequency up-conversion effect. Both prescalers draw about 2.4mA from a 1V power supply.

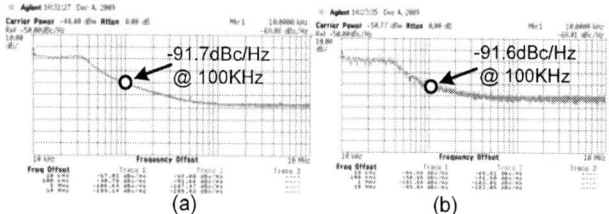

Fig. 7. Measured output phase noises of (a) first divide, (b) second divider.

A chip photo is shown in Fig. 8 with the core chip area 0.12mm x 0.09mm. Both prescalers possess the same area with the only difference being the inductor size.

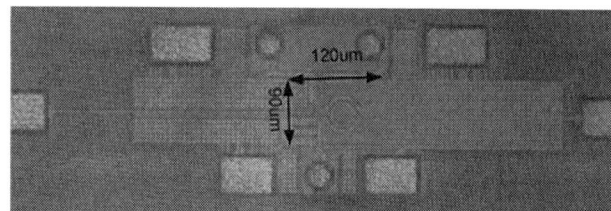

Fig. 8. Die photo of CMOS prescaler with time-interleaved dual injection locking.

Table 1 summarizes key performance measured from both prescalers and indicates that their performance figure-of-merits have exceeded prior arts substantially in terms of dividing frequency, locking range, and power consumption. We compare their performance according to a defined F.O.M. of

$$FOM = Center\ Frequency \times Locking\ Range\ /\ Power$$

where the center frequency and locking range are in Giga-Hertz (i.e. GHz) and power consumption in milli-Watt (i.e. mW). The measured F.O.M.s of our two prescalers are 2721 and 2188 GHz^2/mW, respectively, which are in either case almost 10 times higher than that of prior arts.

V. CONCLUSION

In summary, this paper successfully demonstrates an unique time-interleaved injection locking scheme for CMOS prescalers to achieve the highest dividing frequency (195GHz/208GHz) ever reported for any semiconductor technology (versus the highest dividing frequencies reported by SiGe [5] and InP HBT [6]), simultaneously having wide locking range (37GHz/27GHz), high input sensitivity (< -1dBm/0dBm across the bands), low phase noise (< -91dBc/Hz @100KHz offset), as well as low power consumption (2.4mW). The combined F.O.M. (2721/2188 GHz^2/mW) in either case has exceeded that of prior arts by almost 10 times [1-6]. The demonstrated time-interleaved dual injection scheme has paved the road to implement prescalers in commercial CMOS technology for future integrated mm-Wave and Terahertz communication/imaging systems.

Table 1   Performance comparison with CMOS state-of-the-arts.

	[1]	[2]	[3]	This Work	
Technology	65nm	90nm	65nm	65nm	65nm
Center Frequency	89GHz	121GHz	133GHz	176.5GHz	194.5GHz
Locking Range	82~94GHz (12GHz)	117~125GHz (8GHz)	128~137GHz (8.76GHz)	158~195GHz (37GHz)	181~208GHz (27GHz)
Input Power	0 dBm	N/A	N/A	0dBm	-1dBm
Phase Noise	N/A	N/A	-78.8dBc/Hz @400KHz	-91.7dBc/Hz @100KHz	-91.6dBc/Hz @100KHz
Power Consumption	3.92 mW	10.5 mW	5.5 mW	2.4mW	2.4mW
Chip Area	N/A	0.3mm* 0.14mm	0.32mm* 0.16mm	0.12mm* 0.09mm	0.12mm* 0.09mm
FOM (GHz^2/mW)	272	92	212	2721	2188

REFERENCES

[1] P. Mayr, C. Weyers, and U. Langmann, "A 90GHz 65nm CMOS Injection-Locked Frequency Divider," *ISSCC Dig. Tech. Papers*, pp. 198-199, Feb., 2007.

[2] B. Razavi, "A Millimeter-Wave Circuit Technique," *IEEE J. Solid-State Circuits*, vol. 43, no. 9, pp. 2090-2098, Sep., 2008.

[3] B.-Y. Lin, K.-H. Tsai and S.-I. Liu, "A 128.24-to-137.00GHz Injection Locked Frequency Divider in 65nm CMOS," *ISSCC Dig. Tech. Papers*, pp. 282-283, Feb., 2009

[4] Q. Gu, Z. Xu, D. Huang, T. La Rocca, N. Y. Wang, W. Hant, and M. C. F. Chang, "A low power V-band frequency divider with wide locking range and accurate quadrature output phases," *IEEE J. Solid-State Circuits*, vol. 43, no. 4, pp. 991–998, Apr. 2008.

[5] S. Trotta, H. Li, V.P. Trivedi and J. John, "A Tunable Flipflop-Based Frequency Divider up to 113GHz and a Fully Differential 77GHz push-push VCO in SiGe BiCMOS Technology," *IEEE Radio Frequency Integrated Circuits Symposium*, June 2009.

[6] Z. Griffith, M. Dahlstrom, M. Rodwell, "Ultra High Frequency Static Dividers >150 CHz in a Narrow Mesa InGaAs/InP DHBT Technology," 2004 Proceedings of the Bipolar/BiCMOS Circuits and Technology, 2004

[7] T.-N. Luo, Y. -J. E. Chen, "0.8mW 55GHz Dual-Injection-Locked CMOS Frequency Divider," *IEEE Trans. Microw. Theory Tech.*, vol. 56, no. 3, pp. 620-625, Mar. 2008.

Transmitter Chipset for 24/77-GHz Automotive Radar Sensors

Vittorio Giammello, Egidio Ragonese, and Giuseppe Palmisano

Università di Catania, Facoltà di Ingegneria, DIEES, Catania, 95125, Italy

Abstract — This paper presents a SiGe BiCMOS transmitter chipset for 24/77-GHz automotive radar sensors. The chipset adopts a dual-band architecture consisting of a 24-GHz section for ultra-wideband short-range radar operation, which is able to drive the 77-GHz long-range radar transmitter front-end. The proposed solution allows using a single 24-GHz frequency synthesizer to implement both operation modes. The 77-GHz transmitter demonstrates an output power of 12 dBm, a power gain of 20 dB and an output-referred 1-dB compression point of 11 dBm, while drawing 155 mA from a 2.5-V supply voltage.

Index Terms — BiCMOS integrated circuits, millimeter wave transmitters, mixers, phase-locked loops, power amplifiers, transformers.

I. INTRODUCTION

During last years automotive companies have pushed the development of intelligent safety systems, which allow the vehicle to perceive the surrounding environment and avoid road accidents. The adoption of both 24-GHz short-range radar (SRR) and 77-GHz long-range radar (LRR) sensors has different advantages in comparison with other safety systems (e.g. lidar, video, infrared, acoustic, etc.). Moreover, the development of mm-wave dual-band transceivers [1]-[3] in advanced silicon-based technologies [4], [5], guarantees low cost and can improve the effectiveness of such safety systems. In this scenario, the choice of a proper architecture is of utmost importance to take advantage of well-established mm-wave building blocks developed for 24-GHz SRR sensors.

This paper presents the architecture, the design and the experimental measurements of a SiGe BiCMOS transmitter chipset implemented for a dual-band 24/77-GHz automotive radar sensor.

II. TRANSMITTER ARCHITECTURE

The simplified block diagram of the 24/77-GHz transmitter chipset is shown in Fig. 1. The front-end architecture is compliant to both SRR and LRR requirements. The SRR transmitter is able to generate a ultra-wideband (UWB) signal by exploiting a 24-GHz phase-locked loop (PLL). The sub-ns switch is designed to produce a 24 GHz modulated pulse train, with proper time repetition and pulse width, according to main SRR

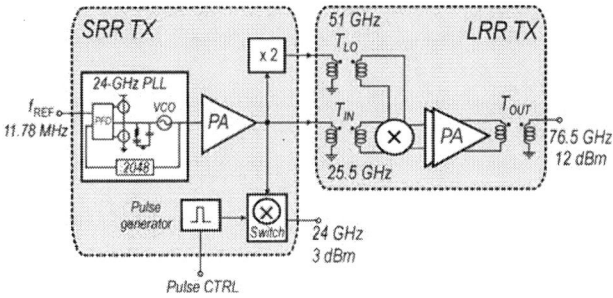

Fig. 1. Simplified block diagram of the dual-band 24/77-GHz transmitter chipset for automotive radar sensors.

requirements (i.e. minimum resolution, maximum unambiguous range, etc.). In the proposed architecture, the SRR TX can be also used to drive the LRR transmitter by using the 25.5-GHz continuous-wave (CW) signal at the PA output. The LRR transmitter takes advantage of a mm-wave mixer, which up-converts the 25.5-GHz IF input signal to 76.5 GHz by using the 51-GHz LO. The adopted architecture and the selected frequency plan allow sharing the same PLL of the SRR section to produce both IF (directly) and LO (multiplying by two) signals. A two-stage PA provides the required output power level for long-range radar operation. On-chip transformers (T_{IN}, T_{LO}, and T_{OUT}) are used at circuit block interfaces to provide high isolation, electro-static discharge protection, impedance/power matching, and single-ended-to-differential conversion.

The 24-GHz SRR section has been demonstrated in [6] and [7], whereas the 77-GHz LRR transmitter is here described in details.

III. CIRCUIT DESCRIPTION AND DESIGN

The transmitter chipset was designed and implemented in a low-cost 0.13-μm SiGe:C BiCMOS technology featuring npn bipolar transistors with f_T/f_{max} of 166/175 GHz and 1.8-V BV_{CEO} [8]. This choice arises from the need of high performance SiGe HBTs for the 24-GHz SRR front-end. Although this process was profitably used for 24-GHz ICs, its back-end-of-line is not well optimized for W-band applications, due to inadequate thickness of both top metal layers and inter-metal oxides.

978-1-4244-6240-7/10 $26.00 © 2010 IEEE

Relative low f_T/f_{max} and large RLC parasitics make the design of the 77-GHz circuits very challenging. The adoption of proper circuit topologies and an electromagnetic/circuital co-design approach, allow overcoming these technology limitations. Moreover, this process makes compatible the 24-GHz TX front-end with the 77-GHz section, thus implementing a low-cost dual-band transmitter.

A. 77-GHz Transmitter

The simplified schematic of the 77-GHz LRR transmitter front-end is shown in Fig. 2. The up-conversion mixer exploits a fully differential topology and consists of a voltage-to-current (V-I) converter and a double-balanced Gilbert quad. On-chip center-tapped transformers T_{IN} and T_{LO} perform both single-ended-to-differential conversion and 50- conjugate matching at the IF and LO ports, respectively. Degeneration folded microstrip inductors L_{EM} improve the linearity of the V-I converter and also adjust the input impedance. A resonant load (i.e. transformer T_1 along with MIM capacitors C_{SM}) is used to obtain the required conversion gain and drive the PA into saturation, while providing impedance matching at the PA interface.

The two-stage 77-GHz amplifier adopts a pseudo-differential architecture to take advantage of AC grounding at Q_7-Q_8 and Q_{11}-Q_{12} emitters. A cascode topology is preferred to common source to achieve excellent reverse isolation, higher stable gain and breakdown voltage. Transformer T_2, together with series MIM capacitors C_S, provides optimum load for the first stage and maximizes the power transfer, whereas T_{OUT} is designed for maximum delivered power at the PA output. Bias currents and HBT sizes are scaled by a factor of 4

from the driver to the power stage. PA transistors are biased at the peak f_T current density (around 7 mA/μm^2) to maximize both gain and output power. Furthermore, to avoid reduction of the actual f_T due to layout connections in the elementary power cell, only single multi-finger transistors with minimum emitter width (i.e. 0.17 μm) are used in this design. In particular, transistors with five and three emitter fingers are adopted for the PA and mixer, respectively. The bias networks of the common base transistors (Q_9-Q_{10} and Q_{13}-Q_{14}) are designed to provide a low base impedance (of about 150), which boosts the effective breakdown voltage from 1.8 V (BV_{CEO}) to 3.1 V (BV_{CER}), thus increasing the PA output power.

Finally, the layout design is a crucial point to avoid unexpected instability and/or coupling effects. For these reasons, RLC parasitics must be minimized by using an optimized layout approach (i.e. high symmetry, low-resistance/low-inductance metal ground planes, isolation guard rings, etc.) and electromagnetic (EM) post-layout simulations are essential for their evaluation.

B. On-chip Transformers

The optimization of a silicon-integrated mm-wave circuit requires an accurate design of the inductive components. Fig. 3 shows a simplified 3-D view of the adopted transformers with a summary of their electrical and geometrical parameters. In particular, T_{IN} exploits an interleaved configuration with 1:2 turn ratio, which helps obtaining the 50- impedance matching at the IF port. Each transformer coil adopts a three-layer structure consisting of two Cu metals (metal 6 and metal 5) plus a top alucap layer in order to reduce the series losses. The other adopted transformers (T_{LO}, T_1, T_2, and T_{OUT}) take advantage of stacked configurations to maximize the

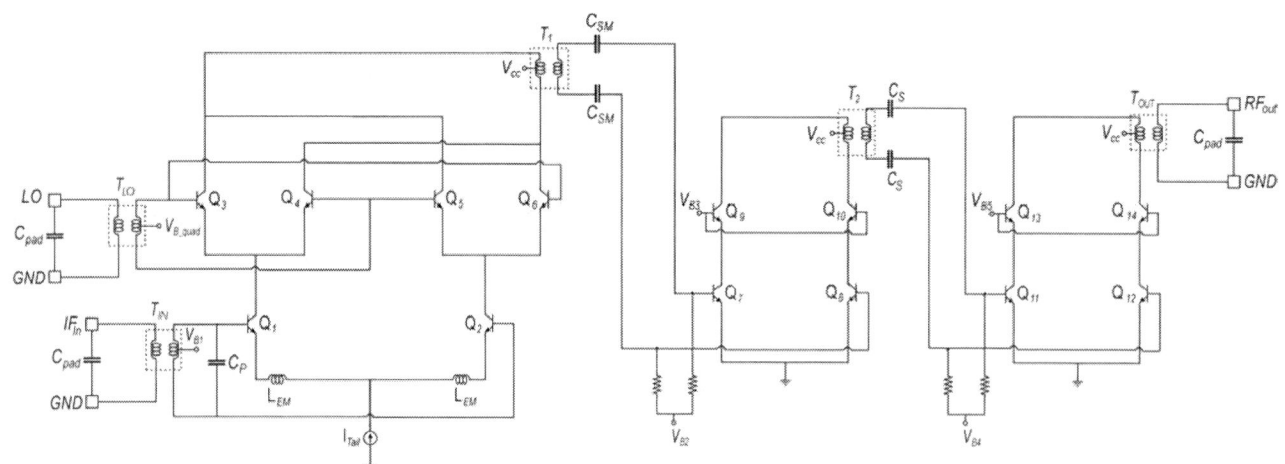

Fig. 2. Simplified schematic of the 77-GHz LRR TX.

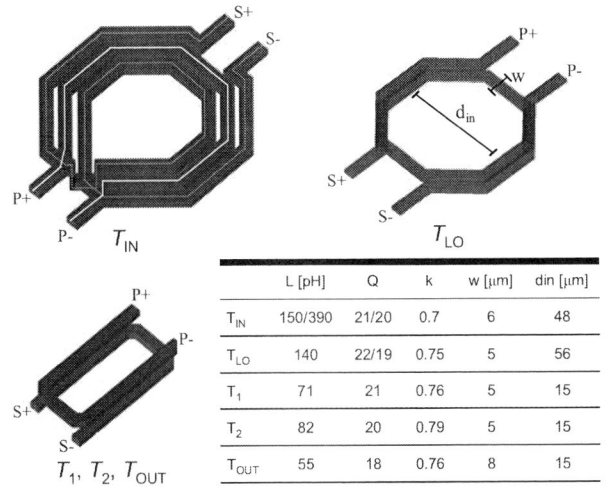

	L [pH]	Q	k	w [μm]	din [μm]
T_{IN}	150/390	21/20	0.7	6	48
T_{LO}	140	22/19	0.75	5	56
T_1	71	21	0.76	5	15
T_2	82	20	0.79	5	15
T_{OUT}	55	18	0.76	8	15

Fig. 3. 3-D view of transformers and summary of electrical and geometrical parameters.

coupling factor, k. Primary and secondary windings exploit a two-layer structure, consisting of alucap / metal 6 and metal 5 / metal 4, respectively. Coil widths were properly optimized to tradeoff series resistance and self-resonance frequency (f_{SR}). Moreover, both transformers operating at 25.5 GHz and 51 GHz (i.e. T_{IN} and T_{LO}) adopt a polysilicon patterned ground shield (PGS) to reduce the substrate losses. The 77-GHz transformers (i.e. T_1, T_2, and T_{OUT}) do not exploit this layout arrangement since the benefits at such high frequency are negligible, while the PGS considerably increases the capacitive parasitic and hence reduces the actual transformer f_{SR}. Transformer design was carried out by means of extensive EM simulations, taking into account the coupling effects due to the ground plane.

IV. EXPERIMENTAL RESULTS

The die micrograph of the fabricated 77-GHz LRR transmitter front-end is shown in Fig. 4. The die size is 800 μm x 725 μm, including on-chip baluns.

All measurements have been performed on-wafer at a standard 2.5-V supply voltage. Fig. 5 shows the S-parameters at the IF, RF and LO ports. The input return loss (S_{11}) is −18 dB at 25.5 GHz, whereas the return losses at the output and LO ports are better than −11 dB around 51 GHz and 77 GHz, respectively. The port-to-port isolation is higher than 40 dB up to 110 GHz. The conversion gain and the output-referred 1-dB compression point are 20 dB and 11 dBm, respectively.

Large-signal measurements have been also performed for the 77 GHz PA stand alone. As shown in Fig. 6, the PA achieves a maximum gain of 18.5 dB and a saturated output power of 12 dBm, with a power added efficiency

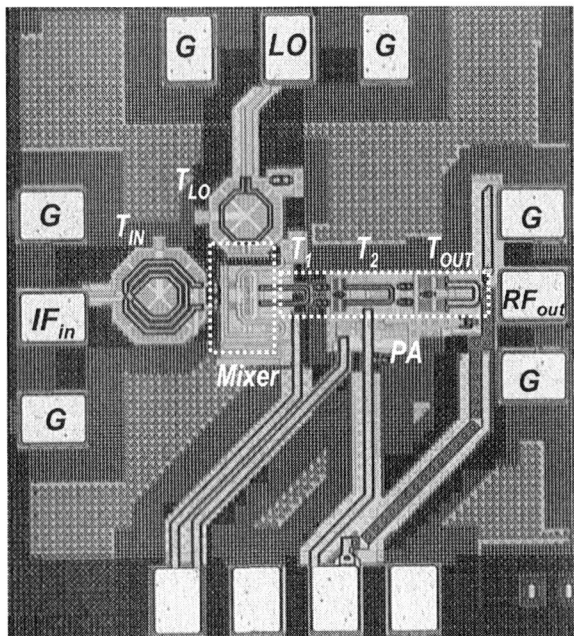

Fig. 4. Die micrograph of the 77-GHz LRR TX.

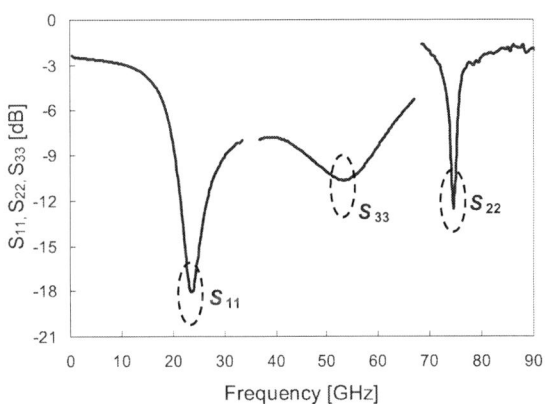

Fig. 5. Measured return losses at the IF (S_{11}), RF (S_{22}) and LO (S_{33}) ports.

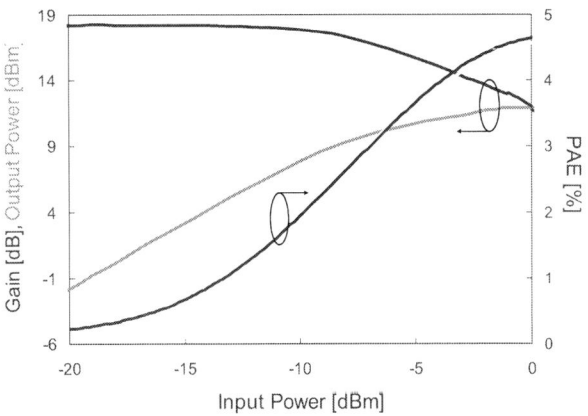

Fig. 6. Measured PA large-signal parameters at 77 GHz.

978-1-4244-6240-7/10 $26.00 © 2010 IEEE

of 4.7% at 77 GHz. The 77-GHz transmitter draws 155 mA from 2.5-V supply voltage.

Table I reports a summary of the measured performance of the 24/77-GHz transmitter chipset. Despite the relative-low f_T/f_{max} process, the 77-GHz transmitter demonstrates state-of-the-art performance in terms of both power gain and output power. Moreover, the adoption of a single PLL for dual-band operation allows considerable saving of both current consumption and silicon area.

TABLE I
SUMMARY OF THE MEASURED PERFORMANCE OF THE 24/77-GHZ TRANSMITTER CHIPSET

	24 GHz SRR TX	77 GHz LRR TX
Technology	0.13-µm SiGe BiCMOS 160/175 GHz f_T/f_{max}	
Radar operation mode	UWB	CW
Phase Noise @ 1MHz offset [dBc/Hz]	-104.3	-
Tuning Range [GHz]	4.7	-
TX Gain [dB]	-	20
TX OP$_{1dB}$ [dBm]	-	11
P$_{out}$ [dBm]	3	12
Silicon Area [mm^2]	1.56	0.58
I$_{CC}$[mA]	46	155
V$_{CC}$ [V]	2.5	

IV. CONCLUSIONS

In this paper, a SiGe BiCMOS 24/77-GHz transmitter chipset has been presented. The adopted dual-band architecture along with the use of a standard silicon BiCMOS process allows implementing a cost-effective solution for automotive SRR/LRR radar. The experimental measurements of the 77-GHz LRR transmitter confirm the soundness of circuit/layout co-design methodologies up to the W-band.

ACKNOWLEDGEMENT

The authors would like to thank STMicroelectronics for chip fabrication, the Microwave Characterization Center, Lille, France and A. Castorina, STMicroelectronics, Catania for the valuable support in carrying out measurements.

REFERENCES

[1] E. Laskin, P. Chevalier, A. Chantre, B. Sautreuil, and S. Voinigescu, "80/160-GHz transceiver and 140-GHz amplifier in SiGe technology," in *IEEE Radio Frequency Integrated Circuits Symp. Dig.*, Jun. 2007, pp. 153-156.

[2] V. Jain, F. Tzeng, L. Zhou, and P. Heydari, "A single-chip dual-band 22-29-GHz/77-81-GHz BiCMOS transceiver for automotive radars," *IEEE J. of Solid-State Circuits*, vol. 44, pp. 3469-3485, Dec. 2009.

[3] K.-H. Chen, C. Lee, and S.-I. Liu, "A dual-band 61.4-63 GHz/75.5–77.5 GHz CMOS receiver in a 90 nm technology," in *VLSI Circuits Symp. Dig.*, Jun. 2008, pp. 160-161.

[4] B. A. Orner *et al.*, "A BiCMOS technology featuring a 300/330 GHz f_T/f_{max} SiGe HBT for millimeter wave applications," in *Proc. BCTM*, Oct. 2006, pp. 49–52.

[5] G. Avenier et al. "0.13 µm SiGe BiCMOS technology fully dedicated to mm-wave applications," *IEEE J. of Solid-State Circuits*, vol. 44, pp. 2312-2321, Sept. 2009.

[6] E. Ragonese, A. Scuderi, V. Giammello, E. Messina, and G. Palmisano, "A fully integrated 24GHz UWB radar sensor for automotive applications," in *ISSCC Dig. Tech. Papers*, Feb. 2009, pp. 306-307.

[7] A. Scuderi, E. Ragonese, and G. Palmisano, "24-GHz ultra-wideband transmitter for vehicular short-range radar applications," *IET Circuits Devices Syst.*, vol. 3, pp. 313-321, Dec. 2009.

[8] M. Laurens *et al.*, "A 150 GHz f$_t$/fmax 0.13 µm SiGe: BiCMOS technology," *IEEE BCTM*, 2003, pp. 199-202.

RMO1D-2

A 94-GHz Passive Imaging Receiver using a Balanced LNA with Embedded Dicke Switch

Leland Gilreath[1], Vipul Jain[2], Hsin-Cheng Yao[1], Le Zheng[1], and Payam Heydari[1]

[1]University of California, Irvine, CA, 92697 [2]SaberTek, Irvine, CA 92614

Abstract — A fully-integrated silicon-based 94-GHz direct-detection imaging receiver with on-chip Dicke switch and baseband circuitry is demonstrated. Fabricated in a 0.18-μm SiGe BiCMOS technology (f_T/f_{MAX} = 200 GHz), the receiver chip achieves a peak imager responsivity of 43 MV/W with a 3-dB bandwidth of 26 GHz. A balanced LNA topology with an embedded Dicke switch provides 30-dB gain and enables a temperature resolution of 0.3-0.4 K. The imager chip consumes 200 mW from a 1.8-V supply.

Index Terms — SiGe, BiCMOS, millimeter-wave, W-band, passive imaging, low-noise amplifiers, power detectors.

I. INTRODUCTION

Passive mm-wave (PMMW) imaging provides the unique capability to create high-resolution images in low-visibility conditions (e.g., through clothing, clouds, and fog) and is therefore useful for such applications as concealed-weapon detection and airplane landing [1]. A low-attenuation atmospheric window from 80-110 GHz (W-band) makes this band an ideal candidate for PMMW systems. The cost of current PMMW cameras is dominated by the W-band compound-semiconductor electronics. Typical III-V-based imaging receivers (RX) are not fully-integrated and require at least two MMW chips [2]. Advanced SiGe BiCMOS technologies with transistor f_T/f_{MAX} beyond 200 GHz can potentially reduce the cost and size of PMMW systems by integrating the MMW RX circuits with imager readout electronics on a single chip.

Passive imagers operate by detecting naturally-emitted thermal (black-body) radiation from an object. Imager performance is measured by the minimum achievable thermal resolution, referred to as the noise-equivalent temperature difference (NETD). NETD depends on the performance of the RX circuits as [2]

$$NETD = T_{sys}\sqrt{\frac{1}{BW\cdot\tau}+\left(\frac{\Delta G}{G}\right)^2}, \qquad (1)$$

where T_{sys} is the noise temperature at the RX input, BW denotes the MMW bandwidth, τ is the integration time, G is the pre-detection gain, and ΔG is the rms variation of G. A camera frame rate of 25 frames/s limits the maximum integration time to 40 ms, and further reductions in NETD can only be achieved by improvements in the front-end noise figure (NF) and bandwidth. Typical indoor applications such as concealed-weapon detection require an NETD of less than 0.5 K [2].

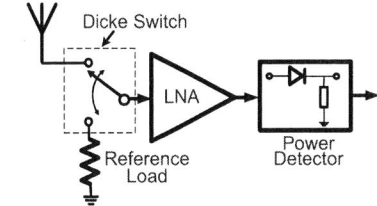

Fig. 1. Conventional direct-detection imaging architecture.

This paper presents the first reported integration of a silicon-based 94-GHz passive imaging receiver with on-chip baseband circuitry. The SiGe BiCMOS RX employs a balanced LNA (using 90° hybrid couplers) and reflection-type binary phase-shifters (RTPS) to realize Dicke-switch functionality with minimal impact of switching loss on NF and NETD performance.

II. RECEIVER ARCHITECTURE

The imaging RX in this work is based on the direct-detection architecture, shown in Fig. 1, as it does not require mixers and complex LO generation associated with a heterodyne detection scheme [3]. In a direct-detection imager, the received MMW radiation is amplified by the LNA and then converted to a DC voltage (proportional to the input power) by the square-law power detector. Since the detector output is a DC voltage, low-frequency phenomena, such as 1/f noise and LNA gain fluctuations (ΔG), severely degrade the signal-to-noise ratio at the detector output, and thus become the dominant contributor to NETD. In order to alleviate this problem, an SPDT switch (referred to as a Dicke switch [4]) is typically used to modulate the MMW input signal such that the detector output signal is shifted to a frequency above the 1/f noise corner. In the classical architecture of Fig. 1, the SPDT switch is placed directly in front of the LNA and periodically switches the LNA input to either

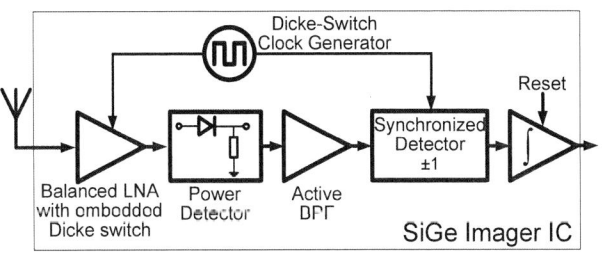

Fig. 2. Block diagram of the proposed SiGe imaging RX.

978-1-4244-6240-7/10 $26.00 © 2010 IEEE

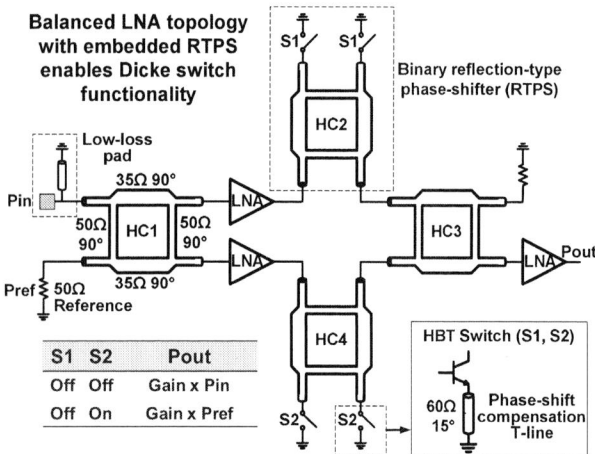

Fig. 3. Balanced LNA with embedded Dicke switch.

the antenna or to a calibrated reference load. By continuously subtracting the two detected signals (antenna and reference), the detector output voltage contributions due to low-frequency fluctuations are canceled (i.e, $\Delta G \rightarrow 0$). The NETD in this case is given by

$$NETD = 2 \cdot T_{sys} / \sqrt{BW \cdot \tau} \,, \qquad (2)$$

where the factor of 2 appears because the Dicke switch is connected to the antenna only for a half cycle. The insertion loss of the switch should be the same in both states, so that the antenna and reference inputs experience the same RF gain. The insertion loss of the switch in Fig. 1 directly adds to the front-end NF, thereby degrading the NETD. Since low-loss switches are available in III-V technologies, this simple architecture has been widely used in practical imagers. On the other hand, silicon-based MMW switches exhibit unacceptably high insertion loss (~5 dB for an HBT switch in our SiGe technology). System-level analysis shows that this 5-dB loss before the LNA will degrade the RX NETD by a factor of 3.

The proposed architecture, shown in Fig. 2, eliminates the above problem by embedding the Dicke switch within a balanced LNA such that it does not degrade the RX noise figure (*cf.* Section III), allowing the system to achieve the lowest possible NETD for a given technology.

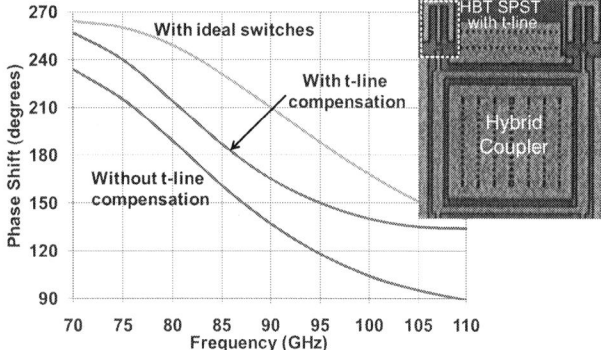

Fig. 4. Simulated RTPS phase shift. T-lines compensate for the HBT SPST.

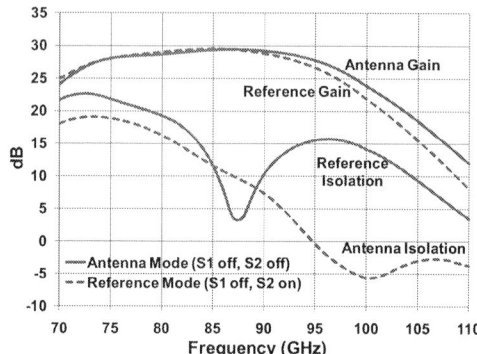

Fig. 5. Measured pre-detection gain and isolation in the two switch states.

The highly non-linear operation of the detector usually results in a high detector NF, necessitating the use of a high-gain LNA (30 dB in this work, *cf.* Section IV).

III. BALANCED LNA WITH EMBEDDED DICKE SWITCH

Fig. 3 shows the schematic of the balanced LNA incorporating the embedded Dicke switch. Inspired by the GaAs topology in [5], the circuit is comprised of a balanced LNA with the addition of a reflection-type binary phase shifter in each branch. When the phase shifters are in the same state (S_1 off, S_2 off), the circuit reduces to a standard balanced LNA, and the amplified signal from the antenna input appears at the output port. However, when the phase shifters are in opposite states (S_1 off, S_2 on), there is an extra 180° phase shift in the reference branch. This results in a constructive amplification of the noise power from the reference input, while the signal from the antenna input is suppressed. By toggling between these two states, the desired chopping operation of the Dicke switch is achieved. System analysis shows that implementing the Dicke switch with this architecture yields a 2x improvement in NETD, as compared with the conventional architecture of Fig. 1.

The switches S_1 and S_2 in the RTPS are implemented using HBTs (Fig. 3). As mentioned in Section II, silicon

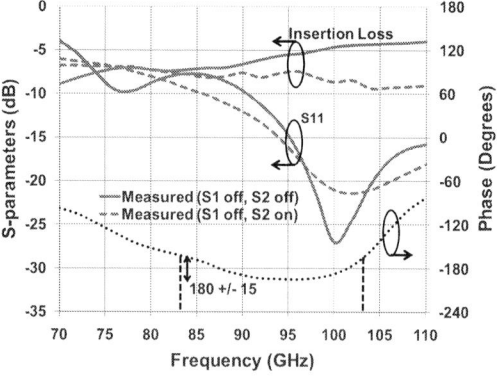

Fig. 6. Measured RTPS S-parameters and output phase difference between the two states.

978-1-4244-6240-7/10 $26.00 © 2010 IEEE

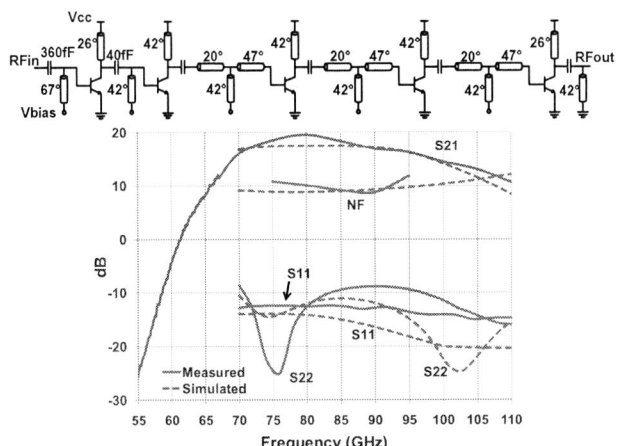

Fig. 7. LNA schematic and its measured performance.

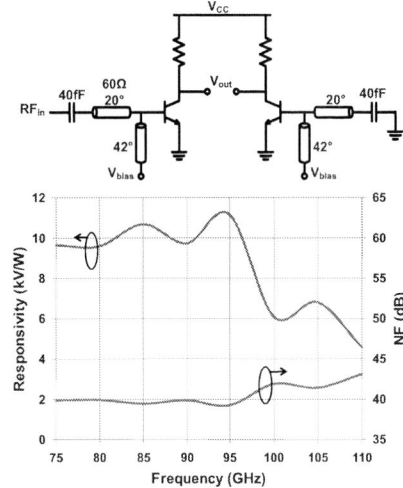

Fig. 8. Detector schematic and its measured responsivity and NF.

high-speed HBTs make poor switches. Aside from the high insertion loss, they introduce additional phase shift in the on-state, thereby corrupting the phase response of the RTPS. 15° t-line stubs have been used to compensate for the HBT switches and recover the RTPS phase response. As shown in Fig. 4, the t-lines shift the RTPS phase response such that it is closer to that of an ideal switch.

Fig. 5 shows the measured gain and isolation from the antenna and reference ports for the two different phase-shifter states. As expected, in the antenna mode, the signal from the antenna is amplified while the reference signal is suppressed. Similarly, in the reference mode, the reference input is amplified while the antenna signal is suppressed. Note that the balanced structure ensures equal gains in both the antenna and reference modes. An additional LNA is used after the balanced structure in order to achieve a total pre-detection gain of 30 dB. The measured performance of the phase shifter is shown in Fig. 6. The coupler-based RTPS topology provides good input and output return losses over a wide bandwidth.

Each of the three identical LNAs in Fig. 3 is designed as a five-stage amplifier shown in Fig. 7. The HBTs are biased at the optimum-NF current density and exhibit 3.9-dB maximum available gain (MAG) and 7.2-dB NFmin at 90 GHz. The theoretical maximum gain and minimum NF of the 5-stage LNA at 90 GHz are 19.5 dB and 9 dB, respectively. The first two stages have been designed for optimum noise match, while the remaining stages are conjugate-matched. In order to minimize matching-network loss, t-lines are implemented as slow-wave coplanar waveguides (CPWs). EM simulations show that a slow-wave CPW achieves 40% higher phase-shift than a conductor-backed CPW for a given length. This translates to 40% reduction in the loss of the matching networks.

The measured LNA gain and NF at 90 GHz are 17 dB and 9 dB respectively (Fig. 7), in good agreement with the calculated predictions. The NF was not measured above

95 GHz due to test setup constraints. The LNA draws 35 mA from a 1.8-V supply.

IV. W-BAND DETECTOR

The W-band power detector employs a common-emitter HBT as a square-law device (Fig. 8). The HBT is biased at a low emitter current of 70 µA in order to minimize shot noise. A replica common-emitter device is used in order to provide differential outputs to the baseband circuitry. The input matching network provides maximum power transfer and increases the detector responsivity (defined as the ratio of the detector output DC voltage to the input power) by a factor of 3. The detector measurements exhibit a peak responsivity of 12 kV/W and a minimum NF of 39 dB (Fig. 8). The detector NETD is calculated from the measured responsivity and NF as 17 K at 94 GHz. A high-gain LNA is necessary in order to suppress the contribution of the detector NF to the RX NETD. To achieve the target NETD of 0.5 K, 25-to-30-dB pre-detection gain is required.

V. SYSTEM PERFORMANCE

A 4-stage differential ring oscillator generates the 1-MHz Dicke switch clock. Note that, unlike [4], the 1/f noise corner in this work is determined by the MOSFETs

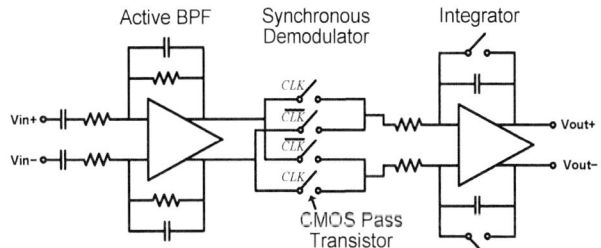

Fig. 9. Simplified schematic of the baseband chain.

978-1-4244-6240-7/10 $26.00 © 2010 IEEE

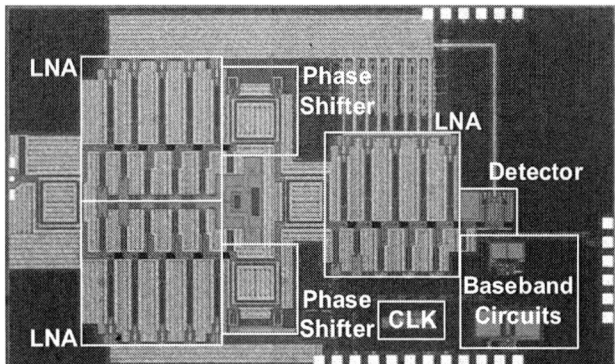

Fig. 10. Die micrograph of the SiGe imaging RX.

(100 kHz) in the baseband circuits. An active bandpass filter (Fig. 9) with an in-band gain of 20 dB and bandwidth of 0.1-10 MHz captures the first 9 harmonics of the detector square-wave output. The synchronous detector and the integrator demodulate the voltage signal, providing a true DC signal at the RX output.

The RX chip has been implemented in a 200-GHz SiGe BiCMOS process offering six metal layers and HBTs with 0.15-μm emitter width. Fig. 10 shows the PMMW RX chip micrograph. The imager chip achieves an NETD of 0.3/0.4 K with integration times of 40/30 ms, meeting the typical specifications for indoor imaging applications. System measurements show a peak responsivity of 43 MV/W and a minimum NF of 12 dB (Fig. 11). The RX draws 110 mA from a 1.8-V supply. Table I summarizes the imaging RX performance, and Table II compares this work with other published W-band imagers. The design in [4] achieves a similar NETD performance using the conventional architecture of Fig. 1. However, it employs a faster 0.12-μm SiGe technology. For comparison, at 90 GHz, the SiGe technology in [4] has a MAG of 6 dB and an NFmin of 5 dB, whereas our technology has a MAG of 3.9 dB and an NFmin of 7.2 dB.

VI. CONCLUSION

A passive mm-wave imaging system operating in the 80-110-GHz atmospheric window has been demonstrated in a 0.18-μm SiGe BiCMOS process. The imager features

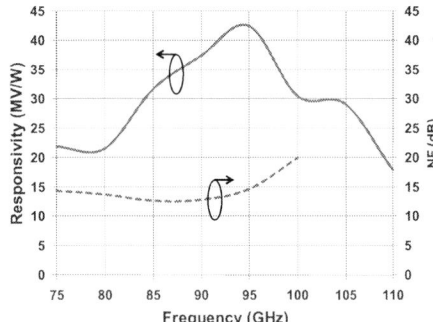

Fig. 11. Measured imager responsivity (τ = 30 ms) and NF.

TABLE I
SUMMARY OF THE RECEIVER PERFORMANCE

Pre-detection Gain	30 dB
3-dB Bandwidth	26 GHz
Noise Figure	12 dB
Responsivity	19–43 MV/W
NETD	0.3/0.4 K (τ = 40/30 ms)
Supply Voltage	1.8 V
Power Dissipation	200 mW
Technology	0.18-μm SiGe BiCMOS (fT/fmax = 200 GHz) MAG = 3.9 dB, NFmin = 7.2 dB @ 90GHz
Die Area	5x2.5 mm²

TABLE II
PERFORMANCE COMPARISON OF 94-GHz IMAGERS

Ref.	Technology	Integration	NETD	Responsivity
[3]	InP HEMT	LNA+Detector chipset	0.45 K[1]	0.5 MV/W
[6]	GaAs HEMT	LNA + Detector + Dicke Switch	1.6 K[2]	-
[4]	0.13-μm SiGe	LNA+Detector	0.6-0.8 K[3]	4 MV/W
[7]	65nm CMOS	LNA + Detector + Dicke Switch	10 K[4]	0.09 MV/W
This work	0.18-μm SiGe	LNA+Detector+Dicke Switch+Baseband	0.4 K	43 MV/W

[1]τ = 3.125 ms. [2]τ = 10 ms. (τ = 30 ms for others)
[3]Accounting for external Dicke switching. [4]Calculated from data.

an RTPS-based Dicke switch embedded in a balanced LNA to achieve the lowest possible NETD for a given technology. To the authors' knowledge, the 0.4 K NETD achieved in this work is the lowest reported for a silicon-based imaging RX.

ACKNOWLEDGEMENT

The authors thank Jazz Semiconductor for chip fabrication and Northrop-Grumman Corp. for providing test equipment. This work was supported in part by SRC under contract 2009-VJ-1962.

REFERENCES

[1] L. Yujiri, M. Shoucri, and P. Moffa, "Passive millimeter-wave imaging," *IEEE Microwave Magazine*, vol. 4, pp. 39-50, Sep. 2003.

[2] J. J. Lynch, et al., "Passive millimeter-wave imaging module with preamplified zero-bias detection," *IEEE Trans. Microwave Theory Tech.*, vol. 56, no. 7, pp. 1592–1600, July 2008.

[3] D.C.W. Lo, et al., "A monolithic W-band high gain LNA/Detector for millimeter-wave radiometric imaging applications," *IEEE MTT-S International Symposium*, pp. 1117-1120, June 1995.

[4] R. H. Dicke, "The measurement of thermal radiation at microwave frequencies," *The Review of Scientific Instruments*, vol. 17, no. 7, pp. 268-275, July 1946.

[5] J. W. May and G. M. Rebeiz, "High-performance W-Band SiGe RFICs for passive millimeter-wave imaging," *IEEE RFIC Symp.*, pp. 437-440, July 2009.

[6] D.C.W. Lo, et al., "Novel monolithic millimeter wave multi-functional balanced switching low noise amplifiers," *IEEE Trans. Microwave Theory and Techniques*, Vol. 42, No. 12, pp. 2629-2634, Dec. 1994.

[7] A. Tomkins, P. Garcia, and S. P. Voinigescu, "A passive W-band imager in 65nm bulk CMOS," *IEEE CSICS*, pp. 1-4, Oct. 2009.

94 GHz silicon co-integrated LNA and Antenna in a mm-wave dedicated BiCMOS technology

R. Pilard[1,2], D. Gloria[1], F. Gianesello[1], F. Le Pennec[2], C. Person[2]

[1]STMicroelectronics, Technology R&D – STD – TPS, 850 avenue Jean Monnet, 38926 Crolles, France
[2]Lab-STICC/MOM, UMR CNRS 3192, Telecom Bretagne-UBO, CS 83818, 29238 Brest, France
romain.pilard@st.com

Abstract - A co-integrated Low Noise Amplifier (LNA) with a dipole antenna is designed considering a millimeter-wave dedicated BiCMOS technology. The targeted application is a 94 GHz passive imaging for security applications. The LNA is based on a high-speed SiGe:C 130 nm HBT. The interest of the co-integration on a common silicon substrate is demonstrated through the decrease of insertion losses between the antenna and the amplifier. The capability of the BiCMOS9MW technology is illustrated to achieve this co-integration reaching a total gain of 3.0 dB ($G_{antenna}$ + G_{LNA}) for a power consumption of 11 mW, in a single-stage LNA configuration. A two-stage configuration achieves a total gain of 8.5 dB with a power consumption of 21 mW.

I. INTRODUCTION

Transition frequencies of transistors in silicon-based technologies have reached performances that enable designers to address millimeter-wave applications such as 60 GHz W-HDMI, 77 GHz automotive radar, 94 GHz passive imaging... With the increase of the frequency, and especially at millimeter-wave frequencies, losses are limiting factors in the design of high performance circuits and especially in Radio-Frequency Front-Ends (RF FE). These losses mostly come from passive devices (transmission lines, inductors), but are also due to packaging or assembly techniques (wire-bonds, flip-chip bumps).

Hence, high performance passive devices are expected by process designers from STMicroelectronics to develop a BiCMOS technology compatible with millimeter-wave application requirements. BiCMOS9MW is a dedicated millimeter-wave technology featuring a high-speed 130 nm bipolar transistor with 230GHz/280GHz f_T/f_{MAX}, and a Back-End-Of-Line (BEOL) containing two thick metal layers (3 μm) [1].

With the previous concerns about losses, one can understand the benefit of the integration of the antenna to the RF FE on the same substrate, in a standard silicon technology. Indeed, losses in the assembly of an external antenna can therefore be suppressed, or at least significantly reduced (as well as additional interconnection costs). Furthermore, if the antenna is directly matched to

the input - respectively output - impedance of a low noise amplifier (LNA) - respectively a power amplifier (PA) -, these losses can be completely neglected. Figure 1 illustrates the concept of such a co-integration technique for a receiving mode configuration.

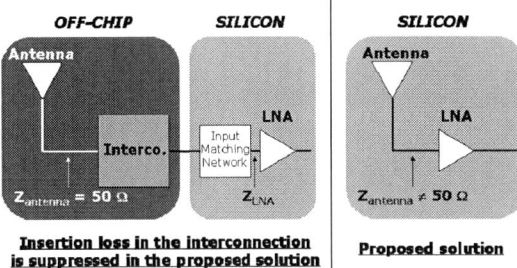

Figure 1 – Co-integration technique

The other benefit of the integrated antenna lies in the reduction of the size of the whole system, provided that the application is addressed to a short range low gain application with reduced gain expectation for both the LNA and the antenna. The proposed co-integration technique could be applied to a typical application such as 94 GHz passive imaging for security applications.

In this paper we demonstrate the potentiality of the BiCMOS9MW process for the integration on the same chip of the fundamental building blocks of an RF FE (only the receiving operating mode being addressed in this paper), i.e. the antenna and the LNA at 94 GHz. The antenna is designed using a 3D full-wave electromagnetic simulation software. We first describe the technology used to design the LNA and the antenna. Secondly, results of the LNA co-designed with the antenna impedance are presented and discussed.

II. TECHNOLOGICAL FEATURES

In this part, we describe the main features of the dedicated BiCMOS technology used for the integration of the LNA and the antenna.

A. Heterojunction Bipolar Transistor (HBT)

For the design of the LNA, we choose a high-speed NPN 130 nm SiGe:C HBT with a CBEBC structure. This

978-1-4244-6240-7/10 $26.00 © 2010 IEEE

choice is motivated by its symmetrical configuration and its RF performances. The device exhibits attractive performances for the design of a 94 GHz LNA. Indeed, with an emitter width $W_E = 0.27$ μm, and an emitter length $L_E = 3$ μm, we can reach a maximum transition frequency f_T of 243 GHz for $V_{BE} = 0.92$ V (at $V_{BC} = -0.5$ V). The f_T values are reported on Figure 2 as a function of the Base-Emitter voltage, V_{BE}. The current consumption of this transistor is $I_C = 7$ mA at peak f_T.

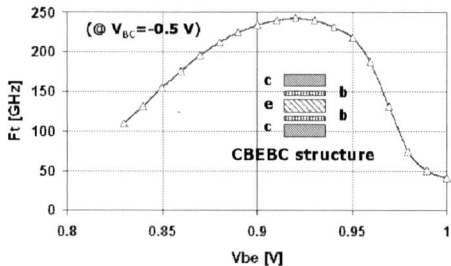

Figure 2 – f_T as a function of V_{BE} for an emitter size of 0.27 x 3.0 μm² in a CBEBC structure configuration

In addition, the noise parameters (NF_{min}, Z_{opt}, and R_n) are given below at 94 GHz:

$$NF_{min} = 5.5 \text{ dB} \qquad (1)$$
$$Z_{opt} = 76 + j*40 \text{ } \Omega \qquad (2)$$
$$R_n = 80 \text{ } \Omega \qquad (3)$$

The technology provides good gain characteristics at this frequency and the Maximum Available Gain (MAG) at 94 GHz is:

$$MAG = 8 \text{ dB} \qquad (4)$$

B. Back-End-Of-Line

The BiCMOS9MW BEOL (cf. Figure 3) is made of 6 metal layers with two upper most thick ones (Metal 5 and Metal 6), stacked on a standard silicon substrate (resistivity: 12 Ω.cm).

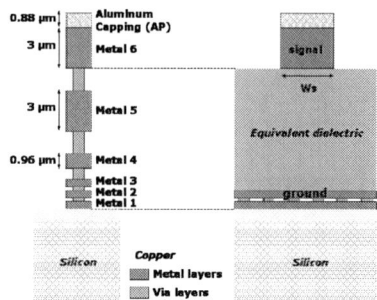

Figure 3 – Back-End-Of-Line of BiCMOS9MW and microstrip line cross-section configuration

Passive components such as microstrip lines suffer from high loss when they are integrated using a standard BEOL.

The transmission line can be built using Metal 6 and AP, while the ground plane can be built with interconnected Metal 1 and Metal 2 layers (interconnection with vias). For a same technology node (130 nm), but with a millimeter-wave dedicated BEOL, a 50 Ω microstrip line exhibits only 0.7 dB/mm at 94 GHz. As summarized in [2], microstrip lines in BiCMOS9MW are even the less lossy transmission lines amongst the latest topology currently used today in CMOS or BiCMOS processes.

Transmission lines are considered for connecting the different sub-functions of the RF FE and also for feeding each radiating structure. The less energy is dissipated along these feeding lines, the more energy will be provided to the antenna and radiated.

At millimeter-wave, antenna dimensions are comparable to integrated circuits ones. Thus, they can be integrated on a same silicon substrate without additional costs. As an example, the resonant length of the antenna designed and described in the following section is 700 μm on the considered technology BEOL, to be compared to a 1.2 mm² area of a classical single-stage LNA.

III. 94 GHZ ANTENNA AND CO-INTEGRATED LNA DESIGNS

In this section, a dipole antenna is designed using the millimeter-wave dedicated BEOL. Then a 94 GHz LNA is co-designed with the antenna impedance.

A. Dipole antenna with integrated balun

A schematic of a dipole antenna with integrated balun, fully described in [3] is presented in Figure 4. The antenna is fed by a 50 Ω characteristic impedance microstrip line ($W_s = 12.5$ μm). The antenna parameters L, W and S are adjusted such as the input impedance of the antenna $Z_{antenna}$ is matched to 50 Ω in the frequency band of interest. The ground plane of the access line and the antenna is made of Metal 1 and Metal 2 connected to each other with vias. The signal line is carried by Metal 6 and AP. The connection on the balun end is made of vias from Metal 6 down to Metal 2. The silicon substrate has a standard thickness of 375 μm.

The antenna is designed using Ansoft HFSS™, a 3D full-wave electromagnetic software based on finite element numerical method (FEM). The antenna is simulated taking into account the surrounding environment, which is a metallic plane under the silicon, and for which the impedance of the antenna is highly impacted because of the reflection of the electromagnetic field. This metal plane corresponds to the chuck of the probe-based test-bench used for S-parameter measurements up to 110 GHz.

The antenna is matched to 50 Ω ($S_{11} < -10$ dB) over a large frequency band, from 70 GHz to beyond our measurement capability (cf. Figure 5). Center frequency is 92.5 GHz. The simulated maximum gain of the antenna in this configuration is -2.0 dBi, and the radiation efficiency is 20%. The radiation characteristics of the antenna will be extracted in a dedicated antenna measurement test-bench previously presented and validated up to 60 GHz [4].

In the next section, the LNA is designed using the antenna impedance.

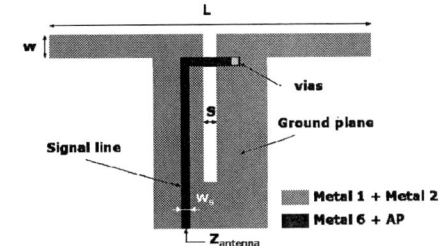

Figure 4 – Dipole antenna with integrated balun

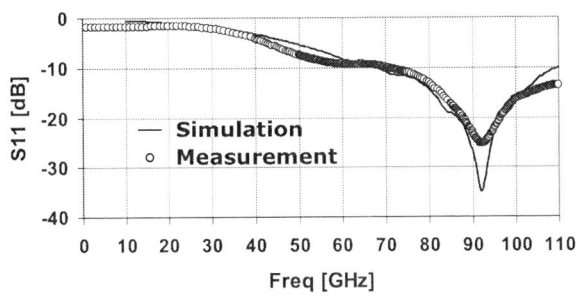

Figure 5 – Antenna S_{11} in dB as a function of frequency

B. 94 GHz LNA design

A single stage LNA is designed to evaluate the performances of such a circuit in order to address applications at 94 GHz. The microstrip topology is preferred to take benefit of the low loss into the transmission lines built on the dedicated BEOL depicted previously. And as a matter of fact, this topology is compatible with the dipole antenna with integrated balun.

The transistor described in section II.A. is biased to work at its maximum transition frequency (f_T = 243 GHz) with V_{BE} = 0.92 V and V_{BC} = -0.5 V. The LNA is a standard common-emitter topology using microstrip transmission lines (TL) in the input and output matching networks and biasing networks. The schematic of the LNA is depicted on Figure 6. The transistor is naturally stable from 50 GHz to 130 GHz. Below 25 GHz, both $S_{11}*$ and $S_{22}*$ are in their respective unstable regions. Source and load stability circles are plotted on Figure 7 from 50 GHz up to 130 GHz and $S_{11}*$ and $S_{22}*$ at 94 GHz are also placed on the Smith chart. To stabilize the transistor at lower frequencies (f < 25 GHz), a compensation network is placed in the input biasing network. This network is composed of an N+ silicided polysilicon resistor and a

Metal-Insulator-Metal (MIM) capacitor. In the design, all capacitors are MIM.

As commonly known, the conditions for matching the input for maximum gain and minimum noise figure are conflicting and a trade-off is required. This fact is illustrated in Figure 7 with different positions held by $S_{11}*$ and S_{opt} at 94 GHz. As already mentioned, the LNA is designed from the impedance of the antenna $Z_{antenna}$.

The output return loss is better than 10 dB in the [90 GHz – 105 GHz] frequency band (cf. Figure 8).

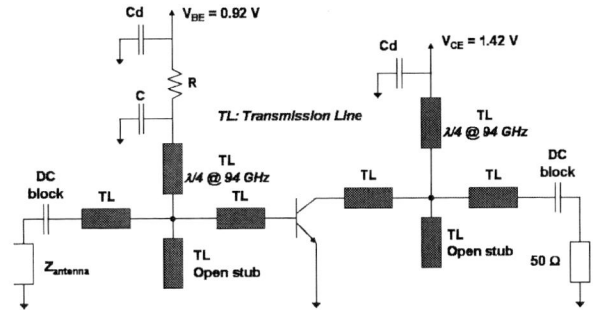

Figure 6 – LNA schematic

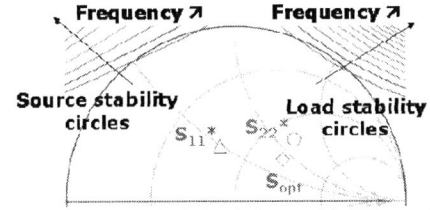

Figure 7 – Smith chart: $S_{11}*$, $S_{22}*$, S_{opt} at 94 GHz, and stability circles from 50 GHz to 130 GHz (V_{BE} = 0.92 V, V_{BC} = -0.5 V)

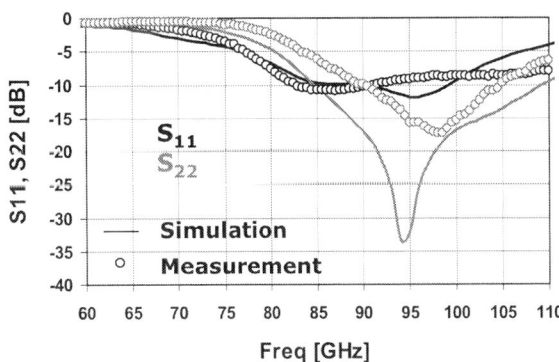

Figure 8 – Input matching under $Z_{antenna}$ and 50 Ω output matching of the LNA (V_{BE} = 0.92 V, V_{BC} = -0.5 V)

The trade-off between gain and noise leads to the following performances at 94 GHz (cf. Figure 9):

$$S_{21} = 5.0 \text{ dB} \quad (5)$$
$$NF = 7.0 \text{ dB} \quad (6)$$

The 94 GHz SiGe:C LNA exhibits state-of-the-art noise performances [3]. Furthermore, the noise figure remains under 7.2 dB over a large frequency band from 92 GHz to 104 GHz. A promising value of maximum gain is achieved and remains above 4 dB over a large frequency band from 80 GHz to 100 GHz. The interest of the integration of the antenna on a same silicon substrate is obvious, since we are able to relax the critical constraint of assembly loss. With an antenna gain of -2 dBi, we can calculate the total gain:

$$\text{Total gain} = 3.0 \text{ dB} \qquad (7)$$

Our study is extended to the design of a 2-stage LNA, with the first stage co-designed considering the antenna impedance. The LNA exhibits very good performances with a higher value of the S_{21} and a slight increase of the NF as shown on Figure 10:

$$S_{21} = 10.5 \text{ dB} \qquad (8)$$
$$NF = 7.7 \text{ dB} \qquad (9)$$

In this case, the total gain of the co-integrated block at 94 GHz is:

$$\text{Total gain} = 8.5 \text{ dB} \qquad (10)$$

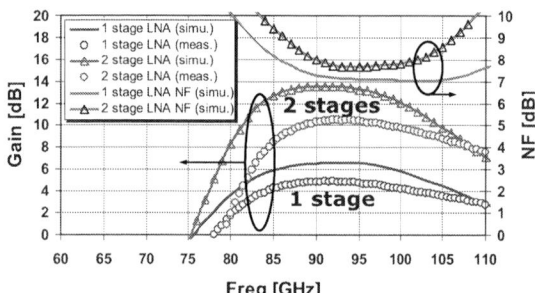

Figure 9 – S_{21} and noise figure of the LNA under $Z_{antenna}$
($V_{BE} = 0.92$ V, $V_{BC} = -0.5$ V)

The main results and performances of the LNAs are summarized in the following Table 1. Table 2 summarizes the performances of the LNAs taking into account the antenna gain into the total gain value. Figure 10 shows the small signal measurement setup on the 2-stage LNA. Noise figure measurements are in progress and results up to 110 GHz will be presented in the final paper as the full radiation pattern LNA+Antenna.

V CONCLUSION

A fully millimeter-wave dedicated BiCMOS technology is used to demonstrate the interest of the co-integration of an amplifier with a dipole antenna on a same silicon substrate for 94 GHz passive imaging applications. State-of-the-art performances are demonstrated with a single-stage and a two-stage LNA. In both cases, the LNA is based on a SiGe:C 130 nm HBT and directly matched to the antenna impedance which is built on the BEOL of the technology specifically processed for millimeter-wave high performance passive components such as transmission lines. With an antenna gain of -2.0 dBi, a

total gain of 3.0 dB is achieved for the building block [single-stage LNA co-integrated with the antenna]. A gain of 8.5 dB is achieved for the block [two-stage LNA with antenna]. These performances appear compatible with conventional millimeter-wave RF FE features.

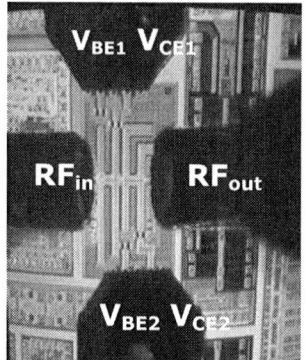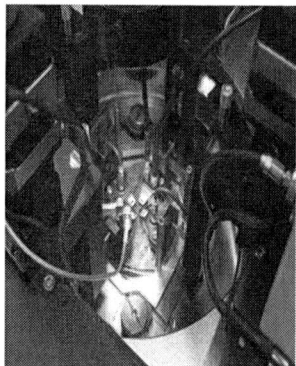

Figure 10 – Measurement setup (2-stage LNA)

Measured parameters	Co-integrated LNA (1 stage)	Co-integrated LNA (2 stages)
Frequency [GHz]	94	94
S_{11} **[dB]**	-9	-9
S_{22} **[dB]**	-14	-14
S_{21} **[dB]**	5.0	10.5
NF [dB]	7	7.7
Power [mW]	11.2	21.3

Table 1 - 94 GHz LNA and LNA summary

Measured parameters	Co-integrated LNA+Antenna (1 stage)	Co-integrated LNA+Antenna (2 stages)
Frequency	94 GHz	94 GHz
S_{22}	-14 dB	-14 dB
Total gain	3.0 dB	8.5 dB
Power	11.2 mW	21.3 mW

Table 2 - 94 GHz LNA and LNA + Antenna performance

REFERENCES

[1] G. Avenier, *et al.*, "0.13 μm SiGe BiCMOS technology for mm-wave applications", *IEEE BCTM*, pp. 89-92, October, 2008

[2] A. Niknedjad, *et al.*, "mm-Wave silicon technology: 60 GHz and beyond", Chapter 2, p. 52, Springer, 2008

[3] H.R. Chuang, *et al.*, "A 60 GHz millimeter-wave CMOS RFIC-on-chip dipole antenna", *Microwave Journal*, vol. 50, No. 1, p. 144, January, 2007

[3] J. Powell, *et al.*, "SiGe receiver front-ends for millimeter-wave passive imaging", *IEEE Trans. on MTT*, vol. 56, No. 11, pp. 2416-2425, November, 2008

[4] R. Pilard, *et al*, "Dedicated measurement setup for millimeter-wave silicon integrated antennas: BiCMOS and CMOS high resistivity SOI process characterization", *EuCAP*, March, 2009

A 3 G-Bit/s W-Band SiGe ASK Receiver with a High-Efficiency On-Chip Electromagnetically-Coupled Antenna

Jason W. May, Ramadan A. Alhalabi, Gabriel M. Rebeiz

University of California, San Diego, CA, 92093, USA

Abstract — A novel high-efficiency on-chip W-Band microstrip antenna is designed in a commercial SiGe process (IBM 8HP). The antenna has a measured gain of 2-4 dB and an efficiency of 50-57% at 92-98 GHz. An ASK receiver including a W-Band SPDT, LNA, and power detector is integrated with the antenna and is used to demonstrate a low-power 94 GHz 3-Gb/s on-chip wireless data link.

Index Terms — ASK, Millimeter-wave antennas, integrated antennas, RFIC, SiGe.

I. INTRODUCTION

The W-Band (75-110 GHz) is commonly utilized for automotive radar, imaging, and high data-rate communications. The trend towards high integration has resulted in several SiGe and CMOS imaging and communications RFICs which typically interface with an off-chip antenna [1],[2]. There has been progress towards fully-integrated IC's including on-chip antennas; the 94 GHz monopulse receiver/phased-array in [3] includes a backside Si lens, and a low-efficiency (< 5%) on-chip dipole antenna is used in the 140 GHz silicon transceiver in [4]. The silicon lens results in an expensive solution and should be avoided in low-cost systems. This paper demonstrates a SiGe W-Band receiver and wireless data link based on a novel high-efficiency electromagnetically coupled microstrip antenna.

II. W-BAND ON-CHIP EM-COUPLED ANTENNA

A. Antenna Design

Silicon substrates introduce special challenges for high efficiency millimeter-wave antennas. First, their low resistivities (0.1 to 10 Ω-cm) results in high dielectric loss and significantly reduces the antenna efficiency. Second, 200-500 μm thick substrates introduce another challenge since TE and TM surface waves, which have undesirable effects on the antenna pattern and efficiency, can be easily triggered and trapped inside the substrate [5].

One way to eliminate these effects is by shielding the antenna from the silicon substrate using a ground plane. However, the thickness of the metal/SiO₂ layers above the silicon substrate is only 10-12 μm in most RF processes

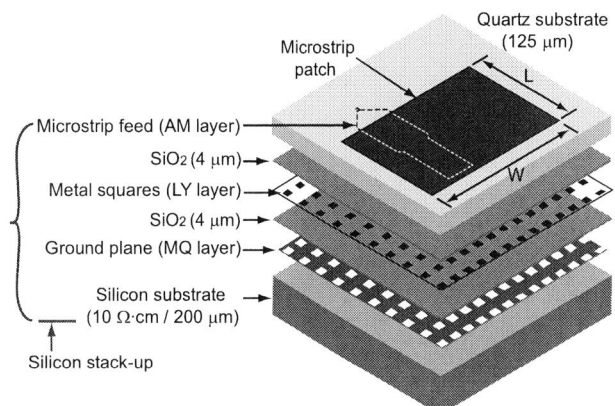

Fig. 1: On-chip EM-coupled microstrip antenna geometry: L=690 μm, W=970 μm.

(IBM 8HP, 9RF, etc.) and it is hard to build efficient antennas with such a ground spacing. As a result, one needs to add another substrate on top of the silicon chip so as to have enough ground-plane spacing for efficient radiation.

Figure 1 presents the layout of the 94 GHz EM-coupled microstrip antenna. The microstrip antenna is integrated on a 125 μm quartz substrate which is placed on top of a silicon RFIC. No via holes are used and the feed between the silicon RFIC top metal layer and the microstrip antenna is achieved entirely using electromagnetic fringing-field coupling. The ground plane of the microstrip antenna and the electromagnetic feed-probe (AM layer – top metal) is fabricated using the MQ layer, and isolates the antenna from silicon substrate. Also, 40x40 μm metal squares are introduced on LY layer to satisfy the required metal rule density and have little effect on the antenna performance.

B. Antenna Measurements

Fig. 2a shows the 94 GHz EM-coupled microstrip patch antenna fabricated using the IBM 8HP process. The quartz substrate is attached to the silicon chip using a small amount of epoxy placed at the corners. The input impedance of the antenna is measured using a CPW probe located 1.1 mm from the antenna, and the measured S_{11}

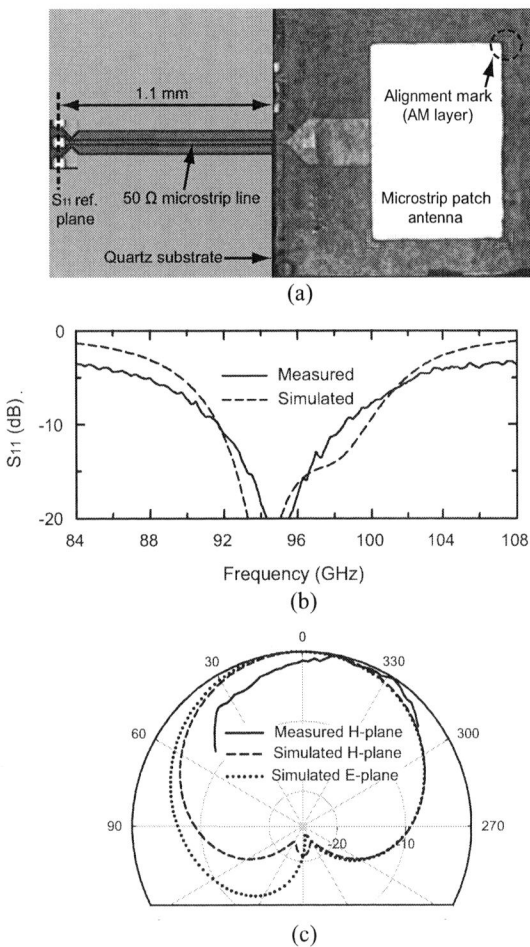

(a)

(b)

(c)

Fig. 2: (a) Fabricated on-chip EM-coupled microstrip antenna, (b) measured and simulated S_{11}, (c) 94 GHz radiation patterns.

agrees well with simulations (Fig. 2b). We have found that the resonant frequency can shift to 86 GHz if there is residual epoxy between the quartz substrate and the silicon die. Therefore, it is essential to mount the quartz microstrip antenna with care on the silicon wafer.

The H-plane radiation patterns are measured in the receive mode using a waveguide mm-wave diode detector. The diode detector was connected to the antenna feed using a CPW probe. A special circular arm set-up is used above the on-chip antenna and allows for a near-perfect radial scan in the H-plane. The transmit antenna is a W-band horn with a gain of 23 dB placed at R $\cong$ 28 cm from the on-chip antenna (well into the far field). Fig. 2c presents the measured and simulated radiation patterns at 94 GHz and with good agreement. Due to the metal-chuck and CPW probe positioner, one can notice some standing waves in the measured patterns.

The absolute gain of the 94 GHz on-chip antenna is measured using a similar set-up but with a calibrated Agilent Power Meter (E4417). The gain of the on-chip

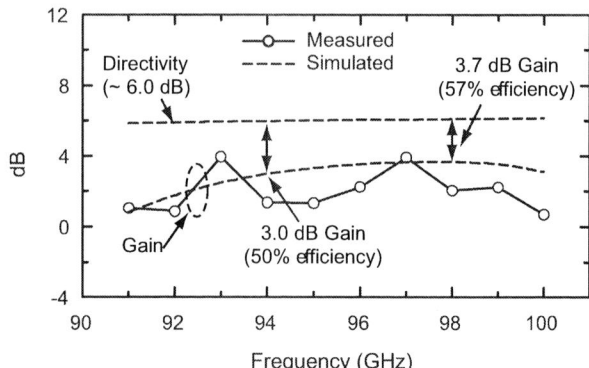

Fig. 3: Measured and simulated gain of the on-chip EM-coupled microstrip antenna.

EM coupled antenna is obtained using the Friis transmission formula and knowing the gain of the transmit horn. The antenna impedance mismatch is not taken out of the measurement. The loss of the on-chip microstrip line and the CPW waveguide probe are normalized out of the measurement (a back-to-back probe-microstrip-probe was measured independently). Fig. 3 presents the measured peak gain of the on-chip antenna at 91-100 GHz and agrees well with simulations. Note that the on-chip antenna has an efficiency of 50-57% (3-2.4 dB loss) at 94-98 GHz, which is excellent for W-band frequencies.

III. W-BAND ON-CHIP ASK RECEIVER

A. Receiver Topology

A W-band ASK receiver (Fig. 4a) with the on-chip microstrip antenna was fabricated in the IBM 8HP process (Fig. 4b). The receiver includes a W-Band SPDT switch, a 5-stage LNA, and a power detector, with a total power consumption of 35 mW (Fig. 5) [1]. The SPDT (2.3 dB insertion loss, 20 dB isolation) is included for use in radiometer applications to reduce the effect of detector 1/f noise and low-frequency variations in the LNA gain, and includes a 50 Ω reference resistor. The LNA uses a 5-stage common emitter design and has a gain of 26 dB and a NF of 8 dB. The detector consists of a single RF transistor biased in class A-B operation with an input matching network and an output 94 GHz notch filter. The receiver performance is described in [1] in terms of thermal-imaging performance and achieves an NEP of 21 fW/Hz$^{1/2}$ over a 85-99 GHz bandwidth (NEDT~0.84 K with a 30 mS integration time using Dicke switching).

B. Responsivity Measurements

The receiver responsivity was tested as shown in Fig. 6a with the input SPDT switched to the antenna port. A variable attenuator is present after the tripler to prevent

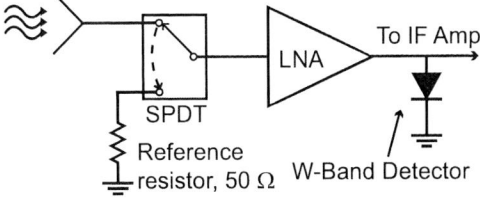

On-chip microstrip antenna

To IF Amp

LNA

SPDT

Reference resistor, 50 Ω

W-Band Detector

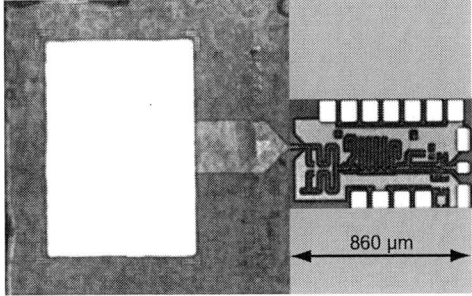

860 µm

Fig. 4: W-Band SiGe receiver with on-chip EM-coupled microstrip antenna. A SPDT with a reference resistor is included for use in radiometers.

receiver saturation (receiver input $P_{1dB} \approx$ -37 dBm). The RF input signal was pulsed at 100 kHz and the resulting output square wave was measured using an oscilloscope after amplification by a low-noise amplifier (SRS 552).

The system responsivity (Resp=Vout/Pin, V/W) was measured using an open-circuit load (Rload >> 750 Ω). The radiated power (P_t) from the W-band horn antenna is -42 dBm at 94 GHz, the horn antenna gain (G_t) is 23 dB, and the distance between the transmitter and the on-chip antenna (R) is 32 cm. In integrated antenna systems, the

responsivity can be defined in different ways since the input power can be defined at several locations. In this case, the input power is calculated as $P_{in}=S \cdot A_{eff}$, where S is the power density in W/m^2 ($S=P_t G_t/4\pi r^2$) and A_{eff} is the effective area of the antenna ($A_{eff}=G_a\lambda^2/4\pi$), where $G_a = 3$ dB at 94 GHz and varies with frequency as in Fig. 3. Note that this is not the standard definition in quasi-optical systems since it normalizes out the antenna efficiency and frequency response. However, it allows us to compare the responsivity with the CPW probe measurements.

Another definition of the input power is $P_{in}=S \cdot A_{eff}$, with $A_{eff} = D_{max}\lambda^2/4\pi$, where D_{max} is the maximum directivity of the antenna and is taken as constant vs. frequency. In this case, the power is referenced to the plane in front of the antenna (air-side) and this definition includes the effect of the antenna efficiency and frequency response. One can think of this definition as the available power at the antenna input (P_{avs}).

Fig. 6b presents a comparison between the responsivity obtained using measurements with the on-chip antenna for different Pin definitions and also using CPW probes. The slight dip at 93 GHz is a result of a standing wave in the measurement setup. The on-chip response is centered at 92 GHz with a 3-dB bandwidth of 89-95 GHz.

C. 3 Gb/s ASK Link Measurements

The receiver was also used to demonstrate a 94 GHz multi-Gb/s data link (Fig. 7). The output of an Agilent N4903A Bit Error Rate Tester (BERT) was applied to the IF port of a 94 GHz mixer. An LO-cancellation circuit

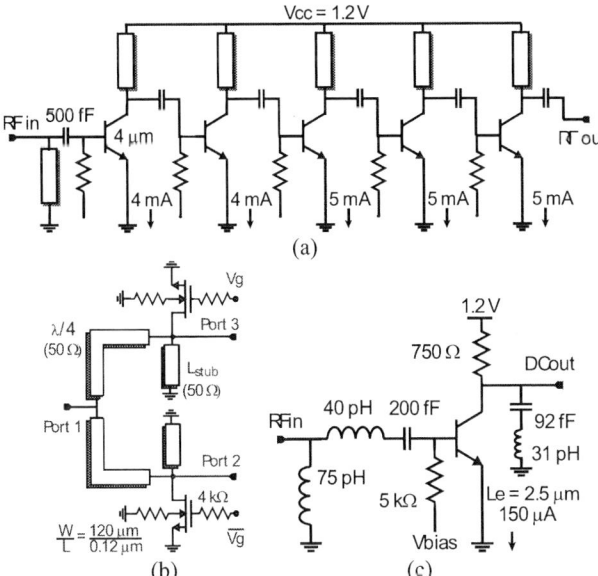

Fig. 5: W-Band (a) LNA, (b) SPDT, (c) ASK detector, all in IBM 8HP.

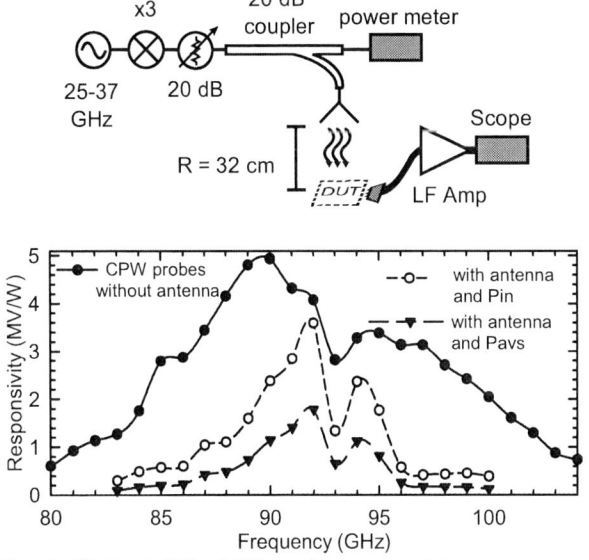

Fig. 6: W-Band SiGe ASK receiver responsivity measurement setup and measured responsivity with and without integrated antenna for two different input power definitions (see text).

978-1-4244-6240-7/10 $26.00 © 2010 IEEE   77

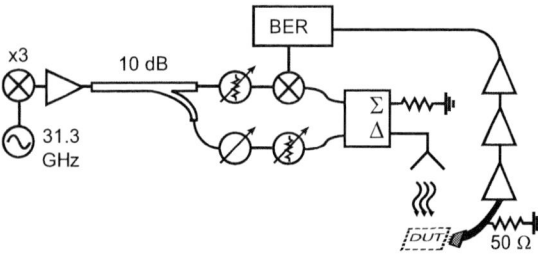

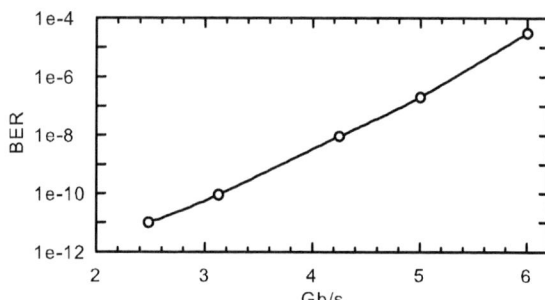

Fig. 8: Measured 94 GHz data link BER, 2^7-1 PRBS

ASK spectrum. A data rate of 3 Gb/s requires a system bandwidth of at least 6 GHz (first spectrum sidelobe is sampled by the ASK detector) and preferably 9-12 GHz (first and second spectrum sidelobes).

VII. CONCLUSION

A novel high-efficiency W-Band electromagnetically-coupled antenna was used to demonstrate an on-chip 94 GHz ASK wireless data link. The planar antenna is compatible with a wafer-scale process, and will enable high performance CMOS and SiGe mm-wave imaging and communications systems at 60 - 300 GHz.

ACKNOWLEDGEMENT

This work was funded in part by the DARPA MIATA program (Thomas Kenny and Alfred Hung) and by Intel Corporation (Ian Young and Jad Rizk).

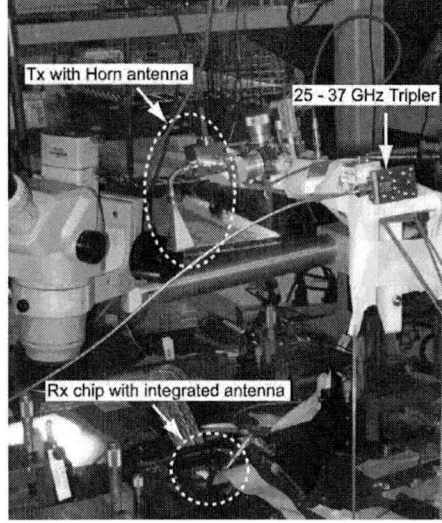

Fig. 7: 94 GHz wireless link BER measurement system.

was used to prevent receiver saturation due to mixer LO feed-through and to maintain maximum receiver responsivity. An IF amplifier chain (~45 dB gain from 20 MHz to 8 GHz) was used after the on-chip ASK receiver to amplify the receive signal before sending it to the BERT. The lower bandwidth of the IF amplifiers (20 MHz) limits the length of the allowable data pattern to 2^7-1 PRBS. An SMA tee and a 50 Ω load were placed at the output port to reduce the receiver output impedance from 750 Ω (useful for high responsivity in imaging applications) to 50 Ω (necessary for Gb/s operation). This reduces the responsivity of the ASK receiver by a factor of 16 (~250 kV/W without antenna effects). RF absorbers were placed around the DUT to reduce reflections from the probes and probe station.

The measured BER for a 2^7-1 PRBS is shown in Fig. 8 for an available power of –27 dBm at the on-chip antenna. The input power to the LNA/Detector is –27 dBm – 3 dB (antenna loss) -3 dB (SPDT switch and t-line loss) and is –33 dBm. Previous work on a 60 GHz SiGe receiver showed that ~-36 dBm is optimal for an ASK detector [6]. The increasing BER with Gb/s modulation is due to the 5-6 GHz bandwidth of the antenna/LNA/Detector and the

REFERENCES

[1] J. W. May and G. M. Rebeiz, "Design and characterization of W-Band SiGe RFICs for passive millimeter-wave imaging," *IEEE Trans. Microwave Theory & Tech.*, accepted for publication, December 2009.

[2] Jeng-Han Tsai, "A 55-64 GHz fully-integrated sub-harmonic wideband transceiver in 130 nm CMOS process," *IEEE MWCL.*, vol. 19, pp.758-760, Nov. 2009.

[3] S. Raman, S. Barker, and G. M. Rebeiz, "A W-Band dielectric-lens-based integrated monopulse radar receiver," *IEEE Trans. Microwave Theory & Tech.*, vol. 46, no. 12, pp. 2283-2288, Dec. 1998.

[4] S. T. Nicolson *et. al.*, "A 1.2 V 140 GHz receiver with on-die antenna in 65nm CMOS," *IEEE RFIC Symp. Dig.*, pp. 229-232, June 2008.

[5] G. M. Rebeiz, "Millimeter-wave and terahertz integrated circuit antennas," *IEEE Proc.*, vol. 80, pp. 1748-1770, Nov. 1992.

[6] W. Shin, M. Uzonkol, and G. M. Rebeiz, "Ultra low power 60 GHz ASK SiGe receiver with 3-6 GBPS capabilities," *IEEE Compound Semiconductor Integ. Circuits Conf.*, pp. 1-4, Oct. 2009.

RMO1D-5

A 325 GHz Frequency Multiplier Chain in a SiGe HBT Technology

Erik Öjefors[1], Bernd Heinemann[2] and Ullrich R. Pfeiffer[1]

[1] University of Wuppertal, Rainer-Gruenter-Str. 21, D-42119 Wuppertal, Germany
[2] IHP GmbH, Im Technologiepark 25, D-15236 Frankfurt (Oder), Germany

Abstract—A single-chip 325-GHz x18 frequency multiplier chain based on two cascaded active differential triplers and a balanced output doubler is presented. The multiplier operates over a 317 to 328 GHz bandwidth with a 0-dBm 18-GHz input signal. A peak output power of -8 dBm is obtained at 325 GHz. The multiplier chain is realized in an evaluation SiGe HBT technology with cut-off frequencies f_T/f_{max} of 250 GHz / 380 GHz.

Index Terms—Frequency conversion, heterojunction bipolar transistors, millimeter wave integrated circuits, power amplifiers, silicon, submillimeter wave generation.

I. INTRODUCTION

Submillimeter-wave or terahertz systems operating in the 0.3 to 3-THz band have potential uses in a wide range of sensor and imaging applications [1]. Radar-based 3D active imaging systems are of particular interest for stand-off security screening [2]. A critical component in such systems is an illumination or local-oscillator signal source with sufficient power. Schottky-diode and heterostructure-barrier-varactor (HBV) frequency multipliers in waveguide technology are capable of generating more than 8 dBm of output power in the 260 to 400-GHz frequency range [3]-[4], but are costly to implement and difficult to integrate in multichannel or power-combining systems. However, III-V MMIC multipliers have recently provided monolithically integrated alternatives capable of 0 dBm output power at 195 GHz [5] as well as -6.4 dBm at 300 GHz [6].

In comparison with Schottky diodes or III-V-based MMICs, silicon technologies offer lower cost in large volumes and the possibility of large-scale integration of the submillimeter-wave source with an on-chip frequency synthesizer and signal-modulation blocks. Submillimeter-wave receiver circuits with sufficient sensitivity for active imaging have been demonstrated using CMOS [7] and SiGe HBT [8] devices. Silicon signal sources in this frequency range, however, have only been able to generate comparatively low output power. A SiGe-HBT based 278-GHz VCO provided -38 dBm output power in [9] and a 410 GHz 45-nm CMOS VCO provided -47 dBm in [10], respectively. A linear superposition technique in CMOS has demonstrated -46 dBm at 324 GHz [11].

In this paper a 325-GHz x18 frequency multiplier with improved output power compared to previously demonstrated silicon submillimeter-wave sources is presented. The monolithic multiplier is implemented in an f_{max}-optimized evaluation SiGe HBT technology using cascaded transistor-based multiplication and amplification stages.

II. MONOLITHIC FREQUENCY MULTIPLIER CHAIN

The frequency multiplier chain consists of a 325-GHz doubler circuit fed differentially by two cascaded triplers as shown in Fig. 1. Each tripler is equipped with a frequency selective

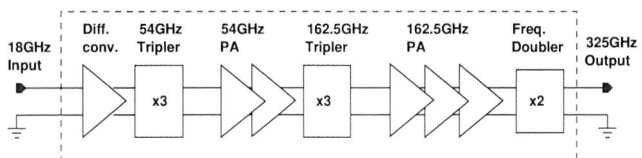

Fig. 1. Block diagram of the monolithically integrated ×18 multiplier chain. Frequency selective power amplifiers are used between the multiplier stages to boost the drive power and to filter out unwanted multiplication products.

power amplifier at the output in order to boost the power entering the next multiplier stage and to suppress spurious frequencies. A differential amplifier is used at the input to convert the the 18-GHz single-ended signal from an external synthesizer to a balanced drive for the 54-GHz tripler.

A. Balanced 325-GHz Frequency Doubler

The balanced 325-GHz frequency doubler consists of a differential pair with a shared collector connection as shown in the circuit schematic (Fig. 2). The two $A_E = 4 \times (0.16 \times$

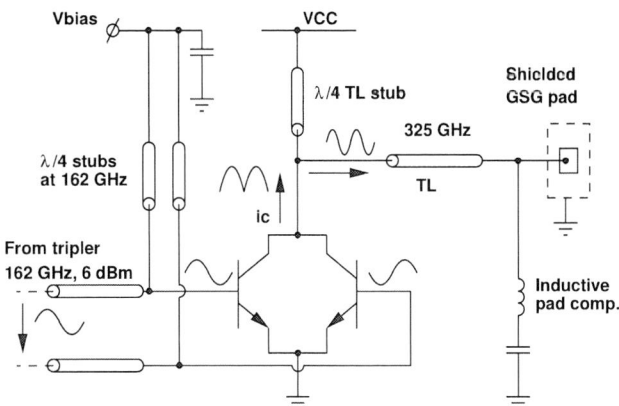

Fig. 2. Schematic of the 325-GHz doubler, which is based on a differential stage driven in class-B to generate combined I_C pulses of twice the frequency.

$0.84)$-μm^2 large devices are driven differentially in class B by a 6-dBm 162.5-GHz signal, thus yielding current pulses of 325 GHz at the shared collector connection. A quarter-wave transmission-line stub connected to V_{CC} is used to filter out the 325-GHz signal and to provide the transistors with a 2-V V_{CE}. The output of the doubler stage is fed through a 50-Ω transmission line to the output pad. On-chip inductive compensation of the pad capacitance is used to minimize the insertion loss of the pad. The simulated output power of the doubler is -5 dBm.

B. 162.5-GHz and 54-GHz Frequency Triplers

The 162.5-GHz frequency tripler, shown in Fig. 3, is based on a differential cascode amplifier. The amplifier is saturated by a strong input signal, which generates square-wave-like collector currents in the $A_E = 4 \times (0.16 \times 0.84)$-$\mu m^2$ large devices Q1/Q2 and Q3/Q4. The differential cascode is powered from the 4-V V_{CC} line with the transistors Q3/Q4 voltage-biased with $V_{base} = 2$ V in order to minimize impact ionization and delay the onset of breakdown. A conventional L-type

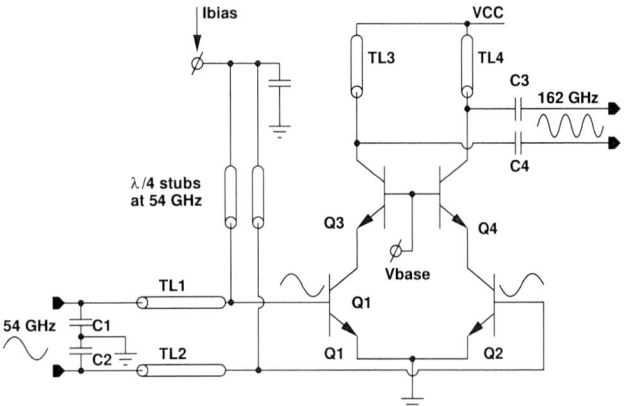

Fig. 3. Simplified schematic of the 162-GHz tripler core. Similar, but differently tuned, differential cascode stages are used in the 54-GHz tripler as well as in the 54-GHz and 162-GHz PA.

impedance-matching network based on the transmission-line inductances TL3/TL4 and the MIM-capacitors C3/C4 is used to filter out the tripled 162-GHz signal. It also provides an impedance match for maximum power delivery in a 100-Ω differential system impedance. The transmission lines TL1/TL2 and the capacitors C1/C2 are used to match for minimum return loss at the tripler input.

The 54-GHz tripler and the multi-stage power amplifiers in the chain use differential cascode stages of similar architecture, but with differently tuned matching elements.

III. TECHNOLOGY AND MANUFACTURING

An HBT-only evaluation version of a BiCMOS technology is used for the circuit fabrication. It represents a combination of the IHP 0.25-μm BiCMOS process SG25H1 [12] and an HBT module similar to that of the recently presented 0.13-μm BiCMOS technology [13]. The HBT architecture is illustrated in Fig. 4. In contrast to the 0.13μm BiCMOS process, the following steps, affecting the HBT performance, were altered: The profile of the SiGe:C base layer was changed resulting in a

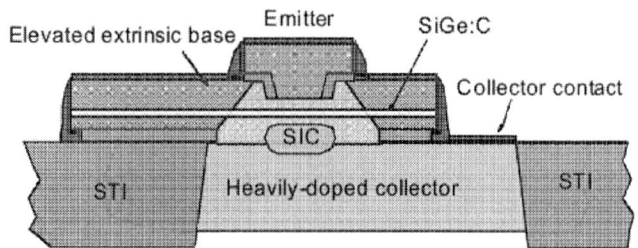

Fig. 4. Schematic cross section of an HBT with elevated extrinsic base. Mono-crystalline and poly-crystalline regions of the base layer are indicated by different hachure.

lower base-sheet resistance while the collector current density was maintained approximately. The temperature of the final spike anneal was reduced. Furthermore, the wafers exhibit a 45° rotated substrate orientation, an enhanced sub-collector and a reduced salicide sheet resistance.

For circuit fabrication, the process offers poly-silicon and silicide resistors. The back-end manufacturing corresponds to the process flow and design rules of the SG25H1 technology with five aluminum metal layers, including a 1fF/μm^2 MIM capacitor. The transistors achieve at $V_{BE} = 0.7$ V a current gain of about 500 and demonstrate an open-base collector-emitter breakdown-voltage BV_{CEO} of 1.7 V. Open and short de-embedded small-signal current-gain h_{21} and unilateral gain U were used for the extrapolation of f_T and f_{max} from 40 GHz with -20 dB per frequency decade. Compared with a 0.13μm BiCMOS reference HBT, the peak f_T/f_{max} values could be increased from 240 GHz / 330 GHz to approximately 250GHz/380GHz in this modified technology (see Fig. 5).

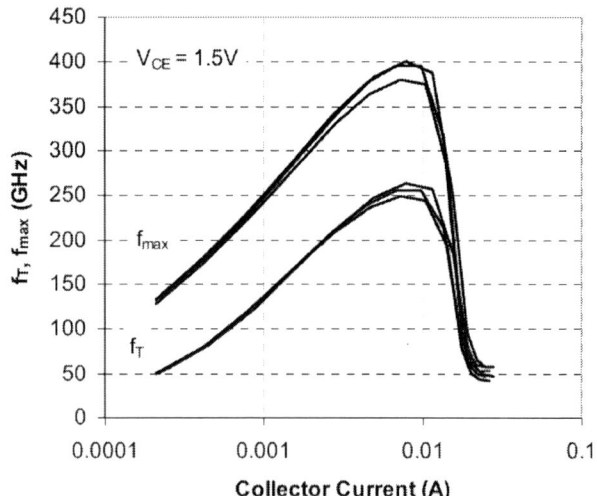

Fig. 5. Transit frequency f_T and maximum oscillation frequency f_{max} as a function of collector current extrapolated from h_{21} and the unilateral gain U at 40 GHz. Multi-emitter HBTs with an effective emitter area of AE = 4×(0.16x0.84) μm^2 were measured on three sites of the wafer at $V_{CE} = 1.5$ V.

978-1-4244-6240-7/10 $26.00 © 2010 IEEE

Fig. 6 shows the chip micrograph with a die size of 2.2×0.43-mm^2. Clearly visible are the 50-Ω shielded microstrip transmission lines used as tuning stubs and interconnects between the stages of the multiplier chain. The

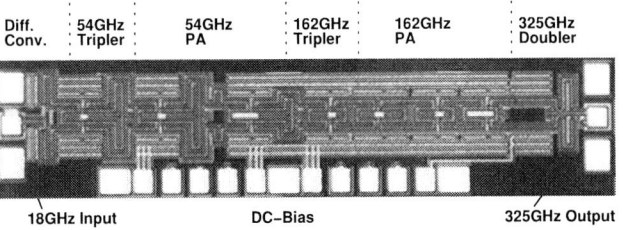

Fig. 6.   Micrograph of the 2.2×0.43-mm^2 large multiplier chip.

transmission-line signal conductors and the side-shields are implemented as strips in the top metal (M5) layer while M3 is used as the microstrip ground plane.

IV. CHARACTERIZATION SETUP

The frequency multiplier (DUT) has been characterized in a wafer prober setup (Fig. 7). The 18-GHz input signal is fed from a synthesizer to the chip through a coaxial wafer probe. A ground-signal-ground (GSG) WR-03-waveguide probe is used to extract the output signal from the frequency multiplier. The output power has been measured using two independent

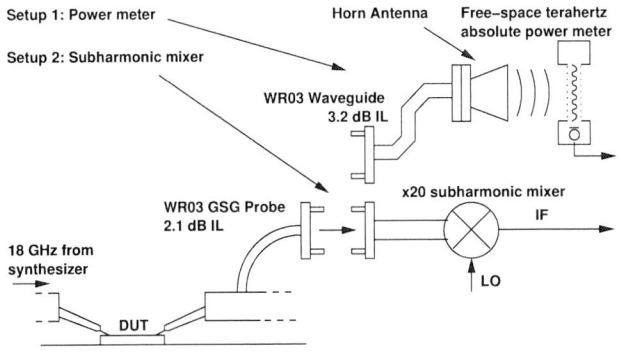

Fig. 7.   Setup used for characterization of the integrated multiplier. The wafer probe is connected to a calibrated mixer for single tone power measurements while the total output power is verified by an absolute power meter.

methods. The first one uses a total output power measurement technique, with a calibrated Thomas Keating free-space terahertz absolute power meter. A waveguide-connected horn antenna illuminates the power meter in this case. The power meter requires a 30-Hz chopped input signal, which is provided by on-off keying of the 18-GHz input signal. The 174-GHz cutoff frequency of the WR-03 waveguide prevents any 162-GHz leakage through the final frequency doubler from reaching the power meter. The second method uses a calibrated ×20 subharmonic down-conversion mixer instead of the power meter. With this additional measurement the output spectrum is observed, and hence, it provides a way to verify that spurious tones are significantly suppressed so that they don't contribute to the the total power measurement.

V. RESULTS

The measured output power, swept over the 315 to 330-GHz range, is shown in Fig. 8 for the two methods described above. The 2.1-dB insertion loss of the probe and the 3.5-dB losses of the waveguide sections were de-embedded from all measurements. Within the 317 to 328-GHz frequency range

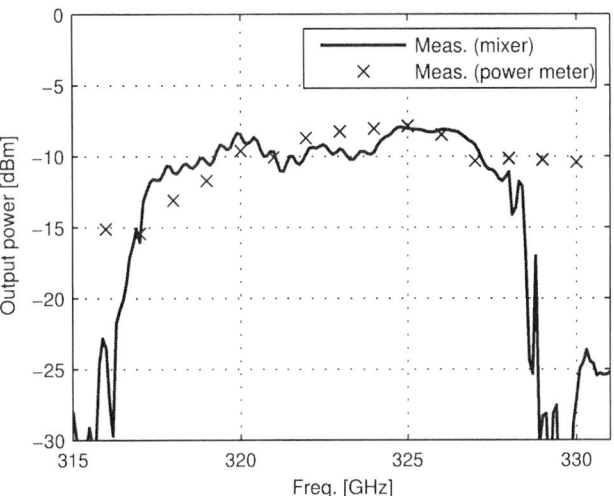

Fig. 8.   Measured output power at the 18'th harmonic (mixer measurement, solid line) as well as the total output power within the waveguide band registered by the absolute power meter (crosses) for the corresponding 1/18-frequency 0-dBm input signal.

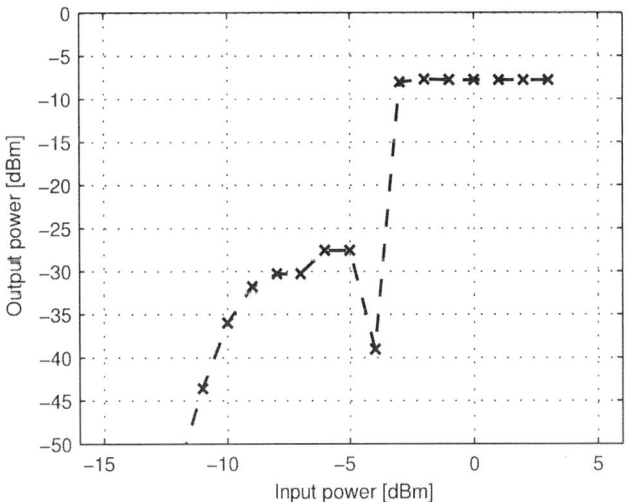

Fig. 9.   Measured output power at the 325-GHz design frequency for an 18-GHz input power swept from -13 dBm to 3 dBm. Below -3.5 dBm input power, the two triplers in the chain are not sufficiently saturated to provide a stable 162-GHz drive to the final doubler.

the mixer measurement of the 18'th harmonic and the total power in the waveguide band show good agreement. The minimum power in this range is -10 dBm and the peak power of -8 dBm is reached at 325 GHz. Note, that outside of this frequency range, significantly larger spurious tones appear

978-1-4244-6240-7/10 $26.00 © 2010 IEEE   81

TABLE I

COMPARISON OF SOLID-STATE SUBMILLIMETER SOURCES

Technology [μm]	Circuit [type/multiplication factor]	BW[1]/Freq. [GHz]	Output Power[2] [dBm]	Monolithic Integration	Reference
III/V-based					
GaAs Schottky	x4 multiplier	352-396	8	no	[3]
InP HBV	tripler	263-278	8.1	no	[4]
mHEMT	x6 multiplier	155-195	0	yes	[5]
mHEMT	doubler	250-310	-6.4	yes	[6]
Silicon-based					
0.13-μm SiGe	x18 multiplier	317-328	-8	yes	This Work
0.14-μm SiGe	VCO	278	-38	yes	[9]
45-nm CMOS	VCO	410	-47	yes	[10]
90-nm CMOS	Linear superposition	324	-46	yes	[11]

[1]3-dB output power bandwidth
[2]Peak output power

at the output because the multiplier stages in the chain are not fully saturated. Hence, the two measurement techniques deviate from each other. The power of the 325-GHz 18'th harmonic with an 18-GHz input power level swept from -12 dBm to 3 dBm is shown in Fig. 9. An input power level of -3 dBm is enough to sufficiently compress the multiplier chain at the 325-GHz design frequency. However, a larger input drive level compresses the stages further so that the operating bandwidth is extended. The measured saturated output power is -8 dBm, which can be compared to the simulated level of -5 dBm. The difference can be explained by measurement uncertainty as well as inaccurate modelling of the output-network losses in the simulation.

The total current consumption of the input differential converter, the two triplers and the five amplifiers is 300 mA at 4 V, while the 325-GHz doubler draws 8.9 mA from the 2-V collector supply voltage.

VI. CONCLUSION

A submillimeter-wave source capable of more than 0.1 mW output power has been realized in a silicon technology. As shown in Table I, the output power is comparable to the results obtained with III-V-based MMIC frequency multipliers, and it is significantly higher than the power levels demonstrated by VCO-based sources in CMOS and SiGe-HBT technology. The implemented multiplier chain can be driven from an external microwave synthesizer or it can be integrated together with a VCO and PLL in order to form a single-chip 325-GHz frequency synthesizer. In combination with recently developed receivers in CMOS [7] and bipolar technology [8] it forms an important building block for silicon-based submillimeter-wave imaging systems.

VII. ACKNOWLEDGMENT

This work was partially funded by the European Commission within the project DOTFIVE (no. 216110).

REFERENCES

[1] P. H. Siegel, "Terahertz technology," IEEE Trans. Microw. Theory Tech., vol. 50, no. 3, pp. 910–928, March 2002.

[2] K. Cooper, R. Dengler, N. Llombart, T. Bryllert, G. Chattopadhyay, E. Schlecht, J. Gill, C. Lee, A. Skalare, I. Mehdi, and P. Siegel, "Penetrating 3-D imaging at 4- and 25-m range using a submillimeter-wave radar," IEEE Trans. Microw. Theory Tech., vol. 56, no. 12, pp. 2771–2778, Dec. 2008.

[3] G. Chattopadhyay, E. Schlecht, J. S. Ward, J. J. Gill, H. H. S. Javadi, F. Maiwald, and I. Mehdi, "An all-solid-state broad-band frequency multiplier chain at 1500 GHz," IEEE Trans. Microw. Theory Tech., vol. 52, no. 5, pp. 1538–1546, May. 2004.

[4] Q. Xiao, J. L. Hesler, T. W. Crowe, B. S. Deaver, Jr., and R. M. Weikle, II, "A 270-GHz tuner-less heterostructure barrier varactor frequency tripler," IEEE Microw. Wireless Compon. Lett., vol. 17, no. 4, pp. 241–243, 2007.

[5] M. Abbasi, R. Kozhuharov, C. Kärnfelt, I. Angelov, I. Kallfass, A. Leuther, and H. Zirath, "Single-chip frequency multiplier chains for millimeter-wave signal generation," IEEE Trans. Microw. Theory Tech., vol. 57, no. 12, pp. 3134–3142, 2009.

[6] I. Kallfass, A. Tessmann, H. Massler, D. Lopez-Diaz, A. Leuther, M. Schlechtweg, and O. Ambacher, "A 300 GHz active frequency doubler and integrated resistive mixer MMIC," in Proc. 4th European Microwave Integrated Circuits Conf., Rome, Italy, September 2009, pp. 200–203.

[7] U. R. Pfeiffer, E. Öjefors, A. Lisauskas, D. Glaab, and H. Roskos, "A CMOS focal-plane array for heterodyne terahertz imaging," in IEEE Radio Frequency Integrated Circuits Symp., July 2009, pp. 437–440.

[8] E. Öjefors and U. R. Pfeiffer, "A 650GHz SiGe receiver front-end for terahertz imaging arrays," in IEEE Intl. Solid-State Circuits Conf., February 2010, pp. 430–431.

[9] R. Wanner, R. Lachner, G. Olbrich, and P. Russer, "A SiGe monolithically integrated 278 GHz push-push oscillator," in IEEE Intl. Microwave Symp., June 2007, pp. 333–336.

[10] E. Seok, C. Cao, D. Shim, D. J. Arenas, D. B. Tanner, C.-M. Hung, and K. K. O, "A 410 GHz CMOS push-push oscillator with an on-chip patch antenna," in IEEE Intl. Solid-State Circuits Conf., 2008, pp. 472–473.

[11] D. Huang, T. R. LaRocca, M.-C. F. Chang, L. Samoska, A. Fung, R. L. Campbell, and M. Andrews, "Terahertz CMOS frequency generator using linear superposition technique," IEEE J. Solid-State Circuits, vol. 43, no. 12, pp. 2730–2738, 2008.

[12] B. Heinemann, R. Barth, D. Knoll, H. Rücker, B. Tillack, and W. Winkler, "High-performance BiCMOS technologies without epitaxially-buried subcollectors and deep trenches," Semicond. Sci. Technol., vol. 22, pp. 153–157, 2007.

[13] H. Rücker, B. Heinemann, W. Winkler, R. Barth, J. Borngräber, J. Drews, G. Fischer, A. Fox, T. Grabolla, U. Haak, D. Knoll, F. Korndörfer, A. Mai, S. Marschmeyer, P. Schley, D. Schmidt, J. Schmidt, K. Schulz, B. Tillack, D. Wolansky, and Y. Yamamoto, "A 0.13um SiGe BiCMOS technology featuring fT/fmax of 240/330 GHz and gate delays below 3ps," in Proc. BCTM Conf., Capri, Italy, Oct. 2009, pp. 166–169.

978-1-4244-6240-7/10 $26.00 © 2010 IEEE

Semi-Active High-Efficient CMOS Rectifier for Wireless Power Transmission

Stephen T. Kim[1], Taejoong Song[1], Jaehyouk Choi[1], Franklin Bien[2], Kyutae Lim[1], and Joy Laskar[1]

[1]Georgia Electric Design Center, Georgia Institute of Technology, Atlanta, Georgia, 30332

[2]Ulsan National Institute of Science and Technology, Ulsan, South Korea

Abstract — A semi-active high-efficient (SA-HE) CMOS rectifier with reverse leakage control has been developed. It employs a cross-coupled NMOS pair and two leakage control comparators to reduce reverse charge leakage current. In addition, the adaptive body bias control technique is utilized to improve the reliability of the rectifier. The SA-HE rectifier has been fabricated in a 0.18um CMOS technology and shows 15% improvement over conventional rectifiers.

Index Terms — Rectifier, wireless sensor, bio-implantable devices, semi-active, wireless power transmission.

I. INTRODUCTION

There has been growing demand for bio-implantable devices and wireless sensors that can operate for extended periods without maintenance, while not disturbing their surroundings either physically or operationally. Wireless power transmission has been a popular solution for these applications because it provides a more reliable power source than energy scavenging scheme [1]. A conventional remotely powered system is described in Fig. 1. The inductive link (or an antenna) catches the RF power generated by the base station. The rectifier then converts AC power into DC power, which is used in the final processing block.

The maximum transmitted power level at the base station should be restricted due to human safety issues. Therefore, given the path loss and the matching loss, the power and voltage conversion efficiency of the rectifier ultimately determines the DC power available at the processing block. Therefore, a high efficiency rectifier is essential for maximizing the DC power, which is closely related to the performance of the processing block.

In this paper, the leakage current issue is analyzed, and

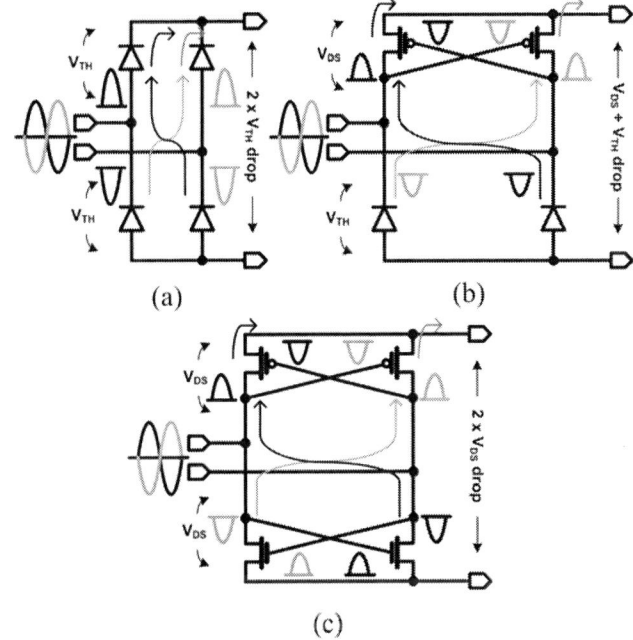

Fig. 2. (a) Full-wave rectifier, (b) PMOS cross-coupled rectifier [2], and (c) NP-cross-coupled rectifier [3].

the semi-active leakage preventing technique and the active body biasing technique are proposed to enhance the conversion efficiency without degrading the system reliability.

II. CMOS RECTIFIER

A diode-bridged rectifier is a simple way to realize a full wave rectifier shown in Fig. 2 (a). Schottky diodes are often employed to enhance the conversion efficiency. However, considering additional costs and the fact that Schottky diodes are not supported in all CMOS technologies, a CMOS-based implementation is considered as practical solutions [3-4].

Moreover, the CMOS implementation offers better compatibility with the following processing block which is usually implemented in a CMOS technology.

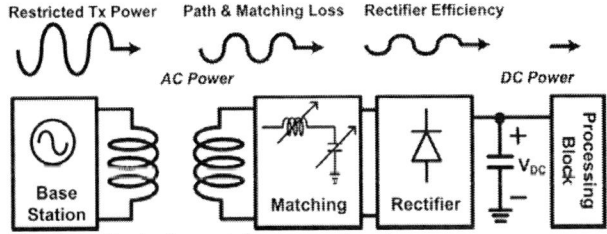

Fig. 1.   Typical remotely powered system

978-1-4244-6240-7/10 $26.00 © 2010 IEEE

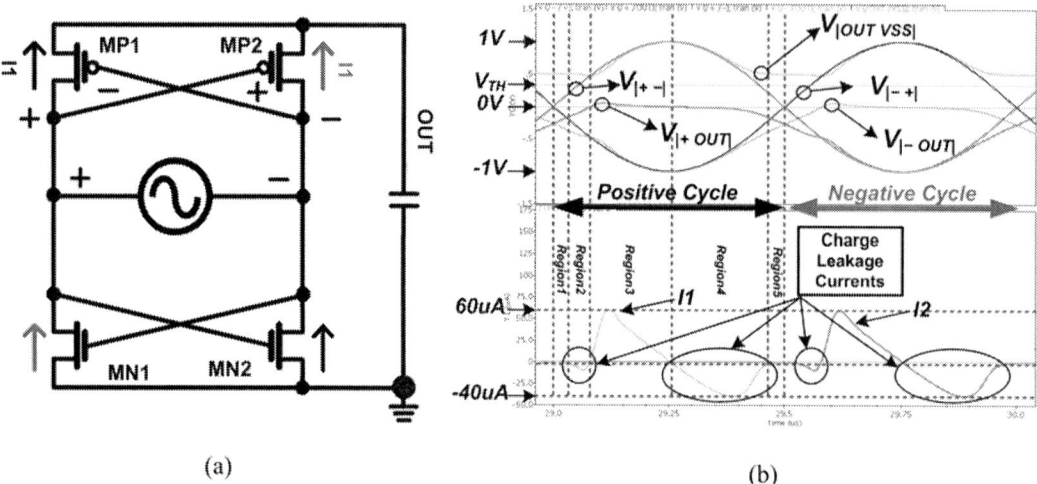

Fig. 3.   (a) The charge leakage test set-up of a single stage rectifier cell, and (b) the simulated waveform of each voltage node.

A CMOS diode-bridged rectifier in Fig. 2 (a) picks up differential RF-input during both input transition cycles - positive and negative. The black line represents the positive cycle and the gray line represents the negative cycle. At each cycle, the corresponding complementary diode pair turns on and delivers the input RF current to the output at the expense of $2 \times V_{TH}$ drop.

Thus, this configuration inherently suffers from $2 \times V_{TH}$ voltage drop. One way to improve this threshold voltage drop is to use a cross-coupled PMOS pair shown in Fig. 2 (b) [2]. During each cycle, the voltage drop across the cross-coupled PMOS pair (V_{DS}) depends on the input current level and if the input current level is sufficient and the transistors are properly sized, V_{DS} can be lower than the V_{TH}. However, the bottom diodes still suffer from the V_{TH} drop.

If a cross-coupled NMOS pair is adopted instead of the two bottom diodes, the remaining V_{TH} drop issue can be solved as shown in Fig. 2 (c) [3]. However, on every cycle, there exists a period where the output node potential is higher than that of the input node. When this happens, the cross coupled pairs cannot be turned off completely, thereby causing charge leakage. The charge leakage then degrades the conversion efficiency. In addition, the rectifier circuit does not have a stable supply voltage that can guarantee the highest potential in the system. Therefore, if PMOS body terminals are statically tied to fixed potentials, the source-body (or drain-body) diodes of PMOS transistors can be forward-biased.

A. Charge Leakage Problem

For the conventional rectifier shown in Fig. 3(a), as the RF differential input is connected to the gate-source nodes of the transistors, and the RF signal fluctuates smoothly,

the cross-coupled PMOS and NMOS pairs cannot turn on and off abruptly. When the output node is charged up to the targeted level, the output voltage can be higher than the input voltage for considerable duration of time. Since the PMOS cross-coupled pair cannot be turned-off completely during this period, the charge stored at the load capacitor discharges through the partially turned-on PMOS pair.

To observe this problem, a sinusoidal wave with 1-V amplitude and 1-MHz frequency is applied to a single stage NP-cross-coupled rectifier shown in Fig. 3 (a). The corresponding voltage and current waveforms are shown in Fig. 3 (b). The rectifying cycle can be divided into 5 regions. As the chosen topology is a full-wave rectifier, the 5 regions repeat every half-input period.

Region 1: The gate drive voltage ($V_{|+-|}$) is smaller than V_{TH}, and the output node voltage is lower than the input voltage. Therefore, there is subthreshold conduction which has minor effects on the leakage problem.

Region 2: The gate drive voltage exceeds V_{TH}. Since the output node ($V_{|OUT\ VSS|}$) is still lower than the input voltage, MP1 starts conducting current and current flows from output to input.

Region 3: As the input voltage increases, it becomes larger than the output voltage. When the gate drive voltage is higher than V_{TH}, the output capacitor is charged.

Region 4: The gate drive voltage is still higher than V_{TH}, but the input voltage is lower than the output node voltage. Thus, the output node is discharged until MP1 turns off.

Region 5: The gate drive becomes smaller than V_{TH}, and there is subthreshold conduction as in Region 2.

We can also observe from the above current waveforms that there are charge leakage currents during both Region2 and Region4.

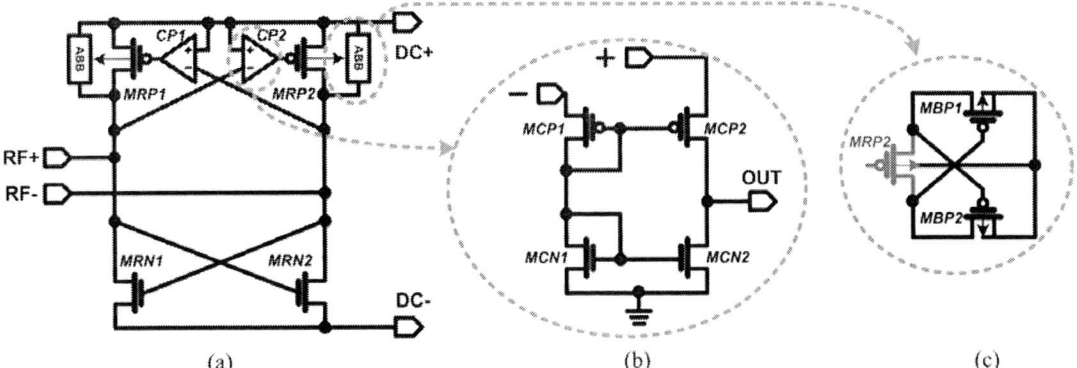

Fig. 4. Diagram of (a) the proposed rectifier cell, (b) the leakage control comparator, and (c) the active body biasing block [7-8].

B. Active Rectifiers

Existing literatures propose utilization of several active circuitries to enhance the conversion efficiency of rectifiers [4-6]. In [6], the V_{TH} cancellation technique was adopted to lower the threshold voltage drop of diode-connected transistors. However, this approach requires external reference and supply voltage. In [4], an active diode was utilized to reduce conduction drop. The active diode was implemented with a 4-input-comparator that does not require external supplies. However, since this comparator does not use an external supply, and it is connected to NMOS transistors, the NMOS transistors cannot be fully turned on. Therefore, if this rectifier cell is multiply stacked to obtain higher voltage conversion gain, the active diodes in the latter stages would suffer from partially high resistive current paths.

III. PROPOSED SEMI-ACTIVE HIGH-EFFICIENT CMOS RECTIFIER

The proposed semi-active high-efficient (SA-HE) CMOS rectifier cell is shown in Fig. 4 (a). It utilizes the NP-cross-coupled rectifier to minimize intrinsic voltage drop. Moreover, the leakage control comparators, shown in Fig. 4 (b), are inserted in the PMOS cross-coupled connection to alleviate the reverse leakage problem. In addition, the active body-biasing (ABB) technique is adopted to prevent the PMOS body-junction diodes from turning on for better reliability of the rectifier [7-8].

A. Semi-Active Leakage Control Comparator

For the NP-cross-coupled rectifier, the transistors can be regarded as variable resistors, of which values vary in a non-linear fashion as the operating regime alternates between saturation and active region. The desired characteristic for improved efficiency is to have a low RC time constant during the charging period when the output

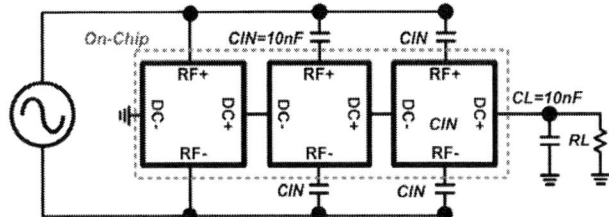

Fig. 5. Measurement setup

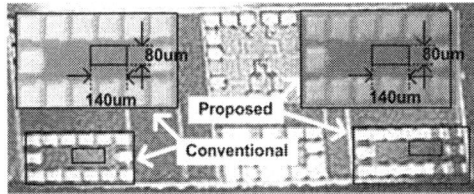

Fig. 6. Die photo of the proposed and conventional rectifier.

voltage is lower than the input voltage. Conversely, a high time constant is desired during the charge leakage period.

To adjust the time constant adaptively, the leakage control comparator is utilized. This comparator takes one of the RF signals (RF+ or RF-) as its positive input and the rectified node as its negative input. Since the negative input also acts as a virtual supply, this comparator does not require an external supply. When RF+ (or RF-) is higher than DC+, the output of CP1 (or CP2) goes low. This reduces the RC time constant associated with MRP1 (or MRP2) and helps the charging process. Conversely, the RC time constant becomes higher when RF+ (or RF-) becomes smaller than DC+. In this case, CP1 (or CP2) raises its output voltage to decrease the gate drive voltage of MRP1 (or MRP2).

B. Active Body Biasing

As mentioned earlier, there is no stable supply voltage to tie up the body terminals of PMOS transistors. For example, when the body of MRP1 and MRP2 is tied to DC+, RF+/RF- can be significantly higher than DC+ at the beginning of the rectifying operation. If this voltage difference beats the threshold voltage of the parasitic

978-1-4244-6240-7/10 $26.00 © 2010 IEEE   85

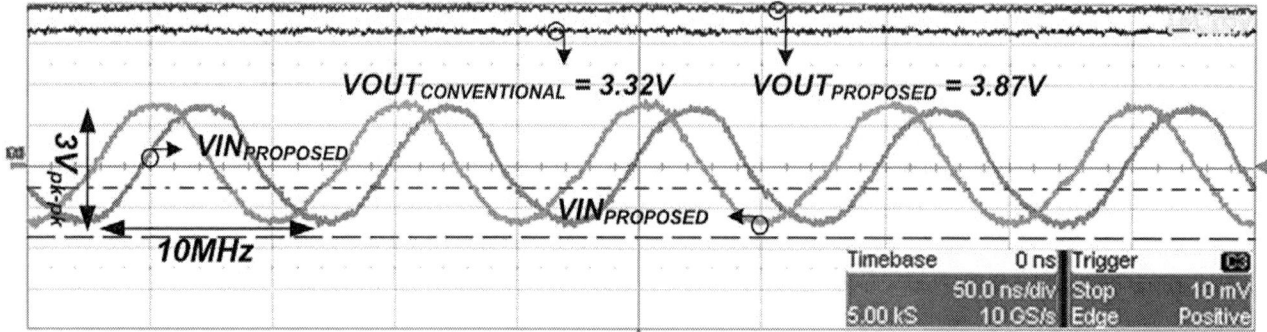

Fig. 7. Measured voltage waveform of the proposed and conventional rectifier when $VRF_{AMP} = 1.5V$ and $f = 10MHz$.

body junction diode, currents can flow into the body and damage the device.

Therefore, the ABB cell is applied to ensure the body of MRP1 and MRP2 is always tied to the higher potential of RF+/RF- or DC+.

IV. EXPERIMENTAL RESULTS

The proposed 3-stage SA-HE rectifier and a 3-stage NP-cross-coupled rectifier were implemented in a 0.18-um CMOS technology. The measurement setup is shown in Fig. 5, and the die photo is shown in Fig. 6. For fair comparison, the cross-coupled pairs in the two rectifiers are sized equally and the same PCB setup was used. Fig. 7 shows a transient waveform of the inputs and outputs of the proposed and the other rectifier when the input frequency and amplitude are 10 MHz and 3 Vpk-pk, respectively. As the proposed SA-HE rectifier prevents leakage currents, the amount of the leakage currents flowing in the conventional rectifier can be saved and this leads to improved conversion efficiency. The output voltages were measured as the input amplitudes were

swept as in Fig. 8. The proposed SA-HE rectifier shows approximately 15% higher voltage conversion efficiency.

V. CONCLUSION

In this paper, a SA-HE CMOS rectifier was presented. Measurements show that the semi-active leakage control comparator effectively prevents the reverse charge leakage, and enhances the conversion efficiency. In addition, by adopting the ABB technique, the reliability of the proposed rectifier system can be improved.

REFERENCES

[1] W. Brown "The history of power transmission by radio waves," *IEEE Trans. Microwave Theory & Tech.*, vol. 32, no. 9, pp. 1230-1242, Sept. 1984.

[2] M. Ghovanloo, et al., "Fully integrated wideband high-current rectifiers for inductively powered devices," *IEEE JSSC*, vol. 39, no. 11, pp. 1976-1984, Nov. 2004.

[3] S. Mandal, et al., "Low-Power CMOS rectifier design for RFID applications," *IEEE Trans. Circuits Systs. I*, vol. 54, no. 6, pp. 1177-1188, Jun. 2007.

[4] Y.-H. Lam, et al., "Integrated low-loss CMOS active rectifier for wirelessly powered devices," *IEEE Trans. Circuits Systs. II*, vol. 53, no. 12, pp. 1378-1382, Dec. 2006.

[5] C. Peters, et al., "CMOS integrated highly efficient full wave rectifier," *Proc. IEEE Int. Symp. circuits Syst*, pp. 2415-2418, May 2007..

[6] T. Umeda, et al., "A 950MHz rectifier circuit for sensor networks with 10M distance," *IEEE JSSC*, vol. 41, no. 1, pp. 35-41, Jan. 2006.

[7] J. Cha, et al., "Analysis and design techniques of CMOS charge-pump-based radio frequency antenna switch controllers," *IEEE Trans. Circuits Systs. I*, vol. 56, no. 5, pp. 1053-1062, May 2009.

[8] J. Shin, et al., "A new charge pump without degradation in threshold voltage due to body effect," *IEEE JSSC*, vol. 35, no. 8, pp. 1227-1230, Aug. 2000

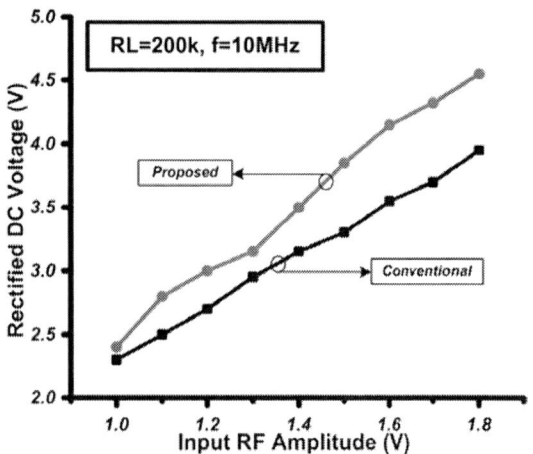

Fig. 8. Measured DC voltages varying input amplitudes

RMO2A-2

A Single-Chip CMOS UHF RFID Reader Transceiver

Runxi Zhang, Chunqi Shi, Yihao Chen, Wei He, Ping Xu, Shuai Xu and Zongsheng Lai

IMCS, East China Normal University, Shanghai, China, Dept. of Electronic Engineering, Suzhou Vocational University, Suzhou, China

Abstract — A novel single-chip 860-960MHz band UHF RFID reader transceiver IC is fabricated in 0.18μm CMOS technology. The transceiver consists of a compact high-linearity low-noise-figure RF front-end, a programmable analog baseband for Rx path; and an image reject filter, a PGA, a switchable up-conversion modulator and a driver amplifier for Tx path. The 3-bit 3rd-order DSM fractional-N frequency synthesizer with optimized all-band phase noise performance is integrated to implement Rx/Tx frequency translation. The transceiver IC has a die area of 4mm×4mm. It achieves IIP3 of 13dBm, sensitivity of -75dBm in listen-before-talk (LBT) mode and -66dBm in normal mode in the presence of -4.4dBm self-jammer for the backscatter modulation while drawing 112mA from 3.3V power supply.

Index Terms — Single-chip, UHF RFID, reader transceiver

I. INTRODUCTION

With the rapid growth of demanding for 860-960MHz band UHF RFID application systems, the significant research effort is dedicated to the fabrication of single-chip UHF RFID reader IC in the past three years [1-6]. Among various solutions, the standard CMOS technology and the direct-conversion architecture is popular in reducing cost and power consumption and improving integration level [2-5]. During each communication rounds, the reader sequentially transmits continuous-wave and useful commands, such as Select and Query, to activate and configure passive tags, which is mainly regulated by EPC C1G2 protocol [7]. Inevitably, comparing with the feeble backscatter signal, a huge in-band self-jammer coupled from the Tx port will deteriorate and even saturate the Rx path. The large power difference between the designed signal and the leakage interferer proposes a prime challenge in the design of monolithic UHF RFID reader transceiver IC, especially in Rx Path.

Some schemes with higher isolation using two antenna [8] or on-chip intelligent canceller [9] is preferred in performance, but bulky in physical size or costly due to needing additional calibration control [4]. Similar with the traditional strategy of accommodating the large leakage interference in Rx Path [1-6], we proposes a novel I/Q direct-conversion UHF RFID reader transceiver. Section II describes system architecture and critical requirements like noise figure, input linearity, setting-up time and phase-noise performance. Sections III and IV discuss circuit details and measurement results, respectively.

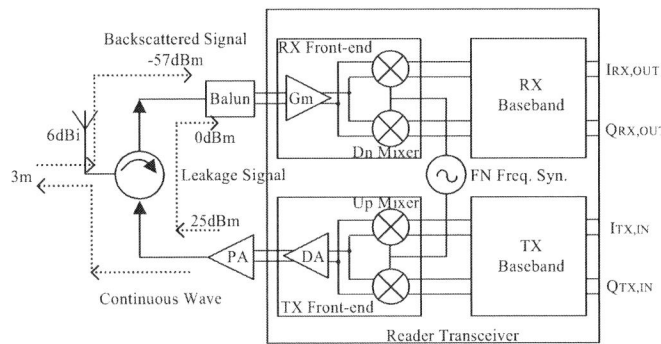

Fig. 1. Block diagram of the UHF RFID system

II. SYSTEM ARCHITECTURE AND REQUIREMENTS

Fig. 1 illustrates the single antenna UHF RFID system block diagram using the proposed direct-conversion reader transceiver. The complete tag identification is finished with some additional discrete components such as antenna, power amplifier and protocol controller. The I/Q Tx structure is designed to achieve the selectable modulations consisting of DSB-, SSB- and PR-ASK. The SSB-ASK is very valuable for compressing bandwidth and improving channel utilization. The I/Q Rx structure is used to eliminate the zero-effect in terms of UHF RFID applications for different phase delay from variable operation distance. In fact, the I/Q Rx down-conversion scheme is also preparing for the future PSK demodulation. After the detection of the anticipant sub-band is finished and the vacant channel is arbitrated, the reader converts from LBT to normal mode. In consequence of that, there are two different linearity and noise figure requirements of the reader Rx path in each mode for distinct operation means. In general, the limited LO phase noise performance would significantly exacerbate Rx input noise floor and the transmit-to-receive turnaround time from reader transmission to tag answer would essentially decide the necessary setting-up time of reader transceiver.

A. Noise Figure

Assuming a minimum SNR of 11dB for 10^{-5} BER [2], the reader sensitivity in LBT mode is calculated as follows

$$NF = 174 - 10\log BW_n - SNR + P_{sensitivity} \quad (1)$$

where BW_n is the channel bandwidth, $P_{sensitivity}$ is only referred in ETSI 302 208-1 standard [10] and up to -83dBm,

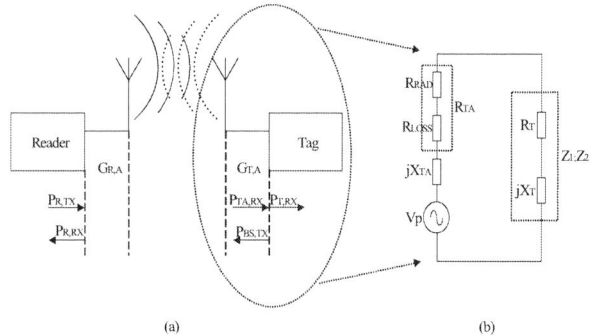

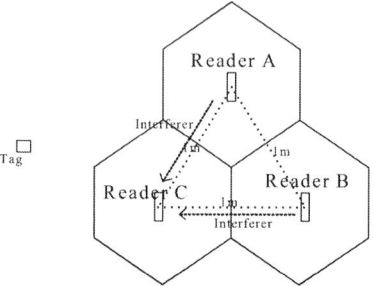

Fig. 2. (a) Reader-tag power transmission model; (b) power transmission model between tag antenna and tag IC

-90dBm and -93dBm for 20dBm, 27dBm and 30dBm maximum PA output power. And then, the LBT mode noise figure for Rx path should be 15dB according to Fig. 1. Fig. 2 illustrates the power transmission model between tag and reader and tag antenna and tag IC. Tag-to-reader backscatter data received by reader can be calculated as

$$P_{R,RX} = P_{R,TX} \times G_{R,A}^2 \times G_{T,A}^2 \times \left(\frac{\lambda}{4\pi r}\right)^4 \times \xi \qquad (2)$$

where $P_{R,TX}$ is output power of the reader, r is operation distance between tag and reader and $G_{R,A}$ and $G_{T,A}$ are the antenna gain of reader and tag. The symbol ξ indicates the modulation efficiency, which is decided by energy distribution scheme of tag IC and related with modulation types [11]. When the ASK1 type is assumed, the tag antenna radiation efficiency is 1, two impedance states emerge with equal probability, the output power is up to 25dBm, $G_{R,A}$ and $G_{T,A}$ are equal to 6dBi and 2dBi and 10μW (-20dBm) is needed by powering on tags, similar with eq. (1), the normal mode reader Rx sensitivity should attain -67dBm. This result is calculated when 10dB margins is included for the undesired reflection and diffraction in practical applications. And then, the normal mode noise figure for reader Rx path should be 35dB.

B. Input Linearity

A multiple reader environment is illustrated in Fig. 3. Reader A and B are communicating with tag group, reader C is entering into this cell environment. The distance between the center to adjacent cells is one meter. The spacing under one meter is costly and inefficient. In LBT mode, the reader C only receives without sending large power continuous-wave to tags. The undesired interferer located in adjacent channel for reader C will degenerate Rx gain as

$$a_1 \cdot V_d + \frac{3 \cdot a_3}{4} V_d^3 + \frac{3 \cdot a_3}{2} V_I^2 \cdot V_d = 0 \qquad (3)$$

where V_d is the desired signal, V_I is the interferer in adjacent channel and a_1, a_2, a_3 are nonlinear coefficients. The interferer power on reader C input can be written as

Fig. 3. LBT interference in multiple reader environment

$$P_C = P_A \times G_{R,A} \times G_{R,C} \times \left(\frac{\lambda}{4\pi r}\right)^2 \qquad (4)$$

where P_A is the output power of reader A, $G_{R,A}$ and $G_{R,C}$ are the antenna gain of reader A and C. From eq. (3), eq. (4) and the tag-reader communication power budget illustrated in Fig. 1, the reader C Rx path input 1-dB compression point in LBT mode is up to -1dBm. The reader C Rx path input 1-dB compression point for normal mode should be up to 0dBm in order to endure the in-band self-jammer. The self-jammer is down-converted to DC for the direct-conversion architecture and removed by DC-offset canceller prior to Rx baseband.

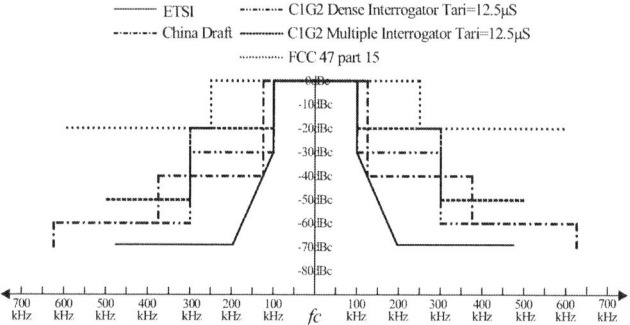

Fig. 4. Tx path transmission spectrum mask

C. Phase Noise

Fig. 4 illustrates the Tx transmission spectrum mask in accordance with protocols [8][10][12][13]. Apart from the ACPR requirements, the LO phase noise performance is also determined by the Rx path capability for withholding adjacent interferer and in-band self-jammer. The former is known as reciprocal-mixing effect and the latter can be described as

$$NF = NF_{fe} + \sin(\Delta\varphi) \cdot \frac{\int_0^B L(\Delta f) \cdot \overline{V_c^2} d(\Delta f)}{4kTR \cdot B} \qquad (5)$$

where NF_{fe} is the native noise figure of Rx path, $\Delta\varphi$ is the phase difference between LO signal and self-jammer, $L(\Delta f)$ is the phase noise property of LO and the 2nd polynomial of eq. (5) describes the deterioration effect of self-jammer. The key phase noise is concluded as -126dBc/Hz at 1MHz

TABLE I A SINGLE TAG REPLY FOR THE MAXIMUM RATE

Type	Command, Symbol or Delay	Maximum Time	Cumulative Time
R-T	Preamble+Query or Frame-sync+Queryrep	12.5µS+4Tari+4Tari	62.5µS
	T1	10Tpri	79µS
T-R	Preamble+RN16	(6+16)Tpri	114µS
	T2	20Tpri	145µS
R-T	FS+ACK	12.5µ+4Tari+18Tari	295µS
	T1	10Tpri	311µS
T-R	P+PC+EPC+CRC5	(6+16+96+5)Tpri	511µS
	T2	20Tpri	542µS

offset from center frequency and the integrated phase noise from 10kHz to 10MHz should be less than -41dBc.

D. Setting-up Time

A single tag reply breakdown has been shown in Table I. When the BLF is 640kHz, the Tpri is 1.56µS, the Tari is 6.25µS, the TRext is 0 and the CRC employs 5bits, the cumulative turnaround time from transmit to receive for the maximum rate and the shortest preamble is about 79µS. Fortunately, the LO channel switching and the Rx path DC-offset cancellation has started simultaneously during the reader transceiver setting-up progress.

III. IMPLEMENTATION OF CRITICAL BUILDING BLOCKS

A. Fractional-N Frequency Synthesizer

The fractional-N frequency synthesizer based on 3-bit 3rd-order delta-sigma modulator is shown in Fig. 5 [5]. The single-loop multiple-feedback architecture with the configured zero is used to suppress middle-band phase noise from DSM. The converged DSM output data distribution is helpful for decreasing coupling noise from substrate. All digital implementation of DSM is able to reduce the cost and size. The noise energy pushed up to higher frequency band is further attenuated by the off-chip passive RC filter for flexible deployment. The double-switch LC-VCO with LDO power supply and current noise filtration is designed to achieve better noise performance which decides the out-of-band phase noise property of frequency synthesizer.

B. Compact Three-stage Rx Front-end

Fig. 6 illustrates the compact Rx front-end consisting of transconductance, switching and transimpedance. The fully differential structure reduces on-chip noise coupling from digital parts. The conventional LNA+Mixer scheme only achieves limited Rx input linearity and the LNA has to be switch-off in normal mode for accommodating the self-jammer [2][3][4]. The complementary transconductance input stage is able to hold circuit with single node nonlinearity which is valuable for Rx path. The passive

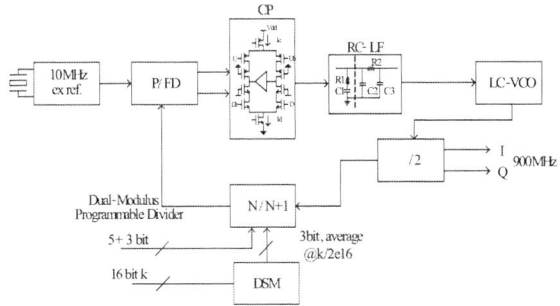

Fig. 5. Fractional-N frequency synthesizer

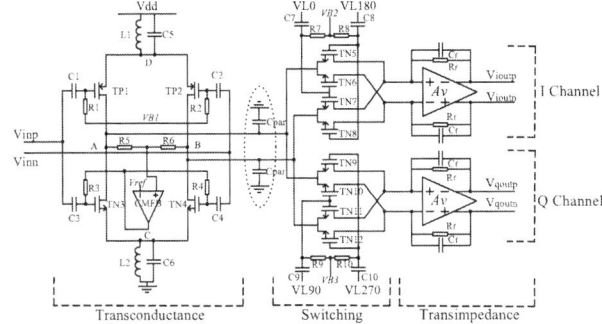

Fig. 6. Compact three-stage Rx front-end

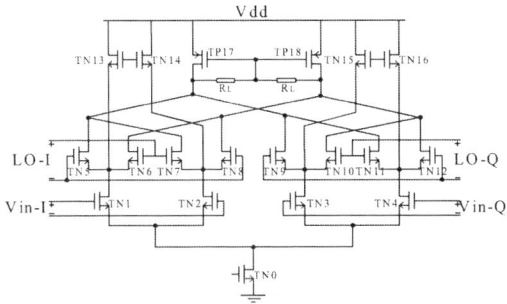

Fig. 7. Switchable I/Q direct-conversion Tx modulator

I/Q mixer biased on "on-overlap" mode benefits the front-end for lower noise power and nonlinearity of mixer. The operational-amplifier-RC transimpedance dissipates lower power for same gain due to lower operation frequency.

C. Switchable I/Q Direct-conversion Tx Modulator

Fig. 7 illustrates the switchable I/Q direct-conversion Tx modulator supporting DSB-, SSB- and PR-ASK modulations. The I/Q signals are generated in digital baseband with Hilbert transformation. These two signals with DC level are passed through Tx path using DC-coupling in order to meet with variable modulations. Current-bleeding is to trade-off between noise and gain [14].

IV. MEASUREMENT RESULTS

The reader transceiver chip is fabricated in IBM 0.18µm CMOS process. The die photograph is illustrated in Fig. 8. The reader transceiver is optimized for 835-930MHz and

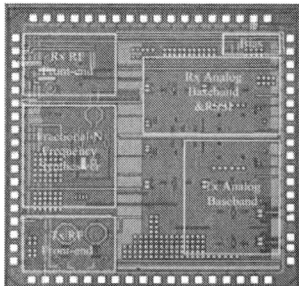

Fig. 8. Die microphotograph

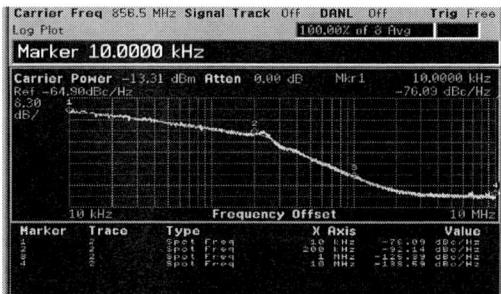

Fig. 9. Frequency synthesizer phase noise performance

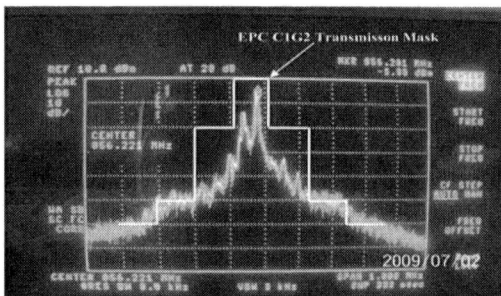

Fig. 10. Reader SSB-ASK modulation transmission spectrum

draws 112mA from 3.3V power supply. The die size is 4mm×4mm and packaged into LQFP64A. An xc4vx35 FGPA is used to supply digital signal processing like coding, checking and protocol controlling. Fig. 9 illustrates the phase noise performance of synthesizer. The spot phase noise is -92 and -125dBc/Hz at 0.2 and 1MHz offset from 856.5MHz operating frequency. The integrated phase noise from 10kHz to 10MHz is -39dBc. The RMS jitter is 6.9pS, the residual FM and PM are 4.5kHz and 2°. The Rx IIP3 and noise figure are 13dBm and 35dB in the presence of -4.4dBm in-band self-jammer. Fig. 10 depicts reader transmission spectrum with SSB-ASK modulation, 40kbps PIE coding and 10dBm output. Fig. 11 is the tag-to-reader response signal when the Select and Query commands are "1010, 000, 000, 01, 0000, 0000, 0000, 0000, 0, CRC16" and "1000, 0, 00, 1, 00, 00, 0, 0001, CRC5". The inset in Fig. 11 shows the RN16 details.

V. CONCLUSION

A single-chip CMOS UHF RFID reader transceiver in

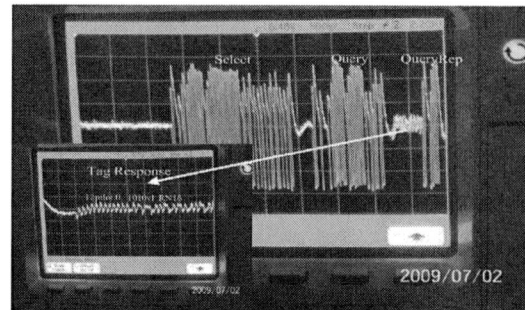

Fig. 11. Tag-to-reader response RN16

accordance with EPC C1G2, ETSI 302 208-1 and China draft is proposed in this paper. The key problems including noise figure, input linearity, phase noise and setting-up time are recapitulated for system planning. It can fulfill requirement of portable low-power low-cost UHF RFID systems.

REFERENCES

[1] S. Chiu, I. Kipnis, M. Loyer, J. Rapp, D. Westberg, J. Johansson, and P. Johansson, "A 900MHz UHF RFID reader transceiver IC," *IEEE J. Solid-State Circuits*, vol. 42, no. 12, pp. 2822–2833, 2007.

[2] I. Kwon, H. Bang, K. Choi, S. Jeon, S. Jung, D. Lee, Y. Eo, H. Lee, and B-Y. Chung, "A single-chip CMOS transceiver for UHF mobile RFID reader," *IEEE J. Solid-State Circuits*, vol. 43, no. 3, pp. 729–738, 2008.

[3] P. B. Khannur, X. S. Chen, D. L. Yan, D. Shen, B. Zhao, M. K. Raja, Y. Wu, R. Sindunda, W. G. Yeoh, and R. Singh, "A universial UHF RFID reader IC in 0.18μm CMOS technology", *IEEE J. Solid-State Circuits*, vol. 43, no. 5, pp. 1146–1155, 2008.

[4] W. T. Wang, S. Z. Lou, K. W. C. Chui, S. J. Rong, C. F. Lok, H. Zheng, H. T. Chan, S. W. Man, H. C. Luong, V. K. Lau, and C. Y. Tsui, "A single-chip UHF RFID reader in 0.18μm CMOS process," *IEEE J. Solid-State Circuits*, vol. 43, no. 8, pp. 1741–1751, 2008.

[5] R. X. Zhang, C. Q. Shi, and Z. S. Lai, "A low-phase-noise frequency synthesizer for single-chip CMOS UHF RFID reader", *IEEE ICMMT2008 Proceeding*, vol.3, pp.1477-1480.

[6] H. H. Roh, and J. S. Park, "Differential colpitts VCO for enhancing tramnsceiver performance in specified UHF mobile RFID environment conditions", *IEEE Trans. Microw. Theory Tech.*, vol. 56, no. 11, pp. 2662–2670, 2008.

[7] EPC radio-frequency identity protocols C1G2 UHF RFID protocol for communications at 860-960MHz. ver. 1.1.0, EPCglobal, 2005.

[8] Intel UHF RFID transceiver R1000: electrical, mechanical, and thermal specification. Intel, Mar., 2007.

[9] A. Safarian, A. Shameli, A. Rofougaran, M. Rofougaran, F. D. Flaviis, "An integrated RFID reader," *IEEE Solid-State Circuits Conference Proceedings*, 2007, pp.218-219.

[10] ETSI 302 208-1 ERM; RFIE operating in the band 865 MHz to 868 MHz with power levels up to 2 W; Part 1: Technical requirements and methods of measurement, ver. 1.1.1, ETSI, 2004.

[11] U. Karthaus, M. Fischer, "Fully integrated passive UHF RFID transponder IC with 16.7-μW minimum RF input power," *IEEE Journal of Solid State Circuits*, vol.38, no.10, pp. 1602-1608, 2003.

[12] 800/900MHz RFID application tech. (Draft).MIIT, no.205, 2007.

[13] FCC 247 CFG Ch.1 Part 15.

[14] S. G. Lee, J. K. Choi, "Current reuse bleeding mixer." *IEE Electronics Letters*, vol. 36, no. 8, pp. 696–697, 2000.

Near Zero Turn-on Voltage High-Efficiency UHF RFID Rectifier in Silicon-on-Sapphire CMOS

Paul T. Theilmann[1], Calogero D. Presti[1], Dylan Kelly[2] and Peter M. Asbeck[1]

[1]University of California, San Diego, La Jolla, CA 92093, USA
[2]Peregrine Semiconductor, San Diego, CA 92121, USA

Abstract — A UHF RFID rectifier which turns on at near zero input voltage is demonstrated. The rectifier is fabricated in 0.25-µm silicon-on-sapphire (SOS) CMOS technology using intrinsic, near zero threshold devices. A novel improved cross-coupled bridge topology is used to minimize the leakage incurred through the use of intrinsic devices while maintaining their low power turn on characteristics. The fabricated rectifier demonstrates a peak power conversion efficiency (PCE) of 71.5% at 915MHz with a RF input of -4 dBm and a 30 kΩ load. More importantly, a PCE > 30% was measured for all RF input powers between -28 and -4 dBm demonstrating state-of-the-art efficiency across a wide range of input powers.

Index Terms — AC-DC power conversion, CMOS integrated circuits, power conversion efficiency (PCE), radio frequency rectifier, RFID, ultra-high frequency (UHF), wireless power transmission.

I. INTRODUCTION

Passive UHF RFID tags utilize a rectifier to convert incoming RF energy to DC energy to power the device. The rectifiers currently used in UHF tags utilize standard threshold voltage (V_{th}) CMOS devices [2] or low V_{th} Schottky diodes ($V_{th} < 200mV$) [3], [4]. Unfortunately if the voltage at the input of the tag is lower than the turn on voltage of the rectifier, which is directly related to the V_{th} of the devices used, the tag will not power up and will not be read. The power and thus voltage seen at the rectifier can be degraded by many factors including distance from the reader to the tag, detuning of the tag antenna due to the proximity of liquids or metals and many other unintended circumstances which can occur in the supply chain [1]. Thus rectifiers which are able to efficiently convert low RF voltages to DC voltages are highly desirable.

In the quest for lower turn on voltages, V_{th} cancellation techniques have been proposed [5]-[7]. In [5] a static technique was utilized which while successful, required a semi-active tag design. The design reported in [6] and [7] uses a passive cross coupled differential circuit configuration which led to a record PCE of 67.5%, but only achieves complete Vth cancellation at a single input power level. State-of-the-art UHF rectifiers generally achieve peak PCE values around 30% [3], [5].

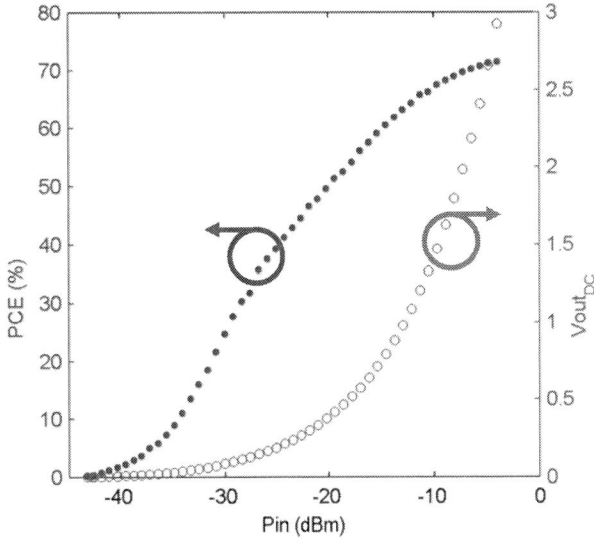

Fig. 1. Measured power conversion efficiency and DC output voltage vs. RF input power at 915 MHz.

In this work, we use near zero threshold transistors available in a 0.25-µm silicon-on-sapphire (SOS) CMOS process. A novel cross-coupled bridge topology is implemented to achieve a low turn-on, high PCE design. The measured PCE and DC output voltage are shown in Fig. 1. Using a 30k load, a peak PCE of 71.5% at 915MHz with a RF input power of -4dBm was obtained. Furthermore, a PCE > 30% was measured for all input powers between -28 and -4dBm. These results not only demonstrate exceptionally high peak efficiency, but also state-of-the-art PCE over a wide dynamic power range which will lead to RFID tag robustness with respect to varying power availability.

II. CROSS-COUPLED CMOS BRIDGE RECTIFIER AND ZERO THRESHOLD DEVICES

Fig 2. shows the schematic of a standard cross coupled CMOS bridge rectifier [2], [6], [7]. A differential input signal is applied across nodes V_{inRF+} and V_{inRF-} and a DC output voltage develops across the load impedance Z_{load}. The operation of this rectifier can be understood by tracing through a single RF cycle. As V_{inRF+} - V_{inRF-} increases beyond the device threshold, M3 and M2 will switch on

978-1-4244-6240-7/10 $26.00 © 2010 IEEE

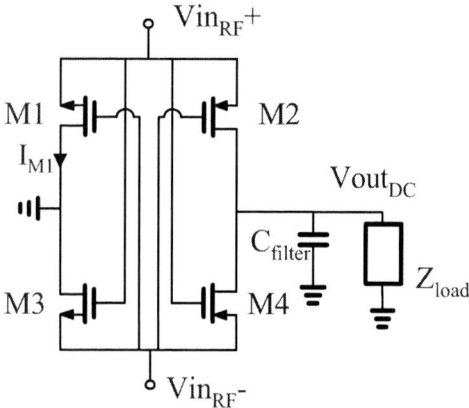

Fig. 2. Standard CMOS cross-coupled bridge rectifier schematic.

allowing current to flow into the load while M1 and M4 will remain off. Continuing through the cycle as V_{inRF+} - V_{inRF-} drops below the device threshold M3 and M2 will turn off and no current will flow to the output until V_{inRF+} - V_{inRF-} becomes more negative than $-V_{th}$, at which point M1 and M4 turn on rectifying the negative half of the incoming RF signal. It is important to note that if $|V_{inRF+} - V_{inRF-}|$ never achieves a value greater than V_{th}, V_{outDC} will remain 0V.

Intrinsic devices, sometimes referred to as native devices, are created by blocking the channel implantation process steps to create a channel with very low doping, which generates devices with near zero thresholds. In twin-tub and silicon-on-insulator processes such as Peregrine's SOS UltraCMOS, both PMOS and NMOS intrinsic devices can be fabricated. The resulting threshold voltages for the devices used in this paper are approximately 80mV for the NMOS and -50mV for the

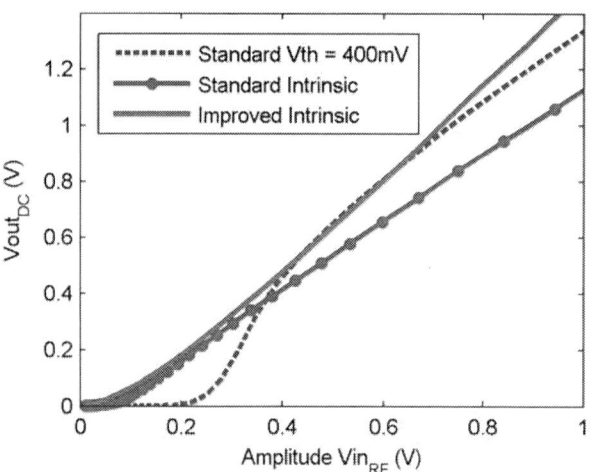

Fig. 3. Simulated DC output voltage vs. RF input voltage of a cross coupled bridge rectifier using standard threshold devices and near zero threshold intrinsic devices. Results for the improved efficiency design with intrinsic devices are also show.

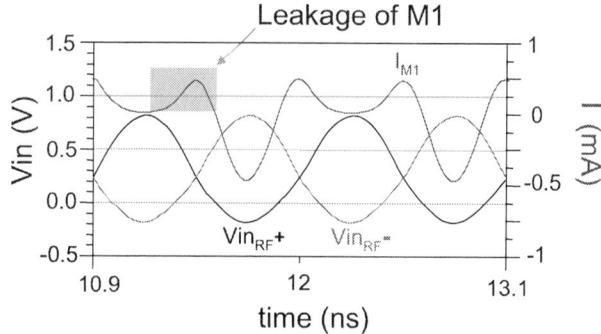

Fig. 4. Input voltage waveforms and current waveform for transistor M1.

PMOS transistors.

In Fig. 3 simulated results of $Vout_{DC}$ versus the amplitude of Vin_{RF} are presented. Compact 0.25-μm SOS models were used. A PMOS width of 45-μm, a NMOS width of 30-μm, a C_{filter} = 0.5pF and a load resistance of 30k were selected. The simulation was performed at 915MHz and results for both standard threshold (V_{th} ~400mV) and intrinsic devices are shown. The poor low voltage performance of the rectifier using standard devices is quite evident in the figure. The zero threshold design performs much better at these lower values, but as the input voltage rises further the zero threshold design performs poorly in comparison to the standard threshold design. The reason for this is illustrated in Fig. 4 which shows the time waveforms from the zero threshold simulation.

As a DC voltage builds across the load, the Vin_{RF}+ and Vin_{RF}- nodes develop a DC voltage with respect to the ground node. Thus when Vin_{RF}+ - Vin_{RF}- = 0, Vin_{RF}+ and Vin_{RF}- • 0. We can look at what effect this has with respect to M1 and the same analysis can be applied to M2-M4. When Vin_{RF}+ - Vin_{RF}- = 0, M1 should be turned off and I_{M1} should be 0 so energy at the load does not leak to ground. Since Vin_{RF}+ - Vin_{RF}- = 0, the gate to source voltage (VGS) is 0. Yet the gate voltage (Vin_{RF}-) > 0, since the drain is connected to ground, the gate to drain voltage, (VGD) > 0. If VGD > V_{th} a channel is formed and current flows to ground, because CMOS devices are symmetric the drain and source have essentially flipped. For devices where V_{th} = 0, this leakage occurs for very low DC offsets and occurs over a greater portion of the RF cycle than standard devices. This current leakage reduces $Vout_{DC}$ for a given load impedance.

III. ENHANCED EFFICIENCY ZERO TURN-ON VOLTAGE RECTIFIER

To suppress this leakage current while retaining the low turn-on levels of the zero-threshold design, the circuit topology has been modified by the addition of four more

978-1-4244-6240-7/10 $26.00 © 2010 IEEE

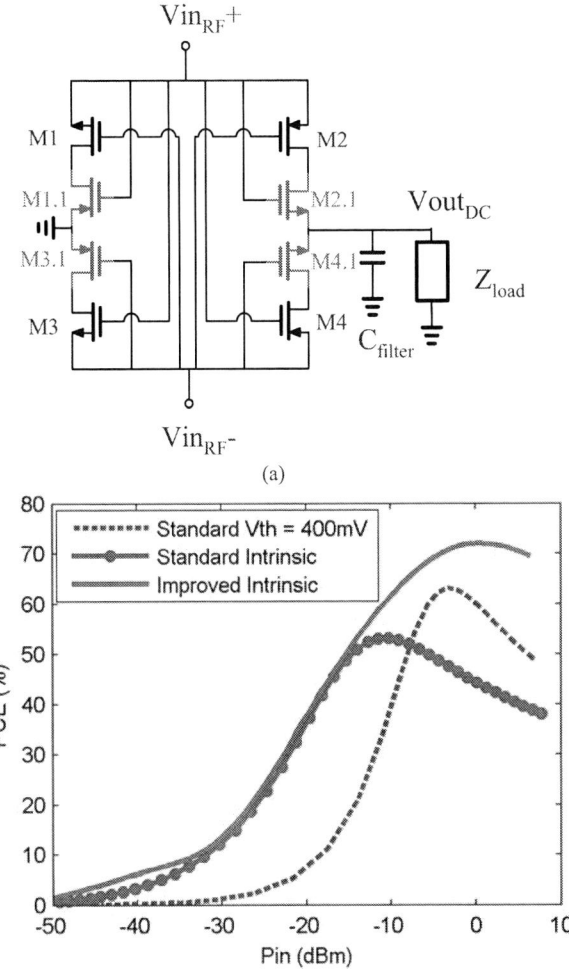

(a)

(b)

Fig. 5. (a) Improved CMOS cross-coupled bridge rectifier with added efficiency enhancing transistors M1.1, M2.1, M3.1 and M4.1. (b) Simulated PCE vs. RF input power for the standard cross coupled bridge rectifier as well as the proposed higher efficiency design at 915MHz.

zero-threshold transistors as shown in Fig. 5a. The function of the new topology can be understood by looking at M1.1 when $Vin_{RF}+ - Vin_{RF}- = 0$ and $Vin_{RF}+$, $Vin_{RF}- > 0$. VGS will be > 0 turning off M1.1, which reduces current flow through M1 to ground, directly improving efficiency. The added transistor does not disrupt the functionality of the circuit because, when M1 should turn on, $Vin_{RF}+$ drops below the ground node turning M1.1 on.

The benefits of this new design can be seen in Fig. 3. The load components and NMOS and PMOS sizes were kept the same as the standard designs. The rectifier retains the improvements provided by the intrinsic devices at low input voltages and does not suffer the losses which result in lower output voltage at higher input levels.

It is also important to look at the power efficiency of a rectifier when quantifying the differences between designs. Rectifier efficiency is most commonly defined by PCE which is computed as follows:

$$PCE = \frac{Pout_{DC}}{Pin_{RF}} = \frac{Vout_{DC} \cdot Iout_{DC}}{(1/2) \cdot Vin_{RF} \cdot Iin_{RF}} \cdot 100(\%) . \quad (1)$$

It should be noted that in a real system the rectifier can only extract all of the power from an antenna and simultaneously obtain this PCE if the antenna is ideally matched to the rectifier.

The PCE results of the simulations performed for Fig. 3 are shown in Fig 5b. The improved efficiency with respect to the standard design with standard devices at low input levels can be attributed to the use of intrinsic devices since both designs with zero threshold devices achieve similar low power PCE. The improvements at higher power levels are due to the novel topology. As the input voltage level rises output leakage in the standard design rises, this causes the decrease in PCE seen in Fig. 5b. These losses are greatly reduced in the improved design.

IV. EXPERIMENTAL VERIFICATION AND DISCUSSION

A test chip was fabricated using Peregrine's 0.25-μm SOS UltraCMOS technology. Intrinsic devices available in the process were used to obtain near zero threshold voltage. A single stage design like the one shown schematically in Fig. 5a was created using NMOS transistors of size 30-μm/0.25-μm and PMOS transistors of size 45-μm/0.25-μm. The relative PMOS to NMOS size was selected for approximately equal on-state resistance and absolute size was optimized through simulation to obtain output powers around the 10μW to 100μW range as is generally required by UHF tags [7]. A load resistance of 30k was used with an on chip MIM ripple filter capacitor of 0.5pF. A die photo of the fabricated chip is shown in Fig. 6, the chip size including pads is 350-μm x 380-μm with an active area of only 52.5-μm x 40.5-μm. The parallel MIM capacitors which form the output filter capacitor are placed directly on top of the transistor structures to minimize area.

A VNA (Agilent PNA-X) was used to drive the rectifier with balanced differential signals while simultaneously measuring balanced S-parameters. Input power was swept while input frequency was held at a constant 915 MHz. The input impedance was measured to be approximately 94-j928 at Pin = -10dBm, it varied slightly with input power level.

DC output voltage versus RF input voltage is shown in Fig. 7. The rectifier turns on and produces a DC output voltage at near zero RF input voltage amplitudes,

978-1-4244-6240-7/10 $26.00 © 2010 IEEE

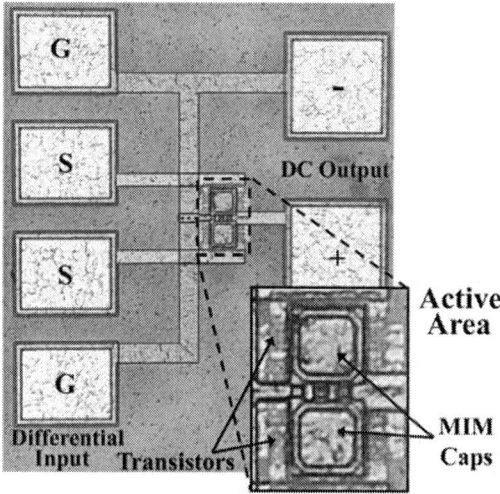

Fig. 6. Die photo of the improved zero threshold voltage rectifier in 0.25um SOS CMOS. Chip size 350μm x 380μm, active area 52.5μm x 40.5μm.

indicating that the rectifier is able to convert very small signals to usable DC power. To quantify PCE, input power to the rectifier must be measured. Pin_{RF} can be calculated using the differential S-parameters measured during the power sweeps used to measure DC output voltage [7].

$$Pin_{RF} = P_A \cdot (1 - |S_{dd11}|^2 - |S_{cd11}|^2) \qquad (2)$$

where P_A is the power available from the source and S_{dd11} and S_{cd11} are the differential-to-differential and differential-to-common mode reflection coefficients.

Fig. 1 displays the PCE and DC output voltage of the rectifier with respect to Pin_{RF}. The fabricated rectifier achieves excellent PCE over a much larger Pin_{RF} range than other recently reported UHF rectifiers [5-7]. The peak measured PCE was 71.5%. Up until very recently a peak PCE of 30% was considered state-of-the-art for CMOS UHF rectifiers [3], [5], [7]. The fabricated rectifier achieves a PCE greater than 30% for $Pin_{RF} = -28$ to -4dBm. In real world situations input power to the rectifier varies significantly, thus achieving high PCE over a wide dynamic range leads to a more robust tag with a longer read range and higher percentage read rate.

VII. CONCLUSION

A high efficiency UHF RFID rectifier has been demonstrated in 0.25-μm SOS CMOS technology. The rectifier uses intrinsic devices to enable conversion of near zero input voltages. The rectifier achieves a maximum PCE of 71.5% at 915MHz with a $Pin_{RF} = -4dBm$. An enhanced cross coupled design is used to achieve state of the art >30% PCE over a wide dynamic range of $Pin_{RF} = -$

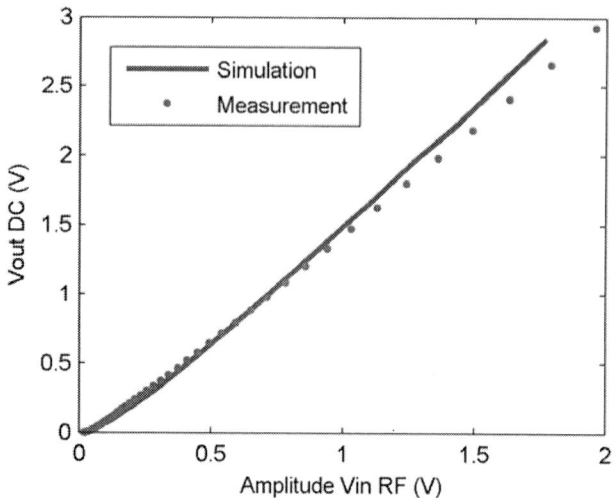

Fig. 7. Measured DC output voltage versus RF input amplitude for the improved zero threshold design at 915MHz.

28 to -4 dBm. The designed rectifier's ability to efficiently convert low input RF power/voltage to usable DC energy will improve the read range and percentage read rates of UHF RFIDs.

ACKNOWLEDGEMENT

The authors would like to thank Peregrine Semiconductor for chip fabrication and support.

REFERENCES

[1] K. Finkenzeller, *RFID Handbook: Fundamentals and Applications in Contactless Smart Cards and Identification*, 2nd ed. Chicester, U.K.: Wiley, 2003.

[2] Z. Zhu, B. Jamali, and P. H. Cole, Brief comparison of different rectifier structures for RFID transponders. Internal Rep. of Auto-ID Lab at University of Adelaide, 2004 [Online].Availible:http://autoidlab.eleceng.adelaide.edu.au/Papers/

[3] U. Karthaus and M. Fisher, "Fully integrated passive UHF RFID transponder IC with 16.7-μW minimum RF input power," *IEEE J. Solid-State Circuits*, vol. 38, no. 10, pp. 1602-1608, Oct. 2003.

[4] J.-P. Curty, N. Joehl, F. Krummenacher, C. Dehollain, and M. j. Declercq, "A model for μ-power rectifier analysis and design," *IEEE Trans. Circuts Syst. I, Reg. Papers*, vol. 52, no. 12, pp. 2771-2779, Dec. 2005

[5] T. Umeda, H. Yoshida, S. Sekine, Y. Fujita, T. Suzuki, and S. Otaka, "A 950-MHz rectifier circuit for sensor network tags with 10-m distance," *IEEE J. Solid-State Circuits*, vol. 41, no. 1, pp. 35-41, Jan. 2006.

[6] S. Mandal, R. Sarpeshkar, "Low-power CMOS rectifier design for RFID applications," *IEEE Trans. Circuts Syst. I, Reg. Papers*, vol. 54, no. 6, pp. 1177-1188, Jun. 2007

[7] K. Kotani, A. Sasaki, and T. Ito, "High-Efficiency Differential-Drive CMOS Rectifier for UHF RFIDs," *IEEE J. Solid-State Circuits*, vol. 44, no. 11, pp. 3011-3018, Nov. 2009

RMO2A-4

A Low Power Low Cost Fully Integrated UHF RFID Reader with 17.6dBm Output P1dB in 0.18 μm CMOS Process

Jingchao Wang, Chun Zhang, Zhihua Wang

Institute of Microelectronics, Tsinghua University, Beijing, 100084, P.R. China

Abstract — **A low power low cost fully integrated single-chip UHF radio frequency identification (RFID) reader for short distance handheld applications is presented in this paper. The IC integrates all building blocks—including an RF transceiver, a PLL frequency synthesizer, a digital baseband and a MCU—in a 0.18 μm CMOS process. A high-linearity RX front-end and a low-phase-noise synthesizer are designed to handle the large self-interferer. A class-E power amplifier with high power efficiency is also integrated to fulfill the function of a UHF passive RFID reader. The measured output P1dB power of the transmitter is 17.6dBm and the measured receiver sensitivity is -60dBm. The digital baseband including MCU core consumes 3.9mW with a clock of 10MHz and the analog part including power amplifier consumes 281.5mW. The chip has a die area of 5.1mm*3.8mm including pads.**

Index Terms — **Radio frequency identification (RFID), reader, single-chip CMOS reader, transceiver.**

I. INTRODUCTION

UHF RFID has been increasingly received considerable attention worldwide these years. Today it is finding its ways into industrial sectors ranging from retail for tracking inventory to manufacturing for tracking product status to airlines for finding lost baggage because it takes numerous of advantages to the traditional bar code, such as capability of programmability, possibility of multiple tag identification, unlimited storage capacity, out sight operation distance, and high data throughput. [1] Currently many different UHF RFID stationary readers are available in the market to meet different demands. Mobile and handheld readers operating in UHF band are also being developed to meet the requirements of different applications such as pharmaceuticals where doctors can carry such readers on their belts. The market of mobile or handheld readers is increasing rapidly these years, much faster than that of stationary readers.

The requirements are very different for handheld readers from stationary readers. The key goal in design of handheld reader is low power consumption to get a longer battery life and low cost to be integrated in cheap handheld equipments while the first goal in design of stationary reader is high sensitivity to operate in a longer range. All the building blocks including power amplifier and protocol processor should be integrated to make the work of high level design easier for handheld readers due

to the limited processing capability of the handheld equipments.

In this paper, a fully integrated single-chip UHF RFID reader for short distance handheld applications in the 0.18 μm CMOS process is reported. The significant improvement of the reader in this paper is integrating a high efficiency power amplifier (PA) with satisfied output power and all the building blocks including a low power MCU and a protocol process module to meet the requirements of handheld applications.

II. SYSTEM ARCHITECTURE

A typical UHF RFID system consists of a reader and several passive tags. The communication between the reader and the tags is half duplex, and the forward data transfer utilizes the ASK modulation scheme while the return data transfer utilizes the back-scattered modulation scheme. It is very convenient for the reader to use I/Q direct-conversion architecture in the receiver to get high integration, low power consumption and to avoid blind zone of the reader. Considering the cost of the chip and the battery lifetime, OOK modulators and high-efficient non-linear PA can be used in the transmitter chain. A MCU with protocol processing ability should also be integrated to release the requirements of high level design.

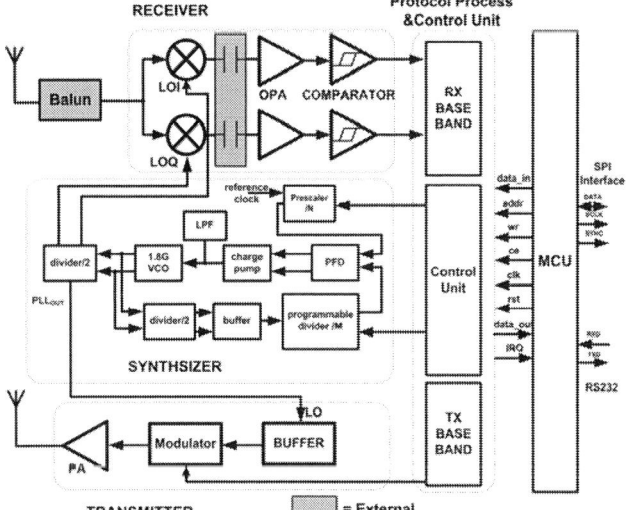

Fig.1. Block Diagram of the Presented RFID Reader

978-1-4244-6240-7/10 $26.00 © 2010 IEEE

The presented reader architecture is shown in Fig. 1. It consists of all the building blocks including frequency synthesizer, I/Q direct-conversion mixer, operational amplifier, comparator, OOK modulator, power amplifier in the transceiver front-end and a MCU with protocol process module in the baseband. The DC offset cancellation is realized by an off-chip capacitor. The transceiver does not use LNA in the receiver. This is because that the target sensitivity is not very high and it is good to avoid the LNA saturation problem induced by carrier leakage in RFID system. The protocol processing module can be fully accessed by the MCU integrated through the address bus and data bus of the MCU.

III. CIRCUITS IMPLEMENTATION

A. Frequency Synthesizer

The carrier frequency is generated by an integer- PLL with a divide-by-8/9 circuit to meet the frequency resolution of different local regulations as depicted in Fig.1. The output frequency can be calculated by $f_{out} = M * f_{ref} / N$, where M is the division ratio of the programmable divider and N is the division ratio of the prescaler. Both M and N can be changed by the integrated MCU through refreshing specified registers. The channel spacing can be changed by rewrite prescaler N. The frequency synthesizer integrates voltage-controlled oscillator (VCO), phase frequency detector (PFD), charge pump, high-frequency dual-modulus divider, and digital programmable divider. A 1.8GHz LO signal is generated by the integrated VCO and then the 900 MHz differential I/Q LO signal are obtained by a divide-by-two circuit to avoid VCO pulling. To meet the phase-noise requirement, the LC VCO (Fig. 2) was designed with the following characteristics. 1) A long-channel pMOS transistor tail current source is used to reduce the 1/f noise. 2) Cross-coupled pMOS with a symmetric inductor using thick top-layer metal transistors are added to improve the symmetry of the differential output for better close-in phase-noise performance. [2]

B. ASK modulator and Power Amplifier

The transmitter front-end is comprised of an OOK modulator and a two-stage class-E power amplifier. The direct-conversion architecture minimizes the off-chip components and provides a low-cost, high-efficient solution. The modulator (Fig. 3) controls the RF signal to PA to generate the amplitude modulation wave. In order to obtain the different RF output signal, two series resistors are connected to the source node of a source follower. Meanwhile, two switch transistors are followed in the RF

signal path, respectively. The gates of the switch transistors are controlled by the transmitted OOK data, fulfilling the different modulation of RF signal by different ratio of R1 and R2. Also an off-chip resistor can be connected in shunt with R2 to adjust the modulation depth to satisfy the different requirements of various tags.

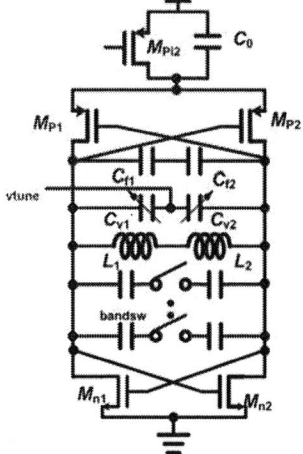

Fig. 2.   Schematic of the VCO

Fig. 3.   Schematic of the Modulator

A high efficiency non-linear PA is chosen to reduce the power consumption. Fig. 4 shows the schematic of the 2-stage class-E PA. The first stage works in class AB mode and the second stage works in class E mode. Each stage has about 11dB power gain. Note that, the second stage can be shut down by control signal through M0. The power transistors are put in deep N-well to release substrate noise coupling impact to other parts.

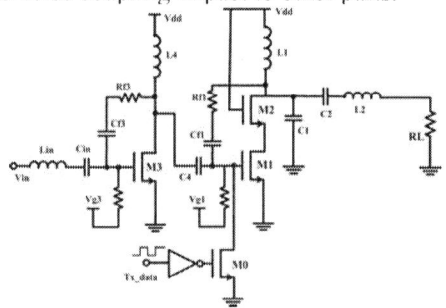

Fig. 4.   Schematic of the class-E PA

C. Down-conversion Mixer and IF Circuits

As the first stage of the receiver, the linearity of the mixer should be high enough to deal with the large LO leakage in passive RFID reader. A passive differential mixer is used due to its high linearity (Fig. 5). An off-chip balun is employed to realize the single-ended to the differential transform. The employed balun provides 6dB voltage gain because its unbalanced impedance is 50ohm while its balanced impedance is 200ohm. The voltage gain of the off-chip matching network not only compensates the $2/\pi$ voltage attenuation caused by the switch of the passive mixer, but also provides the receiver voltage gain to improve the noise performance.

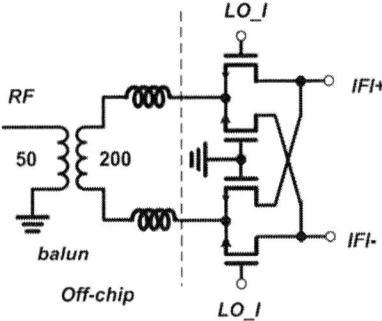

Fig. 5. Schematic of the Down-conversion Mixer

The output signals of the RF front-end are coupled to the input of the IF amplifiers through capacitances of about 100 nF. A 40dB gain, 2MHz bandwidth OPA is designed as IF amplifier. The OPA is designed using cascaded architecture to get high voltage gain and low power consumption. All the bias voltages are got by resisters in series on chip. A hysteretic structure comparator is chosen to avoid the noise interference. A buffer stage is also added at the end of the circuit to increase the output drive capability. To deal with the DC drift in the receiver, one input of the comparator is obtained directly from the OPA output while the other is low-pass filtered signal of the OPA output. The corner of the low-pass filter is carefully designed to achieve a best sensitivity.

D. Digital Baseband

The digital baseband (Fig. 6) is designed based on DW8051 IP core, which has stronger performance than traditional 8051 processor [3]. The associate protocol processing module is designed coincident with the requirement of the MCU data bus and address bus. The accessorial protocol processing module mainly deals with the decoding and encoding task which has strict timing requirements. The module can generate timing signal or decode backscattered signal with different rates and encoding methods under the control of the MCU through control registers. The firmware running in the MCU core mainly deals with protocol flow control as anti-collision arithmetic realization.

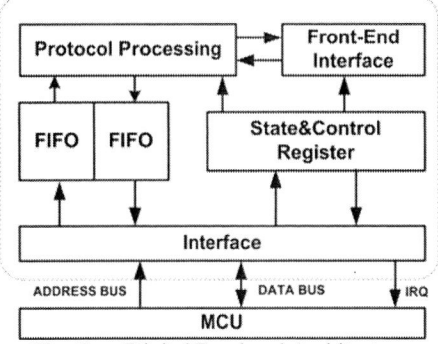

Fig. 6. Digital Baseband Architecture

IV. MEASUREMENT RESULTS

The RFID reader IC was fabricated using a 0.18 μm CMOS process. The chip microphotograph is shown in Fig. 7 with a die area of 5.1mmX3.8mm including pads. The IC was packaged in a 9x9mm body 64-lead QFN package with excellent thermal properties and RF grounding. The IC performance was evaluated on FR-4 test board. To get the sensitivity, we set output of the PLL at a fixed frequency and connected the RF input of the IC with a signal generator which has a 40 kHz difference in frequency from that of the PLL. When the output power of the signal generator is -60dBm, the signal can still be demodulated at the output of the IC. So the sensitivity of the IC can achieve -60dBm@40kbps. The reader performance is summarized in Table I. The measured VCO tuning curve is plotted in Fig. 8. The 8 bands cover frequency from 760MHz to 1005MHz which can cover the whole passive UHF RFID band in the world. The output P1dB power of the integrated PA is about 17.6dBm.

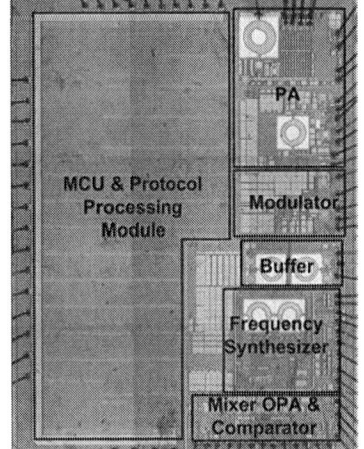

Fig. 7. Die photo of the proposed transceiver

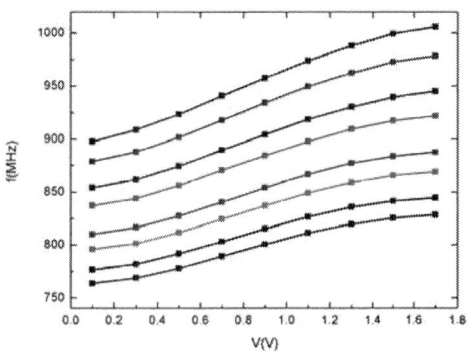

Fig. 8.   Measured VCO tuning curve

The reader performance is summarized in Table I.

TABLE I. SUMMARY OF THE READER IC PERFORMANCE

Parameter		Conditions	Measured Results
Transmitter	Output power(P1dB)	Power Amp	17.6dBm
	Power Consumption	1.8V for modulator and driver stage 2.5V for output stage	245.9mW
PLL+ Receiver	ICP		5dBm
	Sensitivity	40kHz data rate	-60dBm
	Frequency Range	Fref=13 MHz	760MHz~1005MHz
	Power Consumption	1.8V	35.6 mW
Digital Part with MCU	Power Consumption	1.8V CLK=10MHz	3.9 mW

Fig. 9 illuminates the captured communication procedure between the proposed reader and ISO 18000-6 type C tags. The reader first sends the Query command. When the tag receives correct command Query, it returns a random number RN16. The reader gets the RN16 and sends ACK command with the received RN16. After the tag receives the correct acknowledgement, it returns the PC and EPC codes.

Table II shows the comparison results between the UHF RFID reader presented in this paper and with the other recently published works. We can see that the proposed reader achieves higher power output with limited power consumption. To our knowledge it would be the best choice for handheld UHF RFID reader applications.

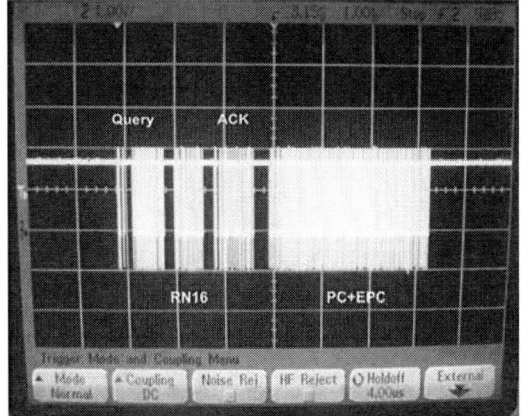

Fig. 9.   Communication procedure between the proposed reader and tags

TABLE II PERFORMANCE SUMMARY OF READER ICs

	[4]	[5]	[6]	[7]	This Work
Process	0.18um SiGe BiCMOS	0.18um CMOS	0.18um CMOS	0.18um CMOS	0.18um CMOS
Integration level	RF+Analog+Digital Baseband	RF+Analog+Digital Baseband	RF+Analog+Data Converter	RF+Analog+Digital Baseband	RF+Analog+Digital Baseband
PA on Chip	YES	NO	NO	NO	YES
Rx Input P-1dB	11dBm	8dBm	-8dBm	3.5dBm	5dBm
Sensitivity	-95dBm	N/A	-85dBm	-90dBm	-60dBm
Output Power	19dBm	4dBm	10dBm	10.4dBm	17.6dBm
Die Size	21mm²	23.9mm²	36mm²	18.3mm²	19.38mm²
Total Power	1.2W	160mW	540mW	<276.4mW	285.4mW

V. CONCLUSION

A single chip UHF passive RFID reader is fabricated in 0.18 μm CMOS process with a die area of 5.1mmX3.8mm including pads. The IC was packaged in a 9x9mm body 64-lead QFN package with excellent thermal properties and RF grounding. The IC integrates all building blocks— including an RF transceiver, a PLL frequency synthesizer, a digital baseband and a MCU. A high-linearity RX front-end and a low-phase-noise synthesizer are designed to handle the large self-interferer. A class-E power amplifier with high power efficiency is also integrated to fulfill the function of a UHF passive RFID reader. The measured output P1dB power of the transmitter is 17.6dBm and the measured receiver sensitivity is -60dBm. The digital baseband including MCU consumes 3.9mW with a clock of 10MHz and the analog part including power amplifier consumes 281.5mW.

REFERENCES

[1] K. Finkenzeller, *RFID Handbook*, 2nd ed., Wiley, 2004, pp. 7-9.

[2] Hajimiri A, Lee T H. "A general theory of phase noise in electrical oscillators," *IEEE J. Solid-State Circuits*, vol. 33, no. 2, pp. 179-194, February 1998.

[3] Synopsys Inc., "DesignWare DW8051 MacroCell Databook", August 2004.

[4] S. Chiu, I. Kipnis, M. Loyer, J. Rapp, D. Westberg, J. Johansson, P. Johansson, "A 900 MHz UHF RFID reader transceiver IC", *IEEE J. Solid-State Circuits*, vol. 42, no. 12, pp. 2822–2833, December 2007.

[5] I. Kwon, Y. Eo, H. Bang, K. Choi, S. Jeon, S. Jung, D. Lee, H. Lee, "A Single-Chip CMOS Transceiver for UHF Mobile RFID Reader", *IEEE J. Solid-State Circuits,* vol. 43, no. 3, pp. 729-738, March 2008.

[6] P.B. Khannur, X. Chen, D.L. Yan, D. Shen, B. Zhao, M.K. Raja, Y. Wu, R. Sindunata, W.G. Yeoh, R.A. Singh, "A Universal UHF RFID Reader IC in 0.18-um CMOS Technology", *IEEE J. Solid-State Circuits*, vol. 43, no. 5, pp. 1146-1155, May 2008.

[7] W. Wang, S. Lou, K.W. Chui, S. Rong, C.F. Lok, H. Zheng, H.T. Chan, S.W. Man, H.C. Luong, V.K. Lau, C.Y. Tsui, "A Single-Chip UHF RFID Reader in 0.18um CMOS Process", *IEEE J. Solid-State Circuits*, vol. 43, no. 8, pp. 1741-1754, August 2008.

978-1-4244-6240-7/10 $26.00 © 2010 IEEE

Far-field RF Powering System for RFID and Implantable Devices with Monolithically Integrated On-Chip Antenna

Soheil Radiom[1], Majid Baghaei-Nejad[2,3], Guy Vandenbosch[1], Li-Rong Zheng[2], Georges Gielen[1]

[1]MICAS-ESAT, Katholieke Universiteit Leuven, Leuven, Belgium
[2]Royal Institute of Technology, Stockholm, Sweden
[3]Sabzevar Tarbiat Moallem University, Sabzevar, Iran

Abstract — **A fully integrated far-field powering system for RFID and implantable devices with monolithically fully integrated on-chip antenna in 0.18μm CMOS is presented. The chip receives power, clock and data wirelessly through RF signal at all the three ISM bands of 915 MHz, 2.45 GHz and 5.8 GHz. Measurements show a minimum input power of -19.41 dBm at 900MHz for chip operation, corresponding to 15.7 meter of operation range with an off-chip 0dB gain antenna. On the other hand, with its on-chip antenna at 5.8 GHz, the chip can be powered-up up to 7.5 cm distance. This is a huge improvement in terms of operation distance compared with other reported similar works with on-chip antenna as well as the off-chip antennas.**

Index Terms — **On-chip Antenna, OCA, RFID, Implantable, Wireless powering, far-field powering**

I. INTRODUCTION

While there have been real initial successes in the use of RFID technologies, the long-term potential benefits are immense. To date, RFID has demonstrated brighter prospects for companies wishing to gain process efficiency and transparency, as well as to identify additional business potential. Few examples of future substantial impact of RFID can be the advanced administration of healthcare services, pharmaceuticals, safety of bank notes and valuable documents. In all these applications, there may be a desire to implant the RFID chip into the target objects where size of the tag becomes very essential and determiner. The goal of this paper is to introduce our recent passive RFID chip with monolithically integrated On-Chip Antenna (OCA) on its top layer for these type of applications, where there is a huge improvement comparing to previously reported commercial and academicals works.

II. OPERATION PRINCIPLE

In general in the passive (battery-less) circuits the required power supply can be provided by any kind of energy sources such as thermal energy, vibration, movement, solar energy and so on. In our case, we use the electromagnetic energy of the incoming UHF wave at ISM frequencies of 900 MHz, 2.45 GHz or 5.8 GHz; for the latter this happens with an on-chip antenna where it enables the chip to be implanted in bank notes or any valuable document. For the uplink and data communication, usually backscattering or load modulation is used. However, the on-chip antenna low-gain reduces the operation distance significantly to a very close proximity. Therefore, an active transmitter can be used in uplink as a solution. However, the available power is too low to provide continuous powering for the circuitry operation; thus the tag captures energy for a longer time, rectifies it and stores it in a storage capacitor, and then uses this energy during the tag's operating period, hence duty cycling the tag's operation. As a result, the amount of available energy is limited and the tag's operation should be done in a time and power efficient approach. So for the uplink the Impulse UWB can be chosen for our system as a power-time efficient solution; this is discussed in more detail in our previous work at [7].In our system the powering-up signal is also used to carry the command and clock to the tags. Data and clock modulate the RF signal using Amplitude-Shift Keying (ASK) modulation with modulation depth from 30% to 100%. To realize the reception simply, binary data are encoded as pulse width modulation of the low-amplitude pulse such as conventional class 1 RFIDs. Time intervals for bit 1 are chosen to be 3 times of bit 0 (e.g. 1.5us for bit 0 and 4.5us for bit 1). A discriminator circuit in the tag will distinguish this difference and extract the 0s and 1s.

III. MONOLITHICALLY INTEGRATED ON-CHIP ANTENNA DESIGN

The most critical point of this design was the antenna as it directly determines the coverage distance. Targeting the standard CMOS technology and considering the very lossy silicon substrate,10 S/m conductivity, make the antenna design and considerations quite different with a normal antenna design. Simulations show that on such lossy substrates the antenna must meet the minimum required surface current path to have better efficiency (less loss) while at the same time needs to be conjugated matched to a capacitive impedance of the rectifier. In order to realize this a multi-turn loop-dipole structure with inductive and resistive stubs is developed and optimized

978-1-4244-6240-7/10 $26.00 © 2010 IEEE

for this work. The antenna, shown in the die microphotographin of *Figure 4*, is less than 3X1.5 mm² in size and its IE3D and HFSS simulated realized-gain is around -29.5 dBi at 5.8 GHz.

IV. CIRCUIT IMPLEMENTATION

To demonstrate the proposed system concept, the module circuitry has been implemented in UMC 0.18μm CMOS.Figure 1 shows the detailed block diagram of the proposed module. It consists of a power management unit, an RF demodulator, a clock generator unit, and an on chip antenna for both short range wireless powering as well as data and clock. Unlike conventional passive tags where the incoming power is used directly for chip operation, here a different powering strategy has been used which reduces the required input power significantly. A power scavenging unit consisting of a chain of CMOS full-wave multiplier rectifies the incoming RF wave to DC voltage in an on-chip storage capacitor. Hence, the incoming power is accumulated in the storage capacitor and a low-power voltage sensor (Vsen) activates the chip when the voltage in the storage capacitor reaches an upper limit (V_H). The operation current is provided by discharging the storage capacitor down to the lower limit (V_L). Based on our simulation a minimum voltage of 1.2V is required for the transmitter to generate the desired pulse. Since the goal is to have the whole system on-chip so the storage capacitor size is an issue as well. Its size depends on the operation time, the required current consumption, and the difference between upper limit and lower limit as :

$$C = \frac{I.time}{\Delta V} = \frac{I\_dc*(Rx\_time + Tx\_time) + I\_Tx*Tx\_time}{V_H - V_L}$$

Where I_dc is the current we estimate for baseband and other receiving blocks, I_Tx is the current of the transmitter and the clock generator (10uA at 1Mbps data rate based on our previous experiences and measurements), Rx_time is the receive time (around 200us i.e. 20 bits at 200Kbps). High upper limit reduces the required storage capacitor size, but it increases the required input power limiting the operational distance. On the other hand, choosing very low upper limit demand a large storage capacitor which occupies large area on-chip which is not feasible.

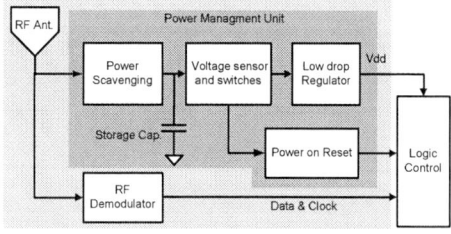

Figure 1: Block diagram of the proposed module

Based on our simulation the upper limit of 1.8V and a 6nF capacitor have been chosen. It can be implemented by MOSCAP which has higher capacity density. A low-drop-out (LDO) regulator provides a regulated voltage of 1.2V for the circuits A Power-on-Reset (PoR) circuit generates a reset signal for the logic control to eliminate the transient response of the Vsen and the LDO.

The operating distance of a chip depends on the efficiency of the power scavenging unit. A cross-connected CMOS full-wave multiplier is chosen here, since this architecture eliminate the threshold voltage of the transistor. Therefore, the required input voltage(power) would be less resulting longer operational distance.. The schematic of the power scavenging unit is shown in Figure 2. It includes a 9-stage full-wave voltage multiplier and a voltage limiter, which keeps the output voltage lower than the breakdown voltage. The goal is to design the rectifier able to produce 1.8V at the output and drive the Vsen with current of 330nA (i.e. 594nW)

The Vsen is the only operating part of the chip in the powering phase, therefore its static current is very critical. Figure 2 shows the schematic of the voltage sensor including a reference voltage generator, a Schmitt trigger comparator and a power switch. Having low power resistor-based bandgap references would need very big resistors, which occupy a large area [1]. Therefore, in this work a fully CMOS reference voltage is used [2]. A CMOS voltage divider divides the voltage across the storage capacitor and the comparator compares it with the reference voltage. The divider is designed to set the switch with 1.8 input voltage. As can be seen, the positive feedback through V2 changes the divider fraction. This changes the comparison voltage to 1.2V during discharging. It generates the 0.6 voltage range which provide the current needed for the chip operation.

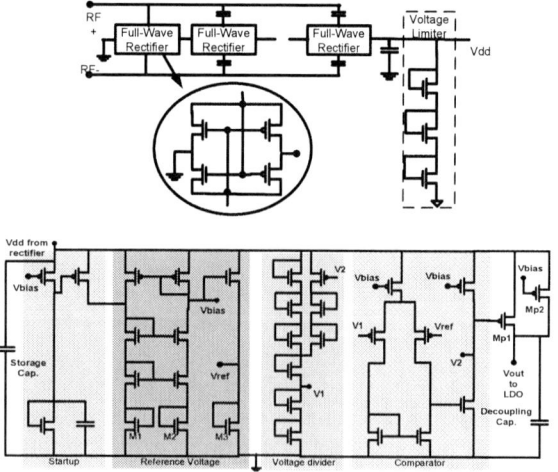

Figure 2: Schematic of the (a) power scavenging unit and (b)the voltage sensor

The schematic of the LDO is shown in Figure 3. The same reference voltage form the voltage sensor is used here. To reduce the power consumption, the bias current needs to be decreased which causes longer transient response for the LDO. In order to decrease the LDO's transient response, transistor MP2 in the voltage sensor provides a small bias current for the LDO circuitry during harvesting priod, which charges the LDO capacitors (parasitic and decoupling capacitors). This bias current can be very low hence does not affect the rectifier performance. In operation period MP1 is switched on and provides enough current for the LDO and the chip operation. Using this technique, can reduces the LDO's transient response significantly. The PoR circuitry is shown in Figure 3. A capacitor charged by a current source generates a pulse which is delayed compared to the signal from the Vsen.

The envelope detector is similar to the power scavenging unit structure but with only 2 stages. Figure 3 shows the block diagram of the RF demodulator including the envelope detector and a comparator circuitry. Q1 acts as an extra load for the envelope detector output which prevents generating high voltage when the incoming RF has higher levels. The extracted clock is used for data sampling and logic control; therefore no local oscillator is needed, which reduces the tag's power consumption significantly. The chip microphotograph is shown in Figure 4 and it occupies 4.5 mm^2.

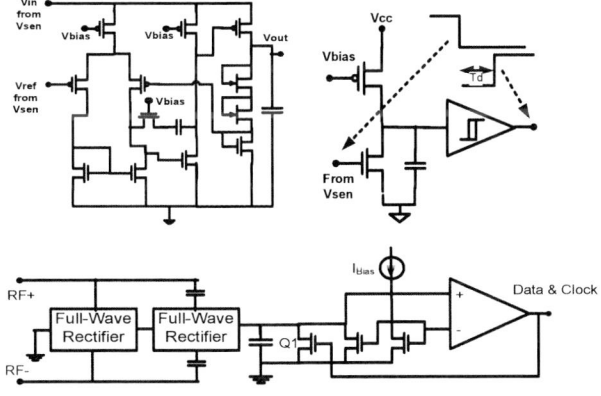

Figure 3: (a) Low-drop-out voltage regulator, (b) Power-on-Reset circuit and (c) RF demodulator

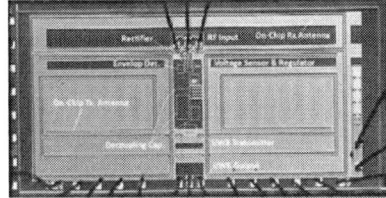

Figure 4: Die microphotograph

V. MEASUREMENT RESULTS AND DISCUSSION

To measure the input sensitivity of the power scavenging unit, the Rhode & Schwarz ZVM Network Analyzer has been used. The voltage sensor, the voltage regulator, and the power-on-reset are considered as the load of the power scavenging.

Measurement results have shown that as it was designed, the rectifier works for the all the 3 ISM bands of 900MHz, 2.45GHz and 5.8GHz while at the latter one the on-chip antenna are being used for short range power scavenging and data receiving. The input sensitivity at 900 MHz without on-chip antenna (having the antenna shoot out by the laser machine) has been measured to be -19.41 dBm by direct probing the chip. Considering 36 dBm EIRP radiation and an external 0 dB gain receive antenna, this corresponds to a 15.7 meter powering distance, which is a great improvement compared to existing passive RFIDs [3]. This improvement is due to the use of the voltage sensor and duty cycling the operation, which reduces the power consumption during harvesting improving the input sensitivity. To occupy a small area, the on-chip antenna has been designed at 5.8 GHz. A wireless operational distance of 7.5cm is achieved under 36 dBm EIRP, which is a huge improvement compared with previously reported works [4-6].

Figure 5.a shows the frequency response of the wireless powering. As can be seen, the chip can operate in a wide range of the frequency band. It can help the chip works when present any frequency shift due to proximity or any unexpected variation. Figure 5.b depicts the rectified voltage versus distance for different extra loads at the rectifier output. It confirms that having circuit with less current consumption longer operational distance can be achieved. Figure 6.a shows the rectifier output voltage across the storage capacitor and the PoR signal. As can be seen the upper and lower limit is measured to be 2V and 1.3V respectively. The deviation from the simulation results is due to the process variation. However this variation does not affect the circuit operation but increment in upper limit reduces the operational distance.

An ASK-modulated RF signal as explained before has been used to measure the data and clock recovery performance. The measurement result at 500 kHz clock is shown Figure 6.b. Higher clock could not be measured because of the instrument limitation in our lab.

Table 1 summarizes and compares the measurement results with three other related works.

VII. CONCLUSION

A fully integrated far-field powering system for RFID and implantable devices with monolithically fully integrated on-chip antenna in 0.18μm CMOS is presented. Measurement results confirm that the chip can be powered up at all the three ISM bands of 915 MHz, 2.45 GHz and 5.8 GHz. In order to minimize the power consumption the chip receives its clock and the commands wirelessly through the modulated RF powering-up signal. Measurement results show that the chip can operate with a minimum input power of -19.41 dBm at 900MHz band, corresponding to 15.7 meter of operation range with assuming an off-chip 0dB gain antenna. On the other hand, with the implemented on-chip antenna at the top layer at the 5.8 GHz band, the chip can be powered up up to 7.5 cm distance. This is a huge improvement in terms of operation distance compared with other reported similar works with on-chip antenna as well as the off-chip antennas.

Table 1: Summary of measurement results

		This work	[4]	[7]
Technology		0.18μm	0.18μm	0.18μm
Die area(mm²)		4.5	0.6	4.5
Frequency		900 MHz, 2.45 &5.8	900MHz	900MHz
Input sensitivity (dBm)	900 MHz	-19.41	-	-18.5
	2.45 GHz	-16.2	-	-
	5.8G Hz	-14.22 dBm	-	-
Vout/I (μA)		1.8 V/0.33	1 V/10	2.7V/1.5
Distance with off-chip antenna	900 MHz	15.7 meter		13.9 m
	2.45 GHz	4 meter	-	-
	5.8 GHz	1.5 meter		-
Distance with OCA		7.5cm (4W EIRP/5.8G	4 mm	-
Need for off-chip component		No off-chip component	-	Antenna & storage capacitor

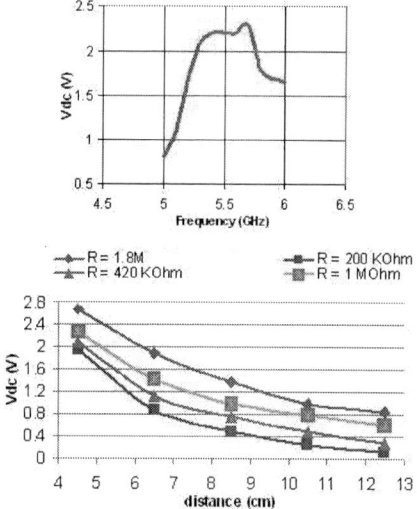

Figure 5: a) Frequency response of wireless powering and b) the wireless powering response versus distance

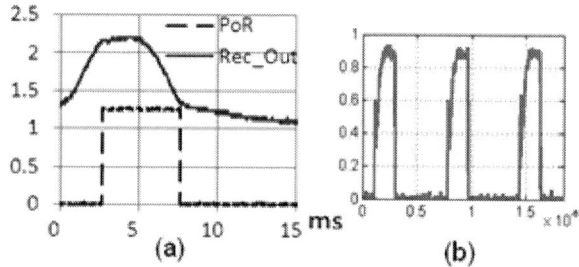

Figure 6: a) Rectifier output charge/discharge and Power-on-Reset output waveforms, b) Measurement result for the ASK demodulator

REFERENCES

[1] T. Preetam, "A CMOS bandgap reference with correction for device-to-device variation," in *ISCAS '04.* pp. I-397-400

[2] G. D. Vita, et al, "A 300 nW, 12 ppm/C Voltage Reference in a Digital 0.35 um CMOS Process," in *VLSI Circuits, 2006.*

[3] R. Barnett, et al, "A Passive UHF RFID Transponder for EPC Gen 2 with -14dBm Sensitivity in 0.13Â¿m CMOS," in *Solid-State Circuits Conference, ISSCC 2007.* pp. 582-623.

[4] A. Shameli, et al, "A UHF Near-Field RFID System With Fully Integrated Transponder," *Microwave Theory and Techniques, IEEE Transactions on,* vol. 56, pp. 1267, 2008.

[5] M. Usami, "An ultra small RFID chip: μ-chip," in *Radio Frequency Integrated Circuits (RFIC) Symposium, 2004. Digest of Papers. 2004 IEEE*, 2004, pp. 241-244.

[6] Y. W. G. C. Xuesong, et al, "A 2.45-GHz Near-Field RFID System With Passive On-Chip Antenna Tags," *Microwave Theory and Techniques, IEEE Transactions on,* vol. 56, pp. 1397-1404, 2008.

[7] Majid Baghaei-Nejad, et al, " A Remote-Powered RFID Tag with 10Mb/s UWB Uplink and -18.5dBm-Sensitivity UHF Downlink in 0.18μm CMOS", IEEE International Solid-State Circuit Conference ISSCC, 2009.

A 2-1100 MHz Wideband Low Noise Amplifier with 1.43 dB minimum Noise Figure

Mohamed El-Nozahi, Ahmed A. Helmy, Edgar Sánchez-Sinencio and Kamran Entesari

Department of Electrical and Computer Engineering, Texas A&M University, College station, TX, 77843, USA

Abstract — A new wideband low noise amplifier (LNA) is proposed in this paper. The LNA utilizes a composite NMOS/PMOS cross-coupled transistor pair to increase the amplification while reducing the noise figure. The introduced approach provides partial cancellation of noise generated by the input transistors, hence, lowering the overall noise figure. An implemented prototype using IBM 90 nm CMOS technology shows a measured conversion gain of 20 dB across 2-1100 MHz frequency range, an IIP3 of -1.5 dBm at 100 MHz, and minimum and maximum noise figure of 1.43 dB and 1.9 dB from 100 MHz to 1.1 GHz. The LNA consumes 18 mW from 1.8 V supply and occupies an area of 0.06 mm².

Index Terms — low noise amplifier, noise cancellation, wideband, broadband.

I. INTRODUCTION

Today, multi-band multi-standard transceivers are the main design trend for compact and low-cost mobile units. These units are covering several standards such as digital video broadcasting 450-850 MHz, FM transceivers 87-108 MHz, satellite communications 950-2150 MHz, and GSM 850-1900 MHz. Stacking several front-ends for the reception of various standards was one of the design trends to realize these wideband receivers. On the other hand, today the design trend is focusing on single wideband front-ends to accommodate all the standards to reduce the area of the front-end. Single wideband front-end faces many challenging problems including very low noise figure, high linearity requirements, and low area consumption.

Inductor-less wideband low noise amplifiers (LNAs) are key building blocks that partially solve the area consumption challenge [1]-[6]. These LNAs usually rely on resistive feedback techniques for wideband input matching, which leads to poor noise figure (sensitivity). In addition, due to the flicker noise, they are not suitable for sub-100 MHz communications. Therefore, noise cancellation techniques have been proposed in the literature to overcome the poor noise figure of these inductor-less wideband LNAs [1]-[2]. These techniques rely on the matching between the devices and a minimum noise figure of 1.9 dB was reported in [1]. Reducing the noise figure below 1.9 dB is still challenging and a

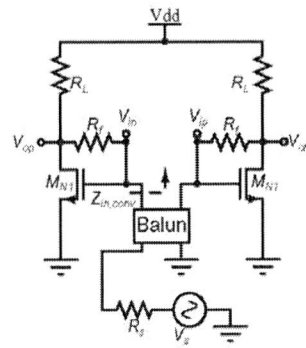

Fig. 1 Conventional boradband LNA with resistive matching

solution is provided in this paper. The paper is organized as follows: In Section II, the conventional wideband LNA with resistive matching is discussed. Section III presents the basic idea of the proposed wideband LNA. Section IV discusses the implementation details, and Section V demonstrates the measured results of the fabricated prototype. Finally, Section VI concludes the paper.

II. BACKGROUND

The conventional broadband LNA with resistive matching is shown in Fig. 1. For this architecture, the differential input impedance, $Z_{in,conv}$, differential voltage gain, $A_{v,conv}$, and noise figure, NF_{conv}, are given as follows

$$Z_{in,conv} = 2 \cdot \frac{R_f}{g_{mn} \cdot R_f \,//\, R_L} = 2 \cdot \frac{R_f + R_L}{g_{mn} \cdot R_L}, \quad (1)$$

$$A_{v,conv} = \frac{V_{op} - V_{on}}{V_{in} - V_{ip}} = -g_{mn} \cdot R_f \,//\, R_L, \quad (2)$$

$$NF_{conv} = 1 + \gamma_n. \quad (3)$$

where g_{mn} is the transconductance of the transistor M_n and γ_n is the thermal noise factor. Eqs. (1)-(3) assumes that the noise is mainly due to the thermal noise of the input transistor, $R_L >> R_f$, and that the LNA is designed with perfect matching. Hence for the conventional LNA with resistive matching, the noise figure is higher than the value defined in (3) and practically a minimum value of 2.3 dB is achievable. Decreasing the noise figure below this value is not possible, and therefore, a new architecture is proposed in this paper to reduce the limit defined by (3).

978-1-4244-6240-7/10 $26.00 © 2010 IEEE

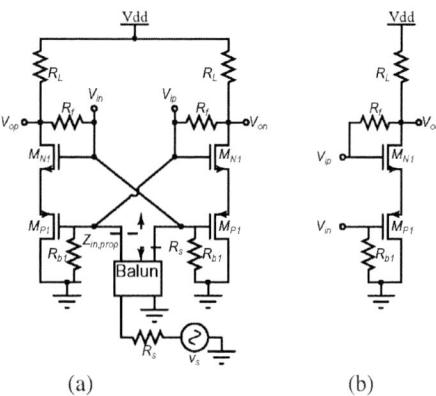

(a)   (b)

Fig. 2 (a) Simplified schematic of the proposed broadband LNA (b) Half circuit model for gain calculation

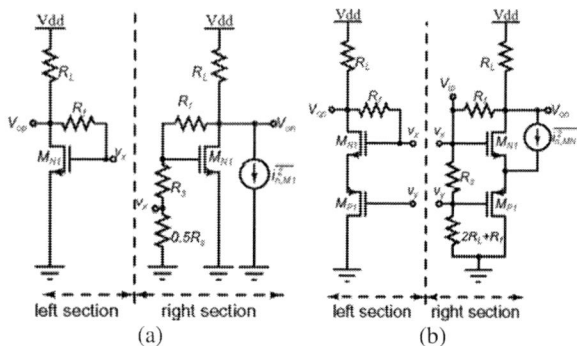

(a)   (b)

Fig. 3 Equivelant circuit model showing the effect of noise current of M_{N1} for (a) conventional and (b) proposed architectures

III. The Proposed LNA

The simplified schematic of the proposed broadband LNA architecture is shown in Fig. 2(a). This architecture is similar to the conventional broadband LNA with resistive matching, however, the overall noise figure is reduced by incorporating the transistor M_{p1} and connecting the gate of M_{p1} to the gate of M_{N1} in a cross-coupled fashion. As shown below, this transistor reduces the output noise by one half leading to a lower output noise. The input matching is adjusted through the resistive feedback resistance R_f and the effective transconductance of the overall LNA, g_{meff}.

The input signal is amplified by considering the half circuit as shown in Fig. 2(b). In this model, input signals to gates of M_{N1} and M_{P1} carry different polarity ($V_{in}=-V_{ip}$) leading to an amplification of the input signal. In case both inputs have the same polarity ($V_{in}=V_{ip}$), the output AC current is zero, leading to common mode noise rejection. This observation is used also to cancel part of the noise as demonstrated below.

The noise cancellation is clarified qualitatively by considering the noise of M_{N1} for both the conventional and proposed architectures, as shown in Fig. 3. In this figure, the noise current due to the right NMOS transistor, $i_{n,MN1}$, is considered, and the input impedance of the left half circuit of the LNA is considered. For the conventional case, the input impedance of the left half circuit is half the value of the source resistance, R_s as shown in Fig. 3(a). The noise current, i_{nMN1}, generates a noise voltage at the gate of the left M_{N1}, V_x, which is a fraction of the output noise at V_{on}. This noise voltage is amplified by the left half circuit with a different polarity (gain is negative) leading to an increase in the overall differential output noise.

On the other hand for the proposed LNA, the input impedance of the left half circuit is $2R_L+R_f$ as shown in Fig. 3(b). With the assumptions of $R_s << R_L$ and $R_s << R_f$, the gates of the NMOS and PMOS transistors can be considered the same ($V_x \cdot V_y$). As a result, the

NMOS/PMOS transistor pair in the left half part of the LNA does not produce an output AC current leading to V_{op} close to V_{on} with the same polarity. The resultant differential output noise voltage is reduced. Similarly the noise generated by M_{P1}, R_f and R_L are reduced.

This qualitative explanation is verified using circuit-level analysis. The resultant differential input impedance, $Z_{in,prop}$, differential voltage gain, $A_{v,prop}$, and noise figure (thermal noise), NF_{prop}, are given as follows

$$Z_{in,prop} = 2 \cdot \frac{R_f}{g_{meff} \cdot R_f \,/\!/\, R_L} = 2 \cdot \frac{R_f + R_L}{g_{meff} \cdot R_L}, \qquad (4)$$

$$A_{v,prop} = \frac{V_{op} - V_{on}}{V_{in} - V_{ip}} = -g_{meff} \cdot R_f \,/\!/\, R_L, \qquad (5)$$

$$NF_{prop} = 1 + \frac{\gamma_n + \gamma_p}{4}, \qquad (6)$$

where g_{meff} is the effective transconductance defined by

$$g_{meff} = \frac{2 \cdot g_{mn} g_{mp}}{g_{mn} + g_{mp}}. \qquad (7)$$

Eqs. (4) and (5) indicate that the input impedance and voltage gain is similar to the conventional LNA with resistive matching as shown in (1) and (2). The main difference between these equations is that the effective transconductance is only g_{mn} for the conventional case, while it is the parallel combination of g_{mn} and g_{mp} for the proposed architecture. Note that setting $g_{mn}=g_{mp}$ sets the value of $g_{meff}=g_{mn}$.

The noise factor of the proposed architecture provides a reduction in the input referred noise as defined by (6) when compared to the conventional case defined in (3). As an example, if the minimum noise figure of the conventional LNA is 2.3 dB and $\gamma_n=\gamma_p$, then the minimum noise figure of the proposed LNA is 1.3 dB. For both cases, LNAs consume the same current consumption and achieve the same gain. However, the proposed architecture requires an increase in the supply by $V_{gs,MP1}$ to accommodate the additional PMOS transistor.

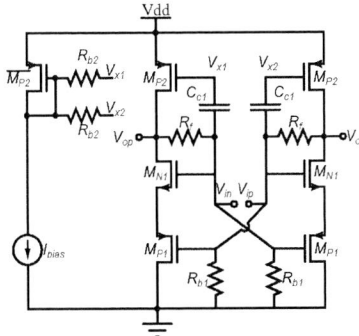

Fig. 4 Complete schematic of the proposed broadband LNA demonstrating the biasing circuit

In summary, the main key difference between the proposed LNA architecture and the conventional one is that the PMOS/NMOS pair provides the required transconductance for the signal, while it cancels part of the noise of M_{N1} and M_{P1} at the output.

IV. IMPLEMENTATION

The actual implementation of the proposed LNA is shown in Fig. 4. In this implementation, the load, R_L, is replaced by a PMOS transistor, M_{P2}. This transistor serves for two main purposes: (1) to provide the DC biasing, and (2) to provide an additional gain to increase the overall gain of the LNA. The DC biasing is adjusted with the current source I_{bias}, which is mirrored through the current mirror M_{P2}. This current also determines the gate-source voltage of M_{N1} and M_{P1}, and therefore no additional DC biasing circuit is required. The DC voltage of the output node is determined from the gate-source voltages of M_{P1} and M_{N1}, i.e. $V_{on,DC}=V_{op,DC}=V_{SG,MP1}+V_{GS,MN1}$. The gate of M_{P1} is biased to ground through the resistance R_{b1}, which is much higher than the value of the source resistance, R_s.

M_{P2} also provides an additional transconductance to increase the overall gain of the LNA. Increasing the overall gain helps to reduce the noise contribution of the load and feedback resistances, and therefore lowering the overall noise figure. The capacitor, C_{c1}, and the resistance, R_{b2}, act as a bias-T such that M_{P2} can provide both the DC biasing and the amplification for the input signal, simultaneously. The cut-off frequency of C_{c1} and R_{b2} is adjusted at 500 kHz.

Widths of the transistors are increased to reduce the flicker noise at lower frequencies. Increasing the size of these transistors helps in lowering the overall noise figure at frequencies lower than 100 MHz. However, this increase limits the operating bandwidth at higher frequencies because of the additional parasitic capacitance at the gate, which forms a low-pass filter with the source resistance for the input signal.

Fig. 5 Die micrograph of the proposed broadband LNA

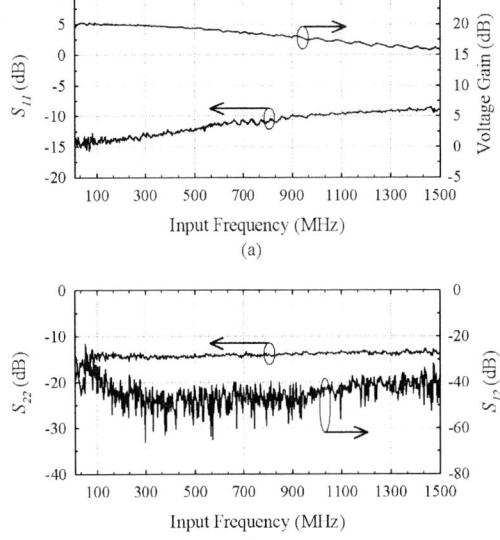

Fig. 6 Measured (a) S_{11} and voltage gain, (b) S_{22} and S_{12}

V. EXPERIMENTAL RESULTS

The broadband LNA is fabricated using 90 nm CMOS technology provided by IBM. The die micrograph is shown in Fig. 5, where the area of LNA core is 0.2x0.3 mm². The core LNA consumes 10 mA from 1.8 V supply. The chip is encapsulated in an MLP package for applying/monitoring the DC biasing and input/output signals. The effect of the output buffer is de-embedded from the LNA+Buffer measurements. The buffer is added at the output of the LNA to drive the 50 Ω input impedance of the network analyzer.

Fig. 6(a) shows the measured S_{11} of the packaged LNA. Measured S_{11} is lower than -10 dB for the entire 2 MHz to 1.1 GHz frequency range. The measured voltage gain after de-embedding the buffer effect is also shown in Fig. 6(a). The buffer was designed to drive the 50 Ω impedance of network analyzer and a measured S_{22} better than -13 dB across the band of interest is obtained as shown in Fig. 6(b). The measured voltage gain is 20 dB with a 3-dB bandwidth of 1.1 GHz (Fig. 6(a)). The measured reverse isolation, S_{12}, is less than -35 dB over the entire band. The bandwidth of the amplifier is limited to 1.1 GHz because

978-1-4244-6240-7/10 $26.00 © 2010 IEEE

TABLE I

PERFORMANCE SUMMARY AND COMPARISON

Ref	Gain (dB)	Freq-Range (GHz)	NF$_{min}$ (dB)	NF@100MHz (dB)	IIP3 (dBm)	P$_{DC}$ (mW)	A. Area (mm^2)	Tech (CMOS)	Package	Topology
[1] JSSC-04	13.7	0.002-1.6	1.9	2.4[b]	0	35	0.075	0.25 μm	On-wafer	Single-Ended
[2] RFIC-07	12.5[a]	0.1-1.6[b]	2.5	> 4.5[b]	16	11.6	0.1	0.13 μm	On-wafer	Single-Ended
[3] ISSCC-07	17.4	0-6	2.5	>2.6[b]	-8	9.8	0.002	90 nm	On-wafer	Single-Ended
[4] ISSCC-07	17	1-7	2.4	N.A.	-4.1	25	0.019	0.13 μm	Chip-on-board	Differential
[5] ESSCIRC-07	15.6	0.2-5.2	3	N.A.	0	14	0.75[c]	65 nm	Chip-on-board	Differential
[6] ISSCC-09	21	0.3-0.92	2	N.A.	-3.2	3.6	0.33	0.18 μm	N.A.	Differential
This Work	20	0.002-1.1	1.43	1.9	-1.5	18	0.06	90 nm	MLP	Differential

[a] Power gain,   [b] Estimated from data provided in papers,   [c] total area including pads

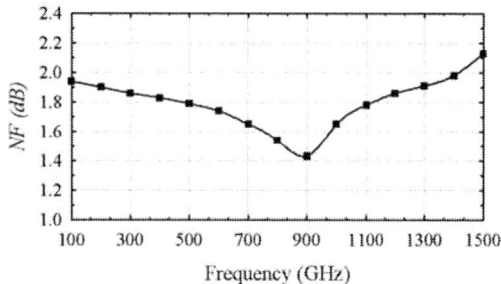

Fig. 7 Measured noise figure versus operating frequency

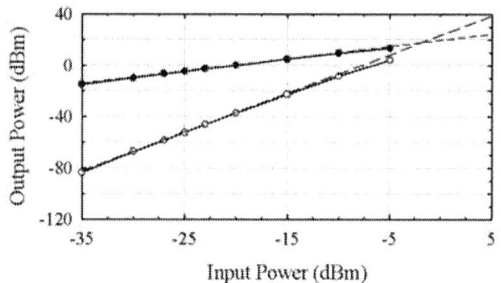

Fig. 8 IIP3 measurements (Tones at 100 and 101 MHz)

of the parasitic capacitance introduced by the package. Without this capacitance, the bandwidth can increase to 2.5 GHz as indicated by post-layout simulations.

The measured noise figure versus the operating frequency is shown in Fig. 7. It has a minimum value of 1.43 dB at 900 MHz. Below this frequency, the noise figure increases because of the flicker noise reaching 1.9 dB at 100 MHz. Above 900 MHz the noise increases because of the bandwidth of the LNA that is limited by the package parasitics. If the bandwidth increases, the NF remains almost at 1.5 dB up to 2.5 GHz as indicated from post-layout simulations. A two-tone IIP3 measurement is performed for the LNA and the results are shown in Fig. 8 for 100 MHz operating frequency. The two tones are applied with the same amplitude and a frequency offset of 1 MHz. A measured IIP3 of -1.5 dBm is obtained. The total current consumption of the broadband LNA is 10 mA

from 1.8V supply. Finally, the performance of the proposed wideband LNA and comparison with existing state-of-the-art inductor-less broadband LNAs around the same frequency range are summarized in Table I.

VI. CONCLUSION

A broadband LNA employing a new technique for noise reduction is proposed in this paper. The LNA relies on a composite NMOS/PMOS transistor pair for implementing the noise cancellation technique. Measurements of a fabricated prototype using 90 nm CMOS technology show a voltage gain of 20 dB with a 3-dB bandwidth of 1.1 GHz. A minimum noise figure of 1.43 dB is also measured with an IIP3 of -1.5 dBm. This measured noise figure is lower than the best reported noise figure by 0.5 dB. The LNA consumes 18 mW from a 1.8 V supply.

REFERENCES

[1] F. Bruccoleri, E.A.M. Klumperink, and B. Nauta, "Wide-Band CMOS Low-Noise Amplifier Exploiting Thermal Noise Cancellation," *IEEE JSSC*, pp. 275-282, Feb. 2004.

[2] W.-H. Chen, G. Liu, B. Zdravko, A.M. Niknejad, "A Highly Linear Broadband CMOS LNA Employing Noise and Distortion Cancellation," *IEEE RFIC Digs.*, pp. 61-64, 2007.

[3] J. Borremants, P. Wambacq, and D. Linten, "An ESD-Protected DC-to-6GHz 9.7mW LNA in 90nm Digital CMOS," *IEEE ISSCC Digs.*, pp. 422-423, 2007.

[4] R. Ramzan, S. Andersson, and J. Dabrowski, "A 1.4V 25mW Inductorless Wideband LNA in 0.13μm CMOS," *IEEE ISSCC Digs.*, pp. 424-425, 2007.

[5] S.C. Baakmeer, E.A.M. Klumperink, B. Nauta, and D.M.W. Leenaerts, "An inductorless wideband balun-LNA in 65nm CMOS with balanced output," *IEEE ESSCIRC Digs.*, pp. 364-367, 2007.

[6] S. Woo, W. Kim, C.-H. Lee, K. Lim, and J. Laskar, "A 3.6mW Differential Common-Gate CMOS LNA with Positive-Negative Feedback," *IEEE ISSCC Digs.*, pp. 218-219, 2009.

RMO2B-2

A 0.045mm² 0.1-6GHz reconfigurable multi-band, multi-gain LNA for SDR

Arnd Geis*,†, Yves Rolain†, Gerd Vandersteen† and Jan Craninckx*

*IMEC, Kapeldreef 75, 3001 Leuven, Belgium

†Vrije Universiteit Brussel, Pleinlaan 2, 1000 Brussels, Belgium

Abstract—A low area fully reconfigurable multi-band LNA array based on active feedback amplifiers with mixed resistive and switched inductor loads is presented. The 90nm baseline digital CMOS implementation covers the entire frequency range of interest for SDR from 0.1 to 6GHz with a dynamic gain range from 0dB to 22dB. A noise figure as low as 2.7dB and an input-referred linearity IIP3 of -4dBm at 16dB gain is achieved. An IIP3 of +9dBm is reached in low gain mode to allow for high signal and interferer power at the antenna input. The LNA draws between 10 and 26mA from a 1.2V supply. The active area of the array is only 0.045mm².

Index Terms—LNA, multi-band, multi-mode, switched inductor, active feedback.

I. INTRODUCTION

Multi-standard and software defined radios (SDR) rely on widely gain and frequency tunable linear amplifiers for low noise amplification of the RF-signal. Miniaturization and cost considerations further demand for full integration of RF-front-end and digital baseband logic in one system-on-chip (SoC) despite reduced voltage headroom and limited availability of RF options such as ultra-thick-metal (UTM) in deep scaled CMOS nodes.

Multiple front-end solutions have been proposed for SDR ranging from wide-band solutions [1], [2] to multi-band approaches [3], [4]. Wide-band amplification typically results in demanding linearity specifications and strong antenna filter requirements to prevent desensitization, while area requirements are increased when using multiple tuned amplifiers due to an increased number of planar-inductors.

The multi-mode environment sets stringent linearity requirements on the RF-front-end that must provide sufficient flexibility in linearity-gain trade-off to accomodate the large input power range of the signal. The linearity requirements are further aggravated by challenging blocking tests such as the out-of-band blocking test specified in the GSM/DCS standard [5]. In order to prevent desensitization in LNA and following blocks, an LNA by-pass mode is essential.

In this paper we introduce the concept, design and measurement results of a gain programmable multi-mode LNA array in standard digital 90-nm technology. We demonstrate full coverage of the entire frequency range of interest to SDR for DVB-H, GSM, LTE, 802.11a/b/g/n and WiMAX. Area reduction and RF selectivity is achieved by using a combination of low-band and tunable switched inductor based high-band LNA. LNA by-pass operation was implemented in order to maximize linearity and to prevent front-end desensitization for

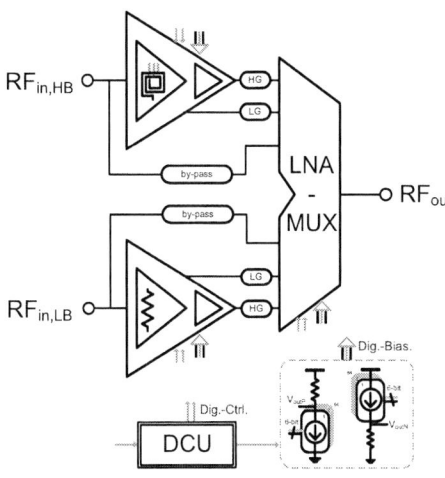

Fig. 1.   Simplified diagram of the proposed LNA array architecture.

strong signal or blocker scenarios. Digitally controlled biasing allows for additional flexible gain/linearity/NF trade-offs.

The following section will introduce the implemented architecture and its individual building blocks. Measurement results on a prototype are shown in Section III. In Section IV conclusions are drawn.

II. LNA ARCHITECTURE

In order to provide seamless coverage of a frequency range of interest, spanning from the UHF band starting as low as 300MHz and up to 6GHz for 802.11, the receive band was split in low-band (LB) ranging from 0.1 to 2.5GHz and high-band (HB) covering frequencies from 2.5GHz to above 6GHz. The separation of LB and HB has multiple advantages such as separate antenna inputs which eases antenna design and optimization. Also the footprint of the LNAs can be optimized by avoiding large inductors necessary for the low frequency range.

The two bands are amplified in individual LNAs. A feedback LNA with resistive load provides flat amplification and input matching throughout the LB as depicted in Fig. 1. The HB is served by an inductively peaked LNA which establishes input matching through a feedback similar to the LB-LNA. The HB-LNA features a programmable resonant load in which coarse frequency selection is achieved by switching between different taps of the load inductor [6]

978-1-4244-6240-7/10 $26.00 © 2010 IEEE

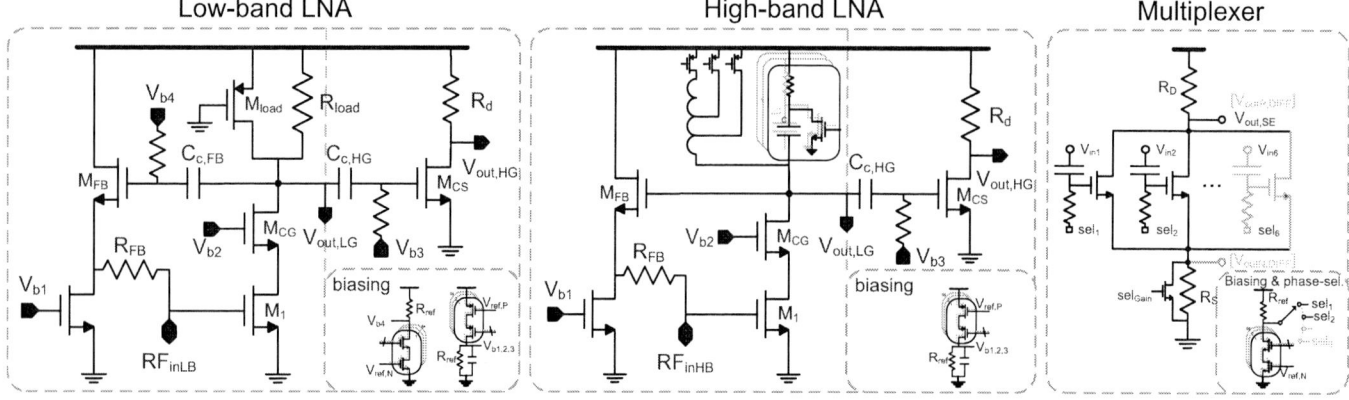

Fig. 2. Schematic of the proposed multi-gain multi-band LNA including digitally controlled biasing.

while frequency fine-tuning is performed with a switched capacitor bank. A multi-tap inductor allows for band switching without sacrificing area which would otherwise be required in order to implement separate individually tuned inductor coils. Gain programming is implemented by additional second gain stages and by-pass modes for both LB and HB inputs. LNA and gain selection is performed in a multiplexing stage which can also be exploited for balun operation in case a balanced RF-output is needed.

A. Low-band LNA

The first stage of the LB-LNA adopts a shunt-shunt feedback LNA topology similar to [7] which was modified to improve linearity and to allow for gain flexibility. High linearity is crucial in order to achieve interference robust front-ends, particularly important in multi-standard environments.

The first stage employs a resistively loaded NMOS cascode amplifier. An additional PMOS in the load operating in triode region (M_{load}) acts as a parallel resistance to the load resistor and allows for decoupling of gain and output potential without degrading the linear performance. An NMOS source follower (M_{FB}) provides feedback to the input of the LNA and establishes 50Ω matching when (1) is satisfied.

$$g_{m,FB} = \frac{1}{50\Omega(g_{m1}Z_{load} + 1) - R_{FB}} \quad (1)$$

Since the input of the source follower is subject to the large signal swing from the output of the cascode amplifier, significant linearity improvements were achieved by biasing the source follower with high overdrive through an RC-network. The overall linearity of the first stage was further boosted by biasing the common-gate in the cascode amplifier sufficiently below V_{dd}. This results in a reduced overdrive and therefore higher swing of the output node without deteriorating the noise performance. A common-source second gain stage provides roughly $7.5dB$ of extra gain. AC-coupling between the first and second stage allows for power-down of the common-source amplifier in medium-gain settings. The LNAs can be entirely switched off by pulling V_{b1} thru V_{b3}

to ground in the DAC sources, which turns off the feedback network and the first and second stage, respectively.

B. High-band LNA

Frequency bands beyond $2.5GHz$ are amplified in the HB-LNA, which, similar to the LB-LNA, is based on NMOS cascode amplifier with a source follower in the feedback path for impedance matching.

At high frequencies, typically above 2GHz, on-chip spiral inductors are readily available with inductance values suitable for inductive peaking in LNAs. The use of resonant loads is desirable since RF-selectivity is provided by the bandpass characteristic of the load. At the same time the voltage drop over the load is reduced when compared to resistive or active loads which improves linearity and operation at low supply voltages.

Various methods of band switching for multi-band LNAs have been proposed. In [8] an approach which is based on a common input device and multiple individually peaked resonant loads multiplexed through the cascode device was employed. The topology has the advantage of having a shared input transistor and therefore a single LNA input. However, it requires as many individual inductors as individual bands are sought. The area overhead is aggravated when a tunable varactor bank for frequency fine tuning is deployed as this bank will have to be multiplied together with the inductors. For the HB-LNA presented in this paper band-switching was introduced through a multi-tap inductor which significantly relaxes area requirements and circuit complexity. The combination of cascode amplifier and multi-tap inductor allows for two different switching configurations as depicted in Fig. 3. Tap-switching can be absorbed into the cascoding devices which improves the tank quality when compared to switching at V_{dd} since no additional switch on-resistance is introduced in series to the inductor. Cascode-switching, however, increases circuit complexity in the same way as the aforementioned switched multi-inductor approach since frequency tuning varactor banks and feedback networks must be repeated for each tap. At the same time multiple outputs are provided which require additional multiplexing. For these reasons a V_{dd}-switched

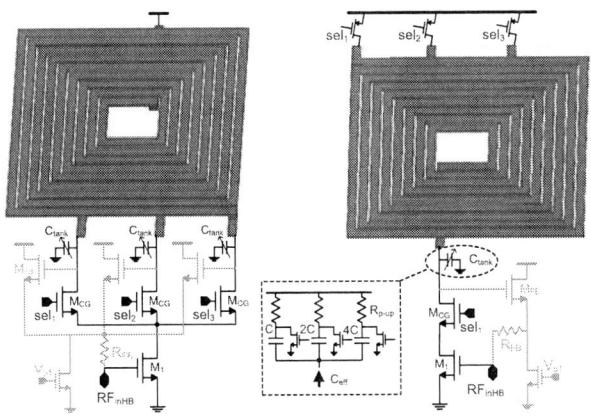

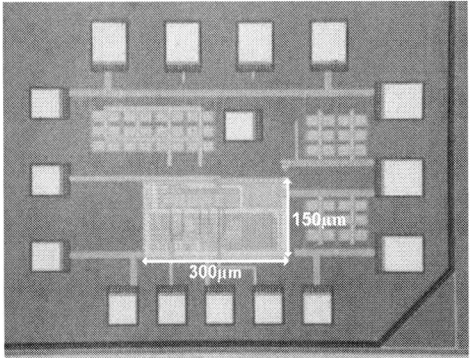

Fig. 4. Chip micrograph.

Fig. 3. Two possible topologies for inductor tap-switching, showing cascode switching (left) and the actual implemented V_{dd}-switching (right).

approach was chosen and sufficiently large switches were used in order to avoid excessive Q degradation.

1) Multi-tap inductor design: The load inductor resonates with the lumped output capacitance of the LNA's first stage. The individual contributors to this capacitance are the stray ground capacitances of the AC-coupling capacitors, the gate capacitance of the feedback NMOS (M_{FB}) and the peripheral capacitance of the cascoding NMOS (M_{CG}). Circuit simulation shows that this capacitance is roughly 1pF. In order to position the resonant peaks evenly throughout the receive band at 2.9, 4.4 and $6.5 GHz$, the inductor taps were placed to provide effective inductance values of 3, 1.3 and $0.6 nH$ according to $f_c = (2\pi\sqrt{LC})^{-1}$.

2) Switched varactor design: Frequency fine tuning is implemented through a bank of 3-bit binary weighed switched MOMCAPs $\{C_{tank}\}$ as shown in Fig. 3. In the off-state the MOMCAP is pulled to V_{dd} by the large resistor R_{p-up} which creates a high impedance path to AC-ground and minimizes the capacitive loading. In the on-state the switch shorts the MOMCAP to ground which maximizes the loading. The effective capacitive tank loading C_{eff} is given by (2) where R_{par} is the parallel resistance of R_{p-up} and the switch resistance.

$$C_{eff} = \frac{\{C_{tank}\}}{1 + (\omega \cdot R_{par} \cdot \{C_{tank}\})^2} \qquad (2)$$

C. Multiplexer

A source degenerated common source amplifier with multiple AC-coupled input devices, as shown in Fig. 2, serves as multiplexer for the different LNA outputs. The source degeneration can be turned off with the NMOS shunt in parallel to R_s which allows for additional gain flexibility. In case a balanced RF-output is needed the multiplexer topology can easily be extended for balun operation simply by sizing the source degeneration resistor (R_s) and drain load resistor (R_d) equally and removing the R_s-shunt NMOS. In this case the differential output is obtained from the drain and source node of the multiplexing transistor and an additional 6dB of

small signal gain is provided from single ended to differential conversion when compared to the un-degenerated single ended version of the multiplexer. The multiplexer was designed for an output driving strength of 700fF in order to drive the output buffer required for measurement. As this buffer is not required when the LNA is embedded in a receiver system, significant power savings of a factor of 4 can be achieved over the dissipation values reported in Table I without NF degradation.

III. EXPERIMENTAL RESULTS

A prototype was fabricated in 1.2V 90nm 1P9M digital CMOS technology without additional analog options. The active area of the LNA array is $0.045mm^2$. The implementation includes a digital control unit for biasing and gain/mode selection which is programmed through a serial interface. The chip was measured with on-wafer probing and the measured power figures reported in Table I include all biasing sources needed for operation.

S-parameter analysis shows the available forward gain (S_{21}) depicted in Fig. 5 and summarized in Table I which is in good agreement with simulation. A very wide dynamic gain-range between 0 and 22dB is achieved by exploiting the available gain settings of the LNAs and the multiplexer. An even finer gain grid is available if the digital-controlled biasing is exploited for gain selection.

Fig. 7 shows the input matching of both RF-inputs which is typically below -10dB within the pass-band.

Two-tone intermodulation tests were carried out in high-gain (HG), low-gain (LG) and by-pass (BP) mode with two sinusoidal tones in the passband and a spectral spacing of 1MHz. Circuit simulation indicates that the linearity in the LNA is dominated by the multiplexing unit which deteriorates the intrinsic LNA linearity by 4 to 8dB. An input compression test (ICP) in by-pass mode shows the saturation point and therefore maximum allowed input power. It is a measure for the LNA's ability to cope with strong interfering signals and a measured ICP of about -3dBm was reached, which is in compliance with the blocking tests of most wireless standards. The LNA's noise figure (NF) was measured in low-gain and high-gain mode and a NF of 2.7dB is reached. The NF deteriorates by 1dB in the high-frequency (HF) inductor

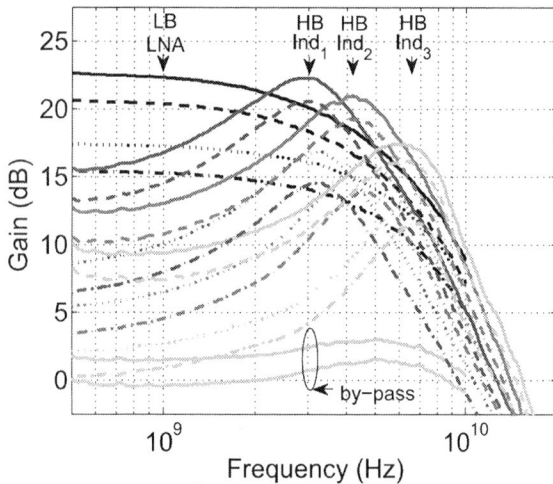

Fig. 5. Measured system gain for various band and gain settings.

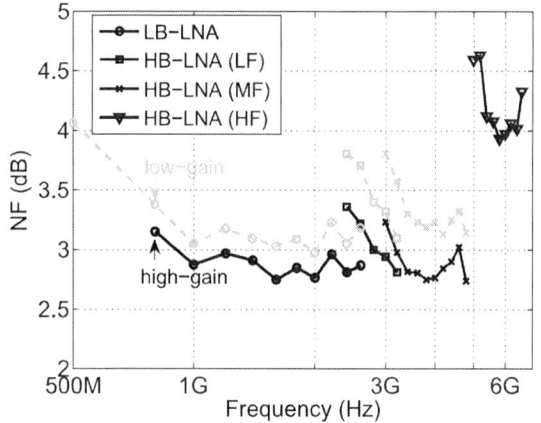

Fig. 6. Measured system NF for various band and gain settings.

setting of the HB-LNA which, related to the gain drop, could be compensated for by optimizing the switch of the HF-coil.

IV. CONCLUSION

A reconfigurabe multi-band LNA array with a wide dynamic gain range in 90nm CMOS has been presented. The design covers the entire frequency range of interest for flexible receivers and in particular SDR applications. Two RF-inputs split the input frequency range which simplifies antenna optimization and complements the LNA's intrinsic RF-selectivity. A by-pass mode was implemented in order to prevent strong interfering signals from saturating the LNA. A combination of LB and HB-LNA with resistive and switched inductor load respectively was employed which effectively reduced the die footprint without compromising on performance and power consumption.

REFERENCES

[1] R. Bagheri et al., "An 800MHz to 5GHz Software-Defined Radio Receiver in 90nm CMOS," in IEEE Int. Solid-State Circuits Conf. (ISSCC) Dig. Tech. Papers, Feb. 2006, pp. 1932–1941.

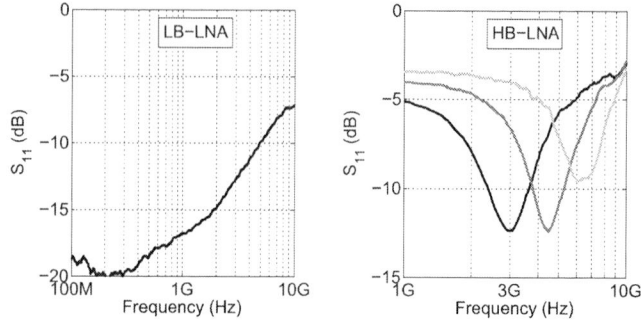

Fig. 7. Measured input matching of the LB and HB antenna inputs.

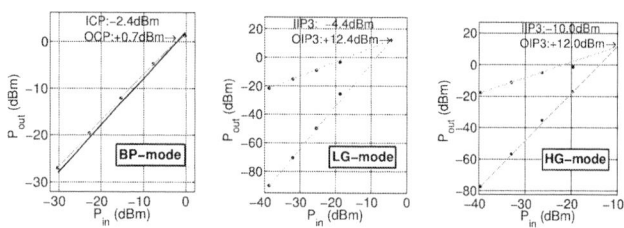

Fig. 8. Measured -1dB compression point (ICP) in by-pass (BP) mode and third order intermodulation tests for low-gain (LG) and high-gain (HG) setting.

TABLE I
MEASURED PERFORMANCE SUMMARY

Ant. input		LB-LNA			HB-LNA		
Freq. (GHz)		0.5	1.0	2.0	3.0	4.5	6.0
Gain (dB)	BP	0-5dB					
	LG	15.4	15.3	14.7	14.5	14.2	12.0
	HG	23	22.5	21.5	22.5	21	18
IIP3 (dBm)	BP	+8.6	+7.0	+9.4	+10.5	+7.5	+7.8
	LG	-4.0	-3.0	-3.6	-4.4	-4.9	-5.0
	HG	-10	-9.8	-8.6	-11	-10	-8.6
ICP (BP)		-3.3	-2.5	-2.9	-5.0	-3.0	-2.4
S$_{11}$ (dB)		-18.5	-17	-15	-12.4	-12.3	-9.6
NF (dB)		4.0	2.8	2.7	2.8	2.7	3.9
P$_{diss}$ (mW) @1.2V	LNA	BP: 0 LG: 12.4 HG: 17.0			BP: 0 LG: 10.8 HG: 18.2		
	Mux	12..14					
	Total	12..32.2					
Technology		UMC 90nm logic/digital CMOS					
Active area		0.045mm^2					

[2] R. van de Beek et al., "A 0.6-to-10GHz Receiver Front-End in 45nm CMOS," in IEEE Int. Solid-State Circuits Conf. (ISSCC) Dig. Tech. Papers, 2008, pp. 128–601.

[3] J. Craninckx et al., "A Fully Reconfigurable Software-Defined Radio Transceiver in 0.13μm CMOS," in IEEE Int. Solid-State Circuits Conf. (ISSCC) Dig. Tech. Papers, 2007, pp. 346–607.

[4] V. Giannini et al., "A 2mm^2 0.1-to-5GHz SDR receiver in 45nm digital CMOS," in IEEE Int. Solid-State Circuits Conf. (ISSCC) Dig. Tech. Papers, 2009, pp. 408–409.

[5] T. S. G. G. R. A. Network, "Radio transmission and reception," 3GPP TS 45.005 V7.11.0 (2007-08).

[6] C.-T. Fu et al., "A 2.45.4-GHz Wide Tuning-Range CMOS Reconfigurable Low-Noise Amplifier," IEEE Transactions on Microwave Theory and Techniques, vol. 56, no. 12, pp. 2754–2763, Dec. 2008.

[7] J. Borremans et al., "An ESD-Protected DC-to-6GHz 9.7mW LNA in 90nm Digital CMOS," in IEEE Int. Solid-State Circuits Conf. (ISSCC) Dig. Tech. Papers, 2007, pp. 422–613.

[8] ———, "Low-Area Active-Feedback Low-Noise Amplifier Design in Scaled Digital CMOS," IEEE J. Solid-State Circuits, vol. 43, no. 11, pp. 2422–2433, Nov. 2008.

A Linearity-Enhanced Wideband Low-Noise Amplifier

Kihwa Choi [1,2], Tamal Mukherjee [2] and Jeyanandh Paramesh [2]

[1] Telecommunication Systems Division, Samsung Electronics, Suwon, Gyeonggi-do, 443-742, Korea

[2] Dept. of ECE, Carnegie Mellon University, Pittsburgh, PA 15213, USA

Abstract- **Techniques are proposed to enhance linearity in a low-voltage wideband LNA for use in a multi-standard wideband receiver. To achieve high linearity over wide frequency range, two previous IMD₃ cancellation techniques are merged and modified to obtain IIP3 peaks at different frequencies, while minimizing component count. A self-biasing current reuse technique is developed to enhance low-voltage operation. Two LNAs are designed in 0.13 μm CMOS to demonstrate these techniques: the (Chebyshev, transformer) LNAs achieved (+10.6, +12.0) dBm IIP3 over (2.3–6.0, 2.0–5.3) GHz, (12.7, 12.4) dB gain, (4.8, 4.9) dB noise figure, while consuming 6.9mA from 1.2V.**

I. INTRODUCTION

An IIP3 peak can be achieved by controlling a FET's gate bias so that the third-order derivative of its dc transfer characteristic is zero. This peak is very sensitive to the gate bias voltage, making it difficult to accurately achieve a high IIP3 in the face of PVT variations. Derivative Superposition (DS) technique reduces this sensitivity by extending the zero crossing points. But its linearity improvement is limited by the contribution of the 2nd-order intermodulation distortion (IMD₂) products that are fed back to the input and mix with the fundamental term, resulting in additional generation of 3rd-order IMD₃ terms. To achieve high linearity at high frequencies, extensions to DS have been proposed [1] [2], so that the IMD₂ and IMD₃ cancel each other in narrow-band LNAs.

While numerous techniques have been proposed for increasing LNA bandwidth (e.g., [3], [4]), there have been relatively few studies of linearity enhancement in wideband LNAs. For example, [5] and [6] employ noise and distortion cancellation techniques to achieve an IIP3 of about 0 dBm at 0.8–2.1 GHz and 0.2–5.2 GHz while consuming 17.4 mW and 21 mW, respectively.

In linearity-enhanced wideband LNAs, linearity enhancement and wideband circuit techniques must be used simultaneously. This paper starts with a source-degenerated topology with input and output matching networks. Two techniques, namely the wideband DS (WBDS) and self-biasing current reuse (SBCR) techniques are proposed to achieve high linearity over a wide frequency range. In the wideband DS method, two IIP3 peaks are realized by merging and modifying two linearity enhancement techniques [1], [2]. SBCR is then used to increase the headroom required by a shunt-peaking load for low voltage operation.

The proposed linearity-enhanced wideband LNA is analyzed and its distortion cancellation is visualized using

Volterra series analysis, leading to a 2–6 GHz design in a 0.13-μm CMOS process. Measurement results show that linearity is enhanced as much as 10 dB, compared to precedent designs.

II. PROPOSED LINEARITY-ENHANCED WIDEBAND LNA

A. Wideband Derivative Superposition Method (WBDS)

Fig. 1 illustrates the principle used to achieve high linearity over wide bandwidth. The linearity enhancement techniques of [1] and [2] are employed and modified to obtain IIP3 peaks at two frequencies, as shown in Fig. 1 (a) and (b), resulting in two independently-set IIP3 peaks (Fig. 1 (c)).

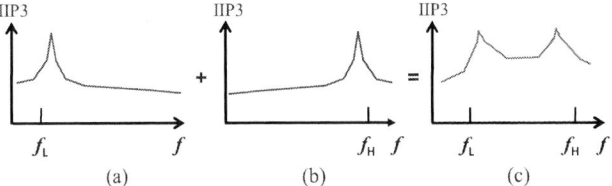

Fig. 1. Conceptual diagram of linearity behavior over an operating frequency: (a) with the low-frequency optimized topology, (b) with the high-frequency optimized topology, and (c) with the composite topology.

Fig. 2 (a) shows a schematic of the WBDS method. The modified DS method of [1], shown inside box (A), is used to set the low frequency IIP3 peak. A modified version of the DS technique of [2], shown inside box (B), is used to set the high frequency IIP3 peak since this technique requires large source inductances to obtain distortion cancellation and good input match. To minimize the number of circuit components, M_B, L_A, and L_B are shared by the two topologies while adjusting the IMD₂ and IMD₃ terms out of phase at the two frequencies.

To understand how the proposed topology can achieve high linearity over a wide bandwidth, an expression for IIP3 is derived using a conventional Volterra series with the equivalent circuit of Fig. 3. Assuming that the drain current is related to the gate-source voltage by a Taylor series, $i_D(V_{GS}+v_{gs}) = I_D + g_1 v_{gs} + g_2 v_{gs}^2 + g_3 v_{gs}^3 + \cdots$, the IIP3 is given by (1)-(2). In the derivation of (1)-(2), several assumptions described in [1] and additional approximations are introduced: a) $g_{1A} \approx 0$, $g_{1F} \approx 0$, $g_{2A} \approx 0$, $g_{2F} \approx 0$, b) two sinusoidal excitation signals are located closely in frequency, c) $\omega_2 C_B L_B \ll 1/4$, d) $\omega_2 C_F L_B \ll 1$. After tedious derivation and approximation, the equations (1)-(2) can be obtained.

$$IIP_3 = \frac{1}{6 \operatorname{Re}[Z_1(s_a)]} \left| \frac{C_1(s_a)}{C_3(s_b, s_b, -s_a)} \right| \cdot \frac{4 g_{1B}^2 \omega^3 [L_A(C_A + C_B) + C_B L_B]}{3 |\varepsilon|} \quad (1)$$

$$\varepsilon = g_{3B} - \frac{g_{2B}^2}{3g_{1B}} + \left[1 + \frac{L_B\left(C_B + j\omega\, g_{1B} C_A L_A\right)}{L_A\left(C_B + C_B\right) + C_B L_B}\right] \cdot \{g_{3A}[1 + (\omega\, g_{1B} L_B)^2]$$
$$\cdot [1 + j\omega\, g_{1B} L_B] + j\omega\, g_{1B} L_B [g_{3A} + \omega^2\, g_{1B}^2 g_{3F} L_B^2]\} \tag{2}$$

Assuming a conjugate input impedance match $Z^*_{in}(s)=Z_1(s)$, the input impedance $Z_{in}(s)$ is given by

$$Z_{in}(s) = sL_A + \frac{1 + sL_A\left[g_{1A} + g_{1B} + (g_{1A} + g_{1F})\cdot(g_{1B} + sC_B)Z_m\right] + (g_{1B} + sC_B)Z_m}{s\left[C_A + C_B + C_A(g_{1B} + sC_B)Z_m\right]} \tag{3}$$

where $Z_m(s)=sL_B/(1+s^2C_F L_B)$.

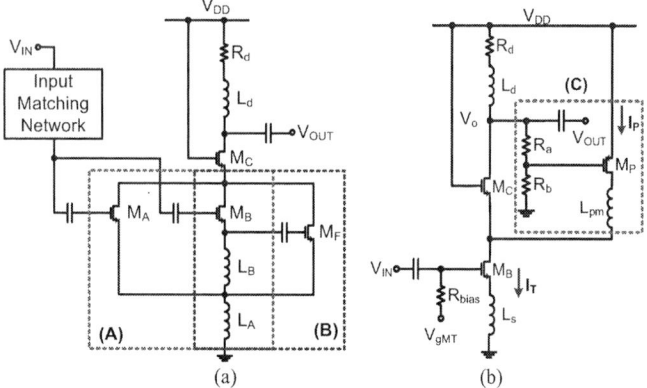

(a)

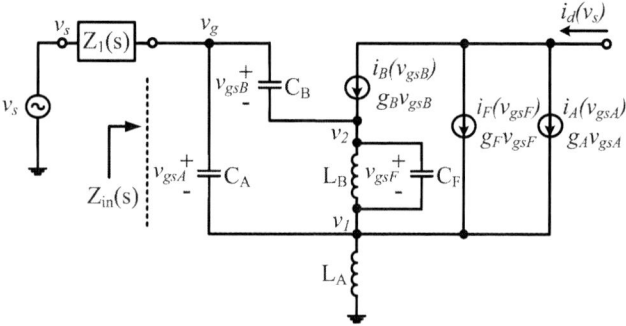

(b)

Fig. 2. LNA employing (a) WBDS and (b) SBCR techniques.

Fig. 3. Small-signal equivalent circuit used for Volterra analysis.

Although the IIP3 expression with dc nonlinearity coefficients (e.g., g_{1B}, g_{2B}, g_{3A}) provides insight about how the coefficients or circuit components (e.g., L_A, L_B) may be adjusted to achieve an IIP3 peak, there are still two remaining limitations of this approach, as addressed below.

First, the IIP3 expression with the dc nonlinearity coefficients accounts for only linear and nonlinear terms (g_1v_{gs}, $g_2v_{gs}^2$, $g_3v_{gs}^3$) in Fig. 4. At higher frequencies, the device capacitances (shown in gray in Fig. 4) affect the $i_d - v_{gs}$ relationship greatly. Thus, the frequency dependence of the device nonlinearity itself, is not accounted adequately, thus hurting the ability of (1)-(2) to accurately predict IIP3. Specifically, the IIP3 calculated by (1)-(2) using the dc nonlinearity coefficients does not match well with the simulated IIP3 as frequency increases since the dc nonlinearity coefficients cannot capture memory effects in the transistors.

Second, the g_{2B} and g_{3B} terms in (2) are independent of the other design parameters, making it difficult to obtain two IIP3 peaks, assuming that g_{2B} and g_{3B} are real numbers. Therefore, the IIP3 expression with the dc nonlinearity coefficients doesn't support the conceptual approach of the WBDS technique shown in Fig. 1.

To address these two issues, a new set of coefficients, called the *extended Volterra coefficients* are introduced. These coefficients are extracted from Harmonic Balance (HB) simulation of the transistors. These coefficients are classified at each output frequency where IMD products are generated. The 1st- and 3rd-order extended Volterra coefficients are defined as the ratio of the *ac* drain current to the *ac* gate-source voltage at each output frequency:

$$g_{1,HB,f1} = \frac{i_{d,HB,f1}}{v_{gs,HB,f1}} \quad g_{3,HB,2f1-f2} = \frac{i_{d,HB,2f1-f2}}{v_{gs,HB,f1}} \tag{4}$$

where $i_{d,HB,f1}$ is an ac drain current at f_1 and $v_{gs,HB,f1}$ the ac gate-source voltage at f_1. The subscripts HB are used to indicate the fact that the coefficients are derived using HB simulation. The 2nd-order coefficient is defined likewise. Using all the extracted coefficients, the 1st-, 2nd-, and 3rd-order composite coefficients are calculated by summing the coefficients and dividing the sum by the number of the coefficients as in (5)-(6).

$$g_{1,HB} = \left(g_{1,HB,f1} + g_{1,HB,f2}\right)/2 \tag{5}$$

$$g_{3,HB} = \left(g_{3,HB,2f1-f2} + g_{3,HB,2f2-f1}\right)/2 \tag{6}$$

With the newly extracted extended Volterra coefficients, the drain current may now be written as

$$i_{d,HB}(v_{gs}) = g_{1,HB}(s)\circ v_{gs} + g_{2,HB}(s_1,s_2)\circ v_{gs}^2 + g_{3,HB}(s_1,s_2,s_3)\circ v_{gs}^3. \tag{7}$$

Fig. 4. Small-signal nonlinear equivalent circuit including the parasitics.

In the practical step of circuit design, the extended Volterra coefficients are applied to analytically estimate the IIP3 of the WBDS topology, and to set a reasonably accurate starting point for component size before optimization through circuit simulation. The extended Volterra coefficients will replace dc nonlinearity coefficients (e.g., g_1, g_2, g_3) in (1)-(3), with the capacitances (C_A, C_B, C_F) set to extremely small values because they are included in the extraction of the extended Volterra coefficients. The IIP3, thus calculated, matches well to the simulated IIP3.

The extended Volterra coefficients are also valuable in understanding the WBDS technique. These are extracted for the transistors in Fig. 2 (a) and are plotted in a polar plot. The polar plot shows that 1st-, 2nd-, and 3rd-order coefficients (e.g., $g_{1,HB}$, $g_{2,HB}$, $g_{3,HB}$) revolve in the clockwise direction because of their frequency dependence on amplitude and phase, with the amplitude shrinking as frequency increases. Using the fact that all the coefficients revolve clockwise as frequency increases, two independent topologies having the IMD$_2$ and IMD$_3$

cancellation techniques can be employed to obtain two IIP3 peaks at two frequencies at which the composite 3rd-order coefficient has the same amplitude and opposite phase with respect to the 2nd-order coefficient. The mechanism of distortion cancellation can be demonstrated by the conceptual vector diagram shown in Fig. 5 (a)-(b). f_L and f_H represent the low- and high-corner frequencies and "∘" is the Volterra operator in the frequency domain.

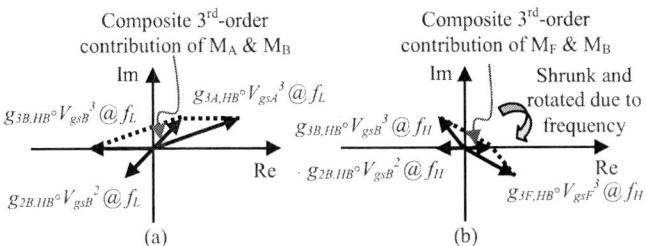

Fig. 5. Conceptual vector diagram with the WBDS method (a) at low-corner frequency and (b) at high-corner frequency.

B. Self-Biasing Current Reuse Technique (SBCR)

A shunt-peaking output load network is often employed in wideband LNAs to extend bandwidth. A large peaking resistance is necessary for high gain, but degrades headroom and linearity, especially at low V_{DD}. When the gate bias voltage of the main transistor M_B in Fig. 2 (b) is increased, the increased drain current I_T of M_B flows through the load resistor R_d. The larger load current increases the voltage drop across the load resistor R_d and results in decreased drain-source voltages of M_B and M_C. This causes difficulties in keeping M_B and M_C in saturation and thus degrades the RF performance of the LNA. The current-reuse technique [7], originally proposed for low-V_{DD} Gilbert cell mixers, is adopted and modified herein to improve the headroom available in the proposed LNA, as shown inside box (C) of Fig. 2(b). The current I_P flowing through the PMOS M_P is controlled by the feedback voltage from the drain node of the cascode transistor M_C to the gate of M_P. This is called the self-biasing current reuse technique because I_P self-adjusts to keep the voltage V_o relatively constant. The current I_P is set by the resistance ratio of the voltage dividing resistors, R_a and R_b. With SBCR, large voltage headroom can be guaranteed to obtain high gain and linearity even at low V_{DD}. The RF choke L_{pm} prevents ac current from flowing into M_P.

C. LNA Design

Two LNA prototypes are designed in a 0.13 μm CMOS process to verify the effectiveness of the proposed technique. One LNA uses a Chebyshev matching network (Fig. 6 (a)) [3] while the other uses a transformer-based matching network (Fig. 6 (b)) [4]. The simplified LNA is shown in Fig. 7 with a buffer amplifier added for test purposes. The initial size of the transistors (M_A, M_B, M_F) is determined so that the composite 3rd-order dc characteristic is set to near zero as done in the narrow band DS topologies. Then the LNAs are optimized in *MATLAB* using the IIP3 expressions (1) (2) and the extended Volterra coefficients extracted from HB simulation.

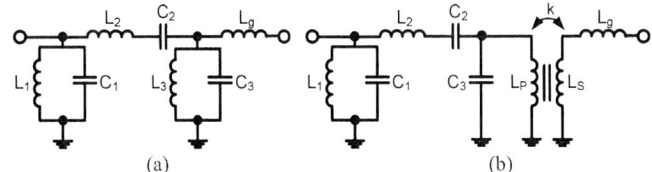

Fig. 6. Input matching networks (a) Chebyshev BPF (b) transformer-based.

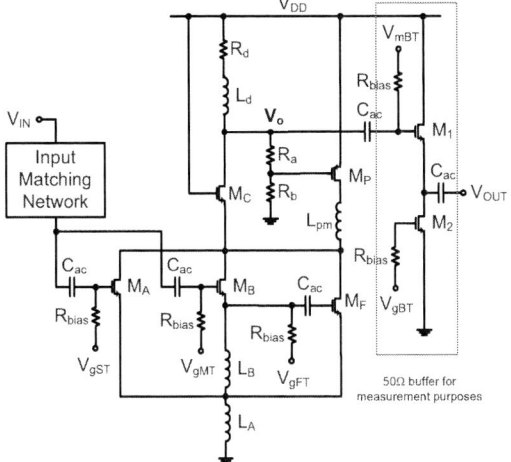

Fig. 7. Schematic of the proposed linearity-enhanced wideband LNA.

III. SIMULATED AND MEASURED RESULTS

Fig. 8 (a) and (b) show the simulated and measured S11, gain, and NF of the two prototypes. The dotted and solid lines represent the simulated and measured performance, respectively. The LNA with the Chebyshev matching network shows the 3 dB bandwidth from 2.3 GHz to 6 GHz with the maximum gain of 12.7 dB and the minimum NF of 4.8 dB. The LNA with the transformer-based matching network shows the 3 dB bandwidth from 2 GHz to 5.3 GHz with the maximum gain of 12.4 dB and the minimum NF of 4.9 dB. Both LNAs show S11>-10dB at some frequency bands. The poor S11 comes from the complexity of the linearity-enhanced topology and design procedure to achieve high linearity, *i.e.*, first the size of the three transistors is determined and then two degeneration inductors are chosen.

The IIP3 is simulated and measured over the operating frequency range to confirm the effectiveness of the proposed linearity-enhanced wideband topology with two the LNAs. The results in Fig. 9 show high linearity over the wide frequency range. The LNA with the Chebyshev matching network has a maximum IIP3 of +21.5 dBm at 4 GHz and shows an average IIP3 of +10.6 dBm from 2.3 GHz to 6 GHz. The LNA with the transformer-based matching network shows a maximum IIP3 of +15.6 dBm and relatively flat IIP3 of +12 dBm on average from 2 GHz to 5.3 GHz.

An interesting issue in a wideband LNA is the IIP3 performance when differently spaced interferers exist within the multi-standard frequency bands. To observe this important measure for a multi-standard wideband LNA, two test tones with different frequency spacing are applied. The resulting IIP3 performance is shown in Fig. 10. The IIP3 holds steady for small frequency spacing and starts to degrade for larger frequency spacing since the linearity enhancement techniques

978-1-4244-6240-7/10 $26.00 © 2010 IEEE

lose their effectiveness. This shows that the highest IIP3 can be obtained in the proposed linearity enhancement technique when two test tones are closely placed. For multi-standard wideband LNAs, the IIP3 measurement setup should be carefully considered depending on a field environment having multiple coexisting in-band interferers (which would be out-of-band interferers in a narrowband LNA).

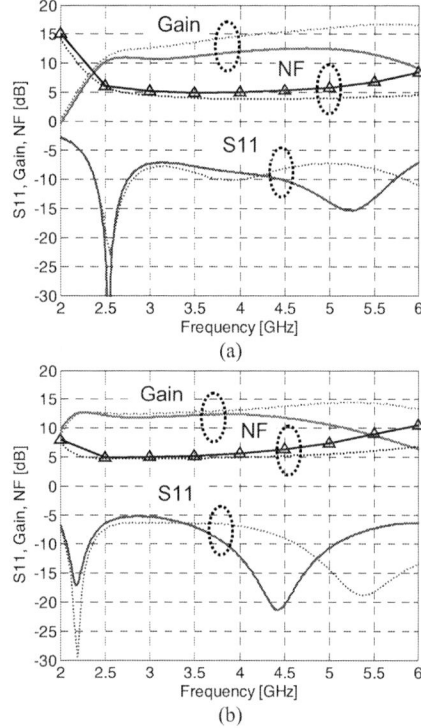

(a)

(b)

Fig. 8. Comparison of simulated and measured results in the LNA with (a) Chebyshev BPF and (b) transformer-based input matching network.

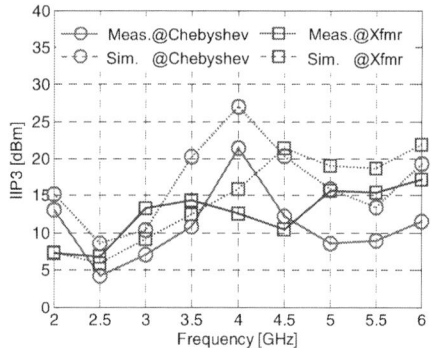

Fig. 9. IIP3 over operating frequency range.

The overall performance is summarized in Table 1 and compared to the current state-of-art designs of linearity-enhanced narrow band topologies [1], [2], wideband topology [3], and wideband topologies with noise and distortion cancellation [5], [6].

IV. CONCLUSION

To achieve high linearity in a wideband LNA, two linearity enhancement techniques are proposed: the wideband derivative superposition method to obtain two IIP3 peaks over a wide frequency range and the self-biasing current reuse technique to provide sufficient voltage headroom. The

effectiveness of the proposed topology is analyzed by Volterra series and supported by conceptual diagram inspired from the behavior of the extended Volterra coefficients. The simulated and measured results in two LNA topologies show relatively high IIP3 over the wide frequency range. Furthermore, the IIP3 performance as a function of frequency spacing of two test tones is demonstrated to estimate IIP3 performance variation when differently spaced interferers are received.

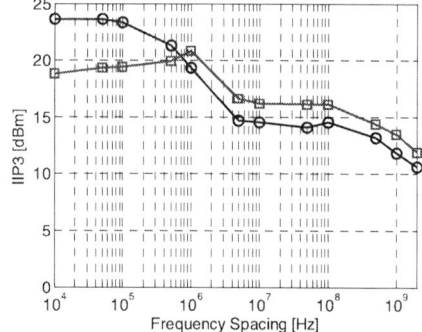

Fig. 10. IIP3 vs frequency spacing plot at 4GHz in the LNA with Chebyshev BPF and transformer-based input matching network.

Table 1. Performance summary and comparison with precedent designs.

Ref.	Freq. [GHz]	Gain [dB]	NF [dB]	IIP3 [dBm]	Supply [V]	Power [mW]	Tech.
[1]	0.9	15.5 [1]	1.65	+22	2.6	24.2	0.25um
[2]	0.9	18.5 [1]	1.76	+21	2.5	22.5	0.35um
[3]	2.3–9.2	9.3 [1]	4.0	-6.7	1.8	9	0.18um
[5]	0.8–2.1	14.5	2.6	16 [2] ~ 0 [3]	1.5	17.4	0.13um
[6]	0.2–5.2	15.6	< 3.5	> 0	1.2	21	65 nm
This work (Chebyshev)	**2.3–6.0**	**12.7**	**4.8**	**10.6 [4] 21.5 [5]**	**1.2**	**8.3**	**0.13um**
This work (Transformer)	**2.0–5.3**	**12.4**	**4.9**	**12.0 [4] 15.6 [5]**	**1.2**	**8.3**	**0.13um**

[1] S21 [2] at large freq. spacing [3] at small freq. spacing [4] average [5] peak

ACKNOWLEDGMENT

We would like to thank Samsung (Korea) for supporting Kihwa Choi with a fellowship.

REFERENCES

[1] V. Aparin and L. E. Larson, "Modified derivative superposition method for linearizing FET low-noise amplifiers," *IEEE Trans. Microwave Theory and Techniques*, pp. 571-581, Feb 2005.

[2] S. Ganesan, E. Sanchez-Sinencio, and J. Silva-Martinez, "A highly linear low-noise amplifier," *IEEE Trans. Microwave Theory and Techniques*, pp. 4079-4085, Dec 2006.

[3] A. Bevilacqua and A. M. Niknejad, "An ultrawideband CMOS low-noise amplifier for 3.1–10.6-GHz wireless receivers," *IEEE J. Solid-State Circuits*, vol. 39, p. 2259–2268, Dec 2004.

[4] K. Choi, D. H. Shin, and C. P. Yue, "An ultra-wideband RF front-end receiver with an active balun in a 0.13-μm CMOS process," Carnegie Mellon University, Pittsburgh, Technical Report 2008.

[5] W. -H. Chen, G. Liu, B. Zdravko, and A. M. Niknejad, "A highly linear broadband CMOS LNA employing noise and distortion cancellation," *IEEE J. Solid-State Circuits*, vol. 43, no. 5, pp. 1164-1176, May 2008.

[6] S. C. Blaakmeer *et. al.*, "Wideband balun-LNA with simultaneous output balancing, noise-cancelling and distortion-cancelling," *IEEE J. Solid-State Circuits*, vol. 43, no. 6, pp. 1341-1350, June 2008.

[7] S. -G. Lee and J. -K. Choi, "Current-reuse bleeding mixer," *Electronics Letters*, vol. 36, no. 8, pp. 696-697, Apr 2000

Power Efficient Distributed Low-Noise Amplifier in 90 nm CMOS

Brecht Machiels, Patrick Reynaert, and Michiel Steyaert

K.U.Leuven, ESAT-MICAS, Kasteelpark Arenberg 10, 3001 Heverlee, Belgium

Abstract — A low-power wideband distributed low-noise amplifier (DLNA) in 90 nm CMOS is presented. Various techniques have been combined in the design to increase the distributed amplifier's power efficiency. These techniques range from moderate inversion biasing to transmission line tapering. The measured gain of the 12.5 mW DLNA is larger than 15 dB from DC to 21 GHz. The average noise figure in the pass-band is 5.4 dB, the IIP3 at 5 GHz is -6.6 dBm and the total die area is 0.41 mm^2.

Index Terms — broadband, CMOS, distributed amplifier, low-noise amplifier (LNA), low-power, tapered.

I. INTRODUCTION

Distributed amplifiers (DA) are a class of amplifiers that do not suffer from the classic gain-bandwidth trade-off. Instead, it can be said that signal delay can be exchanged for gain by distributing gain cells along artificial transmission lines. This makes them well suited for broadband applications such as medical imaging and high-frequency instrumentation. The distributed amplifier was originally conceived in the 1940's using vacuum tubes as gain cells [1]. In the past decade, this type of amplifier has resurfaced in the form of monolithic circuits.

While distributed amplifiers are capable of very wide bandwidths, they are rather power inefficient. This makes them less suited for low-power applications such as wireless frontends in mobile devices.

In this paper, the design of a distributed wideband low-noise amplifier (LNA) with very low DC power consumption is presented. In section II, a number of techniques to improve a distributed amplifier's power efficiency are explored. Section III discusses the design of the low power distributed LNA. Finally, section IV reports about the measurement results of the DLNA.

II. LOW-POWER TECHNIQUES

There are several opportunities for cutting back power consumption in a distributed amplifier. A number of techniques are discussed in this section. These have been combined in the design presented in section III.

A. Output Line Tapering

In a distributed amplifier, each gain cell (typically a common source or cascode amplifier) injects current into

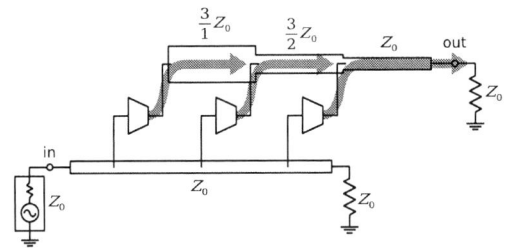

Fig. 1.   Splitting of the currents in right- and left-going waves.

Fig. 2.   A distributed amplifier with a tapered output line.

the output transmission line. This current splits equally into a right-going and a left-going wave, as is shown in Fig. 1. Only the right-going waves add in phase and are dissipated in the load. The left-going waves interfere constructively or destructively depending on the frequency and are dissipated in an internal termination resistor, as shown on the left in Fig. 1. This means that half of the generated output power of the distributed amplifier is lost.

However, it is possible to taper the output line so that all of the output power ends up in the load [1], as pictured in Fig. 2. The tapering of the transmission line characteristic impedances is chosen such that there are no left-going current waves. As can be seen from the figure, the required range of the lines' characteristic impedances is quite large.

The transmission lines are typically approximated by means of LC ladders [1], in which the capacitors exist in the form of gain cell (transistor) parasitics. The following equations express the relations between the characteristic impedance Z_0 and the values of the inductors L and capacitors C in an LC ladder.

$$Z_0 = \sqrt{\frac{L}{C}} \qquad (1)$$

978-1-4244-6240-7/10 $26.00 © 2010 IEEE   115

The range of Z_0 in the tapered transmission line translates to a range of required values for the inductors ($L \sim Z_0^2$) and capacitors ($C \sim 1/Z_0^2$) making up the LC ladder. In order to ensure that signals add in phase, it is important to keep the delay T_D constant across the LC ladder sections. The delay of an LC ladder section is given by

$$T_D = \sqrt{LC} = Z_0 C = \frac{L}{Z_0} \qquad (2)$$

This means that when the characteristic impedance increases, the capacitance needs to decrease by the same factor. Also, the inductance needs to increase by this factor.

B. Cascading

One disadvantage of the distributed amplifier is that the gain of the cells *add* instead of *multiply*, as in cascaded amplifiers. This is a result from the fact that the cells are connected in parallel, as opposed to in series. The additive gain implies that a the ratio of the total gain to DC power consumption will typically be less than that of a classic cascade topology. It is however possible to place a number of DA's in series in order to increase the gain and still maintain the wideband behaviour. Unfortunately, this implies that a very long transmission line – the series combination of an output line and an input line – is formed. Due to the losses in the active and passive components making up this transmission line, a signal traveling over it will be strongly attenuated. Therefore, it is advantageous to cascade the DA's in a "matrix" structure [2], so that the output line of the first stage is also used as the input line of the second stage. However, this topology suffers from a frequency-dependent gain due to the left-going current waves in the shared transmission line. We therefore propose to use a tapered transmission line in which currents flow in one direction only, as shown in Fig. 3.

An interesting property of a distributed amplifier with a tapered output line is that the voltage at the output of each gain cell is equal to the total output voltage of the amplifier.

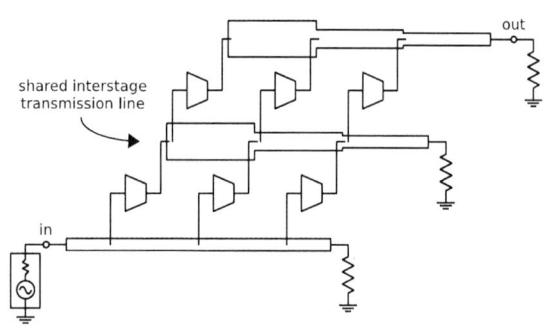

Fig. 3. A tapered matrix distributed amplifier.

This is due to the fact that the characteristic impedance increases towards the left. Thanks to this property, the total gain of the two stages is the same whether cascading the two stages the classic way or in the proposed matrix configuration (Fig. 3).

C. High-Z_0 Interstage Transmission Line

In a cascaded distributed amplifier, the gain can be increased by choosing a higher characteristic impedance Z_0 for the interstage transmission line [3]. The amount by which Z_0 can be increased is however limited by the values of the inductances and (parasitic) capacitances which make up the artificial transmission lines (2).

D. Input Mismatching

Nearly all input power is dissipated in the termination resistor of the input transmission line. Only a fraction of the input power is dissipated in the gates of the MOSFETs and thus amplified. Increasing the characteristic impedance of the input transmission line lowers the total amount of RF input power the circuit absorbs due to mismatch with the source. On the other hand, more power is dissipated in the MOSFETs due to the higher gate voltage, leading to a higher gain.

E. MOSFET Biasing

Another measure to reduce an amplifier's power consumption is to bias the MOSFETs toward weak inversion [4]. Biasing the transistors towards weak inversion increases the ratio of the transconductance g_m to the DC drain current. However, larger (W) transistors are required to preserve g_m. The corresponding larger parasitic capacitances limit the achievable gain-bandwidth (GBW) of the amplifier. The amount of power that can be saved by this means is thus limited by the application's gain and bandwidth requirements. Therefore, the transistors' gate bias voltage is chosen as low as possible for given bandwidth (C_{gs} and thus W) and gain (g_m) requirements.

F. Common-Source Gain Cells

Many distributed amplifier designs adopt cascode gain cells, as they improve the reverse isolation of the amplifier. However, a cascode cell requires a higher output bias voltage in order to ensure proper operation, and thus is less power efficient than a common-source cell. In addition, the internal node of a cascode cell is typically highly capacitive which limits the bandwidth of the amplifier [5], [2]. For these reasons, we choose to use common-source gain cells instead of cascode cells, provided the circuit's stability can still be guarantueed.

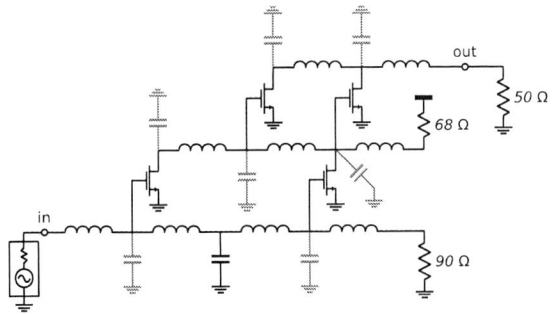

Fig. 4. Circuit schematic. Capacitors in gray represent parasitic MOSFET capacitances.

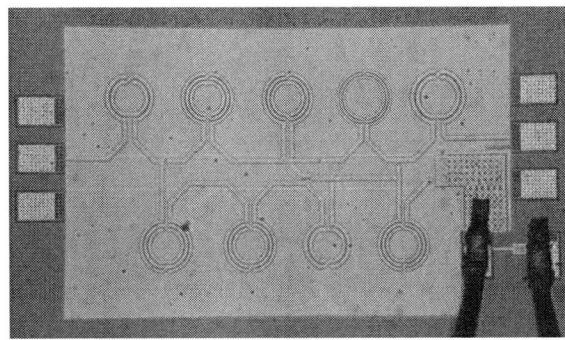

Fig. 5. Chip photograph of the distributed low-noise amplifier. The occupied die size is 0.9×0.46 mm^2.

III. DESIGN

A low-power distributed amplifier was designed in a 90 nm bulk CMOS technology. The circuit shown in Fig. 4 represents a compromise between the design techniques listed in the previous section:

- The amplifier consist of two distributed amplifiers with tapered transmission lines in a matrix configuration. This exploits multiplicative gain and avoids losses in internal termination resistors (sections II-A and II-B).
- The amount of gain cells in each of these distributed amplifiers is limited to two. This limits the range of required coil values (due to the tapering) and keeps losses in the transmission lines to a minimum.
- The characteristic impedance of the input and inter-stage transmission lines are 90 Ω and 68 Ω respectively in order to obtain a higher gain to DC power consumption ratio (section II-C and II-D).
- The four common-source MOSFETs were arranged and sized such as to satisfy the required capacitance in each of the LC ladder sections, dictated by the desired bandwidth. The biasing follows from the desired total gain (section II-E and II-F).
- The inductors were optimized to provide equal signal delays in each of the LC ladder sections.

A micrograph of the fabricated circuit is shown in Fig. 5.

IV. MEASUREMENT RESULTS

Fig. 6 and 7 shows the amplifier's S-parameters. Measurements and simulations of the forward gain S_{21} show good agreement. The average measured gain of 15.4 dB is only slightly less than the average simulated gain of 16 dB. The peaking around 1 GHz is due to a bondwire. Gain variation in the pass band (not considering the bond wire peaking) is less than 1 dB. Reverse isolation is shown to be adequate, being smaller than -30 dB across the whole frequency range. The μ stability factors calculated from the

S-parameters indicate the circuit is unconditionally stable (not shown). S_{11} and S_{22} show maxima of about -6.5 dB. This is due to the intentional mismatching at the input and the output line tapering respectively (see section II).

The amplifier's noise figure (NF) and input-referred third-order intercept point (IIP3) are plotted in Fig. 8. The measured noise figure is within 1 dB of the simulations. It drops from about 6 dB at the low end of the band

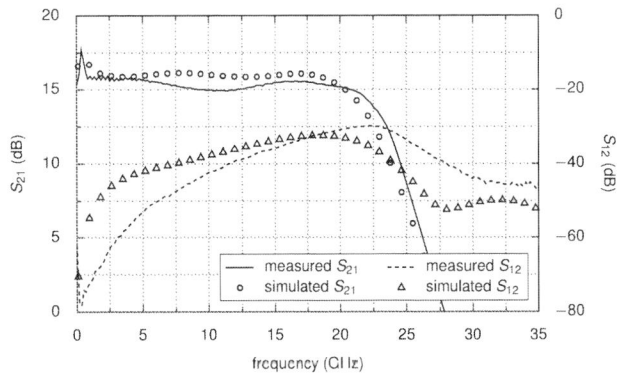

Fig. 6. Forward and reverse gain.

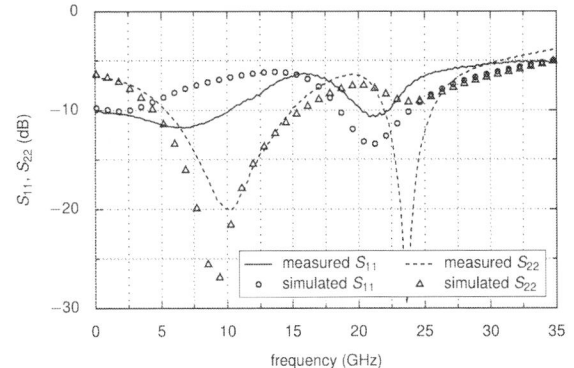

Fig. 7. Reflection coefficients.

	CMOS [μm]	flat-gain BW [GHz]	avg S_{21} [dB]	spot NF [dB]	IIP3 [dBm]	P_{DC} [mW]	Area [mm^2]
[2]	0.18	33.4	17			260	2.3
[3]	0.18	22	7.3	4.3 - 6.1 (to 18 GHz)	8.7	52	1.35
[4]	0.18	6.2	8	4.2 - 6.2	3 @ 2 GHz	9	1.16
[5]	SiGe 0.18	12	8.5	2.9 - 4	-3.8 @ 5 GHz	21.6	0.76
[6]	0.18	23	9		-1 @ 10 GHz	60	0.36
[7]	0.18	2.3 - 9.2	< 9.3	4 - 7.5	-6.7 @ 6 GHz	9 + 9 (buffer)	1.1
[8]	0.13	3 - 9.4	12	1.8 - 4.7		30	0.83
This Work	0.09	21	15.4	4.4 - 6	-6.6 @ 5 GHz	12.5	0.41

Table I
BROADBAND LNA PERFORMANCE COMPARISON

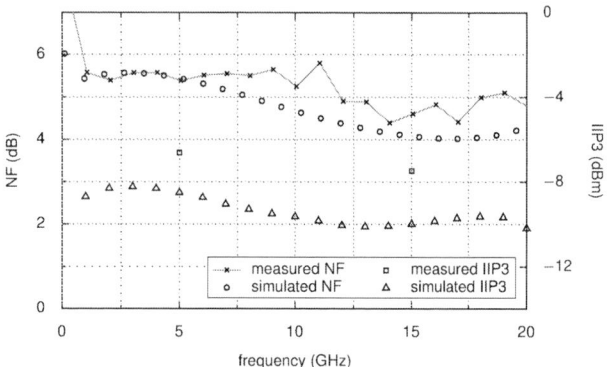

Fig. 8. Measured versus simulated noise figure and IIP3.

to date, making it an attractive option for low-power broadband applications. The low power consumption comes partly at the cost of linearity however. In addition, the LNA demonstrates a good noise figure and very small die area.

ACKNOWLEDGMENT

The authors would like to thank Frederik Daenen (ESAT-MICAS/IMEC), Noëlla Gaethofs (ESAT-MICAS) and Ilja Ocket (ESAT-TELEMIC) for bonding and technical support during measurements. The authors also thank Elie Maricau (ESAT-MICAS) for the enlightening discussions concerning distributed amplifiers. Brecht Machiels is supported by an IMEC research scholarship.

REFERENCES

[1] E. Ginzton, W. Hewlett, J. Jasberg, and J. Noe, "Distributed amplification," *Proceedings of the IRE*, vol. 36, no. 8, pp. 956–969, aug 1948.

[2] J. Chien and L. Lu, "40-Gb/s high-gain distributed amplifiers with cascaded gain stages in 0.18-um CMOS," *Solid-State Circuits, IEEE Journal of*, vol. 42, no. 12, pp. 2715–2725, 2007.

[3] R. Liu, C. Lin, K. Deng, and H. Wang, "Design and analysis of DC-to-14-GHz and 22-GHz CMOS cascode distributed amplifiers," *Solid-State Circuits, IEEE Journal of*, vol. 39, no. 8, pp. 1370–1374, 2004.

[4] F. Zhang and P. Kinget, "Low-power programmable gain CMOS distributed LNA," *Solid-State Circuits, IEEE Journal of*, vol. 41, no. 6, pp. 1333–1343, 2006.

[5] P. Heydari, "Design and analysis of a performance-optimized CMOS UWB distributed LNA," *Solid-State Circuits, IEEE Journal of*, vol. 42, no. 9, pp. 1892–1905, 2007.

[6] M.-D. Tsai, K.-L. Deng, H. Wang, C.-H. Chen, C.-S. Chang, and J. Chern, "A miniature 25-GHz 9-dB CMOS cascaded single-stage distributed amplifier," *Microwave and Wireless Components Letters, IEEE*, vol. 14, no. 12, 2004.

[7] A. Bevilacqua and A. Niknejad, "An ultrawideband CMOS low-noise amplifier for 3.1-10.6-GHz wireless receivers," *Solid-State Circuits, IEEE Journal of*, vol. 39, no. 12, pp. 2259–2268, 2004.

[8] K. Moez and M. Elmasry, "A low-noise CMOS distributed amplifier for ultra-wide-band applications," *Circuits and Systems II: Express Briefs, IEEE Transactions on*, vol. 55, no. 2, pp. 126–130, 2008.

[9] R. Baki, T. Tsang, and M. El-Gamal, "Distortion in RF CMOS short-channel low-noise amplifiers," *Microwave Theory and Techniques, IEEE Transactions on*, vol. 54, no. 1, pp. 46–56, Jan. 2006.

down to 4.4 dB around 17 GHz. IIP3 measurements were performed at 5 GHz (124.3 MHz tone spacing) and 15 GHz (10 MHz spacing). The IIP3 values are strongly linked with the relatively low gate bias voltage of the MOSFETs [9] resulting from the biasing strategy discussed in section II-E and should be weighted against the power consumption of the circuit. The total measured DC power dissipation equals 12.5 mW, whereas the simulation reports 12.8 mW.

The circuit's performance is summarized in table I together with specifications of other published circuits featuring a comparable bandwidth. The presented DLNA demonstrates a record gain-bandwidth to power consumption ratio. Its noise figure is comparable to that of the other broadband amplifiers. Linearity on the other hand suffers from the low-power design. Finally, the circuit's die area is among the smallest published.

V. CONCLUSION

This paper discussed a number of techniques for drastically increasing a distributed amplifier's power efficiency. These have been combined in the design of a low-power wideband LNA. Measurements of the LNA show the highest published gain-bandwidth to power consumption ratio

A Wide-Band RF Front-End with Linear Active Notch Filter for Mobile TV Applications

Seung Hwan Jung, Kang Hyuk Lee, Young Jae Lee[1], Hyun Kyu Yu[1], Yun Seong Eo

Radio Frequency Circuits and Systems Laboratory Kwangwoon University, Seoul Korea

[1]Electronics and Telecommunications Research Institute (ETRI), Daejeon Korea

Abstract — This paper presents a wide-band RF front-end with linear active notch filter covering both T-DMB and DVB-H. A single to differential converter with the low amplitude/phase error and 6dB step RF VGA using the capacitor are implemented. Also, highly linear and Q-enhanced tunable active inductor is proposed. The linear active notch filter rejects GSM band up to 23dB and achieves 20dB linearity improvement. The RF front-end is fabricated on 90nm CMOS technology and consumes 29.7mW.

Index Terms — Wide-band LNA, Single to differential converter, High linear active inductor, Notch filter, CMOS.

I. INTRODUCTION

Nowadays, the mobile TV markets such as Terrestrial-Digital Multimedia Broadcasting (T-DMB) in Korea and Digital Broadcasting Handheld (DVB-H) in Europe have been growing rapidly in the aid of explosive interest in watching TV with very small handheld device. Up to date, the conventional CMOS receiver chips are realized in the direct conversion or the low IF receiver architectures and each band occupies a separated narrow band RF receiver chain [1]. Contrary to the previous works, a single chain wide-band RF front-end applicable to digital RF receiver architecture is designed in this work for the dual mobile TV bands. Recently, to handle wide-band RF input signal, many literatures have been addressing to the noise canceling LNA [2].

In the digital RF receiver, the gain control range (more than 100dB) of receiver should be covered only in RF front-end and ADC stage due to the absence of very large gain control BBA. In our digital RF receiver, the allocated dynamic range of ADC is 43dB, which corresponds to about 7bits, therefore, the required gain dynamic range of RF front-end is more than 57dB. In order to achieve large dynamic range and accurate gain step, the RF VGA is realized using capacitor divider and switched g_m stage. Apart from the dynamic range issue, the strong in-band or out-band interference may come into the receiver front-end due to wide-band characteristics. Therefore, a tunable RF notch filter with the active inductor is adopted to reject the strong interferer such as the undesired GSM signal around 900MHz. However, the conventional active

inductor has a poor linearity and cannot withstand the strong GSM interferer even with the external GSM band rejection SAW filter. To overcome these drawbacks, a newly proposed linear Q-enhanced active notch filter is presented in this paper. In order to obtain a high linearity of the active inductor, the MGTR (Multiple Gated Transistor) topology is adopted to the gyrator's trans-conductor cell and push-pull type negative resistor circuit is also used, which maintain the constant inductance and quality factor in spite of very strong GSM input interferer.

II. THE PROPOSED ARCHITECTURE

Fig. 1 illustrates the proposed wide-band RF front-end with linear active notch filter. The RF front-end consists of the wide-band noise canceling LNA, the low amplitude/phase error single to differential (S/D) converter, the 6dB step RF VGA, and the linear active notch filter. The Integrated S/D converter acts as an off-chip transformer with 3bit gain control of 2dB step. Additionally, the RF buffer which is not shown in Fig. 1 is also included for the measurement.

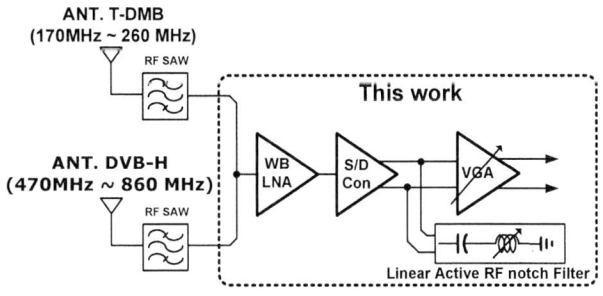

Fig. 1. Block diagram of the RF front-end.

III. THE DESIGN OF RF FRONT-END

A. Wide-band noise canceling LNA

Fig. 2 shows the schematic of the designed wide-band CMOS noise canceling LNA and how to achieve noise cancellation, simultaneously.

978-1-4244-6240-7/10 $26.00 © 2010 IEEE

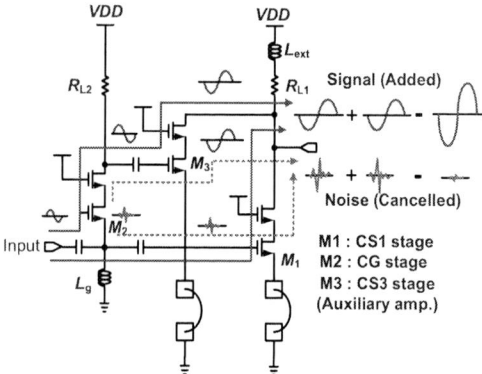

Fig. 2.   Schematic of the noise canceling LNA

Similar to a previously reported the noise canceling LNA [2], the input signal (solid line) goes through two paths of common-gate (CG) and common-source (CS) amplifier, and is combined constructively at the output. Whereas, the channel thermal noise occurring at CG stage would be subtracted at the output node, since the output noise signal through two paths (dotted line) is out of phase to each other. The noise factor of designed LNA can be obtained as follows with the assumption of low frequency operation.

$$F = 1 + \frac{4\gamma g_{do2}(g_{m1}R_s - g_{m3}R_{L2})^2 + 4\gamma(g_{do1} + g_{do3})}{G_s(g_{m1}R_s + g_{m3}R_{L2})^2} . \quad (1)$$

where γ is a coefficient of channel current noise, g_{do1} and g_{do3} are drain-source conductance of M1 and M3 at zero V_{DS}, R_{L1} and R_{L2} are load resistor of CG and CS stage and G_s is the source admittance. If we set $g_{m1}R_s = g_{m3}R_{L2}$ for noise cancellation, (1) can be minimized simultaneously with wide-band input matching property.

In order to obtain a broad-band response, the shunt peaked load is employed for the output load using a resistor R_{L1} and an off-chip inductor. Only with the help of resistor load, the gain cannot be sufficient to meet the receiver NF requirement in spite of resistor's wide-band characteristic. Finally, in order to obtain low gain mode, there is an additional through path switched by MOS transistor between input and output of LNA, which is not shown in Fig. 2.

B. Wide-band Single to differential converter

The proposed S/D converter consists of two amplifier stage. The first stage converts a single ended signal to differential one and the following fully differential stage with feedback capacitor, C_F, improve frequency response of S/D converter. Also, 3 bit gain control with 2dB step is included at the second stage load, R_L with switched resistor array.

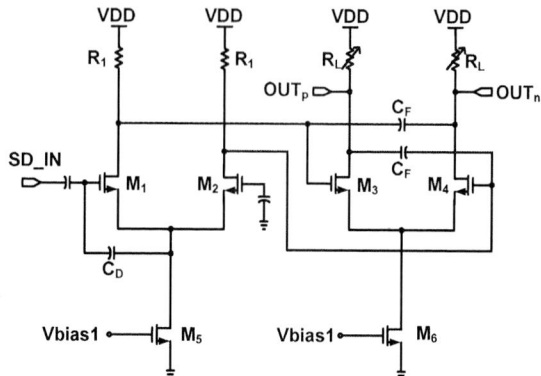

Fig. 3.   Wide-band single to differential converter

As shown in Fig. 3, the capacitor C_D compensates parasitic capacitance between drain and source of M5. Therefore, the input small signal can be divided equally through the parasitic capacitances (C_{gs}) of M1 and M2 at high frequency region and amplitude/phase imbalance caused parasitic mismatches can be minimized. In the wide-band circuit design, the parasitic drain-gate feedback capacitance is the most significant limitation such as a second CS differential stage. This parasitic capacitance can be reduced significantly by inserting feedback capacitor, C_F through M3's gate to M4's drain vice versa [3]. In second stage, the frequency response will be enlarged since the zero pole of output impedance can be boosted as large as C_F and the S/D converter has wide-band property from low frequency region to 1GHz with the aid of adjusting C_F properly.

C. RF VGA with capacitor divider

R-2R ladder attenuation is a well known circuit topology to implement the RF VGA. However, R-2R ladder has drawbacks such as noise degradation due to the resistors [4]. In this paper, the designed RF VGA using capacitor divider consists of six differential trans-conductance stages and capacitive voltage divider instead of resistive one. A capacitor divider which substitutes 2C-C ladder of RF VGA provides 6dB gain control per one G_m stage switching. Total 30dB gain range is achieved with six identical G_m stages and only one G_m stage is enabled at a time. The resistive load is used to provide flat frequency response from 170MHz to 860MHz (T-DMB and DVB-H band). A digital control section decodes a 6bit gain control from 12dB to -18dB with 6dB gain step.

D. Linear Q-enhanced active notch filter

In order to reject the strong interferer such as GSM band, the RF front-end requires a linear LC notch filter. The LC notch filter which can be applicable to GSM band should be implemented around 900MHz and the inductance of inductor should be as large as hundreds nH to obtain a reasonable rejection property. It is difficult to realize hundreds nH inductance as on-chip inductor, therefore, an active inductor which can obtain high inductance and quality factor is needed. Generally, the active inductor is realized as the gyrator-C structure in RF frequency region [6]. The G_m cell of gyrator-C and its small signal equivalent circuit are illustrated in Fig. 4.

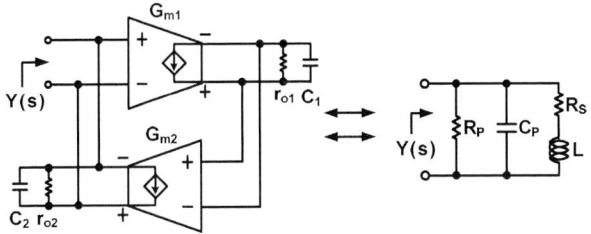

Fig. 4.   Gyrator-C and its small signal circuit

The gyrator-C can be represented equivalently by RLC network shown in Fig. 4. The equivalent inductance of the active inductor and other parameters are as follows;

$$R_p = r_{o2}, \quad C_p = C_2, \quad L = \frac{C_1}{g_{m1}g_{m2}}, \quad R_s = \frac{1}{g_{m1}g_{m2}r_{o1}}. \quad (2)$$

where $g_{m1,2}$ is trans-conductance of the $G_{m1,2}$ cell and R_s is parasitic resistance of the active inductance.

Previously reported active inductors [5], [6] by gyrator-C and Q-enhanced negative g_m circuits have defects that the g_m and L can be varied drastically as the input signal increases. Thus, a resonant frequency $(1/\sqrt{LC})$ of the notch filter using the active inductor is shifted unexpectedly. Meanwhile, the rejection of the notch filter which is associated with the loss of active inductor, resistive components in (2), can be improved by adopting negative g_m (Q-enhanced) circuit. The conventional negative g_m circuit which consists of only NMOS pair has negative input impedance, $-2/g_m$ and also has nonlinearity due to input power variation. To complement these detrimental, a newly proposed active inductor adopts MGTR [7] in the gyrator input G_m cell and push-pull structure at negative g_m stage while remaining class AB bias to improve linearity. Thus the proposed notch filter including gyrator-C of MGTR G_m cell and push-pull negative g_m circuit preserves the quality factor as well as constant notch frequency even when the large input power. The circuit schematic of the proposed highly linear Q-enhanced gyrator G_m cell and conventional one [6] are illustrated in Fig. 5.

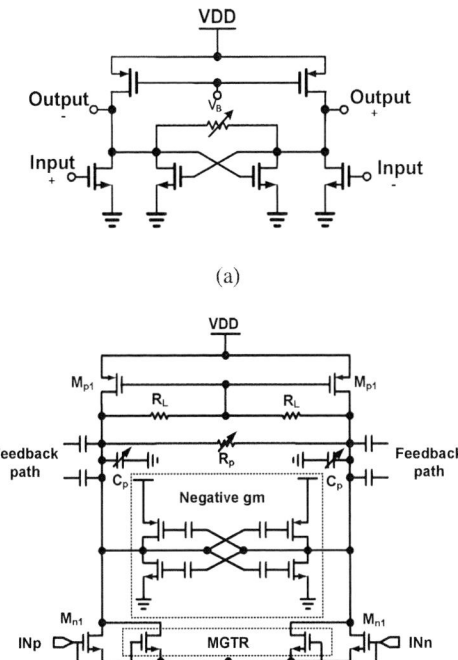

Fig. 5.   (a) Conventional g_m cell [6] (b) Proposed linear Q-enhanced g_m cell

IV. EXPERIMENT RESULTS

The designed RF front-end was implemented in a standard 90nm RF CMOS technology. Fig. 6 shows the chip micro-photograph, which has an active area of 2270um x 650um including pad frame.

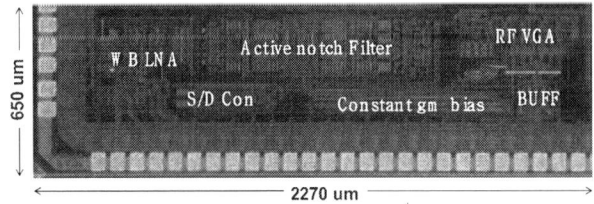

Fig. 6.   Chip micro-photograph

As shown in Fig. 7, the input reflection coefficient is below -10dB from T-DMB to DVB-H band regardless of high/low gain mode. The power gain is 41.0dB and 3-dB frequency is 850MHz, also the notch frequency is 900MHz with 23dB rejection. In this frequency range of T-DMB and DVB-H, the noise figure is from 3.1dB to 3.8 dB. The gain control range of the RF front-end is from -28dBm to 41dBm as shown in Fig. 8. The improved linearity of GSM band rejection using the push-pull structure and that of the notch filter resonant frequency employing MGTR are both around 20dB as shown in Fig. 9 and Fig. 10. The OIP3 is 4.3dBm at high gain mode in the 500MHz region. The wide-band RF front-end employing active notch filter consumes 14.6mW or 29.7mW with the active notch filter on or off.

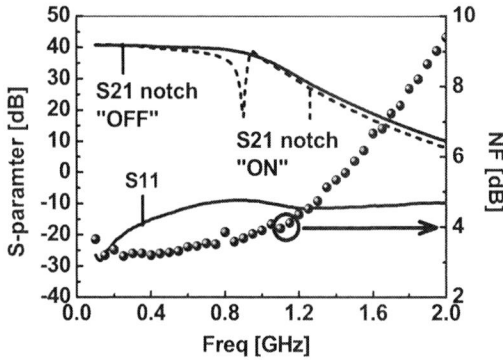

Fig. 7.   Measured s-parameter and Noise figure

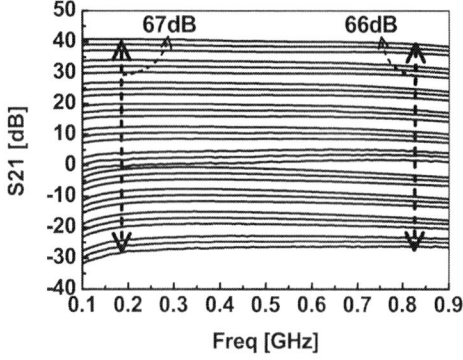

Fig. 8.   Measured RF front-end Gain control

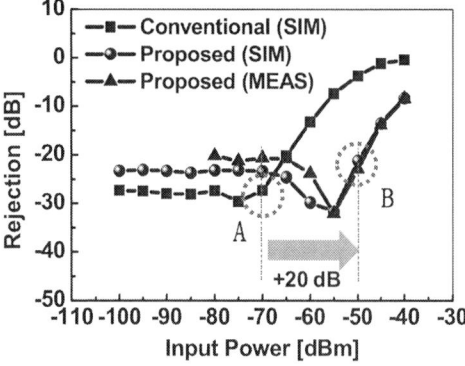

Fig. 9.   Improved GSM band rejection of the linear Q-enhanced active notch filter

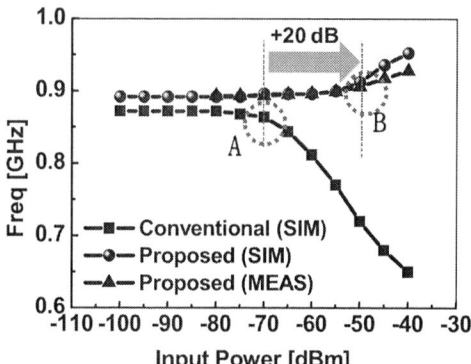

Fig. 10.   Improved linearity of the notch filter resonant frequency

VII. CONCLUSION

A wide-band RF front-end which exploits the wide-band LNA, the low amplitude/phase error S/D converter, the 6dB step RF VGA using capacitor divider, and the linear active notch filter is presented. The RF front-end achieves a 66dB gain control range and 3.8dB noise figure. The RF front-end also has newly proposed Q-enhanced active notch filter that improves linearity about +20dB to sufficiently reject the GSM interferer with the only help of poor external SAW filter.

ACKNOWLEDGEMENT

The authors would like thank for supporting from the IT R&D program of MKE/KEIT [2008-F-008-01].

REFERENCES

[1]   I. Vassiliu et al., "A 65nm CMOS Multistandard, Multiband TV Tuner for Mobile and Multimedia Applications," *IEEE J. Solid-State Circuits*, vol. 43, no. 7, pp. 1522-1533, Jul. 2008.

[2]   Chih-Fan Liao et al., "A Broadband Noise-Canceling CMOS LNA for 3.1-10.6GHz UWB Receiver," *IEEE Custom Integrated Circuits Conf.*, pp. 161-164, Sep. 2005.

[3]   Alen B. Grebene, *Bipolar and MOS Analog Integrated Circuit Design,* Wiley INTESCIENCE 2003, pp. 415-416.

[4]   Xiao, J et al., "A High Dynamic Range CMOS Variable Gain Amplifier for Mobile DTV Tuner", *IEEE J. Solid-State Circuits*, vol. 42, pp. 292-301, Feb 2007.

[5]   Fei Yuan, *CMOS Active Inductor and Transformers: Principle, Implementation and Applications,* Springer, 2008.

[6]   Farsheed Mahmoudi et al., "8GHz Tunable CMOS Quadrature Generator using Differential Active Inductors", *IEEE International Symposium on Circuits and Systems 2005*, vol. 3, pp. 2112-2115, May 2005.

[7]   Bonkee Kim et al., "A New Linearization Technique for MOSFET RF Amplifier Using Multiple Gated Transistors," *IEEE Microwave and Guided Wave Letters*, vol. 10, no. 9, pp.371-373.

RMO2C-1

A mm-Wave Arbitrary 2^N Band Oscillator Based on Even-Odd Mode Technique

Alvin Hsing-Ting Yu, Sai-Wang Tam, David Murphy, Tatsuo Itoh, M.C. Frank Chang

University of California, Los Angeles, CA 90095

Abstract — **A technique to build mm-wave arbitrary 2^N band oscillators is presented. Based on even-odd mode operation, the technique breaks the fundamental tradeoff between frequency switching range and tank quality factor, Q, which exists in classical switched-capacitor and switched-inductor methods. As a result, this technique achieves multi-band operation with FOMs comparable to single band oscillators. To verify the theory, a quadruple band oscillator with 4 arbitrary chosen frequencies (43, 49, 58 and 75 GHz) is implemented in 65-nm CMOS technology. The phase noise measurements taking at 1 MHz offset are -100.3, -95.3, -93.8 and -86.2 dBc/Hz, respectively. The power consumption of the oscillator core is 12mW. The presented technique would enable the development of mm-wave software-defined multi-standard radios.**

Index Terms — **mm-wave, multiband, quadruple band, even-odd mode, oscillator, VCO, 60 GHz, software-defined radio, CMOS.**

I. INTRODUCTION

Driven by software-defined multi-standard radios, for frequency below Ku band, many VCO multiband techniques have been presented [1]-[4]. However, such techniques have not been applied to mm-wave frequencies. Recently FCC released several mm-wave licensed and unlicensed bands to fulfill increasing demand for multi-Gb/sec data transmission. For example, there are licensed bands at 71-76, 81-86 and 92-95 GHz, which are open for high-speed, point-to-point wireless local area networks. Furthermore, an unlicensed band at 57-64 GHz has caught attention from both academia and industry. In addition, the 40.5-43.5 GHz band is used for local multipoint distribution services. Therefore, it is expected that a need for a frequency source that can cover all or a substantial part of these bands will emerge in the future. Classically, implementing such a source is done by multiplexing several sources on-chip [5], [6]. At mm-wave frequencies, however, the multiplexer (MUX) itself is required to operate over an ultra-wide frequency range. Such a design consumes excessive area and power. The MUX can be avoided if the oscillator employs a single tank with multiple resonant modes.

For frequencies below Ku band, the switched-capacitor and switched-inductor methods are the most common [1]. However when using these methods at mm-wave frequencies, a large tuning range corresponds to excessive

switch loss, and a resultant degradation in the Q of the LC tank. An alternative is the coupled inductor method [2], [3], however, for N bands, there are N-choose-2 (C_2^N) inductor coupling factors, k, that need to be carefully specified. For example, a 4-band oscillator requires 4 coupled inductors and needs 6 individual k-factors to be set. Designing such a passive structure with a limited number of good conducting layers in CMOS is difficult. In [4], Goel and Hashemi present a dual-resonance oscillator composed of two LC tanks in series. While the topology cleverly avoids Q-degradation due to switch loss, the resistance and parasitic capacitance of series non-oscillating tank can be problematic; as mentioned in [4], asymmetric waveforms, increased flicker noise and a small reduction in the Q of the LC tank can be expected. As the number of bands increases, the number of series LC tanks also increases, and these issues become more serious. Another approach is the dual-band oscillator that exploits left-handed material (LH) [7]. But, to expand this approach to N bands, 2·(N-1) inductors are required, i.e., 6 inductors for 4 band. Furthermore, for N larger than two, in N-cell LH ring resonator, it is very challenging to switch between various equal phase node sets with 2nπ/N phases to deliver the desired modes, where n=0, -1 … -(N-1). To resolve the problems, we present the simple even-odd mode 2^N band technique without the previous issues. And the technique requires only one inductor per tone.

II. EVEN-ODD MODE MULTIBAND TECHNIQUE

A. Fundamental Even-Odd Mode Block

Fig.1 shows a coupled transmission line and its equivalent capacitance network [8]. With the assumption of TEM propagation, two special types of excitation, even- and odd-mode, have been studied in literature as shown in Fig. 2. With even-mode excitation, the voltages on the strip conductors are in phase and equal in amplitude, which leads to the equivalent capacitor

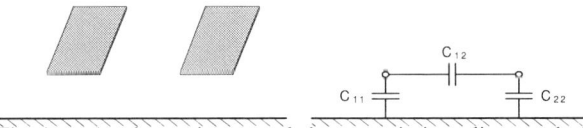

Fig. 1. A three-wire coupled transmission line and its equivalent capacitance network.

978-1-4244-6240-7/10 $26.00 © 2010 IEEE

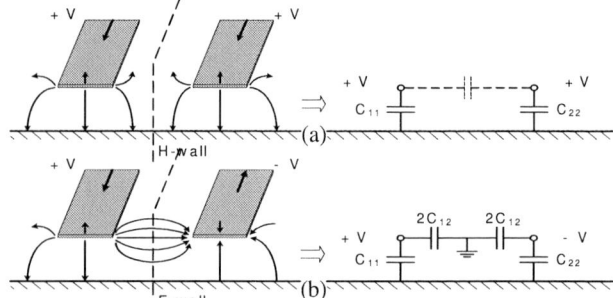

Fig. 2. Even- and odd-mode excitations for a coupled line, and the resulting equivalent capacitance networks. (a) Even-mode excitation. (b) Odd-mode excitation.

network shown in Fig. 2(a), where C_{12} is effectively open-circuited. On the other hand, for the odd mode, the voltages on the strip conductors are 180° out of phase and equal in amplitude, which leads to a virtual ground plane in between the two strip conductors. Hence each of the strips is effectively connected to the virtual ground by $2C_{12}$.

Following this concept, we can create the fundamental even-odd mode building block: a differential dual band even-odd mode resonator, as shown in Fig. 3. One can understand Fig. 3(a) as two LC tanks connected to each other by two C_1 capacitors. Fig. 3(b) and 3(c) show the differential even- and odd-mode resonance equivalent circuits, respectively, where "differential" refer to the voltages across each of the LC tank resonate differentially. Since the two LC tanks are identical, the resonant voltages are equal in amplitude. For even mode, the two LC tanks resonate in phase and C_1 is effectively open-circuited. On the other hand, for odd mode, the two LC tanks resonate 180° out of phase. Hence, each end of each LC tanks is effectively connected to the virtual ground by $2C_1$. The two resonant frequencies are $1/\sqrt{LC}$ and $1/\sqrt{L(C + C_1)}$. These are the only two modes that the resonator can support.

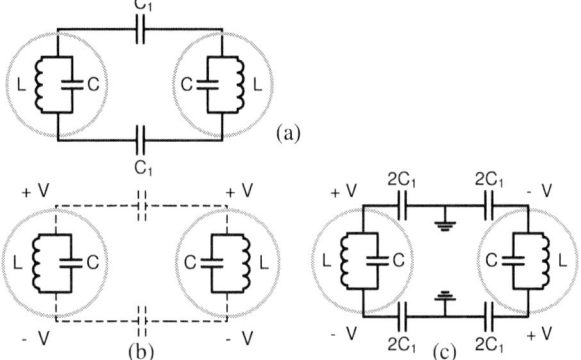

Fig. 3. Differential dual band even-odd mode resonator (a) Schematic. (b) Differential even-mode resonance equivalent circuit. (c) Differential odd-mode resonance equivalent circuit.

B. 2^N band Even-Odd Mode Resonator

To build a 2^N band differential even-odd mode

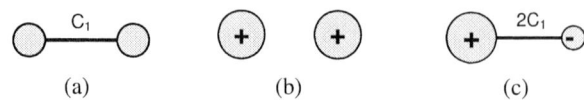

Fig. 4. Simplified schematic of Fig. 3 (a) Schematic. (b) Differential even-mode resonance equivalent circuit. (c) Differential odd-mode resonance equivalent circuit.

resonator, first simplify the schematic of Fig. 3 as Fig. 4, with each circle representing a LC tank. Since only two phases can exist, let us mark one as "plus" and the other, which is 180° out of phase with "plus", as "minus". The thick line in schematic 4(a) represents the two capacitors connected between two LC tanks in Fig. 3(a). The thick line in equivalent circuit 4(c) represents that each end of the LC tanks effectively sees $2C_1$ to virtual ground.

To expand the resonator from 2^{N-1} band to 2^N band, first, make a replica of a 2^{N-1} band resonator. Than connect the corresponding LC tanks of the two resonators by C_N capacitor pairs. By this method, Fig. 5 and Fig. 6 show a exemplary differential 4-band and 8-band even-odd mode resonators, respectively. The 4 resonant frequencies of the 4-band resonator are

$$1/\sqrt{LC}, \ 1/\sqrt{L(C + C_1)}, \ 1/\sqrt{L(C + C_2)}, \ 1/\sqrt{L(C + C_1 + C_2)}$$

The 8 Resonant frequencies of the 8-band resonator are

$$1/\sqrt{LC}, \ 1/\sqrt{L(C + C_1)}, \ 1/\sqrt{L(C + C_2)}, \ 1/\sqrt{L(C + C_3)}$$
$$1/\sqrt{L(C + C_1 + C_2)}, \ 1/\sqrt{L(C + C_2 + C_3)},$$
$$1/\sqrt{L(C + C_1 + C_3)}, \ 1/\sqrt{L(C + C_1 + C_2 + C_3)}$$

Fig. 5. Differential 4 band even-odd mode resonator (a) Schematic. (b) Equivalent circuits of 4 different modes.

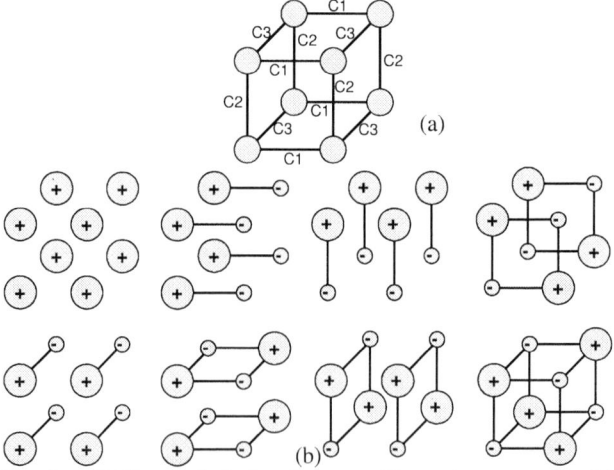

Fig. 6. Differential 8 band even-odd mode resonator (a) Schematic. (b) Equivalent circuits of 8 different modes.

978-1-4244-6240-7/10 $26.00 © 2010 IEEE   124

To prove the 2^N inductor resonator has 2^N differential resonant frequencies, notice that each LC tank is connected to N other LC tanks through $C_1 \ldots C_N$ capacitor pairs. From even-odd mode analysis, each capacitor pair can either be seen or act as equivalent opened circuit. Thus the total number of combinations will be

$$C_1^N + C_2^N + \cdots + C_N^N = (1+1)^N = 2^N$$

An alternative way to think this is as follow. First, assume a 2^{N-1} inductor resonator have 2^{N-1} resonant frequencies, which is true for N=2. When we add additional C_N pairs to expand the resonator to 2^N inductor structure, all the combination of resonant frequencies of the 2^{N-1} resonator get another degree of freedom from C_N pairs, which offer 2 extra choice, even or odd mode. Thus the total number of resonant frequencies equals to $2^{N-1} \times 2 = 2^N$.

C. Even-Odd Mode Selecting Switch

To select the wanted mode, even-odd mode selecting switches are added to each of the C_N capacitor pairs, as shown in Fig.7. The switch set is compose of a pair of parallel switches, S_e, to select even mode and a pair of cross switches, S_o, to select odd mode. The switches S_e and S_o are operated complimentarily. Notice that once the wanted mode is selected, the voltages at the two ends of each "on" switch are the same in both phase and amplitude. Thus the switches are effectively open-circuited. Hence, they do not consume power, do not load the tank, and do not contribute noise. That is, the switches degrade the Q of unwanted modes, but do not affect the Q of the wanted mode.

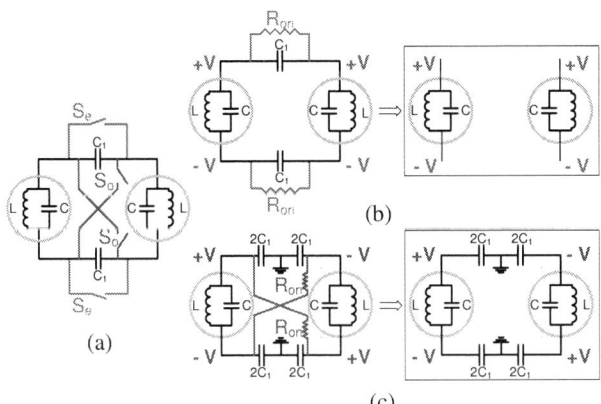

Fig. 7. Differential dual band even-odd mode resonator with complimentary mode selecting switches. (a) Schematic. (b) Differential even-mode equivalent circuit. (c) Differential odd-mode equivalent circuit.

III. mm-Wave Quadruple Band Oscillator

A quadruple band oscillator with the resonator in Fig. 5 and mode selecting units in Fig. 7 is implemented to verify the even-odd mode multiband technique, as shown in Fig. 8. The total parasitic capacitance each inductor sees is

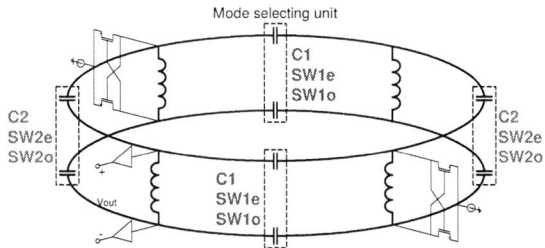

Fig. 8. Schematic of the quadruple band oscillator.

designed to match each other. Fig. 9 shows the large signal tank impedance (when driven with a 1.6V sine-wave) versus frequency under 4 different switch-set configurations. The small signal impedance of the switchless passive tank is also plotted for comparison. Notice that, even with this small switch size W/L=1/0.06μm, the quality factor of the wanted mode is well maintained in each plot. The impedance of the other tree unwanted modes is suppressed.

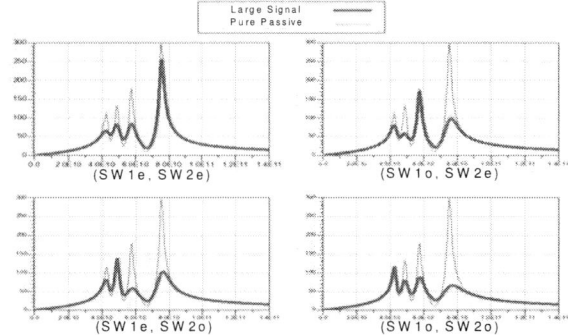

Fig. 9. Impedance of switchless passive tank vs large signal 1st harmonic impedance of the tank with W/L=1/0.06μm mode selecting switches over different turned-on switch set.

IV. Measurements

The proposed quadruple band oscillator is implemented in a 65 nm CMOS process. The 4 arbitrary chosen frequencies were 43, 49, 58 and 75 GHz. The size of C_1, C_2 and L are 20fF, 38fF and 160pH, respectively. The sizes of the mode selection PMOS switch, negative resistance pair and buffer are 3/0.06, 8/0.06 and 16/0.06 μm. The total power consumption of the oscillator core is 12mW. The measured four oscillating frequencies are 42.9, 48.1, 56.4 and 73.7 GHz (Fig 10), and the phase noise at 1MHz offset are -100.3, -95.3, -93.8 and -86.2 dBc/Hz, respectively. Fig. 11 shows the phase noise measurement at 48.1 and 56.4 GHz.

The Figure of Merit (FOM) of a current biased oscillator is given as

$$FOM = \frac{1}{P_{SUPPLY} \, \mathcal{L}\{\Delta\omega\}} \left(\frac{\omega_0}{\Delta\omega}\right)^2 = 2 \left(\frac{Q^2}{kT}\right) \left(\frac{A}{\pi F V_{DD}}\right)$$

where $P_{SUPPLY} = V_{DD} I_{BIAS}$ and $A = 2 \cdot I_{BIAS} R_P / \pi$. Neglecting the lesser importance of F and assuming that the amplitude is maximized, the FOM is determined by Q. Since in our

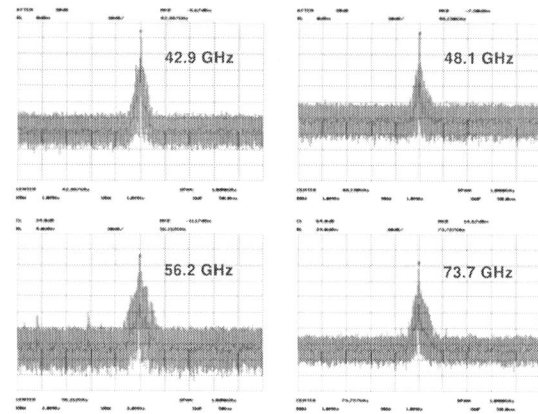

Fig. 10. Measured output spectrums

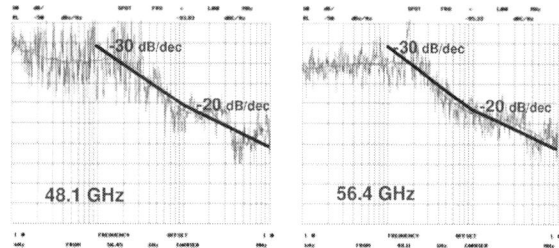

Fig. 11. Measured phase noise

topology the Q is well maintained for all the 4 oscillating frequencies, the proposed quadruple band oscillator has FOM comparable to a state-of-the-art single band oscillator. This is demonstrated in the comparison table and plot in Fig. 12. To ensure a fair comparison, only works with a frequency tuning range less than 5% are included (large continuous tuning typically requires a large varactor, which has a deleterious effect on the Q of mm-wave oscillator.) Fig. 13 shows the chip photograph. The core active area is 200μm×200μm.

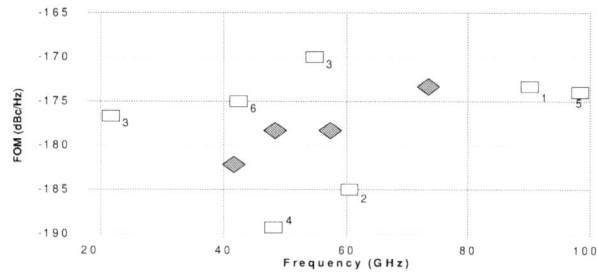

Reference	Process CMOS nm	# of band	f_o (GHz)	FTR	P_{DC} (mW)	Phase Noise @1M (dBc/Hz)	FOM (dBc/Hz)
This Work	65	4	42.8	n/a	12	-100.3	-182
			48.1	n/a		-95.3	-178
			56.4	n/a		-93.8	-178
			73.7	n/a		-86.2	-173
1 D. Kim, ISSCC 2009	SOI 32	1	89.9	4.45	7.7	-102.7 @10M	-173
2 L. Li, JSSC 2009	90	1	61.7	4.81	1.2	-90	-185
3 S. Tam, RFIC 2009	90	2	21.3	n/a	14	-100.8	-175.8
			55.3	n/a		-86.7	-170
4 Y. Lin, VLSI 2009	130	1	47.9	1.59	5.6	-102.5	-189
5 C. Cao, JSSC 2006	130	1	98.5	2.5	7	-102 @10M	-174
6 A. van der wel, JSSC 2004	130	1	43	1.7	7	-91	-175

Fig. 12. FOM comparison plot and table.

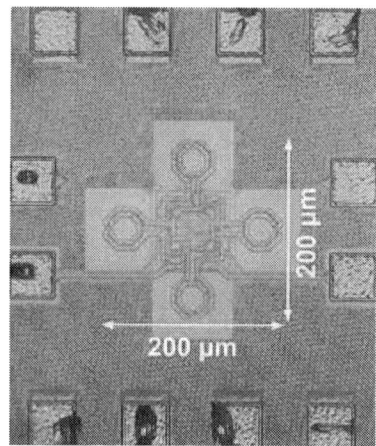

Fig. 13. Die micrograph.

V. CONCLUSION

The developed multiband technique can build an oscillator that can switch between arbitrary 2^N mm-wave bands without degrading Q. The technique breaks the fundamental tradeoff between Q and switching range that exists in classical switched-capacitor and switched-inductor methods. As a result, this even-odd mode 2^N band technique achieves FOMs comparable to single band oscillators.

ACKNOWLEDGEMENT

The Authors would like to thank TSMC for foundry support.

REFERENCES

[1] Z. Li, K.K. O, "A low-phase-noise and low-power multiband CMOS voltage-controlled oscillator," *IEEE J. Solid State Circuits*, vol.40, no.6, pp. 1296-1302, June 2005.

[2] B. Catli, M. M. Hella, "A 1.94 to 2.55 GHz, 3.6 to 4.77 GHz Tunable CMOS VCO Based on Double-Tuned, Double-Driven Coupled Resonators," *IEEE J. Solid State Circuits*, vol.44, no.9, pp.2463-2477, Sept. 2009.

[3] Z. Safarian, H. Hashemi, "Wideband Multi-Mode CMOS VCO Design Using Coupled Inductors," *IEEE Trans Circuits and Systems I: Regular Papers,* , vol.56, no.8, pp.1830-1843, Aug. 2009.

[4] A. Goel, H. Hashemi, "Frequency Switching in Dual-Resonance Oscillators," *IEEE J. Solid State Circuits*, vol.42, no.3, pp.571-582, March 2007.

[5] V. Jain, B. Javid, P. Heydari, "A BiCMOS Dual-Band Millimeter-Wave Frequency Synthesizer for Automotive Radars," *IEEE J. Solid State Circuits*, vol.44, no.8, pp.2100-2113, Aug. 2009.

[6] D.D. Kim, J. Kim, C. Cho, J.-O. Plouchart, M. Kumar, W.-H. Lee, K. Rim, "An array of 4 complementary LC-VCOs with 51.4% W-Band coverage in 32nm SOI CMOS," *ISSCC Dig. Tech. Papers*, pp.278-279, 8-12 Feb. 2009.

[7] S.-W. Tam; H.-T. Yu; Y. Kim; E. Socher, M.C. F. Chang, T. Itoh, "A dual band mm-wave CMOS oscillator with left-handed resonator," *IEEE RFIC Symp. Dig. Papers.*, pp.477-480, 7-9 June 2009.

[8] D. M. Pozar, "Microwave Engineering," 3rd ed., Wiley 2005.

978-1-4244-6240-7/10 $26.00 © 2010 IEEE

RMO2C-2

A 24-GHz and 60-GHz Dual-Band Standing-Wave VCO in 0.13µm CMOS Process

Liang Wu[1], Alan W. L. Ng[1], Lincoln L. K. Leung[2] and Howard C. Luong[1]

[1] Hong Kong University of Science and Technology, Clear Water Bay, Hong Kong

[2] Qualcomm, San Diego, USA

Abstract — By exploiting the intrinsic multiple oscillation modes of a standing-wave oscillator, a dual-band millimeter-wave VCO is designed. Implemented in 0.13µm CMOS with an area of 0.05mm^2, the VCO prototype measures a dual-band operation at 24 GHz and 60 GHz with tuning range of 10.8% and 7.2%, phase noise of -120dBc/Hz and -114dBc/Hz at 10MHz offset, power consumption of 11mW and 24mW, corresponding to FoM of -177dB and -176dB, respectively.

Index Terms — Voltage controlled oscillator, millimeter-wave, standing-wave, dual-band, CMOS, 24 GHz, 60 GHz.

I. INTRODUCTION

With recent advances in CMOS technology, millimeter-wave (MMW) circuits become more and more attractive for many interesting applications such as vehicle radars at 77 GHz and 24 GHz and communication systems at 60 GHz and 24 GHz. Moreover, to support multiple-band systems and to increase the level of integration without increasing the chip area and thus the fabrication costs, multiple-band MMW VCOs are highly desired. However, existing multiple-band VCO techniques, including tapped-inductor based [1] and transformer-based [2], are not suitable for MMW frequency bands. Prediction of their performance is difficult due to the inaccurate modeling of both passive and active devices at the MMW frequencies. In addition, the performance of these existing VCOs is typically not good because of the low quality factor of inductors and transformers. Wave-based oscillators such as standing-wave oscillators (SWO) can reach operation frequencies approaching the devices' transition frequency f_T with proper distribution of active gain elements along a transmission line. Given the quasi-transverse electromagnetic mode of propagation, transmission lines are capable of realizing precise values of small reactance and are inherently scalable in length. Also, the well-defined ground return path significantly reduces magnetic and electric field coupling to adjacent structures [3].

In this paper, a dual-band MMW SWO at 24 GHz and 60 GHz is demonstrated by exploiting and switching the intrinsic multiple standing-wave modes. Employing only one differential transmission line, this dual-band VCO occupies an area of only 0.05mm^2 and achieves figures of

merit (FOMs) in the two frequency bands comparable with that of existing single-band MMW VCOs.

II. MULTIPLE-MODE OSCILLATION OF SWO

In a λ/4 SWO, the boundary conditions allow several standing-wave modes at $l_0 = λ/4 \times n$ (n=1, 3, 5 …) where l_0 is the length of transmission line. The corresponding frequency can be expressed as:

$$f_n \approx \frac{n}{4 l_0 \sqrt{LC}} \cdot \tag{1}$$

where L and C are the inductance and capacitance per unit length.

In the fundamental mode (n=1), the voltage amplitude exhibits monotonic variations as a function of the position with an oscillation frequency f_L as depicted in Fig. 1(a). On the other hand, in the third-order mode (n=3), the voltage amplitude exhibits periodic variations along the transmission line with an oscillation frequency f_H as shown in Fig. 1(b). Oscillations in the fifth mode and higher modes are insignificant and negligible due to the substantial high-frequency loss.

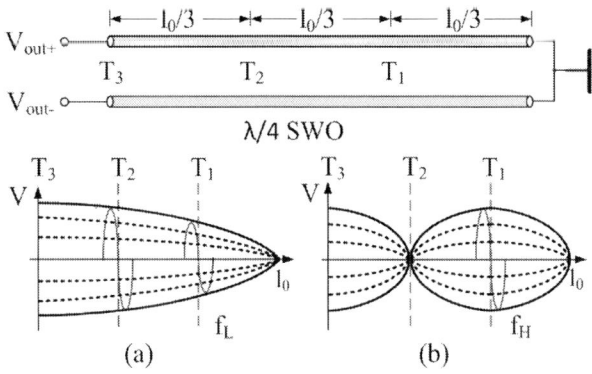

Fig. 1 Voltage amplitudes of a λ/4 SWO operating in (a) the fundamental mode and (b) the third-order mode

The multiple modes of SWO imply multiple impedance peaks of the transmission line. Using the S-parameters of the transmission line obtained from EM simulations in

978-1-4244-6240-7/10 $26.00 © 2010 IEEE

ADS Momentum and the capacitor models from the foundry, the small-signal model is built and simulated as shown in Fig. 2, where C_p is the parasitic capacitance. The impedance seen from Node T_3 exhibits two distinctive peaks with $|Z_{3L}| > |Z_{3H}|$, while the impedance seen from Node T_1 also exhibits two distinctive peaks: $|Z_{1H}| > |Z_{1L}|$, where $f_H \approx 3f_L$. As a result, mode-switching for dual-band operation can be achieved by injecting energy at different nodes, i.e. low-band and high-band oscillations are excited by injecting energy at Nodes T_3 and T_1, respectively.

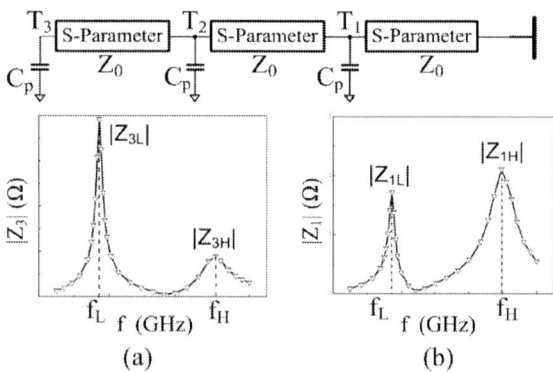

Fig. 2 Impedance seen at Nodes (a) T_3 and (b) T_1

III. ANALYSIS ON THE STABILITY ISSUE

In order to achieve stable oscillation of the VCO, it is critical to ensure that the impedance at the desired frequency is significantly larger than that at the other frequencies. If energy is injected at Node T_3, the low-band oscillation is excited and can work stably since $|Z_{3L}|$ is much larger than $|Z_{3H}|$ as shown in Fig. 2(a). However, if energy is injected at Node T_1, the high-band oscillation is potentially unstable and may jump to the low-band in the presence of process variations since $|Z_{1H}|$ is just slightly larger than $|Z_{1L}|$ as shown in Fig. 2(b). Therefore, a stabilization technique is indispensable to achieve stable and proper oscillation.

Analysis on SWO is performed with transmission line theory in the situation when the high-band oscillation is desired by injecting energy at Node T_1. To simplify the calculation, it is assumed that parasitic capacitance is absorbed into the transmission line. By treating the SWO as a distributed oscillator, the start-up requirement can be derived. Fig. 3 shows a simplified model of the SWO [4] with a negative-conductance stage injecting energy to Node T_1.

Starting from Node T_1, the forward wave V_1 is amplified by the gain stage and travels toward the right end where it is completely reflected. The first reflected

wave travels along the transmission line to the left until it arrives at Node T_3 where it is partially reflected if Z_s is neither infinitely large nor 0. The second reflected wave propagates along the transmission line to the right and finally reaches T_1 where it is amplified to V_1'. The amplitude of V_1' can be expressed as:

$$V_1' = -\left(\frac{1}{2} g_m Z_0\right)^2 e^{-(\alpha+j\beta)2l_0} \Gamma_3 V_1. \qquad (2)$$

where α is the attenuation constant and β is the phase constant of the transmission line. The reflection coefficient at Node T_3 can be express as:

$$\Gamma_3 = |\Gamma_3| e^{j\theta} = \frac{Z_s - Z_0}{Z_s + Z_0} = 1 - \frac{2Z_0}{Z_s + Z_0}. \qquad (3)$$

To satisfy the phase condition for oscillation:

$$e^{-j\beta \cdot 2l_0} e^{j\theta} = -1. \qquad (4)$$

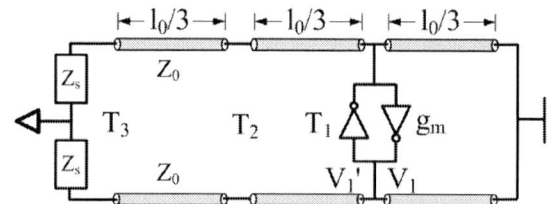

Fig. 3 Simplified model of SWO operating in the high-band mode

The required normalized gain to sustain oscillation can be derived by setting $|V_1| = |V_1'|$, that is:

$$\left(\frac{1}{2} g_m Z_0\right)^2 = \frac{1}{e^{-\alpha 2l_0} |\Gamma_3|}. \qquad (5)$$

As the attenuation α increases with frequency, the high-band oscillation would need more energy than the low-band oscillation. However, the gain condition is relaxed and the high-band oscillation is easily achieved if $|\Gamma_3|$ is increased with frequency, which can be implemented by adding capacitor C_s at Node T_3 in such a way that $C_s > l_0 C$, where C is the capacitance per unit length. This result is further verified by simulation. As shown in Fig. 4(b), the difference between $|Z_{1L}'|$ and $|Z_{1H}'|$ becomes more distinct if C_s is added. However, since C_s is in parallel with the transmission line, $|Z_{3L}|$ decreases to $|Z_{3L}'|$ as shown in Fig. 4(a), and thus more power is needed to sustain the low-band oscillation. The oscillation frequencies f_L and f_H are also shifted down due to the loading effect of C_s.

978-1-4244-6240-7/10 $26.00 © 2010 IEEE

The mode-switching SWO can also be approximated by a lumped model which is simply a second-order LC tank. After calculating the impedance seen from Nodes T_1 and T_3 and utilizing the notch-peak cancellation concept [2], it can be proved that adding capacitance at Node T_3 can help suppress unwanted oscillation and hence stabilize desired oscillation, which is consistent with the results from the analysis above with a distributed model.

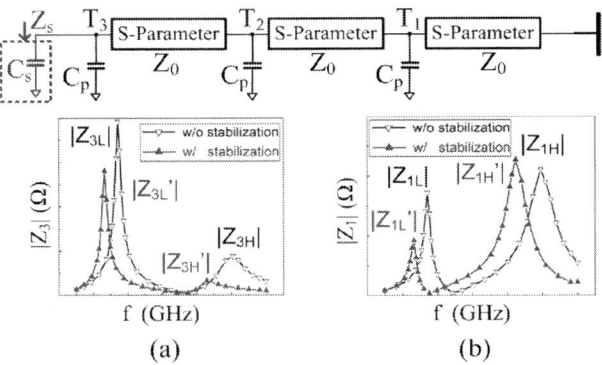

(a)   (b)

Fig. 4 Impedance with and without stabilization technique seen at Nodes (a) T_3 and (b) T_1

IV. CIRCUIT IMPLEMENTATION

Fig. 5 shows the schematic of the proposed dual-band standing-wave VCO with the stabilization technique. The differential transmission line is implemented by a differential micro-strip line with the topmost and thickest Metal 8 as the signal line and Metal 1 as the ground plane. The width is optimized to be 16 μm to maximize the quality factor Q at the frequency in the middle of the range. Frequency tuning is done by two varactors C_{v1} (~40fF) and C_{v3} (~300fF) located at Node T_1 and Node T_3, respectively. Large C_{v3} is used to ensure no frequency jumping while small C_{v1} and no varactor at Node T_2 are employed to reduce the effective loading capacitance at Node T_1. Three equally-spaced NMOS cross-coupled negative gm cells g_{m1}, g_{m2}, and g_{m3}, with an additional g_{m1}' at Node T_1 are utilized to compensate the losses of the transmission lines and varactors. Each of the gm cells is a current-biased gm cell which can be fully turned on/off by switching the bias current. Series switches are avoided in the high-frequency signal path to minimize attenuation. In the presence of the capacitance contributed from varactors and parasitics, the length of the transmission line is chosen to be 260 μm to obtain the desired output frequency.

Dual-band operation is achieved by switching different combinations of negative gm cells at different nodes. When g_{m1}, g_{m2}, and g_{m3} are activated by turning on their associated bias currents I_{b1}, I_{b2}, and I_{b3}, energy is injected

to all the three nodes T_1, T_2, and T_3. As a result, the fundamental-mode standing-wave is excited, and the low-band oscillation is obtained. In this mode, the three negative gm cells are distributed to reduce the effective parasitic capacitance at the output. They are also scaled to be $g_{m3} = 2g_{m2} = 2g_{m1}$ to save the power consumption [4]. On the other hand, when g_{m1} and g_{m1}' are activated by turning on the bias currents I_{b1} and I_{b1}', energy is only injected to Node T_1. As such, the third-mode standing-wave is excited, and the high-band oscillation is obtained. The loading capacitance at Node T_3 looks smaller at Node T_1 due to the loss of the transmission line [5], but the impedance is still low that both g_{m1} and g_{m1}' are needed for high-band oscillation.

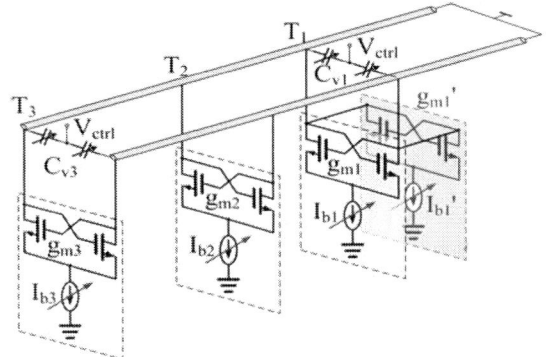

Fig. 5 Proposed dual-band standing-wave VCO

V. EXPERIMENTAL RESULTS

A dual-band standing-wave VCO prototype operating at 24 GHz and 60 GHz in a 0.13μm CMOS process for wireless communication systems is implemented with proposed stabilization technique. With 0.8V and 1.2V supply voltages, the frequency tuning curves are measured as shown in Fig. 6.

Fig. 6 Measured frequency tuning curves with 0.8V and 1.2V supply voltages

978-1-4244-6240-7/10 $26.00 © 2010 IEEE

For low-band phase noise measurement, the output is measured directly with Agilent E4440A spectrum analyzer, and the phase noise is -120dBc/Hz at 10MHz offset while drawing a current of 14mA from a 0.8V supply, as shown in Fig. 7(a).

For high-band phase noise measurement, the output signal is first down-converted by Ducommun V-band balanced mixer with LO input at 50 GHz and then measured with spectrum analyzer. With current consumption of 20mA from a 1.2V supply, the phase noise is -114dBc/Hz at 10MHz offset, as shown in Fig. 7(b).

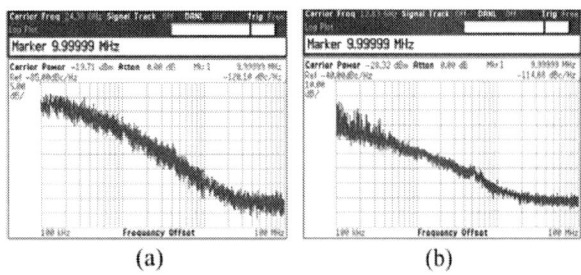

(a)   (b)

Fig. 7 Measured phase noise plots in (a) the low-band mode at 24 GHz and (b) the high-band mode at 60 GHz

The performance and comparison with recent published state-of-art dual-band VCOs and MMW VCOs are summarized in Table 1. The core area of the implemented VCO is 0.31mm×0.16mm, as shown in Fig. 8.

VI. CONCLUSION

In this paper, a MMW mode-switching technique with standing-wave architecture is presented. Based on the detailed stability analysis with a distributed model, a stabilization technique is proposed and demonstrated for a dual-band MMW VCO with good performance. The oscillation frequencies are much higher than existing dual-band VCOs while the performance is comparable to existing single-band MMW VCOs.

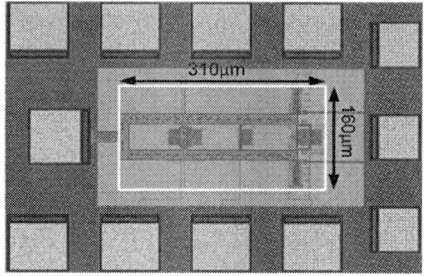

Fig. 8 Die photo

REFERENCES

[1] J. Borremans, et al., "A Single-Inductor Dual-Band VCO in a 0.06mm² 5.6GHz Multi-Band Front-End in 90nm Digital CMOS," *ISSCC Digest of Technical Papers*, pp. 324-325, Feb. 2008.

[2] R. Sujiang and H. C. Luong, "A 1V 4 GHz-and-10 GHz transformer-based dual-band quadrature VCO in 0.18 μm CMOS," *Custom Integrated Circuits Conference*, pp. 817-820, Sept. 2007.

[3] C. H. Doan, et al., "Millimeter-wave CMOS design," *IEEE Journal of Solid-State Circuits*, vol. 40, no. 1, pp. 144-155, Jan. 2005.

[4] C. Jun-Chau and L. Liang-Hung, "Design of Wide-Tuning-Range Millimeter-Wave CMOS VCO With a Standing-Wave Architecture," *IEEE Journal of Solid-State Circuits*, vol. 42, no. 9, pp. 1942-1952, Sept. 2007.

[5] L. Jri, et al., "A 75-GHz Phase-Locked Loop in 90-nm CMOS Technology," *IEEE Journal of Solid-State Circuits*, vol. 43, no. 6, pp. 1414-1426, Jun. 2008

[6] L. Lianming, et al., "Design and Analysis of a 90 nm mm-Wave Oscillator Using Inductive-Division LC Tank," *IEEE Journal of Solid-State Circuits*, vol. 44, no. 7, pp. 1950-1958, Jul. 2009.

TABLE 1 PERFORMANCE SUMMARY AND COMPARISON

Ref.	This Work		[2]		[4]	[5]	[6]
Technology	0.13 μm CMOS		0.18 μm CMOS		0.18 μm CMOS	90 nm CMOS	90 nm CMOS
f_{osc} [GHz]	24	60	4.2	10	40	75	58.4
FTR	10.80%	7.20%	42.00%	18.00%	20.00%	1.30%	9.32%
PN [dBc/Hz]	-120@10M	-114@10M	-116@1M	-112@1M	-100@1M	-108@10M	-91@1M
Supply [V]	0.8	1.2	1.0	1.0	1.5	1.45	0.7
P_{diss} [mW]	11	24	6	10	27	8	8.1
FoM [dB]	-177	-176	-181	-182	-178	-176	-177

978-1-4244-6240-7/10 $26.00 © 2010 IEEE   130

Multi-band Multi-standard Local Oscillator Generation for Direct up/down Conversion Transceiver Architectures Supporting WiFi and WiMax Bands in Standard 45nm CMOS Process

Ram Sadhwani[1], Assaf Ben Bassat[1], Adil A. Kidwai[1], Shahar Rivel[2]

[1]Intel Corporation, Hillsboro, OR; [2]Intel Corporation, Petah Tikva, Israel

Abstract — A highly linear LO generation architecture for direct up/down conversion transceiver is designed in standard 45nm 1.15V digital CMOS process. The spectral purity is better than 50dBc for WiFi and 45dBc for WiMax. The frequency plan is targeted to overcome the VCO pulling due to on-chip transmit power amplifiers. This architecture generates LO frequencies for WiFi 802.11b/g and WiMax 802.16e bands covering 2.3-2.7GHz, 3.3-3.8GHz and 4.8-5.8GHz from narrow input VCO range of 8-9.6GHz. Integrated (1k-40MHz) phase noise contribution is lower than 60dBc across all three bands.

Index Terms — LO, VCO, PA, LB, MB, HB, CMOS.

I. INTRODUCTION

The fast growth and demand of wireless communication has enabled new market segments along with multiple standards. Until recently WiFi 802.11abg was the predominant form of wireless communication modem present in home and office environment and used to implement local area network (LAN). The emerging 4G technology such as WiMax (802.16e) will expand this network and provide the direct wireless connectivity from service provider to subscriber. With the increasing number and complexity of these wireless standards there is ever increasing desire to enable new circuits capable of supporting multiple standards on single die. This enables cost reduction and higher levels of integration. One of the more critical candidates for such shared functionality is the frequency synthesizer and local oscillator (LO) generation block. In most direct up-conversion transceiver architecture, the synthesizer frequency plan is chosen to make VCO frequency of operation a non-integer multiple of the transmitter central frequency. This is to avoid the VCO pulling due to power amplifier which is a more severe issue with integrated power amplifiers becoming more and more common in CMOS technology [1]. Thus a LO generation block is used to convert the VCO frequency to the desired up/down conversion frequency.

For the handheld devices, compact form factor along with cost and power are the critical requirements. Recently, CMOS technology has demonstrated the ability to integrate the entire transceiver along with digital PHY/MAC baseband into a single-silicon substrate for even the, most aggressive performance standards [2]. When it comes to implementing multi-standards onto single die, the digital baseband starts to dominate the cost and area for die. Thus there is strong desire to go with shorter channel length processes. However, the modern nm-length CMOS devices having low intrinsic device gain ($g_m r_o$) are presenting a new set of challenges for RFIC designers [3]. In a standard 45nm digital CMOS process, the combination of thinner oxide stack-up, thinner metals and the lower supply voltage creates a significant challenge in the design of high frequency gain blocks.

In this paper, we present one of the first multi-standard LO generation block that covers the WiFi bands 2.4-2.5GHz (802.11b/g), 4.8-5.8GHz (802.11a), WiMax 2.3-2.7GHz (802.16e) and 3.3-3.8GHz (802.16e) while requiring single VCO frequency range of 8.05-9.6GHz. In a low cost flip-chip four layer substrate package, the chip is tested and validated for performance as standalone block to accurately measure the spectral purity. A novel signal-shaper block is implemented for internal mixer that reduces the output spurs from this LO generation blocks by as much as 17dB. Special attention is paid to the bias block and ground sharing and will be discussed later in this paper.

II. MULTI-BAND LO GENERATION ARCHITECTURE

A direct up/down conversion transceiver requires LO signal generation for the mixer block. The spectral purity of this LO signal determines the signal to interfere ratio. The interfering signal, if it falls within the filter pass-band would degrade the SNR. Fig.1 graphically shows that the frequency of the LO spur and its spacing to desired RF signal is most critical factor in architecture selection. If the LO spur is close to the desired RF signal frequency, there would be very limited filtering from the Rx LNA or transmitter block. This would increase the power of the interferer for Rx (or spurs for Tx, which can lead to regulatory issues) [4]. Also, if the LO spur falls in critical frequency bands used by handheld devices such as

GSM/Cellular/WiMax, the risk of high power interferers increases, which needs to be taken into account during architecture selection.

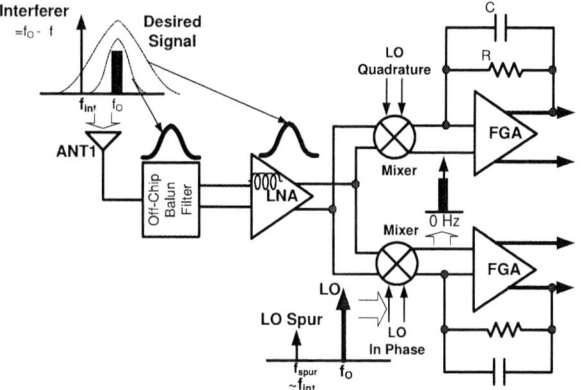

Figure 1 Spur impact on receiver performance

Fig. 2 shows the architecture that is used to generate WiFi and WiMax bands that balances design complexity for performance and block reuse to minimize the silicon area. The regenerative loops are used to generate Low Band (LB) (2.3GHz-2.7GHz) and Mid Band (MB) (3.3GHz – 3.8GHz) LO signals and a feed-forward architecture is used to generate the High Band (HB) (4.8GHz-5.8GHz) LO signal. The architecture selection is done such that the VCO frequency tuning range is within 20% and realizable with single VCO design. This reduces the complexity of the synthesizer design and validation.

1. LOW BAND AND MID BAND REGENERATIVE LOOP

The LB and MB LO frequencies are generated in a single regenerative loop based architecture. As shown in Fig 2, a switch is used to either select the division by four for MB or division by eight for LB configuration. This way the loop dividers and mixer core are shared for generating either band. Regenerative loop with division ratio of N has two possible output solution frequencies (eq. 1)

$$F_{out} = \frac{N}{N \pm 1} F_{in} \qquad ...eq\,(1)$$

where F_{in} is the frequency of the signal from VCO to the regenerative mixer block and F_{out} is the output frequency at mixer block. For the MB, the desired mode of operation is such that the loop locks at lower frequency solution, whereas for the LB, loop locks at upper frequency solution. In order to ensure the lock at the correct mode, a single side band (SSB) mixer is employed to select the upper/lower band. In order to further assist gain selectivity, inductive loading is used at the mixer and the load is designed separately for LB and MB operation. This

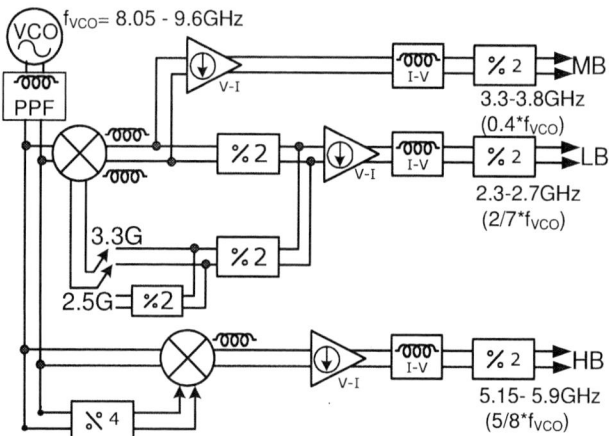

Figure 2 LO generation architecture

is done by having separate LO switching devices and inductive loads for LB and MB operation (Fig. 3). Once the correct lock condition is satisfied, the feedback of the regenerative loop based architecture has the advantage of suppressing the undesired sideband and spurs in the loop.

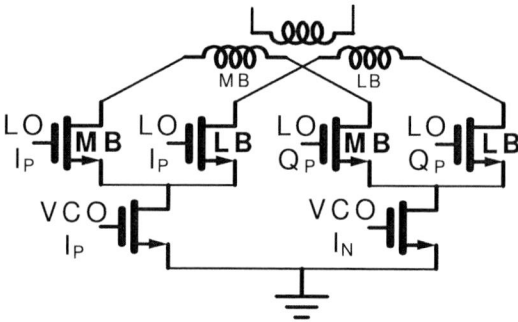

Figure 3 Mixer half block diagram

For the LB operation, the mixer load is tuned at 10GHz (and providing at least 8dB lower gain at 7GHz) and for MB the mixer load is tuned at 7GHz (while providing at least 10dB lower gain at 12GHz). Thus it is not possible to share the same load since for both bands the requirements are almost opposite. However this load switching mechanism also implements the functionality of the loop switch shown in Fig.2. that selects the loop division of four or eight based on MB/LB mode of operation. In the regenerative loop, the common-mode signal can cause stability issues, divider functionality failures and distortions due to high loop gain. In order to suppress the common-mode signals in the loop, a transformer was chosen for the mixer load instead of an inductor. No additional area penalty is created by transformer and it eases the dc-biasing of mixer output signal (through center tap of secondary coil of transformer).

978-1-4244-6240-7/10 $26.00 © 2010 IEEE

2. HIGH BAND FEEDFORWAR LOOP

The HB LO frequency is generated by classic feed-forward mixing of f_{VCO} and $f_{VCO/4}$. In such architecture the most critical spur is the image side band. The desired signal frequency is about 11GHz (for 5.5GHz operation) with image being at 6.6GHz. In order to achieve suppression for this image, we rely on the inductive tuned mixer stage. The SSB mixer is essential and a poly-phase filter (PPF) is also required to generate the quadrature signals from VCO differential output. In order to overcome the 6dB loss in two-stage PPF while maintaining reasonable current consumption, inductive tuned gain stages were used for 8-10GHz VCO signal after PPF. In smaller channel length process such as 45nm, the gain can be achieved with a simple inverter style amplifier biased to optimize the Gm of Pmos and Nmos by appropriately selecting Vp and Vn as shown in Fig. 4 (inductorless PPF buffers). This option of inductor less PPF buffer is implemented for the next version of design and will be tested in future. One important consideration that needs attention is the fact the PPF I and Q quadrature outputs have asymmetric frequency response, one is low pass filter and other is high pass filter. So the input signal to the PPF has to be narrow band sinusoid and it cannot be square wave. With square wave signals, the quadrature outputs of PPF has very asymmetric harmonic contents and degrades the spur performance from LOGEN. Since feed-forward architectures are inherently susceptible to generate higher number of spurs (due to absence of any feedback loop like regenerative ones), achieving good spur performance required additional technique of signal shaping discussed in next section.

3. SIGNAL SHAPING FOR ENAHANCED SPUR PERFORMANCE

The novel technique demonstrated in this paper is the signal shaping at the mixer LO ports to reduce spurs. The LO ports of the mixer contains the lower frequency RF signal from loop dividers as shown in Fig.4. The harmonic content of the square wave at the LO port of the mixer is the dominant source of output spurs. Ideally, we want to have a pure sinusoid at the RF and LO ports to limit these un-desired IF tones. We employed an intermediate tradeoff by using triangle waveform instead of square wave. As shown by eq2 below, we can see that triangle wave would have about $1/3^{rd}$ the 3rd and $1/5^{th}$ the 5th relative harmonic

$$f_t = \frac{8}{\pi^2}(\sin \omega t - \frac{1}{3^2}\sin 3\omega t + \frac{1}{5^2}\sin 5\omega t + ..) \quad ...eq(2)$$

$$f_t = \frac{4}{\pi}(\sin \omega t + \frac{1}{3}\sin 3\omega t + \frac{1}{5}\sin 5\omega t + ..) \quad ...eq(3)$$

content compared to square wave eq3.

The triangle waveform is generated using simple current source feeding current in linear passive capacitance. The slope of triangle is controllable to fine tune it in the silicon. The disadvantage of using triangle waveform is slightly higher noise contribution from mixer block. However for the application of the LOG block, this is not an issue or critical specification. The noise contribution is still dominated by the VCO/Synthesizer block.

4. SUPPLY ISOLATION AND BIAS CONSDERATIONS

Careful attention to biasing scheme is a must for achieving low spur power at the output. The presence of various frequency translation blocks (mixer, divider) operating at near rail to rail signals injects significant noise on supply and ground. This noise can easily appear at the output unless proper attention is paid to supply and ground isolation. Since IO pins are limited, there is always going to be supply/ground sharing among various blocks on silicon. This sharing can be optimized to limit the coupling. Fig. 5 below shows an example of the gain block having large output signal and the impact of having bias ground isolation. Since the output is going to be a large signal, it will couple the $2xF_0$ to the Vcc. Since Vcc and Vss have on-chip decaps, the noise couples to Vss. Having a dedicated bias ground in this case is not necessary a

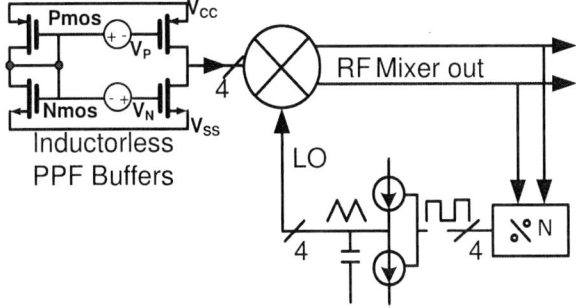

Figure 4 Signal shaping to reduce mixer spurs

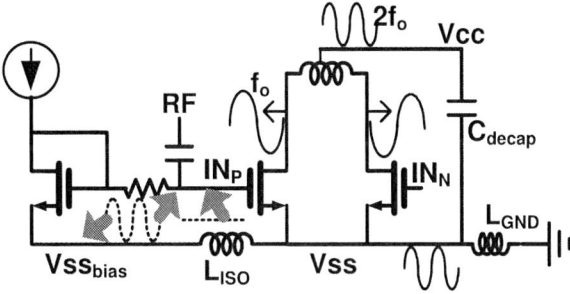

Figure 5 Bias ground selection scheme

good ideal. As shown in Fig. 5 the dedicated quite bias ground Vss_{bias} is going to lead to V_{gs} modulation at the input RF device and hence distortion at the amplifier output. In such cases we want to minimize the isolation between bias ground and active block ground (L_{iso}). This is in-general true for most large signal swing blocks. However such sharing carries the risk of oscillations and in-stability and needs to be examined carefully.

III. MEASUREMENT RESULTS

Packed in flip chip BGA 164 ball package, the testing is carried out using custom designed PCB board (Fig. 6).

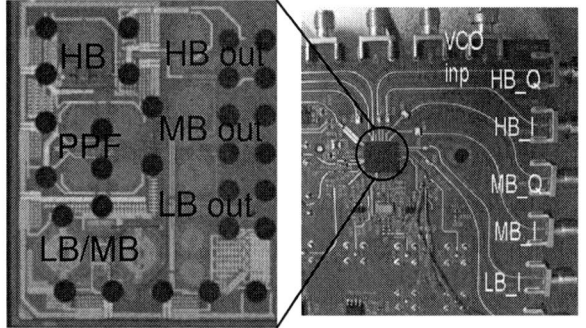

Figure 6 Die photo and test board for LOGEN

Dedicated SMA ports are made available for VCO input and LB/MB/HB outputs. Fig. 7 shows the full LOGEN LB/MB/HB spur measurements. The LB and HB are relatively clean spectrum with better than 50dBc spur number. For HB spur calculation, signal attenuation due to package and board losses is accounted for in quoting numbers. For MB the performance is 47dBc, and problematic spur falls in GSM band 1.6-1.8GHz, However it can be easily filtered out through on chip LNA or off chip filter. Also notice that in Fig. 7, the HB has much more spurs compared to LB/MB which is due to the fact

that HB is classic feed-forward loop whereas LB/MB is implemented using regenerative loop. The high frequency spurs and harmonics are not as critical due to relatively low mixer conversion gain and higher on-chip LO distribution losses. All three bands were characterized using low phase noise source to measure the LOGEN noise contribution. Using a 8-10GHz signal source with integrated (1k-40MHz) phase noise of 65dBc, the measured output integrated phase noise for LB/MB/HB is 67/65/60dBc showing minimal noise addition by LOGEN.

IV. CONCLUSION

We present a novel LOG architecture capable of providing LO signal for both WiFi and WiMAX standards while requiring narrow VCO tuning range (~18%) in standard 45nm CMOS process. The LO shaper architecture when enabled on-chip, reduced the spurs by about 17dB. Overall, the performance for all three bands met the target spur rejection of better than 50dBc for close-in spurs. Table 1. below summarizes the power and die area for each mode of operation from 1.15V voltage source.

Table 1. 45nm LOGEN Area and Power Summary

Band of Operation	Current Consumption	Die Area (including common blocks) *
LB 2.3-2.7GHz	54.6 mW	0.65 mm2
MB 3.3-3.8GHz	56.3 mW	0.65 mm2
HB 4.8-5.8GHz	52.9 mW	0.45 mm2

*Total Area with LB/MB/HB including ESD/Bumps 1.59mm2

VI. REFERENCES

[1]. Adil A. Kidwai et al, "Fully Integrated 23dBm Transmit chain with on-chip Power Amplifier and Balun for 802.11a application in standard 45nm CMOS process", *IEEE Radio Frequency Integrated Circuits Symposium, June-2009.*

[2]. M. Zargari, L. Y. Nathawad et al, "A Dual-Band CMOS MIMO Radio SoC for IEEE 802.11n Wireless LAN", *IEEE J. Solid-State Circuits, vol. 43, No. 12, pp. 2882-2895, Dec 2008.*

[3]. J.C.Rudell, "Low-Voltage Transceiver Design in 45nm CMOS with an emphasis on WIMAX", *IEEE International Solid-State Conf. Giraffe Forum, Feb.2008*

[4]. Chang-Wan Kim et al, "A CMOS Multi-LO Frequency Synthesizer Block for MB-OFDM UWB Systems", *IEEE 64th VTC-2006 Sept. 2006.*

[5]. Rozi Roufoogaran et al, "A Compact and Power Efficient Local Oscillator Generation and Distribution for Complex Multi Radio Systems", *IEEE Radio Frequency Integrated Circuits Symposium June-2008.*

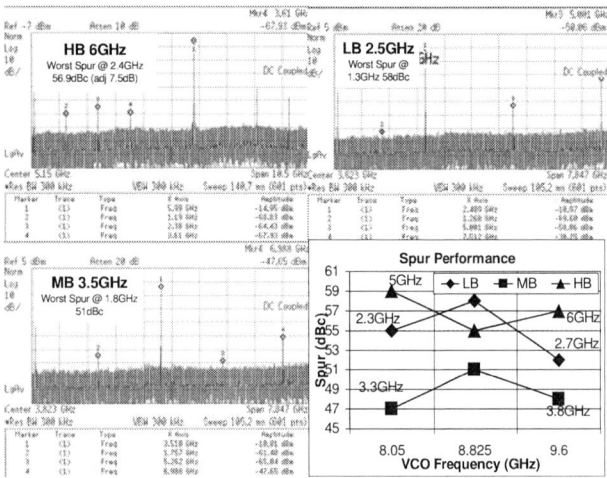

Figure 7 Spur measurement results

A 0.13-μm CMOS Local Oscillator for 60-GHz Applications Based on Push-Push Characteristic of Capacitive Degeneration

Tino Copani, Hyungseok Kim, Bertan Bakkaloglu, Sayfe Kiaei

Electrical, Energy and Computer Engineering, Arizona State University, Tempe, AZ, 85287, USA

Abstract — **A 60-GHz 10mW CMOS VCO is implemented together with a high-speed prescaler in a 130nm CMOS process. Compared to other push-push topologies, capacitive degeneration technique does not impact the resonator and switching transistors are re-used as buffers minimizing noise due to following amplifiers. The measured phase noise at 1MHz offset is -89dBc/Hz and FoM is -174dBc/Hz.**

I. INTRODUCTION

The last few years have shown a growing interest on the development of wireless systems for unlicensed communications in the 60GHz range. The main application area is high data rate wireless links for both audio/video (AV) streaming and bulk data transfer. Proposed system scenarios include not only WPAN and AV streaming over 10m range in multipath environments, but also links over 1 to 3m line-of-sight distance where range and multipath performance are traded in favor of low power and low cost implementation. The possibility of integrating these mm-wave systems in high-speed, sub-micron CMOS technologies, which offer significant cost and mixed-signal advantages, has been extensively explored [1]. However, the design of 60GHz CMOS circuits still faces severe challenges due to poor device model characterization, low quality factor of passive components, and large current consumption to guarantee high unity-gain frequencies (f_{max}). These issues are especially critical to the design of local oscillators (LO), which are key components of wireless transceivers [2], [3]. This work presents a 60GHz LO and high-speed prescaler in a 130nm CMOS technology, with system architecture and circuit topologies optimized to address the process limitations.

New circuit design approaches are required to overcome the lower supply voltage, higher device noise and lower available gain at the desired high frequency band. Moreover besides increasing losses, the conductive CMOS substrate adds parasitic signal paths, which complicate the modeling of active and passive devices. Although the use of metal ground shields under the passive devices helps to mitigate these problems, it increases the capacitive coupling to ground and lowers the self-resonance frequency of monolithic passive components. This requires optimization of the system architecture to define the best frequency planning, device sizing and layout floorplan.

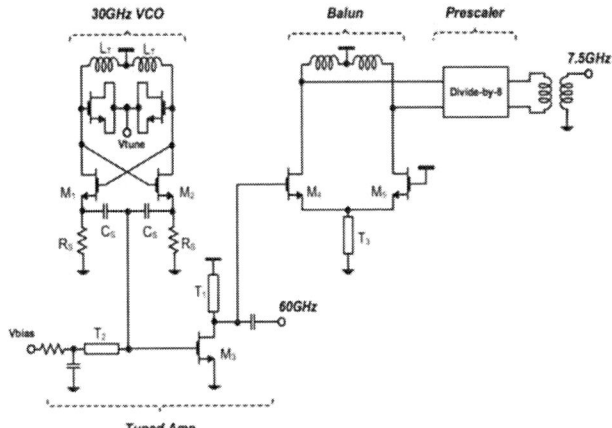

Fig. 1. Architecture of the implemented 60-GHz LO.

The paper is organized as following: The LO architecture is presented in Section II. Section II deals with the detail of the push-pull VCO. The high speed-prescaler is described in Section IV. Finally, Section V presents the main experimental results.

II. SYSTEM ARCHITECTURE

The architecture of the implemented LO is shown in Fig. 1. The system consists of a 30 GHz VCO core with capacitive source degeneration and inherent frequency doubling (push-push), a 60-GHz tuned amplifier (2nd harmonic booster), and an active balun driving a divide-by-8 prescaler.

The voltage-controlled oscillator is based on a cross-coupled CMOS pair and a 30 GHz LC tank resonator. Inversion mode NFETs are used as tuning elements because the provided device model was not fitted with measurements in accumulation mode region. Capacitive degeneration is employed at the source terminals of the M_1-M_2 pair, which avoids the oscillation start-up at undesired frequencies due to parasitic feedback paths. Furthermore, series capacitors' (C_S) center tap is exploited to provide push-push 2nd harmonic generation.

The use of a 30GHz VCO core with embedded frequency doubling instead of a direct 60GHz oscillator is required due to the passive device losses and model characterization limits of the adopted technology.

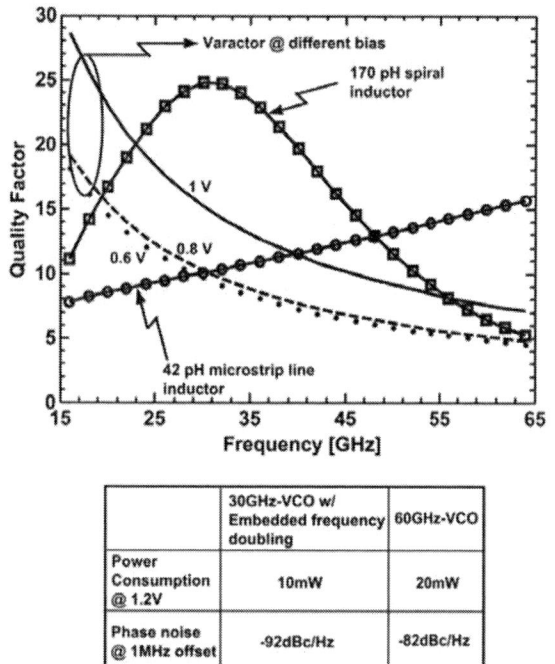

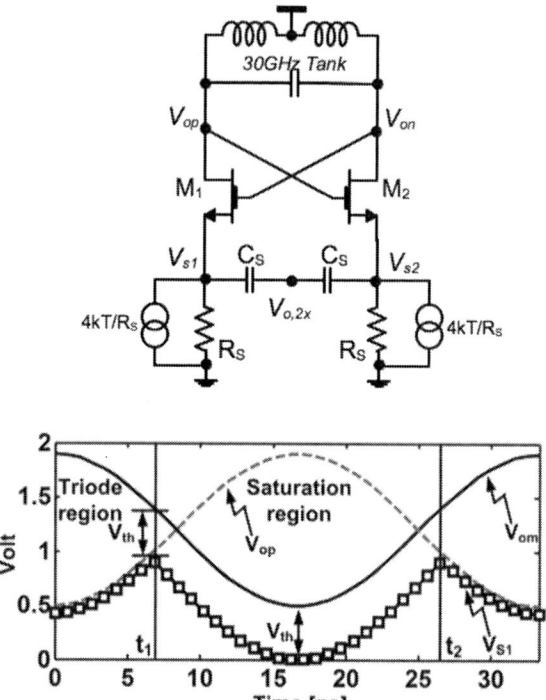

Fig. 2. Comparison of resonator's components and VCOs optimized at 30GHz and 60GHz, respectively.

Fig. 2 shows the quality factor behaviors of the devices commonly used in passive resonators. In the adopted CMOS process, the quality factor of NMOS varactors (Q_V), symmetrical inductors (Q_L) and microstrips (Q_M) are taken into consideration. While the performance of a 60GHz microstrip is comparable to that of a 30 GHz symmetrical inductor, the quality factor of the varactor degrades over frequency by a factor of 3, impacting the overall resonator quality factor (Q_T). When the Leeson's model is considered, oscillator phase noise is inversely proportional to Q_T,

$$ S_\phi(\Delta f) \propto \left(\frac{f_o}{Q_T \Delta f} \right)^2, \tag{1} $$

where f_o and Δf are the oscillation and the offset frequencies respectively. In the adopted technology, combining the high resonator loss with the low gain of the M_1-M_2 pair results in a 60-GHz VCO consuming more power and exhibiting higher phase noise than the proposed 30-GHz VCO core and frequency doubling, despite 6 dB phase noise penalty of the 2nd harmonic content. The simulation results of the two cases are summarized in the table of Fig. 2.

III. PUSH PULL CMOS VCO

Usually, 2nd harmonic generation at the common-source node has been considered a detrimental effect in current-limited oscillators because of folding of high frequency noise components.

Fig. 3. 2nd harmonic generation embedded in a CMOS LC-tank VCO.

However, in the proposed topology shown in Fig. 3, squaring function of the M_1-M_2 pair common source node is exploited to provide frequency doubling through source degeneration capacitors. Compared to other push-push topologies, the proposed technique does not impact the resonator and also minimizes any noise coupling of the subsequent amplifier because M_1 and M_2 are re-used as buffers. Moreover, the up-converted noise contribution of R_S resistors is minimized because when M_1 is on (M_2 off), only one noise current ($4kT/R_S$) flows into the tank while the other contribution is negligible due to C_S capacitors and M_2 switched off state.

The 2nd harmonic generation is qualitatively described in Fig. 3. Assuming that the LC tank filters out all the harmonics but the fundamental, the waveforms at the VCO output nodes (V_{op}, V_{on}), and at the source of M_1 (V_{S1}) are plotted in Fig. 3. Signal V_{S1} is a periodical waveform with period $2\pi/\omega_o$, where ω_o is the oscillation angular frequency. By using Trigonometric Fourier's series, V_{S1} can be approximated by,

$$ V_{S1}(t) \cong \frac{a_0}{2} + a_1 \cos(\omega_0 t) + a_2 \cos(2\omega_0 t) + \\ + a_3 \cos(3\omega_0 t) + ... + a_k \cos(k\omega_0 t) \tag{2} $$

The waveform at M_2 source node (V_{S1}) is the $180°$ phase shifted version of V_{S1}, therefore the common mode voltage, $V_{o,2x} = \dfrac{V_{S1} + V_{S2}}{2}$, contains only even harmonics of the fundamental ω_o. The amplitude of the 2nd harmonic at $V_{o,2x}$ node is then given by

$$
\begin{cases}
a_2 = \dfrac{A_m}{\pi}\big[\sin(\omega_0 t_2) - \sin(\omega_0 t_1)\big] - \dfrac{V_{th}}{2\pi}\big[\sin(2\omega_0 t_2) - \sin(2\omega_0 t_1)\big] + \\[2mm]
\qquad + \dfrac{A_m}{3\pi}\big[\sin(3\omega_0 t_2) - \sin(3\omega_0 t_1)\big] \\[4mm]
t_1 = \dfrac{\cos^{-1}\!\left(V_{th}\big/2A_m\right)}{\omega_0} \\[6mm]
t_2 = \dfrac{2\pi - \cos^{-1}\!\left(V_{th}\big/2A_m\right)}{\omega_0}
\end{cases}
\quad , (3)
$$

where A_m is the amplitude of the oscillation at ω_o and V_{th} is the NMOS threshold voltage.

The signal at C_S capacitors' common node, $V_{o,2x}$, is fed to a boosting amplifier (M_3) with microstrip load (T_1), which also provides impedance matching for the 50Ω external instrumentation. A transmission line, T_2, is used to achieve matching at M_3 input node and decouple the bias source (V_{bias}) of the amplifier. The output of the amplifier is also fed to an active balun driving a static divide-by-8 prescaler. Usually, simple single-ended balun topologies exploit the $180°$ phase difference between drain and source nodes of a MOS device. However any small difference between the drain and source impedances affects the single-ended to differential operation, increasing the conversion of AM noise into phase noise. The adopted balun uses a differential pair (M_4-M_5 in Fig. 1) with microstrip current source (T_3) to increase common mode impedance at high frequency. The fully differential active balun theoretically presents a higher noise floor compared to single-ended topologies because of the extra-device noise. However, circuit simulations showed that the performance of a LO design with single-ended balun rapidly degrades in the $1/(\Delta f)^2$ region of the phase noise characteristic when device and layout mismatches are taken into account.

IV. HIGH-SPEED PRESCALER

In the implementation of a local oscillator for mm-wave applications, the design of high-speed digital prescalers is critical to interface the LO to a PLL circuit. Injection locked dividers offer excellent performance at low power consumption and very high frequency operation. However, the use of such dividers is affected by the limited injection locking range. Although, this problem can be overcome by increasing the power level of the injection tone, large buffers are needed between the LO circuit and the divider. Moreover tracking mechanisms between the resonators used in the VCO and the divider must be arranged for best tuning range performance. The LO circuit employs a VCO and a frequency doubler complicating the tracking of an injection locked divider in this architecture. Static digital dividers show high-speed performance at the expense of high quiescent power consumption. The implemented divide-by-8 prescaler consists of D Flip Flop (DFF) based divide-by-2 stages, as shown in Fig. 4. The first stages of the prescaler use the topology in the inset of Fig. 4 to implement the D latches. Degeneration capacitor, C_D, is used to increase the speed of the latch cell (M_3-M_4). Indeed, at mm-wave frequencies the gate resistance, r_g, of small FET devices impairs the maximum oscillation frequency of the M_3-M_4 cross-coupled pair, $f_{max} = \sqrt{\dfrac{f_T}{r_g C_{gs}}}$,

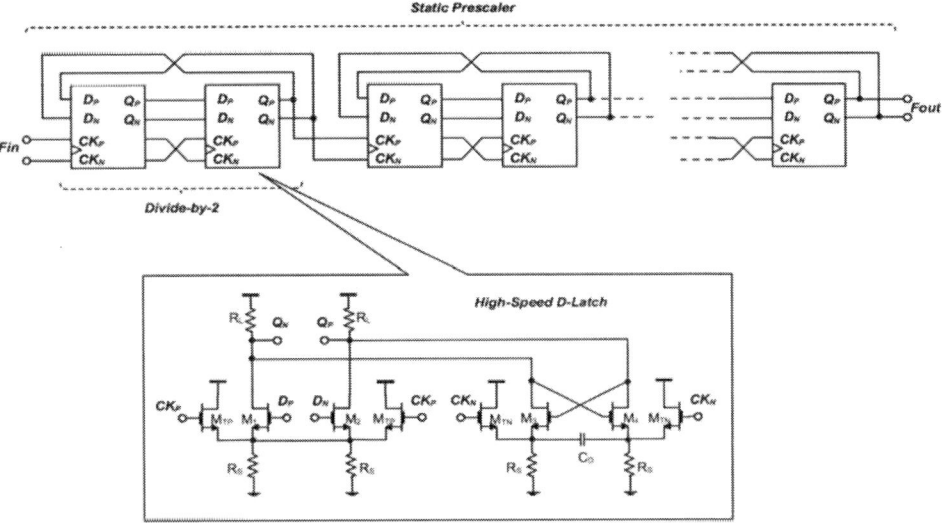

Fig. 4. Prescaler and high-speed D-latch.

where f_t is the device transit frequency and C_{gs} is the gate-source capacitance. When a degeneration capacitor is used, the maximum oscillation frequency is increased up to

$$f_{max} = \sqrt{\frac{f_T\left(\frac{1}{C_{gs}}+\frac{1}{C_D}\right)}{r_g}},$$ thus speeding up the latch phase of the

DFF. Triple tail devices (M_{TP}-M_{TN}) are used to switch between the read and hold phases of the D latch, allowing a low voltage implementation. Furthermore, M_{TP}-M_{TN} perform as source followers thus increasing the impedance at the clock inputs (CK_p, CK_n).

V. IC IMPLEMENTATION

The 60 GHz LO circuit is fabricated in a 130 nm CMOS process and it has been characterized through on-waver micro-probing. The output power is -25dBm for a 56GHz carrier. The broadband emission spectrum of the LO is shown in Fig. 5, which shows leakage of the 30GHz VCO core. The spur at 42GHz results from the mixing of the LO signal with the leakage at 14GHz from the divide-by-4 prescaler stage. The reported tuning range is 6GHz in a [0,1.2V] interval. The phase noise measurement reported in Fig. 6 has been performed at the prescaler output due to instrumentation capability limitations.

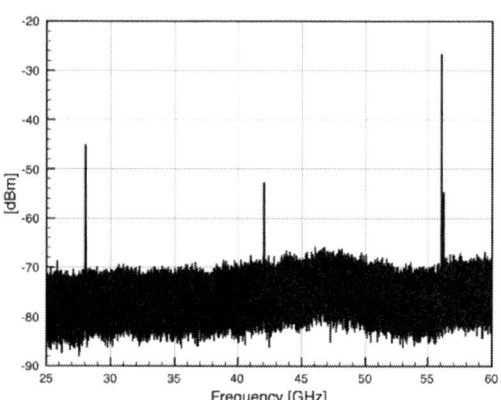

Fig. 5. Measured broadband LO spectrum.

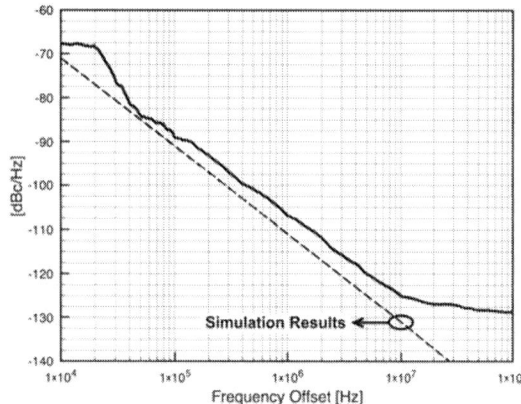

Fig. 6. Phase noise measurement of the divide-by-8 prescaler output signal.

TABLE I

	f_O [GHz]	FTR [%]	PN @ 1MHz [dBc/Hz]	P_{DISS} [mW]	FoM [dBc/Hz]	CMOS Tech.
[3]	59	9.8	-89	9.8	-175	130nm
[4]	40	20	-100	27	-178	180nm
[5]	58	9.3	-90	8.1	-176	90nm
[6]	44.5	8.1	-96	3.6	-184	65nm
This work	55	11	-89	10	-174	130nm

Phase noise at the divide-by-8 output is -107dBc/Hz at 1MHz offset, which corresponds to -89dBc/Hz at the 54.7GHz carrier. The IC draws 47mA from a 1.2 V supply and occupies a $900 \times 550\mu m^2$ area. The LO circuit performance is summarized in Table I. The die micrograph is shown in Fig. 7.

CONCLUSIONS

A 60-GHz CMOS LO circuit has been presented in this work. The system comprises a push-push VCO and a high speed prescaler. The proposed architecture makes a compact and low power solution for next generation high-speed wireless links.

REFERENCES

[1] B. Razavi, "CMOS Transceivers for the 60-GHz Band," *IEEE RFIC Symp.,* Jun. 2006, pp. 231-234.

[2] A. Parsa, B. Razavi, "A 60-GHz CMOS receiver using a 30-GHz LO," *ISSCC Dig. Tech. Papers,* Feb. 2008, pp. 190-191.

[3] C. Cao, K. K. O, "Millimeter-Wave Voltage-Controlled Oscillators in 0.13-μm CMOS Technology," *IEEE J. Solid-State Circuits,* vol.41, no.6, pp. 1297-1304, Jun. 2006.

[4] J.-C. Chien, L.-H. Lu, "Design of Wide-Tuning-Range Millimeter-Wave CMOS VCO With a Standing-Wave Architecture," *IEEE J. Solid-State Circuits,* vol.42, no.9, pp. 1942-1952, Sept. 2007.

[5] L. Li, P. Reynaert, M. Steyaert, "A 90nm CMOS MM-Wave VCO Using an LC Tank with Inductive Division," *ESSCIRC Dig. Tech. Papers,* pp. 238-241, Sept. 2008.

[6] H. M. Cheema, et al., "A 44.5 GHz Differentially Tuned VCO in 65nm Bulk CMOS with 8% Tuning Range," *IEEE RFIC Symp.,* pp. 649-652, Jun. 2008.

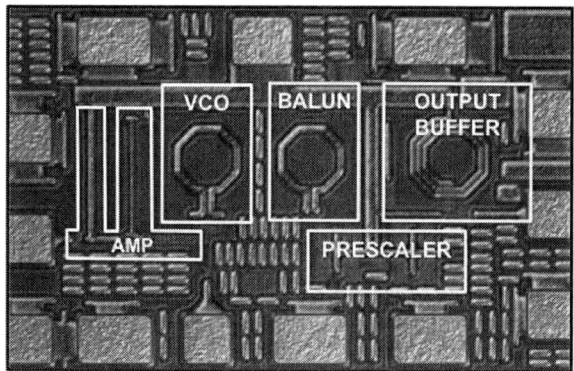

Fig. 7. Die micrograph of the LO system.

A Switched-Capacitor mm-Wave VCO in 65 nm Digital CMOS

Mohammad Nariman[1,2], Reza Rofougaran[2], and Franco De Flaviis[1,2]

[1] University of California at Irvine, [2] Broadcom Corporation, Irvine, CA 92617

Abstract — A 34-40 GHz VCO fabricated in 65 nm digital CMOS technology is demonstrated in this paper. The VCO uses a combination of switched capacitors and varactors for tuning and has a maximum Kvco of 240 MHz/V. It exhibits a phase noise of better than -98 dBc/Hz @ 1-MHz offset across the band while consuming 12 mA from a 1.2-V supply, an FOM_T of -182.1 dBc/Hz. A cascode buffer following the VCO consumes 11 mA to deliver 0 dBm LO signal to a 50Ω load.

Index Terms — CMOS LC-VCO, figure of merit, FOM_T, mm-wave, phase noise, switched capacitor, tuning range.

I. INTRODUCTION

In recent years, the large amount of unlicensed spectrum at the millimeter-wave frequency range has attracted attention for wireless uncompressed video streaming and high performance personal area network applications. Realization of mm-wave radios in digital CMOS makes it possible for the high performance applications to benefit from high integration capacity offered by this technology. Further integration can be achieved by using monolithic mm-wave antennas as they require small footprints.

The high level of integration creates the possibility of realizing *"DC-Boards"* on which interconnections are limited to pure DC routings and their high speed data buses are replaced by chip to chip wireless links, resulting in significant cost and power savings. This work is part of a project with an aim of evaluating the feasibility of small, low power, and high performance mm-wave transceivers in pure digital CMOS technology. As one of the most challenging parts of a mm-wave radio, the VCO often has to meet stringent specifications. This paper demonstrates the design and implementation of a VCO in 65 nm digital CMOS process with f_{max} of 200 GHz and f_T of 140 GHz.

As indicated in (1), phase noise of the VCO degrades 6 dB/octave as the oscillation frequency increases [1]-[2]. At frequencies > 20 GHz non-quasi-static effects for the capacitors and the skin and the proximity effects for the inductors severely degrade their quality factors and pose significant challenges to the design of the VCO [3].

$$L\{\Delta f\} = 10\log\left[\frac{F.k.T}{2P_s} \cdot \left(\frac{f_0}{2Q_T.\Delta f}\right)^2 \cdot \left(1 + \frac{f_c}{\Delta f}\right)\right] \quad (1)$$

Where $L\{\Delta f\}$ is the single sideband power spectral density of the phase noise in dBc/Hz at the offset frequency of Δf from the oscillation frequency of f_0. f_c is the flicker noise corner. Q_T is the loaded quality factor of the tank. P_s is the VCO signal power in Watt. F is the effective noise factor of the oscillator. T is the temperature in °K and k is the Boltzmann's constant.

Section II covers the architecture design and describes the topology. In section III detailed description of each component of the design is provided, with emphasis on the innovative parts. Section IV discusses the experimental results and summarizes the performance characteristics in comparison with the results of other recently published CMOS VCOs. Probing unlocked mm-wave VCOs is extremely challenging and section V is dedicated to discuss that. Section VI concludes this paper by presenting a quick summary of what has been achieved.

II. ARCHITECTURE AND TOPOLOGY

Several architectures have been evaluated including Colpitts in terms of their achievable FOM. As shown in Fig. 1 a cross-coupled VCO architecture with doughnut-shaped grounded-gate NMOS pair has been chosen due to its superior power and area efficiency. A combination of switched capacitors and accumulation-mode varactors comprise the variable capacitance of the tank. A shared-diffusion cascode buffer with a balanced transformer load delivers the LO signal to the probe pads.

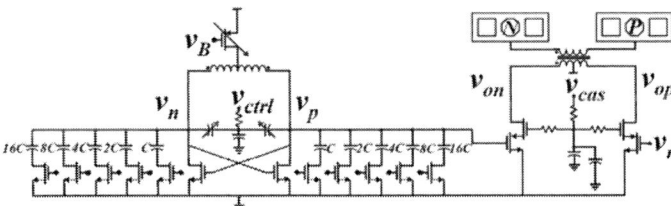

Fig. 1. Simplified schematic of the VCO and its buffer.

Using only varactors to achieve a large tuning range for the VCO results in a high K_{VCO} which as (2) indicates increases the *AM* to *PM* conversion gain which in turn causes high phase noise and spurious signal levels. In this work using the switched-capacitors has helped in reducing the K_{VCO} and increasing the tuning range significantly.

$$\Phi_{PM}(f_0 + f_{AM}) = \left[\frac{K_{VCO}}{2f_{AM}}\right] \cdot V_{AM}(f_{AM}) \quad (2)$$

Where Φ_{PM} is the level of the *PM* signal at the output of the VCO oscillating at the frequency of f_0, in response to an *AM* disturbance voltage with amplitude of V_{AM} and frequency of f_{AM}.

The reliability concerns of deep sub-micron processes make the VCO operate in the current-limited mode. Using a PMOS current source is optimum for biasing the VCO in this mode. This is also to provide the VCO buffer with a proper DC bias needed for an efficient LO power delivery.

III. CIRCUIT DESIGN AND IMPLEMENTATION

Detailed descriptions of the VCO components and explanations of the design are presented in this section.

A. VCO Core

Doughnut transistors comprised of square-shaped poly-silicon gates are used for the VCO cross-coupled pair. This reduces the contribution of the drain junction capacitors to less than 5% of the overall tank capacitance. Size selection for the core devices has been done based on four different factors: gain, Noise Figure (NF), current switching capability, and efficient power delivery by providing a proper bias to the VCO buffer. The optimum current density for the lowest NF depends on the lateral electric field in the channel [4]. For the process of this VCO, a current density of 0.35 mA/μm results in a suitable performance from all of the mentioned aspects.

Fig. 2 shows the unit cell of the VCO core with an optimal layout. The layout of the transistors should be optimized in order to achieve an optimal performance in terms of f_{max}, maximum stable gain, and the noise figure. Wider routings cause larger parasitic capacitances and couplings, while narrower interconnections create larger series resistances and inductances.

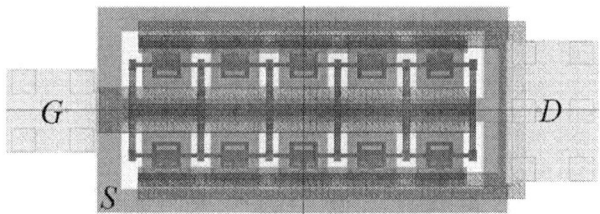

Fig. 2. Layout of the unit cell of the VCO core containing ten doughnut-shaped transistors with optimized routings.

B. Resonant Tank

The resonant tank is composed of a 135 pH differential octagonal-shaped inductor, accumulation-mode varactors and 5-bit binary-weighted switched finger-capacitors. The inductor uses a wall of well-grounded metals to provide a good current return-path and help confinement of the field, at the price of more parasitic capacitance and a little lower overall inductance. To reduce the losses due to the eddy currents, any substrate doping under the inductor and any metal loops in the guard rings and the routings have been avoided. The differential inductor has a quality factor of 24 at 34 GHz and a self resonance frequency of 200 GHz.

The switched capacitor uses doughnut-shaped NMOS switches to achieve a high ON to OFF capacitance ratio. As Fig. 3 demonstrates the capacitor array benefits from an innovative design in which a large finger capacitor splits to five binary-weighted capacitors to be used for the digital frequency tuning. The LSB units are positioned in between the MSB units in order to achieve a better accuracy. Conventional unit-cell based arrays suffer from

significant relative accuracy when used for small capacitor sizes. The proposed compact layout results in a better accuracy, a lower parasitic capacitance, and shorter connections to the switches. Effective parallel resistance of the tank is dominated by the capacitive branches, resulting in an overall tank quality factor of 10 at 40 GHz.

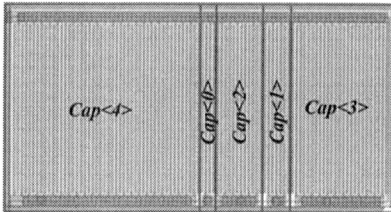

Fig. 3. Layout of the proposed capacitor array consisting of five binary-weighted capacitors in the optimum order.

C. Current Source

The current source is programmable to work in either the linear mode or the saturation mode in order to be able to optimize the overall phase noise. When the current source is in the saturation mode its noise contribution increases, but the contribution of the supply and the cross-coupled pair decrease. In this design, should the current source be set to supply the same amount of current in both modes the linear mode shows a slightly better phase noise.

D. VCO Buffer

A differential cascode buffer has been designed to deliver a 0 dBm LO power to a 50Ω load while isolating the VCO from the load. The VCO buffer also reduces the load pull for the VCO significantly. The operating points are set for an efficient LO power delivery. As shown in Fig. 4 the diffusion regions of the input and the cascode devices have been shared to minimize the parasitic capacitances of the junctions and the routings which create significant loss and current leakage at this common node.

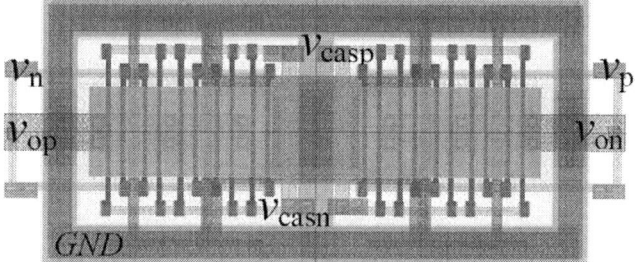

Fig. 4. The unit cell of the differential VCO buffer with five differential pairs of the cascode devices.

A transformer has been used to convert the buffer output impedance of 125Ω, which is optimum for an efficient power delivery, to a 50Ω level at the input of the following stages which can be a mixer and a divider. This feature together with the possibility of having shorter signal feeds to the following stages make the transformer superior to a differential inductor as the VCO buffer load.

978-1-4244-6240-7/10 $26.00 © 2010 IEEE

E. Common-Mode Nodes

A ground grid covers all the layers of the layout, with the exceptions of the parts which introduce excessive eddy currents or couplings to the sensitive nodes. All Common-Mode (CM) nodes of the circuit have their unnecessary bandwidths suppressed by maintaining low impedance levels to the ground in all frequencies. This requires different sizes of decoupling capacitors to be used at all of the CM nodes and to be repeated in short intervals. This also improves CM noise rejection and CM stability.

There are two CM nodes which do not need extra capacitors: The center tab of the inductor and the voltage control node. At the center tab of the inductor higher impedance levels can suppress the noise of the cross-coupled pair. And the voltage control node has to connect to the loop filter when the VCO is employed by a PLL.

IV. EXPERIMENTAL RESULTS

The measurement results of the switched-capacitor VCO prove that it can satisfy stringent specifications for phase noise, tuning range, and LO power. It covers from 34.29 to 39.88 GHz., a tuning range of 15.1% and has a maximum K_{VCO} of 240 MHz/V. It exhibits a worst case phase noise of -98.1 dBc/Hz at 1-MHz offset from the highest oscillation frequency, as shown in Fig. 5.

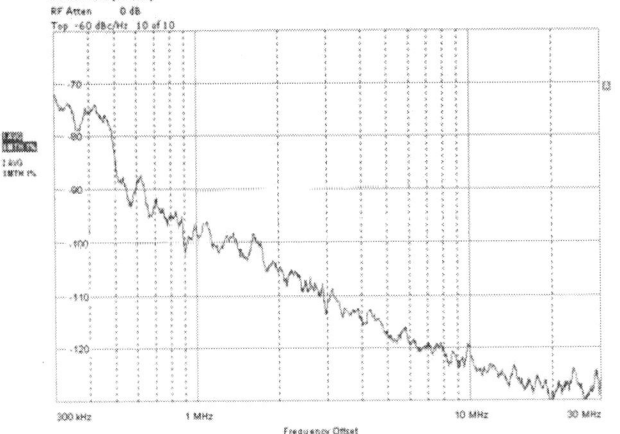

Fig. 5. Phase noise measurement result at 1-MHz offset from 39.88 GHz with 4.5 dB loss in the test equipments.

The VCO consumes 12 mA from a 1.2-V supply and the differential VCO buffer uses 11 mA to deliver a Single-Ended (SE) LO power of 0 dBm to the 50Ω probe. FOM and FOM$_T$, as defined in (3) and (4), are better than -178.5 dBc/Hz and -182.1 dBc/Hz respectively at 1-MHz offset across the band. Since the VCO operates in the current-

limited mode, using a 0.9-V supply to lower the power consumption improves the FOM$_T$ to -183.4 dBc/Hz. In this case the Single-Ended LO power reduces by 2.5 dB.

$$FOM(\Delta f) = L\{\Delta f\} - 20\log\left(\frac{f_0}{\Delta f}\right) + 10\log\left(\frac{P_{DC}}{1mW}\right) \quad (3)$$

$$FOM_T(\Delta f) = FOM(\Delta f) - 20\log\left(\frac{TR}{10}\right) \quad (4)$$

Where the quantities share the descriptions with (1). P_{DC} is the VCO power dissipation in mW, and TR is the tuning range in percentage.

Across the band, the varactor comprises a small portion of the overall capacitance of the tank. Thus, its undesired effects on the delivered power and the phase noise are minimal. However the digital tuning with 15.1% change in the frequency causes 3 dB change in the phase noise. Fig. 6 summarizes the results of the simulations and the measurements of the VCO versus the digital tuning code when the control voltage is set to the mid-rail.

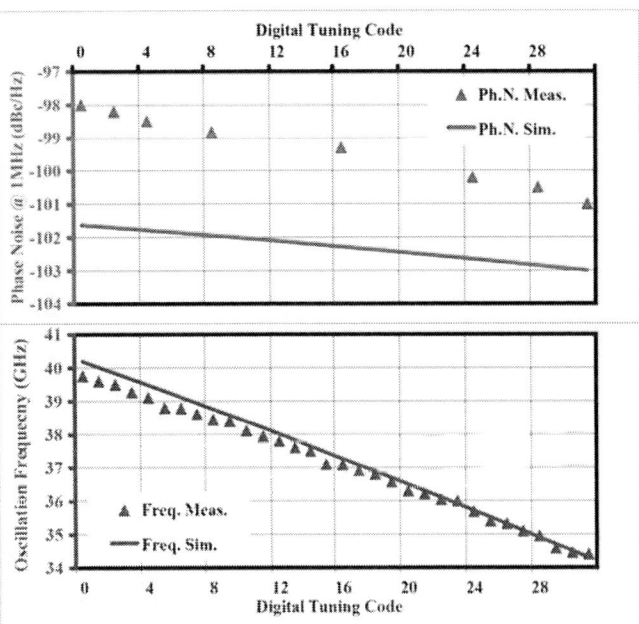

Fig. 6. Summary of the measured and simulated characteristics of the VCO vs. the digital tuning code.

The results of the VCO show significant improvements compared to an earlier version of the design which used only varactors. The varactor-only version measured a tuning range of 5.8% and a K_{VCO} of 2.1 GHz/V.

Table I presents a summary of the characteristics of this switched-capacitor VCO in comparison with other recently published mm-wave CMOS VCOs [5]-[9]. The table reports the worst case results of each work.

TABLE I

SUMMARY OF THE SPECIFICATIONS OF THIS DESIGN IN COMPARISON WITH RECENTLY PUBLISHED MM-WAVE CMOS VCOS

Reference	Year	Publ.	Process Technology	Freq. (GHz)	Ph.N.@1MHz (dBc/Hz)	Tuning Range%	PDC (mW)	FOM$_T$ (dB)	Pout (dBm)
[3]	2006	RFIC	CMOS 0.18 μm	21.6	-101.7	6.25	45	-171.9	-4.2
[5]	2009	VLSI	CMOS 0.13 μm	47.9	-102.5	1.59	5.6	-172.7	-19
[6]	2009	JSSC	CMOS 90 nm	58.4	-91.0	9.32	8.1	-176.7	-14
[7]	2008	ISSCC	CMOS 90 nm	49.9	-87.0	12	10.4	-172.4	-
[8]	2009	CICC	CMOS 65 nm	60	-97.1	16.7	30	-182.3	-
[9]	2009	IMS	CMOS 90 nm	61	-90.1	9.27	10.6	-174.9	-5
This Work	2010	RFIC	CMOS 65 nm	39.9	-98.1	15.1	14.4	-182.1	0

Fig. 8 shows the micrograph of the 0.15 mm^2 VCO, including the differential inductor and the transformer along with their guard rings. The DC pads on the top are to supply ground, supply, and the control signals.

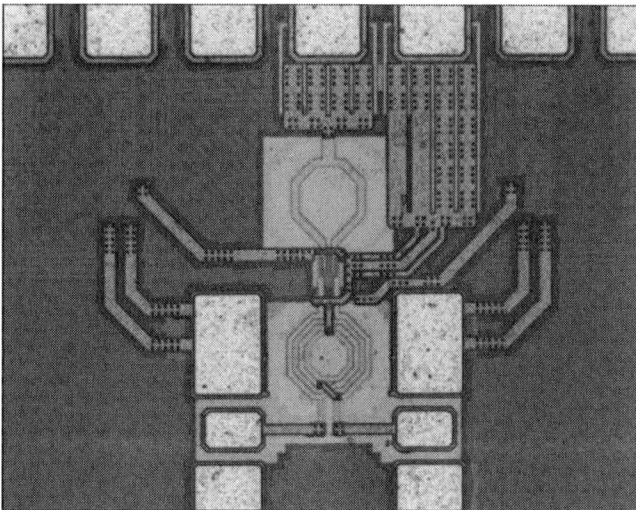

Fig. 8. Micrograph of the 300 μm * 500 μm VCO.

V. MEASUREMENT CHALLENGES

Probing unlocked high performance mm-wave VCOs for the purpose of phase noise measurements is extremely sensitive to their natural frequency drifts. The range of the frequency drift can exceed the offset frequency range of interest which makes it hard to have reliable low-offset phase noise measurements. Elements like the illuminator's light and mechanical stress or strain are destructive to these measurements. Vibration isolation table, wide probe tips, and wafer probing, as opposed to single die probing, significantly improve the reliability of the measurements.

Phase noise analysis mode of Rohde Schwartz FSUP50 has been used to measure this design. There are some settings which are vitally important. In the measurement setup, manual setting with a RBW of 0.1% increases the chance of success. Tracking and verifying the level and the frequency as well as the AFC should be turned off as they are not designed for unlocked VCOs. A combination of mechanical stability of the setup, quiet environment, and proper instrument settings can make low-offset phase noise measurements of the mm-wave VCOs reliable.

VI. CONCLUSION

The design and implementation of a switched-capacitor VCO in 65nm digital CMOS to cover 34 to 40 GHz has been presented. The measurement results demonstrate a tuning range of 15.1%, a phase noise of better than -98.1 dBc/Hz at 1-MHz offset across the band, a maximum K$_{VCO}$ of 240 MHz/V, and a minimum SE LO power of 0 dBm. The 0.15 mm^2 VCO can support a mm-wave transceiver with stringent phase noise, tuning range, and LO power delivery specifications.

ACKNOWLEDGEMENT

The authors wish to acknowledge the invaluable assistance and support of Michael Boers, Ali Parsa, Hsin-Hsing Liao, and Arya Behzad.

REFERENCES

[1] D. Leeson, "A simple model of feedback oscillator noise spectrum," *Proc. IEEE*, vol. 54, pp. 329-330, Feb. 1966.

[2] J. Rogers, et al., "The effect of varactor nonlinearity on the phase noise of completely integrated VCOs," *IEEE J. Solid-State Circuits*, vol. 35, no. 9, pp. 1360–1367, Sep. 2000.

[3] D. Ozis, N. Neihart, D. Allstot, "Differential VCO and passive frequency doubler in 0.18um CMOS for 24GHz applications," *IEEE RFIC Symposium. Dig.*, Jun. 2006.

[4] K. Yau, M. Khanpour, et al., "One-die source-pull for the characterization of the W-band noise performance of 65 nm general purpose (GP) and low power (LP) n-MOSFETs," *IEEE MTT-S IMS Dig.*, pp. 773- 776, Jun. 2009.

[5] Y. Lin, et al., "Low-power 48-GHz CMOS VCO and 60-GHz CMOS LNA for 60-GHz dual-conversion receiver," *IEEE VLSI Circuits Symposium Dig.*, pp. 88–91, Apr. 2009.

[6] L. Li, et al., "Design and analysis of a 90 nm mm-wave oscillator using inductive-division LC tank," *IEEE J. Solid-State Circuits*, vol. 44, no. 7, pp. 1950-1958, Jul. 2009.

[7] K. Scheir, S. Bronckers, J. Borremans, P. Wambacq, Y. Rolain, "A 52GHz phased-array receiver front-end in 90nm digital CMOS," *IEEE ISSCC Dig.*, pp. 184-185, Feb. 2008.

[8] B. Cath, M. Hella, "A 60 GHz CMOS combined mm-wave VCO/divider with 10-GHz tuning range," *IEEE CICC Dig.*, pp. 669–672, Sep. 2009.

[9] T. LaRocca, et al., "CMOS digital controlled oscillator with embedded DiCAD resonator for 58-64GHz linear frequency tuning and low phase noise," *IMS*, pp. 685-688, Jun. 2009.

RMO2D-1

Multi-Mode WCDMA Power Amplifier Module with Improved Low-Power Efficiency using Stage-Bypass

Gary Hau [1], and Mahendra Singh [2]

ANADIGICS, Inc., 300 Potash Hill Road, Tyngsboro, MA 01879, USA. [1]

ANADIGICS, Inc., 141 Mt. Bethel Road, Warren, NJ 07059, USA. [2]

Email: ghau@anadigics.com [1]

Abstract — **This paper presents a multi-mode power amplifier (PA) with very low DC quiescent current and current consumption under large power backoff operation. The PA is optimized to operate in three power modes. A dual-path PA is designed for high- and medium-power-mode operations, while a stage-bypass is applied to the final stage of the medium-power PA for low-power-mode operation. Load impedance for each power mode is individually optimized for best efficiency and linearity, achieving significant current saving compared to two-power-mode PAs. A 1.95GHz WCDMA PA module (PAM) has been developed using GaAs BiFET technology to validate the proposed circuit. The PAM demonstrates a very low quiescent current of 3.5mA under low-power-mode bias. The PAM exhibits 42% PAE and -40dBc ACLR1 at 28.5dBm Pout. At 17dBm and 8dBm backoff Pout, the PAM achieves 22% and 15% PAE, respectively, with -40dBc ACLR1. The current consumption at 8dBm Pout is reduced by 59% with the stage-bypass configuration.**

Index Terms — **wideband code-division multiple access (WCDMA), efficiency enhancement, heterojunction bipolar transistor (HBT), linear amplifiers, monolithic microwave integrated circuit (MMIC), power amplifiers.**

I. INTRODUCTION

Power amplifiers with low current consumption are essential for extending battery life of handheld devices. This requirement becomes even more critical for high-end mobile handsets due to the high level of integrations and functionalities (such as camera, color display, high data rate transmission).

Various techniques for reducing RF PAs' current consumption have been reported extensively in recent years. Special focus is placed on PAs for WCDMA system since the PAs usually transmit at significant power backoff from the maximum linear output power [1].

Among the most effective approach is dynamically adjusting the collector/drain bias voltage of the PA based on the output power level through a DC-DC converter [2].

The main drawbacks are higher cost, increased complexity and potential issue with noise coupling.

Enhanced-backoff-efficiency PAs with switchable load-impedance [3-5] are effective solution and have set the trend for the handset PA industry. Load impedances for such PA are optimized for multi-mode operation over two or three output power ranges, resulting in significant current saving under backoff operation over conventional design.

A multi-mode PA with dual paths optimized for two-power-range operation was reported previously [5]. In order to achieve further current reduction under larger backoff operation, this work investigates the feasibility of a stage-bypass configuration applying to the final stage of the medium-power-mode PA. The aim of this approach is to achieve:

a) a multi-mode PA with optimized current consumption over three output power ranges;

b) very low DC bias current.

Special attention is also paid on minimizing additional circuitry to limit cost impact.

In this paper, the architecture of the proposed PA is first discussed. The MMIC process used for the fabrication of the PA is then described. The design and implementation of the PA MMIC and PA module (PAM) are detailed. The WCDMA power characteristics of the proposed design are illustrated to demonstrate its performance advantage.

II. PA ARCHITECTURE

A dual-path PA, designed to operate in two power modes, is shown in Fig. 1 [6]. Each power path was optimized for different power levels with different load impedance terminations for the PAs. The high power path is designed to deliver the maximum Pout, while the low power path is targeted for backoff operation - for example, 10dB or more below the maximum Pout. Depending on the Pout requirement, the PA is switched between the high- and low power paths via the input and output RF switches, thus achieving higher efficiency under backoff operation over conventional single-path PA.

978-1-4244-6240-7/10 $26.00 © 2010 IEEE

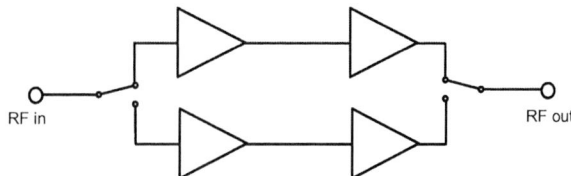

Fig. 1. Block diagram of the dual-path power amplifier [6].

The backoff efficiency of the dual-path PA starts to drop off rapidly if the low-power-path PA is operated at large backoff from its maximum Pout. In order to address this limitation, an additional power mode will therefore be required to further enhance the efficiency of the PA at very low Pout.

Fig. 2 shows the block diagram of the proposed PA with (three) high-, medium- and low-power modes (HPM, MPM, LPM), allowing the PA to operate optimally over three different power ranges. The HPM and MPM employ the dual-path approach similar to our previously reported design shown in Fig. 1 [6]. A different approach is selected for the LPM operation. Instead of adding another parallel signal path, the new design employs a stage-bypass configuration to the final stage of the MPM PA. One advantage of this approach is the re-use of existing active circuitry from the driver stage of the MPM PA and thus minimizes additional components.

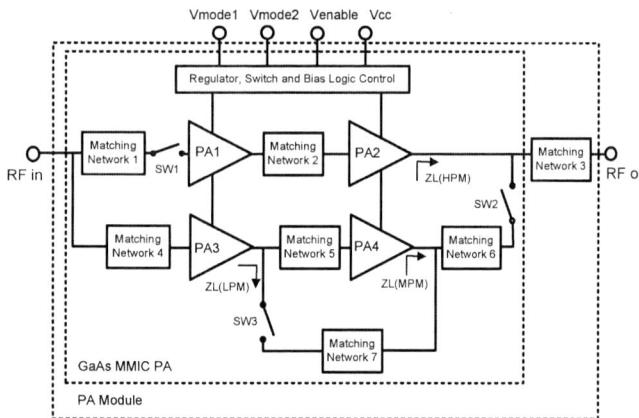

Fig. 2. Block diagram of the proposed power amplifier with stage-bypass.

Under HPM operation, the input signal will be amplified by the upper branch shown in Fig. 2, where PA1,2 will be turned on and PA3,4 will be shut off. The input RF switch SW1 will be closed to allow the signal amplified through the upper path while switches SW2,3 will remain open. The HPM PA2 output is decoupled from the MPM output through switch SW2, minimizing the parasitic effect of matching network 6 on HPM performance.

Similarly, under MPM operation, the input signal will be amplified by the middle branch shown in Fig. 2, where PA3,4 will be turned on while PA1,2 will be turned off. The RF switch SW2 will be closed, and switches SW1,3

will remain open in order to divert the signal through the middle path.

Under LPM operation, the input signal will be diverted through part of the middle path and then through switch SW3 to the output. Only PA3 will be turned on and the RF switches SW2,3 will be closed, while switches SW1 will remain open to support this operation. PA3 will be switched to a lower DC bias current condition compared to the MPM case.

Table 1 summarizes the ON/OFF conditions of the PAs and switches under different mode of operations. Each of these components is controlled individually by the switch and bias logic control circuitry, and the mode selection is determined by the logic combination of Vmode1,2 and Venable PINs.

	Mode of Operation			
	Disable	HPM	MPM	LPM
PA1,2	OFF	ON	OFF	OFF
PA3	OFF	OFF	ON	ON
PA4	OFF	OFF	ON	OFF
SW1	OFF	ON	OFF	OFF
SW2	OFF	OFF	ON	ON
SW3	OFF	OFF	OFF	ON

Table 1. ON/OFF characteristic of individual component of the proposed PA under different operating modes.

With the use of three different amplification paths, it becomes possible to present different load impedance to the output stage of each power mode. RF performances can therefore be optimized for each specific output power range.

Fig. 3 illustrates the RF impedances at various locations of the proposed PA shown in Fig. 2. Matching network 3 transforms the 50Ω output termination to the optimum load impedance, ZL(HPM), for PA2. Matching networks 6 and 7 transform the impedance to ZL(MPM) and ZL(LPM) for MPM and LPM operations, respectively.

Note that the optimum load impedance increases from high- to low-power-mode. The higher load impedance results in lower saturated power and the PA will be able to operate at a less backoff, higher efficiency region.

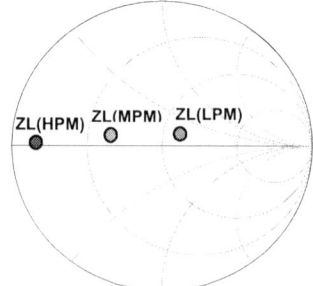

Fig. 3. RF impedances at various matching points of the proposed power amplifier shown in Fig. 2.

Fig. 4 shows the representative DC bias conditions and loadlines for the high-, medium-, and low-power output stages. The bias point and loadline for the PA2/3/4 are selected for optimum efficiency and linearity at maximum output power for the respective power range. As discussed earlier, the shallower slopes for the MPM and LPM loadlines limit the saturated Pout and allow the PA to operate at less backoff, higher efficiency region. Since a single-stage PA (PA3) is sufficient to meet the gain requirement for LPM operation, very low DC bias current can be used to achieve additional current saving.

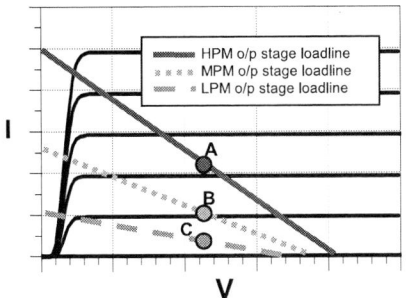

Fig. 4. Representative loadlines and DC bias conditions for the final stage of the proposed PA on a typical transistor I-V curve. Points A to C are the quiescent bias points for PA2 (HPM), PA4 (MPM) and PA3 (LPM), respectively.

III. GaAs-Based BiFET Technology

To realize high level of integration for the PA topology shown in Fig. 2, a MMIC process that offers co-integration of both HBTs (for high efficiency, high linearity PA with low leakage current), and low loss pHEMT switches is required. The ANADIGICS InGaP-*Plus*[TM] process [5-6] was used for the fabrication of the PA MMIC. This process features high performing HBT and pHEMT devices integrated on the same GaAs substrate. Epi layers for pHEMTs (AlGaAs) are grown first on the semi-insulating substrate and then the Epi layers comprising the HBT (InGaP) are grown on top of the pHEMT. The advantage of this process is that the HBT and pHEMT are completely decoupled, allowing pHEMT to be used as low loss RF switches without compromising its performance. The switches also allow on-chip integration of control logic, as well as voltage regulator which eliminates the need of external reference voltage for PA bias.

IV. Circuit Design and Implementation

The proposed PA discussed in Section II was implemented in a 50Ω matched, 3x3mm² multi-layer module with an industry standard footprint. The PAM was designed for WCDMA handset application for the frequency band of 1920-1980MHz. Fig. 5 shows the photograph of the 3x3mm² PAM. The module is 50Ω

matched at the RF input and output ports. The PA output match was implemented with standard 0201 SMD components to minimize circuit loss. The inner layers contained the DC routing and printed inductor line for RF/DC bias isolation. Thermal vias were added underneath the MMIC for improving heat dissipation.

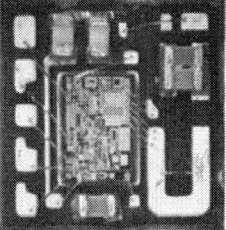

Fig. 5. Photograph of the 3x3mm² PA module.

The RF switches, and HBT peripheries for the drivers and power stages were sized for the power handling requirements. The medium- and low-power stages were considerably smaller than the high-power stages due to the lower Pout requirements. The various matching networks were designed according to the requirements discussed in Section II. The amplifiers were biased under Class AB operations and output stages were terminated with appropriate harmonic impedance for PAE enhancement.

The switch and bias logic control network was designed for modes selection according to Table 1. Enhanced current mirrors with temperature compensated characteristics were employed for the bias circuits. An on-chip voltage regulator was included to provide a stable reference voltage to the current mirrors of the PAs.

IV. Measured Results and Discussions

The proposed PAM was characterized for both DC and RF operations. The PAM was biased at 3.4V Vcc with 0/1.8V logic applied to Venable and Vmode1,2 PINs. Table 2 compares the quiescent currents and power ranges of the PAM under different mode of operations.

Mode of Operation	Quiescent Current	Power range
LPM	3.5 mA	< 8 dBm
MPM	20 mA	8 to 17 dBm
HPM	100 mA	17 to 28.5 dBm

Table 2. Quiescent currents and power ranges of the proposed PAM. Vmode1,2=0/0V for HPM, Vmode1,2=1.8/0V for MPM, and Vmode1,2=1.8/1.8V for LPM operations.

The LPM quiescent current of the proposed PAM is 3.5 mA, which is 83% lower than that of the two-power-mode PAM. This significant bias current reduction offers additional current saving especially when the PAM is under stand-by operation.

The power performances of the PAM were evaluated using a HPSK modulated signal with a chip rate of 3.84Mcps at 1950MHz. Fig. 6(a) shows the measured gain of the PAM from 0-30dBm Pout. The proposed PA achieved 27.5dB gain at 28.5dBm Pout. Under MPM and LPM, the gain of the PA reduced to 24dB and 12dB respectively.

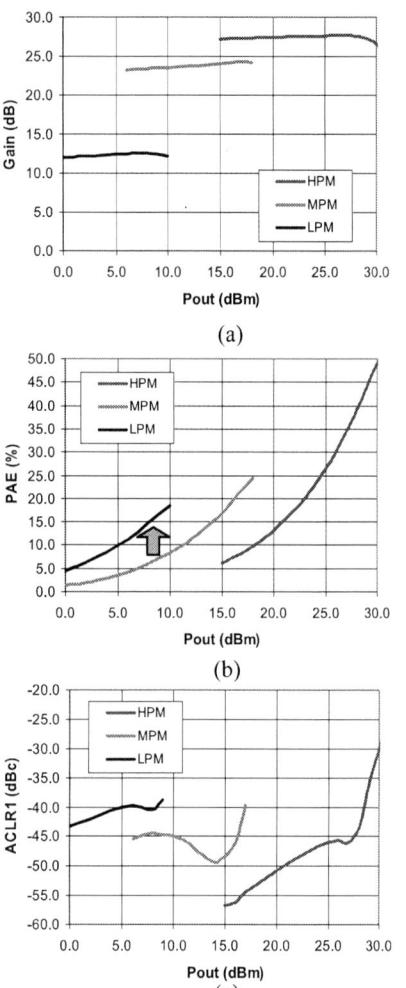

Fig. 6. (a) Gain, (b) PAE, and (c) ACLR1 of the stage-bypass WCDMA PAM measured at 1950MHz with Vcc=3.4V.

Fig. 6(b) and 6(c) compare the PAE and ACLR1 of the PAM respectively. The PAM achieved 42% PAE at 28.5dBm HPM Pout. The MPM and LPM saturated Pout are 12dB and 20dB lower than that of the HPM due to higher load impedance terminations as discussed in Section II. As a result, the PAM operated at less backoff for Pout below 17dBm, and achieved significant PAE improvement. At 17dBm Pout (MPM), the proposed PAM demonstrated 22% PAE. And at 8dBm Pout (LPM), the PAM achieved 15% PAE which represented 59%

reduction in current with this stage-bypass configuration (as compared to MPM at the same Pout). At all three Pout levels, the PAM maintains -40dBc ACLR1, measured at 5MHz offset from the center frequency. The Pout switchover points between LPM, MPM and HPM are determined by the worse case ACLR1 level based upon the system requirements.

V. CONCLUSION

A multi-mode PA with low-power-mode stage-bypass was proposed for reducing DC quiescent current and current consumption under large power backoff operation. The technique was implemented in a WCDMA HBT PAM. The PAM achieves a 3.5mA quiescent current with a 59% reduction in current consumption at 8dBm Pout over conventional design. Excellent high power performance was maintained with 42% PAE at 28.5dBm Pout. The MPM and LPM PAEs are 22% and 15% at 17dBm and 8dBm Pout respectively. This state-of-the-art performance and the low cost, small size implementation are very attractive for WCDMA handset applications.

VI. ACKNOWLEDGEMENT

The authors would like to acknowledge T. Arell, T. Chen, and R. Lertpiriyapong for valuable technical discussions and suggestions. We also like to thank B. MacDonald for technical support.

REFERENCES

[1] D.A. Teeter *et al.*, "Average current reduction in (W)CDMA power amplifiers," *IEEE RFIC Symposium Dig.*, pp. 429-432, June 2006.

[2] G. Hau *et al.*, "A 3x3mm² Embedded-Wafer-Level Packaged WCDMA GaAs HBT Power Amplifier Module with Integrated Si DC Power Management IC," *IEEE RFIC Symposium Dig.*, pp. 409-412, June 2008.

[3] T. Apel *et al.*, "Efficient Three-State WCDMA PA Integrated with High-Performance BiHEMT HBT / E-D pHEMT Process," *IEEE RFIC Symposium Dig.*, pp. 149-152, June 2008.

[4] G. Zhang *et al.*, "Dual Mode Efficiency Enhanced Linear Power Amplifiers Using a New Balanced Structure," *IEEE RFIC Symposium Dig.*, pp. 245-248, June 2009.

[5] Y. Kwon *et al.*, "Low-Power PAE Enhancement for 3G Handset Power Amplifiers," *IEEE RFIC Symposium Workshop*, June 2009.

[6] A. Gupta *et al.*, "InGaP-PlusTM A major advance in GaAs HBT Technology," *IEEE CSICS Dig.*, pp. 179-182, Nov, 2006.

[7] G. De la Rosa *et al.*, "A GSM-EDGE Power Amplifier with a BiFET Current Limiting Bias Circuit," *IEEE RFIC Symposium Dig.*, pp. 595-598, June 2009.

A 3.4 GHz to 4.3 GHz Frequency-Reconfigurable Class E Power Amplifier with an Integrated CMOS-MEMS LC Balun

Leon Wang and Tamal Mukherjee

Department of ECE, Carnegie Mellon University, Pittsburgh, PA, 15213

Abstract — **A monolithically integrated differential class E power amplifier capable of dynamically switching between 3.4 GHz and 4.3 GHz operation has been designed and fabricated in a 0.35 µm BiCMOS process; this power amplifier also includes an integrated CMOS-MEMS variable capacitor enabled LC balun for differential to single-ended conversion. The power amplifier achieves a maximum output power of 19.1 dBm and a maximum power added efficiency of 15.1% with a supply voltage of 3.3 V.**

Index Terms — **Power amplifiers, power combiners, microelectromechanical devices.**

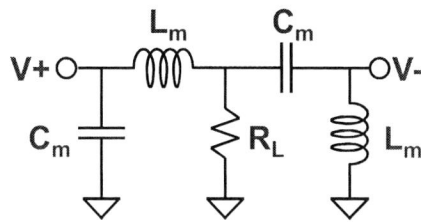

Fig. 1. Lattice-Type LC Balun.

I. INTRODUCTION

With the abundance of wireless standards, there is a desire to make RF front-ends frequency-reconfigurable. Of particular importance is the power amplifier which typically dominates the power consumption of the RF front end and frequency-reconfigurability is needed to maximize output power and efficiency at frequency extremes. Frequency-reconfigurable power amplifiers reported in the past [1]-[2] are not monolithically integrated although the tunable element itself in [2] is integrated and the entire power amplifier holds promise for future integration. This paper presents a frequency-reconfigurable power amplifier that is monolithically integrated.

With monolithic integration comes the strong motivation to use power combining techniques in order to deliver high power from low-breakdown voltage devices [3]-[4]. It is logical that a frequency-reconfigurable power combiner would be needed. Of the several power combining techniques, the LC balun (see Fig. 1) is best suited to be made frequency-reconfigurable with a monolithically integrated CMOS-MEMS variable capacitor. Not only is an LC balun useful for differential to single-ended conversion, but it also has a higher power enhancement ratio than an L-match for achieving high output power with low transformation loss [5]. With the LC balun, this new power amplifier is able to outperform a previously designed monolithically integrated frequency-reconfigurable power amplifier [6].

II. TUNABLE ELEMENTS

A. CMOS-MEMS Variable Capacitor

The CMOS-MEMS variable capacitor (see Fig. 2) is fabricated in a 0.35 µm BiCMOS process along with CMOS circuitry, and the CMOS-MEMS post-process [7] releases the MEMS structures without any additional masks. This CMOS-MEMS post-processing uses the existing metal as a mask by removing any oxide and 50 µm of the silicon substrate that is not covered by metal. In this way, the MEMS variable capacitor is monolithically integrated with CMOS. The CMOS-MEMS variable capacitor has interdigitated capacitance beams consisting of the back-end-of-line metal dielectric stack that form the basis of the variable capacitance. Because these capacitance beams are released by the CMOS-MEMS

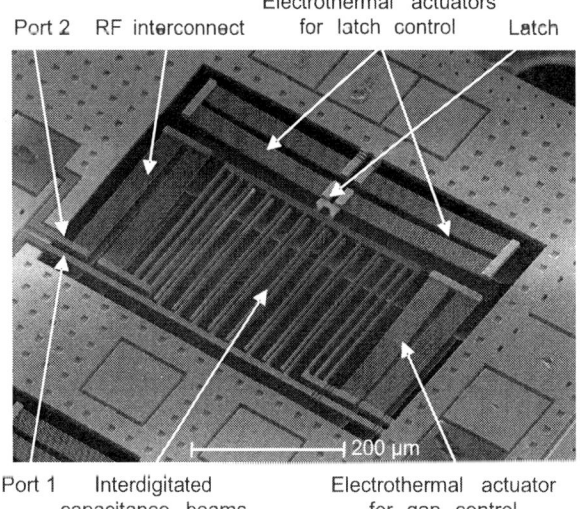

Fig. 2. CMOS-MEMS Variable Capacitor.

post-process, the gap between the beams, and hence the capacitance of the device, can be varied by electrothermal actuators that use embedded polysilicon resistors. A mechanical latch can maintain the capacitance position so that the device does not burn static power. This CMOS-MEMS variable capacitor has a tuning ratio of 6.9:1, a quality factor of 28 at 3 GHz [8].

B. Switched Capacitors

A switched capacitor is simply a capacitor in series with a switch. The main drawback of a switched capacitor is that there is an unfavorable tradeoff between its quality factor and tuning range. The switch would be implemented as a transistor and a large transistor is desired for a high quality factor when the switch is ON. However, when the switch is OFF, the capacitance consists of the series combination of the fixed capacitance and the parasitic capacitance of the transistor, and therefore a small transistor is desired for a large tuning range (see Fig. 3).

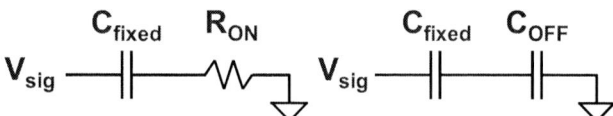

Fig. 3.   Simplified model of a switched capacitor when ON (left) and OFF (right).

A differential switched capacitor can win back a factor of two in either tuning range or quality factor by taking advantage of the fact that the switch can be shared and that a differential ground exists in the middle of the switch (see Fig. 4). For the same sized switch and capacitor (but with two capacitors), the OFF capacitance of the switch is the same as before, but the ON resistance of the switch is half of what it was before because the ground is now half a channel length closer than before, resulting in a doubling of the quality factor (see Fig. 5). Tiny grounding transistors ensure that the source of the switch is grounded so that a maximum gate overdrive voltage can be applied to the gate of the switch to minimize the ON resistance of the switch.

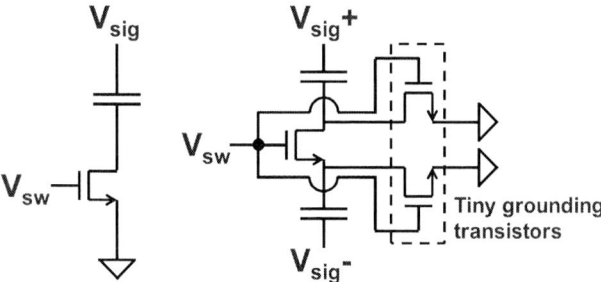

Fig. 4.   Single-ended switched capacitor (left) vs. differential switched capacitor (right).

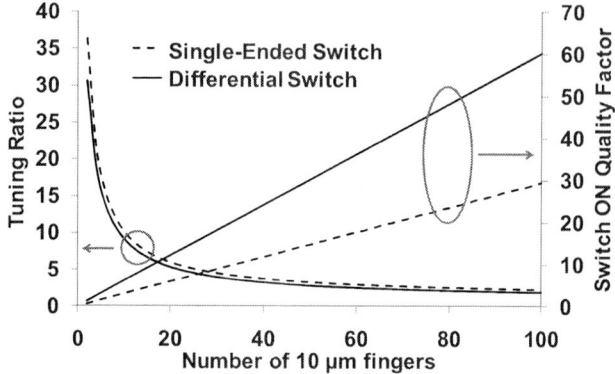

Fig. 5.   Simulated quality factor and tuning ratio of a switched capacitor as a function of switch size: single ended vs. differential. Capacitor value = 1 pF. Frequency = 3 GHz.

C. Use of Tunable Elements

Although the CMOS-MEMS variable capacitor has a quality factor of 28 at 3 GHz and a tuning range of 6.9:1, a single device measures roughly 400 μm by 500 μm but only has a variable capacitance of 54 fF to 372 fF and should only be used where the benefits justify the added area such as in an LC balun. If the LC balun requires a series capacitance of 400 fF at 4 GHz, the impedance of the capacitor is –j*100 Ω, and if an optimistic quality factor of 10 is assumed for a switched capacitor, then there are 10 Ω of series resistance in the capacitive branch of the LC balun. Such a large series resistance would severely degrade the power amplifier's overall efficiency and therefore a CMOS-MEMS variable capacitor is used in the LC balun. The CMOS-MEMS variable capacitor also has the added benefit of having a low parasitic shunt capacitance. The area-efficient switched capacitor is used for all other variable capacitors in the power amplifier.

The use of MOS or diode varactors was considered but switched capacitors were ultimately chosen because of the varactors' signal dependent capacitance. For the tuning ratios that were chosen, the two have similar quality factors so quality factor was not the deciding factor. If varactors must be used, the distortion can be reduced with stacking, and higher quality factor varactors can be fabricated in a silicon-on-glass process [2].

III. POWER AMPLIFIER DESIGN

The output stage with the LC balun is preceded by a pre-amplifier in order to form the input match and to reduce the power amplifier's drive requirement. Since CMOS transistors in a 0.35 μm BiCMOS process are too slow to create a good CMOS inverter above 3.4 GHz, a simple inverter chain as the pre-amplifier was ruled out. Even if the transistors were fast enough, the $f_0 C_{gate} V_{DD}^2$ power would be significant above 3.4 GHz and the gate capacitance of the output stage clearly needs to be resonated with an inductor. A variable capacitor C_1 was added in parallel with this gate capacitance to enable frequency-reconfigurability (see Fig. 6). This frequency-reconfigurable LC tank is driven by a pre-amplifier with an input match that simply consists of a 50 Ω resistor to a differential ground, in parallel with the input transistor (see Fig. 6). This input match works well from DC until slightly beyond the frequencies of interest as long as the input capacitance of the transistor is small enough. To get the most transconductance out of the input transistor for a given input capacitance, a BJT was used instead of a MOS device because of the BJT's higher cutoff frequency. Such a simple input match is appropriate for an already complex power amplifier with many tunable elements.

To save area, several things were done. First, L_1 and L_2 are both differential inductors rather than two single ended inductors (see Fig. 6). Second, C_1 and C_2 are both arrays of 15 identically sized switched capacitors that are controlled by low power 4-bit flash ADCs for testing purposes. As discussed in section II-C, switched capacitors are area-efficient and therefore they are suitable for nodes with a large fixed capacitance. C_1 is in parallel with a large gate capacitance that together resonates with L_1, and C_2 is in parallel with a large drain capacitance that is a part of the output network. Third, the shunt C_m and shunt L_m in the LC balun (see Fig. 1) were omitted in the design. An L_m of 3.5 nH including wire parasitics is large compared to 1.03 nH on one half of L_2, and the two in parallel is still relatively close to L_2; therefore the shunt L_m was omitted. A C_m of 590 fF is also small compared to the single-sided capacitance C_2 of 1.4 pF in the high capacitor setting; therefore the shunt C_m was also omitted.

The chip was tested using a probe station and due to the limited number of probe tips and the abundance of control signals, a low inductance path to ground was provided via bondwires (see Fig. 7). Although only one probe tip is used for the supply in each stage, 60 pF and 140 pF of on-chip decoupling capacitance are provided for the first and second stages respectively. An off-chip DC blocking capacitor was used during testing.

TABLE I

COMPONENT VALUES

Component	Value
Half of L_1	496 pH
Half of L_2	1.03 nH
Half of C_1	1.23 pF - 3.98 pF
Half of C_2	923 fF - 1.84 pF
L_m	2.05 nH
C_m	162 fF - 1.12 pF

IV. MEASUREMENT RESULTS

In Fig. 9 and Table II, the drain efficiency (DE) only includes the power consumption of the output stage while the global efficiency (GE) includes the whole amplifier. Looking at Table II, the power amplifier's performance is comparable to [5] which is another power amplifier with an integrated LC balun. The lower drain efficiency compared to [5] is the result of a higher operating frequency with slower transistors and the fact that the LC balun is missing the shunt L_m and C_m to save area. These high frequencies were chosen for the appropriateness of inductor and CMOS-MEMS variable capacitor values. While a more advanced CMOS process may have been more appropriate for these frequencies, the CMOS-MEMS variable capacitor exists at the 0.35 μm process node. The distortion components (HD_2 and HD_3) are slightly high also because the LC balun is missing the shunt L_m and C_m. Although this work uses a higher supply voltage than [5], the output powers are similar because of the higher loss in

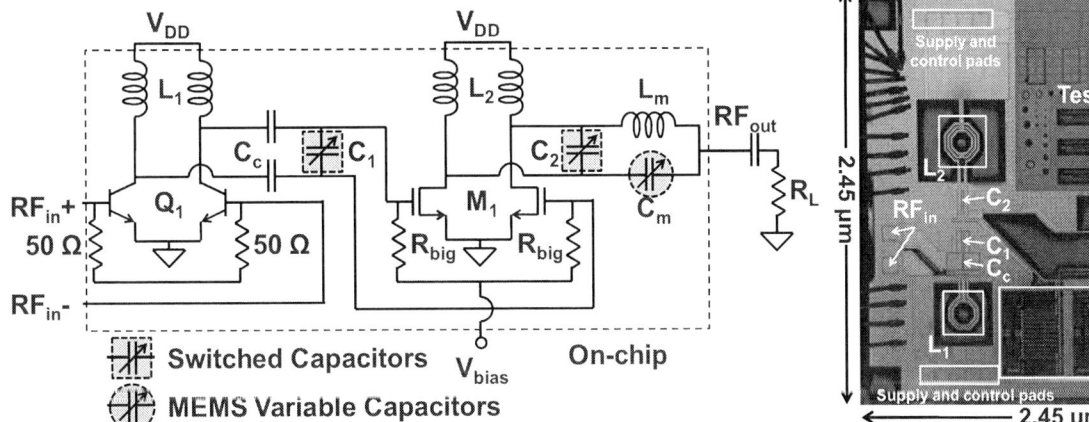

Fig. 6.   Complete power amplifier schematic.

Fig. 7.   Power amplifier die photo.

978-1-4244-6240-7/10 $26.00 © 2010 IEEE   149

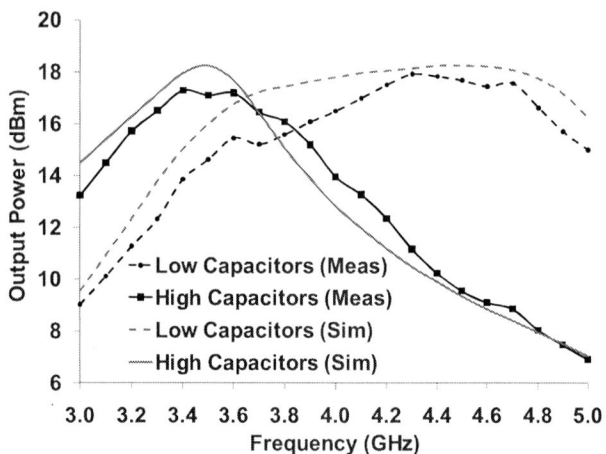

Fig. 8.　Output power versus frequency. Input power = 4 dBm.

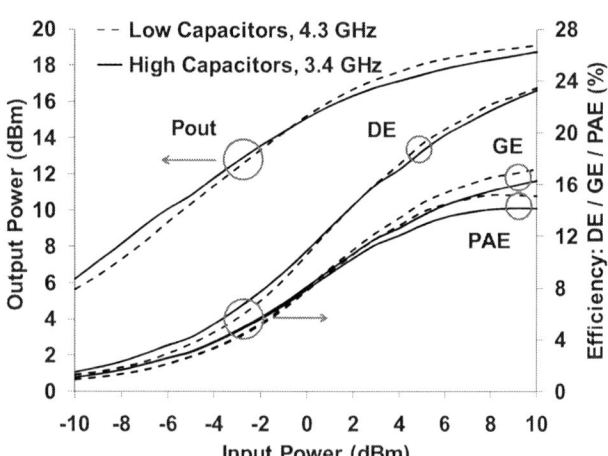

Fig. 9.　Measured output power/efficiency versus input power.

TABLE II

PERFORMANCE SUMMARY AND COMPARISON

	This work		[5]
Process	0.35 µm BiCMOS		0.13 µm CMOS
V_{DD}	3.3 V		1.5 V
Setting	High Caps	Low Caps	1 section–2 PAs
Frequency	3.4 GHz	4.3 GHz	2.45 GHz
Max P_{out}	18.7 dBm	19.1 dBm	18 dBm
Max DE	23.6%	23.5%	32%
Max GE	16.3%	17.2%	21%
Max PAE	14.1%	15.1%	21%
HD_2*	-16.6 dBc	-19.9 dBc	-
HD_3*	-25.3 dBc	-29.6 dBc	-

*At P_{in} = 10 dBm

the passive components at higher frequencies and because [5] uses a higher transformation ratio in the LC balun; this work opted for a lower transformation ratio in order to obtain a lower C_m that is appropriate for the CMOS-

MEMS variable capacitor. Although the tuning range of this amplifier is from 3.4 GHz to 4.3 GHz determined by the output power peaks in Fig. 8, the output power at 4.7 GHz is higher than the peak at 3.4 GHz.

V. CONCLUSION

To the authors' best knowledge, this power amplifier is the first frequency-reconfigurable power amplifier with a power combiner that is monolithically integrated; [1]-[2] are frequency-reconfigurable but not monolithic and [3]-[4] are monolithic and use power combining but are not frequency-reconfigurable. Although the CMOS-MEMS variable capacitor simultaneously has a high quality factor and a wide 6.9:1 tuning ratio, the tuning range of the power amplifier is limited by the weakest link: the area-efficient switched capacitors. Future linearization can be achieved by adding parallel LC balun sections and turning each section on or off as appropriate as discussed in [5].

ACKNOWLEDGEMENT

This research was sponsored by ITRI Labs @ CMU and the SRC. The authors would like to thank S. Santhanam for CMOS-MEMS micromachining, J. Reinke for the CMOS-MEMS variable capacitor design and modeling, and Y. Fang for wirebonding.

REFERENCES

[1] A. Fukuda, et al, "A 900/1500/2000-MHz Triple-Band Reconfigurable Power Amplifier Employing RF-MEMS Switches," *IEEE IMS*, pp. 334-347, June 2005.

[2] W.C.E. Neo, et al, "Adaptive Multi-Band Multi-Mode Power Amplifier Using Integrated Varactor-Based Tunable Matching Networks," *IEEE J. Solid-State Circuits*, vol. 41, no. 9, pp. 2166-2176, Sept. 2006

[3] I. Aoki, S. D. Kee, D. B. Rutledge, and A. Hajimiri, "Fully-Integrated CMOS Power Amplifier Design Using Distributed Active Transformer Architecture", *IEEE J. Solid-State Circuits*, vol. 37, no. 3, pp. 371-383, Mar. 2001.

[4] P. Haldi, et al, "A 5.8-GHz linear power amplifier in a standard 90-nm CMOS process using a 1-V power supply," *IEEE RFIC*, pp. 431–434, Jun. 2007.

[5] P. Reynaert and M. S. J. Steyaert, "A 2.45-GHz 0.13-µm CMOS PA with parallel amplification," *IEEE J. Solid-State Circuits*, vol. 42, no. 3, pp. 551–561, Mar. 2007.

[6] A. Jajoo, L. Wang, and T. Mukherjee, "MEMS Varactor Enabled Frequency-Reconfigurable LNA and PA in the Upper UHF Band," *IEEE IMS*, pp. 1121-1124, June 2009.

[7] G.K. Fedder, et al, "Laminated High-Aspect Ratio Microstructures In A Conventional CMOS Process," *Sensors & Actuators*, vol. A57, no. 2, pp. 103-110, Mar. 1997.

[8] J. Reinke, et al, "CMOS-MEMS Variable Capacitors with Low Parasitic Capacitance for Frequency-Reconfigurable RF Circuits," *IEEE RFIC*, pp. 509-512, June 2009.

A Q-band 6W MMIC Power Amplifier with 3-way Power Combination Circuit

Hiroshi Otsuka, Kazuhisa Yamauchi, Koji Yamanaka, Shin Chaki*, Kazuhiko Nakahara**,
Kunihiro Endo**, Akira Inoue and Yoshihito Hirano

Information Technology R&D Center, Mitsubishi Electric Corp.
5-1-1 Ofuna, Kamakura-city, Kanagawa 247-8501, Japan
E-mail: Otsuka.Hiroshi@bc.MitsubishiElectric.co.jp
*High Frequency & Optical Device Works, Mitsubishi Electric Corp.
**Kamakura Works, Mitsubishi Electric Corp.

Abstract —A Q-band 6W MMIC power amplifier was developed, which employed 3-way power combination circuit to combine 12 FET cells for high output power. In addition, the pi type bias and matching circuit was applied to the inter-stage matching circuit to suppress loop oscillation. 37.9dBm （6.2W） saturated output power was successfully achieved, which is the highest output power for Q-band power amplifiers reported to date.

Index Terms — Power amplifiers, GaAs, MMIC, Q-band, Millimeter-wave.

I. INTRODUCTION

In recent years, with increasing traffics on commercial wireless communication systems, millimeter-wave band has been of much interest [1][2]. High power MMIC amplifiers are the most important components for such systems. Over years, significant efforts to obtain high output power have been expended on the design and development of MMIC power amplifiers [3-9]. So far, MMIC power amplifiers at millimeter-wave band were designed employing 2^N power combination circuit of tournament type. Considering limit of chip size, the number of the most FET cells are 16. Consequently, MMIC power amplifier beyond 3W at millimeter-wave band has not been reported. If an arbitrary number of cells are available, output power more than 3W can be achieved.

In this paper, we presented 3-way power combination circuit which selected a number of combinations freely and a Q-band 6W MMIC power amplifier with the power combination circuit. The 3-way power combination circuit consists of "multi-feed divider" which realizes low-loss power combining of FET cells. By employing the circuit, this amplifier combined 12 FET cells and achieved high output power. In addition, the inter-stage matching circuit employed the pi type bias and matching circuit to suppress loop oscillation. 37.9dBm （6.2W） saturated output

power was successfully achieved, which is the highest output power for Q-band power amplifiers reported to date.

II. MULTI-FEED DIVIDER

In power amplifier design, it is important that power is equally distributed to each FET for attaining high gain and high power. The multi-feed divider was employed for efficient power dividing. A one-stage amplifier with the multi-feed divider and a conventional one-point feed divider [8][9] are shown in Fig. 1. The multi-feed divider consists of a microstrip transmission line with a large width and a two-way divider. The microstrip line acts as a power divider and an impedance transformer. The multi-feed divider reduces phase deviation by supplying RF power to the microstrip line at two points. Moreover, arbitrary divided number is selected, because output paths of this divider are not separated clearly.

Fig. 2 shows calculated current density of the multi-feed divider and the conventional divider. Phase contour of the multi-feed divider is perpendicular to a direction of electromagnetic wave propagation. The divider distributes RF power to each FET in equal phase.

Measured MSG/MAG of the one-stage amplifier with the multi-feed divider is shown in Fig. 3. MAG of the amplifier with the multi-feed divider is approximately 1.5dB higher than that of the amplifier with the conventional divider at Q-band. The multi-feed divider is efficient for power amplifiers.

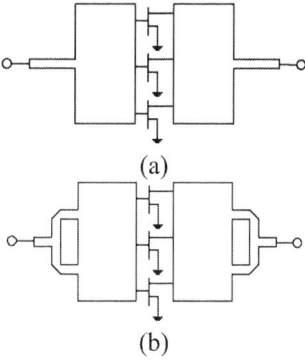

Fig. 1. One-stage amplifier. (a) Conventional divider, (b) Multi-feed divider

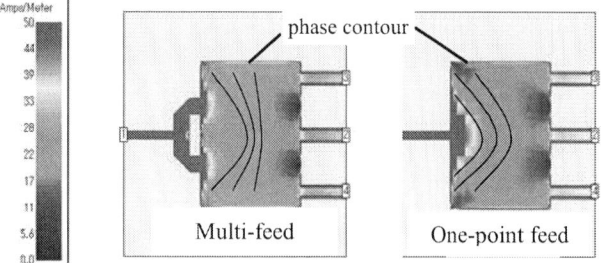

Fig. 2. Calculated current density of the multi-feed divider.

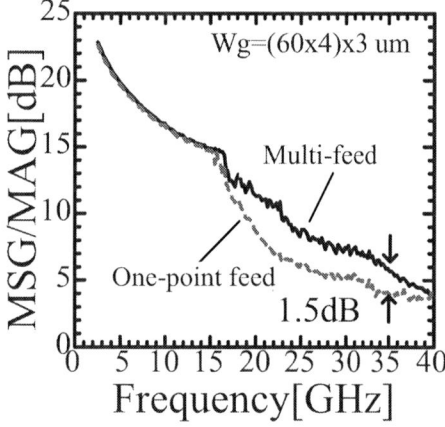

Fig. 3. Measured MSG/MAG of the one-stage amplifier with the multi-feed divider.

III. PI TYPE BIAS AND MATCHING CIRCUIT

The pi type bias and matching circuit is shown in Fig. 4. This circuit is a bias circuit for DC. DC is supplied to the unit amplifier through the quarter-wave transmission line. On the other hand, this circuit is a matching circuit for RF. The impedance Z_A is infinity because the impedance of the node 2 is zero by the capacitor and transformed by the

quarter-wave transmission line. Therefore, this circuit acts as short stub for RF.

Furthermore, this circuit suppresses loop oscillation because the resistor absorbs imbalance power between the node 1 and the node 2. Fig. 5 shows the schematic diagram for loop gain analysis [10] of the amplifier. Fig. 6 shows calculated loop gain of loop A shown in Fig. 5. Loop gain of loop A is higher than 0dB around 12 GHz. The amplifier with the conventional circuit satisfies loop oscillation condition around the frequency. In contrast, loop gain is lower than 0dB for the amplifier with the pi type bias and matching circuit. This circuit reduces loop gain and suppresses loop oscillation.

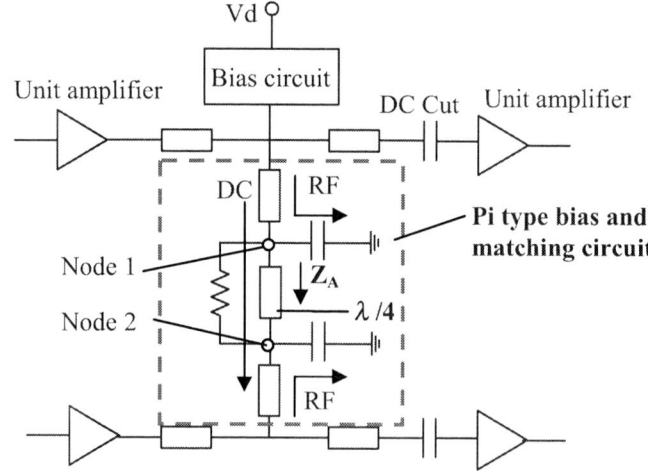

Fig. 4. Pi type bias and matching circuit.

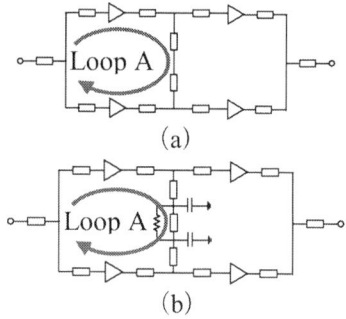

Fig. 5. Schematic diagram for loop gain analysis of the amplifier. (a) Conventional circuit (b) Pi type bias and matching circuit

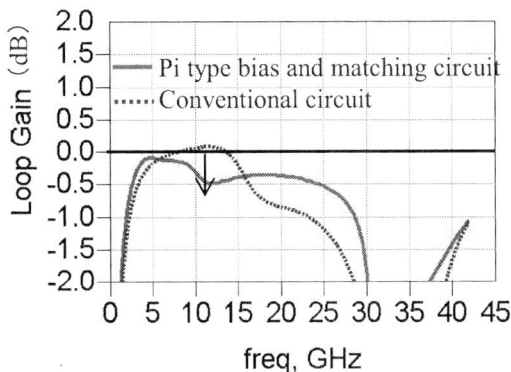

Fig. 6. Calculated loop gain of loop A.

IV. POWER AMPLIFIER CIRCUIT DESIGN

Fig. 7 shows the photograph of 6W MMIC power amplifier chip. The chip size is 5.93x4.93mm.The power amplifier is two-stage amplifier. Each stage consists of four unit amplifiers, which combines three 1mm gate periphery FETs using the multi-feed divider. A high power density TaN/Au T-gate pHEMT[11] was employed. The input and output matching design employed quarter-wave impedance transformers. The inter-stage matching design employed the pi type bias and matching circuit to realize DC power feed and inter-stage matching simultaneously.

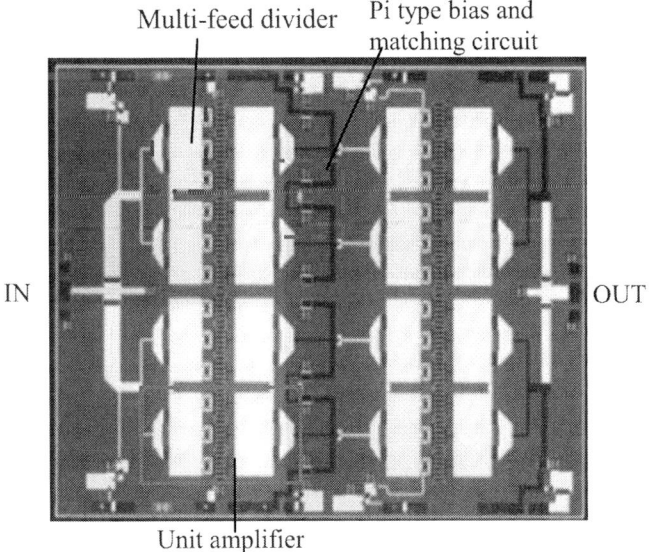

Fig. 7. Photograph of 6W MMIC power amplifier chip.

V. MEASUREMENT RESULTS

Measured and designed small-signal performance of the power amplifier is shown in Fig. 8. The quiescence bias conditions were Vds=6V and Idq=2.4A. Fig. 8 shows good agreement between measured and designed small-signal performance. Fig. 9 shows measured output power, gain and PAE of 6 W MMIC power amplifier. The quiescence bias conditions were Vds=8V and Idq=2.4A. Saturated output power of 37.9dBm(6.2W) and PAE of 17% were achieved. State-of-the-art power performance of Q-band MMIC power amplifiers is shown in Fig. 10. As can be recognized, 6.2W output power is the highest output power for Q-band power amplifiers.

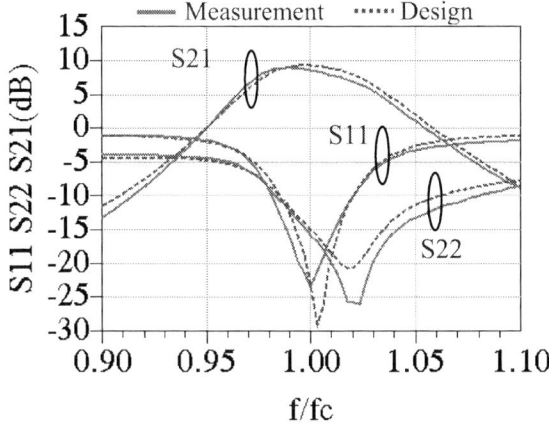

Fig. 8. Measured and designed small-signal performance of 6W MMIC power amplifier at Vds=6V and Idq=2.4A at Q-band.

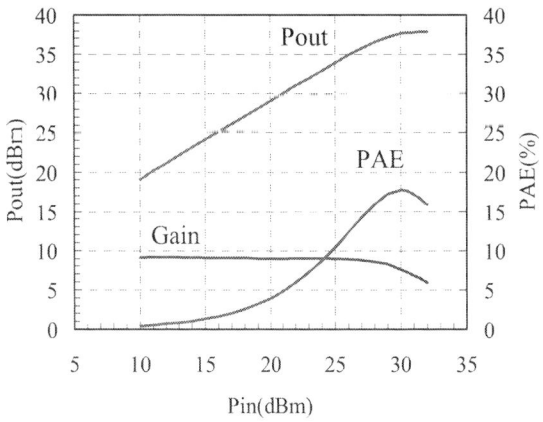

Fig. 9. Measured Pout, Gain and PAE of 6W MMIC power amplifier at Vds=8V and Idq–2.4A at Q-band.

978-1-4244-6240-7/10 $26.00 © 2010 IEEE

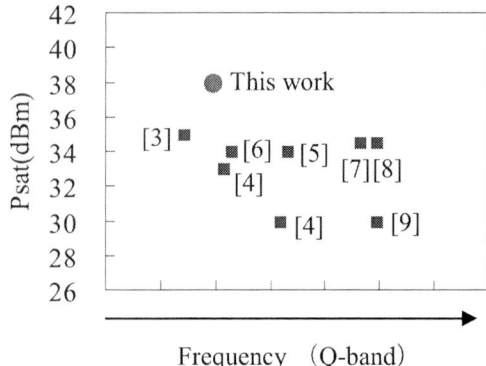

Fig. 10. Power performance of Q-band MMIC power amplifiers.

VI. CONCLUSION

In this paper, a Q-band 6W MMIC power amplifier with 3-way combination circuit was presented. The multi-feed divider and the pi type bias-and-matching circuit were applied to the amplifier. The amplifier demonstrated saturated output power of 37.9dBm(6.2W). This is the state-of-the-art output power for Q-band MMIC power amplifiers.

ACKNOWLEDGEMENT

The authors wish to acknowledge the assistance and support of Masami Saito of Mitsubishi Electric Engineering Corporation who performed all the microwave measurements for this work.

REFERENCES

[1] M. K. Siddiqui, A. K. Sharma, L. G. Callejo, and R. Lai. "A high-power and high- efficiency monolithic power amplifier at 28 GHz for LMDS applications.", *IEEE Trans. Microwave Theory & Tech.,* vol. 46, no. 12, pp. 2226-2232, December 1998.

[2] J. Shu, T. Hwang, D. Nguyen, R. Pumares, P. Chye, and P. Khanna, "Ka-band 2 watt power SSPA for LMDS application.", *IEEE MTT-S Int. Microwave Symp. Dig.,* pp. 573- 576 1998.

[3] P. M. Smith et al, "Progress in GaAs Metamorphic HEMT Technology for Microwave Applications," *2003 IEEE GaAs IC Symp. Dig.,* pp. 21- 24, 2003.

[4] TriQunit data sheet: TGA1073-SCC, TGA1171-SCC, TGA1141-EPU.

[5] Fairchild/Raytheon data sheet: RMPA39200, RMPA39300.

[6] S. Chen et.al., "A Ka/Q-Band 2 Watt MMIC Power Amplifiers Using Dual Recess 0.15um PHEMT Process," *IEEE MTT-S Int Microwave Symp. Dig.,* vol 3, pp. 1669-1672, June 2004.

[7] Q. H. Wang et.al., "A high power Q-band MMIC power amplifier based on dual-recess 0.15 /spl mu/m pHEMT," *IEEE Compound Semiconductor IC Symposium.,* pp. 133-136, October 2004.

[8] M. V. Aust et.al., "A highly efficient Q-band MMIC 2.8 Watt output power amplifier based on 0.15/spl mu/m InGaAs/GaAs pHEMT process technology," *IEEE Compound Semiconductor IC Symposium.,* pp. 228-231, October 2005.

[9] Y. Hwang et.al., "Fully-matched, high-efficiency Q-band 1 watt MMIC solid state power amplifier," *IEEE MTT-S Int Microwave Symp. Dig.,* vol 1, pp. 149-152, June 1996.

[10] T. Takagi, M. Mochizuki, Y. Tarui, Y. Itoh, S. Tsuji, and Y. Mitsui, "Analysis of High Power Amplifier Instability due to f0/2 Loop Oscillation ", IEICE Transactions on Electronics, vol. E78-C, no. 8, pp. 936-943, August 1995

[11] H. Amasuga, et al.,. "A High Power Density TaN/Au T-gate pHEMT with High Humidity Resistance for Ka-Band Applications*", IEEE MTT-S Int. Microwave Symp. Dig.,* pp. 831-834, June 2005.

A Multi-Band Reconfigurable Power Amplifier for UMTS Handset Applications

Unha Kim, Kyutae Kim, Junghyun Kim*, and Youngwoo Kwon

School of Electrical Engineering and Computer Science and Institute of New Media and Communications,
Seoul National University, 599, Gwanak-ro, Gwanak-gu, Seoul, 151-742, Korea
*School of Electrical Engineering and Computer Science, Hanyang University, 1271 Sa 3-dong, Sangrok-gu,
Ansan, Gyeonggi-do, 426-791, Korea

Abstract — **A new practical reconfigurable structure for a multi-band power amplifier (PA) is proposed for UMTS handset applications. The proposed reconfigurable output matching network can reconfigure the output power as well as the frequency. It consists of a path-selection network, a power-reconfigurable network, and a frequency-reconfigurable network. To demonstrate the performance of the proposed structure, a 5 mm × 5 mm prototype reconfigurable PA module is developed for UMTS high-frequency band application. The fabricated PA module can cover any two bands out of three popular high-frequency UMTS bands. The fabricated amplifier module showed adjacent channel leakage ratios (ACLRs) better than –38 dBc up to the rated linear power and power-added efficiencies (PAEs) of higher than 38% at 28 dBm over all high-frequency UMTS bands. Efficiency degradation was limited to 2–3% compared to the single-band PA. Measured RF performance of the reconfigurable PA validates the usefulness of the proposed reconfigurable structure for multi-band UMTS applications.**

Index Terms — **Band switching, concurrent, matching network, multi-band, power amplifier, reconfigurable, UMTS, W-CDMA**

I. INTRODUCTION

As a larger number of frequency bands are allocated for 3-G UMTS communication standards and global roaming has become popular, W-CDMA handsets are required to support the ever increasing number of frequency bands. Simple addition of the dedicated single-band power amplifiers (PA's) for each additional frequency band would result in excessive increase in the overall cost and size of the mobile phones. Researchers have recently started to develop reconfigurable PA's that cover several frequency bands using only one PA core [1]-[4]. Programmable matching networks using varactors and MEMS switches were often employed to cover both low-band and high-band cellular frequencies. However, attempts to cover too wide frequency range using a single PA core would require complicated input/interstage/output matching networks and would result in larger module size and performance degradation due to the losses of the reconfigurable network. For example, the net additional loss of the reconfigurable matching network can be as large as 0.96 dB at 1.6 GHz, which adds to the inherent loss of the output matching network of 1.75 dB. The additional loss of 0.96 dB translates up to 6.5% power-added efficiency (PAE) degradation due to reconfiguration [4]. PAE degradation

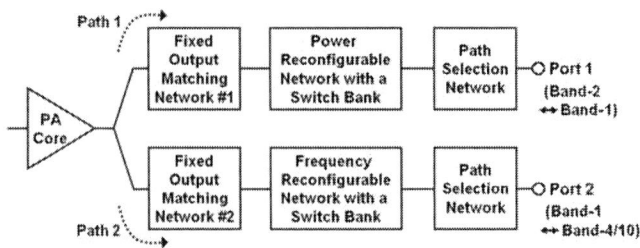

Fig. 1. Block diagram of the proposed multi-band reconfigurable PA for high-band UMTS applications

cannot be easily accepted for mobile phone applications due to the thermal concerns and talk time metric. Moreover, the approach of using a single RF input and output port for multi-band amplifier implementation [1]-[2] will lead to further PAE degradation due to the requirement of post-PA distribution switches, whose losses further degrades the overall efficiency.

In this work, a new approach for multi-band UMTS reconfigurable power amplifier is proposed for practical handset applications with minimal performance degradation. The UMTS transmit frequency bands can be grouped into low-band (0.7–0.9 GHz) and high-band (1.4–2.5 GHz). The complications in covering wide frequency range is mitigated in this work by limiting the band reconfigurability within either low-band or high-band group. The additional losses due to post-PA switches are also avoided by expanding the number of output ports. Besides, the reconfigurable network of this work does not only reconfigure the frequency but also the output power according to the selected UMTS bands, which may have different post-PA duplexer losses; power reconfiguration helps to maximize PAE at the rated max linear output power. In this paper, the operation principle of the reconfigurable network is presented in Sec. II, followed by the design, implementation and measured results of a prototype multi-band reconfigurable PA in Sec. III.

II. OPERATING PRINCIPLE OF THE RECONFIGURABLE OUTPUT NETWORK

The block diagram of the proposed reconfigurable PA is shown in Fig. 1. The PA is designed to support any two bands

TABLE I
UMTS Frequencies and Target Linear Output Powers

	Tx frequency (MHz)	Rx frequency (MHz)	Target P_{out} (dBm)
Band-1	1920 – 1980	2110 – 2170	27.5
Band-2	1850 – 1910	1930 – 1990	28.5
Band-4/10	1710 – 1770	2110 – 2170	27.5

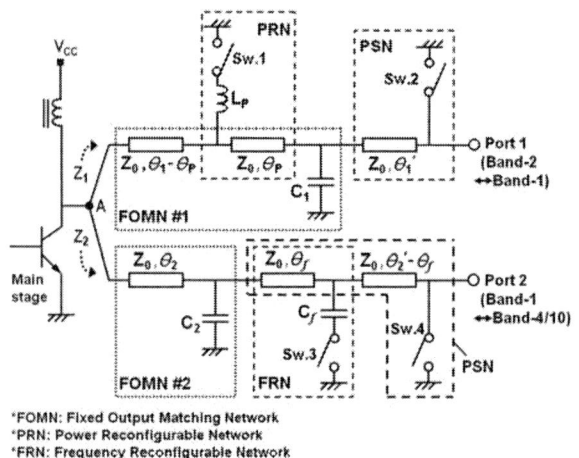

*FOMN: Fixed Output Matching Network
*PRN: Power Reconfigurable Network
*FRN: Frequency Reconfigurable Network
*PSN: Path Selection Network

Fig. 2. Schematic of the proposed reconfigurable output network

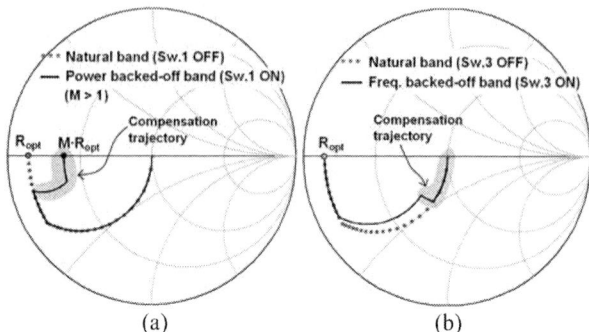

Fig. 3. Load trajectories of (a) the power reconfigurable network and (b) the frequency reconfigurable network

out of three popular high-frequency UMTS bands, band-1, band-2 and band-4/10. The details of the frequency and the target maximum linear output power are summarized in Table I. Most of the phones require only two high-frequency bands out of these three UMTS bands. Tx-frequencies of band-1 (1920–1980 MHz) and band-2 (1850–1910 MHz) are so close to each other that no frequency reconfigurable matching network is required to cover both bands with a single PA. However, the required power level is different due to the differences in the post-PA losses; band-2 has smaller Tx-Rx separation, which causes the subsequent duplexer to have higher Tx insertion loss than band-1. Thus, power reconfigurable network has been inserted in path-1 to decrease the rated linear power for band-1. On the other hand, band-1 and band-4/10 have almost identical output power requirement since the post-PA losses are almost the same. However, the Tx frequency difference is large enough (~12%) to require a frequency reconfigurable network. To this end, the frequency reconfigurable network has been inserted in the path-2 to share band-1 and band-4/10 in the same path. In this way, all following three combinations of dual-band operation can be achieved by selecting the path, frequency and power.

- Combination 1: band-2 and band-1
- Combination 2: band-1 and band-4/10
- Combination 3: band-2 and band-4/10

The detailed schematic of the output network is shown in Fig. 2. The output network has two output paths, each with two available bands. Because the two output paths cannot be operated simultaneously, the unused path is deactivated by the path-selection network (PSN), which is based on the concept

of shunt SPDT switches. By closing the shunt switches (Sw.2 and 4), short-circuit is provided at the unused output port, which is transformed to open-circuit at the junction point (A in Fig. 2) in conjunction with the preceding networks; the artificial quarter-wavelength lines is implemented with the preceding transmission line sections and shunt capacitors. The capacitor-loaded lines help to reduce the physical line lengths ($\theta_1 + \theta_1$' in path-1 and $\theta_2 + \theta_2$' in path-2) well below the quarter-wavelength. The impedance looking into the unused path was designed to be at least 10 ~ 15 times larger than the input impedance of an activated path not to cause performance degradation due to signal leakage.

When a path is selected and not reconfigured ("as is" state), fixed output matching networks (FOMN's) provides the optimum load impedances to the natural bands of each path. Band-2 is selected as the natural band in path-1 while band-1 is the natural band in path-2. Transmission line lengths, θ_1 and θ_2, and shunt capacitors, C_1 and C_2, are designed to transform 50 Ω impedance to the optimum load impedance ($R_{opt} \approx 4$ Ω for band-2 in path-1 and $R_{opt} \approx 5$ Ω for band-1 in path-2). As stated previously, path-1 has a power reconfigurable network so that it can also support band-1 with 1 dB lower target power in the reconfigured state. A shunt network consisting of a switch (Sw.1) and an inductor (L_P) is connected to the transmission line of FOMN (θ_P) to increase R_{opt} from 4 Ω (target $P_{out} = 28.5$ dBm) to 5 Ω (target $P_{out} = 27.5$ dBm).

In the case of path-2, the frequency reconfigurable network has been employed to support also band-4/10 operation in the reconfigured state. To this end, a combination of a series transmission line (θ_f) and a shunt switched capacitor (C_f and Sw.3) is inserted between FOMN and PSN. In the "as is" state, the load impedance is matched to R_{opt} (≈ 5 Ω) for higher frequency (band-1 Tx center frequency = 1950 MHz) by FOMN. The line length (θ_f) and capacitor value (C_f) are designed to present the same R_{opt} at a lower frequency (band-4/10 Tx center frequency = 1740 MHz) when the switch, Sw.3, is turned on. These power- and frequency-reconfiguration mechanisms are described with the load-impedance trajectories in Fig. 3. It should be noted in the case of path-2 that the line length of the path-selection network, θ_2', includes the supplementary line length of the frequency-

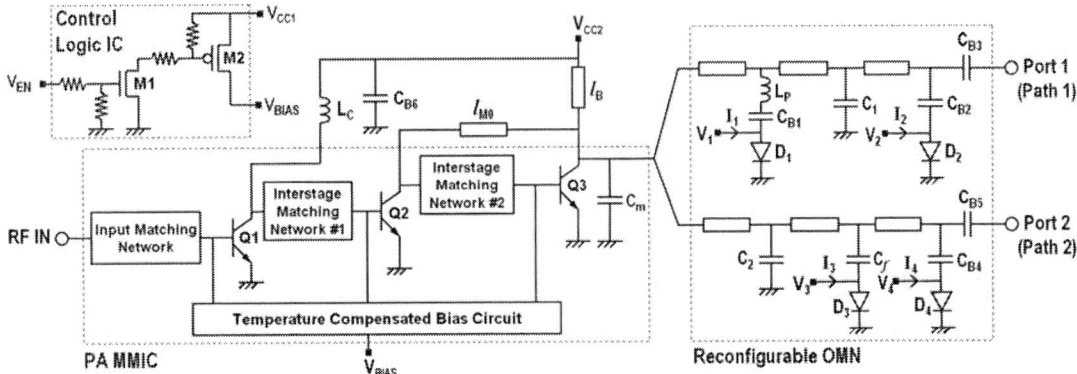

Fig. 4. Schematic of the proposed reconfigurable PA module

TABLE II
LOGIC TABLE AND OPERATION DESCRIPTION OF PIN DIODES

PIN diode control voltage				Operation	
V_1	V_2	V_3	V_4		
Low	Low	Low	High	Path-1	Band-2
High	Low	Low	High	Path-1	Band-1
Low	High	Low	Low	Path-2	Band-1
Low	High	High	Low	Path-2	Band-4/10

reconfigurable network, θ_f.

III. FABRICATION AND MEASUREMENT OF A PROTOTYPE MULTI-BAND RECONFIGURABLE POWER AMPLIFIER

To verify the performance of the proposed reconfigurable matching network, a 5 mm × 5 mm prototype multi-band reconfigurable PA module was designed and fabricated. A schematic of the proposed reconfigurable PA module is shown in Fig. 4. The PA module consists of the reconfigurable output matching network (OMN), an integrated 3-stage PA MMIC, and a control logic IC.

The reconfigurable OMN is implemented using the lumped elements and the transmission lines on a 4-layer substrate with a 100-μm thick dielectric core ($\varepsilon_r \sim 4.7$, tan$\delta = 0.025$). The module size can be further reduced if higher-density substrate is used and higher-level of integration is applied. In Fig. 4, C_1 and C_2 are the lumped-element capacitors of the FOMN's optimized for the natural bands, and L_P and C_f are the lumped elements optimized for the power- and frequency-reconfiguration networks, respectively, as mentioned in Sec II. Five capacitors, C_{B1}, C_{B2}, C_{B3}, C_{B4}, and C_{B5}, are used for DC blocking and C_{B6} is used as a bypass capacitor. Switches are realized with PIN diodes, which have a series resistance of 1.8 Ω with forward on-current of 5 mA and a junction capacitance of 160 fF in the "off" state. They were controlled by the digital control voltages, V_1, V_2, V_3, and V_4, according to the required UMTS band combinations. The mapping between the control voltages and the selected frequency bands is summarized in Table II.

The PA MMIC was designed and fabricated using a 2-μm InGaP/GaAs HBT process. It was based on a 3-stage amplifier

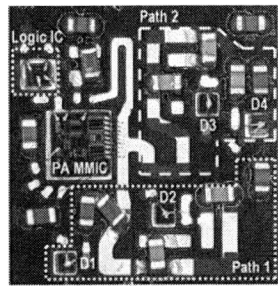

Fig. 5. The photograph of the fabricated PA module

design, and the emitter areas of the pre-stage (Q1), driver stage (Q2), and main stage (Q3) were chosen to be 240 μm^2, 1440 μm^2, and 5760 μm^2, respectively. The low-pass-type input matching network and two high-pass-type interstage matching networks were fully integrated in the MMIC. The PA MMIC die size was 1.1 mm × 1.08 mm.

The fabricated 5 mm × 5 mm PA module is shown in Fig. 5. The PA module works with 3.5 V supply voltage, and the 3GPP uplink W-CDMA signal (Rel'99) was used for the measurement. The measured results of power gain, adjacent channel leakage ratio (ACLR), and PAE are shown in Figure 6, 7 and 8 according to the selected band combinations. In the case of combination 1 (band-2 and -1) shown in Fig. 6, which corresponds to the natural bands of each path, the PA showed linear power gains of 25.6 dB and ACLRs of better than –38 dBc up to the rated maximum output powers (27.5 dBm at band-1 and 28.5 dBm at band-2). PAE at the maximum linear power meeting –38 dBc ACLR, was higher than 37% for both bands (37% at 27.5 dBm for band-1 and 39% at 28.5 dBm for band-2). For combination 2 (band-1 and band-4/10) shown in Fig. 7, where power and frequency reconfiguration was applied to band-1 and band-4/10, respectively, the PA showed linear power gains of 25.0 dB and ACLRs of better than –38 dBc up to the rated linear output powers (27.5 dBm at band-1 and band-4/10). PAE of higher than 38.4% was measured at $P_{out} = 28$ dBm. Finally, for combination 3 (band-2 and band-4/10) in Fig. 8, where frequency reconfiguration was applied to band 4/10, linear power gains of 25.6 dB were measured and ACLRs were maintained better than –38 dBc up to the rated linear output powers (28.5 dBm at band-2 and 27.5 dBm

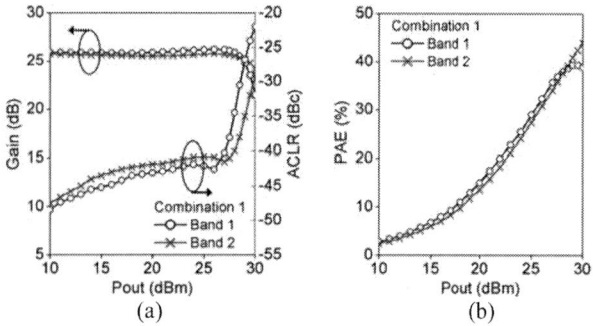

Fig. 6. Measured results for combination 1: (a) gain and ACLR, (b) PAE

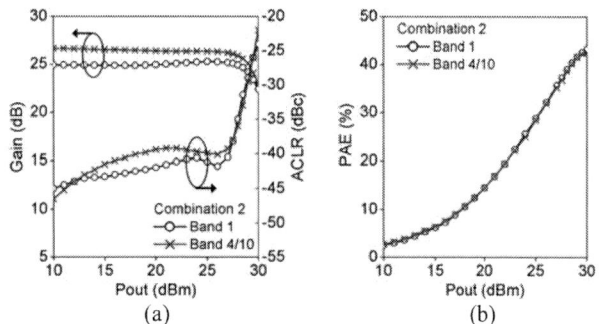

Fig. 7. Measured results for combination 2: (a) gain and ACLR, (b) PAE

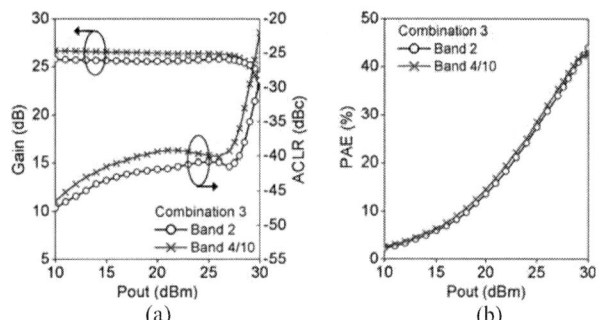

Fig. 8. Measured results for combination 3: (a) gain and ACLR, (b) PAE

at band-4/10). The max linear PAE was higher than 37% for both bands (39% at 28.5 dBm for band-2 and 37% at 27.5 dBm for band-4/10).

In order to compare the performance of the multi-band reconfigurable PA with a single-band PA, a reference PA was also fabricated using the same PA die with a fixed output matching network. The reference PA for each band showed a PAE of 40–41% at the rated linear output power with an ACLR of –38 dBc. Thus, the multi-band reconfigurable PA showed PAE degradation of 2–3%, which is attributed to the additional bias current of the PIN diodes and the losses of the switches. PAE degradation can be further reduced by using either pHEMT or MEMS switches. The total estimated losses of the reconfigurable output matching network are less than 0.7 dB, out of which 0.3 dB arises from the reconfiguration.

This is far less than the loss of the reported reconfigurable structures. To the best of our knowledge, this is the first demonstration of UMTS multi-band reconfigurable PA's meeting the system linearity requirements with minimal PAE degradation.

IV. CONCLUSION

A design methodology to realize reconfigurable output matching network (OMN) for a multi-band UMTS PA is developed. A reconfigurable OMN, which has two output ports and covers any two bands out of three popular UMTS frequency bands, is proposed. To demonstrate the performance of the proposed OMN, a prototype reconfigurable PA module is designed and fabricated in a 5 mm × 5 mm form factor. The measured RF characteristics showed linearity better than –38 dBc for all the selected band combinations. Compared with the single-band PA with a fixed output matching network, the maximum power efficiency was degraded by 2–3% mainly due to the bias current of the PIN diodes and the losses of the switches. With the strong demand for multi-band coverage for global roaming, the proposed reconfigurable PA can be a practical solution for UMTS multi-band Tx applications. It is worthwhile to note that a similar design approach can also be applied to low-frequency UMTS bands to share a PA for UMTS band-5 (Tx: 824–849 MHz) and band-8 (Tx: 880–915 MHz) applications.

ACKNOWLEDGEMENT

This work was supported by the Acceleration Research Program of the Ministry of Education, Science and Technology of the Republic of Korea and the Korea Science and Engineering Foundation.

REFERENCES

[1] A. Fukuda, T. Furuta, H. Okazaki, and S. Narahashi, "A 0.9-5-GHz Wide-Range 1W-Class Reconfigurable Power Amplifier Employing RF-MEMS Switches," *IEEE MTT-S Int. Microw. Symp. Dig.*, pp. 1859-1862, June 2006.

[2] W. Neo *et al.*, "Adaptive Multi-Band Multi-Mode Power Amplifier Using Integrated Varactor-Based Tunable Matching Networks," *IEEE Journal of Solid-State Circuits*, vol. 41, no. 9, pp. 2166-2176, Sep. 2006.

[3] K. Kim, J. Kim, and C. Park, "A Single-Input Single-Chain Dual-Band Power Amplifier for CDMA Mobile Application," *Microw. and Optical Tech. Letters*, vol. 48, no. 5, pp. 981-983, May 2006.

[4] C. Zhang and A. Fathy, "A Novel Reconfigurable Power Amplifier Structure for Multi-Band and Multi-Mode Portable Wireless Applications using a Reconfigurable Die and a Switchable Output Network," *IEEE MTT-S Int. Microw. Symp. Dig.*, pp. 913-916, June 2009.

978-1-4244-6240-7/10 $26.00 © 2010 IEEE

High-Efficiency Reconfigurable RF Transmitter for Wireless Sensor Network Applications

Francesco Carrara and Giuseppe Palmisano

Università di Catania, Facoltà di Ingegneria, DIEES, Viale Andrea Doria 6, 95125 Catania, Italia

Abstract — **In this paper, a 90-nm CMOS 1.2-V ultra-low-power RF transmitter for wireless sensor networks is presented. A wideband topology guarantees continuous frequency coverage from 300 to 960 MHz, thus enabling easy reconfigurability through the most popular sub-GHz bands. At 300 MHz (960 MHz), the transmitter is able to deliver a 10.2-dBm (10.3-dBm) output power with 63% (46%) system efficiency. The proposed circuit features extensive use of dynamic current biasing for improved efficiency in back-off. Substantial power saving in excess of 80% are achieved compared to constant bias operation. Linear operation with variable-envelope input is also demonstrated, since the transmitter exhibits a 8.5-dBm output power, 37% efficiency, and –30-dBc adjacent channel power ratio with π/4-DQPSK 200-ksps input excitation.**

Index Terms — **RF transmitter, power amplifier (PA), efficiency enhancement, dynamic current bias, ultra-low power (ULP), wireless sensor network (WSN).**

I. INTRODUCTION

Wireless sensor networks (WSNs) are expected to find widespread deployment in a variety of applications, including industrial, military, medical, and home automation. The key challenge for the pervasive diffusion of WSNs is the maximization of the node battery lifetime, which mandates for ultra-low-power (ULP) circuit operation. Typically, the transmit section of the RF interface [1]-[8] is the most power-hungry block of a sensor node. Therefore, effective lifetime extension can only be achieved by improving the RF transmitter efficiency. Moreover, this goal shall be pursued all through the output power range rather than at the peak power level only, as the efficiency in back-off operation is usually prone to a substantial drop, which largely affects the average power consumption.

The issue of back-off efficiency enhancement in RF transmitters has been widely addressed in recent years by means of proper system solutions, such as polar, Doherty, or outphasing architectures [9]. As an alternative to system-level approaches, several circuit-level solutions have been suggested (dynamic biasing, load adaptation, stage bypassing), which have been mostly applied to power amplifiers (PAs) in the watt range [10]. Although quite lower power levels are required for WSNs (typically in the milliwatt range [1]-[8]), the problem of transmitter efficiency is still crucial to prolong battery life.

Complex efficiency-oriented system solutions are not suitable for WSN devices since the limited power resources of a network node only enable basic functionality and limited computation capability. On the contrary, smart low-power circuit solutions are mostly indicated. Moreover, such design criteria should be applied to the whole transmitter architecture, since in back-off operation the power consumption of the up-converter and driving stages can compare to the one of the PA itself [11].

In this work, a RF transmitter architecture for WSN applications in the sub-GHz frequency range is proposed. The problem of back-off efficiency is aggressively faced by recourse to a combination of circuit innovations throughout the transmit chain. The reported results offer a quantitative insight into the potential of dynamic current biasing [10] and provide the reader with useful hints for power-aware circuit design.

II. CIRCUIT DESCRIPTION

The proposed ULP RF transmitter adopts a direct conversion architecture, as sketched in Fig. 1, comprising a quadrature up-converter, a power amplifier, and an external matching network. A quadrature generator is also included in order to derive two orthogonal LO waveforms from a double-frequency input. The output matching network is the only frequency-selective block of the transmitter, whereas all other building blocks are designed for wideband operation in the 300–960 MHz range. Therefore, flexible exploitation through several sub-GHz frequency bands (such as ISM, SRD, or MAS) is obtained by simply changing the output matching components.

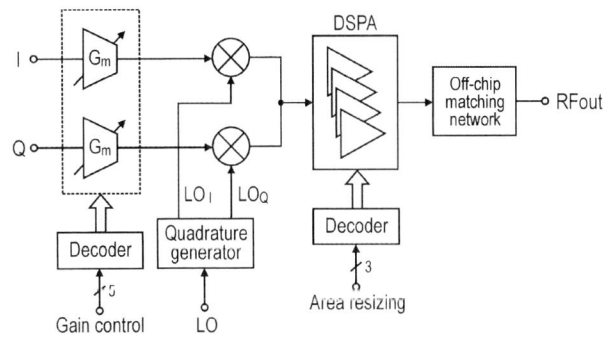

Fig. 1. Block diagram of the ULP RF transmitter.

The concept of dynamic current biasing has been extensively employed to improve the overall system efficiency through the whole transmit power range. To this aim, a digitally scalable power amplifier (DSPA) architecture has been adopted for the final stages of the transmit chain. This solution entails the use of multiple amplifying cells in parallel connection, which can be switched on and off, according to the required output power. At back-off power levels, turning off a portion of the DSPA results in reduced current consumption and hence improved efficiency. To the same purpose, a convenient gain control methodology has been adopted for the variable-gain up-converter, which is able to lower the up-converter bias current while reducing the signal power at the input of the DSPA. Therefore, current saving is guaranteed for all the circuit blocks in back-off operation.

The DSPA comprises 8 identical parallel amplifying stages, whose on/off state is set by a binary-to-thermometric decoder driven by 3 control bits. A simplified schematic of the DSPA is depicted in Fig. 2. The unit cell adopts a two-stage topology with dc coupling between the driver and the power stage. This guarantees broadband operation within the target frequency range. The chosen topology also allows very simple circuit biasing, since it is inherently based on current mirror ratios. Indeed, at dc transistor M_1 implements a current mirror with M_2 and M_3 implements a mirror with M_4. This way, only one current reference (I_B) suffices to bias the entire PA circuit. When a gain cell is to be switched off, the bias is disconnected and the gate-source voltages of both stages are shorted.

For reliability reasons, the final stage of each cell adopts a cascode topology with a thin-gate-oxide common-source transistor (M_4) and a thick-gate-oxide common-gate device (M_5). Actually, under Class-E-like operation the final stage output node suffers severe overvoltage stress, attaining peak voltages as high as 2-3 times the supply level. At a 1.2-V supply voltage, a simple common-source topology with standard thin-gate transistors would fail. Indeed, for the adopted technology, thin-gate transistors can sustain up to 1.2 V, whereas thick-gate ones can tolerate up to 3.3 V (380-nm minimum channel length).

Unlike previously published solutions [12], in the proposed circuit the input impedance of each DSPA cell is disconnected through a series switch when the cell is off. This arrangement is meant to further improve the transmitter efficiency at back-off power levels. Indeed, it lowers the DSPA input admittance, thus requiring a smaller driving power from the up-converter (i.e., lower up-converter current consumption).

As already shown in Fig. 1, the up-converter comprises a couple of variable-gain transconductors and a quadrature mixer. A novel circuit topology has been adopted to im-

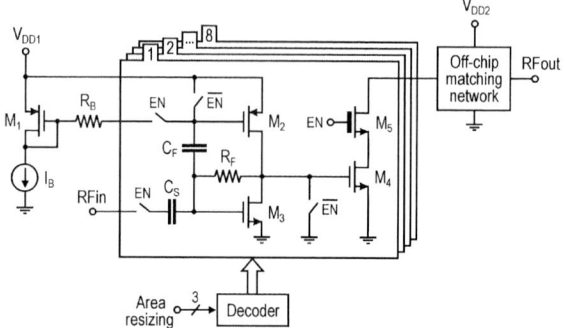

Fig. 2. Simplified schematic of the DSPA.

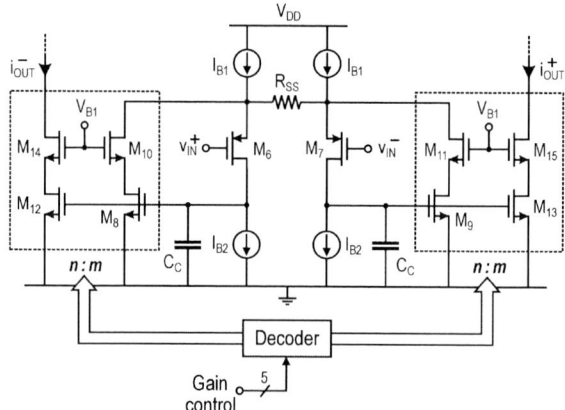

Fig. 3. Schematic of the variable-gain transconductor.

plement the input transconductors, which is shown in Fig. 3. The proposed solution features a feedback loop to linearize the voltage-to-current conversion, hence improving the circuit performance with variable-envelope linear modulations schemes. Capacitors C_C guarantee proper phase margin for the loop.

The transconductance of the proposed circuit is

$$G_m = \frac{i_{OUT}}{v_{IN}} = \frac{i_{OUT}^+ - i_{OUT}^-}{v_{IN}^+ - v_{IN}^-} = \frac{n:m}{R_{SS}} \qquad (1)$$

where $n:m$ is the transfer ratio of the output cascode current mirrors. Therefore, gain variation can be performed by simply changing such a mirror ratio. This is achieved by properly switching on and off a set of parallel mirror cells according to the desired transconductance. The cells are digitally controlled by a decoder converting 5 control bits into 32 gain settings with 1.5-dB steps.

The adopted gain control methodology simultaneously changes both the dc and ac components of the transconductor output currents. Since the transconductors' output branches are fed to the tails of the Gilbert quads in the quardature mixer, changing the dc current in the transconductors proportionally affects the total up-converter bias current. Hence, lower gain settings are

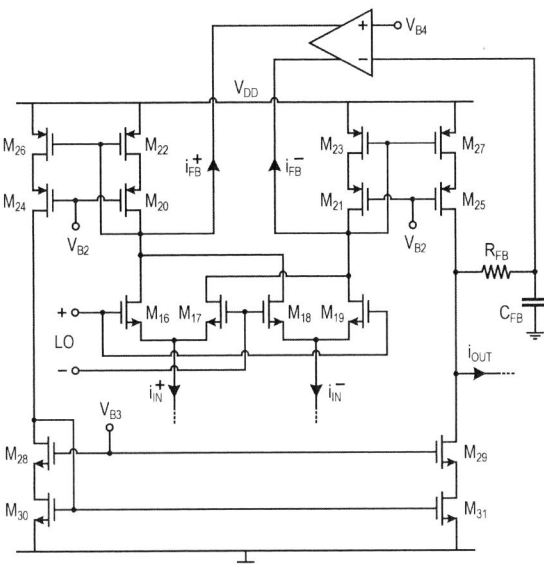

Fig. 4. Simplified schematic of the up-conversion mixer.

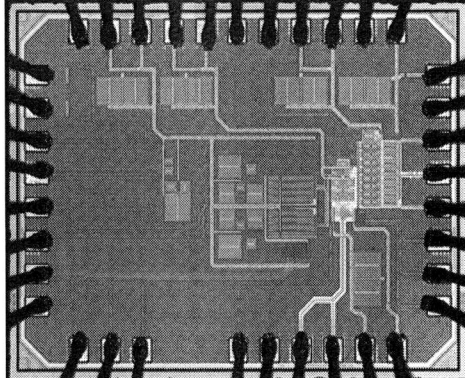

Fig. 5. Micrograph of the ULP RF transmitter.

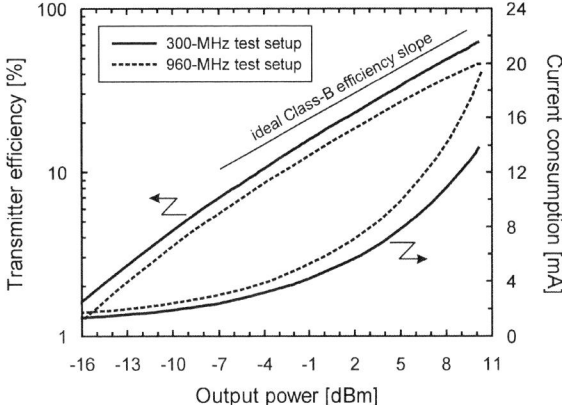

Fig. 6. Transmitter efficiency and current consumption versus output power with constant-envelope continuous-wave output (constant baseband input amplitude, optimized GCW and SCW).

achieved with lower power consumption, which is the specific benefit of the proposed solution.

A simplified schematic of the quadrature up-conversion mixer is shown in Fig. 4. Only one Gilbert quad is shown for simplicity. Currents i_{IN}^{+} and i_{IN}^{-} come from the input transconductor. Therefore, the mixer circuit is subject to substantial bias current variations across the gain control range. The use of cascode current mirrors has been preferred for lower gain sensitivity to such bias drift.

Voltages V_{B2} and V_{B3} in Fig. 4 are adaptively set by *ad-hoc* control loops (not shown) providing proper biasing for each gain setting. The mixer also features a feedback system to conveniently set the output node dc voltage around $V_{DD}/2$.

III. EXPERIMENTAL RESULTS

A micrograph of the proposed transmitter is shown in Fig. 5. The circuit was fabricated in a 90-nm CMOS technology with 7 copper interconnect layers. The die size is 1.2 mm × 1 mm, dimensions being pad-limited because of testing needs. Net silicon area is around 0.54 mm². The device was wire-bonded to a FR4 printed circuit board with lumped matching at the output.

The circuit was tested at 1.2 V with constant-envelope continuous-wave output. Fig. 6 shows the efficiency characteristic of the circuit at the corners of the operating frequency band, i.e. at 300 and 960 MHz, respectively. The plots were obtained by varying the up-converter 5-bit gain control word (GCW) and DSPA 3-bit size control word (SCW), while keeping the baseband input amplitude constant. Both GCW and SCW can be optimized to achieve best transmitter efficiency at any output power.

When operating at 300 MHz, the transmitter is able to deliver a 10.2-dBm output power with a 63% system efficiency (66% for the DSPA alone) and 13.8-mA overall current consumption. At 960 MHz, the circuit can provide a 10.3-dBm output power with a 46% system efficiency (51% for the DSPA alone) and 19.4-mA current consumption. For both test setups, nearly ideal efficiency slope versus output power can be detected (i.e., proportional to the square root of the output power like for Class-B theory) over a larger than 10-dB back-off range.

The beneficial effect of the adopted bias scheme on the transmitter back-off power consumption was tested by comparing the efficiency values in Fig. 6 (constant input, swept GCW/SCW) to the performance obtained with traditional constant bias (constant GCW/SCW, swept input). According to Fig. 7, significant efficiency improvements are demonstrated. The 960-MHz test setup exhibits a higher than 80% power consumption reduction with 1.6×, 3.5×, 5.5× efficiency increase at 10-dB, 20-dB, 30-dB back-off, respectively. Substantial efficiency enhancement is achieved at the 300-MHz corner as well.

978-1-4244-6240-7/10 $26.00 © 2010 IEEE

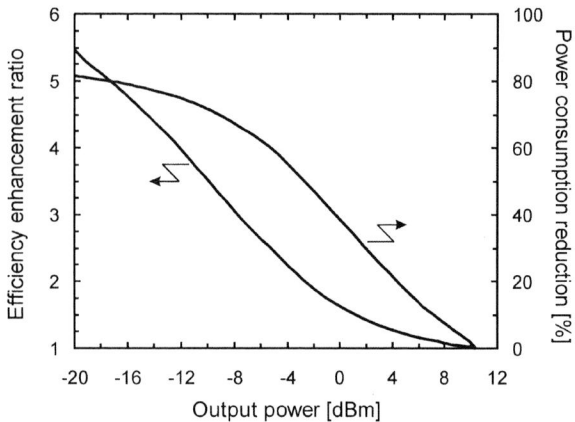

Fig. 7. Transmitter back-off efficiency increase versus output power (960-MHz test setup, constant GCW / SCW as reference).

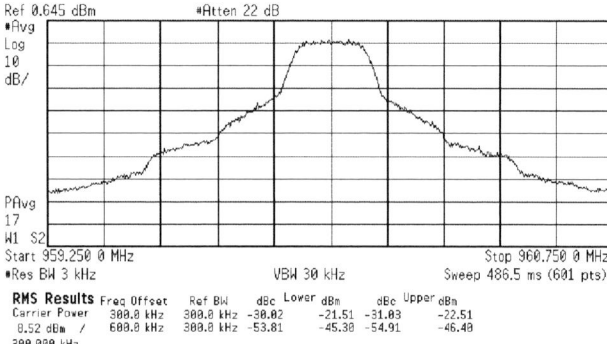

Fig. 8. Output power spectrum with a π/4-DQPSK 200-ksps modulated input signal (Nyquist filter with $\alpha = 0.4$, output matching for 960-MHz operation, maximum GCW and SCW).

The circuit was also tested for linearity with variable-envelope input modulation (no further output matching refinement). Fig. 8 reports the output power spectrum of the 960-MHz test setup with a π/4-DQPSK 200-ksps input signal. Thanks to the linearized input transconductor and careful design of the DSPA bias currents, the transmitter is able to achieve a remarkable 8.5-dBm linear output power with a 37% efficiency, while complying with a –30-dBc adjacent channel power ratio at a 300-kHz frequency offset. At 300 MHz, the same ACPR performance is achieved while delivering a 8.1-dBm output power with a 49% efficiency.

Finally, the transmitter reconfigurability for applications targeting lower output power levels was explored. To this aim, an additional test setup was used in which the SCW was steadily set to 0 (thereby constantly isolating 7 out of 8 DSPA cells) and the output matching was optimized for a 0-dBm nominal output power at 960 MHz. Under this conditions, a fair 28% system efficiency is still obtained (38% for the DSPA) with a 3-mA total consumption.

IV. CONCLUSIONS

In this paper, a RF transmitter architecture featuring intensive use of dynamic current bias has been proposed. The reported results demonstrate that the issue of back-off efficiency enhancement in ULP transmitters can be successfully addressed by recourse to simple yet innovative circuit topologies rather than complex system solutions, the latter being unsuitable for WSN applications.

Though demonstrated at relatively low operating frequencies, the proposed ideas can be profitably applied at higher frequency bands as well [4]. Indeed, the use of adaptive bias on the driving stages of the transmit chain is even more effective when a lower power gain is suffered.

Finally, it should be noted that the proposed architecture claims excellent *flexibility* of use, since it can be easily reconfigured for efficient operation on a wide range of frequencies (from 300 MHz to 960 MHz), output power levels (from 0 dBm to 10 dBm nominal), and modulation schemes (both constant-envelope and linear modulations).

ACKNOWLEDGEMENT

The authors would like to thank Calogero D. Presti, UCSD, La Jolla CA, USA, for many fruitful discussions, and Salvatore Cantella, STMicroelectronics, Catania, Italy, for his help in drawing the circuit layout.

REFERENCES

[1] T. Melly, A.-S. Porret, C. C. Enz, and E. A. Vittoz, "An ultralow-power UHF transceiver integrated in a standard digital CMOS process: Transmitter," *IEEE J. Solid-State Circuits*, vol. 36, pp. 467-472, Mar. 2001.

[2] V. Peiris *et al.*, "A 1V 433/868MHz 25kb/s-FSK 2kb/s-OOK RF transceiver SoC in standard digital 0.18μm CMOS," in *IEEE 2005 Int. Solid-State Circuits Conf. (ISSCC 2005) Dig. Tech. Papers*, Feb. 2005, pp. 258-259.

[3] Y. H. Chee, A. M. Niknejad, and J. M. Rabaey, "An ultra-low-power injection locked transmitter for wireless sensor networks," *IEEE J. Solid-State Circuits*, vol. 41, pp. 1740-1748, Aug. 2006.

[4] B. W. Cook, A. Berny, A. Molnar, S. Lanzisera, and K. S. J. Pister, "Low-power 2.4-GHz transceiver with passive RX front-end and 400-mV supply," *IEEE J. Solid-State Circuits*, vol. 41, pp. 2757-2766, Dec. 2006.

[5] D. C. Daly and A. P. Chandrakasan, "An energy-efficient OOK transceiver for wireless sensor networks," *IEEE J. Solid-State Circuits*, vol. 42, pp. 1003-1011, May 2007.

[6] http://www.semtech.com/images/datasheet/sx1211.pdf

[7] http://focus.ti.com/lit/ds/symlink/cc1101.pdf

[8] http://www.analog.com/static/imported-files/data_sheets/ADF7020.pdf

[9] F. H. Raab *et al.*, "Power amplifiers and transmitters for RF and microwave," *IEEE Trans. Microw. Theory Tech.*, vol. 53, pp. 814-826, Mar. 2002.

[10] J. Deng, P. S. Gudem, L. E. Larson, and P. M. Asbeck, "A high average-efficiency SiGe HBT power amplifier for WCDMA handset applications," *IEEE Trans. Microw. Theory Tech.*, vol. 52, pp. 529-537, Feb. 2005.

[11] V. W. Leung, L. E. Larson, and P. S. Gudem, "Digital-IF WCDMA handset transmitter IC in 0.25-μm SiGe BiCMOS," *IEEE J. Solid-State Circuits*, vol. 39, pp. 2215-2225, Dec. 2004.

[12] C. D. Presti, F. Carrara, A. Scuderi, P. Asbeck, and G. Palmisano, "A 25 dBm digitally modulated CMOS power amplifier for WCDMA/EDGE/OFDM with adaptive digital predistortion and efficient TX power control," *IEEE J. Solid-State Circuits*, vol. 44, pp. 1883-1896, July 2009.

A Triband 65nm CMOS Tuner for ATSC Mobile DTV SoC

Sanghoon Kang, Huijung Kim, Jeong-Hyun Choi, Jae-Hong Chang, Jong-Dae Bae, Wooseung Choo and Byeong-ha Park

Samsung Electronics, Korea

Abstract — **A triband 65nm CMOS tuner is designed and implemented for first ATSC mobile DTV SoC. It supports VHF-I, VHF-III, and UHF bands. The tuner achieves a 2.5/3.0/4.0dB NF at VHF-I, VHF-III and UHF band respectively, while consuming less than 100mW. By using narrowband LNA architecture with input and load tuning ability, it can meet linearity requirement for ATSC mobile with small power consumption. It also automatically calibrates baseband low-pass filter cut-off frequency and LNA LC load resonant frequency.**

Index Terms —**ATSC-M/H, DTV, tuner, receiver, SoC**

I. INTRODUCTION

Many digital TV applications such as DVH-H/T, T-DMB/DAB and ISDB-T are competing in world market. A/153 ATSC Mobile DTV Standard is approved recently by Advanced Television Systems Committee (ATSC). ATSC mobile DTV is designed to provide new services to mobile and handheld devices with current DTV services without any adverse impact on legacy receiving equipment. It is based on 8-VSB and uses same frequency and channel bandwidth with ATSC DTV.

Increasing integration level of mobile devices requires small physical size and less power consumption to chip developers. Many single chip tuner are reported [2][3] and two chip solution of analog tuner chip and baseband digital chip is main stream at this time. [1] But requirement for higher integration level is increasing. System-in-Package (SiP) can be another option, but it also requires more packaging cost and loose cost merit when analog part area is much smaller than that of digital part. It is clear that single chip or system on chip (SoC) receiver

should be the target to be competitive considering the market trend. For SoC receiver, it is difficult to isolate sensitive analog part from digital part and to reduce digital noise and spurious signal. To get a sufficient isolation level, it should be considered in all developing process - circuit design, power and ground planning, layout and package design.

II. CIRCUIT DESIGN

ATSC mobile supports VHF-I, VHF-III and UHF bands. Analog tuner part consists of three LNA for each band, one pair of mixer, trans-impedance amplifier, analog baseband path and LO signal synthesizer. Fig. 1 shows the block diagram of the implemented SoC receiver. Though LNA input has each pin for three bands, single antenna can be used with adequate RF switch and RF filters.

RF tuner is implemented as direct conversion receiver architecture to lower power consumption and image rejection requirement. Down converted signal is filtered by channel selection filter. Analog tuner gain is controlled by automatic gain control (AGC) loop and regulated signal is applied to ADC input.

A. Low Noise Amplifier and Single-to-differential Amplifier

In many works, wideband LNAs were used for mobile TV application because they can provide wideband input matching and relatively low noise figure. But they suffers from relatively higher power consumption and far-off blocker linearity. In contrast, narrow band LNA can have high gain and low noise figure with low power consumption and even it can have far-off blocker rejection. In this work, we use inductor degenerated narrow band LNA with input matching and load tuning circuit as shown in Fig. 2. For input matching, we tuned the gate-source capacitance with MIM capacitor bank.

For wide dynamic range, LNA has four gain steps, high gain, high-mid gain, mid-gain and low gain. In high gain mode, input signal is amplified by M1 and balanced by M2 and 2nd stage operates as buffer amplifier. To improve linearity at high-mid gain and mid-gain mode 2nd stage is bypassed and LNA output is connected directly to mixer. For low gain mode, LNA is bypassed and input signal is

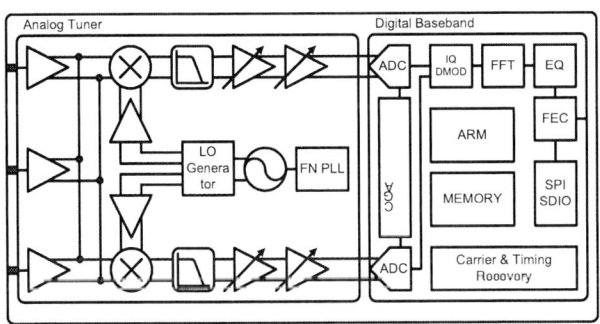

Fig 1. ATSC mobile receiver bock diagram

978-1-4244-6240-7/10 $26.00 © 2010 IEEE

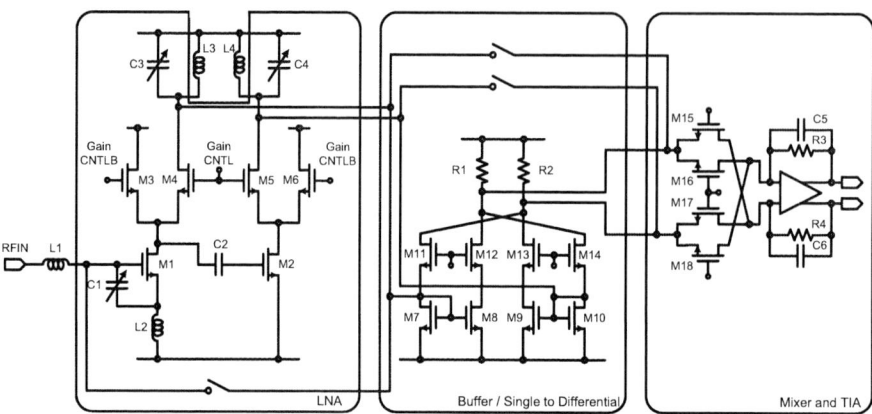

Fig 2. Low noise amplifier schematic

converted to single-to-differential converter implemented at mixer input and feed to mixer switching cores.

We used LC tuned circuit for LNA load. LC tuned load is very useful to increase far off blocker linearity without additional power consumption. But for wideband applications like UHF band receiver, LC load must track each channel to maintain blocker rejection and other performance such as gain, noise figure, etc. But LC load with high quality factor is sensitive to process variation and can not be used without tuning. Automatic calibration circuit is preferred for gain flatness.

In this work, we make oscillation signal with LC load and tune the capacitor bank to match its frequency with LO signal frequency. Cross-coupled negative gm cell is connected to LC load as in Fig. 3 for generating oscillation signal with frequency of LC tune. To change the RF channel, PLL should be locked to RF channel frequency. After PLL lock LNA is turned off and negative gm cell is turned on to generate oscillation signal. It is compared with the LO signal generated by synthesizer. Automatic calibration circuit uses binary search algorithm to find out the tuning code of load capacitor bank.

LC resonator has different quality factor and makes

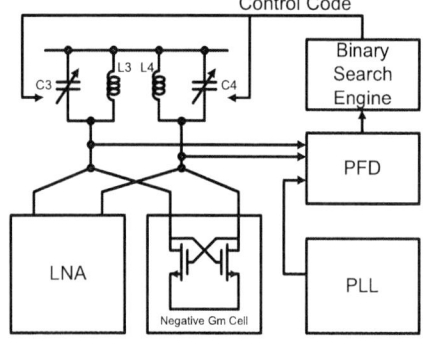

Fig 3. LNA load automatic calibrator block diagram

different gain for each channel. So, we added resistor bank in parallel with LC resonant load and control its value to flatten gain response and load quality factor.

B. Mixer and Trans-Impedance Amplifier

To reduce power consumption and low frequency noise contribution, passive mixer is used for frequency conversion. Also, passive mixer has better linearity than active mixer with low supply voltage. Since passive mixer performance is very sensitive to gate bias voltage, it is generated by bandgap referenced current and diode connected MOS. This type of bias circuit can make the mixer performance more insensitive to process and supply voltage variation than constant voltage bias circuit. Gate voltage where the transistor channel is slightly turned on is best in most applications because mixer can be switched on and off rapidly with small LO signal. It is also very useful technique to make trimming circuit for mixer bias because mixer is the key component of noise-linearity trade-off of receiver.

Trans-impedance amplifier converts mixer output signal to voltage. Input impedance of the trans-impedance amplifier should be lower than when it is used with active mixer, because output impedance of the passive mixer is lower than that of active mixer,

C. Baseband Analog Path

The analog baseband path consists of channel selection filter, offset gain amplifier and output buffer.

The channel selection filter is implemented in seventh order Chebychev low-pass filter to suppress adjacent channel signal sufficiently. It has 0 ~ 54dB voltage gain with 6dB gain step.

To compensate PVT variation of channel selection filter, an on-chip automatic calibration loop using biquad oscillator to adjust RC pole is used.[4] LPF cut-off

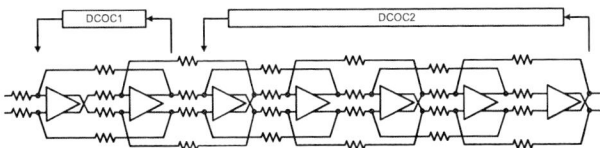

Fig 4. Analog filter block diagram

frequency is adjusted to 2.7MHz with trade-off between signal to noise ratio (SNR) and adjacent channel rejection performance.

To remove DC offset component, dc offset cancelling (DCOC) loop with very low cut-off frequency is adopted which provide high pass response around DC at signal path. For dc offset loop stability and dynamic range, baseband analog filter is divided into two parts with signal gain 24dB and 30dB each. DCOC loop is implemented from output to input of each part as in Fig. 4.

Long DCOC settling time can limit AGC update frequency and make AGC unstable. To reduce dc offset loop settling time, high pass cut off frequency should be increased, but it degrades SNR. ATSC mobile is more sensitive to group delay ripple than other DTV standards which use OFDM, because of 8-VSB method. To meet settling time and SNR requirement simultaneously, DCOC bandwidth adaptation technique is used. At start of DCOC operation, DCOC cut-off frequency is set to high for fast settling and changed to lower value for SNR. DCOC function is turned off at low baseband gain setting to increase SNR. At low baseband gain, DC offset voltage at ADC input is small and can be removed at digital circuit.

Output buffer has gain control range from 0dB to 5.5dB with 0.5dB gain step. Offset gain amplifier is used for trimming gain hysterisis among four LNA gain steps.

D. Frequency Synthesizer and LO signal generation

To support a wide frequency range of LO signal and small area of VCO, the LO generation block consists of several frequency divider blocks which has selectable dividing ratio. With the VCO of the oscillation range from 2.2 to 4GHz, the LO generation block can make all required LO signal for ATSC mobile from 45MHz to 700MHz.

To achieve low in-band phase noise, $\Delta\Sigma$ fractional-N synthesizer which consists of 8/9 prescaler, 3bit 4th order $\Delta\Sigma$ modulator, PFD, CP and divider are implemented. A 6 bit adaptive frequency calibration (AFC) block is implemented and used for tuning the VCO prior to the start of phase lock.[4]

Quadrature LO signal is generated from 1/2 frequency divider. To adjust IQ mismatch a phase trimming block is implemented in UHF band and LO amplifier is designed to be trimmed for LO amplitude trimming.

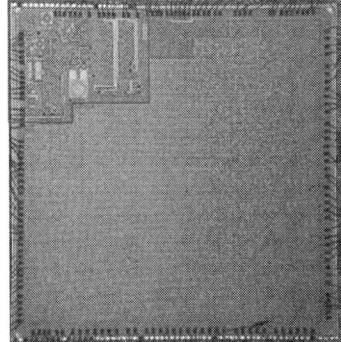

Fig 5. Die photograph. Chip size is 6000 x 6000 um².
Analog tuner part is at upper-left corner

E. ADC and Digital baseband

10 bit differential ADC is used to convert analog baseband output signal to digital code. It has -300mV to +300mV maximum input range. Analog baseband output signal level is controlled to around -10dBm by the automatic gain control (AGC) loop to avoid ADC input saturation considering peak-to-average ratio of 8-VSB, dc offset voltage and implementation margin.

Digital signal from ADC output is used for signal power estimation and averaged power is converted to log scale. Then it is compared to target level and AGC loop controls RF front-end and analog baseband path gain to minimize the difference.

Digital baseband path has its own DCOC function, AGC loop and many other function blocks such as channel filter, equalizer, demodulator, etc. But detailed description of digital part is out of the scope of this paper.

III. MEASUREMENTS

Designed circuit is implemented in a 65nm CMOS process with MIM capacitor and thick metal option. Total chip size is 6mm x 6mm including 130um scribe lane and analog tuner part occupies 6 mm². A die photograph is shown in Fig. 5. We placed 100um isolating space between analog tuner and digital part to reduce digital noise coupling to sensitive analog part. Fabricated chip is packaged in 7.5mm x 7.5mm 144-ball FBGA with careful ball selection and routing placement to avoid cross talk and excessive parasitic components.

The measured performance is summarized in Table I. Power consumption of analog tuner part is 100mW from 1.8V supply at maximum. This power consumption is measured in continuous operation mode (PRC=1) and time slicing operation with PRC=4 can reduce these power to about 1/3 and PRC=7 to about 1/4. Digital baseband part uses 1.2V power supply.

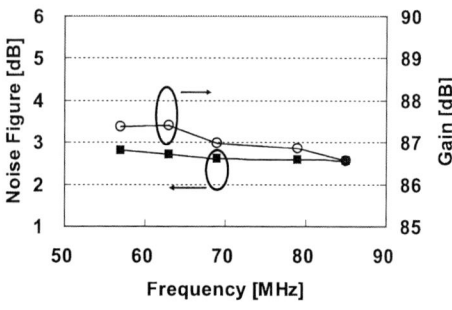

(a) VHF-I band

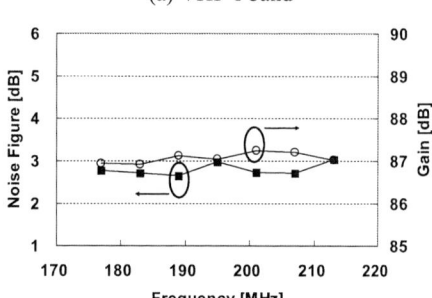

(b) VHF-III Band

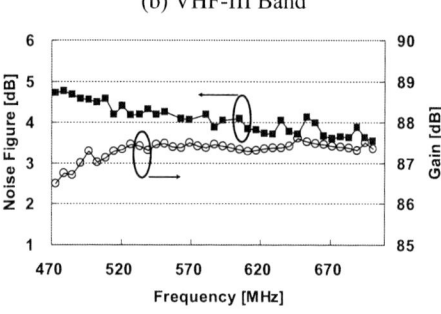

(c) UHF Band

Fig. 6.   Measured gain and noise figure for (a) VHF-I, (b) VHF-III and (c) UHF

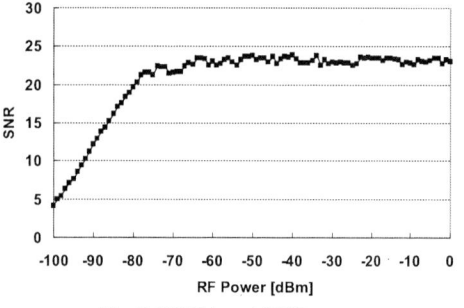

Fig 7. UHF band SNR curve

Fig 6. shows measured analog tuner gain and noise figure for VHF I, VHF III and UHF band. With the LNA load quality factor trimming code, receiver shows flat gain response for all channels. Also, three band gains are trimmed to be similar in each gain mode. These make same gain table can be used for all bands and channels and

Die Size		6000 um x 6000 um			
Package		7.5 mm x7.5 mm 144 Ball FBGA			
Frequency Range		54 ~ 88 / 174 ~ 216 / 470 ~ 698 MHz			
Supply Voltage		1.8/1.2/1.8 V (VDDRF/VDDBB/VDDIO)			
Channel Bandwidth		3 MHz			
		SPEC	VHF I	VHF III	UHF
Power Consumption			100	100	100
Sensitivity	1/4 Code Rate		-98	-97	-97
	1/2 Code Rate		-93	-93	-93
Selectivity	ACS @ -68dBm	-28	-38	-40	-38
	ACS @ -53dBm	-28	-39	-38	-41
	ACS @ -28dBm	-20	-26	-22	-28

a. Power Consumptions are measured for analog tuner part
b. Sensitivity and selectivity are measured for SoC.

TABLE I
SUMMARY OF MEASURED PERFORMANCE

reduce performance variations such as noise figure and linearity

SNR curve in Fig 7. is determined by tuner noise figure, EVM and gain table. In low SNR area, which means weak input signal, SNR is dominated by NF of analog tuner, but in high SNR area, SNR is determined by complex parameters such as group delay ripple, phase noise, filter bandwidth and equalizer performance. Maximum SNR is about 24dB.

LNA input matching is measured for all channel and S11 is less than -15dB, -12dB and -10dB for VHF-I, VHF-III and UHF band respectively.

Sensitivity of receiver is -98/97/97dBm for 1/4 code rate and -93/-93/-93dBm for 1/2 code rate. ACS with digital modulated adjacent signal meets ATSC A74 requirements and detailed performances are in Table I.

IV. CONCLUSION

Triband 65nm CMOS SoC receiver for ATSC mobile is implemented. It meets requirements for ATSC mobile and consumes only 100mW in analog tuner part. Sensitivity level is -97dBm for 1/4 code rate and -93dBm for 1/2 code rate. This is the first report of ATSC mobile receiver.

REFERENCES

[1]   Takae Sakai, et. Al., "A Digital TV Receiver RF and BB Chipset with Adaptive Bias-Current Control for Mobile Applications" *ISSCC Dig. Tech. Papers, pp. 212-213*, Feb., 2007

[2]   Iason Vassiliou, et. Al., "A 65nm CMOS Multistandard, Multiband TV tuner for Mobile and Multimedia Applications" *IEEE J. Solid-State Circuits*, vol. 43, no. 7, pp. 1522-1533, July 2008.

[3]   Ming-Ching, et. Al., "A 1.2V 114mW Dual-Band Direct-Conversion DVB-H Tuner in 0.13um CMOS" *IEEE J. Solid-State Circuits*, vol. 44, no. 3, pp.740-750, March 2009.

[4]   Jae-Hong Chang, et. Al., "A Multistandard Multiband mobile TV RF SoC in 65nm CMOS," *accepted for ISSCC 2010*

RMO3A-2

A Multi-standard Multi-band Tuner for Mobile TV SoC with GSM Interoperability

Huijung Kim, Sanghoon Kang, Jae-Hong Chang, Jeong-Hyun Choi, Hangun Chung, Jungwook Heo, Jong-Dae Bae, Wooseung Choo, and Byeong-ha Park,

Samsung Electronics Co. LTD., Yongin-City, Korea

Abstract — A multi-standard multi-band tuner for mobile TV SoC satisfying the GSM850 & GSM900 Interoperability (IOP) is presented. The single-chip SoC satisfies all requirements of DVB-H/T, ISDB-T, and DAB application with margin. Moreover, this SoC meets not only the GSM900 IOP which is described in MBRAI 2.0, but also GSM850 IOP, in DVB-T/H mode. To suppress GSM transmitter signal in UHF band, LC-tuned load and tunable input matching scheme are adopted in UHF LNA. The SoC consists of three RF LNAs covering the 174 to 248MHz (VHF), 470 to 862MHz (UHF), and 1450 to 1490MHz (L-BAND) bands, dual mode analog baseband filter supporting low pass filtering for direct conversion in DVB-H/T mode and complex band pass filtering for low-IF conversion in ISDB-T 1-segmentation and DAB. The measured sensitivity at UHF for the QPSK 1/2 DVB-T mode is under -95.5dBm with GSM rejection filter.

Index Terms — LC tuned LNA, DVB-T, DVB-H, ISDB-T, DAB, Mobile TV, tuner, GSM IOP, receiver, SoC.

I. INTRODUCTION

As technology grown up, the wireless communication market is moving inevitably toward multi-standard and multi-band mobile terminals. These mobile terminals will integrate multiple functions not only receiving phone call, but also watching mobile TV or high speed internet connectivity e.t.c. This demand leads to change in the system architecture design and the technologies to deploy to achieve the high level integration while minimizing the interferences between the various radio standards [1].

This paper deals with the coexistence of a digital TV receiver and GSM system in a mobile terminal. To support various applications with small form factor and size, the multi-standard and multi-band mobile TV RF tuner is developed as SoC technology [2]-[3]. In addition to multi-standard multi-mode function, DVB-T/H receiver and GSM transmitter (Tx) are able to operate same time without limitations in cellular phone or other mobile terminals. To record TV-channel while receiving phone call or watch TV-channel while WEB-browsing / making a phone call, GSM IOP performance is need to the mobile TV tuner.

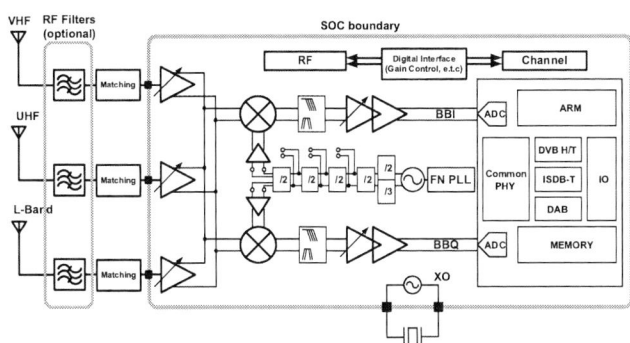

Fig. 1. Multi-standard Multi-band SoC Block Diagram.

II. CIRCUIT DESIGN

As shown in the frequency plan in Fig. 2.(a), the GSM Tx band is very close to the UHF band. Due to this proximity, the received TV signal can be drastically degraded if the interfering GSM signal is not properly rejected. According to the MBRAI 2.0 [4], the sensitivity degradation is less than 1.5dB caused by the blocking effect i.e. the sensitivity is better than –94.1 dBm up to 746MHz in 8 MHz channel when GSM Tx signal is present. The GSM Tx power level assumes worst case antenna isolation value of 15dB i.e. the signal level is +33 dBm - 15dB = +18 dBm at DVB-T/H antenna, as shown in Fig. 2 (b). Typical GSM rejection filter can suppress the GSM band signal up to 40dB, GSM Tx signal level at the UHF LNA input is about -23dBm. In case of UHF 746MHz channel, DVB-T/H signal is 80MHz from the lowest GSM850 frequency.

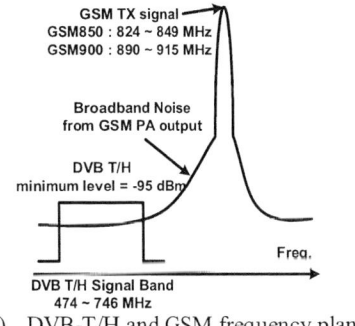

(a)   DVB-T/H and GSM frequency plan.

978-1-4244-6240-7/10 $26.00 © 2010 IEEE   167

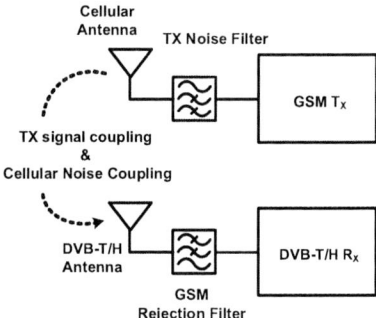

(b) Antenna coupling in a mobile terminal

Fig. 2.   GSM Tx Coupling Mechanism

A. LNA design

To suppress GSM Tx signal in UHF band, a high Q LC-tuned load and tunable input matching scheme are adopted. As shown in Fig. 3, programmable internal C_{LOAD} and external L_{LOAD} make resonance frequency to maximize the LNA gain and suppress the GSM Tx signal. The tunable C_{GS} reflects the GSM Tx signal with good input matching in desired frequency. Simulated S-parameters of LNA is shown in Fig. 4. According to the desired frequency, C_{LOAD} and C_{GS} are tuned to maximize the LNA gain and the suppression of the GSM Tx signal.

The VHF and L-band LNA is designed using typical LC load without frequency tuning.   All three LNAs have 30dB dynamic range using current diverting scheme which provide high gain with low noise figure to meet minimum sensitivity and high linearity to meet the blocking and linearity specifications according to RF input power.

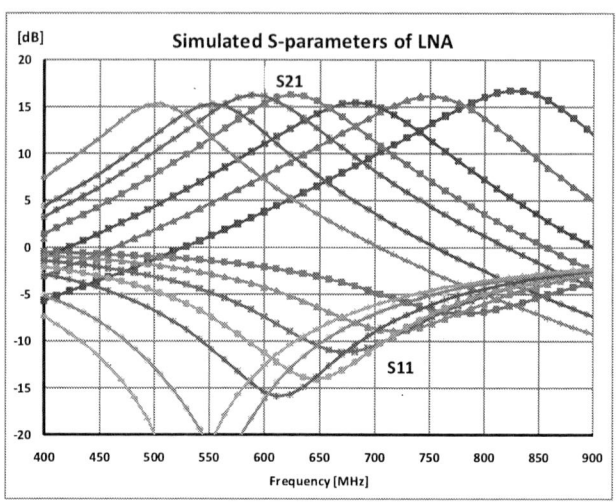

Fig. 4. Simulated S-parameters of UHF LNA.

B. Mixer design

The high loaded-Q of LNA load is essential to suppressing the GSM Tx signal effectively. To maintain the high loaded-Q of LNA load, the Gilbert cell mixer with high-impedance AC-coupled g_m stage is used as shown in Fig. 3.

In high-Q mode, bypass SW is off and G_M stage operates. G_M stage amplifies RF signal and maintain the high loaded-Q of LNA load with high input impedance using common-source differential stage.

In low-Q mode, G_M stage is off and bypass switch is turned on. Low input impedance of the Gilbert cell mixer makes the loaded-Q of LNA load low and relaxes the voltage swing at the LNA load. This mode scheme effectively reduces the distortions caused by voltage modulation at the N-MOS gate [5].

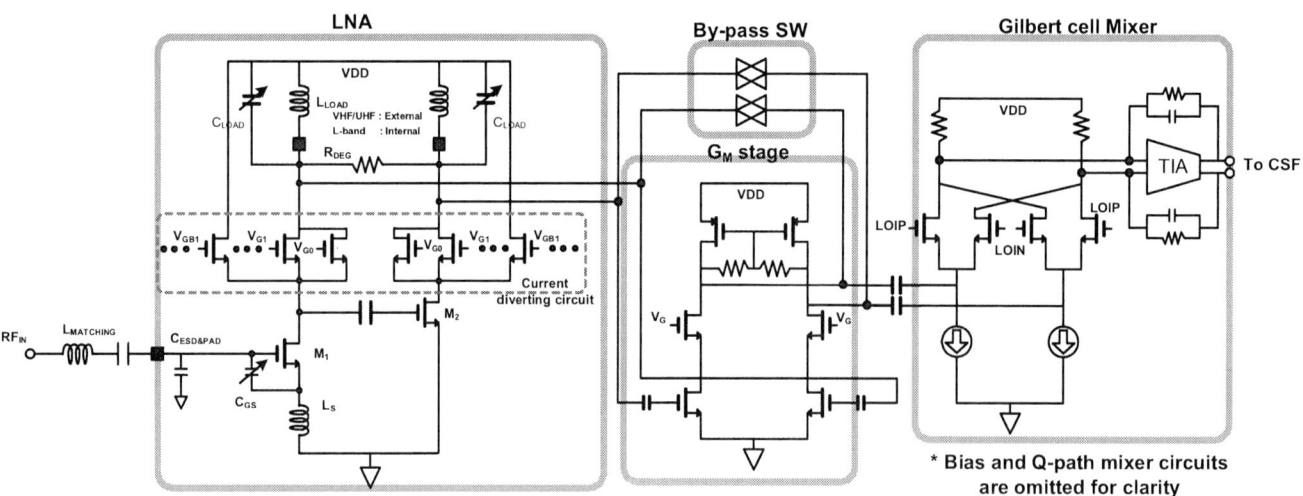

Fig. 3. Simplified schematic of LNA and Mixer.

978-1-4244-6240-7/10 $26.00 © 2010 IEEE

The trans-impedance amplifier (TIA) with 1-pole RC relaxes the linearity requirement of Channel Selection Filter (CSF) for far-off interfere.

C. Other circuits design

The 7th order Chebychev LPF is implemented using ladder type active RC filter. In DVB-T/H direct conversion system, channel bandwidth is tuned from 5MHz to 8MHz. For narrow channel bandwidth system such as ISDB-T 1-segment and T-DMB, low IF mode is also implemented using complex band pass filter [6]. To compensate PVT variation of the different bandwidth, an on-chip auto-calibration by comparing its RC time constant with a reference clock is used.

To support a wide frequency range of LO signal and small area of VCO, the LO generation block consists of three selectable divide-by-2, UHF divider-by-2, LBAND divide-by-2 and divide-by-3 followed by VCO. With the VCO of the oscillation range from 2.2 to 4.0GHz, the LO generation block creates the LO frequency range from 45MHz to 1GHz and from 1.2GHz to 2000MHz. To achieve low in-band phase noise, $\Delta\Sigma$ fractional-N synthesizer which consists of 8/9 prescaler, 3bit 4th order $\Delta\Sigma$ modulator, PFD, CP, divider and XO is implemented. The synthesizer can handle the reference signals of 10~38.4MHz. The baseband demodulator of this SoC consists of data converts, demodulator, FEC, hardwired MPE-FEC, ARM CPU and SRAM.

Fully differential structures are used to avoid digital noise and substrate coupling in SoC environment. To reduce the substrate coupling, a resistive region which blocks well implants, creating regions with higher substrate resistance is inserted on the space between sensitive block such as LNA and other noisy block such as LO generator and digital part. Using on-chip regulator and bypass capacitor, reduction of supply crosstalk is accomplished.

III. EXPERIMENTAL RESULTS

Without GSM-reject filter, the sensitivity at UHF for the QPSK 1/2 DVB-T mode is around -97.5dBm, as shown in Fig. 5. In this mode, operation band is extended to 858MHz as described in MBRAI 2.0.

In GSM-rejection mode which includes GSM-rejection filter, the sensitivity at same condition is around -96 dBm up to 746MHz, as shown in Fig. 5.

As discussed above, GSM IOP test measures the DVB-T/H signal sensitivity coexisting with GSM Tx signal level, +18dBm at the RF input port. Fig. 6 shows the sensitivity measurement results for full channel GSM850 signal (a), and GSM900 (b) signal with all GSM 124-channel which has 200KHz channel spacing.

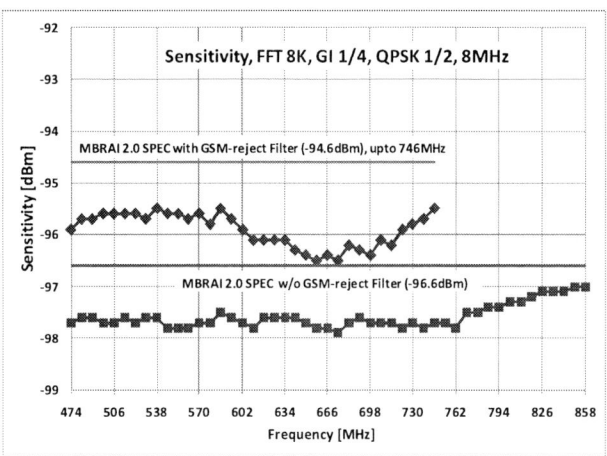

Fig. 5. Sensitivity in DVB-T mode, 2x10⁻⁴ BER.

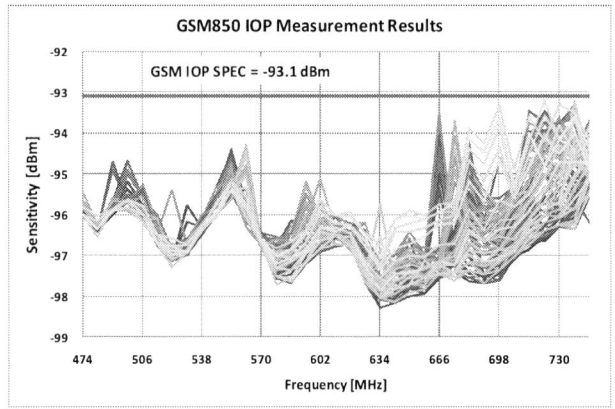

(a) GSM850 (824~849MHz, all GSM signal)

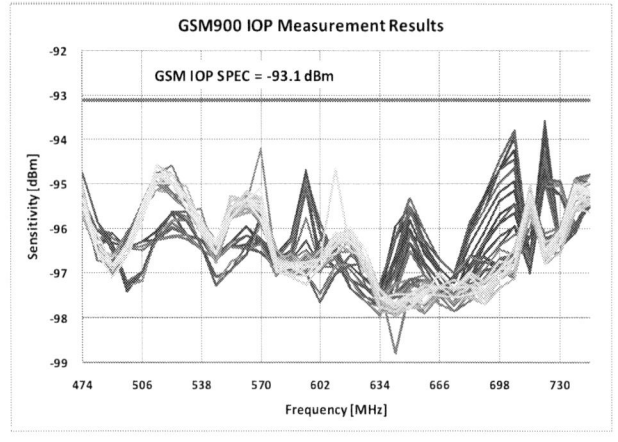

(b) GSM900 (890~915MHz, all GSM signal)

Fig. 6. GSM IOP Measurement Results

TABLE I

MEASURED PERFORMANCE SUMMARY

Die Size [um²]	5000 x 5000 (RF : 2300 x 1700)	
Package	119 Ball WFP	
Frequency range [MHz]	174~248, 470~862, 1450~1490	
Supply Voltage [V]	1.8/1.2/1.8 (VDDRF/VDDBB/VDDIO)	
Phase noise integrated (40 Hz ~ 4 MHz) [Degree]	0.35~0.5(VHF) 0.6~0.9(UHF)	
Channel bandwidth	ISDB-T 1-seg(0.43MHz)	LIF/ZIF
	ISDB-T 3-seg(1.29MHz)	LIF/ZIF
	DMB (1.536MHz)	LIF/ZIF
	DVB-T/H(5, 6, 7, 8MHz)	ZIF
System	Parameter	Value
DVB-T/H UHF, 666MHz With GSM Filter Test Condition < 2x10⁻⁴ BER	Power [mW]	171
	Sensitivity (QPSK ½) [dBm]	-96.4 (-94.6)
	Sensitivity (16QAM ⅔) [dBm]	-89.1 (-86.5)
	Sensitivity (64QAM ¾) [dBm]	-82 (-79.4)
	S1 N±1 / N±2 (16QAM ⅔) [dB]	45 (38) / 54 (48)
	S2 N±1 / N±2 (16QAM ⅔) [dB]	44 (29) / 49 (40)
	L1 (16QAM ⅔) [dB]	48 (45)
	L2 (16QAM ⅔) [dB]	48 (45)
	L3 (16QAM ⅔) [dB]	44 (40)
	L4 (16QAM ⅔) [dB]	48 (45)
ISDB-T 665.143 MHz With GSM Filter Test Condition < 2x10⁻⁴ BER	Power [mW]	140
	Sensitivity (QPSK ½) [dBm]	-97.4 (-95.5)
	Sensitivity (16QAM ½) [dBm]	-92.1 (-88.9)
	Sensitivity (64QAM ½) [dBm]	-86 (-83.5)
	S1 N-1 / N+1 (16QAM 7/8) [dB]	39 (33) / 42.5 (35)
	S1 N-1 / N+1 (16QAM 7/8) [dB]	31.5 (20) / 32.5 (21)
DAB w/o system Filter (IEC-62104) Test Condition DQPSK ½ < 1x10⁻⁴ BER	Power(VHF/L-band) [mW]	140 / 152
	Sensitivity of VHF band [dBm]	-101 ~ -104 (-95)
	Sensitivity of L- band [dBm]	-100 ~ -101 (-95)
	S1 N±1 VHF / L-band [dB]	37 (30) / 36 (30)
	Far off Selectivity (N ± >5Mhz) VHF / L-band [dB]	52~55.5 (40) / 52~55 (40)

*note : ( ) means system spec.

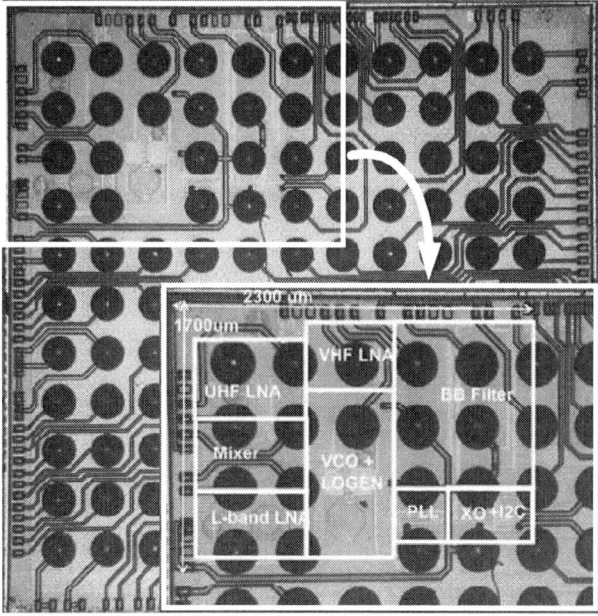

Fig. 7. Die Photo.

A microphotograph is shown in Fig. 7. It occupies an area of 25 mm² while about 16% of SoC is RF area. The IC is packaged in a 119-ball Wafer-level Fabricated Package (WFP) with careful ball and routing placement to avoid RF inductors, as shown in Fig. 7. The measured performance is summarized in table I.

IV. CONCLUSION

A multi-standard multi-mode tuner for mobile TV RF SoC in 65nm CMOS is presented. The SoC satisfies all requirements of DVB-T/H, ISDB-T, and DAB application with margin as shown in table I. Moreover, this SoC meets not only the GSM900 IOP, but also GSM850 IOP in DVB-T/H mode. It occupies an area of 25 mm², while about 16% of SoC is RF area.

ACKNOWLEDGEMENT

The authors wish to thank Woosung Choe for chip measurement support and HyungSeok Seo for chip layout support.

REFERENCES

[1] Michael Flash, "Address challenges in DVB-H receiver design," *EE Times-Asia*, June 1-15, 2007..

[2] Iason Vassiliou, et.al., "A 65nm CMOS Multistandard, Multiband TV tuner for Mobile and Mulitmedia Applications," *IEEE J. Solid-State Circuits*, vol. 43, no. 7, pp. 1522-1533, July 2008.

[3] Jae-Hong Chang, et.al., "A Multistandard Multiband Mobile TV RF SoC in 65nm," *IEEE ISSCC Dig. Tech. Papers*, pp. 462-463, Feb. 2010.

[4] Mobile and Portable DVB-T/H Radio Access – Part 1: Interface Specification. EICTA, 2007.

[5] Ming-Ching Kuo, et.al., "A 1.2V 114mW Dual-Band Direct-Conversion DVB-H Tuner in 0.13um CMOS," *IEEE J. Solid-State Circuits*, vol. 44, no. 3, pp. 740-750, March 2009.

[6] Thomas Byunghak Cho, et.al., "A 2.4 GHz Dual Mode 0.18um CMOS Transceiver for Bluetooth and 802.11b," *IEEE J. Solid-State Circuits*, vol.39, no. 11, pp. 1916 -1926, Nov. 2004.

RMO3A-3

A 2.2 mW Regenerative FM-UWB Receiver in 65 nm CMOS

Nitz Saputra, John R. Long, and John J. Pekarik*

ERL/DIMES, Delft University of Technology, 2628CD Delft, the Netherlands
*IBM Microelectronics, Essex Junction, VT 05452

Abstract — A 4-4.5 GHz receiver front-end consisting of a 35 dB voltage gain regenerative amplifier, ultra-narrowband RF filter and an envelope detector demodulator for FM-UWB communication is described in this paper. Implemented in 65 nm CMOS, the measured receiver sensitivity is -83 dBm at 100 kbps data rate with 15 dB output SNR (10^{-6} BER). The 0.3 mm² test chip includes a 50 Ohm buffer amplifier to facilitate testing and consumes 2.2 mW (excluding buffer) from a 1 V supply.

Index Terms — FM-UWB, FM Demodulator, RF-CMOS, RF receiver front-end.

I. INTRODUCTION

Low power wireless communication is enabling home/office automation, personal-area networking (PAN), and wireless sensor networks (WSN). These systems aim at low-cost wireless access anywhere in the world in the (unlicensed) ISM bands or the ultra-wideband (UWB) spectrum (3.1-10.6 GHz in the US). Radios operating in unlicensed bands must be robust enough to handle co-channel interference (e.g., co-existence with WiFi in the 2.4 GHz ISM band), while UWB links are also constrained by potential wideband interferers and a span-limiting maximum transmit power density of -41 dBm/Hz.

Ultrawideband frequency modulation (FM-UWB) has been proposed for low-complexity wireless transceiver implementations, as it resists frequency selective (multipath) fading by spreading transmit energy across a wide spectrum [1]. Wideband frequency modulation of a sinusoidal carrier by a triangle-wave sub-carrier generates an ultra-wideband FM signal. Data is FSK-modulated onto the sub-carrier, resulting in a uniform spectral density transmit signal that can easily conform to regulatory requirements. The high gain RF amplifier and wideband delay line/multiplier demodulator reported in [2] and [3] demonstrated the simplicity inherent in FM-UWB receivers, and their low cost potential.

The goals of this work are reduced power consumption, integration in deep submicron CMOS, and improved co-existence on a targeted frequency band of 4-4.5 GHz. An FM-UWB receiver that is portable and powered by a small battery or a power scavenging unit should consume few mW. Chip area should be minimized with integration in CMOS, so that low cost/high volume applications can be served. Improved co-existence with other systems enhances reliability and makes operation in the UWB more practical.

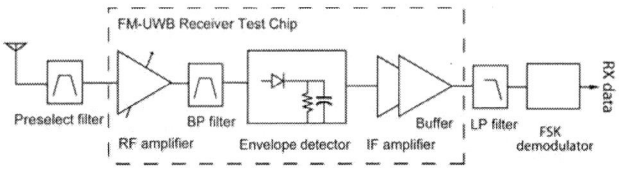

Fig. 1. FM-UWB data receiver architecture.

In contrast to UWB receivers which receive the entire transmitted signal, including noise and unwanted interference, the receiver developed in this work (see Fig. 1) is designed according to the cognitive radio paradigm [4] so that it can select a sub-band of the RF signal. Potential interferers may be avoided in this way and the received signal-to-noise ratio (SNR) can be optimized to improve robustness.

The receiver proposed in this work (Fig. 1) consists of a transconductance RF amplifier and pulse-shaping RF filter followed by an envelope detector. An IF amplifier and an output buffer are included on-chip to facilitate characterization. The antenna bandwidth is 500 MHz to 1 GHz rather than the entire UWB spectrum, and selectivity may be improved further by adding preselect filtering between the antenna and RF input as shown in Fig. 1.

The prototype FM-UWB receiver and details of its design and operation are described in Section II of this paper. Measurements of the prototype realized in 65 nm CMOS are presented in Section III, followed by a brief discussion and a comparison with other low-power FM-UWB receivers reported recently in the literature.

II. FM-UWB RECEIVER FRONT-END

In FM-UWB, the carrier sweeps back and forth across the band (as shown in Fig. 2a) at a rate proportional to the modulating sub-carrier signal (see Fig. 2b). The bandpass RF amplifier in Fig. 1 detects and amplifies the carrier present in its passband, otherwise, the carrier signal is attenuated. The narrowband response of the front-end therefore shapes the envelope of the received FM into an AM signal, which follows the amplitude response of the bandpass filter. The envelope detector then detects the resulting AM (a pulse train, as shown in Fig. 2c), and any

978-1-4244-6240-7/10 $26.00 © 2010 IEEE

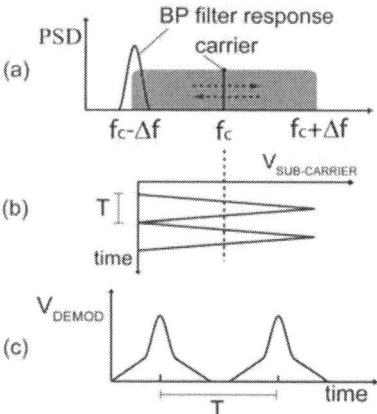

Fig. 2. Illustration of FM-UWB demodulation process, (a) power spectrum density of FM-UWB signal with BP filter response, (b) sub-carrier wave (c) demodulated signal.

unwanted harmonics may be removed from the envelope by a lowpass filter to recover the FSK-modulated data on the subcarrier.

A conventional FM-UWB demodulator as described in [2] and [3] first transforms the FM to a phase-modulated (PM) signal via a delay line, and then multiplies it with the original FM signal to yield an amplitude-modulated (AM) output suitable for envelope detection. In the receiver developed in this work, the FM-AM transformation is realized directly via a bandpass filter. Selecting a sub-band does not (ideally) affect the SNR of the received signal compared to processing the entire FM-UWB transmit signal. In fact, it will be seen from measurement of the prototype that the received SNR and sensitivity for the sub-band FM-UWB receiver is on par with a conventional wideband receiver. However, the receiver RF amplifier can be made more power efficient by processing a sub-band rather than the entire band. One disadvantage of this concept is that a receiver calibration is required to optimize the SNR for a selected sub-band.

The bandpass filter center frequency may be tuned anywhere in the FM-UWB signal bandwidth. However, it can be shown that the output amplitude from the bandpass filter is proportional to the difference between the center frequency of the bandpass filter and the FM carrier frequency (assuming that the bandpass filter overlaps the frequency deviation of the FM-UWB). Thus, setting the center of the bandpass as shown in Fig. 2a (i.e., close to the maximum deviation) maximizes the output amplitude.

Tuning the bandpass filter center frequency adds complexity to the receiver, however, it also permits optimization of the SNR when interference or noise is present. An automated tuning scheme has not been implemented for this prototype, although calibration circuits that optimize the SNR and bit-error performance of the receiver could be devised and implemented on-chip. The tuning range must be wide enough to compensate for

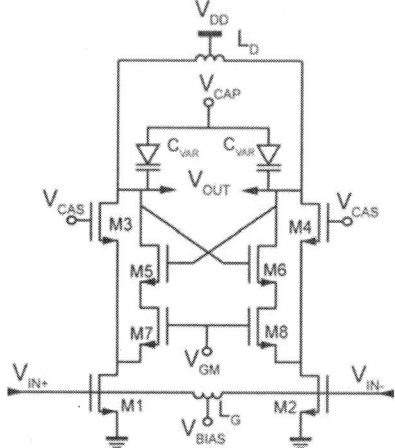

Fig. 3. Simplified schematic of the narrow band preamplifier.

any process, voltage and temperature (PVT) variations that could not be compensated by conventional bias compensation schemes (e.g., PTAT biasing).

When an envelope detector is used to demodulate the received signal, the demodulated signal amplitude is determined by the maximum output from the RF bandpass filter (which occurs when the FM-UWB carrier is in the center of the filter passband). Minimum output is reached when the separation between the FM-UWB and bandpass filter center frequencies is greatest, as previously noted (assuming that this level is above the circuit noise floor).

A. Regenerative Preamplifier and Filter

Since MOSFET transconductance is small at low bias current, a Q-enhanced amplifier is developed to realize >30 dB RF gain and a receiver sensitivity better than -80 dBm at low DC power. The amplifier utilizes positive feedback to increase RF amplification from a low-gain device [5], and consumes 55% less power than the conventional BP-LNA described in [3].

In this topology, the filter Q-factor is increased by negative resistance (derived from positive feedback) that partially enhances the selectivity of the passive LC tank. Positive feedback increases the voltage gain of the preamplifier from 16 dB to 35 dB at the expense of less bandwidth (550 MHz to 45 MHz) and linearity (input P_{1dB} from –20 dBm to -41 dBm). A slight degradation of 2 dB is observed on the linearity given the change in gain. The preamplifier consumes 1.6 mW from the 1 V supply.

The schematic of the preamplifier is shown in Fig. 3. M1 and M2 form a differential transconductance amplifier, which drives the LC tank formed by L_D, C_{VAR} and the parasitic capacitances at the drains of M3, M4, M5 and M6. Cascode transistors M3, M4, M7 and M8 improve isolation between the input and output. Cross-coupled transistors M5 and M6 form a positive feedback loop at the preamplifier output. V_{CAS} and V_{GM} control the

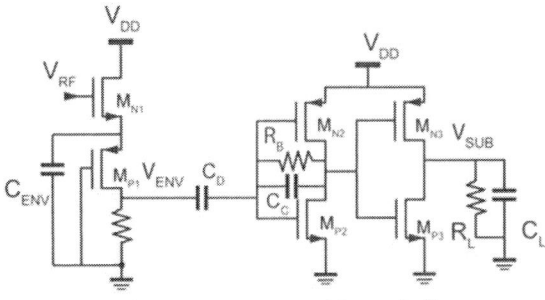

Fig. 4. Schematic of envelope detector, IF amplifier and test buffer.

proportion of signal current flowing through M5 and M6, thereby controlling the amount of positive feedback. V_{CAP} tunes the filter center frequency via varactors in parallel with the LC tank at L_D. The preamplifier input is matched to 50 Ohm in the 4-4.5 GHz band by shunt input inductor L_G. L_G also provides a DC bias path (V_{BIAS}) at the input. L_D and L_G are fully-symmetric on-chip inductors that provide the highest Q-factor in the smallest possible chip area.

B. Envelope detector and IF amplifier

The envelope detector (M_{N1} in Fig. 4) removes the carrier from the filtered RF signal. It is followed by an intermediate frequency (IF) amplifier and output buffer stages, as shown in Fig. 4. C_{ENV} suppress any carrier signal that passes through the source follower detector. Common-gate transistor M_{P1} amplifies and buffers the detected signal to the IF amplifier input. The envelope detector is biased at 200 μA and is sensitive to PVT variation because its bias current depends on the threshold voltages of M_{N1} and M_{P1}, but a compensation circuit can be added [6].

The envelope is amplified 20 dB further by CMOS IF amplifier M_{N2} and M_{P2}, which has a bandwidth of 100 kHz to 10 MHz. R_B provides DC feedback path to bias the amplifier. The output test buffer stage (M_{N3} and M_{P3}) is capable of driving the 50 Ohm impedance of the test instrumentation used for chip characterization. M_{N3} and

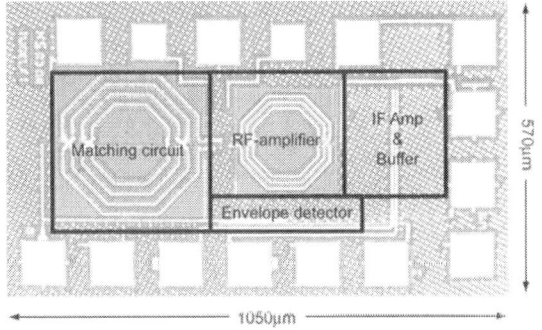

Fig. 5. Chip micrograph of the 65 nm CMOS receiver prototype.

M_{P3} are scaled six times larger than M_{N2} and M_{P2}, and the symmetry in their physical layout is used to promote good matching of the device parameters. The output buffer consumes 2.5 mW from 1 V supply voltage.

III. EXPERIMENTAL MEASUREMENTS

The 0.3 mm^2 receiver prototype (see Fig. 5) was realized in 65 nm RF-CMOS technology [7]. On-wafer probing and a 4-port network analyzer were used to characterize the RF input reflection coefficient (S_{11}). The measured S_{11} of the receiver from 4-4.5 GHz is plotted in Fig. 6a. When the positive feedback turned off, the measured S_{11} showed good agreement with post-layout simulation. The center frequency of the filter cannot be measured directly at the preamplifier output, so it was observed indirectly by measuring S_{11} when the positive feedback was applied in the RF preamplifier. The null normally seen in S_{11} without feedback (see plots of Fig. 6b) was disturbed, and can be controlled by changing V_{CAP}.

The receiver was tuned to optimum sensitivity by adjusting V_{GM} and V_{CAP} (see Fig. 3) to change Q and the sub-band frequency, respectively. Firstly, V_{GM} was increased until an abrupt change in the output DC level occurs, which was caused by oscillation in the RF amplifier. Then, V_{GM} was slowly decreased until oscillation stops as the DC level drops to its original value. The receiver was operating near the point of critical

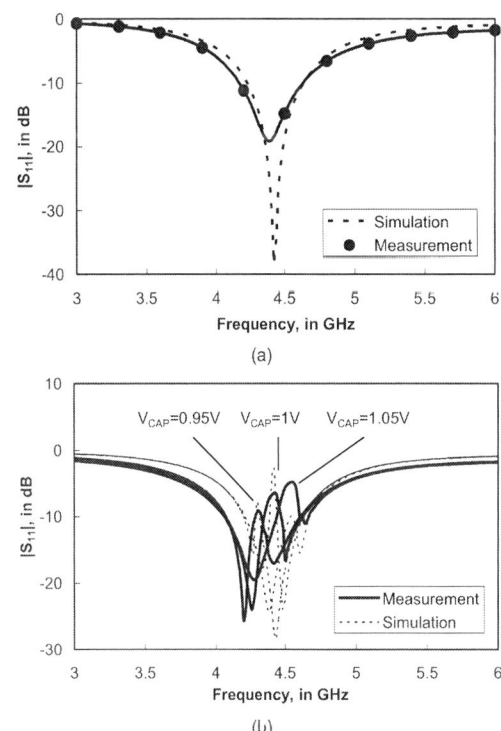

Fig. 6. (a) Measured and simulated S_{11} of the receiver input (positive feedback not applied) and (b) when positive feedback is applied for different settings of V_{CAP}.

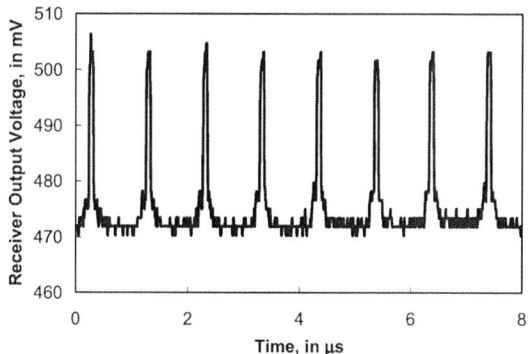

Fig. 7. Measured pulses at receiver output for 1 MHz sub-carrier.

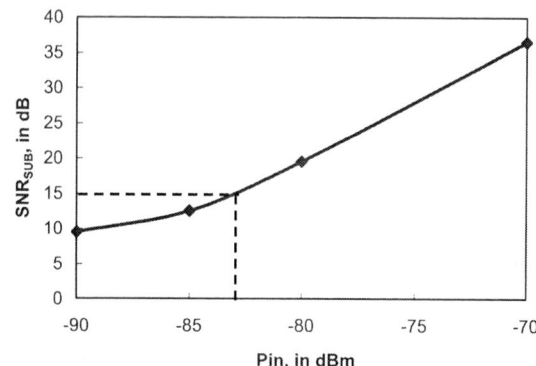

Fig. 8. Measured SNR of sub-carrier vs. RF input power.

stability, where it had the highest voltage gain and selectivity.

Fig. 7 shows the measured received sub-carrier output waveform at 1 MHz for an input power of -50 dBm, proving that the receiver functions as expected. The measured signal-to-noise ratio of the subcarrier (SNR_{SUB}) at the output of the receiver as a function of input power is plotted in Fig. 8. The SNR_{SUB} required to achieve a bit-error rate (BER) of 10^{-6} for FSK modulated data at 100 kbps is 15 dB. The result implies that for ideal baseband processing the whole receiver would have a sensitivity of -83 dBm. At 100 kbps, the energy efficiency of the front-end receiver is 22 nJ/bit. The receiver performance is summarized and compared with other FM-UWB receivers in Table I. This receiver consumes the least power among FM-UWB receivers where FM-UWB has a better co-existence compared to other low power schemes such as OOK.

IV. CONCLUSIONS

The circuit presented in this paper implements a new method for detecting FM-UWB signals by employing positive feedback in a controlled fashion to lower power consumption for a given RF input sensitivity. The receiver prototype was realized in 65 nm RF-CMOS technology and has an active area of 0.3 mm². The measured receiver front-end sensitivity is -83 dBm when operating in the 4-4.5 GHz band and the receiver front-end consumes 2.2 mW, excluding buffer, from a 1 V supply.

ACKNOWLEDGEMENTS

The authors would like to thank A. Akhnoukh and M. Spirito for measurement support. IC fabrication was provided by IBM and facilitated by MOSIS.

TABLE I
SUMMARY OF FM-UWB PERFORMANCE

Parameters	This work	[2]*	[3]
Technology	CMOS 65 nm	SiGe BiCMOS 180 nm	SiGe BiCMOS 250 nm
RF frequency (GHz)	4-4.5	3.1-4.9	7.2-7.7
Sub-carrier frequency (MHz)	1	1	0.5
Power Consumption (mW)	2.2	10	9.1
Receiver sensitivity (dBm)	-83	-76	-86.8
Active area (mm²)	0.3	0.72	0.88

* Uses external LNA

REFERENCES

[1] IEEE P802.15, CSEM FM-UWB Proposal, IEEE P802.15-09-0276-00-0006, May 2009.

[2] J.F.M Gerrits, J.R. Farserotu, J.R. Long, "A Wideband FM Demodulator for a Low-Complexity FM-UWB Receiver", *Proceeding of the 9th European Conference on Wireless Technology*, September 2006.

[3] Y. Dong, Y. Zhao, J.F.M. Gerrits, G. van Veenendaal, J.R.Long, "A 9mW High Band FM-UWB Receiver Front-end", *Proceeding of ESSCIRC*, pp.302-305, Sep. 2008.

[4] A. Sahai, R. Tandra, S.M. Mishra, N. Hoven, "Fundamental Design Tradeoffs in Cognitive Radio Systems", *proceeding of TAPAS*, Aug. 2006.

[5] E.H. Armstrong, "Some Recent Developments in the Audion Receiver". *Proceedings of the IRE* Vol. 3, No.9, pp. 215–247, Sept. 1915.

[6] Y. Tsugita, K. Ueno, T. Hirose, T. Asai, Y. Amemiya, "On-chip PVT Compensation Techniques for Low-voltage CMOS Digital LSIs," *Proc. International Symposium on Circuits and Systems*, May, 2009.

[7] Z. Luo et al., "High Performance and Low Power Transistors Integrated in 65 nm Bulk CMOS Technology", *IEDM Technical Digest*, pp. 661-664, Dec. 2004.

RMO3A-4

A $75\,pJ/bit$ All-Digital Quadrature Coherent IR-UWB Transceiver in $0.18\,\mu m$ CMOS

Enrique Barajas, Didac Gómez, Diego Mateo, José Luis González

Universitat Politècnica Catalunya, SPAIN

Abstract— In this paper a $75\,pJ/b$ all-digital quadrature coherent impulse radio ultra-wideband transceiver in $0.18\,\mu m$ CMOS is presented. It consumes $42\,mW$ operating at a $560\,Mbps$ datarate. The receiver and transmitter share most of the components reducing the area. This design is optimal for low-power low-cost short-range high-speed communications.

Index Terms— Ultra-wideband (UWB), Impulse Radio (IR), Short Range, Low Power, All digital

I. INTRODUCTION

Although ultra-wideband technology (UWB) has lately been prone to criticism, it still has its niche applications: pulse-based UWB can be successfully used in short-range high datarate communications [1]–[3]. The compromise in terms of power consumption, datarate and distance can be conveniently exploited to achieve at the same time low power consumption and high datarate at the expense of range [1], [4]. The transceiver presented in this paper was designed taking into account these bound conditions (datarate, power and range), resulting in a IR-UWB transceiver with the best performance compared with previous works, to best of the knowledge of the authors.

The receiver and transmitter architectures are described in section II while the measurements are presented in section III. In section IV a comparative with previous works is carried out. Finally the conclusions are summarized in section V.

II. TRANSCEIVER ARCHITECTURE

IR-UWB transceivers can be implemented with a coherent or non-coherent architecture. The coherent architecture, such as that shown in Fig. 1, was chosen for the transceiver because it's more appropriate for noisy or interference-rich environments. The FCC part 15 [5] defines several frequency bands. The 3–$5\,GHz$ band was selected as a compromise between available bandwidth and center frequency.

In order to reduce area occupancy as well as complexity the transceiver architecture presented in this work shares

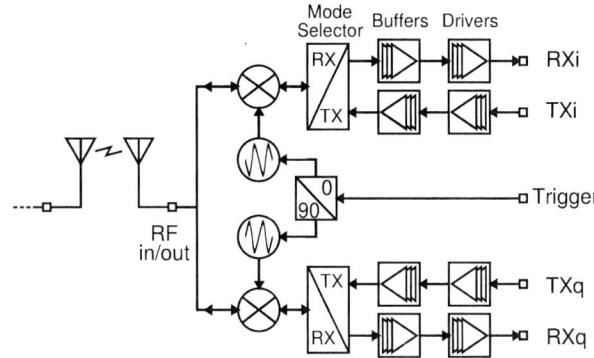

Fig. 1. Quadrature transceiver architecture.

as much components as possible between the receiver and the transmitter. To achieve this functionality the unilateral blocks of the RF front-end are suppressed based on two inherent properties of impulse-radio ultra-wideband (IR-UWB). In the first place, receivers are not noise limited but rather interference limited, thus the LNA is not required [4]. In the second place, to conform to the tight power spectral density specifications, the transmitted power is so limited that it can be achieved without the need of a PA. Besides suppressing these blocks, a quadrature topology is used, only possible in coherent receivers, not only to improve spectral efficiency but also to ease signal tracking and synchronization [6]. The whole transceiver is implemented in a pure digital CMOS technology; this yields to reduced cost and area (no inductances are used), at the expense of some performance lose, though.

A. Receiver Mode

IR-UWB coherent receivers are usually built upon a correlator or a matched filter [7]. This architecture can be implemented in analog receivers by combining a mixer, a template generator and an integrator [7]. The receiver full differential topology proposed in this work is depicted in Fig. 2. The common gate input stage (M1–M2), shared between the I and Q branches, is not matched to $50\,\Omega$

978-1-4244-6240-7/10 $26.00 © 2010 IEEE

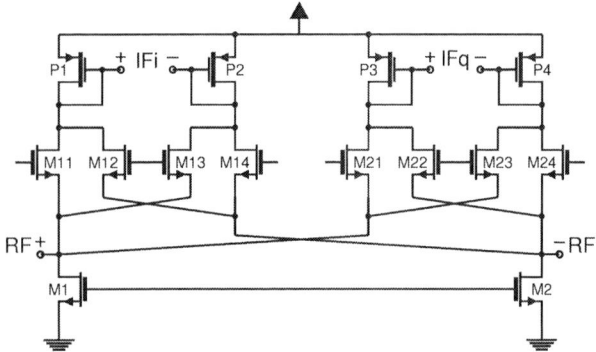

Fig. 2. Quadrature receiver mode schematic.

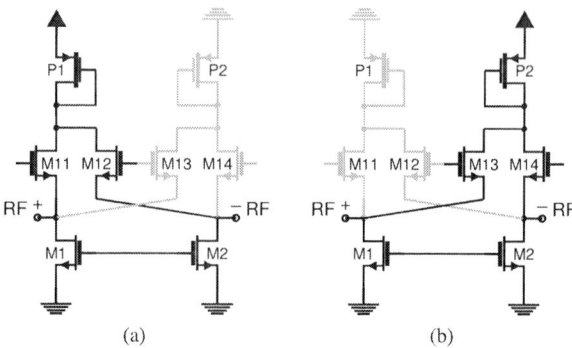

Fig. 4. Simplified transmitter mode schematics for amplitude modulation. (a) Transmission of "0", (b) transmission of "1".

to obtain UWB operation [4]. The Gilbert cell of the I branch (M11–M14) is connected to active loads (P1–P2) to obtain the demodulated IF signal. This signal is then amplified by the baseband buffers shown in Fig. 1 and finally connected to a driver to drive the 50 Ω input of the oscilloscope. For the sake of simplicity the connections to the template generator (gates of transistors M11–M14) are not shown. The Q branch (M21–M24 and P3–P4) is connected in a similar way.

Post-layout transistor level simulations of the signals involved in the receiver are depicted in Fig. 3. The upper panel represents the RF input signal. The signal in the middle panel is the output at the active load in Fig. 2. Finally the last panel shows the signal amplified by the baseband buffers in Fig. 1, just before the driver.

B. Transmitter Mode

There are different modulations schemes for IR-UWB communications (PPM, PAM, ...). An amplitude modulation (PAM) is easily implemented in this circuit unbalanc-

ing the Gilbert cell by connecting one active load source to the positive supply and the other to ground for a "0" pulse, or vice versa for a "1" pulse. A simplified non-quadrature version of the transceiver is shown in Fig. 4a while generating a "0" pulse and alternatively generating a "1" signal in Fig. 4b. It's worth noting that to achieve dynamic operation an inverter (with complementary outputs) must be connected to the sources of transistors P1–P2. Also a driver to adapt the signals to the 50 Ω signal generator output is implemented, as shown in Fig. 1. The IR-UWB transmitted signal for the IQ transceiver operating as a transmitter is represented in Fig. 5.

C. Mode selection, frequency control and synchronization

We have introduced a transceiver that can operate with the same basic blocks either as receiver or transmitter. However, this can be only achieved in half-duplex, as the receiver and transmitter share most of the components and area.

The receiver synchronization is performed by measuring the detected energy at the outputs of the IQ branches. The frequency of the template generator and the delay between each IQ branch are controlled externally. These two controls together with the external trigger are used in coarse synchronization and tracking.

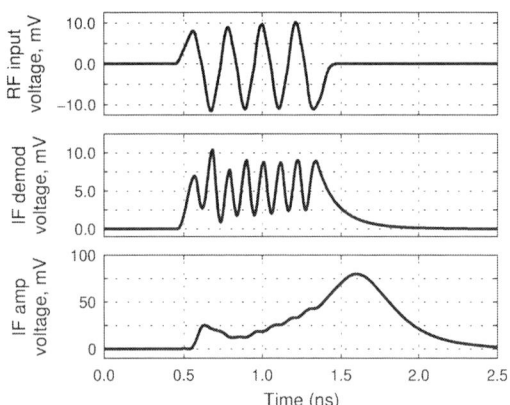

Fig. 3. Receiver transistor level simulation.

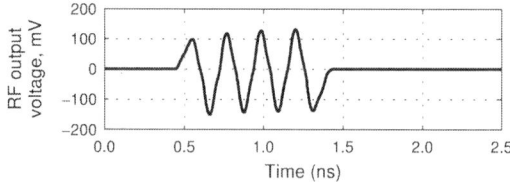

Fig. 5. Transmitter transistor level simulation.

978-1-4244-6240-7/10 $26.00 © 2010 IEEE

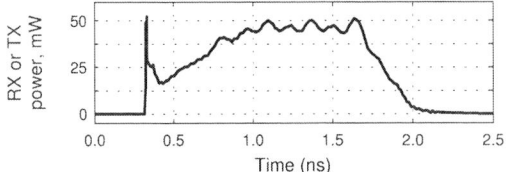

Fig. 6. Receiver or transmitter power consumption transistor level simulation.

D. Power consumption

IR-UWB transceivers can operate in a very low duty cycle, thus only the instantaneous power consumption during the transmitter and receiver operation time window is shown in Fig. 6. Non optimized systems usually have RF front-end blocks such as the LNA always on, which yields to higher power consumption specially in low duty cycle operation. There exist some techniques to switch these blocks off [8], but they need extra control. Due to the fact that the actual transmitted power is very low and the power consumption is dominated by the template generator and buffers, the power consumption in both the receiver and transmitter mode is the same. The power consumption of the transceiver for either receiving or transmitting mode is represented in Fig. 6.

III. MEASUREMENTS

The transceiver was fabricated in a $0.18\mu m$ 1P6M CMOS technology. A microphotography of the chip is shown in Fig. 7a. The layout of the transceiver core is depicted in Fig. 7b, covering an area of $0.11\,mm^2$. The complete system was characterized at different datarates with a pair of UWB antennas at $30\,cm$ distance.

Measurements of the power spectral density and time domain measurements of the transmitted and received

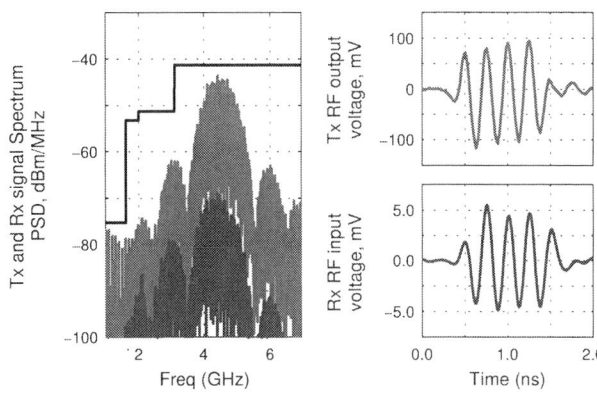

Fig. 8. Measurements of transmitted and received (a) PSD and (b) time domain waveforms.

pulses —at a center frequency of $4.48\,GHz$— are represented in Fig. 8a and Fig. 8b, respectively. The link attenuation is measured to be around $28\,dB$ in the whole band.

The datagram of the transmitted bits and demodulated pulses is shown in Fig. 9. The datarate is $560\,Mbps$ in IQ modulation, with an overall power consumption for the receiver and transmitter of $42\,mW$. This yields to an energy per bit of $75\,pJ/s$. The BER is better than 10^{-4}. The system was tested up to $1.2\,Gbps$ using a cable and a $12\,dB$ attenuator between the transmitter and the receiver. The power consumption of the transceiver was also measured for different datarates, as shown in Fig. 10. The center frequency can be swept from 3–$5\,GHz$ to accommodate different bands of operation and compensate for PVT effects.

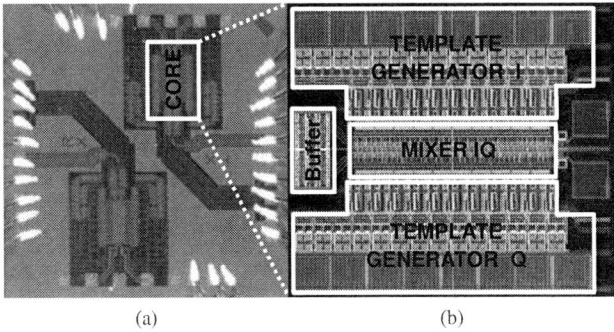

(a)   (b)

Fig. 7. Chip microphotography and core layout.

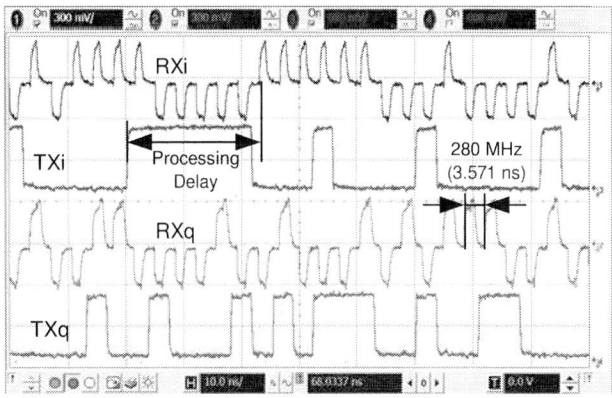

Fig. 9. Datagram measurements of the transmitted bits and demodulated signals for IQ modulation.

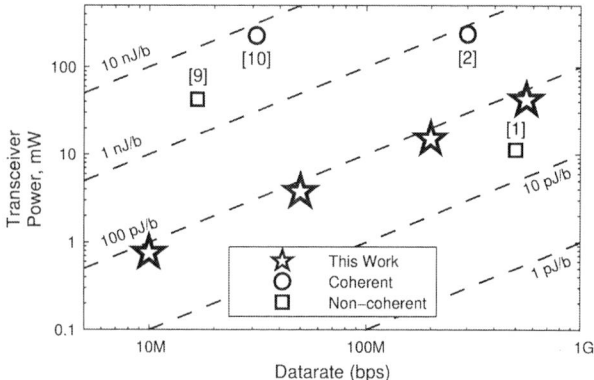

Fig. 10. Power, datarate and energy per bit comparative

IV. COMPARATIVE

As explained in section II there are several IR-UWB architectures proposed in the literature with significant differences. When compared to low speed transceivers the circuit proposed in this work performs much better in terms of efficiency even whether they're coherent [2] or not [9]. When compared to medium to high speed IR-UWB transceivers it's clearly better than any other coherent systems [10] and only a non-coherent system such as the one presented in [1] has better energy per bit figures. This comparative is illustrated in Fig. 10.

Besides a lower power, the presented transceiver has also a smaller area, mostly due to the fact that all the other transceiver use inductances. Moreover, the system in [1] lacks any frequency tuning control, making it difficult to calibrate or compensate for PVT effects. This comparative is summarized in Table I.

TABLE I
COMPARATIVE

Reference	[10]	[9]	[2]	[1]	This
Energy/bit (pJ/b)	7240	2540	787	22.6	**75.0**
Datarate ($Mbps$)	31.2	16.7	300	500	**560**
Power (mW)^a	226	42.5	236	11.3	**42**
Freq. (GHz)	3–9	3–5	4&8	1–2	**3–5**
Coherent	Yes	No	Yes	No	**Yes**
Area (mm^2)^b	4.50	2.28	16.7	0.36	**0.11**
Inductors	Yes	Yes	Yes	Yes	**No**
Freq Tuning	Yes	Yes	Yes	No	**Yes**
CMOS (nm)	180	90	180	180	**180**
Year	2008	2008	2007	2008	**2009**

^{a} Excluding buffers and post processing logic.

^{b} For half-duplex operation whenever possible.

V. CONCLUSION

We have presented a quadrature all-digital coherent IR-UWB transceiver in $0.18\,\mu m$ CMOS which can operate up to $560\,Mbps$ with an energy efficiency of just $75\,pJ/s$. It is based on a novel architecture that shares most of the RF front-end between the receiver and the transmitter resulting in a smaller area. The system performs better than any IR-UWB coherent transceiver found in the literature and only some non-coherent systems have lower energy per bit, but at the expense of performance.

ACKNOWLEDGMENTS

The authors would like to thank the Centre Tecnologic de Telecomunicacions de Catalunya for their support during the measurements and Agilent Technologies for providing measurement equipment.

This work has been partially supported by EU-FEDER funds, TEC2008-01856 project and AGAUR SGR 1497 funds.

REFERENCES

[1] M. Sasaki, "A 12-mW 500-Mb/s 1.8 μm CMOS pulsed UWB transceiver suitable for sub-meter short-range wireless communication," in *Proc. IEEE Radio Frequency Integrated Circuits Symposium RFIC 2008*, 2008, pp. 593–596.

[2] Y. Zheng, K.-W. Wong, M. Annamalai Asaru, D. Shen, W. H. Zhao, Y. J. The, P. Andrew, F. Lin, W. G. Yeoh, and R. Singh, "A 0.18 μm CMOS Dual-Band UWB Transceiver," in *Proc. Digest of Technical Papers. IEEE International Solid-State Circuits Conference ISSCC 2007*, 2007, pp. 114–590.

[3] "TransferJet White Paper 2009." [Online]. Available: http://www.transferjet.org/en/tj/transferjet_whitepaper.pdf

[4] W. Vereecken and M. Steyaert, "An I/Q based CMOS Pulsed Ultra Wideband Receiver Front End for the 3.1 to 10.6 GHz Band," in *Proc. IEEE 2006 International Conference on Ultra-Wideband*, 2006, pp. 191–194.

[5] FCC, "First Report and Order in The Matter of Revision of Part 15 of the Commission's Rules Regarding Ultra wideband Transmission Systems," *ET-Docket, FCC 02-48*, pp. 98–153, April 2002.

[6] H. Huang, H. Yin, G. Wei, and J. Zhu, "The structure and performance on an orthogonal sinusoidal correlation receiver of impulse radio," in *Vehicular Technology Conference, 2004. VTC2004-Fall. 2004 IEEE 60th*, vol. 2, Sept. 2004, pp. 1192–1196 Vol. 2.

[7] P. Heydari, "A study of low-power ultra wideband radio transceiver architectures," in *Wireless Communications and Networking Conference, 2005 IEEE*, vol. 2, March 2005, pp. 758–763 Vol. 2.

[8] E. Barajas, R. Cosculluela, D. Coutinho, M. Molina, D. Mateo, J. Gonzalez, I. Cairo, S. Banda, and M. Ikeda, "Low Noise Amplifiers for Low-Power Impulse-Radio Ultra Wide-Band Receivers," in *Ultra-Wideband, The 2006 IEEE 2006 International Conference on*, Sept. 2006, pp. 471–476.

[9] P. Mercier, D. Daly, M. Bhardwaj, D. Wentzloff, F. Lee, and A. Chandrakasan, "Ultra-low-power UWB for sensor network applications," in *Proc. IEEE International Symposium on Circuits and Systems ISCAS 2008*, 2008, pp. 2562–2565.

[10] Y. Zheng, M. Annamalai Arasu, K.-W. Wong, Y. J. The, A. Suan, D. D. Tran, W. G. Yeoh, and D.-L. Kwong, "A 0.18 μm CMOS 802.15.4a UWB Transceiver for Communication and Localization," in *Proc. Digest of Technical Papers. IEEE International Solid-State Circuits Conference ISSCC 2008*, 2008, pp. 118–600.

RMO3A-5

A 90nm-CMOS, 500Mbps, Fully-Integrated IR-UWB Transceiver Using Pulse Injection-Locking for Receiver Phase Synchronization

Changhui Hu,* Patrick Y. Chiang,* Kangmin Hu,* Huaping Liu,* Rahul Khanna,[†] and Jay Nejedlo [†]

*Oregon State University, Corvallis, OR 97330, USA

[†] Intel DEG group, Hillsboro, OR 97124, USA

Abstract— A fully-integrated, 3.1-5GHz Impulse-Radio UWB transceiver with on-chip flash ADC is designed in 90nm-CMOS. A new scheme for receiver phase acquisition is proposed that uses pulse injection-locking to synchronize the receive clock with the transmitted data, eliminating the need for clock/data recovery (CDR). Occupying $2mm^2$ die area, the transceiver achieves a maximum data rate of 500 Mbps, energy efficiency of 0.18nJ/b at 500Mbps, and a RX-BER of 10^{-3} across a distance of 10cm at 125Mbps.

Index Terms— Impulse radio (IR), UWB, equalization, injection-locking, transceiver.

I. INTRODUCTION AND MOTIVATION

Impulse-radio systems have received much attention as a possible architecture for UWB transceivers due to the large data bandwidth, low spectral interference with nearby channels. Due to the low emitted power spectral density of -41.3dBm/MHz, impulse radio is especially well suited for low cost, low power and short-range wireless communications[1-6].

A. Conventional Impulse-Radio Receiver Architectures

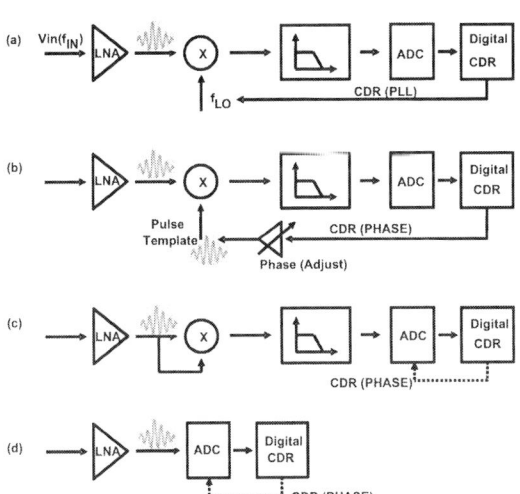

Fig. 1. RX architecture overview

While recent research has demonstrated energy-efficiency of impulse-based UWB transmitters [3], [4], the more critical problem lies within the receiver. Conventional IR-UWB receiver architectures are summarized

in Fig. 1: a) direct-conversion receiver using local oscillator multiplier; b) Direct-conversion receiver using pulse template multiplier [6]; c) non-coherent receiver with self-correlation [5]; d) direct over-sampling analog-digital converter (ADC)[7]. The LO direct-conversion architecture is the most common, and is similar to narrowband receiver systems. This approach typically consumes the most power, due to the high frequency of the local oscillator. In addition, due to the large data bandwidth, the low-pass filter is difficult to design, and ADC oversampling is still required to recover the optimal ADC sampling phase. In addition, a CDR is still required, as the local receiver clock may be plesiochronous from the transmitter clock. The pulse-template direct-conversion architecture is well used in coherent IR-UWB RX, but synchronization will be difficult and power consuming. The non-coherent, self-correlating receiver is an attractive option, as it simplifies the pulse-template synchronization. Unfortunately, bit-error rate will increase as the receiver will not be able to discriminate between noise and transmitted data. In addition, the design of a CDR loop is still required, as the demodulated data needs to be phase locked with the local receiver clock. The direct over-sampling ADC method is the most straightforward, as the pulse input is directly quantized by the ADC, moving the demodulation and CDR requirements to the digital baseband. Unfortunately, the power overhead for the over-sampling ADC is extremely expensive, as a multi-gigahertz, medium resolution ADC is necessary for a 3.1-10GHz transmitter bandwidth.

One overarching constraint of all of these conventional structures is that some mechanism for synchronizing the receiver sampling clock with the incoming transmitted data is required. Because the eventual goal for IR-UWB systems is several hundred Mbps, the design of an over-sampling CDR loop adds both system complexity as well as additional power consumption.

B. Proposed Architecture: Receiver Pulse Injection-Locking Phase Synchronization

In this work, we present a new receiver phase synchronization method using pulse injection-locking, as shown in Fig. 2. This technique provides several advantages over the previously described architectures. First, no CDR is necessary, as the received local oscillator is injection-locked

978-1-4244-6240-7/10 $26.00 © 2010 IEEE

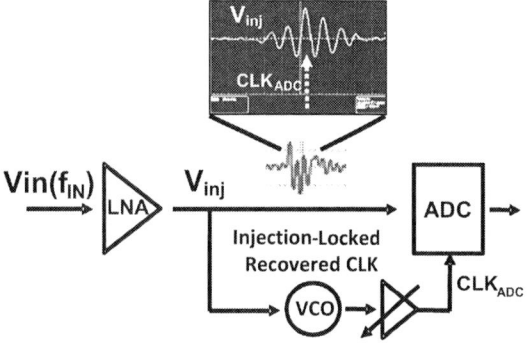

Fig. 2. RX_IL_VCO

to the incoming pulses, and hence is automatically phase-aligned with the transmitted clock. Second, the architecture is inherently a feed-forward system, with no issues with feedback loop stability. The proposed system is similar to a 'forwarded clock' receiver approach used for high-speed links which have been shown to be extremely energy-efficient[10], but here the receiver sampling clock is locked to the actual incoming transmitted pulses, eliminating any requirement for a separate clock channel. Third, since the receiver clock is now injection-locked and synchronized with the transmitter, the ADC sampling requirements can be severely relaxed and now can run at the actual data rate. This is a significant advantage for power reduction, as a multi-gigahertz, over-sampling ADC is no longer necessary.

II. SYSTEM OVERVIEW

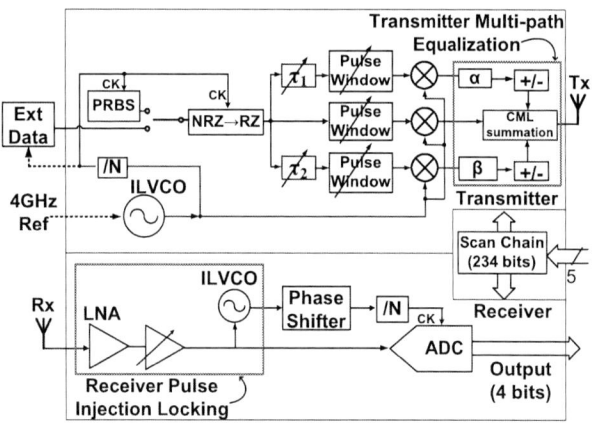

Fig. 3. IR-UWB transceiver architecture

The proposed IR-UWB transceiver is shown in Fig. 3, consisting of a multi-path equalization transmitter, a pulse-injection-locking receiver with an integrated ADC, an on-chip PRBS TX-generator and RX-checker, and a 234-bit scan chain for low-frequency calibration of DC control

bits such as current sources and resonant tank tuning. In the transmitter, OOK modulation is generated from a passive modulator, including a $2^{15}-1$ bit Pseudo-Random-Bit-Sequence (PRBS) selectable for testing operation. An on-die, 3-5GHz LC-VCO is clock-gated to generate the transmitted pulses, where a pulse shaping control block is integrated to enable tunable pulse widths of 0.4-10ns.

In the receiver, the received pulse is amplified by a 2-stage LNA and then directly injected into both the 5 level flash ADC and a 3.4-4.5GHz, injection-locked VCO (IL-VCO). After the receiver VCO is injection-locked and phase synchronized with the transmitted pulses, it is phase shifted and divided down to provide the ADC sampling clock. After the ADC sampling clock is divided down to the same frequency as the incoming data rate, the sampling clock is phase locked and aligned to the peak of the received input pulse, eliminating any requirements for baseband clock/data recovery. Optimally setting the phase position of the ADC sampling clock can be achieved by measuring the BER and building a bath-tub curve, sweeping through all possible phase positions. The 5-level flash ADC is designed using dynamic sense amplifiers with threshold-adjustable, current-steering DACs. The phase-shifter, which enables tunable phase delay of the ADC sampling clock, uses a Glibert-cell, current-summing DAC with a minimum step size of 7ps.

A. Multi-path equalization

For some serious multi-path environment, such as computer chassis, multi-path reflections can degrade the receiver BER. In order to reduce the interference from nearby reflection, a multi-path transmitter equalizer is designed that can reduce the two most severe multi-path reflections [8]. Tap1 and Tap2 are delayed version of the main signal, with sign and coefficient control, depending on the actual multi-path channel environment [8].

B. Receiver Pulse Injection-Locking

Receiver clock phase acquisition with the received UWB pulses is very critical for achieving low power consumption as discussed in the introduction. It consists of a 2-stage LNA and an ILVCO as in Fig. 4.

The proposed receiver clock recovery by using pulse injection-locking from the transmitted pulses, is analogous to sub-harmonic injection-locking. The effective division ratio N can be expressed as:

$$
N = \frac{f_{out}}{\alpha \cdot f_{inj} \cdot n} = \frac{f_{out}}{\alpha \cdot f_{inj} \cdot (W_{pulse}/T_{out})}
$$
$$
= \frac{1}{\alpha \cdot f_{inj} \cdot W_{pulse}} \tag{1}
$$

Where α is the possibility when data is "1"; f_{inj} is the data rate; f_{out} and T_{out} are the the ILVCO output

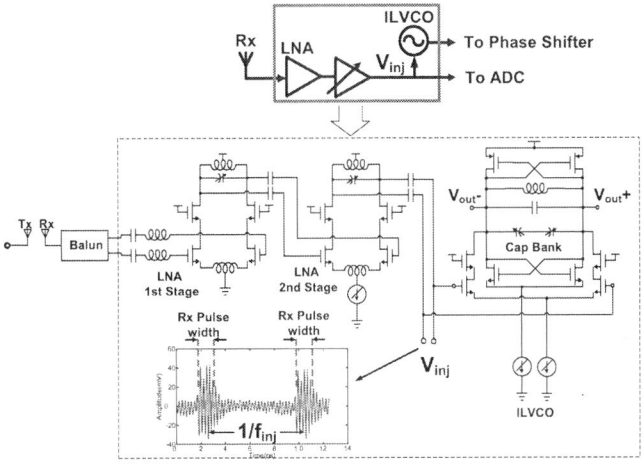

Fig. 4.   Receiver injection locking

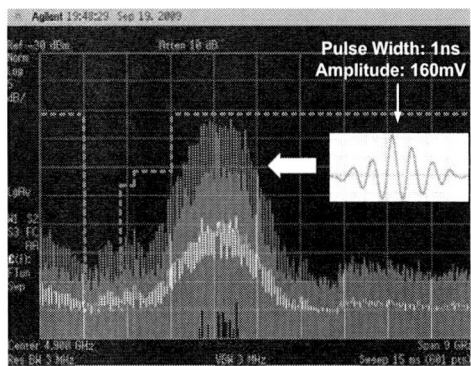

Fig. 6.   Transmitted signal and power spectrum

frequency and period; and W_{pulse} is the pulse width as in Fig. 4. Similar to [9], the phase noise will degrade $20logN$ dB compared to the injected signal. From Eq.(1), we can see that increase injection pulse rate or pulse width, will reduce the the phase noise of ILVCO output, because more energy injected.

III. MEASUREMENT RESULTS

The $2mm^2$ IR-UWB transceiver is built in 90nm-CMOS, 1.2V mixed-signal technology as shown in Fig. 5. The chip is mounted on a PCB using chip-on-board (COB) assembly with an off-chip, low-speed scan interface through a NIDAQ/Labview module.

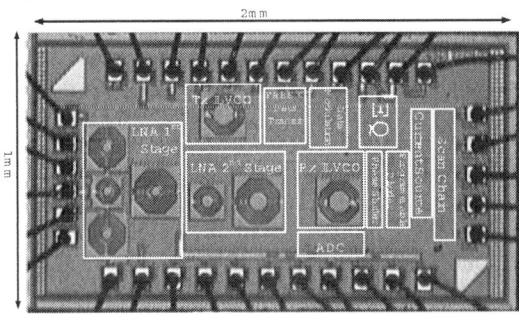

Fig. 5.   Die photo

The measured transmitted signal and its spectrum are shown in Fig. 6. The amplitude of the pulse is 160mVpp, with a nominal pulse width of 1ns. Its frequency spectrum fulfills the FCC UWB spectral mask except for the GPS band, which can be achieved by the addition of a high pass filter. The maximum transmission data rate is 500Mbps.

Fig. 7 shows the recovered IL-VCO clock locked to the LNA output, and after phase/data alignment of the pulse

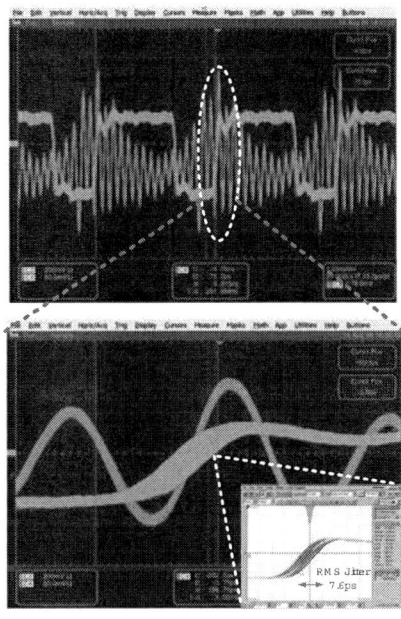

Fig. 7.   ADC clock and received signal alignment

zero crossing with the ADC sampling clock. With a 1ns pulse width, the recovered clock jitter is 7.6ps-RMS.

Fig. 8(a) shows the measured injection-locking range versus varying pulse width and pulse repetition rate. As can be seen, wider pulse width and higher data rate will improve the locking range, as more transmitted pulse energy synchronizes the receiver IL-VCO. Fig. 8(b) shows the measured close-in phase noise, from free running without injection, to pulse repetition frequencies (f_{inj}) 125Mbps, 500Mbps, and sine wave injection. Lower phase noise is exhibited at higher injection rates, as the phase updates occur at a higher frequency, similar to the dynamics in a 1st-order phase-locked loop (PLL). The results also verify Eq.(1), 12dB phase noise difference between 125Mbps and 500Mbps pulse injection is observed. Without pulse injection, the free running VCO shows very large

978-1-4244-6240-7/10 $26.00 © 2010 IEEE   181

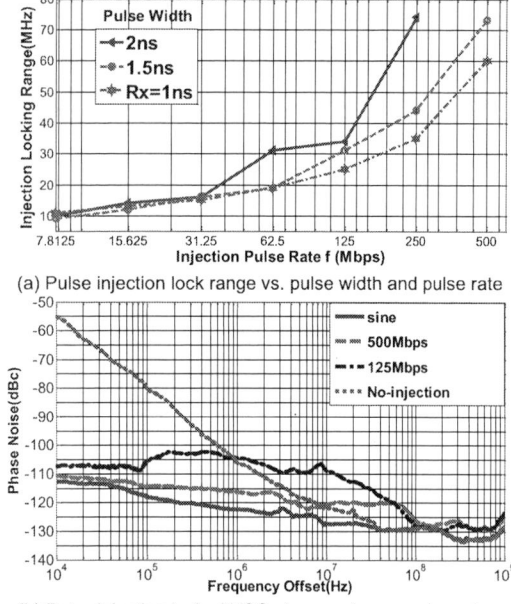

(a) Pulse injection lock range vs. pulse width and pulse rate

(b) Pulse injection locked VCO phase noise vs. pulse rate

Fig. 8. Injection locking measurement

phase noise at low frequency offset.

While a long-string of empty data transitions would result in loss of phase synchronization, conventional DC-balanced codes such as 8b/10b can limit maximum run length. Transmission using the on-chip PRBS-15 modulator, exhibiting a maximum string of fourteen zeros, showed no loss in receiver phase synchronization.

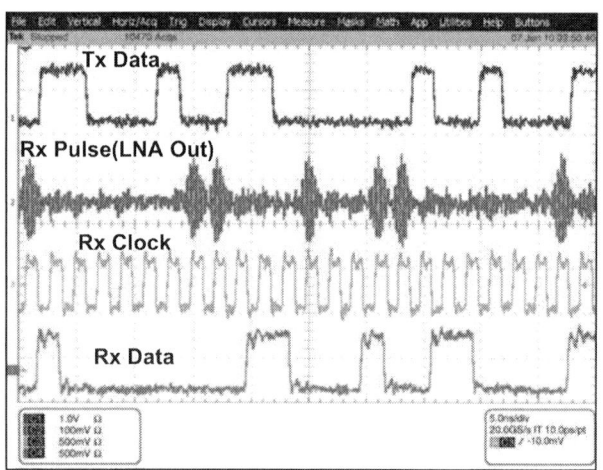

Fig. 9. Tx data, Rx clock, received pulse and recovered data(500Mbps)

The free-space measurement setup uses two UWB antennas that are placed 10cm away. Fig. 9 shows the transmitted digital data, received pulses after LNA gain, recovered Rx clock, and finally the received demodulated data at 500Mbps.

IV. CONCLUSION

A fully integrated single-chip low-power IR-UWB transceiver with ADC in 90nm CMOS is presented. A novel pulse-injection-locking method is used for receiver clock synchronization in receiver demodulation, leading to significant power reduction by eliminating the high power ADC and mixer. The complete transceiver achieves a maximum data rate of 500Mbps, through a 10cm distance, consuming 0.18nJ/bit. And measured BER achieved 10^{-3} at 125Mbps through 10cm.

TABLE I

PERFORMANCE SUMMARY AND COMPARISON

Work	CMOS	Frequency	Energy(pJ/b)		Data Rate	Size
	(nm)	(GHz)	Tx	Rx	(Mbps)	(mm^2)
[1]	180	3.1-9	740	6500	1000	4.5
[2]	130	0-0.96	-	110	40	4.52
[3]	130	3.1-5	1100		31	8
[4]	90	3.1-5	47	-	16.7	0.08(w/o pads)
[5]	90	3.1-5	-	2500	16.7	2.2
This	90	3.1-5	90	90	500	2

ACKNOWLEDGMENT

This work was funded by NSF-GOALI award #0901883. We thank Intel for gift donations, MOSIS for chip fabrication and Will Bettie for Labview support.

REFERENCES

[1] Y. Zheng, et al., "A 0.18μm CMOS 802.15.4a UWB Transceiver for Communication and Localization," *ISSCC Dig. Tech. Papers*, pp. 118-119, Feb. 2008.

[2] M. Verhelst, et al., "A Reconfigurable, 0.13μm CMOS 110pJ/pulse, Fully Integrated IR-UWB Receiver for Communication and Sub-cm Ranging," *ISSCC Dig. Tech. Papers*, pp. 250-251, Feb. 2009.

[3] D. Lachartre, et al., "A 1.1nJ/b 802.15.4a-Compliant Fully Integrated UWB Transceiver in 0.13μm CMOS," *ISSCC Dig. Tech. Papers*, pp. 312-313, Feb. 2009.

[4] D. Wentzloff, et al.,"A 47pJ/pulse 3.1-to-5GHz All-Digital UWB Transmitter in 90nm CMOS," *ISSCC Dig. Tech. Papers*, pp. 118-119, Feb. 2007.

[5] F. Lee, et al.,"A 2.5nJ/b 0.65V 3-to-5GHz Subbanded UWB receiver in 90nm CMOS," *ISSCC Dig. Tech. Papers*, pp. 116-117, Feb. 2007.

[6] L. Zhou, et al., "A 2Gbps RF-Correlation-Based Impulse-Radio UWB transceiver Front-End in 130nm CMOS," *IEEE RFIC Symp. Dig.,*, pp.65-68, June 2009.

[7] I.D.O'Donnell, et al., "A 23mW baseband impulse-uwb transceiver front-end,"*VLSI Circuits Symposium*, pp. 200, Jun. 2009.

[8] C. Hu, et al., "Transmitter Equalization for Multipath Interference Cancellation in Impulse Radio Ultra-Wideband(IR-UWB) Transceivers," *VLSI-DAT International Symposium*,pp.307-310, April 2009.

[9] J. Lee, et al., "Subharmonically Injection-Locked PLLs for Ultra-Low-Noise Clock Generation" *ISSCC Dig. Tech. Papers*, pp. 92-93, Feb. 2009.

[10] K.Hu, et al., "A 0.6mW/Gbps, 6.4-8.0Gbps Serial Link Receiver Using Local, Injection-Locked Ring Oscillators in 90nm CMOS"*VLSI Circuits Symposium*, pp. 46-47, Jun. 2009.

RMO3B-1

The "Load-Thru" (LT) De-embedding Technique for the Measurements of mm-Wave Balanced 4-Port Devices

Zhiming Deng, Ali M. Niknejad

Berkeley Wireless Research Center, University of California at Berkeley, CA

Abstract—**The differential-mode behavior of a balanced 4-port device can be characterized by simple 2-port measurements if baluns are placed at both the input and the output. But the traditional insertion loss technique is not able to fully de-embed the baluns. Therefore, we propose the "load-thru" de-embedding technique which uses the differential-mode characteristics of a balun to fully extract the complete differential-mode behavior of the DUT. Theoretical analysis and mm-wave measurement verifications are provided.**

Index Terms—**De-embedding, mixed-mode, differential, balun, balanced, 4-Port.**

I. INTRODUCTION

In SOC designs, differential signaling has well-known advantages such as superior noise immunity. Unfortunately, characterization of a 4-port device, especially at mm-wave frequencies, is not simple. An accurate and complete characterization requires a dual-source VNA that can generate puremode drives [1], and this is very costly. However, if only the differential-mode (DM) behavior is of interest, on-chip baluns can be placed at both the input and the output of a device-under-test (DUT) as common-mode (CM) blockers and the differential-mode behavior can be investigated by 2-port measurements. The traditional "insertion loss" (IL) technique is usually used to de-embed the baluns, or more precisely, to compensate the insertion loss introduced by the two baluns. The IL technique needs only one extra de-embedding structure wherein two baluns are back-to-back connected. Though simple, the IL technique is strictly restricted to the scenario where all connection ports are power matched.

In this work, we will propose the "load-thru" (LT) de-embedding technique that can fully characterize a balun so that the complete differential-mode 2-port parameters of a symmetric 4-port DUT can be extracted. The paper is organized as follows. In section II, we introduce the procedure of the LT de-embedding technique and give a theoretical analysis. In section III, we discuss practical design considerations that can make the de-embedded results more accurate. Finally in section IV, measurement results of a mm-wave differential amplifier are used to verify the theory.

II. DE-EMBEDDING THEORY

A. DM and CM Separation

The signaling in all of our measurement setups is represented in mixed-mode parameters [2]. The proposed LT de-embedding technique relies on the condition that the DM and the CM operations are completely separated.

The most critical element, a balun, is a 3-port device. Let port "1" denote the unbalanced port connected to the pad and port "2" and "3" denote the balanced ports connected to the DUT. Then the signals at port "2" and "3" can be separated to DM and CM signals by introducing two linear transformation matrices, $\mathbf{K}_V^B$ and $\mathbf{K}_I^B$:

$$
\begin{bmatrix} V_1 \\ V_d \\ V_c \end{bmatrix} = \underbrace{\begin{bmatrix} 1 & 0 & 0 \\ 0 & 1 & -1 \\ 0 & \frac{1}{2} & \frac{1}{2} \end{bmatrix}}_{\mathbf{K}_V^B} \begin{bmatrix} V_1 \\ V_2 \\ V_3 \end{bmatrix}, \tag{1}
$$

$$
\begin{bmatrix} I_1 \\ I_d \\ I_c \end{bmatrix} = \underbrace{\begin{bmatrix} 1 & 0 & 0 \\ 0 & \frac{1}{2} & -\frac{1}{2} \\ 0 & 1 & 1 \end{bmatrix}}_{\mathbf{K}_I^B} \begin{bmatrix} I_1 \\ I_2 \\ I_3 \end{bmatrix}. \tag{2}
$$

Correspondingly, the balun can be characterized by the mixed-mode parameters. If Y-parameters are used, one can derive $\mathbf{Y}^{B(m)} = \mathbf{K}_I^B \mathbf{Y}^B (\mathbf{K}_V^B)^{-1}$:

$$
\begin{aligned}
\mathbf{Y}^{B(m)} &= \begin{bmatrix} Y_{11}^R & Y_{1d}^R & \boxed{Y_{1c}^R} \\ Y_{d1}^B & Y_{dd}^B & \boxed{Y_{dc}^B} \\ \boxed{Y_{c1}^B} & \boxed{Y_{cd}^B} & Y_{cc}^B \end{bmatrix} \\
&= \begin{bmatrix} \mathbf{Y}^{B(d)} & \\ & Y_{cc}^B \end{bmatrix}. \tag{3}
\end{aligned}
$$

For an ideal balun, the 4 boxed terms of $\mathbf{Y}^{B(m)}$ will vanish and $\mathbf{Y}^{B(m)}$ becomes block-diagonal. The CM port is then independent of other ports. As shown in Fig. 1 (a), a mixed-mode model of an ideal balun consists of a 2-port network and an isolated 1-port network

A similar transformation can also be applied to a 4-port device. Both the input signals at port "1" and "2", and the output signals at port "3" and "4", are separated to DM and CM signals. The voltage and current transformation matrices

978-1-4244-6240-7/10 $26.00 © 2010 IEEE   183

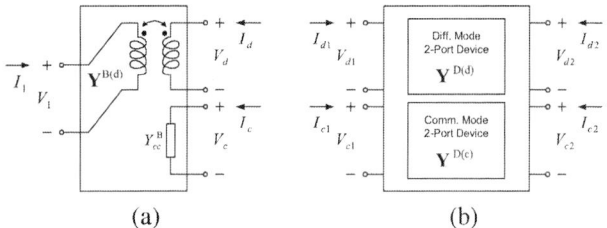

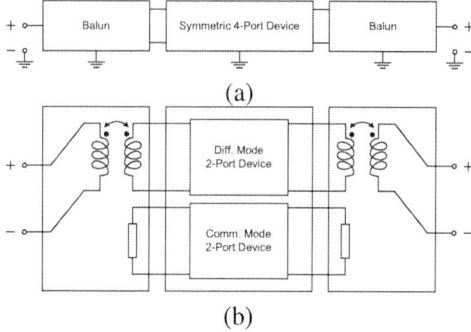

Fig. 1. The mixed-mode models of (a) an ideal balun and (b) a balanced 4-port device.

Fig. 2. (a) The measurement setup of a symmetric 4-port device using ideal baluns. (b) The equivalent mixed-mode model of the setup.

are $\mathbf{K}_V^{\mathrm{D}}$ and $\mathbf{K}_I^{\mathrm{D}}$ respectively.

$$
\begin{bmatrix} V_{d1} \\ V_{d2} \\ V_{c1} \\ V_{c2} \end{bmatrix} = \underbrace{\begin{bmatrix} 1 & -1 & 0 & 0 \\ 0 & 0 & 1 & -1 \\ \frac{1}{2} & \frac{1}{2} & 0 & 0 \\ 0 & 0 & \frac{1}{2} & \frac{1}{2} \end{bmatrix}}_{\mathbf{K}_V^{\mathrm{D}}} \begin{bmatrix} V_1 \\ V_2 \\ V_3 \\ V_4 \end{bmatrix}. \tag{4}
$$

$$
\begin{bmatrix} I_{d1} \\ I_{d2} \\ I_{c1} \\ I_{c2} \end{bmatrix} = \underbrace{\begin{bmatrix} \frac{1}{2} & -\frac{1}{2} & 0 & 0 \\ 0 & 0 & \frac{1}{2} & -\frac{1}{2} \\ 1 & 1 & 0 & 0 \\ 0 & 0 & 1 & 1 \end{bmatrix}}_{\mathbf{K}_I^{\mathrm{D}}} \begin{bmatrix} I_1 \\ I_2 \\ I_3 \\ I_4 \end{bmatrix}. \tag{5}
$$

The original 4-port parameters, $\mathbf{Y}^{\mathrm{D}}$, can be converted to the mixed-mode parameters by $\mathbf{Y}^{\mathrm{D(m)}} = \mathbf{K}_I^{\mathrm{D}} \mathbf{Y}^{\mathrm{D}} (\mathbf{K}_V^{\mathrm{D}})^{-1}$ which gives

$$
\mathbf{Y}^{\mathrm{D(m)}} = \begin{bmatrix} Y_{d1d1}^{\mathrm{D}} & Y_{d1d2}^{\mathrm{D}} & \boxed{Y_{d1c1}^{\mathrm{D}}} & \boxed{Y_{d1c2}^{\mathrm{D}}} \\ Y_{d2d1}^{\mathrm{D}} & Y_{d2d2}^{\mathrm{D}} & \boxed{Y_{d2c1}^{\mathrm{D}}} & \boxed{Y_{d2c2}^{\mathrm{D}}} \\ \boxed{Y_{c1d1}^{\mathrm{D}}} & \boxed{Y_{c1d2}^{\mathrm{D}}} & Y_{c1c1}^{\mathrm{D}} & Y_{c1c2}^{\mathrm{D}} \\ \boxed{Y_{c2d1}^{\mathrm{D}}} & \boxed{Y_{c2d2}^{\mathrm{D}}} & Y_{c2c1}^{\mathrm{D}} & Y_{c2c2}^{\mathrm{D}} \end{bmatrix}
$$
$$
= \begin{bmatrix} \mathbf{Y}^{\mathrm{D(d)}} & \\ & \mathbf{Y}^{\mathrm{D(c)}} \end{bmatrix}. \tag{6}
$$

For a perfectly symmetric 4-port device, port "1" is symmetric with port "2" and port "3" is symmetric with port "4". The 8 boxed terms of $\mathbf{Y}^{\mathrm{D(m)}}$ equal zero which means the DM and the CM signals are separable. In Fig. 1 (b), we show the mixed-mode model of a balanced 4-port device comprised of one DM 2-port network and one CM 2-port network.

B. De-embedding Formula

Under the condition that the two modes are separable for both the balun and the DUT, we consider the 2-port measurement setup in Fig. 2 (a) and depict its mixed-mode model in Fig. 2 (b). From the model, one can see that the CM operation of the DUT is completely suppressed since all external stimuli are converted to DM signals only. The setup is simplified to three cascaded 2-port networks. Suppose we have known the DM 2-port parameters of a balun, the DM $ABCD$-parameters $\mathbf{A}^{\mathrm{D(d)}}$ of the DUT can be de-embedded

from the measured data $\mathbf{A}^{\mathrm{M}}$ using

$$
\mathbf{A}^{\mathrm{D(d)}} = (\mathbf{A}^{\mathrm{B(d)}})^{-1} \mathbf{A}^{\mathrm{M}} \begin{bmatrix} 1 & 0 \\ 0 & -1 \end{bmatrix} \mathbf{A}^{\mathrm{B(d)}} \begin{bmatrix} 1 & 0 \\ 0 & -1 \end{bmatrix}. \tag{7}
$$

$\mathbf{A}^{\mathrm{B(d)}}$ is the DM $ABCD$-parameters of a balun and we have used the fact that the second balun is mirror symmetric with the first one.

C. Characterization of the Balun

The de-embedding equation (7) requires that $\mathbf{A}^{\mathrm{B(d)}}$ is known. Shown in Fig. 3, we propose a measurement setup to characterize the DM behavior of a balun wherein Z_T is supposed to be a known impedance. $\mathbf{A}^{\mathrm{B(d)}}$ includes three unknowns, but one measure only provides two independent equations considering the two ports are symmetric. Therefore, we need at least two measurements with different values of Z_T.

The first measurement uses a finite Z_T and it is called the "load" structure. The measured data are

$$
\mathbf{Y}^{\mathrm{L}} = \begin{bmatrix} Y_{11}^{\mathrm{L}} & Y_{12}^{\mathrm{L}} \\ Y_{21}^{\mathrm{L}} & Y_{22}^{\mathrm{L}} \end{bmatrix}. \tag{8}
$$

The second measurement uses a trivial setup, $Z_T = \infty$. In fact, the structure degenerates to two back-to-back connected baluns and it is called the "thru" structure. The measured data are

$$
\mathbf{Y}^{\mathrm{T}} = \begin{bmatrix} Y_{11}^{\mathrm{T}} & Y_{12}^{\mathrm{T}} \\ Y_{21}^{\mathrm{T}} & Y_{22}^{\mathrm{T}} \end{bmatrix}. \tag{9}
$$

With (8) and (9), we are able to derive analytical expressions for the elements of the DM Y-parameters of a balun, $\mathbf{Y}^{\mathrm{B(d)}}$.

$$
Y_{11}^{\mathrm{B}} = Y_{11}^{\mathrm{L}} - Y_{12}^{\mathrm{L}}. \tag{10}
$$

$$
Y_{dd}^{\mathrm{B}} = \frac{Y_{12}^{\mathrm{L}}}{4 Z_T (Y_{12}^{\mathrm{T}} - Y_{12}^{\mathrm{L}})}. \tag{11}
$$

$$
Y_{d1}^{\mathrm{B}} = Y_{1d}^{\mathrm{B}} = -\sqrt{\frac{Y_{12}^{\mathrm{L}} Y_{12}^{\mathrm{T}}}{2 Z_T (Y_{12}^{\mathrm{T}} - Y_{12}^{\mathrm{L}})}}. \tag{12}
$$

In (12), the square root operation selects the root that has a non-negative real part. Therefore, $\mathbf{A}^{\mathrm{B(d)}}$ can be obtained by converting the calculated Y-parameters to $ABCD$-parameters.

978-1-4244-6240-7/10 $26.00 © 2010 IEEE

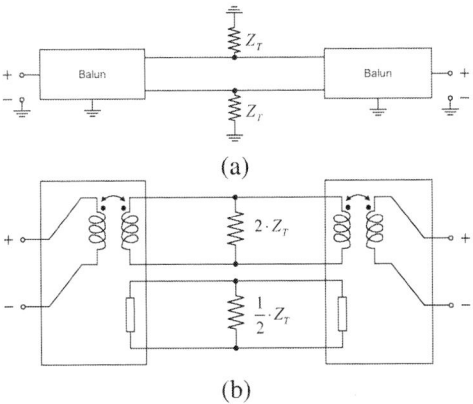

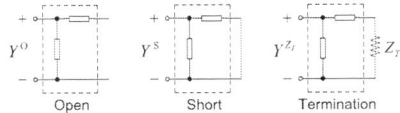

Fig. 3. (a) The measurement setup of two back-to-back baluns with termination loads Z_T in the middle. (b) The equivalent mixed-mode model of the setup.

Fig. 4. The setup for the characterization of Z_T using 1-port open-short de-embedding. The lumped-element model for the de-embedding parasitics are depicted inside the dashed-line box.

D. Characterization of Z_T

The last question is how to characterize Z_T. This can be accomplished by using the 1-port version of the "open-short" de-embedding technique [3]. The three required structures are shown in Fig. 4 and the 1-port measurement results for the "open", "short" and "termination" structures are denoted by Y^{O}, Y^{S} and Y^{Z_T} respectively. The formula to calculate Z_T is

$$Z_T = \frac{1}{Y^{Z_T} - Y^{\mathrm{O}}} - \frac{1}{Y^{\mathrm{S}} - Y^{\mathrm{O}}}. \tag{13}$$

III. DESIGN CONSIDERATION

The above analysis has assumed an ideal balun whose CM port is isolated. For a real balun, signals at the unbalanced port and the DM port can leak to the CM port. This is called mode-conversion. however we can optimize the balun design so that mode-conversion is limited to a satisfying level. The requirement for mode-conversion depends on the CM gain of the DUT. To investigate the mode-conversion level of a balun, we need to study its mixed-mode S-parameters $\mathbf{S}^{\mathrm{B(m)}}$:

$$\mathbf{S}^{\mathrm{B(m)}} = \begin{bmatrix} S_{11}^{\mathrm{B}} & S_{1d}^{\mathrm{B}} & S_{1c}^{\mathrm{B}} \\ S_{d1}^{\mathrm{B}} & S_{dd}^{\mathrm{B}} & S_{dc}^{\mathrm{B}} \\ S_{c1}^{\mathrm{B}} & S_{cd}^{\mathrm{B}} & S_{cc}^{\mathrm{B}} \end{bmatrix}. \tag{14}$$

The mode-conversion level is evaluated by calculating $\left| \frac{S_{c1}^{\mathrm{B}}}{S_{d1}^{\mathrm{B}}} \right|$ and $\left| \frac{S_{cd}^{\mathrm{B}}}{S_{d1}^{\mathrm{B}}} \right|$. Notice that the computation of the S-parameters relies on the selection of port impedance [4]. Therefore, the port impedance needs to reflect the real port connection situation which can be different from that of a standard 50 Ω system.

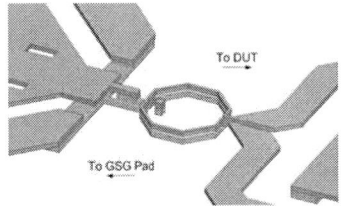

Fig. 5. The 3-D balun structure.

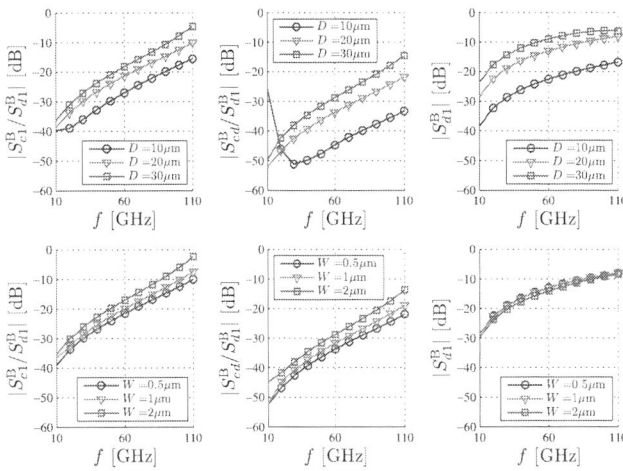

Fig. 6. The simulated mode-conversion level $\left| \frac{S_{c1}^{\mathrm{B}}}{S_{d1}^{\mathrm{B}}} \right|$, $\left| \frac{S_{cd}^{\mathrm{B}}}{S_{d1}^{\mathrm{B}}} \right|$ and the insertion gain S_{d1}^{B} of baluns with different diameters (fixed $W = 0.5\mu$m) and different widths (fixed $D = 20\mu$m).

Fig. 5 shows the 3-D structure of the balun that is used in our prototype design. The two coils use different thick metal layers and they completely overlap to maximize magnetic coupling. The metal overlapping also results in undesirable capacitive coupling, a source of mode conversion. There are two adjustable geometric parameters for the balun: the coil diameter and the trace width. To select the best design, various baluns are simulated using a 3-D EM simulation tool. Their the mode conversion levels are plotted and compared in Fig. 6. In general, a balun with smaller diameter and narrower width introduces less mode conversion but it suffers from higher insertion loss, equivalently, lower $|S_{d1}^{\mathrm{B}}|$. High insertion loss can make the whole structure sensitive to noise either from measurement operations or instruments. Hence, a good balun design is required to balance the two requests.

IV. MEASUREMENT VERIFICATION

A 60 GHz differential amplifier has been fabricated in a 65nm LP CMOS technology to demonstrate the LT de-embedding technique. The chip micrograph is shown in Fig. 7 (a). The input and output matching networks applies single-stub matching method using conventional 75 Ω CPW transmission lines. Two baluns in mirror symmetric are placed at the input and output of the amplifier. The geometric parameters of the baluns are: $D = 20\mu$m and $W = 0.5\mu$m. According to the simulation data in Fig. 6, $\left| \frac{S_{c1}^{\mathrm{B}}}{S_{d1}^{\mathrm{B}}} \right|$ and $\left| \frac{S_{c1}^{\mathrm{B}}}{S_{d1}^{\mathrm{B}}} \right|$ are -20 dB and -35

978-1-4244-6240-7/10 $26.00 © 2010 IEEE   185

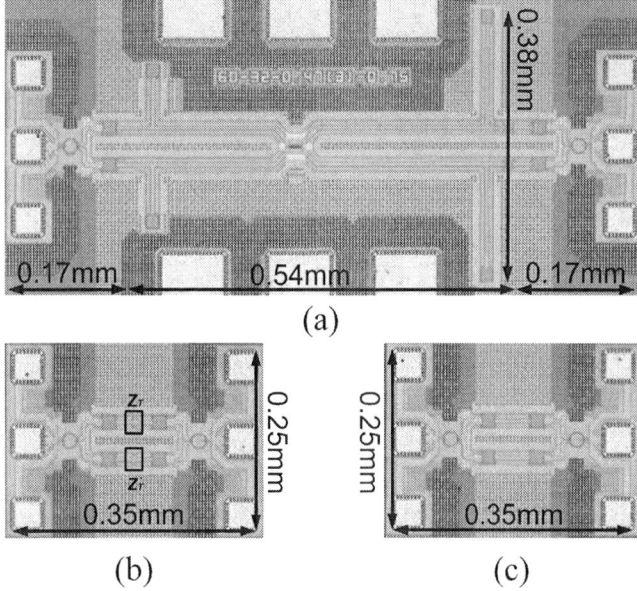

Fig. 7. The micrograph of (a) the 60 GHz differential amplifier with input and output baluns, (b) the "load" de-embedding structure and (c) the "thru" de-embedding structure.

dB respectively, and $|S_{d1}^{B}|$ is -12 dB. The chip micrograph of the "load" and the "thru" de-embedding structures are shown in Fig. 7 (b) and (c). The connecting points from a balun to the DUT also use the same kind of CPW lines to avoid discontinuity. The load impedance Z_T is implemented using a polyresistor with a value of about 80 Ω.

In order to verify the validity of the LT de-embedding technique, the same amplifier with the input and the output port connected to GSSG pads is also fabricated on the same wafer. The pads are modeled and embedded into the matching networks. It is measured by a balun probe to obtain its DM behavior directly. These direct measurement data are compared to the de-bedded data. In Fig. 8, the S-parameters obtained from both methods are plotted together. Both magnitudes and phases show good agreement over wide frequency range.

V. CONCLUSION

We have proposed the LT de-embedding technique for the characterization of the DM performance of balanced 4-port devices. It requires total of 5 de-embedding structures besides the DUT structure. The de-embedding procedure can be summarized in 3 steps:

1) : Measure the 1-port de-embedding structures and extract Z_T using (13).

2) : Measure the "load" and the "thru" structure. Compute $\mathbf{Y}^{B(d)}$ using (10)-(12) and convert it to $\mathbf{A}^{B(d)}$.

3) : Measure the DUT structure to obtain $\mathbf{A}^{M}$, then compute $\mathbf{A}^{D(d)}$ according to (7).

ACKNOWLEDGMENT

The authors acknowledge ST Microelectronics for the chip fabrication. In particular, the authors would like to thank Andreia Cathelin and Daniel Gloria of ST Microelectronics,

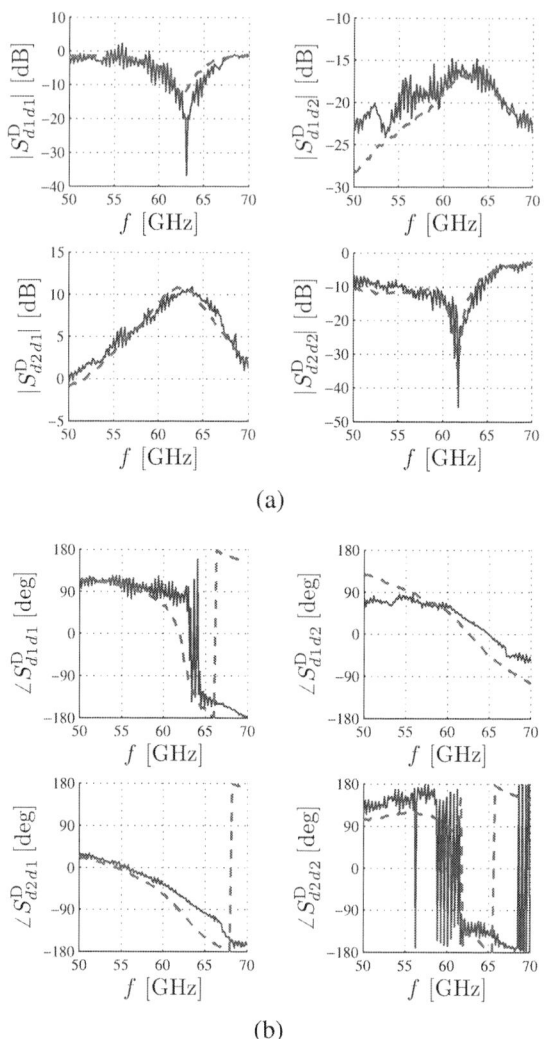

Fig. 8. The de-embedded differential-mode S-parameters of a differential amplifier using the LT method (solid line) are compared with the directly measured data by using a balun probe (dashed line). (a) Magnitudes and (b) phases.

Joel Dunsmore and Suren Singh of Agilent Technologies for their support and guidance, and Peter Hannaway of Cascade Microtech for access to balun probes.

REFERENCES

[1] J. Dunsmore, "New methods & non-linear measurements for active differential devices," *Microwave Sym. Digest, 2003 IEEE MTT-S*, vol. 3, pp. 1655-1658, June 2003.
[2] D. E. Bockelman, *et al*, "Combined differential and common-mode scattering parameters," *IEEE Trans. Microwave Theory Tech.*, vol. 43, pp. 1530-1539, July 1995.
[3] M. C. A. M. Koolen *et al*, "An improved de-embedding technique for on-wafer high-frequency characterization," *Proc. IEEE Bipolar/BiCMOS Circuits and Technology Meeting*, pp. 188-191, September 1991
[4] K. Kurokawa, "Power Waves and the Scattering Matrix," *IEEE Trans. Microwave Theory Tech.*, vol. 13, issue 2, pp. 194-202, March 1965

RMO3B-2

RF-pad, Transmission Lines and Balun Optimization for 60GHz 65nm CMOS Power Amplifier

Sofiane Aloui [#], Eric Kerherve [#], Robert Plana [*] and Didier Belot [**]

[#]University of Bordeaux, IMS Laboratory, bordeaux, France, sofiane.aloui, eric.kerherve@ims-bordeaux.fr
[*]LAAS-CNRS, Toulouse, France, plana@laas.fr
[**]STMicroelectronics, Central R&D, Crolles, France, didier.belot@st.com

Abstract—**Design and optimization of** $65nm$ **CMOS passive devices which are used in the implementation of a** $60GHz$ **Power Amplifier (PA) are presented. The targeted application is the low cost Wireless Personal Area Network (WPAN). A new optimized Radio Frequency (RF)-pad is used to minimize the losses of the PA access. The PA is matched via balun and Transmission Lines (T-Lines). T-Lines ensure both broadband inter-stage matching and biasing. S-parameters and large signal measurement results are demonstrated and compared with electromagnetic simulations. The PA achieves a maximum output power** P_{sat} **of** $7.3dBm$ **with a gain of** $8.5dB$ **while consuming** $96mA$ **from a** $1.2V$ **supply. The active die area of the chip is** $0.065mm^2$**. Additionally, innovative technique is adopted in the balun design to improve the balanced-to-unbalanced mode conversion and PA performances.**

Index Terms— **60GHz ,RF-pads, transmission lines, millimeter wave power amplifiers, baluns.**

I. INTRODUCTION

Standars based on Bluetooth or WiFi have reached the maturity in terms of performances and marketplace. Moreover, they are not able to ensure gigabit wireless communication data rate. IEEE 802.15.3C standard occupies a free $7GHz$ unlicensed band around $60GHz$ to support faster and safer link [1]. This application targets a large scale market and a large volume production. Therefore, a cost constraint is imposed. In this context, a $65nm$ bulk CMOS technology ($\rho = 20\Omega.cm$) is chosen to design the PA.

STMicroelectronics provides the $65nm$ CMOS technology with 7 metal layers and an Alucap layer. Low power transistors are the standard active devices. They offer a high unity current gain frequency (f_t=160GHz) and a high maximum oscillation frequency (f_{max}=200GHz).

The challenge in CMOS technology design is to take into account a maximum of high frequency considerations as metal and substrate losses, return current losses, tee and L junctions, access to transistor, pad capacitor. This work focuses on passive devices characterization and their optimization. This paper is organized in three sections.

1) Section 1 details the impact of RF-pad at millimeter-wave (mmW) frequencies. Both traditional and optimized RF-pads are presented and measured. Electrical characteristics of Co-Planar Waveguide (CPW) T-lines and tee junction are highlighted.

2) Section 2 presents the experimental results of a $65nm$ CMOS $60GHz$ fully integrated PA. Lumped and distributed passive elements are used. The PA reaches high gain and P_{sat}. The PA performances are compared with PAs [2]-[3] designed with the same technology.

3) Section 3 discusses about the balun design which is used as a Distributed Active Transformer (DAT) at the input and the output of the PA. Generally, three-port mixed-mode S-parameters characterize the balun performances. In this paper, electromagnetic simulations are performed to visualize the electric and magnetic fields amount which are responsible for the voltage and current crossing over the balun. Finally, a high performance balun using an innovative methodology design is presented.

II. PASSIVE DEVICES

A. RF-Pad Optimization

Most of today's pad realizations are composed by the top metals (Metal 6 or Metal 7 and Alucap) and by the bottom metals (Metal 1 and Metal 2) representing the solid metal shield. The main motivation of this design is the reduction of the substrate coupling and the improvement of the quality factor. This technique suffers from capacitive coupling losses because of the low thickness of the Back End Of Line (BEOL) in the emerging CMOS technology [4].

Here, an investigation is brought to determine the impact of the substrate in the RF-pad design. The bottom metals are

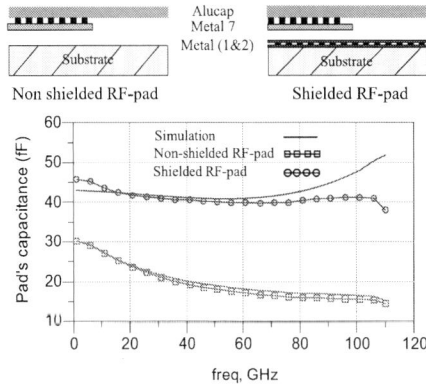

Fig. 1. Measured and simulated capacitance of RF-pads

suppressed and the RF-pad is exposed to the silicon substrate to support that characterization. Additionally, only Metal 7 and Alucap are used as a top plate to obtain the minimum capacitance. Fig. 1 depicts the two octagonal contact pads. The pad pitch is $100\mu m$.

The two pads are fabricated and measured from DC to $110GHz$. The intrinsic capacitance (C_p) of the shielded (Sh) RF-pad varies from $45fF$ to $40fF$ (11%). Whereas, the C_p of the non-shielded (NSh) RF-pad varies from $30fF$ to $15fF$ (50%). Nevertheless, for frequencies beyond $55GHz$ the NSh RF-pad presents almost the same C_p value and is equal to $16fF$ (cf. Fig. 1). This pad has a quasi-constant capacitance and can be used for the most important mmW applications which are centred at $60GHz$, $77GHz$ and $94GHz$.

Investigating to lower C_p has an important effect in the design of PA, LNA or VCO. For instance, at $60GHz$, the input and the output impedances are deviated from 50Ω to $Z_{1pad} = (42 - j*9)\Omega$ for the NSh RF-pad and is drastically deviated to $Z_{2pad} = (15 - j*13)\Omega$ for the Sh RF-pad (cf. Fig. 2). In fact, the impedance deviated by the Sh RF-pad is closer to the input impedance of the transistor (around 10Ω) than the impedance deviated by the NSh RF-pad. This behavior is advantageous for the matching in the design of single-function circuit. Nevertheless, the ultimate goal of mmW applications is to design fully integrated transceivers. Consequently, high-loss passive devices are required and must be added for each circuit to match every stage to 50Ω in the transceiver design. That drops the total gain of the transceiver.

The insertion loss is shown in Fig. 2 to quantify the effect of the mismatching caused by the RF-pad. At $80GHz$, when the input impedance $Z_{in} = 50\Omega$, the attenuation reaches $1.3dB$ for the Sh RF-pad and only $0.7dB$ for the NSh RF-pad. If $Z_{in} = Z_{in}^*$ (conjugate impedance), the NSh RF-pad causes more attenuation ($0.5dB$ vs. $0.3dB$). This result is expected because of the substrate losses which increase more rapidly than the conductor losses at high frequencies.

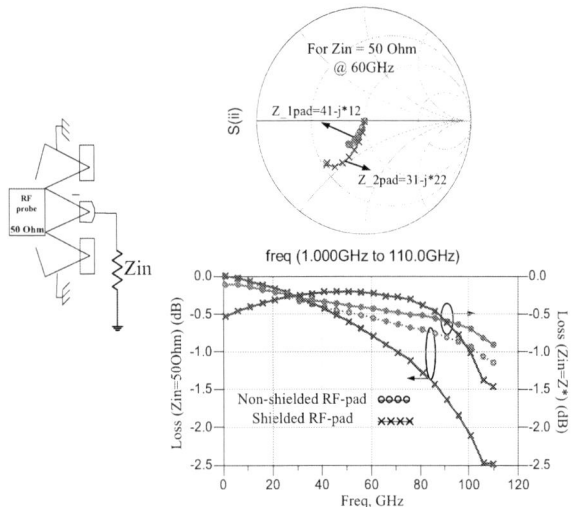

Fig. 2.   Insertion loss of the SH and NSh RF-pads

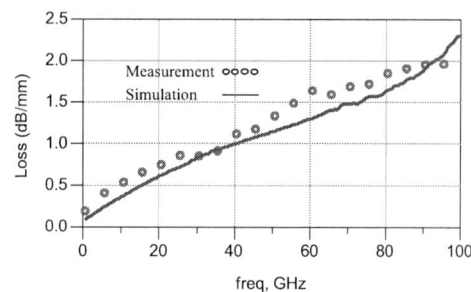

Fig. 3.   Measured losses of a 50Ω CPW line

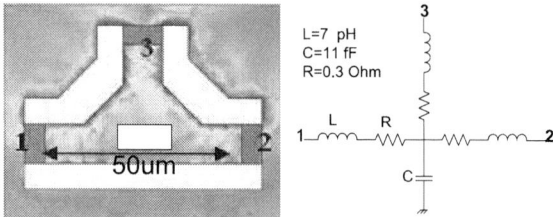

Fig. 4.   Tee junction current distribution and lumped model

B. CPW Characterization

The thin **BEOL** of CMOS technology makes hard a realization of high characteristic impedance (Z_c) line with microstrip line. Consequently, CPW line is chosen for our design. To obtain a $Z_c = 50\Omega$, the width (w) and the gap (g) separating the RF line and the adjacent ground plan are set to $(w, g) = (10\mu m, 6.5\mu m)$. The CPW T-Line exhibits an attenuation of $1.6dB/mm$ at $60GHz$ (cf. Fig. 3).

C. Tee Junction Characterization

he use of T-Line for biasing and matching leads the RF signal crossing over this line. Consequently, the tee junction is considered in the design of the PA to avoid frequency shift due to tee junction parasitic capacitance. Fig. 4 confirms that the current flows mainly on the corners. This phenomenon is amplified by the skin effect which is responsible of the high current density concentrated on the edges of the line. A part of the metal in the center of the tee junction is cut in order to decrease the parallel capacitance. The tee junction has an inductive effect with serial resistive losses and a capacitance in parallel at the middle of the junction. This model is taken into account in the simulation schematic.

III. PA DESIGN AND MEASUREMENT RESULTS

A. PA Design Description

The schematic and the die photography of the PA are depicted in Fig. 5. Two different stacked baluns are placed at the input and the output of the circuit to ensure simultaneously matching and single-to-differential conversion. The common source structure is chosen for its high voltage excursion. The power transistor has a width of $90\mu m$ and a width of $48\mu m$ for the driver stage. The biasing is performed by the CPW transmission line which ensures a broadband inter-stage matching. The transistors are driven to operate in class AB

978-1-4244-6240-7/10 $26.00 © 2010 IEEE   188

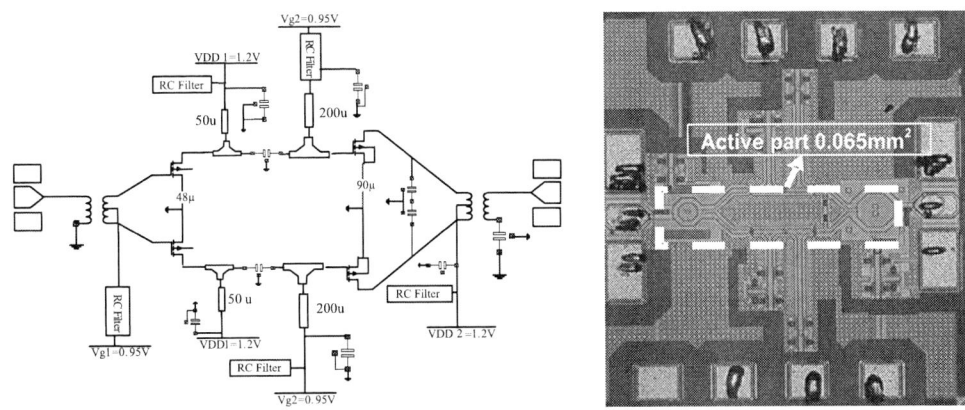

Fig. 5. Schematic and die photography of the fabricated $60GHz$

TABLE I

COMPARISON OF PAS IN 65NM CMOS TECHNOLOGY FOR $60GHz$ WPAN STANDARD

Reference Frequency	[Tech(nm)] [Foundry]	Number of stages	Gain (dB)	P_{sat} (dBm)	OCP_1 (dBm)	PAE (%)	Power Consumption (mA@V)
[2]-60GHz	[65][IBM]	1	4.5	8.5	6	8.5	23@1.2
[3]-58GHz	[65][IBM]	3	15	11.5	2.5	11	43@1
[This work]-61GHz	[65][STM]	2	8.5	7.2	4.2	2.3	96@1.2

to reach good trade-off between gain and linearity. The NSh RF-pad is used. The active area is $0.065mm^2$.

B. PA Measurement Results

The fully integrated $65nm$ CMOS PA is designed and fabricated. An Agilent E83612 vector network analyzer is used to measure S-parameters from 10MHz to $110GHz$. The small signal measurements are plotted in Fig. 6. It operates at the desired frequency band because of the accurate modeling of each interconnect junction and accesses to the transistor. At a supply of $1.2V$, the PA reaches a gain of $8.5dB$ at $61GHz$ in spite of losses caused by power dividing and power combining performed by the baluns. The input and output reflection coefficients are lower than $-10dB$ at $60GHz$. The isolation is more than $20dB$ from DC to $110GHz$. The output power as function of the input power is measured to carry out the large signal performances. The PA reaches a P_{sat} of $7.2dBm$, an Output $1dB$ Compression Point (OCP_1) of $4.2dBm$ with a weak Power Added Efficiency (PAE). Tab 1 compares performances of PAs found in literature designed with 65nm CMOS technology. We are requesting that you follow these guidelines as closely as possible so that the Digest has a professional look and resembles the MTT Transactions. An easy way to comply with the IMS 2009 paper formatting requirements is to use this document as a template and simply type your text into it.

C. Interpretation of Measurement Results

The PA was expected to exhibit better linearity performance. The PA does not demonstrate linearity or gain improvement even after performing load and source pull measurements. The two differential stages are biased in the same conditions and

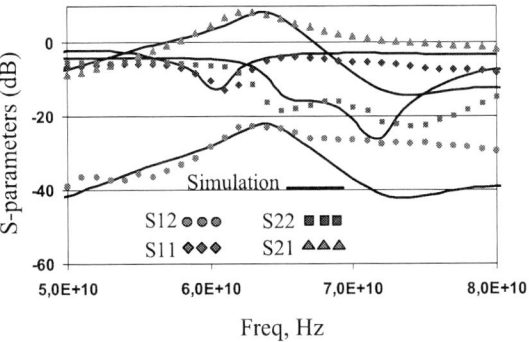

Fig. 6. Measured S-parameters at $(V_{gs}, V_{ds}) = (0.95V, 1.2V)$

have the same power consumption. Retro-simulations enable identifying that the two differential stages are not acting in a balanced mode at $60GHz$. Simulations exhibit a magnitude difference of $2.3dB$ (instead of $0dB$) and a phase difference of $153deg$ (instead of $180deg$) between the two differential stages. In view of the aforementioned comments, the first half stage operates in the compression region while the second half stage remains in the linear region.

IV. BALUN CHARACTERIZATION AND OPTIMIZATION

The balun is measured using only a general two-port network analyzer. The end of the primary and the secondary are connected to ground. The transformer exhibits high coupling (more than $0.8dB$). Due to the limit of this measurement setup, the single-to-differential conversion performances are not extracted.

To understand the reason for this unbalance, 3D-

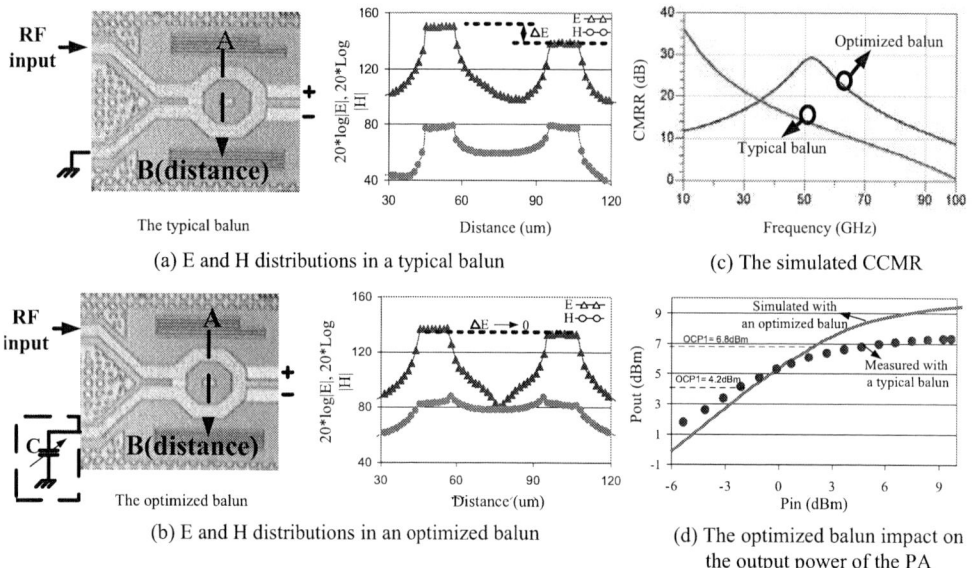

Fig. 7. Comparison between a typical and an optimized baluns

electromagnetic simulations are performed. Therefore, the magnitude of electric field and the magnetic field are probed at the cross section AB (cf. Fig. 7.a). A single RF-mode is applied at the input of the balun.

According to Fig. 7.a, the electric field is the responsible for this imbalance. It is due to high capacitive coupling between the primary and the secondary and between the primary and the substrate. The retrieved energy by each half secondary is unequal. This phenomenon is always located in one turn balun and can be lowered:

1) by decreasing the capacitive coupling to the detriment of the coupling coefficient.
2) by interleaving the primary and the secondary to improve the symmetry coupling [5]. This method lowers the resonance frequency and increases the coupling area with the high-loss substrate.

Unlike the electric field, the magnetic field is distributed symmetrically to the secondary because of the constant current crossing over the primary.

An original idea to improve the balance performances of this balun is proposed. It consists in adding a serial capacitor at the end of the primary (cf. Fig. 7.b). This capacitor compensates the coupling (previously discussed). It also forces a non-zero voltage at the end of the primary to be coupled with the secondary at the resonance. Hence, a symmetric distribution of the electric and the magnetic fields in the secondary is set.

Typically, the Common Mode Rejection Ration ($CMRR$), defined as the ratio between the differential mode and common mode, is calculated to evaluate the single-to-differential mode conversion balun performance. Fig. 7.c depicts the simulated $CMRR$ of the balun which was used in the PA and the optimized balun. Up to $15GHz$, the first balun has a good $CMRR$. Furthermore, getting higher in frequency induces more losses due to the capacitive coupling and a considerably

drop of the $CMRR$. For the optimized balun, the added capacitor of $87fF$ allows the balun reaching a $CMRR$ of $30dB$ at $60GHz$. A frequency reconfigurable balun van be designed by tuning the added capacitor.

The optimized balun is integrated instead of the first balun. The PA is able to deliver up to $9.5dBm$ of output power, and has an OCP_1 of $6.8dBm$ instead of $4.2dBm$ (cf. Fig. 7.d).

V. CONCLUSION

An optimized RF-pad, a CPW T-lines, a tee junction and baluns are customized and considered in the design of a fully integrated $60GHz$ PA . $65nm$ CMOS technology from STMicroelectronics is used for the design. Small signal and large signal analysis of the PA are demonstrated. This PA offers a gain of $8.5dB$ in spite of using baluns at the input and the output of the PA. A critical point of view of the fabricated stacked balun is given concerning balanced-to-unbalanced conversion performances. Finally, an optimized high-performance balun is proposed contributing to the design of mmW-frequency circuits.

ACKNOWLEDGEMENT

The authors acknowledge the foundry support provided by STMicroelectronics.

REFERENCES

[1] C.H Doan, S. Emami, A.M. Niknejad and R.W. Brodersen, "Design of CMOS for 60GHz application", *IEEE ISSCC Digest of Technical Papers*, pp. 440-53, Feb 2004.
[2] A.V. Garcia, S. Reynolds, J.O. Plouchart, "60GHz Transmitter Circuits in 65nm CMOS", *IEEE RFIC Symposium*, pp. 1187-1190, June 2008.
[3] W.L. Chan, J.R Long, M. Spirito, J.J Pekarik , "A 60GHz-band 1V 11.5dBm Power Amplifier with 11% PAE in 65nm CMOS", *IEEE ISSCC Digest of Technical Papers*, pp. 380-381, Feb 2009.
[4] S. Aloui, E. Kerherve, J.B. Begueret, R. Plana, D. Belot, "Optimized pad design for millimeter-wave applications with a 65nm CMOS RF technology", *IEEE EUMC*, pp. 641-644, Sept 2009.
[5] H. Gan, S. Simon Wong "Integrated transformer baluns for RF low noise and power amplifiers", *IEEE RFIC Symposium*, pp. 11-13, June 2006.

RMO3B-3

200GHz f_T SiGe HBT Load Pull Characterization at mm–Wave Frequencies

Luciano Boglione and Richard T. Webster

University of Massachusetts, Lowell, MA; and
Air Force Research Laboratory (AFRL/RYHA), Hanscom AFB, MA

Abstract— **The load pull measurement of a commercially available SiGe HBT device has been performed at Q band over frequency and bias. Measured mm–wave results for the SiGe process under test have never been made available to the general public before and no comparable information on similar SiGe devices is available in the public domain. The goal of this paper is to begin to fill this gap: load pull results along with a discussion of the characterization setup and procedure are presented.**
Index Terms— **mm–wave SiGe HBT device, measurement, load–pull.**

I. INTRODUCTION

Load pull characterization is a well established procedure, crucial in understanding and optimizing the power performance of a device. The load pull procedure allows designers to optimize the performance of an amplifier. The output power and power added efficiency of a transistor are strong functions of the complex load impedance presented to the device output. Load pull characterization consists of measuring the transistor output power and efficiency at a large number of complex loads over a range of frequencies, bias points, and input powers, then fitting an empirical function to the data to provide the information required for an optimized amplifier design.

While load pull systems, both active and passive, in the low GHz range have become fairly sophisticated [1], the challenges that this measurement procedure must overcome at mm–wave frequency are not trivial [2]. Load pull characterization results of compound semiconductor devices at mm–wave frequencies are scarce in the public domain [3], [4]. Despite the growing interest in mm–wave applications based on Si technology, load pull data demonstrating the optimum power performance of silicon devices in the mm–wave range is virtually nonexistent in the public domain.

While the load pull procedure is valid at any power level, it is especially important under the nonlinear conditions where power amplifiers typically operate. Nonlinear be-

The views expressed in this article are those of the authors and do not reflect the official policy or position of the United States Air Force, Department of Defense, or the U.S. Government.

havior is synonymous with harmonic generation and the load seen by the device at the harmonic frequencies has proven to be crucial in optimizing the power performance of the device under test (DUT) [5]. Recent advances in this field at microwave frequencies have focused on computer–driven load pull setups capable of controlling loads, Γ_L, up to the 3rd harmonic [6] with a high degree of repeatability. However, a device operating at $f_0 = 43$GHz would generate a 2nd harmonic in W band ($2 \times f_0 = 86$GHz) and the 3rd harmonic above that ($3 \times f_0 = 129$GHz). Currently, no measurement technology can access this wide bandwidth simultaneously. Furthermore, propagation losses in the connections between the device and the load generally increase with frequency. Since the signal must travel from the device output terminals to the load and reflect back to the device, each dB of loss in the connection reduces the effective load reflection coefficient by 2 dB. Highly reflective loads ($|\Gamma_L| \to 1$) are difficult to transfer to the device terminals, particularly with passive setups. Because of these limitations, only power at the fundamental frequency can be controlled and measured effectively at Q band.

This paper will present for the first time, the results of a load pull characterization of standard 200GHz f_T SiGe HBT SBC18H2 devices manufactured by Jazz Semiconductors at two Q band frequencies and two bias levels. Full de–embedding up to the device terminals – including the GSG pad structure – is carried out on the measured raw data.

The paper is organized as follows: Section II will describe the device and highlight its small signal performance; Section III will discuss the load pull setup and how the measurement challenges have been addressed; Section IV will present the load pull results and Section V will conclude the paper.

II. THE DEVICE UNDER TEST

The DUT is a $0.15\mu m \times 10.16\mu m$ SiGe HBT available commercially in Jazz Semiconductor's SBC18H2 process [7], [8]. Full DC characterization of the DUT has been executed with an Agilent 4156C Precision Semiconductor

978-1-4244-6240-7/10 $26.00 © 2010 IEEE

Parameter Analyzer (PSPA) prior to performing the load pull measurement. Current-Voltage (IV) characteristics and Gummel–Poon measurements have been executed routinely after developing appropriate SCPI based code in an Excel/Visual Basic environment.

Scattering parameters at a given bias point have been measured with an Agilent E8364B Precision Network Analyzer (PNA) over the range 30–50GHz. A Thru–Reflect–Line (TRL) calibration [9] with a Cascade Impedance Standard Substrate model 101–190 has been used, as on–chip calibration structures were not available.

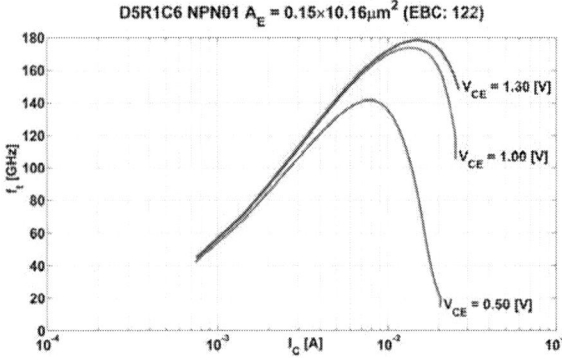

Fig. 1. Measured f_T vs. I_C for 3 different V_{CE} values. The device has 1 emitter, 2 base and 2 collector fingers (EBC=122).

Deembedding the pad structure surrounding the device and moving the measurement planes to the base and collector nodes is key to correctly evaluating the performance of the device. A pad model has successfully been developed and applied to both small and large signal measurement data. The transition frequency, f_T, at which the magnitude of the small signal current gain, $|h_{21}|$, is equal to 1 (0dB) has been extracted (Fig. 1) to confirm the validity of the pad model.

The results obtained from our DC and small signal measurements are consistent with the manufacturer's data and have provided the necessary confidence to move on to the unchartered territory of the load pull characterization at Q band.

III. LOAD PULL SETUP

A load pull setup has been assembled at the Air Force Research Laboratory's Hanscom facility and it is shown in Fig. 2. The source is a Wiltron 10MHz – 65GHz synthesizer which drives a custom–made Q band power amplifier and a Hughes WR22 precision variable attenuator. Hewlett Packard Q281B and Wiltron 35WR22K adaptors provide transitions from coax to WR22 waveguide. A Hughes 45342H–1110 directional coupler is used to sample incident power to, and reflected power from the DUT with Agilent's HP8487A power sensors. Cascade Infinity probes

Fig. 2. Main components of the load pull setup: the output tuner connected between probe and power sensor is clearly visible on the right hand side of the picture.

are used to contact the SBC18H2 device at the probe pads. A low loss semirigid cable is used to connect the output probe to the input port of a 12–50GHz Maury 7941A manual tuner.

Each linear component of the setup has been characterized in terms of scattering parameters in the range 30–50GHz after appropriate calibration of the PNA. The scattering parameters associated with each Infinity probe have been determined with a second–tier calibration following a coax calibration of the network analyzer. The second–tier calibration is based on the standard TRL procedure [9], for which Matlab routines have been developed to determine the probes' scattering parameters.

The tuner has been characterized for 110 settings, each corresponding to one individual load Γ_L available for the load pull measurement. Repeatability of the tuner settings for 10 loads has been found to be within about ± 0.5dB in magnitude and within $5°$ in phase.

A power sensor is connected to the tuner's output port. The power it detects is a function of the tuner setting, since highly reflective loads ($|\Gamma_L| \to 1$) will reflect most power flowing into the tuner's input port, thereby reducing the amount of power reaching the power head. An alternative setup making use of a directional coupler has been investigated in order to reduce the dependence of the detected power on tuner settings; however, it was ruled out due to the additional loss associated with it.

Two constraints limit the setup to operate in the 30–50GHz range:

1) the WR22 waveguide's cut–off frequency of the dominant TE_{10} mode (about 27GHz); and
2) increasing loss in passive components

It should be noted that the passive components in use (e.g.

TABLE I

NOMINAL BIAS POINT AND FREQUENCY AT WHICH LOAD PULL CHARACTERIZATION TESTS HAVE BEEN EXECUTED.

Test #	Bias Point $(I_B; V_{CE})$	Frequency GHz
1	$(180\mu A; 1.3V)$	43.275
2	$(60\mu A; 1.0V)$	44.800

probes, cables, etc.) are not designed for operation above ~ 50GHz. For instance, standard waveguide operation calls for single mode operation and losses dominate passive components outside their specified bandwidth. Furthermore, the HP8487A power sensors are specified for the 50MHz – 50GHz range in frequency and -30dBm to +20dBm range in power. It is therefore assumed that power levels associated with any harmonics of the fundamental frequency at the output power sensor plane are negligible due to the loss of the setup's passive components and undetectable because of the power sensor's inherent limitations. Finally, the passive components are extremely long compared signal wavelengths. Correct determination of the associated delays imposes an upper limit on the PNA frequency step to avoid computational errors.

IV. LOAD PULL RESULTS

A C++ program has been developed to control the source and collect the power readings from the sensors as the source power is swept at constant nominal bias and frequency. The Agilent PSPA is used to record changes in base and collector currents and related voltages as the input RF power is increased. The setup is calibrated prior to executing the load pull test in order to determine the input available power, P_S, from the directional coupler. The previously measured scattering parameters of all the connecting components are used to correct for losses and phase shifts at the frequency of operation and move the input and output measurement planes to the base and collector nodes within the pad structure. The power going into the base and the power coming out of the collector for a given load are designated P_{in} and P_o respectively.

A number of load pull measurements have been carried out. Table I describes two: test #1 corresponds to the bias point for peak f_T; test #2 biases the device slightly below the peak f_T, at lower collector voltage.

Fig. 3 shows the measured output power, power gain, and power added efficiency at 43.275GHz (test #1). Maximum output power of 8.12 dBm is obtained when $P_{in} \approx 3.72$dBm is applied at the device's base terminal; load pull contours for this input power level P_{in} are shown in Fig. 4. A similar plot is shown in Fig. 5, test #2, for an input power level $P_{in} \approx 2.74$dBm. Table II collects the results of

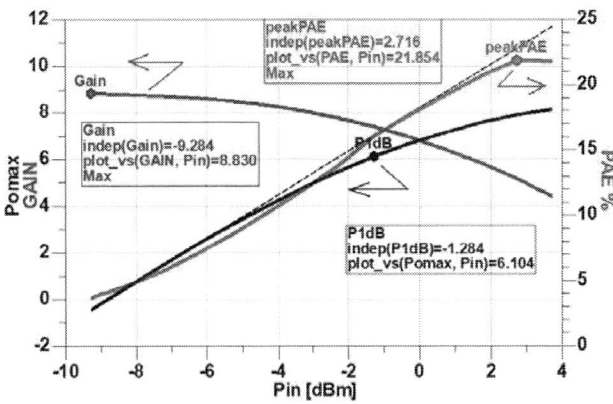

Fig. 3. Maximum output power $P_{o(max)}$ [dBm] at optimum load; power gain [dB]; and corresponding power added efficiency PAE $= (P_o - P_{in})/P_{DC}$ vs. input power P_{in}, test #1.

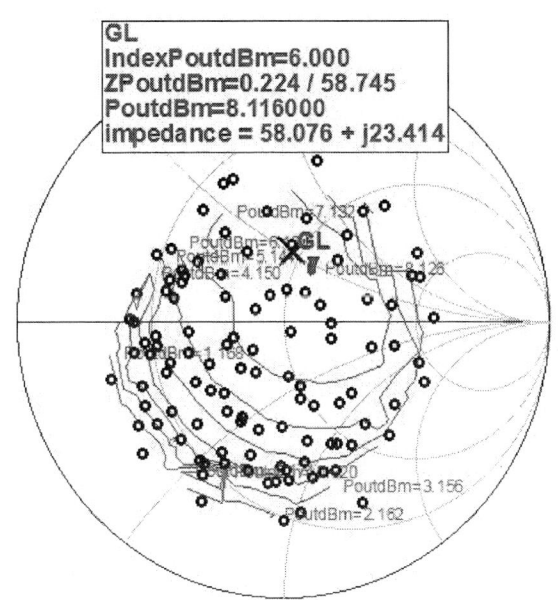

Fig. 4. Load pull contours referred to the device's collector node for test #1. The black cross shows the optimum load $S_{22}^\star$ determined from small signal scattering parameter measurement at the same nominal bias point, red traces show the contours converging to the optimum load Γ_L at ~ 3.7dBm input power level; black circles show the 110 tuner positions at 43.275GHz.

978-1-4244-6240-7/10 $26.00 © 2010 IEEE

TABLE II
OPTIMUM Γ_L FOR MAXIMUM OUTPUT POWER $P_{\max}$ AND MAXIMUM POWER ADDED EFFICIENCY PAE.

Test #	P_{in} [dBm]	$P_{\text{o(max)}}$ [dBm]	$|\Gamma_L|$ -	$\angle\Gamma_L$ [deg]	$\text{PAE}_{\max}$ %
1	+3.72	+8.12	0.224	58.745	21.854
2	+2.74	+3.83	0.554	47.713	20.544

the load pull measurements.

V. CONCLUSIONS

A load pull setup has been assembled to identify the optimum load Γ_L of a $0.15\mu m \times 10.16\mu m$ SiGe HBT device for a number of bias points and frequencies. The limitations associated with the setup have been discussed, particularly the harmonics the device generates when operating non–linearly. To the authors' knowledge, this is the first time that this type of design information has been made available for a SiGe process at mm–wave frequency.

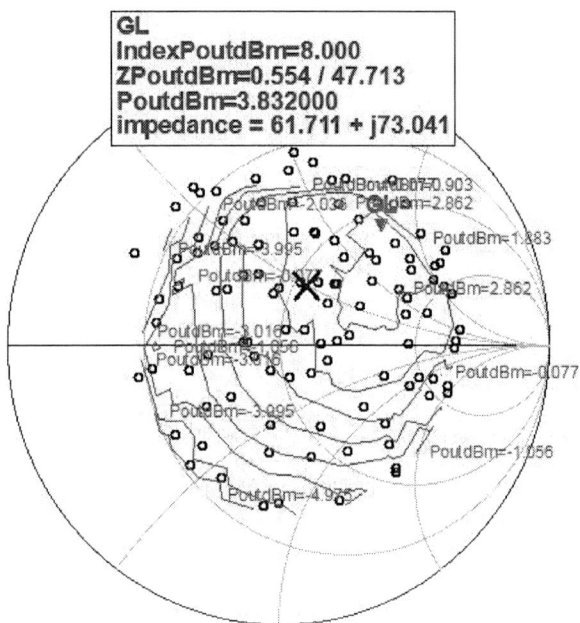

Fig. 5. Load pull contours referred to the device's collector node for test #2. The black cross shows the optimum load S_{22}^* determined from small signal scattering parameter measurement at the same nominal bias point; red traces show the contours converging to the optimum load Γ_L at ~ 2.7dBm input power level; black circles show the 110 tuner positions at 44.800GHz.

ACKNOWLEDGEMENT

The authors wish to thank Samir Chaudhry and his team at Jazz Semiconductor, Newport Beach, CA and the Air Force Office of Scientific Research Summer Faculty Fellowship Program.

REFERENCES

[1] F. DeGroote, P. Roblin, Y.-S. Ko, C.-K. Yang, S. J. Doo, M. V. Bossche, and J.-P. Teyssier, "Pulsed multi–tone measurements for time domain load pull characterizations of power transistors," *73rd ARFTG Microwave Measurement Conference*, pp. 1–4, 12 June 2009.

[2] D. J. Williams and P. J. Tasker, "An automated active source and load pull measurement system," *The 6th IEEE High Frequency Postgraduate Student Colloquium*, pp. 7–12, 9–10 September 2001, Institute of Microwaves and Photonics, School of Electronic and Electrical Engineering, The University of Leeds, UK.

[3] C. Gaquiere, E. Bourcier, B. Bonte, and Y. Crosnier, "A novel 26–40 GHz active load pull system," *25th European Microwave Conference*, pp. 339–342, October 1995.

[4] D. W. Baker, R. S. Robertson, R. T. Kihm, M. Matloubian, M. Yu, and R. Bowen, "On–wafer load pull characterization of W–band InP HEMT unit cells for CPW MMIC medium power amplifiers," vol. 4, pp. 1743–1746, 13–19 June 1999.

[5] B. Noori, A. Sheikh, G. Bigny, P. Tasker, S. Cripps, and J. Benedikt, "The effect of harmonic terminations variation on the accuracy of load pull measurements," *European Microwave Conference*, Rome, Italy, 28–30 September 2009, Workshop WHMO4 Notes.

[6] Focus Microwave Inc., *High Resolution Tuners Eliminate Load Pull Performance Errors.* www.focus–microwaves.com, Application Note AN–15.

[7] Jazz Semiconductors, *Electrical Parameters of the SBC18 Process Family.* Jazz Semiconductor, Newport Beach, CA, NPB–PS–0267, rev. 17.

[8] ———, *SBC18 Design Manual.* Jazz Semiconductor, Newport Beach, CA, NPB–PS–0288, rev. 10.

[9] G. F. Engen and C. A. Hoer, "Thru-Reflect-Line: An improved technique for calibrating the dual six port automatic network analyzer," vol. 27, No. 12, pp. 987–993, December 1979.

RMO3B-4

A Miniature 26-/77-GHz Dual-band Branch-line Coupler Using Standard 0.18-μm CMOS Technology

[*]Yu-Sheng Lin, [*]Cheng-Ying Hsu, [*]Huey-Ru Chuang and [#]Chu-Yu Chen

[*]Institute of Computer and Communication Engineering, Department of Electrical Engineering, National Cheng Kung University, Tainan, Taiwan, R.O.C

[#]Department of Electronic Engineering, National University of Tainan, Tainan, Taiwan, R.O.C.

Abstract — A 26-/77-GHz dual-band branch-line coupler fabricated using standard 0.18-μm CMOS technology is presented. The dual-band coupler is with a compact size of 1.0 × 1.0 mm². The simulated frequency response is to 110 GHz. The measured frequency response shows a good performance with amplitude and phase imbalnce of ± 0.5 dB and ± 5°, respectively, in the low band. The S_{21} in the high band is about 7.5 dB The measured isolation and the return loss are better than 15 dB within the two passbands. The on-chip coupler will be useful for the integrated design of a 26/77-GHz CMOS single-chip RF transceiver.

Index Terms — 26-/77-GHz, Branch-line coupler, CMOS, Dual-band, Millimeter-waver

I. INTRODUCTION

In recent years, the main applications for radar systems are vehicular systems at 24 GHz and 77 GHz. At present, the Federal Communication Commission (FCC) has allocated 22-29 GHz for such devices However, the FCC has limited the center frequency to larger than 24.075 GHz. In addition, the European Union (EU) will prohibit the band near 24 GHz for vehicular radar systems after 2013. For this reason, we decided to set the operating frequency at 26 GHz for this coupler. In addition, the spectrum above 70 GHz is of increasing interest to service providers and systems designers due to its unique propagation characteristics and the wide bandwidth available for carrying communications traffic. For example, the European Telecommunications Standards Institute (ETSI) discovered that the 77 GHz band is suitable for short range communication .

Complementary metal oxide semi-conductors (CMOS) have the advantages of highly-integrated capability and low cost for mass production. CMOS integrated circuits have been reported in millimetre-wave frequencies. The CMOS branch-line coupler is one of the most commonly used elemental components used in wireless communication systems, such as power dividers, balanced mixers, frequency discriminators and phase shifters. It provides equal amplitude and quadrature phase outputs at the operating frequency. The dual-band brand-line coupler based on lumped elements was reported in [1]. In [2]-[3],

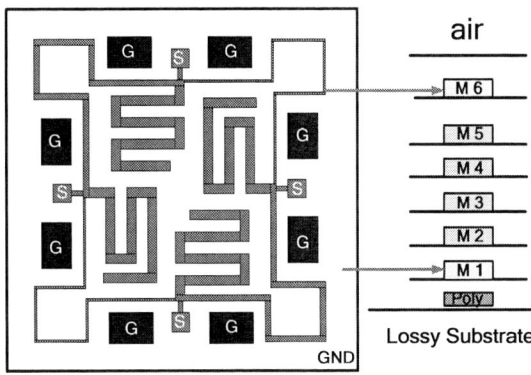

Fig. 1. Schematics and cross-sectional view of the proposed 26-/77-GHz dual-band branch-line coupler.

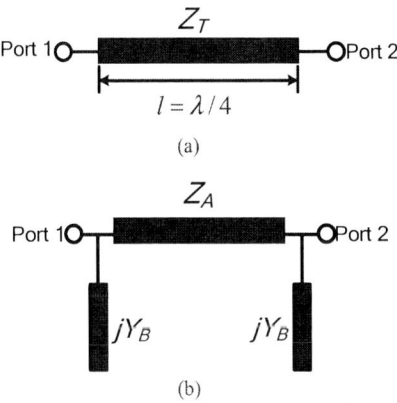

Fig. 2. (a) Quarter-wavelength branch-line. (b) Equivalent branch-line structure.

the dual-band brand-line couplers for planar structure were presented. However, the size of couplers is large and the bandwidths are narrow in these earlier works. In [4], the additional sections of a coupler are introduced to improve the above shortcoming points.

This paper provides a design for a more compact 26-/77-GHz CMOS dual-band branch-line coupler to be used in the integrated design of the 26/77-GHz CMOS single-chip RF transceiver. The coupler is fabricated with the standard0.18-μm CMOS process. To avoid an additional transmission

978-1-4244-6240-7/10 $26.00 © 2010 IEEE

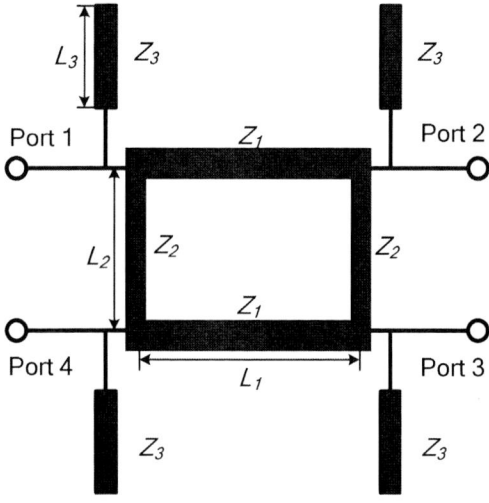

Fig. 3. The dual-band branch-line coupler structure.

$$Z_A = \frac{Z_T}{|\cos(\delta\pi/2)|} \qquad (1)$$

$$Y_B = \begin{cases} \sin(\delta\pi/2)/Z_T, & f = f_1 \\ -\sin(\delta\pi/2)/Z_T, & f = f_2 \end{cases} \qquad (2)$$

where $\delta = (f_2 - f_1)/(f_2 + f_1)$, and the operating frequencies f_1 and f_2 are the lower and upper frequency, respectively. Fig. 3 shows the schematic of the dual-band branch-line coupler. The characteristic impedance of each transmission line can be expressed by the following:

$$Z_1 = \frac{Z_0}{\sqrt{2}} \cdot \frac{1}{\cos(\delta\pi/2)} \qquad (3)$$

$$Z_2 = Z_0 \cdot \frac{1}{\cos(\delta\pi/2)} \qquad (4)$$

$$Z_3 = \frac{Z_0}{1+\sqrt{2}} \cdot \frac{1}{\sin(\delta\pi/2)\tan(\delta\pi/2)} \qquad (5)$$

Based on the above formulas, the CMOS dual-band branch-line coupler can be designed. In order to obtain a better matching between the contact pads and the branch-line coupler, the stepped impedance matching network is used in the design. Based on the transmission line theory, the equivalent circuit for a small θ and large characteristic impedance transmission line is equal to an inductor, whereas the equivalent circuit for a small θ and small characteristic impedance transmission line is equal to a capacitor [7]. Therefore, the contact pad has a larger dimension, which may introduce a parasitic capacitance, and the feeding line has a smaller width, which may introduce an inductor. The electric length and characteristic impedance of the contact-pads and feeding line can be easily obtained and optimized by using an EM simulator. Finally, two section transmission lines can be modelled as a matching network and be easily integrated with the dual-band branch-line coupler.

Since the coupler is fabricated on a 0.18-μm CMOS multi-layered structure, a way to reduce the simulation complexity is to evaluate the effective dielectric constant and simplify it to a single equivalent homogeneous substrate. Based on the formulation developed in [6], the equivalent dielectric constant can be calculated accurately for the structure and dimension design of the coupler. The IE3D is then used for the fine tuning in the design simulation.

loss from mismatch, a step impedance transmission line matching network is considered between the contact pad and feeding port [5]. Finally, both the simulated and measured results are shown in section III.

II. DESIGN METHODOLOGY

Microstrip dual-band branch-line couplers are widely and usefully applied in planar phase shifters. The layout of the proposed CMOS dual-band branch-line coupler is shown in Fig. 1. It is composed of four quarter wavelength transmission lines and four open stubs. The meander open stubs are used to reduce the overall dimension. The dual-band branch-line coupler presented in this work is more compact than an ordinary square dual-band branch-line operated at the same frequency. To operate at millimeter wave frequency, it is crucial to prevent currents from leaking into the substrate and also to reduce insertion loss in a CMOS device. This can be achieved by placing the ground plane far away from the coupler structure. In this case, the ground plane is placed at the bottom metal (M1) and the coupler is placed at the top metal (M6).

Fig. 2 shows a quarter-wavelength branch line can be equivalent to a transmission line cascade with two open stubs. By calculating the ABCD-parameters, the characteristic impedance of new transmission line and the open stubs can be expressed as [3]:

III. SIMULATION AND MEASUREMENT RESULTS

Fig. 4 shows the chip micrograph.The chip size is 1.0×1.0 mm2, including all contact-pads and dummy metals. From (3)-(5), the characteristic impedances of Z_1, Z_2, and Z_3 were found to be approximately 38.3 Ω, 54.1 Ω, 130.6 Ω, respectively. The length of the couplers are L1 = 1160 μm, L2 = 1160 μm, and L3 = 990 μm. Figs. 5 shows the simulation results of the frequency response of the coupler to 110 GHz.. Figs. 6 to 8 show the measurement results together with simulated data. Two measurement setups are used The first setup is a 40-GHz on-wafer measurement system which can measure the S-parameters of each two port. The measured results by using the first setup are shown in 6. At 26 GHz, the isolation property between the port 1 and port 4 is greater than 18 dB. The return loss is better than 18 dB within the passband. At the port 2 and port 3, the measured bandwidth of the output amplitude balance is about ± 0.5 dB. The phase variation of the quadrature phase balance is maintained at ± 5° over the measured bandwidth. The simulated return loss of the passband is better than 15 dB, and the magnitude of coupling is about 5 dB.

The second setup is a 110-GHz on-wafer measurement system which can only measure the S-parameters of the two ports which are arranged in a face-to-face layout. Therefore only port-1 and 2 are measured with S_{21} and S_{11} are plotted in Fig. 7. The simulated and measured results are all in good agreement The S_{21} in the high band is about 7.5 dB and the S_{11} is better than 17 dB. The insertion loss in the high band is about 2.5 dB higher than that in the low band. Table I shows the summarized performance and comparison with reported works [9]-[10] (lower than 40 GHz). The presented CMOS coupler not only has a dual-band operation and also has a better performance and smaller chip size in the compared low band .

IV. CONCLUSION

A 26-/77-GHz dual-band branch-line coupler with a mender line structure fabricated using standard 0.18-μm CMOS technology is presented. The IE3D is used for the design simulatuion. Dummy metals and parasitic effects are also considered in the design. The simulated and measured results are in good agreement. The measured insertion loss in the two passbands are about 5 dB and 7.5 dB (include 3 dB split loss), respectively. The return losses are all better than 15 dB within the two passbands. The phase imbalnce is about ± 0.5 dB and ± 5°, respectively, in the low band. The S_{21} in the high band is about 7.5 dB The measured isolation and the return loss are better than 15 dB within the two passbands.The on chip coupler designed in this work will be useful for the integrated design of a 26/77-GHz CMOS single-chip RF transceiver.

Fig. 4. Chip micrograph.

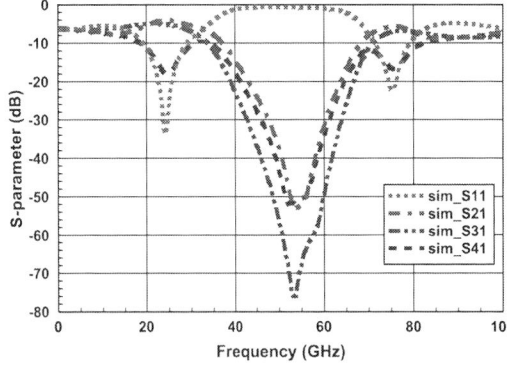

Fig. 5. Simulated frequency response of the 26-/77-GHz dual-band branch-line coupler.

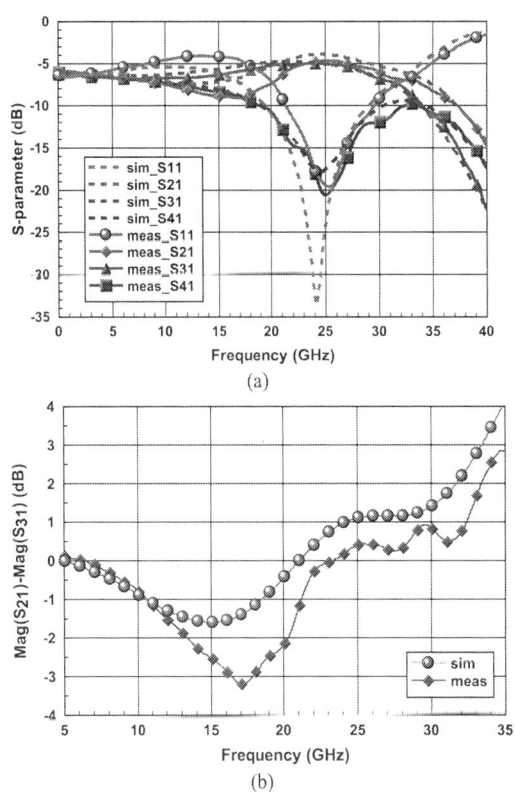

978-1-4244-6240-7/10 $26.00 © 2010 IEEE

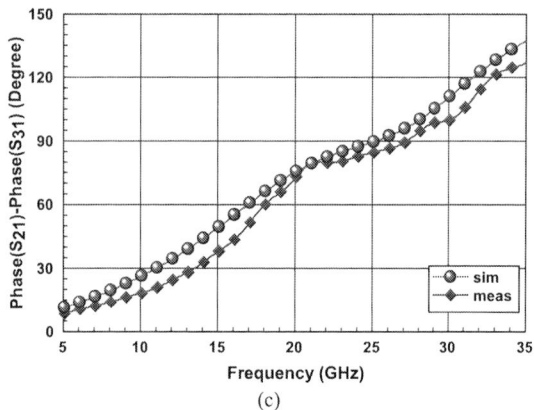

(c)

Fig. 6. Simulation and measurement result to 40 Ghz. (a) S-parametrers., (b) output amplitude imbalance, (c) output phase difference.

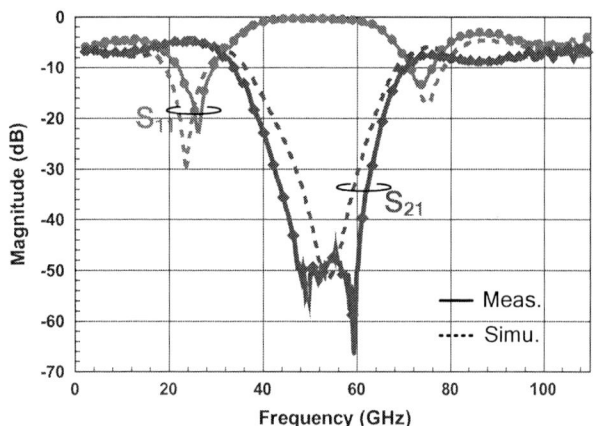

Fig. 7. Simulation and measurement result of the magnitude of S_{11} and S_{21} to 110 GHz.

Acknowledgement

The authors would like to thank the Chip Implementation Center (CIC) of National Science Council, Taiwan, ROC, for supporting the TSMC CMOS process.

References

[1] [1] I. -H. Lin, C. Caloz and T. Itoh, "A branch-line coupler with two arbitrary operating frequencies using left-handed transmission lines," in *IEEE MTT-S Int. Microwave Symposium Dig.*, vol. 1, pp. 325-328, Jun. 2003.

[2] F. L. Wong and K. K. M. Cheng, "A novel planar branch-line coupler design for dual-band applications," in *IEEE MTT-S Int. Microwave Symposium Dig.*, vol. 1, pp. 903-906, Jun. 2004.

[3] K. K. M. Cheng and F. L. Wong, "A novel approach to the design and implementation of dual-band compact planar 90° branch-line coupler," *IEEE Trans. Microw. Theory Tech.*, vol. 52, no. 11, Nov. 2004.

[4] R. Levy and L. F. Lind, "Synthesis of symmetrical branch-guide directional couplers," *IEEE Trans. Microwave Theory Tech.*, vol. 19, no. 2, pp. 80-89, Feb. 1968.

[5] C. Y. Hsu, C. Y. Chen and H. R. Chuang, "A 60- GHz millimeter-wave bandpass filter using 0.18-um CMOS technology," *IEEE Electron Device Letters.* vol. 29, no. 3, pp. 246-248, Mar. 2008.

[6] W. K. W. Ali' and S. H. AI-Charchafchi, "Using equivalent dielectric constant to simplify the analysis of patch microstrip antenna with multi layer substrates." in *Proc. IEEE AP-S Int. Symposium*, vol. 2, pp. 676-679, Jun. 1998.

[7] D. M. Pozar, *Microwave Engineering*, Wiley 3rd ed, 2005.

[8] Dicle Ozis and David J. Allstot, "A CMOS 5 GHz phase-compensated quadrature coupler," in *IEEE Radio and Wireless Symposium*, pp. 51-54, Jan. 2006.

[9] Y. Zhu and H. Wu, "A 10-40 GHz 7 dB directional coupler in digital CMOS technology," in *IEEE MTT-S Int. Microwave Symposium Dig.*, pp. 1551-1554, Jun. 2006.

[10] M Ozgur, U.C Kozat, M.E Zaghloul and M Gaitan, "Micromachined branch line coupler in CMOS technology," *Microwave Symposium Digest., IEEE MTT-S International*, vol.1, no., pp.291-294 vol.1, 2000

TABLE I

MERASURED ERFORMANCE COMPARISON WITH PREVIOUS WORKS

Ref.	Tech.	Frequency	Coupler structure	Insertion loss	Return loss	Isolation	Magnitude imbalance	Phase imbalance	Size (mm × mm)
[9]	0.18-μm CMOS	10-40 GHz	Directional coupler	7-10 dB	> 11 dB	> 16 dB	> 3 dB	~ 90°	1.4 × 0.58
[10]	1.2-μm CMOS	25-30 GHz	Branch-line coupler	> 5 dB	> 10 dB	-	± 1 dB	-	-
This work	0.18-μm CMOS	24/77 GHz	Branch-line coupler	4.9 dB@24 GHz 7.5 dB@77 GHz	> 15 dB	> 18 dB	± 0.46 dB (first band)	90 ± 6.2° (first band)	1.0 × 1.0

RMO3B-5

Compact Transformer Power Combiners for Millimeter-wave Wireless Applications

Yi Zhao, John R. Long, and Marco Spirito

ERL/DIMES, Delft University of Technology, Mekelweg 4, 2628CD Delft, The Netherlands

Abstract — Two current-summing transformer combiners for 60GHz-band power amplification in millimeter-wave wireless applications are characterized. The parasitic-compensated balun and fully-differential combiners mitigate imbalances caused by interwinding capacitance, while self-shielded output windings inhibit substrate coupling. Excellent agreement is seen between measurement and electromagnetic simulation. Power loss for both prototypes at 60GHz is <1.0dB and chip area is <0.015mm². Reflected impedance uniformity between ports at 60GHz for the balun and fully-differential combiners is better than 2.4% and 4.5%, respectively. A lumped-element model for large-signal circuit design with <5.0% error in power loss and reflected port impedance across the 55-65GHz band is also described.

Index Terms — Current-summing transformer, power combiner, millimeter-wave, SiGe-BiCMOS, power amplifier.

I. Introduction

Silicon-based power amplifiers (PAs) rarely deliver peak output powers larger than 15dBm at millimeter-wave (mm-wave) frequencies [1], primarily because the trade-off between gain-bandwidth product and breakdown voltage limits the output power from a single-stage (e.g., peak-f_T =166GHz and BV_{CEO} <2V in 130nm SiGe-BiCMOS [2]). Greater output power may be realized using passive power combining (PC) (e.g., 21dBm in [3]) if the insertion loss of the combiner is lower than the power added by summing multiple PA outputs (e.g., combiner loss <3dB for a 2:1 combiner). A compact power combiner on a silicon chip (i.e., a combiner with outer dimensions << wavelength) can increase mm-wave power output and preserve power-added efficiency (PAE) without incurring a large increase in chip area if metal and substrate losses are small.

The goals for the monolithic combiners developed in this work are insertion loss <1.0dB, uniform load impedance scaling required to realize maximum output power from each PA cell (e.g., 50Ω to 10-20Ω transformation), and operation over >5GHz bandwidth (e.g., 59-64GHz) for Gbit/s mm-wave wireless links. Identical reflected impedances at each input port are required to drive all PA cells to their maximum output powers, as mismatch would drive one PA cell into early breakdown and limit the total output power.

In this paper, a compensated 50Ω output, 4:1 balun (see Fig. 1, top) and a fully-differential (Fig. 1, bottom) 100Ω

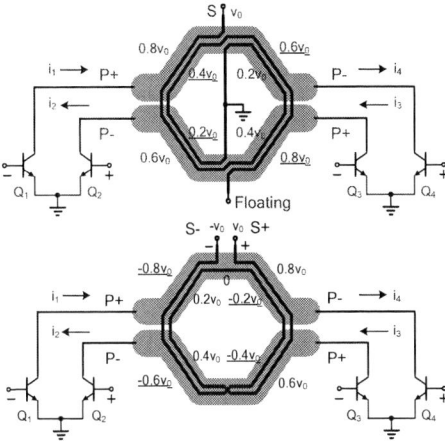

Fig. 1 Parasitic-compensated balun (top) and fully-differential (bottom) transformer power combiner layouts.

output, 16:1 transformer power combiners with <1.0dB insertion loss, <4.5% variation in reflected port-to-port impedance at 60GHz and consuming <0.015mm² chip area are described. The new balun topology is amenable to physical scaling as the layout minimizes undesired coupling between bondpad, secondary and primary transformer windings, while minimizing power loss through optimal use of top (i.e., thick) metal layers.

Details of the power combiners designed in the backend of a 130nm SiGe-BiCMOS technology with thick metal options are reported in the following section. Measurement results of the combiner prototypes are presented in Section III, and a lumped-element model developed for large-signal circuit simulation is described in Section IV. Agilent's Momentum 2.5-D EM simulator was used for all passive component simulations that are presented in this paper.

II. Transformer Power Combiner Design

Imperfect magnetic coupling causes energy leakage and can lead to losses of 1.5dB or more in transformer power combiners at 60GHz. Moreover, interwinding capacitance introduces uneven voltage coupling between primary and secondary windings, and imbalance in the impedances reflected from the load to the PA outputs. Impedance imbalance can be reduced by splitting the secondary into 2 identical windings to balance out uneven voltage coupling [3], as indicated by the voltage swing annotated alongside each winding in Fig. 2. However, the secondary winding

978-1-4244-6240-7/10 $26.00 © 2010 IEEE

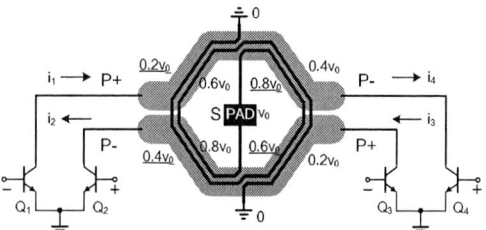

Fig. 2 Multifilament transformer combiner layout [3].

encloses the output bondpad S, which constrains scalability and RF performance. An upper frequency limit of just 40GHz is predicted from EM simulation if a $45 \times 45 um^2$ output pad and $80 \times 80 um^2$ winding inner dimension (ID) are assumed. Also, the output must be flip-chip connected when the I/O bondpad is placed at the center of the transformer layout.

Fig. 3 (top) shows the reflected port impedances simulated over the 50-70GHz range when the multifilament design of Fig. 2 is scaled to operate in the 60GHz band (3-levels of metal and $45 \times 45 um^2$ bondpad). More than 20% variation is observed in the real part of the impedances at 65GHz. Moreover, the two groups of data (seen at collectors of $Q_{1,3}$ and $Q_{2,4}$) resonate at different frequencies (i.e., 57.5GHz vs. 63GHz), due to the uneven capacitive loading and coupling between the bondpad at layout center and balun windings.

New transformer power combiner layouts which do not suffer from this problem are shown in Fig. 1. These combiners can operate beyond the 60GHz band, and have output(s) at the transformer periphery rather than the center. In each of the new designs, differential pairs drive the second-metal primary winding push-pull, thereby creating a virtual ground along the vertical line of symmetry in the layout. The 2-turn topmetal secondary gives an impedance ratio of 1:4 (Fig. 1, top) and 1:16 (Fig. 1, bottom) between primary and secondary windings. The dummy secondary turn connected to a floating output node of the balun (Fig. 1, top) replicates the voltage swing seen at the actual output (S), thereby providing balanced parasitic loads at each primary terminal. The new balun equalizes not only the capacitive loads of one differential pair (e.g., the trifilar in [4]), but also those of all the inverting and non-inverting ports (2 +/- pairs in Fig. 1). Interwinding capacitance does not affect impedance balance of the fully-differential combiner in Fig. 1 (bottom), as the layout is fully-symmetric and all ports are differential. The winding voltages annotated in Fig. 1 indicate the capacitive load equalization in both designs.

Self-shielding [3] is employed to reduce the magnetic flux leakage, as the (relatively) low-voltage primary winding is placed beneath the secondary, thereby shielding it from the substrate. Thus, only the primary couples energy to the substrate, but at a relatively low level due to the lower voltage swing. The compact layouts shown in Fig. 1

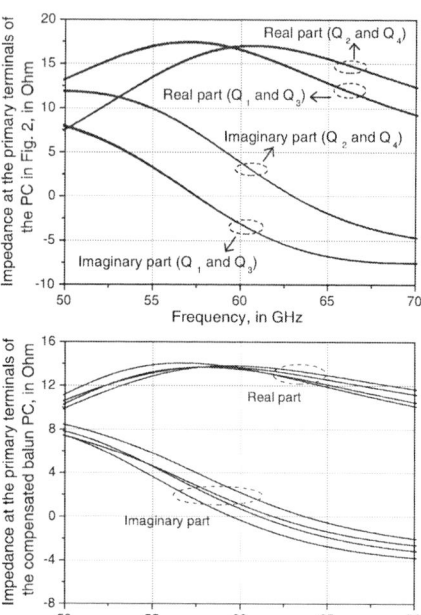

Fig. 3 Simulated reflected port impedances of scaled multi-filament (top) and compensated balun (bottom) combiners.

($<150 \times 150 um^2$ for 60GHz-band operation) are a tradeoff between substrate dissipation and operating bandwidth.

The effectiveness of parasitic compensation in the transformer balun is shown in Fig. 3 (bottom). The real and imaginary parts of the reflected impedances at each of the 4 primary ports are plotted from 50-70GHz. The 50Ω output load is transformed to four 13.5Ω (real) at 60GHz (within 2.4% of each other). The fully-differential power combiner converts the 100Ω load impedance to 11.5Ω (real) at 60GHz within 4.5% error. Deviation from the ideal impedance transformation is caused mainly by non-ideal magnetic coupling (e.g., leakage inductance interaction with parasitic capacitance). By comparison, the impedance imbalance is >20% for the balun layout of Fig. 2 when it is scaled to the 60GHz band as described previously.

Power loss of the combiner is defined as the ratio of the power delivered to the 50Ω load to the total power entering the four primary terminals. With ~300fF loading at each primary terminal, EM simulation predicts the loss of the scaled multifilament (Fig. 2), parasitic-compensated balun (Fig. 1, top) and fully-differential (Fig. 1, bottom) combiners at 0.81dB, 0.56dB and 0.65dB, respectively.

III. Experimental Results

The prototype 60GHz-band power combiners were implemented in a 130nm SiGe-BiCMOS technology with 6 layer copper and 2 thick metal interconnect layers [2]. The multifilament combiner (layout as in Fig. 2; scaled to operate at 60GHz in 3 metal layers) was also implemented for comparison. A 40GHz-band compensated balun was

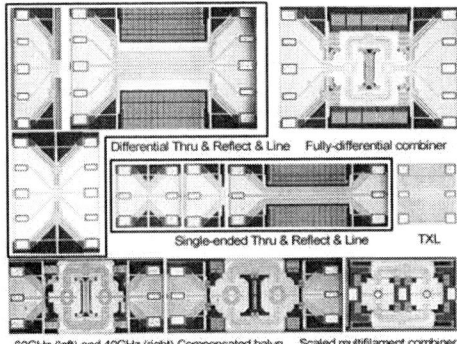

Fig. 4 Die photo of the prototypes and de-embedding standards.

also designed in order to verify frequency scalability and check simulation vs. measurement accuracy. Fig. 4 shows the prototypes and de-embedding structures.

A. Measurement Setup and De-embedding

Each power combiner has 6 or 7 test ports (depending on the design). In order to simplify testing, identical power combiners are connected back-to-back by two pairs of 400um coupled transmission lines, creating a 2-port network. A fully-symmetric cross-over [5] in the transmission lines maintains symmetry of the test structures. The minimum insertion loss is determined from measurement of the back-to-back test structures, thereby verifying and validating predictions from simulation.

The compensated balun and scaled multifilament combiners were characterized using a 2-port network analyzer directly, while the fully-differential combiner was measured 6 times as a 2-port network (other ports terminated in 50Ω), and then renormalized to determine the S-parameters of the 4-port test structure [6].

A thru-reflect-line (TRL) de-embedding was adopted to extract the frequency dependent behavior of pad and interconnect parasitics. A separate transmission line is used to extract the characteristic impedance of line standard. The characteristic impedance extracted from measurement of this test structure is 41Ω, versus 42Ω from EM simulation.

B. Measured Performance

Fig. 5 shows the measured S-parameters of the four combiners compared to the EM-simulation results. Excellent agreement is observed between measurement and simulation for all prototypes. The measured data at 40GHz and 60GHz show that the layouts are easily scalable and that the transformer design principles are applicable to multiple frequency bands.

Maximum available gain (MAG) is used to determine the minimum insertion loss of the power combiners. The back-to-back MAG is twice the value of a single power combiner, including losses from transmission lines in the back-to-back test structure that connect the 4 primary ports to each other. Fig. 6 compares the back-to-back measured

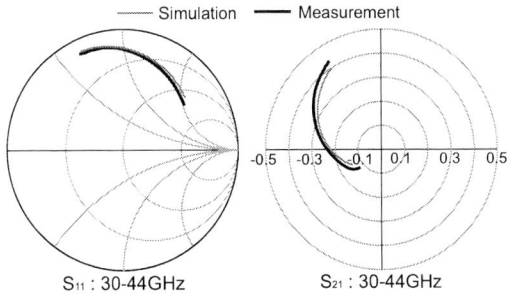

(a) 40GHz compensated balun combiner

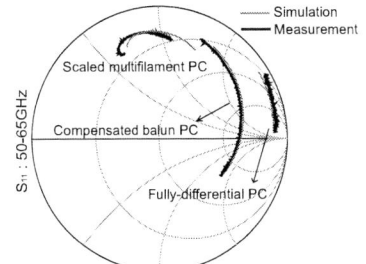

(b) S_{11} of three 60GHz combiners

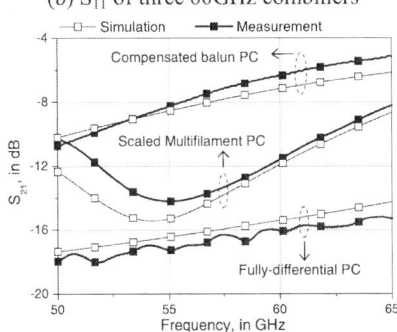

(c) S_{21} of three 60GHz combiners

Fig. 5 Measured and simulated S_{11} and S_{21} in back-to-back testing (note that $S_{22}=S_{11}$, and $S_{12}=S_{21}$).

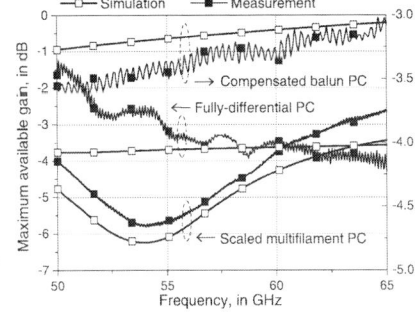

Fig. 6 Measured and simulated MAG in back-to-back testing.

and simulated MAG for the three 60GHz combiners. Good agreement is achieved in the 50-65GHz band, while the discrepancy seen across 50-55GHz for the fully-differential combiner likely arises from accumulated inaccuracy in the multi-step S-parameter measurement procedure described previously. Measurement errors in the 6 measured 2-port datasets are magnified by the sensitivity of MAG to errors in S-parameters. The single compensated balun and fully-

978-1-4244-6240-7/10 $26.00 © 2010 IEEE   201

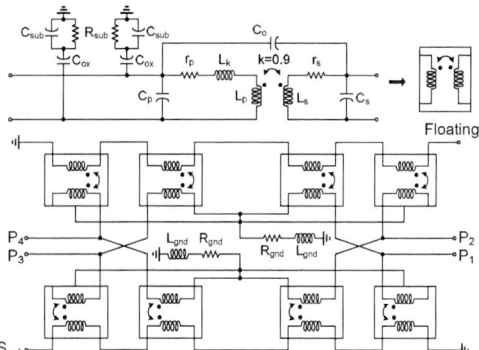

Fig. 7 Simplified model for the compensated balun combiner.

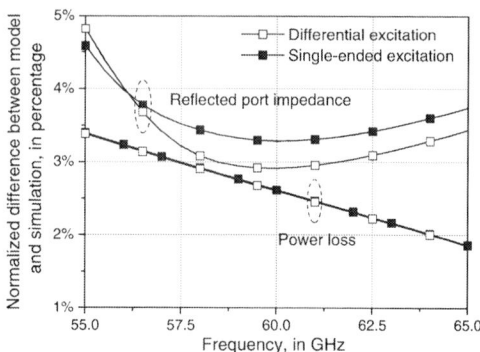

Fig. 8 Difference between model and simulation in the reflected port impedance and power loss.

differential combiners achieve an insertion loss of 1.6dB and 1.8dB at 60GHz, respectively (including the loss of a 200um transmission line). The fully-differential combiner has a higher loss, as the secondary winding is implemented using both top and second (thick) metal layers, leaving the primary in relatively thin and lossy third metal. The scaled multifilament combiner measured insertion loss is 2.0dB at 60GHz, which is the highest among the prototypes. The difference in measured insertion loss of the 3 combiners agrees very well with the simulated results from Section II. The single 40GHz compensated balun has a measured insertion loss of 2.1dB (also including the loss of a 200um transmission line), versus 1.96dB from simulation.

The agreement seen between measurement and simulation validates the EM simulator settings, thereby allowing prediction of the loss of a single combiner from EM simulation. The power loss of both combiners from simulation is <1.0dB. This simulation includes a 50um long transmission line connection at each primary terminal, which would be used in an actual PA layout to connect to the 4 driver transistors (see Fig. 1). Given this insertion loss, mm-wave PA output power can (ideally) be increased by 5dB over that of a single PA cell using either power combiner developed in this work.

IV. Lumped-element Model

A lumped-element circuit model was built to capture the electrical characteristics of the power combiner for large-signal circuit simulation and design in the 55-65GHz range. The simplified model for the compensated balun combiner, which consists of eight 1:1 transformers, is shown in Fig. 7. The model for the fully-differential design is similar.

Optimization in Agilent-ADS simulator is used to extract all the equivalent circuit parameters. Behaviors for both the differential and single-ended excitations are captured by the model. Fig. 8 shows the difference between model and EM simulation in the reflected port impedance and power loss from 55-65GHz. The mismatch between the two is better than 5.0% across the entire band for both single-ended and differential excitations.

V. Conclusions

A current-summing parasitic-compensated balun $(120 \times 120um^2)$ and a fully-differential $(110 \times 90um^2)$ transformer combiners have been implemented in a 130nm SiGe-BiCMOS technology and characterized. Power loss of both prototypes is <1.0dB at 60GHz. Reflected port-to-port impedance uniformity in the 60GHz band within 2.4% and 4.5% is achieved for the compensated balun and fully-differential combiners, respectively. A transformer-based lumped-element model captures the combiner characteristics with <5.0% error in the 55-65GHz band in power loss and reflected port impedance. The combiners reported in this work outperform previous designs in both power loss and uniformity of the transformed impedances, and promise fully-monolithic mm-wave amplifiers with greater output power for wireless communication applications in future.

Acknowledgement

This work was supported by the MEDEA+ project SIAM.

References

[1] U. Pfeiffer, S. Reynolds and B. Floyd, "A 77 GHz SiGe power amplifier for potential applications in automotive radar systems," RFIC Symposium, pp. 91-94, June 2004.

[2] M. Laurens et al, "A 150GHz f_T/f_{max} 0.13μm SiGe:C BiCMOS technology," Proc. of the BCTM, pp. 199-202, Sept. 2003.

[3] T.S.D. Cheung and J.R. Long, "A 21-26-GHz SiGe bipolar power amplifier MMIC," IEEE Journal of Solid-State Circuits, vol. 40, no. 12, pp.2583-2597, Dec. 2005.

[4] G.G. Rabjohn, "Balanced Planar Transformers," US patent, no. 4,816,784, Mar. 1989.

[5] T.Y. Lin, Y.Z. Juang, H.Y. Wang, and C.F. Chiu, "A Low Power 2.2-2.6GHz CMOS VCO with a Symmetrical Spiral Inductor," Proc. of ISCAS, pp. 1641-1644, May 2003.

[6] J.C. Tippet and R.A. Speciale, "A Rigorous Technique for Measuring the Scattering Matrix of a Multiport Device with a 2-port Network Analyer," Trans. on MTT, vol. 30, no. 5, pp. 661-666, May 1982.

RMO3C-1

Integration of multi-standard Front End Modules SOCs on High Resistivity SOI RF CMOS Technology

F. Gianesello[1], S. Boret[1], B. Martineau[1], C. Durand[1], R. Pilard[1], D. Gloria[1], B. Rauber[1] and C. Raynaud[2]

[1]STMicroelectronics, TR&D, STD, 850 avenue Jean Monnet, 38926 Crolles
[2]CEA Leti, avenue des martyrs, 38000 Grenoble

Abstract — RF front end modules (FEMs) are currently realized using a variety of technologies. However, since integration drives wireless business in order to achieve the appropriate cost and form factor, we see significant research concerning FEM integration on silicon [1]. In this quest, SOI technology has already addressed two key blocks, the antenna switch and the power amplifier. In this paper, we will focus our investigation on high performance passive functions in order to demonstrate the capability of SOI CMOS technology to integrate the whole FEM. To do so, balun, harmonic filter, diplexer and directional coupler have been achieved in a 130 nm SOI CMOS technology. Measured performances are clearly competitive with most commercially available Integrated Passive Device (IPD) solutions, which paves the way of FEM silicon SOCs.

Index Terms — SOI, Front End Module, High Resistivity, integrated passive, balun, harmonic filter, diplexer, coupler.

I. INTRODUCTION

CMOS dominates the digital/baseband arena, and thanks to advances in the design and manufacture of integrated circuits (ICs) it has also taken over most highly integrated transceivers such as cellular phones and WLAN. However, CMOS is used only in support roles in the RF front end of modern cell phones. The front end of the cell phone is today served by an assortment of technologies such as GaAs HBT and CMOS.

Today we see significant research in silicon front ends, both bipolar and CMOS [1]. FEM have traditionally integrated multiple technologies on the surface of the Low Temperature Co-fired Ceramic (LTCC) to achieve the overall functionality, but in the same time an important trend has been observed in FEM design, namely, ever-increasing levels of system integration, which has continuously driven FEM manufacturers to reduce component count added to LTCC. The reasons for this trend—especially in consumer electronics—come from the need for lower costs, lower power consumption (especially in mobile and portable products), and smaller product size.

IPDs have been the key enablers to lowering overall component count in the RF section, especially within FEM. For a high volume market such as WLAN and cellular phones, passive devices account for 75 to 85 % of all components used (as illustrated Fig. 1). Several different integrated passive technologies exist today, but the most widely available includes LTCC, Thin Film Silicon, Conventional thin dielectric materials used as PCB, and packaging substrate technology [2].

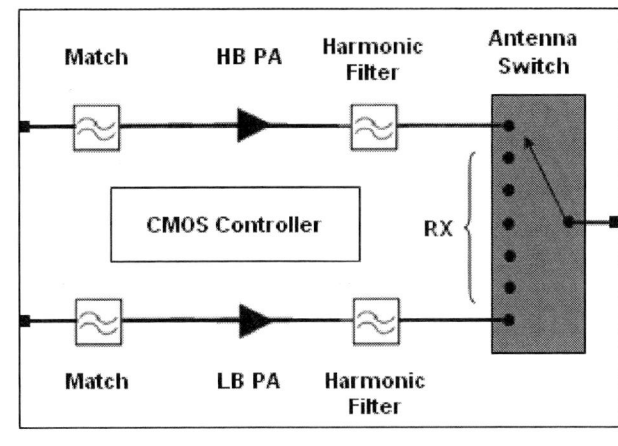

Fig. 1. Architecture of a typical quad band GSM Font End Module

Because SOI CMOS process utilizes an insulating silicon substrate, high Q passives - comparable to those found on LTCC and thin-film silicon solutions - can be integrated into a monolithic die; therefore eliminating costly packaging and die-to-die interconnect issues.

By reviewing the performances offered by STMicroelectronics 0.13 μm HR CMOS SOI technology platform we will discuss here the suitability of SOI technology to integrate the whole FEM on silicon. Since high performance antenna switches have already been integrated on SOI [3] and as the feasibility of SOI power amplifiers has also been reported very recently [4], we focus here our work on the implementation of passive functions (WLAN balun, DCS harmonic filter, WLAN diplexer and GSM directional coupler) on SOI in order to review the achievable performance on CMOS SOI and then conclude concerning the suitability of this technology to integrate complete FEM on a same die.

II. HR SOI CMOS TECHNOLOGY DESCRIPTION

FEMs require a combination of good isolation, high power/high voltage handling, high efficiency, good passive elements and complementary devices for digital integration. We will quickly review the performances offered by STMicroelectronics 130 nm SOI CMOS Technology in the context of FEM.

978-1-4244-6240-7/10 $26.00 © 2010 IEEE

A. Antenna Switch Integration in HR SOI CMOS

In recent years SOI switches have received widespread acceptance in the cellular handset market [4], [5], [6]. Antenna switches in bulk CMOS have been researched, but suffer from lower power handling and increased insertion loss. SOI switches remove the bulk effects, which allow high voltage handling through device stacking. They also have lower capacitance, reduced parasitics, and better isolation than bulk CMOS switches.

B. Thin Film SOI NLDMOS

Integration of high voltage devices in advanced technology is mandatory to enable the integration of PA. Drain extension MOS transistors have been successfully integrated in thin film HR SOI CMOS technology [7]. In STMicroelectronics 130 nm HR SOI CMOS technology, this device has a breakdown voltage of 17 V and Ft/Fmax of 25/50 GHz. Achieved large signal performance are clearly suitable for the integration of PA dedicated to WLAN and cellular product. Moreover, equivalent results have already been achieved in 65 nm SOI CMOS technology [8], which is promising for the integration of the PAs with the RF transceiver and digital processor on the same die in an advanced SOI technology.

C. High Performance Passive Components Integrated on HR SOI CMOS Technology

The integration of high quality passive components in standard bulk silicon technology is not obvious. It is well known that passive components integrated in standard silicon technologies suffer from high substrate losses. On the contrary, since SOI technology is fully compatible with HR substrate without inducing any impact on latch up or isolation, high performance passive components can be achieved on SOI. We can note that the insulating behavior of HR SOI material used here has been confirmed up to 200 GHz [9].

Then high quality factor inductor (Q factor exceeding 30 cf. Fig. 2) can be achieved on SOI using a standard digital BEOL, which are mandatory for FEM design. Those results can also be improved by the use of a thick copper module [10].

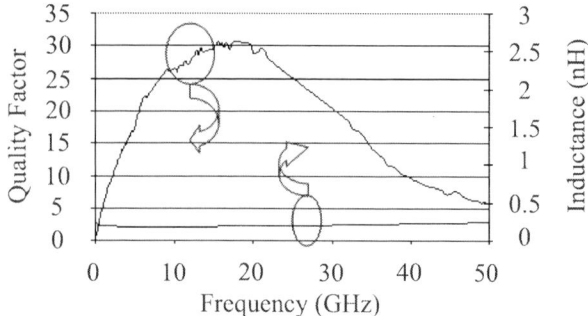

Fig. 2. Quality factor of a 0.183 nH inductor achieved in STMicroelectronics 0.13 µm HR SOI CMOS technology [11]

III. REVIEW OF FEM PASSIVE FUNCTIONS INTEGRATED IN HR SOI CMOS

In order to investigate the potentiality of FEM passive devices integration on SOI, we have designed WLAN balun, DCS harmonic filter, WLAN diplexer and GSM directional coupler using STMicroelectronics 0.13 µm HR SOI CMOS technology.

A. WLAN Balun Integrated in HR SOI

Recently, integrated baluns have demonstrated to become a key component in the design of advanced RF transceivers since most of them have now single ended outputs toward the FEM (while the design of the RF transceiver itself remains differential).

Designers have been used to employ very low loss baluns (< 1 dB) because achieved using IPD technologies. Integrating the balun on CMOS silicon technology generally degrades the situation since it is very difficult to integrate a balun (at RF frequency) in bulk CMOS with insertion losses lower than 1.5 dB.

But as for inductor, SOI CMOS technology has here an advantage. To investigate this point, a classical 2.5 GHz stacked balun (50 Ohms single ended towards 100 Ohms differential) has been achieved. Measured insertion loss is 0.9 dB (cf. Fig. 3) and an excellent agreement has been obtained with electromagnetic simulation.

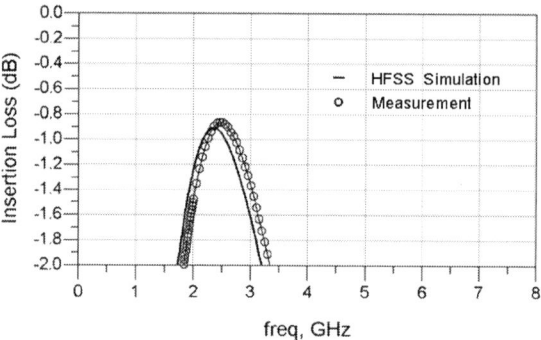

Fig. 3 Insertion loss of a 2.5 GHz stacked balun achieved on SOI

B. DCS Harmonic Filter Integrated in HR SOI

Harmonic filter is another key block of FEM design. During this work, a very compact DCS harmonic filter has been designed on SOI (cf. Fig. 4).

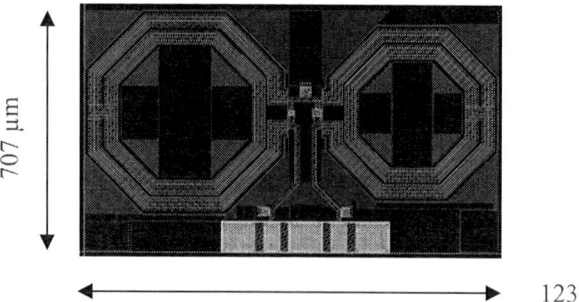

Fig. 4 Die photography of the DCS harmonic filter

978-1-4244-6240-7/10 $26.00 © 2010 IEEE

Measurements have been performed on wafer up to 10 GHz using an Agilent 50 GHz HP8510C vector network analyzer and cascade microtech GSGSG RF probes.

As we can see on Fig. 5 & 6, good agreement has been obtained between measurement and simulation (demonstrating Design Kit inductor scalable models accuracy). Obtained insertion losses are 0.88 dB @ 1710 MHz and 1.03 dB @ 1980 MHz (cf. Fig. 6), which is suitable for the design of PA.

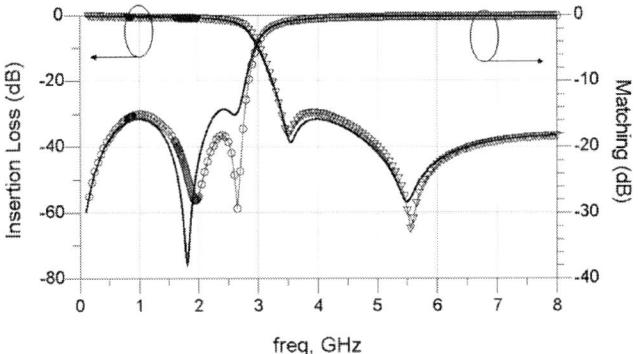

Fig. 5 Comparison of DCS harmonic filter simulated S-Parameter using electrical DK models (solid line) and measured ones (symbols)

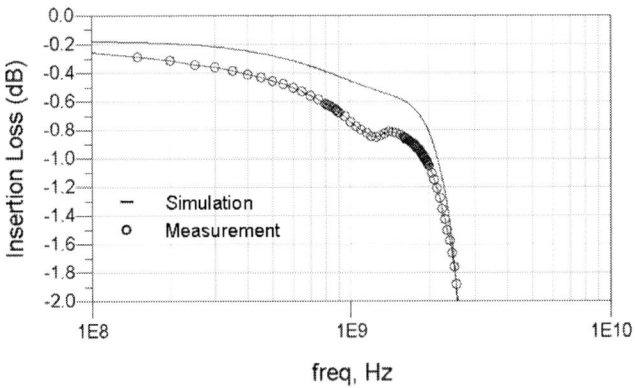

Fig. 6   Comparison of DCS harmonic filter simulated insertion loss using electrical DK models (solid line) and measured one (symbols)

C. WLAN Diplexer Integrated in HR SOI CMOS

Diplexer is the following next key passive function required in the design of modern FEMs. Especially if considering MIMO application such as the popular 802.11n standard.

To illustrate the capability of the SOI a very compact WLAN diplexer has been designed on SOI (cf. Fig. 7) [12]. As we can see Fig. 8, excellent agreement has been obtained between measurement and simulation (demonstrating both DK models accuracy and design methodology workability). Obtained insertion losses are in the order of 1 dB (cf. Fig. 9) and isolation ~ -18 dB for both bands.

We can note here the very compact size achieved for this diplexer (0.325 mm2), which is from the author knowledge the smallest one reported in the literature up to date.

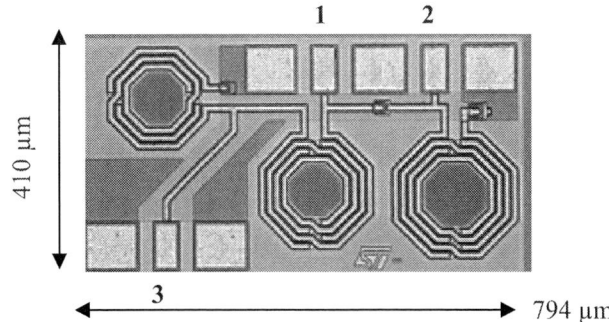

Fig. 7   WLAN diplexer Die photography [12]

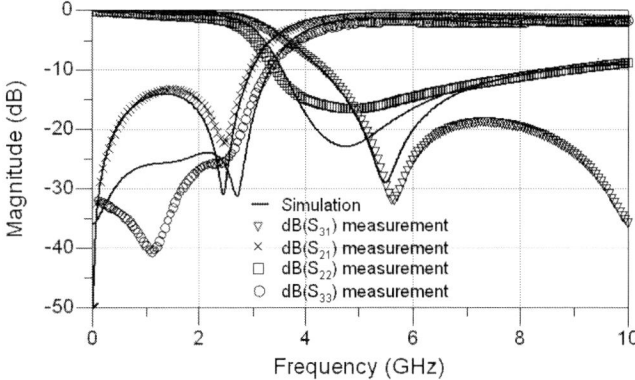

Fig. 8 Comparison of WLAN diplexer simulated S-Parameter using electrical DK models (solid line) and measured ones (symbols) [12]

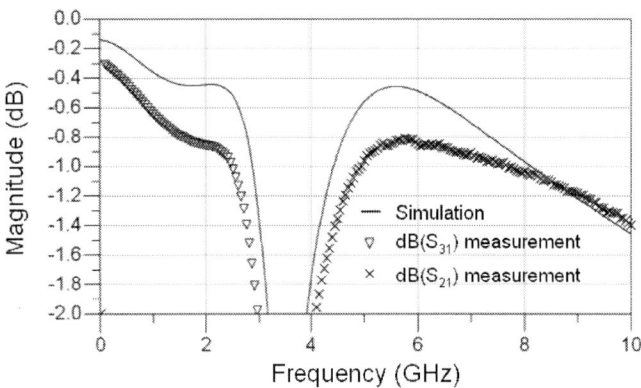

Fig. 9   Comparison of DCS harmonic filter simulated insertion loss (solid line) and measured one (symbols) [12]

D. GSM Directional Couplers Integrated in HR SOI CMOS

Directional coupler is the last key passive function required in the design of an advanced FEM design. This circuit is used to sense the power level at output of the PA.

Since the energy management circuit is already achieved in CMOS, if we can integrate the directional coupler on silicon with the appropriate performances this will enable the integration of the PA and the energy management on the same silicon die. To investigate this point a lumped backward directional coupler has been achieved on SOI (cf. Fig. 10).

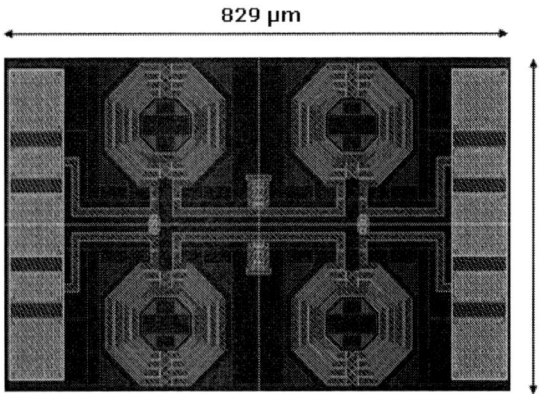

829 µm

520 µm

Fig. 10   Die photography of the GSM directional coupler

Measurements have been performed on wafer up to 10 GHz using an Agilent 26.5 GHz PNA-X four ports vector network analyzer and cascade microtech GSGSG RF probes.

As we can see Fig. 11 & 12, good agreement has been obtained between measurement and simulation (demonstrating Design Kit inductor scalable models accuracy). Measured insertion losses are 0.9 dB @ 800 MHz and 1. 4 dB @ 950 MHz with a coupling of 20 dB (cf. Fig. 6), directivity is ~ -40 dB and isolation -30 dB, which is suitable for the design of a GSM PA.

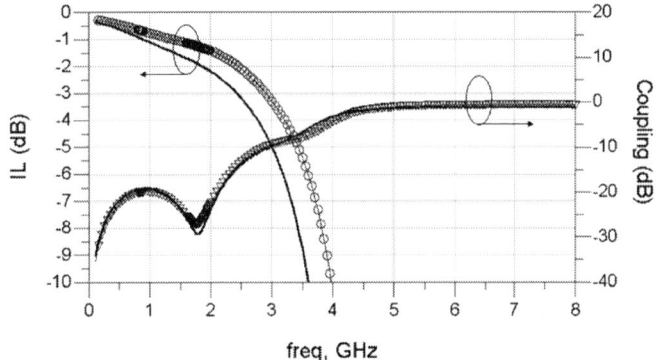

Fig. 11   Comparison of GSM directional coupler simulated insertion loss and coupling (solid line) and measured one (symbols)

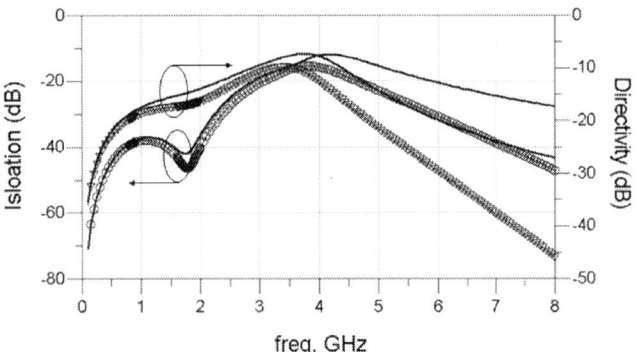

Fig. 12   Comparison of GSM directional coupler simulated isolation and directivity (solid line) and measured one (symbols)

IV. CONCLUSION

In this paper, FEM passive functions have been achieved in an HR SOI CMOS technology demonstrating performances competitive with most commercially available IPD solutions. Those results are promising for the integration of ultra low cost FEM on HR SOI CMOS technology.

CMOS HR SOI technology has then clearly to be considered has a promising one for the integration on silicon of multi standard RF FEM since it offers the highest possible degree of integration and a cost effective approach.

Moreover, since equivalent results are expected in STMicroelectronics 65 nm HR SOI CMOS technology, it opens the way for the whole integration with the base band processor, which would take advantage of an advanced Low Power CMOS SOI technology in order to reduce its power consumption and lower the cost of the global chipset solution.

REFERENCES

[1]   M. Apostolidou et al., "A 65nm CMOS 30dBm Class-E RF Power Amplifier with 60% Power Added Efficiency", IEEE RFIC, June 2008, Atlanta, pp. 141-144.

[2]   T. Kamgaing et al., "A Compact 802.11 a/b/g/n WLAN Front-End Module Using Passives Embedded in a Flip-Chip BGA Organic Package Substrate," IEEE MTT-S International 7-12 June 2009, Boston, pp. 213-216.

[3]   C. Tinella et al., "0.13um CMOS SOI SP6T Antenna Switch for Multi-standard Handsets," Topical Meeting on Silicon Monolithic Integrated Circuits in RF Systems, Jan 2006, San Diego, pp. 58-61.

[4]   A. Tombak et al., "Integration of a Cellular Handset Power Amplifier and a DC/DC Converter in a Silicon-On-Insulator (SOI) Technology", IEEE RFIC 2008, Atlanta, pp. 413-416.

[5]   J. Costa et. al, "A Silicon RFCMOS SOI Technology for Integrated Cellular/WLAN RF TX Modules," 2007 IEEE Microwave Symposium, 3-8 June 2007, pp. 445-448.

[6]   D. Kelly et al., "The State-of-the-Art of Silicon-on-Sapphire CMOS RF Switches," Compound Semiconductor Integrated Circuit Symposium, Oct 30-Nov. 2, 2005.

[7]   O. Bon et al., "RF Power NLDMOS Technology Transfer Strategy from the 130nm to the 65nm node on thin SOI", IEEE International SOI Conference, Palm Spring, 1-4 Oct. 2007, pp. 61 - 62

[8]   O. Bon et al., "RF Power NLDMOS Technology Transfer Strategy from the 130nm to the 65nm node on thin SOI", IEEE SOI Conference, Palm Spring, October 2007, PP. 61-62.

[9]   F. Gianesello et al., "1,8 dB insertion loss 200 GHz CPW band pass filter integrated in HR SOI CMOS Technology", IEEE MTT-S, Hawaï, June 2007, pp. 453-456.

[10] C. Pastore et al., "Double Thick Copper BEOL in Advanced HR SOI RF CMOS Technology: Integration of High Performance Inductors for RF Front-End Module", IEEE SOI Conference, New York, October 2008, pp. 137-138.

[11] F. Gianesello et al., "Opportunity and Perspectives of using Advanced High Resistivity SOI CMOS technology for the Integration of Multi-Standard RF Front-End", IEEE Power Amplifier Symposium, San Diego, January 2009.

[12] F. Gianesello et al., "Small-Size Low Losses WLAN and GSM/DCS Diplexers integrated in a Low Cost 130 nm High Resistivity SOI CMOS Technology", IEEE EuMW, Roma, October 2009, pp. 590-593.

RMO3C-2

Vast-Fast Low-Triggering LTdSCR ESD Protection Structure for RF ICs in CMOS

Jian Liu[1], Lin Lin[1], Xin Wang[1], Hui Zhao[1], He Tang[1], Qiang Fang[1], Albert Wang[1], Hongyi Chen[2], Haolu Xie[3], Siqiang Fan[4], Bin Zhao[4] and Gary Zhang[5]

[1] Department of Electrical Engineering, University of California, Riverside, CA 92521, USA, aw@ee.ucr.edu, [2] Tsinghua University, [3] Fudan University, [4] Freescale Semi., [5] Skyworks

Abstract — This paper reports design of a novel low-parasitic ultra-low-triggering voltage dual-directional LTdSCR ESD protection structure in foundry CMOS. It features programmable low triggering voltage of 4.7~6V, low discharging resistance of ~0.77Ω, low leakage of ~0.1nA, extremely low parasitic capacitance of ~10fF and ultra fast response of ~100ps. it achieves ESD protection of >7.8kV HBM and ~500V CDM for a 90μm device. Measurement matches simulation very well. This low-parasitic low-triggering ESD protection structure is suitable for high data rate and low-voltage RF ICs in CMOS.

Index Terms — ESD, low-parasitic, RF IC, CMOS, SCR.

I. INTRODUCTION

The International Technology Roadmap for Semiconductor (ITRS) has cited low-voltage low-parasitic ESD (electrostatic discharge) protection design as a key emerging reliability challenge to very-deep-submicron (VDSM) ICs, particularly for parasitic-sensitive mixed-signal and RF ICs. Silicon-controlled-rectifier (SCR) ESD structure is superior over other ESD protection devices due to its deep snapback I-V, hence, smaller size and lower parasitic for even higher ESD protection [1]. However, ESD triggering voltage (V_{t1}) of SCR structure, dictated by the high Nwell/Pwell junction breakdown (Fig. 1a), is too high for most ICs. Modifications were done to reduce V_{t1} of SCR for relatively lower V_{t1} by the N+/Pwell junction breakdown in middle-voltage MVSCR structure (Fig. 1b) [2], and for even-lower V_{t1} in a low-voltage LVSCR structure (Fig. 1c) by use of ggNMOS triggering (Fig. 1d) [3]. However, such low triggering V_{t1} is still too high for advanced low-voltage ICs. On the other hand, up to four ESD devices may be needed for each I/O pin if using traditional asymmetrical one-directional ESD structure to ensure complete ESD protection at whole chip, resulting in substantial ESD size and ESD-induced parasitics (capacitance, C_{ESD}, and noise figure, NF, etc) that are intolerable to advanced RF ICs [4].

This paper presents a novel very-low-parasitic ultra-low-triggering dual-directional LTdSCR ESD protection structure designed and fabricated in 0.18μm foundry CMOS to address the above challenges, featuring lowest reported $V_{t1} \leq 4.76V$, ultra fast response of $t_1 \sim 100ps$, full symmetrical I-V snapback, 7kV+ HBM and 500V+ CDM ESD protection, and a record low $C_{ESD} \sim 10fF$.

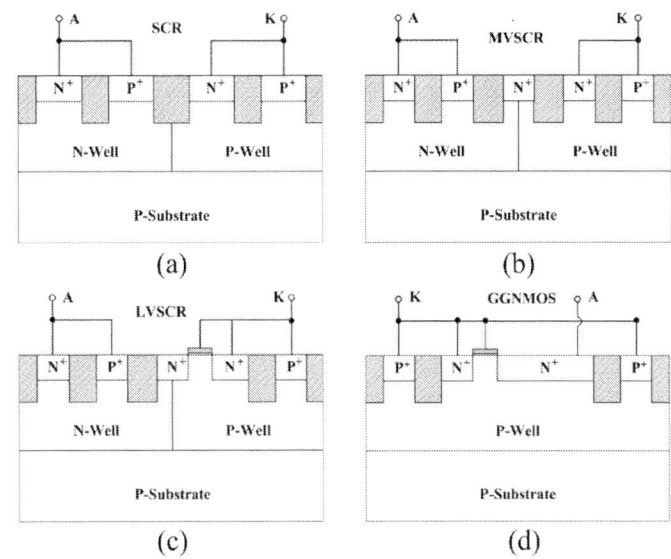

Fig. 1 Traditional SCR ESD structures: (a) SCR uses Nwell/Pwell breakdown for triggering; (b) MVSCR reduces V_{t1} by inserting an N+ across Nwell/Pwell boundary; (c & d) LVSCR uses ggMOS (grounded-gate) to further reduce V_{t1}.

II. LTdSCR ESD STRUCTURE AND DESIGN

Fig. 2 shows a cross-section for the new LTdSCR ESD structure in standard CMOS, which is a symmetrical two-terminal (Anode, A, and Cathode, K) five-layer device ($N_1P_2N_3P_4N_5$) consisting of one PNP transistor (Q_1), two NPN transistors (Q_2 and Q_3), series resistors (R_1, R_2, R_3 & R_4), and two floating-gate NMOS transistors (M_1 and M_2). Fig. 3 gives its equivalent circuit. For general ESD protection, an LTdSCR is connected to an I/O pin (A) and Ground (GND, K). For the core LTdSCR without M_1/M_2, as a positive ESD transient occurs at I/O with respect to GND, it reverse-biases $P_2N_3P_4$ collector of Q_1 (*i.e.*, deep-Nwell/Pwell, N_3/P_4) to its breakdown. The avalanche current is collected by GND via the series R_2 (P_4), which biases $N_3P_4N_5$ emitter of Q_3, turns on the Q_3 and Q_1, consequently triggers the SCR of Q_1+Q_3 at a given V_{t1}

978-1-4244-6240-7/10 $26.00 © 2010 IEEE

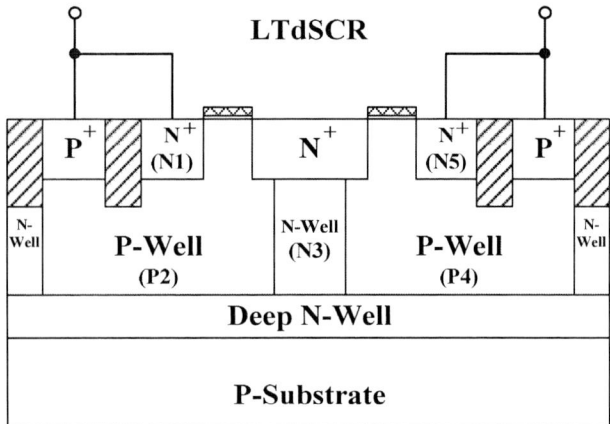

Fig. 2 Cross-section for new LTdSCR ESD protection structure.

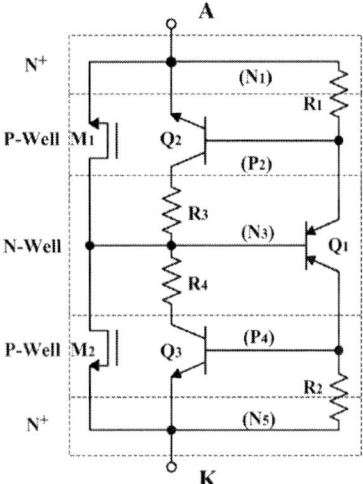

Fig. 3 Equivalent circuit for the new LTdSCR ESD structure.

and forms a very-low-resistance (R_{on}) conduction path to discharge the large ESD pulse. The deep snapback I-V, inherent to SCR, serves to clamp the I/O pad to a very low holding voltage (V_h) to prevent dielectric breakdown in CMOS. A similar ESD discharging takes place when a negative ESD transient occurs at I/O with respect to GND, in which a SCR of Q_1+Q_2 is triggered off to provide ESD protection. Therefore, the desired symmetrical dual-directional dSCR ESD operation with snapback I-V is realized. The integrated M_1/M_2 devices serve to substantially further reduce the V_{t1} for SCR via punch-through breakdown. Unlike in the existing LVSCR where V_{t1} starts by BV_{DS} breakdown of ggNMOS, the M_1/M_2 are designed for channel punch-through by fine-tuning its channel length (L) for varying breakdown voltage, which in turn, determines V_{t1} of LTdSCR. Hence, new LTdSCR achieves much lower V_{t1}, which is otherwise limited by a relatively higher ggNMOS breakdown in LVSCR. A novel LTdSCR ESD protection structure is therefore

realized with adjustable ultra low triggering voltage and full symmetrical snapback I-V characteristics for low-voltage ICs. The very high current-handling capability ensures that the new LTdSCR can realize super compact, extremely low parasitic ESD protection highly desired for RF ICs.

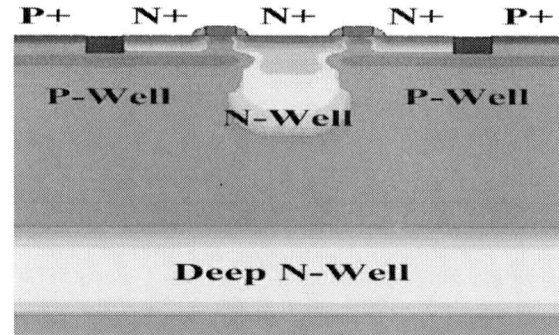

Fig. 4 An LTdSCR structure created by ESD simulation.

III. DESIGN AND MEASUREMENT

Various LTdSCR structures with different channel length (L) were designed and fabricated in a foundry 0.18μm CMOS. To verify the concept and optimize its performance, mixed-mode ESD simulation technique [5] was used to design the new LTdSCR. Fig. 4 shows an LTdSCR device created by simulation and Fig. 5 depicts its current discharging path under ESD stressing where the SCR ESD discharging behavior is readily observed. Fig. 6 shows a perfect I-V characteristic for one exemplary LTdSCR device under ESD stressing by mixed-mode ESD simulation. Design splits were used to optimize ESD operation by changing critical dimensions for LTdSCR including varying L = 0.18μm~0.45μm to adjust V_{t1} and a device width of W=90μm.

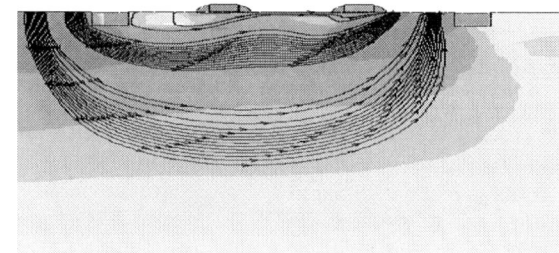

Fig. 5 Mixed-mode ESD simulation shows SCR-type ESD discharging current flows for LTdSCR under ESD stressing.

Comprehensive ESD testing was conducted for all LTdSCR devices including TLP (transmission line pulsing) for ESD performance, s-parameter measurement for ESD

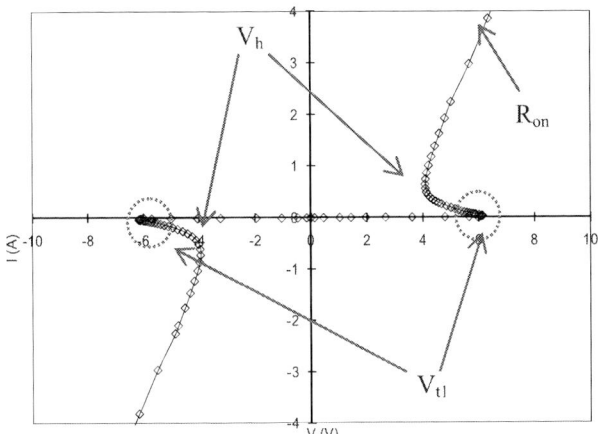

Fig. 6 Mixed-mode ESD simulation shows symmetrical ultra-low-V dual-directional snapback I-V behavior for LTdSCR. Critical ESD operation parameters for design prediction by ESD simulation are triggering (V_{t1}), holding (V_h), discharging resistance (R_{on}) and thermal breakdown (I_{t2}, Fig. 8), etc.

parasitic C_{ESD} extraction and noise figure test, etc. Fig. 7 shows measured I-V curve for LTdSCR by TLP with a pulse rise time (t_r) of 10ns and pulse width of 100ns for HBM ESD testing. It clearly shows the expected symmetrical dual-directional SCR-type I-V snapback behavior required for ESD protection with a desired very low triggering voltage of V_{t1}~ 5.62V achieved. Fig. 8 gives a complete I-V characteristic by TLP in one-direction ESD stressing until thermal breakdown, showing excellent low-voltage triggering V_{t1}~5.62V, snapback holding voltage V_h ~ 3.19V, very low ESD discharging resistance R_{on}~0.77Ω and high ESD failure current I_{t2}~5.25A for a 90μm device. Table I summarizes all measured critical ESD operation parameters for the new LTdSCR structures. It also shows highly desired very low leakage current for LTdSCR, I_{leak} ~ 0.1nA. Table I shows that TLP test agrees well with simulation except for I_{t2}. The discrepancy between the simulated and measured I_{t2} values may be explained as following: In ESD simulation, the I_{t2} corresponds to purely thermal breakdown current; however, the I_{t2} point is obtained much more conservatively in TLP testing that reflects real ESD failure threshold corresponding to where the I_{leak} jumps over by several orders. The so-identified high I_{t2}~5.25A translates into a HBM ESD protection level of >7.8kV, or ESDV≅88V/μm width, which is a very high ESD protection to size ratio for any ESD protection structure. Other minor discrepancies are attributed to more simulation calibration needed. To further characterize the LTdSCR for more challenging CDM ESD protection model that features extremely fast ESD pulse rise time of t_r~400ps compared to t_r~10ns for HBM model, very-fast TLP (VF-TLP) test was conducted with the obtained I-V

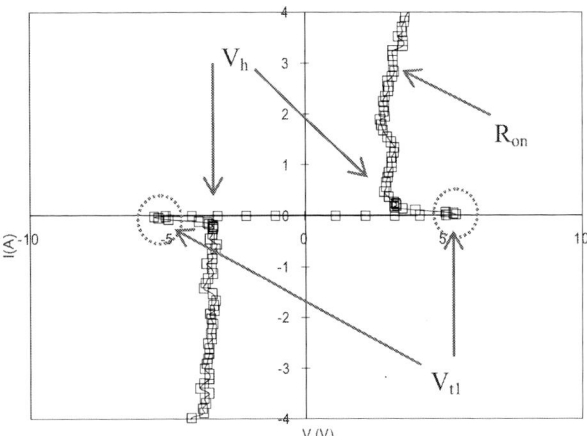

Fig. 7 TLP test reveals a symmetrical ultra-low-V dual-directional snapback I-V curve for LTdSCR structure. All ESD-critical parameters match simulation results quite well.

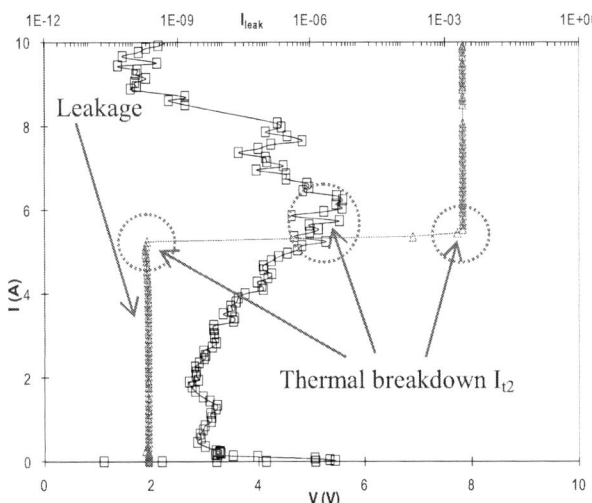

Fig. 8 Full 10ns TLP test shows the desired I-V snapback curve measured all the way to its thermal breakdown and the desired very low leakage current for LTdSCR structure.

Table I.   Critical ESD parameters for LTdSCR.

ESD Parameters	Characterization Methods		
	Simulation	TLP	VF-TLP
V_{t1} (V)	6.16	5.62	5.57
V_h (V)	4.09	3.19	3.60
I_{t2} (A)	8.60	5.25	4.51
R_{on} (Ω)	1.24	0.77	0.21

characteristics shown in Fig. 9 and the key ESD parameters listed in Table I. VF-TLP test results clearly show that the new LTdSCR readily passed CDM ESD testing of I_{t2}~4.51A, or ~500V CDM ESD protection level. Fig. 10 shows the tested V_{t1}~L relation designed to adjust the ESD triggering voltage for the new LTdSCR structures for multi-supply applications, with the key parameters

978-1-4244-6240-7/10 $26.00 © 2010 IEEE

given in Table II. The results clearly show that ESD triggering V_{t1} can be adjusted by L, resulting in a programmable V_{t1} ranging from 4.76V to 6V in this exemplary LTdSCR. A wider V_{t1} tuning range can be readily achieved by carefully designing the LTdSCR structures, including L, hence, making the new LTdSCR highly flexible to various low-voltage ICs. Table III gives the tested $V_{t1} \sim t_r$ results for various TLP pulses, which show good triggering for very fast ESD stressing down to 100ps and confirms that the new LTdSCR ESD protection structure can work for both HBM and ultra fast CDM ESD protection, highly desired for advanced mixed-signal ICs. Table IV compares the measured V_{t1} values for the new LTdSCR and other ESD structures, which readily confirms that the LTdSCR features exceptionally low triggering voltage over other ESD structures. Full S-parameter and noise testing was conducted across a 10GHz spectrum. The testing result shows that the LTdSCR structure has a record-low parasitic capacitance of $C_{ESD} \sim 10fF$ at GHz operation, which suggests that the novel LTdSCR structure is ideal for ESD protection of advanced mixed-signal and RF ICs

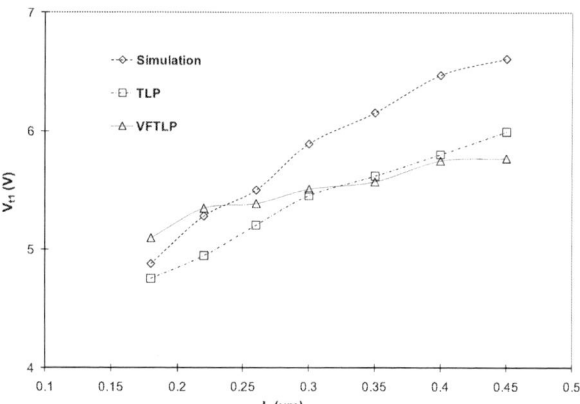

Fig. 10 $V_{t1} \sim L$ curves by simulation and measurement confirm V_{t1} reduction and programmability for LTdSCR.

Table IV. V_{t1} for different ESD devices

ESD Structures	V_{t1} (V)	
	TLP	VF-TLP
SCR	16.71	17.5
MVSCR	13.61	13.7
LVSCR	6.743	6.61
GGNMOS	6.584	6.20
LTdSCR	**<4.76**	**<5.10**

IV. CONCLUSION

We report a novel low-parasitic, ultra-fast, ultra low triggering, dual-directional LTdSCR ESD protection in CMOS. LTdSCR features ultra-low and adjustable $V_{t1} \cong 4.7 \sim 6V$, $R_{on} \cong 0.77\Omega$ and $I_{leak} \cong 0.1nA$, $C_{ESD} \cong 10fF$ and ultra fast response time of ~100ps. It achieves ESD protection levels of >7.8kV HBM and ~500V CDM for a 90μm device. LTdSCR is ideal for RFIC ESD protection.

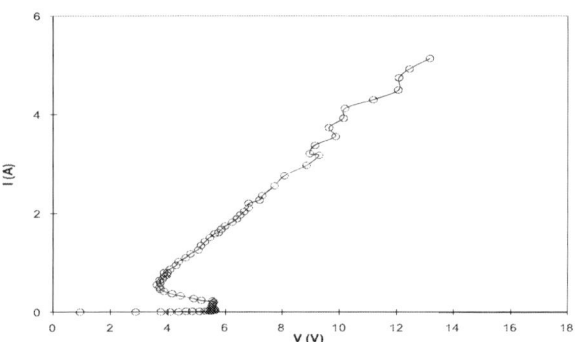

Fig. 9 Ultra fast 100-400ps VF-TLP test for LTdCSR confirms very fast response time (t_r•100ps).

Table II. $V_{t1} \sim L$ controllability for LTdSCR.

Gate Length L (μm)	V_{t1} (V)		
	Simulation	TLP	VF-TLP
0.18	4.88	4.76	5.10
0.22	5.28	4.95	5.35
0.26	5.50	5.21	5.39
0.30	5.89	5.45	5.51
0.35	6.16	5.62	5.57
0.40	6.47	5.80	5.75
0.45	6.61	6.00	5.77

Table III. $V_{t1} \sim t_r$ for LTdSCR in TLP tests.

	t_r			
	10ns	400ps	200ps	100ps
V_{t1} (V)	5.62	5.76	5.78	5.77

REFERENCES

[1] A. Wang and C. Tsay, "On a dual-direction on-chip ESD protection structure", *IEEE Trans. Elec. Dev.*, V48, N5, p978, May 2001.
[2] L. Avery, "Low trigger voltage SCR protection device and structure", *US Patent No. 5,072,273*, Dec. 1991.
[3] A. Chatterjee and T. Polgreen, "A low-voltage triggering SCR for on-chip ESD protection at output and input pads", IEEE *Elec. Dev. Lett.*, V12, N1, p21, Jan. 1991.
[4] A. Wang, H. Feng, R. Zhan, H. Xie, G. Chen, Q. Wu, X. Guan, Z. Wang and C. Zhang, "A review on RF ESD protection design", *IEEE Trans. Elec. Dev.*, V52, N7, p1304, July 2005.
[5] H. Feng, G. Chen, R. Zhan, Q. Wu, X. Guan, H. Xie, Z. Wang and R. Gafiteanu, "A mixed-mode ESD protection circuit simulation-design methodology", *IEEE J. Solid-State Circ.*, V38, N6, p995, July 2003.

RMO3C-3

A Cost-Competitive High Performance Junction-FET (JFET) in CMOS Process for RF & Analog Applications

Yun Shi, Robert M. Rassel, Richard A. Phelps, Panglijen Candra, Douglas B. Hershberger, Xiaowei Tian, Susan L. Sweeney, Jay Rascoe, BethAnn Rainey, Jim Dunn, and David Harame

IBM Microelectronics, Essex Junction, Vermont, 05452, USA

Abstract — in this paper, we present a cost-effective JFET integrated in 0.18μm RFCMOS process. The design is highly compatible with standard CMOS process, therefore can be easily scaled and implemented in advanced technology nodes. The design impact on R_{on} and V_{off} is further discussed, providing the insights and guidelines for JFET optimization. Besides the superior flicker noise (*1/f* noise) characteristics, this JFET device also demonstrates promising RF characteristics such as maximum frequency, linearity, power handling capability, power-added efficiency, indicating a good candidate for RF designs.

Index Terms — JFET, R_{on}, V_{off}, *1/f* noise, RF characteristics, optimization.

I. INTRODUCTION

We are living in an era of information and technologies. Wireless communication, which emerged as one of the major media, has evolved tremendously from conventional voice service to high speed data and multimedia applications. The emerging generation is able to provide a comprehensive IP solution, where voice, data and streamed multimedia can be given to users on an "Anytime, Anywhere" basis. Over the time, Silicon technology has continuously been the backbone and key contributor for the successful wireless evolution. Facing today's challenge in system-on-chip (SoC), technology innovations are being driven to create various semiconductor devices in Silicon substrate. However, Junction-FETs (JFETs), featuring low flicker noise and robust breakdown, are long being investigated using compound materials for RF and millimeter-wave IC (MMIC) applications [1]-[3]. Our effort was to explore a Silicon JFET solution with adequate *ac* and RF characteristics to assist SoC designs.

In this paper, we introduce a cost effective, implanted JFET in a standard CMOS process. The device demonstrates low ON resistance (R_{on} = 1.3 Ω-cm), low pinch off voltage (V_{off} = -2V), 20X lower *1/f* noise than MOSFET, and good power capability with peak f_{max} at 23 GHz. The load pull data from the discrete JFET shows the maximum power-added efficiency (PAE) can achieve 40% at 1.9 GHz. All data suggest very competitive performance comparing to SiC and GaAs JFETs. This JFET feature makes the existing RFCMOS technology more appealing to advanced RF and MMIC designs.

II. DEVICE INTEGRATION

This Silicon JFET is an implanted, symmetric device, integrated into 0.18μm RFCMOS technology. A schematic cross section is shown in Fig. 1. Unlike the conventional designs, this JFET is bounded by shallow trench isolation (STI). The STI enables high drain-to-source operation voltage, while still maintaining aggressive pitch and low ON resistance. As shown, the n-channel is sandwiched between top and bottom gates, which are tied together via layout. The dual gate design enables lower pinch off voltages since the channel is being depleted from both top and bottom junctions. The channel and bottom gate doping profile is optimized to fulfill the following criteria: (1) competitive ON resistance vs. pinch off voltage trade off, (2) the same "top gate to channel capacitance" and "bottom gate to channel capacitance", and (3) avoiding pinch off the channel underneath STI. The JFET utilizes the standard CMOS triple well isolation, deep N-well, to isolate the bottom gate from underlying substrate. The "bottom gate/deep N-well junction" is designed for low capacitance and high breakdown voltage. Overall, this JFET design is highly compatible with CMOS process, very cost-effective, and scalable to advanced CMOS nodes.

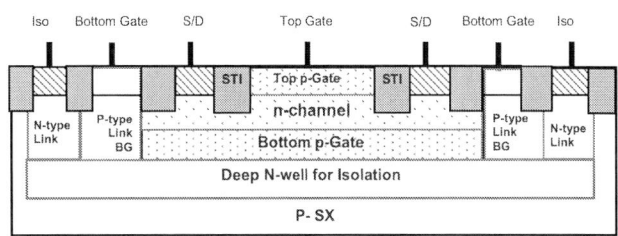

Fig. 1 Schematic cross section of the implanted n-channel JFET, integrated in a standard CMOS process flow.

978-1-4244-6240-7/10 $26.00 © 2010 IEEE

III. DEVICE CHARACTERISTICS

A. DC characteristics

In the first order, a JFET is characterized by the "R_{on} vs. V_{off}" benchmark. A high channel doping is favorable for low R_{on}, however, makes it difficult to pinch off. Therefore, a fundamental trade-off between these two parameters is shown in Fig. 2, when the channel design is varied. R_{on} is calculated as the ratio of V_{ds}/I_d, where $V_{ds} = 0.05V$ and $V_{gs} = 0V$. To measure V_{off}, V_{gs} is swept from 0V towards negative values, and V_{off} is defined as the V_{gs}, at which the drain current I_d is lowered by four order of magnitude (10^4) than I_d at $V_{gs} = 0V$. The red diamond population in Fig.2 represents the nominal JFET. The channel length (L_{ch}) is 0.4µm, R_{on} is 1.3 Ω-cm, and V_{off} is -2V at $V_{ds} = 1.8V$.

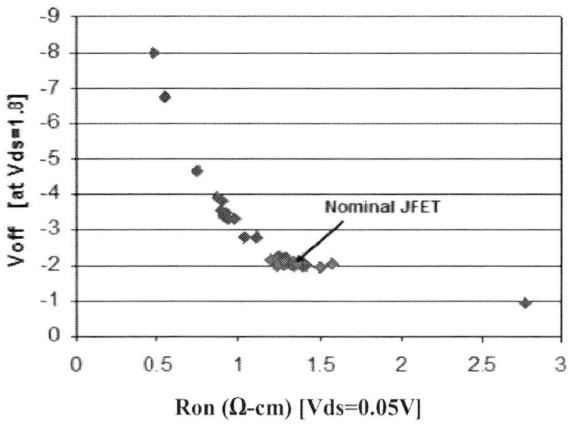

Fig. 2 The R_{on} vs. V_{off} benchmark.

To gain insights of the device, we examined the JFET with different channel length (L_{ch}) at 0.3µm, 0.4µm, 0.5µm, and 1.2µm, respectively. The R_{on} data are plotted against L_{ch}, as shown in Fig. 3a. Due to the intrinsic channel resistance reduction, R_{on} is shown to scale linearly with L_{ch}. When we perform a linear extrapolation, the interception with *y-axis* is the parasitic link resistance of the JFET. This allows us to separate the R_{on} contribution between intrinsic channel resistance (R_{ch}) and extrinsic link resistance (R_{link}). The fitting result shows that R_{ch} is 0.3 Ω-cm, accounts for 23% of total R_{on}, and the remaining 77% is due to R_{link}. Accordingly, pinch off (V_{off}) vs channel length characteristic is plotted in Fig. 3b. Data shows L_{ch} scaling does not impact V_{off}, prove the intrinsic channel pinch off, and a well designed JFET without premature extrinsic pinch off, nor leaky junctions.

Though reducing R_{link} can effectively lower R_{on}, the risk remains for pinch off and breakdown performances. When doping is increased, the field at the junction also increases. As a result, the device leakage current is high; eventually, V_{off} is no longer controlled by gates. Secondly, a more abrupt junction also risks a lower breakdown voltage V_{bd}. It requires extensive design works to ensure neither V_{off} nor V_{bd} will degrade. We utilize the Technology CAD (TCAD) capabilities to assist the optimization. A wide range of processes were simulated to gain the knowledge, which enables us to shift R_{on} vs. V_{off} to a more competitive curve without sacrificing V_{bd}.

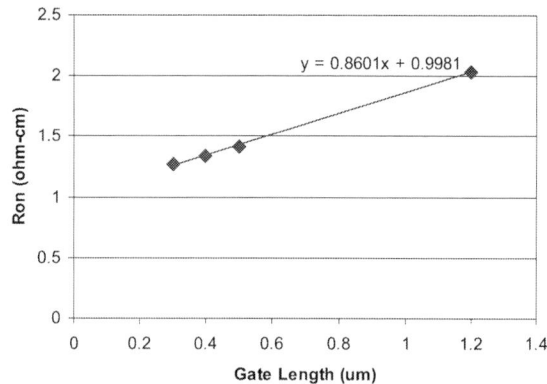

Fig. 3a The R_{on} vs. L_{ch}. The linear fitting of the data can extract R_{on} different components, R_{ch} and R_{link}

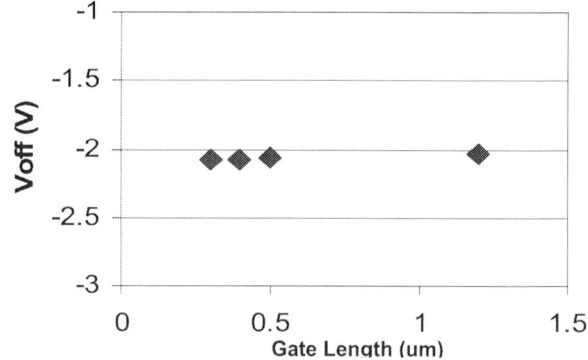

Fig. 3b. Pinch off Voltages vs. L_{ch}.

Another interesting discovery is the trade off in process control. While reducing R_{link}, intrinsic region becomes more dominant, the misalignment and STI module control have less impact on R_{on} variation, meaning tighter R_{on} control. However, V_{off} variation increases, because the junction abruptness is directly controlled by implants. A sharper junction is more sensitive to implant control.

978-1-4244-6240-7/10 $26.00 © 2010 IEEE   212

Next, a 0.4μmX1mm JFET is measured to characterize the output family curves as shown in Fig. 4. Gate voltage, V_{gs}, sweeps from 0 V to -1.8 V; drain voltage, V_{ds}, sweeps from 0 V to 8 V. Data show the JFET has the sufficient drive current, good output impedance, and is very robust with avalanche breakdown > 8V.

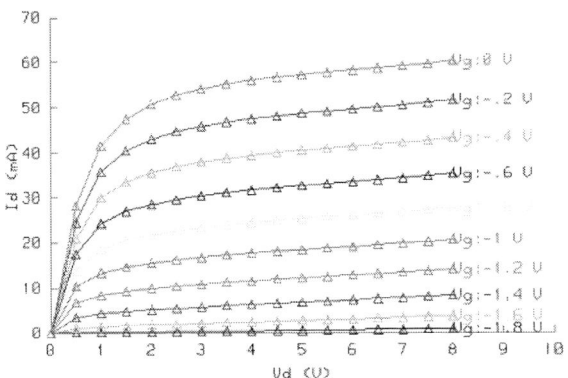

Fig. 4 output $I_d - V_{ds}$ family curves for JFET W = 1 mm.

B. 1/f noise

JFET is well known to have superior 1/f noise performance than conventional MOSFET [4]. Noise is reduced since the current conducting path is pushed into the single crystal silicon body, rather than along the Silicon-Oxide interface. In our design, the intrinsic JFET is bounded by STI, therefore a carefully designed junction depth is needed to avoid current crowding along the bottom STI, which could potentially degrade the 1/f noise.

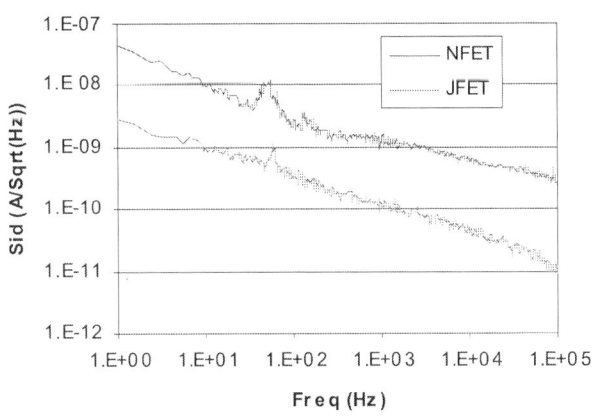

Fig. 5 1/f noise comparison of MOSFET vs. JFET.

A comparison of 1/f noise between the integrated JFET (L_{ch} = 0.4μm) and 1.8V NFET (L_{ch} = 0.18μm) is shown in Fig. 5. V_{ds} = 1.8V for both, while V_{gs} = 1.8V for regular

FET and V_{gs} = 0V for JFET. The drain current noise (S_{id}) is normalized to the same area of 1 μm² for a fair comparison. JFET shows 20X lower noise data. The bias dependence is shown in Fig. 6, with V_{gs} sweeping from 0V, towards pinch off. As expected, with less conducting current, the 1/f noise drops dramatically.

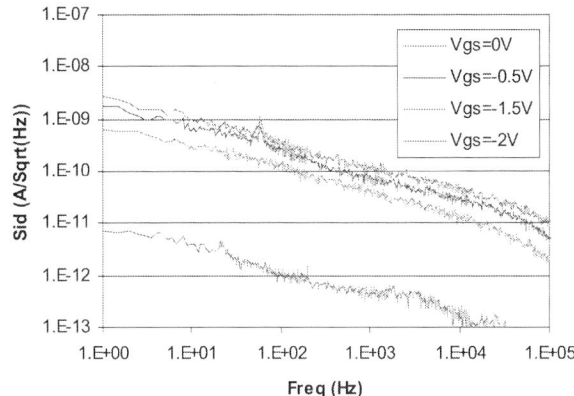

Fig. 6 The V_{gs} dependence of 1/f noise. L_{ch} = 0.4μm, data are normalized to 1μm²

C. ac and RF Characteristics

Small signal ac analysis is conducted to characterize the cut off frequency and maximum frequency of the JFET. Two-port open/short de-embedding method is used to extract parasitic components. Using -20db/decade extrapolation of |H21|, peak f_T is 3.5GHz. The maximum frequency f_{max} is extracted using maximum gain method. As an indicator of the power gain, f_{max} peaks at 23 GHz, and the V_{gs} dependence is shown in Fig. 7. It is suitable for RF and mm-wave applications.

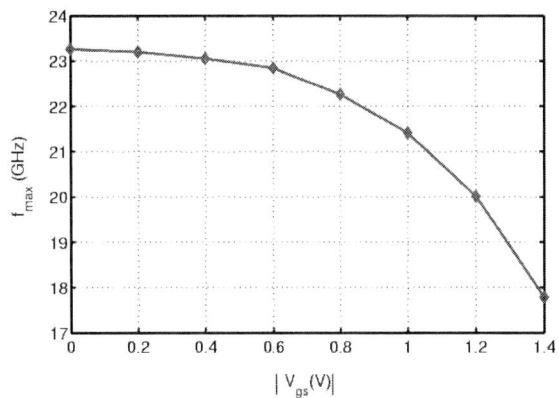

Fig. 7 Maximum frequency f_{max} vs. gate bias $|V_{gs}|$

978-1-4244-6240-7/10 $26.00 © 2010 IEEE

To support RF front-end module designs, the device linearity and power handing capability are very important. The load pull measurement is performed on a discrete 0.4μmX1mm JFET at fundamental frequency f_o = 1.9 GHz. Fig. 8 shows the P-1db compression and 3rd order intermodulation (IIP3) characteristics. Fig. 9 shows the 2nd and 3rd order harmonic components, both below -30 dBc up to P_{out} = 12 dBm. Fig. 10 shows the power gain and power-added efficiency (PAE) for a Class-AB configuration. The plots demonstrate the maximum 40% PAE, competitive with SiC, GaAs JFET for high voltage RF applications, yet with much lower cost.

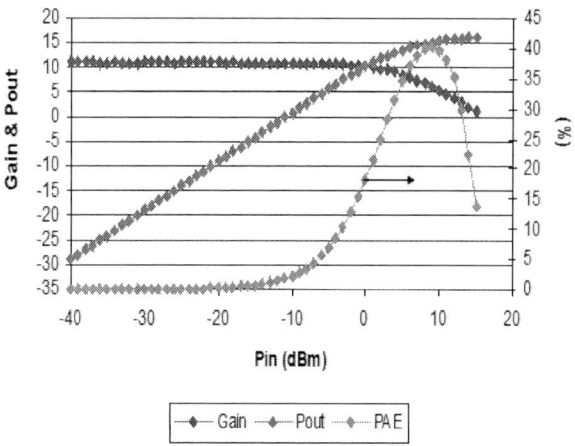

Fig. 10 Pout, Power Gain & PAE at Fo = 1.9 GHz

IV. CONCLUSION

We have demonstrated a Si CMOS compatible, cost effective JFET design. The design details are discussed on Ron vs. Voff optimization and an improved process has been found to improve the device performance. Both *ac* and RF data show very competitive characteristics. Currently, this JFET is qualified and enabled in IBM's 0.18 μm RFCMOS technology, as a feature well suited for RF and analog designs.

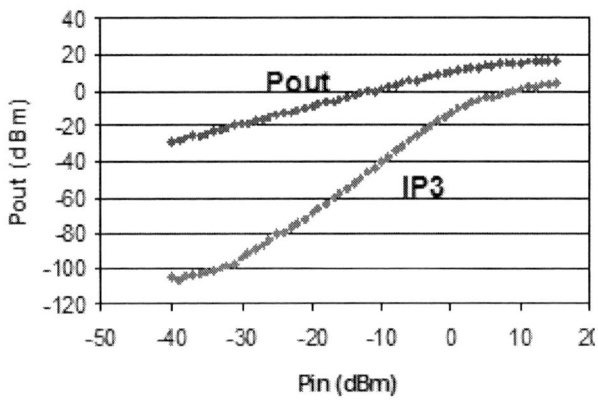

Fig.8 P1-db compression and third order intermodulation (IIP3) characteristics of a stand alone JFET.

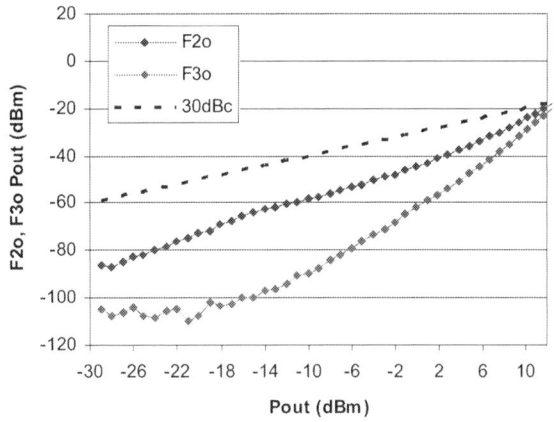

Fig. 9 2nd order and 3rd order harmonics of the stand alone JFET. A 30dBc line is drawn for reference.

REFERENCES

[1] C. W. Hatfield,, G. L. Bilbro S. T. Allen, and J. W. Palmour, "DC I-V Characteristics and RF Performance of a 4H-SiC JFET at 773K," *IEEE Transactions on Electron Deivces*, Vol. 45, No. 9, pp. 2072 – 2074, September, 1988.

[2] C. E. Weitzel, J. W. Palmour, C. H. Carter Jr., K. J. Nordquist, K. Moore, and S. Allen, "SiC microwave power MESFET's and JFET's," *Compound Semiconductors 1994, Inst. Phys. Conference*. Series no. 141, pp. 389 – 394, 1995.

[3] T. Ohgihara, S. Kusunoki, M. Wada, and Y. Murakami, "GaAs JFET Front-End MMICs for L-Band Personal Communications," *IEEE Microwave and Millimeter-Wave Monolithic Circuits Symposium*, 1993, Digest of Papers., pp. 9 – 12, 1993.

[4] F. A. Levinzon, "Noise of the JFET Amplifier," IEEE Transactions on Circuits and Systems – I : Fundamental Theory and Applications, Vol. 47, No. 7, pp. 981 – 985, July 2000.

Tunability of Bulk Acoustic Wave Filters Using CMOS Transistors: Concept, Design and Implementation

M. El Hassan[1], E. Kerherve[2], Y. Deval[2], J.B. David[3], D. Belot[4]

[1]University of Balamand – Al kura, Lebanon/ e-mail: moustapha.hassan@balamand.edu.lb
[2]IMS Laboratory – UMR 5218 CNRS – ENSEIRB – University of Bordeaux
[3]CEA-Leti – Minatec – Grenoble
[4]STMicroelectronics Crolles

Abstract — **This paper presents the feasibility of a new method to tune the Bulk Acoustic Wave - Solidly Mounted Resonator (BAW-SMR) filters by adding capacitors to the shunt resonators and by controlling these capacitors using CMOS transistors that act as switches. The tunable BAW-SMR filter is realized in a ladder topology. It is used for the 802.11b/g standard (2.40 - 2.48 GHz). Mainly, the filter fulfills the requirements for the WLAN 802.11 b/g standard, presenting a measured -3.3 dB of insertion loss, -12.7 dB of return loss and selectivity higher than 33 dB @ ±30 MHz of the bandwidth. Moreover, a measured shift of 0.5% of the centre frequency (2.44 GHz) towards higher frequencies is obtained. This tunable BAW-SMR filter has reduced dimensions (1035*1075 μm²).**

Index Terms — **Tunable BAW filters, SMR devices, passive elements.**

I. INTRODUCTION

In telecommunication systems, all filters and resonators that constitute the RF part have the tendency to be integrated on the same chip that contains the information treatment. The need to reduce the cost and size of electronic equipments has led to a continuing need for ever smaller filter elements. Hence, there has been a progressing effort to provide inexpensive and compact filter units. Many such devices utilize filters that must be tuned to precise frequencies. In this context, BAW (Bulk Acoustic Wave) filters are expected to replace the traditional RF filters technologies: SAW and ceramic filters. The BAW filters offer unique advantages: low losses (Q up to 1000), temperature stability, large signal handling capability (up to 3W) and present reduced dimensions [1].

Therefore, having a manufacturing process compatible with VLSI-CMOS procedures, BAW filters can be directly integrated above active circuits, making possible the design of fully integrated RF front-ends at very competitive costs [2-3]. These factors added to the increasing number of mobile systems radio interface from 500 MHz to 6 GHz make the BAW filters very adapted to the mobile front-end applications.

In this paper, we will propose the feasibility of a new technique to tune BAW-SMR filters. This technique consists of adding capacitors to the BAW-SMR filter and digitally controls these capacitors by CMOS transistors that act as switches. To validate our new method, a tunable bandpass filter for the WLAN standard (802.11b/g) has been designed and realized in a first step in a modular way to prove its functionality. Then, the measured results have been associated with the "bonding wires", the capacitors added in series with the shunt resonators and the CMOS transistors. The combination of experimental results and modeling of other components constitute the final response of the tunable BAW-SMR filter. The tunable BAW filter will be flip chipped with IC die that contains the CMOS transistor and the capacitors. The on-wafer measured results of the filter present a -3.3 dB of insertion loss, -12.7 dB of return loss and selectivity higher than 33 dB @ ±30 MHz of the bandwidth. In addition, a measured shift of 0.5% of the centre frequency (2.44 GHz) towards higher frequencies is obtained.

II. BAW IMPEDANCE BEHAVIOR

In the solidly mounted resonator (SMR), the piezoelectric is solidly mounted to the substrate. Some means must be used to acoustically isolate the piezoelectric from the substrate if a high quality factor (Q) resonance is to be obtained [4-5].

The Butterworth-Van-Dyke (BVD) model is an electrical schematic around resonance (Fig.1). The elements Ra, La, Ca present the series resonance and the insertion losses. The capacity C_0 represents the piezoelectric material between the two electrodes.

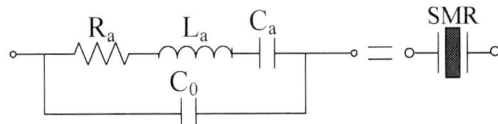

Fig. 1. SMR-BVD model.

The measured impedance characteristic of a BAW-SMR resonator is shown in Fig. 2. In this graph, it can be observed that the SMR presents mainly two resonance pulsations: the series resonance (f_s), when the electrical impedance approaches to zero, and the parallel resonance (f_p), when the electrical impedance approaches to infinity.

For all other frequencies far from the resonances, the SMR presents static capacitor behavior. f_s is adjusted according to the thickness of the piezoelectric layer and it is spaced by the parallel resonance f_p. The instantaneous frequency deviation between the two resonances is determined by the electromechanical coupling coefficient of the piezoelectric layer. The quality factor of the measured resonator is 192.5 and its active area is 16.800 μm².

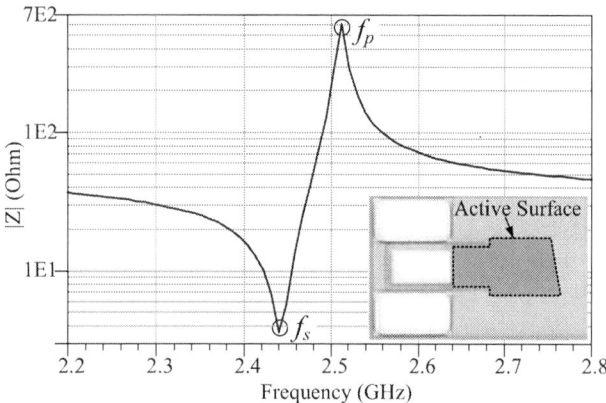

Fig. 2.   Impedance characteristic of a measured BAW-SMR resonator.

The electrical impedance of an SMR is obtained by solving the acoustic boundary problem and applying the transmission line theory [6]. The electrical SMR impedance can be simplified and expressed by the following equation:

$$Z_{SMR} = \frac{1}{j2\pi f C_0} \frac{(f^2 - f_s^2)}{(f^2 - f_p^2)} \qquad (1)$$

III. BAW FILTER DESIGN

The ladder filter is an association of resonators in series and in parallel. The shunt resonators are loaded and their resonance frequencies are smaller than the series resonators.

The addition of the capacitors (Q > 140) to the SMR circuitry doesn't affect severely the quality factor of the overall design. The electromechanical coupling resonator is described indirectly by the capacitor ratio C_0/C_a as determined by the resonator physical configuration and piezoelectric material properties [7]. When changing the electromechanical coefficient of the piezoelectric material, the bandwidth changes. Our goal is to tune the capacitor ratio C_0/C_a by adding series or parallel capacitors to the resonator. Thus, controlling this ratio will enable us to control the electromechanical coefficient, and as a sequence the bandwidth of the resonator.

Ladder BAW topology presents a good selectivity which enables to block undesired signals near the pass band.

In this study, a BAW-SMR filter was designed for the 802.11b/g standard. This filter is composed of five SMRs, associated in ladder topology. Fig. 3 shows the microphotograph of the BAW-SMR filter. The filter occupies a small area and has reduced dimensions (1035*1075 μm²).

The filter stack can be divided into resonators' layers and Bragg reflector's layers. The resonator's layers are composed by the classical couple AlN-Mo [8]. However, in contrast to [8] the Bragg reflector was implemented using an exclusive dielectric stack composed by SiOC:H and Si_xN_y [9]. The acoustical performance of the fully dielectric stack is comparable to the traditional SiO2-W reflectors; however, it strongly reduces the coupling between resonators through the Bragg reflector. Furthermore, the filter stack was realized on a high resistive silicon substrate in order to reduce losses due to the capacitive coupling [10].

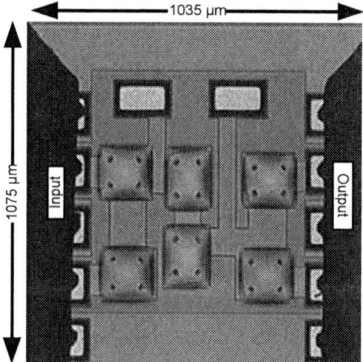

Fig. 3.   Microphotograph of the ladder BAW-SMR filter designed for the 802.11b/g standard.

In order to optimize the filter performance, we have used a double resonator and apodized geometries, and we have taken in consideration the effect of bonding wires. Indeed, double resonators present large electrodes' areas, which results in lower resistive losses. Also, the filter resonators present apodized geometries in order to avoid spurious resonances caused by the parasitic lateral acoustic modes [11].

Fig. 4 shows the comparison between the measured and simulated results of the ladder BAW-SMR filter. Electromagnetic simulation of the overall filter structure has been performed using the ADS-Momentum software. Next, the acoustic effects have been considered using the Mason Model [12] and included in the simulations.

The filter design was realized for implementation in SiP context. The performances of the BAW-SMR filter are in concordance with the simulation results. Mainly, the filter fulfills the requirements for the WLAN 802.11 b/g standard, presenting -3.3 dB of insertion loss, -12.7 dB of

978-1-4244-6240-7/10 $26.00 © 2010 IEEE

return loss and a selectivity higher than 33 dB at ±30 MHz of the bandwidth.

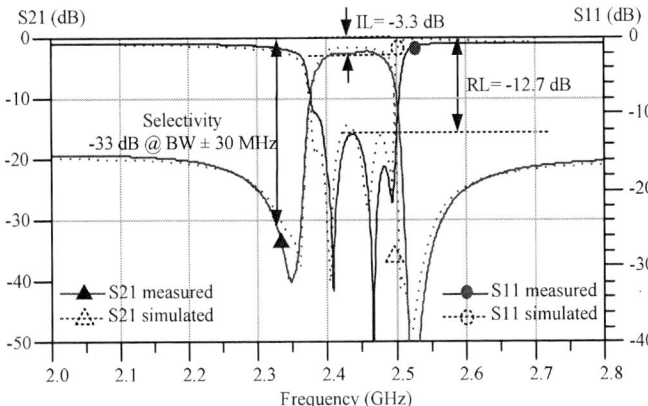

Fig. 4.   Comparison between measured and simulated results of the 802.11b/g ladder BAW-SMR filter.

The filter high insertion losses are mainly due to the low resonators quality factor obtained in the fabrication (Q = 200). Therefore, these losses can be strongly reduced using mechanical energy concentration techniques in the resonator acoustical cavity [13].

IV. LADDER FILTER TUNABILITY

The shunt resonators of the ladder filter determine the position of the zeroes at the left of the center frequency and the series resonators determine the position of the zeroes at the right of the center frequency. Thus, changing the impedance of the parallel and series resonators leads to a change in the zeroes' positions [14].

Based on this theory and in order to tune the BAW-SMR filter, we can add passive elements to the shunt and series resonators that constitute the filter.

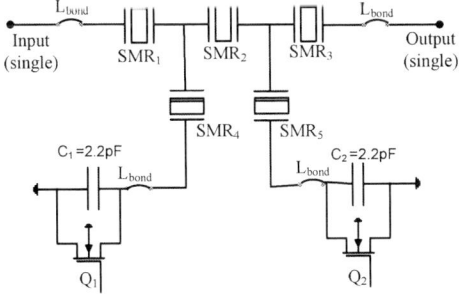

Fig. 5.   Tunable BAW-SMR filter using CMOS transistors.

To validate the concept of digitally tuning BAW filters using passive elements controlled by CMOS transistors, we present in this part the use of CMOS switches at the terminals of capacitors added in series with the shunt resonators (Fig.5).

When a transistor is ON, the capacitor is short circuited, and when a transistor is OFF, the capacitor will be considered in series with the shunt resonator. Thus, the bandwidth and the characteristics of the filter will be modified. The simulation results of the filter, capacitors, transistors and the associated "bonding wires" are shown in Fig.6.

The MOS transistors used during co-simulation are from STMicroelectronics technology (CMOS 65 nm). The width and length of the gate dimensions are: W = 45µm and L = 0.25 µm. The main parasitic elements are taken into account in the co-simulation (C_{gs} = 90 fF, C_{gd} = 70 fF, R_{on} = 65). The length of "bonding wire" is 2 mm, which represents an inductive effect of approximately 2 nH.

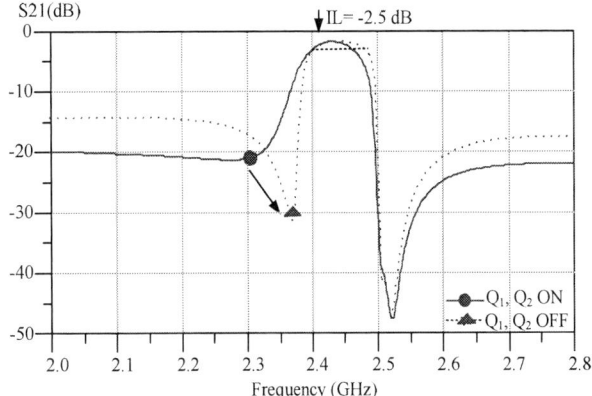

Fig.6.   Transmission responses of the digitally tunable BAW-SMR filter.

Fig.6 shows that when the transistors Q_1, Q_2 are OFF, capacitors affect the performance of BAW-SMR filter. Thus, we see the displacement of zero situated to the left of the center frequency to higher frequencies. Due to the effect of the capacitors (C_1, C_2) and the added "bonding wires" (2nH), the central frequency (2.44 GHz) has moved 0.5% towards higher frequencies (shift of 12 MHz). The insertion losses of filter are -2.5 dB and the out-band rejection is 15 dB.

The measurement results of the tunable BAW filter was made in two parts. First, the experimental results of the filter have been carried out using on-wafer measurements. Then, these results have been associated with the transistors, capacitors, and the bonding wires. The combination of experimental results and modeling of the other elements constitutes the final response of the digitally tunable filter.

978-1-4244-6240-7/10 $26.00 © 2010 IEEE   217

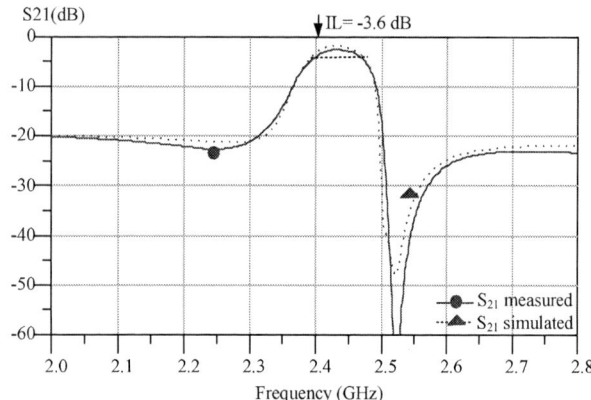

Fig.7. Comparison between the measured and the simulated results of the digitally tunable BAW-SMR filter.

Fig.7 shows the comparison between the measured and the simulated results. When the transistors Q_1, Q_2 are ON, we note a -3.6 dB insertion loss which is higher than the estimated value in simulations (I.L. = -2.5 dB) and the filter has an out of band rejection of 20 dB.

When the transistors (Q_1, Q_2) are OFF, the insertion losses obtained are -3.7 dB and the out-band rejection is 15 dB. These losses are due to the high number of resonators in the direct path and to the low quality factor of the resonators ($Q_{measured}$ = 200 while $Q_{simulated}$ = 500), and the length of bonding associated with the filter. Also, the central frequency (2.44 GHz) has moved 0.5% towards higher frequencies (shift of 12 MHz).

In conclusion, the MOS transistors acting as switches control the passive elements associated with the BAW-SMR filter. When the transistors (Q_1, Q_2) are in the saturation region (ON-state), the capacitors do not affect the transfer function of the filter. The only influence will be from the parasitic capacitors of the transistors. When transistors are in the triode region (OFF-state), the added capacitors change the position of zeros situated to the left of center frequency towards higher frequencies.

V. CONCLUSION

In this paper, we have shown a tunable BAW-SMR filter realized in a ladder topology and used for the 802.11b/g standard (2.40 - 2.48 GHz). The filter fulfilled the requirements for the WLAN 802.11 b/g standard, presenting a measured -3.3 dB of insertion loss, -12.7 dB of return loss and selectivity higher than 33 dB @ ±30 MHz of the bandwidth. The BAW-SMR filter has reduced dimensions (1035*1075 μm^2). Moreover, we have presented a new method to digitally shift the centre frequency of this tunable filter towards higher frequencies by adding capacitors and controlling these capacitors by

CMOS transistors that act as switches. Measured shifts of +0.5% of the centre frequency (2.44 GHz) towards higher frequency is obtained.

ACKNOWLEDGEMENT

IMS laboratory is acknowledged for all facilities offered and the access to obtain the filter measurements. Also, CEA-LETI (Grenoble, France) and STMicroelectronics (Crolles, France) are acknowledged for the technology access and filter fabrication.

REFERENCES

[1] P. Bradley et al. "A Film acoustic bulk resonator (SMR) duplexer for USPCS Handset Applications", *IEEE MTT-S*, pp.367–370, 2001.

[2] J. F. Carpentier et al. "A SiGe:C BICMOS WCDMA zero-IF RF front-end using an above-IC BAW filter", *IEEE ISSCC*, pp. 394-395, 2005.

[3] A. A. Shirakawa, J-M. Pham, P. Jarry, E. Kerherve, C. P. Moreira, "SMR performance evaluation: applications on microwave filters and circuits", *APMC*, pp. 1051, 2004.

[4] Newell, "Face-mounted piezoelectric resonators", *Proc. IEEE*, Vol 53, pp.575-581, June 1965.

[5] K.M. Lakin, K.T. McCarron, and R.E. Rose, "Solidly mounted resonators and filters", *IEEE Ultrasonic Sumposium*, pp. 905-908, 1995.

[6] K. Lakin, G. Kline, and K. McCarron, "High-Q microwave acoustic resonators and filter", *IEEE Trans. On Microwave Theory and Techniques*, vol. 41, pp. 2139-2146, Dec. 1993.

[7] K.M. Lakin, J. Belsick, J.F. McDonald, and K.T. McCarron, "Improved bulk wave resonator coupling coefficient for wide bandwidth filters", *IEEE Ultrasonics Symposium*, Paper 3E-5, October 9, 2001.

[8] G. G. Fattinger et al. "Thin Film Bulk Wave Devices for Applications at 5.2 GHz", *IEEE UFFC Symposium*, pp. 174-177, 2003.

[9] P. Ancey, "Above IC RF MEMS and BAW filters: fact or fiction", *IEEE BCTM Proceedings*, pp. 186-190, 2006.

[10] A. A. Shirakawa, J-M. Pham, P. Jarry, E. Kerherve, F. Dumont, J-B. David and A. Cathelin, "A High Isolation and High Selectivity Ladder-Lattice BAW-SMR Filter", 36th *European Microwave Conference, Manchester*, UK, 10-15 September 2006.

[11] K. M. Lakin, K. G. Lakin, "Numerical Analysis of Thin Film BAW Resonators", *Proceedings of Ultrasonics Symposium*, Vol. 1, pp. 74-79, 2003.

[12] W. P. Mason, "Electromechanical Transducers and Wave Filters", Princeton, New Jersey, Van Nostrand, 1948.

[13] J. Tsutsumi, M. Iwaki, Y. Iwamoto, T. Yokoyama, T. Sakashita, T. Nishihara, M. Ueda and Y. Satoh, "A Miniaturized FBAR Duplexer with Reduced Acoustic Loss for the W-CDMA Application", *IEEE Ultrasonics Symposium Proceedings*, pp. 93-96, 2005.

[14] M. El Hassan, E. Kerhervé, Y. Deval, D. Belot, "A New Method to Reconfigure BAW-SMR Filters Using CMOS Transistors", *IEEE MTT-S International Microwave Symposium*, Honolulu, Hawaii, 3-8 June 2007.

A Layout Efficient, Vertically Stacked, Resonator-Coupled Bandpass Filter in LTCC for 60 GHz SOP Transceivers

Rony E. Amaya

IMMC Research Group, Terrestrial Wireless Systems Research Branch,
Communications Research Centre, Ottawa, Canada, K2H 8S2
e-mail: rony.amaya@crc.ca

ABSTRACT—This paper describes the design and implementation of a layout efficient bandpass filter implemented in Low-Temperature Co-fired Ceramic (LTCC) substrates. Applications for this filter include band select filters for 60 GHz System-On-Package transceivers. Two bandpass filters based on a quasi-elliptic configuration were implemented here using four half-wavelength resonators. The first filter uses all planar resonators to achieved a measured insertion loss of 3.7 dB and a return loss > 10 dB and with a layout area of 1.263mm². The second filter uses vertical stacking of two of its resonators to reduce the layout area and achieved higher immunity to process variations. Measurements for the second filter show an IL of 3.4 dB while maintaining a RL > 10 dB. The vertically stacked filter consumes a layout area of 0.811mm², corresponding to a layout reduction of 36% compared to the planar filter with performance less prone for process variations.

Index Terms — Quasi-Elliptic Bandpass Filter, Low-Temperature Co-fired Ceramics, System-On-Package.

I. INTRODUCTION

The recently allocated 7 GHz of available bandwidth around 60 GHz [1] opens up the possibility of implementing millimeter wave transceivers with high-bandwidth and high-data rate. Applications include personal area networks, point-to-point data links and radio-over-fiber. The high attenuation due to oxygen and atmospheric absorption at this frequency limits the usable range for wireless devices, but it is a feature that can be exploited to enhance security during data transfer. An increase in the frequency of operation also translates into a size reduction for passive circuits reducing the cost of fabrication. Low-Temperature Co-fired Ceramics (LTCC) offer a low cost and versatile solution for System-in-Package millimeter wave systems, with up to twelve layers of connectivity and dielectric properties suitable for millimeter wave frequencies [2]-[3]. Resonator based Quasi-Elliptic bandpass filters offer the best tradeoff between filter performance and layout area at millimeter-wave frequencies when compared to other filter topologies: resonator cavity filters, dielectric waveguide filters and edge-couple microstrip filters [5]-[10].

II. QUASI-ELLIPTIC FILTER THEORY

Quasi-elliptic bandpass filters (QEBPF) offer better selectivity than Chevyshev filters due to the finite location of their poles. Analytical work for these filters is available in the literature [4]. The transfer function for a bandpass filter can be expressed as follows

$$|S_{21}(\Omega)| = \frac{1}{1 + \epsilon^2 F_N^2(\Omega)} \qquad (1)$$

where Ω is the frequency variable which is normalized to the passband cutoff frequency of a low-pass prototype filter and ε is a ripple constant related to a given return loss L_R in decibels by

$$\epsilon = \frac{1}{\sqrt{10^{-(L_R/10)} - 1}} \qquad (2)$$

Also, the form of $F_N(\Omega)$ for the selective filters can be expressed as a function of the pair of attenuation poles, so that $\Omega = \pm\Omega_a \ (\Omega_a > 1)$ are the frequency locations of a pair of attenuation poles. The locations of two finite frequency attenuation poles of the bandpass filters, namely ω_{a1} and ω_{a2}, are given by

$$\omega_{a1} = \omega_0 \frac{-\Omega_a \cdot FBW + \sqrt{(\Omega_a \cdot FBW)^2 + 4}}{2} \qquad (3)$$

$$\omega_{a2} = \omega_0 \frac{\Omega_a \cdot FBW + \sqrt{(\Omega_a \cdot FBW)^2 + 4}}{2} \qquad (4)$$

where FBW corresponds to the fractional bandwidth of the filter. The closer the attenuation poles are to the cutoff frequency $(\Omega = 1)$, the sharper the filter skirt at the expense of lowering the out of band filter selectivity (see Figure 1).

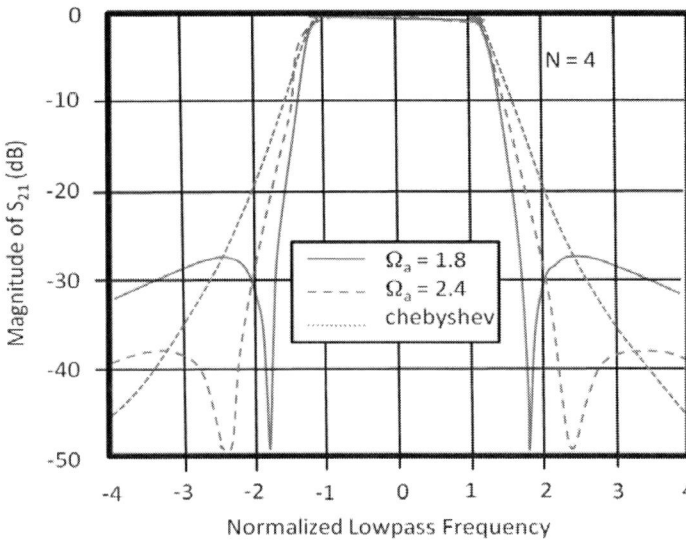

Figure 1. Normalized low-pass response for a quasi-elliptic filter [4].

Table 1. Quasi-Elliptic Filter coefficients [4]

Ω	C_1	C_2	J_1	J_2
1.80	0.95974	1.42192	-0.21083	1.11769
1.90	0.95691	1.39927	-0.18429	1.08548
2.00	0.95449	1.38235	-0.16271	1.06062
2.10	0.95242	1.36934	-0.14487	1.04094
2.20	0.95063	1.35908	-0.12992	1.02499
2.30	0.94908	1.35084	-0.11726	1.01187
2.40	0.94772	1.34408	-0.10642	1.00086

From Table 1, we can use filter coefficients to calculate the required coupling between all resonators (see Figure 2).

$$Q_{e,i} = Q_{e,o} = \frac{C_1}{FBW} \quad (5)$$

$$M_{1,2} = \frac{FBW}{\sqrt{C_1 C_2}} = M_{3,4} \quad (6)$$

$$M_{2,3} = \frac{FBW \cdot J_2}{C_2} \quad (7)$$

$$M_{1,4} = \frac{FBW \cdot J_1}{C_1} \quad (8)$$

where $Q_{e,i}$ and $Q_{e,o}$ correspond to input and output loaded quality factor of resonators 1 and 4 and $M_{i,j}$ represents the magnetic coupling between the resonators.

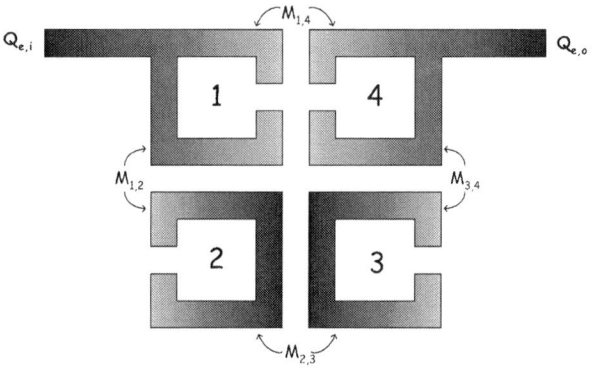

Figure 2. Bandpass filter using a four-pole coupled resonator configuration.

The system specifications for the bandpass filter are as follows:

- Centre frequency, f_c = 61 GHz
- 3-dB Bandwidth = 3 GHz (FBW = 4.9%)
- 20-dB Rejection Bandwidth = 4 GHz
- Passband return loss > 10 dB

Based on the filter response shown in Figure 1, the above specifications can be achieved by using a 4-pole filter with Ω = 1.8. Filter coefficients can be obtained from Table 1 and using the above design equations Eq. (5) - Eq. (8). The calculated magnetic coefficients between the resonators are as follows:

- $Q_{e,i} = Q_{e,o} = 28.82$
- $M_{1,2} = M_{3,4} = 0.0285$
- $M_{1,4} = -0.0073$
- $M_{2,3} = 0.0261$

III. BANDPASS RESONATOR FILTER DESIGN IN LTCC

A. Resonator Design

The resonators used in this work were modeled using commercial EM software, HFSS [11]. Two separate resonators were designed for use in a planar and vertical QEBPF filter configuration as shown in Figure 3. The length of each resonator corresponds to a half wavelength at the resonant frequency.

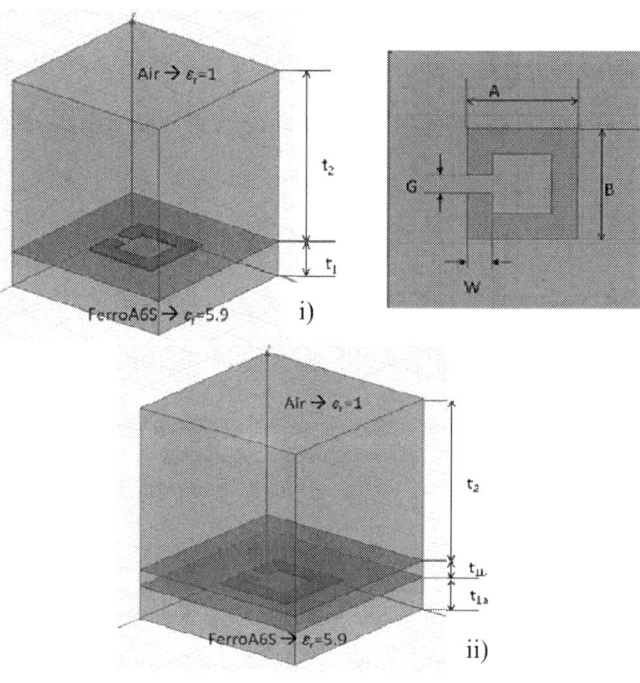

Figure 3. Resonator design in HFSS: i) Planar resonator A ii) Vertically stacked resonator B.

Using the eigenmode solver in HFSS [11], layout dimensions as well as resonator self-resonant frequencies (f_{res}) and unloaded quality factors (Q_U) were extracted for both cases (Table 2).

Table 2. Initial resonator design in HFSS (Figure 3).

Parameter	Resonator A	Resonator B
A (µm)	462.5	390
B (µm)	462.5	390
G (µm)	100	100
W (µm)	100	100
t_{1a} (µm)	198	99
t_{1b} (µm)	0	99
t_2 (µm)	1000	1000
f_{res} (GHz)	60.4	62.4
Q_U	175	162

B. QEBPF Planar filter Design

The design of the planar version of the Quasi-Elliptic BF was implemented using four resonators on the top LTC metal layer (all four resonators were based on Resonator A shown in Figure 3i). A similar design is available in the literature [5] using a Dupont LTCC process [3]. One of the limitations of a QEBPF based on planar resonators in LTC is the minimum spacing of 50μm to 100μm between conductors for LTCC processes [2]-[3], which limits the minimum spacing between resonators 1, 4 and resonators 2 as well as the gap of each resonator ring. The design dimensions for the planar QEBPF were achieved by using commercial EM software [11], and are shown in Figure . The size and spacing of the resonators were designed to me the required electric and magnetic coupling coefficients based on the design equations Eq. (5) - Eq. (8).

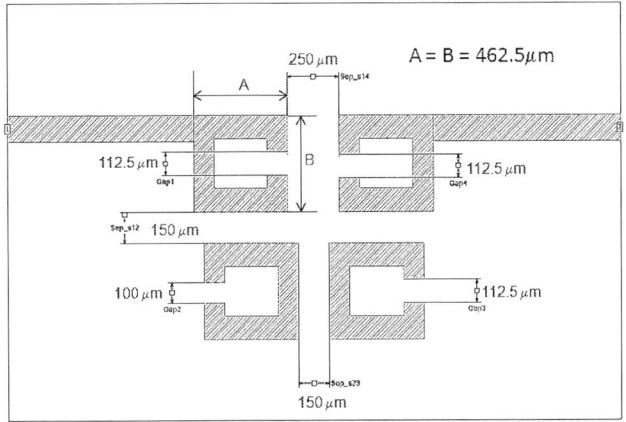

Figure 4. Planar filter design showing final design values.

The layout area of this filter, excluding the microstrip feed lines, is 1175μm by 1075μm corresponding to 1.263mm².

C. QEBPF Vertical filter Design

A second version of the Quasi-Elliptic BPF was implemented by vertically stacking resonators 2 and 3. As shown in Figure 5, resonators 1 and 4 overlap with vertically stacked resonators 2 and 3 by 100μm. Vertically stacked resonators are no longer limited by the minimum conductor to conductor spacing in LTCC, allowing for increased design flexibility in the separation between resonators 1,4 and resonators 2,3. The dimensions (excluding microstrip feed lines) for the vertically stacked filter are 1100μm by 737.5μm corresponding to a layout area of 0.811mm². This represents a reduction in the layout area of the filter of 36% compared to the planar version implemented here.

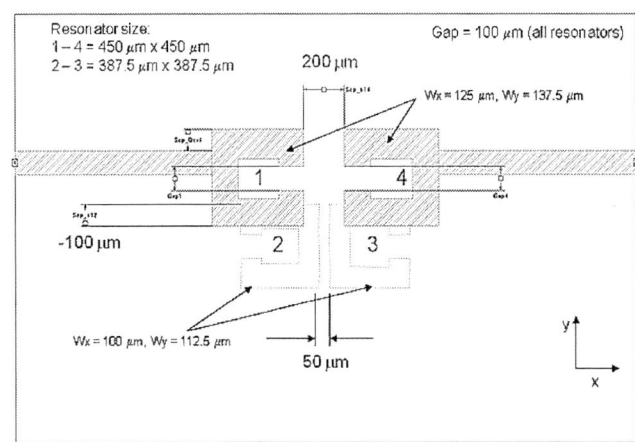

Figure 5. Vertical filter design showing final design values.

IV. LTCC TECHNOLOGY

The resonator filters used in this work were implemented using the Ferro A6S Low-Temperature Co-fired Ceramic process with $\varepsilon_r = 5.9$ and $\tan \delta = 0.002$ [2]. Each LTCC tape layer has a post-fired thickness of 99 μm and uses silver metallization with $t = 9\mu m$ and $\sigma = 5.5 \times 10^7$ S/m. Filter structures were fabricated by the VTT Technical Institute in Oulu, Finland.

V. MEASUREMENTS

Measurements were performed using an Agilent N5250A PNA and characterized for frequencies of up to 67 GHz. A Karl Suss probe station PA 200 with programmable probe heads was used. TRL de-embedding structures were included in order to remove the effect of probes, pads and other external components. GSG probes with 500μm pitch were used. The final layout for the filters implemented here can be seen in Figure 6.

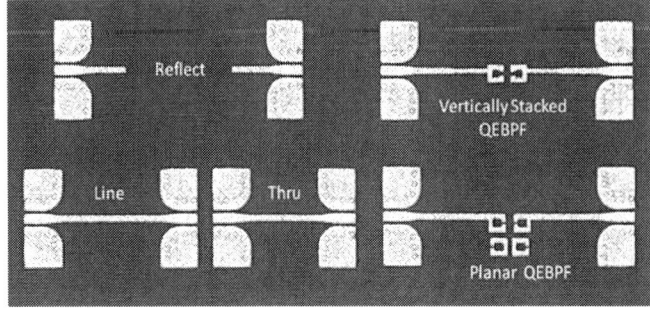

Figure 6. Fabricated filters in Ferro A6S LTCC process.

Results for the planar and vertical versions of the Quasi-Elliptic Bandpass filters are shown in Figure 7 and Figure 8. A total of four filters of each design were measured and showed similar performance.

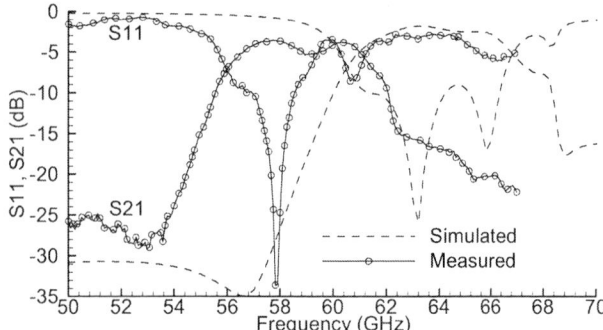

Figure 7. VNA Measurements for a planar filter in LTCC

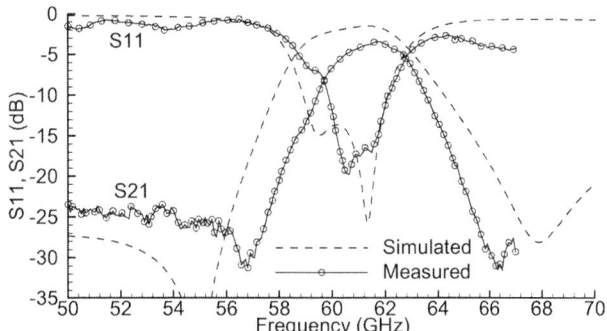

Figure 8. VNA Measurements for a vertically stacked filter in LTCC

As seen in Figure 7, process variations during the firing process in the LTCC tape layers shifted the centre frequency of the planar version of the filter from a simulated/expected centre frequency of 62.5 GHz to its measured value of 58 GHz. Similarly the return loss was affected for this filter, narrowing the useful bandwidth of operation (return loss > 10 dB) to less than 1 GHz (simulated value is 3 GHz). Upon inspection of the fabricated physical dimensions of the planar QEBPF it was found that the measured gap of the resonators was 20% lower than its design value contributing to the reduction in the centre frequency of the planar filter. In contrast, Figure 8 shows the measured results for the vertically-stacked Quasi-Elliptic Bandpass filter. Both the measured insertion loss (S_{21}) and return loss (S_{11}) matched simulated values closer in the vertically-stacked version than for the planar version of the QEBPF. The simulated/measured insertion loss for the planar and vertical version of the Quasi-Elliptic BPF were 1.86dB/3.7dB and 1.6dB/3.4dB respectively.

VI. CONCLUSIONS

A layout efficient bandpass filter was designed for 60 GHz applications and with better process variation immunity to standard LTCC processes. Results indicate that the size of the resonator coupled QEBPF can be reduced by vertically stacking one of the resonator pairs, allowing for a reduction in the layout area of 36% while improving filter performance and achieving higher immunity to LTCC process variations. This Quasi-Elliptic BPF offers a compact solution for the implementation of band-select BPF for radio transceiver integrated System-in-Package systems operating at millimeter-wave frequencies.

Table 3. Measured comparison of 60 GHz bandpass filters results.

Ref	f_c (GHz)	FBW (%)	IL (dB)	RL (dB)	Tech.	Area (mm x mm)
[5]	62.3	5.46	3.48	>10	LTCC [3]	1.02 x 1.05
[6]	59.5	1.4	2.14	>10	LTCC [3]	1.95 x 1.32
[7]	60	4.9	4.0	>10	LTCC [3]	2.0 x 3.0
[8]	60.4	3.5	4.0	>10	LTCC [3]	1.8 x 3.5
[9]	59.5	5.0	3.2	>10	Alumina	4.7 x 3.2
[10]	61	8.3	4.0	>10	LTCC [2]	1.0 x 4.0
This Work (Planar)	61	3.4	3.7	>10	LTCC [2]	1.18 x 1.08
This Work (Vertical)	61	4.8	3.4	>10	LTCC [2]	1.1 x 0.74

To the author's knowledge, the vertically stacked filter reported in this work shows the smallest layout area of all bandpass filters at 60 GHz available in the literature with insertion loss of 4 dB or lower (Table 3).

ACKNOWLEDGEMENT

The authors wish to thank Kari Kautio of the VTT Technical Laboratory in Oulu, Finland for providing access to fabrication and process expertise in LTCC packaging technologies.

REFERENCES

[1] IEEE 802.15.3C, *http://www.ieee802.org/15/pub/TG3c.html*

[2] Ferro A6S LTCC process, *http://www.ferro.com*

[3] Dupont LTCC Green Tape process, *http://www2.dupont.com/Packaging_and_Circuits/en_US/products_services/mems/index.html*

[4] Jia-Sheng Hong *et. al*, *"Design of Highly Selective Microstrip Bandpass Filters with a Single Pair of Attenuation Poles at Finite Frequencies,"* IEEE Trans. MTT, Vol. 48, No. 7, Jul. 2000.

[5] J. H. Lee *et. al*, *"V-band Integrated Filter and Antenna for LTCC Front-End modules,"* Proc. of IEEE IMS, pp. 978-981, Jun. 2006.

[6] J. H. Lee *et. al*, *"Low-Loss LTCC Cavity Filters Using System-on-Package Technology at 60 GHz,"* IEEE Trans. MTT, Vol. 53, No. 12, pp. 3817-3824, Dec. 2005.

[7] Tomohiro Seki *et. al*, *"60GHz Monolithic LTCC Module for Wireless Communication Systems,"* Proc. 36th. IEEE European Microwave Conf., Manchester UK, pp. 1671-1674, Sep. 2006.

[8] Young Chul Lee et. al, "A Fully Embedded 60-GHz Novel BPF for LTCC System-in-Package Applications," IEEE Trans. Adv. Packaging, Vol. 29, No. 4, pp. 804-809, Nov. 2006.

[9] Masaharu Ito *et. al*, *"A 60-GHz-Band Planar Dielectric Waveguide Filter for Flip-Chip Modules,"* IEEE Trans. MTT, Vol. 49, No. 12, Dec. 2001.

[10] Isabel Ferrer *et. al*, *"A 60 GHz Image Rejection Filter Manufactured Using a High Resolution LTCC Screen Printing Process,"* Proc. 33th. IEEE European Microwave Conf., Munich GER, pp. 423-425, Oct. 2003.

[11] Ansoft HFSS, *http://www.ansoft.com/products/hf/hfss/*

Co-Design Considerations for Frequency Drift Compensation in BAW-based Time Reference Application

S. Razafimandimby[1], D. Petit[1], P. Bar[1], S. Joblot[1], J.-F. Carpentier[1], J. Morelle[1],
C. Arnaud[1], G. Parat[2], P. Garcia[1], C. Garnier[1].

[1] STMicroelectronics, Crolles, France, [2] CEA-LETI/MINATEC, Grenoble, France.

Abstract — **In order to take up the challenge of BAW–based time reference, this paper presents new BAW/Integrated Circuits (IC) co-integration considerations. For the demonstration, a SiP approach is proposed where the Solidly Mounted Resonator (SMR) has been directly flip-chipped on the top of the IC. This 2.5GHz oscillator reaches a -93dBc/Hz phase noise at a 2kHz carrier offset for a 7.3mW power consumption. A 5bit switched capacitor bank permits to correct process deviations with a 12.5kHz accuracy while a varactor capacitance allows compensating a SMR with a -4.2ppm/°C Temperature Coefficient of Frequency (TCF) in a [-40°C,85°C] temperature range.**

Index Terms — **Time reference, SMR, BAW, low phase noise, thermal stability, frequency accuracy.**

I. INTRODUCTION

Time reference solutions are today dominated by Crystal-based oscillators (XO). This type of oscillators is a very mature technology and is declined over a wide range of products. It starts from simple XO for classical low performance applications up to Temperature Compensated Crystal-X Oscillator (TCXO) for high performance. Primary markets where BAW technology could bring advantage are the ones where Si content is more important such as TCXO or multiple clock generation. Applications requiring accuracy, stability and low cost circuits can profit from this integrated solution which allow the reduction of the BOM (Bill of Materials) in RF transceivers. In this demonstration, the BAW resonator is realized using an SMR technology with a silicon substrate thinned down to 150µm. Then, it is assembled by flip-chip on top of a 0.13µm BiCMOS die from STMicroelectronics. Finally, divider stages associated with the proposed BAW oscillator provide several cost-efficient reference frequencies thanks to one resonator whose temperature and process drifts need nevertheless to be compensated.

After presenting BAW resonators and their models in section II, the circuit principle of a BAW oscillator will be introduced in section III in which the required BAW resonator performances will be discussed for time reference application. Section IV describes the used VCO architecture and its implementations taking into account BAW and IC constraints. Here, the electrical compensations of temperature and process will be presented. Finally, measurement and analysis will be dealt with in section V before some conclusion and perspectives.

II. BAW RESONATOR AND ITS MODEL

BAW resonators are typically composed of 3 parts: electrodes, a piezoelectric layer and an isolation part (a Bragg reflector for SMR). Thus, thanks to the isolation part, most of the mechanical energy is confined in the piezoelectric material. To minimize TCF, SMR is fabricated taking advantage of silicon dioxide (positive thermal velocity coefficient) as a thermal compensation layer. Our design flow includes a 1-D Mason thermal model to reach optimal stack layers thicknesses [1]. Moreover, the well-known Modified Butterworth Van Dycke (MBVD) model is used to optimize the electrical part. It represents the BAW resonator's electrical behavior by a network of lumped components where the access resistance to the SMR (called R_s), turns out to be one of the key points of high performance and low power VCO.

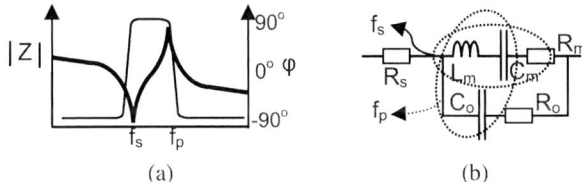

Fig. 1. (a) BAW resonator impedance. (b) MBVD model.

BAW resonators are characterized by a series (f_s) and a parallel (f_p) resonance frequencies. Out of f_s-f_p band, it is seen as a capacitor and in this band, it is equivalent to an inductor, which we use to generate oscillation.

III. BAW CONSIDERATIONS FOR BAW VCO

The BAW VCO principle is to load the BAW resonator by a capacitor C_L. This tank resonates at the frequency at which the inductive part of the resonator is equal to the impedance of C_L. Thus, changing C_L tunes the oscillation frequency. This way, by considering R_m (inversely proportional to C_0) and k_t^2, the coupling coefficient of the BAW resonator correlated to the gap between f_s and f_p. if C_L is ideal, the losses we have to compensate are given by:

$$R_{losses} = R_s + R_m(1+\frac{C_0}{C_L})^2 \qquad (1)$$

Therefore, in order to increase Q of the tank, the access resistance to the BAW has to be decreased. Thus, a good

---

978-1-4244-6240-7/10 $26.00 © 2010 IEEE

way of designing a low power oscillator is to increase k_t^2 and decrease C_o as well. Nonetheless, a high k_t^2 is not compatible with a small frequency step unless having a more complex capacitor matrix, while a high k_t^2 contributes to reach a better phase noise. Furthermore, extra spurious modes can be generated for a lower C_0. Finally, trade-offs between current consumption, phase noise performances and process compensation emerge. Compensating for larger process dispersions implies a larger capacitor bank, hence increasing resistive losses, which in turn need to be corrected, burning extra power.

Frequency sensitivity of each layer of the stack cumulated with deposition dispersions implies trimming steps. Today, optimized STM/CEA-LETI process permits to reach a +/-2MHz accuracy on f_p at wafer scale. Like the thickness non-uniformity of layers, substrate resonances prevent from achieving electrical parameters reproducibility at wafer scale. Indeed, when SMR acoustic isolation is not fully performed, a significant amount of leaking acoustic energy can be reflected from the bottom substrate surface leading to harmful Q inflections (see Fig. 2a). Unlike to spurious mode, this parasitic substrate resonance harmonics disturb resonator thermal response.

Furthermore, in order to be efficient in the fine tuning, it is important that the temperature variation follows a known shape. The use of an appropriate stack for processing SMRs allows obtaining monotone (almost linear) TCF (see Fig. 2b). A negative TCF is required to obtain an optimized Q-factor with this thermal strategy. Note the agreement between simulation (including substrate influence) and measurement of TCF on this -4.2ppm/°C SMR. Improving longitudinal reflections of Bragg mirror is required to prevent thermal response disturbances at wafer scale. Finally, temperature parameters are introduced in the MBVD model to simulate thermal effect at circuit level. Therefore, according to the BAW VCO principle, the response of oscillation frequency versus C_L could not be linear in the vicinity of a substrate harmonic resonance due to a variation of the BAW resonator imaginary part. Operating in the vicinity of substrate parasitic harmonic resonance prevents an accurate calibration of the BAW VCO for the process compensation. Furthermore, a slope change in the vicinity of the oscillation frequency will also make the step frequency non constant. Working in a more linear frequency domain that is close to $(f_s+f_p)/2$ optimizes the accuracy calibration. Nevertheless, at this frequency, the phase noise will be not as good as near f_p.

In this context of designing a low-power time reference, the VCO core has to compensate losses by means of an active negative resistance for reasonable power consumption and good phase noise performances. A coarse tuning corrects process deviations via a switched capacitor matrix (annoted C_{SW}) in order to calibrate the BAW VCO at 27°C before its operating mode in the RF transceiver. Finally, a varactor (C_{var}) will provide a continuous fine tuning to compensate for the temperature deviation during its operating mode.

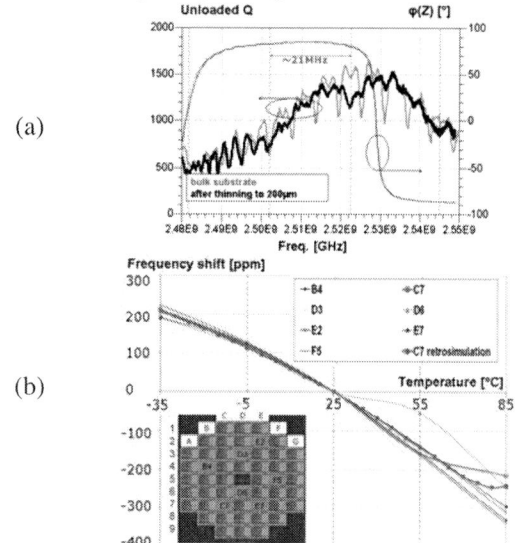

Fig. 2. (a) Measured SMR Q-factors and phases before and after substrate thining. (b) Example of measured frequency drift from -35°C to 85°C of compensated SMRs.

III. . DESIGN CONSIDERATIONS

A. VCO Core

The oscillator uses the well-known Colpitts technique and is presented in figure 3, where is defined the global loading capacitance C_L versus the capacitances of the VCO core (C_1, C_2 and the base-emitter capacitor of Q_1, C_π). By resonating with C_L, the oscillation frequency and the sensitivity of this BAW-based VCO are:

$$f_{osc} = f_s \sqrt{1 + \frac{2C_m}{2C_0 + C_L}} \qquad (2)$$

$$S = -\frac{f_s^2}{2 f_{osc}} \frac{2C_m}{(2C_0 + C_L)^2} \qquad (3)$$

Optimizing the oscillator phase noise is done by a maximization of the signal amplitude, thus decreasing the C_2 capacitor. However, it is convenient to limit the signal amplitude on the BAW terminals in order not to increase BAW 1/f noise near the carrier. Nevertheless, C_2 cannot be reduced as low as wanted because of possible bipolar transistor saturation. Thus, a 0.5 ratio for $C_1/(2.C_2)$ appears as a rule of thumb to optimize the phase noise performances of a Colpitts oscillator. To maintain good phase noise performances, C_2 tuning range has to be limited. Indeed, a trade-off between phase noise and tunability emerges.

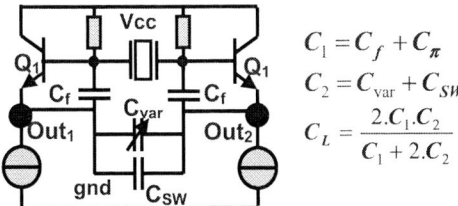

Fig. 3.   Scheme of the implemented Colpitts VCO.

$$C_1 = C_f + C_\pi$$
$$C_2 = C_{var} + C_{SW}$$
$$C_L = \frac{2.C_1.C_2}{C_1 + 2.C_2}$$

By combining (1), (2) and (3) and by considering $C_1/(2.C_2)=0.5$, reducing C_o allows optimizing both the sensitivity and the power consumption for a given oscillation frequency, a given SMR Q-factor and a given k_t^2 (see Fig. 4).

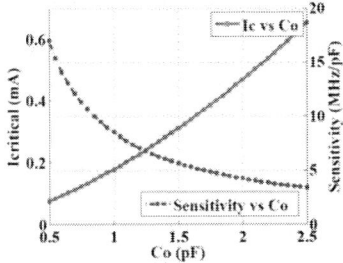

Fig. 4.   Co-integration optimizations.

For all these reasons, an SMR with a 4.5% k_t^2, a 1.35pF C_o and a 2.526GHz f_p has been designed. It presents a measured loaded Q of around 950 and a measured -4.2ppm/°C TCF.

B. Process and Temperature Compensations

The oscillator frequency can be precisely adjusted over the tuning range by means of a matrix capacitors tuning bank. Ideally, in order to increase the yield of the processed BAW resonators, the capacitor matrix has to compensate for all the process frequency deviations. This will be obtained at the expense of phase noise performances. Indeed, adding a bit corresponds to doubling the capacitor matrix size and the higher the capacitor matrix, the higher the insertion losses mainly coming from the interconnect. Moreover, a large capacitor tuning range is not compliant with an optimized phase noise because of a non ideal ratio $C_1/(2.C_2)$. For those reasons, a 5bit switched capacitor matrix has been implemented and allows compensating a 1.5MHz process frequency drift. The minimum, the middle and the maximum achievable capacitance of the switched capacitor matrix will be noted C_{min}, C_{mid} and C_{max}. In order to increase the accuracy of the capacitor step, elementary capacitors attached to a given bit are spread over the whole matrix. It prevents from a change in the value of the switched capacitor reduced by the inductive part of interconnect. Thus, statistically, a constant capacitor step

is obtained whatever the binary word. Simulations with parasitic show the coarse bank tuning step is 25fF and its Q-factor varies between 890 (for the LSB) and 40 (for the MSB) for a 200µmx130µm size while the global capacitor matrix varies between 386fF and 1.16pF. The varactor block capacitance varies from 0.8pF to 2pF for a Q varying between 33 and 40. Finally, the implemented BAW VCO is able to compensate for a 1.5MHz process dispersion and a -4.2ppm/°C TCF in a [-40°C, 85°C] temperature range (for $C_{SW}=C_{min}$) for a 3.65mA current under 2V supply voltage. For the demonstration, a 1.5mA divider by 2 has been used with this VCO while a 50Ω buffer has been designed for test purposes. Just notice no ACL (Amplitude Control Loop) has been implemented in order to simplify the calibration. Indeed, an ACL changes the biasing current, on which C_π (thus C_L) depends. This would lead to an extra uncontrollable shift in the oscillation frequency. Finally, electrical oscillator core presents a simulated -1ppm/°C temperature deviation.

V. MEASUREMENT RESULTS

The SMR die and the IC have been assembled in the assembly product line of STM (Tours, France)(see Fig. 5).

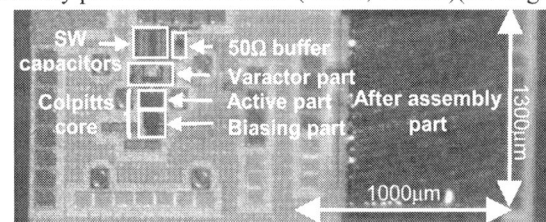

Fig. 5.   Photo-micrograph of VCO without and with its flip-chip mounted BAW resonator.

Good agreements between simulation and measurement is obtained as long as the MBVD model is close to the flip-chipped BAW resonator (see an example in Fig. 6a for $C_{SW}=C_{min}$ at 85°C). At 27°C and for $C_{SW}=C_{min}$, a -99dBc/Hz phase noise at 2kHz offset frequency has been achieved. At 27°C and for $C_{SW}=C_{min}$ and $C_{SW}=C_{mid}$, a 100ppm/V pushing has been measured while phase noise performances remains constant down to 5mW consumption power. As discussed before, the presence of substrate harmonic resonance in the operating frequency domain induces a frequency drift and a shift in the frequency step size (see Fig. 6b: SMR2 curve). It results in a large degradation of the phase noise performances due to a reduced BAW Q-factor because of higher acoustic energy losses at a parasitic mode. Now, all the discussed results will concern the VCO with SMR1 having no substrate harmonic resonance (see Fig. 6b) near the operating frequency.

978-1-4244-6240-7/10 $26.00 © 2010 IEEE

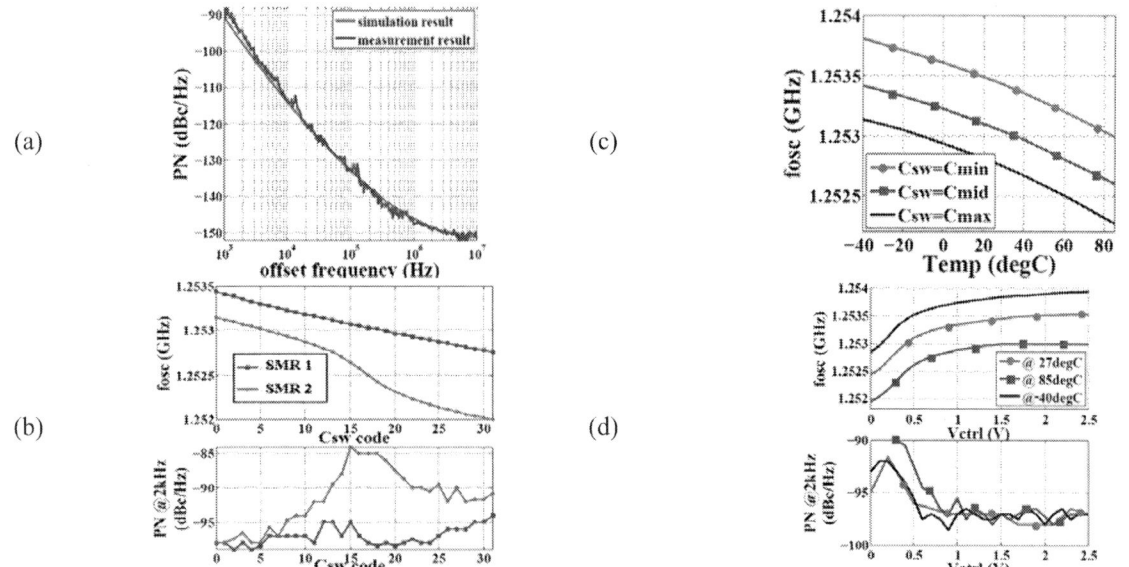

Fig.6. Measurement results: (a) measured PN vs simulated PN for $C_{SW}=C_{min}$ @85°C, (b) f_{osc} and PN @2KHz vs C_{SW} for 2 SMR, (c) BAW VCO TCF for $C_{SW}=\{C_{min},C_{mid},C_{max}\}$, (d) f_{osc} and PN@2kHz vs V_{ctrl} for $C_{SW}=C_{min}$.

TABLE I

SUMMARY OF HIGH PERFORMANCE BAW-BASED VCOs.

Ref.	Technology	Fosc (GHz)	PN (dBc/Hz)	Fm (kHz)	Process correction	Temp. (*) correction	I_{dc}/V_{supply} (mA/V)
[3]	0.35µm CMOS TFBAR	1.5	-125	100	No	200ppm	0.975/1V
[4]	90nm CMOS TFBAR	1.7	-116	100	3820ppm		1.67/0.9(VCO)+2.1/1.1(tuning)
[5]	0.13µm CMOS TFBAR	1.575	-114	100	No	130ppm	0.75/1
This work	0.13µm BiCMOS SMR	2.505	-132	100	560ppm	600/840ppm	3.65/2

(*) over around a 100°C temperature range.

As simulated, process compensation allows tuning the oscillation frequency over a 1.4MHz frequency range with a 12.5kHz accuracy. Finally, the global TCF of the VCO (see Fig. 6c) is around -5.2ppm/°C. The shape of the BAW VCO TCF is almost entirely defined by the shape of the SMR TCF. Nevertheless, the TCF slightly varies according to the applied C_{sw} codes. It varies from -5.2ppm/°C for $C_{sw}=C_{min}$ to -5.53ppm/°C for $C_{sw}=C_{max}$. Finally, the temperature compensation is effective for $C_{sw}=C_{min}$. Actually, the higher the C_{sw} value, the lower the ability to compensate temperature deviations. Indeed, the higher C_{sw}, the lower the sensitivity. In the worse case that is for $C_{sw}=C_{max}$, the maximum corrigible BAW resonator TCF has been estimated around -3ppm/°C (see Fig. 6d). Finally, Table I compares the presented work to other state-of-the-art BAW-based oscillators.

VI. CONCLUSION AND PERSPECTIVES

This work shows for the very first time the impact of the BAW resonator behaviour on the VCO performances. Some extra efforts on the next SMR generation will permit to optimize both power consumption and phase noise performances. Namely, sorting out +/-1MHz process-shifted resonator and a reduced access resistance (R_s) to a substrate resonance free SMR will allow reaching the stringent GSM requirements. Now, the BAW based VCO thermal calibration in industrial environment remains the main issue to solve out. In order to perform the correction, a temperature sensor with high coefficient temperature will be put on the BAW chip [2].

REFERENCES

[1] D. Petit et al., "Temperature Compensated Bulk Acoustic Wave Resonator and its Predictive 1D Acoustic Tool for RF Filtering", Ultrasonics Symp. Dig., pp. 1243-1246, October 2007.

[2] D. Petit et al., "Temperature Compensated BAW Resonator and its integrated Thermistor for a 2.5GHz Electrical Thermally Compensated Oscillator", RFIC Symp. Dig., pp. 339-342, June 2009.

[3] S. Rai et al., "A 1.5GHz CMOS/FBAR Frequency Reference with ±10ppm Temperature Stability", Freq. Control Symp. Dig., pp.385-387, April 2009.

[4] H. Hito et al., "A 1.7GHz 1.5mW Digitally Controlled FBAR Oscillator with 0.03ppb Resolution", ESSCIRC Dig., pp. 98-101, September 2008.

[5] Julie R. Hu et al., "A 750µW 1.575GHz Temperature-Stable FBAR-Based PLL", RFIC Symp., pp. 317-320, June 2009.

High Efficiency and Wideband Envelope Tracking Power Amplifier with Sweet Spot Tracking

Dongsu Kim, Jinsung Choi, Daehyun Kang, and Bumman Kim

Department of Electrical Engineering, Pohang University of Science and Technology, Pohang, Gyeongbuk, 790-784, Republic of Korea

Abstract — This paper describes the implementation of a high efficiency and wideband envelope tracking power amplifier with sweet spot tracking. By modulating supply voltage of power amplifier (PA), efficiency can be increased significantly. And linearity is improved by envelope shaping and sweet spot tracking. The supply modulator has a combined structure of a switching amplifier and a linear amplifier to achieve high efficiency as well as wide bandwidth. The measurement results show efficiencies of 36.4/34.1 % for 10/20 MHz long term evolution (LTE) signals with peak to average power ratio (PAPR) of 7.5/7.42 dB.

Index Terms — Boost converter, envelope tracking, linear amplifier, long term evolution (LTE), power amplifier (PA), sweet spot, switching amplifier.

I. INTRODUCTION

As wireless communication systems provide high data rate services, the channel bandwidth and PAPR of the signals are increased and the efficiency for the power amplifier is decreased. In the case of conventional PA with fixed supply voltage (Fig. 1a), the PA should be operated in the back off power region to linearly amplify the modulated signal with high PAPR and its efficiency is much lower than its peak value as shown in Fig. 2. On the other hand, the envelope tracking PA (Fig. 1b) operates under modulated supply voltage according to its output power level and its efficiency is degraded slightly.

Because overall efficiency of the envelope tracking PA is proportional to efficiency of the supply modulator and its linearity is affected by linearity of the supply modulator, a realization of the supply modulator with a high efficiency and good linearity is very important. In [1], they have designed a low dropout (LDO) regulator as a supply modulator, but its efficiency is very low for high PAPR signals. In [2], a switching amplifier is used as a supply modulator. Although it achieves high efficiency, it requires high order passive filter and its bandwidth is too narrow to use for wide bandwidth signals such as LTE and WiMAX. To achieve high efficiency and wide bandwidth, we use hybrid switching supply modulator combining the advantages of two supply modulators [3]-[7]. To improve the performance of power amplifier, a boost converter is added to the supply modulator as shown in Fig. 3. By boosting the supply voltage of the linear amplifier from 3.4V to 5V, the output voltage of the supply modulator is increased up to 4.5V and the power amplifier shows higher gain, efficiency, output power and wider bandwidth.

In [8], they analyze nonlinear distortion of envelope tracking PA. Because of knee voltage and nonlinear capacitance, AM-AM and AM-PM distortion are generated. By adopting envelope shaping and sweet spot tracking, linearity can be improved.

In this paper, we implement a high efficiency and wideband envelope tracking PA for LTE applications using a hybrid switching supply modulator, HBT PA, envelope shaping, and sweet spot tracking.

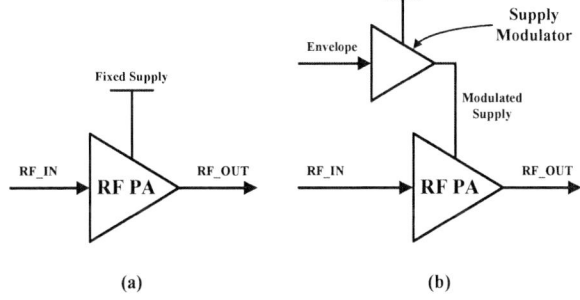

Fig. 1. (a) Conventional PA with fixed supply voltage. (b) Envelope tracking PA with modulated supply voltage.

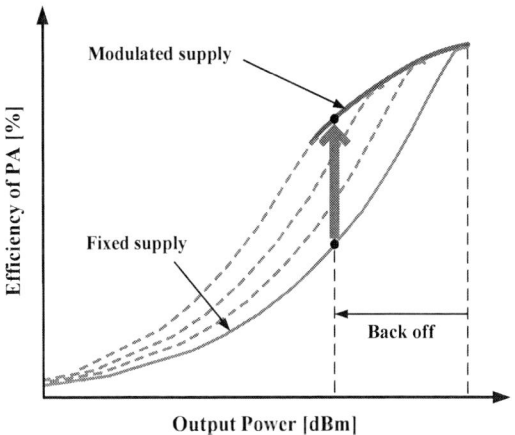

Fig. 2. PA's efficiency curves with fixed supply voltage and modulated supply voltage.

978-1-4244-6240-7/10 $26.00 © 2010 IEEE

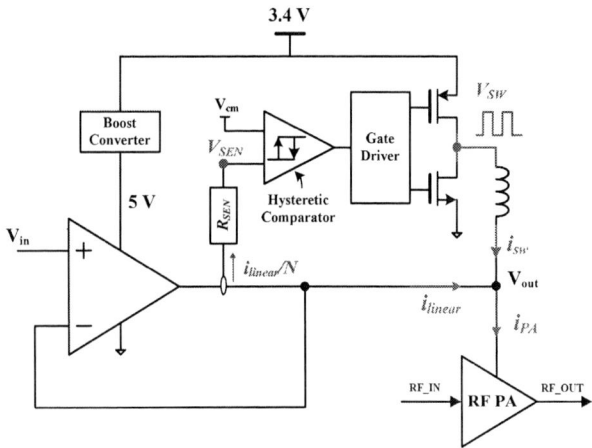

Fig. 3. Block diagram of the hybrid switching supply modulator with boost converter.

II. DESIGN OF HIGH EFFICIENCY AND WIDEBAND SUPPLY MODULATOR

The proposed supply modulator consists of a wideband linear amplifier, a high efficiency low speed switching amplifier, and a boost converter. Usually the switching amplifier supplies low frequency component of the envelope signal with high efficiency and the linear amplifier supplies other high frequency component with high speed. Because most of the power of the envelope signal is located at a low frequency, this structure is suitable for an operation with high efficiency and wideband.

In Fig. 3, the wideband linear amplifier operates as a voltage-controlled voltage source (VCVS). It means the output voltage of the linear amplifier is the same with its input voltage up to tens of MHz due to its high gain, wide bandwidth, and negative feedback. As shown in Fig. 4, we use folded-cascode OTA as a gain stage to achieve a large bandwidth and high DC gain. For large current driving capability and rail-to-rail operation, the output buffer has a common source configuration and it is biased as class-AB for linearity and efficiency.

The high efficiency, low speed switching amplifier operates as a dependent current source. It senses the direction of the linear amplifier's current and controls the switching amplifier using a hysteretic comparator. Generally, the average switching frequency is dependent on the hysteresis width, inductor value, and some other parameters for a narrow-band signal. For a wideband signal, the average switching frequency is mainly determined by its bandwidth. The sizes of the power switches are determined by considering the conduction loss and switching loss at the specific load resistance,

switching frequency, and duty ratio. For the protection, high efficiency, and low switching noise of the switches, anti-shoot-through circuit and divided switches with current control technique are employed (Fig. 5) [9]. Gate driver for the divided switches, which is shown in Fig. 6, turns on / off the 4 switches with a little delay. It can be designed easily using 4 MUXs and inverter chains.

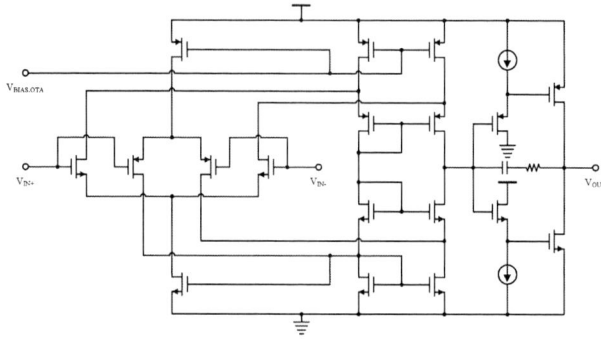

Fig. 4 Wideband linear amplifier.

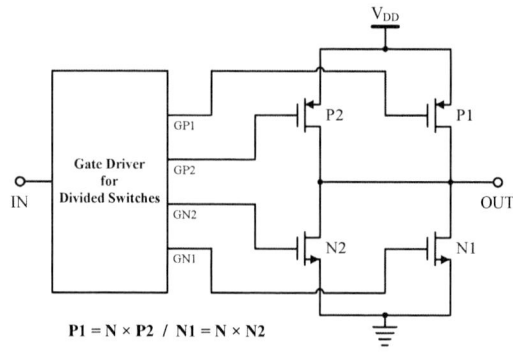

Fig. 5 High efficiency switching amplifier.

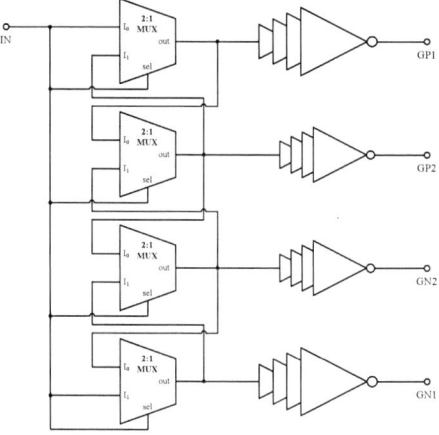

Fig. 6. Gate driver for divide switches with current control technique.

III. ENVELOPE SHAPING AND SWEET SPOT TRACKING

The modulated PA operates differently according to the supply voltage level. Especially at a low supply voltage, a power amplifier shows severe nonlinear characteristics such as AM-AM and AM-PM distortions because of knee voltage effect and nonlinear capacitance. To compensate these effects, an envelope shaping method should be used [7].

In addition to this basic method, a sweet spot tracking is proposed in this work. Fig. 7 is third-order and fifth-order intermodulation distortions (IMD) of PA in two-tone analysis. In this figure, there are sweet spots which are local minimums of IMD and are occurred by cancellation of the harmonics. As supply voltage decreases, the sweet spot also moves to lower power. By adjusting the supply voltage to minimize the distortions at each power level, the linearity of the envelope tracking PA can be improved significantly.

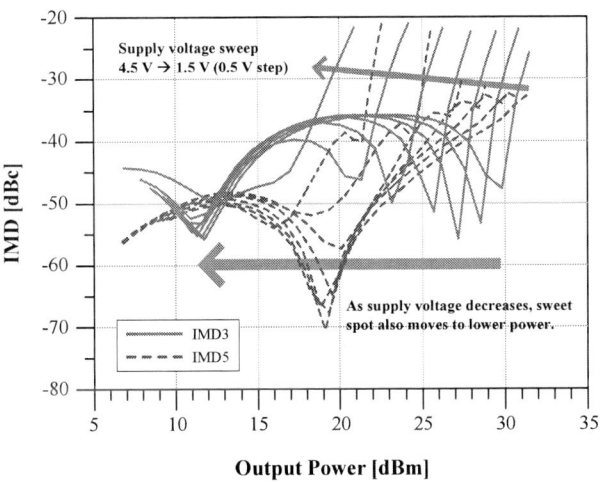

Fig. 7. Simulated third-order and fifth-order intermodulation distortions of PA in two-tone analysis.

IV. MEASUREMENT RESULTS

The designed supply modulator is fabricated using 65nm CMOS process and it uses thick oxide I/O devices for a high voltage operation. Chip photograph is shown in Fig. 8 and its size is 2.6 mm × 1.7 mm. The supply voltage for the supply modulator is 3.4 V (the battery voltage) and the boost converter generates 5 V for supply of the linear amplifier. In this configuration, output voltage range of the supply modulator is 0.5 to 4.5 V regardless of the battery voltage fluctuation, replacing the DC-DC converter [10]. The linear amplifier shows over 100 MHz bandwidth and over 55 dB DC gain. The average switching frequency of the switching amplifier is varied

from 3 MHz to 6 MHz according to the bandwidth of an input signal.

To implement the envelope tracking PA, 2.535GHz class-AB PA, which is fabricated using InGaP/GaAs 2um HBT process, is used. It has about 30 dBm peak output power at 3 V supply voltage. By boosting the supply voltage of the linear amplifier, PA's supply voltage increases up to 4.5 V and peak output power of the PA also increases to 33.4 dBm. Performance of the envelope tracking PA is measured using 10/20 MHz LTE signals with 7.5/7.42 dB of PAPR.

Fig. 9 shows the measured efficiency and gain of the envelope tracking PA. For the 10 MHz LTE signal, the envelope tracking PA has efficiency of 36.4 % at output power of 27.2 dBm. For the 20 MHz LTE signal, its efficiency is 34.1 % at output power of 26.1 dBm. Estimated efficiencies of the supply modulator are about 75/71 % for 10/20 MHz LTE signals. These values can be calculated from PA's efficiency curve at each supply voltage. Fig. 10 is measured output spectra of the envelope tracking PA at the peak output powers without any linearization technique.

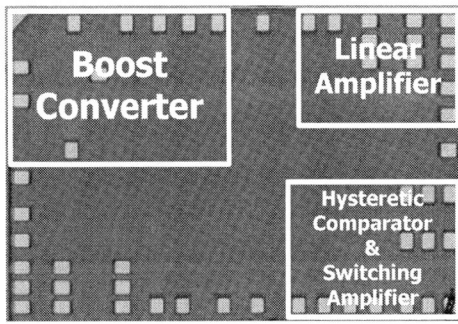

Fig. 8. Fabricated chip photograph of the supply modulator.

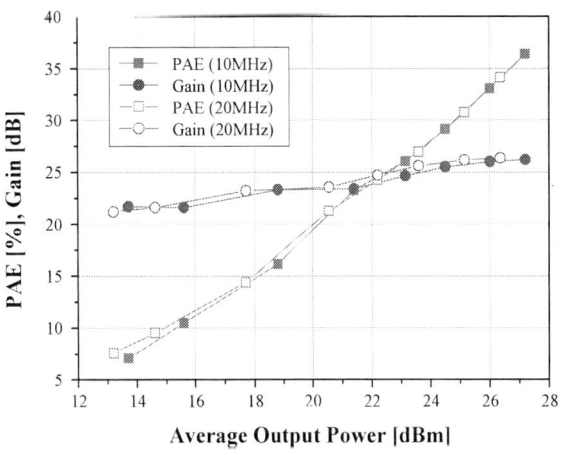

Fig. 9. Measured efficiency and gain of the envelope tracking PA.

978-1-4244-6240-7/10 $26.00 © 2010 IEEE

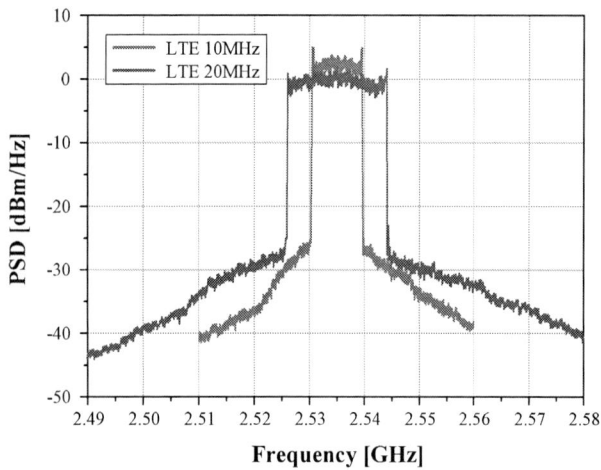

Fig. 10. Measured output spectra of the envelope tracking PA at peak output power.

TABLE I

PERFORMANCE SUMMARY OF ENVELOPE TRACKING POWER AMPLIFIER FOR LTE APPLICATIONS

Signal bandwidth	10 MHz	20 MHz
PAPR	7.5 dB	7.42 dB
Supply voltage	3.4 V	
Peak output power	27.2 dBm	26.1 dBm
Peak efficiency	36.4 %	34.1 %
Estimated efficiency of supply modulator @ peak output power	75 %	71 %

V. CONCLUSIONS

A high efficiency and wideband envelope tracking PA with sweet spot tracking technique is proposed and implemented for LTE applications. For the supply modulation, a hybrid switching supply modulator with boost converter is fabricated using 65nm CMOS process. An envelope shaping with sweet spot tracking is adopted to compensate AM-AM and AM-PM distortions. Efficiencies of the implemented envelope tracking PA are 36.4/34.1 % at output power of 27.2/26.1 dBm for 10/20 MHz LTE signals, respectively. Measured results show the proposed supply modulator with sweet spot tracking is a suitable structure to achieve a high efficiency and wideband envelope tracking PA.

ACKNOWLEDGEMENT

This work was supported by WCU (World Class University) program through the Korea Science and Engineering Foundation funded by the Ministry of Education, Science and Technology (Project No. R31-2008-000-10100-0), and by the MKE (The Ministry of Knowledge Economy), Korea, under the ITRC (Information Technology Research Center) support program supervised by the NIPA (National IT Industry Promotion Agency) (NIPA-2009-C1090-0902-0037).

REFERENCES

[1] P. Reynaert and M. Steyaert, "A 1.75-GHz polar modulated CMOS RF power ampli•er for GSM-EDGE," *IEEE J. Solid-State Circuits*, vol. 40, no. 12, pp. 2598–2608, Dec. 2005.

[2] V. Pinon, F. Hasbani, A. Giry, D. Pache, and C. Garnier, "A single-chip WCDMA envelope reconstruction LDMOS PA with 130MHz switched-mode power supply," *IEEE Int'l Solid State Circ. Conf. Dig. Tech. Papers*, Feb. 2008, pp. 564–565.

[3] T. Kwak, M. Lee, B. Choi, H. Le, and G. Cho, "A 2W CMOS hybrid switching amplitude modulator for EDGE polar transmitter," *IEEE Int'l Solid State Circ. Conf. Dig. Tech. Papers*, Feb. 2007, pp. 518–519.

[4] F. Wang, D. F. Kimball, D. Y. Lie, P. M. Asbeck, and L. E. Larson, "A monolithic high-efficiency 2.4-GHz 20-dBm SiGe BiCMOS envelope-tracking OFDM power amplifier," *IEEE J. Solid-State Circuits*, vol. 42, no. 6, pp. 1271–1281, June 2007.

[5] J. Kitchen, W. Chu, I. Deligoz, S. Kiaei, and B. Bakkaloglu, "Combined linear and Δ-modulated switched-mode PA supply modulator for polar transmitters," *IEEE Int'l Solid State Circ. Conf. Dig. Tech. Papers*, Feb. 2007, pp. 82–83.

[6] W. Chu, B. Bakkaloglu, and S. Kiaei, "A 10MHz-bandwidth 2mV-ripple PA-supply regulator for CDMA transmitters," *IEEE Int'l Solid State Circ. Conf. Dig. Tech. Papers*, Feb. 2008, pp. 448–449.

[7] J. Choi, D. Kim, D. Kang, and B. Kim, "A polar transmitter with CMOS programmable hysteretic-controlled hybrid switching supply modulator for multistandard applications," *IEEE Trans. Microw. Theory Tech.*, vol. 57, no. 7, pp. 1675-1686, July 2009.

[8] J. C. Pedro, J. A. Garcia, and P. M. Cabral, "Nonlinear Distortion Analysis of Polar Transmitters," *IEEE Trans. Microw. Theory Tech.*, vol. 55, no. 12, pp. 2757–2765, Dec. 2007.

[9] S. Sakiyama, J. Kajiwara, M. Kinoshita, K. Satomi, K. Ohtani, and A. Matsuzawa, "An on-chip high-efficiency and low-noise dc/dc converter using divided switches with current control technique," *IEEE Int'l Solid State Circ. Conf. Dig. Tech. Papers*, 1999, pp. 156-157.

[10] J. Choi, D. Kim, D. Kang, J. Park, B. Jin, and B. Kim, "Envelope Tracking Power Amplifier Robust to Battery Depletion," *in IEEE MTT-S Int. Microw. Symp. Dig.*, May 2010.

978-1-4244-6240-7/10 $26.00 © 2010 IEEE

A 150MHz, 84% efficiency, Two Phase Interleaved DC-DC Converter in AlGaAs/GaAs P-HEMT Technology for Integrated Power Amplifier Modules

Han Peng, V. Pala, T. P. Chow, and Mona Hella

Rensselaer Polytechnic Institute, ECSE Department, 110 8th Street, Troy, NY 12180, USA.

Email:pengh2,palav,chowt,hellam@rpi.edu

Abstract—**This paper presents a high efficiency, high switching speed, two-stage interleaved DC-DC buck converter with negatively-coupled inductors in AlGaAs/GaAs technology, targeting integrated power amplifier modules. The flip chip DC-DC converter is implemented in 0.5 μm GaAs pHEMT process and occupies $2 \times 2.1 mm^2$ without the output network. The inductors in the output network are implemented in 65 μm thick top copper metal layer and have a quality factor of 25 at 150 MHz. The interleaved DC-DC converter achieves 84% efficiency when operating at 150MHz switching frequency with 4.5V/3.3V conversion ratio and 1A load current.**

Index Terms—**Gallium Arsenide Technology, p-HEMT, interleaved DC-DC converters, coupled inductors, conversion efficiency, supply modulators.**

I. INTRODUCTION

Mobile communication systems utilizing non-constant envelope modulation, require highly linear power amplifiers. To meet the stringent linearity requirements, the power amplifier (PA) typically operates in "back-off" power mode, leading to lower efficiency. Transmitter architectures such as envelope tracking and polar modulation have been developed to enhance the efficiency at back-off power, by modulating the supply voltage of the PA. High efficiency, small size, high modulation bandwidth, and multi-mode operation are the main requirements of supply modulators in mobile applications. Different types of supply modulators include switched-mode DC-DC converters, linear regulators, and hybrid-solutions that combine both functionalities. While switched-mode DC-DC converters can provide the highest efficiency, their reported bandwidth in silicon technologies have been limited due to the gate charging and switching losses of the employed field effect transistors (FETs).

This paper proposes the use of GaAs technology for the implementation of switched-mode, high efficiency, DC-DC converters targeting GSM/EDGE applications with their relatively high voltage, high current requirements. Given that GaAs is the most popular technology for power amplifier implementations, it is natural to integrate the supply modulator with the PA in the same technology. It has been previously shown that GaAs P-HEMT based switching transistors have a lower switching figure of merit compared to silicon MOSFETs for the same voltage range [1]. In this paper, we demonstrate a flip-chip integrated 150MHz, two phase interleaved DC-DC converter with negatively coupled inductors in the output

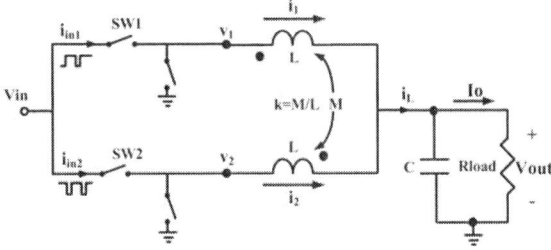

Fig. 1. Ideal interleaved topology with negatively coupled inductors.

filter network. The DC-DC converter achieves 84% maximum efficiency for 1A load current and 4.5/3.3V ratings. Section II discusses the advantages of coupled inductors in terms of the steady state and transient performance of the DC-DC converter. Section III provides the details of the circuit implementation. Measurement results are presented in section IV, while conclusions are drawn in Section V.

II. ANALYSIS OF INTERLEAVED TOPOLOGY WITH COUPLED INDUCTORS

Increasing the switching speed and the use of interleaved architectures in DC-DC converters have been shown to reduce the values of filter inductors by more than 50% [2], [3], [4], [5]. In this section, we study the effect of introducing coupling between the filter inductors in two phase interleaved converters on the steady state current ripple and the system bandwidth.

A. Steady State Analysis

Fig. 1 shows conceptually the core interleaved structure with coupled inductors, where M is the mutual inductance between the two phases. Here, we assume that the two branches have equal inductances L and $k = M/L$, where k is the coupling factor. Following the analysis presented in [5], [6], the steady state current ripple per phase in the inductor for duty ratios larger than 0.5 can be given by:

$$\Delta i_1 = \frac{V_{in}(1-D)(\frac{D}{1-D} + k)}{L(1-k^2)} \cdot (1-D)T \quad (1)$$

Where V_{in} is the input voltage and D is the duty cycle. Fig. 2 shows Δi_1 as function of coupling factor for different duty cycles assuming an input voltage V_{in} of 4.5V, a switching frequency of 150MHz and a filter inductor of 8nH. As can be seen from the figure, the coupling factor that results

978-1-4244-6240-7/10 $26.00 © 2010 IEEE

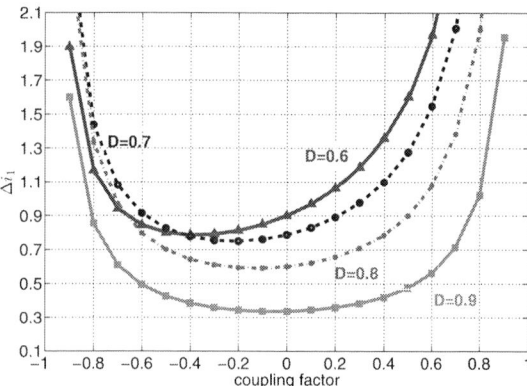

Fig. 2. Current ripple versus inductors' coupling factor for different duty cycles.

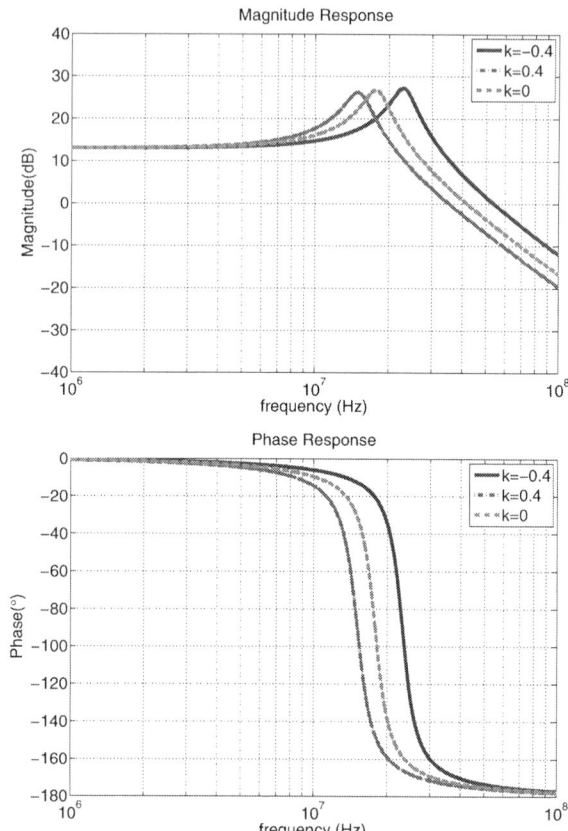

Fig. 3. Open loop transfer function for different coupling factors.

in minimum current ripple is a function of the duty ratio. However, one can make the statement that a negative coupling factor between -0.2∼-0.4 would generally result in lower current ripple, while positive coupling increases the ripple beyond the uncoupled case. A higher current ripple results in larger inductor loss, which lowers the efficiency of the DC-DC converter. In the given example of 4.5/3.3V conversion ratio, the optimum coupling factor for minimum current ripple is around -0.3.

B. System Stability and Transient Response

Considering the transient response, the open loop transfer function $\frac{v_o(s)}{d(s)}$, where $d(s)$ is the small signal function of duty ratio D, can be derived from the inductor current and capacitor voltage as in (2), following the analysis described in [7], while including the dc resistance of the inductors.

The small signal transfer function for different coupling factors is plotted in Fig. 3. The figure compares the bandwidth and phase margin of non-coupled, positively coupled and negatively coupled inductors in two phase interleaved converters. Negatively coupled inductors increase the bandwidth by 29.2% compared to non-coupled inductors, and 52.7% compared to positively coupled inductors for the case of $k = \pm 0.4$. Thus, it is evident from the discussion above that negative coupling can improve both the steady state and transient response, by reducing the current ripple and increasing the bandwidth of the DC-DC converter. However, optimum ripple cancellation depends on the selected coupling factor, which varies for different duty cycles. To achieve the best ripple cancellation, the coupling factor should be selected according to Fig. 2.

III. CIRCUIT IMPLEMENTATION

A prototype of the interleaved DC-DC converter is designed for 4.5V to 3.3V conversion, 1A load current, in a three metal

layer 0.5 μm GaAs p-HEMT process with both depletion and enhancement mode p-HEMT devices. The coupled inductors are implemented in 65 μm top copper layer. The circuit diagram of the converter is shown in Fig. 4.

A. Output Stage and Gate Driver Design

Two loss mechanisms are encountered in the switching stage; the switching loss and the conduction loss. Fig. 5 shows the variations of both losses as a function of the switching transistor width for 4.5V to 3.3V conversion ratio. The sizes of the high side switches are selected at the point where the conduction loss equals to the switching loss to achieve maximum efficiency. It is important to include the gate driver losses with the overall losses when sizing the switching transistor. For 1A output current and the required voltage conversion ratio, the widths of SW1 and SW2 are chosen as 10mm. M3 and M4 provide a path for the current when SW1 and SW2 are off and they are sized at the same width as SW1 and SW2.

Given that there are no complementary devices in the used technology, the supply voltage of the gate driver needs to

$$\frac{v_o(s)}{d(s)} = \frac{\frac{2V_{in}}{C(L+M)}\left(s + \frac{r_L}{L-M}\right)}{s^3 + \left(\frac{1}{R_{load}C} + \frac{2Lr_L}{L^2-M^2}\right)s^2 + \left(\frac{2Lr_L}{R_{load}C(L^2-M^2)} + \frac{r_L^2}{L^2-M^2} + \frac{2}{C(L+M)}\right)s + \left(\frac{r_L^2}{R_{load}C(L^2-M^2)} + \frac{2r_L}{C(L^2-M^2)}\right)} \quad (2)$$

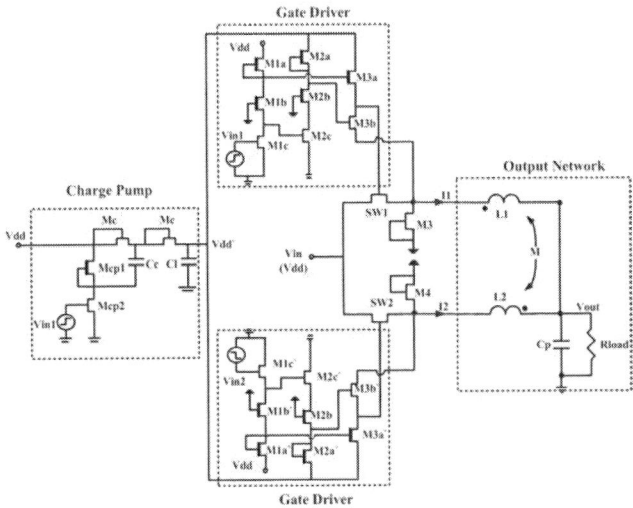

Fig. 4.   Circuit topology.

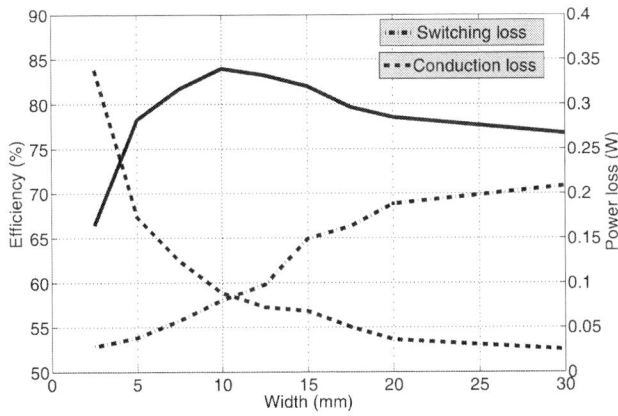

Fig. 5.   Transistor loss and efficiency versus transistor size.

be higher than the supply voltage of the converter to drive the high side switches. The minimum value of gate driver supply voltage is $V_{dd} + V_p$, where V_p is the pinch off voltage of the enhancement mode p-HEMT. A single stage Dickson charge pump is adopted to increase the supply voltage. The gate driver stage is a two-stage active inverter with the second stage referenced to the source of the high side switches. The second inverter stage is designed as pseudo complementary switches with high side depletion mode HEMT and low side enhancement mode HEMT. The first two stages generate the gate control signals of the M_{3a} and M_{3b} separately. The sizing of the gate driver stages is a tradeoff between reducing the gate driver losses and enhancing the driving capability for the high side switches, which affects the switching loss of SW1 and SW2. The enhancement mode pHEMT M_{3b} is the main switching transistor, and its width is chosen as $1/10$ of the high side switch SW1, while M_{3a} is sized as 1/3 of M_{3b}.

B. Inductor Design

The coupled inductors must be properly selected to achieve an optimal balance between the required inductance value

Fig. 6.   Die photo of coupled inductors.

at the given switching speed and a low series resistance to minimize the losses. The size of inductors is set according to the switching frequency, output current level and current ripple requirements. The maximum current ripple can be defined at the boundary of continuous conduction mode (CCM) and discontinuous conduction mode (DCM), which is the point at which the steady state current ripple Δi_1 equals to the load current I_o [6]. Hence, using (1), the minimum required inductor can be defined as:

$$L_{min} = \frac{V_{in}(\frac{D}{1-D} + k)(1-D)^2}{(1-k^2)f_{sw}I_o} \tag{3}$$

The minimum inductor for the given circuit specifications is 6.28nH. Fig.6 shows the die photo of the coupled inductors, implemented using the interleaved topology and fabricated in 65 μm copper layer. The electromagnetic simulation results of the coupled inductors are shown in Table I.

TABLE I
INDUCTOR PARAMETERS

Area	$2.3 * 2.7 mm^2$
Width	60 μm
Turns	2.75
Spacing	140 μm
L	6.28nH
R_{dc}	0.055 ohm
k	0.3
$Q_{at 150MHz}$	25

IV.   MEASUREMENT RESULTS

The circuit shown in Fig. 4 is designed at 150MHz with 6.28nH coupled inductors and 20nF load capacitor. The circuit converts 4.5V input to 3.3V output with 1A output current. The duty cycle is 0.65 and the coupling factor is -0.3. The die micrograph is shown in Fig. 7. The converter die and coupled inductors are flip chip bonded to a PCB board. The area of the converter is $2 \times 2.1 mm^2$. The interleaved DC-DC converter is tested using an external pulsed source, provided by Agilent B1110A. The transient response is measured using HP Infinium 1.5GHz Oscilloscope. Fig. 8 shows the transient response of the output voltage. It is worth noting that the relatively high measured output voltage ripple of 112 mV, which is 2 times higher than the simulated value, is due to the deviation of implemented inductors from the target values of inductance and coupling factors for maximum ripple

978-1-4244-6240-7/10 $26.00 © 2010 IEEE

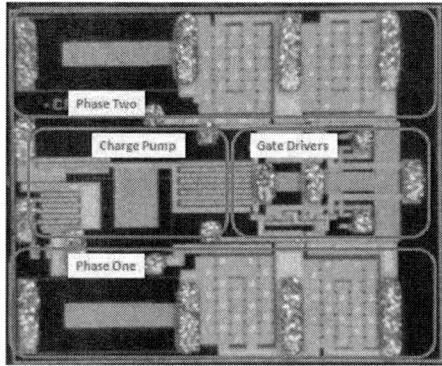

Fig. 7.   Die photo of DC-DC converter.

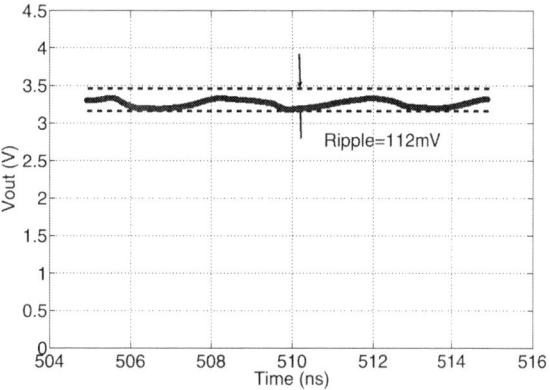

Fig. 8.   Output transient response.

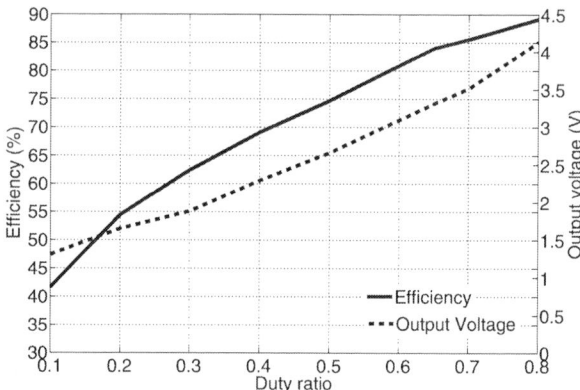

Fig. 9.   Efficiency and output voltage at varying duty cycle.

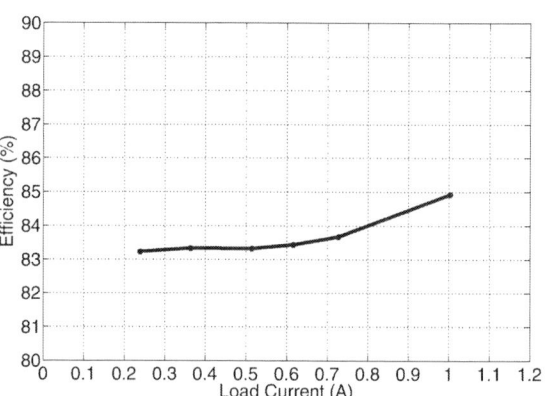

Fig. 10.   Efficiency at different load currents.

cancellation (the implemented inductors have a value of 8.7nH and a coupling factor of 0.46 compared to the target values of 6.28nH inductance and 0.3 coupling factor).

The measured efficiency for 4.5V input, 1A output current at 150MHz switching frequency with varying duty cycle is plotted in Fig. 9. The efficiency at the target conversion of 4.5V/3.3V is 83.8%. For duty ratios between 0.2 to 0.8, the efficiency variations are around 30%, which can be improved by using synchronous rectifiers and adding control techniques such as adaptive dead-time control. The efficiency is maintained over a wide range of output current as seen in Fig. 10.

V. CONCLUSION

A high efficiency, high switching frequency interleaved DC-DC converter with negatively coupled inductors has been demonstrated in 0.5μm, p-HEMT GaAs technology. GaAs technology provides a faster switch with lower on-resistance and smaller parasitic capacitance compared to CMOS technology. The interleaved converter achieves a peak efficiency of 84% at 150MHz with 4.5V/3.3V conversion ratio and 1A load current. This work demonstrates the potential of GaAs technology to extend the efficient operation of switching supply modulators to the hundreds of MHz, ultimately enabling fast tracking and multi-mode operation of integrated GaAs power amplifier modules.

VI. ACKNOWLEDGEMENT

The authors would like to acknowledge TriQuint Semiconductor for fabrication.

REFERENCES

[1] V. Pala, K. Varadarajan, and T. Chow, "GaAs pseudomorphic HEMTs for low voltage high frequency DC-DC converters," *Power Semiconductor Devices and IC's, 2009, 21st International Symposium on*, pp. 120–123, June 2009.

[2] P. Hazucha, G. Schrom, J. Hahn, B. Bloechel, P. Hack, G. Dermer, S. Narendra, D. Gardner, T. Karnik, V. De, and S. Borkar, "A 233-MHz 80%-87% efficient four-phase DC-DC converter utilizing air-core inductors on package," *IEEE J. Solid-State Circuits*, vol. 40, pp. 838–845, April 2005.

[3] G. Schrom, P. Hazucha, J. Hahn, D. Gardner, B.A.Bloechel, G. Dermer, S. Narendra, T. Karnik, and V. De, "A 480-MHz, multi-phase interleaved buck DC-DC converter with hysteretic control," *Power Electronics Specialists Conference, 2004 IEEE 35th Annual*, pp. 4702–4707, June 2004.

[4] S. Abedinpour, B. Bakkaloglu, and S. Kiaei, "A Multistage Interleaved Synchronous Buck Converter With Integrated Output Filter in 0.18 μm SiGe Process," *IEEE Trans. Power Electron.*, vol. 22, no. 6, pp. 2164–2175, Nov. 2007.

[5] P.-L. Wong, P. Xu, P. Yang, and F. Lee, "Performance improvements of interleaving VRMs with coupling inductors," *IEEE Trans. Power Electron.*, vol. 16, no. 5, pp. 499–507, Jul. 2001.

[6] R. W. Erickson and D. Maksimovic, *Fundamentals of Power Electronics (Second Edition)*.   Springer, 2001.

[7] J. Abu-Qahouq, M. Batarseh, L. Huang, and I. Batarseh, "Analysis and Small Signal Modeling of a Non-Uniform Multiphase Buck Converter," *Power Electronics Specialists Conference, 2007. IEEE*, pp. 961–967, June 2007.

A 0dBm 10Mbps 2.4GHz Ultra-Low Power ASK/OOK Transmitter with Digital Pulse-Shaping

Xiongchuan Huang, Pieter Harpe, Xiaoyan Wang, Guido Dolmans, Harmke de Groot

Holst Centre – imec, High Tech Campus 31, Eindhoven, 5656AE, The Netherlands

Abstract — **This paper presents an ultra-low power transmitter used for wireless sensor network (WSN) and wireless body area network (WBAN) applications. The proposed 2.4GHz direct modulation transmitter radiates 1mW with 3.88mW power consumption, and it supports OOK and ASK modulation up to 10Mbps. The novel power amplifier structure enables digital pulse-shaping to improve spectrum efficiency of OOK transmission. When applied with OOK modulation with equal probability of 1's and 0's, it consumes 2.3mW with an energy efficiency of 0.23nJ/bit/mW. The transmitter is implemented in a 90nm CMOS technology and packaged in a QFN56 package.**

Index Terms — **Low power, radio transmitter, amplitude shift keying, body area network, sensor network, power amplifiers.**

I. INTRODUCTION

Autonomous operation of wireless transducer nodes is one major challenge in today's WSN and WBAN applications. For truly autonomous networks, the sensor nodes must be self-powered. The radio part of the sensor nodes draws considerable amount of power, and therefore its power consumption must be minimized [1]. Unlike wake-up receivers (WuRx) which are optimized for active power consumption [2], the radio transmitter in WSN and WBAN nodes must have high energy efficiency since it is only activated when there are data to send. In order to achieve high efficiency, these transmitters are usually confined to simple structures, low data-rate, and limited options on modulation schemes [3-6].

This work implements an ultra-low power 2.4GHz radio transmitter with high data-rate and enhanced spectrum efficiency. It supports amplitude modulation in the form of ASK and OOK with pulse-shaping, and its 10Mbps data-rate makes it capable of handling data intensive applications such as WBAN ECG/EMG/EEG waveform transmission or high quality personal audio/video streaming. The transmitter outputs a nominal 1mW of radiated power while consuming 3.88mA from a 1V supply, resulting in an overall efficiency of 26 %.

II. TRANSMITTER ARCHITECTURE

At 2.4GHz, just a small amount of output power (~0dBm) is sufficient to cover the link distance (3 ~ 5m)

of most WSN and WBAN applications. In order to deliver high power efficiency at this output level, not only should the PA efficiency be optimized, but also the pre-PA power consumption has to be minimized [4]. The direct modulation architecture simplifies the stages preceding the PA, thus reducing the power overhead from the pre-PA stages.

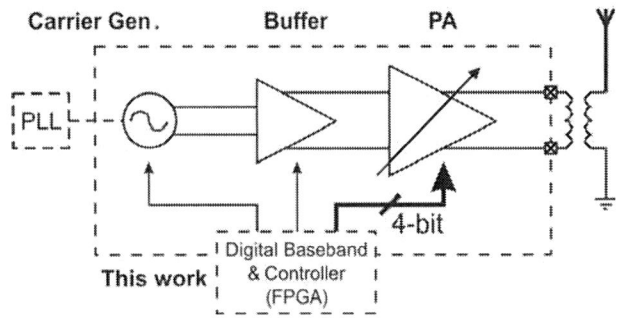

Fig. 1.   Proposed transmitter architecture.

Fig. 1 shows the proposed transmitter block diagram. The 2.4GHz carrier is generated by the integrated LC VCO, and amplitude modulation is applied on the PA. In between the PA and VCO, a low power buffer is inserted to avoid the dynamic loading effect from the PA, thus improving the frequency stability of the VCO. Unlike other ultra-low power OOK transmitters [3 5], here only the PA is switched on and off to avoid spectrum artifacts during the oscillator startup period. This also increases the data-rate since the start-up time of an LC oscillator with reasonable Q-factor is much longer than the time needed to toggle on/off the PA. In this work, no phase locked loop (PLL) is implemented; depending on the application environment and required frequency stability, a PLL can be adopted to regulate the carrier frequency either continuously, or periodically.

III. CIRCUIT IMPLEMENTATION

A. Low Power Integrated LC VCO

The schematic of the integrated VCO is shown in Fig. 2. It is a cross-coupled oscillator with an integrated LC tank. PMOS tail current source is adopted to improve the phase noise performance of the VCO [7]. The complementary cross-coupled pair provides twice the negative conductance compared to a single PMOS or NMOS cross-coupled pair, therefore reducing the current consumption by half for a given LC tank. M1-M4 are biased in weak or moderate inversion, to further reduce power consumption.

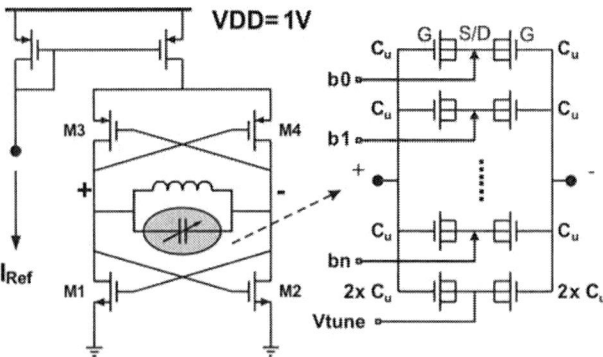

Fig. 2. Complementary cross-coupled LC VCO.

An on-chip inductor is chosen instead of bondwire or off-chip inductors for its immunity to package parasitics. In order to compensate for its low Q-factor, a large inductance is used. A 7.3nH differential spiral inductor with Q-factor of 17 is chosen, which results in a tank impedance of 1.87kΩ at resonance. This is comparable with a 4nH bondwire inductor with a Q-factor of 30. The tank capacitance consists of an array of switched varactors in parallel with a pair of varactors controlled by an analog tuning voltage. This ensures a large tuning range while reducing the phase noise originating from the control voltage. The switched varactor array is implemented with 31 unit varactor pairs with their S/D switched by 5 digital control bits, while the analog controlled part are in fact 2 unit varactor pairs with S/D connected to an analog tuning voltage. This combination provides redundancy against mismatch in the 31 unit pairs, such that any frequency spot within the tuning range can be selected. In this technology, the switched varactor bank occupies less area, and offers higher Q-factor than switched MIM banks since it does not suffer from the on-state resistance and off-state capacitance of the switches in the signal path.

Due to its high parallel loss resistance and power-efficient active part, the oscillation starts with just 112µA of current consumption at 1V supply.

B. Direct Modulation Power Amplifier

The proposed PA topology illustrated in Fig. 3 takes direct modulation beyond on/off keying. The PA consists of multiple identical pseudo-differential NMOS pairs connected in parallel. The LO signal is driving the input

NMOS pairs with a fixed amplitude, and the output of each unit cell is combined by summing up the RF current. Each unit PA cell can be switched on and off by controlling the gate voltage at the cascode transistors. In this work there are 15 unit cells in parallel, controlled by 4 digital bits. An off-chip matching network is implemented to transform the 100Ω differential antenna impedance up to 500Ω to boost up the voltage swing at the drain nodes. Also the input NMOS pairs are operating in class-AB for better power efficiency. At 0dBm output level, this PA achieves a drain efficiency of 33%.

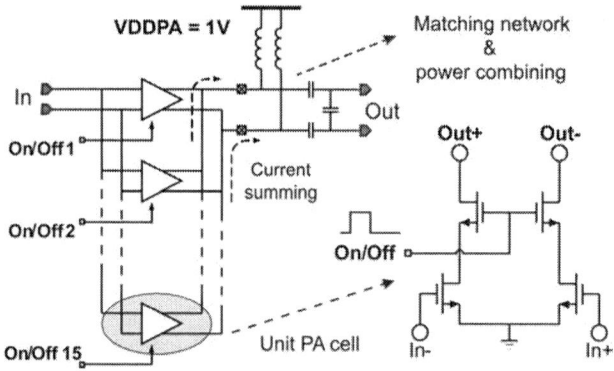

Fig. 3. Proposed OOK/ASK power amplifier.

By selectively activating certain unit cells, the output power level can be modulated either statically or dynamically. For example, higher-order ASK modulation can be applied to this PA, or OOK modulation with pulse-shaping can be realized. Both techniques offer better spectrum efficiency than traditional OOK modulation.

C. Low Power VCO Buffer

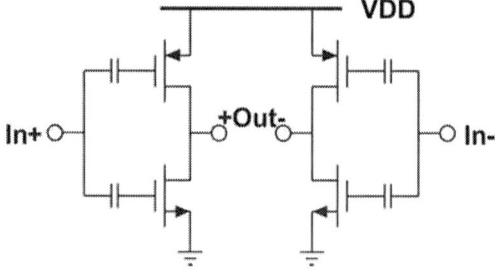

Fig. 4. Complementary VCO buffer (biasing not shown).

The buffer isolates the VCO from external perturbation, while amplifying the LO signal to drive the PA. The schematic of the VCO buffer is shown in Fig. 4. It also utilizes current-reuse between the N- and P- type transistors to improve power efficiency. A passive feedback network with large resistors sets the output quiescent level, while not loading the output substantially.

In order to provide the desired input swing for the PA (400mV differentially), the VCO is biased at 250uA to provide 300mV input to the buffer, and the buffer boosts it up to 400mV with 500uA current consumption.

D. Digital modulator and control logic

For testing and demonstration, digital baseband logic is implemented on FPGA for the transmitter. It is able to choose OOK modulation with or without pulse-shaping, as well as ASK modulation, all at variable data-rate. An ASIC version of the modulator will be realized together with other control logic, such as TX enable/disable, power control, bias adjustment, and frequency calibration.

E. Implementation

The transmitter front-end is fabricated in a 1P9M 90nm CMOS technology with RF/mixed-signal option. The chip micrograph is shown in Fig. 5. The chip measures 0.564 by 1.564 mm excluding pad-ring, and is bonded inside a QFN56 air-cavity package for testing.

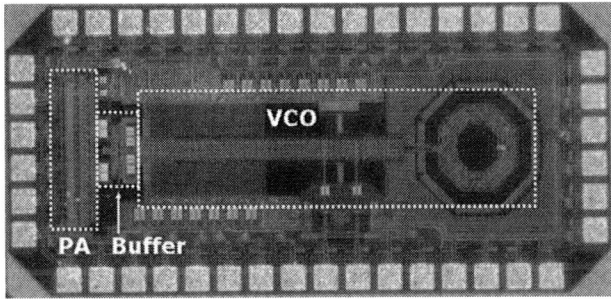

Fig. 5. Die photo of the transmitter.

IV. MEASUREMENT RESULTS

A. Measurements with single-tone carrier

The single tone carrier is measured at the transmitter output. The VCO is able to tune from 2.11 to 2.78 GHz, and all frequency spots are covered by the 5-bit digital plus 1 analog control pin. The harmonic components at 4.8 and 7.2 GHz are -51 and -38 dBc respectively. When the oscillator is biased at minimal (112µW) and nominal power consumption (250µW), its phase noise at 1MHz offset is -109 and -112 dBc/Hz respectively.

The output power control is shown in Fig. 7. As more unit PA cells are switched on, the carrier level grows linearly in voltage swing. When all 15 unit PA cells are turned on, the output power has not reached the 1dB

compression point, indicating a good linearity with respect to the PA control bits.

When 13 out of 15 PA cells are on, the transmitter outputs 0dBm with 3.88mW power consumption under 1V V_{DD} (PA: 3.14mA; pre-PA: 740µA). This results in an overall efficiency of 26%. When all the unit PA cells are on, the overall efficiency becomes 32% with 1.2dBm output power.

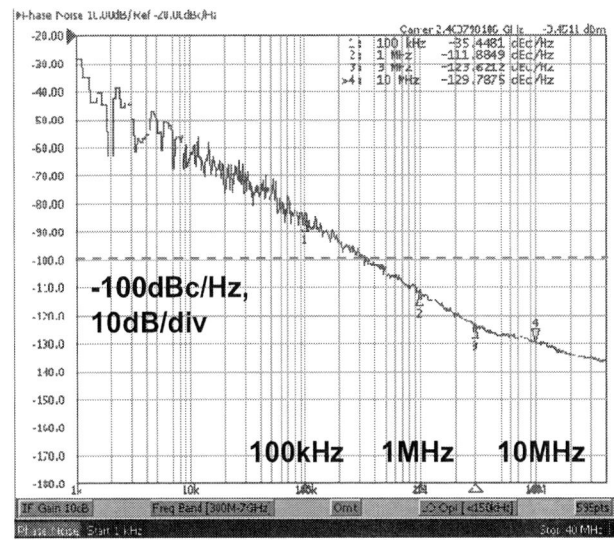

Fig. 6. VCO phase noise at nominal power consumption (250µW).

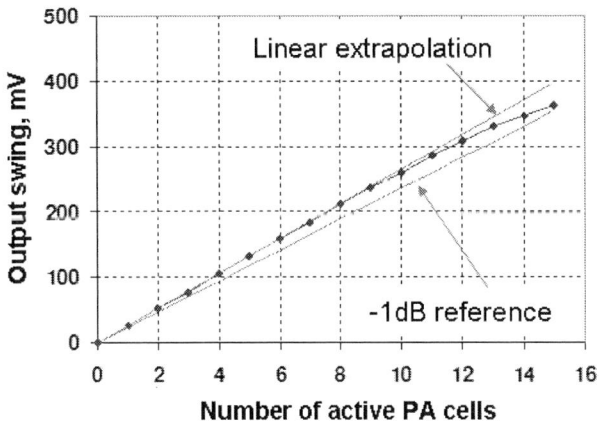

Fig. 7. TX output swing vs. number of active unit PA cells.

B. Measurements with modulated carrier

A consecutive '0101' series is applied to the PA on/off control to generate OOK output. Fig. 8 shows the transient waveform of a 10Mbps OOK output. The rise- and fall-time of the carrier envelope is 20 and 10 ns respectively; therefore the data-rate can go up to 20Mbps.

By gradually activating certain PA cells within one bit period, we can achieve pulse-shaping. Raised-cosine pulse-shaping with 8X oversampling and a roll-off factor of 0.2 is applied to a 3.125Mbps OOK '1' impulse. The resulting output waveform is shown in Fig. 9, with the FIR coefficients overlaid onto the waveform. By comparing the RF spectrum of the modulated output, the advantage of pulse-shaping is clearly shown in the spectrum efficiency: the -20dBc bandwidth of the traditional and pulse-shaped OOK transmission is 15.6MHz and 5.2MHz respectively. The side-lobes of the pulse-shaped OOK originate from the 8X oversampling clock, thus residing at +/- 25MHz away from the carrier, 32dB below the main-lobe.

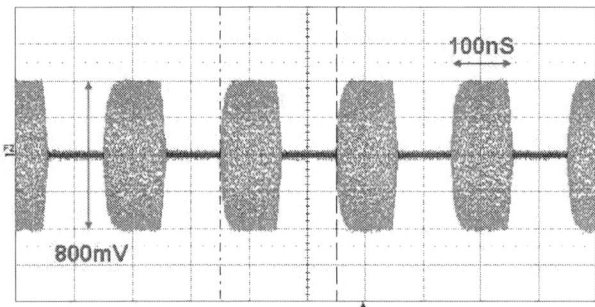

Fig. 8. Transient output of 10Mbps OOK signal.

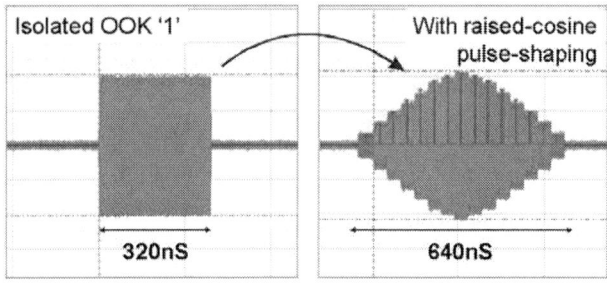

Fig. 9. Transient output of pulse-shaped OOK.

Table I: Comparison with state-of-the-art OOK transmitters

Ref.	Freq. (GHz)	P_out (mW)	η_{TX} (%)	Data-rate (Mbps)	FOM (nJ/bit/mW)
This work	**2.4**	**1**	**26**	**10**	**0.23**
[3]	0.433	0.054	4.5	10	0.97
[4]	1.9	1	28	0.156	11.54
[5]	1.9	1.2	46	0.33	3.41
[6]	0.9	0.6	6.9	1	14.4

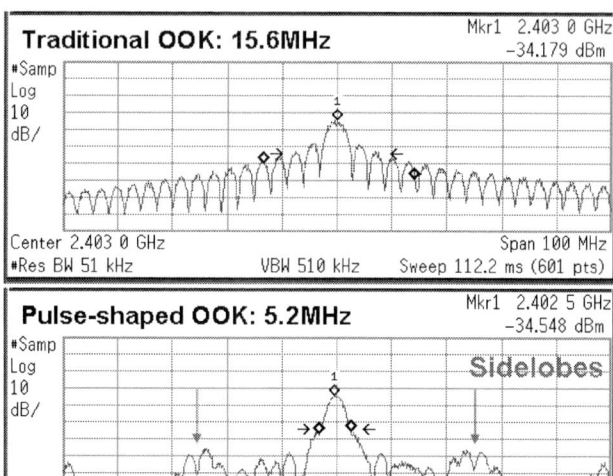

Fig. 10. Output spectrum of 3.125Mbps OOK modulation.

V. CONCLUSION

An ultra-low power 2.4GHz direct digital modulating ASK transmitter is presented. It achieves 26% efficiency while delivering 0dBm to a 100Ω differential load. With 50% OOK, it consumes 2.3mW, resulting in an FOM [3] of 0.23nJ/bit/mW at 10Mbps data-rate. In addition to OOK modulation, high-order ASK and pulse-shaped OOK are also supported with better spectrum efficiency. As shown in Table I, this transmitter compares favorably with other state-of-the-art low power transmitters.

REFERENCES

[1] J. Rabaey et al., "PicoRadios for Wireless Sensor Networks: The Next Challenge in Ultra-Low Power Design", ISSCC Dig. Tech. Papers, pp. 200-201, Feb 2002;

[2] X. Huang et al., "A 2.4GHz/915MHz 51 W Wake-Up Receiver with Offset and Noise Suppression", to appear in ISSCC 2010;

[3] M.K. Raja et al., "A 52 pJ/bit OOK transmitter with adaptable data rate," ASSCC 2008, pp.341-344, Nov 2008;

[4] Y.H. Chee et al., "An Ultra-Low-Power Injection Locked Transmitter for Wireless Sensor Networks," JSSC, vol.41, no.8, pp.1740-1748, Aug 2006;

[5] Y.H. Chee et al., "A 46% Efficient 0.8dBm Transmitter for Wireless Sensor Networks," 2006 VLSI Symposium Dig. Tech. Papers. pp.43-44;

[6] D.C. Daly et al., "An Energy-Efficient OOK Transceiver for Wireless Sensor Networks," JSSC, vol.42, no.5, pp.1003-1011, May 2007;

[7] D.J. Young et al., "2 GHz CMOS Voltage-Controlled Oscillator with Optimal Design of Phase Noise and Power Dissipation," 2007 RFIC, pp.131-134, Jun 2007;

A Linear-in-dB SiGe HBT Wideband High Dynamic Range RF Envelope Detector

Hsuan-yu Marcus Pan and Lawrence E. Larson

University of California, San Diego, 9500 Gilman Drive, La Jolla, CA, 92093, USA

Abstract—**A linear-in-dB wideband, SiGe HBT high dynamic range RF envelope detector is presented. The detector operates from 200MHz to 2.5GHz with 50dB maximum dynamic range.**

Index Terms—**Power detector, Gilbert multiplier, Variable Gain Amplifier, Multi-tanh, Linear-in-dB.**

I. INTRODUCTION

RF envelope detectors [1] are used in many applications, such as instrument interfaces, polar and envelope tracking transmitters [2], receive signal strength indication (RSSI) [3], power control [4], voltage standing-wave ratio (VSWR) measurements and automatic gain control (AGC) circuits.

When detecting wide dynamic range RF signals, "linear-in-dB" detectors generate an output proportional to the logarithm of the input envelope [5] and can be classified into two types: logarithmic amplifiers and AGC-based detectors [4]. Logarithmic amplifiers rely on the progressive compression of cascading gain cells, with summing/filtering each stage to produce a rectified voltage [6][7]. For high peak-to-average ratio (PAR) signals, AGC-based detectors (also referred to as "true RMS" detectors) operate in a closed-loop manner. The output of the VGA/attenuator is a constant when the loop is locked, and the envelope information is extracted from the control voltage [8][9].

In this paper, an improved AGC-based linear-in-dB envelope detector is proposed. The paper is organized as follows: Section II reviews the AGC-based power detector. Section III covers the implementation of sub-blocks. Section IV shows the measurement results.

II. AGC-BASED LINEAR-IN-DB ENVELOPE DETECTOR OPERATION

The basic AGC-based linear-in-dB envelope detector is shown in Fig. 1. The output of the VGA can be expressed as:

$$V_{out}(t) = V_{in}(t) \times \alpha \times 10^{V_{env}(t)/V_C} \qquad (1)$$

where α is the VGA gain constant and V_C is the exponential slope. If $V_{in}(t)$ is an envelope modulated signal:

$$V_{in}(t) = V_{in,env}(t) \times cos\omega_c t \qquad (2)$$

where $V_{in,env}(t)$ is the envelope and ω_c is the carrier frequency.

An ideal peak detector will extract the envelope $V_{in,env}(t)$ in (2). i.e.

$$V_{pd}(t) = V_{in,env}(t) \times \alpha \times 10^{V_{env}(t)/V_C} \qquad (3)$$

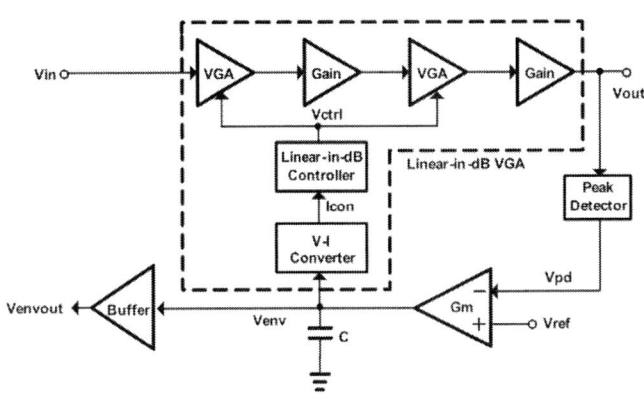

Fig. 1.   Block diagram of linear-in-dB envelope detector.

When the bandwidth of $V_{in,env}(t)$ is relatively small compared with $\frac{V_{ref}}{V_C} \times \frac{g_m}{C}$ [10], the loop forces $V_{pd}(t)$ to the reference voltage V_{ref}. Hence $V_{env}(t)$ is

$$V_{env}(t) \approx -\frac{V_C}{20} \times 20log(V_{in,env}(t)) + V_K \qquad (4)$$

where $V_K = V_C \times log(V_{ref}/\alpha)$.

III. CIRCUIT DESIGN

The linear-in-dB VGA function is realized by a chain of cascaded VGAs and gain stages, a V-I converter and a linear-in-dB controller.

A. Gilbert Multiplier Type VGA

Fig. 2 shows the variable gain stage, which keeps the output of the VGA constant when the input signal varies [11]. The voltage gain is approximately

$$A_{VGA} \approx \frac{I_{EE}R_L}{I_{EE}R_{EE} + V_T} \times \frac{1 - e^{V_{ctrl}/V_T}}{1 + e^{V_{ctrl}/V_T}} \qquad (5)$$

where V_T is KT/q. Equation (5) shows that a linear-in-dB control is needed for the VGA.

B. Gain Stages with Bandwidth Broadening

Fig. 3 shows the fixed gain stages with bandwidth broadening. The second gain stage input is in parallel with a differential pair $Q_9 - Q_{10}$ with a degeneration capacitor $C_{EE}/2$, which serves to broaden the bandwidth. The high frequency voltage gain of two gain stages can be approximated as

$$A \approx \frac{(\frac{g'_{m1}g_{m3}}{C_L^2})(s + \frac{g'_{m2}g_{m3}}{C_{EE}(g'_{m2}+g_{m3})})}{(s + \frac{g_{m3}}{C_{EE}})(s + \frac{1}{R_{L1}C_L})(s + \frac{1}{R_{L2}C_L})} \qquad (6)$$

978-1-4244-6240-7/10 $26.00 © 2010 IEEE

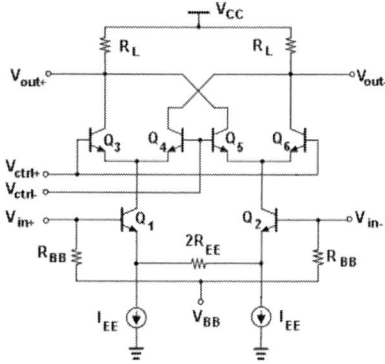

Fig. 2. Variable gain stage. The control voltage is provided by the linear-in-dB controller.

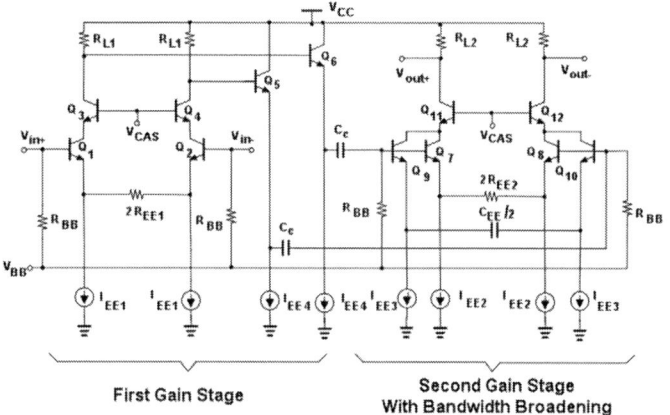

First Gain Stage

**Second Gain Stage
With Bandwidth Broadening**

Fig. 3. Fixed gain stages with gain broadening. The second gain stage input is in parallel with a differential pair with a degenerated capacitor $C_{EE}/2$ to cancel the dominant pole due to the load.

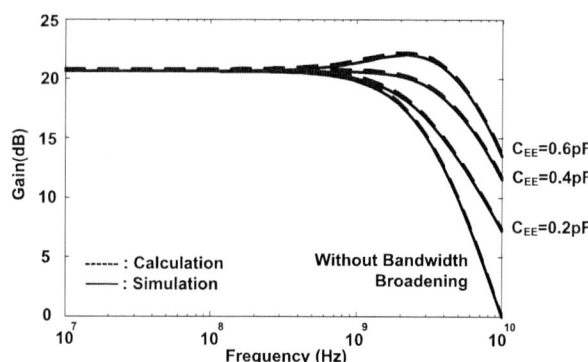

Fig. 4. Bandwidth broadening simulation/calculation results. $I_{EE1}=I_{EE2}=I_{EE3}=I_{EE4}=1\text{mA}$, $V_{CC}=3\text{V}$, $R_{L1}=R_{L2}=0.5\text{K}$, $R_{EE1}=R_{EE2}=0.125\text{K}$, $C_c=10\text{pF}$, $C_L=0.1\text{pF}$.

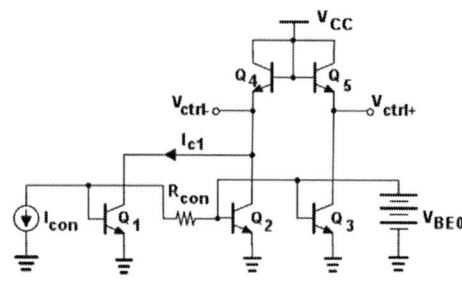

Fig. 5. Linear-in-dB controller for Gilbert multiplier VGA. The difference between I_{C4} and I_{C5} will be proportional to $e^{-I_{con}R_{con}/V_T}$.

where $g'_{m1} \approx 1/R_{EE1}$, $g'_{m2} \approx 1/R_{EE2}$ and $g_{m3} = I_{EE3}/V_T$. The circuit generates a zero and mitigates the roll-off due to dominant pole $1/R_{L2}C_L$ and $1/R_{L3}C_L$ to extend the bandwidth. Fig. 4 shows the simulation/calculation results of the bandwidth broadening.

C. Linear-in-dB Controller for Gilbert Multiplier Type VGA

Fig. 5 shows the linear-in-dB controller for Gilbert multiplier type VGA, where I_{con} is the control current proportional to V_{env} from the output of V-I converter (in Fig. 1). When I_{con} flows through R_{con}, I_{C1} will be

$$I_{C1} = I_{C2,3} \times e^{-I_{con}R_{con}/V_T}. \tag{7}$$

Therefore the controller's output voltage is

$$V_{ctrl} = V_T \times ln(1 + e^{-I_{con}R_{con}/V_T}) \tag{8}$$

Combining (5) and (8), A_{VGA} is

$$dB(A_{VGA}) \approx \frac{-8.7 \times I_{con}R_{con}}{V_T} + dB\left(\frac{I_{EE}R_L}{I_{EE}R_{EE} + V_T}\right) \tag{9}$$

Equation (9) shows the gain of the VGA is linear-in-dB to the control current. The temperature dependance of V_T is

cancelled by setting I_{con} proportional to absolute temperature (PTAT).

D. V-I Converter

Fig. 6 shows the V-I converter to transform V_{env} to I_{con}. The differential input voltage V_{env} is linearized by a PMOS degenerated differential pair M_1-M_2. Based on the translinear loop Q_1-Q_4, the output current I_{con} can be approximated as

$$I_{con} \approx \frac{V_{env}}{2R_{SS}}\frac{I_{PTAT}}{I_{BG}} + \frac{I_{SS}I_{PTAT}}{I_{BG}} \tag{10}$$

where I_{PTAT} is a PTAT current source and I_{BG} is a constant current source. The current mirror M_3-M_4 and Q_5-Q_6 are

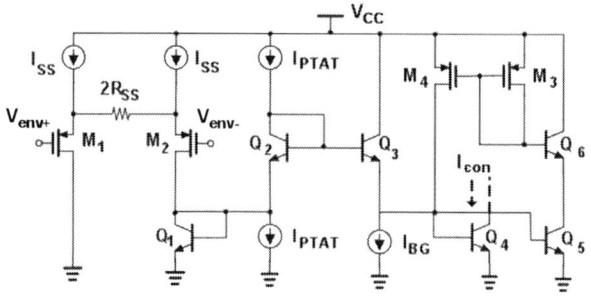

Fig. 6. V-I converter to generate I_{con}. The output current I_{con} is proportional to the differential voltage V_{ctrl} and I_{PTAT}.

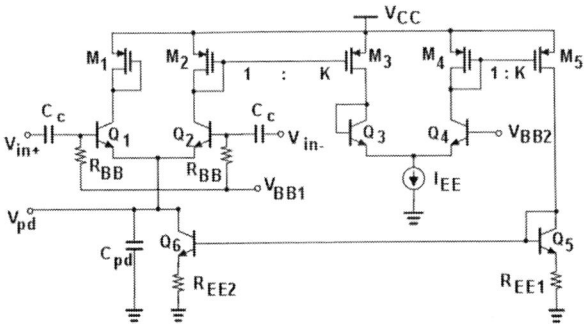

Fig. 7. Adaptive biasing peak detector. I_{C6} varies with the input signal to speed charging/discharging.

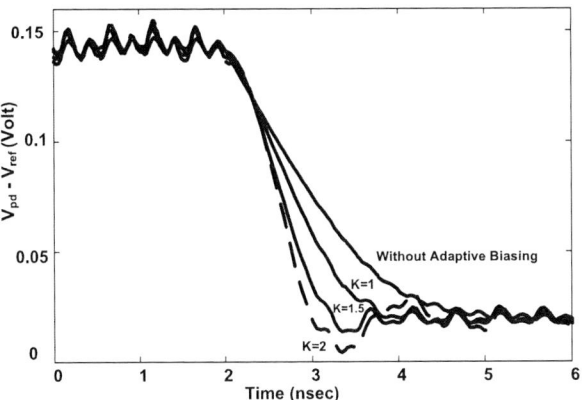

Fig. 8. Simulation results of $V_{pd} - V_{ref}$ with/without adaptive biasing techniques where K is the current mirror ratio. Input carrier frequency=1.7GHz.

used to compensate the base current of Q_4.

E. Adaptive Biasing Peak Detector

Fig. 7 shows the peak detector [12] with adaptive discharging current source I_{C6}. When the input V_{in} drops, I_{C6} will increase to speed up the discharging process of C_{pd}. When V_{in} increases, I_{C6} will become smaller to assist C_{pd}'s charging process.

Fig. 8 shows the $V_{pd} - V_{ref}$ settling behavior with/without adaptive biasing techniques. The discharging time can be significantly improved by selecting the appropriate K value.

F. Multi-tanh Differential g_m-C Filter

Fig. 9 shows a multi-tanh differential g_m-C filter composed of four asymmetric differential pairs. In order to speed the settling, a modified multi-tanh g_m profile is applied shown in Fig. 10 [13]. When $|V_{ref} - V_{pd}|$ is large, the transconductance is enhanced to speed the charging/discharging process without sacrificing stability.

IV. MEASUREMENT RESULTS

The detector chip is fabricated in TowerJazz Semiconductor's 0.18μm SiGe SBC18 BiCMOS process [14], and is shown in Fig. 11. The chip occupies $1.8 \times 1\ mm^2$ die area including the pads.

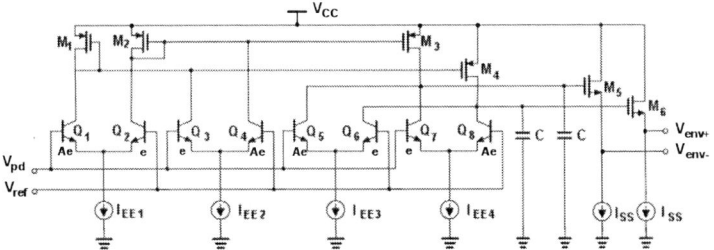

Fig. 9. Multi-tanh differential g_m-C filter. The differential pairs are asymmetric to improve the AGC loop settling time without sacrificing stability.

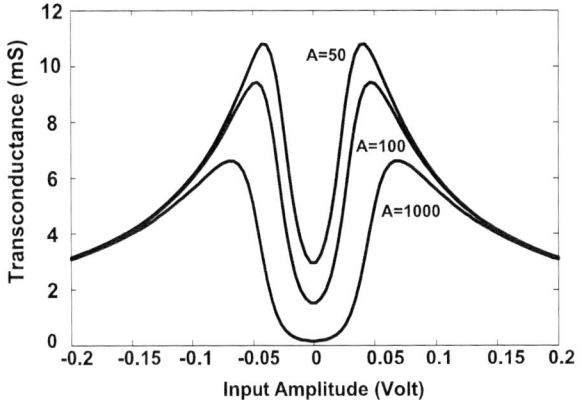

Fig. 10. Multi-tanh simulated g_m profile. The transconductance is offset to speed the AGC settling.

The measurement results are shown in Fig. 12. Fig. 12(b) shows linear-in-dB error for different input power and frequencies. When the input frequency is close to 1GHz, the optimal dynamic range is 50dB with +-2dB error range. The error degrades at higher frequency – the dynamic range reduces to 30dB when the frequency increases to 2.2GHz. When the input signal is small, the linear-in-dB error is limited by the isolation between the VGA and gain stages. When the input signal is large, the linear-in-dB error is mainly dominated by

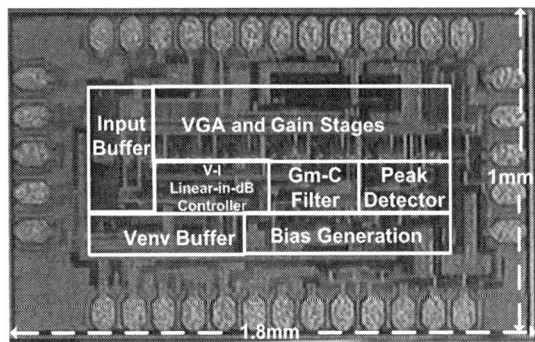

Fig. 11. AGC-based envelope detector chip microphotograph. The chip occupies $1.8 \times 1\ mm^2$ die area including the pads.

978-1-4244-6240-7/10 $26.00 © 2010 IEEE

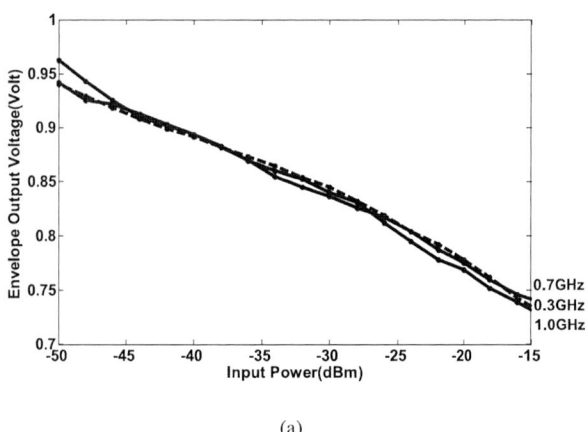

(a)

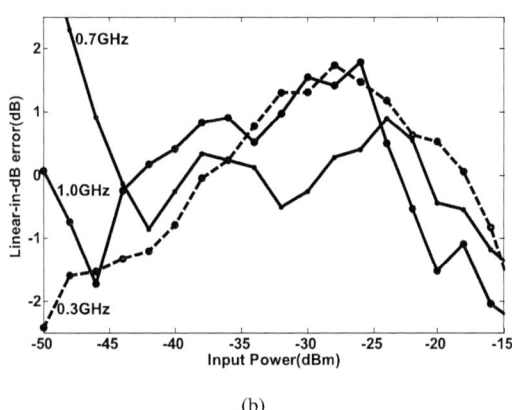

(b)

Fig. 12.   Linear-in-dB measurement results (a) V_{env} vs. input power of the envelope detector. (b)linear-in-dB error vs. input power of the envelope detector.

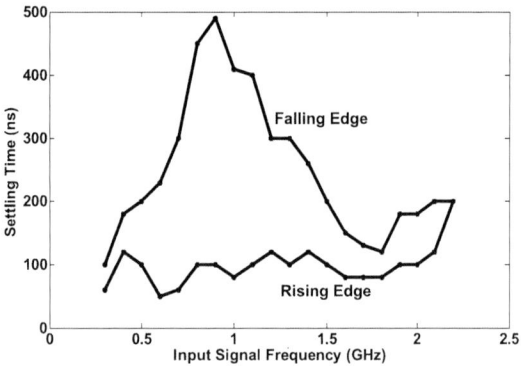

Fig. 13.   Envelope settling time for different input frequencies. Input power varies from -10dBm to -60dBm.

the mismatches of Gilbert type VGA devices [15].

The -10dBm to -60dBm transient impulse response is shown in Fig. 13 and a 10%-90% settling time is adopted here. The envelope detector's rising/falling edge settling time ranges from 60ns to 200ns and 100ns to 490ns, respectively. Table I compares existing linear-in-dB and linear envelope detectors.

TABLE I
TABLE OF COMPARISON: AGC-BASED LINEAR-IN-dB RF POWER DETECTORS

	AD8362[8]	HMC614LP4[9]	Cha[1]	This Work
Type	Linear-in-dB	Linear-in-dB	Linear	Linear-in-dB
Min. Vcc	4.5V	5V	1.8V	3V
Min. I_{DC}	24mA	65mA	1mA	34mA
RF BW.	0-3.8GHz	0.1-3.9GHz	1-2.4GHz	0.2-2.5GHz
Max. DR.	65dB	70dB	35dB	50dB
Slope	50mV/dB	40mV/dB	1V/V	5.7mV/dB
Accuracy	+-1dB	+-1dB	N/A	+-2dB
Min. T_{Rise}	45ns	34ns	>200ns*	60ns
Min. T_{Fall}	400ns	620ns	>200ns*	100ns

*:Estimate from 5MHz sine-wave envelope bandwidth.

V. CONCLUSION

A 2.5GHz 50dB dynamic range linear-in-dB envelope detector operating with 3V power supply is presented. The detector consumes 34mA supply current with 1.8 mm^2 die area.

ACKNOWLEDGMENT

The authors would like to thank TowerJazz Semiconductor for fabricating the chip, UCSD Center of Wireless Communications for supporting this work, and Donald Kimball's assistance on measurements.

REFERENCES

[1] J. Cha et al, "A Highly-Linear Radio-Frequency Envelope Detector for Multi-Standard Operation," in *IEEE Radio Frequency Integrated Circuits Symposium*, 2009, pp. 149–152.

[2] F. Wang et al, "Design of wide-bandwidth envelope-tracking power amplifiers for OFDM applications," *IEEE Transactions on Microwave Theory and Techniques*, vol. 53, no. 4, pp. 1244–1255, 2005.

[3] P. C. Huang et al, "A 2-V 10.7-MHz CMOS Limiting Amplifier/RSSI," *IEEE Journal of Solid-State Circuits*, vol. 35, no. 10, pp. 1474–1480, 2000.

[4] J. Cowles, "The evolution of integrated RF power measurement and control," in *Proceedings of the 12th IEEE Mediterranean Electrotechnical Conference*, vol. 1, 2004.

[5] J. Cowles et al, "Accurate gain/phase measurement at radio frequencies up to 2.5 GHz," *Analog Dialogue*, vol. 35, no. 5, pp. 1–4, 2001.

[6] B. Gilbert, "Noise Figure and Logarithmic Amplifiers," *Analog Dialogue 42-06*, p. 1, 2008.

[7] Y. J. Chuang et al, "A Wideband InP DHBT True Logarithmic Amplifier," *IEEE Transactions on Microwave Theory and Techniques*, vol. 54, pp. 3843–3847, Nov 2006.

[8] "AD8362 Data Sheet," Analog Devices, Inc., 2009. [Online]. Available: http://www.analog.com

[9] "HMC614LP4 Data Sheet," Hittie Microwave Corporation, 2009. [Online]. Available: http://www.hittite.com/

[10] H. Y. M. Pan et al, "Improved Dynamic Model of Fast-Settling Linear-in-dB Automatic Gain Control Circuit," in *IEEE International Symposium on Circuits and Systems*, 2007, pp. 681–684.

[11] J. W. Lai et al, "Design of variable gain amplifier with gain-bandwidth product up to 354 GHz implemented in InP-InGaAs DHBT technology," *IEEE Transactions on Microwave Theory and Techniques*, vol. 54, no. 2 Part 1, pp. 599–607, 2006.

[12] R. G. Meyer, "Low-power monolithic RF peak detector analysis," *IEEE Journal of Solid-State Circuits*, vol. 30, no. 1, pp. 65–67, 1995.

[13] B. Gilbert, "The multi-tanh principle: A tutorial overview," *IEEE Journal of Solid-State Circuits*, vol. 33, no. 1, pp. 2–17, 1998.

[14] "TowerJazz Semiconductor SBC18 Process." [Online]. Available: http://www.jazzsemi.com

[15] W. M. C. Sansen et al, "An integrated wide-band variable-gain amplifier with maximum dynamic range," *IEEE Journal of Solid-State Circuits*, vol. 9, no. 4, pp. 159–166, 1974.

Cellular Antenna Switches for Multimode Applications Based on a Silicon-on-Insulator Technology

Ali Tombak[1], Christian Iversen[2], Jean-Blaise Pierres[3], Dan Kerr[1], Mike Carroll[1], Phil Mason[1], Eddie Spears[1] and Todd Gillenwater[1]

[1] RFMD Inc., 7628 Thorndike Rd., Greensboro, NC 27409 USA
[2] RFMD Denmark Design Center, Aalborg, DENMARK
[3] RFMD Toulouse Design Center, Toulouse, FRANCE

Abstract — A Silicon-on-Insulator (SOI) CMOS technology on high resistivity silicon substrates is presented for the design of cellular antenna switches. The design and measurement results for an SP9T cellular antenna switch based on this technology are presented. To the best of our knowledge, this is the first demonstration of an SP9T cellular antenna switch with adequate intermodulation and harmonic distortion performance on a high resistivity SOI CMOS technology.

Index Terms — SOI, silicon-on-insulator, high resistivity silicon, multimode, SP9T, intermodulation distortion, harmonics, blocker, RF switch, antenna switch.

I. INTRODUCTION

Tremendous changes are happening in cellular communications. The ever increasing need for higher data rates and connectivity are driving service providers and OEMs to evolve the standards. One of the major hurdles facing the suppliers of RF components for the cellular markets is the fact that the evolution of RF system standards (such as the evolution to the 3G and 4G cellular standards) creates the need for increasingly more complex RF system architectures. For 3G, for example, RF system suppliers are being asked to provide integrated RF TX modules capable of supporting both the existing GSM/EDGE standard and the multi-band UMTS WCDMA 3G standard. Fig. 1 depicts a typical TX module architecture for a 3G system. It supports dual-band GSM/EDGE standard and 3 UMTS bands. Hence, an SP7T switch is needed to connect the GSM/EDGE transmitters/receivers and UMTS duplexers to the antenna. Future system specifications beyond 3G require an even larger numbers of multi-standard TX functions, hence higher throw count switches with more stringent requirements. It is obvious that such complex RF TX modules could greatly benefit from higher degrees of integration and more modular solutions [1]-[3]. As a result, switch duplexer modules (SDMs), antenna switch modules (ASMs), and switch filter modules (SFMs) have gained popularity in the market.

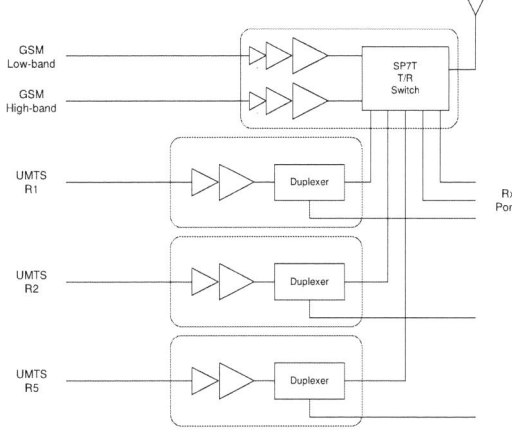

Fig. 1. A 3G RF TX module block diagram illustrating the need to parallel 4 different PA blocks to serve typical GSM/EDGE and WCDMA system specifications.

Since post-PA losses are very difficult to recover, cellular antenna switches must exhibit low insertion loss. In addition, they must be capable of handling high RF power levels (up to +35 dBm) along with widely varying antenna impedances. This translates into very high RF voltages across the switch during operation. The switch must also maintain high levels of linearity, generate low harmonics, and be reliable for long term operation. Since conventional CMOS devices cannot handle large RF voltages, multiple devices must be stacked to divide the RF voltage across many devices [4]. GaAs pHEMT and Silicon-on-Sapphire (SOS) based switches are widely used in the industry today since they offer an insulating substrate for stacking devices [5]-[6]. On the other hand, SOI CMOS technology on high resistivity silicon substrates has been studied over the years for the design of cellular antenna switches which offers device stacking on a CMOS based technology [7]-[12]. However, since the high resistivity handle wafer was still not an ideal insulator compared to GaAs and SOS based technologies, notable levels of harmonic and intermodulation distortion from the handle wafer was observed which prevented the design of high throw count switches, especially with intermodulation distortion specifications [13]. With the advances in the high resistivity SOI CMOS technology, these effects have been reduced to make SOI a more competitive technology.

In this paper, the design/technology details and measurement results for an SP9T cellular antenna switch based on a 2.5 V high resistivity SOI CMOS technology are presented. To the best of our knowledge, this is the first demonstration of an SP9T cellular antenna switch with adequate intermodulation and harmonic distortion performance on a high resistivity SOI CMOS technology.

II. TECHNOLOGY OVERVIEW

The high resistivity SOI CMOS technology used in this work offers 2.5 V NFETs with gate oxide thickness of 52 Å and gate length of 0.32 μm. A single NFET provides a typical on-resistance of 0.80 ohm-mm and off-capacitance of 310 fF/mm, resulting in an Ron*Coff product of about 250 fsec. Ron*Coff product is an important figure of merit in the design of RF switches. Ron of a device can be decreased for a lower insertion loss by increasing the device size, however Coff increases at a similar rate, reducing the usable bandwidth and isolation of the switch and increasing losses due to mismatch. A lower Ron*Coff product overall allows better insertion loss and/or isolation for the switches designed in the technology, which is desired for high throw count switches since OFF state switch branches contribute to mismatch losses. The SOI technology described in this work also offers 0.18 um devices (although not used for the switch branches), hence 0.18 μm design rules allow much tighter contacted gate pitch compared to a GaAs pHEMT based switch technology, resulting in very compact switch branches. Table I shows a comparison of the key performance parameters of a SOI NFET switch branch from this work with a switch branch from a GaAs pHEMT based technology already in production. Both of the switch branches in Table I are capable of handling up to +35 dBm RF power under worst-case antenna impedance mismatch and temperature. The ON state insertion loss and resistance of the particular SOI switch branch for this comparison is slightly higher than the pHEMT switch branch, whereas the OFF state isolation and capacitance is superior compared to the pHEMT switch branch. Overall, SOI offers lower Ron*Coff product, enabling the design of RF switches with lower insertion loss and higher isolation. It should also be noted that GaAs pHEMT switches usually require an external control circuitry on a CMOS die, whereas the control circuitry can be integrated in the same SOI die as the RF switch, hence it is easier to realize highly integrated complex switch circuits. Being a CMOS technology, high-resistivity SOI also offers the manufacturing capability that has been developed over many years to support silicon ICs.

TABLE I
COMPARISON OF RF NFET SWITCH BRANCH
PARAMETERS FOR SOI AND pHEMT TECHNOLOGIES

Parameter	SOI	pHEMT
2 GHz Insertion Loss, dB	0.34	0.24
2 GHz Isolation, dB	20.2	15.0
R_{ON}, Ohm	2.7	1.9
C_{OFF}, fF	92	147
$R_{ON} * C_{OFF}$, fsec	250	280

Cellular antenna switches have stringent harmonic and intermodulation distortion specifications [14]. 2nd and 3rd harmonic distortion of -40 dBm at a forward power of +35 dBm, and intermodulation distortion of less than -105 dBm at the RX bands (with the presence of a +20 dBm forward power at the TX port and -15 dBm blocker present at the antenna port [14]) are typical specifications. A high resistivity handle wafer (>750 ohm-cm) and thick buried oxide (1 μm) are used in the SOI CMOS technology used in this work. Notable levels of intermodulation and harmonic distortion of RF signals was observed by THRU structures over high-resistivity silicon substrates [13]. The source for this distortion was identified as the fact that since the doping density is low in the handle wafer, a small amount of fixed charges in the SiO_2 near the $Si-SiO_2$ interface can invert the Si interface which is then modulated with the presence of high RF voltages on metal structures above the handle wafer, hence distortion occurs [13]. Several techniques for reducing this distortion have been proposed, and significant improvements have been observed for metal transmission lines over the high resistivity silicon substrate [11], [13]. With the availability of these techniques in production SOI processes, a more adequate harmonic and intermodulation distortion can be expected.

III. DESIGN AND MEASUREMENT RESULTS

An SP9T switch was designed using the high resistivity SOI CMOS technology described in Section II. It provides two TX ports for GSM/EDGE low/high band transmitters, four RX ports to GSM/EDGE receivers, and three TRX ports for UMTS mode. A photo of the SP9T switch while being measured on an evaluation board is shown in Fig. 2. Each switch throw is composed of series and shunt switch branches. TX and TRX port series branches utilize the same device size, whereas the RX series branches utilize a smaller device size. The switch was controlled using an on-chip controller which demultiplexes the 4-bit external

978-1-4244-6240-7/10 $26.00 © 2010 IEEE

control signals and provides the necessary bias voltages to the switch branches for the given state.

Fig. 2.   A photo of the SP9T switch being measured on an evaluation board.

Both small and large signal measurements were performed on the SP9T switch. Fig. 3 and Fig. 4 show the typical corrected small signal insertion loss and return loss measured on the TX1, TRX1, and RX1 modes of the switch as a function of frequency. The rest of the ports achieves a similar performance within the ports' family and was not included to simplify the figures. The measured results show that TX and TRX ports have 0.45 dB and RX ports have 0.55 dB insertion loss on average at 915 MHz. The measured return loss was better than 23 dB. At 1990 and 2170 MHz, TX/TRX/RX port insertion losses increase by about 0.2 and 0.25 dB on average, respectively. The return loss stays better than 15 dB up to 2170 MHz. The reason for the slightly higher insertion loss on RX ports is due to the utilization of a narrower series branch for the RX port as compared to TX and TRX ports. Port to port isolation range was also measured as summarized in Table II, which specifies the measured isolation range within the isolation port family (such as TX1-2 to RX1-4). Since shunt branches were used, the switch achieves favorable isolation performance. It should however be noted that the reported isolation numbers on Table II also include bond wire coupling between the switch ports, hence the inherent isolation of the switch should be better than the reported numbers. Fig. 5 and Fig. 6 show the measured 2nd and 3rd harmonic distortion at the TX1 and TRX1 modes of the SP9T switch at fundamental frequencies of 900 and 1780 MHz, respectively. The rest of the ports achieve a similar performance and were not included. At 900 MHz, the measured 2nd and 3rd harmonic distortion were better than -58 and -47 dBm up to 37 dBm input power, respectively, whereas at 1780 MHz, the

measured 2nd and 3rd harmonic distortion were better than -49 and -46 dBm up to 36 dBm input power, respectively. Overall, measured harmonics were considerably better than most switch specifications. The measured intermodulation distortion levels are also given in Table III for the TRX ports. The measured IMD3 and IMD2high (IMD 791.5/1760 and IMD 1718/4090) were significantly better than the typical -105 dBm specification [14]. On the other hand, IMD2low (IMD 45 and 190) was marginal, but considering the low frequency attenuation contributed by the commonly used shunt inductors for IEC ESD protection at the antenna, the system requirements can easily be met.

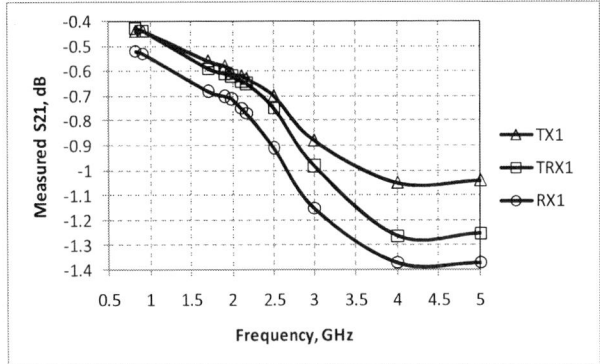

Fig. 3.   Measured insertion loss (corrected) for the TX1, TRX1, and RX1 modes as a function of frequency.

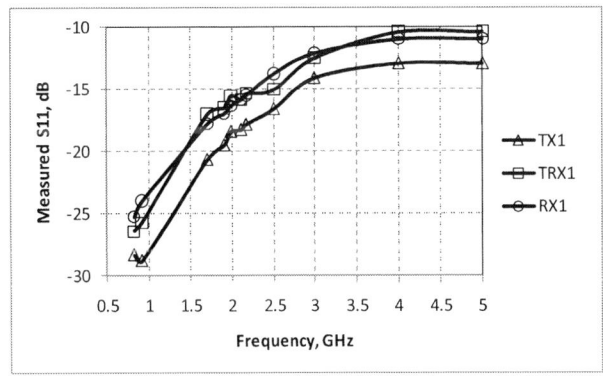

Fig. 4.   Measured return loss for the TX1, TRX1, and RX1 modes as a function of frequency.

TABLE II
SUMMARY OF THE MEASURED PORT TO PORT
ISOLATION FOR THE SOI SP9T SWITCH

Isolation Ports	Isolation Range, dB		
	915 MHz	1990 MHz	2170 MHz
TX to RX	41-58	36-45	35-44
TRX to RX	42-58	33-45	32-45
TX to TRX	37-54	30-41	29-40
TX to TX	50 55	38 42	37 41
RX to RX	35-58	29-44	28-43

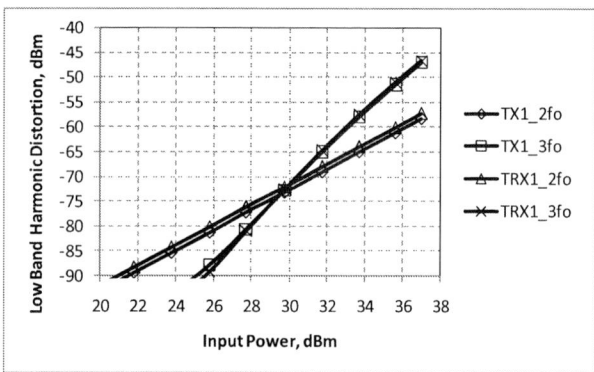

Fig. 5. Measured 2nd and 3rd harmonic distortion on the TX1 and TRX1 modes at a fundamental frequency of 900 MHz.

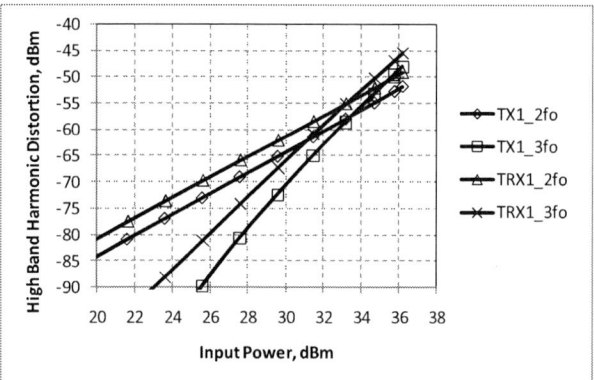

Fig. 6. Measured 2nd and 3rd harmonic distortion on the TX1 and TRX1 modes at a fundamental frequency of 1780 MHz.

TABLE III
SUMMARY OF THE MEASURED INTERMODULATION
DISTORTION LEVEL FOR THE SOI SP9T SWITCH

IMD Port	Measured Intermodulation Distortion Level, dBm					
	IMD 45	IMD 791.5	IMD 1718	IMD 190	IMD 1760	IMD 4090
TRX 1-3	-102	-122	-120	-106	-119	-110

IV. CONCLUSION

An SOI CMOS technology on high resistivity silicon substrates was presented for the design of cellular antenna switches. The design and measurement results for an SP9T cellular antenna switch based on this technology were presented. The measured small signal, intermodulation and harmonic distortion performance of the switch were competitive to other technologies and meet most cellular antenna switch specifications.

REFERENCES

[1] A. Upton and V. Steel, "The Current State of Technology and Future Trends in Wireless Communications and Applications", Microwave Journal, vol. 49, no. 9, Sep. 2006.

[2] R. Jos, "Technology Developments Driving an Evolution of Cellular Power Amplifiers to Integrated RF Front-End Modules", IEEE Journal of Solid-State Circuits, vol. 36, no. 9, pp. 1382-1389, Sept. 2001.

[3] P. V. Wright, "Integrated front-end modules for cell phones", 2005 IEEE Ultrasonics Symposium, vol. 1, pp. 564-572, Sept. 2005.

[4] M. Shifrin, P. Katzin, and Y. Ayasli, "Monolithic FET Structures for High-Power Control Component Applications," IEEE Trans. on Microwave Theory and Techniques, pp. 2134-2141, Dec. 1989.

[5] H. C. Chiu, T. J. Yeh, Y. Y. Hsieh, T. Hwang, P. Yeh, C. S. Wu, "Low Insertion Loss Switch Technology Using 6-inch InGaP/AIGaAs/InGaAs pHEMT Production Process", Proceedings of the 2004 Compound Semiconductor Symposium.

[6] D. Kelly, C. Brindle, C. Kemerling, M. Suber, "The state of the art of silicon-on-sapphire CMOS RF switches", Proceedings of the 2005 IEEE Compound Semiconductor Symposium, pp 200-205.

[7] C. Tinella et al., "0.13 CMOS SOI SP6T antenna switch for multi standard handsets," Silicon Monolithic Circuits in RF Systems, 2006 Topical Meeting.

[8] J. Costa, M. Carroll, J. Jorgenson, T. McKay, T. Ivanov, T. Dinh, D. Kozuch, G. Remoundos, D. Kerr, A. Tombak, J. McMacken, M. Zybura, "A Silicon RFCMOS SOI Technology for Integrated Cellular/WLAN RF TX Modules", 2007 IEEE International Microwave Symposium, June 2007.

[9] T. McKay, M. Carroll, C. Iversen, D. Kerr, G. Remoundos, "Linear Cellular Antenna Switch for Highly Integrated SOI Front-End," 2007 IEEE SOI Conference, Oct. 2007, pp. 126-126.

[10] T. McKay, M. Carroll, D. Kerr, J. Costa, "Advances in Silicon-on-Insulator Cellular Antenna Switch Technology," 2009 IEEE Topical Meeting on Meeting on Silicon Monolithic Integrated Circuits in RF Systems.

[11] A. Botula et al., "A Thin-film SOI 180nm CMOS RF Switch Technology," 2009 IEEE Topical Meeting on Meeting on Silicon Monolithic Integrated Circuits in RF Systems.

[12] M. Carroll, D. Kerr, C. Iversen, A. Tombak, J.-B. Pierres, P. Mason, J. Costa, "High-Resistivity SOI CMOS Cellular Antenna Switches", 2009 IEEE Compound Semiconductor Integrated Circuit Symposium, Oct. 2009, pp. 1-4.

[13] D. Kerr, J. Gering, T. McKay, M. Carroll, C. Neve, J.-P. Raskin, "Identification of RF Harmonic Distortion on Si Substrates and its Reduction Using a Trap-Rich Layer," 2008 IEEE Topical Meeting on Meeting on Silicon Monolithic Integrated Circuits in RF Systems, pp. 151-154.

[14] T. Ranta, J. Ella, H. Pohjonen, "Antenna switch linearity requirements for GSM/WCDMA mobile phone front-ends", 2005 European Conference on Wireless Technology, Oct. 2005, pp. 23-26.

978-1-4244-6240-7/10 $26.00 © 2010 IEEE

RMO4A-1

A 120μW Fully-Integrated BPSK Receiver in 90nm CMOS

Han Yan, Jose Gabriel Macias-Montero, Atef Akhnoukh, Leo C. N. de Vreede, John R. Long,
John J. Pekarik* and Joachim N. Burghartz

ERL/DIMES, Delft University of Technology, 2628CT, Delft, the Netherlands

*IBM Microelectronics, Essex Junction VT

Abstract — **In this work a highly integrated, ultra-low-power BPSK receiver for short-range wireless communications is presented. The receiver consists of a power divider, two injection-locked RC oscillators with limiting buffers and an XOR output stage. The demodulation principle is based on the dynamic phase response of the two BPSK signal injected oscillators. As proof of concept, a 300 MHz receiver was implemented in a 90nm CMOS technology. The whole receiver has an active die area of 0.04 mm², a sensitivity of -34 dBm at 1Mbps and consumes only 120 μW from 1V supply, which relates to an energy per bit of only 0.12 nJ/bit, a value which is among the best reported up-to-date for low-transmission rate systems.**

Index Terms — **Low-power RF, CMOS integrated circuit, BPSK, injection-locked oscillator.**

I. INTRODUCTION

Low-power short-range wireless systems have enjoyed wide popularity in recent years. They find application in wireless sensor networks [1], RFID systems and implantable medical devices. A small-size, highly-integrated and inexpensive RF receiver is an essential component to all these wireless nodes. In order to minimize the power consumption, amplitude modulation schemes, such as on-off keying (OOK), have been proposed and demonstrated [2][3]. However, it is well-known that AM has poorer noise immunity and bandwidth efficiency compared to phase-shift keying (PSK) in data communication applications [4]. However, to ensure error-free detection of a PSK signal, synchronization must be established between the received signal and the local frequency reference. Costa's loop is often used for synchronous detection of PSK, but its complexity and power consumption makes it unsuitable for wireless devices aimed at energy efficiencies on the order of pJ/bit. Recently, new low-power, low-complexity BPSK demodulators were proposed [5][6][7] that utilize the phase response of two super-harmonic injection-locked LC oscillators to phase changes in their input signal. In [7] a 2.5mW receiver with two LC injection-locked oscillators and output mixer demodulates a 19GHz input signal, resulting in a measured performance of 250pJ/bit at a 10MB/s data rate. In this work we aim for a BPSK

receiver implementation that has one order of magnitude smaller power consumption and 10 times smaller chip area.

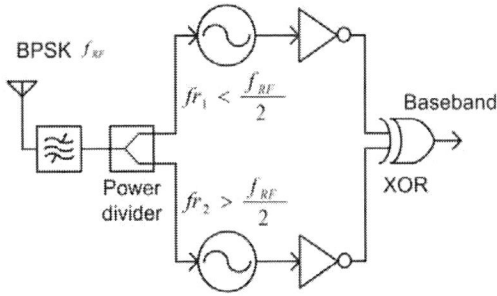

Fig. 1. Block diagram of the low-power BPSK receiver based on two injection-locked RC-oscillators.

To reach this goal, a power divider, two RC based oscillators and an XOR output stage are used, as shown in Fig. 1. This configuration provides the original phase modulation of the BPSK signal directly as a digital output at baseband, while the higher impedance levels used in the RC oscillators significantly reduce the required DC power. The proposed ultra-low-power BPSK receiver structure, can be made extremely compact (<0.05 mm²), with a wake-up time <100 ns (determined by the start-up time of the oscillators). Due to the inherent frequency selection of injection-locked oscillators, the need for any external or on-chip filtering is relaxed. A 300 MHz proof-of-concept receiver is implemented in a standard CMOS 90nm technology. The measured energy efficiency is 0.12 nJ/bit, making it one of the most efficient receivers for ultra-low-power wireless communication systems reported [3][12].

II. PRINCIPLE OF THE BPSK RECEIVER

In the proposed BPSK receiver, the injection-locked oscillators must be operated close to one-half of the receiver RF input frequency, while their free running frequencies fr_1 and fr_2 (indicated in Fig. 1) are tuned as:

$$fr_1 < f_{RF} / 2 < fr_2 \qquad (1)$$

where f_{RF} is the frequency of the received signal. When both oscillators are locked to the received signal, a 180°

978-1-4244-6240-7/10 $26.00 © 2010 IEEE   247

phase change at the input signal will result in a -90° phase change at the output of oscillator #1 (due to its lower free running frequency) and +90° phase change at the output of oscillator #2 (due to its higher free running frequency). Therefore, the digital information carried on the BPSK signal can be directly reproduced by performing an XOR operation on the two oscillators' outputs after limiting. To ensure correct demodulation, the oscillators must be properly locked to the RF input. Therefore, the RF signal amplitude must exceed the critical locking level of the oscillators. Fig. 2 shows the minimum detectable signal levels at different frequencies. Proper demodulation is ensured when the RF signal falls into the shadowed region of Fig. 2.

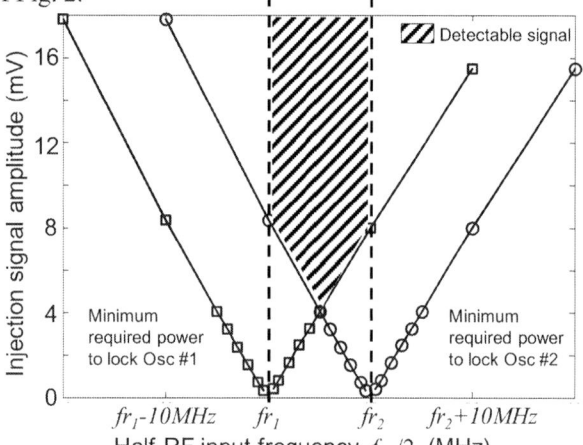

Fig. 2.   Required signal level to ensure proper demodulation of the BPSK receiver.

III. Circuit Implementation

A proof-of-concept BPSK receiver at 300MHz was implemented in a standard 90nm RF-CMOS technology [8], which offers 5 Cu metal layers, 2 thick Cu metal layers and a top aluminum layer.

A. Second Harmonic RC Injection Locked Oscillator

Fig. 3 shows the schematic of the implemented RC injection-locked oscillator. The oscillator core is a typical RC oscillator topology in which M3 and M4 form the gain stage and M5 and M6 are PMOS varactors in inversion mode, whose voltage-dependent capacitance dominates the injection locking process [7]. The free-running frequency of the oscillator can be controlled from an external pad connected to the common connection between the varactors. In order to minimize power consumption, the smallest varactor value in combination with the highest possible resistance (R1 and R2) is preferred. However, the varactor capacitance is chosen so that the capacitance of M5 and M6 are at least 10 times

larger than all the parasitic capacitance from the devices and wiring. Consequently, a resistance value of 30k-Ohm is chosen for this $f_r/2$=150MHz oscillator implementation. Transistor aspect ratios of 360nm/100nm are used in the switching pair to reduce the parasitic capacitance. The current source transistors M1 and M2 draw 7 μA each to maintain the oscillation. The 2nd harmonic injection of the oscillator core is realized by actively driving current source transistors M1 and M2.

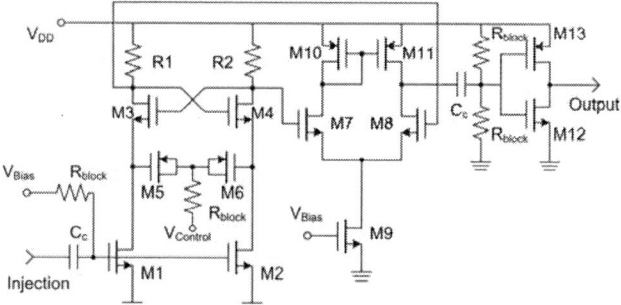

Fig. 3.   The RC injection-locked oscillator, including buffer and digitizer.

An active load stage (M10 and M11) amplifies and buffers the oscillator output, and converts the differential to a single-ended signal. The signal swing at the output of this buffer is typically 800mV peak-peak, and is AC coupled to a CMOS inverter (M12's W/L=360nm/100nm, M13's W/L=1000nm/100nm), where the oscillator output is finally limited.

B. Power Divider and XOR gate

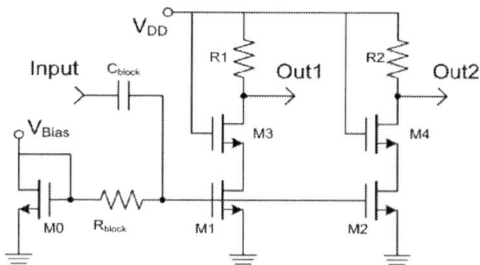

Fig. 4.Schematic of the power divider. Cascode stages have been used to increase isolation between the two outputs

A difference between two injection-locked oscillator's free-running frequencies is required for correct operation of the proposed receiver. Consequently locking or pulling of one oscillator by the other must be avoided at all times. This requires high-isolation from the input power divider to block even harmonic leakage between the two oscillators. Fig. 4 shows the schematic of the power divider. A cascode stage increases isolation between the two outputs. Since the load impedance for this stage is dominated by the gate-source capacitance of the oscillator bias transistors (M1 and M3 in Fig. 3), about 5 dB voltage gain can be achieved with less than 50 μA DC current by choosing R1 and R2 of 20k Ohm. Here we must make a balance between the minimum detectable signal level and the power budget. The use of on-chip inductors at the RF input is avoided to keep chip size to an absolute minimum. The XOR gate was realized by a conventional CMOS logic topology (Fig. 5).

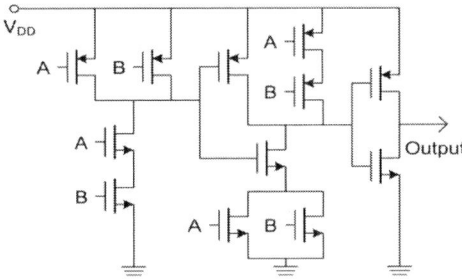

Fig. 5. Schematic of the XOR gate.

C. Layout Considerations

The physical layout of the circuits is optimized to reduce parasitic wiring capacitance. The supply lines are decoupled by 10 pF MIM capacitors to minimize the risk of mutual locking of the oscillators through poor power supply isolation. The target power divider isolation is 40dB.

IV. EXPERIMENTAL RESULTS

A photograph of the fabricated die is shown in Fig. 6, which occupies 520 x 540 μm^2 die area including pads. Currently, many of these pads are DC pads for testing individual circuit blocks, and can be omitted in future implementations. The actual active IC area is about 200 x 200 μm^2. The circuit is wire bonded to a test PCB, on which all the DC control voltages are generated.

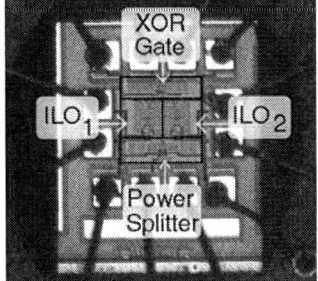

Fig. 6. Photograph of the assembled die.

A. Performance of the 2nd harmonic-locked oscillator

The RC oscillator has a measured frequency tuning range between 120MHz and 170MHz, as shown in Fig. 7. Frequency shift caused by temperature and process variations can be compensated by tuning the bias current.

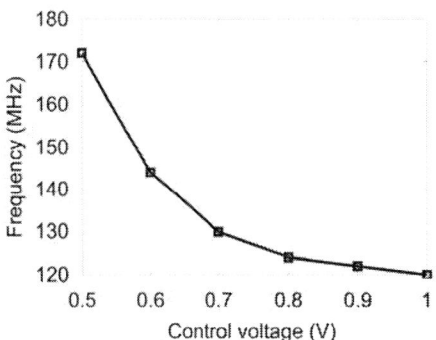

Fig. 7. Measured tuning range of the RC injection-locked oscillator.

The oscillator including buffers consumes 27 μA from a 1V supply. The locking range of the oscillator in combination with the power divider is measured using a continuous RF signal at the receiver input, while the oscillator is set to a free running frequency of 139.4MHz. The resulting minimum RF power at the receiver input required to lock the oscillator versus the injection frequency is plotted in Fig. 8. A minimum RF input power of -60dBm is required to lock the oscillators when they are offset by 0.5 MHz (i.e. the two oscillators' free running frequencies are separated by 1 MHz).

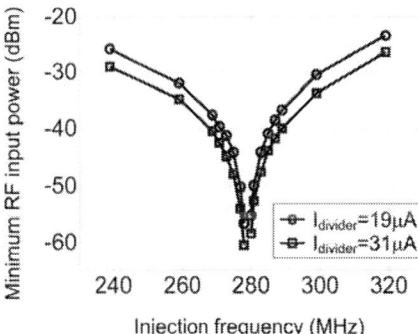

Fig. 8. Measured locking range of the RC injection-locked oscillator.

B. Performance of the complete BPSK receiver

For the measurement a BPSK modulated RF source signal was connected directly to the receiver input and the demodulated output monitored on an oscilloscope. The amplitude of the RF signal was set to -25dBm to ensure locking of both oscillators. Fig. 9 gives the original signal generator data and the receiver output when the oscillators' free-running frequency offset Δf is 10MHz and the data rate is 1M bps. All the original bits are correctly recovered. The pulse slopes indicate that the 9pF input capacitance of the oscilloscope form the limiting factor in this measurement. Higher data rate can be achieved when a buffer is connected. Reducing Δf can improve input

978-1-4244-6240-7/10 $26.00 © 2010 IEEE   249

TABLE I
COMPARISON OF RECENT LOW POWER RECEIVERS

	Technology	Frequency (MHz)	Modulation	Data rate (kb/s)	Power consumption (μW)	Sensitivity (dBm)	Energy per bit (nJ)
[9]	0.5 μm CMOS	868/915	PWM	N/A	2.25	-17.7	N/A
[10]	0.18 μm CMOS	433/868	FSK/OOK	25/2[a]	2100	-111/-117 [b, c]	84/1100[a]
[11]	BAW resonator and CMOS	1900	PWM	5	400	-100.5 [c]	80
[12]	0.13 μm CMOS	2400	FSK	1000	330	N/A	0.33
[2]	0.18 μm CMOS	916.5	OOK	1000	500/2500	-37/-65 [c, d]	0.5/2.5[d]
[3]	BAW resonator and 90 nm CMOS	2000	OOK	100	52	-72[c]	0.52
[7]	90 nm CMOS	19000	BPSK	10000	2500	-31	0.250
This work	90 nm CMOS	300	BPSK	1000	120	-34 [e]	0.12

[a] FSK and OOK respectively
[b] At 433 MHz
[c] At BER = 10^{-3}
[d] With 500 μW and 2500 μW power consumption respectively .
[e] At 1M bps

sensitivity. When Δf is 4MHz, a sensitivity of -34dBm is measured while keeping the data rate 1M bps. Table II summarizes the realized receiver performance.

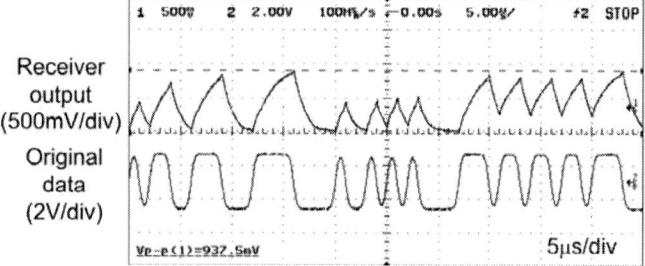

Fig. 9. The waveforms of receiver baseband output and the original data.

V. CONCLUSIONS

An ultra low-power BPSK receiver based on injection-locked RC-oscillators has been introduced. The proof-of-concept receiver circuit was implemented in a 90nm CMOS technology. The receiver consumes 120 μW from a 1V supply. It has an RF input sensitivity of -34dBm at a data rate of 1Mb/s, for a measured energy efficiency of 12nJ/bit. The active circuit area of the receiver is only 0.04mm². When comparing these results with other published low-data rate, low-power, short-range RF receivers (Table I), the proposed BPSK receiver architecture compares very favorably, making it a promising candidate for low-power short-range wireless applications.

ACKNOWLEDGEMENTS

This work is supported by the Dutch BSIK project Freeband-WiComm. IC fabrication via IBM was facilitated by MOSIS. Thanks to Ing. W.G.M. Straver at TU Delft for test fixture assembly.

TABLE II
BPSK RECEIVER PERFORMANCE SUMMARY

Technology	IBM CMOS 90nm
Supply Voltage	1 V
DC Power Dissipation	120 μW
Power Divider DC Current	40 μA
Injection-locked Oscillator DC Current	30 μA
Frequency	300 MHz
Sensitivity	-34dBm at 1Mbps
Chip dimensions	520 μm x 540 μm

REFERENCES

[1] J. Rabaey et al., "PicoRadios for Wireless Sensor Networks: The Next Challenge in Ultra-Low Power Design," ISSCC Dig. Tech. Papers, pp. 200-201, Feb., 2002.

[2] Denis C. Daly et al., "An Energy-Efficient OOK Transceiver for Wireless Sensor Networks", IEEE J. Solid-State Circuits, vol. 42, pp. 1003-1011, May, 2007.

[3] N. M. Pletcher et al., "A 2GHz 52uW Wake-Up Receiver with -72dBm Sensitivity Using Uncertain-IF Architecture", ISSCC Dig. Tech. Papers, pp. 524-525, Feb., 2008.

[4] B. Sklar, "Digital Communication Fundamentals and Applications," NJ: Prentice Hall, 2001.

[5] J. M. López-Villegas, et al., "BPSK to ASK Signal Conversion Using Injection-Locked Oscillators—Part I: Theory," IEEE Trans. on MTT, vol. 53, pp. 3757-3766, Dec., 2005.

[6] J. M. López-Villegas, et al., "BPSK to ASK Signal Conversion Using Injection-Locked Oscillators—Part II: Experiment," IEEE Trans. on MTT, vol. 54, pp. 226-234, Jan ., 2006.

[7] J. G. Macias-Montero et al., "A 19 GHz, 250pJ/bit Non-linear BPSK Demodulator in 90nm CMOS," Proc. ESSCIRC, pp. 304-307, Sept., 2009.

[8] Foundry technologies 90-nm CMOS IBM product brief IBM, 2004.

[9] Udo Karthaus et al., "Fully Integrated Passive UHF RFID Transponder IC with 16.7-μW Minimum RF Input Power," IEEE J. Solid-State Circuits, vol. 38, pp. 1602-1607, Oct., 2003.

[10] Vincent Peiris et al., "A 1V 433/868MHz 25kb/s-FSK 2kb/s-OOK RF Transceiver SoC in Standard Digital 0.18μm CMOS", ISSCC Dig. Tech. Papers, pp. 258-259, Feb., 2006.

[11] B. Otis et al., "A 400μW-RX, 1.6mW-TX Super-Regenerative Transceiver for Wireless Sensor Networks," ISSCC Dig. Tech. Papers, pp. 396-397, Feb., 2005.

[12] Ben W. Cook et al., "Low-Power 24-GHz Transceiver With Passive RX Front-End and 400-mV Supply," IEEE J. Solid-State Circuits, vol. 41, pp. 2757-2766, Dec., 2006.

RMO4A-2

A Fully Integrated 2.4-GHz CMOS Diversity Receiver with a Novel Antenna Selection

Yong-IL Kwon*, Sang-Ku Park*, T.J.Park*, and Hai-Young Lee**

*UC solution team, SAMSUNG Electro-Mechanics, Suwon, 443-743, Korea, **Department of Electronics Engineering, Ajou University, Suwon, 443-749, Korea

Abstract — A new low-complexity antenna diversity architecture, using a 2.4-GHz single low-IF receiver chain with a novel antenna selection scheme, is exploited by using 0.18-μm CMOS technology. The receiver has been developed for the IEEE standard 802.15.4 radio system and two RF input channels are selected through an efficient analog-type antenna selection scheme for achieving the diversity. Compared to conventional receivers without diversity, 10~15 dB improvement of the received signal strength (RSS) has been measured for non-line-of-sight (NLOS) channels. By incorporating a wake-up function for the baseband blocks, the receiver operates at a very low power of 8.5 mW, with a 1.8 V power supply in the standby mode for receiving. The antenna selection error is negligible (<1 %) and the antenna selection time is very fast (<20 μs).

Index Terms — CMOS, LNA, Receiver, antenna diversity, wake-up, wireless communications

I. INTRODUCTION

A new type of wireless communication, with higher density of nodes and simple protocol, is emerging for low-data-rate distributed sensor network applications, such as home automation and industrial systems. Therefore, their sensors must have effective and sensitive operation in multi-path fading circumstances. Diversity reception is a technique of radio communication to provide better received signal strength (RSS) with the cost of an extra set of antennas. The diversity not only enhances the signal strength but also reduces the co-channel interference, leading to the increased system capacity. Many proposals [1]-[2] have been published with respect to the diversity reception.

Fig. 1(a) shows a typical signal processing chain of a conventional receiver with the antenna diversity, where the base-band processor, using a diversity algorithm, selects an antenna receiving the strongest signal. The receiver chain requires high power consumption because the programmable gain amplifier (PGA), analog-to-digital (ADC), modulator-demodulator (Modem), and Micro Controller Unit (MCU) remain in their active modes in order to continually analyze the signal strength. The complexity of the digital part is also increased to incorporate the antenna selection algorithm and the external switches are required for the antenna selection.

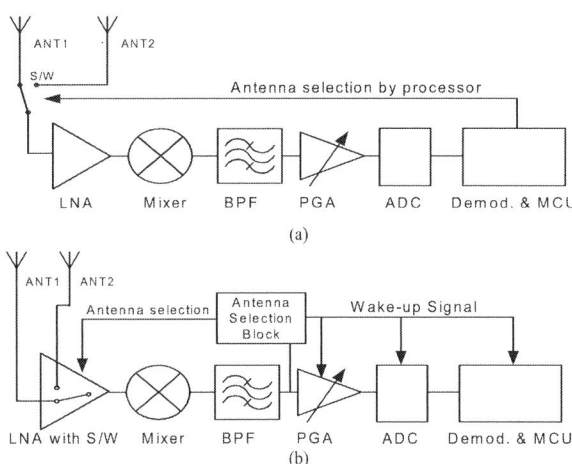

Fig. 1. Receiver architectures : (a) Conventional and (b) Proposed.

This work proposes a low-IF receiver architecture, which uses a novel antenna selection scheme to realize efficient antenna diversity, as shown in Fig. 1(b). The analog antenna selection block selects an antenna after detecting and comparing the signal strength received at the PGA input. Hence, operating blocks are only the low-noise amplifier (LNA) with analog switches, the mixer, the band-pass filter (BPF), and the antenna selection block before the antenna selection block wakes up the PGA, the ADC, the Modem, and the MCU, after selecting an antenna with the strongest signal. Therefore, the wake-up scheme significantly reduces the power consumptions of the PGA, the ADC, the Modem, and the MCU. The power consumption of the whole receiver is 8.5 mW, with a 1.8 V power supply in the standby mode, and a 10~15 dB improvement of the RSS has been measured using the proposed scheme. It is expected that the proposed antenna selection scheme be very effectively used for developing various low-power diversity receivers.

II. DESIGN OF THE RECEIVER ARCHITECTURE

A. Low-noise amplifier and down-conversion mixer

Fig. 2 shows the schematic description of the CMOS receiver with the antenna diversity and the wake-up function. Fig. 3 represents the detailed LNA circuit

978-1-4244-6240-7/10 $26.00 © 2010 IEEE   251

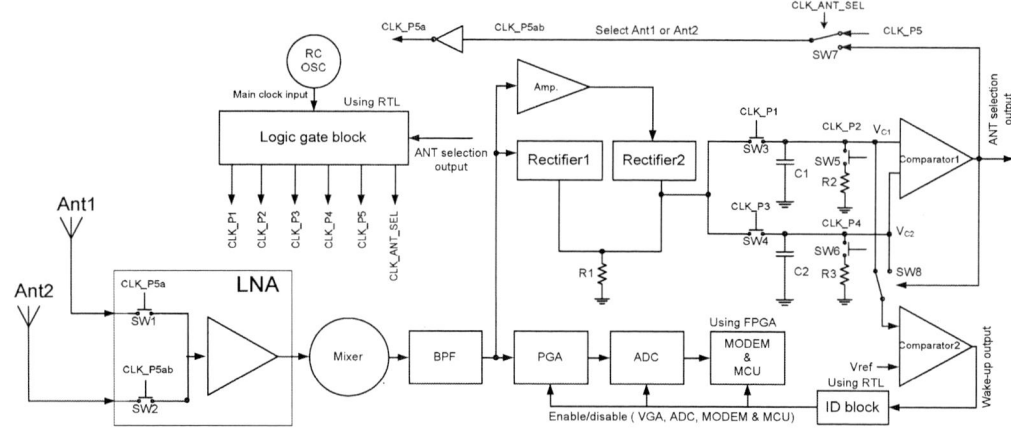

Fig. 2.   The detailed circuit description of the CMOS RF receiver with antenna diversity.

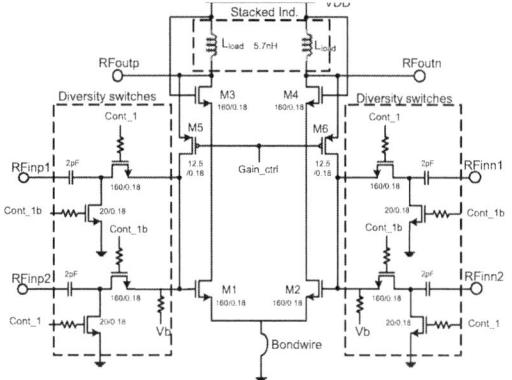

Fig. 3. The detailed low noise amplifier.

including the diversity switching function; the LNA topology uses a common-source amplifier tuned to 2.4 GHz. The diversity switches employ series and shunt devices that are adopted to obtain the required isolation between ANT1 and ANT2. The diversity switches yield an insertion loss of 1.2 dB and a consequent noise figure (NF) penalty compensated by the diversity gain. The LNA diversity switches are controlled by the antenna selection block. Fig. 3 shows that the transistor connected to the strongest signal channel is switched ON (and the shunt transistor is OFF) and the other transistors are switched OFF (and the shunt transistor is ON). An adequate isolation(23dB) is achieved between the ANT1 and the ANT2 terminals. The down-conversion mixer employs the Gilbert cell mixer topology. The total gain of the LNA and the Mixer is programmable between 33dB and 1dB [3].

B. BPF and PGA

The IF block consists of the BPF, the PGA, and the ADC. Building a BPF for the channel selection is difficult because it operates over a wide range of the input voltage at a low supply voltage (1.8V). This is a 6th order

Butterworth BPF that provides 3rd order base-band response. The BPF is implemented by the leap-frog configuration. The PGA consists of two operational amplifiers. The BPF gain (20dB) and the PGA gain (42dB) are programmable between 0dB and 62dB by 1dB step [4].

C. Antenna selection block

Three kinds of sub-blocks (the rectifiers, the RC stages with switches, and the comparators) are involved in the antenna selection block. The LNA+Mixer+BPF gain is 53 dB and the required sensitivity is approximately -95 dBm and hence, the detection range of the rectifier should be within the range from -45 dBm to 0dBm. Two rectifiers are used for the detection range. The rectifier schematic and the simulated dynamic range [5] are shown in Fig. 4.

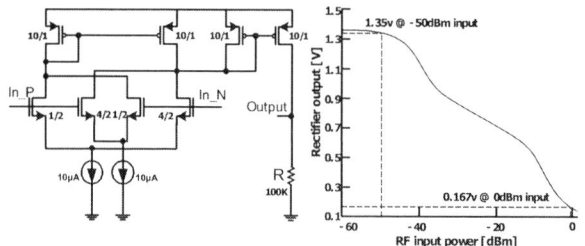

Fig. 4. The simplified rectifier and the dynamic range.

The On-Off Keying (OOK) is applied for the simplification of the identification (ID) block. The Eb/No (Energy per bit to Noise power spectral density ratio) of 12.5 dB is needed for the 10^{-4} bit error probability in non-coherent detection of the OOK signals. An additional 4 dB of the Eb/No ratio will be obtained if the repetition technique is applied to the OOK signaling [6]. Therefore, -95 dBm sensitivity can be achieved by the consequent 8.5 dB signal- to-noise ratio (SNR) at the input of the

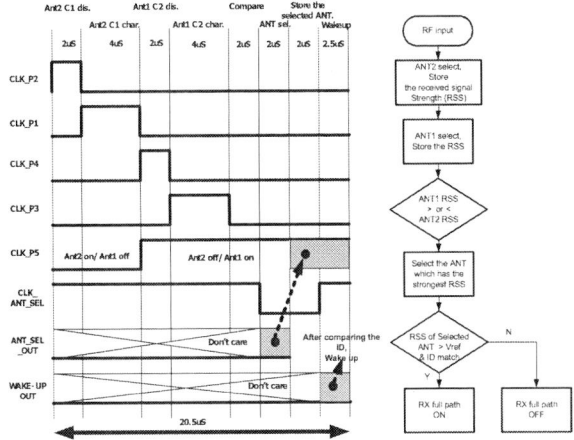

Fig. 5. The timing waveform and the flow chart.

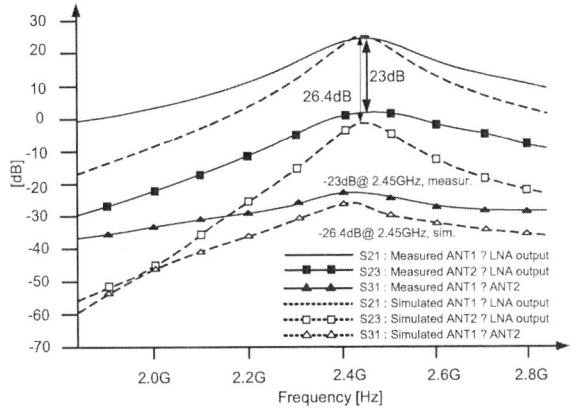

Fig. 6. Simulated and measured S-parameters of the LNA (showing 23dB isolation).

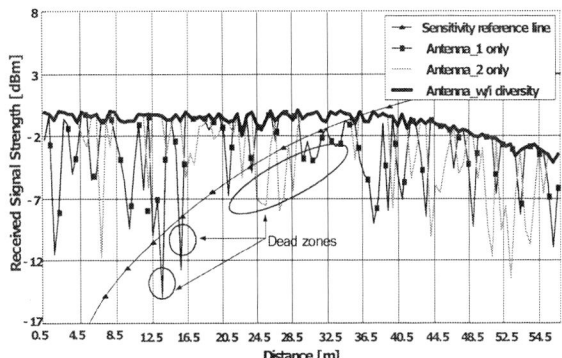

Fig.7. Measured signal strength to the distance.

comparator. The comparator is designed to operate at 12.5 dB input SNR considering the 4 dB noise margin.

The ring oscillator produces a clock frequency for the antenna selection block that is divided by a logic gate block. Fig. 5 shows the timing waveform and the flow chart, respectively. The ANT2 is selected first and then, the capacitor C1 is discharged using R2. The signal comes from the ANT2 and then, the DC current from the rectifier charges C1 until the SW3 is OFF to preserve the voltage on C1. In the next step, the ANT1 is selected and the capacitor C2 is discharged using R3. A similar process is carried out to charge the C2.

The voltage of V_{C1} is compared with V_{C2} using the Comparator1, and then the ANT2 is selected if V_{C1} is higher than V_{C2}; otherwise the ANT1 is selected. The resistors (R2, R3) are used to eliminate the voltage offset of the capacitors. In order to reduce the power consumption, a wake-up function is adopted. If the signal from the selected antenna is higher than the reference

voltage (Vref), then the digital parts (PGA, ADC, Modem & MCU) are enabled, otherwise the digital parts are disabled using the Comparator2. The wake-up input level can be controlled using the Vref. The ID block is used to stop the wake-up function when the interference is present at the input. The transmitter sends the OOK modulation signal to the receiver three times in a preamble for the wake-up. This block consists of less than 3,000 gates using the register transfer level (RTL) code, resulting in a negligible power consumption (under 0.1 mA). Namely, the wake-up functions are enabled if the ID code is matched. Therefore, the in-band or close-to-band interference can be avoided even if the antenna is selected by the interference.

III. EXPERIMENTAL RESULTS

The single-chip integrated circuit has been fabricated at the TSMC foundry, using a 0.18-μm gate length and six metal layers. The fabricated IC of 9.7 mm² die size, including the TX and PLL, is packaged and mounted on a FR-4 PCB test board. Two diversity antennas are separated by a one-quarter wavelength on the test board, which has been designed using the ADS tool [7].

Fig. 6 represents the simulated and measured S-parameters of the LNA, showing a large isolation (23 dB) between the antenna ports. The received signal strength is measured as a function of the distance to the transmitter, and compared with that of a previous work [4], as shown in Fig. 7. A 10~15 dB increment of the RSS has been observed in the non-line-of-sight (NLOS) channels. Wireless dead zones, possible for receivers without the antenna diversity, have not been observed in this diversity receiver. The antenna selection error has been found to be negligible (<1 %), and the antenna selection time is very fast (<20 μs) if the RSS difference between the antennas is over 3 dB, at -93 dBm input power. 46% power saving is also attained by powering down the PGA block (1 mA),

TABLE I
PERFORMANCE COMPARISON OF CMOS RECEIVERS

Reference	[4]	[8]	[9]	This work
Operating frequency	2.4 GHz band	2.4 GHz band	2.4 GHz band	2.4 GHz band
Power supply	1.8 V	1.8 V	1.8 V	1.8 V
RX current	25 mA @ including PLL & MCU	6 mA @ not including PLL & MCU	18 mA @ not including MCU	10 mA @ not including PLL
Stand-by current	No function	No function	No function	4.7 mA (= 8.5 mW)
NF	-	< 10 dB	6 dB (simulation)	< 10 dB
IIP3	10 dBm @ low gain	> -15 dBm	-13.5 dBm @ high gain	-11 dBm @ high gain
RX Architecture	Low-IF	Low-IF	Low-IF	Low-IF
Sensitivity	-94.7 dBm	-	-	-93 ~ -95 dBm @ PER=1%
Communication distance @ NLOS	12 m	-	-	37 m
Die size	5.76 mm² @ including RX, TX, PLL, and MCU	3.61 mm² @ not including PLL & MCU	6.5 mm² @ not including MCU	9.7 mm² @ including RX, TX, PLL, and MCU
Technology	0.18-μm CMOS	0.18-μm CMOS	0.18-μm CMOS	0.18-μm CMOS

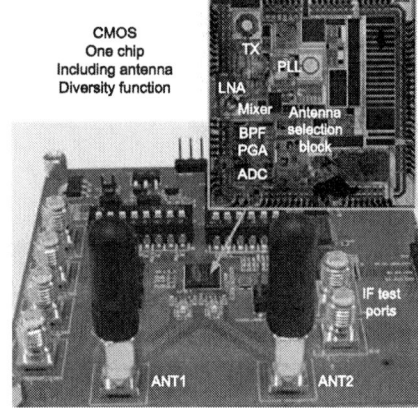

Fig. 8. The fabricated chip and the test board micrographes.

the ADC block (1 mA) and the MCU block (3 mA). The proposed whole receiver consumes 8.5 mW with a 1.8 V power supply in the standby mode for receiving. The current consumption of the LNA, the mixer, the BPF, and the antenna selection block are 1.2 mA, 1.5 mA, 1 mA, and 1 mA, respectively.

Table I summarizes a comparison of the proposed receiver with those in literatures. The proposed receiver has 3 times longer communication distance for the NLOS channels and the current consumption is less than other works. Hence, the proposed CMOS diversity receiver system, using the analog method, achieves enhanced performance compared to other available CMOS receivers. A chip micrograph and the test-bed are shown in Fig. 8.

IV. CONCLUSION

This work has presented a fully integrated diversity low-IF receiver, using a novel antenna selection scheme, for the IEEE standard 802.15.4 radio system. The proposed receiver is designed to operate in the 2.4-GHz band and fabricated using 0.18-μm CMOS technology. Compared to a single antenna system, 10~15 dB increment of the RSS has been measured for the NLOS channels and the antenna selection speed has been found to be very fast (under 20 μs) with a negligible antenna selection error (< 1%). The power consumption of the proposed receiver is 8.5 mW with a 1.8 V power supply in the standby mode for receiving, by virtue of the wake-up function.

REFERENCES

[1] C.N.Zhang, C.Ling, "A Low-Complexity Antenna Diversity Receiver Suitable for TDMA Handset Implementation," *Proc.Veh. Tech. Conf.*, may 1997

[2] K.Tsunekawa, "Diversity antennas for portable telephones," *in Proc. IEEE Veh.Tech.Conf.*, 1989, pp. 50-56.

[3] Y.I. Kwon, T.J.Park, H. Y. Lee, "A low power 2.4GHz CMOS RF front-end with temperature compensaton, " *in Proc. APMC2007*, vol. 3, pp. 1817-1820, Dec. 2007.

[4] J.H.Lim, K.S.Cho, Y.I.Kwon, " A fully integrated 2.4 GHz IEEE 802.15.4 transceiver for Zigbee applications," *in Proc. APMC2006*, vol.3, pp.1779-1782, Dec. 2006.

[5] Kimura,K, "A CMOS logarithmic IF amplifier with unbalanced source-coupled pairs," *IEEE Journal of Solid-State Circuits*, vol.28, pp 78-83, Jan 1993.

[6] S.Weilian, "RFID Reliability Technique for Single Tag Interrogation," *in poster presentations, Netted sensors community workshop.*, May 3.2007.

[7] ADS tools (ver 2005A) : The Agilent Technologies.

[8] I.Nam, K.Choi, .J.H,Lee, H.K.Cha, B.I.Seo, K.Kwon, K.Lee, " A 2.4-GHz Low-Power Low-IF Receiver and Direct-Conversion Transmitter in 0.18-μm CMOS for IEEE 802.15.4 WPAN Applications," *IEEE Transactions on Microwave Theory and Techniques*, vol.55, no.4, pp.682-689, April 2007.

[9] Y.S. Eo, J.Y. Hyun, S.S. Song, Y.J. Ko, and J.Y. Kim, "A fully integrated 2.4GHz low IF CMOS transceiver for 802.15.4 ZigBee applications," *IEEE Asian. Solid-State Circuits Conf.*, Nov. 2007, pp. 164-167.

A 90 μW MICS/ISM Band Transmitter with 22% Global Efficiency

Jagdish Pandey and Brian Otis

Department of Electrical Engineering, University of Washington, Seattle, WA

Abstract—**For fully autonomous implantable or body-worn devices running on harvested energy, the peak and average power dissipation of the radio transmitter must be minimized. We propose a highly integrated 90 μW 400 MHz MICS band transmitter with an output power of 20 μW leading to a 22% global efficiency — the highest reported to date for such systems. We introduce a new transmitter architecture based on cascaded multi-phase injection locking and frequency multiplication to enable low power operation and high global efficiency. Our architecture eliminates slow phase/delay-locked loops for frequency synthesis and uses injection locking to achieve a settling time < 250 ns permitting very aggressive duty cycling of the transmitter to conserve energy. At a data-rate of 200 kbps, the transmitter achieves an energy efficiency of 450 pJ/bit. Our 400 MHz local oscillator topology demonstrates a figure-of-merit of 204 dB.**

I. INTRODUCTION

In 1999, the FCC established the Medical Implant Communications Service (MICS) band to address the need for ubiquitous body area networks (BAN) comprising body-worn or implanted devices that continually sense vital body parameters [1]. Battery replacement for these devices may not be feasible or desirable, placing severe constraints on the power dissipation of the radio transceiver that tends to consume the bulk of total power. Other interesting applications enabled by ultra-low power radios are the animal tracking systems such as a moth flight recorder that requires payloads less than one gram [2]. High power dissipation in the radio will lead to a reduced recording time.

The typical power harvested from common surroundings is on the order of 100 μW [3]. Since the transmitter tends to dominate the total power budget, its power consumption needs to be below 100 μW. We report the first highly integrated 90 μW MICS band transmitter, realizing a significant improvement in the state-of-the-art. To achieve low power consumption without sacrificing performance, we introduce techniques both at the architecture and circuit level. Using cascacaded multi-phase injection locking and frequency multiplication, our local oscillator achieves a high figure-of-merit of 204 dB. Our architecture also helps eliminate the slow phase/delay locked loops used in carrier generation, and therefore permits agile duty cycling of the transmitter to conserve energy.

One significant challenge in the design of ultra-low power transmitters is the loss of global efficiency as the PA output power drops (≤ 25 μW in the MICS band). The power dissipation in the carrier generation and data modulation circuitry constitutes a large percentage of total power consumption of the transmitter, leaving the PA efficiency as a secondary concern. As a result, the best reported global efficiency for MICS band transmitters is <6.5% [5]– [6]. This work presents a greater than $3\times$ improvement in the global transmitter efficiency at 20 μW of transmit power.

II. BRIEF LITERATURE REVIEW

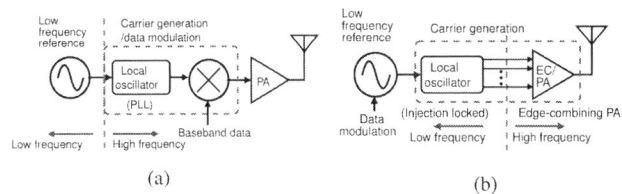

Fig. 1. (a) The conventional transmitter architecture (b) Proposed ultra-low power (ULP) transmitter architecture.

A number of integrated MICS band ultra-low power (ULP) transmitters have appeared in recent literature [4] [5] [6] [7]. In order to reduce the power consumption of the transmitter, the performance burden is shifted to a power hungry receiver. For example, to reduce power dissipation in carrier generation that typically dominates the total power budget in ULP transmitters, an open loop oscillator is used in [4] and [5]. The necessary frequency stability is obtained using a frequency correction/calibration loop running at the receiver. Others have adopted an RFID-style passive tag to perform medical sensing [8]. This again requires a power hungry RFID receiver. In summary, all the low-power transmitters described above shift the performance burden to a complex receiver resulting in high energy per *transceived* bit. This results in a highly asymmetrical link unsuitable for autonomous peer-to-peer body area network applications running on scavenged energy. Authors in [7] attempt to reduce the energy per bit by operating at higher data rates. However, small implantable batteries and scavengers typically exhibit a large source resistance that may not be able to supply the large peak currents needed for such systems. It is therefore desirable to achieve a sub-100 μW of power dissipation in the transmitter *without* introducing asymmetry in the radio link to enable the true autonomous medical sensing and telemetry.

978-1-4244-6240-7/10 $26.00 © 2010 IEEE

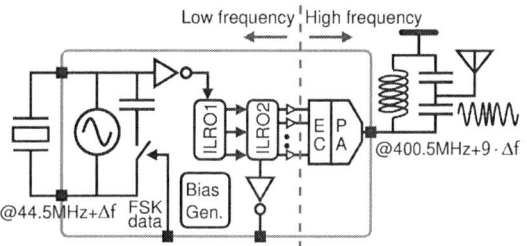

Fig. 2. Detailed block diagram of the proposed transmitter architecture for ultra-low power operation.

III. PROPOSED ARCHITECTURE

Conventional ULP transmitters perform frequency synthesis and data modulation at the carrier frequency leading to poor global efficiency and high power consumption (Figure 1(a)). Our proposed transmitter achieves very low power dissipation by performing these operations at a much reduced frequency and employing an edge-combiner merged into the power amplifier (PA) (Figure 1(b)). However, in order to obtain the equally spaced edges necessary for the edge-combiner, a delay chain or a ring oscillator locked in a PLL/DLL is needed. The additional components in the loop such as charge-pump and loop filter present significant area/power overhead, while the settling time of the loop constrains the maximum possible duty cycling of the transmitter [6]. Any attempt to directly injection-lock the multi-phase low-frequency ring oscillator using the single phase crystal reference will introduce significant mismatches in the delayed waveforms of the RO. We address this issue by using cascaded injection-locking (Figure 2). The proposed architecture is an example of digitally-assisted RF design that benefits from CMOS and power supply scaling. The key building blocks of our proposed architecture — the edge-combining PA, the injection-locked oscillator and data modulation — are described below:

A. Edge-combiner

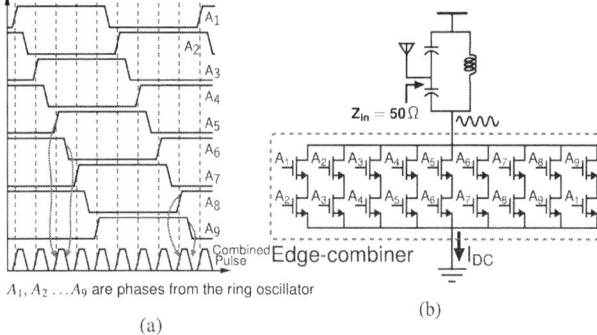

$A_1, A_2 \ldots A_9$ are phases from the ring oscillator

(a)   (b)

Fig. 3. (a) The principle of edge-combining (b) Schematic of the edge-combiner.

The low-power frequency multiplier is based on the principle of edge-combining. Let $A_1, A_2 \ldots A_N$ be the waveforms from a digital ring oscillator running at frequency f_{RO}. The waveform $\Sigma(A_1 A_2 + A_2 A_3 \ldots A_N A_1)$ is a square wave of frequency $N f_{RO}$ where N is both the factor of multiplication and the number of stages in the ring oscillator. For our 9-stage ring oscillator, the waveforms $A_1, A_2 \ldots A_9$ are spaced apart by a period of $T/18$, where T is the time period of the reference input at 44.5 MHz (Figure 3(a)). We use MOS transistor switches to perform an AND operation and sum the switched currents to realize an OR operation (Figure 3(b)).

We take advantage of low output power requirement of the MICS standard to combine the PA functionality with the edge-combiner which behaves like a non-linear power amplifier (Figure 3(b)). The load impedance of the edge-combiner is transformed using a tapped-capacitor matching network to match a 50 Ω antenna. This high-Q load at the edge-combiner output also attenuates the out-of-band spurs resulting from the mismatches in the low-frequency ring oscillator.

B. Injection-locked LO Design

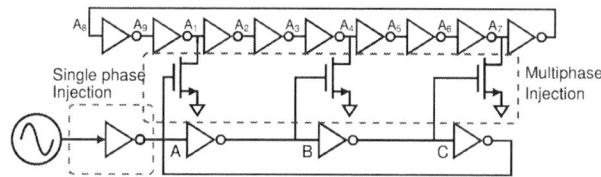

Fig. 4. Schematic of the two-stage multi-phase injecton-locked ring oscillator.

To ensure frequency stability across PVT variations, we injection-lock the low-frequency ring oscillator to an on-chip crystal reference, thereby eliminating the need for a PLL. The fast lock-time (on the order of 250 ns) of the LO allows aggressive duty cycling of the transmitter to further save power. Figure 4 shows the schematic of the 3-phase 2-stage injection-locked oscillator. The first stage of injection locking by the crystal oscillator ensures the correct frequency and low phase noise. However, the single-phase injection introduces asymmetry in otherwise equally spaced phases (A, B and C) of the ring oscillator. This asymmetry will lead to large reference spurs in the frequency multiplied output. The second stage of injection-locking attenuates this phase imbalance by using 3-stage symmetrical injection in a 9-stage ring oscillator. As shown in [9], cascaded multiphase injection-locked oscillators can be used to correct the phase and amplitude mismatches. The phase mismatches in phases $A_1, A_2, \ldots A_9$ are approximately an order of magnitude smaller than those in A, B and C. Multiphase-injection locking also increases the locking bandwidth ensuring reliable operation across PVT variation.

C. Data Modulation

On-chip FSK modulation is accomplished by pulling the quartz reference clock. The resulting frequency deviation is multiplied by 9×. Presence of crystal shunt capacitance C_0 and parasitic capacitance C_p in the circuitry limits the maximum fractional pulling to about a few hundred ppm. For a 44.5 MHz crystal, we obtained >20 kHz of frequency pulling, resulting in a 180 kHz frequency deviation about the carrier frequency.

978-1-4244-6240-7/10 $26.00 © 2010 IEEE

IV. MEASUREMENT RESULTS

This section presents the measured results from our 400.5 MHz transmitter prototype, implemented in 130 nm CMOS, based on the architecture described above. The entire system is integrated except for the crystal and the matching network. Figure 5 shows the frequency multiplied output at -17 dBm output power. The carrier-to-spur ratio (CSR) is 42.77 dB indicating excellent matching in phases $A_1, A_2, \ldots A_9$.

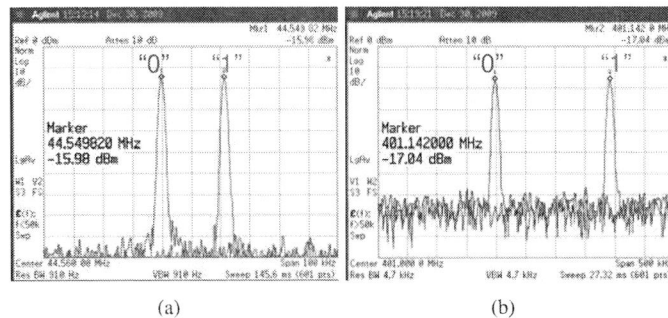

(a)   (b)

Fig. 7. (a) FSK modulated 44.5 MHz injection-locked ring oscillator. $\Delta f_{RO} = 20$ kHz (b) FSK modulated frequency multiplied output at 400.5 MHz. $\Delta f_{RO} = 180$ kHz

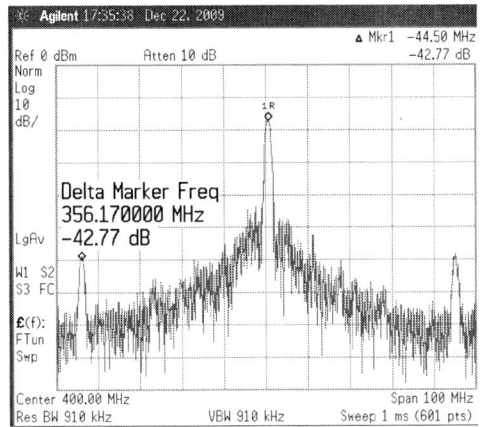

Fig. 5. Measured carrier-to-spur ratio of 42.77 dB was achieved.

Figure 6(a) shows the overlaid spectrum of the free running and the injection-locked ring oscillator. As shown in Figure 6(b), the close-in phase noise of the injection-locked ring oscillator is vastly improved. The locking range of the 45 MHz free-running ring oscillator extends from 32–52 MHz (44% locking bandwidth). This allows operation in the 433 MHz ISM band using a 48 MHz crystal. The FSK modulated waveforms of the locked oscillator and the frequency multiplied output are captured in Figure 7. An FSK modulated pseudo-random data sequence was successfully detected using a commercial off-the-shelf receiver at a data rate of 200 kbps.

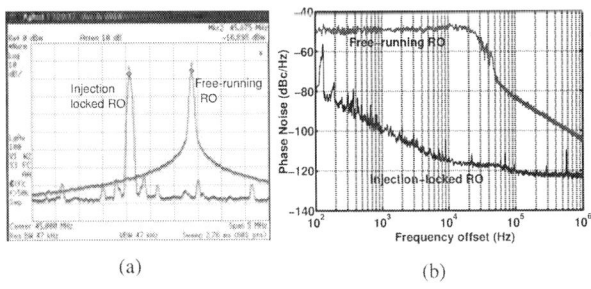

(a)   (b)

Fig. 6. (a) The spectrum and (b) the measured phase noise of the free running and the injection-locked ring oscillator.

Figure 8 presents the phase noise of the injection-locked ring oscillator and the frequency multiplied output. At a given offset, the phase noise of the frequency multiplied output is higher than that of the 44.5 MHz ring oscillator by $20 \log_{10}(9) \approx 19$ dB. At 300 kHz offset, the frequency-multiplied output achieves a phase noise of -105.2 dBc/Hz.

The oscillator figure-of-merit (FoM) is given as

$$\text{FoM(dB)} = -\mathcal{L}(\Delta\omega) + 20\log\left(\frac{f_0}{\Delta f}\right) - 10\log(\text{P(mW)})$$

For our 400.5 MHz frequency multiplied oscillator, the FoM is 204 dB.

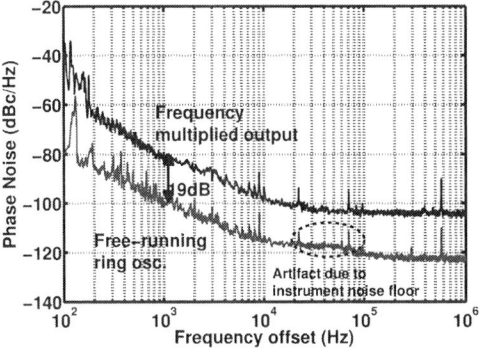

Fig. 8. Measured phase noise of the injection-locked LO and the frequency multiplied output using an Agilent E4446A spectrum analyzer.

Using our proposed techniques of high efficiency frequency multiplication and multi-phase cascaded injection locking, we were able to reduce the total power dissipation in carrier generation and data modulation to less than $24\,\mu$W. The measured PA drain efficiency is higher than 30% for an output power greater than $20\,\mu$W (Figure 9).

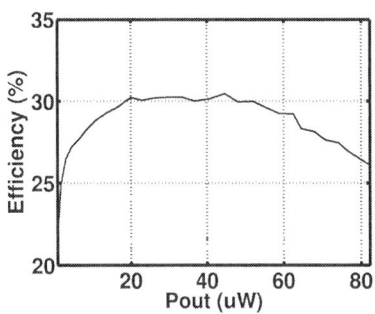

Fig. 9. Measured PA drain efficiency

Figure 10 presents the measured input return loss for the transmitter. Excellent matching with a measured $|S_{11}| <$ -22 dB was achieved.

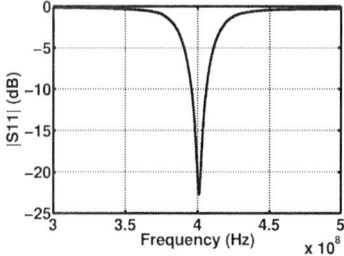

Fig. 10. Measured input return loss. $|S_{11}| <$ -10 dB.

Figure 11 shows the chip micrograph of the die, implemented in 130 nm CMOS. Due to the highly digital architecture of the proposed transmitter, and the absence of the synthesizer loop along with its large loop filter capacitors, the active area is less than $200 \, \mu m \times 200 \, \mu m$.

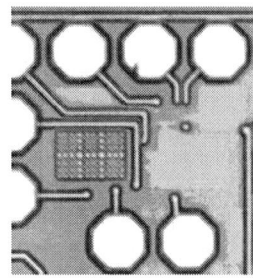

Fig. 11. The chip micrograph of the ultra-low power transmitter. The active area of the transmitter is $\approx 200 \, \mu m \times 200 \, \mu m$.

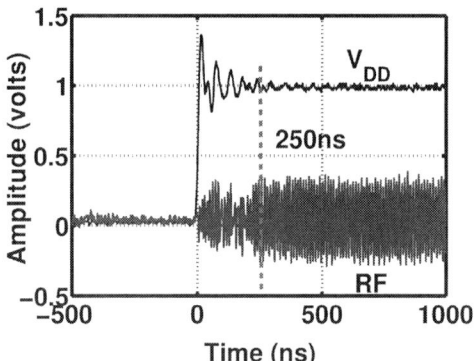

Fig. 12. A settling time <250 ns was measured.

Due to the absence of a PLL, Figure 12 shows a transmitter start-up time less than 250 ns. Table I captures the summary of the latest reported transmitters for the MICS band, along with our proposed work. Our transmitter has a global efficiency of 22% and energy efficiency of 450 pJ/bit which is a 3× improvement in the state-of-the-art.

TABLE I
PERFORMANCE SUMMARY OF THE ULP TRANSMITTERS AND THE
PROPOSED TRANSMITTER

	[4]	[7]	[5]	[6]	This Work
Power dissipation	350 μW	5 mW	400 μW	400 μW	**90 μW**
Data-rate	120 kbps	800 kbps	250 kbps	100 kbps	**200 kbps**
Transmit power	NA	-4 – -17 dBm	-16 dBm	-16 dBm	**-17 dBm**
Energy per bit	2.9 nJ/bit	6.3 nJ/bit	1.4 nJ/bit	4 nJ/bit	**0.45 nJ/bit**
Process	90 nm	180 nm	130 nm	130 nm	**130 nm**
Modulation	MSK	FSK	FSK	FSK	**FSK**

V. CONCLUSION

We report the first sub-100μw MICS band transmitter with 22% global efficiency and 450 pJ/bit energy efficiency at a data-rate of 200 kbps. Using the techniques of cascaded multiphase injection-locking and frequency multiplication, we perform carrier generation without the slow phase/delay locked loops which allow a settling time <250 ns. Except for the crystal and the matching network, the entire transmitter is integrated with an active area less than $200 \, \mu m \times 200 \, \mu m$. The 400 MHz LO is locked to a crystal reference and achieves a FoM of 204 dB.

ACKNOWLEDGMENT

The authors thank the Center for the Design of Analog and Digital Integrated Circuits (CDADIC).

REFERENCES

[1] http://wireless.fcc.gov/services/index.htm?job=service_home&idmedical_implant

[2] B. Otis, C. Moritz, J. Holleman, A. Mishra, J. Pandey, S. Rai, D. Yeager, F. Zhang, "Circuit techniques for wireless brain interfaces, " *IEEE Conf. on Engineering in Medicine and Biology Society, 2009* pp. 3213–3216, Sept. 2009

[3] S. Roundy, E.S. Leland, J. Baker, E. Carleton, E. Reilly, E. Lai, B. Otis, J.M. Rabaey, P.K. Wright, V. Sundararajan, "Improving power output for vibration based energy scavengers," *IEEE Pervasive Computing*, vol. 4, no. 1, pp. 28–36, Jan. 2005

[4] J. Bohorquez, A. Chandrakasan, J. Dawson,"A 350 μW CMOS MSK transmitter and 400 μW OOK super-regenerative receiver for medical implant communications," *IEEE Journal of Solid-State Circuits,* vol. 44,no. 4,April 2009 pp. 1248–1259

[5] J. Bae, Namjun Cho, H.-J. Yoo, "A 490 μW fully MICS compatible FSK transceiver for implantable devices," *IEEE symposium on VLSI Circuits,* June 2009

[6] S. Rai, J. Holleman, J. Pandey, F. Zhang, B. Otis, "A 500 μW neural tag with $2 \mu V_{rms}$ AFE and frequency-multiplying MICS/ISM FSK transmitter," *IEEE International Solid-State Circuits Conference*, Feb. 2009

[7] P. Bradley, "An ultra low power, high performance medical implant communication system (MICS) transceiver for implantable devices," *IEEE Biomedical Circuits and Systems (BioCAS)*, 2006

[8] D.J. Yeager, J. Holleman, R. Prasad, J.R. Smith, B.P. Otis,"NeuralWISP: a wirelessly powered neural interface with 1 m range," *IEEE Trans. on Biomedical Circuits and Systems*, vol. 3, no. 6, Dec. 2009 pp. 379–387

[9] P. Kinget, R. Melville, D. Long, V. Gopinathan, "An injection-locking scheme for precision quadrature generation," *IEEE Journal of Solid-State Circuits,* vol. 37, no. 7, July 2002 pp. 845–851

A 2mW CMOS MICS-Band BFSK Transceiver with Reconfigurable Antenna Interface

Seungkee Min[1], Sridhar Shashidharan[1], Mark Stevens[1], Tino Copani[1],
Sayfe Kiaei[1], Bertan Bakkaloglu[1] and Sudipto Chakraborty[2]

[1]Electrical, Energy and Computer Engineering, Arizona State University, Tempe, AZ, 85287, USA

[2]Texas Instruments Inc., High Performance Analog Design, Dallas, TX, 75243, USA

Abstract — A 0.18µm CMOS MICS-band transceiver with a reconfigurable RF front-end is presented, reusing the same circuit core for super-regenerative wake-up receiver, receive-mode LNA, and transmit power amplifier, eliminating the need for an external T/R switch. The transceiver uses an All Digital Frequency Locked-Loop (ADFLL) for LO signal generation and transmitter's modulation. The OOK wake-up receiver sensitivity is -80dBm @ 50kbps, while the BFSK receiver's sensitivity is -97dBm for a 75kbps signal and 2mW power consumption. The nominal output power of the transmitter is -5dBm.

Index Terms — Medical Implantable Communication Services (MICS), implanted wireless sensor, ultra low power transceiver.

I. INTRODUCTION

Low power transceivers using the Medical Implantable Communication Services (MICS) standard are becoming key devices in the modern health care system such as patient's remote monitoring or drug release control [1][2]. In addition to small size, low unit cost and long battery life are the most critical aspects of MICS devices. To maximize battery life and reduce the bill of material (BOM), simple radio architectures and modulation schemes are usually adopted. Different solutions have been presented in the past years. On the receiver side, super-regenerative architecture using On-Off Keying (OOK) modulation is reported in [3], while Frequency Shift Keying (FSK) implementations using direct conversion or low-IF systems are presented in [4-5]. A dual-receiver system supporting both OOK and FSK has been also reported [2].

Usually, MICS transmitters adopt FSK modulation with high modulation index, which allow low power implementations and minimize flicker noise impact. Direct modulation Binary FSK (BFSK) schemes using fractional-N synthesizers have been used due to their high digital content [3-5].

A MICS transceiver IC that addresses ultra low power consumption requirements, and minimize the number of external components is presented in this work. Circuit re-use and digital intensive techniques are presented, which optimize the analog RF front-end and the LO frequency synthesizer.

The remainder of the paper is organized as follows: Section II explains and justifies the proposed architecture, Section III describes the circuit design of the implemented blocks, and finally Section IV concludes with measured results.

II. SYSTEM ARCHITECTURE

The system level schematic of the implemented half-duplex transceiver is shown in Fig. 1. The system has three modes of operation: wakeup receive (RXW), normal receive (RX), and transmit (TX). Majority of the time the transceiver operates in the RXW mode waiting in standby for any base-station wakeup signal. In this case, the input reconfigurable front-end is a super-regenerative oscillator triggered by a low data rate OOK modulation in the MICS band. Demodulation is performed through an envelope detector and comparator, down-converting the signal to baseband. Once a wakeup signal is detected, the input front-end is reconfigured from SRO to a g_m-boosted LNA topology. The normal reception is then performed with direct conversion architecture. An All Digital Frequency Locked Loop (ADFLL) is adopted as I/Q LO frequency synthesizer. Finally, in the TX mode, the reconfigurable front-end converts to a power amplifier, driven by the frequency modulated ADFLL. The power amplification is achieved by configuring the front-end into a class-AB stage for high data rate or into a power oscillator with injection locking modulation in the case of low bit rates.

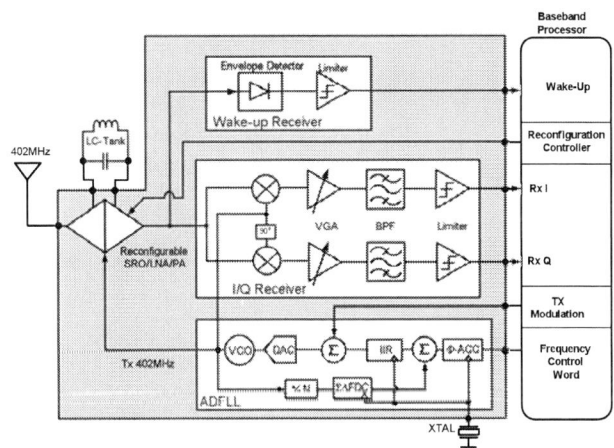

Fig. 1. Block diagram of the proposed MICS transceiver.

OOK and BFSK modulation schemes are commonly adopted in low power radio architectures. The OOK architecture eliminates the need for LO generation, enabling power saving. However, any unwanted interferer in the RX bandwidth will be superimposed with the intended signal at the envelope detector output, degrading the receiver's sensitivity. In contrast, direct-conversion BFSK receivers provide high interference rejection and low BER at high data-rates. In the proposed transceiver, OOK is adopted to wake-up the IC from stand-by, and BFSK is used for normal receive functionality.

The maximum transmit power allowed by the MICS standard is -16dBm [1]. The base-station signal is attenuated due to free space loss (i.e, 30dB for 2m distance), human body loss (i.e, 20dB) and poor implant antenna gain (i.e., -20dB). Thus the maximum signal strength at the receiver is as low as -86dBm. Taking into account a 10dB margin, the targeted sensitivity of the receiver is set at -96dBm. The system sensitivity, P_s, is defined by

$$P_s = -174 + 10\log_{10}(BW) + NF_{SYS} + SNR_{RX}, \quad (1)$$

where SNR_{RX} is the minimum signal to noise ratio required at the digital demodulator input. Using BFSK modulation @ 75kbps with 10^{-3} BER, a 10.5dB SNR_{RX} is needed. Using a channel bandwidth of 90kHz and an input sensitivity of -96dBm, the required system noise figure is 17dB.

Low power transceiver modules require low complexity modulation schemes in order to reduce the decoder power. Direct modulation of the LO frequency synthesizer is a low power solution in the case of low bitrate FM applications where a wide PLL bandwidth is not required. The modulated PLL output signal can be just amplified and fed to the antenna without the need of I/Q mixer upconversion. The maximum allowed output power implanted device is -10dBm.

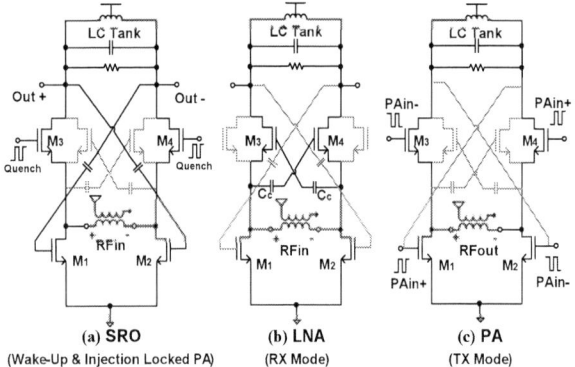

Fig. 2. Reconfigurable RF Front-End in (a) wake-up, (b) normal reception and (c) TX PA modes

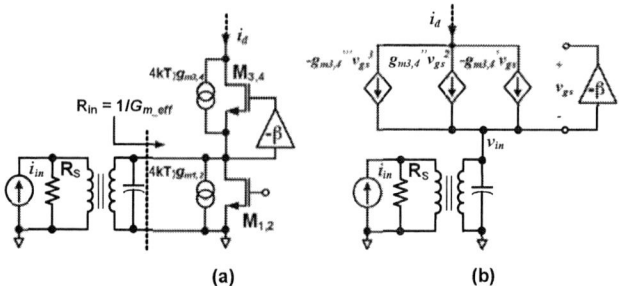

Fig 3. Half circuit approximation of the g_m-boosted LNA (a) for gain, noise figure and (b) linearity analysis.

III. CIRCUIT DESIGN

A. Reconfigurable RF Front-End

The schematics of the reconfigurable front-end in the three different operation modes are shown in Fig. 2. The front-end operates in fully differential mode with a matched antenna interface using the same pair of pins and a tuned load in all cases.

I – Super-regenerative wake-up mode: in the super-regenerative mode (Fig. 2a), the cross-coupled M_1-M_2 pair operates on a positive feedback. In the presence of an RF signal at the SRO input, full-scale oscillations rapidly build up. A quench signal is used to open the positive feedback loop periodically. When the amplitude of the input OOK signal is low, the quench signal dampens the SRO oscillations before they can reach full amplitude, while a strong RF signal allows oscillations to reach steady state before the Quench opens the loop.

II – Receive mode: in the LNA mode (Fig. 2b), the M_1-M_2 devices are fixed current sources, while a capacitive cross-coupling path provides extra gain in the common-gate (CG) devices M_3-M_4 [6]. The half-circuit approximation of the LNA to analyze the gain and noise figure is shown in Fig. 3(a). The effective transconductance of this topology, G_{m_eff}, is given by

$$G_{m\_eff} = (1 + \beta)g_{m3,4} \quad (2)$$

$$\beta = C_C / (C_C + C_{GS\,3,4}), \quad (3)$$

where $g_{m3,4}$ and $C_{GS\,3,4}$ are the transconductance and gate-source capacitance of the CG MOS devices. The transconductance of the LNA is then increased by the capacitive feedback ratio, β. The input impedance of the LNA is set by $1/G_{m\_eff}$, therefore the g_m-boosted approach allows achieving input matching with lower DC consumption

The Noise Factor of the LNA is given by,

$$NF = 1 + \frac{\gamma}{\alpha G_{m\_eff} R_S} + \gamma g_{m1,2} R_S, \quad (4)$$

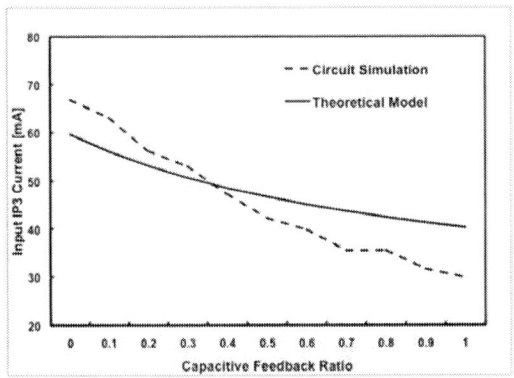

Fig. 4. Comparison of LNA IP3 model against simulations.

where, γ is the MOS excess noise factor, α is the self-gain g_m/g_{d0}, R_S is the source impedance (antenna) and $g_{m1,2}$ is the transconductance of the MOS current sources. The half-circuit equivalent model of the LNA used to analyze the linearity performance is shown in Fig. 3b. Under ideal matching conditions, it can be assumed that any higher order harmonics are filtered out at node v_{in}. In this case, the input referred IP3 of the g_m-boosted LNA can be approximated by

$$IIP3 = \sqrt{\frac{4}{3}\left(\frac{g_{m3,4}'}{g_{m3,4}'''}\right) \cdot \left[\frac{1}{(1+\beta)R_S} + g_{m3,4}'\right]}, \quad (5)$$

where $g_{m3,4}'$ and $g_{m3,4}'''$ are the 1st and 3rd order transconductance coefficients of the M_3-M_4 devices, respectively. The theoretical IIP3 derived from (5) is compared against circuit simulations in Fig. 4. The simulation closely matches the model, with small deviation due to the limited filtering of higher order harmonics at node v_{in} in the real circuit.

III – Transmit PA mode: in the transmitter mode, the differential MOS pair features a push pull configuration as shown in Fig. 2c. The pseudo-differential output provides twice the voltage swing with reduced risk of oxide voltage break down.

In the case where the signal bandwidth is lower than the injection locking bandwidth of the oscillator mode, the power amplification is achieved by injecting the ADFLL BFSK output to the SRO configuration of Fig. 2a. The required injection locking power is lower than the class-AB PA, achieving further power saving in the PA drivers.

However in this approach, the transmitter data rate is limited by the trade-off between the LC-tank Q-factor and the oscillator power consumption. A lower Q-factor leads to a wider injection locking bandwidth, but it also requires higher current consumption to guarantee stable oscillations.

B. All Digital Frequency Locked Loop (ADFLL)

An all-digital frequency locked loop is used for LO generation and transmit modulation as shown in Fig. 5. All-digital implementation of the synthesizer enables a programmable loop bandwidth and fast locking. Due to the use of non-coherent frequency modulation, a frequency locked loop (or a Type-I PLL) rather than a phase locked loop can be used for the receiver LO. The modulation injection for transmit channel is performed inside the loop, which decouples modulation bandwidth from loop bandwidth.

The ADFLL reference frequency is noise shaped via a single-bit first order $\Sigma\Delta$ modulated to obtain phase accumulation similar to direct digital frequency synthesizers (DDFS). In the feedback path, the ring oscillator (VCO) output is divided and passed through a frequency to digital converter resulting also in a single-bit first order $\Sigma\Delta$ noise shaped signal representing the instantaneous VCO frequency. The difference between reference and feedback signals is applied to the Direct Form I IIR loop filter, attenuating the out-of-band quantization noise. The 12-bit IIR filter output is re-modulated to a 3-bit, 7-level noise shaped signal through a second order $\Sigma\Delta$ noise shaper clocked at 50.4MHz signal divided from the VCO output. A 7-level current steering DAC with PMOS-only current source array is used to generate the control voltage for the VCO. The current mode DAC output is fed to a second-order passive RC filter, and a first-order hold response providing reconstruction filtering. The VCO is a four-stage differential ring oscillator with Maneatis delay cells and replica bias circuit with reduced AM/PM conversion.

Using frequency as the loop variable, the ADFLL loop resembles a Type-I PLL, with a single integrator in the loop. The frequency-based approach minimizes the complexity associated with phase domain PLLs. The worst-case near-integer spur of -55dBc and a phase noise of -83dBc/Hz at 300kHz offset is measured.

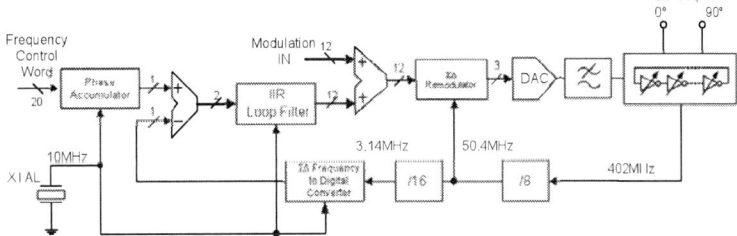

Fig. 5. Block Diagram of the proposed ADFLL.

IV. Measurement Results

The MICS band transceiver IC has been fabricated in a bulk $0.18\mu m$ CMOS process and directly assembled on a FR4 test board. The wakeup receiver achieves a sensitivity of -80dBm for a data rate of 50kbps with measured BER of 10^{-3} and $280\mu W$ consumption from a 1.5V supply. Time domain measurements for a -80dBm input signal are shown in Fig. 6. The BFSK receiver achieves a sensitivity of -97dBm at 75kbps for a power of 2mW, which is equivalent to 24nJ/b. The measured BER at different RX power consumptions is shown in Fig. 7.

The RX measurements show no significant BER degradation up to a +30dB interferer level. The transmitter was tested both in the class-AB operation and power oscillator modes, respectively. The measured spectra for a 100kbps BFSK signal are shown in Fig. 8. The class-AB amplifier fulfills the requirement for a 100kbps signal, whereas the power oscillator mode shows spectral re-growth at this data rate because of the limited injection bandwidth. The measured return loss at the antenna interface is higher than 10dB in all the operation modes. Finally, the chip layout, which occupies an active area of $3.8mm^2$, is shown in Fig. 9.

Conclusions

A low power MICS transceiver has been presented, which adopts a reconfigurable RF front-end and all digital FLL. The proposed system and circuit techniques enable ultra low power operation and also minimize the use of external components, which are the key factors in the implementation of radios for medical implants.

References

[1] MICS Band Plan, *FCC Rules and Regulations*, Part 95, Jan. 2003.

[2] P. D. Bradley, "An Ultra Low Power, High Performance Medical Implant Communication System (MICS) Transceiver for Implantable Devices", *IEEE BIOCAS 2006*, Nov 29 2006-Dec 1 2006, pp. 158-161.

[3] J.L. Bohorquez, et al., "A 350 µW CMOS MSK Transmitter and 400 µW OOK Super-Regenerative Receiver for Medical Implant Communications", *IEEE Journal of Solid-State Circuits*, vol. 44, Issue 4, Apr. 2009, pp. 1248-1259.

[4] A. Tekin, et al., "A low power MICS band transceiver architecture for implantable devices", *IEEE Wireless and Microwave Technology Conference 2005*, pp. 55-58.

[5] N. Cho, et al., "A 10.8 mW Body Channel Communication/MICS Dual-Band Transceiver for a Unified Body Sensor Network Controller", *IEEE Journal of Solid State Circuits, Vol 44*, Issue 12, Dec 2009, pp. 3459-3468.

[6] W. Zhuo, et al., "A capacitor cross-coupled common-gate low-noise amplifier," *IEEE Trans. on Circuits and Systems II*, vol. 52, no. 12, Dec. 2005, pp. 875-879.

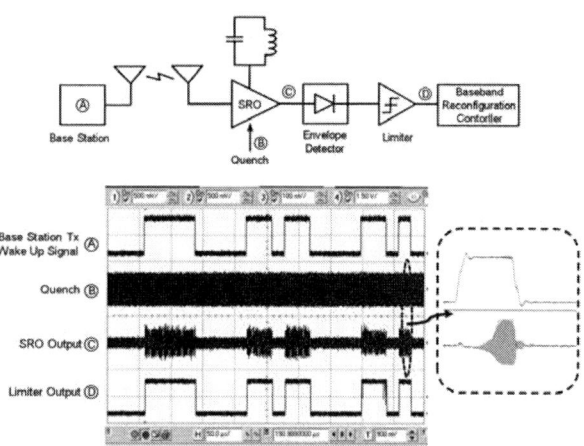

Fig. 6. Transient measurement results of the Wake-up Receiver.

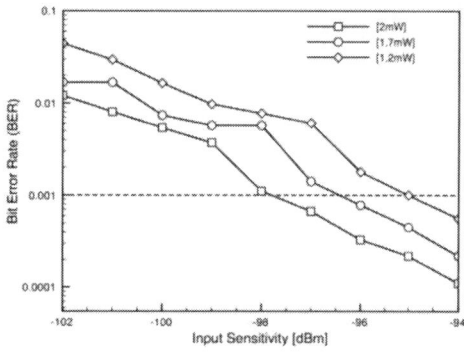

Fig. 7. BER measurement results of the receiver at different power consumption levels.

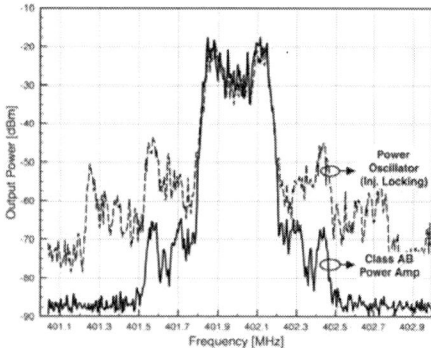

Fig. 8. Trasmitter's output spectra for a 100kbps BFSK signal.

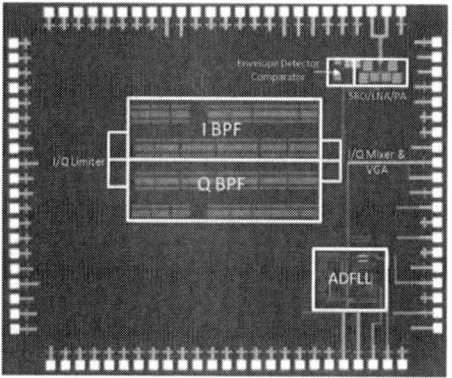

Fig. 9. Die micrograph of the MICS transceiver.

RMO4A-5

A 1.8 to 2.4-GHz 20mW Digital-Intensive RF Sampling Receiver with a Noise-Canceling Bandpass Low-Noise Amplifier in 90nm CMOS

Joonhee Lee, Jaewook Kim, and SeongHwan Cho
Department of Electrical Engineering, KAIST, Daejeon, Republic of Korea
Email: ljh0616@gmail.com, jaeuk83@gmail.com, chosta@ee.kaist.ac.kr

Abstract— **This paper presents a digital-intensive RF sampling receiver composed of a noise-canceling bandpass low-noise amplifier (LNA) and an RF analog-to-digital converter (ADC) for multi-band multi-mode wireless communication. The proposed LNA employs an on-chip transformer to combine the outputs of a common-gate and a common-source LNA to reduce the noise figure and enhance the linearity, while providing tunable bandpass filtering from 1.8 to 2.4-GHz. The RF ADC employs a time-based architecture that uses time-interleaved VCOs with 1st order noise shaping property, which benefits from enhanced time resolution of advanced CMOS process. A prototype chip implemented in 90 nm CMOS process has an area of 0.3 mm^2 and achieves SNR of 50 dB for 1-MHz signal bandwidth at 1.8 to 2.4-GHz carrier frequency, while consuming 20 mW from 1.2 V supply.**

Index Terms— **RF sampling receiver, direct RF sampling, LNA, on-chip transformer, time-based ADC, time-interleaved**

I. INTRODUCTION

The rapid increase in the number of communication standards has intensified the research effort for a multi-band multi-mode wireless receiver. Employing a high-performance ADC near the antenna has been considered an attractive architecture for multi-band multi-mode wireless communication, since it can take advantage of signal processing power and reconfigurability of DSPs. Unfortunately, designing a high performance ADC at radio frequency (RF) is not an easy task. Although there have been several attempts for direct RF sampling $\Delta\Sigma$ bandpass ADCs at GHz range [1], [2], ,[3], their power consumption and area are very large due to the high speed DACs and Gm-LC filters that use multiple on-chip spiral inductors. Moreover, frequency range of the ADC is very limited as the stability of the ADC is highly sensitive to coefficients of the $\Delta\Sigma$ loop, thus making them unsuitable for multi-band applications. While a time-based ADC employing time-interleaved VCOs [4] requires an anti-alias bandpass filter at its input, it is a promising architecture for multi-band multi-mode applications as it consumes low-power, small area and has widely tunable frequency without suffering from any stability issue.

In this paper, we propose a digital-intensive RF sampling receiver that consists of a noise-canceling bandpass LNA and a time-based ADC using time-interleaved VCOs as shown in Fig. 1. The RF signal is bandpass filtered and amplified by the LNA and digitized by the ADC. Next, the digital output

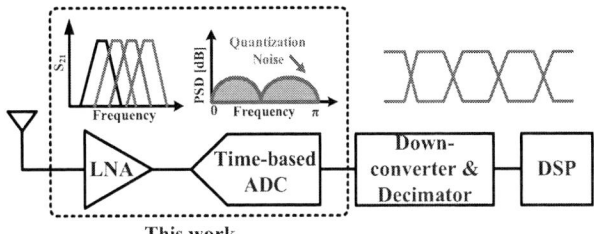

Fig. 1.   Block diagram of the digital-intensive RF sampling receiver.

of the ADC is decimated and filtered in the digital domain. Since the signal transfer function of the time-based ADC [4] is all pass and hence susceptible to aliasing, the LNA provides anti-alias bandpass filtering by employing a digitally tuned LC load with an on-chip transformer that is also used for noise and distortion cancelation. The sampling frequency of the ADC can be chosen so that the ADC either performs direct RF sampling or mirror image sampling [2]. In addition, the ADC is structured so that there is little overhead in decimation, which is in contrast to [4] that requires complex algorithms with heavy computational overhead in decimation.

II. NOISE-CANCELING BANDPASS LOW-NOISE AMPLIFIER FOR MULTI-BAND APPLICATIONS

A. Proposed LNA employing On-chip Transformer

A common-gate (CG) LNA has been widely used in wide-band or multi-band receivers due to its wide-band input matching characteristic. However, the CG LNA suffers from the poor noise figure which degrades the receiver sensitivity. Recently, a noise cancelation technique which combines the outputs of a CG and a common-source (CS) LNA to cancel the thermal noise of the CG LNA has been reported [5], [6], [7]. Unfortunately, these circuit techniques cannot be used in the proposed RF sampling receiver because a PMOS based current mirror or a resistor load do not provide bandpass characteristic necessary for the ADC.

The schematic of the proposed LNA is shown in Fig. 2, which consists of an on-chip transformer, a CS and a CG LNA. 50 Ω input matching is realized by using the transconductance of M_1 and the DC current path is provided by an off-chip inductor (L_s). The cascode transistors (M_3 and M_4) improve the reverse isolation. A 2-bit switched capacitor array is

978-1-4244-6240-7/10 $26.00 © 2010 IEEE   263

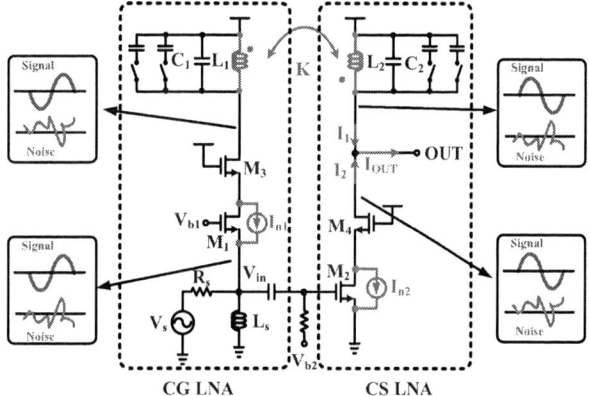

Fig. 2. Schematic of the proposed LNA employing the on-chip transformer.

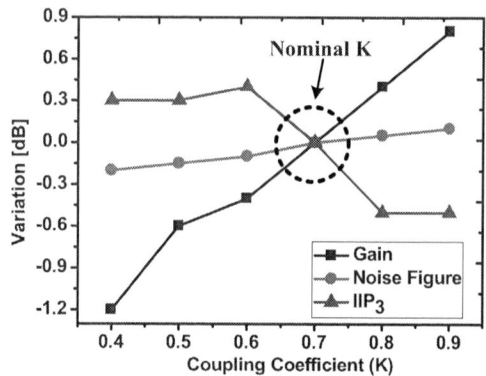

Fig. 3. Performance variation according to coupling coefficient (K).

employed to cover the center frequency from 1.8-GHz to 2.4-GHz. The on-chip transformer which can perform the role of the current mirror is used to combine the outputs of the CS and the CG LNA for noise cancelation. The on-chip transformer has the primary winding of a four-turn symmetric inductor and the secondary winding of a two-turn symmetric inductor. The transformer is designed using EM simulator and its size is 300 μm x 250 μm.

B. Noise and Distortion Cancelation in the Proposed LNA

Assuming that M_1 is the dominant source of distortion and noise, the output current (I_{OUT}) can be expressed as

$$
\begin{aligned}
I_{OUT} &= I_1 + I_2 = N_G \cdot I_{D1} + I_{D2} \\
&= N_G \cdot g_{m1} \cdot (V_{in} + V_n + V_{NL,1}) \\
&\quad + g_{m2} \cdot (V_{in} - V_n - V_{NL,1})
\end{aligned}
\tag{1}
$$

,where V_n and $V_{NL,1}$ represent the input referred noise of I_{n1} and all non-linear high order terms, respectively, and N_G is the current gain of the transformer. When $g_{m2} = N_G \cdot g_{m1}$, noise and distortion terms are removed and I_{OUT} can be described as

$$
I_{OUT} = 2 \cdot N_G \cdot g_{m1} \cdot V_{in}.
\tag{2}
$$

Therefore, the proposed LNA can cancel the thermal noise and the distortion generated by M_1.

C. Noise Figure and Effect of Coupling Coefficient on the Performance of the Proposed LNA

When the condition for noise cancelation ($g_{m2} = N_G \cdot g_{m1}$) is met, the noise factor of the proposed LNA can be described as

$$
F \approx 1 + \frac{\gamma}{4 \cdot R_s} \cdot \frac{(1 + g_{m1} \cdot R_s)^2}{g_{m1} \cdot N_G}
\tag{3}
$$

,where γ is the thermal noise factor due to the transistor channel, R_s is the source impedance.

It can be seen that increasing N_G is desirable for low noise figure (NF). However, increasing N_G results in larger power consumption and area since g_{m2} and L_1/L_2 must be increased.

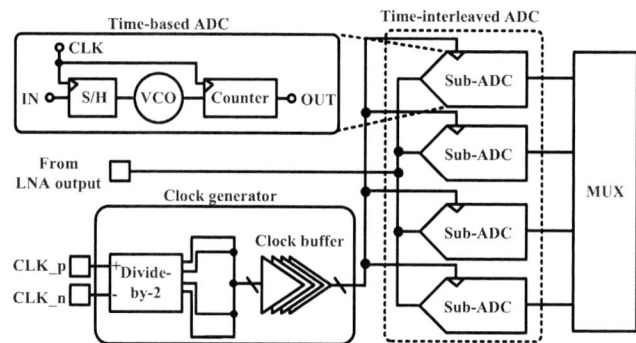

Fig. 4. Block diagram of the time-interleaved time-based ADC.

Considering this trade-off, we chose N_G as 2, which best optimizes the NF, power consumption, and area.

It is noteworthy that the proposed LNA is resilient to variation of the coupling coefficient. This is important as it is difficult to accurately predict K in the design stage. Based on the circuit simulation where nominal K is 0.7, it can be seen from Fig. 3, the variation of the voltage gain, the NF, and IIP_3 is 0.8 dB, 0.15 dB, and 0.9 dB, respectively, when K is changed by ± 14 %, from 0.6 to 0.8.

III. TIME-INTERLEAVED TIME-BASED ADC

The block diagram of the proposed time-based ADC is shown in Fig. 4. Each of sub-ADC consists of a sample and hold (S/H) circuit, a differential ring VCO, and a digital counter. As the linearity of the integrated output of the S/H is important, a passive S/H that consists of MOS switches and capacitor is used [4]. The ring VCO is based on six cascaded differential inverters with a single-ended delay control. Note that the ADC is highly digital as most building blocks are based on digital logics and switches. Moreover, opamps and voltage comparators are not used. The clock generator receives differential clock (F_{clk}) and generates 4 phase clocks ($F_{clk}/2$) for the sub-ADCs. Each interleaved channel operates at $F_{clk}/2$, thereby resulting in an effective sampling frequency of $2 \cdot F_{clk}$ ($F_s = 2 \cdot F_{clk}$).

In the proposed receiver, a direct RF sampling or mirror image sampling can be employed depending on the choice of sampling frequency. The number of zeros in the noise transfer

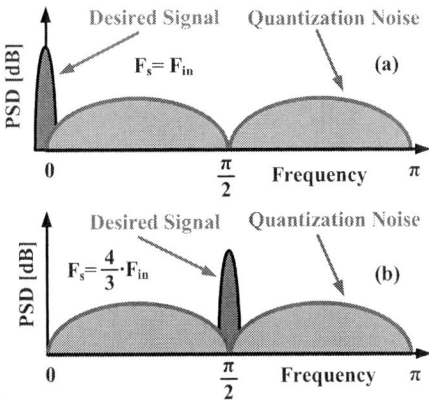

Fig. 5. (a) Output spectrum of the ADC when $F_s = F_{in}$ (b) Output spectrum of the ADC when $F_s = \frac{4}{3} \cdot F_{in}$.

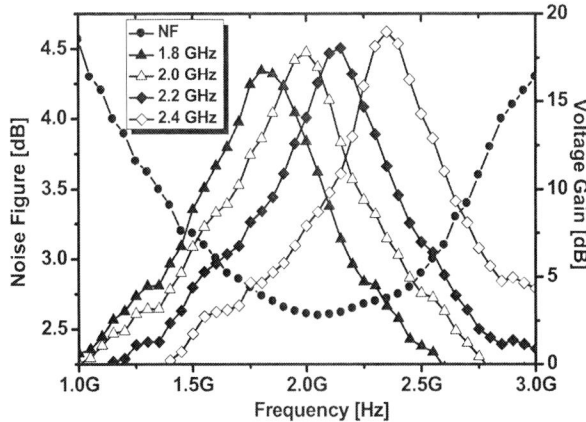

Fig. 7. Measured noise figure and voltage gain of the proposed LNA.

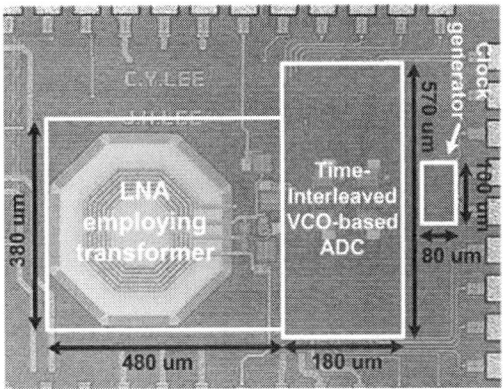

Fig. 6. Chip microphotograph of the proposed RF sampling receiver.

function (NTF) of the time-based ADC is determined by the number of the sub-ADCs [8]. In the proposed ADC, since the number of sub-ADCs is four, the NTF of the ADC has zeros at the discrete-time frequencies of 0, $\pi/2$, and π as shown in Fig. 5. Suppose that the input RF signal is at 2.4-GHz. When $F_s = F_{in} = 2.4$-GHz, the proposed receiver performs direct RF sampling and the output spectrum of the ADC is shown in Fig. 5-(a), where the desired signal is down converted to DC. In this case, additional ADC is needed for quadrature demodulation. When $F_s = 4/3 \cdot F_{in} = 3.2$-GHz, the receiver performs mirror image sampling and the output spectrum of the ADC is shown in Fig. 5-(b), where the desired signal exists around $\pi/2$ that corresponds to 800-MHz. Unlike the case of the direct RF sampling, the digital down-conversion is required and additional ADC for quadrature demodulation is unnecessary. Note that the digital down-conversion can be easily performed as the signal is located at $\pi/2$. This is in contrast to [4] where the signal is at $3\pi/8$, requiring heavy computation for the down-conversion.

IV. EXPERIMENTAL RESULTS

The proposed digital-intensive RF receiver is implemented in 90 nm CMOS process where the die photo is shown in Fig. 6. The active area of the proposed RF receiver is 0.3 mm^2. The measured voltage gain of the proposed LNA is shown in Fig. 7. It can be seen that the center frequency of the LNA can be changed from 1.8-GHz to 2.4-GHz according to the 2-bit control and the voltage gain varies from 16 dB to 18 dB. Fig. 7 shows the measured NF of the proposed LNA when the center frequency is 2.4-GHz. The difference among measured NF for different center frequencies was less than 0.1 dB and worst case is plotted in Fig. 7. It can be seen that the NF is less than 2.8 dB for the desired frequency range from 1.8-GHz to 2.4-GHz. The proposed LNA achieves IIP_3 of 4 dBm and OIP_3 of 22 dBm when two signals of 2.400-GHz and 2.403-GHz are applied for two tone test. The performance of the proposed LNA is summarized in Table I together with other LNAs that employ the noise-cancelation technique. The figure of merit is defined as $10 \log \frac{OIP_3(mW)}{F \cdot P_{DC}(mW)}$ [9], where OIP_3, F, and P_{DC} are the output referred IP_3, the noise factor, and DC power consumption, respectively. It can be seen that the proposed LNA has the largest FOM, implying that the on-chip transformer is very effective in improving the NF and linearity.

The measured output spectrum of the proposed RF sampling receiver with an input frequency of 2.401-GHz and a sampling frequency of 2.4-GHz is shown in Fig. 8 where nulls at DC and 600-MHz can be seen. Due to the non-linearity of the S/H, the ring VCO and the LNA, spurious tones are seen, which degrade the SNDR. The peak SNR and the SNDR at 1-MHz of bandwidth are 50 dB and 38 dB, respectively. The SNR and the SNDR vs. input power when the input bandwidth is 1-MHz and 10-MHz are shown in Fig. 9. It can be seen that the input sensitivity of the proposed RF receiver is -70 dBm when it is assumed that the required minimum SNR in the receiver is 0 dB at 1-MHz bandwidth. Fig. 10 shows the SNR and the SNDR vs. input power according to various sampling frequencies, where it can be seen that the peak SNR at 1-MHz of bandwidth is around $47 \sim 50$ dB. The performance of the proposed RF sampling receiver is summarized in Table II.

V. CONCLUSION

This paper presents a digital-intensive RF sampling receiver which consists of a noise-canceling bandpass LNA and a time-based ADC for multi-band wireless communication. The proposed LNA improves the noise figure and the linearity by using

978-1-4244-6240-7/10 $26.00 © 2010 IEEE

TABLE I

COMPARISON OF LNAS WITH NOISE CANCELATION TECHNIQUE

	[5]	[6]	[7]	This work
Frequency (MHz)	2 ~ 1600	20 ~ 1175	800 ~ 2100	1800 ~ 2400
Gain (dB)	13.7	20.5	14.5	16 ~ 18
NF (dB)	2.5	3.3	2.6	2.8
IIP_3 (dBm)	0	2.7	0	4
OIP_3 (dBm)	13.7	23.2	14.5	22
Power Consumption	14 mA @2.5V	18 mA @1.8V	11.6 mA @1.5V	4.5 mA @1.2V
Differential	NO	YES	NO	NO
F.O.M	-4.2	4.8	-0.5	11.9
Process Technology	0.25 μm CMOS	0.18 μm CMOS	0.13 μm CMOS	90 nm CMOS

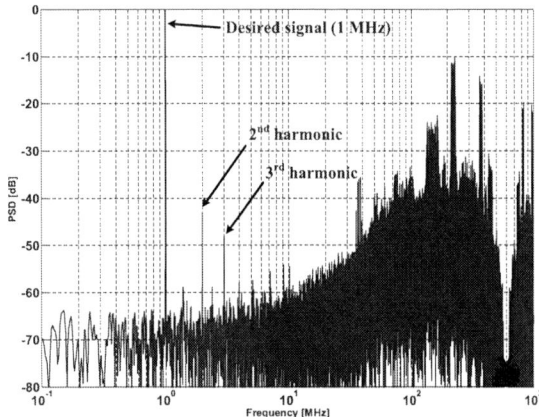

Fig. 8. Output spectrum of the proposed RF sampling receiver

TABLE II

PERFORMANCE SUMMARY OF PROPOSED RF SAMPLING RECEIVER

RF Frequency	1.8 GHz ~ 2.4 GHz
Sampling Frequency	1.8 GHz ~ 2.4 GHz
Peak SNR	50 dB @ 1 MHz, 37 dB @ 10 MHz
Peak SNDR	38 dB @ 1 MHz, 35 dB @ 10 MHz
IIP_3	0 dBm
Power Consumption	20 mW @ 1.2 V (Only I path)
Core Area	0.3 mm^2
Process Technology	90 nm CMOS

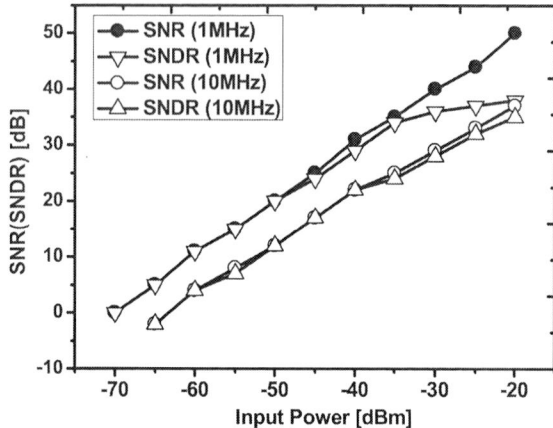

Fig. 9. SNR and SNDR vs. input power When F_{in}=2.401 GHz and F_s=2.4 GHz.

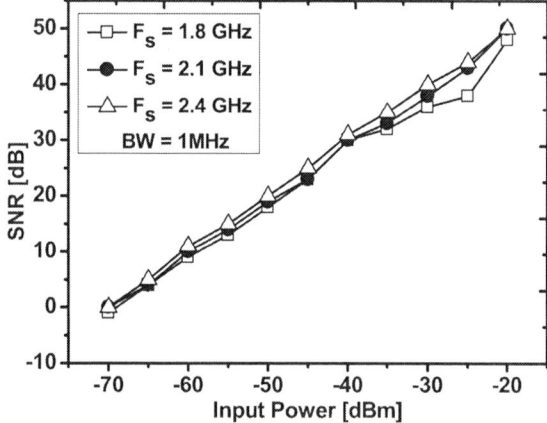

Fig. 10. SNR and SNDR vs. input power according to the sampling frequency

an on-chip transformer to combine the outputs of a CG and a CS LNA. The proposed ADC using time-interleaved VCOs significantly relaxes requirement of the down-conversion and decimation and offers direct RF sampling and mirror image sampling. The proposed RF sampling receiver can be a promising candidate for a multi-band multi-mode wireless receiver.

VI. ACKNOWLEDGMENTS

This work was supported by the Korea Science and Engineering Foundation (KOSEF) grant funded by the Korea government (MEST) under R11-2005-029-04-005-0.

REFERENCES

[1] B. K. Thandri and J. S. Martinez, "A 63 dB SNR, 75-mW Bandpass RF ADC at 950MHz using 3.8-GHz Clock in 0.25-um SiGe BiCMOS Technology," *IEEE J. Solid-State Circuits*, vol. 42, pp. 269–279, Feb. 2007.

[2] J. Ryckaert, J. Borremans, B. Verbruggen, L. Bos, C. Armiento, J. Craninckx, and G. V. der Plas, "A 2.4 GHz Low-Power Sixth-Order RF Bandpass $\Delta\Sigma$ Converter in CMOS," *IEEE J. Solid-State Circuits*, vol. 44, pp. 2873–2880, Nov. 2009.

[3] T. Chalvatzis, M. Repeta, and P. Voinigescu, "A Low-Noise 40-GS/s Continuous-Time Bandpass ADC Centered at 2 GHz for Direct Sampling Receivers," *IEEE J. Solid-State Circuits*, vol. 42, pp. 1065–1075, May 2007.

[4] Y.-G. Yoon and S. H. Cho, "A 1.5-GHz 63dB SNR 20mW Direct RF Sampling Bandpass VCO-based ADC in 65nm CMOS," in *Proc. Symposium on VLSI Circuits*, 2009, pp. 270–271.

[5] F. Bruccoleri, E. Klumperink, and B. Nauta, "Wide-band CMOS Low-Noise Amplifier Exploiting Thermal Noise Canceling," *IEEE J. Solid-State Circuits*, vol. 39, pp. 275–282, Feb. 2004.

[6] S. Song, D. Im, H.-T. Kim, and K. Lee, "A Highly Linear Wideband CMOS Low-Noise Amplifier Based on Current Amplification for Digital TV Tuner Applications," *IEEE Microwave Wireless Compon. Lett.*, vol. 18, pp. 118–120, Feb. 2008.

[7] W.-H. Chen, G. Liu, B. Zdravko, and A. M. Niknejad, "A Highly Linear Broadband CMOS LNA Employing Noise and Distortion Cancellation," *IEEE J. Solid-State Circuits*, vol. 43, pp. 1164–1176, May 2008.

[8] Y.-G. Yoon, J. Kim, T.-K. Jang, and S. H. Cho, "A time-based bandpass ADC using time-interleaved voltage-controlled oscillators," *IEEE Trans. Circuits Syst. I*, vol. 55, pp. 3571–3581, Dec. 2008.

[9] D. Linten, S. Thijs, M. I. Natarajan, P. Wambacq, W. Jeamsaksiri, J. Ramos, A. Mercha, S. Jenei, S. Donnay, and S. Decoutere, "A 5-GHz Fully Integrated ESD-Protected Low-Noise Amplifier in 90-nm RF CMOS," *IEEE J. Solid-State Circuits*, vol. 40, pp. 1434–1442, July 2005.

RMO4B-1

A Differential 4-Path Highly Linear Widely Tunable On-Chip Band-Pass Filter

Amir Ghaffari, Eric A.M. Klumperink, Bram Nauta

University of Twente, CTIT Institute, IC Design Group, Enschede, The Netherlands

Abstract — **A passive switched capacitor RF bandpass filter with clock controlled center frequency is realized in 65nm CMOS. An off-chip transformer which acts as a balun, improves filter-Q and realizes impedance matching. The differential architecture reduces clock-leakage and suppresses selectivity around even harmonics of the clock. The filter has a constant -3dB bandwidth of 35MHz and can be tuned from 100MHz up to 1GHz. IIP3 is better than 19dBm, P_{1dB}=2dBm and NF<5.5dB at P_{diss}=2mW to 16mW.**

Index Terms — **N-path filters, commutated capacitor, CMOS bandpass filter, inductorless, cognitive radio, software-defined radio.**

I. INTRODUCTION

Tunable RF filters have many applications in receivers, transmitters and synthesizers, e.g. to reject out-of-band interference, harmonics, and spurious tones. Although off-chip passive filters provide high rejection, low insertion loss and high linearity, integrated CMOS alternatives are highly desired for reasons of size, cost and programmability. Moreover, especially below 1GHz, filters based on on-chip inductors have limited Q resulting in significant insertion loss, while also consuming large die area. Q-enhanced techniques [1]-[3] can improve filter quality factor but degrade linearity and noise.

Tunable filters based on periodically time variant networks have been addressed in literature under different names such as N-path filters, sampled data filters, commutated capacitors, etc. [4]-[6]. These filters basically transfer the lowpass/notch characteristic of a network to a bandpass/bandstop one by means of frequency mixers. An N-path filter can realize an inductor-less tunable band-pass or band-stop filter [4] in which the center frequency is determined by the mixing frequency. These features make N-path filters interesting for software defined and cognitive radios in which tunable filters with a large frequency tuning range are highly wanted. While the concept of passive N-path filters and commutating networks is known for a long time and has been used for low frequencies [4], the concept seems to be somewhat forgotten. Recently, the concept was applied to high frequencies [7] where an 8-path filter with on-chip clocking was realized in CMOS.

In this paper we propose a 4-path differential bandpass filter. Compared to [7], the differential architecture suppresses selectivity around even harmonics of the mixing frequency. Moreover, input matching is provided by employing a wide-band RF transformer with termination resistor. The effects of switch resistance on the stop-band rejection and the tradeoffs between input matching, noise and achievable Q are addressed as well.

II. N-PATH FILTER

The basic block diagram of an N-path filter with bandpass characteristic is presented in Fig. 1 [4]. It is composed of N similar time invariant lowpass networks and 2N frequency mixers driven by time/phase shifted versions of clock p(t) and q(t). This architecture transfers a lowpass characteristic to a bandpass with the center frequency determined by the mixing frequency. In fact the input signal is downconverted to baseband, filtered and then upconverted again to the same band as V_{in}.

Now suppose that, as in Fig. 2a, we realize mixers with switches driven by multiphase clocks, while implementing the lowpass filters with simple RC networks. Since a resistor is a memory-less element, it can be shared by all paths and shifted in front of them. Moreover, the first set of switches can also implement the function of the second set of switches, if V_{out} is tapped between the resistor and switches (see Fig. 2b).

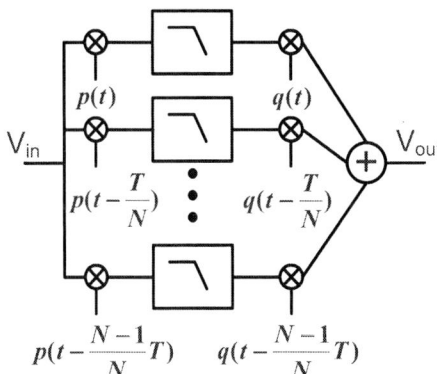

Fig. 1. Architecture of an N-path filter [4] (p and q are the mixing functions and T is the period of the mixing frequency).

978-1-4244-6240-7/10 $26.00 © 2010 IEEE

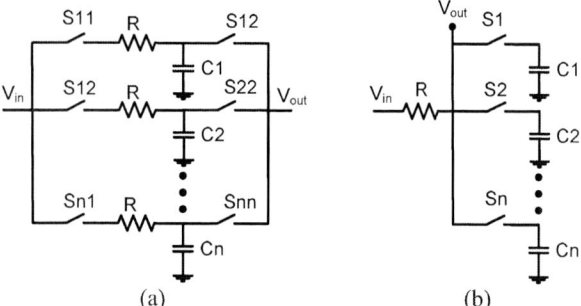

Fig. 2.   (a) Switched-RC N-path filter. (b) Simplified version.

As a matter of fact Fig. 2b not only represents a simple form of an N-path filter, but it can also be utilized as multiphase passive mixer, if the voltages on the capacitors are considered as output [8]-[9]. The RC time-constant determines the -3dB bandwidth and thus the quality factor (Q) of the filter for a given center frequency. Larger RC results in less bandwidth, and hence renders a higher Q. In Fig. 2b, R can be considered as source impedance, but it can also serve as an "auxiliary resistance" to increase Q as in [7]. However, adding resistance results in noise degradation. In this paper we will increase source impedance by using a transformer which (ideally) doesn't add noise.

III. PROPOSED ARCHITECTURE

The architecture of the proposed differential 4-path bandpass filter is illustrated in Fig. 3. Each path is differential-in and differential-out, but contains one grounded capacitor connected to two anti-phase driven NMOS switches. A 4-phase 25%-duty-cycle clock provides all required clocks (see Fig.3) and is derived from an external clock by dividers and logic. Considering the comb-like characteristic of N-path filters [4], i.e. its repetitive selectivity around harmonics of the switching frequency, the architecture of Fig. 3 cancels the even harmonics of the switching frequency due to anti-phase switching of the differential input. In fact for input signals around even harmonics of the clock frequency, no charge is stored on the capacitors in steady state and no upconverted signal appears at the output.

An off-chip wide-band (50-1000MHz) RF transformer serves as a balun for single to differential conversion. Moreover, it increases the impedance level seen by the switched-capacitor circuit, increasing Q of the filter without degrading its noise. Neglecting switch resistance, the input impedance R_{in} in Fig. 3, seen looking into the input of the IC before the switches at the switching frequency f_s, can be written as [8]:

$$R_{in} = \frac{8R_{out}}{\pi^2 - 8} \quad (1)$$

As shown in Fig. 3, R_{out} is the driving impedance seen by the switched-capacitor circuit. Using (1), the required value for R_M to provide matching can be found as:

$$R_M = \frac{4\pi^2 R_s}{16 - \pi^2} \quad (2)$$

For Rs=50Ω the value for R_M becomes 322Ω. In practice, the insertion loss of the transformer is usually non-negligible and in our case it was actually sufficient to implement R_M. In the general case an equivalent total resistance R_M according to (2) is needed for good S_{11}.

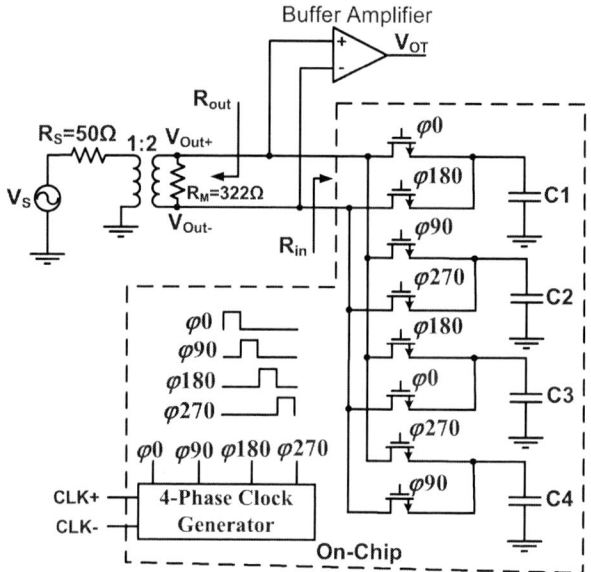

Fig. 3.   Filter architecture including a balun and buffer amplifier for measurements.

IV. MEASUREMENT RESULTS AND COMPARISON

A prototype IC has been fabricated in 65nm standard CMOS technology (Fig. 4). A differential high input impedance buffer amplifier is added to the circuit in Fig. 3 to be able to measure with 50Ω equipment without loading the output of the filter. Measurement results show that the tunable filter in Fig. 3 works from 100MHz up to at least 1GHz. Fig. 5 shows measurement results and compares them to a simulation employing an ideal transformer and R_M=322Ω for f_s=400MHz, while also including a bond-wire inductance estimate. The buffer amplifier gain is de-embedded in all experiments, but the transformer effects are included. Close to f_s the input is matched to 50Ω and the -3dB bandwidth is 35MHz rendering a Q ranging from 3 to 29 (0.1 to 1GHz).

978-1-4244-6240-7/10 $26.00 © 2010 IEEE   268

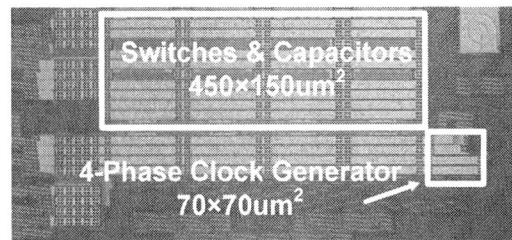

Fig. 4.   Micrograph of the 65nm CMOS chip.

The maximum filter rejection is limited by non-zero switch resistance and impedance R_{out}. For frequencies far away from the switching frequency, input impedance R_{in} can be approximated as two times the switch resistance R_{SW} [8]. The maximum rejection, α, can coarsely be estimated as:

$$\alpha \approx 20\log(\frac{\pi^2}{8}\frac{2R_{SW}}{R_{out}+2R_{SW}}) \qquad (3)$$

Thus increasing R_{out} not only results in less bandwidth and hence an increased Q, but also renders a larger maximum achievable filter rejection. More attenuation can also be achieved using wider switches at the cost of clock driver power. In the implemented architecture, $R_{SW}\approx5\Omega$, $R_{out}=123\Omega$, resulting in α=-20.6dB. Measurement results render -16dB (Fig. 5). The error is mainly due to the effect of the non-zero rise and fall times of the clock which is not considered in deriving (3).

Fig. 5 also illustrates the frequency selectivity around odd harmonics of the switching frequency. A rejection of 10dB is found around the 3rd harmonic of the clock, while downconversion from the 3rd harmonic to the desired band also represents 10dB attenuation. Even order harmonics are rejected due to the differential nature of the circuit. The differential architecture also reduces the power leakage from the switching clock to the RF input. In Fig. 3, the rising and falling edges of the clock mainly produce a common mode signal, which is suppressed by the common mode rejection of the transformer. At the RF input -62dBm was measured. The flexible tuning capability of the filter is illustrated in Fig. 6 for f_s swept from 100MHz up to 1GHz. In-band S_{11} proves to be better than -10dB for the whole tuning range and the voltage transfer characteristic exhibits a maximum of 2dB passband attenuation. Due to parasitics of the transformer and printed circuit board some peaking occurs at 100 and 200MHz center frequencies. The main frequency limitations of the current design are related to the clocking circuit and transformer. Wider frequency ranges are possible by improving the clocking circuit and removing the transformer for on-chip applications. The implemented

4-phase clock generator consumes between 2mW and 16mW (f_s=0.1-1GHz, f_{CLK}=0.4 to 4GHz).

Noise figure measurements have been done using a low noise buffer amplifier. As de-embedding of the transformer and buffer amplifier noise contribution is non-trivial, raw measured data are shown in Fig. 7. The NF including transformer and buffer amplifier is below 5.5dB. Two simulated curves are also shown in Fig. 7 both for an ideal transformer and ideal buffer amplifier and: 1) R_M=322Ω (matched); 2) no R_M (unmatched). The latter case renders about 1dB noise figure. The measured IIP3 is 19dBm and is rather constant over frequency (Fig. 8). In table I the design is compared with two other on-chip filters, one using Q-enhancement [3] and the other an 8-path filter [7], clearly illustrating benefits in tuning-range, linearity and noise. In [7] the achieved Q is increased significantly by increasing source resistance without providing matching and also increasing the number of paths to 8. Inserting a resistor deteriorates the NF significantly. While reactive impedance transformation which is employed in this paper ensures a low NF.

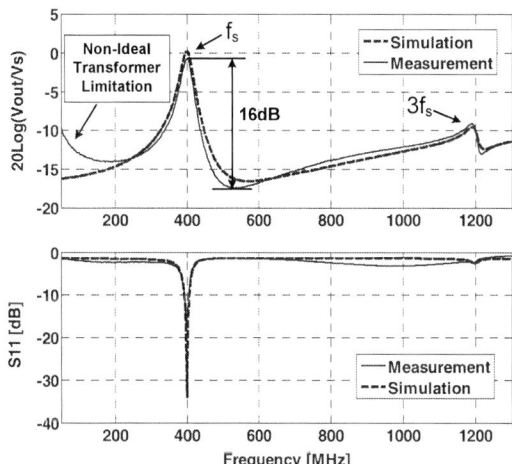

Fig. 5.   Frequency transfer and S_{11} at f_s=400MHz.

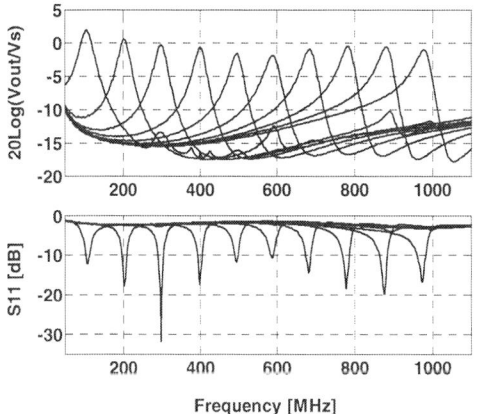

Fig. 6.   Frequency transfer and S_{11} at f_s between 0.1 and 1GHz.

978-1-4244-6240-7/10 $26.00 © 2010 IEEE

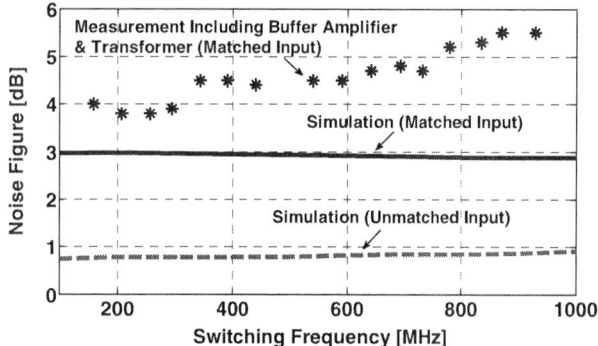

Fig. 7.   Measured and simulated noise figure.

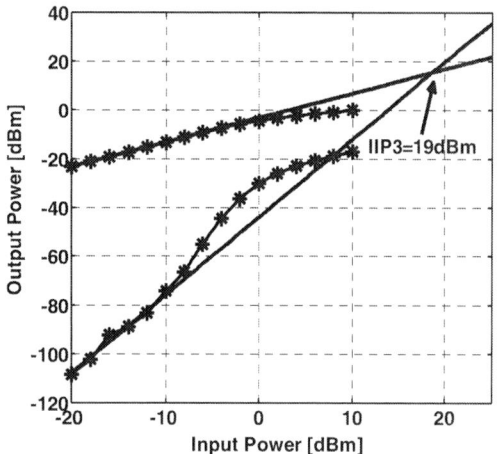

Fig. 8.   Measured IIP3.

TABLE I
COMPARISON WITH OTHER DESIGNS

Performance	This Work	[3]	[7]
Process	65nm CMOS	0.18um CMOS	0.35um CMOS
Active Area	0.07 mm^2	0.81mm^2	1.9mm^2
Power Consumption	2 to 16mW	17mW	63mW
Frequency Tuning Range	0.1 to 1GHz	2 to 2.06GHz	240 to 530MHz
-3dB Band Width	35MHz	130MHz	1.75 to 4.6MHz
Voltage Gain	-2dB	0dB	-2dB
Quality Factor (Q)	3 to 29	15.4 to 15.8	301 to 114
P$_{1dB}$	2dBm	-6.6dBm	-8dBm
IIP3	19dBm	2.5dBm	NA
Noise Figure	<5.5dB	15dB	9dB

V. CONCLUSION

In this paper an integrated tunable filter based on N-path periodically time variant networks is implemented and measured. The proposed differential 4-path architecture provides an inductor-less filter with a decade tuning range. The availability of high quality switches in CMOS technology offers high linearity while the simulation and measurement results confirm a low noise operation as well. Although the filter rejection is currently limited to 16dB, the flexible tunability and high linearity are attractive assets for software-defined or cognitive radio applications.

ACKNOWLEDGEMENTS

This work is funded by STW and we would like to thank M. Soer, G. Wienk and H. de Vries for their helpful contribution.

REFERENCES

[1] W. B. Kuhn, A. Nobbe, D. Kelly, A. W. Orsborn, "Dynamic range performance of on-chip RF bandpass filters," *IEEE Trans. Circuits Syst. II*, vol. 50, no. 10, pp. 685–694, Oct. 2003.

[2] X. He, W. B. Kuhn, "A 2.5-GHz low-power, high dynamic range, self-tuned Q-enhanced LC filter in SOI," *IEEE J. Solid-State Circuits*, vol. 40, no. 8, pp 1618-1628, Aug. 2005.

[3] B. Georgescu, I. G. Finvers, F. Ghannouchi, "2 GHz Q-enhanced active filter with low passband distortion and high dynamic range," *IEEE J. Solid-State Circuits*, vol. 41, no. 9, pp 2029-2039, Sep. 2006.

[4] L. E. Franks and I. W. Sandberg, "An alternative approach to the realization of network transfer functions: The N-path Filters," *Bell Sys. Tech. J.*, vol. 39, pp. 1321-1350, Sep. 1960.

[5] L. E. Franks and F. J. Witt, "Solid-state sampled data band-pass filters," *Proc. Solid-Slate Circuits Conf.* (Philadelphia, Pa.), Feb. 1960.

[6] R. Fischl, "Analysis of a commutated network," *IEEE Trans. Aerospace and Navigational Electronics*, vol. ANE-10, pp. 114-123, June 1963.

[7] A. El Qualkadi, M. El Kaamouchi, J. M. Paillot, D. V. Janvier, D. Flandre, "Fully integrated high-Q switched capacitor bandpass filter with center frequency and bandwidth tuning", *IEEE RFIC Sym.*, pp. 681-684, 2007.

[8] B. W. Cook, A. Berny, A. Molnar, "Low-power 2.4-GHz transceiver with passive RX front-end and 400-mV supply," *IEEE J. Solid-State Circuits*, vol. 41, no. 12, pp. 2757-2766, Dec. 2006.

[9] M.C.M. Soer, E.A.M. Klumperink, Z. Ru, F.E van Vliet, B. Nauta "A 0.2-to-2.0GHz CMOS receiver without LNA achieving >11dBm IIP3 and <6.5 dB NF," *ISSCC Dig. Tech. Papers*, pp. 222-223, Feb. 2009.

A CMOS Wide-Bandwidth High-Power Linear-in-dB Variable Attenuator Using Body Voltage Distribution Method

Yan-Yu Huang[1], Wangmyong Woo[2], Chang-Ho Lee[2], and Joy Laskar[1]

[1]Georgia Electronic Design Center, Georgia Institute of Technology, Atlanta, GA 30308, USA

[2]Samsung Design Center, Atlanta, GA 30308, USA

Abstract — **A wide bandwidth, highly linear variable attenuator designed in 0.18μm triple-well CMOS process is presented. This attenuator is based on three cascade π-networks with body voltage distribution scheme to minimize the effects of the input power levels. Measurements show it achieves minimum 1-dB gain compression of 7.5 dBm. The mid-band insertion loss is 1.6 dB and the maximum attenuation is 34.8 dB. This attenuator has a linear-in-dB controllability from 400 MHz to 3.7 GHz with input return loss better than 9 dB. To our knowledge, this is the highest linear CMOS variable attenuator with a wide bandwidth of 3.3 GHz.**

Index Terms — **attenuator, body voltage swing distribution, body-floating technique, CMOS, linear-in-dB**

I. INTRODUCTION

Circuits that are capable of varying the signal strength uniformly have been widely used in communication systems either for adjusting the input power of receivers or as a part of a correction/pre-distortion loop of transmitters. Although variable gain amplifiers (VGA) used to take over the rule, VGA's gain-varying mechanism makes it inferior in applications requiring high linearity, small power consumption, and wide operation frequency. Attenuators, on the other hand, are proved to have superior performance in such situations [1]. A recent work showed a complementary metal oxide semiconductor (CMOS) attenuator has an input 1-dB gain compression point (IP_{1dB}) of up to 2.5 dBm with a frequency range of 2.5 GHz [2]. However, this is still insufficient for many control operations in mobile system applications such as GSM, EDGE, and WCDMA, etc.

The demands on highly linear CMOS pre-distortion and correction circuits rise with fully integrated CMOS power amplifier (PA) modules, which gradually become feasible. Pre-distortion/correction circuits are often placed right in front of a PA, and a relatively high IP_{1dB} is typically required. The input power they need to handle may be more than 5 dBm, depending on the system.

Conventional CMOS attenuators are very linear in their minimum or maximum attenuation point, but are not in the transition range. The worst case linearity usually happens at the place close to maximum attenuation, where an attenuator is supposed to carry very high input power in a pre-distortion and correction circuits. This, therefore, defeats the design goal of using attenuators in mobile communication systems.

It is known that non-linear effects in CMOS attenuator result from parasitic capacitance and/or diodes in the physical structure of transistors [3].The proposed circuit employs a new body voltage distribution technique as well as adapts parasitic-suppressing methods of high power CMOS switches to a conventional linear-in-dB attenuator to maximize the linearity and bandwidth.

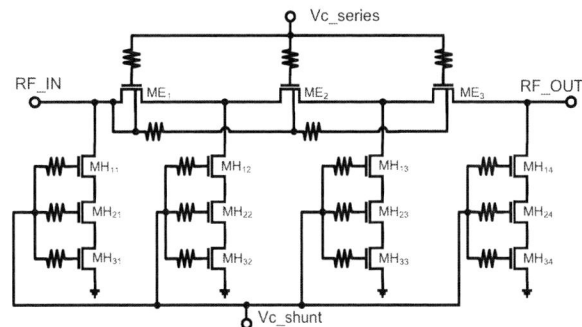

Fig. 1.   Circuit topology of the attenuator

II. ATTENUATOR DESIGN FOR POWER PERFORMANCE

Fig.1 shows the schematic diagram of the proposed attenuator topology. The main target of this design is to minimize the influences of input signals strength onto the impedance value of variable resistors, which commonly are composed of transistors and are the basic components in π-, T-, or bridged attenuators. Ideally, no matter how strong a signal it carries, a resistor should have a constant resistance value. However, the impedance of a transistor from its drain to source is not an independent function of the input voltage swing, resulting in nonlinearity.

This nonlinearity on drain to source impedance is most obvious when the gate voltage (V_G) is operating near threshold. Switching from low to high attenuation, an attenuator has three different power performance regions: (a) transistors are fully ON or OFF, (b) V_G of shunt branches are close to the threshold and (c) V_G of series transistors are close to the threshold. In region (a), attenuators can be seen as simple switches. Transistors in this region are very close to ON or OFF state, and have good linearity up to watt level [4]. The bottleneck is on (b) and (c), where transistors are operating in between cut off and triode region. A large voltage swing exacerbates this transition, making a rapid parasitic value change within a signal cycle.

Multi-stack shunt branches and series transistor with body voltage distribution are incorporated in this design to reduce the parasitics as well as their dependency on input signal, so that the power handling capability of attenuators can be significantly increased.

A. Multi-Stack CMOS with Floated Body

Transistors of shunt branches shown in Fig. 1 are body floated and fabricated using a deep n-well CMOS process. With the multi-stack connection, the voltage swing will be equally divided into whatever number of identical transistors stacked in a branch, and therefore relieve the voltage stress across each drain and source (V_{DS}) [5]. Reducing the swing of V_{DS} helps reduce the variation of drain-to-source resistance (R_{DS}) of a transistor during the signal cycle. With a relatively fixed R_{DS}, attenuators can maintain their voltage/current gain and has better linearity.

Input impedance is not merely determined by R_{DS} but also by various parasitic capacitors/diodes that exist between junctions of different types of doping, and non-ideal effects like latch up. A large transistor with an independent body bias can avoid latch-up problem results from the closely integrated n-MOS and p-MOS; while a VDD-connected deep n-well keeps junction diodes between p-well, deep n-well and substrate reversely biased and induces very small amount of unwanted current [5]. Besides, parasitics such as C_{BD} and C_{BS} are basically a function of V_{BD} or V_{BS}; if a large resistor is put to bias the body, as shown in Fig. 2, then most parts of the voltage signal that originally across C_{BD} and C_{BS} will now carried

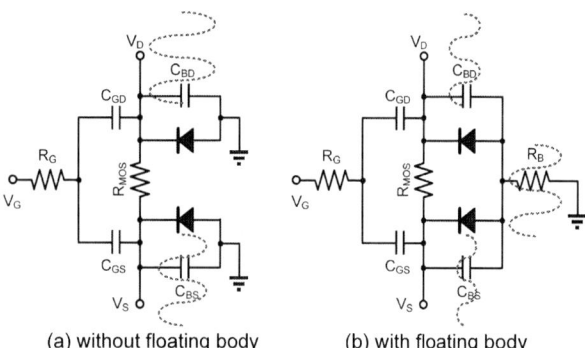

(a) without floating body   (b) with floating body

Fig. 2.   Voltage swing across parasitic capacitors and diodes

by R_B. This configuration makes C_{BD} and C_{BS} less susceptible to input signal strength.

Methods describe above are widely used in switch designs. According to the simulation results, they are also effective on improving power linearity especially when an attenuator operates in region (b). However, in order to keep the minimum insertion loss of an attenuator as small as possible, the multi-stack topology is not preferable in designing series variable resistors. Multiple transistors in series not only increase the ON resistance on an attenuator, but also add extra parasitic capacitance to ground. Since both of them contribute to extra losses, a linearization method with no extra transistors is required.

B. Body Voltage Distribution

The main concept of body voltage distribution is to build a secondary signal path to the body of series transistors, so that a large voltage swing could be evenly distributed across the doping junctions of a transistor.

For an attenuator operating in region (c), its series transistors have very high drain to source resistance and are close to OFF state. Most of the input signal will, therefore, go through the secondary path. Because this path is paved with discrete components, the body voltage swing will have less dependency on parasitics, for example, junction capacitors C_{DB} and C_{SB}. Based on a Metal Oxide Semiconductor Field Effect Transistor (MOSFET) having equally sized drain and source, the parasitic related to C_{DB} and C_{SB} in triode and cut off region can be obtained as:

$$C_{SB} + C_{DB} = K \times \left[(1 + \frac{V_{SB}}{\Phi_0})^m + (1 + \frac{V_{DB}}{\Phi_0})^m \right] \quad (1)$$

In equation (1), K represents a constant depends on the process technology, V_{SB} is the voltage difference between source and body, Φ_0 is the bulk junction potential and m is the bulk junction grading coefficient. For common m values (1/2 or 1/3), $C_{SB}+C_{DB}$ has the minimum value when $V_{SB}=V_{DB}$. That is, if the body voltage level can be kept in the middle of V_D and V_S, the body parasitic capacitance value and its consequent effect on degrading the linearity will be minimized. Similarly, the parasitic diodes of a MOSFET have the smallest forward conduction cycle with the same body bias point. According to this idea, the impedance values of secondary path will then be optimized to achieve maximum linearity.

Fig. 3 shows three different methods that can be applied to series branch transistors. In (a), a large resistor is used to bias the body, making it AC floated. Topology (b) is widely used in CMOS switch designs to increase the linearity at ON state; while in the proposed topology (c), a resistor path is added to make V_B close to a half of V_{DS}.

978-1-4244-6240-7/10 $26.00 © 2010 IEEE   272

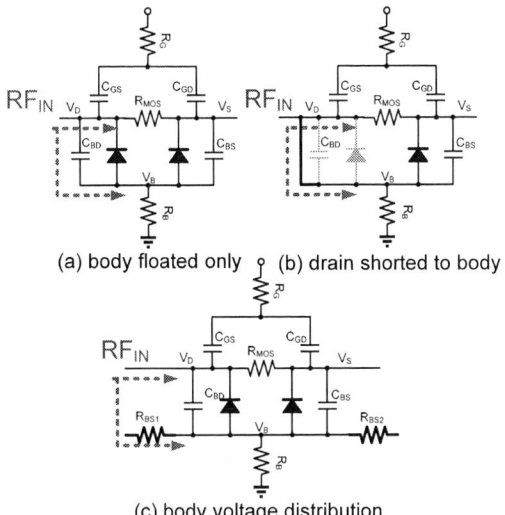

(a) body floated only   (b) drain shorted to body

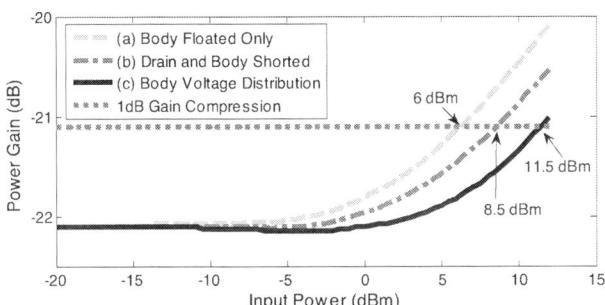

(c) body voltage distribution

Fig. 3.   Equivalent schematic diagrams of transistors with different body connections

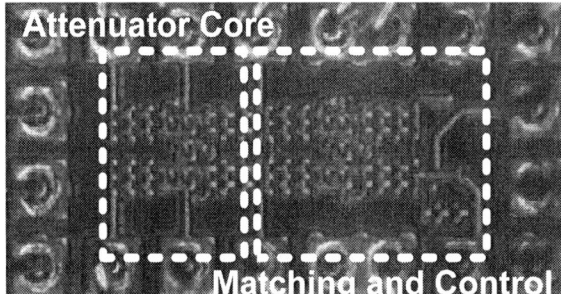

Fig. 5.   Die photograph of the attenuator

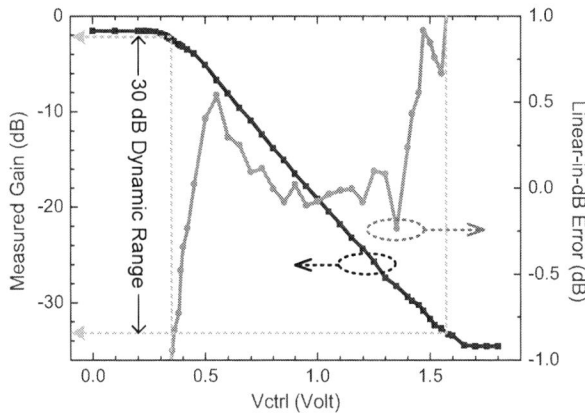

Fig. 6.   Measured gain-control curve and its linear-in-dB error

Fig. 4.   Performance comparison of three different topologies in terms of gain compression points

Their performance in terms of power gain compression points are compared, as shown in Fig. 4. Transistors in all three topologies have equal size and threshold voltage. The simulation results show that with a shorted drain-body connection (b), the input referred 1-dB power gain compression point (IP_{1dB}) of the attenuator is about 2.5 dB better than the body floated topology (a), while the body voltage distribution method (c) can increase the IP_{1dB} value with another 3 dB.

III. MEASUREMENT RESULTS

The attenuator core along with control circuits and I/O pads were fabricated on a $750 \times 375\ \mu m^2$ rectangle, as shown in Fig. 5. The attenuation control mechanism is based on the constant reference voltage method [6], which appears to have the best linear-in-dB nature. The die was mounted on a printed circuit board for a chip-on-board measurement. The control signal (V_{ctrl}) was applied externally from 0 (GND) up to 1.8 volts (VDD).

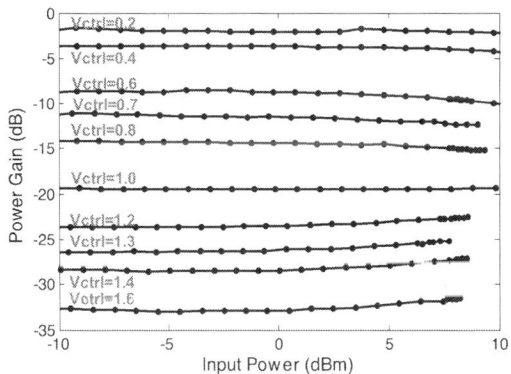

Fig. 6 shows the gain-control curve measured at 1.95 GHz. This attenuator displays 30 dB of dynamic range with ±1 dB linear-in-dB error. Fig.7 shows how input power affects the gain at different control voltages. The worst case IP_{1dB} occurs when the control voltage is between 1.2 and 1.3 volts. The measured IIP3 value at this point was 17.8 dBm.

Fig. 8 shows the worst case S_{11} and S_{22} values over the control voltages. It was measured within a frequency range of 400 MHz to 3.7 GHz. Both the worst S_{11} and S_{22} values occur at a low attenuation setting with a high signal frequency. S_{22} and S_{11} values are different due to the asymmetrical body connection.

TABLE I
COMPARISON BETWEEN CMOS ATTENUATORS

	Kaunisto [1]	Dogan [2]	This work
Technology	0.8-μm CMOS	0.13-μm CMOS	0.18-μm CMOS
Frequency	DC-900 MHz	DC-2.5 GHz	400 MHz-3.7 GHz (range with linear-in-dB control curve)
Minimum Attenuation	3.3 dB	0.9-3.5 dB	0.96~2.91 dB
Max. Attenuation Range	28 dB	42 dB	33 dB (30 dB with ±1 dB linear-in-dB error)
Return Loss	>-12 dB	>-8.2 dB	>-9 dB
Insertion loss flatness	N/A	2.6 dB (DC-2.5 GHz)	2.6 dB (400 MHz-3.7 GHz)
1dB Compression point	5 dBm	2.5 dBm	7.5 dBm

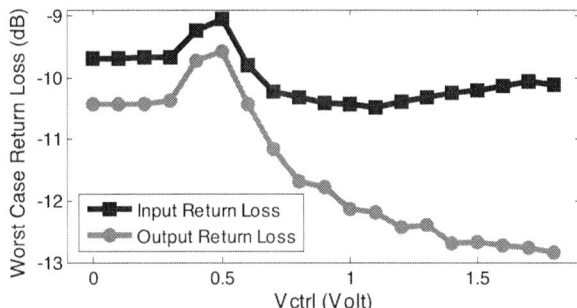

Fig. 8.   Return loss vs. input power at different control voltage

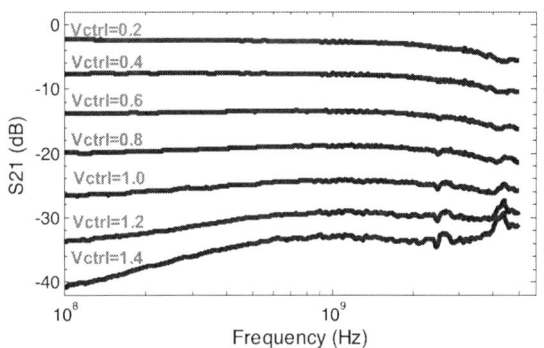

Fig. 9.   Frequency response at different control voltage

Fig. 9 is the frequency response of S_{21} at different control voltages. At the high attenuation range, the resistance values from input to output are large, but the parasitic capacitance between body and source of the last series transistor opens another door for high frequency signals. That makes the attenuator has a "high-pass like" frequency response at the high attenuation region.

The overall current consumption of this circuit was less than 2 mA with a supply voltage of 1.8 V. All of the power was consumed by the matching and control blocks. This current value is determined by the desired slew-rate of an operational amplifier and can be optimized based on the bandwidth requirement of the control signal.

IV. CONCLUSION

In this paper, a very high power linear-in-dB voltage-controlled attenuator was demonstrated. Body voltage distribution and multi-stack methods are applied to body floated transistors to suppress parasitic effects. It was fabricated in a standard CMOS 0.18μm deep n-well process and exhibited an attenuation dynamic range of 33 dB, a bandwidth of 3.3 GHz and a minimum IP_{1dB} of 7.5 dBm. As the measurement results indicate, the proposed methods effectively improve the high power performance as well as the operating frequency, and hence make this attenuator applicable to high power pre-distortion and correction circuits in CMOS-based highly linear transmitter systems.

REFERENCES

[1] R. Kaunisto, P. Korpi, J. Kiraly, and K. Halonen, "A linear-control wideband CMOSattenuator," in *Proc. IEEE ISCAS 2001*, Sydney, Australia, 2001, vol. 4, pp. 458–461.

[2] H. Dogan, R. G. Meyer, and A. M. Niknejad, "Analysis and Design of RF CMOS Attenuators," *IEEE J. Solid-State Circuits*, vol. 43, no. 10, pp. 2269–2283, Oct. 2008.

[3] H. Dogan and R. G. Meyer, "Intermodulation distortion in CMOS attenuators and switches," *IEEE J. Solid-State Circuits*, vol. 42, no. 3, pp. 529–539, Mar. 2007.

[4] M. Ahn, C.H. Lee, B.-S. Kim, and J. Laskar, "A High-Power CMOS Switch Using A Novel Adaptive Voltage Swing Distribution Method in Multistack FETs," *IEEE Trans. Microw. Theory Tech*, vol. 56, no. 4, pp. 849–858, April. 2008.

[5] M.-C. Yeh, Z.-M. Tsai, R.-C. Liu, K. Y. Lin, Y.-T. Chang, and H.Wang, "Design and analysis for a miniature CMOS SPDT switch usingbody-floating technique to improve power performance," *IEEE Trans.Microw. Theory Tech.*, vol. 54, no. 1, pp. 31–39, Jan. 2006.

[6] Y. Araki, T. Hashimoto, and S. Otaka, "A 0.13μm CMOS 90dB Variable Gain Pre-power Amplifier using Robust Linear-in-dB Attenuator," *IEEE RFIC. Symp. Dig.*, pp. 673–676, Jun. 2008.

RMO4B-3

A 17 GHz Transformer-neutralized Current Re-use LNA and Its Application to a Low-power RF Front-end

Sandipan Kundu, Jeyanandh Paramesh

Carnegie Mellon University, Pittsburgh, PA 15213, USA

Abstract — **A 17 GHz current re-use low noise amplifier (LNA) is designed in 0.13 µm CMOS for low power applications such as wireless sensor networks. The LNA also employs transformer based feedback to neutralize the gate-drain capacitance of a MOSFET. The LNA achieves 15.4 dB gain into a 50 Ω load along with 1.9 GHz bandwidth. It features 4.5dB NF and −12 dBm IIP3 while consuming 7.8 mW of power. A 17 GHz receiver frontend using a similar two-stage LNA and a mixer is also demonstrated which achieves 25 dB of voltage conversion gain at 70 MHz IF, 7 dB NF, −18 dBm IIP3 and consumes 8 mW from a 1.2 V supply.**

Index Terms — **CMOS RF frontends, low-noise amplifier (LNA), neutralization.**

I. INTRODUCTION

Two popular approaches to the design of ultra low-power RF front-ends with high gain are: (1) horizontally cascading gain stages with reduced supply voltage and (2) vertically cascading complementary gain stages [1] with bias current re-use at the nominal supply voltage. In both cases, the limited voltage headroom per gain stage limits the stack in each gain stage to a single gain transistor. Unfortunately, this approach suffers from stability and design issues due to the reverse signal flow through the gate-drain overlap capacitance (C_{gd}) of the MOS device. Since C_{gd} becomes large and comparable to the gate-source capacitance in nanoscale devices, the above issues become increasingly severe at high frequencies.

This paper demonstrates the effectiveness of transformer feedback [2] to neutralize C_{gd} in complimentary vertically stacked 17 GHz LNA. Originally proposed in [2], this technique is developed further here by deriving additional design guidelines for simultaneous noise and conjugate matching while maintaining good reverse isolation, based on rigorous mathematical analysis. This permits a better understanding of the design trade-offs. Note that this technique is equally applicable to the horizontal cascade approach. This work targets mm-wave radar, wireless sensor networks and other low-power wireless applications.

II. CURRENT RE-USE IN LNA

A. Circuit Considerations

To reduce power it is desirable to share bias current between different stages of a circuit. A conventional cascode LNA (Fig. 1(a)) can be viewed as a cascaded common source common gate (CS-CG) two-stage amplifier, where the two stages share the same bias current. Since the input impedance of a CG stage is inherently low compared to a CS stage, the load impedance and thus the gain of the CS stage are reduced. One way to increase the overall gain of the LNA is to replace the second CG stage with another CS stage as shown in Fig. 1(b). A virtual ground is created by placing a large bypass capacitor at the node A. This circuit consumes the same power as the cascode but offers higher gain. Alternatively, for same gain, the circuit of Fig. 1(b) uses less power than that of Fig. 1(a).

Another variable which affects power is the supply voltage. Low voltage operation is not possible in Fig. 1(b) due to the stack of NMOS transistors. A possible solution is to use the topology shown in Fig. 1(c) [1] where the second NMOS CS stage is replaced with a PMOS CS stage. In this CMOS amplifier, the minimum required supply voltage is reduced by one transistor overdrive compared to the two previous topologies. Alternatively, for similar supply voltage Fig. 1(c) can accommodate larger output voltage swing. The overall gain can be further increased by adding an additional PMOS CS stage [1] as shown in Fig. 1(d).

In Fig. 1(c) and Fig. 1(d), if the PMOS and the NMOS stages are independently biased, the bias voltage at node A

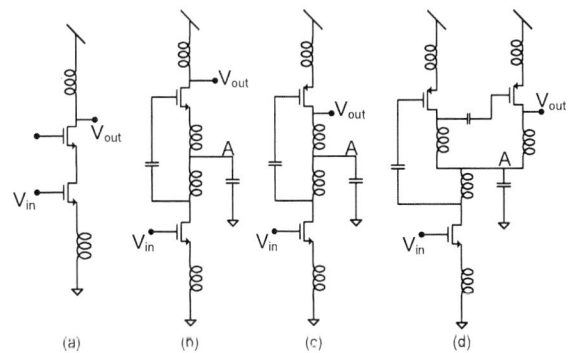

Fig. 1. LNA configurations: (a) Cascode (b) Two NMOS CS stages (c) Two (NMOS-PMOS) stage (d) Three (NMOS-PMOS-PMOS) stage. Biasing not shown.

978-1-4244-6240-7/10 $26.00 © 2010 IEEE   275

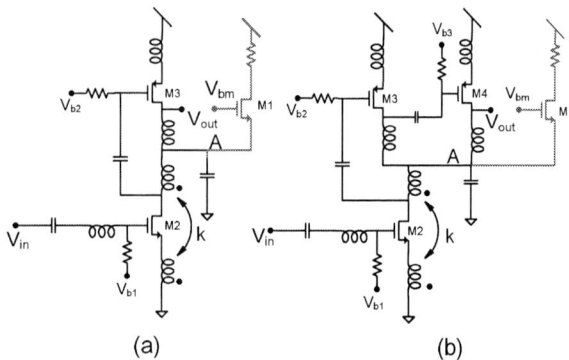

Fig. 2.   (a) LNA used in the receiver, (b) Standalone LNA

would suffer from high sensitivity. To alleviate this, an alternate path for current is inserted as shown in gray in Fig. 2. Any excess current flows through M1, thereby reducing the sensitivity of the node A with respect to PVT and bias variations. To evaluate the effect of these variations, Monte Carlo simulations were performed with a sample size of 200. The standard deviation of the DC voltage at A is found to be 1.5% of VDD with all transistors operating in saturation with adequate margin.

B. Reverse Signal Flow

The high gain in the first stage degrades the input matching of the LNA since no unilateralization or neutralization is used in Fig. 1(c) or Fig. 1(d). The gate-drain capacitance (C_{gd}) provides an additional path between the input and the output of each stage, which allows reverse signal flow. In terms of two port Y parameters, this implies $Y_{12} \neq 0$. The input admittance Y_{in} of such a network can be written as

$$Y_{in} = Y_{11} - \frac{Y_{12} Y_{21}}{Y_{22} + Y_L} \qquad (1)$$

where $Y_L = 1/Z_L$ is the load admittance. For a non-cascoded source degenerated CS amplifier (as in Fig. 3(a) with $M=0$), neglecting r_o, $Y_{in} = 1/Z_{in}$ can be written as

$$Y_{in} = Y_{in,0} + \frac{1 + \left(g_m / Y_{L,a}\right)K}{1/Y_{L,a} + 1/Y_F} \qquad (2)$$

where $Y_{L,a} = 1/(Z_L \| sL_d) = Y_L + (1/sL_d)$ is the admittance of the output tank at the drain, Y_F is the admittance of C_{gd}, $Y_{in,0}$ is the input admittance ignoring C_{gd}, and is given by

$$Y_{in,0} = \frac{1}{Z_{in,0}} = \frac{1}{1/j\omega C_{gs} + j\omega L_s + g_m L_s / C_{gs}} \qquad (3)$$

The factor K is defined as

$$K = \frac{Y_{in,0}}{j\omega C_{gs}} = \frac{1}{1 - \omega^2 L_s C_{gs} + j\omega g_m L_s} \qquad (4)$$

Since Y_F is often small compared to of $Y_{L,a}$, we can rewrite (2) as

$$Y_{in} \approx Y_{in,0} + \frac{1 + \left(g_m / Y_{L,a}\right)K}{1/Y_F} = Y_{in,0} + Y_F + \frac{g_m Y_F}{Y_{L,a}}K \quad (5)$$

The implication of (5) is two-fold. First, since the value of $Y_{L,a}$ is determined by output resonant frequency, and since $Y_{L,a}$ affects the last term of (5), it is seen that the input admittance, and hence input tuning (frequency at which input is best matched) is dependent on output tuning (reverse signal flow). Similarly deriving the expression for output admittance (Y_{out}), it can be shown that output tuning is also dependent on input tuning. So, it becomes difficult to tune both input and output simultaneously. Had there been no C_{gd} ($Y_F = 0$), the last term in (5) would vanish and the input tuning would be independent of output tuning.

The second implication of (5) arises from the fact that $Y_{L,a}$ is inductive and thus has a negative imaginary part below the output resonant frequency. Therefore, the ratio $Y_F / Y_{L,a}$ will be negative real. Since the input and output resonant frequencies are generally close by design, K will have positive real part below resonance (since $1 - \omega^2 L_s C_{gs} > 0$). Thus, below resonance, the last term of (5) will have a negative real part which counteracts the positive real part of $Y_{in,0}$ – this can potentially cause the total real part of Y_{in} to become negative thus rendering the amplifier unstable. Furthermore, the real part of Z_{in} will be frequency dependent as opposed to real part of $Z_{in,0}$ ($= g_m L_s / C_{gs}$).

The gray curves in Fig. 4 show the deleterious effects of C_{gd} on the input matching. For most of the lower half of the band, S_{11} is seen to be poor and then sharply improves.

III. TRANSFORMER FEEDBACK

To alleviate problems arising from reverse signal flow, the amplifier must be neutralized or unilateralized. One way of doing this is to couple the drain and the source inductors of the first stage; this was introduced in [2] as shown in Fig. 2. This coupling introduces an additional feedback path from the drain to the input of the first stage and can be used to reduce or nullify the reverse signal flow. This method is particularly well-suited in the current re-use LNA since it does not necessitate extra voltage headroom (unlike cascade unilateralization), a large inductor (unlike resonant neutralization) or a differential topology (unlike cross-coupled capacitive neutralization).

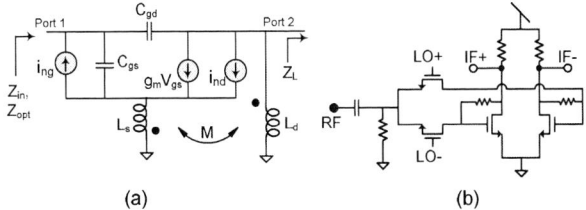
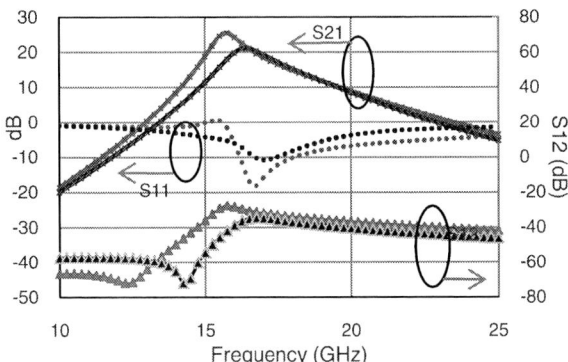

Fig. 3. (a) Small signal model of transformer feedback CS stage with MOSFET noise sources (b) Mixer circuit

Unlike [2], explicit equations will be derived here for simultaneous noise and input matching while maintaining good reverse isolation. From (1), it is clear that for Y_{in} to be independent of Y_L, Y_{12} must be zero. The necessary condition for this can be written using the small-signal model of Fig. 3(a):

$$Y_{12} = sC_{gd} - \frac{sC_{gs}n}{s^2 C_{gs} L_s \left(1-k^2\right) + sg_m L_s \left(1-k^2\right)+1} = 0 \quad (6)$$

where $n = M/L_d$ and $k = M/\sqrt{L_s L_d}$ is the coupling factor of the transformer. (6) is satisfied when

$$C_{gd}/C_{gs} = n \quad (7)$$

and

$$k = 1 . \quad (8)$$

The ratio of C_{gd} to C_{gs} is approximately 1/3 in the process used here. (7) is the same as the conclusion found in [2], (8) is, however, an additional constraint and needs further discussion. On the other hand, substituting $Y_{12}=0$ in (1) and using (6), the input impedance Z_{in} (see Fig. 3(a)) is

$$Z_{in} = 1/Y_{11} = \frac{1}{sC_{gs}} \frac{1}{1+n} + sL_s \frac{1}{1+n}\left(1-k^2\right) \\ + \frac{g_m L_s}{C_{gs}} \frac{1}{1+n}\left(1-k^2\right) \quad (9)$$

(9) shows that for perfect coupling as in (8), Z_{in} will not have any real part hence input matching is not possible. So, k has to be less than 1. To keep expressions simple, we will assume (9) is approximately true when $Y_{12} \neq 0$. For conjugate match it can be shown that

$$L_s = \frac{\mathrm{Re}[Z_s]C_{gs}\left(1+n\right)}{g_m\left(1-k^2\right)} \quad (10)$$

where Z_s is the impedance looking back from the gate towards the input. (10) shows that required L_s increases with increase in k, everything else remaining constant. Higher valued inductors however show poorer Q which

Fig. 4. Simulated effect of transformer feedback on S-parameters of the three-stage stand-alone LNA, gray: without transformer black: with transformer

gives rise to higher resistive source degeneration thereby decreasing gain.

We now turn to the noise analysis of the LNA. Assuming that most of the noise comes from the transformer neutralized first stage, and using the small-signal noise equivalent circuits of Fig. 3(a), the optimum source impedance $(R_{opt}+jX_{opt})$ for minimum noise figure can be calculated as

$$X_{opt} \approx -j\omega L_s \frac{\left(1-k^2\right)}{1+n} \frac{(1+n)+\alpha^2 \chi^2 - |c|\alpha\chi(2+n)}{(1+n)^2 + \alpha^2\chi^2 - 2|c|\alpha\chi(1+n)} \\ - \frac{1}{j\omega C_{gs}} \frac{1+n-|c|\alpha\chi}{(1+n)^2 + \alpha^2\chi^2 - 2|c|\alpha\chi(1+n)} \quad (11)$$

$$R_{opt} \approx \frac{1}{\omega C_{gs}} \frac{\alpha\chi\sqrt{1-|c|^2}}{(1+n)^2 + \alpha^2\chi^2 - 2|c|\alpha\chi(1+n)} \\ \times \left(1-\omega^2 n C_{gs} L_s \frac{1}{1+n}\left(1-k^2\right)\right) \quad (12)$$

where c, the correlation coefficient between gate and drain noise current, is purely imaginary and negative. $\alpha = g_m/g_{do}$ and $\chi = \sqrt{\delta/5\gamma}$. δ and γ are the gate and drain noise coefficients respectively. F_{min} now can be written as

$$F_{min} \\ = 1 + \frac{2}{\sqrt{5}} \frac{\omega C_{gs}}{g_m} \sqrt{\gamma\delta} \left[\sqrt{1-|c|^2} \left(1 - \omega^2 n C_{gs} L_s \frac{1-k^2}{1+n}\right) \\ + \frac{\omega C_{gs} n}{g_m(1+n)}\left(|c|(1+n)-\alpha\chi\right) \right] \quad (13)$$

Several approximations used in deriving (11), (12) and (13) including $|\omega C_{gd}/g_m| \ll 1$. Inspection of (9) and (11) reveals that X_{opt} is approximately equal to complex

978-1-4244-6240-7/10 $26.00 © 2010 IEEE   277

conjugate of the imaginary part of Z_{in}, a step closer to achieving simultaneous power and noise match.

(13) shows that minimum noise figure is achieved when $k = 0$. From (10), we can see for minimum L_s, $k = 0$ is needed. However, for achieving best reverse isolation we need $k = 1$ as in (8). As a compromise with these conflicting requirements, k is chosen around 0.5.

Fig. 4 shows a simulated ~10dB improvement in reverse isolation along with a well behaved input matching due to the use of transformer. This allowed the removal of source inductors from subsequent stages to save area. Fig. 2 also shows the final circuit diagram of the LNAs.

IV. IMPLEMENTATION AND MEASUREMENT RESULTS

Two versions of the transformer-neutralized current re-use LNA are designed in 0.13 μm CMOS technology: (a) a three-stage stand-alone LNA, with the third stage directly driving the 50Ω measurement load, and (b) a 17 GHz receiver front-end with a two-stage LNA. A stand-alone first stage of the LNA along with a 50Ω buffer is also implemented. The downconversion mixer uses a single-balanced "near-passive" topology followed by an IF amplifier (Fig. 3(b)). Resistive feedback, used to bias the IF amplifier, also limits the signal swing across the nonlinear parasitic capacitance of the LO devices (switches), thus improving linearity.

The primary of the transformer is realized as a 2.5 turn inductor in the top metal layer (M8) and the secondary as a 1.5 turn inductor in the next metal layer (M7). Half turn inductors are chosen for the transformer to get the terminals on opposite sides, thus reducing the length of any un-modeled RF path for the drain-source AC current. The coupling factor is set by adjusting the distance between the two centers of the inductors. EM simulation is used to extract all RF interconnects and passives including the transformer. This ensures all the parasitics in AC return paths are accounted for.

Fig. 5 shows that the three-stage LNA achieved reverse isolation of 34 dB or better. It also shows the IIP3 inferred

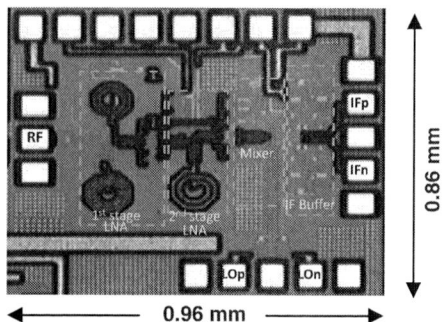

Fig. 6.   Chip micrograph of the receiver

from a two tone test on the receiver. Table 1 shows measured performance summary of the receiver and the LNAs.

TABLE I

MEASURED PERFORMANCE SUMMARY

	2-stage LNA + Mixer	3-stage LNA w/o Buffer	1-stage LNA w Buffer
Centre freq. (GHz)	16.4	16.5	14.5
Bandwidth (GHz)		1.9	5.2
Gain (dB)	>25 [1]	15.4 [2]	12.5 [2]
Noise Figure (dB)	7 [3]	4.5	3.3
S_{11} (dB)	<-10	-12	<-11
IIP3 (dBm)	-18	-12	-3
Supply Voltage (V)	1.2	1.3	0.55
Power (mW)	8	7.8	2.9

[1]Voltage conversion gain, [2]S_{21}, [3]DSB

V. CONCLUSION

In this paper, a current-reuse multistage 17GHz LNA with a transformer-neutralized first stage is presented. The inductors are coupled to neutralize the first stage. A variant of this LNA is employed in a low-power receiver front-end. Measurements on a 0.13 μm CMOS prototype showed 25 dB of voltage conversion gain with 7 dB noise figure (DSB) while sinking 6 mA current from a 1.2 V supply. A stand-alone version of the LNA achieves 15.4 dB gain and 4.5dB noise figure.

ACKNOWLEDGEMENT

This work was supported in part by a gift from Intel.

REFERENCES

[1] H. Hseih and L. Lu, "Design of Ultra-Low-Voltage RF Frontends With Complementary Current-Reuse Architectures," *IEEE Trans. Microw. Theory & Tech.*, vol. 55, no. 7, pp. 1445-1458, Jul. 2007.

[2] D. Cassan and J. Long, "A 1-V Transformer-Feedback Low-Noise Amplifier for 5-GHz Wireless LAN in 0.18-μm CMOS," *IEEE J. Solid-State Circuits*, vol. 38, no. 3, pp. 427-435, Mar. 2003.

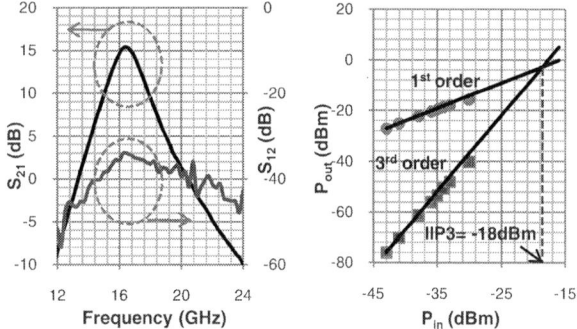

Fig. 5.   Measurement results, *left*: S_{21} and S_{12} of the 3-stage stand-alone LNA, (note: 3rd stage drives 50Ω measurement load), *right*: IIP3 measurement of the receiver

RMO4B-4

A Self-Healing 2.4GHz LNA with On-Chip S_{11}/S_{21} Measurement/Calibration for In-Situ PVT Compensation

Karthik Jayaraman, Qadeer Khan, *Baoyong Chi, William Beattie, *Zhihua Wang, and Patrick Chiang
School Of Electrical Engineering and Computer Science, Oregon State University, Corvallis,OR,USA
*Institute of Microelectronics, Tsinghua University, China

Abstract— **This paper presents a 2.4GHz, reconfigurable RF LNA using on-chip peak detection and calibration to measure and optimize its input impedance (S_{11}) and gain (S_{21}) in-situ, compensating for the unpredictable effects of process, voltage and temperature (PVT) variations. Measurement results show that the calibration of the LNA across PVT corners improves the S_{11} by 5.1dB, S_{21} by 3dB, while not significantly degrading the Noise Figure (0.22dB degradation) and linearity (1.7dBm degradation).**

Index Terms— **LNA, PVT, peak detector, on-chip calibration**

I. INTRODUCTION

As CMOS technology migrates to deep submicron processes, the sensitivity of circuits towards process, voltage and temperature (PVT) variations increasingly degrades the RF circuit performance, making the design of sensitive analog/RF circuits in these technologies extremely difficult. In addition, chip manufacturers are becoming increasingly challenged in accurately predicting the process skew, mismatch, and variation of devices, as the technology exhibits little time and low manufacturing volume to help improve silicon predictability. As a result, the limited modeling information about process characteristics and uncertainties often lead to over-design of the RF section. The low noise amplifier (LNA) is the first component in a radio receiver, and therefore has very stringent requirements such as low noise, high forward gain, high linearity, a well-matched input impedance (for interfacing with the preselect filter that precedes the LNA) and low power consumption. One of the most critical requirements for the LNA is accurate impedance matching, especially of the input reflection coefficient (S_{11}). S_{11} is critical as it important for determining the Noise Figure (NF) and the forward gain (S_{21}). Possible factors that can affect the LNA performance are 1) Variation of on-chip inductance and capacitance (min,typ,max); 2) Variation of off-chip inductance due to variable bond wire length; 3) Temperature variation (-50°C to 70°C) 4) Transistor corner and supply voltage variation (SS,TT,FF). If all these process uncertainties are considered together, the performance can be extremely degraded, as shown in Fig. 1 for example. The degradation of the input match is most severe, showing a worst case 8dB decrease at the frequency of interest.

In this work, a narrowband 2.4 GHz LNA is designed that measures PVT/packaging variations and automatically adapts its performance (i.e. impedances, gain) to the optimal operating condition. This is in contrast to the conventional approach

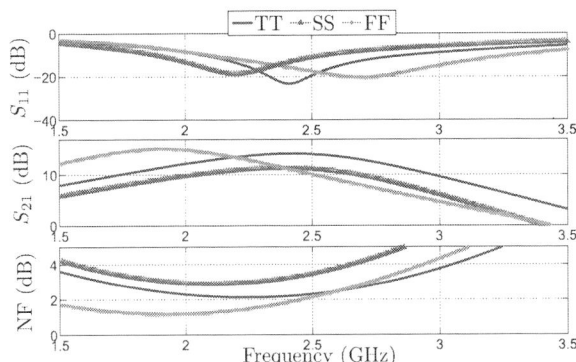

Fig. 1. PVT Variations on a conventional 2.4 GHz LNA

of trying to design a-priori a wideband LNA design that can cover the required RF bandwidth specifications across extreme process variations. The advantage of this tunable narrowband LNA over a wideband LNA are: lower Noise Figure and lower power; reduced pick-up of out-of-band interferers in nearby frequencies; relaxed requirements for simultaneous input and noise matching (SINM); reduced circuit complexity due to the elimination of a broadband, bandpass filter structure. The theme of this work is to build a reconfigurable LNA, enabling the ability to tune the LNA to the desired frequency of interest (2.4 GHz) to compensate for any resonant frequency shift due to PVT variations. The on-die detection of this frequency shift non-invasively is a critical requirement for in-situ calibration of the LNA. A proposed amplitude detector used in this work is area/power efficient, accurate (i.e. matches the optimal S_{11} and S_{21} locations), and capable of translating the RF frequency match information into DC for further digital baseband processing. This detection method is aided by the presence of an on-chip frequency synthesizer that is tunable across 2GHz-2.7GHz. Note that the overhead of this frequency synthesizer is low, as it typically is needed for the entire RF signal chain.

II. LNA DETECTION AND CALIBRATION

In previous works, the sensing of the input match is performed through a small resistor at the source of the input transistor [1]. However, the addition of this current-sensing resistor is invasive, as it is in series with the source inductor degrading the Q_{input} (though not by much as it is a 7Ω resistor). In addition, this process-varying resistor will create

978-1-4244-6240-7/10 $26.00 © 2010 IEEE   279

a parasitic capacitor from the inductor(L_s) node to ground, thereby modifying the resonant frequency. Further, two independent schemes must be used for tuning the input and output resonant frequency shift, adding overhead to the design. In another work [2], the transmitter and receiver chains are linked using a loop back connection with a peak detector used for the on-die detection. Their proposed peak detector topology introduces a large input capacitance and loads the LNA. The calibration is performed using an ADC and digital signal processing blocks. However, this work only reports simulation results and does not give measured results on the linearity, performance improvement, and power consumption of the LNA.

The proposed detection and calibration scheme here overcomes the limitations of the previous schemes by reducing the detector complexity, die area, and power consumption. By using a robust calibration algorithm, this work proposes minimal overhead and degradation of the LNA. An amplitude detector (common drain topology) is used to convert the RF signal to DC. The compensated input and output tank matching schemes help compensate for the frequency shift, with minimal effect on noise figure (NF) and linearity (IIP_3). The baseband, feedback calibration loop is performed off-chip using a NIDAQ and LabView interface that provides the system-level control of the optimal tuning conditions. Note that this baseband digital calibration system can easily be integrated on-chip with a low-frequency ADC and simple digital synthesized controller, but is ignored here due to time/system complexity and minimal intellectual value. Figure 2 and the section below highlight the detection/calibration scheme from a system-level perspective.

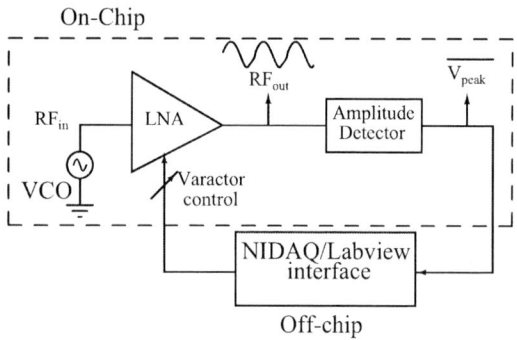

Fig. 2. Proposed detection and calibration for reconfigurable LNA

A. Calibration Algorithm

When the output load is tuned to the best possible frequency matching, configuring the input match results in the maximum power gain, S_{21}. The idea here is to use a single peak detector to detect the variations in the resonant frequencies of both the input and the output tanks. The calibration scheme is performed in the order outlined below:

1) The LNA input is applied a 2.4 GHz sine wave (frequency of interest) using the on-chip LC VCO.

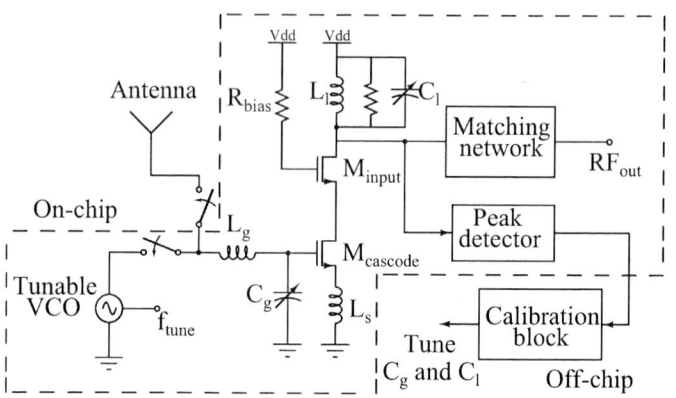

Fig. 3. Reconfigurable LNA topology

2) RF to DC conversion is performed by the peak detector.
3) An off-chip NIDAQ/LabView interface converts the DC analog signal to digital, and provides the digital control signals to adjust the tunable capacitor banks.
4) Tune C_l and measure PD_{out}. Set C_l for maximum PD_{out}.
5) Tune C_g and measure PD_{out}. Set C_g for maximum PD_{out}.
6) Terminate offline calibration, once optimal performance has been reached.

B. Maximum voltage gain correlation with best input match

Maximum voltage gain of the peak detector at the output tank load is used to detect input resonant frequency variations. S parameter analysis below shows that the best input match results in the maximum voltage gain [3].

$$S'_{11} = \frac{b_1}{a_1} = \frac{S_{11}(1 - S_{22}\Gamma_L) + S_{21}S_{12}\Gamma_L}{1 - S_{22}\Gamma_L} \quad (1)$$

$$S'_{11} = S_{11} + \frac{S_{12}S_{21}\Gamma_L}{1 - S_{22}\Gamma_L} \quad (2)$$

Voltage gain with arbitrary source and load impedances,

$$A_v = \frac{V_2}{V_1} = \frac{S_{21}(1 + \Gamma_L)}{1 - S_{22}\Gamma_L(1 + S'_{11})} \quad (3)$$

Under matched load conditions, (i.e) output matching $\Gamma_L = 0$,

$$A_v = \frac{S_{21}}{(1 + S'_{11})} = S_{21} \ when \ S_{11} \ \rightarrow \ 0. \quad (4)$$

Hence, the voltage gain is at a maximum when the input match is optimal, when the LNA is provided with the input frequency of interest.

III. RECONFIGURABLE LNA INPUT MATCH

A conventional LNA input match is determined by both the active and passive devices of the input front-end. It can be seen that at resonance the real part of the input impedance is determined by the g_m of the input MOS , C_{gs} and L_s, and the resonant frequency is determined by L_g , L_s and C_{gs}.

978-1-4244-6240-7/10 $26.00 © 2010 IEEE   280

A. Compensated Input Match

To correct for the input resonant frequency shift, three parameters can be altered : L_g , L_s or C_{gs}. Since the value of L_s is usually very small, using it to tune the input frequency match is difficult. Tuning the input resonant frequency using a variable L_g suffers from the disadvantages of a series transistor switch, with the switch thermal noise causing increased noise figure (NF) [1]. Therefore, modifying C_{gs} to perform tunability is simplest and least invasive, as proposed in [2]. In this work, the input match is modified by adding a shunt varactor from V_g to ground [4]. The choice of adding the varactor C_g improves the tuning of the resonant frequency shift, by modifying the effective value of L_g. While this tuning of L_g modifies both its real and imaginary parts, this is not necessarily a problem as varying C_g adjusts the real part of the input match (S_{11}), calibrating the impedance by counteracting the imaginary portion [4]. The varactor C_g alters the source resistance, R_s, thereby affecting R_b and modifying L_g to L_b, as shown in Fig. 4.

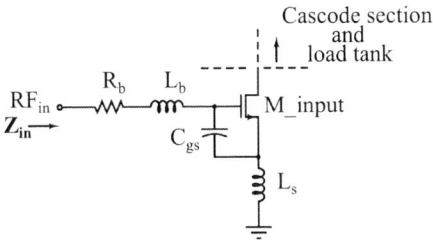

Fig. 4.   Modified Input Match

$$R_b = \frac{R_s}{\omega_0^2 C_g^2 R_s^2 + (1 - \omega_0^2 C_g L_g)^2} \tag{5}$$

$$L_b = \frac{L_g - C_g(\omega^2 L_g^2 + R_s^2)}{\omega_0^2 C_g^2 R_s^2 + (1 - \omega_0^2 C_g L_g)^2} \tag{6}$$

$$\omega_0 - \frac{1}{\sqrt{(L_g + L_s)C_{gs} + L_g C_g}}, Q_{input} - \frac{1}{(R_b C_{gs} + g_m L_s)\omega_0} \tag{7}$$

The initial value of R_b is determined by considering the Noise Figure requirements as follows:

$$NF = 1 + (\frac{\omega_0}{\omega_T})^2 \frac{\gamma}{\alpha} g_m R_b + (\frac{\omega_0}{\omega_T})^2 \frac{2\gamma}{\alpha\kappa} + \frac{\alpha\delta}{\kappa g_m R_b} \tag{8}$$

$$PCC = 4(\frac{\omega_T}{\omega})^2 \frac{R_b}{(R_b + R_{in})^2} \tag{9}$$

The value of R_b should be set to the value at the inflection point where the NF is a minimum. For the case where $R_b \geq R_{in}$, the power-to-current gain (PCC) improves with Q_{input} and the noise will degrade, though not significantly.

$$R_b = (\frac{\alpha\omega_T}{g_m\omega_0})\sqrt{\frac{\delta}{\gamma\kappa}} \tag{10}$$

The capacitor C_g that is added for tunability modifies the noise figure of the LNA, as the quality factor of the input matching circuit (Q_{input}) is reduced. The largest reduction in the channel current noise is obtained by increasing Q_{input} of the input tank circuit. However, this LNA topology is sensitive to gate-induced current noise, which is proportional to Q_{input}. Hence, reducing the Q_{input} using the C_g varactor will increase the channel current noise, but also help reduce the gate induced noise, thereby not degrading the noise figure.

IV. MEASUREMENT RESULTS

The peak detection and calibration scheme is implemented in a 1.8V, 0.18μm mixed-signal/RF process. The die photo of the reconfigurable RF LNA is shown in Fig. 5, along with the off-chip detection and measurement/calibration setup. The LNA, VCO and the peak detectors are implemented on-chip while the calibration routine is implemented off-chip using a NIDAQ and LabView interface operating at 1MHz. The output varactor value is statically set once the maximum peak detector output is reached, followed by statically setting the optimal input resonant frequency.

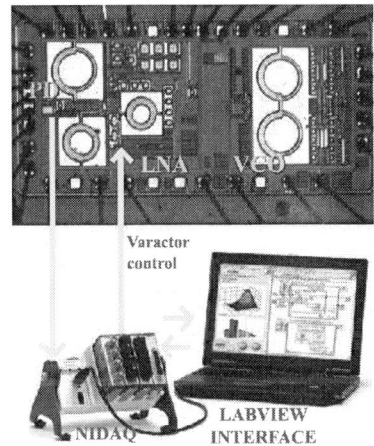

Fig. 5.   Measurement setup of the reconfigurable RF LNA.

A. Peak Detector Performance Summary

The peak detector used in this work [5] is area/power efficient and relatively non-invasive.It uses a common drain topology with a diode-tied transistor providing the current. The peak detector shown in Fig. 6 (a) is used in this work and exhibits a gain of 8.3mV/dBm, input capacitance of 88fF, overall die area of 0.026mm^2, and power consumption of 3.4mW with VDD=1.8V. Note that the peak detector can be shut-down during normal operation, as the calibration procedure is performed offline.

B. Process Compensation

It can be seen from Table I that the worst affected parameter is the input transistor's transconductance (g_m) and intrinsic capacitance (C_{gs}). The variation in g_m affects the real part of the input impedance and the variation in C_{gs} affects the resonant frequency of interest. The LNA performance, input

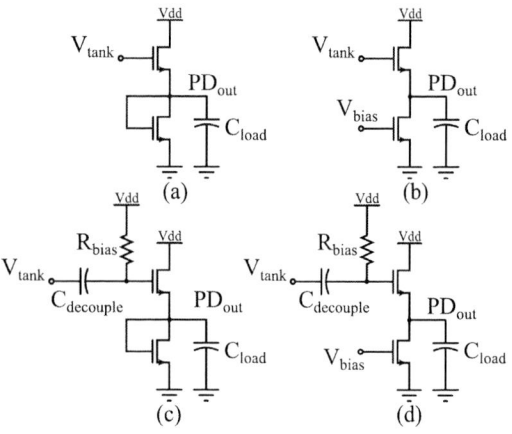

Fig. 6. Peak detector topologies [5]

TABLE I

LNA PERFORMANCE DEGRADATION DUE TO PVT VARIATIONS

PVT Corner	g_m	C_{gs}	i_d
TT, V_{dd} = 1.8 V @ + 27°	65.09 mS	308.04 fF	6.36 mA
SS, V_{dd} = 1.62V @ + 70 °	52.55 mS	325.98 fF	5.87 mA
FF, V_{dd} = 1.98V @ - 50 °	87.65 mS	302.31 fF	6.87 mA

matching and the power gain, with/without the detection and calibration scheme is shown in Fig. 7.

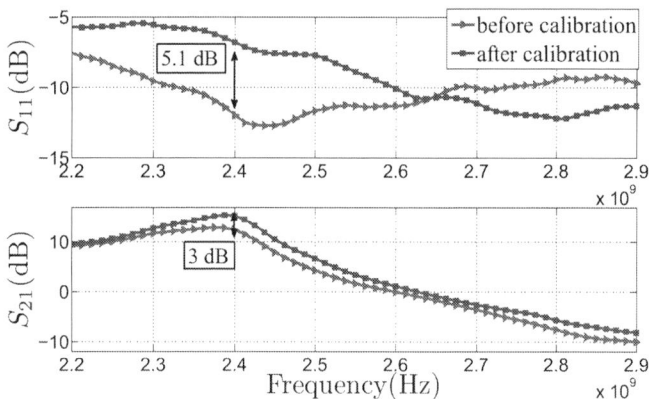

Fig. 7. Measured S_{11} and S_{21} of the LNA with/without the calibration.

Before calibration, the degradation in the input match decreases the gain to 12.4dB (initially designed for 15dB). From Fig. 7 it can be seen that while the output tank resonant frequency is still well oriented at 2.4GHz and does not suffer a frequency shift, while the major problem is with the input match. The proposed algorithm corrects for the input resonant frequency shift, where after calibration, the S_{11} is improved by 5.1dB and S_{21} is improved by 3dB. From post-layout simulations, a conventional LNA without peak detectors had worst case NF of 2.42dB, as opposed

to 2.64dB for the reconfigurable LNA (post layout), which translates to a higher value on silicon. The reconfigurable LNA experimentally shows a NF of 2.92dB and P_{1dB} of - 23.6dBm (1.7dBm degradation). Therefore linearity and noise performance of the LNA are not degraded by much due to the detection and calibration. The post-layout numbers are compared, as only the reconfigurable LNA was fabricated and measured as opposed to the conventional LNA which was post-layout extracted. Figure 8 shows that the peak detector voltage maximum correlates with the S_{11}, ensuring robust calibration.

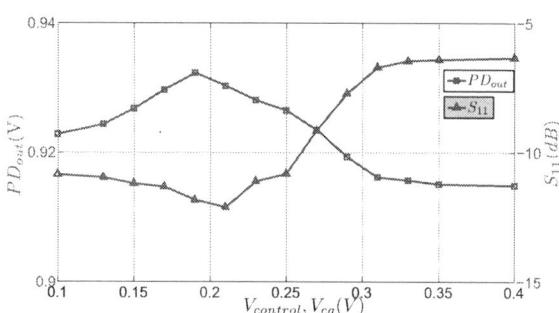

Fig. 8. Peak detector voltage peak in comparison to the input reflection coefficient S_{11} at Vdd = 1.8 V

V. CONCLUSION

A novel peak detection and calibration scheme is proposed that integrates on-die with the RF LNA, enabling reconfigurability and improved performance across process variations. The tunable 2.4 GHz narrowband LNA has been designed using on-chip detection and calibration of the input match and power gain. Measurement results show that the 2.4GHz LNA can improve the input match by 5dB as well as power gain by 3dB at the desired frequency, with minimal degradation of the noise performance and linearity. This integrated measurement and calibration technique can potentially be used for next generation software defined radios (SDRs) in future deep submicron CMOS processes.

REFERENCES

[1] T.Das, A.Gopalan, C.Washburn and P.R.Mukund "Self-calibration of input-match in RF front-end circuitry," IEEE Transactions on Circuits and Systems II: Express Briefs, vol.52, no.12, Dec. 2005, pp. 821-825.

[2] J.Wilson and M.Ismail,"Input match and load tank digital calibration of an inductively degenerated CMOS LNA," Integration, the VLSI Journal, Vol. 42, No. 1, Jan 2009, pp 3-9.

[3] D.Anderson,L.Smith and J.Gruszynski,S-parameter techniques, Test & Measurement Application Note 95-I, Hewlett Packard, Jan. 1996.

[4] Y.Cui, B.Chi, M.Liu, Y.Zhang, Y.Li, P.Chiang and Z.Wang, "Process Variation Compensation of a 2.4GHz LNA in 0.18um CMOS Using Digitally Switchable Capacitance," International Symposium on Circuits and Systems, 2007, pp. 2562-2565.

[5] K.Jayaraman, Q.A.Khan, P.Chiang and B.Chi, "Design and analysis of 160GHz, RF CMOS peak detectors for LNA calibration," International Symposium on VLSI Design, Automation and Test, 28-30 Apr. 2009, pp.311-314.

RMO4B-5

A Low Power LNA using Miniature 3D Inductor without Area Penalty of Passive Components

Akira Tanabe, Ken'ichiro Hijioka, Hirokazu Nagase, and Yoshihiro Hayashi

LSI Fundamental Research Lab., NEC Electronics Corporation
1120 Shimokuzawa, Sagamihara, Kanagawa 229-1198, Japan

Abstract — A low power 5GHz LNA without area penalty of inductors has been fabricated. Because of a miniature 3D vertical solenoid inductor, a chip area of this LNA is as small as that of feedback type LNAs which do not need passive components. A noise and a power consumption are smaller than those LNAs. Because of a small parasitic capacitance of the 3D inductor and a controlled series resistance considering skin and proximity effects, a 15.7dB power gain and a 2.0dB noise factor at 5GHz have been achieved with only 3.6mW power consumption. This LNA with the miniature 3D solenoid inductor is preferable for low power and low cost RF/mixed-signal SoCs.

Index Terms — CMOS, LNA, Inductor, Noise.

I. INTRODUCTION

In spite of a miniaturization of MOSFETs, a chip area of analog circuits was not minimized as small as digital circuits because of large area passive components. To minimize chip area, RF circuits without passive components or partly using passive components, such as active inductors and feedback circuits, have been proposed [1]-[3]. However, the performances, such as a noise and a power consumption, were insufficient because of an additional noise and power arise from MOSFETs. If the LNA circuits using inductors becomes as small as that without inductors, these additional problems are solved.

In this paper, we demonstrate a miniature low power 5GHz LNA using a 3D vertical solenoid inductor [4]-[6]. Owing to a small area and a small parasitic capacitance of this inductor, the area of this LNA is as small as that without passive components. In spite of the small area, this LNA has the low noise and low power performances owing to the passive components.

II. MINIATURE 3D INDUCTOR

The 3-dimentional (3D) vertical solenoid inductor [4]-[6] is suitable to minimize chip area and reduce parasitic capacitance. Fig.1 compares structures of the 3D vertical solenoid inductor and a conventional 3D stack inductor. Both inductors are composed of two metal layers (Top, Bottom) and each layer has two inductor wires (Inside, Outside). The difference between these inductors is the order of the connection of these wires as shown in Fig.2.

While the 3D stack inductor is a series connection of planer spiral inductors, the 3D vertical solenoid inductor is a series connection of concentric solenoid shape inductors [4]-[6], which has different configuration from a horizontal 3D solenoid inductor [7].

In Fig.1, the dominant parasitic capacitance is inter-layer capacitance Cv of the outside wire. In Fig.2, while Cv of 3D stack is connected at IN and OUT terminals of the inductor, Cv of 3D solenoid is connected at the middle node of the inductor. Therefore, effective parasitic capacitance observed from IN and OUT terminals becomes small. Moreover, while the 3D stack inductor has asymmetric parasitic capacitance arise from a capacitance between the inductor and a substrate, the 3D solenoid inductor has a symmetric property. A magnetic field of this inductor is distributed in a smaller volume than that of planer inductors, suppressing coupling between adjacent inductors [4].

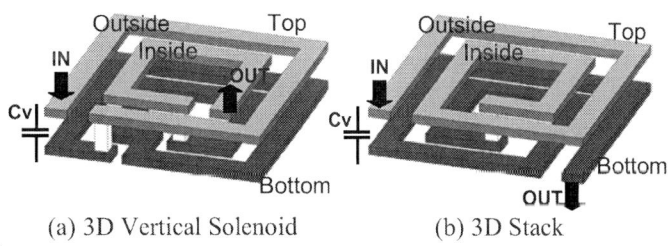

Fig.1 Structure of 3D inductors

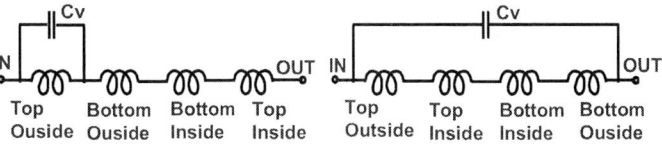

(a) 3D Vertical Solenoid   (b) 3D Stack

Fig.2 Parasitic Capacitance of 3D inductors

III. CHARACTERISTICS OF INDUCTORS

To make the 3D inductor in a small chip area, local metal layers as well as global metal layers are used. Fig. 3 shows cross sections of fabricated inductors in 90nm Cu

978-1-4244-6240-7/10 $26.00 © 2010 IEEE

wire digital CMOS process. This process has 5 local Cu, 1 global Cu and 1 Al layers. The 3D vertical solenoid inductor is composed of M3-M6 and Al layers, in which global M6 and Al layers are used as a single inductor wire and local M3-M5 layers are strapped as a combined inductor wire to reduce series resistance. On the other hand, the conventional planer inductor composed of strapped M6 and Al layers.

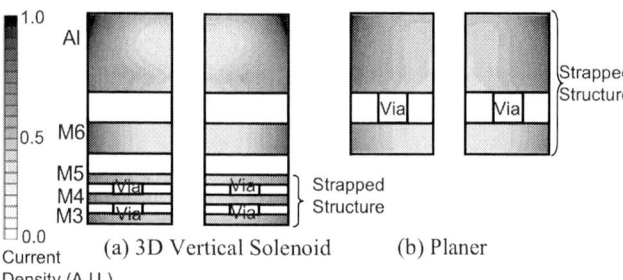

(a) 3D Vertical Solenoid   (b) Planer

Fig.3 Cross-section of inductors. Current density considering skin and proximity effect is also shown.

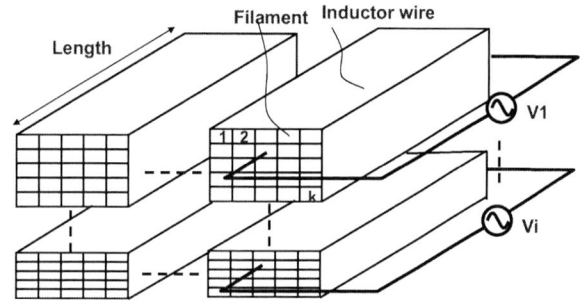

Fig.4 Volume filament model of skin and proximity effects

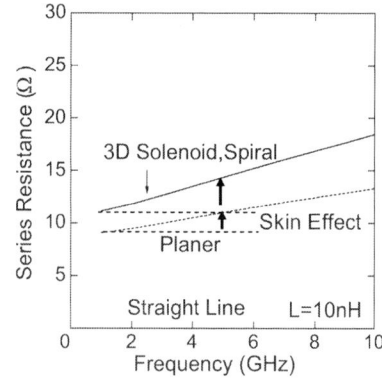

Fig.5 Comparison of series resistance arises from skin and proximity effects. Two straight wires of 5μm width are placed in parallel

Because of the 3D structure, the skin effect and the proximity effect between upper and lower wires occur. These effects are analytically investigated using volume filament model [8] shown in Fig.4. In this model, an

inductor wire is assumed to be a bundle of fine filament inductors. By calculating a self and mutual inductance of each filament, a total series inductance and a total series resistance are obtained. Fig.5 shows calculated series resistance of 3D solenoid, 3D stack and planer inductors which have 10nH inductance and a straight line shape. The series resistance ratio between high and low frequency indicates the skin and proximity effects. It shows that these effects of 3D solenoid and 3D stack inductors are the same and their impact is comparable with that of the planer inductor. Therefore, low noise operation of 3D inductor is promising at high frequency region. Fig. 3 compares a current density of the 3D solenoid and planer inductors. It shows that because of the strapped structure, planer inductor also suffer from these effects because of its large cross section area.

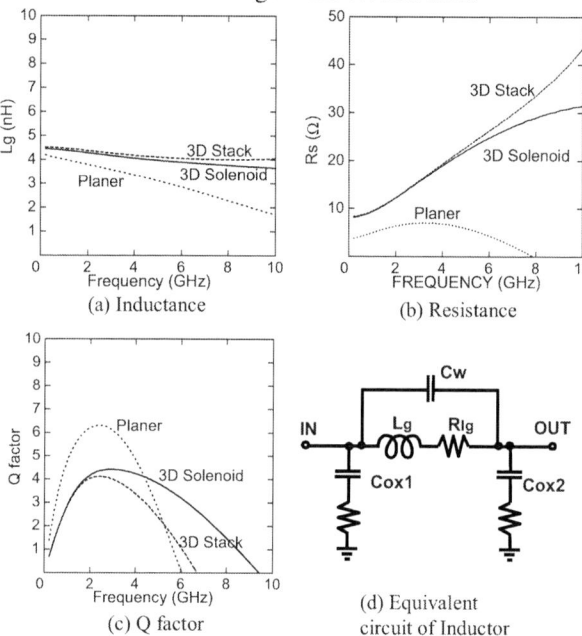

(a) Inductance   (b) Resistance

(c) Q factor   (d) Equivalent circuit of Inductor

Fig.6 Characteristics of inductors

TABLE I Parameter of inductors

	3D Solenoid	3D Stack	Planer
Lg (nH)	4.36	4.45	3.99
Rlg (Ω)	9.14	9.33	4.90
Cox1 (fF)	66.0	98.9	224
Cox2 (fF)	66.6	30.5	241
Cw (fF)	5.5	34.3	19.9

(Parameters are extracted at 1GHz)

Fig.6 shows measured series inductance L_g, series resistance R_{lg} and Q factor of the 3D solenoid, the 3D stack and the planer inductors designed for a 5GHz LNA. An equivalent circuit is also shown. Here, R_{lg} is a series resistance, C_w is a inter wire capacitance, C_{ox1} and C_{ox2}

are capacitances between the inductor and the substrate, R_{sub1} and R_{sub2} are substrate resistances. A line width of the 3D inductors is 5µm and that of the planer inductor is 10µm. Table I shows parameters of these inductors at 1GHz. It shows that peak Q factor of the 3D solenoid inductor is smaller than that of the planer inductor because of a large R_{lg}. However, at 5GHz, it is as large as that of the planer inductor and greater than that of the 3D stack inductor because of small parasitic capacitances.

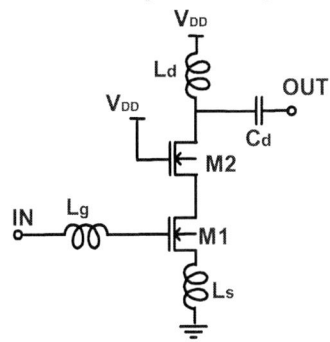

Fig.7 Circuit Diagram of LNA

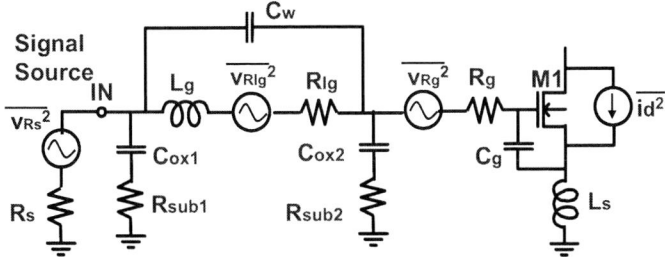

Fig.8 Equivalent circuit of the input stage.

IV. ESTIMATION OF THE NOISE OF LNA CIRCUITS

LNAs using these three types of inductors have been fabricated. Fig.7 shows circuit diagram of the LNA. This LNA is a cascode type and has three inductors L_g, L_s and L_d. Fig.8 shows an equivalent circuit model of the input stage of the LNA. From this circuit, the noise figure NF of this LNA at resonant frequency f0 ($=\omega_0/2\pi$) is simply given by [9].

$$F \cong 1 + \frac{R_{lg}}{R_S(1 - \omega_0^2 L_g C_W)} + \frac{R_g}{R_S} + \gamma g_{d0} R_S \left(\frac{\omega_0 (C_g + C_{ox2})}{g_m} \right)^2 \quad (1)$$

Here, R_S is a signal source resistance, R_g is a gate resistance, C_g is a gate capacitance, γ is a coefficient of the channel thermal noise, g_{d0} is a drain conductance at $V_d=0$, and g_m is a transconductance. Here, C_{ox2} is approximated to be parallel with C_g and a gate induced noise, C_{ox1}, R_{sub1} and R_{sub2} is neglected. The effect of R_{sub1} and R_{sub2} on 3D LNAs is small because of the small C_{ox1} and C_{ox2}. At feedback type LNA which omits inductor for input

matching, a noise from a feedback circuit is added instead of the second term. As a result, a large power is needed to enhance the gain and suppress the noise. Because of the small inter wire capacitance C_w, second term of the 3D solenoid LNA is smaller than that of the 3D stack LNA and because of the large series resistance R_{lg}, second term of the 3D solenoid and 3D stack LNA becomes larger than that of the planer LNA. However, the forth term, arise from the channel noise, indicate that small C_{ox2} of 3D solenoid LNA is effective to reduce noise. Using measured γ of the MOSFET (~2.5) and measured parameters of the inductors shown in Fig.6, estimated noise figure NF(dB) of these LNAs are 3D solenoid = 2.0dB, 3D spiral = 2.3dB, and planer = 1.8dB. It indicates that 3D vertical solenoid inductor has a potential to realize LNA whose noise is smaller than that using 3D stack inductor.

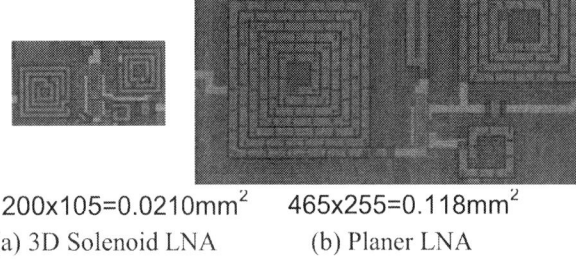

$200 \times 105 = 0.0210\text{mm}^2$ $465 \times 255 = 0.118\text{mm}^2$
(a) 3D Solenoid LNA (b) Planer LNA
Fig.9 Chip Micrograph of LNA

V. FABRICATION AND MEASUREMENT

Fig.9 shows chip micrograph of the 3D solenoid LNA and the planer LNA. The chip area of the 3D solenoid LNA is about 1/6 of that of the planer LNA. This area is comparable to the area of LNAs without inductors. Since this circuit has a good input and output matching, no additional buffer circuits, which require additional area and power, are needed.

Fig.10 shows a measured power gain (S21) and an input matching (S11) of these circuits. A supply voltage is 1.0V. The gain at 5GHz of the 3D solenoid LNA is 15.7dB and the power consumption is only 3.6mW. This gain is greater than that of the planer LNA and as large as that of the 3D stack LNA. This large gain is because of the small parasitic capacitance of the 3D vertical solenoid inductor.

Fig.11 shows the noise figure NF of these LNAs. The NF of 3D solenoid LNA at 5GHz is only 2.0dB which is as small as that of the planer LNA and smaller than that of the 3D stack LNA. While series resistance of the 3D solenoid inductor is larger than that of the planer inductor, its small parasitic capacitance realized the small NF. Because of the noise from substrate through C_{ox1}, C_{ox2} and a mismatch of the noise, NF of the planer LNA is

978-1-4244-6240-7/10 $26.00 © 2010 IEEE

larger than that of the estimated value. Table II compares major characteristics of fabricated and reported LNAs. The 3D solenoid LNA achieved chip area and gain comparable to that of feedback type LNAs and its power consumption and noise is smaller than those.

VI. CONCLUSION

A low power, small area 5GHz low noise amplifier (LNA) using miniature 3D vertical solenoid inductors has been fabricated. The chip area of this LNA is as small as the feedback type LNAs which omit passive components for input matching. The noise and the power consumption are smaller than those LNAs because of the advantage of the passive components. Because of the small parasitic capacitance of 3D inductor and the controlled series resistance considering skin and proximity effects, the 15.7dB power gain and the 2.0dB noise factor at 5GHz has been achieved with only the 3.6mW power consumption. This small area LNA is preferable for low power and low cost RF/mixed-signal SoCs.

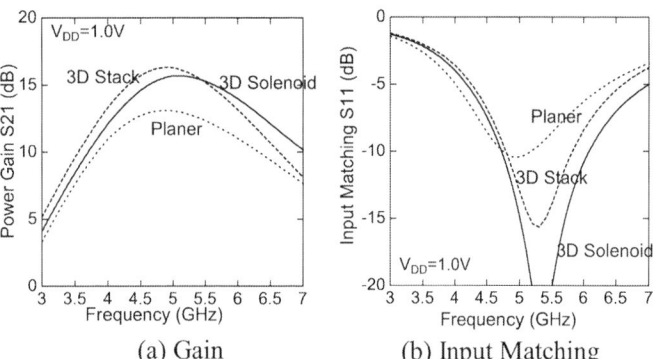

(a) Gain (b) Input Matching

Fig.10 Gain and Matching of LNAs

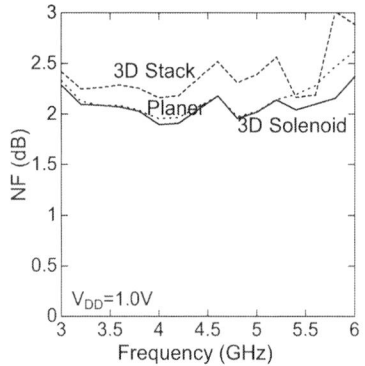

Fig.11 Noise figure of LNAs

ACKNOWLEDGEMENT

The authors thank Y.Mochizuki of NEC and H.Watanabe of NEC Electronics for their research support. The authors also thank Y.Nakashiba and T.Kuramoto of NEC Electronics for their discussions and measurement supports.

REFERENCES

[1] J-H.C.Zhan and S.S.Taylor, "A 5GHz Resistive-Feedback CMOS LNA for Low-Cost Multi-Standard Applications", ISSCC Dig. Tech. Papers, pp200-201, 2006.
[2] J.Borremans, P.Wambacq, G.Van der Plas, Y.Rolain, and M.Kuijk, "A Bondpad-Size Narrowband LNA for Digital CMOS", IEEE RFIC Symp. Dig., pp677-680, 2007.
[3] M.Okushima, J.Borremans, D.Linten, and G.Groeseneken, "A DC-to-22 GHz 8.4mW compact dual-feedback wideband LNA in 90 nm digital CMOS", IEEE RFIC Symp. Dig., pp295-298, 2009.
[4] K.Hijioka, A.Tanabe, Y.Amamiya and Y.Hayashi, "Crosstalk Analysis Method of 3-D Solenoid On-chip Inductors for High-speed CMOS SoCs", IEEE IITC, pp186-188, 2008.
[5] A.Tanabe, K.Hijioka, H.Nagase, and Y.Hayashi, "A Low-Power, Small Area Quadrature LC-VCO using miniature 3D Solenoid shaped Inductor", IEEE RFIC Symp. Dig., pp263-266, 2009.
[6] CC Tang, C-H Wu, and S-I Liu, "Miniature 3-D Inductors in Standard CMOS Process", IEEE J. Solid-State Circuits, Vol.37, No.4, pp471-480, Apr.2002.
[7] J.B.Yoon,B.K.Kim,C.H.Han,E.Yoon,K.Lee,C.K.Kim,"High-Performance Electroplated Solenoid-Type Integrated Inductor(SI^2)for FW Applications Using Simple 3D Surface Micromachining Technology", IEDM, pp544-547, 1998.
[8] S.Mei and Y.I.Ismail, "Modeling Skin and Proximity Effects With Reduced Realizable RL Circuits", IEEE Trans. VLSI Systems, VOl.12, No.4, pp437-447, Apr.2004.
[9] D.K.Shaeffer and T.H.Lee, "A 1.5-V, 1.5-GHz CMOS Low Noise Amplifier", IEEE J. Solid-State Circuits, Vol.32, No.5, pp745-759, May.1997.

TABLE II Characteristics of LNAs

	This Work 3D Solenoid	Conv. Planer LNA	Feedback LNA Narrow Band [2]	Feedback LNA Wide Band [3]	Feedback LNA without Inductor [1]
Supply Voltage(V)	**1.0**	1.0	1.0	1.2	2.7
Area (mm²)	**0.021**	0.118	**0.004 +Buffer**	**0.017 +Inductor**	**0.025**
Power (mW)	**3.6**	3.6	7.3*	8.4*	42
Frequency (GHz)	**5.0**	5.0	3.4	0-22.1	0.5-8.2
Power Gain (dB)	**15.7**	13.1	16.8*	10.7	18*
NF (dB)	**2.0**	2.0	2.2	4.3	1.9-2.6
IIP3 (dBm)	**-8.6**	-5.0	-11.5	-2.67	-14

*Include output buffer

RMO4C-1

A 65nm CMOS DCXO System for Generating 38.4MHz and a Real Time Clock from a Single Crystal in 0.09mm^2

Danielle Griffith, Fikret Dülger, Gennady Feygin, Ahmed Nader Mohieldin, and Prasanth Vallur

Texas Instruments, Dallas, TX, 75243, USA

Abstract — An integrated digitally-controlled crystal oscillator (DCXO) is presented that generates both 38.4MHz and also a 32.768kHz real time clock (RTC) from a single 38.4MHz crystal. The DCXO can startup independently and transition seamlessly in and out of software control. The tuning range is 280ppm with 2ppb/step and guaranteed monotonicity. The phase noise is -135dBc/Hz at 1kHz offset and -146dBc/Hz at 10kHz offset. The current consumption is 5mA from a 1.4V supply in full power mode and 234μA in low power mode, including the LDO and all clock buffers. The DCXO is implemented in standard 65nm digital CMOS with a die area of 0.09mm^2.

Index Terms — CMOS, Crystal oscillators, DCXO, digital control, GSM, phase noise, VCXO.

I. INTRODUCTION

A stable reference frequency is critical for wireless applications. The GSM standard requires the cellular handset to have a frequency accuracy within 0.1ppm of the basestation frequency. Crystal oscillators are used to generate the PLL reference frequency for cellular handset transceivers and reference clocks for phone peripherals such as GPS, WLAN, and Bluetooth. An automatic frequency control loop (AFC) synchronizes the oscillator frequency to within 0.1ppm of the basestation frequency. A digitally controlled crystal oscillator (DCXO) has several advantages over a VCXO for such applications. These include lower cost, smaller board area, potentially higher tuning range, better noise immunity, easier implementation of AFC loop in digital domain, and digital predistortion of the frequency control signal to achieve a linear tuning curve [1].

Cellular phones also require a 32.768kHz clock source to be used as real time clock (RTC). The RTC is the first clock to start on power up and enables phone boot up. Low power consumption is required because it runs continuously as long as battery is connected, even when the rest of the system is placed in a power-saving sleep mode. Usually the RTC is generated with a separate 32.768kHz crystal oscillator, but this architecture has cost and PCB area drawbacks because two crystals are needed. The DCXO clock system presented here in 65nm digital CMOS for a GSM/GPRS/EDGE transceiver eliminates the 32.768kHz crystal by deriving 32.768kHz from the 38.4MHz oscillator. This DCXO can operate in either a low power mode when only the RTC is required or a full power mode when both 38.4MHz and 32.768kHz are needed. It is also capable of autonomous startup and seamless transition to digital control.

II. DCXO ARCHITECTURE AND DESIGN

A. Frequency Tuning

Fig. 1 shows a block diagram of the DCXO clock system. The crystal oscillator oscillation frequency is

$$f_{osc} \approx \frac{1}{2\pi\sqrt{L_s C_s}}\left(1+\frac{C_s}{2C_L}\right) \qquad (1)$$

where L_s and C_s are the crystal's motional inductance and capacitance and C_L is the equivalent load capacitance. Frequency tuning is done by adjusting the tuning capacitors C1 and C2 in Fig. 1, which changes C_L and the oscillation frequency.

A Pierce oscillator was chosen because the tuning capacitors C1 and C2 are referenced to ground. This allows them to be implemented with MOS capacitors instead of the lower density MIM capacitors needed in a Colpitts oscillator. Capacitors C1 and C2 are nearly identical and each consists of a coarse frequency adjust (CFA) array and a fine frequency adjust (FFA) array. The CFA 6 bit binary weighted arrays are used to compensate for process variation and are adjusted only during factory calibration, stored in non-volatile memory and loaded into registers on power-up. The FFA arrays are thermometer-coded arrays arranged in a two dimensional matrix with row/column decoding co-located with each individual capacitive element. In this design, each array is composed of 480 (32x15) unit sized cells and 64 (32x2) half sized cells, as shown in Fig. 2. The FFA employs serpentine encoding to minimize glitches on major binary code transitions and to ensure the best DNL. The size of capacitors in each row is weighted to linearize the tuning curve for easier AFC loop design.

Each whole-sized capacitor in the FFA array changes the frequency by 130-200ppb, depending on process corner. The capacitors are alternatively switched in between C1 and C2 FFA arrays. At any given tuning

978-1-4244-6240-7/10 $26.00 © 2010 IEEE

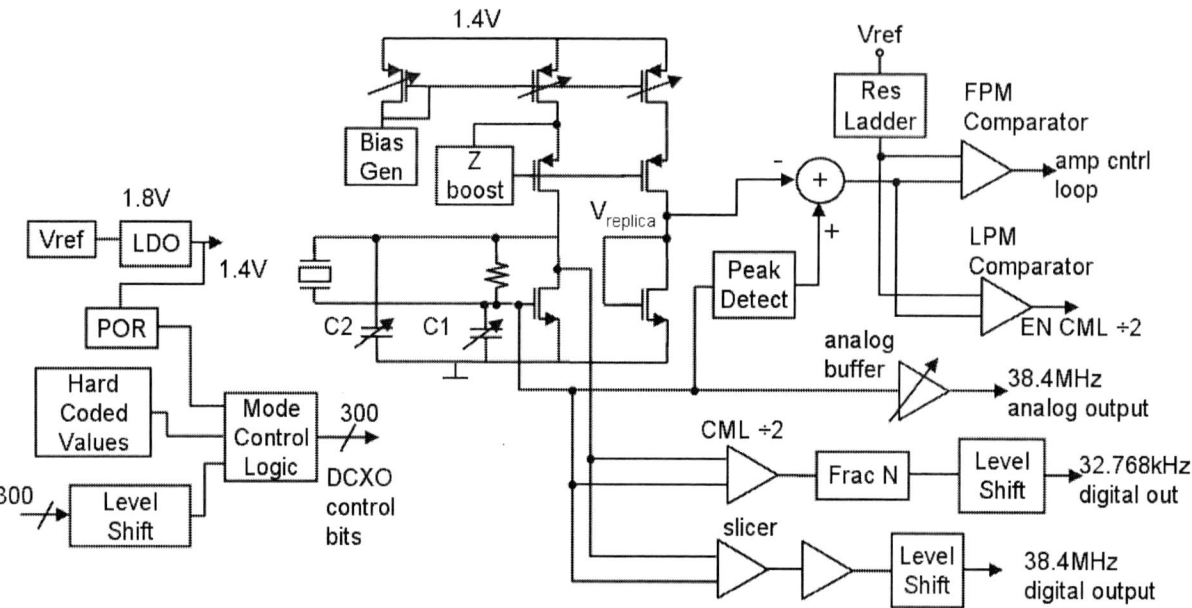

Fig. 1.   Simplified schematic of the DCXO system.

code, the capacitances on the oscillator gate and drain are nearly identical. This minimizes the effective step size while increasing the minimum required capacitor unit [2]-[3]. The half capacitors can be independently selected and also fractionally ΣΔ modulated creating a frequency resolution of 2ppb. For GSM, the AFC loop must keep the frequency accurate within 100ppb to maintain lock with the basestation, and 50ppb of this range is allocated to the DCXO in this design. This is easily achievable with the measured frequency resolution shown in Fig. 3. The total tuning range with only the FFA adjusted is measured to be 173ppm for a mid-code CFA setting. Adjusting the CFA capacitors adds another 107ppm to tuning range, for a total tuning range of 280ppm.

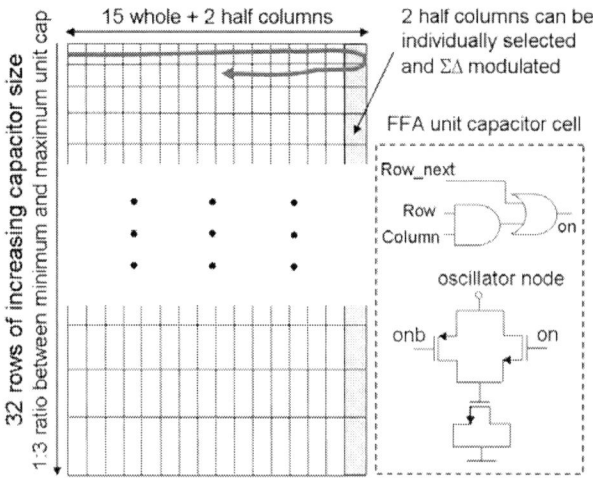

Fig. 2.   Conceptual view of an FFA capacitor array.

B. Amplitude Tuning

Temperature, crystal resistance, loading capacitance, and transistor transconductance all effect oscillator amplitude. If the amplitude is too low, the phase noise is degraded. If it is too high, it stresses the crystal and causes premature aging and frequency drift. The oscillator amplitude is sensed with a peak detector. The peak detector output is converted to a digital word with a SAR ADC. The dc point from the replica bias is also converted to a digital signal. Subtracting the two gives the oscillator amplitude which is used by the amplitude control loop to adjust the oscillator bias current, and therefore amplitude, to the optimum value.

Adjusting the bias current to change the amplitude also changes the frequency. The maximum frequency error is 50ppb, so the bias current minimum step size must cause less than 50ppb frequency shift. The amplitude control loop adjusts the bias current, then AFC adjusts frequency, and the process repeats until amplitude is correct. To allow small changes in current, the bias is provided through a programmable current mirror. The output transistors in the current mirror are binary weighted and used for large changes on startup. One LSB change in the output transistors causes a 90ppb frequency shift. The input transistors are a predistorted row and column encoded array to minimize DNL and guarantee monotonicity, similar to the FFA array. One unit change in the input transistor array causes only an 18ppb frequency shift. If a lower step size is needed, the final unit selected can also be dithered.

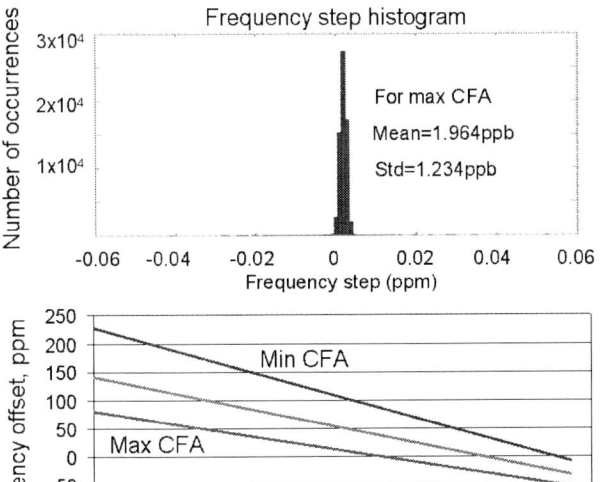

Fig. 3. Measured histogram of FFA step size and tuning range for minimum and maximum CFA values.

III. 32.768kHz Real Time Clock Generation

To generate the RTC, the DCXO must start up independently from digital control because the RTC must be present before the phone's digital circuitry can boot up. All DCXO control bits are generated in the analog power domain as "hard coded values". These values set the DCXO system to a lower power state, which includes the following settings. First, the LDO is set to about 10% of its full power mode (FPM) current. The DCXO capacitor array is set to the minimum value to allow reliable oscillation at 38.4MHz with low current. The oscillator core bias current is set to 20% of its typical full power mode value. The output buffers for the 38.4MHz clock are disabled. The 38.4MHz is then divided internally to the DCXO system with CML to 19.2MHz which drives the fractional divider to get approximately 32.768kHz. This clock is used as the RTC to boot the phone. After the digital VDD and then the digital baseband come up, software synchronizes the control bits in the digital baseband to existing low power mode (LPM) values so that when control is handed over to software there is no frequency jump causing a loss of lock with basestation. Software then transitions the DCXO smoothly from LPM to FPM. This transition includes increasing the LDO operating current, increasing the oscillator core bias current, adjusting C1 and C2 to get exactly 38.4MHz, enabling the 38.4MHz clock buffers, and adjusting the fractional divide ratio to get exactly 32.768kHz. Example waveforms are shown in Fig. 4. When the phone enters the sleep mode, the procedure is done in reverse, handing DCXO control back to the analog domain.

IV. Measured Results

Table 1 contains a performance summary of the DCXO system. Total current consumption in LPM includes margin to ensure reliable startup over process corners and crystal variation. On subsequent sleep cycles, the current can be reduced to approximately 100µA for nominal process corners. In FPM, a pseudo-differential slicer generates a digital signal for the PLL reference clock with low phase noise (-124dBc/Hz at 1kHz worst-case) and a duty cycle of ~49% worst case. Programmable strength buffers to tradeoff rise time and noise immunity with reference spurs on supply/ground were also implemented. Also in FPM, an analog buffer can be used for driving phone peripherals. It has a programmable drive strength from 5pF to 25pF and a sine wave output with amplitude >400mV. Measured phase noise is shown in Fig. 5. Table 2 shows a comparison to recently published crystal oscillators. The Figure of Merit (FOM) used is

$$FOM = 10 \cdot \log\left(\frac{f_{osc}^2}{f_m^2 \cdot L(f_m)} \cdot \frac{1}{P_{diss}} \right) \quad (2)$$

where f_{osc} is the oscillation frequency, $L(f_m)$ is the phase noise at offset frequency f_m, and P_{diss} is the power dissipation. A die photo is shown in Fig. 6.

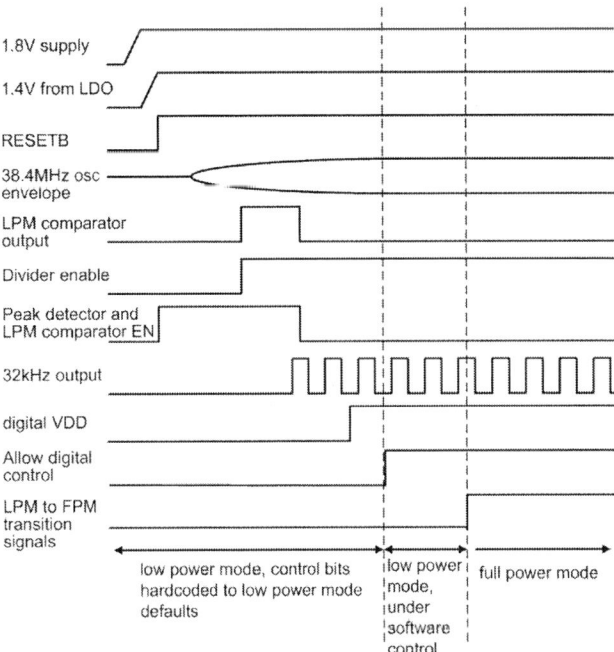

Fig. 4. Example 38.4MHz and 32kHz startup waveforms.

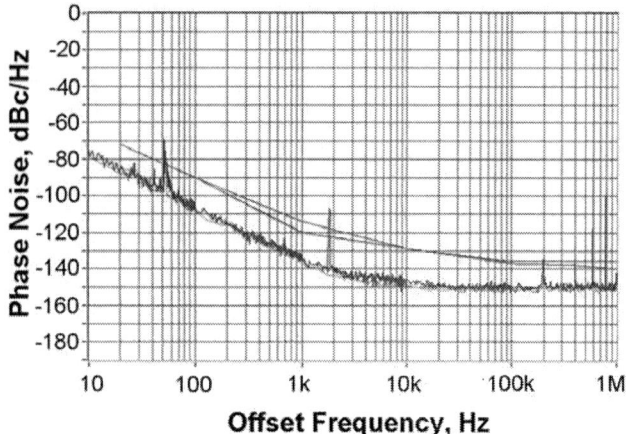

Fig. 5. Measured phase noise at the 38.4MHz analog buffer output.

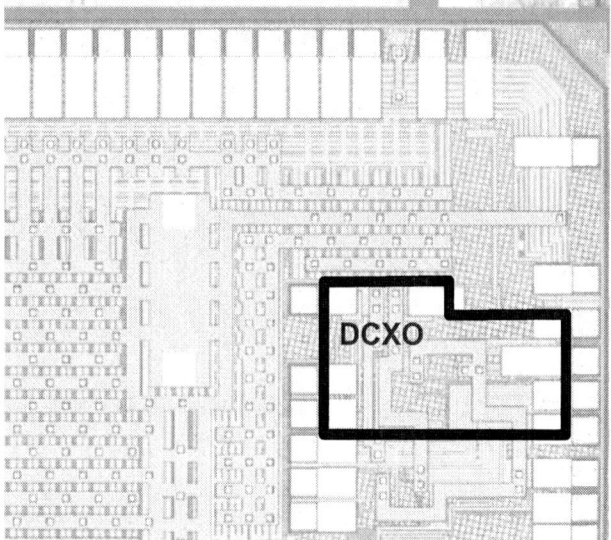

Fig. 6. Die photo of the DCXO in the transceiver.

V. CONCLUSION

In conclusion, a DCXO system that can generate both a 38.4MHz reference frequency and a 32.768kHz real time clock from a single 38.4MHz crystal has been implemented. This allows the 32.768kHz crystal to be removed, reducing cost and board area. The DCXO can startup independently from software control and transition seamlessly in and out of software control. The DCXO has low phase noise of -135dBc/Hz at 1kHz offset and small die area of 0.09mm[2] in 65nm digital CMOS.

ACKNOWLEDGEMENTS

The authors wish to acknowledge John Wallberg, Steve Dondershine and Vishwanath Venkataraman.

TABLE I: DCXO System Performance Summary

Voltage	1.8V, internal LDO generates 1.4V
Current	5mA in FPM (including LDO and all output buffers) 234µA in LPM on initial startup, including LDO
Frequencies	38.4MHz off chip to peripherals 38.4MHz as transceiver PLL reference 32.768kHz for system RTC
Measured 38.4MHz tuning range	280ppm for CFA + FFA 173ppm for FFA only
Measured 32.768kHz performance	6805ppm tuning range, adjusting only divide ratio, 2.57ns-10.1ns rms jitter over possible divide ratios
Measured 38.4MHz frequency step size	2ppb with 6-bit $\Sigma\Delta$ modulation on half sized capacitors
Measured phase noise for peripheral clock (over 5 parts and temperature)	-135dBc/Hz nominal, -132dBc/Hz worst-case @ 1kHz offset -146dBc/Hz nominal, -144dBc/Hz worst-case @ 10kHz offset
Measured frequency settling time to 0.1ppm	1.25ms at room temperature
Frequency stability (-40°C to 85°C)	+/- 10ppm (Crystal and DCXO combined)
Die area	0.09mm[2]
Process technology	65nm standard digital CMOS process

TABLE 2: Comparison to Published Results

	This work	[1]	[2]	[3]
P_{diss}, mW	9[1,2]	3[3]	8[1]	6.5
FOM, dB	247[1,2]	255[3]	227[1]	245
Tuning range, ppm	280	70	200	-
Step size, ppb	2	4	450	30
Technology	65nm CMOS	90nm CMOS	0.5um CMOS	0.35um SiGe BiCMOS
Die area, mm[2]	0.09	0.18	0.27	0.75

[1]Includes LDO, [2]Half of power consumption in analog output buffer, [3]No LDO, no analog output buffer

REFERENCES

[1] J. Lin, "A low-phase-noise 0.004ppm/step DCXO with guaranteed monotonicity in 90 nm CMOS," *IEEE Int. Solid-State Circuits Conf. (ISSCC) Dig. Tech. Papers*, San Francisco, CA, 2005, pp. 418-419.

[2] V. Balan and T. Pan, "A crystal oscillator with automatic amplitude control and digitally controlled pulling range of +/-100ppm," *Proc. IEEE Int. Symp. Circuits and Systems (ISCAS)*, vol.5, Scottsdale, AZ, May 2002, pp.461-464.

[3] S. Farahvash, C. Quek, and M. Mak, "A temperature-compensated digitally-controlled crystal pierce oscillator for wireless applications," *IEEE Int. Solid-State Circuits Conf. (ISSCC) Dig. Tech. Papers*, San Francisco, CA, 2008, pp. 352-353.

RMO4C-2

A 50ppm 600MHz Frequency Reference
Utilizing the Series Resonance of an FBAR

Julie Hu[*], Lori Callaghan[†], Richard Ruby[†], Brian Otis[*]

[*]Department of Electrical Engineering, University of Washington, Seattle, WA

[†] Avago Technologies, Inc., San Jose, CA

Abstract—**A 600MHz thin film bulk-acoustic wave resonator (FBAR)-based differential oscillator fabricated in a 0.13μm CMOS process is presented. The oscillator employs a cross-coupled pair with an FBAR resonator tank providing high Q source degeneration to realize frequency oscillation at the series resonance. The measured phase noise is -126 and -150dBc/Hz at 10kHz and 1MHz frequency offsets respectively; the integrated RMS jitter from 10kHz to 20MHz is 50fs. The oscillator achieves a frequency drift of 50ppm over the temperature range from 25 to 110 °C, providing the potential for quartz replacement in some applications. The figure-of-merit (FOM) of the oscillator is 214dB.**

I. INTRODUCTION

Thin film bulk-acoustic wave resonators (FBARs) have become critical building blocks for miniaturizing mobile phone transceivers. The need for high quality factor (Q) RF duplexers and filters has driven the FBAR manufacturing process to maturity; this includes well-controlled resonator frequency, quality factor, and temperature dependence with continued cost and size reduction [1].

FBAR-based oscillators have demonstrated superb phase noise performance due to their high Qs (>2000) [2][3]. An uncompensated FBAR resonator is sensitive to temperature variation with a temperature coefficient (TC) of about -25ppm/°C [4]. Physical compensation, wherein a layer of positive TC material is added to the FBAR stack to cancel the TC of the piezoelectric material in the first order, significantly reduces its TC. Further electronic temperature compensation can be be implemented as in [5].

As the size of the resonator (typically $100 \times 100 \mu m^2$) shrinks, the Q near the parallel resonance becomes poor. In contrast, Q near series resonance is less sensitive to shrinking area. Therefore, further size reduction of the resonator requires the oscillator to operate near the series resonance to utilize the maximal Q and maintain the phase noise performance. The motivation for this work is to present an efficient oscillator topology allowing operation near the series resonance of an FBAR.

Few previous works on series FBAR/BAW oscillators have been reported. Work in [6] demonstrated a series FBAR resonator in a SiGe BiCMOS process which achieved an inconsistent phase noise performance over a tuning range of 1.2% at a large power consumption.

This paper presents a CMOS differential oscillator that operates near the series resonance of an FBAR resonator (Fig. 1). Section I introduces the proposed oscillator. Section II

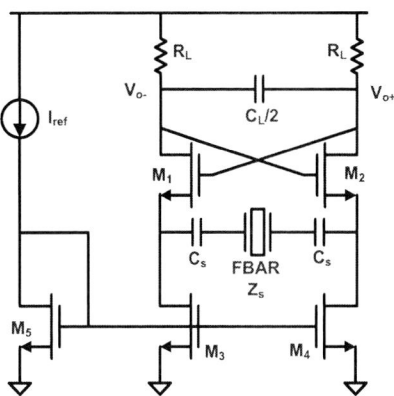

Fig. 1. The proposed FBAR oscillator operating at the series resonance.

discusses design challenges and proposes solutions. Section III presents the test results of our prototype.

II. THE PROPOSED FBAR SERIES OSCILLATOR

FBAR resonators are fabricated in a planar process which involves a minimum of three layers on a silicon substrate: two metal electrode layers surrounding one piezoelectric layer. The electrical behavior near the resonances can be accurately modeled using a modified Butterworth Van-Dyke (mBVD) model (Fig. 2). At the series resonance f_s, the resonator tank impedance reaches the minimum value R_s (several Ω), and at parallel resonance f_p, it reaches the maximum value R_p (> 2000Ω). At both high and low frequencies, the resonator behaves capacitively. The maximum Q value (> 2000) of the resonator appears at a frequency between the two resonances. The FBAR tank characteristic is modified by the additional impedance presented by the oscillator circuitry. A parallel capacitor C_p across the tank pulls the parallel resonance f_p and the maximum impedance R_p to a lower value. Likewise, a capacitor C_s in series with the tank pushes the series resonance f_s and the minimum tank impedance value R_s to a higher value. We will use this property to manipulate the series impedance of the FBAR, allowing efficient series-mode oscillation.

Like most LC oscillators, parallel resonance-mode FBAR oscillators drive the resonator tank with a current and sense the return voltage from the tank to the input of a transconductor. In contrast, a series oscillator drives the resonator with a voltage and senses the current. Using duality, a straightforward oscillator implementation operating at the series resonance of

978-1-4244-6240-7/10 $26.00 © 2010 IEEE

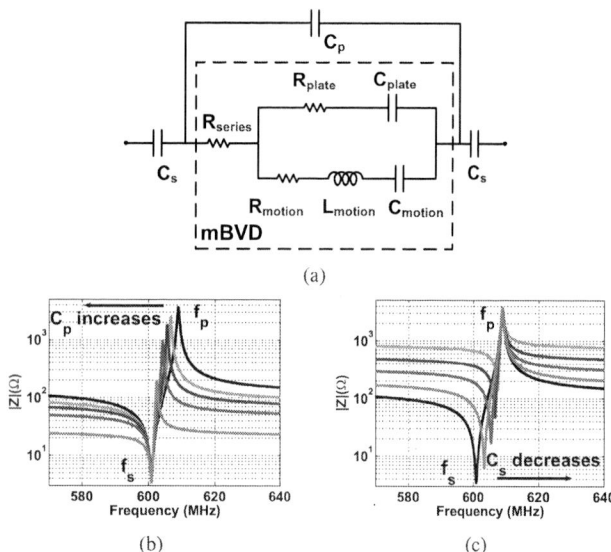

(a)

(b)   (c)

Fig. 2. FBAR characteristic. (a) The mBVD model with load capacitors. (b) Impact of parallel load C_p on frequency response. (c) Impact of series load C_s on frequency response.

the impedance tank is shown in Fig. 3 [6]. This oscillator uses a source follower to isolate the gain stage, which degrades power efficiency. We propose a new differential series oscillator topology that uses two-stage amplification realized with cross-coupled devices (Fig. 1). The equivalent circuit of our

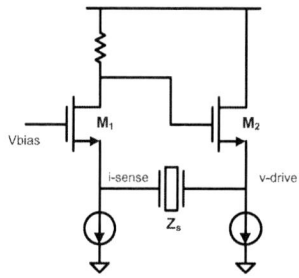

Fig. 3. Traditional implementation of series oscillators.

proposed oscillator is shown in Fig. 4, which consists of two identical source degeneration amplifiers in a positive feedback configuration. The small signal loop gain is

$$Gain \approx \left(\frac{g_m Z_L}{1 + g_m Z_s'} \right)^2, \qquad (1)$$

where $Z_L = R_L \parallel C_L$, and $Z_s' = Z_s/2 + 1/(j\omega C_s)$; here Z_s is the FBAR tank impedance.

At the series resonance, the source degeneration of the cross-coupled pair reaches a sharp local minimum, allowing super-unity loop gain and thus oscillation at this frequency.

In the proposed oscillator circuit (Fig. 1), we have introduced capacitors C_L and C_s which are of critical importance. The next section is dedicated to describing their functions.

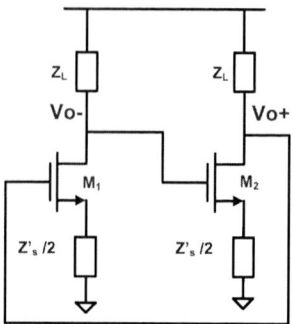

Fig. 4. The equivalent circuit of our proposed series oscillator in Fig. 1.

III. DESIGN CONSIDERATIONS

A. Prevention of parasitic oscillation

The proposed differential series FBAR oscillator in Fig. 1 has a potential parasitic oscillation at a frequency higher than the FBAR resonances f_s and f_p. The impedance looking down into the drains of the cross-coupled pair is

$$Z = \frac{-2}{g_m} \left(1 + \frac{g_m Z_s'}{2} \right). \qquad (2)$$

Eq. 2 indicates that impedance Z is inductive with a capacitive source degeneration impedance Z_s'. Z_L contains a capacitive component contributed by parasitics and explicit load capacitance. These two impedances in parallel form a parasitic resonance that can be excited at a high frequency. This can be intuitively understood by recognizing that, in addition to the series resonance frequency, the cross-coupled pair gain becomes large at high frequencies due to the drop in FBAR impedance. Fig. 5 shows a simulated result of the oscillator loop gain. In addition to the desired oscillation at the series resonance, the condition for oscillation ($|G(\omega)| > 0dB$ and $\angle G(\omega) = 0^o$) also occurs at an undesired frequency >4GHz.

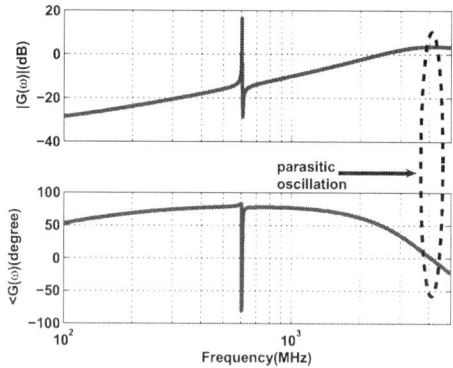

Fig. 5. Simulated loop gain showing parasitic oscillation.

Fortunately, the load impedance tank Z_L provides an opportunity for low-pass filtering the loop gain transfer function to suppress the parasitic oscillation by the inclusion of an explicit C_L. To accommodate variations from both CMOS and FBAR processes, we have built a programmable on-chip load capacitor C_L which value can be modified through a serial IO interface.

B. Trade-off of power and phase noise performance

For the moment, assume $C_s = \infty$. The series load impedance R_{osc} experienced by the FBAR tank is the summation of the two input impedances looking into the sources of the cross-coupled pair,

$$R_{osc} \approx \frac{2}{g_m}. \qquad (3)$$

We must avoid Q degradation of the FBAR tank to maximize phase noise performance. This requires that R_{osc} be small compared to the tank impedance at the series resonance R_s, i.e.,

$$R_{osc} << R_s. \qquad (4)$$

For R_s typically less than 10Ω, it is extremely difficult to avoid de-Qing the FBAR tank in a series oscillator configuration. For typical FBAR resonators with impedances (defined as the impedance magnitude of C_{plate} at resonance) on the order of 50Ω, a typical R_s will be sub-1Ω. Even with smaller resonators, it is difficult to achieve an R_s above 5 or 6Ω. Only extremely small resonators with very high impedance can achieve $R_s > 10\Omega$. This is an area of future study as it is unknown what other trade-offs there will be when going to extremely small area resonators.

We propose to address this limitation by boosting the intrinsic FBAR motional resistance to a higher value using a capacitive transformer as shown in Fig. 2(c). The simulated result of R_s and f_s versus C_s is given in Fig. 6. Both R_s and f_s increase as C_s decreases. With a C_s-tuned resonator, we

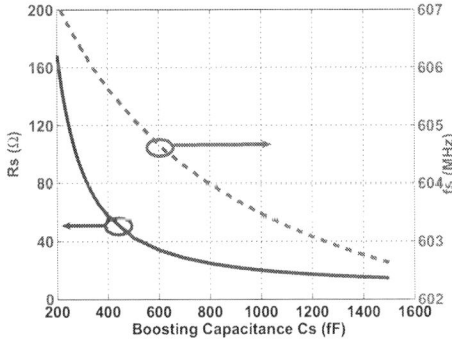

Fig. 6. R_s boosting in resonant tank $Z_s^{'}$ through a series capacitor C_s.

create an extra degree of freedom C_s to mitigate the large current requirement resulted from the need of the minimum loaded resonator Q.

In our implementation, we design a C_s, so that R_s of the resonator compound is approximately 50Ω. We then relax the constraint imposed by Eq. 4 so that the total loaded Q is 50% of the unloaded Q of the resonator. This requires that the nFET cross-coupled pair consume approximately 8mA current, which is about 10 times lower than without the R_s boost.

A tunable C_s was desired to allow for assembly of different FBAR samples. Unfortunately, since series resistance must be

avoided at all costs, transistor switching was not feasible. Thus, we used multiple on-chip C_s configurations that are selected through wirebonding at the packaging level. The complete oscillator schematic is shown in Fig. 7.

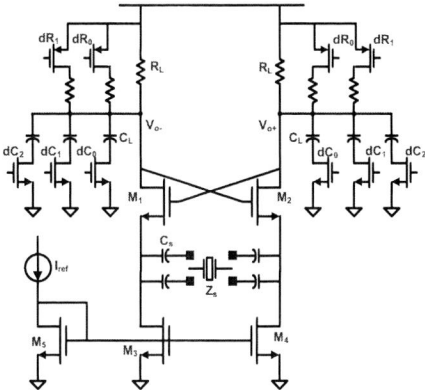

Fig. 7. Final schematic of the oscillator.

IV. EXPERIMENTAL RESULTS

The oscillator was fabricated in a $0.13\mu m$ CMOS process, occupying $(500\times750)\mu m^2$ of area including the pads. The die photo is shown in Fig. 8. The FBAR is epoxied directly on the CMOS chip, allowing in-package assembly of the entire frequency reference.

Fig. 8. Chip micrograph of the oscillator.

The oscillator operates over a supply range of 1 to 1.5V. The measured power-supply-rejection is around 730ppm/V throughout this range.

Fig. 9 shows the measured frequency spectrum of the oscillator at a supply of 1.25V. The current drawn from the supply is 4.5mA. The phase noise of Agilent PSA E4440 is typically -116 and -124dBc/Hz at 30kHz and 100kHz frequency offsets. It can be seen that the measured spectrum noise floor at offset 30kHz and above is dominated by the instrument noise. The phase noise of the oscillator measured with an Agilent 5052B (Fig. 10) is -126dBc/Hz and -150dBc/Hz at 10kHz and 1MHz offsets respectively. The integrated RMS jitter from 10kHz to 20MHz is 50fs.

Fig. 11 shows the results of the temperature sweep measurement of the oscillator. Using a compensated 600MHz FBAR

978-1-4244-6240-7/10 $26.00 © 2010 IEEE

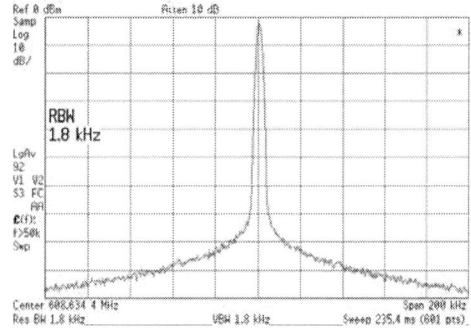

Fig. 9. The measured frequency spectrum of the oscillator with Agilent PSA E4440.

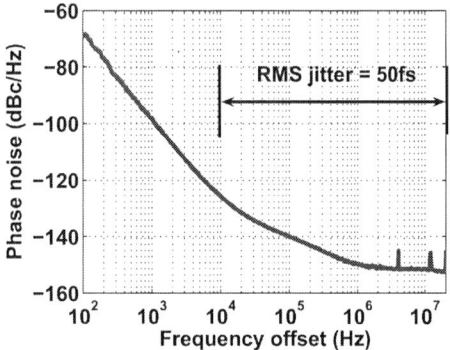

Fig. 10. The oscillator phase noise measured with Agilent SSA 5052B.

resonator, the oscillator drifts only 50ppm over a temperature range of 25 to 110°C. A conventional crystal oscillator has a frequency stability in the range of ±10ppm, which includes temperature stability, aging and initial calibration.

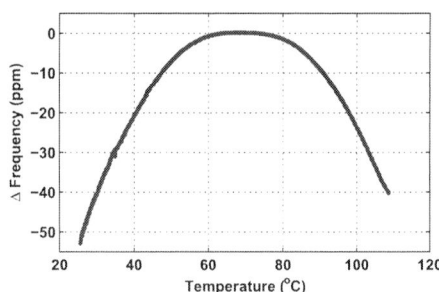

Fig. 11. Measured frequency dependence of the oscillator over temperature.

The measured frequency variation over time at a normally-fluctuating "room temperature" is shown in Fig. 12. The oscillator drifted by less than 6ppm over a 40-hour period.

Table I provides a performance summary and comparison to previously published work.

V. CONCLUSION

We have presented a novel differential oscillator topology operating at the series resonance of an FBAR. The oscillator uses a cross-coupled pair with R_s-boosted FBAR

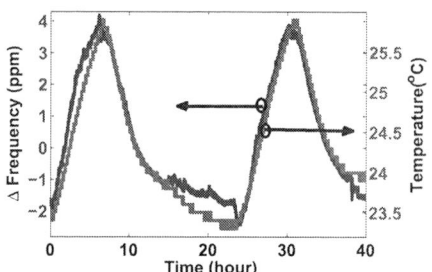

Fig. 12. Frequency stability at room temperature.

Table I
PERFORMANCE SUMMARY

	this work	[7]	[2]	[4]
CMOS process	$0.13\mu m$	$0.5\mu m$	$0.18\mu m$	$0.35\mu m$
f_c (MHz)	600	223	1917	600
V_{dd} (V)	1.25	5	1	3.3
Power (mW)	5.6	10	0.3	17.5
$\mathcal{L}(f)$@100Hz (dBc/Hz)	-69	-60	–	–
$\mathcal{L}(f)$@1kHz (dBc/Hz)	-98	-88	–	-102
$\mathcal{L}(f)$@10kHz (dBc/Hz)	-126	-121	-100	-130
$\mathcal{L}(f)$@100kHz (dBc/Hz)	-140	-148	-120	-149
$\mathcal{L}(f)$@1MHz (dBc/Hz)	-150	-160	-130	–
Int. RMS jitter (10k-20MHz) (fs)	50	–	–	–
Temperature drift (ppm)	$\sim$50 ($25-110^oC$)	–	2125 ($25-110^oC$)	80 ($-35-85^oC$)
FOM(dB)	214	205	211	213

source degeneration to realize a high Q series resonance. Our oscillator FOM is 214dB. The measured frequency stability is 50ppm over a temperature range of 25 to 100°C. This oscillator provides the potential to replace quartz oscillators in applications where size and cost are critical.

REFERENCES

[1] R. Ruby, P. Bradley, I. Larson, J., Y. Oshmyansky, and D. Figueredo, "Ultra-miniature high-Q filters and duplexers using FBAR technology," in *Solid-State Circuits Conference, 2001. Digest of Technical Papers. ISSCC. 2001 IEEE International*, pp. 120–121, 438, 2001.

[2] B. Otis and J. Rabaey, "A 300μW 1.9GHz CMOS Oscillator Utilizing Micromachined Resonators," *Solid-State Circuits, IEEE Journal of*, vol. 38, no. 7, pp. 1271–1274, 2003.

[3] S. Rai and B. Otis, "A 600 μW BAW-Tuned Quadrature VCO Using Source Degenerated Coupling," *Solid-State Circuits, IEEE Journal of*, vol. 43, pp. 300–305, Jan. 2008.

[4] W. Pang, R. Ruby, R. Parker, P. Fisher, M. Unkrich, and J. Larson, "A Temperature-Stable Film Bulk Acoustic Wave Oscillator," *Electron Device Letters, IEEE*, vol. 29, pp. 315–318, April 2008.

[5] S. Rai, Y. Su, W. Pang, R. Ruby, and B. Otis, "A Digitally Compensated 1.5GHz CMOS/FBAR Frequency Reference," *Ultrasonics, Ferroelectrics and Frequency Control, IEEE Transactions on*, vol. 57, March 2010.

[6] K. Ostman, S. Sipila, I. Uzunov, and N. Tchamov, "Novel VCO Architecture Using Series Above-IC FBAR and Parallel LC Resonance," *Solid-State Circuits, IEEE Journal of*, vol. 41, pp. 2248–2256, Oct. 2006.

[7] C. Zuo, N. Sinha, J. Van der Spiegel, and G. Piazza, "Multi-frequency pierce oscillators based on piezoelectric AlN contour-mode MEMS resonators," in *Frequency Control Symposium, 2008 IEEE International*, pp. 402–407, May 2008.

RMO4C-3

An Electronically Temperature-Compensated 427MHz Low Phase-Noise AlN-on-Si Micromechanical Reference Oscillator

Hossein Miri Lavasani, Wanling Pan, and Farrokh Ayazi

School of Electrical and Computer Engineering, Georgia Institute of Technology, Atlanta, GA 30332

Abstract — **This paper reports on the first demonstration of series tuning for lateral micromechanical oscillators and its application in a temperature-compensated 427MHz AlN-on-Si reference oscillator. The sustaining amplifier is a 13mW tunable TIA implemented in 0.18μm CMOS that uses shunt-parasitic cancellation to increase the tuning by 12× to 810ppm. The tunable oscillator along with a 2mW on-chip temperature compensation circuit has reduced the overall frequency drift to 70ppm in -10°C to 70°C. The phase-noise of the oscillator reaches -82dBc/Hz at 1kHz offset with floor below -147dBc/Hz.**

Index Terms — **Micromechanical oscillator, MEMS, temperature compensated, phase-noise.**

I. INTRODUCTION

Frequency reference oscillators play a critical role in determining the performance of modern radio transceivers. With the advent of integrated silicon micromechanical resonators [1], designers are actively considering micromechanical oscillators to deliver highly-stable and low-jitter clock signals with smaller form-factor and lower power than quartz crystal [2].

Despite their excellent short-term and long-term stability, silicon micromechanical oscillators suffer from inferior frequency accuracy compared to quartz crystals, both in terms of temperature stability and manufacturing tolerance [3]. The large native temperature coefficient of frequency (TCF) of these oscillators (~-28ppm/°C) [3] causes more than 2200ppm drift across the commercial temperature range, which is unacceptable for many applications, the majority of which require ±50ppm accuracy. This calls for inclusion of temperature compensation techniques in micromechanical reference oscillators [4]. Several temperature compensation methods are used for micromechanical oscillators; among them, material compensation [5] and electrical compensation [4] are more popular. While material compensation offers the least complex solution with minimal power consumption, near-zero TCF resonators require thick SiO_2 layer that negatively impacts the power handling and Q [5]. On the other hand, electronic compensation has minimal effect on the phase-noise performance but the tuning range is not sufficient for full compensation over -10°C to 70°C. As such, a combination of material and electronic compensation is preferred for near-zero TCF oscillators.

Laterally-excited micromechanical resonators are modeled as a series RLC tank with large shunt parasitic capacitances to ground (2 to 4pF) at the input and output [1]. While the series-resonant structure calls for placing a tuning network in series with the resonator as the most efficient way to adjust the frequency, the presence of large shunt parasitic capacitance drastically reduces the tuning range. Tuning enhancement techniques are therefore key to improving the oscillator tuning range.

This paper reports on the first demonstration of series tuning for lateral micromechanical oscillators. Tuning enhancement techniques, through cancellation of shunt parasitic capacitance has increased the tuning range by 12× to 810ppm and is used to reduce the temperature drift of a 427MHz oscillator from 780ppm to 70ppm in -10°C to 70°C range. The measured phase-noise of the oscillator is better than -82dBc/Hz at 1kHz offset with floor extending below -147dBc/Hz. The trans-impedance amplifier (TIA) and temperature compensation circuitry are fabricated in a 1P6M 0.18μm CMOS and consume 7.2mA from 1.8V supply.

II. BLOCK DIAGRAM

The reference oscillator consists of tunable TIA, MEMS resonator and temperature compensation circuitry (Fig. 1).

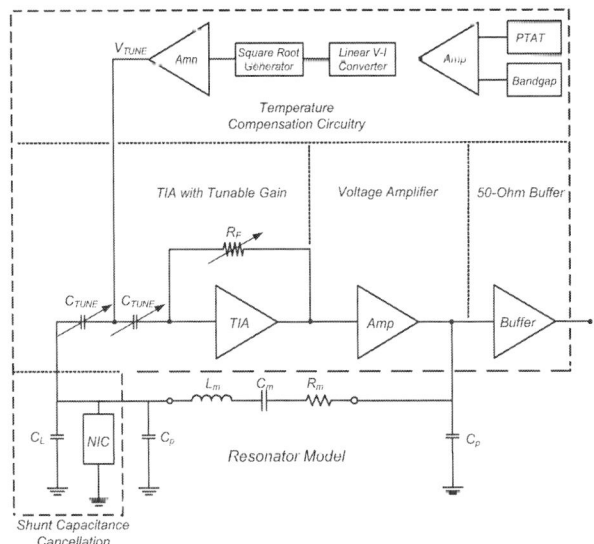

Fig. 1.   Block diagram of the 427MHz reference oscillator.

978-1-4244-6240-7/10 $26.00 © 2010 IEEE   295

The sustaining amplifier is comprised of three sections: a trans-impedance stage that employs tunable shunt-shunt feedback for gain tuning, frequency tuning network that consist of MOS varactors connected to a parasitic cancellation network, and voltage gain amplifiers for additional gain and phase-shift.

III. FREQUENCY TUNING ENHANCEMENT CONCEPT

The 427MHz AlN-on-Si resonator can be modeled as a series RLC with shunt parasitic capacitances to ground at both terminals. The concept of series tuning combined with parasitic capacitance cancellation is shown in Fig. 2.b. Cancellation is performed by placing a negative capacitor in parallel with the shunt parasitic capacitor of the resonator. A bank of varactors (C_{TUNE}) is then placed in series with the resonator to tune the frequency. The tuning voltage can be independently supplied from a temperature and/or process compensation circuitry.

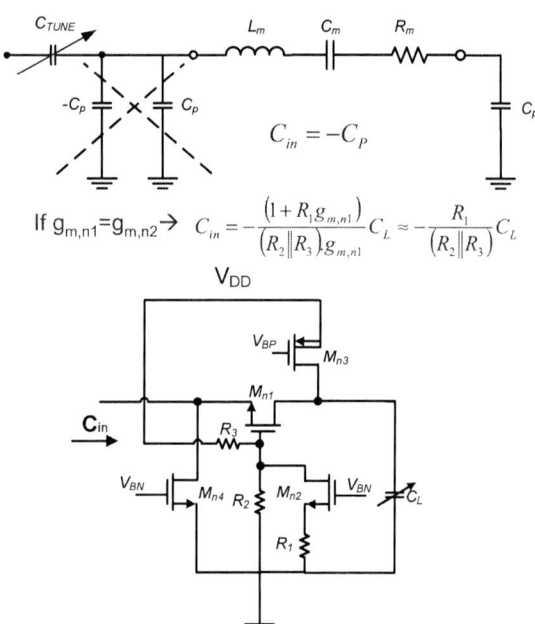

Fig. 2. Tuning enhancement concept for series RLC resonator.

The negative capacitance is realized using a single-port negative impedance converter (NIC). Assuming $g_{m,n1} \sim g_{m,n2} = g_m$, the equivalent negative capacitance is:

$$C_{in} = -\frac{(1+R_1 g_m)}{(R_2 \| R_3) g_m} C_L \approx -\frac{R_1}{(R_2 \| R_3)} C_L \quad (1)$$

where g_m is the transconductance of M_{n1} and M_{n2} transistors. Addition of R_3 allows additional degree of freedom that can be used for biasing. Since the signal is significantly attenuated by resonator before reaching the

input of the amplifier, the varactors are placed at the input. Consequently, the negative capacitor is put in parallel with the input terminal of the resonator (Fig. 2.b). C_L appears as a capacitive load to the input of the TIA and naturally, it has to be minimized such that it does not affect the 3-dB bandwidth (BW). This can be done by increasing the $R_1/(R_2\|R_3)$ ratio. Additionally, C_L can be made tunable for fine adjustment of negative capacitance value and a more accurate capacitance cancellation.

The phase-noise degradation due to tuning enhancement network is inherently small as the close-to-carrier phase-noise performance mainly determined by the Q of the resonating tank, loading from the amplifier, and the loss of varactors [1]. Given the high-Q of varactors in CMOS (>50), reducing the amplifier loading by lowering the input/output resistance and sizing the transistors for lower 1/f noise can minimize the impact on the phase-noise performance. The difference between ratio of varactor capacitance and parasitic capacitance from negative capacitance circuit at the two ends of tuning range causes slightly different loading for resonator Q that shows up in the close-to-carrier phase-noise performance.

Applying another negative capacitor generator and a bank of series varactors to the output help increase the tuning range at the expense of higher attenuation (which forces higher power consumption) and reduced dynamic range (due to large signal nonlinearity of the varactor).

IV. TRANSIMPEDANCE AMPLIFIER DESIGN

The sustaining amplifier is a three-stage TIA comprised of an inverter with tunable shunt-shunt feedback in the first stage for current-to-voltage conversion followed by two voltage gain stages that form a modified cherry-hooper amplifier (Fig. 3). An on-chip 50 buffer is included to interface with the measurement equipment. The choice of inverter over a common source topology is made due to its potential for higher gain when both MOS transistors are in saturation. The availability of gain tuning makes the TIA capable of interfacing with a wide variety of resonators with different loss. It also helps with phase-noise optimization. The gain tuning is incorporated in the first stage and is realized by a tunable NMOS resistor. The third stage employs a shunt-shunt feedback for bandwidth enhancement. This feedback can be in the form of a "T" network to introduce a left-half plane zero.

The tuning network that consists of two MOS varactors connected in series is placed between the input of the TIA and the shunt capacitance cancellation network (Fig. 3). The TIA (including the negative capacitor) consumes 7.2mA from 1.8V and is capable of providing 68dB-Ohm gain up to 700MHz with 3pF input/output load.

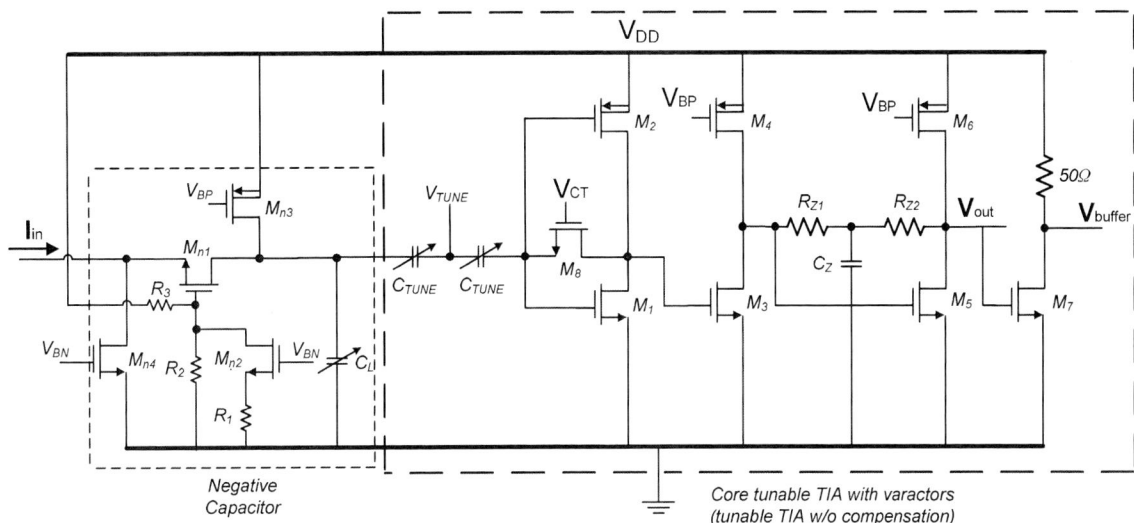

Fig. 3. Schematic of the TIA with frequency tuning and negative capacitor cancellation network connected to the input.

IV. TEMPERATURE COMPENSATION

Temperature compensation is performed in two steps; first the TCF of the resonator is reduced from ~-28ppm/°C to ~-10ppm/°C. This is achieved by including a 0.7μm SiO_2 layer in the resonator stack. This is a relatively thin layer and does not have a major impact on the resonator performance. The next step is the application of electronic temperature compensation to achieve sub-ppm/°C.

The integrated electronic temperature compensation circuitry includes temperature sensing and control signal generation units. The temperature sensing unit relies on a proportional-to-absolute temperature (PTAT) reference whose output is compared against the output of a bandgap reference to accurately sense the temperature variation.

Due to the nonlinear change in varactor capacitance with the tune voltage, the frequency variation is not linear. For most cases, this frequency vs. tune voltage relationship can be approximated by a parabolic function [4]. Therefore, a square-root generator block is necessary to convert a 2nd order parabolic function back to linear output. To this end, the output of the temperature sensing unit is first passed through a square-root generation circuit and then scaled and applied to the frequency tuning network. The linear V-to-I converter is added to increase the accuracy. The circuit consumes less than 2mW.

V. MEASUREMENT RESULTS

The sustaining amplifier and temperature compensation circuitry are fabricated in a 1P6M 0.18μm CMOS process and occupy 1.35mm×0.4mm (Fig. 4). For the purpose of performance comparison, another TIA without parasitic cancellation network is also fabricated on the same chip.

To demonstrate the effectiveness of negative capacitance cancellation technique, a 427MHz AlN-on-Si resonator (R_m~180Ω, $Q_{unloaded}$~1400) is interfaced with both TIAs. The TIA with negative capacitance achieves tuning range beyond 810ppm, 12× larger than the 64ppm tuning achieved by the TIA without tuning enhancement, and is in good agreement with simulated results (Fig. 5). The small difference between measured and simulated data is due to the parasitic interconnects.

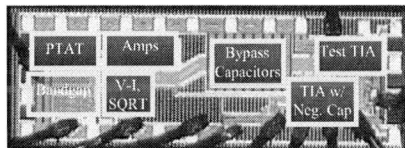

Fig. 4. Micrograph of the TIA with capacitive cancellation and integrated temperature compensation.

The oscillator phase-noise is measured with an Agilent E5500 phase-noise system. Careful optimization of ratio of varactor capacitance to parasitic capacitance from negative capacitance circuit has confined the total close-to-carrier phase-noise variation to ~5dB (Fig. 6). For best case, the phase-noise measures -82dBc/Hz at 1kHz offset with floor reaching below -147dBc/Hz (Fig. 6).

Finally, this extended tuning range along with electronic temperature compensation circuitry is used to fully compensate for frequency drift in -10°C to 70°C range. The temperature drift of the 427MHz oscillator is reduced from 780ppm to 70ppm (<0.88ppm/°C) in -10°C to 70°C range (Fig. 7). The oscillation frequency is monitored in the span of 50kHz with 801 point resolution. This setting enables 0.15ppm resolution, which is ~6× better than the required accuracy.

978-1-4244-6240-7/10 $26.00 © 2010 IEEE

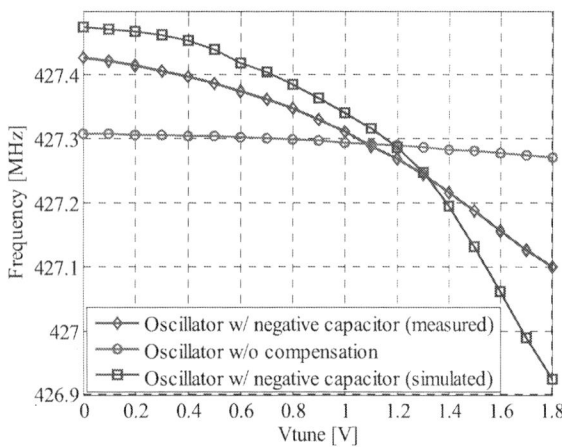

Fig. 5. Comparison of tuning range for oscillators with and without shunt parasitic cancellation.

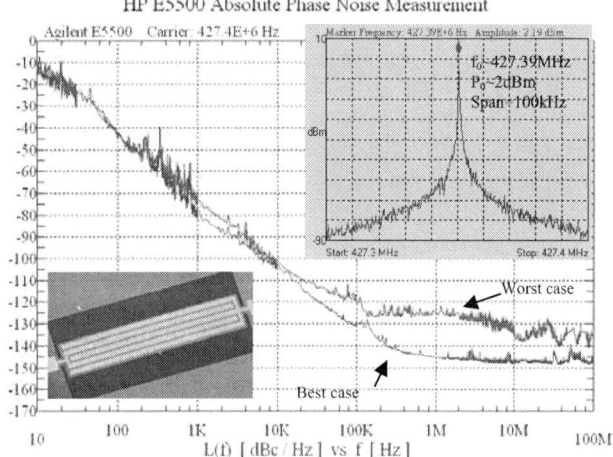

Fig. 6. SEM view of resonator along with the measured output spectrum and phase-noise of the 427MHz tunable oscillator.

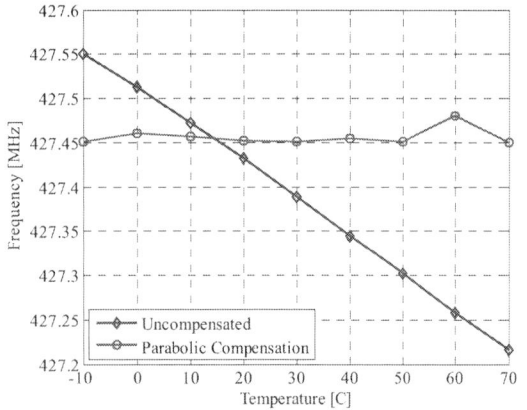

Fig. 7. Frequency drift over temperature for 427MHz temperature-compensated oscillators.

TABLE I. COMPARISON OF TEMP.-STABLE OSCILLATORS

Spec		AlN [6]	ZnO [5]	FBAR [7]	FBAR [8]	XTL[9]	This work
PN [dBc/Hz]	@1kHz	-69*	-81*	-83*	N/A	-113*	-82
	floor	N/A	-148*	N/A	-136*	-134*	-147
Unloaded Q		850	1,850	700	~1,500	~30,000	1.400
f_{osc} (MHz)		482	467	2000	1500	26	427
R_m (Ω)		N/A	600	~5	~5	~50	180
P_{DC} (mW)		10	9.4	~5	> 100	4.2	13
Comp. Method		N/A	Mat.	Mat.	Elec.	Elec.	Mat.+Elec.
Electronic Comp.		N/A	N/A	N/A	Digital	Digital	Analog
Range (ºC)		N/A	-5 to 90	-30 to 80	0 to 100	-10 to 55	-10 to 70
Total Drift (ppm)		N/A	375	530	20	2	70
Improvement		N/A	3×	4×	5×	14×	32×
IC Process		0.5μm CMOS	0.18μm CMOS	BiCMOS (100GHz)	0.35μm CMOS	0.13μm CMOS	0.18μm CMOS

* referred to 427MHz

VI. CONCLUSION

An electronically temperature-compensated AlN-on-Si 427MHz oscillator with phase-noise of -82dBc/Hz at 1kHz offset is reported. The oscillator uses, for the first time, shunt parasitic cancellation in a laterally-excited resonator for tuning enhancing with minimal impact on the phase-noise. Using this extended tuning, the on-chip electronic temperature compensation reduces the total drift to 70ppm in -10ºC to 70ºC.

REFERENCES

[1] H. M. Lavasani, et al., "A 500MHz low phase-noise AlN-on-silicon reference oscillator", CICC, pp. 599-602, 2007.

[2] F. Nabki et al., "A Highly Integrated 1.8 GHz Frequency Synthesizer Based on a MEMS Resonator," IEEE J. Solid-State Circuits, vol. 44, no. 8, pp. 2154-2168, March 2009.

[3] F. Ayazi, "MEMS for Integrated Timing and Spectral Processing," CICC, pp. 65-72.

[4] K. Sundaresan et al., "A Low Phase Noise 100MHz Silicon BAW Reference Oscillator", CICC, pp. 841-844, 2006.

[5] H. M. Lavasani, R. Abdolvand, and F. Ayazi, "Low Phase-Noise UHF Thin-Film Piezoelectric-On-Substrate LBAR Oscillators", Proc. IEEE MEMS, pp. 1012-1015, 2008.

[6] C. Zuo, et al., "Multi-Frequency Pierce Oscillators Based on Piezoelectric AlN Contour-Mode MEMS Resonators", IEEE Int. Freq. Cont. Symp., pp. 402-407, 2008.

[7] F. Vanhelmont, et al., "A 2 GHz reference oscillator incorporating a temperature compensated BAW resonator," IEEE Ultrasonics Symp., pp. 333-336, 2006.

[8] R. Shailesh, et al., "A 1.5GHz CMOS/FBAR Frequency Reference with ±10ppm Temperature Stability," IEEE Int. Freq. Cont. Symp., pp. 385-387, 2009.

[9] M.-D. Tsai, et al., "A Temperature-Compensated Low-Noise Digitally-Controlled Crystal Oscillator for Multi-Standard Applications," RFIC, pp. 533-536, 2008.

RMO4C-4

A Wide Tuning 1.3 GHz LC VCO with Fast Settling Noise Filtering Voltage Regulator in 0.18 μm CMOS Process

Hiroshi Akima, Aleksander Dec, and Ken Suyama

Epoch Microelectronics, Inc., 220 White Plains Road, Suite 330, Tarrytown, NY 10591, USA

Abstract — This paper presents a wide tuning LC voltage controlled oscillator (VCO) with integrated voltage regulator to minimize supply pushing. The integrated voltage regulator utilizes on-chip low-corner frequency noise filters to minimize the impact of regulator noise on VCO phase noise. One shot circuit is used with low-corner frequency noise filter to ensure fast settling. The VCO achieves current consumption of 15.9 mA with single integrated low-Q LC tank, tuning range of 620.8 MHz to 1384.5 MHz, phase noise of -128 dBc/Hz at 1 MHz offset from 1.3 GHz carrier at room temperature. With the noise filter, 9 dB improvement in phase noise is demonstrated in measurement. The frequency range can be extended to 38.8 MHz - 1384.5 MHz with integrated frequency dividers.

Index Terms — voltage regulator, voltage controlled oscillator, wide tuning range.

I. INTRODUCTION

Phase noise of local oscillators is a crucial design parameter in RF systems. For example, a transmitted signal with substantial phase noise may corrupt adjacent channel or it may lead to increased bit error rates in digital communication systems [1]. LC VCOs, instead of ring oscillators, are often used in RF transceivers for their superior phase noise characteristics [2][3]. In addition to phase noise, supply pushing can be sensitive to frequency modulation of any noise or interference present on the voltage supply of the VCO. To minimize the supply pushing, today's VCOs often use voltage regulators [4][5][6]. Although voltage regulator is helpful for reducing VCO supply pushing, its low-frequency noise can be modulated with the VCO frequency and negatively impact its phase noise. A common technique to reduce regulator noise is to use large off-chip noise-filtering capacitors. Although this technique is very effective at reducing regulator noise, in applications where chip pin count is important, it may be desirable to use a regulator which does not require any external pins.

This paper presents a noise filtering technique with fast settling time for a voltage regulator, which offers a small layout area, no external components, and sufficient noise suppression, implemented in a 1.3 GHz wideband LC VCO design. The implemented prototype also integrates frequency dividers and buffers.

II. PROPOSED TECHNIQUE

Figure 1 (a) shows a conventional method of implementing a low noise and high power supply rejection voltage regulator [7][8]. The bandgap voltage reference ensures constant voltage over supply and temperature variations, and voltage regulating amplifier scales up the 1.2 V bandgap voltage to desired regulated voltage by the ratio of resistors R2 and R1. External capacitors, C2 and C3, help achieving low noise and good power supply rejection.

Figure 1 (b) shows the proposed approach. To implement low-corner frequencies (<1kHz) needed to filter out regulator noise, noise filters using moderate-size on-chip capacitors and long channel MOS transistors are proposed. In addition, to minimize the noise impact of the bandgap voltage scaling amplifier noise, the VCO supply voltage is regulated using a unity gain regulator. Although it's possible to regulate the VCO directly using bandgap voltage scaling amplifier, the proposed architecture achieves better regulator noise. Since these noise filters have low-frequency cut-offs, the proposed regulator is very slow to settle. To realize fast settling time, one shot circuit is used to momentarily bypass the long-channel triode resistors and quickly pre-charge filtering capacitors to their steady-state voltages. After one shot signal turns OFF, the noise filters are enabled, and the regulator operates at its normal low noise mode.

III. CIRCUIT DESIGN

A. Architecture

The LC VCO is designed to oscillate from 650 MHz to 1300 MHz. A tuning range of 2:1 is chosen so that frequency dividers can be used to extend the tuning range down to 650MHz/16.

The two buffers are connected to the VCO output in order to impose equal loading to the VCO. This is done to minimize VCO frequency pulling when stepping through different divider modes. Frequency dividers are implemented with conventional CML divider circuits [9].

978-1-4244-6240-7/10 $26.00 © 2010 IEEE

B. Bandgap Reference Design

Bandgap voltage reference circuit used in this design is shown in Figure 3. The bandgap core consists of transistors Q1 and Q2 and resistors R1, R2, and R3. The error amplifier formed by M1, M2, M3, and M4 and the output transistor M5 creates negative feedback needed to regulate the bandgap. A long channel transistor M10 and Miller capacitor C1 form a noise filter to minimize bandgap noise. This noise filter is also bypassed using one shot circuit to speed-up bandgap voltage itself via M6 and R4. A start-up circuit is used to ensure proper bandgap start-up. Diodes D1 and D2 are used to short the long-channel MOS in case of large unexpected transition to speed up recovery time.

C. VCO Design

A wide tuning LC VCO is designed to demonstrate the effectiveness of the noise filtering voltage regulator on the VCO phase noise performance. The schematic of the VCO is shown in Figure 4. The wideband LC VCO is implemented in a complementary cross-coupled differential topology to achieve low phase noise with high power efficiency. To minimize VCO phase noise, low VCO gain is chosen [10]. To provide large frequency coverage, 7-bit binary-weighted MIM capacitor array is used. The complementary cross-coupled pair consists with seven switchable pair and one fixed pair. The seven switchable pairs are sized in binary-weighted fashion and track with the 7-bit MIM capacitor coarse tuning for high power efficiency. The number of turned ON pair decreases as the number of MIM capacitor array decreases because the quality factor looking into the MIM capacitor array increases in this case. A 3-bit programmable resistor array is used for post fabrication current and phase noise optimization. Accumulation varactors are used for the fine frequency tuning and implemented in differential fashion to maximize the quality factor [11]. Only one integrated inductor is used.

IV. MEASUREMENT RESULTS

The noise filtering voltage regulator, wideband VCO, divider, and buffer circuits were fabricated in a 0.18 um CMOS process. Figure 5 shows the chip layout. The chip has been packaged in QFN24 package and was mounted on a conventional FR4 device evaluation board. All measurements were done using Agilent 5052B Signal Source Analyzer.

Figure 6 shows the measured tuning characteristic for all 128 bands. The measured tuning range is from 620.5 MHz to 1384.5 MHz for bands 0 to 127, respectively. The

tuning range can be extended to from 38.8 MHz to 1384.5 MHz using the four divide-by-two circuits.

The peak VCO gain varies from 4.2 MHz/V to 41.4 MHz/V for band 0 to 127, respectively. The low VCO gain at low band setting is caused by the cross-coupled CMOS pair parasitic, which constitutes a significant portion of the tank capacitance [12]. The switchable cross-coupled CMOS pair parasitic contribution increases as the coarse tuning band decreases as well as the MIM capacitor contribution increases.

Figure 8 shows the measured phase noise when the noise filtering is turned ON and OFF, and the control voltage was fixed at 1.4 V. The phase noise is -102 dBc/Hz at 100 kHz offset and -128 dBc/Hz at 1 MHz offset from 1.3 GHz carrier with the noise filtering is turned ON. On the other hand, the phase noise where the noise filtering is turned OFF is -93 dBc/Hz at 100 kHz offset and -119 dBc/Hz at 1 MHz offset from the carrier. The measured data shows an improvement of 9 dB in VCO phase noise at 1 MHz offset from 1.3 GHz carrier.

Figure 9 shows the measured temperature drift of the VCO. The measured temperature drift is 14 MHz/115C, 15 MHz/115C, and 13 MHz/115C for bands 0, 63, and 127, respectively. The temperature drift is larger than the frequency tuning coverage at lower bands. Although for TDMA type application this may not be a problem since the synthesizer is periodically re-locked, it is a critical problem for CDMA type applications where synthesizer must stay locked over entire temperature range and cannot be re-locked. To reduced temperature drift, temperature compensation circuit is necessary [13][14].

The measured supply pushing is less than 0.5 MHz/0.3V for band 0 and 63. However, at band 127 supply pushing is 1.5 MHz/0.3V, which is higher than expected.

The VCO current consumption, including the regulator, varies from 9.9 mA to 15.9 mA, depending on the bands selected, from the 3.3 V supply. In order to achieve the tuning range of 2:1, the relatively high current consumption was necessary to compensate for the low Q-factor of the single integrated inductor (Q=8) whereas typical external inductors have Q-factor as high as 40.

V. CONCLUSION

A wide tuning LC VCO with integrated voltage regulator, which uses on-chip noise filters to achieve low noise and one shot circuit to achieve fast settling, has been demonstrated in 0.18 um CMOS process. The effectiveness of the proposed regulator has been confirmed by measurements where 9 dB of improvement in phase noise at 1 MHz offset from carrier of 1.3 GHz was observed when on-chip noise filter functionality was enabled.

ACKNOWLEDGEMENT

The authors would like to thank Roman Skrada for many helpful measurement contributions.

REFERENCES

[1] B. Razavi, RF Microelectronics, Prentice Hall PTR, 1997, pp. 215-217.

[2] N. Nguyen, et. al., "A 16Gb/s differential I/O cell with 380fs RJ in an emulated 40nm DRAM process," *Symposium on VLSI Circuit Digest of Technical Papers*, pp. 128-129, June 2008.

[3] M. H. Perrott, et. al., "A 2.5-Gb/s multi-rate 0.25-um CMOS clock and data recovery circuit utilizing a hybrid analog/digital loop filter and all digital referenceless frequency acquisition," *IEEE Journal of Solid-State Circuits*, Vol. 41, no. 12, pp. 2930-2944, December 2006.

[4] Y. Wu, and V. Aparin, "A monolithic low phase noise 1.7GHz CMOS VCO fro zero-IF cellular CDMA receivers," *IEEE ISSCC Dig. Tech. Papers*, pp. 396-535, February 2004.

[5] A. Maxim, and C. Turinici, "9.953-12.5GHz 0.13um CMOS LC VCO using a high resolution calibration and a constant gain varactor," *IEEE CICC Symp. Dig,*, pp. 545-548, September 2005.

[6] K. Manetakis, D. Jessie, and C. Narathong, "A CMOS VCO with 48% tuning range for modern broadband systems," *IEEE CICC Symp. Dig.*, pp. 265-268, October 2004.

[7] G. W. den Besten and B Nauta, "Embedded 5 V-to3.3 V voltage regulator for supplying digital IC's in 3.3 V CMOS technology," *IEEE Journal of Solid-State Circuits*, Vol. 33, no. 7, pp. 956-962, July 1998.

[8] G. Mora and P Allen, "A low-voltage, low quiescent current low drop-out regulator," *IEEE Journal of Solid-State Circuits*, Vol. 33, no. 1, pp. 36-44, January 1998.

[9] U. Singh, and M. Green, "High-frequency CML clock dividers in 0.13um CMOS operating up to 38GHz," *IEEE Journal of Solid-State Clrcults*, Vol. 40, no. 8, pp. 1658-1661, August 2005.

[10] J. W. M. Rogers, J. A. Macedo, and C. Plett, "The effect of varactor nonlinearity on the phase noise of completely integrated VCOs," *IEEE Journal of Solid-State Circuits*, Vol. 35, no. 9, pp. 1360-1367, September 2000.

[11] A. Porret, T. Melly, C. Enz, and E. Vittoz, "Design of high-Q varactors for low-power wireless applications using a standard CMOS process," *IEEE Journal of Solid-State Circuits*, Vol. 35, no.3 , pp. 337-345, March 2000.

[12] B. Razavi, "A 1.8GHz CMOS voltage-controlled oscillator," *ISSCC Dig. Tech. Papers*, pp. 388-389, February 1997.

[13] Y. Wu, and V. Aparin, "A temperature stabilized CMOS VCO for zero-IF cellular CDMA receiver," *IEEE VLSI Symp. Dig.*, pp. 398-401, June 2005.

[14] A. Kachouri, D. B. Issa, N. Boughanmi, and M. Samet, "A new temperature compensation method for a 2.5 GHz integrated VCO," *International Journal of Computer*

Science and Network Security, Vol. 7, no. 5, pp. 78-85, May 2007.

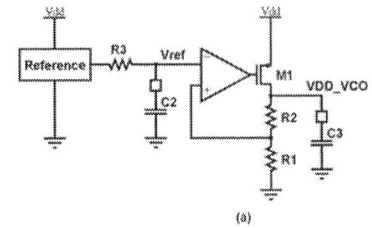

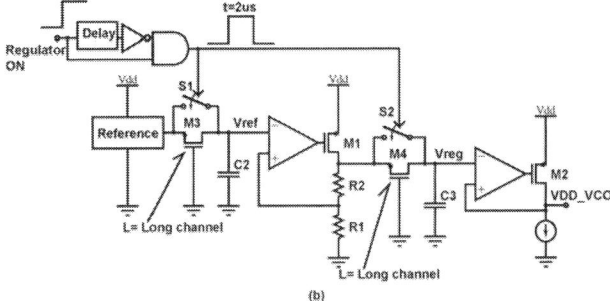

Fig. 1.   Conceptual diagram of (a) conventional regulator, and (b) proposed regulator.

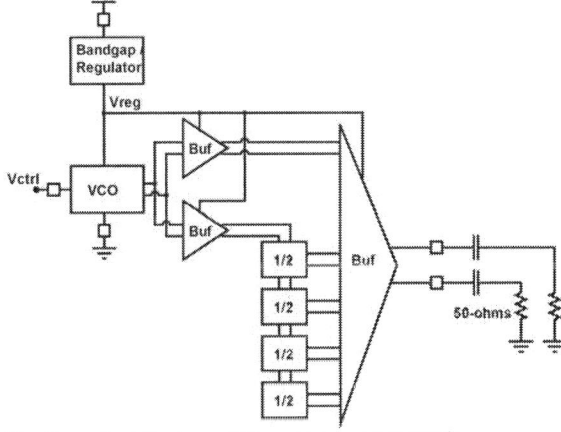

Fig. 2.   Architecture of the wideband LC VCO.

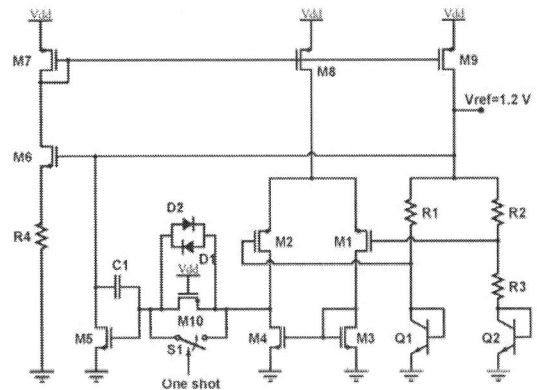

Fig. 3.   Schematic of the bandgap reference generation.

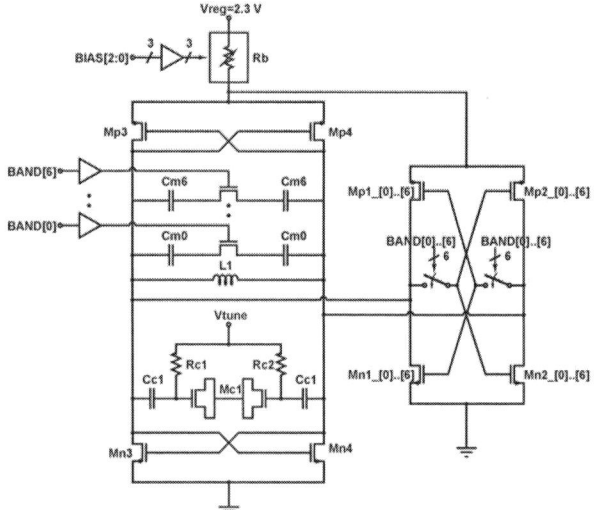

Fig. 4.   Schematic of the wideband LC VCO.

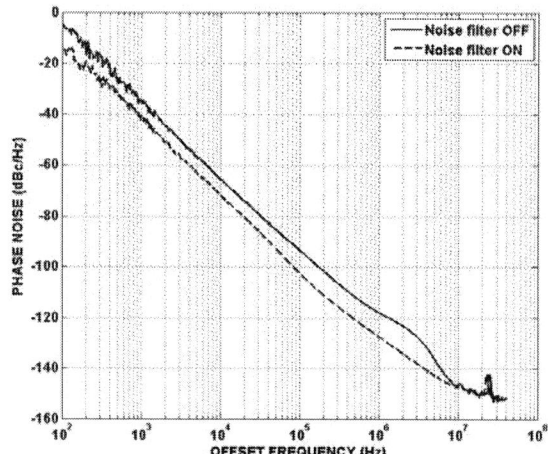

Fig. 8.   Measured phase noise (1.3 GHz carrier).

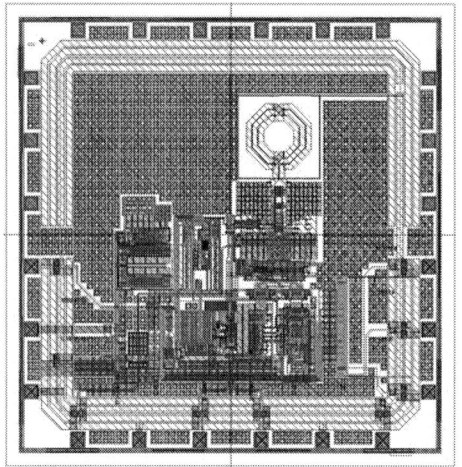

Fig. 5.   Layout diagram of the wideband LC VCO chip.

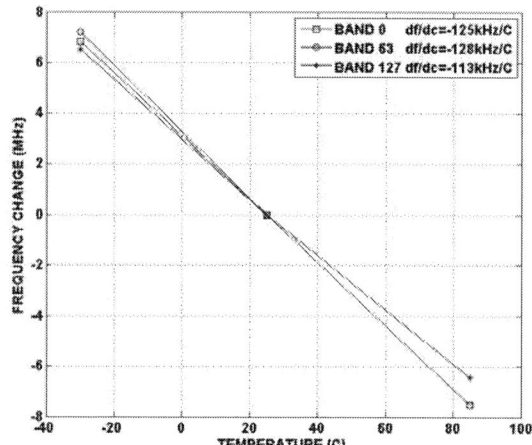

Fig. 9.   Measured temperature drift.

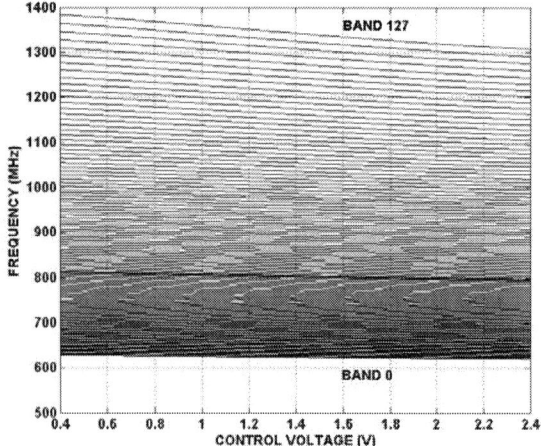

Fig. 7.   Measured frequency tuning.

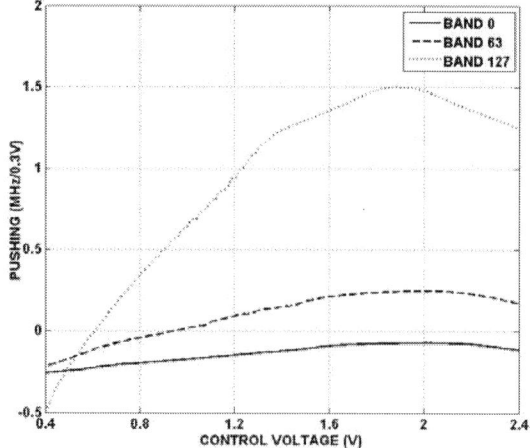

Fig. 10.   Measured supply pushing.

RMO4C-5

A Wide-Range VCO with Optimum Temperature Adaptive Tuning

Behzad Saeidi[1], Joshua Cho[2], Georgi Taskov[2], and Aaron Paff[2]

[1]Marvell Semiconductor, Aliso Viejo, CA 92656;   [2]Skyworks Solutions, Inc., Irvine, CA 92617

Abstract — **This paper presents an integrated wide-range VCO with a modified tuning scheme to deal with VCO frequency drift over temperature. In this approach, during the coarse-tune operation, VCO tune voltage is a function of temperature such that it resembles the inverse function of VCO fine-tune characteristic. Without degrading VCO performance, the proposed temperature adaptive tuning optimizes the maximum tolerable VCO temperature frequency drift over which PLL remains locked. As a result, VCO gain can be reduced significantly, making VCO less sensitive to PLL tune voltage noise. Integrated in a multi-standard multi-band transceiver with a small VCO gain of 50MHz/V at 3.90GHz, PLL remains locked despite 45MHz frequency drift of VCO over [-30°C, 85°C]. Using an on-chip inductor, VCO covers from 3.15GHz to 4.60GHz, achieving -138.0dBc/Hz phase noise at 3.0MHz at 3.90GHz by drawing just 8.5mA from 1.60V supply in 0.13u CMOS process.**

Index Terms — **Wide-range VCO, VCO temperature frequency drift, VCO fine-tune , on-chip inductor, PLL**

I. INTRODUCTION

With the industry trend towards single chip implementation of multi-standard multi-band transceiver [1], there is a growing demand for wide-range VCO design which satisfies very stringent requirements over frequency and temperature. In wide-range VCO design, besides low Q of on-chip inductor, loss of switchable capacitor arrays further degrades tank loaded-Q which consequently results in higher frequency drift over temperature [2]. For many applications such as full duplex systems, once the tuning is finished and the coarse-tune word is selected, PLL must remain locked over the entire temperature range without coarse-tune recalibration [3]–[5]. To guarantee the locked condition, several solutions have been proposed which can be categorized into two main approaches: (I) Extending VCO fine-tune frequency range by increasing VCO gain [3] and/or PLL tune voltage range [4]; (II) Reducing VCO temperature frequency drift by adding redundancy to VCO circuit [5]. High VCO gain causes high sensitivity to PLL tune voltage noise which increases the spur level and degrades the phase noise. Moreover, as power supply and size shrink, PLL tune voltage range decreases too. On the other hand, the proposed circuitries to reduce VCO temperature frequency drift degrade phase noise and make VCO prone to power supply noise and pushing.

This paper presents a modified temperature adaptive tuning scheme. Here, during coarse-tune, VCO tune voltage is a function of temperature which resembles the inverse function of VCO fine-tune characteristic. This guarantees that, regardless of coarse-tune temperature, PLL can remain locked despite as high VCO temperature frequency drift as VCO fine-tune range. Since this is the maximum frequency range VCO can cover without coarse-tune recalibration, the proposed approach is the optimum tuning scheme to deal with VCO temperature frequency drift. It paves the way to design a wide-range VCO with a small gain, making VCO robust to PLL tune voltage noise and spurs. The proposed scheme is implemented in the RX VCO of [1].

This note is organized as follows. The tuning process of a wide-range VCO is briefly reviewed in Section II. The proposed optimum temperature adaptive tuning, the main contribution of this paper, is presented in Section III. Design, circuit implementation and Lab measurement results are described in Section IV. The conclusions are summarized in Section V.

II. WIDE-RANGE VCO FREQUENCY TUNING

To support a wide frequency range, VCO frequency range is first covered in discrete steps by digitally switchable coarse-tune capacitor arrays. Then, VCO fine-tune varactor(s) continuously covers each sub-range by analog tune voltage [4]. To guarantee the coverage, the fine-tune range should be no smaller than the coarse-tune resolution. Figure 1 illustrates block diagram of VCO frequency tuning circuit which involves two loops: DFC (Digital Frequency Converter)/VCO loop in coarse-tune and PLL/VCO loop in fine-tune. Figure 2 provides the details of the frequency tuning process of a VCO with a single PN-Junction varactor. It depicts the corresponding fine-tune curves of two consecutive coarse-tune codes: n and n-1, which the target frequency lies in between. For the illustration purpose, a very large coarse-tune resolution is chosen.

As illustrated, VCO frequency tuning is a two-step procedure: coarse-tune and fine-tune. Based on an M-bit target frequency word, f_{target}, VCO frequency is first coarse-tuned to the closest discrete frequency available by the coarse-tune capacitor arrays, coarse-tune curve number 'n'. The N-bit digital coarse-tune code, 'n', is calculated by the DFC/VCO closed-loop, using the successive approximation method and a reference clock, f_{ref}. During coarse-tune calibration, VCO tune voltage, dfc_vtune, has

978-1-4244-6240-7/10 $26.00 © 2010 IEEE

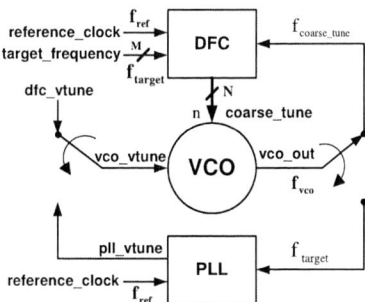

Fig. 1. Block diagram of frequency tuning circuit of a wide-range VCO

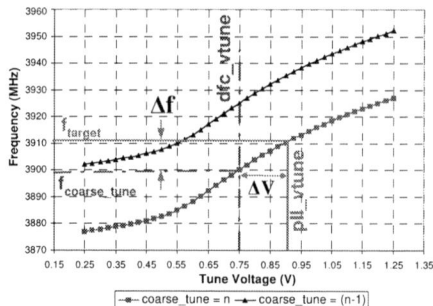

Fig. 2. VCO frequency tuning, two consecutive fine-tune curves which the target frequency lies in between

been so far kept constant [2]-[5]. Once VCO is coarse-tuned, there exists a difference between the target frequency and the coarse-tuned VCO frequency, which is called coarse-tune frequency error, Δf. Now VCO tune voltage is connected to PLL to close the PLL/VCO loop. Once locked, PLL corrects the tune voltage so that VCO frequency equals the target frequency, removing the coarse-tune error, Δf. The amount of the tune voltage correction by PLL, ΔV, is a function of both the coarse-tune frequency error, Δf, and VCO fine-tune characteristic.

III. OPTIMUM TEMPERATURE ADAPTIVE TUNING

In many applications such as 3G transceivers, once the tuning is finished and the coarse-tune word is selected, PLL must remain locked over the entire temperature range. Figure 3 shows the drift of VCO fine-tune curve over temperature which, for the illustration purpose, is equal to the fine-tune frequency range. After the completion of the tuning at f_{nom}, if it gets hot (cold), VCO frequency drifts down (up) linearly over temperature [5]. To maintain the same target frequency of f_{nom}, PLL has to increase (decrease) the tune voltage to make up for VCO frequency drift. The required tune voltage adjustment is a function of both VCO frequency drift over temperature and VCO fine-tune characteristic. The maximum tolerable VCO temperature frequency drift is set by DFC tune voltage. As illustrated in Fig. 3, if it is a fixed voltage [2]-[5], then to guarantee PLL locked condition, VCO temperature frequency drift has to be smaller than half the VCO fine-tune frequency range. Note that, if VCO is tuned at f_{cold} (f_{hot}), it becomes out of lock once temperature rises (drops) to nominal.

Optimally PLL/VCO can remain locked in the presence of as big VCO temperature frequency drift as VCO fine-tune frequency range if DFC tune voltage is set properly. Contemplating on Fig. 3, if DFC tune voltage, 'dfc_vtune', is a function of temperature such that it changes from the minimum PLL tune voltage at cold to the maximum PLL tune voltage at hot, then once PLL/VCO is locked at hot/cold, it remains locked despite as high VCO temperature

frequency drift as VCO fine-tune frequency range. On the other hand, if PLL/VCO is locked at f_{amb} at T_{amb} (ambient temperature), VCO frequency can drop/rise by (f_{amb} - f_{hot})/(f_{cold} - f_{amb}) at hot/cold. Figure 4 illustrates VCO temperature frequency drift, optimum DFC tune voltage and VCO fine-tune characteristic at T_{amb}. Temperature frequency drift of VCO is assumed to be as big as VCO fine-tune frequency range. As shown, the VCO fine-tune characteristic of Fig. 4 suggests that at T_{amb}, there is an optimum VCO tune voltage, V_{amb}, at which if VCO is coarse-tuned, PLL can compensate for VCO frequency drop/rise of (f_{amb}-f_{hot})/(f_{cold}-f_{amb}) by increasing/decreasing tune voltage by (V_{hot}-V_{amb})/(V_{amb}-V_{cold}). As depicted in Fig. 4, the optimum DFC tune voltage is a function of temperature so that it resembles the inverse function of VCO fine-tune characteristic.

IV. DESIGN AND CIRCUIT IMPLEMENTATION

Optimum DFC tune voltage would be a linear function of temperature only and only if VCO fine-tune characteristic is linear. Linear VCO fine-tune characteristic results in a constant VCO gain (K_{VCO}) over entire tune voltage range, making PLL bandwidth independent of tune voltage. Moreover, in circuit level, a linear DFC tune voltage can be implemented more accurately. These motivated us to design a fairly linear VCO fine-tune characteristic using identical MOS varacors biased at different voltages [6].

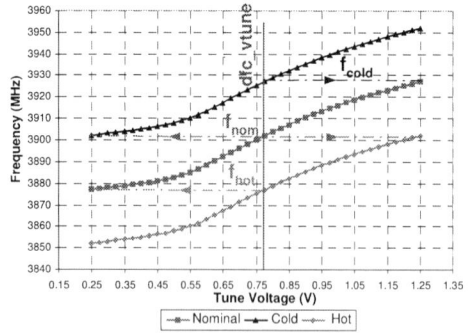

Fig. 3. Drift of VCO fine-tune curve over temperature

978-1-4244-6240-7/10 $26.00 © 2010 IEEE

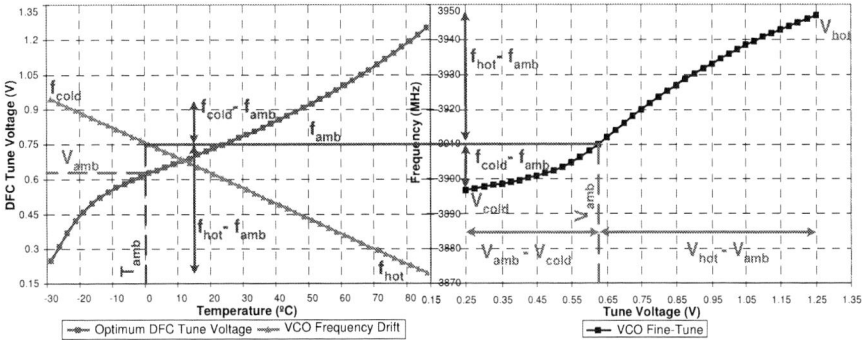

Fig. 4.   VCO temperature frequency drift, VCO fine-tune curve at ambient temperature, T_{amb}, and the optimum DFC tune voltage

The idea is implemented in the RX VCO of [1]. With an 8-bit coarse-tune capacitor array, VCO covers from 3.15GHz to 4.60GHz to guarantee the required frequency coverage of 3.45GHz to 4.35GHz over PVT. Using linear DFC tune voltage vs. temperature, VCO remains locked despite temperature frequency drift of about VCO gain. For instance, while exhibiting gain of 50MHz/V at 3.90GHz, VCO remains locked despite 45MHz frequency drift over [-30°C, 85°C].

In Section III, for illustration purpose, we neglected the coarse-tune resolution and the uncertainty of the DFC tune voltage which both call for a smaller tolerable temperature frequency drift than VCO fine-tune frequency range. The tune voltage equivalent of the coarse-tune resolution of the design is about ±100mV across the frequency range. On the other hand, due to mainly mismatch, though linear with temperature, the DFC tune voltage would fall within a ±100mV (±3σ) wide channel, further reducing effective PLL tune voltage range. Both VCO and PLL operate at 1.60V supply voltage, enabling 1.3V (0.15V to 1.45V) PLL tune voltage range. However, the effective PLL tune voltage range would be shrunk to 0.9V (0.35V to 1.25V) by the explained non-idealities. Therefore the design should be targeted to construct a DFC tune voltage which linearly varies from 0.35V to 1.25V over [-30°C, 85°C]. If PLL is locked at -30°C (85°C), due to the coarse-tune resolution of ±100mV and the DFC tune voltage uncertainty of ±100mV, the PLL tune voltage can be as low as 0.15V (1.05V) and as high as 0.55V (1.45V), leaving 0.9V (45MHz at 3.90GHz) for the VCO temperature frequency shift.

A. DFC Tune Voltage Circuit

Figure 5 shows the proposed circuit to construct such a linear DFC tune voltage. The proportional to absolute temperature (PTAT) current, I_{ptat}, of the already existing BandGap voltage generator, provides the linearity vs. temperature component. However, since the PTAT current is proportional to absolute temperature, a constant current, I_{const}, should be subtracted to compensate for the excessive positive offset of the PTAT current.

B. Design Considerations and Measurement Results

Percentage wise, the uncertainty of the PTAT current is very small. However the excessive offset of the PTAT current which is canceled out by constant current, would elevate the absolute uncertainty of the DFC tune voltage considerably. The uncertainty of the constant current would contribute to that as well. For the designed circuit, although 3σ of PTAT current is only 3%, that of the DFC tune voltage is less than 100mV (smaller in lower voltage).

The tune voltage equivalent of the coarse-tune resolution has to be relatively small over the entire frequency and temperature ranges. Figure 6 shows the PLL tune voltage at -30°C, 27°C and 85°C when the target frequency sweeps the entire frequency range in 2MHz steps. As illustrated, the measured coarse-tune error is limited to ±100mV. Figure 6 also confirms the linearity and the correct coverage of the DFC tune voltage over temperature. Figure 7 depicts the PLL tune voltage deviation due to temperature after VCO is tuned at 3.90GHz, first tuned at -30°C and then at 85°C. To Maintain the locked condition over [-30°C, 85°C], PLL adjusts the tune voltage to make up for the VCO temperature frequency drift. Note that at both -30°C and 85°C, VCO is coarse-tuned to the same coarse-tune curve (code) but at two different DFC tune voltage values which is the indicative of VCO fine-tune characteristic being compensated by DFC tune voltage. Figure 7 also resembles the inverse function of a fairly linear

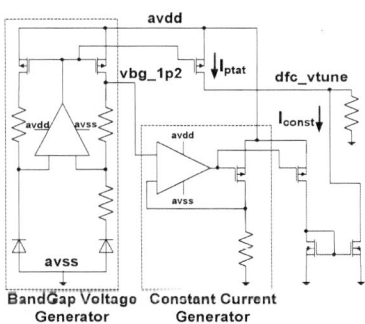

Fig. 5.   The proposed circuit to construct DFC tune voltage

978-1-4244-6240-7/10 $26.00 © 2010 IEEE

VCO fine-tune characteristic at 3.90GHz.

Figure 8 shows VCO phase noise, at 3.9GHz, after passing through a driver. Covering from 3.15GHz to 4.16GHz, VCO draws 8.5mA from 1.60V supply to achieve phase noise of -138dBc/Hz at 3.0MHz at 3.90GHz. The RX VCO location in chip photograph is shown in Fig. 9.

Table 1, summarizes the VCO measurement results. Utilizing an on-chip inductor of 0.13u CMOS process, it comfortably meets the receiver path requirements of the backward compatible 3G SAW-less transceiver [1].

V. CONCLUSION

A modified tuning scheme for wide-range VCO is presented. It optimizes the maximum tolerable VCO temperature frequency drift by allowing it to be as big as VCO fine-tune frequency range. It facilitates the design of a wide-range VCO with a small gain which reduces VCO sensitivity to noise of PLL tune voltage and supply. The method is implemented in the RX wide-range VCO of a multi-band multi-standard transceiver.

ACKNOWLEDGEMENT

The authors gratefully acknowledge the support of the entire Skyworks Solutions, Inc, Mobile Transceiver team.

REFERENCES

[1] T. Sowlati et al.,"Single-chip multi-band WCDMA/HSDPA /HSUPA/EGPRS transceiver with diversity receiver and 3G DigRF interface without SAW filter in transmitter/3G receiver paths," *IEEE ISSCC Dig. Tech. Papers,* pp. 116-117, 117a, February 2009

[2] L. S. L. Loke et al., "A versatile 90-nm CMOS charge-pump PLL for SerDes transmitter clocking," *IEEE J. Solid-State Circuits,* vol. 41, no. 8, pp. 1894-1907, August 2006

[3] L. Daniel et al., "A single-chip tri-band (2100, 1900, 850/800 MHz) WCDMA/HSDPA cellular transceiver," *IEEE J. Solid-State Circuits,* vol. 41, no. 5, pp. 1122-1132, May 2006

[4] Kun-Seok Lee et al., "A 0.13-μm CMOS Σ-Δ frequency synthesizer with an area optimizing LPF, fast AFC time, and a wideband VCO for WCDMA/GSM/GPRS/EDGE applications," *IEEE RFIC Symp. Dig.*, pp. 299-302, June 2008

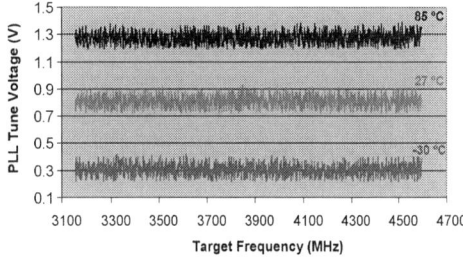

Fig. 6.   PLL tune voltage across frequency and temperature

[5] T. Tanzawa et al., "A temperature-compensated CMOS LC-VCO enabling the direct modulation architecture in 2.4GHz GFSK transmitter," *IEEE CICC Dig.*, pp. 273-276, Oct. 2004

[6] J. Mira et al., "Distributed MOS varactor biasing for VCO gain equalization in 0.13 μm CMOS technology," IEEE RFIC Symp. Dig., pp. 131-134, June 2004

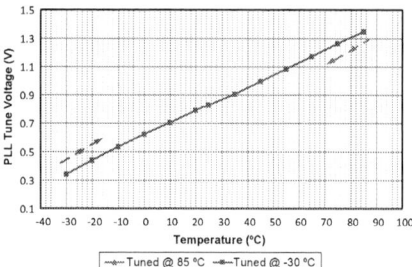

Fig. 7.   PLL tune voltage adjustment due to temperature, after VCO is tuned at 3.90GHz first at -30°C and then at 85°C

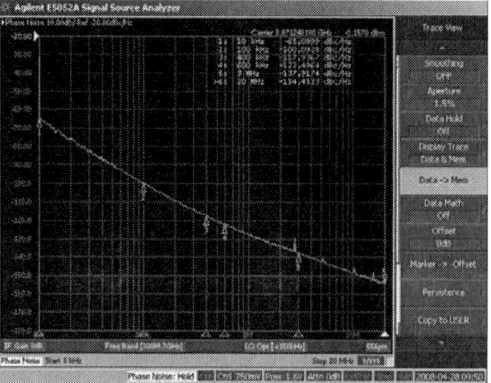

Fig. 8.   VCO phase noise after passing through a driver

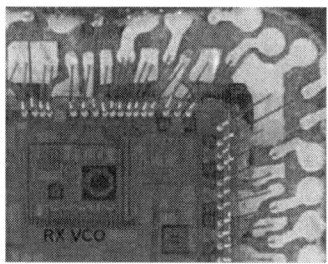

Fig. 9.   RX VCO in chip photograph

Table 1. VCO Measurement Summary

Power Supply / Current	1.60V / 8.5mA
Frequency Range	3.15GHz ~ 4.60GHz
VCO Gain (tune voltage: [0.15V, 1.35V])	50MHz/V ± 15% @ 3.9GHz
Frequency Drift over [-30°C,85°C]	45MHz @ 3.90GHz
Phase Noise	-138dBc/Hz @ 3.9GHz @ 3 MHz
Power Supply Pushing (no LDO)	3MH/V @ 3.9GHz @ DC

RMO4D-1

A 60GHz Transformer Coupled Amplifier in 65nm Digital CMOS

Michael Boers

Broadcom Corporation, Irvine, CA, 92617, USA

Abstract — A three stage transformer coupled amplifier for operation in the 57-64GHz band is presented. The amplifier uses differential capacitive neutralization and low loss transformers to achieve a gain of 30dB at 61GHz. The amplifier has an output compression point of 7.5dBm and a power added efficiency at 1dB compression of 9% at 57GHz. The amplifier has been fabricated in digital CMOS and occupies an area of 0.055mm^2.

Index Terms — CMOS MMIC, Millimeter-wave amplifier, power added efficiency, transformer coupled.

I. INTRODUCTION

Over the last decade interest in the 7GHz of unlicensed spectrum centered at 60GHz has grown extremely rapidly. The proliferation of wireless enabled devices, a continual trend towards higher data rates and the congestion occurring in popular unlicensed bands (such as 2.4 and 5GHz) has fuelled this interest. Applications for 60GHz systems include uncompressed video streaming, wireless personal area networks as well as low energy per bit communication links.

The power amplifier is a critical component in any communication system and typically consumes a large percentage of transmitter power. In millimeter-wave systems, a phased array can be employed to spatially combine the power from multiple antennas and increase the antenna gain. In such a system, achieving high gain and efficiency in the amplifier is crucial as it may be implemented multiple times depending on the number of antennas the solution requires.

Published 60GHz amplifiers typically fall into three different categories; transmission line based [1-3], lumped element based [4, 5] and transformer based [6, 7]. Transmission line amplifiers use typical MMIC design strategies and benefit from a well defined current return path. This aids accuracy; the tradeoff is the large area which these designs generally require. Lumped element and transformer based designs facilitate smaller layouts with the requirement of better modeling and design methodology to get close agreement between simulated and measured performance. In the case of differential designs, transformers are preferred as they allow a much simpler layout.

With a large number of antennas in a phased array, the output 1dB compression point for each individual amplifier can be relaxed. In this type of system, the most important parameters to consider are power added efficiency (PAE) and the gain provided by each stage. High PAE is desired in order to decrease the system power consumption. Gain is required to amplify the low power output from the up-conversion mixer. Having a high gain per stage reduces the number of stages needed in the amplifier.

II. METHODOLOGY AND IMPLEMENTATION

The most important aspect of millimeter-wave design is a structured design flow and a methodology that enables repeatable, first-pass correct design success. At the heart of the design flow is a requirement for accurate transistor and passive models along with an accurate parasitic extraction engine.

The amplifier presented in this paper was designed using the BSIM4 transistor model with RF enhancements based on [8]. Passive devices were simulated in HFSS and mapped to a custom wide-band equivalent circuit model using Microwave office and Matlab. Post-layout parasitic extraction was done to capture interconnect and via parasitics. As the interconnect is very small compared to the wavelength at 60GHz this lumped approach approximates the parasitics well. Care must be taken with the layout of the transistors to reduce effects that are difficult to extract and cause problems with the design such as routing induced gate inductance.

A. Amplifier topology

A differential transformer based topology was chosen for a number of reasons. Compared to the transmission line approach, it saves a lot of area. Compared to the lumped element approach, it makes for a much easier and compact layout. The schematic is shown in Fig 1.

In general differential circuits are preferred over single-ended circuits as they possess greater impunity to common mode noise. In a power amplifier a differential topology also has the benefit of splitting the output between two transistors. If a single-ended output is desired, this power can be efficiently combined using a BALUN like in this design.

978-1-4244-6240-7/10 $26.00 © 2010 IEEE

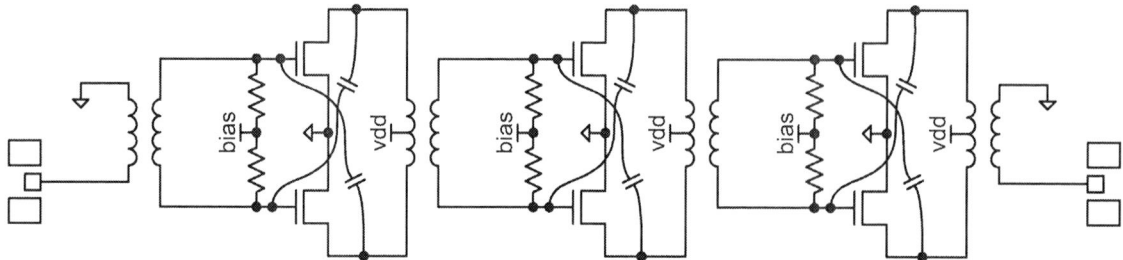

Fig. 1. Three stage amplifier schematic (simplified). The input and output are coupled through baluns. The inter-stage matching and coupling is achieved using single-loop transformers.

B. Transistor size and layout

Sizing the transistors for a millimeter-wave amplifier is a tradeoff between power handling capability and f_{max}. Below a certain width, the layout parasitics associated with the routing dominate; at larger widths the gate resistance is dominant. Simulation including the parasitic components leads to an optimal gate width of 1um.

Transistors are placed using custom P-Cells which include routing up to the highest metal. The P-Cell is designed to reduce the gate resistance and inductance, reduce C_{GD} and resistance between the channel and tie-downs. A 20x1um finger NMOS placed in an island surrounded by substrate contacts has been used as the unit cell in this amplifier. Transistors are sized 2 x 20 x 1um, 3 x 20 x 1um, 4 x 20 x 1um for the first, second and final stages.

C. Transformer design

Transformers are used throughout this amplifier. At the input and output the primary is terminated to provide a single-ended to differential transformation. At the input stage the secondary turns (connected to the transistor gates) center-tap is decoupled to ground through an RC filter to reduce the possibility of common-mode oscillation. Between stages, the transformers are used as matching networks, to provide VDD to the drain and couple the signal from the output of the previous stage to the input of the next.

The transformers are modeled using HFSS. Parameterized multi-port s-parameters are extracted for several loop widths, radius and overlaps. These s-parameters are input into a custom software module that creates equivalent circuit models for Cadence, ADS and AWR MWO. No substrate shielding techniques are used however a multi-metal ground ring is implemented around each transformer. This ground ring ensures that any metal outside the transformer perimeter does not affect the

performance; it also allows ground to be distributed throughout the circuit with low impedance. The Gmax for this transformer is extremely low, -0.5dB at 60GHz. The transformers are implemented using the top 2 copper layers (0.9um thick).

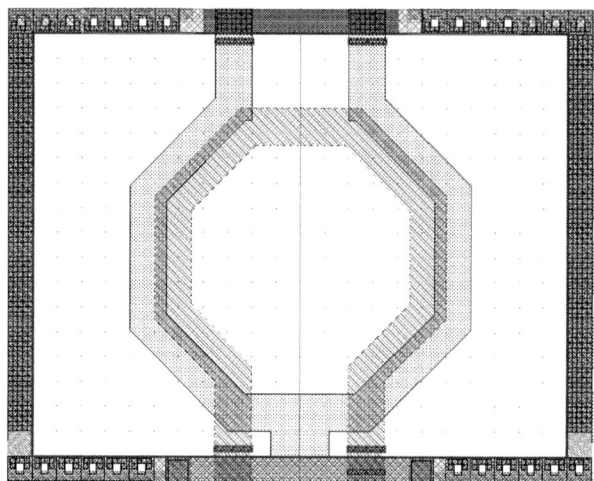

Fig. 2. Layout of an inter-stage transformer. The primary turn is used to provide VDD to the transistors through a center-tap.

D. Neutralization

Common-source amplifiers at millimeter-wave frequencies often suffer from gain degradation due to drain-gate capacitive feedback. This capacitance across a voltage gain node is magnified by the gain of the stage. In addition, this capacitance can de-stabilize the amplifier.

At lower frequencies a cascode topology is often employed to reduce the effect of the miller capacitance. At millimeter-wave frequencies, the capacitance associated with the drain and source nodes often degrades the performance and needs to be removed with a series inductor. This inductance adds area and in an amplifier with limited headroom the cascode topology is not practical.

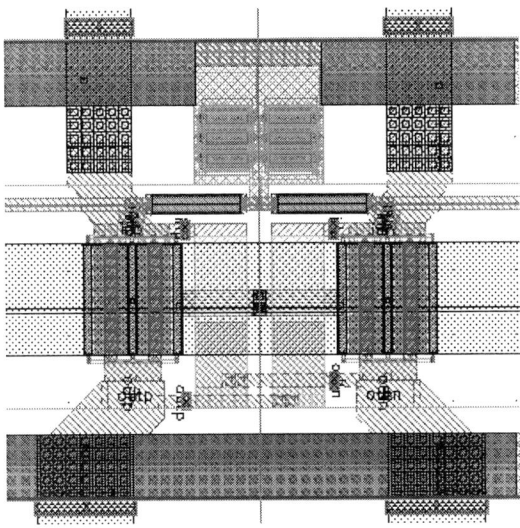

Fig. 3. Layout for the first differential transistor stage. Each side has 2 x 20um transistors, totaling 40um per side and 80um for the entire input. Neutralization capacitors are shown in the middle and are implemented using metal plate capacitors.

Neutralization reduces the effect of the drain-gate capacitance and aids stability. It has been demonstrated recently at 60GHz in [6]. It is easily achievable in a differential amplifier as out of phase signals are available. The neutralization is accomplished by capacitive feedback from the output of one phase to the input of the other. The feedback is implemented using a metal plate capacitor and the size varies depending on the width of the transistor from 15-30fF. The neutralization is only useful across a narrow frequency range and needs to be done with care as too much capacitance will lead to an un-stable design.

To ensure stability at low frequencies where the transistors have a large amount of gain, high pass elements such as transformers or series capacitors need to be used. At high frequencies the quickly rolling off frequency response of the amplifier aids stability.

Compared to other stabilization techniques such as series RC networks the neutralization technique does not degrade the performance of the amplifier.

E. Amplifier layout

The core of the amplifier (not including input balun which is only integrated for testing) takes an area of 350um x 150um. The compact layout is due to the use of transformers through-out the design. Apart from the reduced cost, the small size also keeps routing between the transistors to a minimum improving performance and reducing the need for additional modeling steps.

Fig. 4. Micrograph showing layout of the power amplifier.

III. MEASUREMENT RESULTS

The amplifier presented in this paper achieves a peak efficiency of 18% and a saturated output power of 10.5dBm (both limited by measurement). The gain is greater than 30dB from 61 to 65 GHz demonstrating the highest achieved gain per stage for any CMOS amplifier. The PAE at 1dB compression and 61GHz is 7.7%

Figure 5 shows the s21 and s22 of the amplifier across frequency for different temperature settings. The gates are biased with a constant voltage across temperature. With a temperature dependent bias network, the gain will be more constant across temperature.

Figure 6 shows the output 1dB compression point of the amplifier across frequency and temperature. Figure 7 shows the power added efficiency of the amplifier vs. input power.

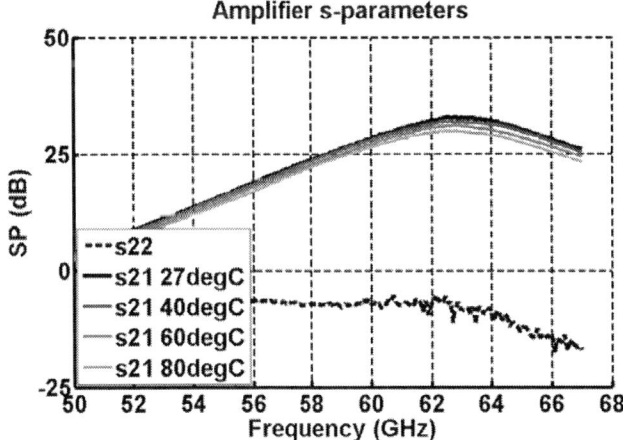

Fig. 5. s21 and s22 of the amplifier. s11 is not included as the input isn't matched. The input's match is differential and designed to be matched to the previous stage. The balun at the input is only integrated to aid testing.

TABLE I
SUMMARY OF RECENTLY PUBLISHED MM-WAVE AMPLIFIERS

Ref	Year	Process	Stages	Topology	Pg	Pg / stage	P1dB dBm	Psat dBm	Vdc (V)	Pdc (mW)	PAE p1dB
[4]	2006	90nm	3	L/C	5.2	1.73	6.4	9.3	1.5	39.75	6.05
[7]	2008	90nm	2	xfmr	7.7	3.85	9.0	12.3	1	88	7.09
[2]	2008	90nm	4	tline	8.3	2.075	8.2	10.6	1.2	220	2.7
[3]	2008	90nm	3	tline	14.3	4.77	10.0	11.0	1.0	150	6.35
[1]	2008	90nm	3	tline	17	5.67	5.1	8.4	1.5	54	5.8
[6]	2009	65nm	3	xfmr	15.8	5.3	2.5	11.5	1	43.5	4.0
[5]	2009	45nm	3	L/C	19	6.3		7.9	1.2		6.4
This work	2009	65nm	3	xfmr	30	10	6.8	10.6	1	65	7.7

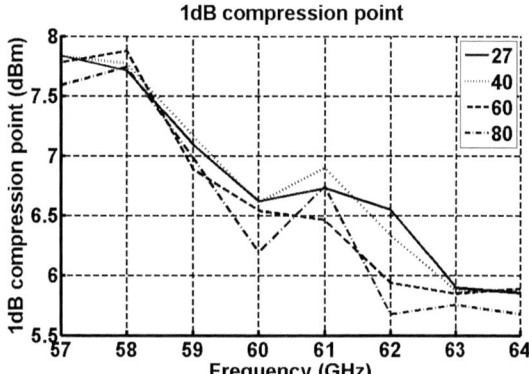

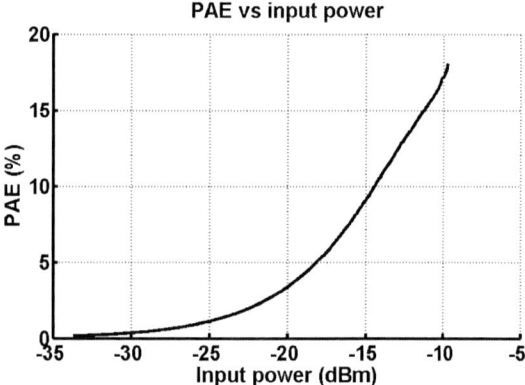

Fig. 6 (top). 1dB vs. freq across temperature.

Fig. 7 (bot). PAE vs. input power.

IV. CONCLUSION

Using a structured design methodology with high quality core active and passive models, first pass right 60GHz design is achievable. This paper demonstrates a amplifier for use in a multi-element 60GHz phased array front-end. The high efficiency reported is critical for reducing power consumption in the front-end where multiple antennas and amplifiers will be used together in order to increase antenna gain and spatially combine the power.

ACKNOWLEDGMENTS

The author would like to acknowledge the invaluable assistance and useful discussion from Reza Rofougaran, Arya Behzad, Mohammad Nariman, Ali Parsa and Hsin-Hsing Liao.

REFERENCES

[1] D. Dawn, S. Sarkar, P. Sen, B. Perumana, D. Yeh, S. Pinel, and J. Laskar: "17-dB-gain CMOS power amplifier at 60GHz". MTT-S, 2008

[2] T. Suzuki, Y. Kawano, M. Sato, T. Hirose, and K. Joshin: "60 and 77GHz Power Amplifiers in Standard 90nm CMOS". Solid-State Circuits Conference, 2008.

[3] M. Tanomura, Y. Hamada, S. Kishimoto, M. Ito, N. Orihashi, K. Maruhashi, and H. Shimawaki: "TX and RX Front-Ends for 60GHz Band in 90nm Standard Bulk CMOS". Solid-State Circuits Conference, 2008.

[4] T. Yao, M. Gordon, K. Yau, M.T. Yang, and S.P. Voinigescu: "60-GHz PA and LNA in 90-nm RF-CMOS". Radio Frequency Integrated Circuits (RFIC) Symposium, 2006.

[5] E. Cohen, S. Ravid, and D. Ritter: "60GHz 45nm PA for linear OFDM signal with predistortion correction achieving 6.1% PAE and −28dB EVM". Radio Frequency Integrated Circuits Symposium, 2009.

[6] W.L. Chan, J.R. Long, M. Spirito, and J.J. Pekarik: "A 60GHz-band 1V 11.5dBm power amplifier with 11% PAE in 65nm CMOS". International Solid-State Circuits Conference, 2009

[7] D. Chowdhury, P. Reynaert, and A.M. Niknejad: "A 60GHz 1V + 12.3dBm Transformer-Coupled Wideband PA in 90nm CMOS". International Solid-State Circuits Conference, 2008.

[8] T. Suet Fong, A.A. Osman, K. Mayaram, and H. Chenming: "A simple subcircuit extension of the BSIM3v3 model for CMOS RF design", Solid-State Circuits, IEEE Journal of, 2000.

A Stage-Scaled Distributed Power Amplifier Achieving 110GHz Bandwidth and 17.5dBm Peak Output Power

Jiashu Chen, Ali M. Niknejad

Berkeley Wireless Research Center, University of California at Berkeley, Berkeley, CA 94702

Abstract — This paper presents the design of a pseudo-differential distributed power amplifier in a 0.13μm SiGe BiCMOS process. Based on the newly proposed efficiency enhancing stage-scaling technique, the distributed power amplifier achieves a small-signal bandwidth of 110GHz, a peak saturated output power of 17.5dBm and a peak PAE of 13.2%. The measured 3dB output power bandwidth is greater than 77GHz. The amplifier consumes 119mA from a 3V supply.

Index Terms — Stage scaling, distributed power amplifier.

I. INTRODUCTION

There is growing interest in mm-wave electronics for communication, imaging and sensing applications. In particular ultra-wideband mm-wave imaging has the potential to complement many existing modalities due to high resolution (sub-mm-wave in the body), high contrast, and the low cost and non-invasive nature. To achieve fine spatial resolution, narrow pulses need to be transmitted and this calls for extremely wideband power generation and amplification. Conventional tuned amplifiers cannot satisfy the stringent system requirements due to its inherent tradeoff between gain and bandwidth. In addition, parasitic resistance and capacitance associated with large power devices significantly reduce the *maximum-stable-gain* and increases the sensitivity to load variations at mm-wave frequencies. Distributed amplification is an effective solution to these problems since transistor parasitics are absorbed into transmission lines with relatively high cut-off frequency and large power devices are inherently split into multiple smaller ones. A number of distributed amplifiers have been reported in silicon technologies with bandwidth approaching 100GHz [1-6]. However, one major deficiency of previously reported distributed amplifiers is their relative low efficiency, preventing them to be used for power amplification.

In this paper, we present the design methodology for extremely wideband distributed power amplifiers. In particular, we propose a stage-scaling technique to improve the collector efficiency as well as the PAE. In a stage-scaled distributed amplifier, the loaded collector-line impedance and the transistor size are scaled simultaneously from stage to stage to reduce the total DC power consumption without affecting the AC power delivered to the load. An 8-stage pseudo-differential stage-

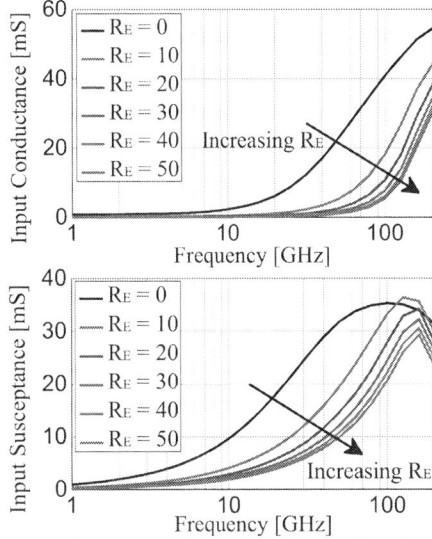

Fig. 1. Input conductance and susceptance of a 9μm transistor as a function of emitter degeneration resistor R_E.

scaled distributed power amplifier has been implemented based on the proposed idea. Measurement results will be discussed.

II. DESIGN CONSIDERATIONS FOR GAIN AND BANDWIDTH

The operation of distributed amplifier is based on a filter-like structure in which the transistor parasitic capacitors are absorbed into base and collector transmission lines to form two wideband multi-section low-pass filters while the gain stages are embedded in between them. In theory, the bandwidth of distributed amplifiers is determined by the cut-off frequencies of the two loaded transmission lines. In practice, however, it is very difficult to approach this theoretical limit due to imperfect on-chip passives. In particular, the transmission line loss dictates the achievable bandwidth in these high f_T processes. Two loss mechanisms exist. The series loss is caused by the metal resistance of the transmission line on the signal path and it is proportional to the transmission line length which is set by the susceptance of the gain stages. The shunt loss is mainly caused by the conductance of the gain stages which load the transmission line. Therefore, it is desirable to minimize both the conductance and the susceptance of the gain stages. Fig. 1 shows the input conductance and input

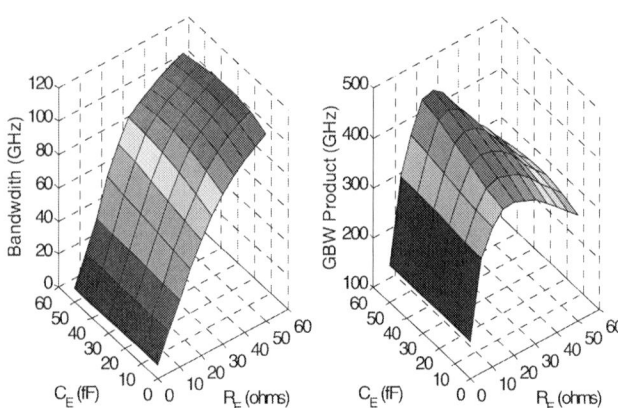

Fig. 2. Simulated bandwidth and GBW product as a function of emitter degeneration resistance R_E and capacitance C_E.

susceptance of a 9μm transistor. Due to the large bipolar base resistance, both parameters are significant at mm-wave frequencies. To remedy this problem, emitter degeneration resistors R_E are added to form a series feedback and both input conductance and susceptance reduce with increasing R_E. Since the collector capacitance is much smaller than the base capacitance, the problem is less severe at the output. In addition, the output conductance can be effectively reduced by using cascode topology. To further improve the bandwidth, small capacitors C_E are added in parallel with the degeneration resistor to create high frequency zeros in the voltage transfer function. Fig. 2 shows the simulated bandwidth and GBW product of the 8-stage distributed power amplifier with cascode gain stages as a function of emitter degeneration resistance R_E and capacitance C_E. As evident in the figure, the bandwidth is very limited without degeneration resistor and increasing R_E can significantly improve the bandwidth. Optimal GBW product can be achieved when R_E is in the vicinity of 30Ω. However, the use of RC combination requires extra care since large C_E values result in negative input impedance and can potentially cause stability problems. Given the aforementioned tradeoffs, R_E and C_E are chosen to be 35Ω and $20fF$ respectively resulting in 11dB of gain and 110GHz of bandwidth.

III. EFFICIENCY ENHANCEMENT USING STAGE-SCALING

In a conventional distributed amplifier, the input signal travels along the base-line and gets amplified by the gain stages. The amplified signals then add up constructively when they travel towards the load as illustrated in Fig. 3a. Due to distributed summation, the largest voltage swing occurs at the last stage and as a result, only the last stage experiences maximum allowed voltage swing when output

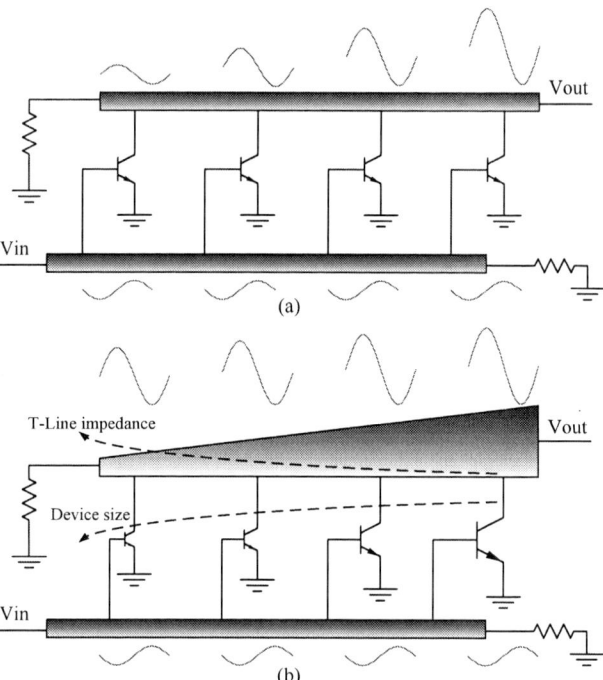

Fig. 3. (a) A conventional uniform distributed amplifier versus (b) a stage-scaled distributed amplifier.

power saturates while preceding stages have sub-optimal voltage swings. Therefore if the voltage swing can be increased in the preceding stages, less current swing is needed to produce the same power. Under Class-A operation, this directly translates into reduction of bias current and DC power consumption. To realize this concept, the loaded collector-line impedance needs to scale up from the last stage to the first stage while the transistor size and the bias current need to scale down in the same direction. The proposed stage-scaled structure is illustrated in Fig. 3b. In order to optimize the efficiency, it's crucial to determine the correct scaling coefficients between the stages. Assuming the bipolar device size and the bias current scales by a factor of k between two stages in the direction from output to input while the loaded collector-line impedance scales by a factor of z, the peak power delivered by the ith stage is $(k^2 z)^{N-i} P_0$ where P_0 is the peak power delivered by each stage in a non-scaled distributed power amplifier and N is the total number of stages. To maintain the same output power as the non-scaled counterpart, the scaling factor $k^2 z$ must be unity. This first order analysis guides us on how to simultaneously scale the collector-line impedance and the transistor size without losing output power, but it neglects the fact that large signal reflection occurs on the loaded transmission line when the impedance scaling becomes significant and one cannot keep reducing the k value. In addition, the length of the transmission line increases with higher impedance values which also introduces more loss.

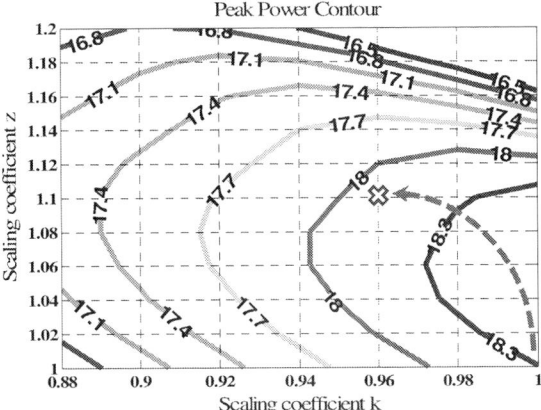

Fig. 4. Peak power contour.

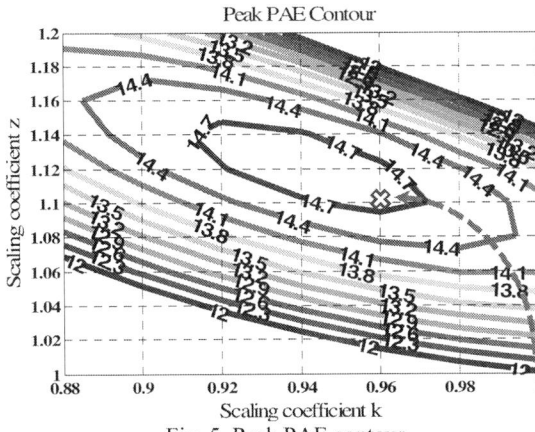

Fig. 5. Peak PAE contour.

To obtain a more accurate scaling prediction, large-signal simulation has been performed with varied k and z values and the results are plotted in Fig. 4 and Fig. 5 in the form of contours. The lower right corners of the figures where $k = z = 1$ represent a non-scaled amplifier and clearly it is far from the optimal PAE region. On the other hand, the peak output power degrades significantly when either the k becomes too small or the z becomes too large. Given the tradeoffs, we choose a k value of 0.96 and a z value of 1.1 which give a simulated peak output power of 18.1dBm and a PAE of 14.7%.

IV. IMPLEMENTATION

Due to the highest loaded transmission line impedance achievable, the first two stages of the amplifier are not scaled. This will slightly degrade the efficiency performance since the DC power consumption becomes larger, but careful analysis and simulation show that it can be compensated by slightly reducing the k value to 0.956. Fig. 6 shows the single-ended schematic of the 8-stage distributed power amplifier incorporating stage-scaling. The complete amplifier is made differential to double the

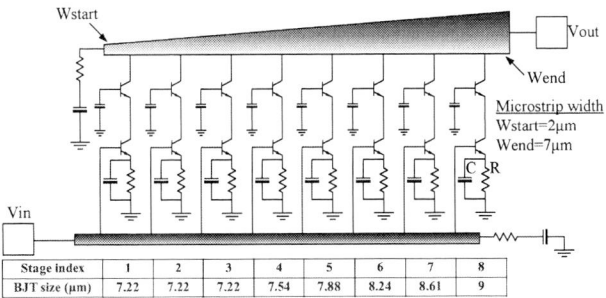

Stage index	1	2	3	4	5	6	7	8
BJT size (µm)	7.22	7.22	7.22	7.54	7.88	8.24	8.61	9

Fig. 6. Schematic of the stage-scaled distributed power amplifier.

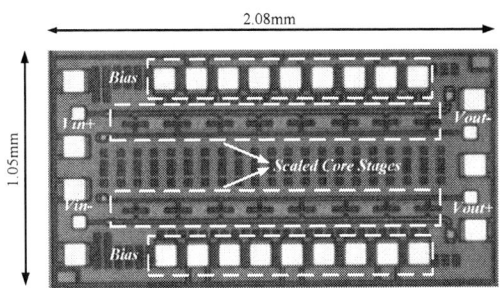

Fig. 7. Chip micrograph of the stage-scaled distributed power amplifier.

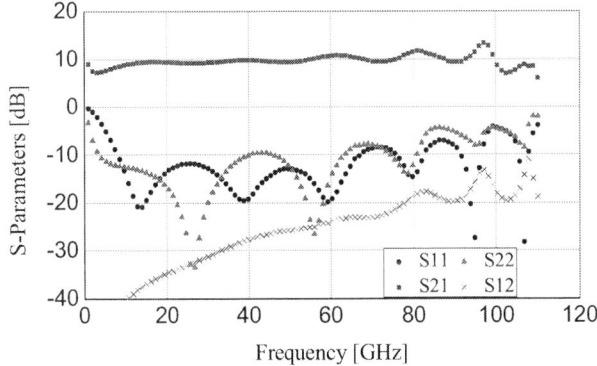

Fig. 8. Measured S-parameters of the stage-scaled distributed power amplifier.

output power. Both base and collector transmission lines are implemented in low-loss microtrip structures and the impedance scaling is achieved by tapering the width of the microstrip. The chip is fabricated in a 0.13µm SiGe BiCMOS process and the chip micrograph is shown in Fig. 7. It measures 2.08mm by 1.05mm.

V. EXPERIMENTAL RESULTS

The measurement is performed by on-wafer probing. The amplifier draws 119mA from a 3V power supply. Fig. 8 shows the measured S-parameters of the distributed power amplifier. It achieves a small-signal gain of 10dB and a 3-dB bandwidth of 110GHz. Fig. 9 and Fig. 10 shows the measured large-signal performance. The measured peak P_{-1dB} and P_{sat} are 16.7dBm at 40GHz and 17.5dBm at 60-

REF	Gain (dB)	Small-Signal BW (GHz)	Peak P_{-1dB} (dBm)	Peak P_{sat} (dBm)	Peak PAE	3dB output power BW (GHz)	Technology
[1]	11	90	12	-	6.8%	-	0.12μm CMOS SOI
[2]	7	70	10	-	6.9%	-	90nm CMOS
[3]	19	74	3.7	-	2.7%	40	90nm CMOS
[4]	14	74	3.2	-	2.4%	50	90nm CMOS
[5]	7.4	80	8	-	4.1%	30	90nm CMOS
[6]	9	90	-	21	3.5%	24	0.13μm SiGe BiCMOS
This Work	**10**	**110**	**16.7**	**17.5**	**13.2%**	**77**	**0.13μm SiGe BiCMOS**

TABLE. I. COMPARISON WITH RECENTLY REPORTED DISTRIBUTED AMPLIFIERS

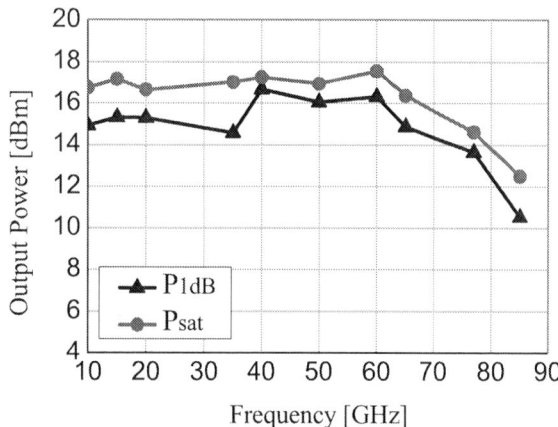

Fig. 9. Measured P_{-1dB} and P_{sat} of the distributed power amplifier.

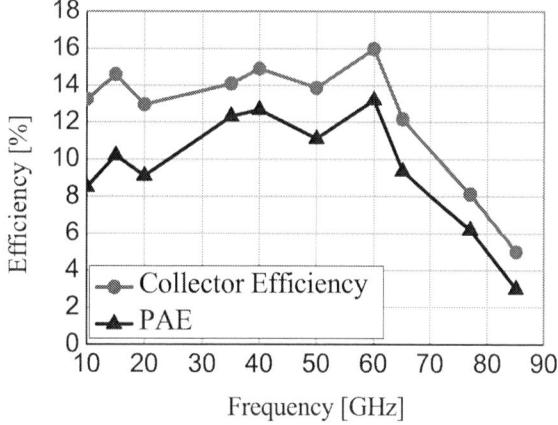

Fig. 10. Measured collector efficiency and PAE of the distributed power amplifier.

GHz respectively. The measured 3-dB output power bandwidth is greater than 77GHz. The output power roll-off above 65GHz is mainly due to the bandwidth limitation of the bias-T used in the experiment. A peak collector efficiency of 16% and a peak PAE of 13.2% are achieved at 60GHz. Compared to measurements on a uniform non-scaled distributed power amplifier designed in the same technology, the normalized efficiency improvement is more than 20% at the same output power

level. Comparison with other recently reported distributed amplifiers are summarized in Table. I.

VI. CONCLUSION

In this paper, we presented the analyses and design procedure for a distributed power amplifier incorporating the proposed stage-scaling concept. By utilizing unused voltage swing of gain stages, the stage-scaling technique can effectively reduce the DC power consumption without affecting the AC power delivered to the load. The fabricated amplifier achieves a record 110GHz small-signal bandwidth and a peak 17.5dBm saturated output power at 13.2% peak PAE.

ACKNOWLEDGEMENT

The authors acknowledge BWRC sponsors, NSF Infrastructure Grant No. 0403427, chip fabrication donation by IBM, MOSIS and Dawn Wang for research support. The authors also want to thank Ehsan Adabi, Jun-Chau Chien and Amin Arbabian for useful discussions.

REFERENCES

[1] J. Kim, J. Plouchart, N. Zamddmer et al., "A 12dBm 320GHz GBW distributed amplifier in a 0.12μm SOI CMOS," *ISSCC Dig. Tech. Papers*, pp. 478-479, Feb. 2004.
[2] M.-D. Tsai et al., "A 70GHz Cascaded Multi-Stage Distributed Amplifier in 90nm CMOS Technology," *ISSCC Dig. Tech. Papers*, pp. 402-403, Feb. 2005.
[3] A. Arbabian and A. M. Niknejad, "A Broadband Distributed Amplifier with Internal Feedback Providing 660GHz GBW in 90nm CMOS," *ISSCC Dig. Tech. Papers*, pp. 196-197, Feb. 2008.
[4] A. Arbabian and A. M. Niknejad, "A Tapered Cascaded Multi-stage Distributed Amplifier with 370 GHz in 90 nm CMOS," in *Proc. Radio Freq. Integr. Circuits Symp.*, pp. 57–60, Jun. 2008.
[5] R.-C. Liu et al., "An 80GHz travelling-wave amplifier in a 90nm CMOS technology," *ISSCC Dig. Tech. Papers*, pp. 154-155, Feb. 2005.
[6] E. Afshari et al., "Electrical Funnel: A Broadband Signal Combining Method," *ISSCC Dig. Tech. Papers*, pp. 206-208, Feb. 2006.

RMO4D-3

DC Hot Carrier Stress Effect on CMOS 65nm 60 GHz Power Amplifiers

T. Quémerais[1,2], L. Moquillon[2], V. Huard[2], J.-M. Fournier[1], P. Benech[1], N. Corrao[1]

[1]IMEP-LHAC, UMR INPG/UJF/US/CNRS, 3 parvis Louis Néel, BP 257, 38016 Grenoble Cedex, France
[2]STMicroelectronics, 850 rue Jean Monnet 38920 Crolles, France
e-mail : thomas.quemerais@st.com

Abstract — The effects of dc hot carrier stress on the characteristics of 60GHz power amplifiers on CMOS 65nm are investigated. The increase in the threshold voltage, the decrease in the transconductance and the output conductance of the MOSFETs caused by hot carriers leads to a loss performances of the PAs. A reliability study is first made on a 1 stage PA to validate the ageing model and the degradation explanation. A drop of 5% the gain, 7% of the OCP_{1dB}, 7% of the P_{sat} are measured at 58GHz after 50 hours of stress under Vdd=1.7V on a 4 stages amplifier.

Index Terms — CMOS mmw circuits, 65nm technology, power amplifier, hot carrier stress, reliability.

I. INTRODUCTION

CMOS technologies are enabling integration of high frequency applications like HDMI, WLAN or WPAN communications (60GHz band). The CMOS power amplifier (PA) is one of the most challenging blocks in a transmitter due to the important reliability constraints existing on MOS transistors and losses in the integrated passive components.
The feasibility of millimeter wave (mmw) PAs has been demonstrated previously in SiGe and CMOS technologies. Reliability hot carrier impact on PA performances has been studied [8] and [9] but never at millimeter wave frequencies. Indeed, during operation, common source class A mmw power amplifiers degradation is due to the hot carrier injection phenomenon.

Reducing the transistor channel length turns the hot carriers into an important reliability issue [4] to [9]. Carriers in the channel can gain high energy (hot carrier) in the pinch off region and cause an avalanche effect very similar to what happens in a reverse-biased p-n junction. The collisions of hot carriers with the atomic bonds at the interface of the substrate and gate oxide leads to the generation of dangling bonds, also known as interface traps [7]. These traps affect different parameters of the MOS.
Effects of hot carriers on the RF characteristics of single NMOS transistor have been before reported ([4] to [9]).

Studying the power amplifier reliability consists in investigating the MOS transistor degradation with time, when stressed under high voltage levels.

In this paper, a dc hot carrier reliability model is demonstrated to be correct at 60 GHz thanks to a 1 stage 60 GHz power amplifier reliability study in CMOS 65nm. An ageing study is also done on a state of the art 4 stages PA [1]. The reliability model is validated thank to comparison with the measurements of the PA characteristic parameters (small and large signals) degradation after 50 hours of stress under Vdd=1.7V power supply. The reliability test protocol is also described.

The section II is devoted to the devices and circuits description. In section III, the experimental procedure is presented. In section IV, simulations and measurements of power gain, compression point and S parameters are compared for the two amplifiers before and after stress and the results are discussed before conclude.

II. DEVICES AND CIRCUIT DESCRIPTION

The developed design methodology of the PAs is described in [1] and their performances are summarized on Table I. The PAs are biased under Vdd=1.2V and are matched to the 50 ohms external impedances.

To design a millimeter wave power amplifier, a thin film microstrip line and an accurate millimeter wave MOS model are developed [1]. For this purpose, an existing PSP (Penn State Philips) model dedicated to RF application is used and improved with an extrinsic model.
This model includes parasitic elements due to drain, source and gate accesses like series resistors, coupling and substrate capacitors and access lines. Electromigration constraints at 125°C are also considered in the design flow. The amplifier power gain is inversely proportional to the gate length [2]. Increasing this dimension to gain life time deteriorates the PA efficiency.

978-1-4244-6240-7/10 $26.00 © 2010 IEEE

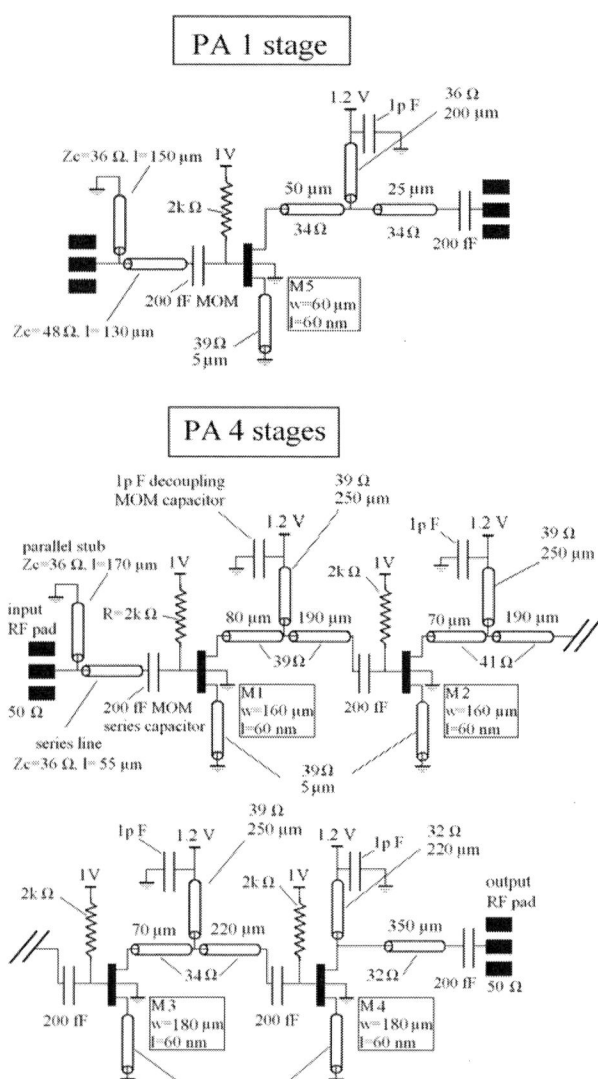

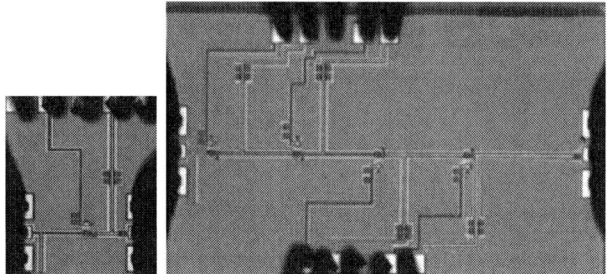

Fig. 1. Schematic of the CMOS 65nm 1 stage (up) and 4 stages (down) power amplifiers including element values.

Fig. 2. Micro-photography of the 1 stage PA (left) and the 4 stages PA (right).

Fig. 2 shows the micro-photography of the PAs implemented in 65nm technology. The die sizes are respectively 0.4×0.6 mm^2 and 1.5×0.8 mm^2 for the 1 stage and the 4 stages amplifiers. The measured performances of the PAs at Vdd=1.2V are summarized in Table I.

TABLE I
CMOS POWER AMPLIFIERS PERFORMANCES

PA	P_{sat} (dBm)	P_{1dB} (dBm)	G (dB)	Cons. (mW)	PAE (%)	FOM
1 stage	9.2	6.4	4.5	20.5	26	22
4 stages	14.2	12.2	13.7	300	8.4	174

$(FOM = f^2 \times G \times P_{sat} \times PAE$, from ITRS$)$

III. EXPERIMENTAL PROCEDURE

When biasing the amplifiers at an operating point higher than Vdd=1.2 V, strong electric field are generated inside the transistor. This reduces the time at which the hot carrier effects can be observed. Nevertheless, too high drain voltage applied to the transistor will make the oxide broken.

A model of degradation under dc hot carrier stress to study the PAs ageing is implemented in Mentor Graphics Eldo simulator. For the simulation of the circuit after stress, the values of carrier mobility, threshold voltage, DIBL (drain induced barrier lowering) coefficient, velocity saturation and channel-length modulation of the MOSFETs take into account the dc hot carrier stress effects. These parameters degradation is only function of the stress duration, the stress voltage Vdd and is modeled using the substrate current model presented in [7]. Furthermore with dc hot carrier stress these parameters degradation are not function of the transistors dimension and biasing current. Then, process parameters are introduced into the model thanks to measurements made on a large quantity of MOSFETs of various widths, length for several stress voltage and stress time durations of stress.

Experimentally, large 65nm MOSFETs width (W>>10 μm) avoid statistical approach of the degradation. Indeed, the degradation of the presented PAs can be measured on few devices.

To study the impact of hot carrier on PAs performances, the power gain, the input and output matching (S_{11} and S_{22}), the output saturated power (P_{sat}) and the output 1dB compression point (OCP1dB) are studied before and after the stress. The measurements (S parameters and power characterization) are performed before ageing and after 1 hour, 10 hours, 20 hours and 50 hours of stress. These measurements are done with Vgs=1V and Vdd=1.56V, 1.7V and 1.83V. These stress voltages are chosen to observe significant degradation (higher than incertitude of measurement) of the PAs parameters after 50 hours.

IV. COMPARISON BETWEEN MEASUREMENT AND SIMULATION

Fig. 3 and 4 respectively present the measured and simulated S_{21} and S_{11} parameter before and after a stress of 44 hours with Vdd=1.83V of the 1 stage PA.

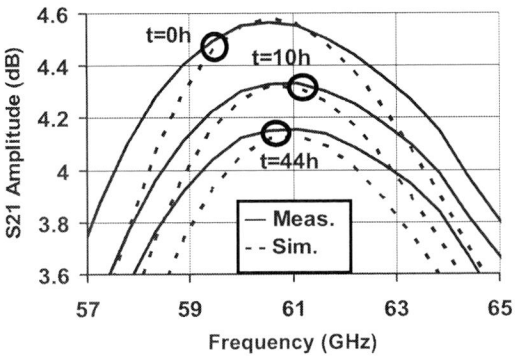

Fig. 3.   Simulated and measured degradation of the 1 stage PA S_{21} parameter stressed with Vdd=1.83 V.

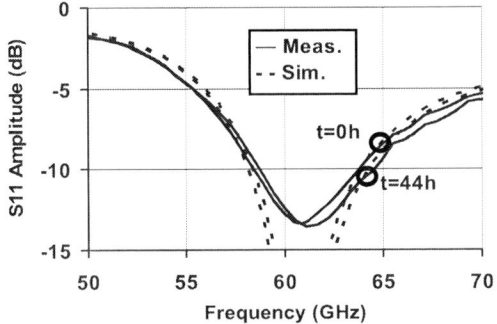

Fig. 4.   Simulated and measured degradation of the 1 stage PA S_{11} and S_{22} parameters stressed with Vdd=1.83V.

The difference between the measured and the simulated bandwidth around 60 GHz shown Fig. 4 and Fig. 5, comes from losses not took into account in the stub models.

A relative gain drift of 8 % is measured at 60 GHz after the amplifier was stressed with 1.83V during 44h, while S_{11} and S_{22} parameters have only a relative drift lower than 3 % at 60 GHz. Moreover measurements and simulations show that MOSFETs intrinsic capacitances are not impacted by hot carrier degradation because no significant frequency shifts are observed on the S parameters after stress. The MOS threshold voltage V_{th}, the transconductance gm and the drain source dynamic resistance R_{ds} are extracted from the transistor static characteristics before and after stress. V_{th} has increased of 3 %, whereas gm has dropped of 3 % and R_{ds} of 2 %.

The analytic equation of the power gain G_p of a 1 stage PA [3]. After dc hot carrier stress the relative drift of the linear gain is only proportional to the relative drift of gm and R_{ds}. In the approximation that $R_{ds}>>R_{load}$:

$$\frac{\Delta G_p}{G_p} \approx 2.\frac{\Delta gm}{gm} \qquad (1)$$

Equation (1) is validated by the measurements presented on Fig. 1 and the MOS transconductance extraction. The transconductance gm relative drift is expressed as:

$$\frac{\Delta gm}{gm} \approx \frac{\Delta V_{th}}{V_{th}} + \frac{\Delta \mu_n}{\mu_n} \qquad (2)$$

where gm is function of the threshold voltage V_{th} and the carrier mobility μ_n. From equation (2) and the measurement of V_{th}, the carrier mobility relative drift can be approximated to 1%.

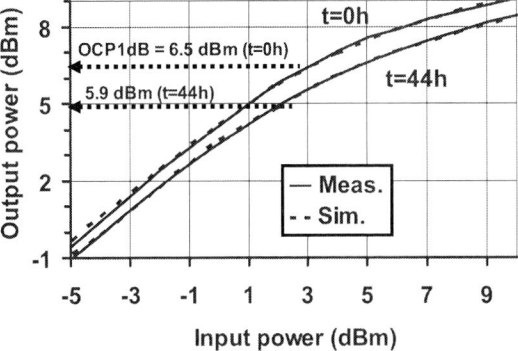

Fig. 5. Simulated and measured degradation of 1 stage PA output power and gain versus input power at 60GHz stressed after 44h of stress at Vdd=1.83V.

Fig. 5 presents the output power, the power gain and the output 1dB compression point before and after a stress of 44 hours under Vdd=1.83 V. The saturated power and the output 1dB compression points have both decreased of 10 %. Furthermore, P_{sat} and $OCP1_{dB}$ [2] are both function of μ_n and $(Vgs-V_{th})^2$. Consequently the relative drift of P_{sat} can be expressed as:

$$\frac{\Delta P_{sat}}{P_{sat}} = \frac{\Delta OCP_{1dB}}{OCP_{1dB}} \approx \frac{\Delta Ids}{Ids} \approx 2.\frac{\Delta V_{th}}{V_{th}} + \frac{\Delta \mu_n}{\mu_n} \qquad (3)$$

Equation (3) is validated through the measurements shown on Fig. 5 and V_{th} and μ_n extracted values. Moreover, Fig. 3 to 5 show an excellent agreement between measurements and simulations at 60 GHz. These results demonstrate that the dc hot carrier stress reliability model is accurate at millimeter wave frequencies to predict the degradation of the small and large signal power amplifier performances.

In a second step, a reliability study is made on a 4 stages state of the art millimeter wave PA. The amplifier is stressed at Vdd=1.7 V during 50 hours. On Fig. 6 and 7 are respectively presented the measured and simulated S_{21}, the output power and OCP_{1dB} parameter before and after stress.

978-1-4244-6240-7/10 $26.00 © 2010 IEEE   317

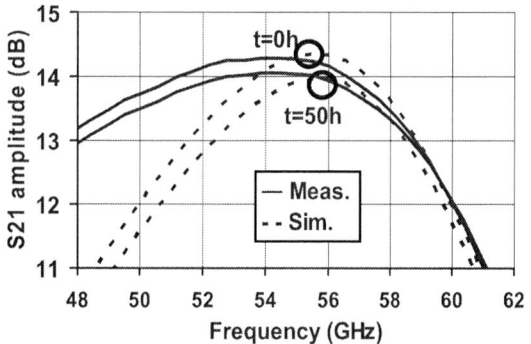

Fig. 6. Simulated and measured degradation of the 4 stages PA S_{21} parameter after 50 hours of stress at Vdd=1.7V.

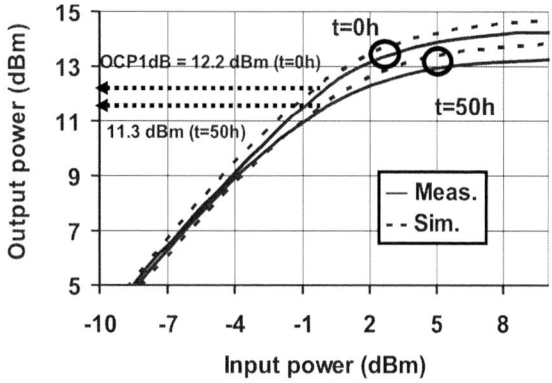

Fig. 7. Simulated and measured degradation of 4 stages PA output power at 58 GHz after 50 hours of stress at Vdd=1.7V.

The difference between the measured and the simulated maximum gain (Fig. 6) comes from parasitic not perfectly took into account in the model.

A linear power gain relative drift of 5 % is measured on Fig 6 after a stress of 50 hours at Vdd=1.7V. A relative drift of 7 % of linear OCP_{1dB} and P_{sat} are measured on Fig. 7 after the same stress at 58 GHz. Consequently, a dc hot carrier stress during 50 hours at Vdd=1.7V leads to a drop of 20 % of the PA figure of merit, which represent a strong loss of performances. In the 4 stages PA, each transistor [1] are much larger than the one used in the 1 stage PA which implies that the source access line acts as a relatively large degenerative inductor. Thus for the MOSFETs used in the 4 stages PA, an equivalent dynamic transconductance gm_0 must be defined for each stage as:

$$gm_0 = \frac{gm_{int\,rinsic}}{1 + gm_{int\,rinsic} \cdot Z_L} \qquad (4)$$

where $gm_{intrinsic}$ and Z_L are respectively the transconductance and the degenerative source access impedance. Consequently, for a relative drift of $gm_{intrinsic}$

due to the dc hot carrier stress, the corresponding relative drift of gm_0 is attenuated according to:

$$\frac{\Delta gm_0}{gm_0} = \frac{\Delta gm_{int\,rinsic}}{gm_{int\,rinsic}} \cdot \frac{1}{\left| 1 + gm_{int\,rinsic} \cdot Z_L \right|} \qquad (5)$$

Furthermore, the stress applied on each MOS in the 4 stages PA (50 hours at Vdd=1.7V) is lower than for the 1 stage PA (44 hours at Vdd=1.83V). Therefore the relative drift of the power gain of the 4 stage PA, proportional to the relative drift of gm_0, is only of 5%.

VI CONCLUSION

In this work, the effects of dc hot carrier stress on the characteristics of 65 nm fully integrated millimeter wave power amplifiers are presented. After stress, the increasing of V_{th}, and the drop of μ_n imply a decreasing of the transconductance and the drain source resistance of the MOS. This leads to a reduction of the biasing current and consequently a reduction of the circuit performances. Measurements show that V_{th} is the largest contributor to the deterioration of the PAs performances. A decreasing of 5% of the power gain and 7% of the OCP_{1dB} are measured on a 4 stages PA at 58 GHz after a stress of 50 hours with 1.7V. That leads to a drop of 20% of the figure of merit of the power amplifier.

REFERENCES

[1] T. Quémerais et al., "A CMOS Class-A 65nm Power Amplifier for 60 GHz Applications", IEEE SIRF, New Orleans, USA, Jan. 2010.

[2] T. Yao, et al., "Algorithmic Design of CMOS LNAs and PAs for 60-GHz Radio," IEEE J. Solid-State Circuits, vol. 42, no. 5, pp. 1044-1057, May 2007.

[3] M.S Gupta et al., "Power gain in feedback amplifiers, a classic revisited", IEEE Trans. on MTT, vol. 40, no. 5, pp. 864-879, 1992.

[4] J. T. Park et al., "RF performance degradation in nMOS transistors due to hot carrier effects", IEEE Trans. Electron Devices, vol. 47, no. 5, pp. 1068-1072, May 2000.

[5] Yehao Shen et al., "Hot Carrier Stress Effect on the Performance of 65 nm CMOS Low Noise Amplifier", IEEE ICICDT, Austin, USA, may 2009, pp. 249-252.

[6] Q. Li, J. Zhang et al., "RF circuit performance degradation due to soft breakdown and hot-carrier effect in deep-submicrometer CMOS technology," IEEE Trans. On MTT., vol. 49, no. 9, pp. 1546-1551, 2001.

[7] Chemming Hu et al. "Hot-electron-induced MOSFET degradation model, monitor, and improvement", IEEE trans. on Electron Devices, 1985, vol. 32, no2, pp. 375-385.

[8] Enjun Xiao et al "Hot Carrier Effect on CMOS RF Amplifier" IEEE IRPS 2005, San Jose, USA, pp. 680-682.

[9] C. D. Presti, et al "Degradation Mechanisms in CMOS Power Amplifier comparison to the DC Case" IEEE IRPS, Phoenix, USA, 2007, pp. 86-92.

RMO4D-4

A Layout-Based Optimal Neutralization Technique for mm-Wave Differential Amplifiers

Zhiming Deng, Ali M. Niknejad

Berkeley Wireless Research Center, University of California at Berkeley, CA

Abstract—**A layout-based optimal neutralization technique is proposed for the designs of mm-wave differential amplifiers. Based on a new layout style which exploits routing signal capacitive coupling, the need for physical neutralization capacitors are obviated which results in compact and robust layout. Experimental prototype designs at 60 GHz and 110 GHz amplifiers demonstrate the utility of the idea by direct comparison with unneutralized designs.**

Index Terms—**Differential amplifier, neutralization, cross-coupled.**

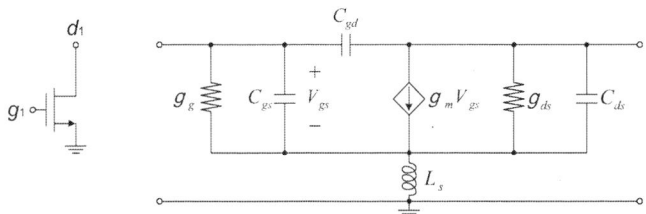

Fig. 1. The single-ended configuration and its small-signal model.

I. INTRODUCTION

CMOS amplifiers are fundamentals for mm-wave IC designs in silicon technologies. As the operating frequency gets close to the $f_{\max}$ of a transistor [1], special care of topology selection and layout optimization must be taken to maximize the available gain. For the simple common-source (CS) configuration whose model is shown in Fig. 1, we can derive the maximal stable gain (MSG) as

$$
MSG_{\mathrm{se}} \approx \frac{g_m}{\omega C_{gd}}\{[1 - \omega^2 L_s(C_{gs} + C_{ds} + \frac{C_{gs}C_{ds}}{C_{gd}}) \\
+ \frac{L_s g_g g_{ds}}{C_{ds}}]^2 + \omega^2 L_s^2[g_m + g_g(1 + \frac{C_{ds}}{C_{gd}}) \\
+ g_{ds}(1 + \frac{C_{gs}}{C_{gd}})]^2\}^{-\frac{1}{2}}.
$$
(1)

Reduction of C_{gd} is the bottleneck of the enhancement of MSG. Additionally, L_s also adds non-dominant poles and accelerates the roll-off of MSG. A differential configuration consisting of two equal single-ended CS transistors uses cross-coupled capacitors between gates and drains to neutralize C_{gd} and improve MSG. Usually the neutralization capacitors are explicitly implemented as MIM or MOM capacitors [2]. This method requires two extra capacitors in the layout. And, it is hard to control accuracy, especially due to the unwanted parasitic inductance associated with the neutralization current path. Moreover, the capacitors occupy extra space and make routing difficult.

In this work, we propose a layout method for differential CS transistors by which neutralization capacitance is intrinsically embedded in the coupling of metal signal wires. Therefore, no extra capacitors are required and the neutralization current path is minimized. The paper is organized as follows. In section II, we provide a description of this method and show critical design equations. In section III, we discuss practical design issues. Finally in section IV, we demonstrate the experimental results of several prototype designs.

II. NEUTRALIZATION TECHNIQUE

In Fig. 2, the proposed layout method is shown by a simplified diagram. Two multi-finger transistors are laid out in an interdigitated style. The complete structure can be regarded as a 1-D array of equal cells and the cells are lined up in the same direction as that of signal transmission. Each cell consists of a unit differential pair whose inputs and outputs are connected to four parallel signal buses accordingly. At the top level, the buses for the input and output of the same transistor are placed apart to avoid coupling, such as "g_1" and "d_1". But those of different transistors are placed close to each other, such as "g_1" and "d_2", and therefore the coupling capacitors can be used as neutralization capacitors.

A model to analyze this layout configuration is shown in Fig. 3. Besides the coupling capacitors, this model also considers the self inductance and the mutual inductance of the signal buses. Moreover there is no parasitic source degeneration inductance for the differential-mode operation in this model arising from the source-sharing configuration in each unit cell. According to this model, the complete MSG expression can be derived as

$$
MSG = \left| \frac{-g_m}{j\omega(C_{gd} - C_n + \Delta \cdot M)} + 1 \right|
$$
(2)

where

$$
\Delta = -\omega^2(C_{gs} + 2C_{gd})(C_{gs} + 2C_n) + g_g g_{ds} \\
+ j\omega g_m(C_{gd} - C_n) \\
+ j\omega(g_g + g_{ds})(C_{gs} + C_{gd} + C_n).
$$
(3)

For physically small structures, $\frac{1}{\sqrt{MC_n}} \gg 2\pi f_{\max}$ is satisfied, which implies the effect of the mutual inductance is negligible. So (2) has a simpler form

$$
MSG = \sqrt{\frac{g_m^2}{\omega^2(C_{gd} - C_n)^2} + 1}.
$$
(4)

978-1-4244-6240-7/10 $26.00 © 2010 IEEE

Furthermore, the stability factor K can also be derived:

$$K = [1 + \frac{2g_g g_{ds}}{\omega^2(C_{gd} - C_n)^2}] \cdot MSG^{-1}. \qquad (5)$$

Both MSG and K monotonically increase to infinity as C_n approaches C_{gd} from either side. But the maximal power gain $G_{\max}$ is different. It simply equals MSG for $K < 1$. But when $K > 1$, it starts to decrease and has a local minimum at $C_n = C_{gd}$ which is actually the invariant U function [3]:

$$U = G_{\max}|_{C_n = C_{gd}} = \frac{g_m^2}{4g_g g_{ds}}. \qquad (6)$$

Therefore, the peak $G_{\max}$ happens when $K = 1$. This includes two possibilities. If $C_n < C_{gd}$, or in other words, C_{gd} is partially neutralized, the ratio between C_n and C_{gd} is

$$n_1 = 1 - \frac{1}{\omega C_{gd}} \sqrt{\frac{g_g g_{ds}}{U - 1}}. \qquad (7)$$

If $C_n > C_{gd}$, or C_{gd} is over neutralized, the ratio is

$$n_2 = 1 + \frac{1}{\omega C_{gd}} \sqrt{\frac{g_g g_{ds}}{U - 1}}. \qquad (8)$$

In both cases, the same peak $G_{\max}$ value is achieved:

$$G_{\max 1} = G_{\max 2} = 2U - 1. \qquad (9)$$

Fig. 4 shows how MSG, $G_{\max}$ and K vary with the capacitor ratio.

In our model, g_g is converted from a physical resistor in series with C_{gs} so $g_g \propto \omega^2$ which implies

$$G_{\max i} = 2U - 1 \propto g_g^{-1} \propto \frac{1}{\omega^2}, \quad i = 1, 2, \qquad (10)$$

$$|n_i - 1| \propto \omega, \quad i = 1, 2. \qquad (11)$$

As the operating frequency becomes higher, the optimal neutralization scheme gets further away from the unilateral design ($C_n = C_{gd}$). Especially, if n_1 calculated from (7) is negative, the optimum can be achieved by using over neutralization only.

In practice, the selection of C_n is shifted from the theoretical optimum ($n_1 C_{gd}$, $n_2 C_{gd}$) in a direction so that the resulted design is even further from the unilateral case. Because $G_{\max}$ drops quickly once K exceeds 1 which can be seen from the slope of $G_{\max}$.

$$\frac{dG_{\max}}{dC_n} = \frac{\frac{dMSG}{dC_n} - \frac{MSG}{\sqrt{K^2 - 1}}\frac{dK}{dC_n}}{K + \sqrt{K^2 - 1}} = \begin{cases} -\infty, & n \to n_{1+} \\ +\infty, & n \to n_{2-} \end{cases}. \quad (12)$$

III. DESIGN APPROACH

From the analysis in section II, it is clear that the design objective is to properly design the layout structure so that U is not lower than a certain level and the capacitor ratio between C_n and C_{gd} is within a desired range. Starting from a simple assumption that the number of cells in line does not affect the performance, we only study the design of a unit cell.

The two critical design variables in a unit cell are the poly-gate finger width, wf, and the spacing between parallel signal buses, s, which are labeled in Fig. 2. U only depends on wf

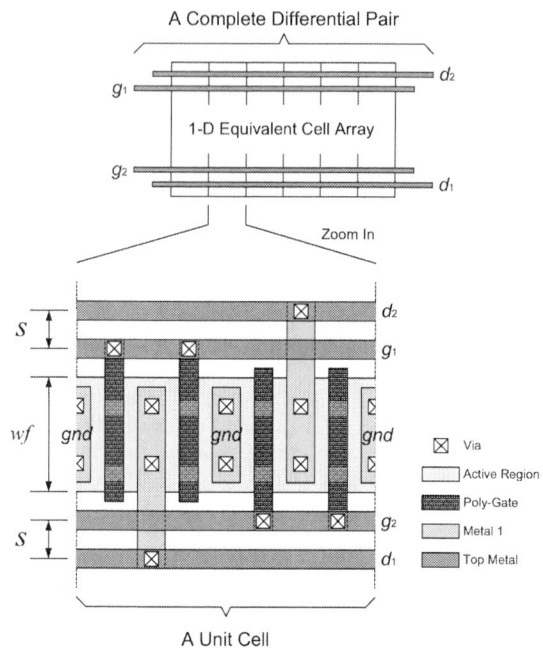

Fig. 2. A simplified diagram of the interdigital layout of a differential pair.

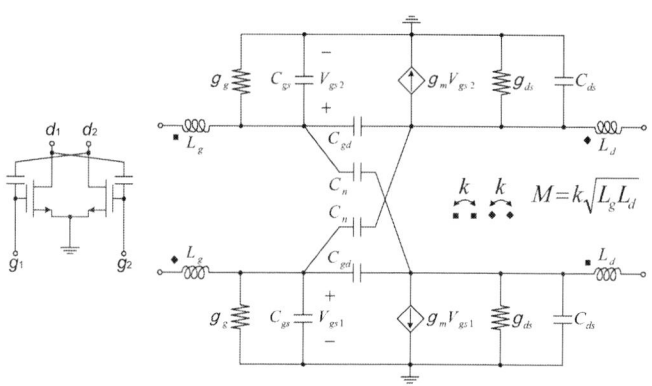

Fig. 3. The differential-mode small-signal model of the proposed layout structure in Fig. 2.

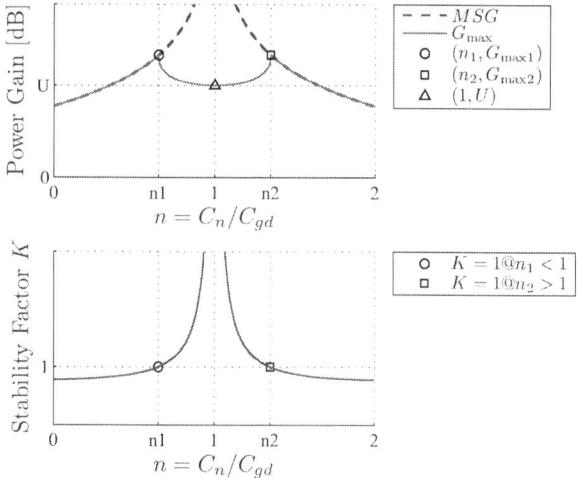

Fig. 4. MSG, $G_{\max}$ and the stability factor K vary with neutralization capacitor C_n (Mutual inductance M ignored).

978-1-4244-6240-7/10 $26.00 © 2010 IEEE   320

by

$$U = \frac{U_0}{1 + \alpha \cdot w f^2}. \tag{13}$$

The quadratic term in (13) reflects the degradation of U caused by the series poly-gate resistors. For the capacitor ratio, we can use simple relations of $C_n \propto s^{-1}$ and $C_{gd} \propto wf$ so

$$n \triangleq \frac{C_n}{C_{gd}} = \frac{\beta}{s \cdot wf}. \tag{14}$$

Eq. (14) ignores fringing effects. Suppose wf has been decided according to (13), then s can be derived from (14). The selection of s must be subject to process design rules so the range of n is limited. β is another parameter that can be adjusted by changing layout styles. For example, the signal buses can be laid out using multiple metal layers in parallel which induces a larger β value.

In Fig. 5 (a) and (b), we show the comparisons of $G_{\max}$ between different layout configurations for the partial neutralization designs ($wf = 1\mu$m) and the over neutralization designs ($wf = 0.75\mu$m) respectively. These figures are generated from post-layout simulation results with extracted RC parasitics. The kinks in the curves results from the fact that the optimum value of s is frequency dependent.

Finally, we make a qualitative study of the effect of the number of cells. In Fig. 5 (c), the layouts that use the same unit cell but have different numbers of cells, or equivalently, different numbers of fingers, nf, are compared. At low frequencies, these curves almost overlap. But the structures with higher nf become unconditionally stable at lower frequencies. This can be explained by the increased series resistance of the signal buses in the direction of signal transmission as more cells are cascaded.

All the comparisons in Fig. 5 do not consider inductive coupling of the signal buses because the design kit does not provide the capability of inductance extractions. However, using the RC extraction only can be good enough as long as the physical size of a structure is not too large.

IV. EXPERIMENTAL RESULTS

To verify the proposed neutralization technique, several prototype mm-wave differential amplifiers are fabricated in a 65nm LP CMOS technology.

The first design is a 60 GHz single-stage amplifier. The unit cell parameters are selected as $wf = 0.75\mu$m and $s = 0.47\mu$m. The signal buses use multi-layer coupling. These correspond to a capacitor ratio of $n = 1.4$. Therefore, this is an over neutralized design. The number of fingers is 32 (16 cells). The input and output matching networks adopts the simple single-stub configuration using 75 Ω conventional CPW transmission lines. The chip micrograph of this design is shown in Fig. 6 (a). This amplifier is designed for direct measurements and the GSSG pads are modeled and included in the matching networks. The differential-mode characteristic is obtained through a 2-port measurement using balun probes. The measured S-parameters are summarized in Fig. 7 (a).

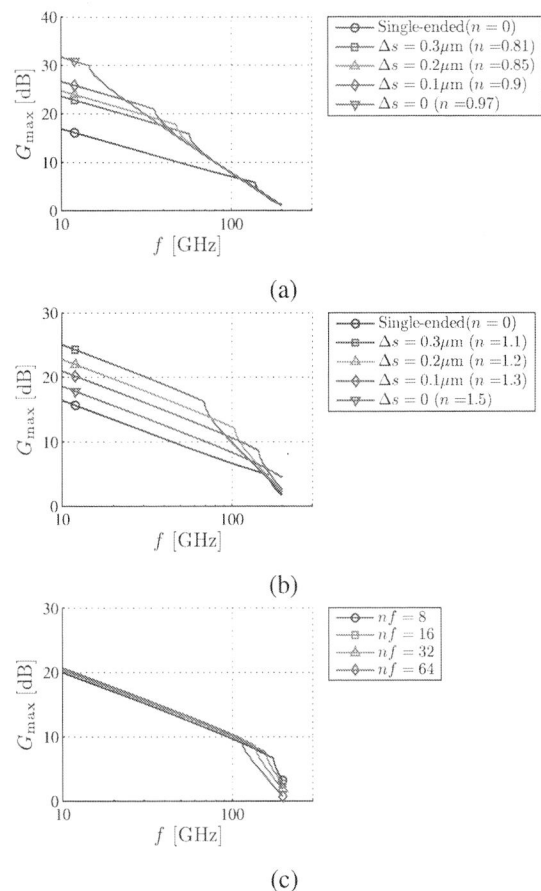

(a)

(b)

(c)

Fig. 5. The comparisons of $G_{\max}$ between different layout configurations. (a) The partial neutralization designs ($wf = 1\mu$m, $nf = 16$ and single-layer coupling). (b) The over neutralization designs ($wf = 0.75\mu$m, $nf = 16$ and multi-layer coupling). (c) Designs use the same unit cell but different numbers of cells ($wf = 0.75\mu$m, $s = 0.47\mu$m and multi-layer coupling). For the neutralized designs ($n > 0$) in (a) and (b), $s = s_{\min} + \Delta s$ and $s_{\min}$ is the minimal metal spacing defined by the process.

The maximal gain is 10.9 dB at 62.2 GHz and the MSG at the frequency is 13.8 dB. The S_{11} and S_{22} at the peak-gain frequency are -11.6 dB and -20.3 dB respectively. The difference between MSG and S_{21} is due to the chosen stability factor, $K = 1.15$. The gate and drain bias voltages are 0.65V and 1.2V. The total power consumption is 13 mW. As a reference, we also fabricate a single-ended CS transistor with the same size ($wf = 0.75\mu$m and $nf = 32$) and measure under the same bias conditions. The measured MSG is also plotted in the same figure which is 9.4 dB, 4.4 dB lower than the neutralized design. These data verify the effectiveness of the proposed neutralization technique because this differential amplifier has superior power gain over all of the unconditionally stable ($K > 1$) single-ended counterparts.

The second design is a 60 GHz two-stage amplifier and each stage is the same as the first design. The chip micrograph is shown in Fig. 6 (b). The two stages are directly connected through AC coupling capacitors. The measurement setup is the same as that of the first design and the measurement results are plotted in Fig. 7 (b). The maximal gain is 18.5 dB at

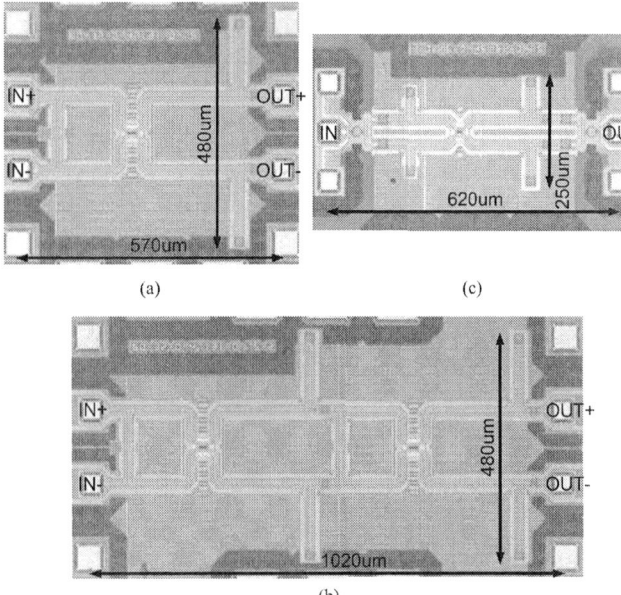

(a)   (c)

(b)

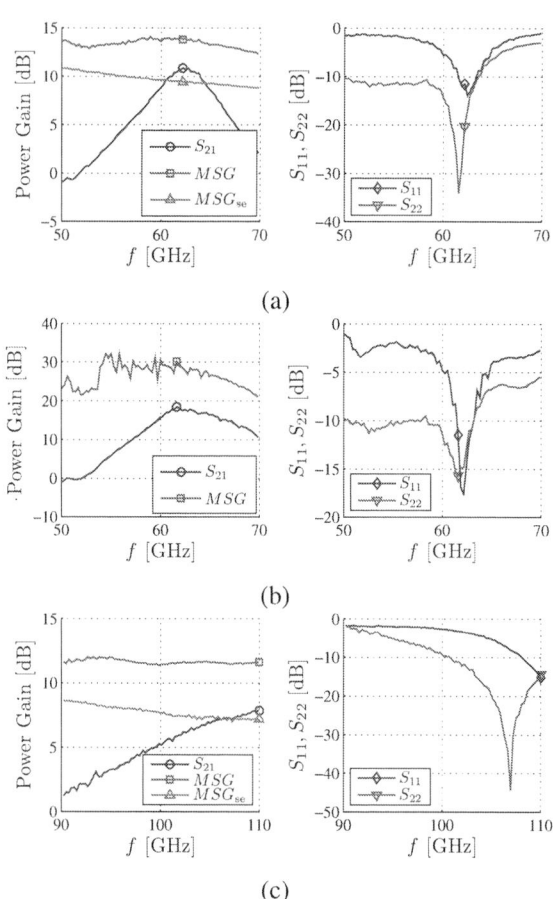

(a)

(b)

(c)

Fig. 6. The chip micrographs of (a) a 60 GHz single-stage amplifier, (b) a 60 GHz two-stage amplifier and (c) a 110 GHz single-stage amplifier with on-chip baluns for de-embedding.

Fig. 7. The S-parameters of (a) a 60 GHz single-stage amplifier obtained from direct measurements, (b) a 60 GHz two-stage amplifier obtained from direct measurements and (c) a 110 GHz single-stage amplifier obtained from de-embedding.

61.6 GHz and the MSG at the frequency is 30.1 dB. S_{11} and S_{22} at the peak-gain frequency are -11.5 dB and -15.7 dB respectively. The stability factor K is 6.6. The bias conditions are the same as that of the first design and the total power consumption doubles.

The third design is a 110 GHz single-stage amplifier. The unit cell parameters are $wf = 0.75\mu$m and $s = 0.45\mu$m. The signal buses use multi-layer coupling. Then the capacitor ratio is about $n = 1.4$ and it is also an over neutralized design. The number of fingers are 16 (8 cells). The chip micrograph is shown in Fig. 6 (c). A de-embedding technique using on-chip baluns are applied to obtain the differential-mode 2-port parameters [4]. The de-embedded S-parameters are summarized in Fig. 7 (c). Limited by the frequency range of the VNA, we can only obtain data up to 110 GHz. At 110 GHz, the gain is 7.8 dB with MSG of 11.6 dB. S_{11} and S_{22} are -15.1 dB and -14.4 dB. K is 1.24. The gate and drain bias voltages are 0.8V and 1.2V. The total power consumption is 11 mW. A reference single-ended transistor with the same size ($wf = 0.75\mu$m and $nf = 16$) and bias conditions has a MSG of 7.2 dB at 110 GHz, 4.4 dB lower than the neutralized design. The measured MSG enhancement by using neutralization agrees with the value predicted by (4).

V. CONCLUSION

A layout-based neutralization technique has been proposed for the designs of mm-wave differential amplifiers. The neutralization capacitors are realized directly from extrinsic transistor signal line coupling, obviating the need for extra capacitors. Several prototype designs demonstrate the proposed technique and measurement results of the amplifiers confirm the theory and effectiveness of the approach. An improve-

ment of 4.4 dB in MSG is observed in the measurements, corresponding exactly with the theoretical value based on the amount of over neutralization.

ACKNOWLEDGMENT

The authors acknowledge ST Microelectronics for the chip fabrication. In particular, the authors would like to thank Andreia Cathelin and Daniel Gloria of ST Microelectronics, Joel Dunsmore and Suren Singh of Agilent Technologies for their support and guidance, and Peter Hannaway of Cascade Microtech for access to balun probes.

REFERENCES

[1] M. Seo *et al*, "A 1.1V 150GHz Amplifier with 8dB Gain and +6dBm Saturated Output Power in Standard Digital 65nm CMOS using Dummy-Prefilled Microstrip Lines," *ISSCC Dig. Tech. Papers*, pp. 484-485, February 2009.

[2] W. L. Chan *et al*, "A 60GHz-Band 1V 11.5dBm Power Amplifier with 11% PAE in 65nm CMOS," *ISSCC Dig. Tech. Papers*, pp. 380-381, February 2009.

[3] S. J. Mason, "Power Gain in Feedback Amplifiers," *IRE Trans. on Circuit Theory*, vol. CT-1, pp. 20-25, June 1954

[4] Z. Deng *et al*, "The "Load-Thru" (LT) De-embedding Technique for the Measurements of mm-Wave Balanced 4-Port Devices," *RFIC Symp. Dig. Paper*, June, 2010.

RMO4D-5

A 100 GHz Transformer-Coupled Fully Differential Amplifier in 90 nm CMOS

Noël Deferm and Patrick Reynaert

K. U. Leuven, ESAT/MICAS, Kasteelpark Arenberg 10, 3001 Heverlee, Belgium

Abstract — **This paper proposes differential design techniques for W-band CMOS applications. Transformers are used as passive matching circuits, which provide numerous advantages compared to traditional matching circuits. Stabilization and gain improvement of the differential pair is achieved by a wideband neutralization technique. These techniques are combined in a fully differential amplifier which is successfully measured. To our knowledge, this is the first fully differential 100 GHz CMOS amplifier.**

Index Terms — **Differential Design, mm-Wave, CMOS, integrated transformer, MOS neutralization, W-band.**

I. INTRODUCTION

Today's wireless communication systems typically operate at relatively low frequencies. Carrier frequencies below 10 GHz are common, resulting in limited channel bandwidth. The demand for wireless 5 to 10 Gbps data rate systems will increase rapidly in the near future. But current wireless systems are not able to support these high data rates due to their limited bandwidth [1]. Today's wireless LAN standards offer data-rates of 54 Mbps up to 600 Mbps if multiple in, multiple out antenna technologies, complex beamforming and channel aggregation are applied [2]. The solution for this bandwidth problem can be found in the field of mm-wave applications. At mm-wave frequencies much higher bandwidths (2-10 GHz bandwidth is not unusual) are available which in turn can support gigabit wireless systems [3]. Other mm-wave applications can be found in the imaging sector for medical or security purposes or in radar and guiding applications like car radar or airplane landing guidance [4]. All these applications require a high performance, low cost implementation. Thanks to the high integration level, this can be achieved in CMOS. Despite the advantages of the CMOS process, the analog performance at mm-wave frequencies is limited. The integration of digital and analog circuits on the same substrate also leads to increased substrate noise. Adequate modeling and simulation of passive circuits is becoming a necessity due to electromagnetic radiation and the lack of a good ground definition.

In this paper design solutions for the high frequency active device performance (Section II) in combination with new high frequency passive devices (Section III) in a 90 nm CMOS technology are proposed. The ground definition and substrate coupling problems can be solved by using fully differential circuits which in turn also creates the opportunity of designing new active and passive device topologies. The design and measurements of a fully differential 100 GHz amplifier in a 90 nm CMOS process are described in section IV. In this amplifier the design techniques proposed in the other sections are adopted to achieve high gain at high frequencies and full range stability.

II. ACTIVE DEVICES

Although the operating frequency of mm-wave circuits approaches the ft and fmax, active circuits can still be designed by using circuit techniques which enhance the gain at these high frequencies. The behavior of a 15 μm 90 nm nMOS transistor is depicted in figure 2. At frequencies lower than the stability break point (75 GHz for 3 μm fingerwidth and 130 GHz for 1 μm fingerwidth in figure 2), the gain of the device is described by the Gmsg or Maximum Stable Gain. In this region the device is conditionally stable. This means that for certain load impedances, such as complex conjugate input and output matching, the device will oscillate. Stabilization is achieved by introducing losses such as mismatch or resistive losses. At frequencies beyond the stability break point the device becomes unconditionally stable which results in a faster roll-off of the gain curve . The position of the stability break point is mainly a function of the parasitic gate resistance. An increase of this resistance, for example by increasing the fingerwidth, results in a more stable device over a wider frequency range. However, this also leads to a lower gain at higher frequencies. A device with a lower gate resistance (smaller fingerwidth) can thus provide more gain at higher frequencies. The first 2 curves of figure 2 show this effect for a transistor with a constant W/L. Decreasing the fingerwidth from 3 μm to 1 μm leads to an increase of the the stability break point from 75 GHz to 130 GHz.

In the conditionally stable region the gain is limited by the internal feedback of the transistor. The problem is the

978-1-4244-6240-7/10 $26.00 © 2010 IEEE

gate-drain parasitic capacitance which creates a feedback loop. This internal feedback can be neutralized by connecting a negative capacitor in parallel with the original one so that the resulting feedback becomes smaller. The negative capacitance can be achieved by cross-coupling a capacitor between gate and drain in a differential pair (Figure 1). This is another benefit of using differential circuits at these high frequencies. Another advantage of this neutralization technique is the increase in stability performance without introducing extra losses, leading to a higher stability over a wider frequency band without losing gain. This also simplifies the design of the input and output matching networks. The effect of this neutralization technique is depicted in figure 2. Clearly, at 100 GHz the gain is improved with about 3 to 4 dB and the stability break point is pushed to an even lower frequency (65 GHz) than the original one (75 GHz).

The cross-coupled capacitors are inactive nMOS transistors (Figure 1). MOS capacitors are chosen here to overcome mismatch problems with the active devices of the differential pair due to process variations. If the cross-coupled capacitor is larger than the parasitic one of the active device, the stability performance of the complete differential pair is decreasing so oscillation can occur again at certain frequencies. To create a save stability margin, the MOS capacitors are approximately 20 % smaller than the active devices.

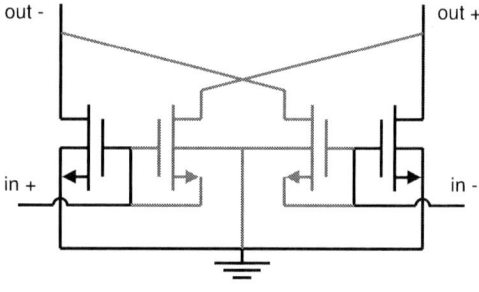

Fig. 1. Neutralized differential pair.

High gain at high frequencies can thus be achieved by reducing the fingerwidth which in turn results in a lower gate resistance, and gate-drain capacitor neutralization. For the 90 nm CMOS process, gain values of approximately 9 dB can be reached for a single nMOS transistor at 100 GHz (Figure 2).

III. PASSIVE DEVICES

Accurate mm-wave impedance matching networks are a must to create a conjugate match at input and output to

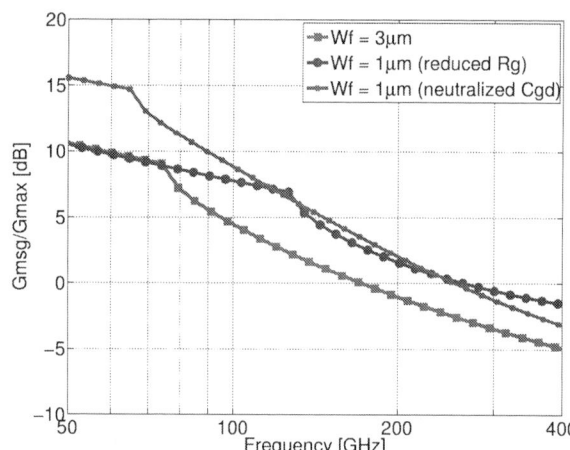

Fig. 2. Gain and stability approval by Rg reduction and Cgd neutralization.

maximize the power gain. Typical matching circuits at 100 GHz are build from bulky transmission lines [5]. Lumped element matching with capacitors and inductors is gaining importance, resulting in a reduced chip area [6], [7]. Using differential circuits, new matching circuit topologies can be obtained. One of the most important differential matching circuits is the transformer [8]. Even at mm-wave frequencies its performance is remarkably good. On top of this, the circuit can act as an impedance transformer and DC-blocker between different stages. DC-biasing can be connected at the center tap of the transformer which takes away the need for large inductors or transmission lines and coupling capacitors. The Gmax of a planar transformer (Figure 3) is depicted in figure 4. At 100 GHz a minimal loss of opproximately 1.1 dB can be reached.

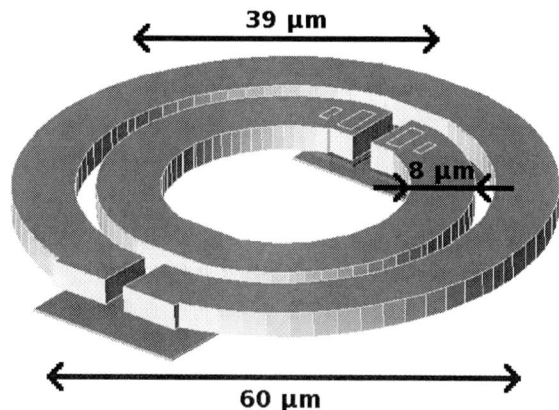

Fig. 3. 3D view of an integrated planar 1:1 transformer.

The loss of the transformer decreases with frequency up to the point where the effect of self resonance becomes important. At this point the loss reaches a minimum. For this particular transformer, the SRF (self resonance frequency) is situated at relatively high frequencies due to the planar design. By altering the dimensions of the transformer, it can be optimized for coupling factor, SRF or impedance matching. The impedance transformation ratio can be adjusted by changing the dimensions and number of inner and outer windings. This also alleviates the need for extra series or parallel matching circuits.

The simulations of the transformer are performed in the 2.5D finite element simulator Momentum. Good agreement between measurements and simulations is achieved (Figure 5).

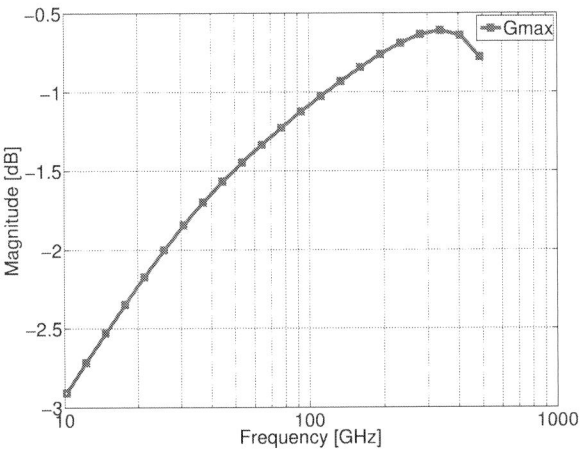

Fig. 4.   Transformer performance.

IV. AMPLIFIER MEASUREMENT RESULTS

A fully differential 6-stage transformer-coupled amplifier was optimized for maximum power gain. One single stage of the amplifier is shown in figure 6. Interstage matching is accomplished by optimized transformers to obtain ideal impedance transformations. This impedance transformation is achieved by increasing the diameter of the primary coil and decreasing the diameter of the secondary coil. In other words, the coupling factor is changed to optimize the leakage and magnetizing inductors for tuning the gate and drain capacitors of the transistors. The input and output matching networks do not only provide impedance matching between the 50 Ohm probes and the circuit, they also act as baluns to convert the differential signal on chip to a single ended 50 Ohm impedance. The use of transformers also creates the possibility to easily connect decouple capacitors and ESD protection for the DC-supply/biasing bondwires. Indeed, large capacitance can be tolerated at the center tap of the transformer, which is a virtual ground for the differential circuit.

Figure 7 shows the chip photograph. The amplifier consumes an area of 1360 μm by 640 μm, including on chip baluns, GSG-probepads, bondpads, decouple capacitors and ESD protection. The actual amplifier only consumes an area of 0.11 mm^2. Each stage consists of a neutralized differential pair with 15 μm active devices and 12 μm crossed capacitors to obtain high gain and stability at 100 GHz.

The proposed 6 stage amplifier provides a small signal gain of 11 dB at 99 GHz. The 3 dB bandwidth is about 11 GHz starting from 93 GHz up to 104 GHz (Figure

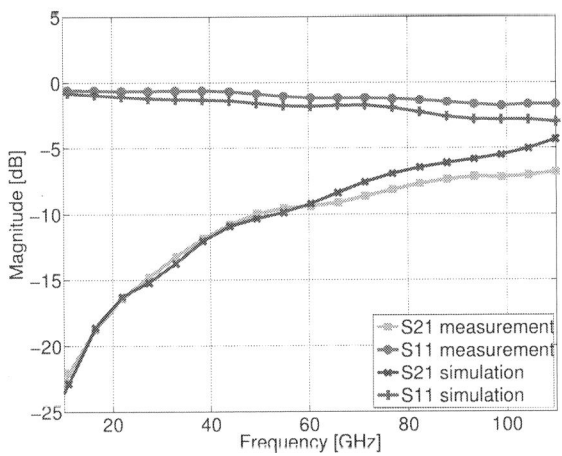

Fig. 5.   Transformer measurements vs. simulations.

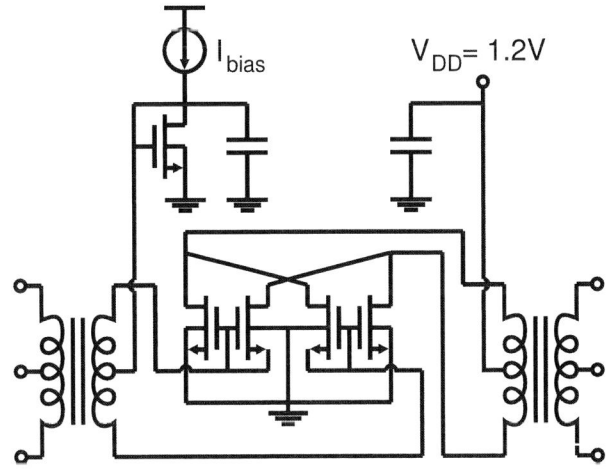

Fig. 6.   Schematic of a single stage of the 100 GHz amplifier.

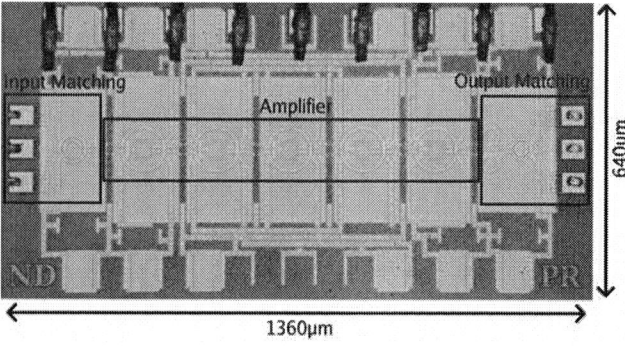

Fig. 7. Chip photograph.

8). The measurements are performed with an Agilent N5250A Vector Network Analyser and Infinity 110 GHz 75 μm pitch GSG probes. The amplifier consumes about 94 mW with a supply voltage of 1.2 V. This results in a current density of approximately 400 μA/μm which is the optimal bias point for maximal fmax [6].

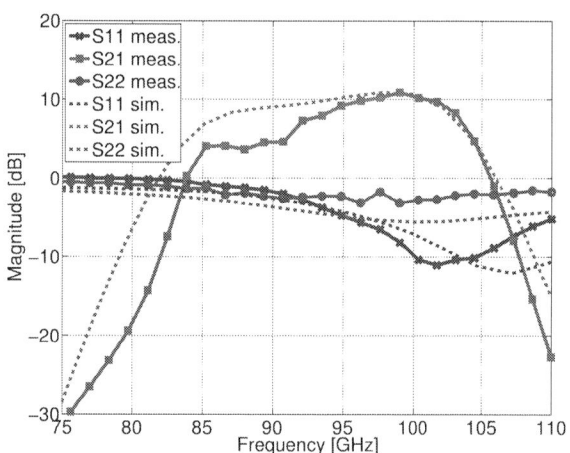

Fig. 8. Measured and simulated S-parameters.

Although the original design was simulated to achieve a gain of 18 dB with the same power consumption, the measurement results show a gain of only 11 dB. This was caused by an error in the port and probepad definitions of the finite-element simulations of the input and output matching networks. The simulations and measurements were matching much better after solving this problem (Figure 8). From this we can conclude that a gain of approximately 20 dB can be achieved when these impedance mismatch problems are solved.

V. CONCLUSION

Differential design techniques for accurate and structured design of W-band mm-wave CMOS circuits were proposed in this paper. Transformers were introduced as passive matching circuits, which provide numerous advantages compared to traditional matching circuits. Stabilization and gain improvement of the differential pair is achieved by a wideband neutralization technique without increasing the power consumption. Compared to traditional mm-wave single ended design techniques, which suffer from poor noise performance and high area consumption due to bulky passive devices, these differential techniques offer innovating solutions. Also, adequate simulation of the active devices, which is less accurate at these high frequencies due to the lack of a virtual ground in single ended circuits, can be achieved with these new techniques. All these techniques were combined in a fully differential 6 stage amplifier which is successfully measured. To our knowledge, this is the first fully differential 100 GHz amplifier in CMOS.

ACKNOWLEDGMENT

The authors would like to thank Prof. D. Schreurs and Dr. Ir. Ilja Ocket of the research group ESAT-TELEMIC for their measuring equipment and support during measurements.

REFERENCES

[1] S. Ohmori, Y. Yamao and N. Nakajima,"The Future Generations of Mobile Communications Based on Broadband Access Technologies," *Communications Magazine, IEEE, vol. 38, Issue 12, pp. 134-142, December 2000.*

[2] J. Wells, "MM-Waves in the Living Room: The Future of Wireless High Definition Multimedia," *Microwave Journal, vol. 52, No. 8, August 2009.*

[3] A. Tomkins, R. A. Aroca, T. Yamamoto, S. T. Nicolson, Y. Doi, S. P. Voinigescu, "A Zero-IF 60GHz Transceiver in 65nm CMOS with > 3.5Gb/s Links," *Custom Intergrated Circuits Conference, CICC 2008. IEEE, pp. 471-474, September 2008.*

[4] K. Mizuno, "Millimeter wave imaging technologies," *International Symposium on Signals, Systems, and Electronics, ISSSE 1998, pp. 289-290, September 1998.*

[5] Y. Jiang, J. Tsai, H. Wang, "A W-Band Medium Power Amplifier in 90 nm CMOS," *Microwave and Wireless Components Letters, IEEE, vol. 18, no. 12, pp. 818-820, December 2008.*

[6] T. Yao, M. Q. Gordon, K. K. W. Tang, K. H. K. Yau, M. Yang, P. Schvan, S. P. Voinigescu, "Algorithmic Design of CMOS LNAs and PAs for 60-GHz Radio," *Journal of Solid-State Circuits, JSSC 2007, vol. 42, Issue 5, pp. 1065-1075, May 2007.*

[7] S. P. Voinigescu, S. T. Nicolson, M. Khanpour, K. K. W. Tang, K. H. K. Yau, N. Seyedfathi, A. Timonov, A. Nachman, G. Eleftheriades, P. Schvan, M. T. Yang, "CMOS SOCs at 100 GHz: System Architectures, Device Characterization, and IC Design Examples," *International Symposium on Circuits and Systems, ISCAS 2007, IEEE, pp. 1971-1974, May 2007.*

[8] D. Chowdhury, P. Reynaert, A. M. Niknejad, "A 60GHz 1V +12.3dBm Transformer-Coupled Wideband PA in 90nm CMOS," *International Solid-State Circuits Conference, ISSCC 2008, IEEE, pp. 560-562, February 2008.*

RTU1B-1

A 68–82 GHz Integrated Wideband Linear Receiver using 0.18 μm SiGe BiCMOS

Austin Ying-Kuang Chen[1,2], Yves Baeyens[1], Young-Kai Chen[1], and Jenshan Lin[2]

[1]Bell Laboratories, Alcatel-Lucent, Murray Hill, NJ 07974, USA
[2]Department of ECE, University of Florida, Gainesville, FL 32611, USA

Abstract — **This paper presents a highly integrated wideband linear receiver with on-chip active frequency doubler implemented in a low-cost 200/180 GHz f_T/f_{max} 0.18 μm SiGe BiCMOS technology. Individual receiver circuit blocks including low-noise amplifier, passive balun, mixer, and frequency doubler have been independently characterized and optimized for wideband, NF, and linearity performance. The receiver highlights a 3 dB RF bandwidth of larger than 14 GHz from 68 GHz to at least 82 GHz. The measured peak power conversion gain is 28.1 dB with an input 1 dB compression point of -23.6 dBm, and NF of 8 dB at 77 GHz. Noise figures of 8–10 dB are achieved over the 3 dB bandwidth. The overall chip size is 1350 x 990 μm^2 and the total power consumption is 413 mW. To the best of authors' knowledge, this receiver reports the highest 3 dB RF bandwidth with excellent linearity performance among all the prior arts in SiGe HBT/BiCMOS technologies to date.**

Index Terms — **Automotive radar, E-band, microstrip line, millimeter-wave, receiver, SiGe BiCMOS, wideband.**

I. INTRODUCTION

Until recently, low-cost millimeter-wave (mm-wave) integrated radio systems have been emerging thanks to the rapid evolution of advanced SiGe and CMOS technologies. In particular, SiGe technology has been identified as a well-suited technology for both active and passive imaging applications such as high-resolution automotive radars and concealed weapon detections due largely to its superior performance in gain, NF, output power, phase noise, and variability over temperature compared even with today's CMOS. To further maintain the low production cost advantage, efficient technology sharing suggests the key components be universal. Therefore, it is an attractive effort to integrate multiple standards that include lower E-band 71–76 GHz high speed point-to-point links, 76–77 GHz long-range and 77–81 GHz short-range automotive radars into a single chip. Moreover, high dynamic range performance is essential as these systems are required to operate under all-weather conditions. To date, numerous research efforts have been undertaken to realize SiGe HBT/BiCMOS automotive phased array radar [1], [2], radar transceivers [3], [4], individual receivers [5]–[9] and down-converter [10] at ~77 GHz. However, no integrated receiver front-end addressing the complete 71–81 GHz band has been reported.

In this paper, we present the first fully integrated universal wideband linear receiver covering the entire 71–81 GHz band that can be used in a direct-conversion zero-IF or low-IF architecture. The simplified block diagram of the wideband receiver is shown in Fig. 1.

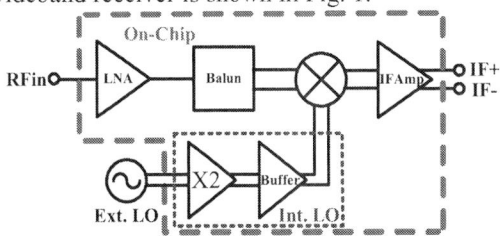

Fig. 1. Simplified block diagram of the wideband receiver with on-chip active frequency doubler.

II. CIRCUIT BLOCKS DESIGN

The architecture chosen for the low-noise amplifier (LNA) and mixer combination is crucial in determining the overall RF bandwidth of the receiver front-end. Between these two important building blocks, the mixer generally gives better bandwidth performance; therefore optimizing the LNA design becomes the remaining key option to simultaneously achieve the wideband operation and low noise upon integration.

A. Low-Noise Amplifier

The schematic of the two-stage LNA is shown in Fig. 1. The amplifier has an input cascode stage for higher power gain and better isolation at the expense of higher NF_{min}. Common emitter was used for the second stage for improved linearity and wider output matching bandwidth. Device layout configurations with two base fingers (CBEBC) were used throughout the LNA design to mitigate the signal loss via device parasitic capacitance (C_{BC} and C_{BE}) to ground, thereby improving the overall power gain and NF. Wideband noise input matching technique with low quality factor Q network (~1.3 after optimization) was employed to achieve high linearity and enhance robustness to PVT simultaneously. Inductive degeneration at the second stage further improves the linearity of the overall LNA. A standalone version of the LNA has also been characterized and has a measured peak S_{21} of 14.5 dB and NF of 6.9 dB at 77 GHz with a 3 dB bandwidth spanning from 69 to 83.5 GHz [11]. The measured NF is lower than 8.4 dB from 64 to 84 GHz while the measured input-referred 1 dB compression point (IP_{1dB}) is -11.4 dBm at 77 GHz. Note the output matching

978-1-4244-6240-7/10 $26.00 © 2010 IEEE

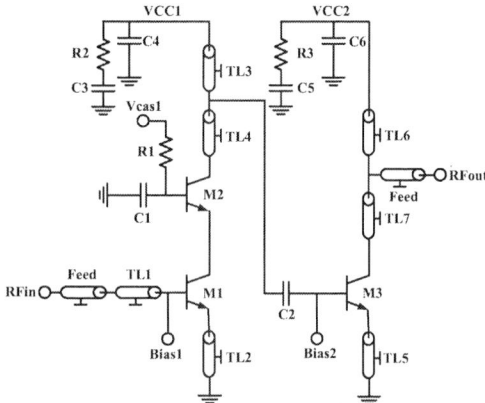

Fig. 2. Circuit schematic of the two-stage LNA [11].

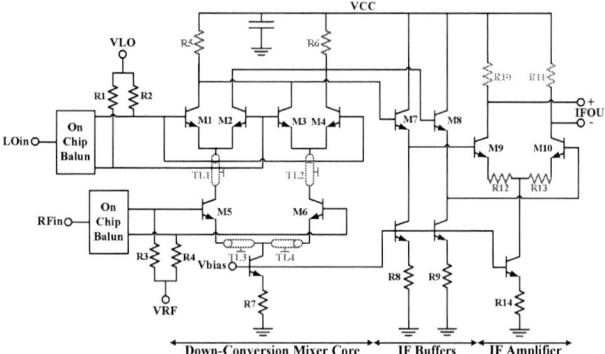

Fig. 3. Circuit schematic of the standalone down-conversion mixer.

of the standalone LNA has been modified slightly to facilitate the integration of the receiver front-end.

B. Down-Conversion Mixer and Integrated Passive Balun

The schematic of the standalone mixer shown in Fig. 3 consists of two integrated wideband passive baluns at LO and RF port, a mixer core, emitter followers, and an IF amplifier. The passive balun features an insertion loss of 1.5–2.2 dB, amplitude imbalance of <0.27 dB, phase imbalance $180° \pm 2°$ from 70–84 GHz, and port return losses better than 10 dB from 60–90 GHz, as reported in [12]. The mixer core design is based on the Gilbert cell for double-balanced structure for broadband performance. The selection of the device sizes involves trade-offs among conversion gain, NF, and linearity. In fact, linearity of the receiver front-end is primarily determined by the performance of the mixer. There are three main factors to consider for linearity design. First, large DC bias current therefore large g_m is usually required from transconductor (g_m) stage to reduce the noise and the 3rd order intermodulation product (IM₃). However, considering the f_{max} of the technology and G_{max} at operating frequency range, device sizes cannot be made arbitrarily large so that the gain of the g_m stage can be conserved. Second, small device geometry (CEBEC) which presents the highest f_T in this technology was chosen for fast switching quad. This configuration reduces the charging and discharging times for complete current steering, thereby reducing the mixer

quad's contribution to the overall nonlinearity. Moreover, it also relaxes the required LO drive and reduce the overall system power budget. Additionally, for a small size device the required DC bias current to achieve peak f_T is smaller, this allows lower current noise arising from the switching quad. Note however that this goes against the large bias current for g_m stage necessary for linearity and NF. As an additional benefit, power conversion gain and LO-to-IF isolation can be improved. During current steering, the peaking transmission line inductors TL1 and TL2 (60 pH) inserted between CE and CB stage resonate out the parasitic capacitance associated with collector node of g_m stage (M5 & M6 with emitter length = 7 µm) and emitter node of switching quad (M1–M4 with emitter length = 2.5 µm), thus improving mixer's power gain, NF, isolation, and bandwidth all at once. The emitter length of M5 and M6 are intentionally scaled more than twice the size of M1–M4 such that the g_m stage can be biased at the optimal current density for high linearity while the switching quad is biased close to peak f_T. The selection shows excellent linearity with minimal performance degradation in gain and NF, with nominal 3.3 V supply. Inductive degeneration transmission lines (TL3 & TL4, 25pH each) further improve the design bandwidth and linearity. The mixer resistive loads (R5 & R6) are 240 Ω, optimized for conversion gain while satisfying the low-voltage headroom requirement. Third, in order to interface with the 50 Ω testing environment and provide adequate gain and output drive to the following circuitry, an IF buffer amplifier is usually unavoidable after the mixer core. The IF differential amplifier core comprises large size devices (M9 & M10 with emitter length = 2 x 6 µm) along with a pair of degeneration resistors (R12 & R13, 27Ω each) to increase the overall linearity. A major part of the current drained in the mixer design comes from the IF amplifier in which the individual device has a current density of 6.5 mA/µm². Two 75 Ω resistors R10 and R11 are used as IF loads for output matching. As a final note, linearity performance (IP₁dB) is usually better without an IF amplifier. The mixer draws a total current of 36 mA.

C. Internal LO Signal Generation

The internal LO signal is generated by an on-chip active by an on-chip active frequency doubler with a power output buffer to relax fundamental signal source phase noise requirement. The frequency doubler based on Gilbert cell was chosen for its fully balanced operation, high conversion gain, high output power, and wide bandwidth. The doubler highlights a wide operable bandwidth from 36–80 GHz with gain and P_{out} of 0.8 dB and -3.9 dBm, respectively, at 80 GHz. More design insights are available in [13]. A power output buffer consisting of a differential cascode stage was added to

978-1-4244-6240-7/10 $26.00 © 2010 IEEE

further increase the output power of the doubler and ensure fast switching of the mixer quad as described in Section II. B.

III. IMPLEMENTATION AND EXPERIMENTAL RESULTS

The receiver was fabricated in a low-cost 200/180 GHz f_T/f_{max} 0.18 μm SiGe BiCMOS process. The emitter width of the HBT is 0.15 μm. All the inductors in the design were implemented using thin-film microstrip transmission lines with full ground plane shield to isolate them from the conductive substrate and maintain continuous ground potential. All the interconnects were modeled and accounted for at both design and layout phases. The microphotograph of the wideband receiver is shown in Fig. 4. The overall chip size is 1350 μm x 990 μm (1.34 mm²). All the measurements were characterized on-wafer. For the receiver conversion gain and linearity test setup, the RF signals were generated with the external multiplier modules (X4 and X6) with synthesized sweeper (HP 83650B). The external LO signal was generated with Agilent vector signal generator (VSG) (E8267D) and external 10–40 GHz 180° hybrid junction (HY1040–180). The on-chip active frequency doubler and LO buffer together provide adequate LO power up to about 80 GHz. The IF outputs were fed through a 50–1000 MHz 180° power combiner. The standalone mixer characterization has a similar setup except that the LO signal is generated by a V- or W-band source module and VSG. Port matching and port-to-port isolation of the mixer were measured with Agilent 110 GHz mm-wave network analyzer (PNA series). The noise figure measurement was done with the calibrated V- and/or W-band noise source in conjunction with spectrum analyzer (PSA E4448). All the off-chip losses were corrected and de-embedded from the measurement results.

As shown in Fig. 5, the receiver achieves a measured power conversion gain of 28.1 dB and NF of 8 dB at 77 GHz with at least 14 GHz 3 dB RF bandwidth from 68 to 82 GHz. The NF of the receiver is between 8 and 10 dB across the 3 dB BW. Note that all the measurements were performed with IF fixed at 100 MHz unless otherwise specified. The measured power conversion gain and single sideband (SSB) NF of the mixer are 13.7 dB and 15.5 dB respectively, at 77 GHz. The SSB NF is ≤ 16.5 dB over the frequency range from 67 to 82 GHz. Both gain and NF measurements of the mixer include the loss of passive balun. The calibrated LO drive level at the input of active frequency doubler is between -9.5 dBm and -4.5 dBm from 67 GHz to 82 GHz. The measurement results beyond 80 GHz are limited by the 10–40 GHz 180° hybrid and available LO drive, therefore the actual 3 dB RF BW of the receiver is expected to be even larger.

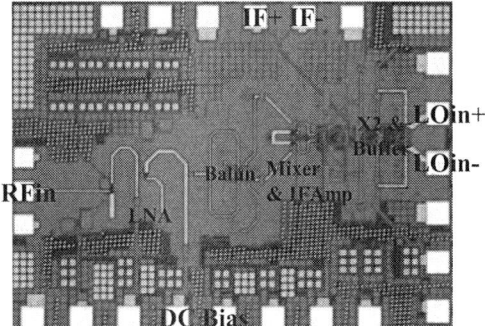

Fig. 4. Microphotograph of the wideband receiver.

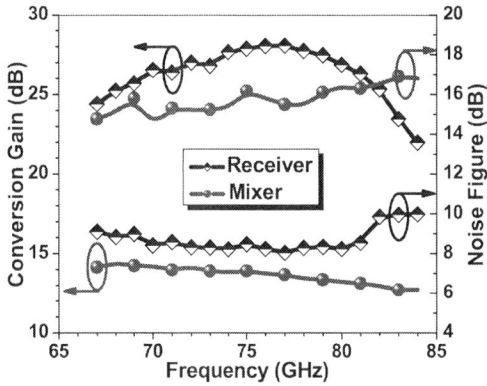

Fig. 5. Measured receiver and down-conversion mixer conversion gain and noise figure (IF fixed at 100 MHz).

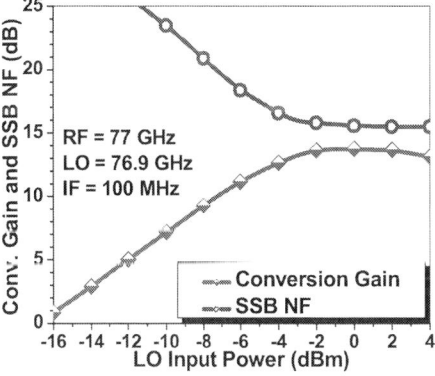

Fig. 6. Measured mixer conversion gain and SSB NF vs. LO input power.

Fig. 6 shows conversion gain and SSB NF of the mixer versus LO input power at 77 GHz. The mixer gain saturates at about -2 dBm LO input drive while the SSB NF reached its minimum at 2 dBm. Degradation in both gain and NF performance can be observed when the LO power level is dropped to below -4 dBm. Both RF and LO port return losses are better than 10 dB while LO-to-RF and RF-to-LO isolation are better than 40 dB and 30 dB, respectively over the target BW as shown in Fig. 7. Fig. 8 depicts the measured linearity characteristics of the mixer and receiver described by input P_{1dB}. At RF = 77 GHz and IF = 100 MHz, the mixer and receiver feature input P_{1dB} of

TABLE 1. PERFORMANCE OF PRIOR PUBLISHED SiGe HBT/BiCMOS ~77 GHz RECEIVERS

Reference	Technology f_T/f_{max} (GHz)	3 dB RF BW (GHz)	RX Gain (dB) @ Freq. (GHz)	RX NF (dB)	Input P_{1dB} (dBm)	P_{diss} (mW)	Chip Area (mm²)	Receiver Integration Level
[1], [2]	0.13 µm SiGe BiCMOS 200/290	76–80	35@ 77 GHz	8–10	-27.5	161	2.25	LNA/Mixer/IF Amp/VCO (4X Array)
[3]	0.13 µm SiGe BiCMOS 170/200	76–81	25.6@ 78 GHz	9	-24	740	1.17	LNA/Mixer/IF Amp/VCO
[4]	0.18 µm SiGe BiCMOS 200/180	76–81	31@ 79 GHz	7.5–9.5	-30.7	601	7.4	LNA/Quadrature Mixer/PLL
[5]	0.18 µm SiGe:C HBT 200/275	75–82	~32@ 77 GHz	11 (SSB)	-16	1073	1.1	LNA/Quadrature Mixer/Branchline Coupler/LO Buffer
[6]	0.13 µm SiGe BiCMOS 200/290	73–81	46@ 77 GHz	7–10.5	-38	195	1.7	LNA/Mixer/IF Amp/VCO
[7]	0.13 µm SiGe BiCMOS 220/250	68–76	24@ 77 GHz	4.8	-21.7	120	0.23	LNA/Mixer/IF Amp/VCO
[8]	0.25 µm SiGe BiCMOS 180/200	79	21.7@ 79 GHz	10.2 (sim)	-35	595	1.26	LNA/Mixer/VCO
[9]	0.14 µm SiGe:C HBT 225/330	75.5–77.5	30@ 77 GHz	11.5 (SSB)	-26	440	1.16	LNA/Active Balun/Mixer
This work	**0.18 µm SiGe BiCMOS 200/180**	**68–82**	**28.1@ 77 GHz**	**8–10**	**-23.6**	**413**	**1.34**	**LNA/Passive Balun/Mixer/ /IF Amp/Doubler/LO Buffer**

-10.3 dBm and -23.6 dBm, respectively. The measured input return loss of the receiver is better than 9 dB across the 3 dB BW. The total power consumption is 413 mW. Table 1 outlines the performance of this work and several recently published state-of-the-art ~77 GHz receivers realized in SiGe HBT/BiCMOS technologies. More than 14 GHz 3 dB RF bandwidth achieved by this receiver is the largest among all.

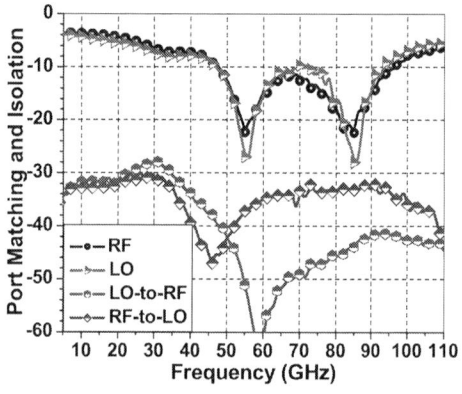

Fig. 7. Measured RF and LO port matching and port-to-port isolation of the mixer.

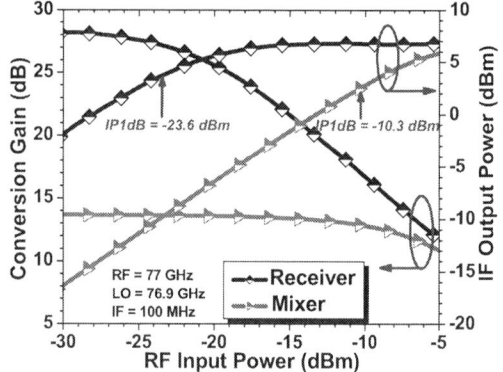

Fig. 8. Measured linearity characteristics of IF output power and conversion gain of the mixer and receiver at RF = 77 GHz.

IV. CONCLUSION

In this paper, a highly integrated wideband linear receiver with a 3 dB RF bandwidth from 68 GHz to 82 GHz was presented. The receiver, fabricated in a low-cost 200/180 GHz f_T/f_{max} SiGe BiCMOS process, achieved a maximum gain of 28.1 dB, NF of 8 dB, and input P_{1dB} of -23.6 dBm at 77 GHz and dissipates 413 mW. The distinct advantage of wideband operation along with excellent linearity performance accomplished in this radio receiver allows applications within the band, such as 71–76 GHz high speed point-to-point links, 76–77 GHz long-range and 77–81 GHz short-range radar sensors to share and reuse the same front-end chip.

REFERENCES

[1] A. Babakhani, X. Guan, A. Komijani, A. Natarajan, and A. Hajimiri, "A 77-GHz phased-array transceiver with on-chip antenna in silicon: Receiver and antennas," *IEEE J. Solid-State Circuits*, vol. 41, no. 12, pp. 2795–2806, Dec 2006.

[2] A. Natarajan, A. Komijani, X. Guan, A. Babakhani, and A. Hajimiri, "A 77-GHz phased-array transceiver with on-chip antenna in silicon: Transmitter and local LO-path phase shifting," *IEEE J. Solid-State Circuits*, vol. 41, no. 12, pp. 2807–2819, Dec 2006.

[3] S. Nicolson, P. Chevalier, A. Chanter, B. Sautreuil, and S. P. Voinigescu, "A 77–79 GHz Dopplar radar transceiver in silicon," in *IEEE Compound Semiconduct. Integr. Circuits Symp.*, Oct. 2007, pp. 1–4.

[4] V. Jain, F. Tzeng, L. Zhou, and P. Heydari, "A single-chip dual-band 22-to-29/77-to-81 GHz BiCMOS transceiver for automotive radars," in *IEEE Int. Solid-state Circuits conf. Dig. Tech. Papers*, 2009, pp. 308–309.

[5] B. Dehlink, H.-D. Wohlmuth, K. Aufiner, F. Weiss, and A. L. Scholtz, "An 80 GHz SiGe quadrature receiver frontend," in *IEEE Compound Semiconduct. Integr. Circuits Symp.*, Nov. 2006, pp. 197–200.

[6] J. Powell *et al.*, "SiGe receiver front ends for millimeter-wave passive imaging," in *IEEE Trans. Microw. Theory Tech.*, vol. 56, no. 11, pp. 2416–2425, Nov. 2008.

[7] S. T. Nicolson *et al.*, "A low-voltage SiGe BiCMOS 77-GHz automotive radar chipset," in *IEEE Trans. Microw. Theory Tech.*, vol. 56, no. 5, pp. 1092–1104, May. 2008.

[8] L. Wang, S. Glisic, J. Borngraeber, W. Winkler, and J. C. Scheytt, "A single-end fully integrated SiGe 77/79 receiver for automotive radar," *IEEE J. Solid-State Circuits*, vol. 43, no. 9, pp. 1897–1908, Sep. 2008.

[9] M. Hartmann *et al.*, "A low-power low-noise single-chip receiver front-end for automotive radar at 77 GHz in silicon-germanium bipolar technology," in *IEEE MTT-S Int. Microw. Symp. Dig.*, Jun. 2007, pp. 149–152.

[10] B. Floyd *et al.*, "Silicon millimeter-wave radio circuits at 60-100 GHz," in *IEEE Silicon Monolithic Integrated Circuits in RF Systems*, Long Beach, CA, Jan. 2007, pp. 213–218.

[11] A. Y.-K. Chen, Y. baeyens, Y.-K. Chen, and J. Lin, "A low-power linear SiGe BiCMOS low-noise amplifier for nillimeter-wave active imaging," *IEEE Microw. Wireless Compon. Lett.*, vol. 20, no. 2, pp. 103–105, Feb. 2010.

[12] A. Y.-K. Chen, H.-B. Liang, Y. Baeyens, Y.-K. Chen, J. Lin, and Y.-S. Lin, "Wideband mixed lumped-distributed-element 90° and 180° power splitters on silicon substrate for millimeter-wave applications," in *IEEE Radio Frequency Integrated Circuit (RFIC) Symposium*, Atlanta, GA, Jun. 2008, pp. 449–452.

[13] A. Y.-K. Chen, Y. Baeyens, Y.-K. Chen, and J. Lin, "A 36–80 GHz high gain millimeter-wave double-balanced active frequency doubler in SiGe BiCMOS," *IEEE Microw. Wireless Compon. Lett.*, vol. 19, no. 9, pp. 572–574, Sep. 2009.

A 24-GHz Low-Power Fully Integrated Receiver with Image-Rejection using Rich-Transformer Direct-Stacked/Coupled Technique

Nobuhiro Shiramizu, Takahiro Nakamura, Toru Masuda, and Katsuyoshi Washio

Central Research Laboratory, Hitachi, Ltd.,
1-280 Higashi-Koigakubo, Kokubunji, Tokyo, 185-8601, Japan

Abstract — **We have developed a low-power, fully integrated receiver for 24-GHz ISM band wireless communication using a Rich-Transformer Direct-Stacked/Coupled (RT-DSC) technique. This technique makes it possible to reduce supply voltage and current without any performance degradation. The 24-GHz receiver circuit was fabricated using 0.18-μm SiGe BiCMOS technology. Receiver gain of 30 dB and noise figure (NF) of 5.6 dB are obtained at low power consumption of 21.5 mW. We utilized the synergistic effect of combining a narrow-band transformer and a notch filter to integrate the circuit's image rejection (IR) function. This resulted in our achieving power consumption only 30 % of that reported previously.**

Index Terms — **Receiver, transformer, image rejection, K-band, quasi-millimeter-wave, ISM band, SiGe BiCMOS, low power, wireless communication.**

I. INTRODUCTION

The rapidly growing demand for high-speed wireless communication is creating strong demand for wide bandwidth radio frequency (RF) systems. Among the promising candidates for such systems are those operating in a quasi-millimeter wave region, such as the 24-GHz ISM band or the K-band [1]-[10]. However, they require lower-power operation and lower-cost fabrication to enable more widespread use of their applications and services.

To reduce the power consumption of RF-ICs, both circuit element design techniques and circuit block connecting techniques are important. In particular, in a quasi-millimeter-wave region, buffer amplifiers dissipate a great deal of power in driving high frequency signals. To address this problem, we previously developed a transformer-based connecting technique and applied it to RF and local oscillator (LO) signal transmission [7]. In the superheterodyne architecture, which has been the most widely accepted approach towards achieving robust and low-power operation in wireless transceivers [1], suppression of image-frequency signals is one of the most fundamental performance factors [6]. Integrating an image rejection function into the receiver is an effective way to reduce cost because it allows external filters to be eliminated. Therefore, an image rejection (IR) technique has been adopted for a low-noise amplifier (LNA) [6]. One of the most power-hungry of receiver circuits is a frequency divider operating at a very high frequency. An injection-locked divider helps to reduce power

consumption; however, locking frequency failures occur when signals with different frequencies are injected. To achieve stable operation in such cases, we previously introduced a direct-coupled transformer connection between the divider and voltage-controlled oscillator (VCO) [8].

To reduce supply voltage and current without performance degradation, we here propose a 24-GHz low-power, fully integrated receiver utilizing a novel rich transformer direct-stacked/coupled (RT-DSC) technique and an image-rejection low-noise amplifier (IR-LNA).

II. RECEIVER DESIGN

Fig. 1 shows a block diagram of the proposed receiver. To achieve low power operation for the first frequency down-conversion, we adopt a superheterodyne architecture, which does not need to generate complicated quadrature-phased LO-signal generation signals at the same frequency as the RF signals. At intermediate frequency (f_{IF}), which alleviates the limitations on LO-signal generation, a direct-conversion architecture is used to extract the baseband (BB) signal. The low power property of this receiver configuration is further enhanced by selecting an f_{IF} of 2.68 GHz, lower than the 5 GHz that has previously been reported in 24-GHz band receivers [4], [5]. Due to the use of super-heterodyne architecture, the receiver essentially requires an image rejection ratio (IRR)

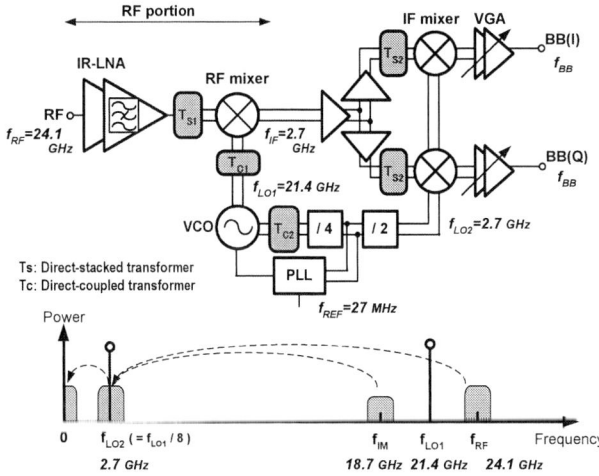

Fig.1. Block diagram of proposed 24-GHz receiver and frequency allocation plan

of more than 30 dB at image frequency (f_{IM}), which is 5.36 GHz lower than f_{RF}. Developing a highly efficient signal transmission technique for use between the core circuits in the receiver chip is one of the key issues for achieving low power operation. With this in mind, we have developed a new RT-DSC technique, which refers to (a) direct-stacked transformers ($T_{S1} \sim T_{S3}$) at the RF port of both the RF mixer and the IF mixers to reduce the supply voltage by alleviating DC-voltage stacking, and (b) usage of direct-coupled transformers (T_{C1} and T_{C2}) that achieve low-power LO transmission from the VCO since the transformer connections make it possible to eliminate power-hungry buffer-amplifiers.

Fig. 2 shows a simplified circuit schematic of the receiver RF portion. The IR-LNA consists of two cascade amplifier stages; a notch feedback circuit (comprising C_{n1}, C_{n2}, and L_n) in the second amplifier stage is used to reduce blocker signals at f_{IM} [6]. The LNA and the RF mixer are connected directly by T_{S1}. This eliminates the need for an emitter-coupled stage as used in conventional Gilbert mixers; thus the T_{S1} allows the RF mixer to operate at 1.5 V even though SiGe heterojunction bipolar transistors (HBTs) are used for the cross-coupled switching stage ($Q_1 \sim Q_4$) [7]. Furthermore, optimizing the output impedance of the IR-LNA allows transmission efficiency between the IR-LNA and the mixer to be maximized at the f_{RF}. The VCO is connected to the RF mixer and frequency divider through direct-coupled transformers (T_{C1} and T_{C2}) [8]. By selecting a proper shape for the transformers and designing the output impedance to have good power-matching with the LO input node of the RF mixer and the input node of the frequency divider, transmission efficiency is improved and this allows the VCO to provide higher driving power at low power consumption.

The advantages of applying the transformer direct-coupling technique were also verified in the proposed VCO configuration. The injection-locked dividers are used as first frequency dividers due to their high-operating-frequency capability at lower power consumption. However, it is easy for the dividers to lock into the wrong frequency. Therefore, finding a robust configuration to suppress the effect of RF-leakage signals on the injection-locked dividers becomes an important issue.

Fig. 3 depicts a case study on power consumption and RF-signal isolation between the LO port of the mixer and the input node of the frequency divider. In conventional case #1 (Conv. #1), the VCO drives both the frequency divider and the mixer through an emitter follower buffer. The buffer has good isolation of 48 dB, but the power consumption of the VCO and the buffer is increased to 17.7 mW. In Conv. #2, the buffer is removed and the VCO drives the RF mixer directly. This decreases the power consumption to 13 mW, but the isolation is degraded to 34 dB. To achieve operation with lower power consumption and higher isolation, we introduced the transformer direct-coupling technique. In the proposed case, two transformers (T_{C1} and T_{C2}) are introduced in the VCO circuits to drive the RF mixer and the frequency divider, respectively. Since the RF-leakage signal goes through both transformers optimized for the LO frequency, the RF frequency signal level is reduced due to attenuation generated by the T_{C1} and T_{C2}. The isolation is consequently improved and reaches 47 dB. The proposed configuration does not require any buffers between the VCO and the mixer, so the power consumption can be a much lower 10.9 mW. Using this transformer direct-coupling technique reduces VCO power consumption to 61 % of that of the conventional case despite having a higher isolation level of 47 dB.

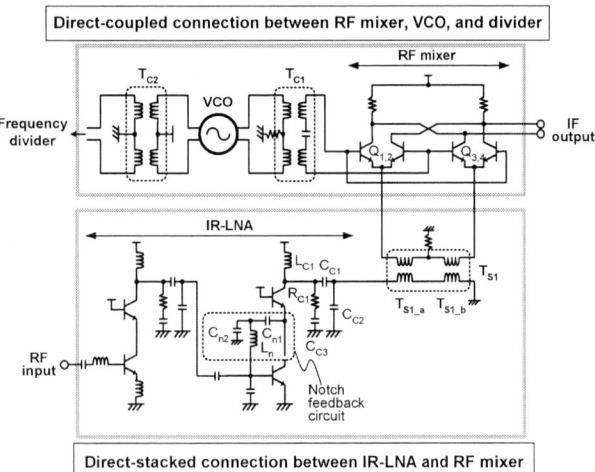

Fig. 2. Simplified schematic of RF portion in proposed 24-GHz receiver

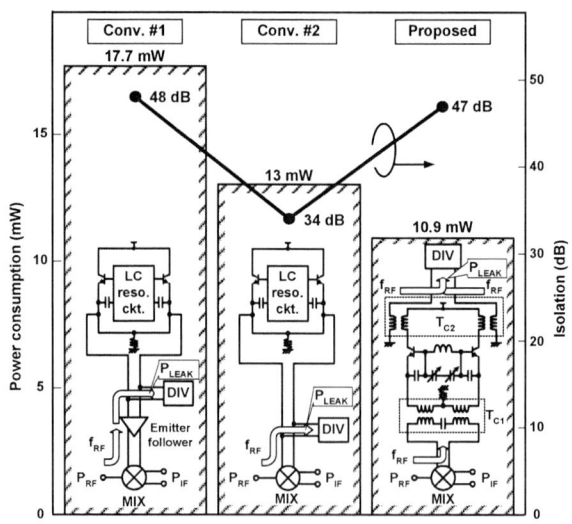

Fig. 3. Simulated characteristics of power consumption and RF-signal isolation dependences on LO-signal distribution circuits

III. Measurement Results

Fig. 4 shows a micrograph of the receiver chip fabricated using 0.18-μm SiGe BiCMOS technology with SiGe HBTs of 140-GHz f_T and 180-GHz f_{max}. The chip occupies an area of 910×2260 μm.

Fig. 5(a) illustrates how the transformer direct-stacking technique improves the image rejection ratio (IRR) of the receiver. The gain characteristics of the RF portion (LNA, RF mixer, and VCO) of the receiver are plotted. As shown in the inset, the LNA has a notch frequency response; the RF mixer with the T_{S1} also shows a gradual band-pass filter characteristic in the frequency response. It is utilized to enhance the overall IRR of the receiver. At f_{IM} of 18.6 GHz, the overall IRR reaches 36.0 dB, a 3.5 dB improvement. The maximum gain of the RF portion is 30 dB at 24.1 GHz. Even though the values of f_{RF} and f_{IM} are close (the frequency ratio is 0.77), a higher IRR of more than 30 dB was obtained in the fully integrated receiver. The transformer direct-stacking configuration also improves the linearity due to the elimination of active devices that generate distortion in signals.

Fig. 5(b) shows the measured two-tone test characteristics of the RF portion. When using a conventional configuration with a transistor (Q_{pre}) located between the LNA and the RF mixer, we obtained an input-referred third-order intercept point (IIP3) of -28.3 dB and a conversion gain of 30 dB. However, eliminating the Q_{pre} and optimizing the power matching between the LNA and the RF mixer improved the IIP3 to -24.5 dBm while maintaining the conversion gain at 30 dB. The proposed transformer connection therefore improves the IIP3 by 3.8 dB. Using the transformer direct-stacking technique also reduced power consumption by 11 % for the RF-signal path from the LNA to the RF mixer.

Fig. 6 shows the overall small signal characteristics of the receiver. The gain observed at the baseband output nodes, shown in Fig. 1, reached a maximum value of 55 dB. This gain should be able to be decreased to at least 35 dB. The 1-dB bandwidth of the gain in the baseband frequency range is more than 455 MHz. As shown in Fig. 6, NF characteristics are stable in the BB frequency range up to 500 MHz. A minimum NF of 5.6 dB is observed at the conversion gain of 55 dB. To the authors' knowledge, this NF is close to the best (minimum) value obtained so far, and the BB bandwidth is clearly the widest reported to date for 24-GHz operation using Si technology [2, 4].

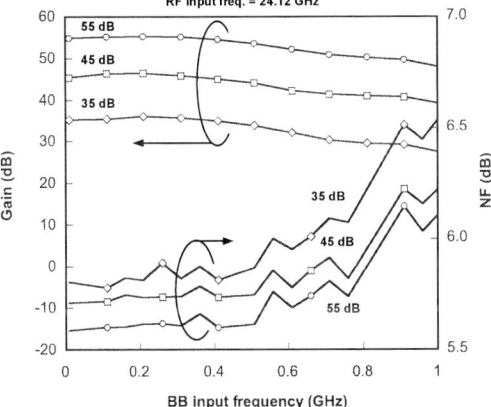

Fig. 6. Measured characteristics of proposed 24-GHz receiver. Baseband frequency dependences of overall conversion gain, and overall DSB noise figure are shown.

The measured performance is summarized in Table I. For the RF portion, a conversion gain of 30 dB, IIP3 of -24.5 dBm, and IRR of 36 dB were obtained at the low power consumption of 21.5 mW. Overall gain of 55 dB and NF of 5.6dB were obtained at the power consumption of 54.7 mW.

Fig. 7 shows the relationship between the power consumption (P_{DC}) and figure of merit (FOM) in terms of this study and previously reported receivers operating in the frequency range from 24 GHz to 60 GHz [1-4, 9-15]. Receiver FOMs are calculated by the equation shown at the top of the figure. In the equation, f_{RF} is the RF input frequency, Gain is the RF front-end gain between the RF input node and the first mixer output node, NF is the receiver noise figure, and BB_BW$_{1dB}$ is the 1-dB

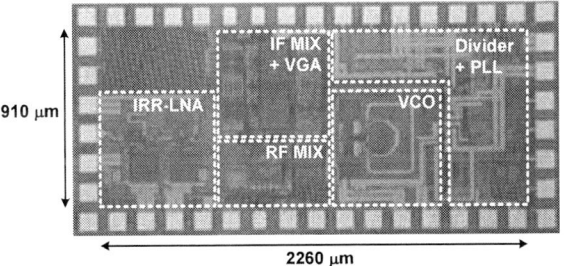

Fig. 4. Photomicrograph of the 24-GHz receiver fabricated with 0.18 μm SiGe BiCMOS technology

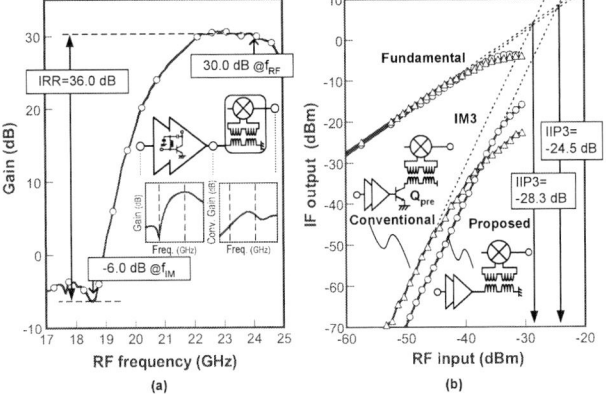

Fig. 5. Measured characteristics of RF portion in proposed 24-GHz receiver. (a) RF frequency dependence on gain, (b) two-tone test characteristics

978-1-4244-6240-7/10 $26.00 © 2010 IEEE

Table 1. Performance summary of proposed 24-GHz receiver

Technology	0.18μm SiGe BiCMOS	
RF freq.	24.1 GHz	
LO freq.	21.4 GHz	
Gain	35 - 55 dB (Overall)	
	30 dB (RF portion)	
NF	5.6 dB	
P1dB	-36.5 dBm (RF portion)	
IIP3	-24.5 dBm (RF portion)	
IRR	36 dB	
BB-BW$_{1dB}$	910 MHz	
I$_{CC}$	IR-LNA:	8.0 mA
	RF mixer:	1.0 mA
	IF mixer+VGA:	9.8 mA
	VCO:	8.7 mA
	DIV:	9.6 mA
P$_{DC}$	54.7 mW (Total)	
	21.5 mW (RF portion)	

bandwidth of the output baseband signal. In the previous-study group, P_{DC} almost always increases as the FOM is increased. This dependency makes it difficult to reduce P_{DC} while maintaining FOM. In contrast, using the RT-DSC technique allows the P_{DC} for this study to be decreased to less than 30 % of that generally obtained in previous studies at the same FOM. These results confirm that the proposed receiver shows the lowest power consumption yet obtained in the quasi-millimeter and millimeter frequency range.

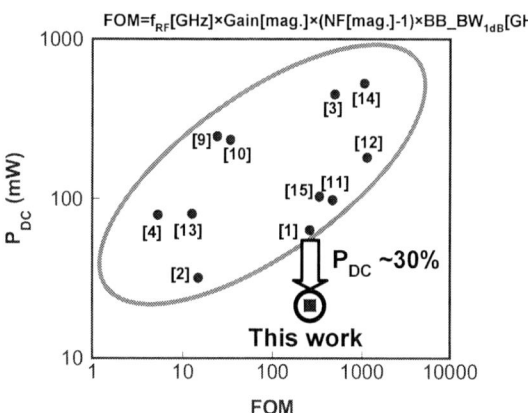

Fig. 7. Figure of merit of receiver studies for 24 GHz and 60 GHz.

IV. SUMMARY

We have proposed a low-power, fully integrated receiver for 24-GHz ISM band wireless communication. The receiver utilizes a Rich-Transformer Direct-Stacked/Coupled (RT-DSC) technique and an image rejection low-noise amplifier (IR-LNA). Along with the low power consumption of 21.5 mW for the RF portion, we obtained conversion gain of 30 dB, input-referred third-order intercept point (IIP3) of -24.5 dBm, and image rejection ratio (IRR) of 36 dB at radio frequency (RF) of

24.1 GHz. The receiver characteristics included a maximum gain of 55 dB and a minimum noise figure (NF) of 5.6 dB. These results demonstrate that for the quasi-millimeter and/or millimeter wave bands, the RT-DSC technique is one of the most promising candidates for providing low-power, single-chip solutions.

ACKNOWLEDGEMENT

This work was supported by "The Research and Development Project for Expansion of Radio Spectrum Resources" of the Ministry of Internal Affairs and Communications, Japan.

REFERENCES

[1] A. Mazzanti, et al., "A 24GHz Sub-Harmonic Receiver Front-End with Integrated Multi-Phase LO Generation in 65nm CMOS," ISSCC 2008, pp. 216-217, Feb. 2008.

[2] C-Y Chu, et al., "A 24GHz Low-Power CMOS Receiver Design," ISCAS, 2008, pp.980-983, May 2008.

[3] K. Takana, et al., "A Highly Integrated Quasi-millimeter Wave Receiver Chip Using 3D-MMIC Technology," Eu-MIC 2007, pp. 12-15, Oct. 2007.

[4] Y-H Chen, et al., "A 24-GHz receiver frontend with an LO signal generator in 0.18-μm CMOS," Trans. on MTT, vol. 56, no. 5, pp. 1043-1051, May 2008.

[5] X. Guan, et al., "A 24-GHz CMOS Front-End," J. of SSC, vol. 39, no. 2, pp. 368-373, Feb. 2004.

[6] T. Masuda, et al., "A 50-dB image-rejection SiGe-HBT based low noise amplifier in 24-GHz band," RFIC Symp., pp. 307-310, Jun. 2009.

[7] N. Shiramizu, et al., "24-GHz 1-V pseudo-stacked mixer with gain-boosting technique," 34th ESSCIRC, pp. 102-105, Sept. 2008.

[8] T. Nakamura, et al, "A Low-Phase-Noise Low-Power 27-GHz SiGe-VCO using Merged-Transformer Matching Circuit Technique," RFIC Symp., pp. 413-416, Jun. 2007.

[9] C. Kienmayer, et al., "17 GHz Receiver in TSLP Package for WLAN/ISM Applications in 0.13 μm CMOS," the 31st ESSCIRC, pp. 133-136, Sept. 2005.

[10] T. Yu, et al., "A 4-Channel 24-27 GHz CMOS Differential Phased-Array Receiver," RFIC Symp., pp. 455-458, Jun. 2009.

[11] C.H. Wang, et al., "A 60GHz Low-Power Six-Port Transceiver for Gigabit Software-Defined Transceiver Applications," ISSCC 2007, pp. 192-193, Feb. 2007.

[12] S Pinel, et al., "A 90nm CMOS 60GHz Radio," ISSCC 2008, pp. 130-131, Feb. 2008.

[13] B. Razzavi, "A Millimeter-Wave CMOS Heterodyne Receiver with On-Chip LO and Divider," J. of SSC, Vol. 43, No. 2, pp. 477-485, Feb. 2008.

[14] S.K. Reynolds, et al., "A Silicon 60-GHz Receiver and Transmitter Chipset for Broadband Communications," J. of SSC, Vol. 41, No. 12, pp. 2820-2831, Dec. 2006.

[15] J. Lee, et al., "A Low-Power Fully Integrated 60 GHz Transceiver System with OOK Modulation and On-Board Antenna Assembly," ISSCC 2009, pp. 316-317, Feb. 2009.

A 60 GHz CMOS Receiver Front-End with Integrated 180° Out-of-Phase Wilkinson Power Divider

Chi-Chen Chen, Yo-Sheng Lin, Jen-How Lee, and Jin-Fa Chang

Department of Electrical Engineering, National Chi Nan University, Puli, Taiwan, ROC
Tel: 886-492912198; Fax: 886-492917810; Email: stephenlin@ncnu.edu.tw

Abstract– A 60-GHz receiver front-end with an integrated 180° out-of-phase Wilkinson power divider using standard 0.13 μm CMOS technology is reported. The receiver front-end comprises a wideband low-noise amplifier (LNA) with 12.4-dB gain, a current-reused bleeding mixer, a baseband amplifier, and a 180° out-of-phase Wilkinson power divider. The receiver front-end consumed 50.2 mW and achieved input return loss at RF port better than −10 dB for frequencies from 52.3 GHz to 62.3 GHz. At IF of 20 MHz, the receiver front-end achieved maximum conversion gain of 18.7 dB at RF of 56 GHz. The corresponding 3-dB bandwidth (ω_{3dB}) of RF is 9.8 GHz (50.8 GHz to 60.6 GHz). The measured minimum noise figure (NF) was 9 dB at 58 GHz, an excellent result for a 60-GHz-band CMOS receiver front-end. In addition, the measured input 1-dB compression point (P_{1dB}) and input third-order inter-modulation point (IIP3) are −20.8 dBm and −12 dBm, respectively, at 60 GHz. These results demonstrate the adopted receiver front-end architecture is very promising for high-performance 60-GHz-band RFIC applications.

Index Terms– CMOS, wideband, receiver front-end, low-noise amplifier, mixer, splitter

I. INTRODUCTION

In USA, Canada, and Japan, there is 7-GHz-wide unlicensed band around 60 GHz for wireless personal area network (WPAN) system applications [1]. A 60 GHz WPAN system can provide short-range (<10 m) and high-speed (>2 Gb/s) multi-media data access to nearby consumer appliances and computer terminals. In the past, III-V semiconductor technologies were adopted in most of the applications for frequencies around 60 GHz and above. Recently, thanks to the rapid development of CMOS/BiCMOS processes, it has become possible to use them to implement 60 GHz WPAN system and even 77 GHz radar system.

In transceiver design, receiver front-end (which includes LNA [2]-[3] and mixer [4]-[5], etc.) is a critical block that receives small signals from antenna over the whole band of interest and amplifies and down-converts them with a good signal-to-noise ratio property. The basic requirements of a receiver front-end include good input impedance matching, good port-to-port isolation, low noise, and high gain over the whole band of interest. Recently, several excellent 60 GHz CMOS receivers front-end have been reported [6]-[7]. For example, in [6], a 60-65 GHz CMOS receiver front-end with

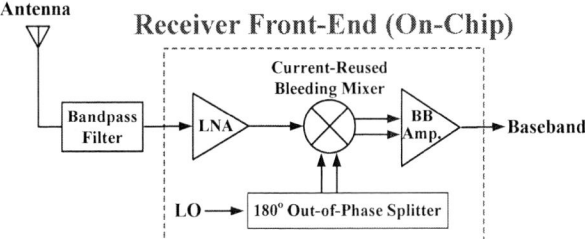

Fig. 1 Adopted receiver front-end architecture.

25-30 dB power gain in 0.13 μm CMOS process was demonstrated. However, its 4 GHz 3-dB bandwidth (ω_{3dB}) and 10-12 dB noise figure (NF) were not satisfactory. In [7], a 57-63 GHz CMOS receiver front-end in 0.13 μm CMOS process was reported. However, its power gain of 11.8 dB and NF of 10.4 dB was not good enough. In this work, a 60 GHz CMOS receiver front-end with excellent 3-dB bandwidth (ω_{3dB}) and noise figure properties using standard 0.13 μm CMOS technology is reported. The receiver front-end comprises a wideband low-noise amplifier (LNA), a current-reused bleeding mixer, a baseband (BB) amplifier, and a Wilkinson-power-ivider-based 180° splitter. The receiver front-end consumed 50.2 mW and achieved high conversion gain of 18.7 dB, wide ω_{3dB} of 9.8 GHz, and low NF of 9 dB. Besides, excellent LO-RF, RF-IF, and LO-IF isolations were also achieved.

II. CIRCUIT DESIGN

The 60-GHz receiver front-end was designed and implemented by a standard 0.13 μm CMOS process (on a p-type silicon substrate with thickness of 300 μm and resistivity of 8-12 Ω·cm) provided by a commercial foundry. This technology offers 8 metal layers, named M_1 to M_8 from bottom to top. The thickness of M_8 was 3.35 μm, and that of M_7, M_6-M_2, and M_1 was 0.83 μm, 0.37 μm, and 0.26 μm, respectively. The IMD (inter-metal dielectric) thickness was 0.695 μm between M_8 and M_7, 0.67 μm between M_7 and M_6, 0.45 μm between other adjacent metal layers, and 0.49 μm between M_1 and the silicon substrate. The interconnection lines as well as the microstrip-line (MSL) inductors were placed on the 3.35-μm-thick topmost metal to minimize the resistive loss. Fig. 1

978-1-4244-6240-7/10 $26.00 © 2010 IEEE

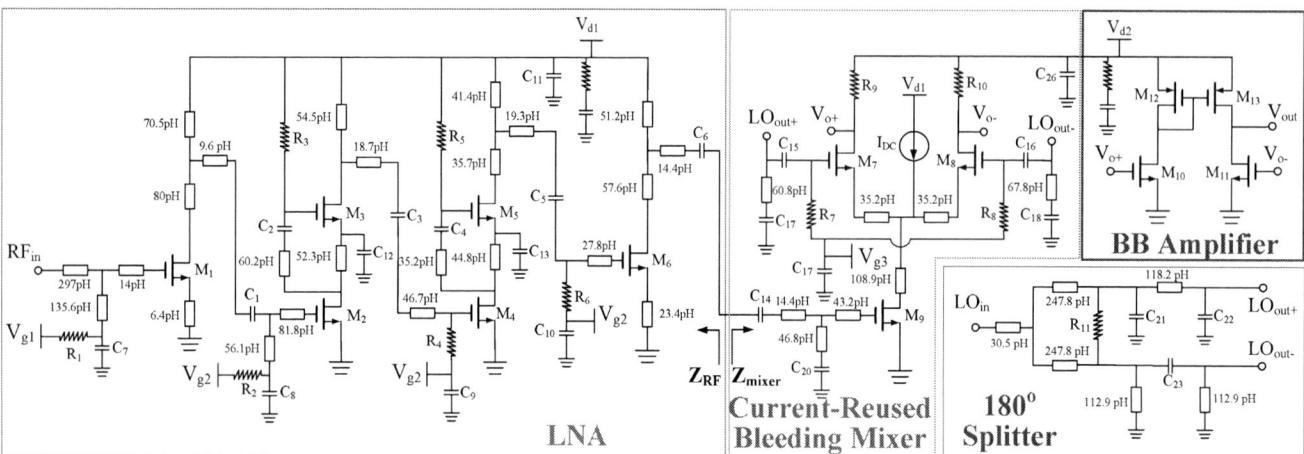

Fig. 2 Schematic f the 60 GHz CMOS receiver front-end.

shows the adopted receiver front-end architecture. Fig. 2 shows the schematic of the 60 GHz CMOS receiver front-end. The receiver front-end comprises a wideband LNA, a current-reused bleeding mixer, a baseband amplifier, and a 180° out-of-phase Wilkinson power divider. The design of the LNA, mixer, baseband amplifier, and divider are described in the following.

II.A. Wideband LNA

To achieve sufficient gain, this LNA was composed of six cascade common-source stages. The output of each stage was equivalently loaded with a bandpass combination of L and C to provide parallel resonance, i.e. to maximize the gain, over the 60-GHz-band of interest. Since the DC current of the third stage and the fifth stage of the LNA was reused in the second and fourth stage, respectively, by adopting the current-sharing technique in [8], no additional driving current was needed for these stages. In this way, low-power consumption was achieved. Based on the method in [9], simultaneous input impedance and noise matching over the 60-GHz-band of interest was achieved by appropriately selecting the values of the T-type input network, the source-degenerative inductor, and the size and bias of transistor M_1, i.e. C_{gs1} and g_{m1}, of the input stage. Besides, a source degenerative inductor was used in the output stage to achieve good output impedance matching and improve stability.

II.B. Current-Reused Bleeding Mixer and Baseband Amplifier

As shown in Fig. 2, the current-reused bleeding topology was adopted in the single-balanced mixer to achieve high gain, low noise, and high linearity [10]. Because of the large bias current (i.e. g_m) of the RF input transistor M_9, low noise and high gain can be achieved simultaneously by appropriately selecting the size of transistor M_9, and the values of its T-type input network. Besides, compared to the traditional single-balanced mixer topology, the bias current of the LO input transistors M_7/M_8 is relatively smaller, so the voltage drop of

the loads (R_9 and R_{10}) is smaller and hence M_7/M_8 can be kept in the saturation region even for small supply voltage (V_{d2} in Fig. 2) of 1 V. This in turn results in better linearity. Moreover, the baseband amplifier is used for single-end-to-differential conversion and providing additional gain.

II.C. 180° Out-of-Phase Wilkinson Power Divider

As shown in Fig. 2, the 180° out-of-phase Wilkinson power divider comprises a traditional Wilkinson power divider with two in-phase outputs, a right-hand π network (RH_π) at port-2 (i.e. port LO_{out+}) for positive phase shift, and a left-hand π network (LH_π) at port-3 (i.e. port LO_{out-}) for negative phase shift. The 180° out-of-phase Wilkinson power divider can convert the single-ended LO input signal (LO_{in}) into differential LO signals (LO_{out+} and LO_{out-}) to provide the needed differential LO inputs of the mixer.

III. RESULTS AND DISCUSSIONS

The chip micrograph of the finished circuit is shown in Fig. 3. The chip area was 0.92×1.3 mm² excluding the test pads. On-wafer measurement was performed by an Agilent's 67 GHz RFIC measurement system, including Agilent's E8361A network analyzer, Agilent's E8257D signal generator, and Agilent's N8975A noise figure analyzer, etc.. The LNA of the receiver front-end was biased at $V_{d1} = 1.8$ V, $V_{g1} = 0.72$ V, and $V_{g2} = 0.75$ V. The total drain current was 20.3 mA, that is, the LNA consumed 36.5 mW power, a relatively low value for a CMOS LNA operated in the 60-GHz band. Besides, the current-reused bleeding mixer and the baseband amplifier of the receiver front-end were biased at $V_{d2} = 1$ V, $V_{g3} = 0.85$ V, and total drain current of 13.7 mA, i.e. they consumed 13.7 mW power. This means the receiver front-end consumed 50.2 mW power.

Fig. 4(a) shows the measured input return-loss of RF port and LO port versus frequency characteristics of the receiver front-end. The RF port achieved input return-loss smaller than

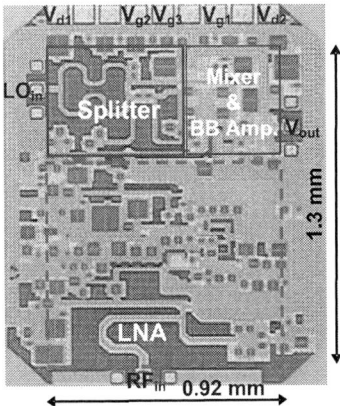

Fig. 3 Chip micrograph of the 60 GHz CMOS receiver front-end.

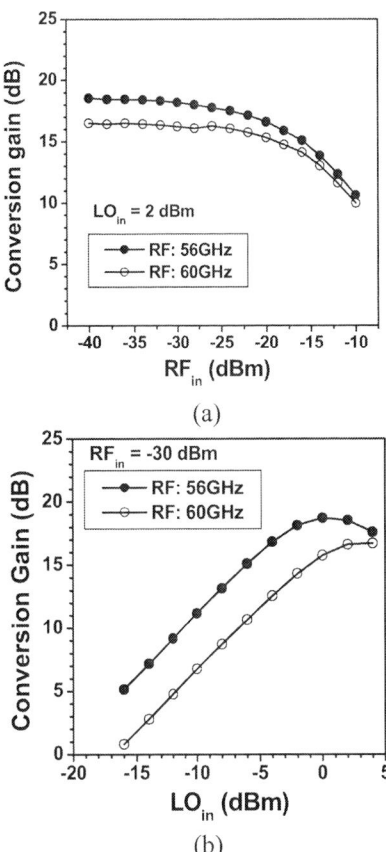

(a)

(b)

Fig. 5 Measured (a) conversion gain versus RF input power characteristics, and (b) conversion gain versus LO input power characteristics of the 60-GHz CMOS receiver front-end.

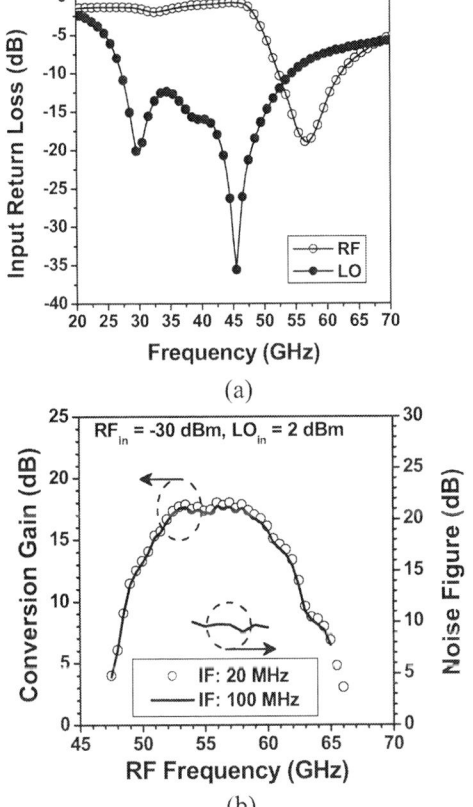

(a)

(b)

Fig. 4 Measured (a) input return-loss of RF and LO ports versus frequency characteristics, and (b) conversion gain and noise figure versus frequency characteristics of the 60-GHz CMOS receiver front-end.

−10 dB for frequencies from 52.3 GHz to 62.3 GHz. Besides, the LO port achieved input return-loss smaller than −10 dB for frequencies from 27.2 GHz to 54.5 GHz.

Fig. 4(b) shows the measured conversion gain and NF versus RF frequency characteristics of the receiver front-end at LO input power (LO$_{in}$) of 2 dBm. At IF of 20 MHz, the

receiver front-end achieved maximum conversion gain of 18 dB at 57 GHz. The corresponding ω_{3dB} is 9.8 GHz, from 50.8 GHz to 60.6 GHz. Besides, at IF of 100 MHz, the measured minimum NF was 9 dB at 58 GHz, an excellent result for a 60-GHz-band CMOS receiver front-end.

Fig. 5(a) shows the measured conversion gain versus RF input power (RF$_{in}$) characteristics of the receiver front end at LO$_{in}$ of 2 dBm. As can be seen, at RF$_{in}$ of −40 dBm, the receiver front-end achieved conversion gain of 18.5 dB and 16.5 dB, respectively, at RF frequency of 56 GHz and 60 GHz. The corresponding input 1-dB compression point (P$_{1dB}$) is −24 dBm and −20.8 dBm, respectively, at RF frequency of 56 GHz and 60 GHz.

Fig. 5(b) shows the measured conversion gain versus LO$_{in}$ characteristics of the receiver front-end at RF$_{in}$ of −30 dBm. For LO$_{in}$ smaller than −6 dBm, conversion gain increases linearly with the increase of LO$_{in}$ since the distortion is not significant. The corresponding maximum conversion gain is 18.7 dB (at LO$_{in}$ of 0 dBm) and 16.7 dBm (at LO$_{in}$ of 4 dBm), respectively, at RF frequency of 56 GHz and 60 GHz.

To characterize the non-linear behavior, two-tone signals with equal power levels and frequency difference of 1 MHz were applied to the receiver front-end. Fig. 6 shows the

978-1-4244-6240-7/10 $26.00 © 2010 IEEE

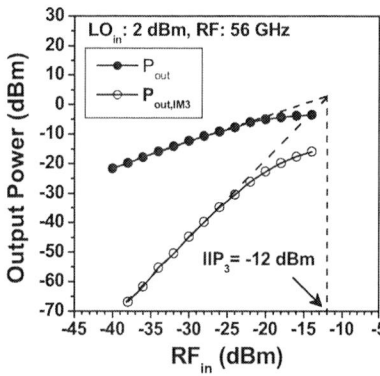

Fig. 6 Measured input third-order inter-modulation point (IIP3) at RF of 56 GHz of the 60 GHz receiver front-end.

Table I Summary of the implemented 60-GHz-band CMOS receiver front-end, and the recently reported state-of-the-art 60-GHz-band CMOS receiver front-ends.

	This Work	[6] 2008 CICC	[7] 2007 ISSCC
Technology	0.13 μm CMOS	0.13 μm CMOS	0.13 μm CMOS
RF Frequency (GHz)	50.8 ~ 60.6	60 ~ 65	57 ~ 63
IF Frequency (MHz)	20	100	2000
RF Power (dBm)	−30	--	--
LO Power (dBm)	2	--	0
3-dB BW (GHz)	9.8	4	6.1
Power Gain (dB)	15 ~ 18	25 ~ 30	11.8 (@ RF = 60 GHz)
Noise Figure (dB)	9 (@RF = 58 GHz)	10 ~ 12	10.4 (@RF = 60 GHz)
IIP3 (dBm)	−12 (@RF = 60 GHz)	--	--
P_{1dB} (dBm)	−20.8 (@RF = 60 GHz)	−27	−15.8 (@RF = 60 GHz)
P_{DC} (mW)	50.2	44	76.8
Chip Area (mm^2)	1.196	0.765	~ 3

measured fundamental and 3-order inter-modulation (IM3) output power versus input power characteristics of the receiver front-end at IF frequency of 20 MHz and two-tone RF frequencies of 56 GHz and 56.001 GHz. The corresponding P_{1dB} and input 3-order intercept point (IIP3) were −24 dBm and −12 dBm, respectively.

Table I is a summary of the implemented 60-GHz-band CMOS receiver front-end, and the recently reported state-of-the-art 60-GHz-band CMOS receiver front-ends. As can be seen, the implemented receiver front-end achieved ω_{3dB} of 9.8 GHz, wider than that (4 GHz) in [6]. Besides, the receiver front-end achieved NF of 9 dB (at RF frequency of 58 GHz), better than those (10~12 dB in [6] and 10.4 dB in [7]) in other works. These results suggest that the proposed receiver front-end with on-chip 180° out-of-phase Wilkinson power divider is very promising for high-performance 60-GHz-band RFIC applications.

IV. CONCLUSION

In this paper, we report a 60-GHz CMOS receiver front-end which comprises a wideband LNA, a current-reused bleeding mixer, a baseband amplifier, and a Wilkinson-power-divider-based 180° splitter. This receiver front-end consumed 50.2 mW and achieved high conversion gain of 18.7 dB, wide ω_{3dB} of 9.8 GHz, and low NF of 9 dB. Besides, excellent input-impedance matching in RF port, and excellent LO-RF, RF-IF, and LO-IF isolations were also achieved. These results indicate that the proposed receiver front-end architecture is very promising for high-performance 60-GHz-band RFIC applications.

ACKNOWLEDGEMENT

The authors are very grateful for the support from CIC, Taiwan, for chip fabrication, and Dr. Guo-Wei Huang, NDL, Taiwan, for high-frequency measurements.

REFERENCES

[1] R. Fisher, "60 GHz WPAN Standardization within IEEE 802.15.3c," *2007 IEEE International Symposium on Signals, Systems, and Electronics*, pp. 103-105.

[2] C. H. Doan, S. Emami, A. M. Niknejad, and R. W. Broderson, "Millimeter-Wave CMOS Design," *IEEE Journal of Solid-State Circuits*, vol. 40, no. 1, pp. 144-155, Jan. 2005.

[3] C. M. Lo, C. S. Lin, and H. Wang, "A miniature V-band 3-stage Cascode LNA in 0.13 μm CMOS," *2006 IEEE International Solid-State Circuits Conference*, pp. 1254-1255.

[4] F. Zhang, E. Skafidas, and W. Shieh, "A 60-GHz double-balanced Gilbert cell down-conversion mixer on 130 nm CMOS," *2007 IEEE RFIC Symposium*, pp. 141-143.

[5] J. H. Tsai and T. W. Huang, "35–65-GHz CMOS broadband modulator and demodulator with sub-harmonic pumping for MMW wireless gigabit applications," *IEEE Transactions on Microwave Theory and Technique*, vol. 55, no. 10, pp. 2075-2085, Oct. 2007.

[6] C. S. Wang, J. W. Huang, K. D. Chu, and C. K. Wang, "A 0.13 μm CMOS fully differential receiver with on-chip baluns for 60 GHz broadband wireless communications," *2008 IEEE CICC*, pp. 479-482.

[7] S. Emami, C. H. Doan, A. M. Niknejad, and R. W. Brodersen, "A Highly Integrated 60GHz CMOS Front-End Receiver," *IEEE International Solid-State Circuits Conference*, pp. 190-191, Feb. 2007.

[8] M. D. Wei, S. F. Chang, Y. C. Liu, "A Low-Power Ultra-Compact CMOS LNA with Shunt- Resonating Current-Reused Topology," *IEEE European Microwave Integrated Circuit Conference*, pp. 350-353, Oct. 2008.

[9] H. W. Chiu, S. S. Lu, and Y. S. Lin, "A 2.17-dB NF 5-GHz-Band Monolithic CMOS LNA with 10-mW DC Power Consumption," *IEEE Trans. Microwave Theory and Techniques*, vol. 53, no. 3, pp. 813-824, Mar. 2005.

[10] S. G. Lee, and J. K. Choi, "Current-reuse bleeding mixer," *Electronics Letters*, vol. 36, no. 8, pp. 696-697, Apr. 2000.

RTU1B-4

Coherent Parametric RF Downconversion in CMOS

Zhixing Zhao, Jean-Francois Bousquet and, Sebastian Magierowski

Department of Electrical and Computer Engineering

Schulich School of Engineering, University of Calgary, Calgary, AB, Canada T2N-1N4

Abstract—**Parametric circuits constitute a longstanding RF technique that has been largely ignored by the RFIC community. Increasing interest in applying CMOS to (sub)millimeter-wave applications plus mounting scaling complexity may combine to revitalize this circuit style. This paper presents basic parametric downconverter structures, their theory of operation, and the benefits to be gained from CMOS implementation. A low-power, sub-1-V, fully integrated mixer in 130-nm CMOS is introduced. It implements two parametric modes and operates on RF signals between 22 and 24 GHz with possible conversion gains in excess of 20 dB.**

Index Terms—**mixer, parametric mixer, MOS varactor, RFIC**

I. INTRODUCTION

Parametric mixers utilize nonlinear or time-varying *reactive* circuit elements to translate the spectrum of an input signal in frequency. Employing a reactance to achieve frequency conversion in wireless communications has been appreciated for nearly a century [1] and its potential as a downconverting technique was discovered in the mid-40s [2]–[4].

Parametric circuits owe such early emergence onto the RF design landscape to their superior gain at higher frequencies relative to contemporaneous technologies based on varistors and transistors. Given the fact that they minimize the presence of resistive components, parametric circuits also generally demonstrate lower noise behaviour, higher power output and lower power requirements for a given technology.

However, with technological advances (e.g. vacuum to solid-state, miniaturization, materials, etc.), transistor circuits managed to attain the performance levels of incumbent parametric circuits for a number of radio applications. And once a raw performance parity was achieved, transistor circuits became favoured for their amenability to unilateralization and hence ability to accommodate more complex designs. As a result, parametric circuits are now largely absent from RF designs aside from a presence in submillimeter-wave frequency multipliers [5]. This absence is especially acute in the case of parametric downconverters which, to the authors' knowledge, are employed in no electronic circuits, discrete or integrated.

However, the growing interest in millimeter and submillimeter-wave signalling for the purpose of high-speed communications and versatile sensing coupled with the increasing difficulty of scaling CMOS transistors to meet such frequency demands may provide a new opportunity for parametric circuits. Not only can parametric techniques boost the raw performance needed to address emerging THz applications, but they can also enhance the utility of mature technologies for existing high-frequency applications.

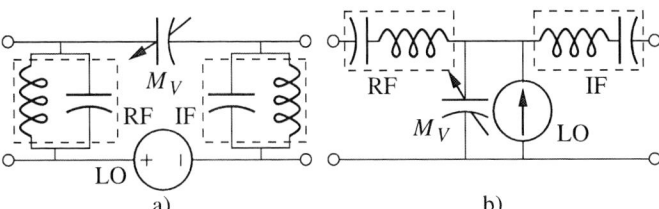

Fig. 1. Basic parametric mixer topologies including a) the Y-mixer and b) the Z-mixer.

Towards this goal, a fully-integrated 130-nm CMOS coherent parametric downconverter is introduced in this paper. Although promising CMOS envelope detectors have been shown [6], [7], coherent mixers provide the most flexibility to systems designers. Operating on RF signals between 22 and 24 GHz, the proposed circuit uses sub-1-V supplies, draws little power, and may be scaled to operate at higher frequencies. In Sections II and III basic circuit topologies, performance characteristics, and MOS varactor utility in parametric mixers are discussed while the IC is detailed in Section IV.

II. PARAMETRIC DOWNCONVERTER STRUCTURES

As with any mixing circuit, parametric downconverters contain a nonlinear reactive component, a capacitive varactor, M_V, whose capacitance is modulated (pumped) by a local-oscillator (LO) signal at frequency, ω_{LO}, while being driven by an RF input centred at ω_{RF}. The intermediate frequency (IF) output signal at ω_{IF} is determined by the difference between the RF and LO frequencies. Fig. 1 shows two basic parametric downconverter topologies, the Y-mixer and the Z-mixer styles [8].

The filter arrangements in Fig. 1 (LC tanks are used to represent the actual filters) are intended to prevent M_V from being driven by out-of-band signals. The Y-mixer configuration does so by shorting-out unwanted signals while the Z-mixer blocks them with open circuits. In many practical situations the Y-mixer is preferred for its relative insensitivity to stray filter parasitics, potential to absorb series matching elements [9], and ability to DC decouple the RF and IF ports.

Besides the filter arrangement, the operation of parametric downconverters is profoundly influenced by the choice of ω_{LO} relative to ω_{RF}. Specifically, in the upper-sideband downconverter (USBDC) configuration, where $\omega_{RF} > \omega_{LO}$ and $\omega_{IF,USB} = \omega_{RF} - \omega_{LO}$, Manley and Rowe [10] predicted that the power gain cannot exceed

$$G_{USB} = \omega_{IF,USB}/\omega_{RF}. \tag{1}$$

978-1-4244-6240-7/10 $26.00 © 2010 IEEE

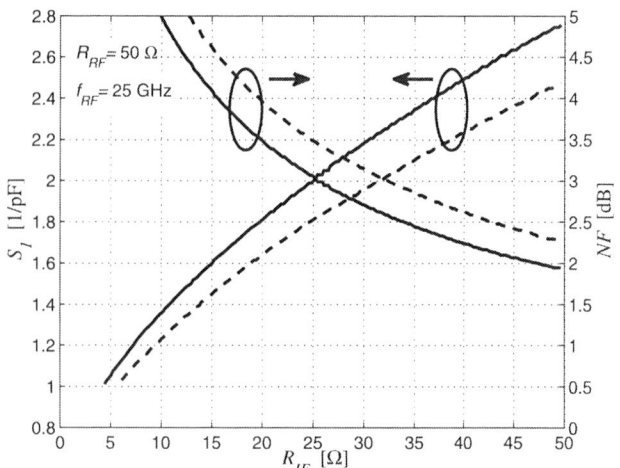

Fig. 2. The S_1 necessary to sustain a 15-dB LSBDC transducer gain as a function of IF road resistance. The noise figure corresponding to the S_1 settings is indicated by the right ordinate.

Since $\omega_{IF} < \omega_{RF}$ we have the prediction that the USBDC can only operate with a conversion loss.

Alternatively, positive power gain can be attained in the lower-sideband downconverter (LSBDC) configuration for which $\omega_{RF} < \omega_{LO}$ so that $\omega_{IF,LSB} = \omega_{LO} - \omega_{RF}$. For a Y-mixer LSBDC driven by a reactively matched RF source with resistance R_{RF} and loaded at its IF port by a series-equivalent resistance R_{IF} (also reactively matched), the transducer gain is [11]

$$G_{LSB} = \frac{4R_{IF}R_{RF}|S_1|^2}{[\omega_{RF}(R_{RF}-R_s)(R_{IF}-R_s) - |S_1|^2/\omega_{IF,LSB}]^2} \tag{2}$$

where R_s is M_V's series loss and S_1 is its fundamental elastance harmonic (i.e. inverse of the modulated capacitance) at ω_{LO} generated by the pumping action of the LO. The acheivable S_1 for a MOS varactor under various LO and bias conditions is discussed in Section III. For the same circuit circumstances, the LSBDC's noise factor is [11]

$$F_{LSB} = 1 + \frac{\overline{e_{n,IF}^2} + \overline{e_{n,RF}^2}[|S_1|/\omega_{RF}(R_{RF}-R_s)]^2}{\overline{e_{n,i}^2}[|S_1|/\omega_{RF}(R_{RF}-R_s)]^2} \tag{3}$$

where $\overline{e_{n,IF}^2} = \overline{e_{n,RF}^2} = 4kTR_s\Delta f$ are the noise sources at IF and RF associated with the varactor loss and where $\overline{e_{n,i}^2} = 4kTR_{RF}\Delta f$ is the RF input noise source.

From (2) and (3), it is apparent that the gain and noise behaviour of the LSBDC are dependent on a myriad of factors. Generally, S_1 is the most important contributor (as discussed in Section III it is regulated by the varactor type and the LO) and reasonable values for this parameter are shown in Fig. 2 which considers the performance of an LSBDC with $R_{RF} = 50\ \Omega$, 25-GHz RF input frequency and 2.5-GHz IF output frequency. The left ordinate indicates the S_1 necessary to maintain a transducer gain of 15-dB as a function of the series-equivalent IF resistance termination.

As shown, increasing R_{IF} requires a greater S_1 to sustain gain. The solid and dashed curves in Fig. 2 indicate that the same gain can be achieved with two different S_1's an outcome of the fact that S_1 controls both the mixer's IF output impedance and open circuit IF voltage gain. Increasing S_1 also reduces the mixer's noise figure (right ordinate of Fig. 2). The low noise figure is characteristic of reactive mixing, which minimizes noise contributions from resistive components.

The LSBDC achieves a part of its gain by presenting a negative resistance at the IF port and hence is conditionally stable. The gain, and therefore stability, of the LSBDC can be managed by the varactor bias and the amount of LO power used to drive it. This issue is an area where the LSBDC stands to benefit from integration with CMOS which affords the opportunity for digital regulation. This was not the case in early parametric designs which preceded the arrival of mixed-signal and system-on-chip technologies.

III. MOS VARACTORS IN PARAMETRIC MIXERS

Another way in which parametric circuits benefit from CMOS is by the availability of accumulation-mode MOS varactors (AMOSVs). Being majority carrier devices, AMOSVs easily provide the frequency response necessary to support high-frequency mixing applications. The key figure of merit in this respect is the *dynamic* cutoff frequency [12],

$$f_{cd} = (S_{max} - S_{min})/2\pi R_s \tag{4}$$

where S_{max} and S_{min} are the maximum and minimum varactor elastances during a period of the LO drive.

A contour plot of the f_{cd} (in THz) for a 0.36-μm gate length varactor model available from the foundry (a 130-nm RF CMOS process is being considered) as a function of LO amplitude, V_m, and AMOSV gate-source bias, V_{DC}, is shown in Fig. 3. With f_{cd} values in excess of 400-GHz for LO voltages below 150 mV, these results, albeit simulation based, hint at the potential utility of AMOSVs in parametric circuits far beyond the microwave regime. As CMOS technologies adopt high-k dielectrics and metal gates this performance is expected to improve.

AMOSVs also allow sufficient S_1 levels to be reached at reasonable LO magnitudes. Fig. 4 is a contour plot of the simulated S_1 of an AMOSV with 0.36-μm gate length and a net width of 40-μm (this device is used in the downconverter IC discussed in Section IV). As shown, under optimal bias, this AMOSV can achieve S_1's near 3 pF^{-1} for LO amplitudes below 200 mV. A reference for the performance achievable at this level is provided in Fig. 2. Indeed, at low R_{IF}'s S_1 settings around only 1.5 pF^{-1} are necessary which can be attained with an LO amplitude of less than 100-mV. CMOS amplifiers can provide such terminations.

IV. A USBDC/LSBDC CMOS IC

The basic circuits shown in Fig. 1 can be ported to an IC in a number of ways. One topology, a realization of the Y-mixer, is shown in Fig. 5. This circuit was implemented in a 130-nm 8-metal layer RF CMOS technology. A die photo is shown in

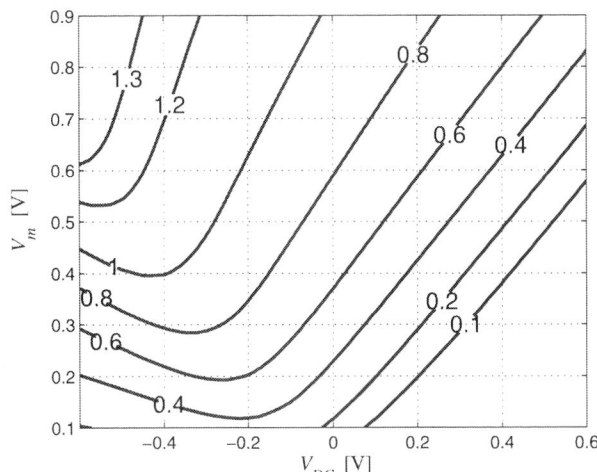

Fig. 3. The dynamic cutoff frequency (in THz) as a function of AMOSV DC bias and LO voltage amplitude for a device with 0.36-μm unit gate length.

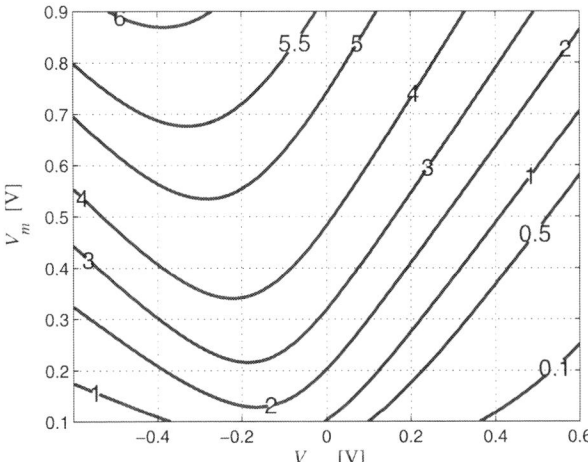

Fig. 4. The simulated S_1 (in pF^{-1}) for a 0.36×40 μm^2 AMOSV as a function of DC bias and LO amplitude.

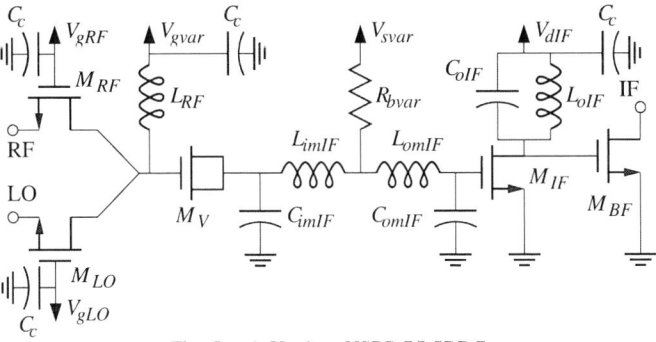

Fig. 5. A Y-mixer USBDC/LSBDC.

Fig. 6. The dimensions of this chip, including pads, are 1 mm by 0.57 mm.

The RF filter consists of L_{RF} and the drain capacitance of transistors M_{RF} and M_{LO}, which couple the RF and LO signals into the circuit, as well as the gate-bulk parasitics of M_V. The RF filter's frequency response is centred such that

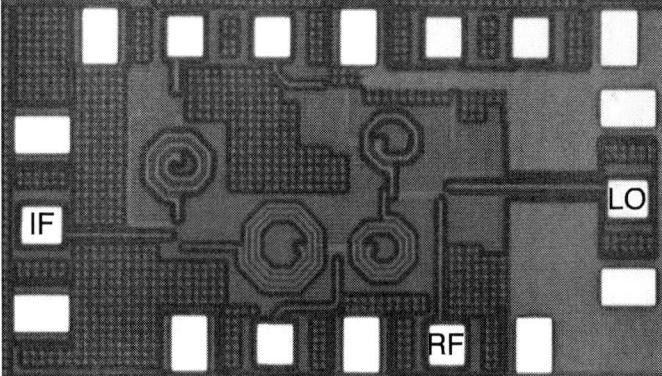

Fig. 6. A microphotograph of the 130-nm CMOS USBDC/LSBDC mixer.

it allows RF and LO signals to excite M_V while acting as a return path for IF. Both high-side and low-side LO signals can be fed into the circuit such that this mixer can function in either LSBDC or USBDC mode.

Minimum gate length transistors are used throughout the design while the varactor dimensions are 0.36 μm by 40 μm. A non-minimum M_V gate length is used in order to prevent capacitive parasitics from degrading S_1; the total area is set such that S_1's in excess of 1 pF^{-1} can be realized at LO voltage amplitudes below 200 mV.

The IF filter consists of C_{imIF}, L_{imIF}, L_{omIF}, and C_{omIF} and realizes both a reactive match between the parametric downconverter and the IF post-amplifier, M_{IF}, as well as a return path for the LO and RF signals. The IF filter also helps realize an LO-to-IF isolation of 24-dB. The average capacitance of M_V is included in the design of the IF filter and also serves to DC-decouple the RF/LO and IF ports therefore allowing a low-voltage design. Also, the source-bulk capacitance of M_V is absorbed into the IF matching filter.

The final circuit is a coarse mixer intended to downconvert inputs to a 9-GHz IF as part of a superheterodyne receiver. Lower output frequencies are possible, but require larger passive components if LC filters are used. In the LSBDC configuration, a 22-GHz RF input is employed along with a 31-GHz LO. For the USBDC, optimal performance is achieved with a 24-GHz RF input and a 15-GHz LO. Only the limitations of available measuring equipment prevented higher RF frequency inputs and hence larger frequency conversion ratios from being explored.

The measured maximum transducer gain, G_{Tmax} and 3-dB bandwidth, f_{3dB}, of the parametric mixer in LSBDC and USBDC mode as a function of M_{IF} gate-source bias, V_{svar}, are shown in Fig. 7. As expected, the mixer can operate with higher gain while in LSBDC-mode because of the lower IF port impedance presented to the IF transistor. The comparison is not entirely fair however, since the LO signals are not conducted equivalently in the two designs.

Since only narrowband matching and IF amplification techniques were used in this design the mixer bandwidth is limited to just over 180 MHz although a positive power gain can be maintained over a 500-MHz span. With a proper IF filter

978-1-4244-6240-7/10 $26.00 © 2010 IEEE

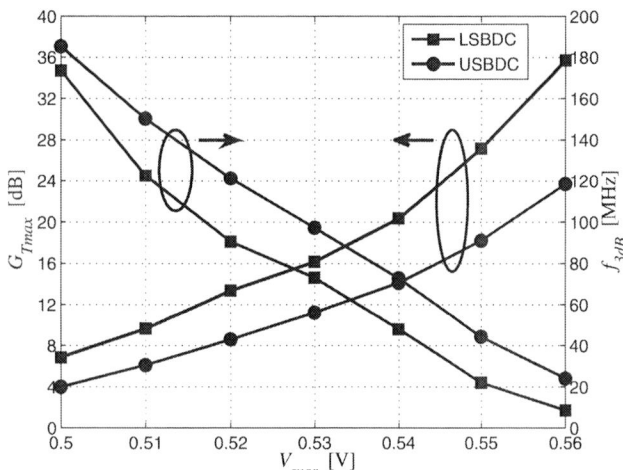

Fig. 7. Measured maximum transducer gain (left ordinate) and 3-dB bandwidth (right ordinate) of the parametric mixer in LSBDC and USBDC-mode respectively as a function of M_{IF} gate bias.

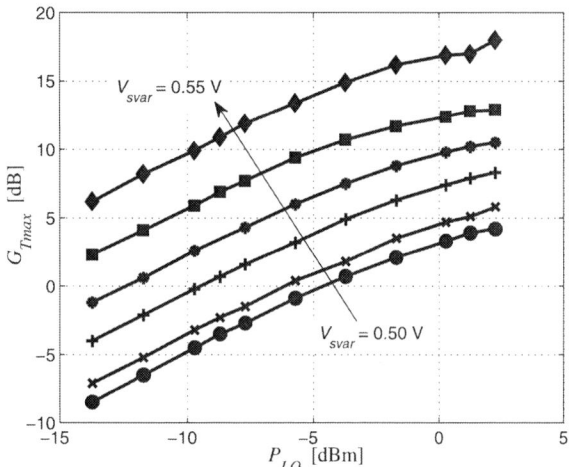

Fig. 8. Measured maximum transducer gain of the USBDC-mode mixer as a function of available LO power for different values of V_{svar} in increments of 100-mV.

design simulations indicate that bandwidths in excess of 1-GHz are possible.

The DC power consumption of the mixer's IF and buffer (M_{BF}) amplifiers varies between 2.2 and 3.2 mW depending on V_{svar}, while the combined power drawn by the RF and LO transistors is 3 mW. It should be recalled however that the mixing component itself (i.e. M_V) consumes no DC power.

A couple of factors are responsible for this low power requirement at the RF/LO port. First, since no transistor stacks are employed in this design, low-voltage supplies are sufficient (not exceeding 0.7 V). Second the M_{LO} can operate in class-AB mode since the varactor elastance can be made to respond primarily to positive LO excitation. This is implied in Fig. 8 which indicates a graceful degradation in the USBDC mixer's transducer gain as a function of available LO power, P_{LO}. The LSBDC follows a similar pattern.

These results indicate that the mixer varactor-core's linearity

is good, however the IF post-amplifier compromises this for the circuit as a whole. For example, as the gain of the US-BDC is varied from around 0-dBm to 15-dBm by increasing V_{svar},o= the input-referred 1-dB compression point of the mixer drops from about -12 dBm to -30 dBm.

V. Conclusion

The basic operation, structure, and performance of parametric downconverters has been discussed. The potential that this circuit design technique has to profit from CMOS technology has also been stressed. Mainly, the benefits of AMOSVs, their dynamic cutoff frequency and high fundamental elastance harmonic, have been highlighted. This analysis indicates that even parametric converters realized in a mature CMOS process should easily be able to operate deep into the millimeter-wave domain.

Also presented was, to the authors' knowledge, the first fully integrated parametric downconverter. Doubling as both a LSBDC and USBDC parametric mixer the circuit operated on RF input signals between 22 and 24 GHz to produce an 9-GHz IF. Although these frequencies are far below the mixer's potential operating point the proposed topology can be retained for higher frequency operation. Besides high gain and intrinsic linearity this approach promises a substantial power savings as low-voltage headroom is sufficient and no DC power is drawn by the varactor mixing component itself .

Acknowledgment

The authors would like to thank CMC Microsystems for chip fabrication support and NSERC for financial support.

References

[1] L. Kuhn, "Uber ein neves radiotelephonische system," *Jahr. fur dracht. Teleg. und Telephone*, vol. 9, pp. 502–534, June 1915.

[2] H. Q. North, "Properties of welded contact germanium rectifiers," *J. Appl. Phys.*, vol. 17, pp. 912–923, Nov. 1946.

[3] H. C. Torrey and C. A. Whitmer, *Crystal Rectifiers* (MIT Rad. Lab. Ser. vol. 15). New York, NY: McGraw-Hill Book Co., Inc., 1948.

[4] R. V. Pound, *Microwave Mixers* (MIT Rad. Lab. Ser. vol. 16). New York, NY: McGraw-Hill Book Co., Inc., 1948.

[5] T. W. Crowe, J. L. Hesler, R. M. Weikle, and S. H. Jones, "GaAs devices and circuits for terahertz applications," *Infrared Physics and Technology*, vol. 40, pp. 175–189, Dec 1999.

[6] E. Seok, C. Cao, S. Sankaran, and K. K. O, "A millimeter-wave Schottky diode detector in 130-nm CMOS technology," in *Symp. on VLSI Circuits Dig. Tech. Papers*, June 2006, pp. 178–179.

[7] E. Ojefors, U. R. Pfeiffer, A. Lisauskas, and H. G. Roskos, "A 0.65 THz focal-plane array in a quarter-micron CMOS process technology," *IEEE J. Solid-State Circuits*, vol. 44, no. 7, pp. 1968–1976, Jul 2009.

[8] A. A. M. Saleh, *Theory of Resistive Mixers*. Cambridge, MA: MIT Press, 1971.

[9] G. L. Matthaei, "A study of the optimum design of wide-band parametric amplifiers and up-converters," *IRE Trans. Microwave Theory and Techniques*, vol. 9, no. 1, pp. 23–38, Jan 1961.

[10] J. M. Manley and H. E. Rowe, "Some general properties of nonlinear elements — Part I. general energy relations," *Proc. IRE*, vol. 44, pp. 904–913, Jul. 1956.

[11] S. Magierowski, T. Zourntos, J.-F. Bousquet, and Z. Zhao, "Compact parametric downconversion using MOS varactors," in *Proc. IEEE International Microwave Symposium*, June 2009, pp. 1377–1380.

[12] H. A. Watson, Ed., *Microwave semiconductor devices and their circuit applications*. New York, NY: McGraw-Hill, 1968.

RTU1B-5

60 GHz Broadband Image Rejection Receiver using Varactor Tuning

Jihoon Kim, Wooyeol Choi, Youngrak Park, and Youngwoo Kwon

School of Electrical Engineering and Computer Science and Institute of New Media and Communications,
Seoul National University, 599, Gwanak-ro, Gwanak-gu, Seoul, 151-742, Korea

Abstract — A pHEMT broadband image rejection receiver with an image rejection ratio (IRR) more than 20 dB from 54 GHz to 66 GHz is presented using varactor tuning topology. Tunable varactors connected in shunt between an RF coupler and mixers are used to control the phase and amplitude of two RF signals. It offers the IRR improvement of 3.1 ~ 21.4 dB in the cost of gain degradation below 1.1 dB from 54 GHz to 66 GHz except for 65 GHz. To the best of authors' knowledge, this work shows the best image rejection performance of V-band receivers. At 61 GHz, this circuit achieves an 18.3 dB conversion gain (C.G) and a 49.3 dB IRR. It shows a noise figure of 5.6 ~ 8.1 dB from 56 GHz to 64 GHz.

Index Terms — Broadband, image rejection receiver, image rejection ratio (IRR), varactor

I. INTRODUCTION

For past few years, V-band transceivers have been actively been developed for various applications including WPAN (Wireless Personal Area Network) [1]-[3]. V-band transceivers mostly employ heterodyne conversion receivers because of good selectivity and sensitivity over broad band. Therefore, the rejection of image signals has become an important requirement for the receiver design. There are generally two solutions for image rejection. One is to use image rejection filters. However, the receiver with low IF frequency requires the filters with steep roll-off characteristics, which are difficult to implement at millimeter-wave frequencies. Recently, dual IF architectures that allow large separation between the image and RF signals have been successfully demonstrated. But these circuits require higher complexity and DC power consumption [3], [7], [10]. Moreover, these require a high IF which could result in many drawbacks. The other solution used in millimeter-wave receivers is Hartley architecture [1]-[2], [8]-[9]. In this architecture, image rejection can be infinite if the conversion gain and phase shift in the mixers are identical, and the phase and amplitude balance of the hybrids is perfect. However, it is virtually impossible to eliminate the phase and amplitude imbalance in the practical circuits, in particular at millimeter-waves due to parasitic effects [4]. Achieving the image rejection higher than 20 dB over wide band becomes very difficult at V-band. To overcome these limits, a new image rejection topology is developed in this work, which adds varactor tuning in the Hartley architecture. Varactors allow one to tune out the phase and amplitude imbalances in the two paths, resulting in the improved image rejection over broad band. The proposed topology requires no

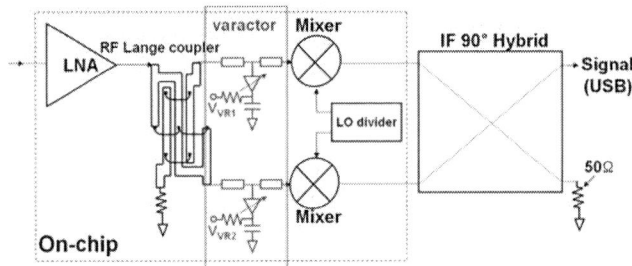

Fig. 1 V-band image rejection receiver system diagram

additional DC power consumption and chip area.

II. VARACTOR TUNING TOPOLOGY

The widely-used equation for image rejection is [4]

$$R_I = -10\log\left[\frac{1-2\sqrt{G}\cos\theta+G}{1+2\sqrt{G}\cos\theta+G}\right] \quad (1)$$

(G: gain imbalance, θ: phase imbalance)

To achieve 20 dB image rejection, the phase error should be kept below 10° and the gain imbalance below 1 dB. High-frequency receivers often result in large phase imbalances [4]. Moreover, unpredicted parasitic and coupling effects make the accurate simulation of the phase imbalance difficult [6]. In addition, for broadband image rejection, the gain imbalance should be also considered. In this work, varactors are inserted in the RF path as the shunt arm of the low-pass-type matching circuit between the coupler and mixer as shown in Fig. 1. Each varactor acts as a tunable capacitor controlled by the external bias. The change of shunt capacitances results in the

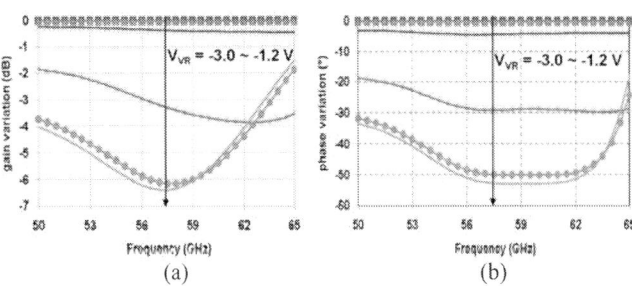

Fig. 2 (a) Gain variation of a shunt varactor (2 × 20 μm) according to bias sweep and (b) phase variation of a shunt varactor (2 × 20 μm) according to bias sweep

978-1-4244-6240-7/10 $26.00 © 2010 IEEE

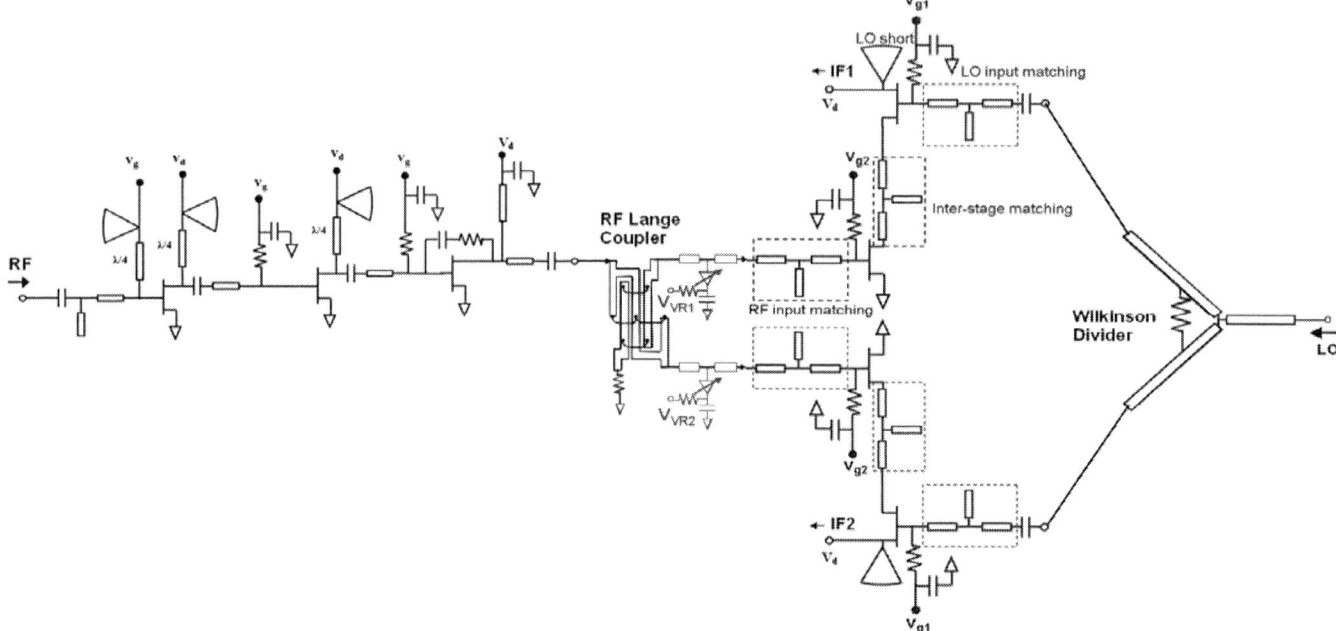

Fig. 3 Circuit schematic of the V-band image rejection receiver MMIC

change of phase of RF paths. It changes the phase of the RF signals entering the mixers and enables one to control the phase imbalance at the IF ports. Besides the phase, the varactors can also control the gain imbalance. The change of shunt capacitances between the coupler and mixers results in the change of RF input impedances of the mixers. It allows variable gain for adjusting gain imbalance. Each of varactor was realized with a source-drain connected HEMT with a size of $2 \times 20~\mu m$. It was connected to bias pads to be supplied by reverse bias voltages using DC grounds of an RF coupler. Figure 2 shows the phase and gain variation of a shunt varactor structure. The phase imbalance up to 50° can be tuned while the gain imbalance up to 6 dB can be controlled from 50 to 65 GHz.

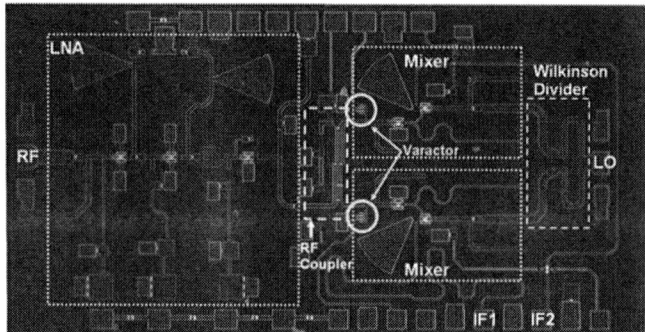

Fig. 4 Photograph of the image rejection receiver MMIC (size = 3.4 mm × 1.8 mm)

III. CIRCUIT DESIGN

The image-rejection receiver consists of an LNA, a quadrature coupler, two cascode mixers and an LO Wilkinson divider. Figure 3 shows the circuit schematic of the image rejection receiver MMIC. The LNA is a 3-stage design, where the first two stages were matched for low noise and the last stage for power gain. For stability and bandwidth, the last stage employed parallel feedback. Lange coupler divides the output signal from the LNA into two mixers with 90° phase difference. The varactors are inserted between the coupler and the mixers for phase and gain control. The chip area occupied by the varactors is only 0.01 mm². Cascode mixers were designed following the methodology presented in our previous work [5], which employed harmonic termination between the common-source and common-gate transistors to improve the conversion gain and linearity. LO signal is

equally split with the same phase through Wilkinson divider and is fed into the gate of the common-gate transistor in the cascode mixer. LO short was provided at the drain port of the common-gate transistor through a radial stub. The size of transistors used in the LNA and mixers is $2 \times 50~\mu m$. The phase error between two ports of a RF coupler predicted by EM simulation is 1.5° at 60 GHz. IF lines are designed to have the phase error of opposite sign for the phase balance. The phase imbalance of an external IF coupler is found 2~3° by measurements. Thus the total phase error in the receiver is expected to be over 2° at least. But various uncontrollable factors such as layout effects can generate gain and phase mismatch [6].

IV. MEASUREMENT

This circuit has been fabricated using a commercial 0.15 μm GaAs pHEMT MMIC process with a f_T of 85 GHz and an

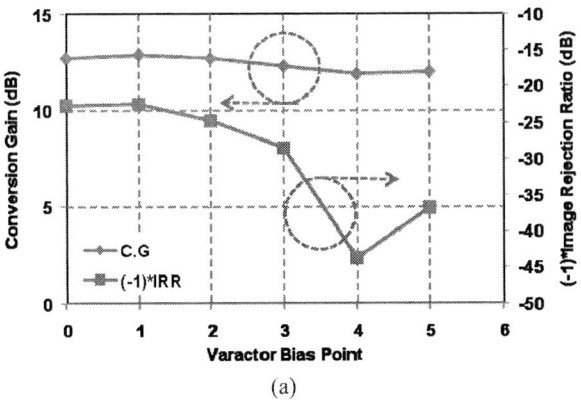

(a)

Bias point	V_{VR1} (V)	V_{VR2} (V)
0	−3	−3
1	−1.5	−1.2
2	−1.5	−1.0
3	−1.5	−0.7
4	−1.5	−0.3
5	−1.3	−0.3

(b)

Fig. 5 (a) Measured conversion gain and image rejection ratio according to varactor biases (RF frequency = 54 GHz, RF power = 31.5 dBm, LO frequency = 53 GHz, LO power = 4.5 dBm), (b) table of varactor bias points

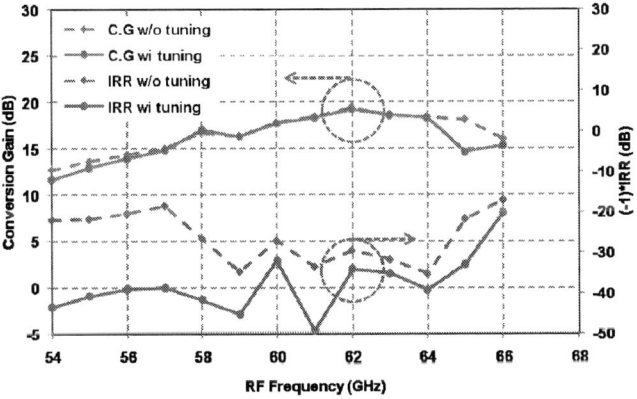

Fig. 6 Measured conversion gain and image rejection ratio according to RF frequency with and without varactor tuning (IF frequency = 1 GHz, RF power = −33 dBm, LO power = 4.5 dBm)

f_{MAX} of 190 GHz. Total chip size is 3400 μm × 1770 μm. Figure 4 shows the photograph of the image rejection receiver MMIC. The RF and LO signals were applied to the circuit using the on-wafer probes. IF ports are connected to an off-chip 90° hybrid (XCO900P-03S) using 40 cm flexible coaxial cables (RG316DS). For testing, the upper side band (USB) output port of the hybrid is connected to a spectrum analyzer while the other port is terminated in 50 Ω. Figure 5(a) shows the measured conversion gain and image-rejection ratio at the RF frequency of 54 GHz and the IF frequency of 1 GHz while

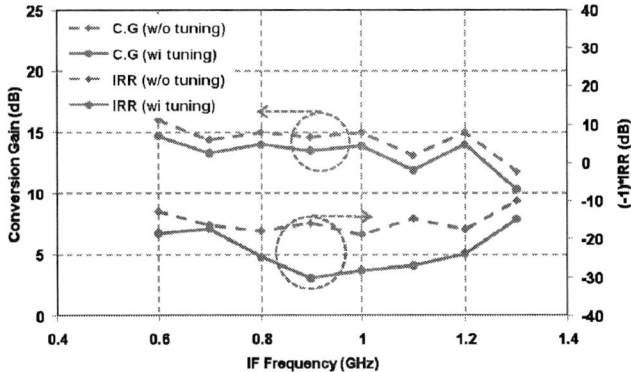

Fig. 7 Measured conversion gain and image rejection ratio according to IF frequency with or without varactor tuning (RF power = −33.5 dBm, fixed RF frequency = 57 GHz, LO power = 4.5 dBm, fixed varactor bias V_{vr1} = −1.5 V, V_{vr2} = −0.3 V)

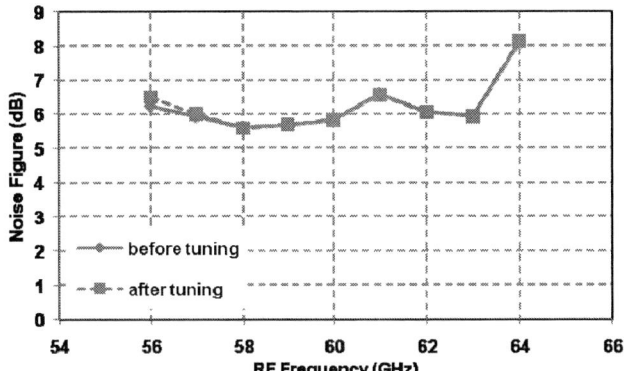

Fig. 8 Measured noise figure (IF frequency = 1 GHz, LO power = 4.5 dBm)

varying the bias voltages of the varactor as seen at the table of fig. 5 (b). As expected from analysis, the bias tuning results in improved image rejection at the cost of slight degradation in the conversion gain. At an RF frequency of 54 GHz in USB mode, when V_{VR1} = −1.5 V, V_{VR2} = −0.3 V is biased, a 43.7 dB IRR which is improved by 21.4 dB, is achieved at the cost of 0.8 dB conversion gain degradation. Likewise, the optimum image rejection points were tested by sweeping RF frequencies from 54 to 66 GHz according to the varactor bias. Figure 6 shows the measured image rejection and conversion gain with and without varactor tuning in USB mode of operation. Over the broad frequency range from 54 to 66 GHz, the varactor tuning offers the image rejection improvement of 3.1 ~ 21.4 dB. With varactor tuning, conversion gain has slight degradation below 1.1 dB except for 65 GHz which suffers largest gain and phase mismatch within measurement band. As a result, the RF frequency bandwidth providing image rejection higher than 20 dB has been extended to 12 GHz with varactor tuning. The image rejection peaks up to 49.3 dB at 61 GHz. Figure 7 shows the measured conversion gain and image rejection according to IF frequencies at the

TABLE I
COMPARISON OF 60 GHz IMAGE REJECTION RECEIVERS

Ref.	Technology	IF freq. (GHz)	C.G (dB)	IRR BW* (GHz)	Max.IRR (dB)	NF (dB)	DC power cons. (mW)
2005 [1]	GaAs pHEMT	2.5	8.5	5	35	9.8	990
2007 [2]	GaAs mHEMT	2.5	12.9	2	23	7.2	450
2006 [3]	SiGe BiCMOS	0	39.8	5	38	6	530
1999 [12]	GaAs pHEMT	n/a	-10	0.14	20	5.9	n/a
1995 [13]	GaAs pHEMT	0.14	-3	0	13	3.3	n/a
2002 [14]	GaAs pHEMT	2.5	-10	0	19	n/a	0
This Work	**GaAs pHEMT**	**1**	**18.3**	**12**	**49.3**	**5.8**	**278**

IRR BW*: Bandwidth of RF frequencies (IRR > 20dB)

fixed RF frequency of 57 GHz. For IF signals from 0.6 to 1.3 GHz, the image rejection is improved by 1 ~ 14.3 dB with gain degradation below 1.5 dB. In order to check the impact of varactor tuning on the receiver noise, the noise figure has been tested across the RF frequency band of 56 to 64 GHz, as shown in Fig. 8. The noise figure showed minimum of 5.6 dB at the RF frequency of 58 GHz, and was less than 7 dB from 56 to 63 GHz. The measured noise figure was almost identical with and without the varactor tuning. The performance of V-band image reject receivers is compared with the previous work in Table I. To the best of authors' knowledge, if varactor bias could be controlled according to frequencies, this work shows the widest band and the best image rejection characteristics (12 GHz for IRR > 20 dB, 49.3 dB max) among the reported V-band receivers, together with comparable or better noise figure and conversion gain. It is worthwhile to note that simple varactor tuning employed in this work does not require additional DC power consumption and complex IF circuitry, resulting in smaller DC power consumption as compared with [3] and other works.

V. CONCLUSION

A V-band image rejection receiver with wideband image rejection has been implemented using on-chip varactor tuning topology. By adjusting the varactor bias voltage, the gain and phase imbalance inherent to the fabricated chip has been corrected effectively, which allowed non-negligible image rejection over wide frequency range. Moreover, the correction was implemented in such a fashion that does not increase the chip size or DC power consumption. No significant noise figure degradation was observed by varactor tuning either. The measured performance of the chip shows the widest frequency image rejection among the reported V-band receivers. This topology is a promising solution for broadband image rejection for millimeter-wave receiver.

ACKNOWLEDGEMENT

This work was supported by the Acceleration Research Program of the Ministry of Education, Science and Technology of the Republic of Korea and the Korea Science and Engineering Foundation.

REFERENCES

[1] S. E. Gunnarsson, C. Kärnfelt, H. Zirath, R. Kozhuharov, D. Kuylenstierna, A. Alping, and C. Fager, "Highly integrated 60 GHz transmitter and receiver MMICs in a GaAs pHEMT technology," IEEE J. Solid-State Circuits, vol. 40, no. 11, pp. 2174–2186, Nov. 2005.

[2] S. E. Gunnarsson, C. Kärnfelt, H. Zirath, R. Kozhuharov, D. Kuylenstierna, C. Fager, M. Ferndahl, B. Hansson, A. Alping, and P. Hallbjörner, "60GHz Single-Chip Front-End MMICs and Systems for Multi-Gb/s Wireless Communication," IEEE J. Solid-State Circuits, vol. 42, no. 5, pp. 1143–1157, May. 2007.

[3] S. Reynolds, B. Floyd, U. Pfeiffer, T. Beukema, J. Grzyb, C. Haymes, B. Gaucher, and M. Soyuer, "A Silicon 60-GHz Receiver and Transmitter Chipset for Broadband Communications," IEEE J. Solid-State Circuits, vol. 41, no. 12, pp. 2820–2831, Dec. 2006.

[4] S. A. Maas, Microwave Mixers. Boston, MA: Artech House, 1993, pp. 280–283.

[5] J. Kim and Y. Kwon, "Intermodulation analysis of dual-gate FET mixers," IEEE Trans. Microwave Theory Tech., vol. 50, no. 6, pp. 1544–1555, June 2002.

[6] S. E. Gunnarsson, D. Kuylenstier, H. Zirath, "Analysis and Design of Millimeter-Wave FET-based Image Reject Mixers," IEEE Trans. Microwave Theory Tech., vol. 55, no. 10, pp. 2065–2074, Oct. 2007.

[7] B. Razavi, "A Millimeter-wave CMOS Heterodyne With On-Chip LO and Divider," IEEE J. Solid-State Circuits, vol. 43, no. 2, pp. 477–485, Feb. 2008.

[8] R. Hartley, "Modulation System," U.S Patent 1,666,206, April 1928.

[9] P. Singh, S. Basu, K. Liao, and Y. Wang, "Highly Integrated Ka-Band Sub-Harmonic Image-Reject Down-converter MMIC," IEEE Microw. Wireless Comp. Lett., vol. 19, no. 5, pp. 305–307, May. 2009.

[10] A. Parsa and B. Razavi, "A New Transceiver Architecture for the 60-GHz Band," IEEE J. Solid-State Circuits, vol. 44, no. 3, pp. 751–762, Mar. 2009.

[11] Gavell, M.; Ferndahl, M.; Gunnarsson, S.E.; Abbasi, M.; Zirath, H., "An Image Reject Mixer for High-Speed E-Band (71-76, 81-86 GHz) Wireless Communication," IEEE Compound Semiconductor Integrated Circuits Symp., pp. 1–4, Oct. 2009.

[12] K. Nishikawa, K. Kamogawa, R. Inoue, K. Onodera, T. Tokumitsu, M. Tanaka, I. Toyoda, and M. Hirano, "Miniaturized millimeter-wave masterslice 3-D MMIC amplifier and mixer," IEEE Trans. Microwave Theory Tech., vol. 47, pp. 1856-1862, 1999.

[13] T. Saito et al., "60-GHz MMIC image-rejection downconverter using InGaP/InGaAs HEMT", IEEE Gallium Arsenide Integrated Circuit (GaAs IC) Symposium, Digest, pp. 222–225, Oct. 1995.

[14] K. Fujii et al., "A 60 GHz MMIC chipset for 1-Gbit/s wireless links," IEEE MTT-S International Microwave Symposium Digest, Vol. 3, pp. 1725–1728, June 2002.

978-1-4244-6240-7/10 $26.00 © 2010 IEEE

A Discrete Resizing and Concurrent Power Combining Structure for Linear CMOS Power Amplifier

Jihwan Kim[1], Hyungwook Kim[1], Youngchang Yoon[1], Kyu Hwan An[1], Woonyun Kim[2],

Chang-Ho Lee[2], Kevin T. Kornegay[1], and Joy Laskar[1]

[1]Georgia Electronic Design Center, Georgia Institute of Technology, Atlanta, GA 30308, USA

[2]Samsung Design Center, Atlanta, GA 30308, USA

Abstract — We propose a new method of power combining for a parallel-combining-transformer (PCT)-based CMOS linear power amplifier (PA). The power cell in parallel paths is divided into three sub-cells to implement device resizing for discrete power control. Concurrent power combining of sub-power-cells utilizes the maximum available transformer efficiency even at the low-power mode, boosting overall PA efficiency. When all sub-power-cells are enabled, the PA exploits output power of 30.7 dBm with PAE of 35.8%. Power back-offs of 6 dB and 12 dB are achieved by discretely turning off sub-cells, showing output power of 25 dBm and 19 dBm with PAE of 19.8% and 10.5%, respectively. With 802.11g WLAN modulated signal used for linearity test, the PA shows 21-dBm output power satisfying -25-dB EVM requirements consuming 560 mA from 3.3 V power supply.

Index Terms — CMOS, parallel amplification, power amplifier, power combining, power control, transformer.

I. INTRODUCTION

With incomparable advantages in terms of cost and integralibility with digital blocks, CMOS PAs have been a brisk research territory pushing up performance limits. To transmit signals with high complexity modulation schemes in mobile environment, high output power capability and good linearity have been the biggest goals in developing CMOS PAs. Low power capability of CMOS PAs due to low break-down voltage has been overcome by using impedance transforming power combiners [1]. Linearity of CMOS PAs has been improved by circuit/system level compensation techniques [2-3].

Moreover, enhancing the efficiency of CMOS PAs at low-output-power levels has been also a challenging issue for extending operation time of mobile devices. Since the PAs are not functioning at the peak output power mode at most of time, waste of energy due to extremely low efficiency at low-power operation mode of the PAs has to be prevented. Dynamic gate-biasing [4] or parallel combining of multiple PAs, in conjunction with implementing multi-mode PAs [5-6], have been proposed to improve the low-power efficiency of the CMOS PAs.

To achieve high output power while maintaining discrete power-controllability, transformer-based power

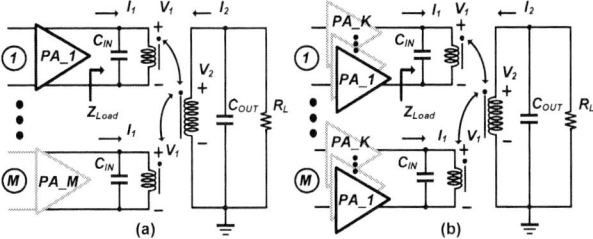

Fig. 1. Diagram of PCT-based PAs (a) conventional structure and (b) proposed structure.

combining techniques for linear PAs have been proposed by Liu *et al.* [7] and An *et al.* [8], overcoming complicated design/layout issues of parallel power combiners based on quarter-wavelength transmission lines [5] or LC-baluns [6]. By turning off individual PAs as the required output power decreases, those transformer-based PAs could effectively increase the efficiency at lower power modes. However, both voltage [7] and current [8] combining techniques could not utilize the maximum efficiency of the transformers when some power cells are disabled due to the effects of idle primary inductors of the transformers.

In this paper, we propose the concurrent power combining of PAs using a PCT in discrete power modes. As shown in Fig. 1 (b), the PAs in each combining path are sub-divided into multiple cells. For low-power mode operation, the PAs are concurrently resized (disabling some PA cells) and combined by a PCT at the output. Using this technique, we can avoid having inactive primary sides of the transformer at different power modes, which occurs inevitably in a conventional structure (Fig. 1(a)). Therefore the efficiency of the transformer can be maximally utilized, boosting the overall PA efficiency even at the low output power mode of operation.

II. CONCURRENT POWER COMBINING STRUCTURE

In the designs of [7] and [8], when some PAs in a combined structure are disabled for low-power operations, the transformers have idle primary inductors that are not

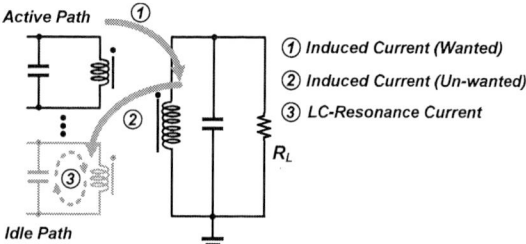

Fig. 2. Diagram of induced currents in the PCT.

participating in the power transfer. Although the inactive PAs are completely turned off, the hanging primary inductors with parasitic junction capacitances at drain nodes of PAs act as lossy loads for the transformers. Shorting the primary inductors to ground is not a perfect solution in terms of the transformer efficiency because induced currents from the secondary side will dissipate energy through the parasitic resistances of the idle primary inductors. This occurrence becomes more obvious if the primary inductors are very closely placed to the secondary inductor, which is true in most cases. Making the idle primary sides open (infinite impedance) may be better solution for the PCT, but the parasitic resistances of the primary inductors will be still wasting energy. In realistic situation, the effect of the idle primary inductors is much severe. If any capacitive reactance at the primary sides is not ignorable, there might be some resonant points where the overall transformer performance is degraded due to circulating RF currents in the LC tank, pulling the energy out of the transferring power in the active paths. The diagram in Fig. 2 presents how RF currents are induced from the active primary side to the secondary side, and back to the inactive primary side in the PCT. The effect of the idle primary inductor on the transformer efficiency in case of a two-pair PCT is shown in Fig. 3. We can observe that the maximum efficiency is attained when all combining paths are active. If one of primary inductors is inactive having open, short or a parallel capacitive load, the overall efficiency is degraded noticeably.

We can avoid issues from idle primary inductors while maintaining the discrete power-controllability by sub-dividing the power cells in each PA branch and combining them in, namely, concurrent manner. The idea is presented in Fig. 1 (b). There are two (or more) PA stages in parallel to be combined by a PCT. Each PA stage is divided into multiple sub-cells (PA_1 to PA_K) and concurrently combined, meaning PA_1s in each PA path are combined when the lowest output power is needed, and all sub-cells in each path are combined when the maximum power is required. Therefore, all primary inductors are participating in the power combining at any operation mode, discarding possible losses in case that the transformer has idle primary inductors.

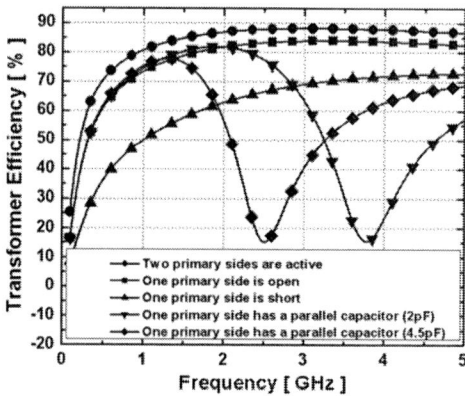

Fig. 3. Simulated efficiency of the two-pair PCT.

In addition, the concurrent combining has another advantage over the conventional technique using the PCT. The structure in [8] has a tendency to have decreasing load impedance when some PAs are turned off, resulting in the efficiency drop at the power back-offs. In contrast, proposed combining scheme maintains constant load impedance (the ratio of the output current (I_1) and the output voltage (V_1) of each PA) by utilizing all combining paths, as long as same-sized PA cells are concurrently combined. With turn ratio of the transformer (n) and the number of combined PA paths (M), the load impedance of each PA in a PCT-based structure (Fig. 1) is given by

$$Z_{Load} = \frac{V_1}{I_1} = \frac{\frac{1}{n} \cdot V_2}{\frac{n}{M} \cdot I_2} = \frac{M \cdot V_2}{n^2 \cdot I_2} = \frac{M}{n^2} \cdot R_L. \quad (1)$$

It is observed that Z_{Load} in the conventional structure decreases as M decreases (some PAs are turned off), whereas it remains constant in the proposed structure. Without disturbing the load impedance, the proposed combining technique can boost low-power efficiency of the PA with maximal utilization of the transformer efficiency at the output. It should be noted that the load impedance of each PA can be further optimized by tunable matching circuitry such as switchable capacitors. Nevertheless, we have fixed output matching network in this design only to investigate the effectiveness of proposed combining scheme along with device resizing.

III. CMOS PA DESIGN AND IMPLEMENTATION

The fully integrated PA has been fabricated in a standard 0.18 μm CMOS technology. As presented in Fig. 4, the PA consists of an input balun, a driver stage, an inter-stage matching network, two parallel power stages, and a PCT. The driver and power stages were implemented in pseudo-differential topology to minimize gain reduction due to source degeneration by grounding bond-wires. They also

978-1-4244-6240-7/10 $26.00 © 2010 IEEE

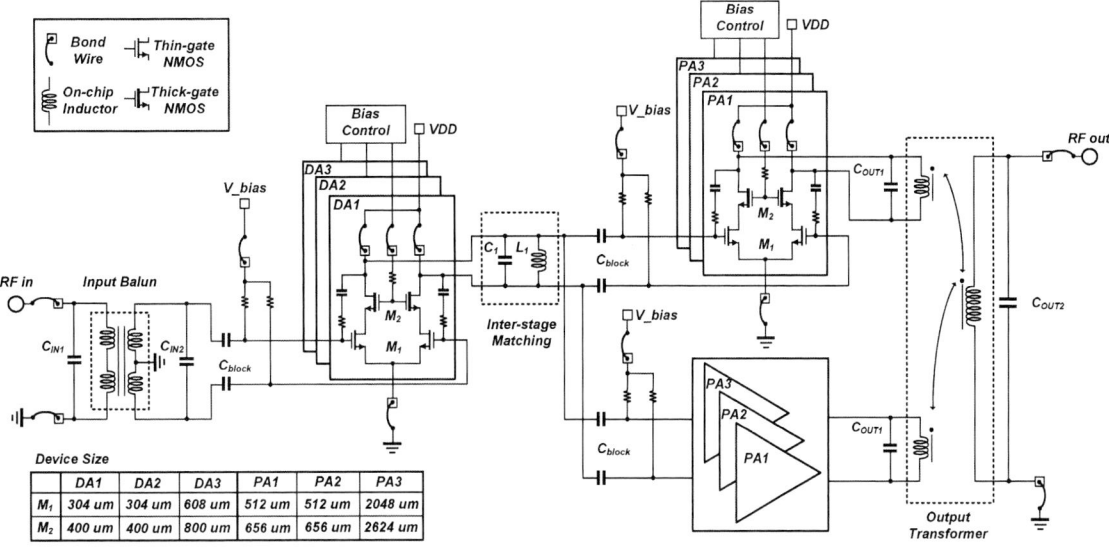

Fig. 4. The schematic diagram of the PA with concurrent power combining structure.

Device Size

	DA1	DA2	DA3	PA1	PA2	PA3
M_1	304 um	304 um	608 um	512 um	512 um	2048 um
M_2	400 um	400 um	800 um	656 um	656 um	2624 um

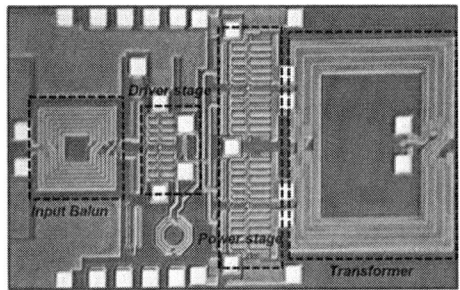

Fig. 5. Micro-photograph of the PA (1.8×1.1 mm²).

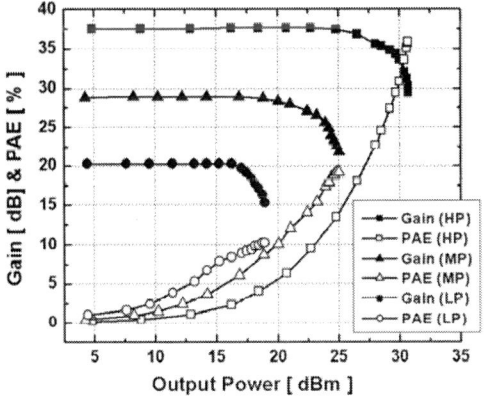

Fig. 6. Measured gains and PAEs in three power modes.

feature cascode configurations to relax voltage stress. Both stages have thin-gate-oxide transistors as common-source (CS) devices and thick-gate-oxide transistors as common-gate (CG) devices. The unit cell of power stage (PA1 in Fig. 4) is composed of a differential pair of a 512-µm width devices (0.18-µm length, 8 µm × 64 fingers) at bottom and a 656-µm width devices (0.4-µm length, 8 µm

× 82 fingers) on top. The PA2 in Fig. 4 is identical with the PA1, and the PA3 is four times larger than PA1. Thus, the cell size ratio of 1:1:4 is achieved and the entire width of the power stage (PA1+PA2+PA3) becomes 3072 µm in CS and 3936 µm in CG. The driver stage has been also sub-divided into three cells (DA1, DA2, and DA3). The DA1 has a differential pair of a 304-µm width (0.18-µm length, 8 µm × 38 fingers) devices in CS and a 400-µm width (0.18-µm length, 8 µm × 50 fingers) devices in CG. The DA2 is same the DA1, and the DA3 is twice larger than the DA1. The size ratio of the driver stage becomes 1:1:2, resulting in the entire width of 1316 µm in CS and 1600 µm in CG. For a maximum output power, the DA1, 2, and 3 with the PA1, 2, and 3 are all enabled. If power back-off is needed, the PA3s in each path and the DA3 are disabled by turning off the gate biases of CG devices. In the lowest power mode, only the PA1s and the DA1 are amplifying a signal. Note that the number of power control modes can be increased by resizing combinations between driver and power stages. Both the driver and the power stages are biased at class AB to compromise good linearity and efficiency. The input balun and output transformer have shunt capacitors (C_{IN1}, C_{IN2}, C_{OUT1} and C_{OUT2}) to cancel out the inductive reactances. An on-chip spiral inductor (L_1) and a capacitor (C_1) compose the inter-stage matching. The bond-wires for power supply of the driver stage are also included in the matching network. The PCT has two primary windings and one secondary winding which are interwoven with 1:2 turn ratio. The transformer has 30-µm width metal lines for signal path which has been implemented in two stacked metal layers. The space between adjacent signal lines is 5 µm. The insertion loss of the transformer at 2.4 GHz is 0.63 dB.

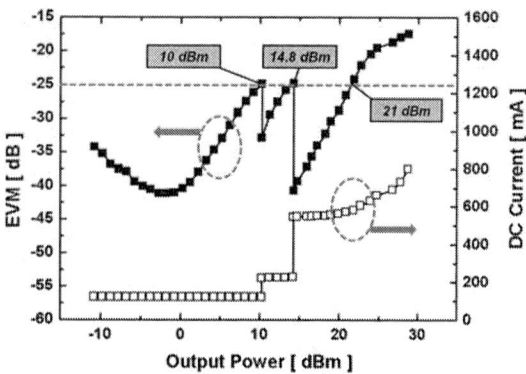

Fig. 7. Measured EVM and DC current with 802.11g WLAN 54 Mbps 64 QAM OFDM signal.

IV. MEASUREMENT RESULTS

The micro-photograph of the fabricated PA is shown in Fig. 5. The die size is 1.8×1.1 mm^2 including wire-bonding pads. A printed circuit board (PCB) was used for measurements, and the loss of signal lines in the PCB was de-embedded. Due to bond-wire effects in the matching networks, optimal operating frequency was a little shifted down from the desired frequency. One-tone power sweep with continuous-wave signal at the RF frequency of 2.1 GHz was performed, and the measurement results are presented in Fig. 6. V_{DD} was 3.3 V and, the gate bias voltages for the low power (LP), the medium power (MP), and the high power (HP) modes were set at 0.5 V, 0.55 V and 0.6 V, respectively. Measured P_{sat}/P_{1dB} were 30.7/27.5 dBm with PAE of 35.8/20% for the HP mode, 25/21.8 dBm with PAE of 19.8/13% for the MP mode (6-dB back-off) and 19/16.5 dBm with PAE of 10.5/9.8% for the LP mode (12-dB back-off). Although we observed that 7% of efficiency enhancement was achieved at 17-dBm P_{out}, this is not as significant enhancement as anticipated in the simulation. We believe that the optimal performance was limited at some point by the fixed matching network. We can improve the performance by adapting tunable capacitors in the output matching. For linearity test, 802.11g WLAN modulated signal was used to evaluate EVM. With 54 Mbps 64 QAM OFDM signal, -25-dB EVM specification was satisfied at output powers of 21 dBm (HP), 14.8 dBm (MP), and 10 dBm (LP), consuming DC quiescent current of 560 mA (HP), 200 mA (MP), and 100 mA (LP) as presented in Fig. 7. We could verify that the linearity of designed PA is comparable to the conventional PA [8] with enhanced low-power efficiency.

V. CONCLUSIONS AND DISCUSSIONS

To fully utilize the maximum available transformer efficiency, the concurrent power combining structure with the sub-divided DA/PA cells was proposed. With discrete power control, along with cell-area resizing, the efficiencies at power back-offs were enhanced by 6~10 %. The proposed PA is compared with other CMOS linear PAs in table I and demonstrates the best performance in terms of efficiency. The proposed structure can further improve the PA performance by having another design degree of freedom in the tunable matching network. It can be also applied to any transformer-based combining PAs.

ACKNOWLEDGEMENTS

The authors wish to acknowledge the support of Samsung Electro-Mechanics Co., Ltd for this work.

REFERENCES

[1] I. Aoki, et al., "Fully Integrated CMOS Power Amplifier Design Using the Distributed Active-Transformer Architecture," IEEE JSSC, vol. 37, no. 3, pp. 371-383, March 2002.

[2] C. Wang, et al., "A Capacitive-Compensation Technique for Improved Linearity in CMOS Class-AB Power Amplifiers," IEEE JSSC, vol. 39, no. 11, pp. 1927-1933, November 2004.

[3] Y. Palaskas, et al., "A 5-GHz 20-dBm Power Amplifier With Digitally Assisted AM-PM Correction in a 90-nm CMOS Process," IEEE JSSC, vol. 41, no. 8, pp. 1757-1763, August 2006.

[4] P.-C. Wang, et al., "A 2.4-GHz +25dBm P1dB Linear Power Amplifier with Dynamic Bias Control in a 65-nm CMOS Process," ESSCIRC, pp. 490-493, September 2008.

[5] A. Shirvani, et al., "A CMOS RF Power Amplifier With Parallel Amplification for Efficient Power Control," IEEE JSSC, vol. 37, no. 6, pp. 684-693, June 2002.

[6] P. Reynaert, et al., "A 2.45-GHz 0.13-μm CMOS PA With Parallel Amplification," IEEE JSSC, vol. 42, no. 3, pp. 551-562, March 2007.

[7] G. Liu, et al., "Fully Integrated CMOS Power Amplifier With Efficiency Enhancement at Power Back-Off," IEEE JSSC, vol. 43, no. 3, pp. 600-609, March 2008.

[8] K. H. An, et al., "A 2.4 GHz Fully Integrated Linear CMOS Power Amplifier With Discrete Power Control," IEEE MWCL, vol. 19, no. 7, pp. 479-481, July 2009.

[9] D. Chowdhury, et al., "A Single-Chip Highly Linear 2.4GHz 30dBm Power Amplifier in 90nm CMOS," IEEE ISSCC, February 2009.

Table I
PERFORMANCE COMPARISON OF CMOS PAs

Ref.	CMOS [nm]	Freq [GHz]	V_{DD} [V]	Size [mm^2]	P_{1dB}/P_{sat} [dBm]	PAE [%] at -12dB/-6dB/P_{sat}
[6]	130	2.45	1.5	5.5	--/23	12/21/29
[7]	130	2.4	1.2	2.0	--/27	12/22/32*
[8]	180	2.4	3.3	2.0	27/31	8/17/33*
[9]	90	2.4	3.3	4.32	27.7/30.1	6/16/33
This work	180	2.1	3.3	1.98	27.5/30.7	10.5/19.8/35.8

*Drain Efficiency

A Single-Chip 2.4GHz Double Cascode Power Amplifier with Switched Programmable Feedback Biasing under Multiple Supply Voltages in 65nm CMOS for WLAN Application

Mingyuan Li, Ali Afsahi, and Arya Behzad

Broadcom Corporation, San Diego, CA, 92127

Abstract — **A 2.4GHz fully integrated power amplifier with an on-chip balun for embedded WLAN applications with direct battery connection (2.3-5.5V) is presented. With a switched programmable feedback bias network, the PA can deliver 23.5dBm to 28.4dBm CW saturated power and 18.2dBm to 23.2dBm OFDM linear power (-25dB EVM) with PAPD when the supply varies from 2.3V to 5.5V. The PA occupies 1.2mm^2 in 65nm CMOS.**

Index Terms — **CMOS, linearization, OFDM, power amplifiers, stacked transistors, WLAN.**

I. INTRODUCTION

Fully integrated power amplifiers with on chip matching have been recently implemented in standard CMOS technologies for different wireless communication systems [1]-[3]. One of the bottlenecks of a SoC WLAN transceiver is an integrated PA providing sufficient linear power with good efficiency while transmitting high peak-to-average ratio OFDM modulated signals. The emerging market of embedding a WLAN IP into portable handheld devices is growing very fast. One of the new design challenges is that the PA has to work properly and reliably with a direct battery connection, which can present supply voltage from 2.3V up to 5.5V. For mobile cellular applications, in the absence of direct "V$_{battery}$" capability for the WLAN PA, a costly and area consuming switching regulator or a power inefficient linear regulator would be required to supply the required voltage to the PA.

With CMOS technologies scaled down, the device reliability problems post a limit on the maximum power supply and the RF signal swing across transistor terminals. Time-dependent dielectric breakdown (TDDB) and hot carriers injection (HCI) are the two critical issues causing degradation of deep sub-micro device and circuit performance. The degradation problem becomes much worse when a higher supply voltage like 5.5V is applied to the PA, which is already under RF stress. The electric fields applied to the device plus the electric field built up by charge injection can exceed the dielectric breakdown threshold in some of the weakest points due to the presence of defects of the gate dielectric. Several approaches have been proposed to relax the voltage stress restrictions. A CMOS PA showing no hot carrier

degradation from a 2.4V supply using a self-bias cascode topology was reported in [4], but it could not withstand such a high supply as 5.5V. In [5]-[6], stacked FET GaAs and SOS PAs were proposed for high voltage applications, but these non-standard CMOS designs imposed the same amount of voltage swing limitations on all the transistors, and the feedback bias network were optimized for a single supply. In this paper, a double cascode PA with switched programmable feedback biasing is designed with considerations of load impedance matching, reliability, and efficiency with a direct battery connection. Under different supply voltages, the PA can deliver the maximum power while each transistor of different channel length experiences different allowable voltage swing across its terminals.

II. DOUBLE CASCODE PA IMPLEMENTATION

Fig. 1 shows the block diagram of the proposed PA. It is a two-stage design of a PA driver and PA with both input and output on-chip matching and a power detector.

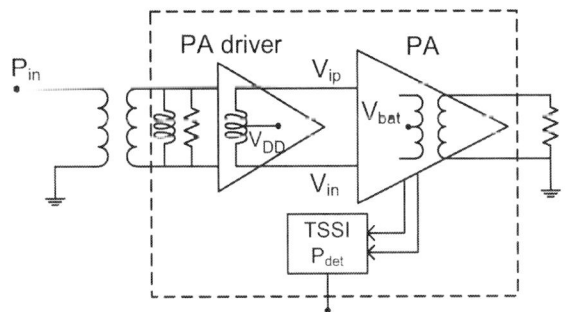

Fig. 1. Block diagram of the two-stage PA design.

The PA stage (M$_1$ to M$_8$) employs a pseudo differential double cascode structure with an on-chip transformer as shown in Fig. 2. The differential configuration is less sensitive to the ground inductance and has better common-mode stability compared with a single-ended version. The transformer is designed to provide the optimal differential load impedance from a 50 Ohm antenna at the 2.4 GHz center frequency based on load-pull simulations. Thin-oxide transistors in 65nm are used for the common-source stage due to their higher transconductance, which provides enough active current without requiring excessive input drive and presenting too much capacitive load for the preceding stage. The G_m linearization method of combining two G_m stages (M_1 and M_2) with different gate bias is adopted for better large signal gain flatness [7]. The middle transistors need not only withstand the large signal swing at saturated power with a 5.5V supply, which is beyond the reliable region of thin-oxide devices, but also provide load impedance about several Ohms to the bottom G_m stage for delivering the maximum power. The real part of the load impedance R_{opt} seen by the bottom stage can be approximately expressed as

$$R_{opt} = \frac{1}{g_m}\left(1 + \frac{C_{gs}}{C_1}\right) \qquad (1)$$

where g_m is the transconductance of the middle transistors, and C_{gs} and C_1 are the intrinsic gate-to-source capacitor and the external capacitor connected to the gate of the middle transistors. The thick-oxide transistors with a particular channel length are selected with both transconductance and reliability considerations. The top transistors experience the largest signal stress; therefore the long channel length transistors are used.

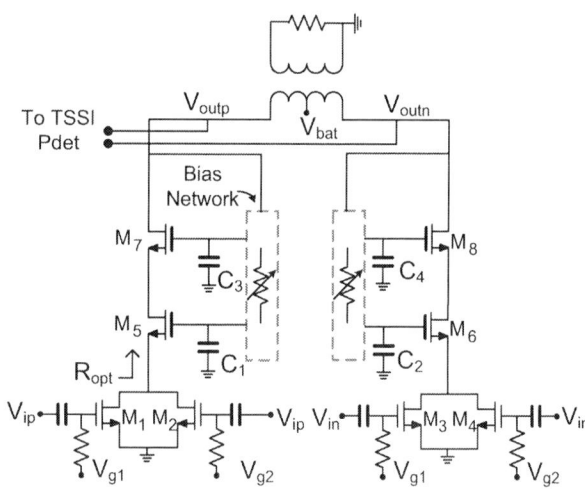

Fig. 2. Schematic of the double cascode power amplifier with switched programmable feedback biasing.

A differential bias network is designed to be able to provide optimal biases under the large supply variations as shown in Fig. 3. When the supply is above 3.3V, the switch SW_1 and SW_3 are on, and a dynamic feedback bias network is employed to keep the voltage swing between drain, gate, and source terminals tracking together to avoid the gate oxide breakdown and the long-term reliability degradation. The programmable resistor array can provide

optimal bias voltages to the top two cascode transistors for delivering maximum power under different supplies. When the supply is below 3.3V, the switch SW_2 and SW_4 are on, and the bias voltages are derived from programmable bias currents. The feedback network is disabled at lower supply voltages to increase the output power and reduce the leakage current.

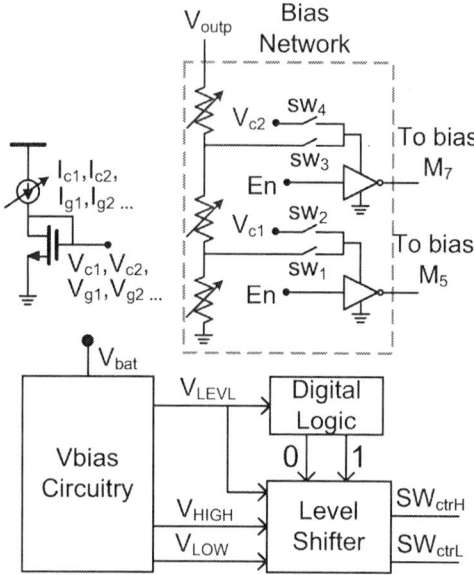

Fig. 3. Schematic of the PA's switched programmable feedback bias network and bias control signal generation circuitry.

In practice, the PA can have multiple branches, which are independently turned on and off to provide additional power control besides previous RF stages in a transmitter chain. The control signals are applied to gates of all the cascode devices instead of the gate of all G_m devices to avoid the off branches to function like Class C operation with a large input signal. For reliability considerations, all the logic control and switch control signals (SW_{ctrH} and SW_{ctrL}) cannot directly use GND and VDD as logic "0" and "1" and must be generated by a dc level shift circuitry with different level reference voltages (V_{LEVL}, V_{HIGH}, and V_{LOW}) provided by a bias circuitry.

The PA driver stage has the same double cascode structure with a differential inductor load. Since the maximum signal swing is much smaller than the PA stage, no feedback bias network is needed for the PA driver. On-chip shunt inductor and resistor combining with an off-chip single-ended to differential balun are used to provide a 50 Ohm input matching. In practice, the PA driver only needs a differential matching, which can be absorbed in the load of the preceding stage.

III. MEASUREMENT RESULTS

This PA is fabricated in 65nm CMOS. The die photo of the PA is shown in Fig. 4 with an occupied active area is about $1.2 \times 1 \text{mm}^2$ including the PA driver input matching and the digital control logics. The chip is packaged with 20-pin QFN package and mounted on an FR4 board. The PA is tested with a CW input signal under four different supply voltages, which are 2.3V, 3.3V, 4.5V, and 5.5V respectively.

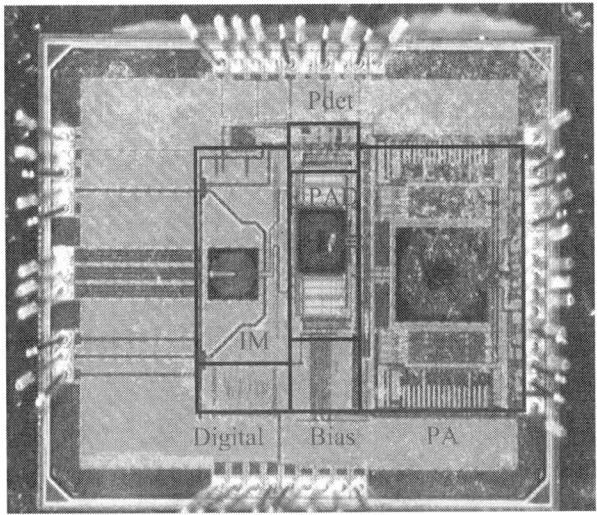

Fig. 4.   Die micrograph of the presented PA chip.

As a representative, the S-parameter plots in Fig. 5 show that the input and the output return loss of the PA with a 3.3V supply is better than -11 dB from 2.4 GHz to 2.5GHz, and the gain 3dB bandwidth is about 400MHz without any tuning.

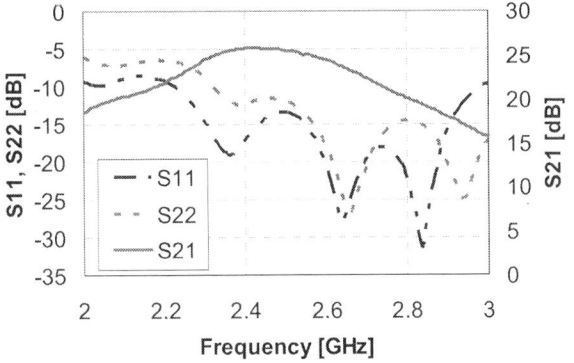

Fig. 5.   S-parameter plots of the PA with 3.3V supply voltage vs. frequency.

Fig. 6 shows that the small signal gain changes from 25dB to 26.6dB and the output saturation power changes from 23.5dBm to 28.4dBm when the supply varies from 2.3V to

5.5V. P_{sat} drops more rapidly than the gain because with a lower supply, the maximum signal swing at the drain of transistors is limited, but the output impedance is largely determinated by the transformed load impedance and does not change significantly as a function of the supply. Fig. 7 shows the drain efficiency of the PA stage vs. the output power. The peak efficiency is relatively constant vs. supply at about 20% including all the on-chip and package losses.

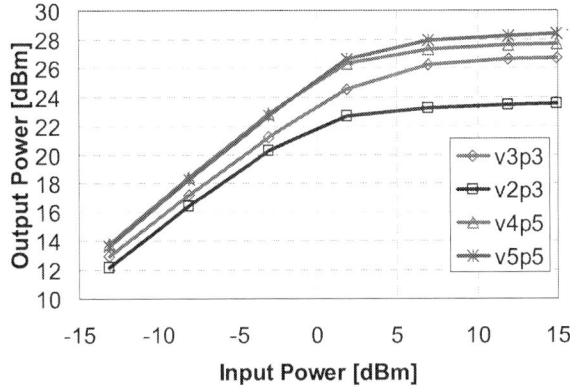

Fig. 6.   Output power of the PA with four different supply voltages at 2.442GHz.

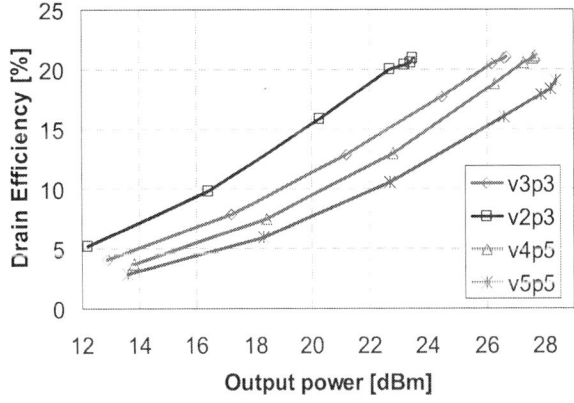

Fig. 7.   Drain efficiency of the PA with four different supply voltages at 2.442GHz.

The PA is tested with a 54Mbps 64QAM OFDM modulated signal. The EVM vs. the output power using a similar power amplifier predistortion (PAPD) linearization method described in [7] is shown in Fig. 8. The -25dB EVM specification is achieved at 18.2dBm to 23.2dBm Pout. The linear power for the PA with a 3.3V supply is improved by about 10dB after PAPD as shown in Fig. 9. Fig. 10 shows the output spectrum without Gaussian filtering after PAPD and spectrum mask for the PA with a 3.3 supply on channel 2.442GHz as a representative.

978-1-4244-6240-7/10 $26.00 © 2010 IEEE

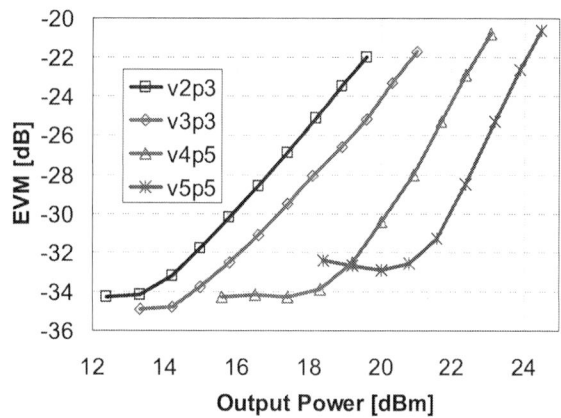

Fig. 8. EVM of the PA after PAPD with four different supply voltages at 2.442GHz.

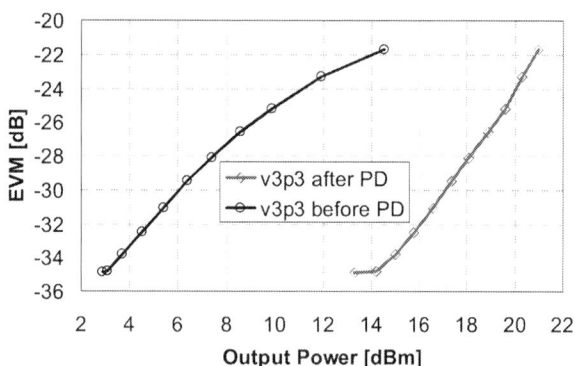

Fig. 9. EVM of the PA before and after PAPD with 3.3V supply voltage at 2.442GHz.

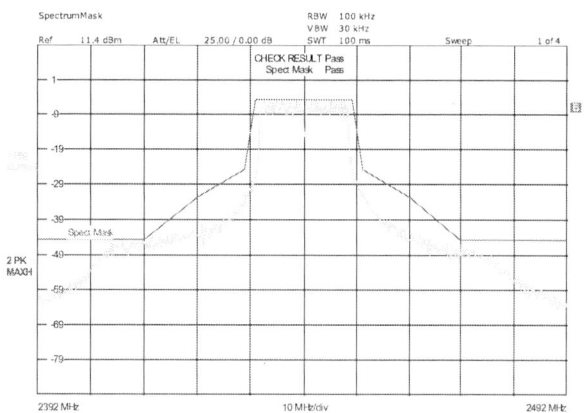

Fig. 10. Spectral mask and output spectrum of the PA after PAPD for 19.5dBm Pout with 3.3V supply at 2.442GHz.

Table I summarizes the PA performance under four different supply voltages. The PA has also been tested under extreme stress conditions for an extended period of

time under both CW and OFDM signals with no appreciable change in output power or EVM.

TABLE I
SUMMARY OF PA PERFORMANCE

	Gain (dB)	P_{sat} (dBm)	Drain Eff @ P_{sat} (%)	Power @ -25dB EVM (dBm)
2.3V	25.3	23.5	20.9	18.2
3.3V	26	26.7	21	19.6
4.5V	26.9	27.7	21.1	21.7
5.5V	26.7	28.4	19	23.2

IV. CONCLUSION

A fully integrated 2.4GHz double cascode power amplifier has been designed and implemented in 65nm CMOS process for direct battery connection embedded WLAN application. Measured results show that the PA can deliver sufficient linear power with good efficiency while transmitting WLAN signals under different supply voltages.

REFERENCES

[1] D. Chowdhury, C. Hull, O. Degani, P. Goyal, Y. Wang, and A. Niknejad, "A Single-Chip Highly Linear 2.4GHz 30dBm Power Amplifier in 90nm CMOS," ISSCC Dig. Tech. Papers, pp. 378-380, Feb., 2009.

[2] I. Aoki, S. Kee, R. Magoon, R. Aparicio, F. Bohn, J. Zachan, G. Hatcher, D. McClymont, and A. Hajimiri, "A Fully Integrated Quad-Band GSM/GPRS CMOS Power Amplifier," ISSCC Dig. Tech. Papers, pp. 570-572, Feb., 2008.

[3] C. Presti, F. Carrara, A. Scuderi, P. Asbeck, G. Palmisano, "A 25 dBm Digitally Modulated CMOS Power Amplifier for WCDMA/EDGE/OFDM With Adaptive Digital Predistortion and Efficient Power Control," IEEE J. Solid-State Circuits, vol. 44, pp. 1883-1896, Jul., 2009.

[4] T. Sowlati and D. Leenaerts, "A 2.4 GHz in 0.18-μm CMOS Self-Biased Cascode Power Amplifier with 23-dBm Output Power," ISSCC Dig. Tech. Papers, pp. 294-296, Feb., 2002.

[5] A. Ezzeddine and H. Huang, "The High Voltage/High Power FET," IEEE RFIC Symp. Dig., pp. 215-218, June 2003.

[6] J. Jeong, S. Pornpromlikit, P. Asbeck, and D. Kelly, "A 20 dBm Linear RF Power Amplifier Using Stacked Silicon-on-Sapphire MOSFETs," IEEE Microwave and Wireless Components Letter, vol. 16, no. 12, pp. 684-686, Dec., 2006.

[7] A. Afsahi, A. Behzad, V. Magoon, and L. Larson, "Fully Integrated Dual-Band Power Amplifiers with on-chip Baluns in 65nm CMOS for an 802.11n MIMO WLAN SoC," IEEE RFIC Symp. Dig., pp. 365-368, June 2009.

RTU1C-3

A 31-dBm, High Ruggedness Power Amplifier in 65-nm Standard CMOS with High-Efficiency Stacked-Cascode Stages

Stephan Leuschner*, Sandro Pinarello[†‡], Uwe Hodel[‡], Jan-Erik Mueller[‡], Heinrich Klar*

*Technical University of Berlin, †Friedrich-Alexander-Universitaet Erlangen-Nuernberg, ‡Infineon Technologies AG

Abstract—A novel, high ruggedness power amplifier topology in a 65-nm CMOS technology is proposed. The proposed stacked cascode topology uses only standard devices available in a modern triple-well CMOS process to achieve breakdown voltages of more than 18V. The power amplifier stage delivers 28 dBm output power at a power-added efficiency (PAE) of 69.9% from a 3.6V supply. The saturation gain is 18 dB. A watt-level power amplifier for GSM low-band operation with 31-dBm output power and 61% PAE is presented.

Index Terms—Breakdown voltage, CMOS, high efficiency, HV device, power amplifier, RF, ruggedness.

I. INTRODUCTION

The implementation of single-chip radios is the ultimate goal in RF CMOS design. This requires the full integration of power amplifiers (PA). The main obstacle is the ruggedness of standard devices (MOSFETs) available in modern CMOS processes. The low device breakdown voltages do not comply with the nominal battery voltage in a mobile handset of 3.6V (even up to 5.5V in the worst case) and the requirement to withstand load mismatch conditions, i.e. high VSWR up to 10:1 without harming the device. Thus, a much lower supply voltage would have to be generated efficiently (e.g. by means of a DC-DC converter) which leads to a reduction of the overall efficiency. Generating high RF power from low supply voltages poses also another problem: the optimum load impedance of the PA output stage transistors becomes very small. A matching network with a high transformation ratio is needed to transform the antenna impedance of 50 to the optimum load impedance. High transformation ratios usually result in small bandwidth and higher losses. Possible solutions are the use of either non-standard high power RF devices or special circuit topologies using only the available standard devices.

The development of non-standard devices (like LDMOS etc. [1], [2]) is expensive and increases technology costs due to additional masks and/or process modifications. Circuit topologies using only standard devices circumvent these additional costs and achieve the required ruggedness by stacking multiple transistors. Simple cascode structures in modern nm-CMOS technologies cannot withstand the high RF voltages of more than 15V occurring at the

output of power devices in GSM PAs under load mismatch conditions. Therefore, more than two standard MOSFETs have to be stacked. This can be done in two ways: either by DC-only coupling of single FETs/cascode circuits [3] with RF power combining at the outputs or by the HiVP (High-Voltage/High-Power device) configuration [4], [5]. DC-only coupled devices require multiple matching networks while the HiVP and similar configurations suffer from reduced efficiency due to increased effective on-resistance because of the in-phase voltage swing at the gates of the stacked transistors.

In this paper, the novel stacked-cascode power amplifier topology is proposed. Section II summarizes the theory of circuit operation. Result from breakdown and load-pull measurements on single-stage test structures are discussed in Section III. Section IV presents a differential 31-dBm Class-AB power amplifier employing the novel stacked-cascode approach. Conclusions are given in Section V.

II. STACKED-CASCODE CIRCUIT

The efficiency is one of the most crucial performance parameters of a power amplifier. An important factor limiting the achievable efficiency is the on-resistance of the power amplifier output stage. From the load-line theory a simple expression for the efficiency can be derived taking the on-resistance into account (here for a class B amplifier):

$$\eta = \frac{\pi}{4} \frac{1}{1 + \frac{2}{V_{max}/(R_{on}I_{max})-1}}.$$

When connecting N multiple MOSFETs in series V_{max} as well as R_{on} increase by a factor of N. Thus V_{max}/R_{on} stays constant and the efficiency will not be compromised by stacking multiple MOSFETs.

The novel stacked cascode structure proposed in this paper is shown in Fig. 1. It consists of two cascode circuits comprising the MOSFETs M1/M2 and M3/M4. The gates of the cascode transistors M2 and M4 are RF shorted by the blocking capacitors C_B to the source nodes of M1 and M3, respectively. This way the transistors M1 and M2

978-1-4244-6240-7/10 $26.00 © 2010 IEEE

act as standard cascode circuit while M3 and M4 form a floating cascode circuit. The cascode behavior results in a lower on-resistance and therefore in a higher efficiency of the structure compared to the HiVP configuration. Proper choice of the tuning capacitor C_{Tune} sets the voltage swing for the top cascode circuit so that the total output voltage is uniformly distributed across all four MOSFETs. This allows the voltages seen by M1-M4 to remain below the oxide breakdown voltage and the maximum voltage ratings imposed by hot carrier effects.

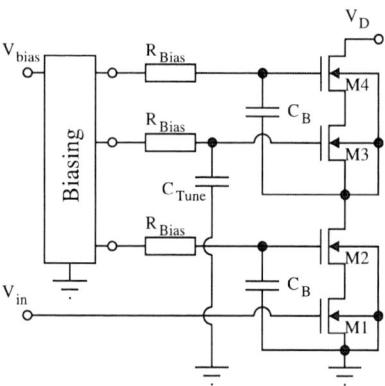

Fig. 1. Stacked cascode circuit comprising four MOSFETs, blocking/tuning capacitors and biasing

Another limiting factor is the breakdown of the drain diode of transistor M4 at high reverse voltages. High RF peak voltages (in the vicinity of the diode breakdown voltage) at the drain of M4 can lead to a current flow through the diode and to a clipping of the output voltage waveform. At higher RF peak voltages the drain diode will be destroyed. In order to avoid breakdown of the drain diode of M4 the cascode circuit formed by M3/M4 is realized in a p-well in n-well with the n-well biased to the supply voltage. The connection to the source of M3 forces the p-well to swing together with the top cascode circuit reducing the voltage across the drain diode of M4 by a factor of 2.

The circuit overall breakdown voltage is limited either by the gate oxide breakdown voltage ($4 \times V_{br,GOX}$) or by the diode and well breakdown voltages ($V_{br,diode} + V_{br,well}$). In the best case stacking of four MOSFETs leads to a four times higher overall breakdown voltage.

III. Measurement of load-pull test structures

A. DC and RF breakdown

To demonstrate the robustness of the stacked cascode topology DC breakdown measurements were performed, as well as RF measurements under optimum load conditions for increasing supply voltages.

The gate oxide of the thick oxide FET and diode/well breakdown voltages in the used technology are 7V and 10V, respectively. Thus, the circuit should be able to sustain a maximum output voltage of about 20V – limited by the breakdown of the N-well and the drain diode of transistor M4.

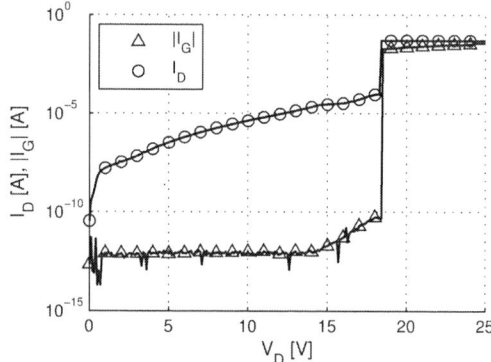

Fig. 2. Measured DC breakdown of the stacked cascode (gate-source voltage of transistor M1: $V_{gs,M1} = 0$V, bias voltage: $V_{bias} = 3/4 \times V_D$)

Fig. 2 shows the results of the DC breakdown measurement. To ensure proper distribution of the total drain voltage V_D across all devices, the bias voltage V_{bias} was swept during measurement along with V_D by setting $V_{bias} = 3/4 \times V_D$. The measured breakdown voltage of $V_{bd} = 18.4$V is slightly below the theoretical maximum.

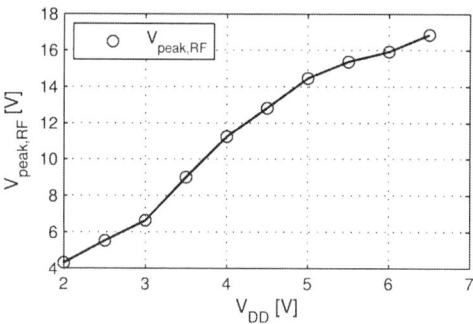

Fig. 3. RF peak voltage at increasing drain bias voltages

Using the possibilities of time-domain load-pull measurements, device breakdown under RF operation was evaluated. For each supply voltage level the amplifiers were operated in saturation at maximum output power. In Fig. 3 the reached peak RF voltage at the drain is plotted against the drain bias voltage V_{DD}. The circuit worked up to a drain bias voltage of 6.5V, which corresponds to a peak RF voltage of 17V; at $V_{DD} = 7$V degradation due to gate oxide breakdown of one of the four transistors was observed. Extrapolating the measured curve the peak RF voltage reached levels of about 18V.

Under DC and RF stress the same breakdown voltages were measured. More importantly it is shown that circuits based on the stacked cascode approach can sustain the high peak voltages occurring in handset power amplifiers under load mismatch conditions.

B. RF performance

Fig. 4 and Table I compare the RF performance of three implemented PA stages. The measured PAE values for all the three PA stages are nearly the same. This result verifies the theoretical considerations in II. The output power increases by 6dB each time the number of FETs is doubled. This is because the drain bias voltage V_{DD} could also be doubled while keeping the stress on each single FET constant.

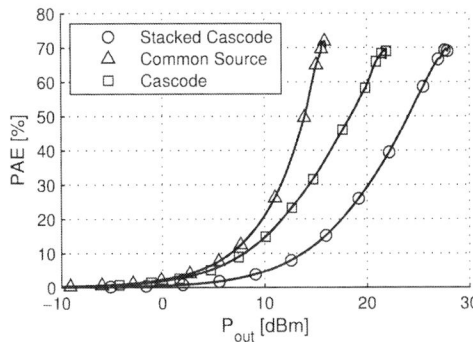

Fig. 4. Comparison of RF performance of a simple common-source stage, a cascode stage and the stacked cascode stage at $f = 900\text{MHz}$

TABLE I
MEASURED PERFORMANCE AT 900MHz

	No. of stacked FETs à ⅛/mm	V_{DD} [V]	P_{out} [dBm]	PAE [%]
Common-Source	1	0.9	15.9	71.9
Cascode	2	1.8	21.9	69.8
Stacked Cascode	4	3.6	28	69.9

The performance vs. frequency of the stacked cascode is plotted in Fig. 5. The circuit shows outstanding performance in the GSM low band at 900MHz and acceptable high efficiency in the GSM high band at 1.8GHz. At higher frequencies the PAE is limited due to the low unity-gain frequency ($\omega_u = 15\text{GHz}$) of the stacked-cascode using 230nm thick-oxide FETs. For higher frequency operation the gate length could be reduced or even thin oxide FETs could be used at the cost of circuit ruggedness.

Table II compares the measured performance of the presented stacked-cascode approach with the state-of-the-art. The circuit shows a performance in CMOS technologies

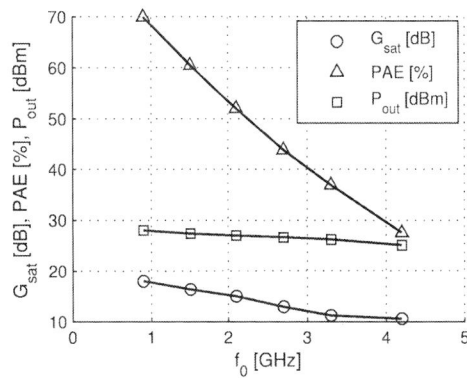

Fig. 5. PAE, gain and output power vs. frequency at $V_{DD} = 3.6\text{V}$ for optimum load conditions for each frequency point

so far only reachable with special devices[1], [2]. Compared to other implementations using only standard devices [5] the efficiency of the stacked cascode is considerably higher.

TABLE II
COMPARISON WITH STATE OF THE ART

Reference	Technology	f [GHz]	P_{out} [dBm]	PAE [%]
[1]	Si-LDMOS	0.9	35	65
[2]	65nm CMOS using special EDMOS device	0.8	30	70
[5]	0.18μm CMOS	1	27.5	36.9
this work	65nm CMOS, thick oxide FETs ($L_{Gate} = 230$nm)	0.9	28	69.9

IV. A 31-DBM, CLASS-AB STACKED-CASCODE POWER AMPLIFIER

To show the feasibility of watt-level power amplifiers for GSM handset application, a differential power amplifier output stage including complete biasing, ESD protection and blocking capacitors in a flip-chip package was fabricated in a 65nm CMOS technology. A simplified block diagram of the chip and the test-board is shown in Fig. 6. The power amplifier occupies a total chip area of 0.6mm². The size of the output stage is 0.2mm².

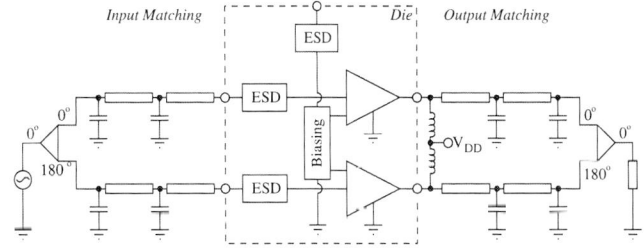

Fig. 6. Block diagram of chip and test-board

978-1-4244-6240-7/10 $26.00 © 2010 IEEE

The matching at the amplifier input and output was realized on-board using grounded coplanar waveguides on a FR-4 substrate and SMD capacitors to form two-stage matching networks. The output matching was optimized for maximum PAE. Fig. 7 shows the complete test-board.

Fig. 7.   Test-board and chip photograph

The power amplifier operates in classical class-AB mode. At 850MHz it delivers 31dBm output power at a power-added efficiency of 60.5% as shown in Fig. 8. The small-signal and saturation gain is 27dB and 16dB, respectively.

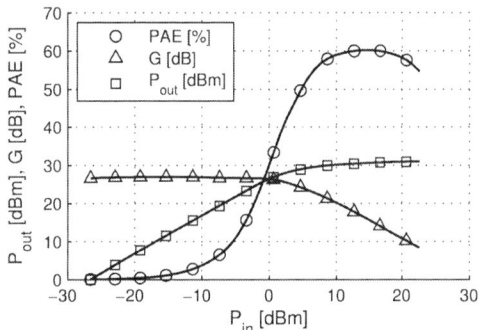

Fig. 8.   Power amplifier performance at $f = 850$MHz

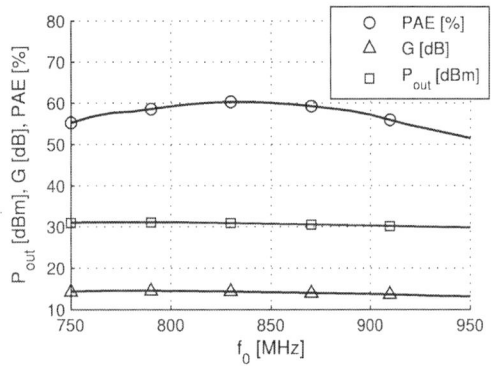

Fig. 9.   Performance vs. frequency at $P_{in} = 14$dBm

The performance plotted versus frequency in Fig. 9 shows the maximum PAE of 61% at 830MHz. The PAE is always higher than 55% at frequencies ranging from 750MHz to 920MHz. In this range the output power is also equal or greater than 30dBm at a minimum saturation gain of 14dB.

The difference in maximum PAE compared to the load-pull measurements shown in Section III results from the losses in the on-board matching networks ($\Delta PAE \approx 5\%$) and a non-optimum ESD structure design ($\Delta PAE \approx 4\%$).

V. Conclusion

The presented stacked-cascode power amplifier topology uses only standard devices in a 65-nm CMOS process. The implemented stacked-cascode stage using 230nm thick-oxide MOSFETs has a breakdown voltage of 18.4V. The on-wafer load-pull measurements show a PAE of 69.9% at an output power of 28dBm at 900MHz. To the knowledge of the authors, this is the first CMOS power amplifier that is able to handle such high voltage levels using only standard devices with an efficiency so far only measured on special devices.

The GSM low-band power amplifier proves the feasibility of power amplifiers using the stacked-cascode topology. It delivers 31dBm output power at an efficiency of 61%.

Acknowledgements

The authors would like to thank Boris Kapfelsperger, Bernhard Sogl, Ronald Thueringer and Norman Wolf for constant support and their help during the measurements.

References

[1] T. Shimizu, Y. Nunogawa, T. Furuya, S. Yamada, I. Yoshida, and H. Masao, "A small GSM power amplifier module using Si-LDMOS driver MMIC," in *Solid-State Circuits Conference, 2004. Digest of Technical Papers. ISSCC. 2004 IEEE International*, 2004, pp. 196–522 Vol.1.

[2] M. Apostolidou, M. van der Heijden, D. Leenaerts, J. Sonsky, A. Heringa, and I. Volokhine, "A 65nm CMOS 30dBm class-E RF power amplifier with 60% power added efficiency," in *Radio Frequency Integrated Circuits Symposium, 2008. RFIC 2008. IEEE*, 2008, pp. 141–144.

[3] I. Aoki, S. Kee, R. Magoon, R. Aparicio, F. Bohn, J. Zachan, G. Hatcher, D. McClymont, and A. Hajimiri, "A Fully-Integrated Quad-Band GSM/GPRS CMOS Power Amplifier," *Solid-State Circuits, IEEE Journal of*, vol. 43, no. 12, pp. 2747–2758, 2008.

[4] A. Ezzeddine and H. Huang, "The high voltage/high power FET (HiVP)," in *Radio Frequency Integrated Circuits (RFIC) Symposium, 2003 IEEE*, 2003, pp. 215–218.

[5] A. K. Ezzeddine, H. C. Huang, R. S. Howell, H. C. Nathanson, and N. G. Paraskevopoulos, "CMOS PA for Wireless Applications," in *IEEE Topical Symposium on Power Amplifiers for Wireless Communications*, 2007. [Online]. Available: http://pasymposium.ucsd.edu/pages/2007papers.htm

RTU1C-4

Analysis and Design of a Wideband High Efficiency CMOS Outphasing Amplifier

M.C.A. van Schie[1], M.P. van der Heijden[2], M. Acar[2], A.J.M. de Graauw[2], and L.C.N. de Vreede[1]

[1]Delft University of Technology, Delft, the Netherlands
[2]NXP Semiconductors, Eindhoven, the Netherlands

Abstract — **This work presents the analysis and design of a novel transformer-based power combining network for an efficient outphasing power amplifier (PA). The proposed power combining network was implemented on PCB together with two 65nm CMOS class-E PA's. Measurements show a peak output power of more than 30dBm over a 29% bandwidth around 700MHz. The peak output power at 700MHz equals 33.9dBm. The 10dB back-off efficiency is larger than 27.3% over the same 29% bandwidth. The 6dB back-off efficiency is larger than 46.4% over this bandwidth. The drain efficiency at 650MHz is larger than 50% over a 10dB power back-off range, which is, to our best knowledge, the highest back-off efficiency reported in a CMOS outphasing PA to date.**

Index Terms — **Outphasing, power amplifiers, CMOS, LINC, Chireix, transformer, power combiner.**

I. INTRODUCTION

Modern wireless communication systems make use of complex digital modulation schemes to achieve high data-rate communication, while using bandwidth efficiently. In these modulation schemes both the phase and the amplitude of a carrier are being modulated. These complex modulation schemes impose challenging demands on the linearity of a transmitter, and therefore on the linearity of its PA. In a wireless transmitter the PA is the most important stage from the perspective of system linearity and energy consumption. For this reason, efficiency is a key performance metric for a PA. Traditional linear PA's are known to have a very low efficiency while efficient switch-mode PA's are known to be very nonlinear.

Outphasing or LInear amplification using Non-linear Components (LINC) is one candidate technique to work around this linearity-efficiency trade-off. In an outphasing transmitter a complex modulated input signal is split into two constant amplitude signals with a relative phase-difference. This relative phase-difference corresponds to the required output power level. Both branch signals are amplified separately by an efficient but non-linear PA. After summation of both branch signals, the original signal results. If a conventional isolating power combiner is used, the overall efficiency still scales with the signal envelope amplitude [1]. However, if a non-isolating power combiner is used, it is possible to enhance the efficiency by partly cancelling the reactive part of the admittance

seen by each PA. This technique is known as Chireix power combining [2]. The most common implementation of a Chireix power combiner is shown in Fig. 1, and can be found in many publications on outphasing transmitters.

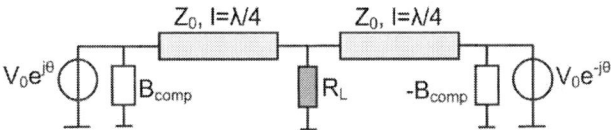

Fig. 1. Most common implementation of a Chireix power combiner

The classical Chireix power combiner from Fig. 1 is elegant in the sense that both voltage sources and the load can be single ended. However, from a bandwidth point-of-view, this combiner is not very elegant. Section II presents an analysis of the frequency-behavior of the classical Chireix power combiner. In section III a wideband transformer-based power combiner is presented. In section IV the design of a class-E outphasing PA with a transformer-based power combiner is presented. Section V shows the measurement results, and section VI presents the conclusions.

II. ANALYSIS OF THE CLASSICAL CHIREIX POWER COMBINER

Referring to Fig. 1, the admittance seen by source 1 and source 2 at the design frequency can be derived as:

$$Y_{1,2} = \frac{2R_L \cos^2(\theta)}{Z_0^2} \mp j \frac{R_L \sin(2\theta)}{Z_0^2} \pm jB_{comp} \qquad (1)$$

Where θ is the outphasing angle, equal to the half of the relative phase difference between the two sources, and B_{comp} is the compensation susceptance. The normalized efficiency can be defined as:

$$\eta_{norm} = \frac{G(\theta)}{|Y(\theta)|} = \frac{G(\theta)}{|G(\theta) + jB(\theta)|} \qquad (2)$$

Such that after addition of compensation susceptances for both sources, the following expression results for the normalized efficiency.

978-1-4244-6240-7/10 $26.00 © 2010 IEEE

$$\eta_{norm1,2} = \frac{G_{1,2}}{|Y_{1,2}|} = \frac{2\cos^2(\theta)}{\sqrt{\left(2\cos^2(\theta)\right)^2 + \left(\sin(2\theta) - \frac{Z_0^2}{R_L}\left|B_{comp}\right|\right)^2}} \qquad (3)$$

This normalized efficiency can be chosen to peak at a certain outphasing angle, corresponding to a certain power back-off (PBO), by choosing the values of the compensation susceptances such that the total susceptances seen by the sources are zero at this angle. At this compensation angle the load seen by both sources is purely ohmic, and therefore the efficiency is highest. The outphasing angle θ relates to (linear) PBO as:

$$\theta = \arccos\left(\frac{1}{\sqrt{PBO}}\right) \qquad (4)$$

For the classical power combiner, the efficiency does not only depend on the outphasing angle, but also on frequency. The contour-plot in Fig. 2 shows the normalized efficiency of source 1 as a function of frequency and outphasing angle for a 50 Ohm load, 100 Ohm transmission lines, a center-frequency f_0 of 700MHz and a compensation angle corresponding to 10dB PBO at the center frequency. The efficiency plot for source 2 looks similar, but mirrored vertically around f_0.

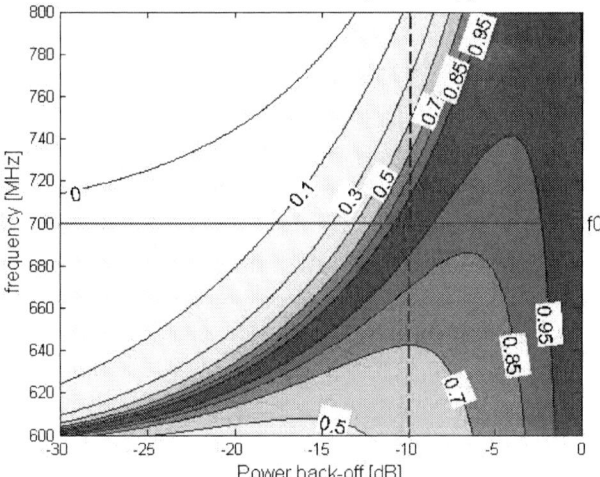

Fig. 2.   Contour plot showing normalized efficiency versus power back-off and frequency for source 1.

From Fig. 2 it can be seen that the back-off efficiency depends heavily on frequency. The efficiency at f_0, (indicated by the blue horizontal line,) peaks at maximum power and at 10dB PBO. The 10dB power back-off line is shown as well. If frequency deviates from f_0, the peak efficiency area, shown in dark red, rapidly moves to a different back-off level. For a wideband implementation of an outphasing system, this is highly undesirable. Apart from the frequency-dependent efficiency, also the

conductance seen by each source depends heavily on frequency. Under some conditions, the conductance seen by the sources can even get negative, with the risk on instable behavior. The negative-conductance region is shown in Fig. 2 as the white region in the top left corner. It can be shown that the frequency-dependence of the efficiency of the classical Chireix combiner can be mainly devoted to the presence of the transmission lines, rather than the compensation elements. When the transmission lines are modeled as perfect quarter wavelength transmission lines at all frequencies, the efficiency contours from Fig. 3 result.

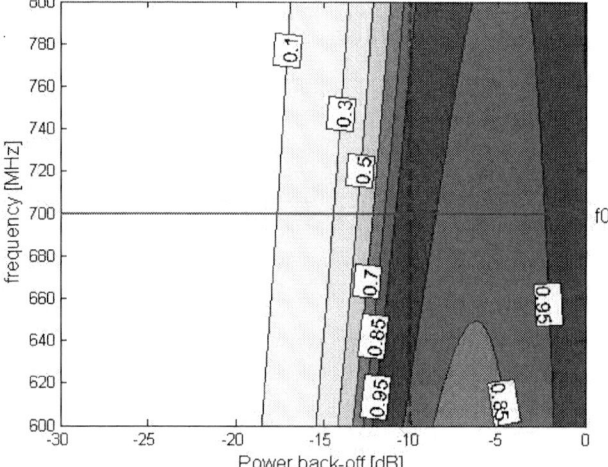

Fig. 3.   Contour plot showing normalized efficiency versus power back-off and frequency for source 1 in case transmission lines are quarter wavelength at all frequencies.

From Fig. 3 it is seen that the normalized efficiency in this nonphysical circuit is way less dependent on frequency than it is in the classical Chireix combiner. The frequency dependence that is left in Fig. 3 can be devoted solely to the frequency dependence of the compensation elements. Also, the conductance seen by source 1 doesn't become negative anywhere. The same holds for source 2. Fig. 3 shows that the outphasing principle can be wideband, as long as the power combining is done with a wideband power combiner.

III. TRANSFORMER-BASED POWER COMBINING NETWORK

As pointed out in the introduction, the classical Chireix power combiner is an elegant circuit from the perspective of having single-ended sources and a single-ended load. The classical way of realizing this convenience causes the back-off efficiency to be very narrow-band, while power combining in an outphasing transmitter is fundamentally wideband. This can also be seen from the basic circuits in Fig. 4. Both circuits do not contain any bandwidth limiting components, and the bandwidth of the sources (PA's) determines the overall bandwidth.

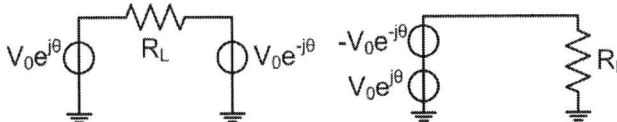

Fig. 4. Fundamental way to add two voltages in a floating load (left) or a single-ended load (right).

Both circuits have the disadvantage of having either a floating load or source. A floating source implies the need for a differential PA, while a floating load can be offered by using a differential antenna. Since transformers are known to be capable of providing large BW [3], it is tried to use a transformer as a balun to avoid the need for differential components, while still offering a large BW. When this is done for the two basic topologies of Fig. 4, the circuits in Fig. 5 result.

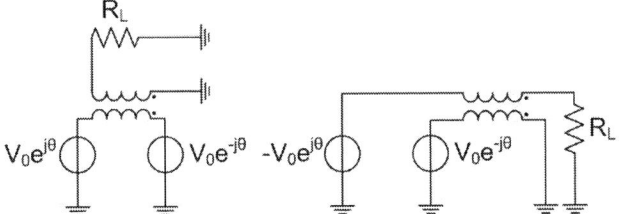

Fig. 5. Resulting circuits after inserting a transformer to solve the floating load issue (left) or floating source issue (right).

Analysis of the left circuit from Fig. 5 shows that the conductance seen by one of the sources can still get negative for some frequencies and outphasing angles. This is caused by the transformer's magnetizing inductance. Since this negative conductance is one of the issues to be avoided, this circuit is not considered as a good candidate.

The right circuit of Fig. 5 turns out to be more useful, even if the transformer parasitics are taken into account. Fig. 6 shows this topology after insertion of the transformer's magnetizing and leakage inductance. Also, a wideband matching network is added at the secondary side of the combiner. The leakage inductance is used as the first element of the matching network. Remember that one of the sources in an outphasing transmitter needs an inductive compensation susceptance. In the topology in Fig. 6, the transformer magnetizing inductance is in the right position to serve as this compensation susceptance. The capacitive compensation element for the other source has to be added separately. It can be shown that the transformer's transformation ratio must be equal to 1 for correct operation. In order to achieve a transformation ratio equal to 1, the primary inductance L_p and secondary inductance L_s must satisfy (5). Assuming k to be fixed for a certain technology, (i.e. the maximum achievable coupling coefficient is limited by the technology,) and L_p to be constrained by the compensation susceptance that is needed, L_s follows automatically. With the relation given by (5), the leakage inductance can be expressed as (6). For

a large bandwidth, L_p must be small, and k must be close to 1.

$$n = \sqrt{\frac{L_s}{L_p}} = \frac{1}{k} \tag{5}$$

$$L_{leak} = L_p \left(\frac{1}{k^2} - 1 \right) \tag{6}$$

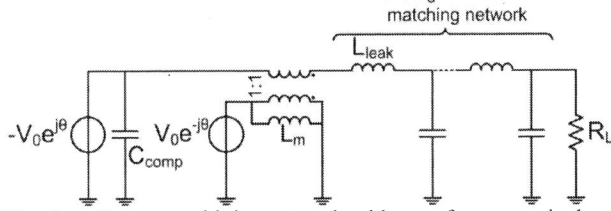

Fig. 6. Power combining network with transformer equivalent circuit, output matching network and compensation elements.

IV. DESIGN OF A CLASS-E OUTPHASING PA

The transformer-based power combining concept is tested with two class-E PA's built with NXP 65nm CMOS extended drain devices [4]. Outphasing systems are usually implemented using class-B, -D or –F PA's, but recently also class-E PA's have been used [5]. The complete schematic is shown in Fig. 7.

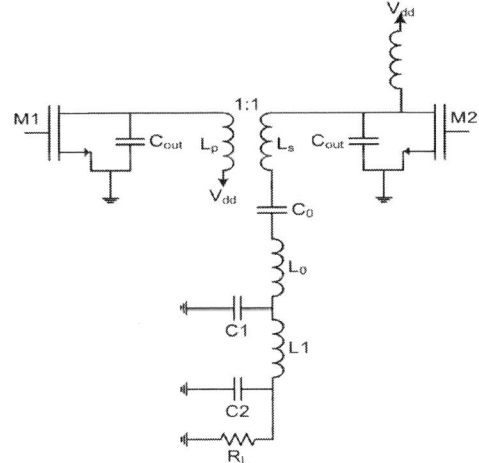

Fig. 7. Complete PA schematic with two CMOS extended drain devices and a transformer-based power combiner.

The series resonators of both class-E PA's are shifted to the secondary side of the combiner, in series with the transformer leakage inductance. The leakage inductance is merged with the resonator inductor and the inductor of the matching network. The compensation susceptance of M2 is created by changing the value of the DC feed inductance to a larger value. The primary transformer winding serves as a DC feed inductance for M1 and a compensation susceptance. The resonator at the secondary side has a low Q_L to enable high BW operation. Furthermore, a 2-stage wideband impedance matching network was designed, to offer the correct load impedance to the devices over a large BW. The transformer was implemented on a three-layer PCB with minimum feature sizes of 100µm. A photograph of the core of the board, containing the transformer, the devices, and the SMD components of the output network is shown in Fig. 8. The primary transformer winding is on the second metal layer, and cannot be seen in this photograph. The transformer coupling coefficient is rather low (only 55%). This low coupling coefficient is the main bandwidth limiting factor.

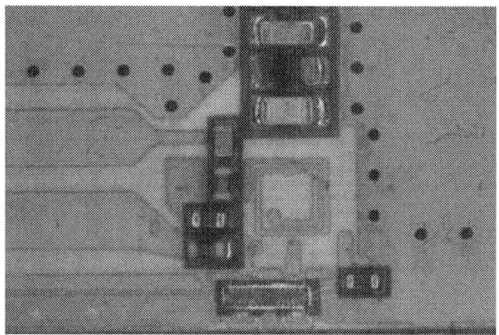

Fig. 8. Core of the PCB. CMOS die is shown in the bottom. Secondary transformer winding is shown in the middle.

V. MEASUREMENT RESULTS

The output power and drain efficiency have been measured under different driving- and supply conditions, over a frequency range of 600MHz – 800MHz. The best 10dB back-off efficiency was measured at 650MHz. The drain-efficiency at 650MHz as a function of output power is shown in Fig. 9. Supply voltage is 3.6V. Apart from a downshift, the curve nicely matches an ideal outphasing efficiency curve. Fig. 10 shows the measured drain efficiency as a function of frequency for different PBO levels. A peak output power of more than 30dBm was measured over a 29% BW around 700MHz (600MHz – 800MHz). The 10dB back-off efficiency is larger than 27.3% over the same BW. The 6dB back-off efficiency is larger than 46.4% over this BW. The peak output power at 700MHz equals 33.9dBm. The maximum operating power gain (G_p) equals 20.1dB and was measured at 700MHz.

VI. CONCLUSION

Most published PA's are either optimized for bandwidth at peak output power, or for back-off efficiency. Traditional outphasing PA's are almost always narrow-band, and use a classical Chireix power combiner, or a lumped equivalent. This work presents a PA with high back-off efficiency over a large bandwidth. The novel power combining technique presented here enables to achieve much larger BW in outphasing PA's compared to the classical Chireix power combiner. The BW can even be further increased by designing a transformer with larger k on a more advanced laminate. To the authors best knowledge, the 10dB back-off efficiency at 650MHz is the highest back-off efficiency reported in a CMOS outphasing PA to date.

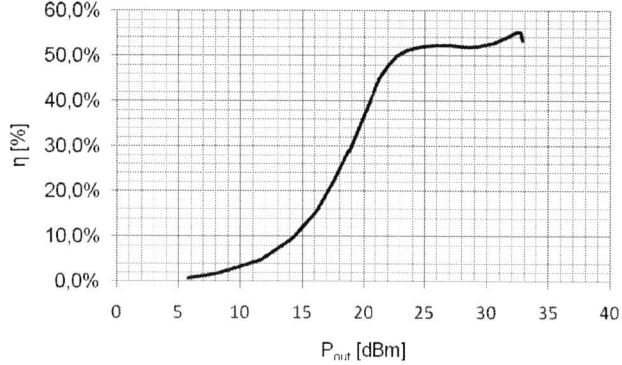

Fig. 9. Drain efficiency versus output power at 650MHz.

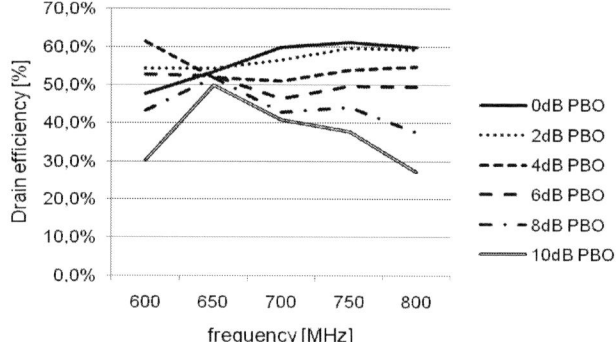

Fig. 10 Measured drain efficiency versus frequency.

REFERENCES

[1] I. Hakala et al., "A 2.14 GHz Chireix Outphasing Transmitter," *IEEE Trans. on Microwave Theory and Techniques*, vol. 53, pp. 2129-2138, Jun. 2005.

[2] H. Chireix, "High Power Outphasing Modulation," *Proc. of the IRE*, pp. 1370-1392, Nov. 1935.

[3] J. R. Long, "Monolothic transformers for silicon RF IC design," *IEEE Journal of Solid-State Circuits*, vol. 53, pp. 1368-1382, Sep. 2000.

[4] J. Sonsky et al., "Innovative High Voltage transistors for complex HV/RF SoCs in baseline CMOS," in *Int. Sym. on VLSI Technology, Syst. and Appl.*, 2008, pp. 115-116.

[5] R. Beltran et al., "HF Outphasing transmitters using class-E PA's," *IEEE MTT-S Int. Microwave Sym. Dig. 2009*, pp. 757-760, Jun. 2009.

RTU1C-5

A Highly Efficient 5.8 GHz CMOS Transmitter IC with Robustness over PVT Variations

[1]Eun-Hee Kim, [2]Jeong-Ki Choi, [2]Seok-Oh Yun, [2]Jinho Ko, [1]Kwyro Lee

[1]Dept. of Electrical Engineering and Computer Science, KAIST, Daejeon, 305-701, Korea

[2]PHYCHIPS Inc., Daejeon, Korea

Abstract — This paper presents a switching PA-based polar transmitter which achieves uniform and robust performances under process, voltage, and temperature variations. A new approach utilizing current driven envelope signal is proposed, leading to much more accurate control of output power of the PA. In addition, the proposed adaptive LO technique extends the available linear control range of the PA. Fabricated in 0.13 um CMOS process, an experimental polar transmitter is designed to fulfill the stringent requirements of the 5.8 GHz Korean DSRC/ETC standards. Delivering output power of + 12 dBm, it consumes average current of 35 mA at 3.3 V supply voltage. Output P-1dB of 15 dBm and ACPR lower than -50 dBc are obtained. The transmitter achieves output power error less than ± 1 dB over the temperature and supply voltage range from -40 to 100 °C and from 3.0 to 3.6 V.

Index Terms — DSRC applications, ETC system, polar transmitter architecture, PVT invariant transmitter, CMOS switching PA, current-mode PA, adaptive LO technique

I. INTRODUCTION

As CMOS technology has evolved, there are many efforts to integrate RF, analog, and digital circuits onto single chip. It leads to reduce the number of external components and to lower the price of chipset. Especially, among all the circuit blocks which compose transceiver integrated circuits (IC), a power amplifier is considered as the most difficult to integrate since CMOS technology inherently has low breakdown and efficiency. In recent years, nevertheless, several researches prove their ability to improve the performances of a CMOS power amplifier, such as in output power, efficiency, and linearity. For mass production, however, not only to achieve high performances of a power amplifier but to get a high yield is the key element. As based on open loop topology in high operating frequency, the power amplifier has vulnerable performances to PVT variations. Considering that the yield of a full transceiver IC is determined by a power amplifier, it is highly needed to raise that of a power amplifier. In this paper, to minimize PVT variations of RF transmitter, polar transmitter with current-mode power amplifier is newly proposed. Section II describes

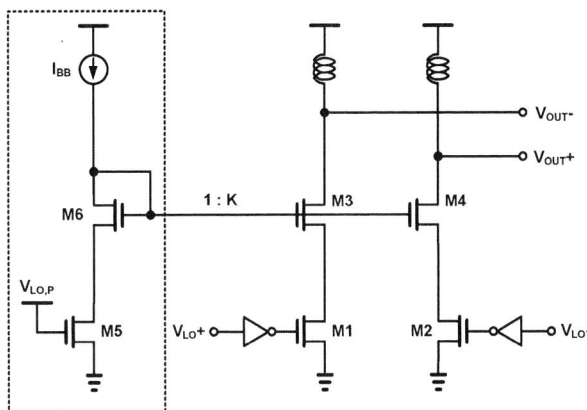

Fig. 1 Proposed current-mode PA

the operation principle of the proposed current-mode power amplifiers and analysis on PVT variation and efficiency compared to linear power amplifiers. In section III, the design of the total system is discussed in detail. The measurement results are given in Section IV and conclusions are discussed in Section V.

II. OPERATION PRINCIPLE AND ANALYSIS OF PROPOSED PA

A. Proposed PA Circuit: Current-Mode Switching PA

Generally, a switching power amplifier for a polar transmitter uses a modulated signal as a supply voltage. Though the current consumption of a power amplifier itself is low, the total current dissipated by a transmitter becomes so large on account of a low dropout voltage (LDO) block to provide modulated supply voltage. In [1], a switching power amplifier directly modulated by a gate voltage of cascode transistors is presented. The power amplifier enables a transmitter to remove a LDO block, and then it results in reduced current consumption for a transmitter. Due to nonlinear relation between an input and output signal, the power amplifier has a drawback in that it is not suitable to N-ary amplitude modulation.

Fig. 1 depicts the concept of a newly proposed power

978-1-4244-6240-7/10 $26.00 © 2010 IEEE

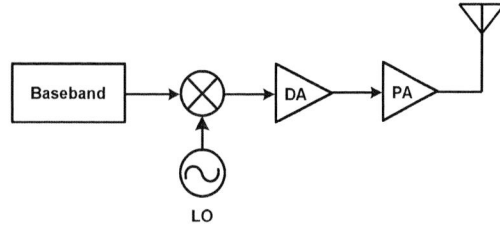

(a) Direct up-conversion transmitter

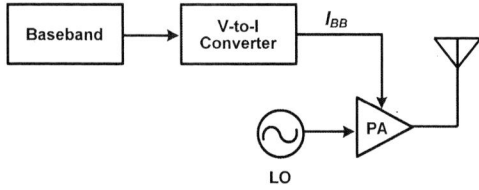

(b) The proposed polar transmitter

Fig. 2 Transmitter architectures

amplifier. The power amplifier is biased by a baseband current IBB, which is provided from a current mirror. In ideal LO, the output current $I_{out}(t)$ of the current-mode power amplifier is given by

$$I_{OUT}(t) = K \cdot \frac{I_{BB}}{2} \cdot \frac{V_{LO}(t)}{V_{LO,P}}$$
$$= K \cdot \frac{I_{BB}}{2} \cdot \left\{ 1 + \frac{4}{\pi} \cdot \sin \omega_{LO} t + \frac{1}{3} \cdot \frac{4}{\pi} \cdot \sin 3 \omega_{LO} t + \cdots \right\} \quad (1)$$

where K is the current mirroring ratio, $V_{LO}(t)$ is the LO voltage, and $V_{LO,P}$ is the peak voltage of the LO signal. When filtered out by output load of the power amplifier, the output current becomes $\{I_{BB} \cdot K \cdot (2/\pi) \cdot sin(\omega_{LO}t)\}$. The current conversion gain of the power amplifier is $(K \cdot 2/\pi)$, and it remains constant. The maximum drain efficiency of the proposed power amplifier is expected as 64 %.

B.PVT Variation Errors in Transmitter Architectures

The function of RF transmitter is classified as three kinds of up-conversion, filtering, and amplification. Above all, with respect to amplification, the PVT variation error of each the transmitter architecture, in Fig. 2, is derived as follows.

Fig. 2 (a) shows a block diagram of a direct up-conversion transmitter architecture, which consists of a mixer, a driver amplifier, and a power amplifier. Assuming that a Gilbert mixer and class A amplifiers are used, for the direct-up conversion transmitter, a simplified equation of total gain $A_{v,Conventioanl}$ is expressed by

$$A_{v,Conventional} = A_{v,Mixer} \cdot A_{v,DA} \cdot A_{v,PA}$$
$$= \left\{ (2/\pi \cdot g_m \cdot Z_L) \big|_{Mixer} \cdot (g_m \cdot Z_L) \big|_{DA} \cdot (g_m \cdot Z_L) \big|_{PA} \right\} \quad (2)$$

where $A_{v,Mixer}$, $A_{v,DA}$ and $A_{v,PA}$ is the gain of a mixer, a driver amplifier, and a power amplifier. To analyze the process variation error, we define a process variation coefficient (PVC) of

$$ProcessVariationCoefficient \triangleq \frac{\Delta X}{X} \bigg|_{(FF \to TT) or (TT \to SS)} \quad (3)$$

Using (2) and (3), the PVC of a conventional transmitter is given by

$$PVC_{conventional} = \sum_{Mixer,DA,PA} \left(\frac{\Delta g_m}{g_m} \right) + \sum_{Mixer,DA,PA} \left(\frac{\Delta Z_L}{Z_L} \right) \quad (4)$$

To derive the temperature variation error, the temperature coefficient (TC) is expressed as

$$TemperatureCoefficient \triangleq \frac{1}{X} \cdot \frac{dX}{dT} \quad (5)$$

From (2) and (5), the TC of a conventional transmitter is found as

$$TC_{conventional} = \sum_{Mixer,DA,PA} \left(\frac{dg_m}{g_m \cdot dT} \right) + \sum_{Mixer,DA,PA} \left(\frac{dZ_L}{Z_L \cdot dT} \right). \quad (6)$$

Fig. 2 (b) shows a block diagram of a polar transmitter with the proposed current-mode power amplifier (CMPA). When using a voltage-to-current converter (VIC) with external resistor, the voltage conversion gain of the polar transmitter is expressed as

$$A_{v,polar} = \left\{ K \cdot (2/\pi) \cdot (Z_L / R_{EXT}) \right\} \quad (7)$$

where R_{EXT} is the resistance of an external resistor. Then, the PVC and TC of the polar transmitter is given by

$$PVC_{polar} = (\Delta K / K) + (\Delta R_{EXT} / R_{EXT}) \quad (8)$$

$$TC_{polar} = (K^{-1} \cdot dK/dT) + (R_{EXT}^{-1} \cdot dR_{EXT}/dT) \quad (9)$$

To expect the numerical value of variation errors, Spectre simulations have been performed with a 0.13 um CMOS technology. Using design parameters of our former work of [3], the PVT variation error in g_m and Z_L is simulated. The PVC and TC of g_m are -7 % and -0.24 %/°C, and those of Z_L are -10 % and -0.13 %/°C, respectively. For the variation parameter of K, the size of transistors is used as the following values; NMOS transistors are (120 um / 0.13 um), (360 um / 0.35 um), (15 um / 0.13 um), and (45 um / 0.35 um) for M1, M3, M5, and M6 in Fig. 1. At then, a current mirror has the PVC of -3 % and TC of -0.18 %/°C in K. The maximum variation of external resistor is assumed to be in ±5%. Over temperature range of (-40~100 °C), total process and temperature variation error in A_v of a conventional and a polar transmitter can be expected as in equations of (4), (6), (8), and (9), being respectively 140.8 % and 17.5 %.

978-1-4244-6240-7/10 $26.00 © 2010 IEEE

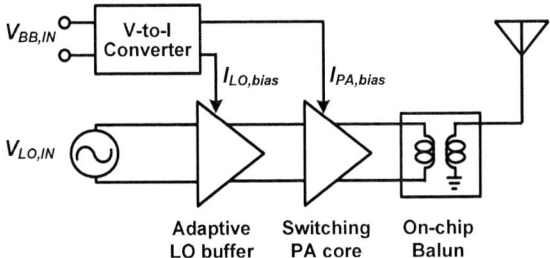

Fig. 3 Architecture of the proposed polar transmitter

TABLE I

COMPARISON OF PT VARIATION ERRORS

Architecture	A_v variation		**Gain variation**
	Block	(%)	**(dB)**
Conventional Transmitter	Mixer	49.3	15
	DA	49.3	
	PA	42.2	
Polar Transmitter	VIC	10.0	1.5
	CMPA	7.5	

The calculated PV variations are summarized in table I, and it is proven that a polar transmitter is superior to a conventional one in PVT variation error as well as in current consumption.

III. CIRCUIT DESIGN OF THE POLAR TRANSMITTER WITH PROPOSED CURRENT-MODE SWITCHING PA

The 5.8 GHz Korean DSRC (dedicated short range communication) system [2] was chosen as a target application due to the following reasons; 1) An ASK-modulated system requires only a magnitude control path in a polar transmitter, and hence the proposed PA circuit is directly applicable to the system without any additional phase control path. 2) The accuracy requirement of the absolute Tx output power, specified in the Korean DSRC standard, is less than ± 1dB over all PVT corners. When using the conventional direct up-conversion architecture, this stringent requirement often results in the need of various complex PVT compensation circuits. It increases silicon area, power consumption, and finally cost [3], [5].

A. 5.8 GHz DSRC Transmitter

The proposed DSRC transmitter, as shown in Fig. 3, incorporates a V-I converter, an adaptive LO buffer, and a current-mode switching PA core. The V-I converter converts differential baseband analog input into two current sources for the adaptive LO buffer bias and the switching PA core bias. The Current fed into PA core linearly controls the output magnitude of the PA. Finally,

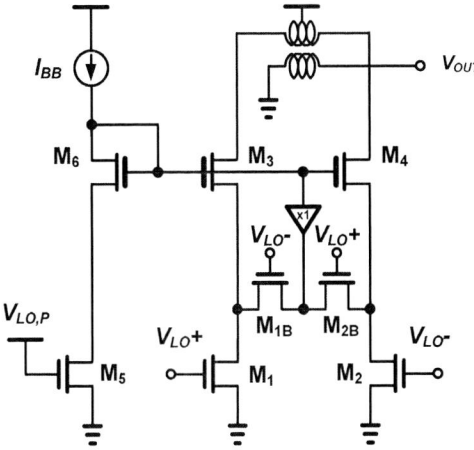

Fig. 4 Final circuit diagram of the proposed PA

the differential output of the PA is converted to a single-ended output via an on-chip balun transformer.

An adaptive LO buffer is introduced to enhance the linearity of the whole PA. When baseband input signal is very low, the output level of the adaptive LO buffer keeps the minimum level which can drive the PA switches. As the baseband input level increases, the LO level also increase linearly. This adaptive LO technique offers two remarkable positive effects on PA performance. First, it can lower the LO leakage level when output power is very low. And secondly, it can compensate nonlinearity caused by the switch loss when output power is very high. In other words, the adaptive LO technique can improve both the lower limit and the upper limit of the dynamic range of the proposed PA.

B. Speed Enhancement Technique of the Proposed PA

In general, to lower current consumption of a PA, while keeping output power constant, it is essential to use higher supply voltage. In this work, a single 3.3V external supply was assumed. So, to fully utilize 3.3V supply while preventing PA from unwanted breakdown, the output part of the PA must use 3.3V NMOS transistors. However, a 3.3V NMOS transistor has even lower fT than that of 1.2V NMOS transistor and further decreases the maxmum operating frequency of the PA. In Fig. 1, M3, M4, and M6 are built by 3.3V NMOS transitor, while the remaining switch transistors, M1, M2, and M5 are built by 1.2V NMOS transistors. Because the fall time determined by $g_{m,M5}$ and $C_{gs,M5}$ is far slower than the rise time determined by $R_{on,M1}$ and $C_{gs,M5}$, the maximum operating frequency is approximately given by,

$$ t_{max} < \frac{1}{T_{LO}/2} \simeq \frac{g_{m,M3}}{C_{gs,M3}} = t_{T,M3} \quad (= t_{T,M4}). \quad (10) $$

TABLE II

PERFORMANCE SUMMARY AND COMPARISON

	Unit	Spec	[3]	[4]	[5]	**This work**
Technology			0.18 um CMOS	0.25 um BiCMOS	0.18 um BiCMOS	0.13 um CMOS
Power supply voltage	V		3.3	3.3	3.3 ± 0.3	3.3 ± 0.3
Current consumption	mA		< 102	< 137	< 37	< 35 @ 75 % MI
Pout,max	dBm	10	10.5	11.7	4	15
Pout,variation	dB	± 1	N/A	N/A	± 1.2	± 1
ACPR	dBc	40	42	34	53.9	50

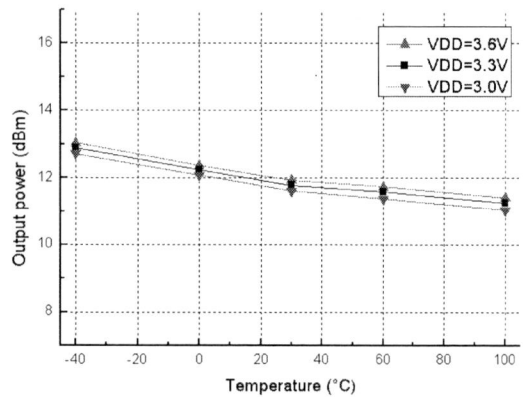

Fig. 5 Measured voltage and temperature variations in P_{out}

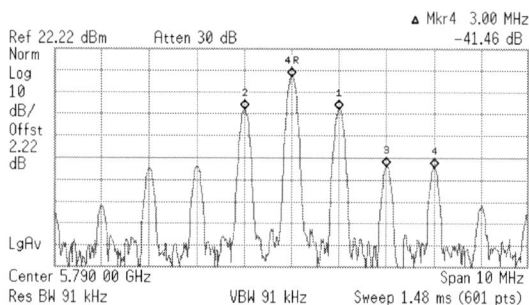

Fig. 6 Output spectrum @ modulation index of 75 %

By just adding switches of M1B and M2B which provide faster discharge path, the maximum operating frequency is calculated as

$$f_{max}' \simeq \frac{g_{m,M1}}{C_{gs,M3}} = \frac{g_{m,M1}}{g_{m,M3}} \cdot f_{T,M3} \quad \left(\sim 3 \cdot f_{T,M3} \right). \quad (11)$$

It can be increased by around three times higher as in (10). In addition, rise and fall time symmetry can be improved.

IV. EXPERIMENTAL RESULTS

The polar transmitter for Korean DSRC/ETC application is implemented with 0.13 um 1-poly 6-metal RF CMOS process allowing 1.2 V and 3.3 V of core and I/O voltage, respectively. The chip area of the proposed transmitter, including bond pads, is 1.52 mm². Maximum output power of 15 dBm, being the same as output P-1dB, is measured. For delivering CW output power of 12 dBm, it consumes 57 mA from 3.3 V supply voltage. As shown in Fig. 5, measured variation error in output power is less than ± 1 dB over temperature range of (-40~100 °C) and supply voltage range of (3.0~3.6 V).

Fig. 6 shows output spectrum of the proposed transmitter when the baseband signal is ASK modulated with a modulation index (MI) of 75 %. The average current consumption of transmitter is 35 mA under 3.3 V supply voltage, and ACPR less than -50 dBc is obtained.

Table II summarizes and compares the performance of the proposed DSRC transmitter against those of other published data.

V. CONCLUSION

A highly efficient polar transmitter, fabricated in 0.13 um CMOS process, is proposed for 5.8 GHz Korean DSRC/ETC applications. It adopts the new PA topology of current-mode switching PA, which uses current mirroring operation to deliver accurate output power over extreme PVT variations. A speed enhancement technique is added to PA. Moreover, for extending linear control range of the PA, an adaptive LO technique is used. The measurement results of the proposed transmitter clearly indicate the potential for high yield transmitter architecture.

REFERENCES

[1] S. Shim, J. Han, and S. Hong, "A CMOS RF polar transmitter of a UHF mobile RFID reader for high power efficiency," *IEEE Microw. Wireless Compon. Lett.*, vol. 18, no. 9, pp. 635-637, Sep. 2008.
[2] TTASI.K0-06.0025, 5.8 GHz Korean DSRC standards
[3] S. Shin, et. al., "0.18 um CMOS integrated chipset for 5.8 GHz DSRC systems with +10 dBm output power," *IEEE Int. Symp. Circuits and Systems*, pp. 1958-1961, May 2008.
[4] N. Sasho et al., "Single-chip 5.8 GHz DSRC transceiver with dual-mode of ASK and pi/4-QPSK," *IEEE Symp. Radio and Wireless*, pp. 799-802, Jan. 2008.
[5] H. Lee et al., "A temperature-independent transmitter IC for 5.8 GHz DSRC aplications," *IEEE Trans. Circuits Syst.*, vol. 55, no. 6, pp. 1733-1741, Jul. 2008.

RTU1D-1

A 700uA, 405MHz Fractional-N All Digital Frequency-Locked Loop for MICS Band Applications

S. Shashidharan[1], W. Khalil[2], S. Chakraborty[3], S. Kiaei[1], T. Copani[1] and B. Bakkaloglu[1]

[1]Electrical, Energy and Computer Engineering, Arizona State University, Tempe, AZ 85287 USA
[2]Electrical and Computer Engineering, Ohio State University, Columbus, OH 43212, USA
[3]Texas Instruments Inc. High Performance Analog Group, Dallas TX 75243, USA

Abstract — An all-digital frequency-locked loop (ADFLL) based frequency synthesizer with a built-in FSK modulator for medical implants communication systems (MICS) band applications is presented. The ADFLL uses a high resolution single-bit digital $\Sigma\Delta$ frequency discriminator in the feedback path and a $\Sigma\Delta$ phase accumulator in the reference path, achieving fractional resolution. The ADFLL uses a digital IIR-based loop filter followed by a digital-intensive $\Sigma\Delta$ current-steering DAC and a first-order-hold filter. The ADFLL achieves 9.5Hz frequency resolution, spanning the ISM 400MHz-410MHz band. The worst-case near-integer spur of -55dBc and a phase noise of -83dBc/Hz at 300kHz offset is measured. The ADFLL is fabricated on a 0.18um CMOS process, occupying 0.14mm2 die area, with a quiescent current consumption of 700uA.

Index Terms — type-I PLLs, digital PLLs, $\Sigma\Delta$ DACs

I. INTRODUCTION

RF transceivers for Medical Implant Communications Systems (MICS) band require long battery life and very low power consumption [1]. An all digital frequency locked loop (ADFLL) designed for a micro-power MICS band transceiver application is presented. The proposed ADFLL is based on a Type-I PLL architecture and it is used for both receiver LO frequency synthesis and transmitter GFSK modulator. In Type-I synthesizers, the loop is locked on the frequency rather than phase, which reduces the loop order and enables wider bandwidth with reduced risk of instability [4][5]. As shown in Fig.1 (a), the proposed digital Type-I loop utilizes a high resolution single-bit all digital $\Sigma\Delta$ frequency to digital converter ($\Sigma\Delta$ FDC) in the feedback path, and a single-bit phase accumulator in the reference path. When the loop is locked, correlated quantization noise and spurious content generated by the FDC and the reference DDFS is cancelled.

In typical digital PLLs, phase detection is performed by a time-to-digital converter (TDC) and frequency synthesis is achieved by a digitally-controlled oscillator (DCO) [2]. The TDC generates a quantized value of VCO phase with respect to the reference phase, and the

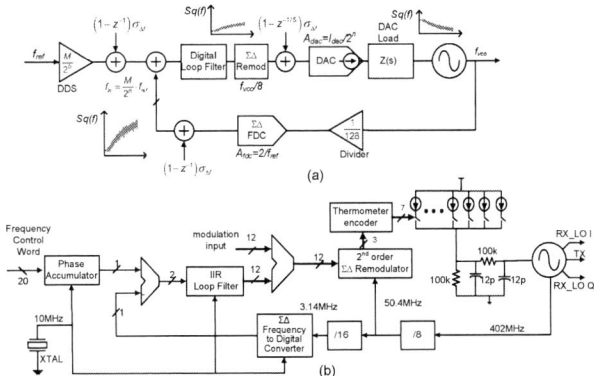

Fig.1 (a) Signal flow diagram of the ADFLL (b) Block diagram of the proposed ADFLL

frequency generation is typically performed by controlling discrete MOS varactors in an *LC* oscillator. However the TDC linearity limits the PLL phase noise performance, because the TDC is in the feedback path [3]. In the proposed approach a noise shaping frequency-to-digital converter is adopted to avoid using nonlinear phase digitization and TDC.

The rest of the paper is organized as follows: Section II describes the proposed synthesizer architecture and overview of the loop dynamics. Section III addresses the implementation of building blocks. Section IV presents the experimental results. Finally, conclusions are drawn in section V.

II. OPERATION OVERVIEW

The block diagram of the proposed ADFLL is shown in Fig.1 (b). The reference frequency control is achieved by a first order phase accumulator typically used in DDFS applications. The overflow bit provides a single-bit first order • noise shaped bitstream representing the fractional reference frequency based on the frequency control word (FCW) input. The phase accumulator is clocked by a crystal-based reference, generating a synchronized output with respect to the feedback signal. In the feedback path, the VCO output is divided down to a nominal 3.14MHz, bringing the instantaneous frequency to the range of the 10MHz reference

978-1-4244-6240-7/10 $26.00 © 2010 IEEE   367

frequency. The divided VCO output signal is passed through the ΣΔ FDC, resulting in a single-bit first order • noise shaped signal representing the instantaneous VCO output frequency [4]. The difference between first-order noise-shaped reference and feedback instantaneous frequency code is applied to a Direct Form I IIR based loop filter. The IIR block filters the residual quantization noise as well. The 12-bit IIR filter output is re-modulated to a 3-bit, 7-level noise shaped signal through a second order • noise shaper clocked at 50.4MHz signal divided from the VCO output. A 7-level current steering DAC with PMOS-only current source array is used to generate the control voltage for the VCO. The current mode DAC output is fed to a second-order passive RC filter, and a first-order hold response providing reconstruction filtering. The VCO is a four-stage differential ring oscillator with Maneatis delay cells and replica bias circuit with reduced AM/PM conversion.

Using frequency as the loop variable, the ADFLL loop resembles a Type-I PLL, with a single integrator in the loop. The closed-loop frequency domain model of the loop along with closed loop and open loop transfer functions is represented as:

$$A_{OL}(s) = K_1 \cdot I_{LSB} \cdot \frac{2 \cdot K_{VCO}}{N \cdot f_{ref}} \cdot \frac{R_1 \left(s + \frac{K_P}{K_I} \right)}{s \left((R_1 R_2 C_1 C_2) s^2 + (R_1 C_1 + R_1 C_2 + R_2 C_2) s + 1 \right)}$$

where R_1, R_2, C_1, and C_2 are the DAC reconstruction filter components, I_{LSB} is the unit current source value of the current-steering DAC, K_P and K_I are proportional and integral gains, K_{VCO} is the VCO gain. With nominal loop parameters, the loop bandwidth is dominated by the digital loop filter and it is around 25kHz and all of the analog poles associated with the DAC is outside the loop bandwidth.

III. CIRCUIT DESIGN AND IMPLEMENTATION

Fundamental building blocks of the architecture are described as follows.

A. ΣΔ Feed-forward Path:

An all-digital circuit to extract instantaneous phase and frequency of an input signal is introduced in [5], and its phase noise characteristics are introduced in [6]. Fig. 2 shows the operation of this digital frequency digitizer operation.

The instantaneous frequency of the input signal can be defined as $f_{in}(t)=f_c+F_n(t)$, where f_c is the carrier frequency, F_n is the instantaneous frequency deviation of the carrier. The ΣΔ FDC digitizes the deviation of $f_{in}(t)$ from its carrier frequency f_c with high-pass quantization noise shaping similar to ΣΔ ADCs.

Assuming that the phase at the time $n \cdot T_s$ is θ_n and defined as

$$\theta_n / 2\pi = p_n + \phi_n \qquad (1)$$

where p_n is an integer representing the received number of rising VCO edges at time $n \cdot T_s$ and $0 < \phi_n \le 1$ is the fractional phase difference between the previous rising VCO edge and the sampling signal edge.

The ΣΔ FDC output can now be expressed as

$$y_n = Q(p_n + \phi_n) - Q(p_{n-1} + \phi_{n-1}) \qquad (2)$$

where $Q(\cdot)$ represents time quantization with respect to a reference clock. By choosing integer quantization thresholds, output y_n can be expressed as

$$y_n = p_n - p_{n-1} + Q(\phi_n) - Q(\phi_{n-1}) \qquad (3)$$

where y_n represents the number of received rising VCO edges during the sampling interval. Since ϕ_n is less than unity and integer quantization thresholds are used, the output of quantization error is always zero, which reduces y_n to $p_n - p_{n-1}$. Input of the frequency discriminator is $T_s(f_c+F_c(t))$. By representing the quantization error as e_n the output can be expressed as

$$y_n \approx T_s (f_c + F_n) + e_n - e_{n-1} \qquad (5)$$

As shown in (5), this circuit discriminates its input modulating signal frequency with first order ΣΔ noise shaping characteristics.

In the proposed architecture, a digital representation of the desired fractional reference frequency is applied to the reference path by using a 20-bit phase accumulator-based DDFS. To cancel the correlated quantization noise of the ΣΔ FDC, the input to the ADFLL must also have a similar noise-shaping in comparison to the feedback path. A phase accumulator is used for this purpose as shown in Fig. 2. A phase accumulator is equivalent to a first order digital noise shaper (ΣΔ-DDFS), where the frequency quantization error can be defined by $\Delta f = f_{ref}/2^{20}$. The spectral content of the SD_0 output signal in Fig. 3 is first order noise shaped.

The input frequency is converted to a single-bit noise shaped digital form, which gives a fine resolution and precise control over the output frequency. Therefore, the

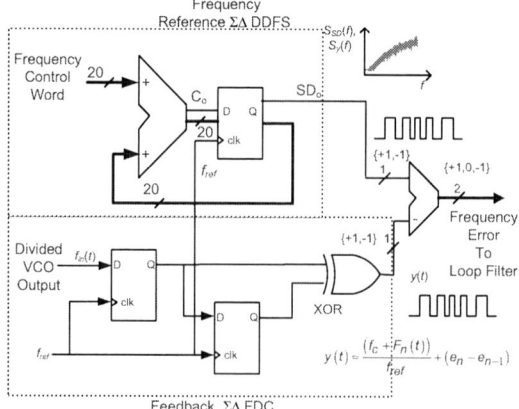

Fig. 2 Digital ΣΔ frequency-to-digital converter and phase accumulator based frequency error generation.

978-1-4244-6240-7/10 $26.00 © 2010 IEEE   368

output frequency resolution is dictated by the bus width of the DDFS accumulator. The ADFLL spans 400MHz to 410MHz MICS band with 9.5Hz frequency resolution.

B. Digital Loop Filter and $\Sigma\Delta$ DAC:

The error signal is applied to the digital loop filter, followed by a $\Sigma\Delta$ DAC. The loop filter is implemented by an IIR Direct-Form-I structure. The input bus width is 2-bits, the output bus-width is truncated to 12 bits. To control the VCO with the required frequency resolution, nominally a 12-bit DAC is required. Instead, an error-feedback $\Sigma\Delta$ digital noise-shaper clocked at 50.4MHz, generated from the VCO output is used to reduce the number of unit elements from 2^{12} down to 7, reducing DAC complexity. When the loop is locked, the error signal into the digital noise-shaper is a constant DC code, therefore a second order re-modulator is used to reduce DC-related idle-tones at the DAC output. The main requirement of the DAC is monotonicity, low glitch energy and speed.

A PMOS cascoded 3-bit, 7-level current steering DAC as shown in Fig. 3 is used for this block. The 8 thermometer output codes are latched with the re-modulator clock, updating the VCO control line at 50.4MHz frame rate. Due to the second order $\Sigma\Delta$ noise-shaped DAC, at least a second order reconstruction filter is needed to remove the shaped quantization noise. An RC based trans-impedance load followed by a first order RC filter is used to filter-out the quantization noise that is located at 25.2MHz, which is three decades wider than the digital loop bandwidth of 23kHz, and do not impact the loop dynamics. The DDFS and $\Sigma\Delta$ FDC consume and IIR loop filter consumes 230uA. The re-modulator and the DAC consume 108uA.

C. Current Controlled Oscillator

A 4-stage differential delay cell based ring oscillator

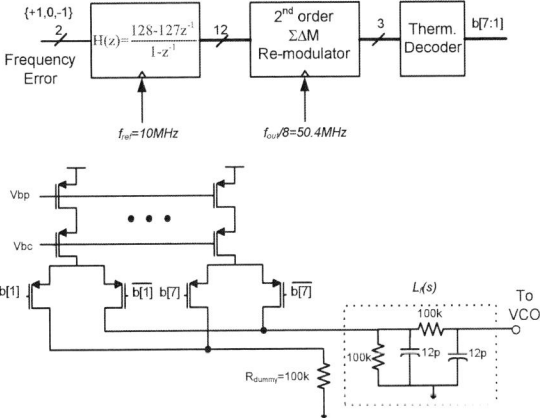

Fig. 3 IIR digital loop-filter, $\Sigma\Delta$ re-modulator and current steering DAC controlling the VCO.

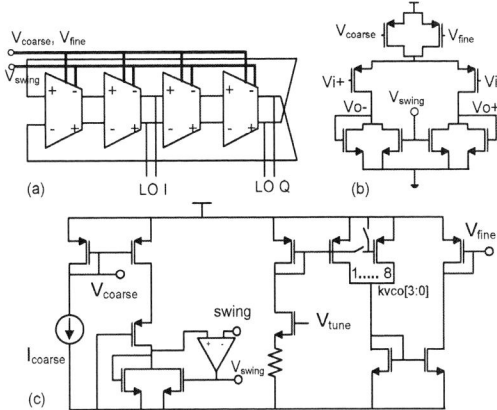

Fig. 4 Replica bias based coarse-fine controlled VCO.

is used for the VCO. The frequency of the oscillator is tuned by the delay current through the unit cells via current control [7]. The in-phase and quadrature LO signals for the LO mode is taken from the second and fourth stage as shown in Fig 4(a). The delay cell has a Maneatis-type load which gives a variable linear resistance with varying input control voltage as shown in Fig. 4(b). In the replica bias circuit, the amplitude of the oscillation is fixed independent of the tuning voltage as shown in Fig. 4(c). This approach minimizes the AM-PM noise conversion. As shown in Fig 6(c), the oscillation amplitude varies by less than 1mV across tuning range. The fine control of frequency tuning is obtained by using a transconductance stage which is controlled by the DAC. The VCO consumes 260uA current.

IV. MEASUREMENT RESULTS

The prototype is fabricated in a 0.18 µm CMOS process. The IC micrograph is shown in Fig. 5. Fig. 6 (a) shows the measured phase noise at 403 MHz operating frequency, with a fractional input. The phase noise at 300kHz offset is measured at -83dBc/Hz. Fig. 6 (b) shows measured ADFLL settling time vs frequency. The ADFLL settles to 1% accuracy in less than 3uSecs. Table I shows the summary of measured performance parameters and Table II provides a comparison of this work with respect to prior art.

IV. CONCLUSION

A low power all digital frequency-locked loop using $\Sigma\Delta$ DDFS in the feed-forward loop and a first order $\Sigma\Delta$ FDC converter in the feedback loop for MICS band applications is presented. The frequency based approach minimizes the complexity associated with phase domain PLLs. The ADFLL is fabricated on a 0.18um CMOS process, occupying 0.14mm2 die area, with a quiescent current consumption of 700uA.

978-1-4244-6240-7/10 $26.00 © 2010 IEEE

Table I. Design parameters

Design Parameter	
Technology	UMC 0.18μm 1P6M CMOS
Reference Frequency	10MHz
Operating output frequency	401-405MHz
Maximum data rate	100kbps
Maximum freq. deviation	150kHz
Loop Bandwidth	23.5kHz
Measurement results	
Phase noise @300kHz offset	-83.6dBc/Hz
Settling time	110μs
Power consumption (average current at 1.8V Vdd)	
VCO	258.4uA
ΣΔ-DAC	108uA
ΣΔ-FDC + IIR Loop Filter + clock buffers	232uA
feedback dividers	100uA

REFERENCES

[1] MICS Band Plan, *FCC Rules and Regulations*, Part 95, Jan. 2003

[2] R. B. Staszewski, *et al*, "A digitally controlled oscillator in a 90 nm digital CMOS process for mobile phones", *IEEE J. Solid-State Circuits*, vol. 40, no. 11, pp. 2203–2211, Nov. 2005.

[3] M. Z. Straayer, M. H. Perrott, "A Multi-Path Gated Ring Oscillator TDC With First-Order Noise Shaping", *IEEE J. Solid-State Circuits*, vol. 44, April pp. 1089-1098, 2009

[4] E. W. McCune, W. B. Sander, "Sigma-delta-based frequency synthesis", *US Patent 6,690,215,* Feb 10 2004.

[5] M. Hovin, A. Olsen, T. S. Lande, "Delta-Sigma modulators using frequency-modulated intermediate values," *IEEE J. Solid-State Circuits*, vol.32, no.1, pp. 13-22, Jan 1997.

[6] J. Kwon, B. Bakkaloglu, "Impact of sampling clock phase noise on Σ−Δ frequency discriminators", *IEEE Trans. on Circuits and Systems – II 54(11)*, Nov 2007.

[7] J. G. Maneatis and M. A. Horowitz, "Precise delay generation using coupled oscillators" *IEEE Journal of Solid-State Circuits, 28(12),* Dec 1993.

[8] Alessandro Italia, Giuseppe Palmisano, "A 1.2-mW CMOS Frequency Synthesizer with Fully-Integrated *LC* VCO for 400-MHz Medical Implantable Transceivers", *RFIC 2009*, pp 333-336, 2009.

[9] V.Peiris,*et al.*,"A1V433/868MHz25kb/s-FSK2kb/s-OOK RF Transceiver SoC in Standard Digital 0.18μm CMOS," *ISSCC 2005*, pp. 258–259, Feb. 2005.

[10] A. Tekin, M. R. Yuce and W. Liu, "Integrated VCO Design for MICS Transceivers," *CICC 2006 ,* pp. 765 – 768, Sept. 2006.

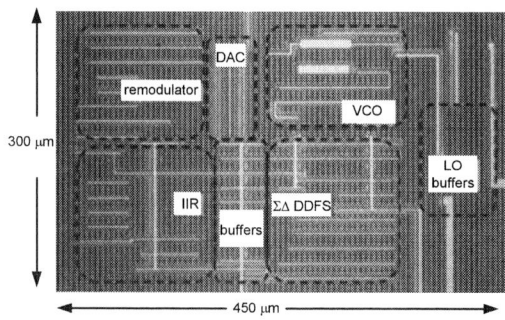

Fig. 5 Die micrograph of the synthesizer

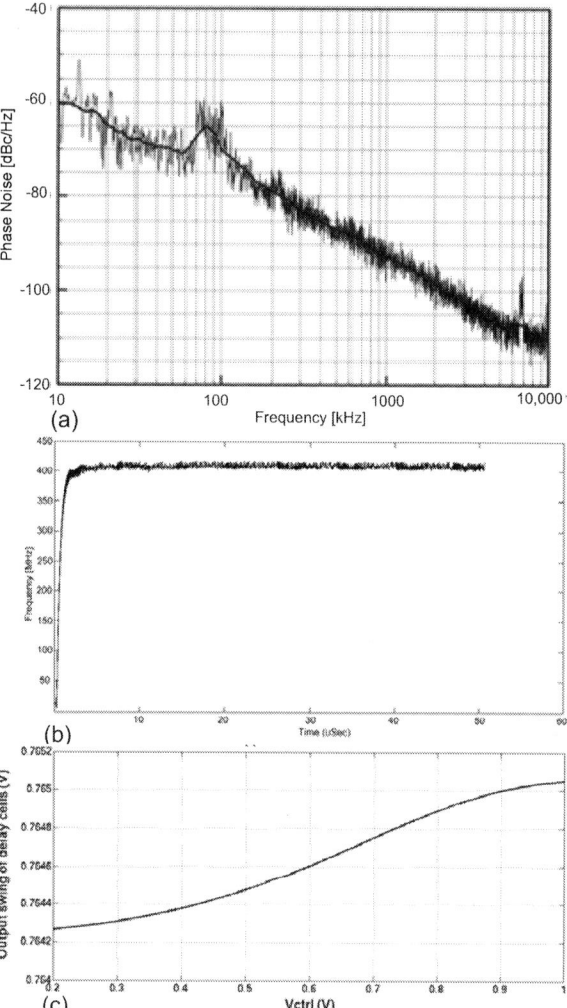

Fig. 6. (a) Measured phase noise of the ΣΔFLL (b) settling time (c) average AM variation with VCO tuning voltage

Table II. Performance comparison

Parameter	This work	[8]	[9]	[10]
Operating Freq (MHz)	400-410	640-650	433/868	402-405
Technology	0.18um	0.13um	0.18um	0.18um
Supply	1.8	1.2	1.8	1.5
Current	700uA	1.2mA	2.1mA*	2.2mA
Phase Noise	83.6 dBc/Hz @300kHz	96 dBc/Hz @100kHz	110 dBc/Hz @600kHz	82d Bc/Hz @160kHz
VCO type	RO	LC	LC	RO
Settling time	110us	500us	200us	--
Bitrate	100kbps	----	100kbps	20kbps
Bandwidth	150kHz	160kHz	100kHz	70-80kHz
Ref Spur	-55dBc	-52dBc	----	----

*RX+PLL combined current

RTU1D-2

A 2-MHz Bandwidth Δ-Σ Fractional-N Synthesizer Based On a Fractional Frequency Divider with Digital Spur Suppression

Pin-En Su and Sudhakar Pamarti

Department of Electrical Engineering, UCLA, Los Angeles, California, 90095, U.S.A.

Abstract — A 2-MHz delta-sigma fractional-N frequency synthesizer based on a staggered switching fractional frequency divider is presented in this paper. The phase generator based fractional frequency divider provides lower instantaneous phase error and hence lowers the delta-sigma quantization noise, so that the synthesizer loop bandwidth can be increased. To suppress fractional spurs due to phase generator phase errors, a digital spurious tone suppression technique is adopted. The frequency synthesizer is implemented in 0.18-μm CMOS process, and it operates at 2.1-GHz carrier frequency with 2-MHz bandwidth. Excluding the output buffer, the synthesizer consumes 33.9-mA and is capable of transmitting 4-Mb/s GFSK signal.

Index Terms — Delta-sigma modulator, fractional spur, frequency synthesizer, Gaussian frequency-shift keying, phase-locked loop.

I. INTRODUCTION

In spite of sophisticated phase noise cancellation techniques, delta-sigma (Δ-Σ) fractional-N phase locked loops (PLLs) are forced to employ low bandwidth (no greater than 1-MHz) in order to suppress Δ-Σ quantization noise [1]-[3]. Using a fractional frequency divider i.e., one that counts ($N+k/M$) voltage controlled oscillator (VCO) cycles each reference period, where N, k, M are integers and $0 \leq k < M$, can lower the quantization noise by $20 \times \log_{10} M$-dB [4]. However, fractional frequency dividers need multiple phases of the PLL output restricting them to noisy ring VCOs or LC VCOs with small M. Furthermore, mismatches between the M phases cause strong spurious tones (*fractional spurs*) at PLL output.

This paper presents (1) a *staggered switching fractional frequency divider* (SSFFD) with a 6-bit fractional resolution ($M=64$) for use with an LC VCO, and (2) a *digital spur suppression technique* to suppress fractional spurs caused by phase mismatches. Together, they enable a 2-MHz bandwidth Δ-Σ fractional-N PLL. The large loop bandwidth allows 4-Mb/s in-loop GFSK modulation, suppresses VCO noise, and reduces the susceptibility of the VCO to pulling. Unlike phase noise cancellation techniques [2, 3], elaborate gain calibration techniques are not required. Furthermore, unlike conventional fractional-N PLLs and Type II PLLs, instantaneous phase errors seen by the phase frequency detector (PFD) are small, resulting in reduced sensitivity to charge pump and PFD non-

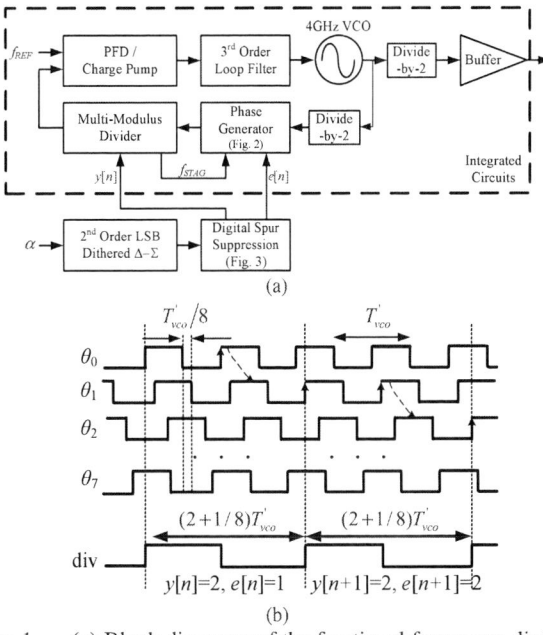

Fig. 1.   (a) Block diagrams of the fractional frequency divider based fractional-N synthesizer, and (b) the operation of fractional frequency divider.

linearity. Unlike [4] the fractional frequency divider is not restricted to ring oscillator based VCOs.

The paper is organized as follows. Section II overviews the fractional frequency divider based synthesizer, and Section III describes the circuit design. Section IV presents the measured synthesizer performance, and conclusion is given in Section V.

II. OVERVIEW OF THE SYNTHESIZER ARCHITECTURE

Fig. 1(a) shows a block diagram of the overall SSFFD synthesizer. The synthesizer is very similar to a conventional Δ-Σ fractional-N synthesizer, except that a phase generator (PG) is inserted before the multi-modulus frequency divider (MMFD). The digital spur suppression technique which uses a 1st order Δ-Σ modulator is also included not only to select the proper divider modulus and the phase for frequency division, but also to suppress the fractional spurs resulting from unevenly spaced phase generator phases. As mentioned, the benefit of using an

978-1-4244-6240-7/10 $26.00 © 2010 IEEE

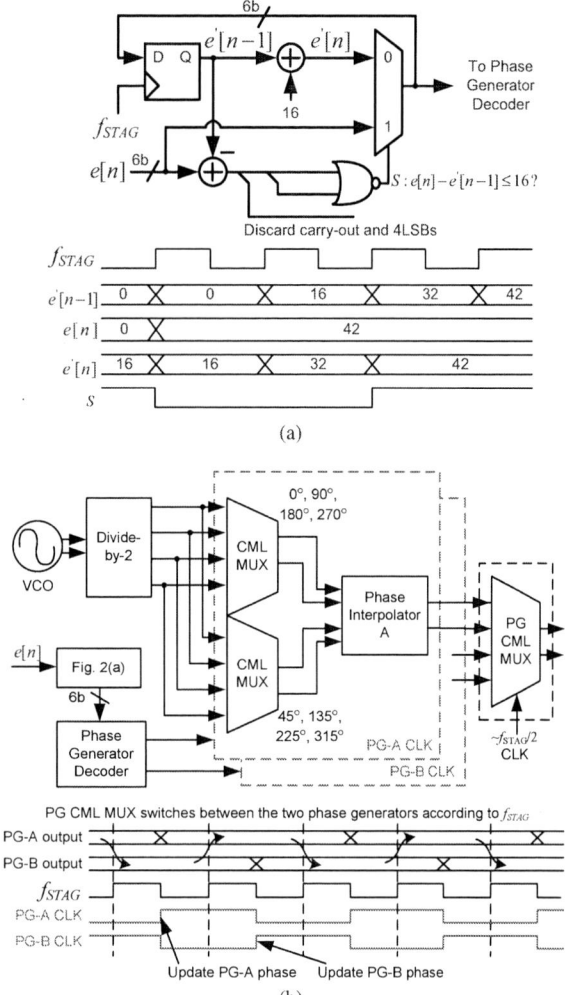

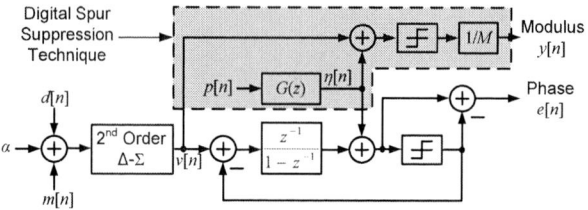

Fig. 3. Control circuitry block diagram for the staggered switching fractional frequency divider including the digital spur suppression technique.

Fig. 2. Detail block diagram of the phase generator block. (a) The control logic for the phase generator, and (b) phase generator architecture and its operation.

M-phase fractional frequency divider is that the phase resolution is enhanced by M, so that the Δ-Σ modulator quantization error is reduced by $20\times\log_{10}M$-dB.

The operation of the fractional frequency divider is illustrated in Fig. 1(b). As an example, assume 8 phases are available from phase generator, θ_0 is the current phase being used for frequency division, and 2 and 1/8 are the integer and fractional modulus, respectively. In this case, the MMFD initially counts the rising edge of θ_0, and during counting, the phase generator output is smoothly switched from θ_0 to θ_1 without glitches. As a result, the total time elapsed when the MMFD finishes counting two rising edges is $(2+1/8)T'_{\text{VCO}}$, where T'_{VCO} is half of the VCO period.

Circuit implementation details and glitch free switching between the phases are described in the following section.

III. CIRCUIT DESIGN

All circuit blocks in the dashed box in Fig. 1 are integrated, whereas the dithered 2nd order Δ-Σ modulator and the digital spur suppression blocks are implemented in software for testing flexibility. The following subsections give the design detail of the SSFFD and the digital spur suppression technique.

A. Staggered Switching Fractional Frequency Divider

The SSFFD in Fig. 1 is composed of the divide-by-2 circuit, the phase generator, and the MMFD. Fig. 2 shows the detail block diagram of the phase generator control logic and phase generator itself. An LC VCO and current mode logic (CML) divider generate the four quadrant phases, 0°, 90°, 180°, and 270°. Two 6-bit phase generator circuits, each of which can generate any one of 64 phases from these quadrant phases, operate in ping-ping fashion. While the output of one phase generator, θ_A, is connected to a conventional MMFD, the other, θ_B, is configured to be used next. Each phase generator uses CML multiplexers and a differential phase interpolator [5]. To count $(N+k/M)$ VCO cycles during the nth reference period, the SSFFD switches the MMFD input gradually in at most 4 steps, none of which advances the phase by more than a quadrant i.e. 90° from its current phase by using the control logic in Fig. 2(a).

For example, suppose $\theta_A = 45°$ is currently connected to the MMFD at the beginning of the nth reference period, and $\{N, k, M\} = \{42, 56, 64\}$ corresponding to 42.875 VCO cycles, which means that the phase should advance to 0° at the end of the reference period. Then, during the nth reference period, the MMFD input is gradually switched forward in the following manner:

$$\theta_A=45° \Rightarrow \theta_B=135° \Rightarrow \theta_A=225° \Rightarrow \theta_B=315° \Rightarrow \theta_A=0°.$$

Note that when moving the phase backwards, the MMFD integer modulus is decremented accordingly. The phase switching is clocked by f_{STAG}, a clock signal whose frequency is roughly $5\times f_{REF}$ so as to accommodate the longest phase jumps, where f_{REF} is the reference frequency. The f_{STAG} signal is generated by the MMFD, but details are omitted for the sake of brevity. The SSFFD relaxes phase interpolator setup time requirements and allows glitch free switching because the slew-rate is small enough to guarantee overlap of phases that are a quadrant apart. A glitch free switching circuitry [6] would otherwise be needed and would consume more than 20-mA at 2-GHz carrier in 0.18-μm CMOS technology.

B. Digital Spur Suppression Technique

Note that when an SSFFD is used in a Δ-Σ fractional-N PLL, the desired division ratio changes potentially every reference period as $(N+v[n]/M)$ where $v[n]$ is an integer generated by a Δ-Σ modulator and $0 \leq v[n] < M$ [4]. Even then, inevitable phase mismatch errors among the M phases cause strong spurs as seen from the measurement results. Note that this behavior is similar to the effect of non-linearity caused by divider modulus dependent delays in a conventional Δ-Σ fractional-N PLL.

The *digital spur suppression technique* uses a 1st order Δ-Σ modulator with *specially filtered dither* as shown in Fig. 3 to reduce these tones. The quantization error $e[n]$

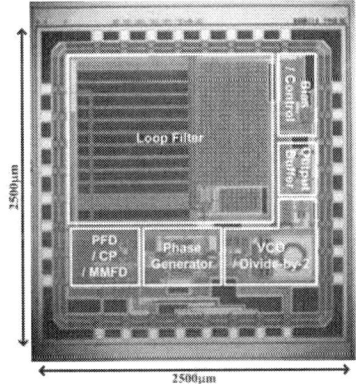

Fig. 4.   Die micrograph.

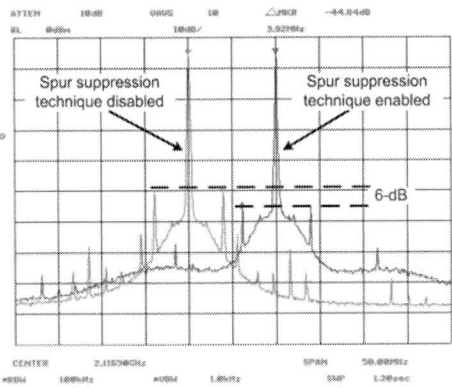

Fig. 5.   Overlaid synthesizer output spectrum showing 6-dB spur reduction at 2.1115-GHz with digital spur suppression technique enabled and disabled.

decides which VCO phase, $\theta_{e[n]}$, the SSFFD should switch to at the end of the reference period. The modulus $y[n]$ is added to the integer part of the division ratio. Note that a one-bit pseudorandom sequence $p[n]$ is filtered by a digital FIR filter, $G(z)$, and added to the quantizer input. In the absence of $p[n]$, the selected phases, $\theta_{e[n]}$, are strongly correlated with $v[n]$ and hence phase errors cause strong fractional spurs. It has been theoretically proven that if $1,2,4,...,M/2$, where M is the number of PG phases and is a power of 2, are the magnitudes of some coefficients of $G(z)$, then random phase mismatch errors do not cause any spurious tones [7]. By choosing an appropriate $G(z)$ the phase noise contribution of $p[n]$ is high-pass shaped and subsequently removed by the PLL's loop filter. Here, $G(z) = 1 - z^{-1} - 2z^{-2} + 4z^{-3} - 2z^{-4}$ is chosen even though $M = 64$, as a compromise between the phase noise contribution of $p[n]$ and residual fractional spur strength.

IV. MEASUREMENT RESULTS

This work was fabricated in TSMC 1P6M 0.18-μm CMOS process with thick top metal layer. The die size is 2.5-mm × 2.5-mm including ESD protected pads. The die micrograph is shown in Fig. 4. It is packaged with TQFN-32 and tested under 1.8-V supply voltage.

Fig. 5 shows measured spectra for the PLL operating as a local oscillator with the digital spur suppression technique enabled and disabled. The bandwidth of the PLL is set to 2-MHz. The center frequencies have been offset to clearly show the spur suppression. It can be seen that the strongest spur is reduced by 6-dB (from −44.8-dBc to −50.8-dBc) with the technique. Fig. 6 shows the measured PLL phase noise and simulated phase noise contributions. As curve (a) shows, the conventional fractional-N PLL with 2-MHz bandwidth has a very high phase noise. Enabling the SSFFD lowers the quantization noise (curve (b)). Further enabling the spur suppression technique reduces the spur by 6-dB while increasing the 10-MHz phase noise by 4.4-dB (curve (c)). The charge pump is the dominant noise source at low frequencies and can be reduced to achieve better noise performance. Note that its contribution increases a little when the suppression technique is on because the technique increases the turn on time of the charge pump by a little. Fig. 7 shows the eye pattern when transmitting a 4-Mb/s GFSK modulation and the measured rms error is 12.4%.

Fig. 8 shows the spur reduction and residual maximum spur versus different fractional channels, α. Only α=0~0.5 is reported in the figure for the spur performance plot is symmetric around α=0.5. This is because for α=α_0, the phases are exercised in the reverse order of the phase exercised when α=1−α_0. As shown in Fig. 8(b), using a

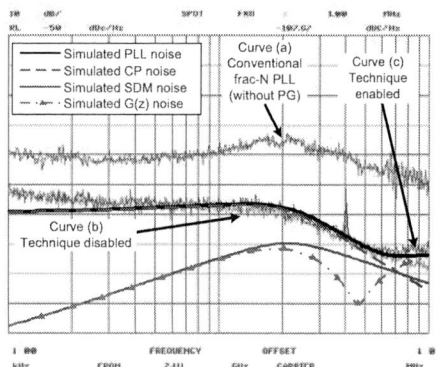

Fig. 6.   PLL's output phase noise, simulation and measured.

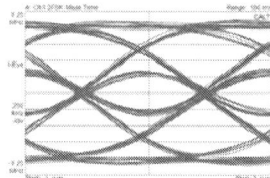

Fig. 7.   4-Mb/s GFSK modulation with 1-MHz deviation.

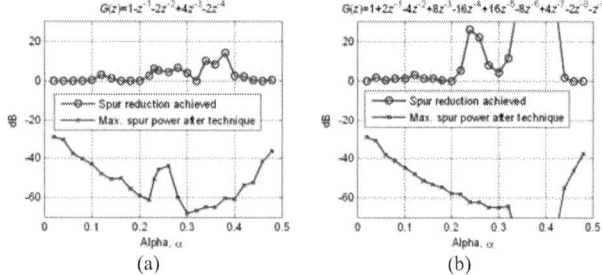

Fig. 8.   Measured spur reduction and maximum spur level after reduction versus different frequency, α, with (a) a small $G(z)$ filter, and (b) a stronger $G(z)$ filter

$G(z)$ with larger coefficients (not yet up to $M/2$) can significantly reduced the spur power. Note that no residual spur is observed for α=0.34~0.42.

For α around 0 and 0.5, although there is no reduction to the strongest spur, reduction is still observed for other spurs. It is suspected that these unsuppressed spurs are caused by nonlinearities other than phase generator errors, such as PFD/CP nonlinearity and multi-modulus divider dependent delay, because they show up at the same level when the synthesizer works as a normal fractional-N PLL.

V. CONCLUSION

A Δ-Σ fractional-N frequency synthesizer that utilizes a staggered switching fractional frequency divider and a digital spur suppression technique is presented. Δ-Σ quantization noise is lowered by SSFFD, enabling 2-MHz bandwidth and 4-Mb/s GFSK modulation. Spur reduction is also observed with the spur suppression technique.

TABLE I
CURRENT BREAKDOWN AND PERFORMANCE SUMMARY

Technology	TSMC 0.18-μm 1P6M CMOS	
Package	5-mm × 5-mm, 32 pin TQFN	
Power consumption	61-mW (excluding output buffer)	
Die area	2.5-mm × 2.5-mm	
Supply voltage	1.8-V	
Current Consumption, by Block		
Digital	1.2-mA	
PLL and bias	14.9-mA	Core circuits: 33.9-mA
VCO	7.8-mA	
VCO buffer and Divide-by-2	4.8-mA	
Phase generation	5.2-mA	
Output buffer	12.0-mA	
Frequency and Bandwidth		
Frequency range	2.1-GHz ± 0.15-GHz	
Reference	50-MHz	
Loop bandwidth	2-MHz	
Measured Phase Noise Performance		
	Technique disabled	Technique enabled
Integrated phase noise (10-kHz to 100-MHz)	1.07°	1.12°
Phase noise @ 1-MHz	-107.8-dBc/Hz	-106.2-dBc/Hz
Phase noise @ 3-MHz	-114.3-dBc/Hz	-112.8-dBc/Hz
Phase noise @ 10-MHz	-126.7-dBc/Hz	-122.3-dBc/Hz

ACKNOWLEDGEMENT

This work was supported in part by the National Science Foundation Award ECCS 0824279.

REFERENCES

[1] S. Pamarti, L. Jansson, and I. Galton, "A wideband 2.4-GHz delta-sigma fractional-*N* PLL with 1-Mb/s in-loop modulation," *IEEE J. Solid-State Circuits*, vol. 39, no. 1, pp. 49-62, Jan. 2004.

[2] A. Swaminathan, K. J. Wang, and I. Galton, "A wide-bandwidth 2.4 GHz ISM band fractional-N PLL with adaptive phase noise cancellation", *IEEE J. Solid-State Circuits*, vol. 42, no. 12, pp. 2639-2650, Dec. 2007.

[3] M. Gupta and B. S. Song, "A 1.8GHz spur cancelled fractional-N frequency synthesizer with LMS based DAC gain calibration," *ISSCC Dig. Tech. Papers*, pp. 478-479, Feb. 2006.

[4] C.-H. Heng and B.-S. Song, "A 1.8-GHz CMOS fractional-*N* frequency synthesizer with randomized multiphase VCO," *IEEE J. Solid-State Circuits*, vol. 38, no. 6, pp. 848-854, Jun. 2003.

[5] S. Sidiropolous and M. Horowitz, "A semi-digital DLL with unlimited phase shift capability and 0.08-400 MHz operating range," *ISSCC Dig. Tech. Papers*, pp. 332-333, Feb. 1997.

[6] N. Krishnapura and P. Kinget, "A 5.3-GHz programmable divider for HiPerLAN in 0.25-μm CMOS," *IEEE J. Solid-State Circuits*, vol. 35, no. 7, pp. 1019-1024, Jul. 2000.

[7] S. Pamarti and S. Delshadpour, "A spur elimination technique for phase interpolation-based fractional-N PLLs," *IEEE Trans. Circuits Syst. I, Reg. Papers*, vol. 55, no. 6, pp. 1639-1647, Jul. 2008.

978-1-4244-6240-7/10 $26.00 © 2010 IEEE

RTU1D-3

A 6fJ/step, 5.5ps Time-to-Digital Converter for a Digital PLL in 40nm Digital LP CMOS

J. Borremans, K. Vengattarmane[1], J. Craninckx

IMEC, Leuven, Belgium
[1] also KUL, Leuven, Belgium

Abstract— **A compact (0.01mm²) coarse-fine time-to-digital converter (TDC) in 40nm LP CMOS achieves 5.5ps resolution using parallel delay lines. A 6fJ/conversion step efficiency is achieved thanks to efficient residue calculation. A 0.8LSB single-shot precision and low DNL are reached thanks to simple calibration which is possible in fractional-N PLLs. Further, metastability avoidance and digital error correction are implemented. This 14-bit architecture operates at a 40MS/s reference clock.**

I. INTRODUCTION

Time resolution scales favorably with CMOS, making Time-to-Digital Converters (TDCs) [5] emerge in disruptive system architectures. TDCs adopt functionality beyond time-of-flight applications, in ADCs [9], PLLs [4], or in mismatch calibration for PAs [6] and fractional dividers [7] to name just a few. In a digital PLL (DPLL), the analog filter is replaced by a digital one, and a TDC is required to sample the phase offset, and thus to replace the phase-frequency detector.

In a DPLL, a TDC with fine resolution (<10ps) accompanied by high dynamic range is required. These requirements favor a coarse-fine architecture, well-known in the ADC world. Unfortunately, the input signal in this case is time, which cannot be stored unlike the voltages in ADCs can be. On top, to achieve sub gate delay resolution (<10ps), typically power-hungry techniques are required. As a result, the design is severely complicated, typically leading to inefficient architectures. As a consequence, TDC based PLLs typically consume significantly more power.

This work presents an area-efficient (0.01mm²), energy-efficient (6fJ/step, 15pJ per-shot or <1mW at 40MS/s) 5.5ps coarse-fine TDC with 14bit dynamic range in 40nm LP CMOS. Only the relevant residue is quantized. A sub-gate delay precision is achieved by using parallel interleaved delay lines. Such TDC offers no power penalty when migrating to a digital rather than an analog PLL architecture.

II. TDC ARCHITECTURE

Time-to-digital converters, akin to voltage converters, have been proposed using different architectures to achieve sub-gate delay resolution. Stochastic- [12], pulse shrinking- [10], passive-interpolation [2], multipath oscillator [3], parallel delay line [8], Vernier line [11] and time amplification-based [1] TDCs are all techniques to achieve sub-gate delay resolution, each with their specific advantages and issues. The main challenge is to create edges on a fine grid (5ps or 200GHz equivalently), and to be able to measure beyond 10ns, the reference period of a PLL. Creating all these closely spaced edges in a thermometer code fashion is very inefficient. From the voltage converter analogy (this would correspond to a >11bit flash ADC), it is well-known that this can be better handled with coarse-fine architectures.

In coarse-fine voltage converters, a coarse estimation of the input is made by a first converter stage which also creates a residue voltage. This residue is stored, often amplified and passed on to the next converter stage. Time converters face the problem that time residues can be created, but not stored. As such, coarse-fine converters typically create all possible residues [1], which is undoing the power benefit of such architecture. Fig. 1 illustrates the issue on a general example. A coarse converter counts on a coarse grid. When the stop edge occurs, the fine converter needs to count the time since the last coarse edge (which occurred in the past) until the *stop* edge, T_F. The fine converter however has difficulties going back in time.

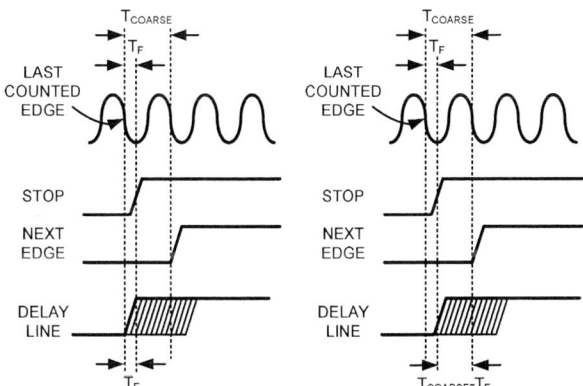

Fig. 1 Coarse-fine TDC operation illustrated. The residue is the time since the last edge to the stop signal (left), or equivalently, the time from the stop edge until the next consecutive falling edge (right).

The issue can be circumvented by sampling *the complement of the residue* T_{COARSE}-T_F, which is the time since the *stop* edge to the *next coarse edge*. Since the coarse period T_{COARSE} is known, T_F is found. Practically, T_F is found by simple inversion of the fine bits. Such TDC has been used in a time-based ADC in [9]. A flip flop finds the next coarse edge. However, this solution will be hard to realize at the targeted 5ps resolution, which involves high speed operation. Metastability in the edge selection – when *stop* occurs almost at the same time as a coarse oscillator edge – will dramatically affect the effective resolution. Two techniques are presented in this work to make the architecture feasible.

Our proposed TDC is depicted in Fig. 2. A *slow* (6.5GHz) gated ring oscillator based coarse TDC (Fig. 2) produces a coarse input estimation (coarse code) while offering low area

978-1-4244-6240-7/10 $26.00 © 2010 IEEE   375

and power consumption. A high-speed counter does so by counting the number of periods of this ring oscillator. This coarse system consumes less than 100µW per 1ns of time input, targeting a jitter below an LSB for the full converter. Since edges need only to be counted and generated on a coarse grid (<<200GHz equivalently), power consumption is low. When the *stop* edge occurs, the counter stops counting, but the oscillator proceeds until the fine TDC is ready too.

fine codes in this measurement is simply the coarse-fine gain.

In simulation, the low-resolution coarse TDC consumes only 100µW per ns of time input, and the high resolution FTDC (which generates all local edges on a fine grid) consumes 300µW once. Digitizing a 5ns input thus consumes 800µW. If the TDC were not to operate as a coarse/fine architecture, this would be approximately 6mW.

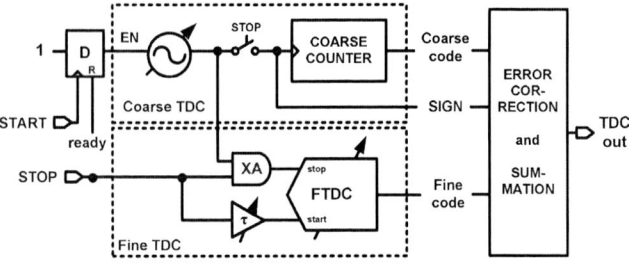

Fig. 2   Coarse-fine TDC simplified architecture.

A fine TDC (FTDC) then digitizes the complement of the residue (Fig. 1), which is the time to the next consecutive rising edge of the coarse oscillator. The *stop* edge starts the FTDC, and the next coarse edge stops the FTDC.

Sub-inverter delay resolution is achieved in the FTDC by using four parallel delay lines (Fig. 3). Buffers rather than inverters are used to overcome rise/fall mismatch, yielding a resolution of $2T_{INV}/4 \approx 5$ps. A Wallace counter adds output words B_1-B_4 to provide the fine code. Time offset between the lines is established by MOSCAPs C_1-C_5, designed to provide linearly spaced delay offset. A fifth dummy delay line serves the purpose of calibration. Ensuring $B_5=B_1$-1 by controlling C_1-C_4, the edges in the delay lines can be equally spaced. Fig. 4 illustrates the edges through the five lines in time, when changing the capacitance C_1-C_5. When the fifth line edge arrives exactly one buffer delay $T_{BUFF}=2T_{INV}$ after the first line edge, the parallel lines are well calibrated.

Proper calibration can be achieved by sending a set of random inputs to the fine TDC, and monitoring the average difference between B_5 and B_1. In a fractional PLL, this occurs automatically, since the input phase is naturally scrambled by the $\Delta\Sigma$-modulator [13]. This technique also allows determining the coarse-fine gain. The number of appearing

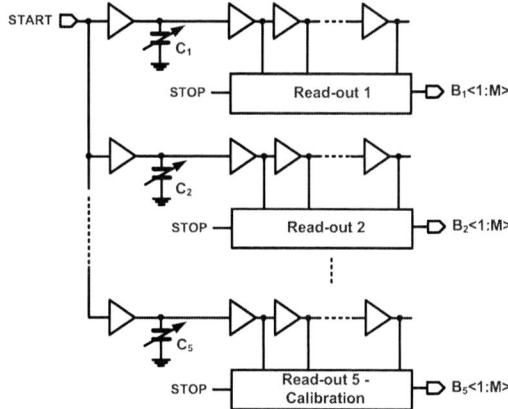

Fig. 3   Fine TDC architecture using parallel delay lines for sub-gate delay resolution.

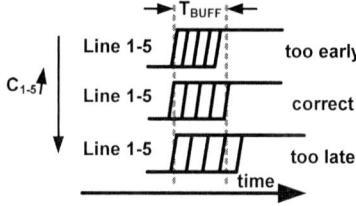

Fig. 4   Fine calibration illustration. Changing the digital value of C1-5 changes the time offset between the parallel delay lines.

III. METASTABILITY AVOIDANCE AND ERROR CORRECTION

For the FTDC to measure the residue, the *next coarse edge and only this edge* needs to be selected, since the STOP edge has occurred. Unfortunately, at a clock speed of 150ps and resolution of 5ps, such edge selector struggles with

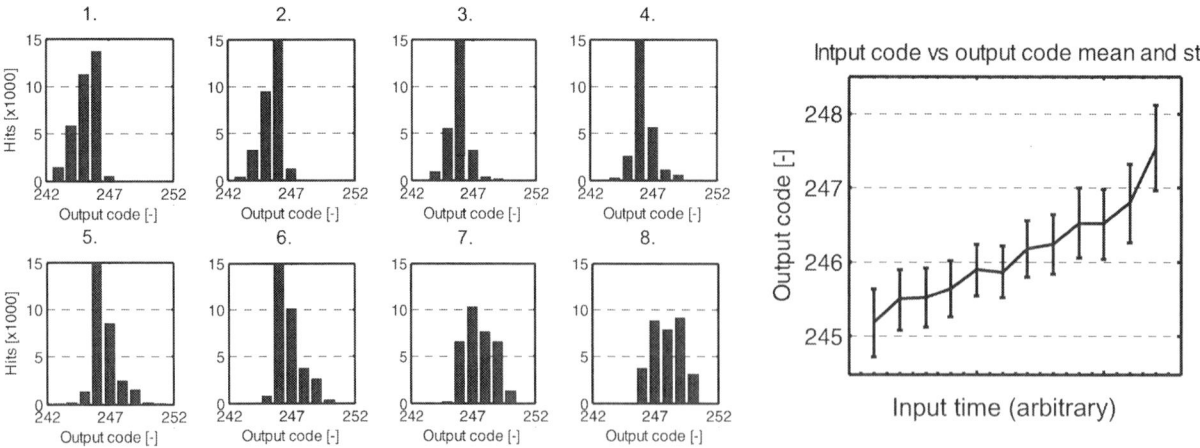

Fig. 5   TDC histogram and single-shot precision for different consecutive input codes (note: input time is arbitrary due to delay in the input testing cables).

metastability when the *STOP* edge occurs almost at the same time as a coarse oscillator edge. Therefore, the coarse code can be wrong, which results in an error of many (fine TDC) LSBs. Metastability cannot be removed, but is alleviated as follows. An AND gate XA in Fig. 2 passes all upcoming coarse periods since *stop* occurred to the FTDC. Only the first period can be short or incomplete yielding a metastable FTDC. To make the FTDC ignore this period, the *FTDC start* edge is delayed (programmable τ in Fig. 2) before being fed into the FTDC.

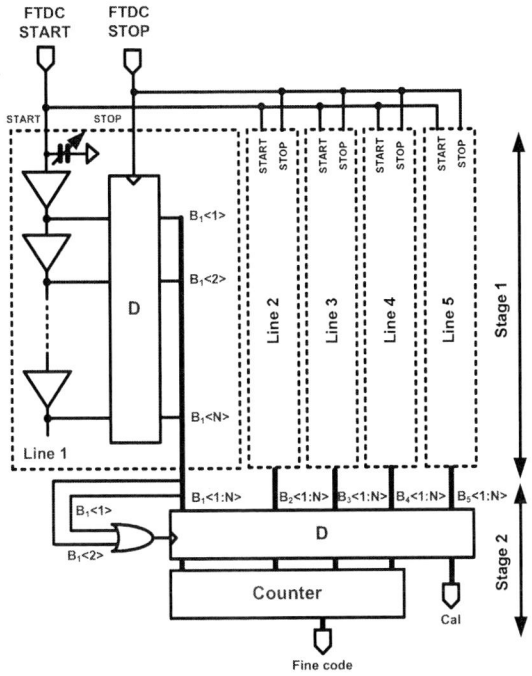

Fig. 6   Fine TDC two-stage readout circuitry to avoid metastability.

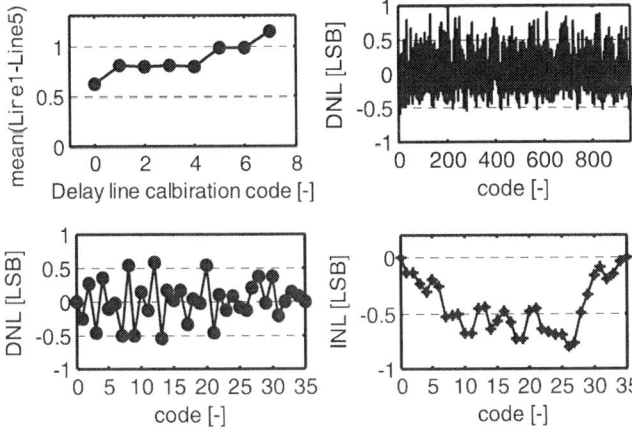

Fig. 7   Fine TDC calibration (top left), fine TDC DNL and INL after calibration (bottom) and DNL of first 10 bit of the full TDC (top right).

Since all periods of the coarse counter since the *stop* edge are fed into the FTDC, a two-stage read-out circuit is used which only latches once (Fig. 6). A first latch operates on the FTDC stop signal. Next, when one of the first two outputs goes high, the second stage latches all FTDC outputs, just once.

A final issue rests in the coarse and fine TDC being independent (unlike in voltage converters where the residue is a product of the coarse converter). Hence, errors can be made. Indeed, while an edge may not yet be detected by the coarse TDC's counter, it may already serve in the residue calculation in the fine TDC. This time offset creates code jumps which can be corrected in the digital domain. To avoid this, we store the last value of the level fed into the counter (which we refer to as the sign bit 'S'). A high value indicates the rising edge is counted less than $LSB_{COARSE}/2$ ago. A zero indicates it's been longer. In the latter case, when the fine TDC outputs a low value (which is not allowed), the code is corrected by a coarse LSB. The estimated power consumption of this simple digital algorithms is a few 10s of μW, including coarse and fine scaling and summation (TDC_{OUT}=fine +coarse*gain).

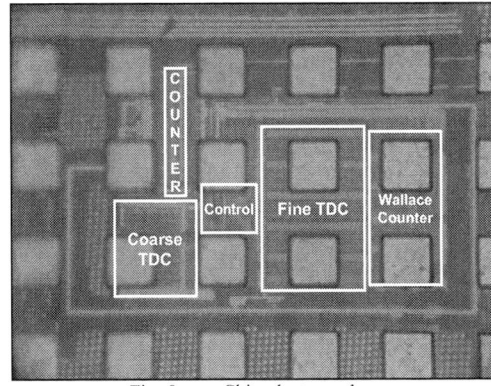

Fig. 8   Chip photograph.

IV. CHIP DESIGN AND MEASUREMENT

The 14-bit circuit has been implemented in 40nm LP digital CMOS, and occupies just 0.01mm^2 (Fig. 8).

First, the fine TDC has been calibrated. Fig. 7 shows the measured average difference between B_5 and B_1 vs. the calibration word. The difference equals 1 for code 5 or 6. The fine gain has been measured to be 36. It also shows the decent INL/DNL of the calibrated fine TDC. Also shown is the DNL of the first 10 bits (limit of measurement setup) of the full TDC, showing proper static operation over the full range. A resolution of 5.5ps has been achieved at 1.1V, and 5ps at 1.3V. Fig. 5 shows the measured histogram over a range of TDC codes. The average single-shot precision is ~0.8LSB (Fig. 5).

To measure dynamic behavior, the supply of an external input buffer has been modulated as in [3], to provide a time sine input. Fig. 9 depicts the spectrum measurement of a 110ps$_{pp}$ input. Harmonic spurs originate from the nonlinear transfer from supply to delay variation of the external buffer. The noise floor of an ideal 5.5ps 40MS/s quantizer with 0.8LSB white noise (single-shot precision) is indicated. This plot confirms the static measurements of Fig. 5. In the full 20MHz bandwidth, the jitter is less than 5ps, leading to a dynamic range of 77dB (or 12.8ENOB) calculated as in [3]. The FOM defined as P/(2^{ENOB}2BW) is only 6fJ/conversion step. This very low value is thanks to coarse-fine approach that only quantizes the relevant residue.

978-1-4244-6240-7/10 $26.00 © 2010 IEEE

TABLE I.   OVERVIEW OF PICOSECOND TDCs IN LITERATURE

	Technique	Tech.	Resolution [ps]	Single shot precision [LSB]	Nr. Bits	Power cons. [mW]	Sample freq [MS/s]	FOM [fJ/step]	Average energy per-shot [pJ]	Area [mm2]
This work	**Parallel lines**	**40nm**	**5.5**	**0.8**	**14**	**0.6-1.8**	**40**	**6**	**15-45**	**0.01**
Henzler [JSSC'08]	Passive interpolation	90nm	4.7	0.7	7	3.6	180	309	19	0.02
Lee [JSSC'08]	Time amplification	90nm	1.25	0.6	9	3	10	994	300	0.6
Straayer [JSSC'09]	GRO with feedforward	130nm	6 (1.2)	-	11	2.2-20	50	200	44-222	0.04
Yu [VLSI'09]	Ring vernier	130nm	8	1	12	7.5	15	345	500	0.26

The FOM has been calculated assuming white noise in all works except [3], and assuming that the reported power consumption was the one at full scale.

An $11ps_{pp}$ time sine has been applied and Fig. 10 depicts the noisy raw output code over time. In a PLL, the loop filter filters off high-frequency noise. Clearly, the TDC achieves a 5.5ps resolution. Fig. 10 illustrates the digitally filtered raw output code, with a filter bandwidth of 300kHz, typically present in a PLL application of the TDC.

The circuit consumes between 0.6-1.8mW at 1.1V at 40MHz rate, or an excellent 6fJ/conversion step. A state-of-the-art overview is indicated in Table 1. Our TDC has a fine resolution of 5.5ps, for the best FOM and lowest area consumption reported.

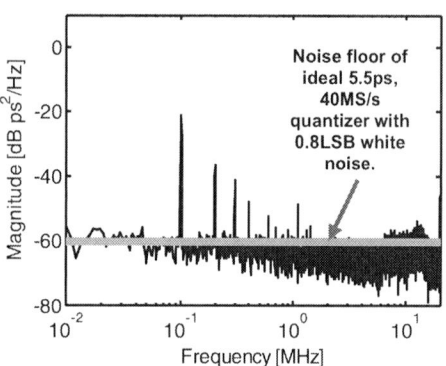

Fig. 9   Spectrum plot of the TDC with a sine input of about $110ps_{pp}$. Harmonic spurs are present due to the modulated nonlinear measurement setup input buffer.

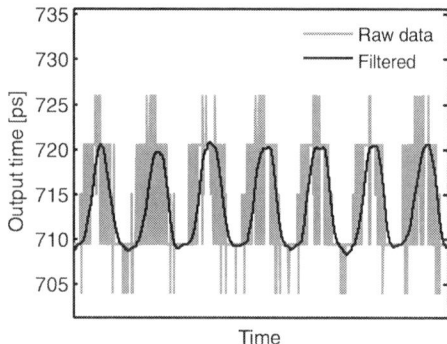

Fig. 10   Raw data and digitally filtered (BW=300kHz) output of the TDC, applying an $11ps_{pp}$ time sine input to the converter.

CONCLUSIONS

A sub-gate delay resolution coarse-fine TDC has been presented using a low-power coarse oscillator, and a fine TDC that samples the complement of the residue. Metastability avoidance techniques, digital error correction and calibration enable the 5.5ps resolution. As a result, an efficiency of 15pJ-per-Shot is achieved, or 6fJ/conversion step on an area of just $0.01mm^2$.

REFERENCES

[1]   M. Lee and A. Abidi, "A 9b, 1.25ps resolution coarse-fine time-to-digital converter in 90nm CMOS that amplifies a time residue," *J. Solid-State Circuits*, vol. 43, no. 4, pp. 769–777, Apr. 2008.

[2]   S. Henzler, et al., "A Local Passive Time Interpolation Concept for Variation-Tolerant High-Resolution Time-to-Digital Conversion", *J. Solid-State Circuits*, vol. 43, no. 7, pp.1666-1676, July 2008.

[3]   M. Z. Straayer, M. H. Perrott, "A Multi-Path Gated Ring Oscillator TDC With First-Order Noise Shaping", J. Solid-State Circuits, Vol. 44, No. 4, pp. 1089 – 1098, April 2009.

[4]   R. Staszewski, et al. ," All-digital PLL and GSM/EDGE transmitter in 90nm CMOS," ISSCC Dig. Tech. Papers, pp.316-317, Feb. 2005.

[5]   R. B. Staszewski, et al., "1.3 V 20 ps time-to-digital converter for frequency synthesis in 90-nm CMOS", Transactions on Circuits and Systems II, Vol. 53, No. 3, pp.220 – 224, March 2006.

[6]   M. E. Heidari, M. Lee; A. A. Abidi, "Digital Outphasing Modulator for a Software-Defined Transmitter", J. Solid-State Circuits, Vol. 44, No. 4, pp.1260 – 1271, April 2009.

[7]   S. Pellerano, P. Madoglio, Y. Palaskas, "A 4.75GHz fractional frequency divider with digital spur calibration in 45nm CMOS", ISSCC Dig. Tech. Paper, pp. 8-12, pp.226 – 227, Feb. 2009.

[8]   J.-P. Jansson, A. Mantyniemi, J. Kostamovaara, "A CMOS time-to-digital converter with better than 10 ps single-shot precision", J. Solid-State Circuits, Vol. 41, No. 6, pp. 1286-1296, June 2006.

[9]   S. Naraghi, M. Courcy, M. Flynn, "A 9b 14µW 0.06mm² PPM ADC in 90nm Digital CMOS", ISSCC Dig. Tech. Papers, pp. 168-169, Feb. 09.

[10]   P. Chen, S. I. Liu, and J.Wu, "A CMOS pulse-shrinking delay element for time interval measurement," IEEE Trans. Circuits Syst. II, Analog Digit. Signal Process., vol. 47, no. 9, pp. 954–958, Sep. 2000.

[11]   J. Yu, F. F. Foster, C. Richard, "A 12-bit vernier ring time-to-digital converter in 0.13µm CMOS technology", Proceedings of VLSI Symposium, pp. 232-233, June 2009.

[12]   V. Gutnik and A. Chandrakasan, "On-Chip Picosecond Time Measurement," IEEE Symp VLSI Circuits., pp. 52 – 53, June 2000.

[13]   J. Borremans, K. Vengattaramane, V. Giannini. J. Craninckx, "A 86MHz-to-12GHz Digital-Intensive Phase-Modulated Fractional-N PLL Using a 15pJ/Shot TDC in 40nm digital CMOS, ISSCC 2010.

RTU1D-4

A 6GHz Direct Digital Synthesizer MMIC with Nonlinear DAC and Wave Correction ROM

Danyu Wu, Gaopeng Chen, Jianwu Chen, Xinyu Liu, Lixin Zhao, Zhi Jin

Institute of Microelectronics, Chinese Academy of Sciences, Beijing, 100029, China

Abstract — **This paper proposes a new DDS architecture combined with Nonlinear DAC and Wave-Correction-ROM (WCR) which shows both high operating speed and accuracy. Based on this architecture, a 6GHz 8-bit DDS MMIC is designed and fabricated in 60GHz GaAs HBT Technology. The DDS MMIC includes 8-bit pipeline accumulator, an 8×8×3bits WCR, two combined DACs and an analog Gilbert Cell for sine-wave generation with 8-bit amplitude resolution. The DDS chip is tested in on-wafer measurement system. The measured spurious free dynamic range (SFDR) is 33.96dBc with 2.367GHz output under a 6GHz maximum clock (FCW=0x65). It shows an average SFDR of 37.5dBc and the worst case SFDR of 31.4dBc (FCW=0x70) within the whole Nyquist band under a 5GHz clock frequency. The whole chip occupies 2.4×2mm² of area consuming 3.27W of power from a single -4.6Vpower supply.**

Index Terms — **direct digital synthesizer (DDS), sine-weighted DAC, wave correction ROM (WCR), emitter coupled logic (ECL), digital-to-analog converter (DAC).**

I. INTRODUCTION

In next generation of radar system, a signal source with faster frequency agility, higher frequency resolution, phase and frequency modulation capability and larger bandwidth has been needed. These requirements are surpassing the performance capabilities of conventional analog phase-locked loops (PLL). In contrast, Direct Digital Synthesizer (DDS) is capable of ultra fast frequency hopping typically within several clock cycles. The recently improvement of semiconductor process enables DDS chips operate at higher frequency such as mm-wave for direct linear frequency modulation (LFM) signal generation.

Recent works show that DDS chips which are implemented in InP or SiGe technologies can operate from 5 to 32GHz frequencies [1]-[5]. The design reported in [1] which can operate at 32GHz is implemented in InP double heterojuction bipolar transistor (DHBT) technology. It eliminates the ROM by using a Sine-Weighted DAC to increase the operating frequency and reduce the power consuming, but it is hard to achieve high worst case spurious free dynamic range (SFDR). To improve SFDR, Current Source Matrix is used to realize the Sine-Weighted DAC with better resolution [2]-[3]. But it calls for a larger area of DAC that also limits its SFDR performance. The design reported in [4] which uses an analog triangle-to-sine wave convertor for sine wave generating achieves high speed and low power consumption but has relatively worse SFDR. A ROM-based DDS reported in [5] shows a better worst case SFDR (30.7dBc) but lower clock rate (24GHz) compared to [1] with similar technologies. In this paper, a DDS MMIC is designed and measured that reaches a maximum operating frequency of 6GHz. It also shows an average SFDR of 37.5dBc and the worst case SFDR of 31.4dBc (FCW=0x70) with output frequency range of 0~2.5GHz under a 5GHz clock.

II. PROPOSED DDS ARCHITECTURE

A DDS typically consists of a phase accumulator, a phase-to-amplitude converter and a digital-to-analog converter (DAC). The width of phase accumulator determines the frequency control resolution. Several LSB's of phase output of the accumulator are generally truncated for a realizable phase-to-amplitude conversion, but this causes generation of spurious signals. The finite word length effects and nonlinearity of DAC also limits its SFDR performance.

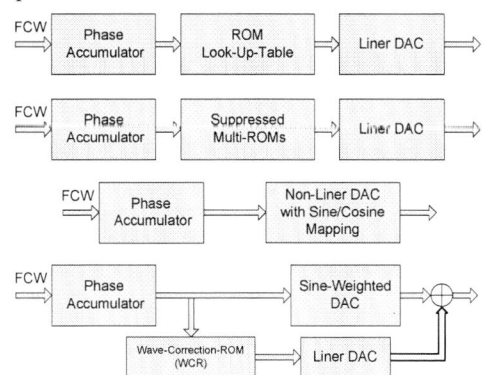

Fig. 1 Conceptual Diagram of different DDS architectures

The phase-to-amplitude converter is usually the slowest and most complex part. Many DDS architecture are proposed such as Sunderland Architecture, Nicholas architecture, CORDIC algorithm etc. to improve resolution and reduce the complexity. But these architectures have a relatively low operating speed. ROM-less architectures are developed in recent years to improve

978-1-4244-6240-7/10 $26.00 © 2010 IEEE   379

the maximum operating frequency of DDS. For best compromise of SFDR and frequency of traditional DDS architecture, this paper propose a new DDS structure combined with sine-weighted DAC and Liner DAC to improve SFDR and maintain high operating frequency simultaneously. Fig. 1 shows the conceptual diagrams of traditional DDS architecture, Multi-ROMs suppressed architecture, ROM-less architecture and proposed architecture.

III. CIRCUITS DESIGN AND IMPLEMENTATION

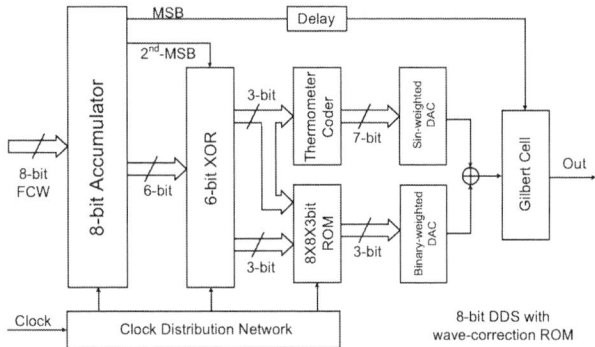

Fig. 2 Detailed Block Diagram of proposed DDS

Fig. 2 shows the detailed diagram of the DDS architecture of this work. It mainly consists of an 8-bit pipeline phase accumulator, a 6-bit XOR complementer, a 3-bit thermometer coder, an 8×8×3 bits WCR, DACs and a Gilbert cell. The MSB of the 8-bit phase accumulator result is used to reverse the output waveform by a Gilbert cell, while the 2nd-MSB is used for XOR complementing the rest 6-bit data. This allows Sin-Weighted DAC and Wave-Correction ROM to contain only a quarter of the sine waveform. An MSB delay circuit is adopted for compensating the delay of DACs. The higher 3-bit of the XOR output are thermometer coded to 7-bit to control the Sin-Weighted DAC, and all 6-bit of the XOR output are used to generate 3-bit Wave Correction data by addressing the WCR. The output of the Sin-Weighted DAC and Wave-Correction DAC is added to form half of the sine waveform. The whole circuit is pipelined to maximize the clock frequency.

For trading off speed, area and power, we split the 8-bit accumulator into 4 stages of pipeline that need 12 registers to make all bits of phase in alignment. All of the digital circuits in this chip are making use of Current-Mode-Logic (CML) and Emitter-Coupled-Logic (ECL). The maximum operating frequency of the accumulator is designed slightly higher than the other circuits which can be estimated by equation $f = 1/(\tau_{c-q} + 2 \times \tau_{carry} + \tau_{setup}) \approx 1/(4 \times \tau)$, where $\tau_{c-q} \approx \tau_{carry} \approx \tau_{setup} \approx \tau$ in ECL logic circuit.

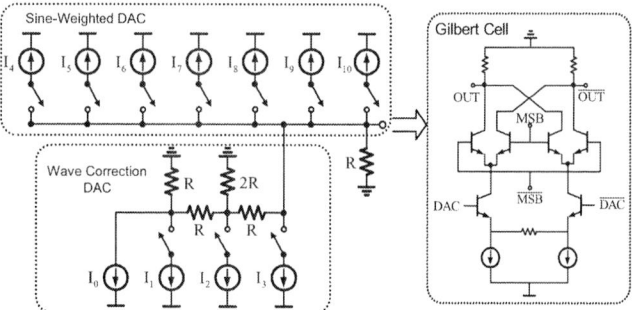

Fig. 3 Sine-Weighted DAC, Wave Correction DAC and Gilbert Cell circuits.

Fig. 3 shows the structure of the Sine-Weighted DAC, Wave Correction DAC and the Gilbert Cell. The Sine-Weighted DAC consists of 7 different sine weighted current sources which are controlled by the thermometer coded X(5:3) of the XOR gates output. The Wave Correction DAC is binary weighted using an R-2R-R Ladder, so the 3 current sources ($I_1 \sim I_3$) are exactly the same. An extra constant current source is employed to make the zero crossing point of output waveform smooth. We can obtain the *Vout* before Gilbert Cell by equation:

$$Vout = \frac{2}{3} \times (I_{10} + I_9 + I_8 + I_7 + I_6 + I_5 + I_4) \times R$$
$$+ (\frac{4}{6} \times I_3 + \frac{2}{6} \times I_2 + \frac{1}{6} \times I_1 + \frac{1}{6} \times I_0) \times R \quad (1)$$

Assuming that $I_1 = I_2 = I_3 = I$, then $I_4 \sim I_{10}$ can be calculated by equation:

$$I_k = \frac{\sin(\frac{k-3}{16}\pi) - \sin(\frac{k-4}{16}\pi)}{\sin(\frac{\pi}{16})} \times 2I \quad (2)$$

I_4 is the largest current source equaling only 3.43 times of I_{10} which is the smallest current source that consumes $0.583 \times I$ of current. This enables all current sources having almost the same switching delays which are one of the most important factors of SFDR performance.

After $I_0 \sim I_{10}$ are determined, data of the WCR can be calculated:

$$WCR(A \times 2^3 + B) = round(\frac{\sin(A \times 2^3 + B) - \sin(A \times 2^3)}{\sin(\frac{\pi}{16})} \times 2^3) \quad (3)$$

A and *B* are MSBs (row) and LSBs (column) of the address of the WCR respectively. The calculated data stored in the 8×8×3-bit WCR are shown in Fig. 4.

The row address X(5:3) and column address X(2:0) are the output of XOR gate that are shown in Fig. 2. The DDS

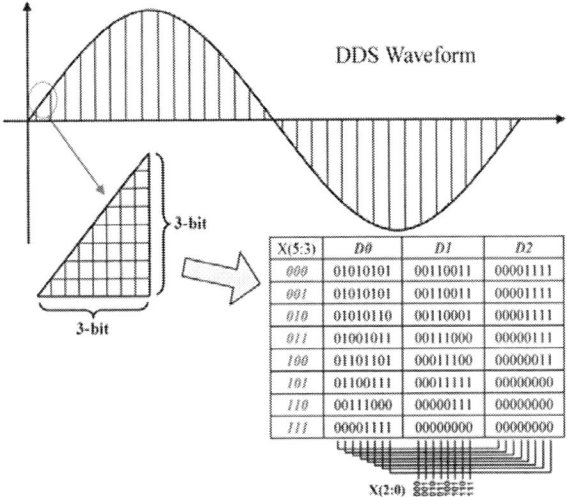

Fig. 4 DDS waveform and the data of Wave Correction ROM

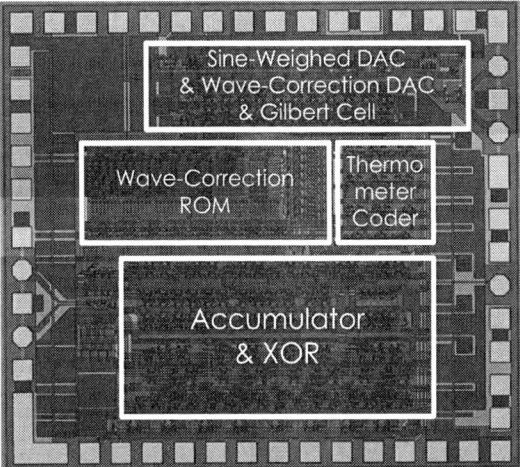

Fig. 5 Die photo of the DDS chip

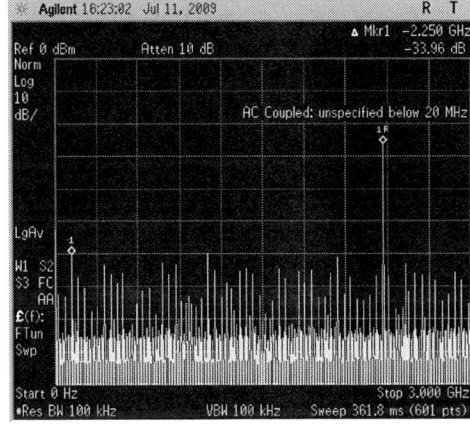

Fig. 6 Measured spectrum of 2.367GHz output frequency with SFDR of 33.96dBc at 6GHz clock (FCW=0x65)

waveform shown in Fig. 4 illuminates the operational principle of proposed DDS.

The WCR only stores the error part between an ideal Sine-Wave and Waveform generated by the Sin-Weighed DAC. So the WCR can be designed smaller to archive a higher operating frequency. The ROM data is truncated to 3-bit width thus making the equivalent DAC resolution being 7-bit which is enough for a DDS with 8-bit phase resolution. We find that data in some adjacent columns are the same and some columns are all zeroes, so we can merge the same-data adjacent columns and ignore the all-zeroes columns. Finally we realized an 8×16=128 bit WCR. In contrast with the traditional DDS architecture, this kind of architecture achieves a ROM compression ratio of 14:1.

The clock distribution circuits are carefully designed to minimize the clock skew of different blocks and prevent the potential instability. Because the delay of ROM is much bigger than thermometer coder, we introduce a fixed clock skew between the registers of ROM and thermometer coder by designing different delay circuits for them in order to maximize the bit rate of the ROM.

IV. MEASUREMENT RESULTS

The DDS MMIC has been fabricated in 1um GaAs HBT technology which has a *ft* of 60GHz. A microphotograph of the fabricated DDS chip is shown in Fig. 5. The DDS is tested on-wafer using two DC probes and two GSG single-ended microwave probes. The measurement results show that it can work stably at 6GHz clock frequency and can generate sine waves of 0~3GHz with frequency resolution of 23.44MHz. It performs better SFDR under 5GHz clock frequency generating sine wave in Nyquist band from 0~2.5GHz.

Fig. 6 shows the measured spectrum of DDS output with FCW=0x65 by Agilent E4440A spectrum analyzer. It shows the DDS has an output frequency of 2.367 GHz with SFDR of 33.96dBc. The chip is internally 50 /100 single/differential terminated, so it can be conveniently matched to external 50 system. Fig. 7 shows the measured SFDR at FCW from 1 to 128. The worst case SFDR is 31.4dBc at FCW=0x70 and the average SFDR over all frequency control words is 37.5dBc, which can be even better if using differential probe for measurement. Based on the measured results, we find that the odd harmonics produced by the Gilbert Cell and output buffer are the largest spurious, so nonlinearity of the analog circuits forms the bottleneck of SFDR performance. The chip has 1978 transistors in die area of 2.4×2.0 mm² and

TABLE I
EXPERIMENTAL PERFORMANCES OF THE PROPOSED DDS IC AND COMPARISON WITH RECENTLY REPORTED DESIGNS

Technology f_t [GHz]	InP HBT 300 [1]	SiGe BiCMOS 120 [2]	SiGe HBT 200 [3]	SiGe HBT 170 [4]	InP HBT 370 [5]	GaAs HBT 60 [This work]
DDS maximal clock frequency f_{clk} [GHz]	32	12	5	15	24	5
DDS phase resolution B [bit]	8	9	24	8	12	8
DDS amplitude resolution A [bit]	5	8	10	6	7.5	8
Worst case DDS $SFDR$ [dBc]	21.56	22	38	21	30.7	31.4
DDS die area [mm^2]	2.7×1.5	2.5×0.7	3.7×3.0	1.1×1.0	5.0×3.3	2.4×2.0
DDS total power consumption P [W]	9.45	1.9	4.7	0.366	19.8	3.27
FOM1=f_{clk}/f_t×100%	10.67%	10.0%	2.50%	8.82%	6.49%	8.33%
FOM2=f_{clk}×A×B×$SFDR/(P$×$f_t)$	9.73	83.4	48.5	243	9.05	51.2

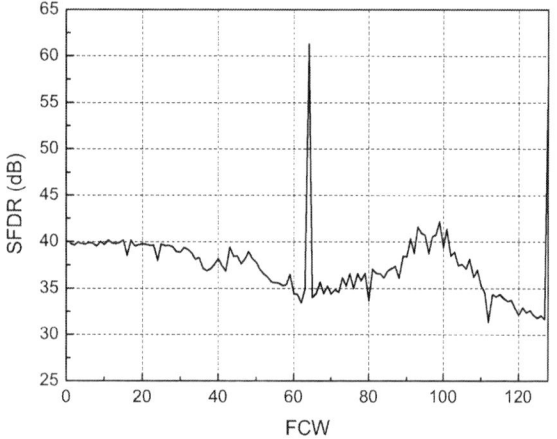

Fig. 7 Measured DDS output SFDR versus FCW at 5GHz clock frequency, average SFDR is 37.5dBc, worst case SFDR is 31.4dBc (FCW=0x70)

consumes 3.27W of power from a single power supply of -4.6V.

Table I summarizes the experimental performance of the proposed DDS along with five recently reported ultra-high-speed designs implemented in InP or SiGe technologies. Two figures of merit (FOM) are calculated for all of the 6 designs for comparison. The proposed DDS shows a good performance of FOM1 and FOM2. The pipeline accumulator of the DDS can be easily extended to 24-bit or 32-bit for actual application without compromising the maximum operating frequency. The amplitude resolution can also be improved by increasing the bit width of Wave-Correction-ROM and corresponding Linear DAC. Considering that fabricating a chip using 60GHz 1um GaAs HBT technology costs much lower than the ones in [1]-[5], the proposed DDS has strong competitiveness and good commercial prospects.

V. CONCLUSION

This paper proposes a new DDS architecture combined with Nonlinear DAC and Wave-Correction-ROM (WCR) to improve SFDR and maintain high operating frequency simultaneously. A 6GHz 8-bit DDS MMIC is designed and fabricated in 60GHz 1um GaAs HBT Technology based on this architecture. The measured SFDR is 33.96dBc with 2.367GHz output under a 6GHz clock. It shows an average SFDR of 37.5dBc and the worst case SFDR of 31.4dBc within the whole Nyquist band under a 5GHz clock.

REFERENCES

[1] S. E. Turner, D. E. Kotecki, "Direct Digital Synthesizer With Sine-Weighted DAC at 32-GHz Clock Frequency in InP DHBT Technology," *IEEE Journal of Solid-State Circuits*, vol. 41, no. 1, pp. 2284-2290, Oct. 2006.

[2] Y. Xuefeng, F. F. Dai, J. D. Irwin, and R. C. Jaeger, "A 12 GHz 1.9 W Direct Digital Synthesizer MMIC Implemented in 0.18um SiGe BiCMOS Technology," *IEEE Journal of Solid-State Circuits*, vol. 43, pp. 1384-1393, June 2008.

[3] G. Xueyang, F. F. Dai, J. D. Irwin, and R. C. Jaeger, "A 5 GHz direct digital synthesizer MMIC with direct modulation and spur randomization," *IEEE Radio Frequency Integrated Circuits Symposium*, pp. 419-422, June 2009.

[4] B. Laemmle, C. Wagner, H. Knapp, L. Maurer, and R. Weigel, "A 366mW direct digital synthesizer at 15GHz clock frequency in SiGe Bipolar technology," *IEEE Radio Frequency Integrated Circuits Symposium*, pp. 415-418, June 2009.

[5] S. E. Turner, R. T. Chan, and J. T. Feng, "ROM-Based Direct Digital Synthesizer at 24GHz Clock Freuqency in InP DHBT Technology," *IEEE Microwave and Wireless Components Letters*, vol. 18, no. 8, pp. 566-568, Aug. 2008.

RTU1D-5

A 10GHz 8-bit Direct Digital Synthesizer Implemented in GaAs HBT Technology

Gaopeng Chen, Danyu Wu, Zhi Jin, Jin Wu and Xinyu Liu

Institute of Microelectronics, Chinese Academy of Sciences, Beijing, 100029, China

Abstract — This paper presents a 10GHz 8-bit Direct Digital Synthesizer (DDS) Microwave Monolithic Integrated Circuit (MMIC) implemented in 1μm GaAs HBT technology. The DDS takes a Double-Edge-Trigger (DET) 8-stage pipeline accumulator with sine-weighted DAC based ROM-less architecture, that can maximize the utilization ratio of GaAs HBT's high-speed potential. With an output frequency up to 5GHz, the DDS gives an average Spurious Free Dynamic Range (SFDR) of 23.24dBc through the first Nyquist band, and consumes 2.4W of DC power from a single -4.6V DC supply. Using 1651 GaAs HBT transistors, the total area of the DDS chip is 2.4×2.0mm².

Index Terms — Accumulator, digital to analog converter (DAC), direct digital synthesizer (DDS), direct digital frequency synthesizer (DDFS), Double-Edge-Trigger, digital-to-analog converter (DAC), Gallium Arsenide (GaAs), heterojunction bipolar transistor (HBT), ROM-less DDS.

I. INTRODUCTION

Compared with other frequency synthesizing techniques such as VCO and PLL, DDS gives much higher frequency resolution, faster frequency channel switching with continuous phase, wider tuning range of frequency and more flexible versatile modulation capability. So that DDS can be widely used in various communication systems, radars and test-measurement instruments. However, the output frequency of DDS is generally much lower than its competitors due to its conventional ROM-based operating mechanism.

With the recent improvement of some advanced semiconductor technologies such as SiGe BiCMOS, InP HBT, several ultra-high-speed DDS integrated circuits with clock frequencies from 8 to 32GHz have been reported. The design reported in [1] uses a traditional DDS architecture with the ROM-based phase to amplitude conversion, while the DDS in [2] utilizes a ROM-less architecture with bipolar differential pairs in saturation to perform triangle to sine wave conversion in analog domain. As similar to the designs in [3]-[5], this proposed DDS employs a ROM-less structure with a sine-weighted DAC combining the sine mapping block and the digital to analog amplitude conversion block together. Besides using advanced technologies such as InP and SiGe, designing new circuit architectures is another way to improve the speed of DDS. Noting the speed of ROM-less DDS is often limited by the speed of the phase

accumulator [4], we developed a Double-Edge-Trigger (DET) technique for 8-stage pipeline accumulator that can finish two accumulating operations in a single clock cycle, thus significantly improved the speed of the DDS by 2 times. So that even implemented in the relatively backward GaAs HBT technology, this proposed DDS can achieve comparable speed with the InP or SiGe technologies based DDSs. It should be noted that the DET technique proposed here is not exclusive for GaAs HBT technology and can be transplanted conveniently to InP or SiGe based designs to achieve even higher performance.

Implemented in a 1μm GaAs HBT technology of which the unit short-circuit current gain frequency (f_t) is about 60GHz, the DDS IC presented here achieves a maximum clock frequency of more than 10GHz, showing the highest semiconductor technology's high-speed potential utilization ratio FOM of 16.67% compared to other reported designs.

II. DDS ARCHITECTURE

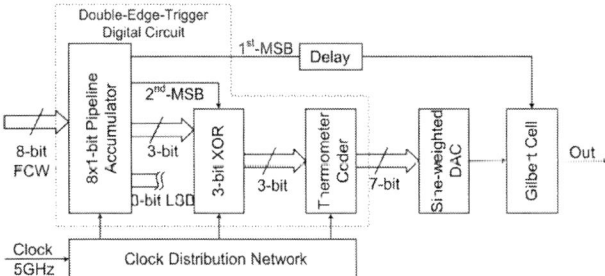

Fig. 1.   Block diagram of the 10GHz 8-bit ROM-less DDS.

Typically a traditional ROM-based DDS consists of an accumulator, a ROM for phase to amplitude conversion and a digital-to-analog converter (DAC). Generally the ROM is the bottleneck of DDS's high-speed performance, so that some ROM-less DDS architectures are used to improve DDS's clock frequency [2]-[5].

The top-level architecture of the proposed DDS is shown in Fig. 1. Similar to the DDS architecture in [3], this proposed DDS mainly consists of an 8-bit pipeline phase accumulator, a 3-bit XOR complementer, a 3-bit thermometer coder, a sine-weighted DAC and a Gilbert cell. The three LSBs of the accumulator's output are

978-1-4244-6240-7/10 $26.00 © 2010 IEEE   383

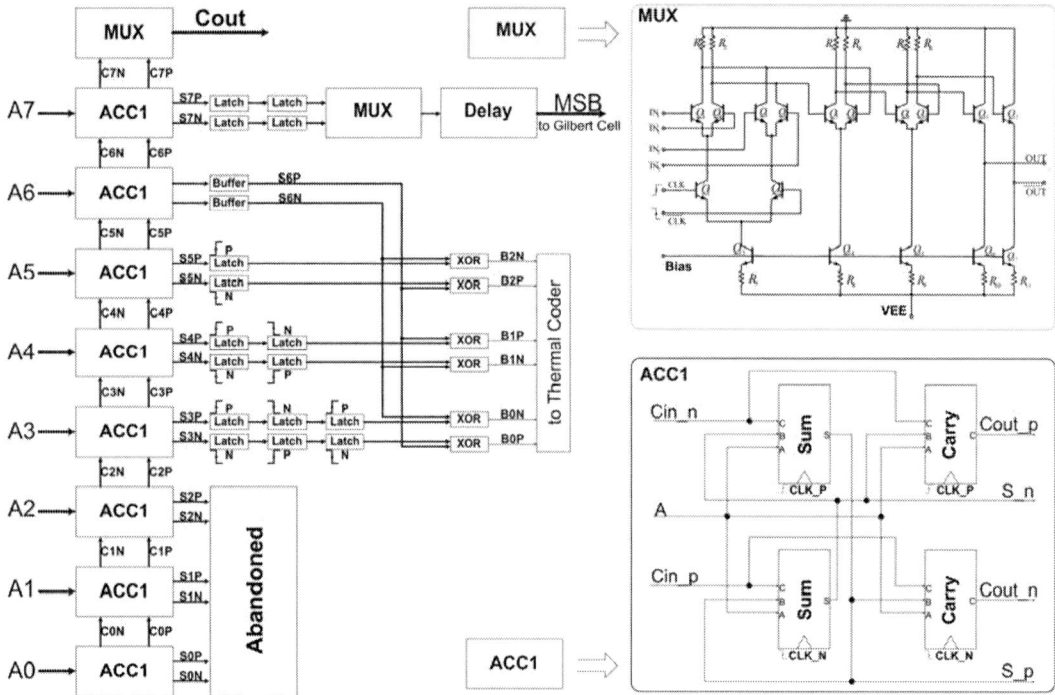

Fig. 2. Block diagram of the DET 8-bit pipeline accumulator and the 1-bit accumulator unit, and the schematic of MUX.

truncated to reduce the size and power consumption of the chip with little spurious penalty. The 1st-MSB output is used as one input of the Gilbert cell to provide the proper mirroring of the sine waveform about the phase point. The 2nd-MSB is used to complement the remaining 3-bit output for the second and fourth quadrants of the sine waveform prior to the thermometer coder. The 7-bit thermometer outputs drive the switches of the sine-weighted tap current sources of the DAC. All the currents are added together and then are converted to a voltage signal as the second input of the Gilbert cell. The output of the Gilbert cell forms the DDS output.

What is different from the DDS architecture in [3] is that we designed a DET 8-stage pipeline accumulator in each stage there is a 1-bit accumulator, rather than a single-edge-trigger 4-stage pipeline accumulator in [3]. By this DET pipelined accumulator, we can improve the speed by at least 2 times.

III. CIRCUIT IMPLEMENTATION

To pursue the highest speed, all of the digital circuits are making use of Current-Mode-Logic (CML) or Emitter-Coupled-Logic (ECL) which are popular in ultra-high-speed bipolar digital designs.

A. 8-bit DET Pipeline Accumulator

Fig. 2 shows the block diagram of the DET 8-bit pipeline accumulator and the 1-bit accumulator unit **ACC1**. Each of the eight **ACC1**'s includes two 1-bit adder **Sum**'s and two 1-bit carrier **Carry**'s. One **Sum-Carry** pair is triggered at the rising edge of clock signal while the other pair is triggered at the sequent clock falling edge. The two corresponding sum results **S_p** and **S_n** are feeding back into the falling-edge-triggered and rising-edge-triggered **Sum-Carry** pairs respectively inside the **ACC1** for accumulating operation. Meanwhile the two corresponding carry results **Cout_p** and **Cout_n** are passing out to the next stage **ACC1**. Thus the pipeline accumulator operates at the speed of **ACC1**, generating two accumulating results in a single clock cycle. There is one **Latch** in the **Sum** block following the sum logic gate, as well as in the **Carry** block. So the critical path propagation delay of the **ACC1** for each accumulating operation is $T_1 = t_{gate} + t_{latch}$, where t_{gate} is the larger one of the propagation delays of sum and carry logic gates, and t_{latch} is the propagation delay of the **Latch**. Compared to the critical path propagation delays of $T_2 = 2t_{gate} + 2t_{latch}$ in the designs of [1,2] and $T_3 = t_{gate} + 2t_{latch}$ in the designs of [3,4], the DET accumulator can achieve higher maximum operating frequency. The latches at the output of each **ACC1** facilitate data alignment for the following operations of the complementer and thermometer coder which are also using DET structure.

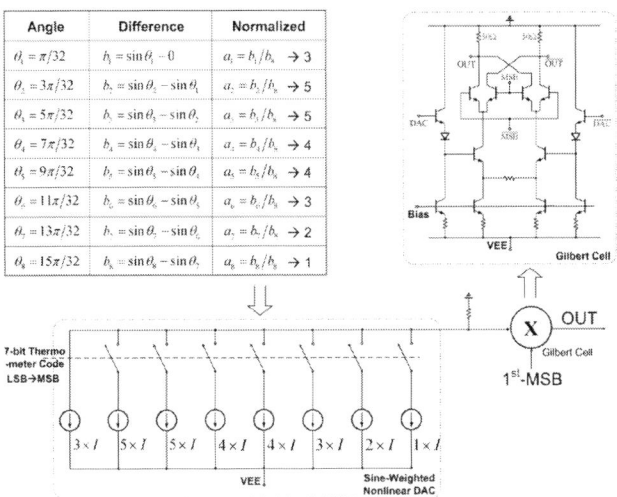

Angle	Difference	Normalized
$\theta_1 = \pi/32$	$b_1 = \sin\theta_1 - 0$	$a_1 = b_1/b_8 \rightarrow 3$
$\theta_2 = 3\pi/32$	$b_2 = \sin\theta_2 - \sin\theta_1$	$a_2 = b_2/b_8 \rightarrow 5$
$\theta_3 = 5\pi/32$	$b_3 = \sin\theta_3 - \sin\theta_2$	$a_3 = b_3/b_8 \rightarrow 5$
$\theta_4 = 7\pi/32$	$b_4 = \sin\theta_4 - \sin\theta_3$	$a_4 = b_4/b_8 \rightarrow 4$
$\theta_5 = 9\pi/32$	$b_5 = \sin\theta_5 - \sin\theta_4$	$a_5 = b_5/b_8 \rightarrow 4$
$\theta_6 = 11\pi/32$	$b_6 = \sin\theta_6 - \sin\theta_5$	$a_6 = b_6/b_8 \rightarrow 3$
$\theta_7 = 13\pi/32$	$b_7 = \sin\theta_7 - \sin\theta_6$	$a_7 = b_7/b_8 \rightarrow 2$
$\theta_8 = 15\pi/32$	$b_8 = \sin\theta_8 - \sin\theta_7$	$a_8 = b_8/b_8 \rightarrow 1$

Fig. 3. The sine-weighted nonlinear DAC and Gilbert cell.

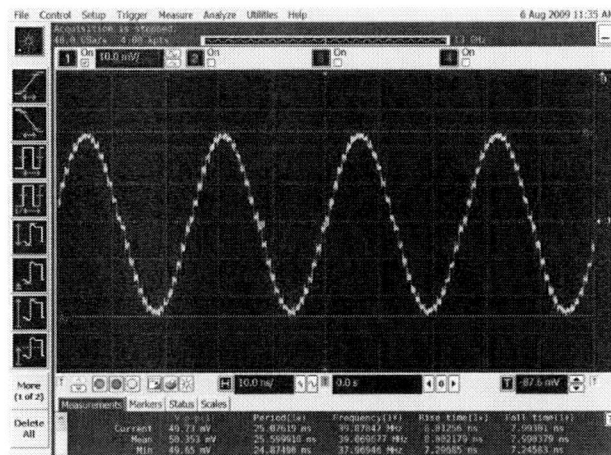

Fig. 5. DDS output waveform of 39.06MHz (FCW=10) with a 5GHz external clock.

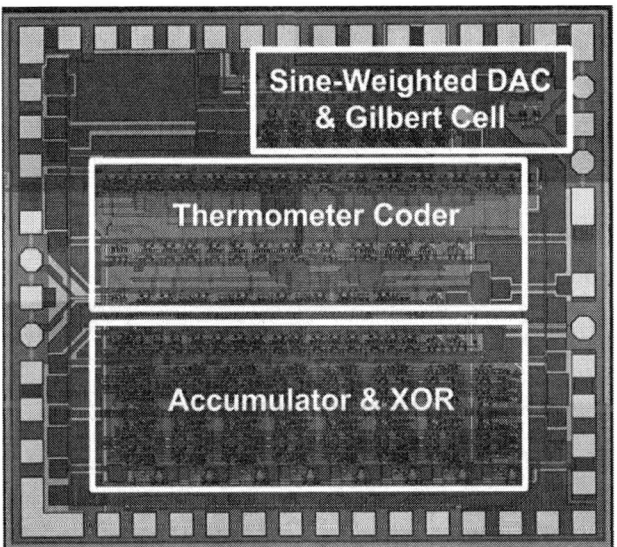

Fig. 4. Die photo of the DDS chip.

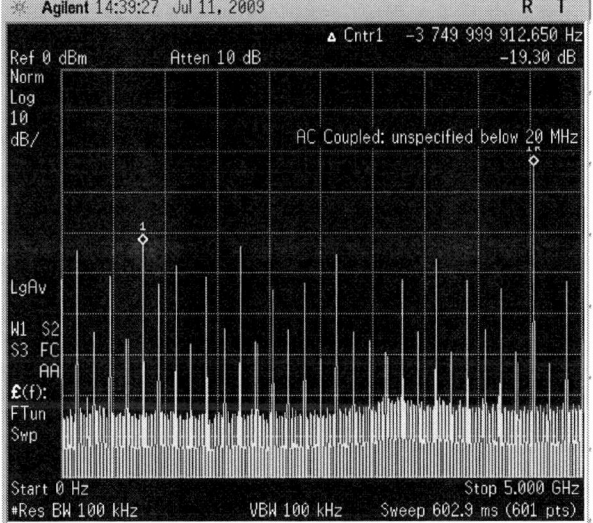

Fig. 6. DDS output spectrum of the worst case SFDR of 19.3dBc at FCW=116. The output frequency is 4.531GHz with an external clock of 5GHz.

B. MUX

For each bit of the 7-bit width thermometer coder output, a MUX circuit shown in Fig. 2 is used to merge the two data streams that generated at the rising and falling edges of the clock respectively. As the same as the other digital blocks in the DDS, the MUX is based on full differential ECL topology, which is well suited for the GaAs HBT devices. Two different signal pairs (IN_1, $\overline{IN_1}$) and (IN_2, $\overline{IN_2}$), representing the two data streams, are gated by the differential clock signal (CLK, $\overline{CLK}$) to form one data stream of which the data rate is doubled. After two stage amplifications, the data stream is buffered out by an emitter follower to drive the switches of the sine-weighted DAC.

C. Sine-weighted DAC

The block diagram of the proposed sine-weighted DAC and the schematic of the Gilbert cell are shown in Fig. 3. The table in Fig. 3 illustrates how the [3 5 5 4 4 3 2 1] tap weighting scheme of the DAC is calculated out. The switches of the current sources in the DAC are controlled by the 7-bit width thermometer-code, thus the weights sum successively to generate a quarter-wave sine output. In order to ensure that the current summing junction has non-zero outputs for all possible states, the first tap weight 3 is always enabled [3]. The summing current is then converted to a voltage signal simply by a pull-up resistor. This differential voltage signal (DAC, $\overline{DAC}$) is

TABLE I
EXPERIMENTAL PERFORMANCE AND COMPARISON WITH RECENTLY REPORTED DESIGNS

Technology f_t [GHz]	InP 370 [1]	SiGe 170 [2]	InP 300 [3]	SiGe 120 [4]	SiGe 200 [5]	GaAs 60 [This work]
Max clock f_{clk} [GHz]	24	15	32	12	5	10
Phase resolution B [bit]	12	8	8	9	24	8
Amplitude resolution A [bit]	7.5	6	5	8	10	5
Worst case DDS $SFDR$ [dBc]	30.7	21	21.56	22	38	19.3
Die area [mm^2]	5.0×3.3	1.1×1.0	2.7×1.5	2.5×0.7	3.7×3.0	2.4×2.0
Power consumption P [W]	19.8	0.366	9.45	1.9	4.7	2.4
FOM1=f_{clk}×B×$SFDR$/P	447	6885	584	1251	970	643
FOM2=f_{clk}/f_t×100%	6.49%	8.82%	10.67%	10.0%	2.50%	16.67%
FOM3=f_{clk}×A×B×$SFDR$/(P×f_t)	9.08	243	9.75	83.44	48.5	53.61

multiplied by the properly delayed 1st-MSB (MSB, $\overline{MSB}$) of the accumulator output in the Gilbert cell, resulting in a full-wave sine output. Linearity of the Gilbert cell is important for quality of the DDS output since nonlinearities reduce the SFDR. There is a resistor between the emitters of the input transistor pair, to improve the linearity of the Gilbert cell by introducing a negative feedback. Two 50 pull-up resistors are used at the output port to perform a wideband impedance match to the measurement system.

IV. EXPERIMENT RESULTS

A microphotograph of the fabricated DDS is shown in Fig. 4. The fabricated DDS is implemented in 1μm GaAs HBT process with an f_t of 60GHz, and has 1651 transistors in die area of 2.4×2mm^2. The DDS is tested on-wafer using GSG single-ended microwave probes. With the Nyquist output, the DDS achieves a maximum internal effective clock frequency of more than 10GHz when consuming 528mA current from a single -4.6V power supply. Fig. 5 shows the measured DDS output waveform of 39.06MHz with 5GHz external clock when the Frequency Control Word (FCW) is set to be 10. The measured SFDR of the DDS through the whole Nyquist band is 23.24dBc on average, with the worst case SFDR of 19.3dBc at an FCW of 116, as shown in Fig. 6. It should be noted that the SFDR performance can be better if measured under differential driven.

Table I summarizes the experimental performance of the proposed DDS, along with five recently reported ultra-high-speed designs. Three figures of merit (FOM) are calculated for all of the 6 designs for comparison. Due to the DET architecture designed, this work shows the best utilization ratio of the technology's f_t (FOM2) of 16.67% which is 1.56 times of the second place. It also

demonstrates a good FOM3 which includes more information on the metrics that are important for DDS performance.

V. CONCLUSION

This paper presented a 10GHz 8-bit DDS implemented in 1μm GaAs HBT process with an f_t of 60GHz. Due to the proposed DET accumulator, this DDS gives a best f_{clk}/f_t ratio compared with the other reported designs.

There are some limitations in this DDS. Future work will improve the linearity of Gilbert cell and extend the bit width of FCW to 32-bit, and reduce the power consumption.

REFERENCES

[1] S. E. Turner, R. T. Chan, and J. T. Feng, "ROM-Based Direct Digital Synthesizer at 24GHz Clock Freuqency in InP DHBT Technology," *IEEE Microwave and Wireless Components Letters,* vol. 18, p. 3, 2008.

[2] B. Laemmle, C. Wagner, H. Knapp, L. Maurer, and R. Weigel, "A 366mW direct digital synthesizer at 15GHz clock frequency in SiGe Bipolar technology," in *Radio Frequency Integrated Circuits Symposium, 2009. RFIC 2009. IEEE,* 2009, pp. 415-418.

[3] S. E. Turner and D. E. Kotecki, "Direct Digital Synthesizer With Sine-Weighted DAC at 32-GHz Clock Frequency in InP DHBT Technology," *Solid-State Circuits, IEEE Journal of,* vol. 41, pp. 2284-2290, 2006.

[4] Y. Xuefeng, F. F. Dai, J. D. Irwin, and R. C. Jaeger, "A 12 GHz 1.9 W Direct Digital Synthesizer MMIC Implemented in 0.18um SiGe BiCMOS Technology," *Solid-State Circuits, IEEE Journal of,* vol. 43, pp. 1384-1393, 2008.

[5] G. Xueyang, F. F. Dai, J. D. Irwin, and R. C. Jaeger, "A 5 GHz direct digital synthesizer MMIC with direct modulation and spur randomization," in *Radio Frequency Integrated Circuits Symposium, 2009. RFIC 2009. IEEE,* 2009, pp. 419-422.

Dual-Band CMOS Transceiver with Highly Integrated Front-End for 450Mb/s 802.11n systems

Shai Gross, Tzvi Maimon, Fabian Cossoy, Mark Ruberto, Georgi Normatov, Alexander Rivkind,
Nikolay Telzhensky, Rotem Banin, Ori Ashckenazi, Assaf Ben-Bassat, Sharon Zaguri, Gabriel Hara,
Mario Zajac, Nir Shahar, Shay Shahaf, Hani Yousef, Eyal Mor, Yishai Eilat, Anna Nazimov,
Zeev Beer, Amir Fridman and Ofir Degani

Intel Corporation, Mobile Wireless Group, Haifa, Israel

Abstract — A 3-stream, 802.11n WLAN MIMO transceiver, with fully integrated PAs and LNAs in both 2.4GHz and 5GHz bands, and a T/R switch in the 2.4GHz band, was implemented in a standard 90nm CMOS technology. The transmitter achieves an EVM of -28dB at output power of 19dBm and 17dBm in the 2.4GHz and 5GHz bands, respectively. The transmitter power consumption per Mb of data, in 3-stream mode, is 3.7mW/Mb and 4.5mW/Mb in the 2.4GHz and 5GHz, respectively. This is four times lower comparing to the single stream (SISO) mode. The receiver NF is 4dB in both bands, and power consumption is 1.6mW/Mb and 1.7mW/Mb, in the 2.4GHz and 5GHz bands, respectively.

Index Terms — IEEE 802.11n, WLAN, MIMO, CMOS, RF transceiver, 3x3, integrated FEM, integrated LNA, integrated PA, Integrated T/R switch, RC filter, frequency synthesizer.

I. INTRODUCTION

The widespread acceptance of 802.11n products has been driven by users seeking to exchange information in a timely manner, stream High Definition (HD) video, and handle voice, video, and data traffic concurrently. From power consumption point of view, the power per Mb of data is significantly lower compare to SISO systems. This has led to the adoption of 802.11n as a key technology component in both consumer and enterprise notebook PCs. While the 802.11n standard supports up to 4 data (spatial) streams, capable of delivering up to 600Mb/s, the most common high throughput products in the market today supports only two data streams, delivering 300Mb/s. Naturally, supporting more data streams, directly impact the silicon and package size, board complexity, and BOM, which eventually translates to higher cost. In recent years, a major effort has been done [1]-[3] in order to reduce the cost of the WiFi radio, by integrating portions of the FEM. Described in this paper, is a highly integrated 3x3, dual band transceiver, capable of delivering a throughput of 450Mb/s, with an integrated PAs and LNAs in both the 2.4GHz (aka LB) and 5GHz (aka HB) bands, and a T/R switch in the LB.

Supporting high MIMO data rates imposes stringent requirements from such a transceiver. Having 3 power amplifiers (PAs) working simultaneously on the same die, introduces a new level of design complexity. The design would need to overcome all reliability issues, keeping the total power consumption low and still be linear enough in order to meet the required EVM and mask requirements. Furthermore, having multiple PAs in close proximity exposes the adjacent chains to a risk of cross talk, which can significantly degrade the MIMO performance. These challenges among others, coupled with the need of keeping the die size and pin count as low as possible, are been addressed in the transceiver presented in this paper.

Section II of this paper describes the high level architecture of this transceiver. Section III discusses the different blocks implementation and challenges, followed by measurements results presented in section IV.

II. ARCHITECTURE

Fig.1 shows high level description of the transceiver block diagram. For brevity, all peripheral content such as LDOs, various calibration features, and biasing systems, were omitted.

The radio consist of three identical, direct conversion transceivers, where each transceiver has two RF chains (one for each band), a LO generation, and a shared BB filter for both TX and RX paths.

The HB and LB chains consist of integrated PAs, and LNA. The LB chain also includes an integrated T/R switch, leaving only a passive balun filter as non-integrated. In the receive path, the signal from the antenna is amplified by a programmable gain, single-stage LNA, passing through a TX/RX selector and down converted using a passive quadrature mixer. The down converted signal is then mux'ed to the 5-pole filter, and a programmable gain amplifier before going to the ADC.

In the transmit path the input signal is mux'ed to the same filter as the receiver and from there, to the TX programmable gain amplifier. It is then up-converted with a passive quadrature mixer and routed through a TX/RX

978-1-4244-6240-7/10 $26.00 © 2010 IEEE

selector to the RF TX chain. The first stage of each chain is a programmable gain, RF amplifier, followed by a fixed gain driver amplifier and finally a power amplifier.

The LO generation is based on a fractional-N synthesizer using a 40MHz reference clock. The VCO output frequency range is between 3.2GHz and 4GHz, which is routed to all three chains and multiplied locally by 3/4 or 3/2 for the LB and HB chains, respectively.

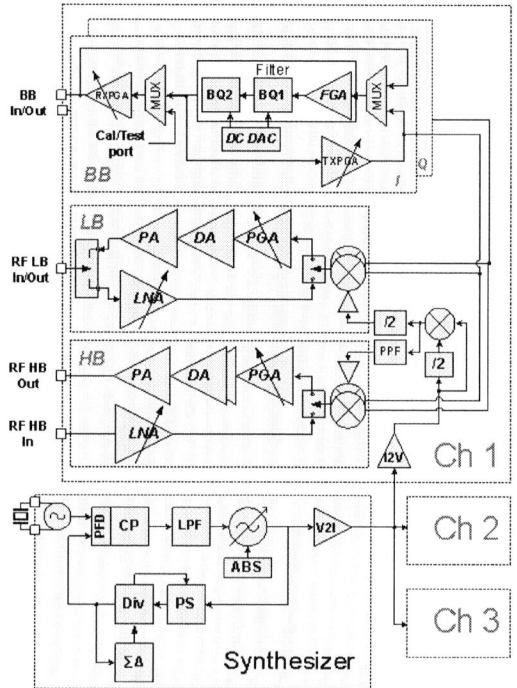

Fig. 1.   Transceiver block diagram.

III. CIRCUITS IMPLEMENTATION

A. Transmitter

Having three, simultaneously working PAs, integrated on the same die, impose a variety of challenges. The main challenge is to design a PA, which on one hand, meets the low-data rate spectral masks in SISO mode with high output power, while on the other hand, can reach -28dB EVM for 3x3 MIMO with good enough efficiency to keep the overall power consumption within the limits. This was mitigated by having two operational modes of the PA. In the first mode, the PA operates at a class-A bias point to meet the CCK spectral mask and band-edge requirements with good linearity for SISO. The second mode is a low linearity mode, where the PA operates at a class-AB bias point for higher efficiency, which enables all three PAs to work simultaneously with lower power consumption. In order to improve the PA nonlinearities inherent in a class-AB mode of operation, a digital pre distortion (DPD)

algorithm is used. Fig. 2 shows the TX chain EVM performance as a function of output power. Implementing the DPD algorithm on the class-AB PA, achieved an EVM of -28dB at output power of 19dBm (PA efficiency of 18%) and 17dBm (PA efficiency of 17.5%) in the LB and HB, respectively. The Psat of the PAs was designed to be 26dBm, in order to meet the requirement for output power without over designing the Psat and ultimately degrading PAE.

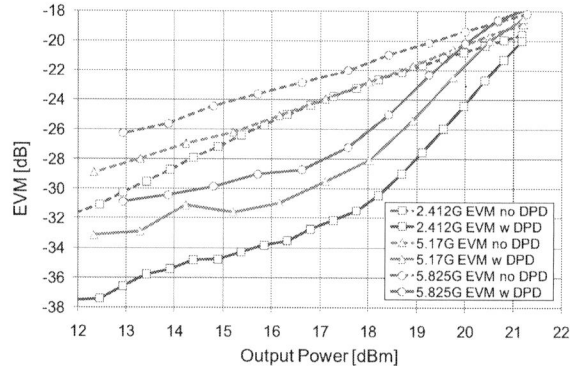

Fig. 2.   EVM vs. Output power, with and without DPD for various channels in both frequency bands.

In order to achieve flat gain and Psat response in the HB, all the matching networks were designed using transformers, allowing very wide-band impedance transformation. Shown in Fig. 3, the simulated gain and P4dB variance across the 700MHz bandwidth is ±0.75dB. The PA employs a standard threshold, thin-gate oxide common-source (CS) device and a thick-gate oxide common-gate device, in a quasi-differential cascade topology for improved reliability and performance. The driver and power amplifiers are working from a 3.3V supply.

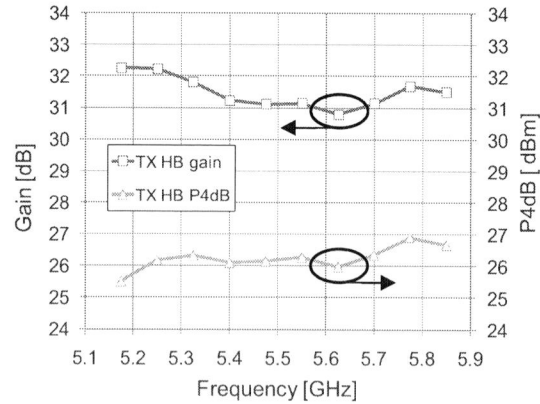

Fig. 3.   HB TX gain and output power across the full band frequency.

B. Receiver

LNAs of both bands have a triple cascade topology [1] resulting a highly unilateral amplifier. It has 4 coarse gain steps of 6dB, using current steering mechanism. By using the triple cascade topology there is a better isolation between the current steering branches and the input of the amplifier, thus reducing the impact of the different gain steps on the input matching. The down conversion is done via passive direct conversion quadrature mixer. The LB receiver has an integrated low loss T/R switch, based on the topology described in [4], were the RX side of the switch is part of the LNA matching network. The receiver dynamic range (in both bands) is 60dB, with gain step resolution of 1.5dB. The receiver NF is 4dB in both bands. Sensitivity is better than -90dBm for the low data rates, and better than -64dBm for high data rate MIMO. Out-of-channel IP1dB of -14dBm provides sufficient immunity to interferes.

C. Analog/BB Circuits

The Rx/Tx shared filter has five poles, implemented by a first pole in a trans-impedance stage, and two bi-quads stages, each having two complex poles. The filter is optimized to improve interferers rejection, and meet the requiring group delay, and in-band flatness.

The filter has two modes of Band Widths (BW) operation, 10MHz and 20MHz, which are configurable with two banks of capacitors with build-in process variations compensations. A DC canceling DAC is connected to the filter input. As shown in Fig. 1, a multiplexer is used before the RXPGA input. The purpose of this MUX is to connect a variety of radio testing signals to the ADC, and enable BB loop back path.

D. LO Generation

A single 3.2GHz - 4GHz LC VCO is used to generate the required frequency source for both the LB and HB. The VCO employ a 5-bit digital switched capacitors bank for coarse tuning, and a varactor for fine tuning. The VCO coarse tuning is done by an automatic tuning algorithm that is PVT (Process, Voltage, Temperature) variation resistant and ultra fast, converging within <10μSec.

The VCO is locked by a fractional-N PLL. The PLL reference is a 40MHz crystal oscillator with integrated programmable load capacitors for frequency calibration. The PFD-CP incorporates a digital lock detect circuit and dynamic charge pump current control to improve transient performance. The PLL feedback uses a custom digital CMOS Prescaler and a digital fractional divider.

The VCO frequency is distributed to each transceiver chain and multiplied locally by 3/2 or 3/4 for the HB or LB, respectively. The VCO signal passes through V to I converter, and the current is distributed by means of differential traces to the three chains. Since the receiver uses a direct conversion architecture, it is extremely important to reduce electromagnetic coupling between the LO and the receiver's signals, in order to minimize DC offset. For this reason, special attention is given to the floor planning of the LO Generation (LOG) inductively loaded mixer and buffer, along with careful design for low currents to minimize the electromagnetic coupling.

IV. MEASUREMENT RESULTS

The transmitter EVM (with DPD) is -28dB@19dBm and -25dB@19.7dBm for LB, and -28dB@17dBm and -25dB@18.5dBm for HB. For an EVM of -28dB, the DPD algorithm improves the output power of the integrated PAs by 3-4dB. Output power of LB PA pass CCK spectral mask requirements at 18dBm and OFDM (HT-20/40) spectral mask at 20.4dBm/18.6dBm. HB PA pass OFDM (HT-20/40) spectral mask at 18.6/17.3dBm.

The PAs reliability was confirmed through accelerated aging testing procedure. All 3 PAs were toggled together, in a multiple units, between ON and OFF repeatedly for 170 hours, under extreme supply voltage and ambient temperature conditions, while monitoring their bias point and 23dBm output power. The measured data showed no noticeable degradation in the output power, and only 2% drop in the current consumption from start to end. This assures proper operation over the life time of the product.

One way to look at the benefit of using multi stream system is the significant reduction in power consumption per (Mega-bit) Mb of data. In the 802.11g mode (SISO) the transmitter discussed in this paper, consumes 15.8mW/Mb and 17.3mW/Mb in the LB and HB, respectively, compare to 3.7mW/Mb and 4.5mW/Mb, in the MIMO 3x3 mode. The receiver consumes, in the 3x3 mode, 1.6mW/Mb and 1.7mW/Mb for the LB and HB, respectively. In both, transmit and receive, the power consumption during 3x3 mode is higher at any given time, however, for a given amount of data, it actually consumes 4 times less power, compare to SISO mode.

Synthesizer lock time, including VCO automated band selection mechanism, takes less than 20uS, and integrated phase noise is better than -46dBc in the LB, and -38.5dBc in the HB, across the whole bands.

The effective throughput in the 5.24GHz channel (3-stream, HT-40, GI=800nS), shown in Fig. 4, reaches more than 260Mbp/s.

TABLE I

PERFORMANCE SUMMARY

Parameter		LB	HB	Unit
RX sensitivity	6Mbps	-92	-90	dBm
	54Mbps	-74	-72	
	450Mbps	-66	-64	
RX NF		4	4	dB
RX power consumption	54Mbps	6.4	7.1	mW/Mb
	450Mbps	1.6	1.7	
TX Psat		26.8	26.6	dBm
TX output power	-19dB EVM	21.6	20.5	dBm
	-25dB EVM	19.7	18.4	
	-28dB EVM	19	17	
	CCK	18	NA	
	OFDM (HT-40)	18.6	17.3	
TX power consumption	54Mbps	15.8	17.3	mW/Mb
	450Mbps	3.7	4.5	mW/Mb
Integrated phase noise		< -46	< 38.5	dBc
Supply	1.8V, 3.3V			
Technology	90nm CMOS			
Die size	25mm²			
Package	QFN dual-row			
Standard	802.11a/g/n, MIMO, 3x3, 3-stream			

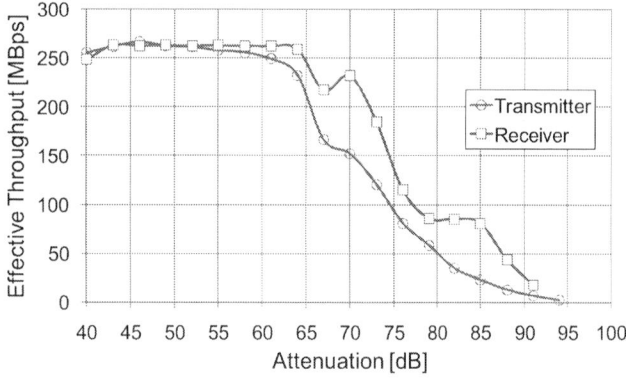

Fig. 4.   Effective throughput (3x3, HT-40, GI=800nS).

V. CONCLUSION

A highly integrated, dual band, 3x3 (3 stream), MIMO transceiver with integrated PAs, LNAs and a T/R switch, implemented in a standard 90nm CMOS technology. Each TX chain can deliver an output power of 19dBm and 17dBm in the LB and HB, respectively, with EVM of -28dB.

The receiver sensitivity, for a 64QAM, 3 stream signals, is -66dBm and -64dBm in the LB, and HB, respectively. In 3x3 mode, the receiver consumes less than 1.7mW/Mb in both bands, and the transmitter consume 3.7mW/Mb and 4.5mW/Mb in the LB and HB, respectively. Fig. 5 shows a die microphotograph of the chip. Manufactured in a 90nm CMOS process with die size of 5mm X 5mm, and packaged in QFN.

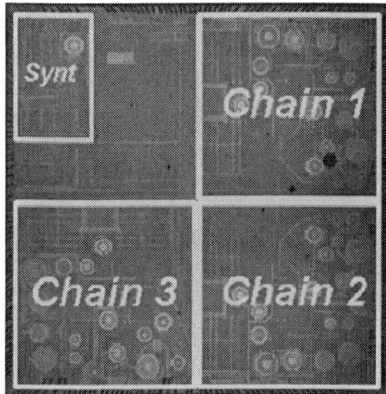

Fig. 5.   Microphotograph of the silicon die.

ACKNOWLEDGEMENT

The authors would like to thank the layout team and silicon validation team, which are an integral part of the design process. The authors would also like to thank the, back-end, validation, integration, Q&R, and production teams for their support.

REFERENCES

[1] Ofir Degani, Mark Ruberto, Emanuel Cohen, Yishai Eilat, Benjamin Jann, Fabian Cossoy, Nikolay Telzhensky, Tzvi Maimon, Gregory Normatov, Rotem Banin, Ori Ashkenazi, Assaf Ben Bassat, Sharon Zaguri, Gabriel Hara, Mario Zajac, Eyal Shaviv, Shay Wail, Amir Fridman, Richard Lin, Shai Gross "A 1x2 MIMO Multi-Band CMOS Transceiver with an Integrated Front-End in 90nm CMOS for 802.11a/g/n WLAN Applications" *ISSCC Dig. Tech. Papers*, pp. 356-357, Feb. 2008.

[2] A. Afsahi, A. Behzad, V. Magoon, L. E. Larson, "Fully Integrated Dual-Band Power Amplifiers with on-chip Baluns in 65nm CMOS for an 802.11n MIMO WLAN SoC" *IEEE RFIC Symp. Dig*, pp.365-368, June 2009.

[3] M. Terrovitis, M. Mack, J. Hwang, B. Kaczynski, G. Tseng, B. Wang, S. Mehta, D. Su, "A 1x1 802.11n WLAN SoC with Fully Integrated RF Front-end Utilizing PA Linearization", *Proceedings of ESSCIRC, 2009.*

[4] A.A. Kidwai, C.T. Fu, R. Sadhwani, D. Chu Chi, J.C. Jensen, S. Taylor, "An Ultra-Low Insertion Loss T/R Switch fully integrated with 802.11b/g/n Transceiver in 90nm CMOS" *IEEE RFIC Symp. Dig*, pp.313-316, 2008.

A CMOS Transceiver with internal PA and Digital Pre-distortion For WLAN 802.11a/b/g/n Applications

Chia-Jun Chang, Po-Chih Wang, Chih-Yu Tsai, Chin-Lung Li, Chiao-Ling Chang, Han-Jung Shih, Meng-Hsun Tsai, Wen-Shan Wang, Ka-Un Chan, and Ying-Hsi Lin

Realtek Semiconductor Corp., Hsinchu, 300, Taiwan

Abstract — A 2.4/5GHz Fully-Integrated Transceiver is implemented in 65nm CMOS technology. To alleviate the cost of external front-end components, the G-mode RF transmit/receive (T/R) switch and a power-efficient linear CMOS PA are fully integrated on-chip. On the other hand, for better performance, only the A-mode PA is integrated on-chip while the external T/R switch is used. It shows 5dB and 5.5dB NF in the G-mode and A-mode receivers respectively. Also, the transmitter delivers an average power of 18dBm OFDM (64QAM, 54MBPS) signal with EVM of –28dB for G-mode application and 16dBm OFDM (64QAM, 54MBPS) signal with EVM of –28dB for A-mode application after digital pre-distortion.

Index Terms — WLAN 802.11A/B/G/N, CMOS PA, CMOS switch, CMOS transceiver, Digital Pre-distortion.

I. INTRODUCTION

The growing demand for wireless applications has spurred the development of high-efficient and low-cost WLAN transceiver. Moreover, the channels of 2.4GHz are not sufficient and the demand for A-band is growing. A 802.11a/b/g/n transceiver in a 65nm CMOS technology is presented in this paper concluding a 2.4GHz-TX chain, a 2.4GHz-RX chain, a 5GHz-TX chain, a 5GHz-RX chain, and a dual-band synthesizer. For low cost and high integration, the CMOS PA and T/R switch for 2.4GHz are also implemented in this chip. With the internal T/R switch, it shows 5dB NF and -73dBm sensitivity (OFDM, 54Mbps) in 2.4GHz RX chain. A 2.4GHz on-chip PA presents 24dBm output P_{1dB}, which only consumes 155mA current. It can deliver 18dBm output power signal with –28dB EVM after our digital pre-distortion. On the other hand, for 5-GHz applications, only the PA is implemented in this chip. The external T/R switch is used for the better performance at high frequency. It shows 5.5dB NF and -72dBm sensitivity (OFDM, 54Mbps) in 5GHz RX chain. A 5GHz on-chip PA presents 23dBm output P_{1dB}, which consumes 180mA current. It can deliver 16dBm output power signal with –28dB EVM also by means of the digital pre-distortion.

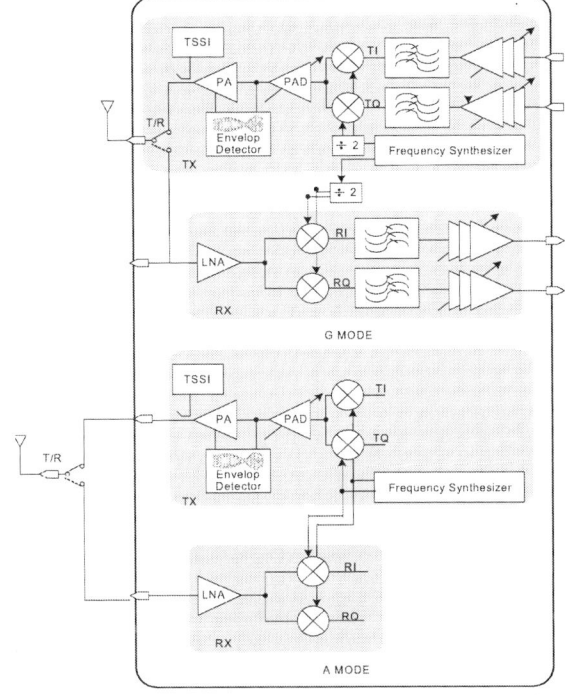

Fig. 1. 2.4GHz/5GHz Transceiver block diagram

II. TRANSCEIVER ARCHITECTURE

Figure 1. shows the block diagram of the direct conversion transceiver. There are two transceivers each for 2.4GHz and 5GHz respectively. In the receive path, the LNA and I/Q mixers ensure the low power, low noise and correct gain. An integer-N synthesizer with on-chip VCO circuit generate the LO signal from 4.8~6GHz. This LO signal can be provided for A-mode application directly while it needs to be divided for G-mode application. Both the LPF and PGA circuits in the RX and TX path can operate with low IQ mismatch and high linearity. To save the area, the LPF and PGA are used for both the 2.4GHz and 5GHz band. In the transmit path, the on-chip PA/PA driver (PAD) boost the signals directly up-converted from baseband. The digital pre-distortion is added here to improve the high power performance. A power control circuit can efficiently decrease the static current of the PA/PAD. A digital serial interface circuit is included to control the transceiver setting and responsible for the logic functions including VCO frequency calibration, filter bandwidth calibration and I/Q imbalance compensation.

The integer-N frequency synthesizer covers the full band for G-mode and A-mode. There is a 40MHz on-chip crystal oscillator with frequency fine-tuning function. The VCO operates at 3.2GHz~4GHz to decrease the effect of PA pulling.

III. RECEIVER

The topology of both G-mode and A-mode RX RF circuits are the same. Each receiver chain consists of a single-stage fully differential cascode LNA followed by I/Q mixers. As shown in Figure 2, the cascode current switches and LNA loading adjustment control the LNA gain setting. In the following mixer, including current bleeding circuits, Gilbert cell is used to decrease the parasitic capacitance and noise sources. The sensitivity can be as low as -73dBm for 2.4GHz band and -72dBm for 5GHz band. In front of the 2.4GHz LNA, there is an internal T/R switch. Even in this CMOS process, we can get isolation as high as 50dB.

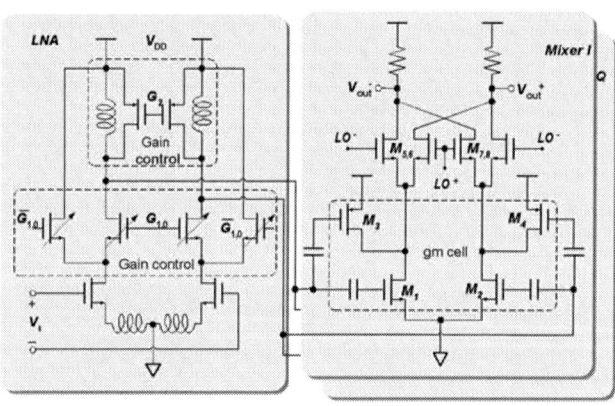

Figure 2. 2.4GHz/5GHz LNA and Mixer block diagram

The baseband circuit consists of 5th order LPF and PGA providing 54dB gain range with 2dB step. The MPF topology is used for the LPF. These OP-based amplifiers can provide precise gain tuning and these blocks are used both for the G-mode and A-mode applications to save the area. The dc offset cancellation circuit provides a high-pass characteristic with 10KHz/20KHz HP corner. Since R, C may vary over ±20%, the filter must be calibrated. A simple calibration loop consisting of comparator and counter is used to program the 5-bit range unit capacitor for proper frequency tuning.

To minimize I/Q imbalance, the IQ calibration technique is used with the digital algorithm. With IQ calibration, the receiver I/Q paths show 42dB IMR as shown in Figure3. The power consumption is 280mW in 2.4GHz receive chain while 281mW in 5GHz receive chain.

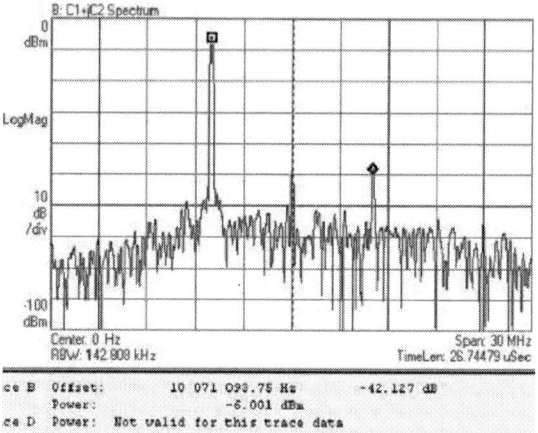

Figure 3. Receiver signal with IMR=42dBc

IV. TRANSMITTER

As shown in Figure 1, a 5th-order anti-alias filter providing 8dB variable gain with 0.5dB step is implemented in MPF architecture with 20/10MHz bandwidth. Both for the G-mode and A-mode applications, an on-chip PA and PA driver (PAD) boost the signals directly up-converted by I/Q modulator. For the 2.4GHz band, the T/R switch shows no observable loss in the transmission path.

The stringent requirement of PA linearity can be accomplished by class-A bias PA, degrading power added efficiency (PAE). To solve this problem, dynamic bias topology, which control gate "dc" bias voltage of CMOS power transistors with varying envelop of RF signal to adapt the drain current dynamically, has been adopted. Based on [2], we can compare the power saving ability from Class-A bias PA to dynamic bias ones. Generally, the peak to average power ratio of an OFDM signal is roughly 9dB, this means:

$$\frac{P_{av}}{P_{peak}} = \frac{V_o^2/(2R_L)}{V_{dd}^2/(2R_L)} \approx \frac{1}{8} \qquad (1)$$

in which V_o and V_{dd} is the output voltage of PA and supply voltage, respectively, and R_L is the optimal load for Class-A bias PA. If an optimal bias level is provided dynamically, the power consumption of the dynamic bias PA is

$$P_{dy} = \frac{V_o}{R_L} * V_{dc} \qquad (2)$$

and the power consumption of Class-A bias PA is

$$P_{classA} = \frac{V_{dc}}{R_L} * V_{dc} \qquad (3)$$

From (1), (2) and (3), the resulting relationship of power consumption between Class-A bias PA and dynamic bias ones in an OFDM system is

$$\frac{P_{dy}}{P_{classA}} = \frac{1}{2\sqrt{2}} \approx 35.4\% \qquad (4)$$

As shown in Figure 4, the PAD and PA are differential and cascode amplifiers both for G-mode and A-mode applications. The pre-PAD stage is common-source stage for lower power supply instead of cascode topology. With the digital pre-distortion[5], the output P_{1dB} of the G-mode transmitter chain is 24dBm, and the transmitter chain consumes 689mW (PA 155mA)at 18dBm output signal power. The output P_{1dB} of the A-mode transmitter chain is 23dBm, and the transmitter chain draws 770mW.

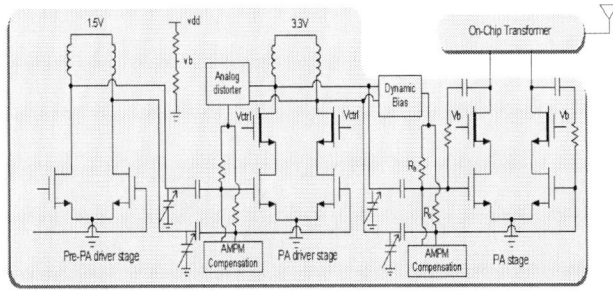

Figure 4.   PAD and PA Topology

V. SYNTHESIZER

The integer-N frequency synthesizer covers the full band from 2412 to 2484MHz for G mode and 4900 to 6000MHz for A band. The VCO oscillates at 3.2~4GHz to prevent from PA pulling. And it is converted by a mixer to produce 4.8GHz ~ 6GHz signal. For the G-mode application, it has to be divided by two, while for the A-mode application, it can be used directly. Also, to compensate the long signal path, the inductive load is used in buffer stage to guarantee the driving ability of the local oscillator.

The synthesizer exhibits -120dBc phase noise at 1MHz offset frequency in 2.4GHz band. That corresponds to 0.46° phase error integrated from 10KHz to 10MHz. And it exhibits -110dBc phase noise at 1MHz offset frequency in 5GHz band. That corresponds to 0.6° phase error integrated from 10KHz to 10MHz as shown in Figure 5.

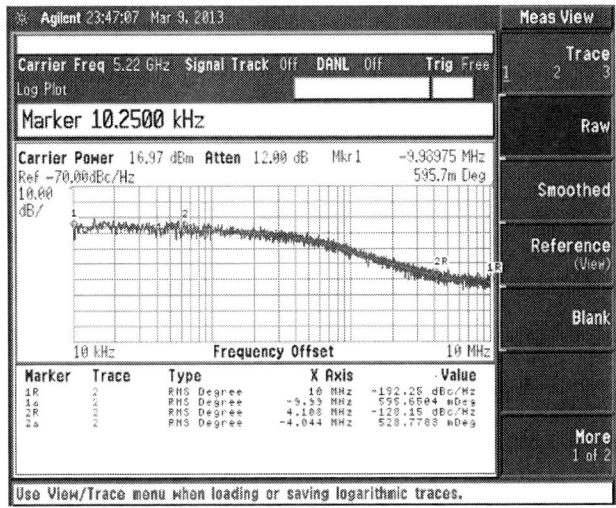

Figure 5.   Measurement result of synthesizer noise

VI. INTEGRATED MEASUREMENT RESULT

In the receive mode, each RX chain can provide signal with EVM (OFDM, 64QAM, 54Mbps) lower than –32dB. The dynamic range of each RX chain is from –72dBm to 0dBm with OFDM (64QAM, 54Mbps) signal.

In the G-band transmit mode, the PA can deliver 18dBm OFDM (64QAM, 54Mbps) signal power with EVM=-28dB, which improves 1dB after digital pre-distortion, as shown in Figure 6. In Figure 7, it reveals the A-mode transmitter can delivers signal power of 16dBm while meeting the stringent EVM of –28dB.

This SOC chip is fabricated in 65nm 1P7M CMOS technology and the size is 5580um X 5080um, including 2 transceivers described above and digital circuits. The measurement result is summarized in TABLE I.

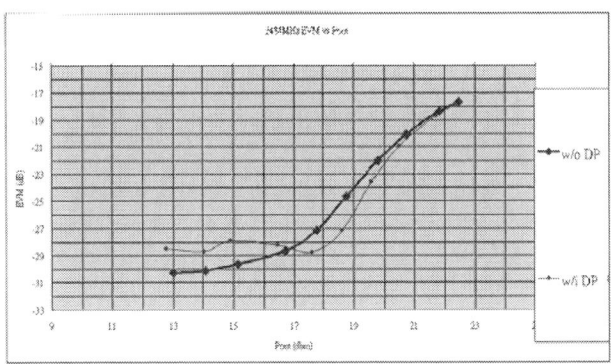

Figure 6.   G-mode TX : @ EVM=-28dB , P_{out}=17dBm w/o PD and P_{out}=18dBm w/i PD

978-1-4244-6240-7/10 $26.00 © 2010 IEEE   393

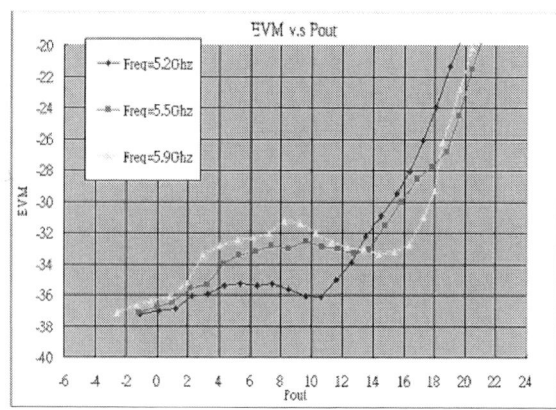

Figure 7. A-mode TX : @ EVM=-28dB and P_{out}=16dBm

Figure 8. Die Photo of SOC chip

TABLE I SUMMARY OF TRANCEIVER PERFORMANCE

Parameter	Results for 2.4/5GHz band
RX NF	5dB @ 2.4GHz
	5.5dB@ 5GHz
TX output P_{1dB}	24 dBm @ 2.4GHz
	23dBm @ 5GHz
TX EVM @P_{out} (54Mb/s)	-28dB @ 18dBm @ 2.4GHz
	-28dB @ 16dBm @ 5GHz

Integrated phase noise	0.46° @2.4GHz
	0.6° @ 5GHz
Technology	65num 1P7M CMOS
Die size	5580um X 5080um
Power Consumption	280 mW @ 2.4G RX
	689 mW @ 2.4G TX
	P_{out}=18dBm
	281 mW @ 5G RX
	770mW @ 5G TX
	P_{out}=16dBm

VII. CONCLUSION

A fully integrated transceiver consisting of 2.4/5GHz TX/RX chain is implemented in 65nm CMOS technology. To alleviate the cost of external front-end component, the 2.4GHz transmit/receive (T/R) switch and two power-efficient linear CMOS Pas for 2.4GHz and 5GHz respectively are fully integrated on-chip. It shows 5dB NF for G-mode RX and 5.5dB NF for A-mode RX respectively. Also, the G-mode transmitter delivers an average power of 18dBm OFDM (64QAM, 54Mbps) signal with EVM of –28dB while the A-mode transmitter delivers an average power of 16dBm OFDM (64QAM, 54Mbps) signal with EVM of –28dB .

REFERENCES

[1] Shih-Chieh Yen, Ying-Yao Lin, Tzung-Ming Chen, Yung-Ming Chiu, Bin-I Chang, Ka-Un Chan, Ying-Hsi-Lin, Ming-Chong Huang, Jiun-Zen Huang, Chao-Hua Lu, Wen-Shan Wang, Che-Sheng Hu and Chao-Cheng Lee "A Low-power Full-band 802.11abg CMOS Transceiver with On-chip PA," IEEE RFIC. Symp. Dig., pp. 103-106, June 2005.

[2] Po-Chih Wang, Chia-Jun Chang, Wei-Ming Chiu, Pei-Ju Chiu, Chun-Cheng Wang, Chao-Hua Lu, Kai-Te Chen, Ming-Chong Huang, Yi-Ming Chang, Shih-Min Lin, Ka-Un Chan, Ying-His Lin, and Chao-Cheng Lee "A 2.4GHz Fully Integrated Transmitter Front End with +26.5-dBm On-Chip CMOS Power Amplifier," IEEE RFIC June 2007.

[3] S.G.Lee and J.K.Choi, "Current Reuse Bleeding Mixer," Electronics Letters, vol. 36, no. 8, pp.696-697, April 2000.

[4] Y.H.Hsieh, et al., "An auto-I/O calibrated CMOS transceiver for 802.11g," in ISSCC Dig. Tech. Paper.Feb.2005,pp.92-93.

[5] Michael Faulkner, Mats Johansson, "Adaptive Linearization Using Predistortion-Experimental Results," in IEEE Transactions On Vehicular Technology, VOL.43, NO2,May 1994,pp.323-332.

Highly Linear SOI Single-Pole, 4-Throw Switch with an Integrated Dual-band LNA and Bypass Attenuators

Chun-Wen Paul Huang, Lui (Ray) Lam, Mark Doherty, and William Vaillancourt

SiGe Semiconductor, Andover, MA 01810, USA

Abstract — **An innovative Silicon-On-Insulator (SOI) SP4T T/R switch is presented. The SP4T switch consists of 2 receive paths with an integrated dual-band LNA and bypass attenuators along with 2 high linearity matched transmit paths. Tx paths feature 0.1 dB compression to 34 dBm input power and 0.5-0.8 dB insertion loss from 1 to 6 GHz with > 20 dB return loss and > 25 dB isolation. Receive paths feature 16 dB gain with 2.3 dB NF for 2.4-2.5 GHz and 14 dB gain with 2.4-2.6 dB NF for 4.9-5.9 GHz. The band selectivity exceeds 40 dB. Cascading with a dual-band WLAN PA, a complex dual-band WLAN/MIMO front-end module (FEM) can be easily constructed with low assembly complexity and post PA losses resulting in dual-band transmit linearity >18 dBm with EVM < 3% and < -50 dBm/MHz harmonic emissions within a 4 x 5 mm QFN package.**

Index Terms — **Dual-Band WLAN/MIMO front-end IC's, switch designs, LNA designs.**

I. INTRODUCTION

Wireless local area network (WLAN) applications have been one of fastest growing areas of data communications. WLAN radios were originally designed for computer networking, but have subsequently been widely implemented in consumer products such as cell phones, security monitoring systems, personal data assistants, gaming systems, multi-media personal entertainment devices, and multimedia (video, data, voice) distribution systems [1]. Furthermore, the demand for more bandwidth and higher throughput rates results in the emergence of multiple-input, multiple-output (MIMO) techniques to increase the data rate from the original 54Mbps to a minimum of 108 Mbps.

Most WiFi and MIMO radios operate at the 2.4-2.5 GHz –'b/g' band, with 3 channels for 54Mbps operations. For b/g band MIMO applications, channel bonding techniques result in only 1 or 2 available channels. Therefore, the demand for more bandwidth is inexorable. One may surmise that dual-band WLAN and MIMO radios will be soon adopted to resolve the bandwidth congestion within b/g band WLAN applications. In this context, to achieve small form factors for compact portable computers and electronics, a front-end module (FEM) is the preferred design implementation, especially when MIMO architecture is adopted in portable electronics or other miniature network cards. One advantage of FEM's is that they simplify both circuit and printed circuit board layout for small form-factored WLAN radio solutions.

In this paper, a novel dual-band, single-pole, 4-throw (SP4T) T/R switch with an integrated dual-band LNA and bypass attenuators is presented as shown in Fig. 1. The design is based on SOI technology, which combines both advantages of CMOS for analogue circuitry and low substrate loss RF FET's for T/R switch and LNA. The funtional block diagram of the switch-LNA is shown in Fig. 2. The SP4T switch-LNA consists of 4 RF switch paths for both dual-band transmit and receive paths. Each Rx path has an integrated LNA and a bypass attenuator path.

In most WLAN radio designs, T/R switch is the last component prior to the antennas. Therefore, high linearity for the transmit switch path is a must to minimize nonlinear distortions. In this design, each transmit path features low insertion loss (IL) of 0.5-0.8 dB between 1 to 6 GHz with >20 dB return loss combined with 25 dB isolation and less than 0.1 dB input compression up to 34 dBm.

Fig. 1.   Die photo for the SP4T T/R switch with dual-band LNA and bypass attenuators. The die size is 1.2x1.6 mm².

In contrast to a conventional FEM architecture shown in Fig. 3 [2], this FEM is based on the proposed switch-LNA IC design and a self-sufficient, dual-band PA presented in [3] can be realized as shown in Fig. 2. The FEM in Fig. 2 has low assembly complexity and low pre-amplifier and post-amplifier losses due to the elimination Tx and Rx diplexers. This leads to improved RF performance including better efficiency, more available linear power, higher gain, and lower noise figure. The proposed FEM can be realized in a 4 x 5 x 0.9mm³ QFN package. Both b/g and a-band transmit paths can provide >18 dBm linear power with EVM < 3% at 54 Mbps and worst-case harmonic emissions less than -50 dBm/MHz up to 20 dBm output power. The Rx path performance of the FEM is dominated by the SP4T switch-LNA design, which has 16 dB gain with 2.3 dB NF between 2.4-2.5 GHz and 14 dB gain with 2.4-2.6 dB NF for 4.9-5.9 GHz band. The band selectivity

978-1-4244-6240-7/10 $26.00 © 2010 IEEE

exceeds 40 dB. In addition, when a WLAN radio is in proximity to a WLAN access points, the Rx path could be saturated by the associated emissions. In this circumstance, the bypass mode attenuators in the switch-LNA can be activated to provide an 18-dB path attenuation to reduce the impact on the receiver. These unique features ensure easy construction of dual-band WLAN and MIMO radios.

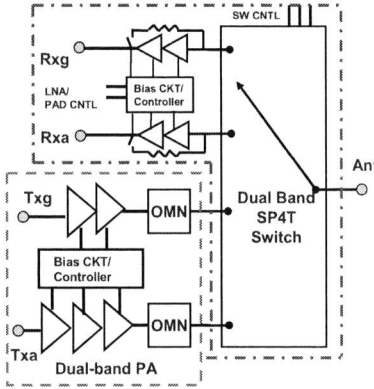

Fig. 2.   Architecture for a dual-band WLAN FEM based on the proposed SP4T switch-LNA and a self-sufficient dual-band PA for a/b/g/n radio applications.

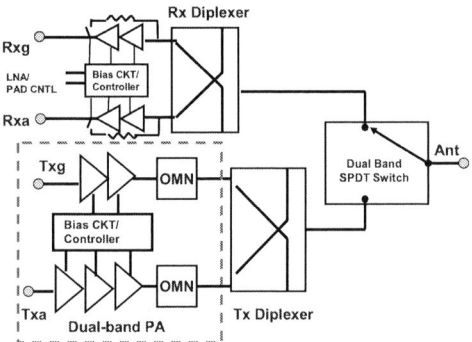

Fig. 3.   Traditional architecture used in dual-band FEM in WLAN a/b/g/n radios.

II. DESIGN

A. SP4T Switch-LNA Architecture

As shown in Fig. 2, the proposed SP4T switch-LNA consists of 2 major functional blocks, a dual-band LNA with bypass mode attenuators and a SP4T T/R switch. Using SOI technology, analogue circuitry such as temperature/voltage compensated bias circuits, enable/disable controls, bypass attenuator controllers, and switch logic decoder can be easily integrated onto the same die. In contrast to the traditional FEM architecture shown in Fig. 3, both Tx and Rx diplexers are eliminated. Therefore, the on-chip low and high band LNA's are directly matched to the 2 Rx paths of the SP4T switch. Similarly, the external dual-band PA will be matched to 2 Tx path of the SP4T switch as shown in Fig. 2. Therefore,

Tx switch paths need to support low insertion and mismatch losses and high linearity. In addition, the Tx switch paths feature fast on/off switching time (< 100 nS) to allow the PA enable/disable controls to drive Tx switch controls without degrading the linearity under pulse mode. This feature also reduces the number of external control lines to the FEM.

B. SP4T Switch

The schematic for a switch path of the SP4T switch is shown in Fig. 4. Since the SP4T switch design has 4 symmetrical switch paths, this description of the switch design can focus on a single switch path. In Fig. 4, two multi-gate SOI switch FET's are used to construct both series and shunt paths in a switch path to minimize insertion loss and maximize high isolation at high frequency. The linearity of a series-shunt switch path is usually constrained by that of the shunt path FET's [4]. Consequently, the maximum transmit power can be calculated by equation (1).

$$P_{\max}(dBm) = 10\log_{10}\left(\frac{[n(Vgs+Vth)\times2]^2}{2\times Z_o}\right) \qquad (1)$$

where Zo is the characteristic impedance of the measurement system, Vgs is the control voltage difference between the gate and source (or drain), Vth is the threshold voltage of the switch FET, and n is the number of cascaded switch FET. When the parasitic capacitance of multiple stacked FET's in a shunt path is well designed and balanced, the RF voltage swing will be evenly distributed across each drain-source junction in the stacked FET's [4]. Therefore, the maximum transmission power can be easily predicted by equation (1).

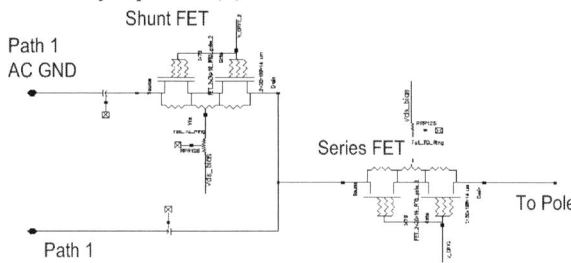

Fig. 4.   Schematics of a series-shunt switch path in the SP4T witch.

C. Dual-band LNA with Bypass Attenuators

Unlike the traditional FEM implementation, the proposed dual-band LNA is directly matched to two of the switch paths. Using a diplexer prior to a dual-band LNA will not only degrade the LNA noise figure but also decrease the gain by the path loss. In addition, when a diplexer does not have sufficient out-of-band isolation, the tuning for one

978-1-4244-6240-7/10 $26.00 © 2010 IEEE   396

LNA will always detunes the second LNA. The proposed SP4T switch features high isolation to each band, which enables each LNA to operate completely independent of each other.

As a single die implementation, both the low and high band LNA's are based on a similar topology. The 'a-band' LNA schematic is illustrated in Fig. 5. To achieve sufficient gain between 4.9 to 5.9 GHz, the cascode topology is used. To reduce the impact of out-of-band interference, two out-of-band traps were implemented to ensure sufficient out-of-band rejections as shown in Fig. 5. In addition, to avoid saturation of the LNA, an 18-dB bypass attenuator is implemented as shown in Fig. 5. The bypass attenuator has the opposite on/off logic to that of LNA.

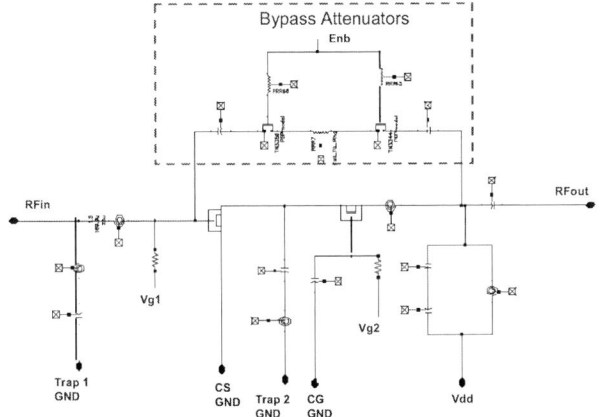

Fig. 5. The schematic for the integrated, low noise amplifier with bypass mode attenuator

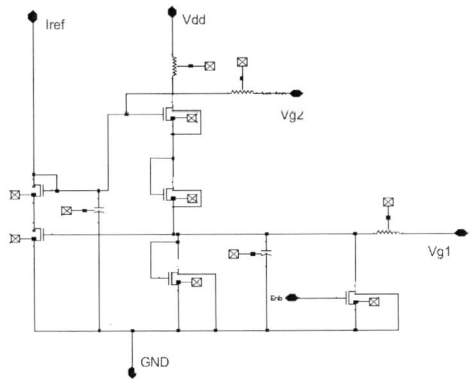

Fig. 6. The schematic for the bias circuit for the LNA.

The bias circuitry is also an important design consideration to ensure the gain response and noise figure of an LNA. The bias circuit for the cascode LNA is shown in Fig. 6. The reference FET devices are chosen to be matched to RF FET's to ensure the proper current scaling. Vg1 and Vg2 are providing the bias currents for both LNA FET's as shown in Figs. 5 and 6. In addition, when the SP4T switch is under transmit mode, the shunt FET at Vg1

will pull down the gate voltages of the cascode LNA to avoid having the LNA unintentionally turn on.

III. PERFORMANCE

Measured results for the design are presented in this section. The Tx path linearity of the SP4T switch-LNA was validated with a high power PA operating at 2.5 GHz. As shown in Fig. 7, the 0.1 dB compression point was found above 34.0 dBm input power with harmonic emission < -50 dBm up to 25 dBm input power.

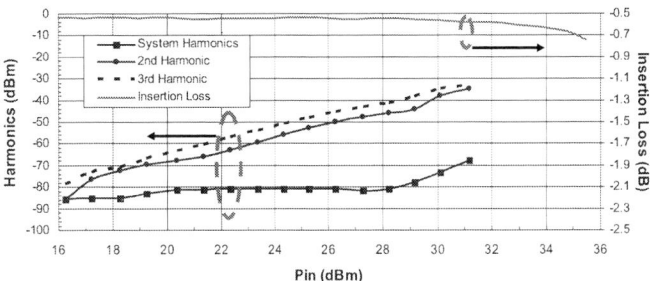

Fig. 7. Transmit characteristic of the SP4T switch-LNA at 2.5 GHz.

The s-parameters of the SP4T switch-LNA are shown in Fig. 8. The transmit switch paths have IL between 0.5-0.85 dB from 1 to 6 GHz. In addition, the return loss for each Tx path is typically >20 dB, providing a good impedance matching condition for the PA. As shown in Figs. 8 and 9, the Rx paths have 16 dB with NF < 2.3 dB between 2.4 to 2.5 GHz and 14 dB gain with NF < 2.6 dB between 4.9 and 5.9 GHz. The band selectivity is measured to be > 40 dB. The bypass attenuators were also measured and shown in Fig 8. The input 1 dB compression of the LNA was measured to be around -5 dBm With current consumption at 15 mA.

In addition to the characterization of the standalone SP4T switch-LNA, the switch-LNA design is also cascaded with a dual-band PA presented in [3] to demonstrate its performance in a FEM configuration as shown in Fig. 2. Since the SP4T switch-LNA is the last component in the FEM design, the Rx path performance remains the same as that previously reported. The s-parameters for the transmit paths of the FEM are shown in Fig. 10. Both Tx paths can provide ~28 dB gain with +/- 0.5 dB and +/- 1 dB flatness for b/g and a-band, respectively, along with 8-10 dB of input return loss. Transceiver spurious rejection at 1.6 GHz, 3.2 GHz, and 3.88 GHz are also addressed with the on-chip PA matching networks.

The modulation quality was tested with OFDM signal at 54 Mbps, having more than 10 dB peak to average ratio, the measured EVM of the FEM is shown in Fig. 11. The linear power is defined as the peak average power with EVM measurement at less than 3%. The power level achieved was roughly 18 dBm in both bands. The EVM at mid power range is less than 2.0%. The quiescent currents for both band is 110 and 140 mA for b/g and a-band PA respectively. The current at peak linear power are approximately 180 and 220 mA with worst-case harmonic emissions below -50 dBm/MHz up to 20 dBm output power. These performance attributes can be partly attributed to the usage of the SOI technology platform for this design. Consequently, a high linearity dual-band FEM can be housed within a 4x5 mm compact package addressing the requirements of dual-band WiFi and MIMO applications.

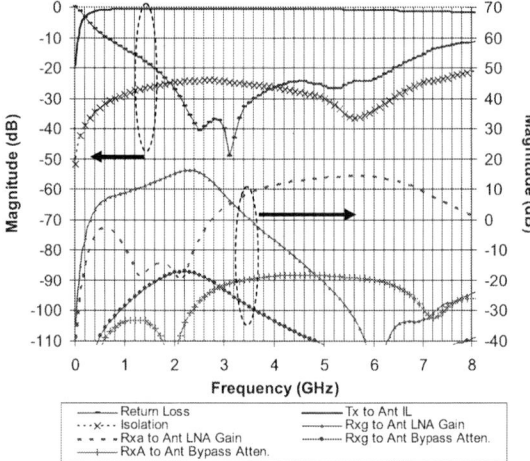

Fig. 8. s-parameters for the SP4T switch-LNA.

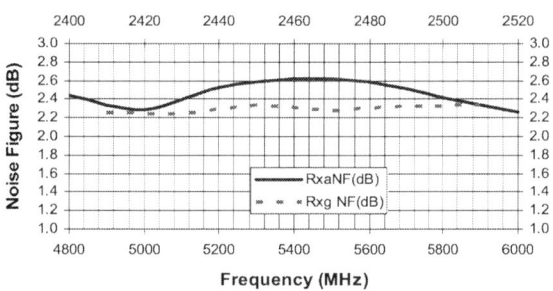

Fig. 9. Measured NF of the Rx paths in the SP4T switch-LNA.

VII. Conclusion

A novel SOI-based SP4T T/R switch with an integrated dual-band LNA and bypass attenuators is presented. The design consists of 2 receive and 2 transmit paths. Each Rx path has an integrated LNA and an 18-dB bypass attenuator. The Tx paths features 0.1 dB compression point above 34 dBm and 0.5-0.8 dB insertion loss between 1 to 6 GHz with > 20 dB return loss and > 25 dB isolation. Receive paths feature 16 dB gain with < 2.3 dB NF for b/g band path and 14 dB gain with 2.4-2.6 dB NF for 'a-band' path. Pairing the switch-LNA with a self-sufficient dual-band WLAN PA, an innovative 2-die dual-band FEM can be realized as shown in Fig. 2 in a 4 x 5 mm compact

package. Moreover, as a result of the low post PA losses, the dual-band transmit linearity at >18 dBm with EVM < 3% and < -50 dBm/MHz harmonic emissions were achieved. The proposed SP4T switch-LNA front-end IC greatly simplifies dual-band FEM designs and enables the reduction of the FEM form factor, which leads to simple constructions of complex dual-band MIMO radios for 802.11n applications.

REFERENCES

[1] W. J. Choi, etc, "Circuit Implications of MIMO Technology for Advanced Wireless Local Area Networks," IEEE RFIC Symp. Dig., June 2006.

[2] C.-W. P. Huang, etc, "A 5 x 5 mm Highly Integrated Dual-band WLAN Front-End Module Simplifies 802.11 a/b/g and 802.11n Radio Designs," 2006 IEEE RFIC Symp. Dig., June 2007.

[3] C.-W. P. Huang, etc, "A Highly Integrated Dual Band SiGe BiCMOS Power Amplifier that Simplifies Dual-band WLAN and MIMO Front-End Circuit Designs," submitted to 2010 IEEE IMS Symp.

[4] C.-W. P. Huang, etc, " Innovative Architecture for Dual-band WLAN and MIMO Front-end Module Based on a Single Pole Three Throw Switch-plexer, " 2009 IEEE RFIC Symp. Dig., June 2009.

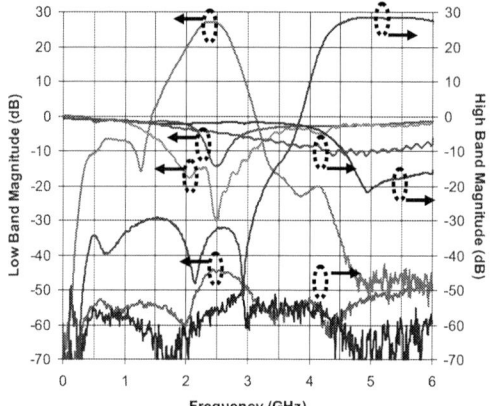

Fig. 10. Transmit characteristics for the proposed FEM.

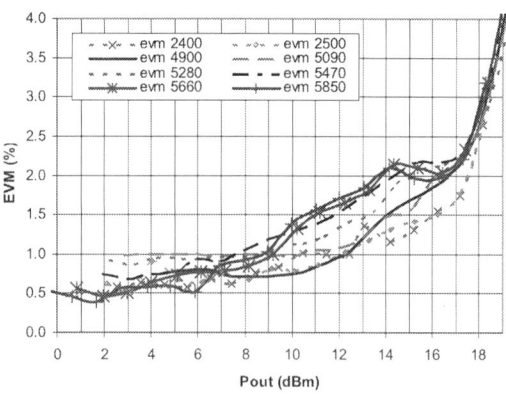

Fig. 11. Measured EVM at 54 Mbps for Tx to antenna path of the proposed 4 x 5 mm FEM.

A 6.1 GS/s 52.8 mW 43 dB DR 80 MHz Bandwidth 2.4 GHz RF Bandpass ΔΣ ADC in 40 nm CMOS

Julien Ryckaert[1], Arnd Geis[1,2], Lynn Bos[1,2], Geert Van der Plas[1] and Jan Craninckx[1]

[1]Interuniversity MicroElectronics Center (IMEC)
[2]Vrije Universiteit Brussel (VUB)

Abstract — **A 2.4 GHz 4th order BP ΔΣ ADC is presented. The feedforward topology uses Gm-LC resonators that can be calibrated in frequency. The quantizer is split in 6 interleaved comparators to relax speed. Clocked at 6.1 GHz, it achieves a DR of 43 dB in 80 MHz consuming 52.8 mW. Implemented in 40 nm CMOS, it achieves a FoM of 3.6 pJ/conv. step, which is to date the lowest published value for RF BP ADCs.**

Index Terms — **ΔΣ ADC, RF bandpass filters, high-speed comparators**

I. INTRODUCTION

The recent growth of circuit techniques that leverage the high speed capabilities of deep submicron CMOS devices is gradually renovating the architectures of RF communication systems. In this evolution, RF bandpass (BP) ΔΣ ADCs have the potential to provide a substantial paradigm shift in RF communication systems towards directly quantized RF signals. These converters could pave the way towards entirely software radios where all signal conditioning would be done in the digital domain. However, the replacement of the entire receiver chain by an ADC poses enormous requirements on the converter mainly in terms of dynamic range, leaving its reality for a long term future. Nevertheless, this work attempts to set a step forward for these systems by demonstrating the capability of integrating such a BP ADC in a digital 40 nm CMOS process at power levels compatible with mobile applications.

The ADC is designed to operate in the 80 MHz wide 2.4 GHz ISM band. The architecture uses a ΔΣ loop where a BP loop filter is used to filter out the oversampled quantization noise created by the internal quantizer.

The paper is structured as follows. Section II describes the architecture of the converter; Section III and IV provide implementation details of the loop filter circuits and the quantizer; Section V introduces the calibration method proposed to adjust the filter center frequency and quantizer thresholds respectively; finally, Section VI provides measurement results of the implementation and Section VII concludes the paper.

II. ADC ARCHITECTURE

Fig. 1 shows a block diagram of the modulator. The 4th order BP loop filter is a feedforward structure with two Gm-LC resonators. The system clock chosen is 6.1 GHz such that the bandwidth of the filter is centered at 0.4xFs = 2.44 GHz. The bandwidth of the modulator is 80MHz leading to an oversampling ratio around 38. The quantizer is split into 6 interleaved quantizers to relax speed requirements mainly limited by comparator metastability.

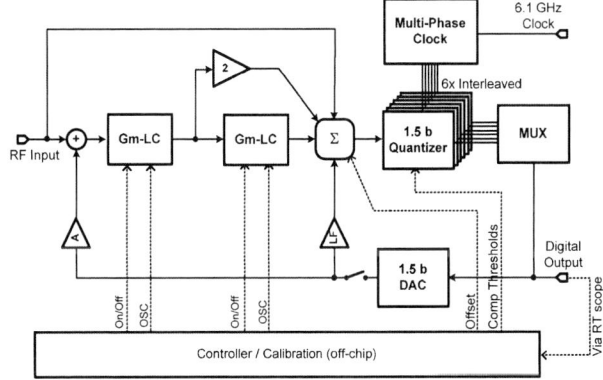

Figure 1: RF Bandpass ΔΣ modulator architecture

The six output streams are then multiplexed to feed a single DAC in order to avoid component mismatches in the feedback path. The signal is quantized into 3 levels (1.5 bit) to extend the dynamic range of the modulator. The center frequencies of the resonators as well as the threshold levels of the comparators are adjusted via a calibration procedure that monitors in open-loop the statistics of the quantized signal. A local feedback path is added to the architecture to accommodate for loop delay in the feedback path.

III. LOOP FILTER CIRCUITS

A resonator circuit is shown in Fig. 2 (left). The input transconductance is a cascoded pseudo-differential pair for improved isolation and linearity. Each resonator tank is made of a center-tap differential 4-turn inductor of 4 nH in parallel with a capacitor. The capacitor is a fixed MoM capacitor in parallel with binary weighted switched NMOS capacitors allowing for center frequency adjustment. To improve the in-band noise shaping by the filter, the quality factor of the tank is enhanced with cross-coupled pairs that compensate for tank losses. This negative resistance is implemented as an array of switchable binary weighted devices. The four-input adder, shown in Fig. 2 (right), is made of 4 cascoded differential pairs loaded by a common resistor. The resistor size of 110 Ω results from a compromise between loop gain and minimum phase shift around 2.4 GHz when loaded by the quantizer. The gain of 2 in the internal feedforward path is emulated by duplicating the corresponding adder stage. All bias voltages and currents are provided by an on-chip bandgap reference circuit, guaranteeing a fixed voltage drop across R_L over PVT variations.

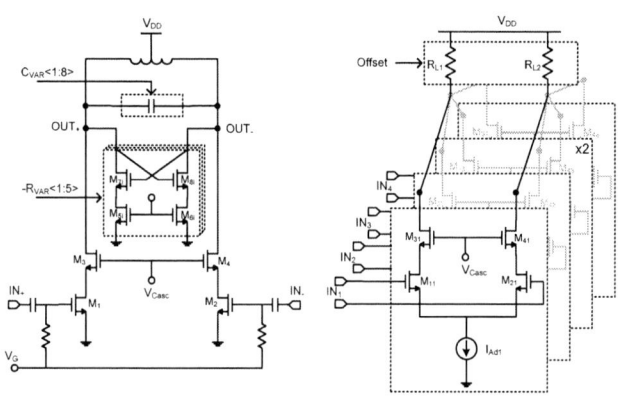

Figure 2: Resonator circuit (left) and adder circuit (right)

IV. QUANTIZER IMPLEMENTATION

The 1.5 bit quantizer is made of two banks of 6 interleaved comparators as used in [1], with opposite threshold levels as shown in Fig. 3. The interleaved clock of the comparators is produced by a synchronous divider made of 12 high speed TSPC flip-flops [2] in a closed loop. The flip-flops are reset to the value '111000111000', which, runs through the flip-flop chain at the rhythm of the 6.1 GHz clock. Each flip-flop output is used as interleaved clock for one single comparator. Although a chain of 6 flip-flops could be sufficient, the use of 12 flip-flops reduces the load on each flip-flop and facilitates the layout. Once a comparator decision is taken, its value is held constant during the reset phase of the comparator by an SR latch. To realize the multiplexing operation, the comparator decision value is NANDed with a pulse of one

clock period, obtained by NANDing two appropriate interleaved clocks. Depending on the decision value, a pull-up pulse or a pull-down pulse is created activating either the PMOS or the NMOS transistor of the tri-state buffer. The short pulse ensures that after the operation, all transistors return to cut-off leaving the line at high impedance. The multiplexed output data are latched before activating the DAC to guarantee a stable loop delay and to reduce jitter in the feedback path. The DAC uses a switched NMOS current source topology with positive/negative/dump-to-supply output for proper 1.5 bit operation.

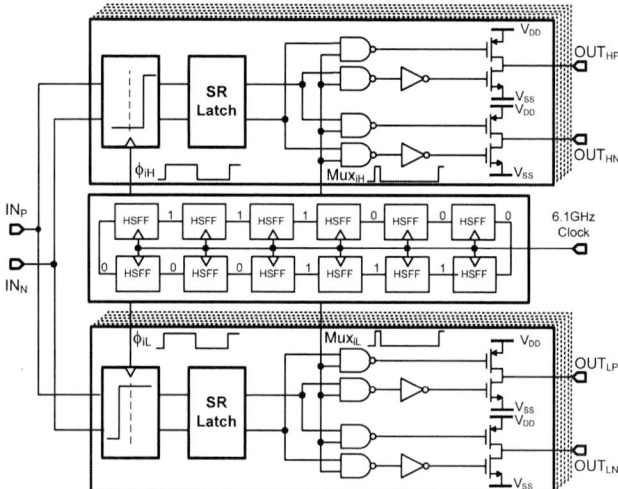

Figure 3: Interleaved quantizer and multiplexing stage

V. ADC CALIBRATION

Before closing the loop, the loop filter and the quantizer thresholds need to be calibrated. The resonators are calibrated by turning them into their oscillating mode using sufficient negative resistance. As shown in Fig 4, the oscillation frequency can be detected digitally with a counter at the output of the quantizer. A frequency locked loop can then adjust the center frequency via a controller, which is positioned off-chip. The two resonance frequencies can be detuned to extend the bandwidth of the modulator. The center frequency is controlled with an accuracy of a few MHz to guarantee the final $\Delta\Sigma$ performance. Once calibrated, the resonator Q is reduced until no more oscillation is observed at the output. The measured calibrated loop filter transfer function is plotted at the bottom of Fig. 3. To calibrate the comparator thresholds, the required offset voltage is applied at their input by simply setting an imbalance in the load resistors of the preceding adder. Each comparator is then calibrated individually using a capacitor array as in [3], of which the setting is selected that flips the output.

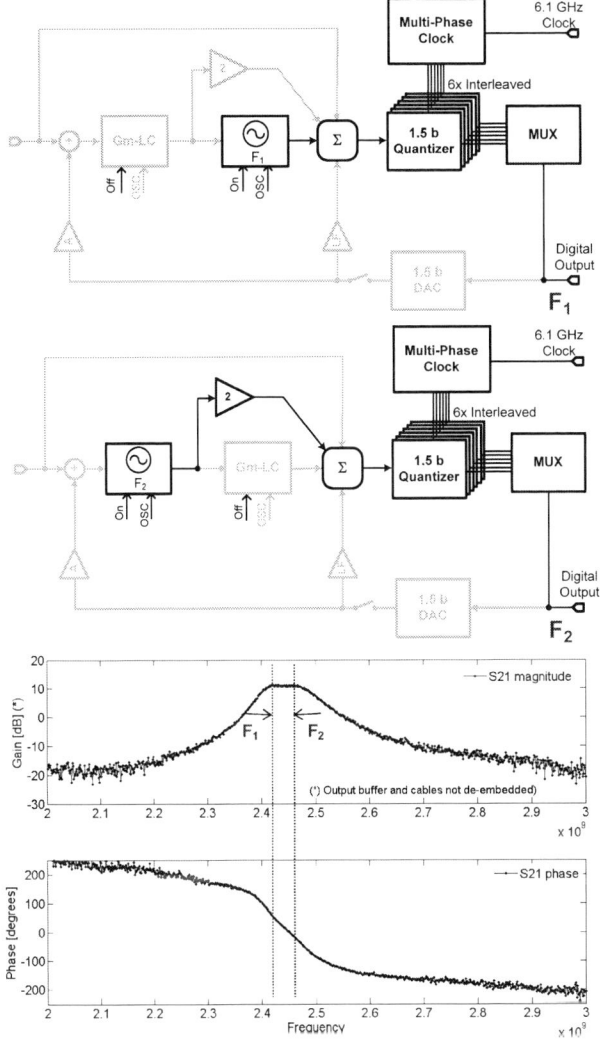

Figure 4: Center frequency calibration and resulting measured loop transfer function

VI. MEASUREMENTS

The ADC was fabricated in a 40 nm CMOS technology and a chip micrograph is shown in Fig. 5. The active area is 0.4 mm² and is substantially reduced with respect to [4] thanks to the use of 2 inductors instead of 4.

Figure 5: Chip micrograph. Quantizer, DAC and clock generation/distribution are highlighted together

The digital output data were measured on a 12GHz real-time oscilloscope. An FFT of the digital bit stream for a 2.42 GHz input tone at -15 dBm input power (-8 dBFS) is given in Fig. 6.

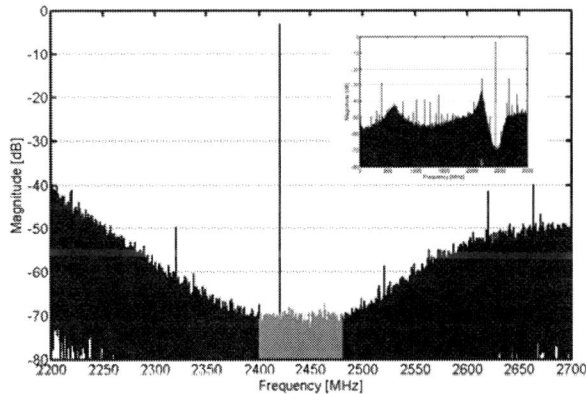

Figure 6: Output spectrum obtained from an FFT on the digital bit stream around the center frequency for an input signal at 2.42GHz

The SNDR and SFDR (single-tone) vs. input signal power are plotted in Fig. 7. Measured on a bandwidth of 80 MHz, a maximum SNDR of 41 dB and SFDR of 65 dB were obtained and a dynamic range higher than 43 dB was measured. The IIP3 of the ADC was measured to be -5 dBm. The current consumption breakdown is 30 mA for the RF filter, 12 mA for the quantizer and clock network, and 6 mA for the DAC including bandgap reference circuit, clock drivers and low-jitter re-timing latches.

978-1-4244-6240-7/10 $26.00 © 2010 IEEE

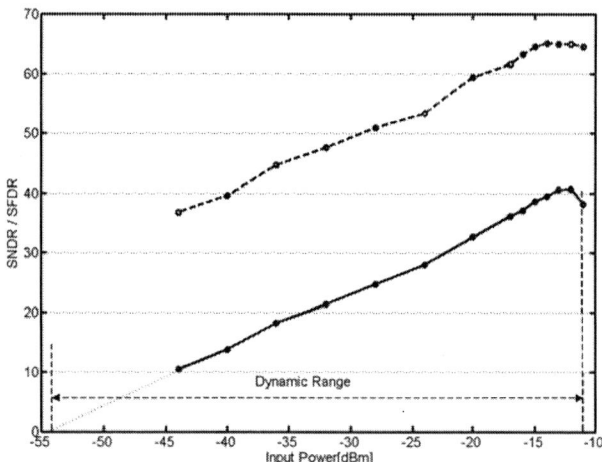

Figure 7: Measured SNDR and SFDR plotted against absolute input power

With a total power consumption of 52.8 mW from a 1.1 V supply, the FoM, is 3.6 pJ/conversion step which is to date the lowest value published for designs in SiGe [5-7] as well as CMOS [4], [8], while implemented in an advanced 40 nm CMOS technology. Also, when compared to [4], the proposed ADC uses 2 inductors instead of 4, which not only reduces the area by more than a factor 2, but also reduces the complexity of the system, simplifying the calibration of the resonators and reducing the effect of coupling between inductors. A comparison of this design with state of art implementations is given in Table 1.

Table 1: Performance summary and comparison to state of art

	[4]	[5]	[6]	[7]	[8]	This work
Center frequency	2.4 GHz	2 GHz	950 MHz	1 GHz	2.442 GHz	2.44 GHz
Clock frequency	3 GHz	40 GHz	3.8 GHz	4 GHz	3.256 GHz	6.1 GHz
SNDR	40 dB	52 dB	59 dB	40 dB	34 dB	41 dB
Bandwidth	60 MHz	120 MHz	1 MHz	20 MHz	25 MHz	80 MHz
Power Consumption	40 mW	1600 mW	75 mW	450 mW	26 mW	52.8 mW
Technology	90 nm CMOS	0.13 µm SiGe	0.25 µm SiGe	0.5 µm SiGe	130 nm CMOS	40 nm CMOS
Active Area	0.8 mm²	~1 mm²	1.08 mm²	1.36 mm²	0.27 mm²	0.4 mm²
FOM [pJ/conv. step.]	4.1	20.4	51.5	142.9	12.7	3.6

VII. CONCLUSION

This paper described the implementation of an RF bandpass ΔΣ ADC in a standard 40 nm CMOS process. Its low power and low complexity demonstrates the feasibility of bringing an ADC next to the antenna of a wireless receiver. A calibration procedure is also proposed to compensate for the inherent process variations in the loop filter and in the quantizer thresholds. Although the performances obtained are not yet sufficient to substitute a complete receiver chain, this design sets a new path towards entirely software receivers.

ACKNOWLEDGEMENT

The authors thank J. Borremans and B. Verbruggen for technical discussions, H. Suys for the PCB design, P Van Wesemael for the design of the network on chip and their colleagues from the Invomec division for their support during digital synthesis and layout.

REFERENCES

[1] M. Miyahara et al. " A low-noise self-calibrating dynamic comparator for high-speed ADCs," *IEEE Asian Solid-State Circuits Conference*, pp. 369–372, Nov. 2008

[2] J. Yuan and C. Svensson. "High-speed CMOS circuit techniques," *IEEE Journal of Solid-State Circuits*, vol. 24, no 1, pp. 62–70, Apr.1989

[3] G. Van der Plas, S. Decoutere, and S. Donnay, "A 0.16pJ/conversion-step 2.5mW 1.25GS/s 4b ADC in a 90nm digital CMOS process," *ISSCC Dig.Tech. Papers*, pp. 566–567, Feb., 2006

[4] J. Ryckaert et al. "A 2.4GHz 40mW 40dB SNDR/62dB SFDR 60MHz bandwidth mirrored-image RF bandpass ΔΣ ADC in 90nm CMOS," *IEEE Asian Solid-State Circuits Conference*, pp. 361–364, Nov. 2008

[5] T. Chalvatzis at al. "A low-noise 40GS/s continuous-time bandpass ΔΣ ADC centered at 2GHz for direct sampling receivers," *IEEE J. Solid-State Circuits*, vol. 42, no. 5, pp. 1065-1074, May 2007

[6] B. K. Thandri and J. Silva-Martinez, "A 63dB 75-mW bandpass RF ADC at 950 MHz using 3.8-GHz clock in 0.25-µm SiGe BiCMOS technology," *IEEE J. Solid-State Circuits*, vol. 42, no. 2, pp. 269-279, Feb. 2007

[7] J. A. Cherry, W. Snelgrove and W. Gao, "On the design of a fourth-order continuous-time LC delta-sigma modulator for UHF A/D conversion," *IEEE T. Circ. Syst.. II, Analog Digital Signal Proc.*, vol. 47, no. 6, pp. 518-530, Jun. 2000

[8] N. Beilleau et al."A 1.3V 26mW 3.2GS/s undersampled LC bandpass ΔΣ ADC for a SDR ISM-band receiver in 130nm CMOS," *RFIC symp. 2009*, pp. 383-386, Jun. 2009

RTU2A-5

Single-Chip WiFi b/g/n 1x2 SoC with Fully Integrated Front-end & PMU in 90nm digital CMOS technology

J. C. Jensen[1], R. Sadhwani[1], A. A. Kidwai[1], B. Jann[1], A. Oster, M. Sharkansky[2], I. Ben-bassat[2], O. Degani[2], S. Porat[2], A. Fridman[2], H. Shang[1], C. Chu[1], A. Ly[1], M. Smith[1]

[1]Intel Corporation, Hillsboro, OR, USA, [2]Intel Corporation, Haifa, Israel

Abstract — **We report a compact 802.11b/g/n MIMO SoC with fully integrated transceiver, on-chip PMU including dc-dc converters, PHY, MAC, PCIe and a non-volatile memory. The transceiver includes on-chip PA, LNA and T/R switch. Fabricated in 90nm standard digital CMOS technology, this IC consumes 663/878mW (Rx/Tx 54Mbps) with an area of approx 33mm2. A peak saturated power of 24dBm is achieved at antenna.**

Index Terms — **Power amplifiers, PAPD, EVM.**

I. INTRODUCTION

In 2002 just over 23 million 802.11b ICs were shipped with an average sales price of $16.50. By 2005, the ASPs dropped to approximately $5.50 and the trend continues downward [1]. Today, over 97% of all notebooks and netbooks are equipped with WiFi. This explosive growth has brought strong competition and strict requirements on the level of integration and associated BOM. Integrating greater functionality into the IC reduces footprint and cost simultaneously but must be accomplished while still achieving the same performance users expect with current generation of products.

In the process of optimizing cost and footprint, single chip solutions have been demonstrated [2,3]; but this paper reports the highest level of integration to date while achieving performance levels equal or better than the existing solutions. This paper describes a single-chip, 802.11b/g/n 1x2 MIMO transceiver with RF, analog, baseband physical layer (PHY), medium access control (MAC) and host functions included. This IC includes on-chip low noise amplifier (LNA), power amplifier (PA), transmit/receive (TR) switch and power management with dc/dc converters. In addition, an on-chip non-volatile memory block was incorporated to replace the off-chip EEPROM to further save the cost.

This IC was fabricated in a 90nm standard digital CMOS process without metal-insulator-metal (MIM) capacitors or ultra thick top layer metal. Standard thick metal was used for RF and power routing including inductors. Aluminum redistribution metal was used to augment the top copper metal layer when required. Custom ESD design and seal rings were designed to reduce coupling around the die. These techniques were employed along with a tiered isolation technique using gradually decreasing p+ taps isolation rings to reduce coupling and improve isolation. Larger taps were placed between high level blocks and smaller taps were placed at the sub levels.

The device reported in this paper achieves the high reliability standards set by the product usage models. The target was 5 years of lifetime with less than $125^{0}C$ junction temperature. These targets are simulated during design phase for device degradation (NBTI and HCI) to be less than 10% and layout meets the electro-migration and self-heat reliability criteria. Utilizing the low cost metal stack-up requires a focused effort to meet the lifetime targets with high current density blocks such as PA and dc-dc converters, hence careful attention is paid to these considerations from the beginning of the design phase.

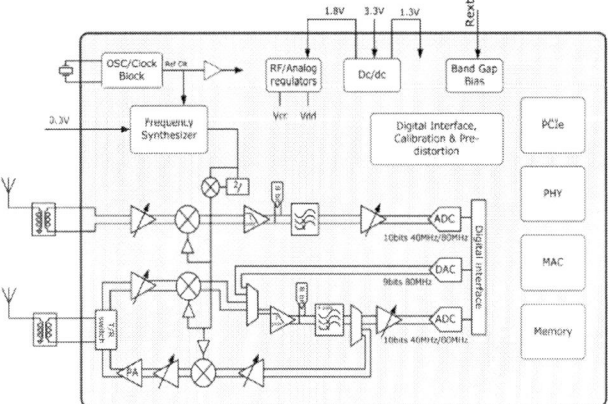

Fig.1. Single-chip MIMO block diagram

II. RECEIVER DESIGN

The receiver with fully integrated ultra-low-insertion loss T/R switch achieves 5.5dB NF measured at the antenna [5]. The T/R switch has 0.3dB insertion loss in the transmit mode and adds a mere 0.1dB of NF in the

978-1-4244-6240-7/10 $26.00 © 2010 IEEE   403

receive mode while occupying 0.02mm2 in total die area. The transmit and the receive side switch employed different architectures. The transmit side switch used a thin gate NMOS series switch with novel high bulk impedance technique which enhances the power handling capability of the transmit side switch and also improves the insertion loss. The switch architecture in the receive side was merged with the series matching inductor of the LNA to achieve the desired switching action.

The LNA was designed using classic common-source architecture. Gain control in the LNA was achieved by current steering to ensure a stable S11 during gain stepping. The input series matching inductor was designed as a custom differential inductor with concentric coils to reduce the area as the mutual coupling contributed to the total inductance required for the series power matching.

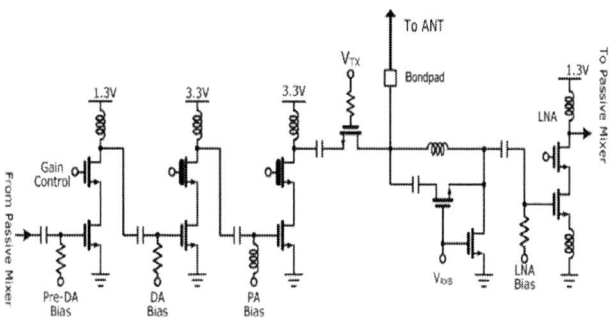

Fig.2. Transistor level diagram of Receiver and Transmitter

III. TRANSMITTER DESIGN

In the transmitter side, the PA used a thin-gate NMOS transistor as the input device and a thick-gate NMOS as cascode device (Fig.2). The PA biasing was found to be challenging and was dealt with utmost care to meet all the performance specifications. Therefore, the bias level of the PA was designed to be digitally programmable so that a variety of levels can be explored. At low bias levels, class-AB action creates AM to AM with significant gain expansion and provides flat gain up until hard clipping resulting in a high OP1dB. At high bias levels, gain rolls off slowly as power is increased and results in a lower OP1dB. Error Vector Magnitude (EVM) is optimized at lower bias levels where gain is flatter for a larger range of powers. However, meeting mask measurements requires the bias to be high to reduce high order non-linearities; thus preferring class-A operation. The EVM however will also improve at sufficiently high biases. Even though high efficiency can be obtained at low currents, variations over process,

temperature and supply vary performance significantly if operated into this highly sensitive class-AB regime (Fig.3). Since the top end of the bias range is limited by thermals, a compromise was made between performance robustness and power requirements. The final bias point was chosen to be 120mA of PA bias current. As can be seen in Fig.3, the TX performance is limited by the regulatory requirements at the band edge and not by the in band EVM performance. The EVM limits the performance for channels other than those at the band edge.

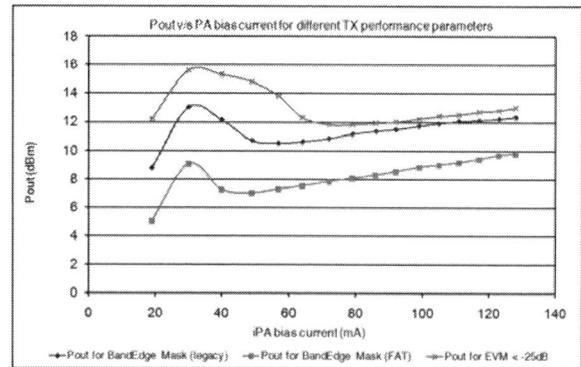

Fig.3. TX performance v/s PA current

To further improve the output power for EVM of -25dB or better, digital pre-distortion is employed. When each unit is calibrated separately, typical EVM improvements are better than 3.5dB. However, to reduce the production testing time, pre-distortion coefficients are kept constant across various skews (slow, typical, fast and temperature variation). It was seen that this reduces the EVM improvement to 2.5dB but reduces production testing time significantly.

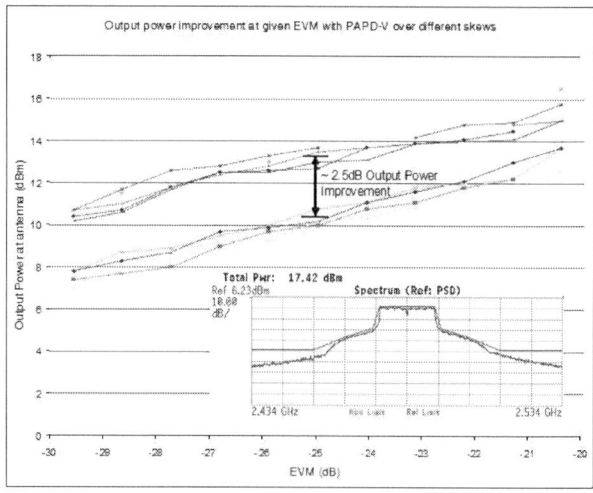

Fig.4. PA Power with pre-distortion at given EVM (HVM data)

Voltage drop from supply loss presents another challenge in this technology. The performance of high current consuming circuits such as PAs will be severely impacted by the IR loss if measures are not taken to reduce metal resistance to a minimum. Metal resistance also leads to self heating and thermals are bound by the package technology used. All inductors and RF lines in the PA were strapped with AL redistribution metal and multiple metal layers were strapped together for underpasses. Multiple wire-bonds were used for the PA supply and the ground.

The final group of components that should be brought on chip is the power management unit (PMU). The PMU can be a single module or a group of discrete devices depending on the functionality, cost, and footprint required. The complex RF design requires multiple supply voltages and separate supply regulators for isolation between digital and sensitive RF/Analog blocks. The power domain is partitioned into 5 main sectors: radio frontend, radio synthesizer, analog (baseband filters and ADC/DAC), PCIe and digital section. The digital circuitry and the PCIe are powered from on-chip dc-dc that uses 3.3V to generate 1.3V with >85% efficiency. For the rest of the sensitive analog and RF blocks, additional dc-dc is used to generate 1.8V which is regulated down to 1.3V supply using multiple linear regulators. This two step supply voltage generation from 3.3V provides good tradeoff between power efficiency and supply isolation. The regulators are optimized to have high PSRR (~40dB) upto dc-dc switching frequency (~1MHz) and each regulator has off-chip decoupling capacitors for stabilization and high frequency PSRR. We integrated all PMU components on chip except for a large inductor used to filter noise on the dc-dc output.

IV. QUALITY AND RELIABILITY

The heat dissipation in single chip is a significant issue. Power Amplifier is the most critical block for meeting heat dissipation target. As seen in Fig.5, the PA met the design target of less than 125^0C junction temperature.

The silicon was validated for reliability and quality by employing accelerated burn-in testing on a large sample of units (~5000) to confirm the low number of defects per million. All the pins passed HBM and CDM ESD requirements. With proper consideration to layout and design for manufacturability rules, the overall yield of this single chip is higher than 99%.

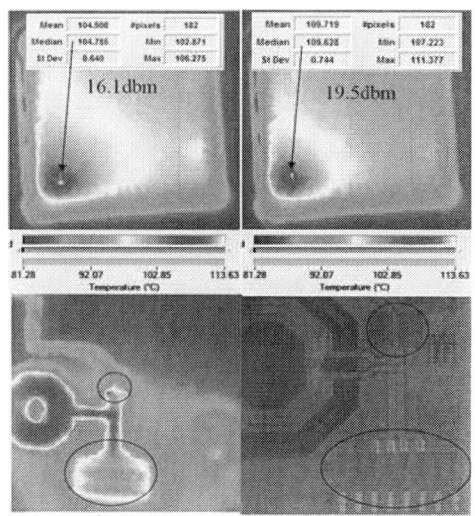

Fig.5. Thermal images showing hot-spots (72^0C ambient)

V. MEASUREMENTS RESULTS

Packaged in 10mm x 10mm dual-row QFN, the IC was mounted on a 29mm x 23mm board which was built using only two layers and components populated on a single side. To the best of the authors' knowledge, this silicon is one of the most compact solutions requiring only a single external supply voltage of 3.3V and highest output transmit power reported till date.

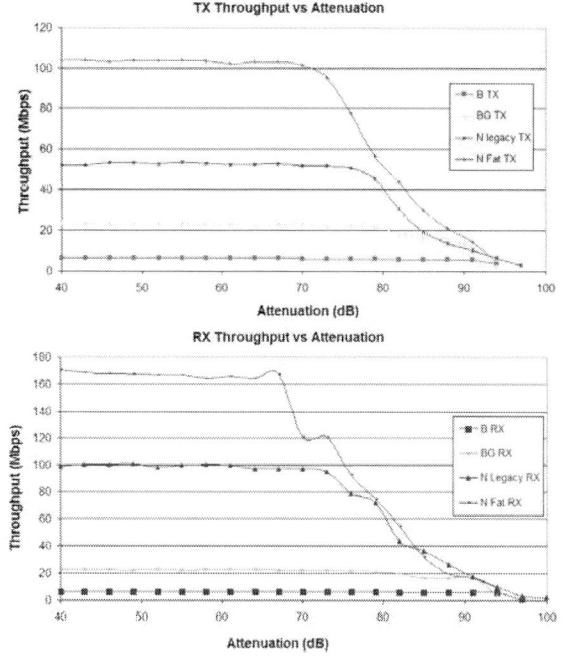

Fig.6. RX/TX throughput performance

Table 1: Performance Specifications

	This Work (measured at the antenna connector)	[2] (measured at the antenna connector)	[3] (Measured at Soc)	[6] (Measured at Soc)
Standard	802.11b/g/n	802.11b	802.11a/b/g/n	802.11a/b/g & BT
Receiver Specifications				
Receiver Sensitivity 6Mbps/54Mbps (dBm)	-89/-73	-93 @ 2Mbps -88 @ 11Mbps	-92/-74	-75 @54Mbps
Receiver Noise Figure (dB)	5.5	6-7	4.0	n/a
Power consumption in 54Mbps (Legacy 802.11g) (mW)	663 (RF Rx/Synth/BB Filter= 264 MAC/PHY/PCIE/ADC =399)	525 mW @ 11Mbps 802.11b	n/a	315 (RFRx/Synth/Filter/ADC) excluding PMU efficiency
Power consumption in 300Mbps (MIMO 802.11n mode) (mW)	818 (RF Rx/Synth/BB Filter= 352 MAC/PHY/PCIE/ADC =466)	n/a	800	n/a
Transmitter Specifications				
Output 1dB Compression point (dBm)	19.3	18	n/a	11
Output power to meet EVM -25dB (dBm)	15.5 @ -25dB EVM (with digital pre-disortion)		-8 @-31dB EVM	0 @ -35dB EVM
Power consumption in 54Mbps transmit rate (Legacy 802.11g) (mW)	878 @ 15dBm output power (RF Tx /Synth/BB Filter= 600 MAC/PHY/PCIE/DAC = 278)	810@13dBm Output power 802.11b 11Mbps		251 (RFTx/Synth/Filter/DAC) excluding PMU efficiency
Power consumption in 150Mbps transmit rate (802.11n mode) (mW)	1007@15dBm output power (RF Tx/Synth/Filter = 600 MAC/PHY/PCIE/DAC = 407)		630 @ -5 dBm output power	n/a
Other Specifications				
Technology	90nm CMOS with 3V I/O	180nm CMOS	130nm, CMOS	90nm CMOS
Die Area	33 sq. mm	32.2 sq. mm	36 sq. mm	n/a
Package	10x10 QFN dual row 86pin	144 pin BGA	88pin leadless	n/a

The elements that remained on the board were power-supply decoupling capacitors, a bias reference resistor and a balun-filter. When benchmarked against similar single-chip solutions [2,3,6] the solution presented here shows similar performance with a greater level of integration.

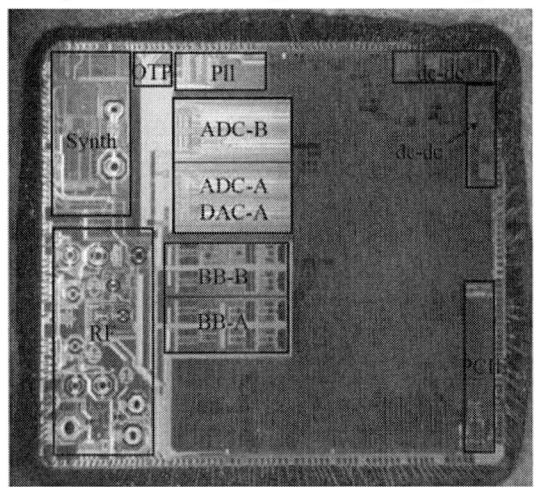

Fig.7. Die Photo

VI. REFERENCES

[1]. Wi-Fi IC Market Data, ABI research, Q3 2007.

[2]. S. Khorram, H. Darabi et al, "A Fully Integrated SOC for 802.11b in 0.18-um CMOS", IEEE J. Solid-States Circuits, vol. 40, No. 12, pp.2492-2501, Dec 2005.

[3]. M. Zargari, L. Y. Nathawad et al, "A Dual-Band CMOS MIMO Radio SoC for IEEE 802.11n Wireless LAN", IEEE J. Solid-State Circuits, vol. 43, No. 12, pp. 2882-2895, Dec 2008.

[4]. A. Shirvani, D. Cheung et al, "A dual-band triple-mode Soc for 802.11a/b/g Embedded WLAN in 90nm CMOS", IEEE CICC Conference, pp89-92, Sep-2006.

[5]. A. A. Kidwai, C. T. Fu et al, "A Fully Integrated Ultra-Low Insertion Loss T/R Switch for 802.11b/g/n Application in 90nm CMOS Process", IEEE J.Solid-States Circuits, vol.44, No. 5, pp1352-1360, May 2009

[6]. P.B. Leong, D. Cheung et al, "World's First 90m CMOS Single-Chip Bluetooth and WLAN with integrated RF", IEEE International Symposium on Integrated Circuits, ISIC OCT 2007

RTU2C-1

A 44-GHz 8-Element Phased-Array SiGe HBT Transmitter RFIC with an Injection-locked Quadrature Frequency Multiplier

Sunghwan Kim[1], Prasad S. Gudem[2], and Lawrence E. Larson[1]

[1]Center for Wireless Communication, University of California San Diego, La Jolla, CA, USA

[2]Qualcomm Inc., San Diego, CA, USA

Abstract—An 8-element 44-GHz phased-array direct up-conversion transmitter, based on a localized injection-locked quadrature oscillator, is fabricated in a SiGe HBT process. The transmitter includes an improved injection-locked quadrature frequency doubler, an LO active phase shifter, I/Q mixers, and an RF PA driver. The transmitter has approximately 20-dB conversion gain per element, continuous $360°$ phase-shift and a baseband I/Q bandwidth of 2.2-GHz. The maximum saturated RF output power is 2-dBm at 45-GHz. Each element consumes 450mW. The chip size, including the pads, is 3×2.4 mm^2.

Index Terms—up-conversion mixer, LO-path phase shifter, SiGe, BiCMOS, phased-array, transmitter, millimeter-wave integrated circuits.

I. INTRODUCTION

Several monolithic millimeter-wave phased-array transmitter IC's have been demonstrated recently for radar and communications applications [1]–[7]. These transmitters often require variable phase shifters in the RF, LO or IF-path, which should exhibit low loss and low dc power consumption. Typically, RF and LO-path phase shifters require a low-loss transmission line structure along with amplifiers to compensate for power loss during signal distribution, which increases at higher working frequencies.

In order to ease the LO distribution at millimeter-wave frequencies, we propose a frequency doubling injection-locked quadrature oscillator, which can be used at each transmitter element to reduce power consumption and LO signal loss during distribution. Specifically, this oscillator functions as a quadrature frequency doubler, which generates differential quadrature outputs from a single differential signal input. This paper describes a millimeter-wave phased-array transmitter with a new injection-locked quadrature frequency doubler.

II. PHASED-ARRAY TRANSMITTER ARCHITECTURE

Fig. 1 shows the architecture of the proposed transmitter. A differential LO signal is converted to differential I/Q signals at twice the LO input frequency in the injection-locked quadrature frequency doubler. An active phase shifter, which is composed of four independently controlled VGA's, uses the I/Q signals to generate phase-shifted I/Q LO signals for the up-conversion mixers. Eight of these proposed phased-array

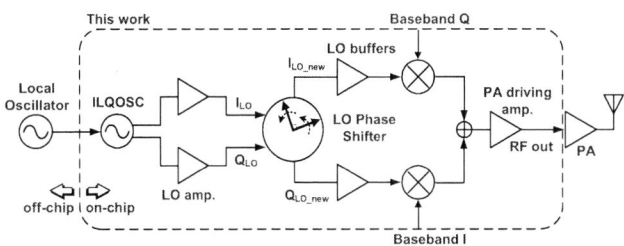

Fig. 1. Architecture of one proposed transmitter element (of eight).

transmitters are integrated on-chip. One differential LO input port and one I/Q baseband input are shared for the phased-array transmitter.

III. CIRCUIT DESIGN

A. Injection-locked Quadrature Oscillator

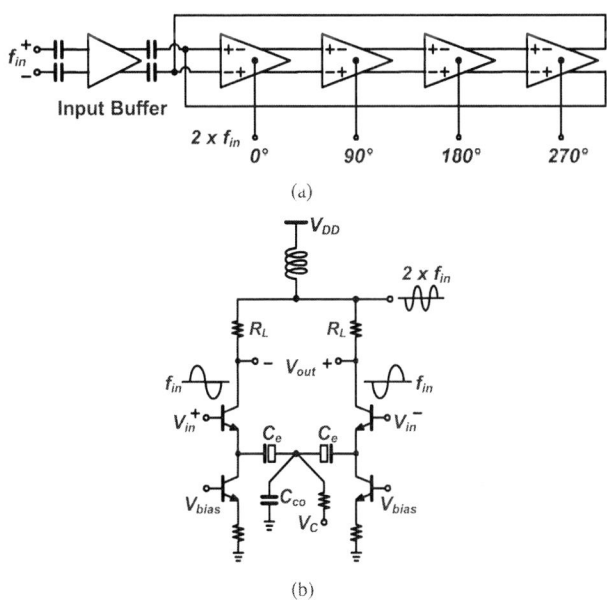

Fig. 2. (a) Block diagram of the injection-locked quadrature ring oscillator, (b) schematic of a differential gain stage with capacitive degeneration.

In the injection-locked quadrature oscillator, resistively loaded differential amplifiers are connected in a ring as shown

978-1-4244-6240-7/10 $26.00 © 2010 IEEE   407

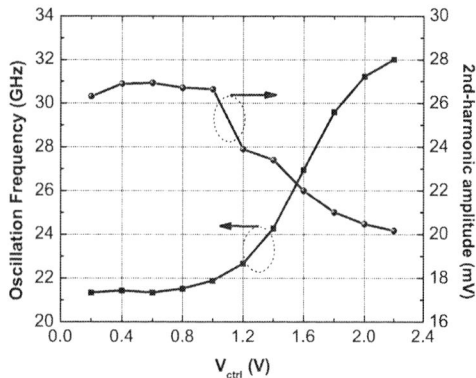

Fig. 3. Simulated oscillation frequency and second-harmonic output amplitude variation with control voltage.

in Fig. 2(a). The first stage is also connected to an LO input buffer to lock the oscillator to the off-chip LO signal.

While each differential amplifier in the ring is operating at f_{in} under injection-locked conditions, the second harmonic tone is extracted at the common-collector node, with an inductive load, as shown in Fig. 2(b). Since the phase shift across each stage is $45°$ under locked conditions, the phase difference across each stage for the *second-harmonic* tone (at $2 \times f_{in}$) is $90°$. Hence, the first (second) and third (fourth) stage's output form a differential output at the doubled frequency and differential I(Q) are obtained. The inductor in the first (second) and the third (fourth) stage's $2 \times f_{in}$ port are merged into one transformer to save chip area. The fact that the injection-locked quadrature oscillator is oscillating at half the LO frequency also mitigates the VCO pulling problem [8].

One potential limitation of this approach is the limited frequency locking range of the oscillator. While a wider locking range can be achieved by increasing the injection power [8] or multiple injection points [9], these techniques involve design challenges such as power loss during distribution of the larger injected signal and more complicated routing when multiple injection-locked oscillators share a common LO input.

A wider input bandwidth for the frequency multiplier can be achieved as shown in Fig. 2(b) by tuning the free-running frequency, by changing the pole location of each gain stage.

The capacitively-degenerated gain stage has a negative input resistance, $R_m \cong -|\beta(\omega)| / (\omega (C_e \| C_{co}))$. The free-running frequency is given by

$$f_o \cong \frac{\left(1 + Q_{in}^2\right)}{2\pi Q_{in}^2 \left[R_L \| (r_b + R_m) \left(1 + Q_{in}^2\right)\right] C_t}, \quad (1)$$

where $Q_{in} = [\omega (r_b + R_m) C_t]^{-1}$, R_L is the load resistance of the gain stage, r_b is the base resistance, and C_t is $(C_\pi \| C_e \| C_{co})$. By adjusting the variable degeneration capacitors (C_e), the total resistance — and hence the free-running frequency — of the oscillator changes. Fig. 3 shows how the free-running oscillation frequency and the amplitude of the second-harmonic output signal changes with MOS varactor tuning.

A small capacitor (C_{co}) is added at the node formed between the two degeneration capacitors (see Fig. 2(b)) to increase second-order harmonic currents at the output node. However, the capacitance must be properly sized to prevent common-mode oscillations.

B. LO Amplifier

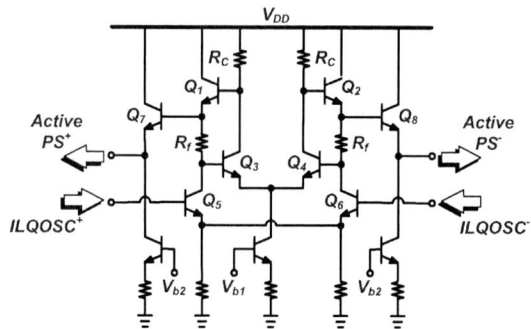

Fig. 4. Schematic of LO boosting amplifier.

An LO amplifier follows the injection-locked quadrature oscillator. A conventional common-emitter amplifier with an inductive load would have a relatively narrow bandwidth. To reduce chip area and power-consumption, a negative feedback Cherry-Hooper topology [10] was adopted, as shown in Fig. 4. Negative feedback formed by Q1 (Q2) and R_f enhances the bandwidth up to the carrier frequency 50GHz. The bias current of this amplifier is 8.8mA from the 3.2V power supply voltage.

C. Active LO Phase Shifter

The active LO phase shifter is implemented using an orthogonal vector summation technique, which is based on the summation of two weighted quadrature vector signals [7]. The preceding injection-locked quadrature oscillator generates differential I/Q signals for this vector summation. The variable gain amplifier (VGA) is implemented with Gilbert cells as shown in Fig. 5(a). Two Gilbert cells share one differential inductor at the output of each I/Q path. Differential voltages V_{conta} and V_{contb} control the dc currents flowing through Q_1-Q_8.

For target phase shifts, both I-VGA and Q-VGA keep the same values of V_{conta} and V_{contb} when the quadrature signal from the injection-locked quadrature oscillator has no I/Q phase mismatches. The moderate I/Q phase mismatch due to PVT variation can be corrected by tuning I-VGA and Q-VGA independently. The control voltages do not change the dc bias current through the RF input devices (Q_9-Q_{12}), so the input impedance stays constant. The bias current of each VGA is 9mA from the 3.2V power supply voltage.

D. Up-conversion Image Rejection Mixer

A double-balanced Gilbert-Cell mixer with an inductive load is implemented, as shown in Fig. 5(b). The switching quad transistors are roughly half the size of the IF transistors for increased f_T to ensure fast switching. The outputs of the mixer are tied together and are connected to the output

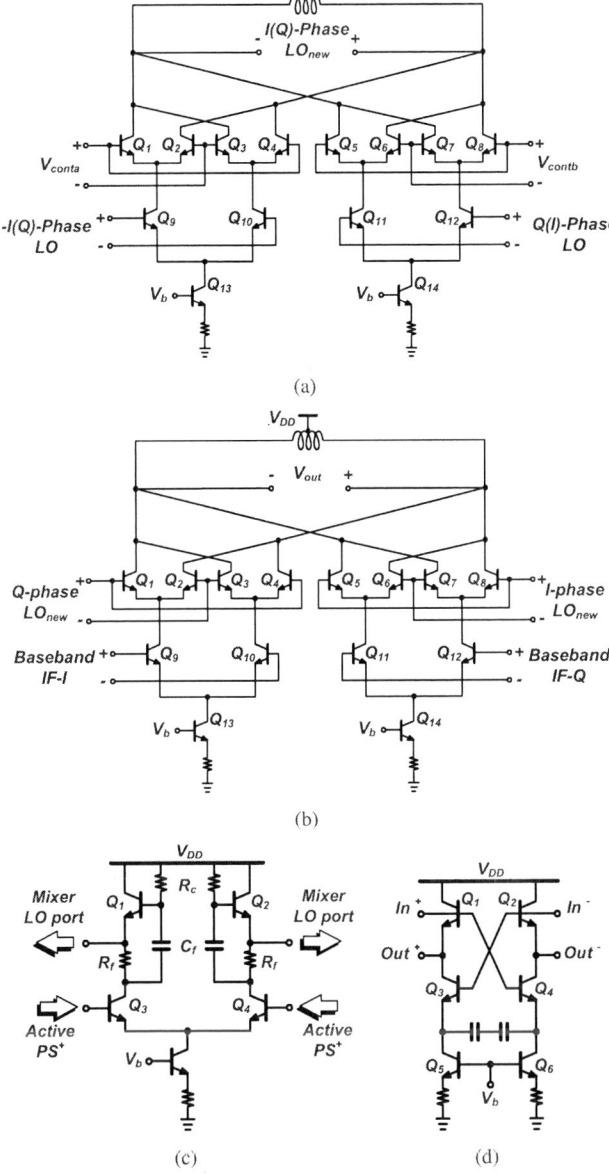

Fig. 5. (a) Active phase shifter (b) double balanced mixer (c) LO Buffer with improved stability and (d) output amplifier.

amplifier. To resonate out the parasitic capacitances, a 150pH differential inductor is connected at the output. The bias current of each mixer is 9.5mA from the 3.2V power supply voltage.

E. LO Buffer and Output Amplifier

A cascade of common-emitter (CE) and common-collector (CC) stage is a widely used topology to drive switches in active current steering mixers. Due to the high quality factor of the passive inductor load of the active phase shifter, a CC stage followed by the active phase shifter would be prone to instability due to the potentially negative real input impedance of the CC stage.

By combining the two topologies into one (see Fig. 5(c)), current can be recycled and better reverse isolation can be

achieved while still maintaining a low output impedance. The finite base resistance of the input transistor in Fig. 5(c) helps to lower the Q factor in the LC tank and its bandwidth is expanded.

A CC stage follows the mixer output to drive a 50Ω off-chip load. For a balanced differential signal and higher gain, current sources in the CC stage are coupled to the opposite phase input signals as shown in Fig. 5(d).

IV. MEASURED RESULTS

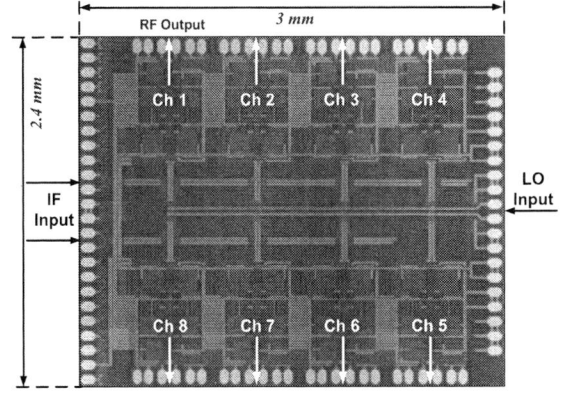

Fig. 6. Microphotograph of the phased-array transmitter.

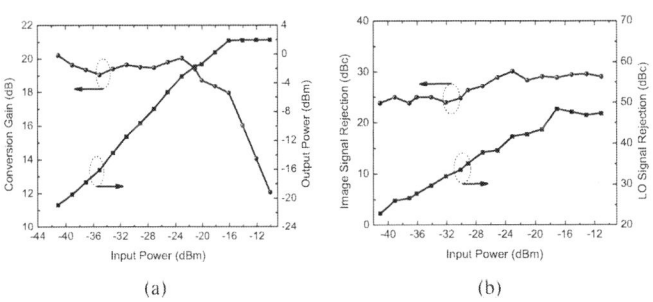

Fig. 7. (a) conversion gain and RF output power and (b) image and LO rejection ratio after calibration with I/Q VGA tuning.

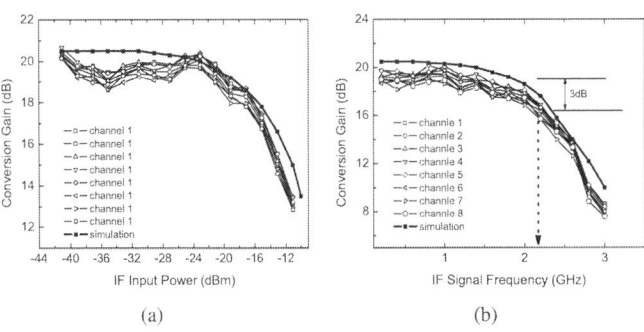

Fig. 8. Simulated and measured conversion gain at 45GHz with IF input (a) power sweep and (b) frequency sweep.

978-1-4244-6240-7/10 $26.00 © 2010 IEEE   409

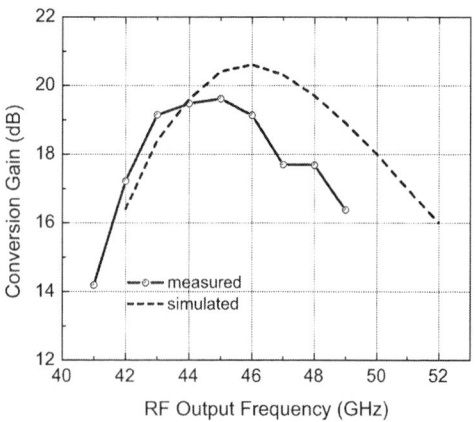

Fig. 9. Conversion gain change with injection-locked LO frequency sweep (f_{if}=1GHz).

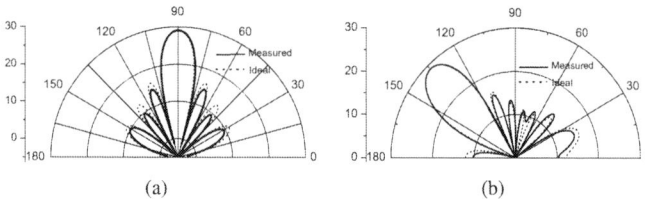

(a)   (b)

Fig. 10. Calculated array beam scanning pattern (a) $0°$ scan angle, (b) $45°$ scan angle.

This phase shifting upconverter was implemented in a 0.18μm BiCMOS process from TowerJazz Semiconductor, which features a peak f_T and f_{MAX} of the SiGe HBT of nearly 150GHz and 200GHz, respectively [11]. The microphotograph of the fabricated die is shown in Fig. 6, and the chip size, including the pads is 3×2.4 mm^2. Eight transmitters are integrated on one chip, while all dc control nodes and LO input ports were tied together. The 8-elements consumes 1.0A from the 3.2V power supply voltage for amplification of LO signals and up-conversion mixers, and 150mA from a 2.4V power supply voltage for the injection-locked quadrature oscillators.

On-wafer measurements were carried out to characterize the performance of the transmitter. After calibration, the conversion gain, output power, image rejection ratio, and LO leakage ratio were as shown in Fig. 7(a) and (b).

The conversion gain is approximately 20dB with maximum output power of 2dBm. Both Image rejection and LO leakage ratio are better than 20dBc. The measured conversion gain of each element is shown in Fig. 8(a) and (b). The conversion gain variation between transmitter elements is less than 2dB. The 3dB bandwidth of the IF input is 2.2GHz.

The bandwidth of the injection-locked quadrature oscillator and LO path were measured for different LO frequencies. The varactor tuning voltage was also adjusted depending on the LO input frequency. Measured and simulated conversion gain over LO frequency change is shown in Fig. 9. The 3dB bandwidth of 7GHz was achieved with minimum LO input power of -12dBm.

In order to construct phased-array patterns, the phase shift of the LO signal at each element was measured. The phased-array

patterns (array factor) were generated using ADS at 45GHz (carrier freq.=44GHz, IF freq.=1GHz) based on the measured phase shift and LO power under an assumption of a standard linear array with isotropic radiators. The resultant $0°$ and $45°$ angle beam were created as shown in Fig. 10(a) and (b).

V. CONCLUSIONS

A 44GHz phase-shifting transmitter based on a direct conversion architecture and LO phase-shifting has been presented in a SiGe HBT technology. This demonstrated an 8-element phased-array transmitter for millimeter-wave applications. The transmitter is based on a new architecture, with an injection-locked quadrature oscillator, which enables simpler LO distribution, wider LO frequency bandwidth, and excellent scalability to larger phased-array applications. The measured lower sideband and carrier suppression exceed 20dB after calibration. 7GHz carrier and 2.2GHz IF signal frequency bandwidth are achieved with a conversion gain of 20dB at 45GHz.

ACKNOWLEDGMENT

The authors would like to acknowledge the support of the DARPA SMART program, as well as the fabrication support of TowerJazz Semiconductor. The authors would also like to acknowledge the support of Dr. Mark Rosker of DARPA and Dr. Jonathan B. Hacker of Teledyne Technologies. The authors would like to thank Professors Gabriel M. Rebeiz, Peter Asbeck, and James Buckwalter of UCSD, and Professor Mark Rodwell of UCSB, for their encouragement and valuable discussions.

REFERENCES

[1] A. Natarajan et al., "A 77-GHz phased-array transceiver with on-chip antennas in silicon: transmitter and local LO-path phase shifting," IEEE Journal of Solid-State Circuits, vol. 41, pp. 2807–2819, Dec. 2006.

[2] B.-W. Min et al., "Single-ended and differential Ka-band BiCMOS phased array front-ends," IEEE Journal of Solid-State Circuits, vol. 43, no. 10, pp. 2239–2250, Oct. 2008.

[3] S. Kim et al., "A low-distortion, low-loss varactor phase-shifter based on a silicon-on-glass technology," IEEE Radio Frequency Integrated Circuits (RFIC) Symposium, pp. 175–178, April 2008.

[4] K.-J. Koh et al., "A millimeter-wave (40-45 GHz) 16-element phased-array transmitter in 0.18-μm SiGe BiCMOS technology," IEEE Journal of Solid-State Circuits, vol. 44, no. 5, pp. 1498–1509, May 2009.

[5] C. Wagner et al., "A phased-array radar transmitter based on 77-GHz cascadable transceivers," IEEE MTT-S International Microwave Symposium Digest, pp. 73–76, June 2009.

[6] S. Kishimoto et al., "A 60-GHz band CMOS phased array transmitter utilizing compact baseband phase shifters," IEEE Radio Frequency Integrated Circuits Symposium, pp. 215–218, June 2009.

[7] S. Kim et al., "A 44-GHz SiGe BiCMOS phase-shifting sub-harmonic up-converter for phased-array transmitters," IEEE Transactions on Microwave Theory and Techniques, May 2009, submitted.

[8] B. Razavi, "A study of injection locking and pulling in oscillators," IEEE Journal of Solid-State Circuits, vol. 39, no. 9, pp. 1415–1424, Sept. 2004.

[9] J.-C. Chien et al., "Analysis and design of wideband injection-locked ring oscillators with multiple-input injection," IEEE Journal of Solid-State Circuits, vol. 42, no. 9, pp. 1906–1915, Sept. 2007.

[10] E. M. Cherry et al., "The design of wide-band transistor feedback amplifiers," Proceedings of the IEEE, vol. 110, pp. 375–389, Feb. 1963.

[11] P. Kempf et al., "Silicon germanium BiCMOS technology," Gallium Arsenide Integrated Circuit Symposium, 24th Annual Technical Digest, pp. 3–6, 2002.

RTU2C-2

A thirty two element phased-array transceiver at 60GHz with RF-IF conversion block in 90nm flip chip CMOS process

Emanuel Cohen[12], Claudio Jakobson[1], Shmuel Ravid[1], and Dan Ritter[2]

[1]Mobile Wireless Group, Intel Haifa, Israel
[2]Electrical Engineering Technion, Haifa, Israel

Abstract — **A 60 GHz 32 element bidirectional phased-array TX/RX chip with a 2 bit phase shifter and IF converter to/from 12GHz, using 90nm CMOS process, is described. The array features 12.5 dB gain, noise figure (NF) of 11 dB, IP1dB of -17dbm for RX, and total output Psat of +8dBm for TX, drawing 390 mA from a 1.3-V supply. The RMS amplitude and phase error of the phase shifter is 0.8dB and 5° max respectively from 57 to 66 GHz. The paper emphasizes the flip-chip assembly technology selected and its impact on performance, and the phase and amplitude errors resulted by physical impairments such as the finite isolation between different chains. Special test structures were designed to measure bump isolation and insertion loss (IL). The designed architecture together with the compact layout results in a die area of 14.5mm² for the full array. To our knowledge, this is the first report on a large bidirectional 60 GHz array, with the lowest reported chip power consumption and size.**

Index Terms — **Phased-array, flip chip, phase shifter, bidirectional, power consumption, 60 GHz**

I. INTRODUCTION

The increasing demand for high data rate combined with today's fast CMOS processes provides new opportunities in the 60GHz frequency range, and creates challenges in designing low cost high performance RF systems. A high quality link requires a phase array steerable antenna with as many as 24-64 elements. Integrated phase array solutions with up to 16 elements for other applications and frequency bands have been reported in [1-3]. However, system calculations indicate that conventional design approaches result in unacceptable chip size and power consumption for a mobile platform. We have recently presented a new approach for designing low power and small size phase array chips using standard 90nm CMOS process, and reported on a 4 element array [4]. Here, we report on a 32 elements array required to support a reliable link for a 5 Gb/s transmission at 10 meter distance. The design challenges associated with a full array can be divided into 3 main challenges. The first is a low power (<1W) architecture that can be realized in a small area IC (<20 mm²) and will meet the system RF requirements. The second challenge is developing a low cost assembly/packaging and testing solution with minimum impact on performance. The third challenge is analyzing

and understanding the potential impact of various impairments, such as coupling and mismatch, on the performance of such a large array. These challenges and proposed solutions will be discussed in this paper.

II. PHASED-ARRAY ARCHITECTURE

Phased array systems are generally composed of an RX and TX array. By sharing the TX and RX elements as phase shifters, combiners, and converters, the array size can be reduced significantly. To reduce size and power, the combining of these array elements is done in the RF domain. Fig. 1 presents the block diagram of the phase array system described in this paper. The array is composed of 32 identical chains with an additional separate chain. The system is divided into 4 sub-arrays of 8 chains each. To achieve the best tradeoff between performance and power consumption only one amplifier is used before the combiner with 4 more after the 8 to 1 combiner. This topology is more sensitive to coupling between chains but this can be managed with optimized phase shifter and combiner designs. The array features independent bias control for each stage to enable performance optimization and compensation between chains. A 12 GHz IF interface is chosen to ease on the external connection, and reduce the coupling from the combining output to the TX/RX inputs.

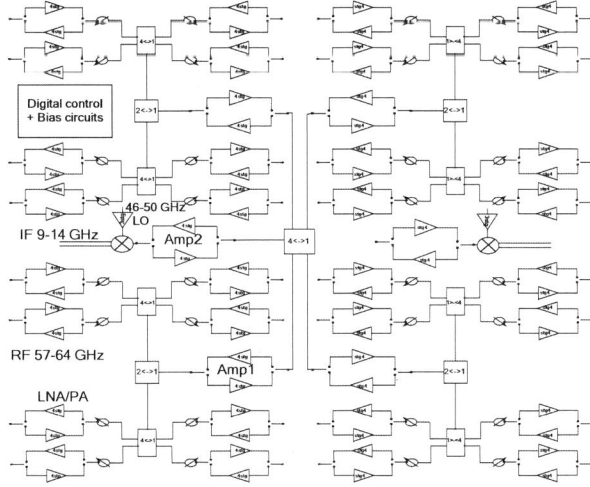

Fig. 1: Block diagram of the 32 element bidirectional array RF-IF

978-1-4244-6240-7/10 $26.00 © 2010 IEEE

III. CIRCUIT IMPLEMENTATION

The design methodology and circuit implementation are described in [4-5]. It is based on dense layout of lumped inductors, repeated layout patterns, passive LC shared structures and a single ended design, all significantly lowering the power consumption and size in a large RF phase-array. Internal coupling between chains is met with a single ended distribution by interleaving between TX and RX chains and routing long lines between ground bumps functioning as shielding.

Fig. 2 shows the schematic of the RF front end. The LNAs/PAs are similar to those reported in [5-6] with 16dB gain at 10mA current supply. Two additional amplifiers indicated as Amp1/2 were added to the full array, exhibiting flat frequency response by employing staggered matching with 11dB gain at 12mA current consumption. The 2 bit phase shifter (PS) implemented was a switched delay passive LC [7-8] with HP/LP configuration. It exhibits wideband phase stability over process variation and has 6dB IL. For combining a Wilkinson LC topology [9] with high isolation was selected. A 4 to 1 and 2 to 1 combiner are used in the array. The combiner exhibits a 2dB IL with 20dB isolation. The converter block uses a complementary differential star type mixer and has 12dB IL. The mixer RF to IF isolation is essential, to reduce the coupling between the IF output and the RF input bumps.

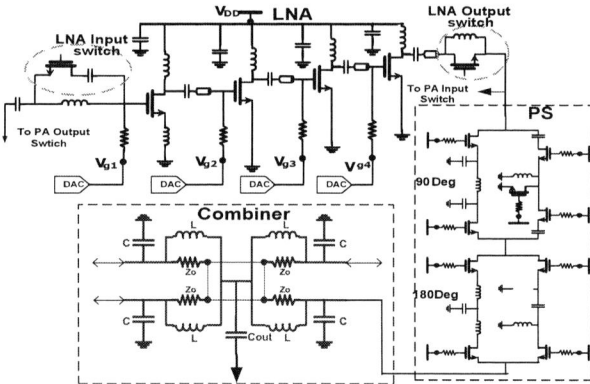

Fig. 2: Schematic of RF main section: LNA + switches, phase shifter and combiner

IV. ARRAY ASSEMBLY AND TEST STRUCTURES

A flip chip interface was chosen to ensure a reduction in the variation between the array chains. The bumps were spread symmetrically across the die with a 200um minimum pitch. The LNA and PA were retuned after accurate 3D EM bumps simulation. The bump impact was minimized by creating a ground ring distant enough from the RF bump. A special alumina test substrate was designed and manufactured to route the RF signals from the die bumps to a measuring point at the edge of the

substrate as shown in Fig.3 allowing the flip chip die to be tested on the probe station. A similar assembly approach will be used to connect the RFIC to a substrate with the antenna array. The interconnect lines were made of a single layer CPW TRL with 5um of electro plated gold to probe pads on the alumina substrate.

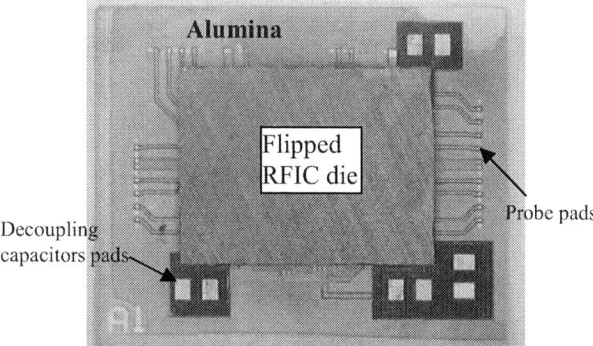

Figure 3: RFIC die flipped on the alumina fan-out structure

The 80um air gap between the die surface and the alumina ground, determined by the bump height, was simulated and found to have minimal impact on the RF performance. To ensure good fit between simulations and measurements, no under-fill material was added to the assembly.

$$IL_{bump} = \left(IL_{tot} - IL_{CPW\_RFIC} - IL_{CPW\_Alumina} \right)/2$$

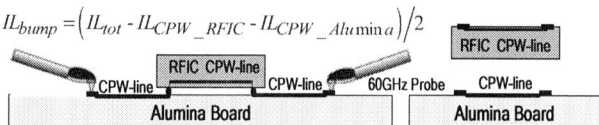

Figure 4: RFIC and alumina test structure for bump IL, isolation, and alumina/RFIC CPW line loss measurement

To measure the bump loss and bump-to-bump isolation, special test structures were created. Fig. 4 shows some of the different structures used and Fig.5 shows the EM simulation configuration. The de-embedded loss of two bumps was measured by subtracting the CPW connecting routing on the alumina and RFIC. The measured bump loss of ~1.5dB at 60 GHz was consistent with 1.3dB predicted by simulation as shown in Fig 6. This loss is hard to improve while maintaining good matching given the current solder bump manufacturing technologies.

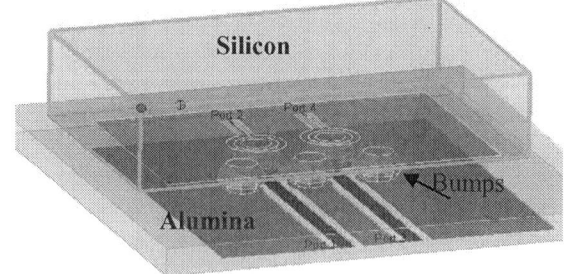

Fig. 5: 3D EM simulation of coupling between bumps and IL

978-1-4244-6240-7/10 $26.00 © 2010 IEEE

Fig. 6 also shows the results of different isolation tests made on the flip chip array and dedicated test structures. For this test, the distance between adjacent bumps is 240um. The 3D EM simulations agree with the measurements showing an isolation of 20dB. This isolation can be improved only by further separating the bumps resulting in increased die size. Since the bump coupling is the limiting factor, a differential circuit design will not help to mitigate this issue. The distance between the two non adjacent bumps is 510um with an additional GND bump between them. In this case, the isolation is around 35dB.

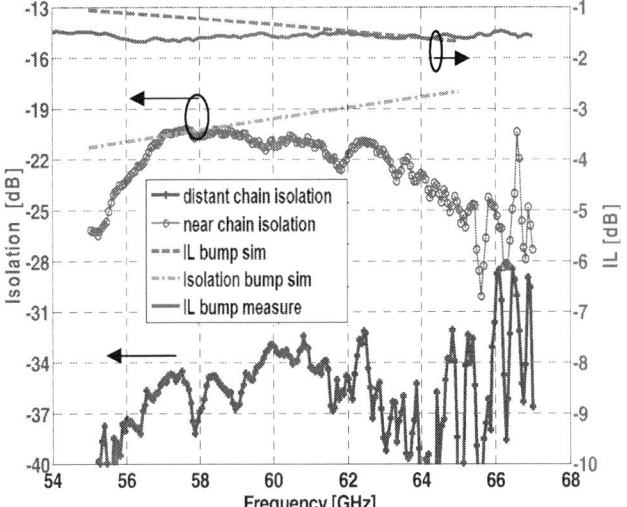

Fig. 6: Simulation and measurement of isolation between bumps in array and a single bump IL

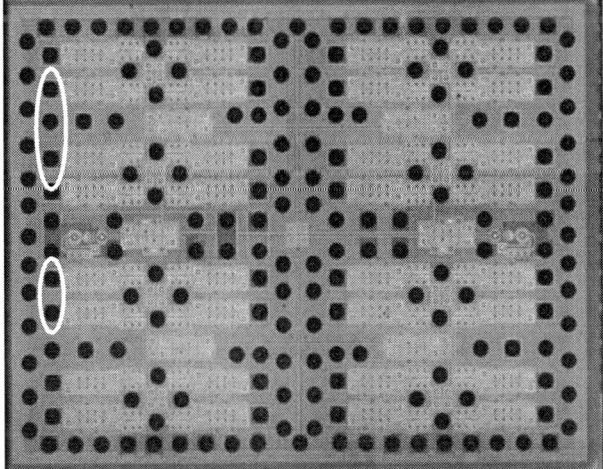

Fig. 7 : 32 elements RF-IF phased-array die 4400 um x 3300 um. Adjacent and non adjacent bumps are marked

V. ARRAY MEASUREMENTS

The array micrograph is shown in Fig.7. The 32 element TX/RX array is 14.5mm^2 in size. S-parameter measurements were made using multiple GSGSG probes with a 67 GHz VNA. De-embedding was implemented by subtracting the alumina CPW routing data. Fig. 8 shows the measured Tx and Rx gain and matching of the array. The transceiver gain is ~12.5dB at 58 GHz and with a total current consumption of 390mA that fits very well with the sum of each block measured separately. A DC probe sense on the alumina is used to keep the Vcc constant and compensate for the IR drop. Different channels were tested showing negligible gain variation between chains. The simulated gain for each block was about 1dB higher than measured. The RF matching for the TX and RX chains are better than -10dB across the band suggesting that bumps behavior was correctly simulated and accounted for. The chain NF is 11dB.

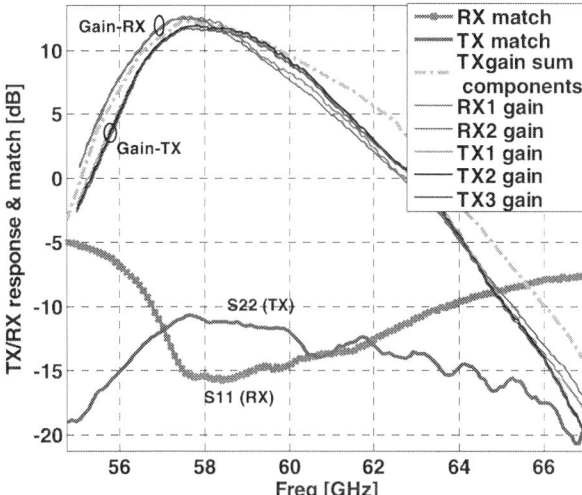

Fig. 8: Measured Gain and matching for TX and RX different chains (LO 48 GHz, -5dBm) on alumina assembly

Fig. 9 shows the phase and amplitude variation over four different phase states measured on the RF-IF array. The phases are measured in TX mode using a reference channel as described in [5]. The phase data is taken per chain with other neighboring chains in different phase states. This measurement is repeated and compared for different chains. In this way the total phase variations of the array can be characterized. The max RMS amplitude error is 0.8dB while the RMS phase error is ~5^0 across the 60GHz band. The variations encountered can be divided into 3 groups. The first is related to process variation that is quite limited due to the broadband PS. The second is due to the finite isolation of the 4 way combiner and the third is due to the coupling between the bumps. The combiner isolation impact is about 2-3^0 and the bump coupling may add as much as 6-7^0. Despite the coupling impact, the RMS error is still tolerable. Moreover, in the full system, there will be additional coupling from the antenna routing and antenna elements that may be a more dominant factor.

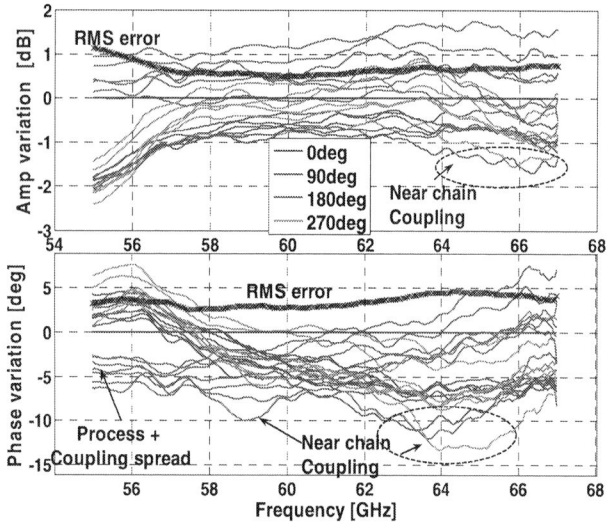

Fig. 9 : Measured amplitude and phase variation for four phases, on different chains and states – TX mode

Fig. 10 shows the measured power sweep for TX and RX at 58GHz. RX IP1dB is -17 dBm and TX Psat of the entire array is 8dBm. The results are close to simulation, and fit well with the calculated results obtained by using the characterization data of each component. In the RX mode, the LNA is the limiting factor while in the TX mode the mixer and amp2 compress first. As the array was optimized for RX performance, the TX compression is being affected. Redesigning the gain and compression of Amp1 and Amp2 will allow an increase of 7dB in TX power with very little effect on power consumption and die size. By changing the bias of the mixer and Amp2, we can improve the linearity significantly at the expense of gain. Changing the gain of the PA will not contribute to linearity as it reduces Psat due to the compression of previous blocks.

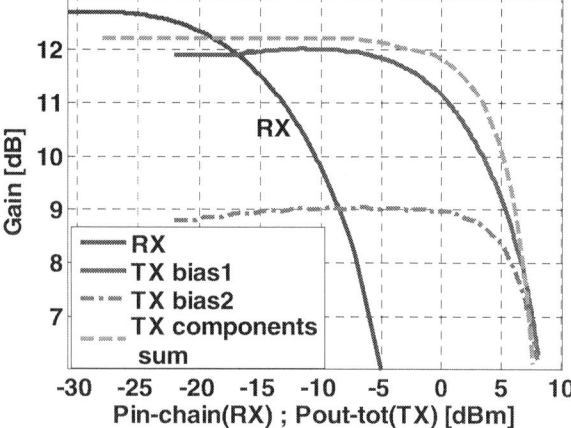

Fig. 10 : Linearity test measured at 58 GHz for TX with mixer bias1 (high gain) and bias2 (high linearity), and RX

VI. CONCLUSION

A 60 GHz fully integrated 32 element phased-array RF system for TX and RX is demonstrated in 90nm CMOS. The entire die occupies an area of 14.5mm² and has a power consumption of 0.5Watt. These numbers show it is possible to integrate such a system in future mobile platforms. The architecture shares the combiner and phase shifters between TX and RX and uses a lumped dense layout for the amplifiers achieving the best known die size compared to arrays at other frequencies. The flip chip design has proven to be very robust with an excellent transition match from the die to the board route. Different tests structures were used to measure the bump loss of 1.5dB and minimal isolation between adjacent bumps of 20dB. The coupling between neighboring chains through the bumps was measured to be a significant contributor to the phase error of 2-3⁰ rms. The total TX power was limited due to pre-compression and can be corrected to reach Psat of 15dBm with very little penalty in size and power consumption. After proving the conductive RF performance, the next research step will be to mount the RFIC to an antenna array and measure the total performance over the air.

REFERENCES

[1] A. Hajimiri, H. Hashemi, A. Natarajan, X. Guan, A. Komijani, "Integrated Phased Array Systems in Silicon", *Proceedings of the IEEE Sept. 2005 Page(s):1637 - 1655*

[2] H. Hashemi, X. Guan, A. Komijani, A. Hajimiri, "A 24-GHz SiGe Phased-Array Receiver—LO Phase-Shifting Approach", *IEEE Trans. Microw. Theory Tech* FEBRUARY 2005

[3] K. Koh, J. W. May and G. M. Rebeiz, "A Millimeter-Wave (40-45 GHz) SiGe BiCMOS 16-Element Phased-Array Transmitter," *IEEE J. Solid State Circuits,* vol. 44, no. 5, pp. 1498-1509, May 2009.

[4] E. Cohen, et. al. "A bidirectional TX/RX four element phased-array at 60GHz with RF-IF conversion in 90nm CMOS" – RFIC09

[5] E. Cohen, S. Ravid, D. Ritter "An ultra low power LNA with 15dB gain and 4.4db NF in 90nm CMOS process for 60 GHz phase array radio" , *RFIC08*

[6] T. Yao, M. Q. Gordon, K. K. W. Tang, K. H. K. Yau, M.-T. Yang, P. Schvan, S. P. Voinigescu, "Algorithmic Design of CMOS LNAs and PAs for 60-GHz Radio" , *IEEE J. of Solid-State Circuits,* pp. 1044 - 1057 May 2007 .

[7] Y. Ayasli, S. W. Miller, R.Mozzi, and L. K. Hanes, "Wide-band monolithic phase shifter,", *IEEE Trans. Microw. Theory Tech.*, vol. MTT-32, no. 12, pp. 1710–1714, Dec. 1984.

[8] D.-W. Kang, H.-D. Lee, C.-H. Kim, and S. Hong, "Ku-band MMIC phase shifter using a parallel resonator with 0.18 um CMOS tehcnology,", *IEEE Trans. Microw. Theory Tech.*, vol. 54, no. 1, pp. 294–301, Jan. 2006.

[9] Jeong-Geun Kim, and Gabriel M. Rebeiz, "Miniature Four-Way and Two-Way 24 GHz Wilkinson Power Dividers in 0.13um CMOS, *IEEE Microw. Wireless Component Letter,* SEPTEMBER 2007

RTU2C-3

A 16-Element Phased-Array Receiver IC for 60-GHz Communications in SiGe BiCMOS

Scott K. Reynolds[1], Arun S. Natarajan[1], Ming-Da Tsai[2], Sean Nicolson[3], Jing-Hong Conan Zhan[2], Duixian Liu[1], Dong G. Kam[1], Oscar Huang[2], Alberto Valdes-Garcia[1], Brian A. Floyd[1]

[1]IBM T. J. Watson Research Center, Yorktown Heights, New York, USA

[2]MediaTek Inc., HsinChu, Taiwan

[3]MediaTek Inc., San Jose, California, USA

Abstract — A 0.12-μm SiGe phased-array Rx IC for beam-steered wireless communication in the 60-GHz band is described. It has 16 RF phase-shifting front-ends with 11° digital phase resolution and hybrid passive-active RF signal combining. It achieves 7.4-7.9 dB NF (not including 12-dB array gain) over the 4 IEEE channels. The IC has a double-conversion superheterodyne Rx core with a maximum of 72 dB of power gain in 1-dB steps, and the on-chip synthesizer achieves < -90 dBc/Hz Rx phase noise at 1MHz offset. The IC draws 1.8 W at 2.7 V with a die area of 38 mm². It has been packaged with 16 antennas in a 288-pin organic BGA and phased-array beamsteering has been demonstrated, along with 5+ Gb/s wireless links using 16-QAM OFDM.

Index Terms — Phased-arrays, beam steering, 60 GHz, millimeter-wave, receiver, SiGe.

I. INTRODUCTION

The 57 to 66-GHz band supports extremely high-rate (1-10 Gb/s) wireless digital communication; however, fixed-antenna 60-GHz systems are sensitive to obstructions in the line-of-sight (LOS). Multi-element phased-array systems can overcome LOS limitations of the 60-GHz band and have recently attracted widespread research interest [1]-[2]. This paper presents a 16-element phased-array receiver (Rx) IC implemented in 0.12-μm SiGe BiCMOS (f_T = 200 GHz). Compared to other reports in the literature, it offers low NF, precision phase shifting, and very high integration level. Combined with a 16-element antenna, the overall phased-array system enables non-LOS communication and can significantly improve either system SNR or link budget.

II. CHIP ARCHITECTURE

Fig. 1 shows a block diagram of the Rx, which employs RF-path phase shifting followed by mostly-passive RF signal combining. Each of the 16 Rx inputs is applied to an RF front-end consisting of a stepped-gain LNA, a digitally-controlled phase shifter, a balun, and a phase-

inverting (0/180) VGA (PIVGA), similar to the phase-shifting front-end in [3]. Fine phase control (11±3° digital resolution, 0° to 180°) is achieved through a reflection-type phase shifter (RTPS), which consists of varactor-adjusted loads on a 90°-hybrid coupler. An additional 180° phase shift is achieved by inverting the output phase in the differential PIVGA following the passive phase shifter. The PIVGA also compensates for the phase-shift dependent loss of the RTPS, ensuring constant front-end gain across phase shift settings. Each of the (N = 16) RF front-ends draws 22 mA from 2.7 V, while the system SNR (in dB) improves as $10 \times \log_{10}(N)$.

Following the front-ends is a 4-stage binary RF power-combining tree, detailed in Fig. 2. The passive power combining uses a modified Gysel combiner. By introducing a cross-coupled t-line between the outputs as shown, the combiner achieves isolation between them, while a) not requiring the outputs to be colocated as in a differential Wilkinson divider, and b) reducing the required t-line length required in a Gysel divider [4]. The reduced signal routing saves 0.5-1.0 dB per combining level, and the overall combining tree area is reduced about 50% compared to a Wilkinson tree. An active combiner provides gain and buffering in the third stage of combining to compensate for the passive losses, and also allows for power down and isolation of groups of 4 front-ends. A final modified Gysel combiner provides the input signals for the RF down-conversion mixer and IF circuits.

In the case of the 16-element array, the input power into the RF mixer can be theoretically 12 dB higher (8-10 dB including losses) than in the case of a single-element Rx [5]; hence, the system required an RF mixer and IF strip with wide dynamic range. The double-balanced Gilbert mixer has iP$_{1dB}$ of -4 dBm and SSB NF of 11 dB. The 50-55.5 GHz LO is provided by a frequency tripler operating from the 16.7-18.5 GHz synthesizer output.

978-1-4244-6240-7/10 $26.00 © 2010 IEEE

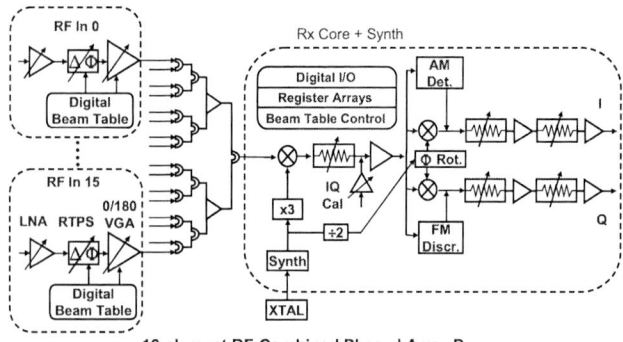

Fig. 1. Block diagram of the 16-Element RF-Combined Phased-Array Receiver. Signals following the 0/180 VGA are differential.

The 1st mixer output passes through a tunable IF filter and a coarse (6-dB step) attenuator before being buffered and converted to a baseband signal by a second set of quadrature (IQ) mixers. The 2nd LO for the IQ mixers is provided by a divide-by-2 operating from the synthesizer output. A phase rotator following the divide-by-2 allows IQ accuracy to be adjusted to within ±1°. An IF loop-back calibration scheme with the companion Tx IC [6] permits finer IQ adjustment in the baseband. The IQ calibration VGA in Fig. 1 allows path gain to be adjusted so calibration can be performed over baseband gain settings.

The baseband signal passes through a cascade of coarse and fine (1-dB) step attenuators and 16-dB fixed gain amplifiers to provide the required gain range. The baseband output buffer has 100 ohm differential output impedance and is designed to drive > 500 mVppd into a 100 ohm differential load in the baseband ADC. Overall, the Rx core (Fig. 1) provides +50 to -10 dB of gain in 1-dB steps. For fast AGC, all Rx gain control bits are grouped into 2 registers, which can be written in two 7.5-ns clock cycles.

III. MEASUREMENT SUMMARY

Fig. 3 shows Rx RF response along with the responses for each of the 4 IEEE channels, measured from a single input at maximum gain to output. RF response is measured by sweeping both the LO and RF frequencies to maintain constant 100MHz baseband output frequency. The Rx channel responses are measured for constant LO and swept RF input frequency. Explicit baseband low-pass filters (for adjacent-channel rejection) have not been included, but IF filters and baseband amplifier bandwidths together produce very nearly the desired ±1 GHz channel responses.

Fig. 4 illustrates performance of the phase-shifting front-ends. Each front-end achieves 360° phase shift in all

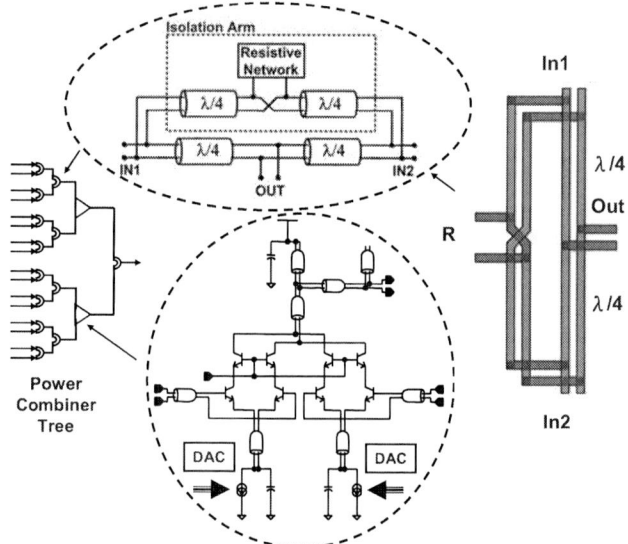

Fig. 2. Architecture of the mostly-passive RF power combining tree, with inset schematics of the active power combiner and modified Gysel combiner, and simplified layout of the Gysel combiner.

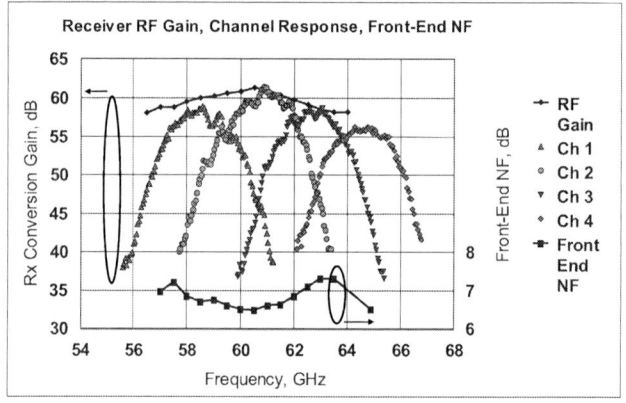

Figure 3: Measured receiver RF gain, along with the responses for each of the four IEEE channels, and front-end NF over frequency.

channels, and gain is equalized over all phase-shifter settings by adjusting the PIVGA. Normalized output power for 8 individual elements across phase shift settings is shown, demonstrating constant gain. Fig. 4 also shows two-element combined output power with the phase settings of one element held constant while the phase shift in the other element is varied. As expected, there is a 6dB increase in output power when the two elements are in phase while minimum output power occurs when the two elements have a relative phase difference of 180°. This measurement was repeated pair-wise across 16 elements with worst-case peak-to-null ratio of 19 dB. Matching of phase shift and gain over phase-shifter settings was

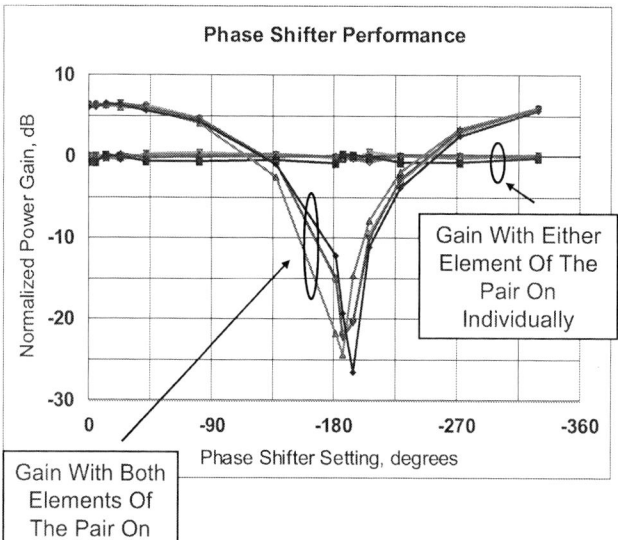

Fig. 4. Two-element phase shifter measurements, showing equalized gain over phase shifter settings for each element individually, as well as the power-combined response with one element swept while the other is held constant.

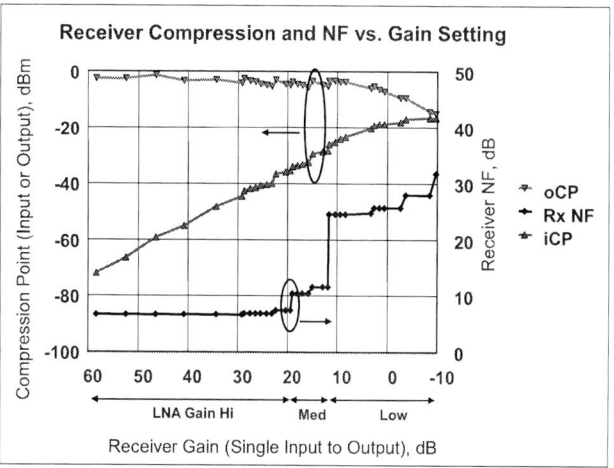

Figure 5: Receiver input compression point (iP$_{1dB}$, referred to a single input), output compression point (oP$_{1dB}$, with combined power from 16 inputs), and NF versus receiver gain, as the digital attenuation settings are increased in 1-dB steps from 0 to 69 dB.

measured across a wafer, with σ = 1.4° and 0.8 dB, respectively.

Fig. 5 combines several measurements to illustrate Rx dynamic range performance, as attenuator settings are increased from 0 to 69 dB. The middle curve is the input 1-dB compression point (iP$_{1dB}$, referred to a single input) at each attenuator step, assuming all 16 inputs are driven at the same power level, and the top curve is the corresponding output 1-dB compression point (oP$_{1dB}$). The attenuation sequence avoids compressing internal stages while maintaining the best possible sensitivity. Since we cannot simultaneously drive all 16 inputs, this data is compiled from measurements on the full Rx, Rx core, and RF front-ends.

For phased-array NF, we follow the approach of Lee [7] and do not include array gain in the Rx NF determination. For link budgeting, array gain is best allocated to the antenna gain. Overall Rx NF is specified when all inputs are driven at the same power level. Since such a measurement is impractical, the NF is computed from front-end NF and single-element Rx NF measurements [7]. The Rx NF versus attenuation setting is shown in the lower curve in Fig. 5 (22°C, Ch. 2, max. phase-shifter loss). The abrupt increases in NF as Rx gain is reduced occur as LNA gain is reduced in 8 and 18-dB steps.

Table 1 summarizes Rx performance. Higher power consumption at 65°C is due primarily to increased front-end bias current to partially counteract the decrease in LNA gain as temperature rises. The 330k FETs are mostly

	Ch 1 58.32 GHz	Ch 2 60.48 GHz	Ch 3 62.64 GHz	Ch 4 64.80 GHz
Maximum Rx Power Gain[1]	70 dB	71 dB	70 dB	68 dB
Front-End NF, max. phase shifter loss[2] 22°C	6.8 dB	6.8 dB	7.3 dB	6.5 dB
65° C	7.4 dB	7.7 dB	7.9 dB	6.9 dB
Rx NF, max phase shifter loss[3] 22° C	7.4 dB	7.4 dB	7.9 dB	7.6 dB
65° C	8.2 dB	8.4 dB	8.7 dB	8.5 dB
iP$_{1dB}$, 0-dB Rx gain, min. LNA gain[1]	-16 dBm			
oP$_{1dB}$, max. Rx gain	-1 dBm			
iIP3, in-channel 300- & 400-MHz tones, 12-dB total Rx gain, max. LNA gain[1]	-23 dBm			
oIP3, in-channel 500- & 600-MHz tones, max. Rx gain	+7.7 dBm			
Phase tuning range and resolution	> 360°, ≈ 11°			
IQ Gain and Phase Error (before fine cal)	± 1 dB, ± 1°			
Channel bandwidth (3 dB), 22° C & 65° C	≥ ±1 GHz	>+1 GHz - 0.9 GHz	≥ ±1 GHz	≥ ±1 GHz
Rx Phase Noise, 1-MHz offset	< -90 dBc/Hz			
Power 22° C, 65° C	1.8 W, 2.0 W (2.7-V supply)			
Size, Device Count	6.08 x 6.2 mm², 1.94k NPNs, 330k FETs			

1. Total output power divided by total input power, assuming all 16 inputs are driven at the same power.
2. Measured on a separate testsite.
3. Overall Rx NF, referred to a single input, assuming all inputs are driven at the same power, not including 12-dB array gain.
4. All data for 22° C unless otherwise stated.

Table 1. Table summarizing Rx performance over the 4 IEEE channels at 22°C and 65°C. Measured by wafer probing.

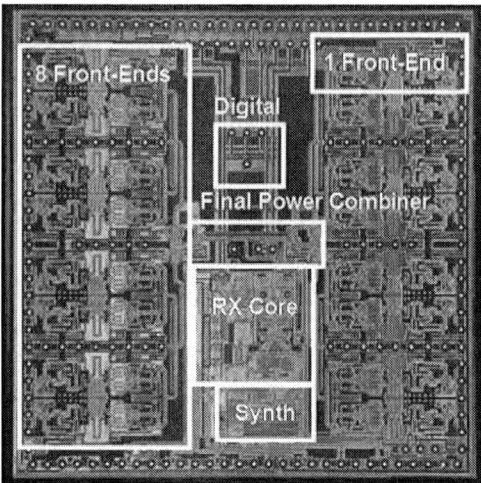

Fig. 6. Chip photo, 6.08 x 6.2 mm² die size.

used for memory to store programmed beam directions; in Fig. 6, the registers are hidden under the front-ends.

In Fig. 7, the Rx IC is shown packaged with 16 planar antennas. Fig. 8 shows phased-array antenna patterns for the packaged IC, recorded in our antenna chamber. A beam direction of 0° is normal to the package plane.

The Rx IC (with the companion Tx IC) has been used in beam-steered, non-line-of-sight, 4.5-m, 5.3 Gb/s, 16-QAM wireless links, in each of the 4 IEEE channels. The links used 6 Tx elements and 12 Rx elements, and occupied the bandwidth of a single 2.16-GHz IEEE channel.

IV. CONCLUSION

This 16-element phased-array Rx uses a novel modified-Gysel power combiner and achieves high integration level and low NF. Packaged ICs with antennas have been used in beam-steered, NLOS wireless links at >5 Gb/s data rates in each of the 4 IEEE channels.

ACKNOWLEDGEMENT

The authors acknowledge John Zhongxuan Zhang, Young Kim, Hsin-Hung Chen, Sammi Chan, Rodel Anonuevo, Lay-Poh Loh, and Doris Lee of MediaTek, and Ben Parker, Sakshi Dhawan, Don Beisser, Sudhir Gowda, and Mehmet Soyuer of IBM for their contributions.

REFERENCES

[1] E. Cohen, C. Jakobson, S. Ravid, and D. Ritter, "A bidirectional TX/RX four element phased-array at 60GHz with RF-IF conversion block in 90nm CMOS process", *IEEE RFIC Symp.*, pp. 207-210, June 2009.

[2] Y. Yu, P. Baltus, A. von Roermund, A. de Graauw, E. van der Heijden, M. Collados, and C. Vaucher, "A 60GHz

Fig. 7. The Rx IC packaged with 16 antennas in a 288-pin (28x28 mm²) organic BGA. The white material is thermally conductive paste.

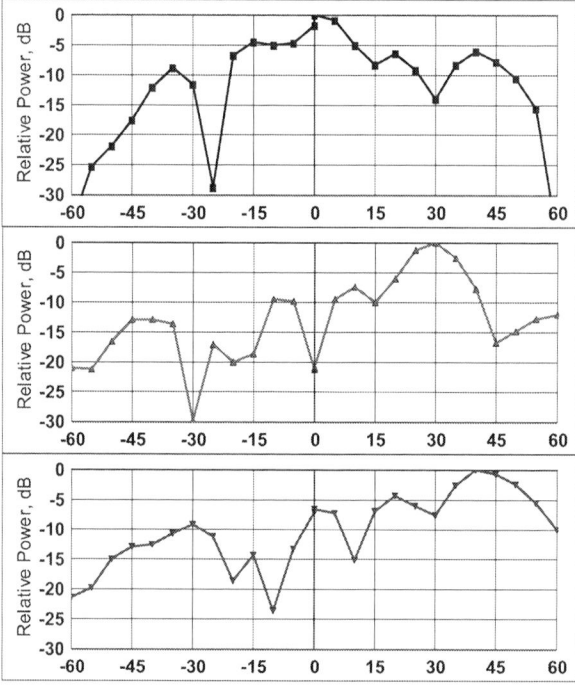

Fig. 8. Normalized 8-element antenna patterns, as the beam is steered from 0° to 30° and 45°.

Digitally Controlled RF-Beamforming Rx Front-end in 65 nm CMOS", *IEEE RFIC Symp.*, pp. 211-214, June 2009.

[3] A. Natarajan, M.-D. Tsai, and B. Floyd, "60GHz RF-path Phase-shifting Two-element Phased-array Front-end in Silicon", *IEEE VLSI Symp.*, pp. 250-251, June 2009.

[4] U. H. Gysel , "A New N-Way Power Divider/Combiner Suitable for High-Power Applications," *IEEE MTT-S IMS Digest*, May 1975, pp. 116-118.

[5] S. Reynolds, B. Floyd, U. Pfeiffer, T. Beukema, J. Grzyb, C. Haymes, B. Gaucher, and M. Soyuer, "A Silicon 60-GHz Receiver and Transmitter Chipset", *IEEE JSSC*, v.41, n. 12, pp. 2820-2831, Dec. 2006.

[6] A. Valdes-Garcia, S. Nicolson, J.-W. Lei, A. Natarajan, P.-Y. Chen, S. Reynolds, J.-H. C. Zhan, and B. Floyd, "A SiGe BiCMOS 16-Element Phased-Array Transmitter for 60-GHz Communications", *ISSCC Dig. Tech. Papers*, pp. 218-219, Feb. 2010.

[7] J. J. Lee, "G/T and Noise Figure of Active Array Antennas", *IEEE Trans. Ant. Prop.*, v. 41, n. 2, pp. 241-244, Feb. 1993.

RTU2C-4

A 24-GHz Phased-Array Receiver in 0.13-μm CMOS using an 8-GHz LO

Satwik Patnaik and Ramesh Harjani

Department of Electrical and Computer Engineering

University of Minnesota, Minneapolis, MN 55455,USA

Abstract—**This paper presents a 24-GHz two-channel phased-array receiver. The receiver adopts the LO-phase-shifting approach and employs a sub-harmonically injection-locked phase-shifter. A CMOS-only prototype, fabricated in a 130-nm SiGe BiCMOS technology, draws 16-mA of current from a 1.5-V supply and consists of a injection-locked oscillator (operating as a phase-shifter, LO-buffer and frequency multiplier), a down-conversion mixer and an IF-buffer. The worst-case measured amplitude and phase errors are 1.5-dB and 4°. The two-channel receiver occupies an active area of 0.23-mm^2.**

Index Terms—**CMOS, injection-locked oscillator (ILO), mm-wave, phased-array, sub-harmonic injection-locking.**

I. INTRODUCTION

To meet the ever-increasing demands for higher data-rates in wireless communication, modern wireless standards are moving to multi-antenna architectures. There is renewed interest in operating at the millimeter frequencies as the congestion at the lower ISM bands are increasing. Commercial systems prefer silicon-based technologies (CMOS/BiCMOS) for cost-related advantages and integration levels. With improvements in technology, silicon is rapidly closing its frequency gap with III-V technologies. Modern 65-nm CMOS and 130-nm BiCMOS processes offer f_Ts beyond 200-GHz. This has led to extensive research in building silicon-based solutions for mm-wave commercial and defense applications [1]–[5]. Millimeter frequencies of commercial interest include 24GHz (for automotive radar and wireless communication), 60GHz (for wireless communication), 77 GHz (for automotive radar) and certain sub-bands in the W-band (for passive imaging).

Phased-array techniques generate a highly directional signal that increases signal-to-noise ratio, reduces interference to other non-collocated communication links allowing for higher data rates in wireless communication and longer operating ranges in radars. Fig. 1 shows the basic concept of a phased-array receiver. The receiver provides spatial filtering as a function of the beam pattern including strong nulls that reduce interference in undesired directions.

In this work, we present part of a two-channel 24-GHz phased-array receiver front-end. The basic block diagram of the receiver is shown in Fig. 2. The primary focus of this design has been on the phase-shifter which is a critical block in any phased-array system. By employing a sub-harmonically injection-locked phase-shifter configuration, the receiver achieves low DC power consumption and small area, making it an ideal candidate for portable mm-wave applications. The VCO of the injection-locked oscillator (ILO) is locked to the third harmonic (of the injection frequency) created by the non-linearity of the G_m-cell. The next section describes the ILO as a phase-shifter block. Section III discusses the design of the various circuit blocks. Measurement

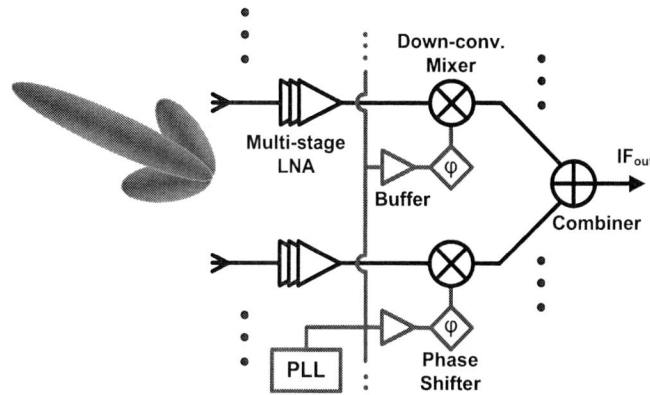

Fig. 1: A typical mm-wave phased-array receiver architecture

results from a CMOS prototype fabricated in a 130nm BiC-MOS process are presented in Section IV.

II. THE ILO AS A PHASE-SHIFTER

Injection-locking is a phenomenon in which an oscillator can be locked to an injected frequency, provided the injected frequency is close to the natural oscillation frequency. The transient behavior of an ILO is well-approximated by Adler's equation [6]. In the steady-state locked condition, the ILO output is phase-shifted from the injection signal by a constant value given by (1), where ω_0 is the natural frequency of the oscillator, ω_{inj} is the injection frequency and ω_L is the lock-range of the ILO, *i.e.*, the ILO locks if ω_{inj} is in the range $(\omega_0 - \omega_L, \omega_0 + \omega_L)$.

$$\phi_{ss} = \sin^{-1}\left(\frac{\omega_0 - \omega_{inj}}{\omega_L}\right) \tag{1}$$

This phase-shift can be controlled by tuning the natural frequency of the oscillator. By doing so, the phase-shift ϕ_{ss} can be tuned within the range $(-\pi/2, \pi/2)$. For differential

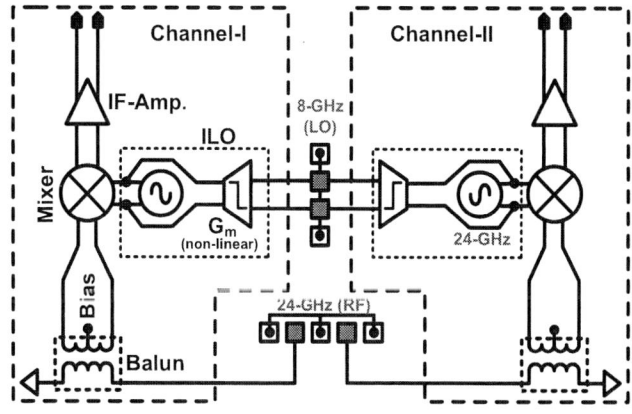

Fig. 2: Proposed phased-array receiver

978-1-4244-6240-7/10 $26.00 © 2010 IEEE

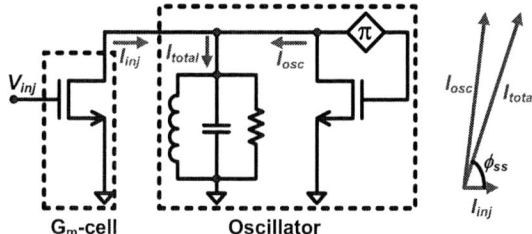

Fig. 3: Simplified schematic of an injection-locked LC-oscillator

implementations, inverting the signals results in an additional phase-shift of π bringing the total phase range to 2π.

Fig. 3 shows the RF model of a simple LC-based ILO. The signal is injected into the oscillator as a current (I_{inj}). The single-sided lock-range of this ILO is given by (2) [6]

$$\omega_L = \frac{\omega_0}{2Q}\left(\frac{I_{inj}}{I_{osc}}\right) \qquad (2)$$

where Q is the quality factor of the LC-tank of the oscillator and I_{osc} is the current dissipated by the oscillator (as depicted in Fig. 3). Further, in order to achieve lock, the injection-signal only needs to possess a frequency component that is within the ILO lock-range. All other components are suppressed by the LC-tank [7]. For example, a square-wave may be used to lock an ILO to any odd-harmonic of the square wave. This is called sub-harmonic injection-locking.

Benefits and design trade-offs: At mm-wave frequencies, frequency synthesis and LO distribution consumes significant power [8]. The exponential increase in signal losses with frequency [9] has to be compensated with additional buffers. By applying sub-harmonic injection-locking, this power consumption can be greatly reduced [10], [11]. Firstly, the LO-distribution is done at a sub-harmonic frequency and hence, lower signal losses. Secondly, with a reduced divider ratio and potential removal of some divider stages, the overall power of the frequency synthesizer (PLL) reduces. In general, a large portion of the PLL power is spent on the initial stages of the dividers. Essentially, the ILO would perform three important functions - frequency multiplication, phase-shifter and signal buffering. All three functions are done utilizing the highly power efficient positive feedback inherent in an LC ILO. This multi-functionality allows the receiver to operate at lower power making phased-arrays, which typically increase circuit complexity, a viable option for portable systems.

Sub-harmonic injection-locking allows for lower power spent on LO distribution. However, there are a number of other system tradeoffs that need to be also considered. Sub-harmonic injection-locking results in the leakage of unwanted harmonics at the ILO output. There is a trade-off between the choice of the sub-harmonic and spurious tones. A lower sub-harmonic (ω_{ILO}/ω_{LO}) value results in higher suppression of spurs, but at the same time, requires that the LO (ω_{LO}) generation and distribution is at a higher frequency, and vice-versa. Though these unwanted harmonics are suppressed by the oscillator tank, they can reduce the SNR at the mixer outputs. In order to further mitigate this problem, techniques like pulse-slimming and narrow-band filtering can be utilized [7].

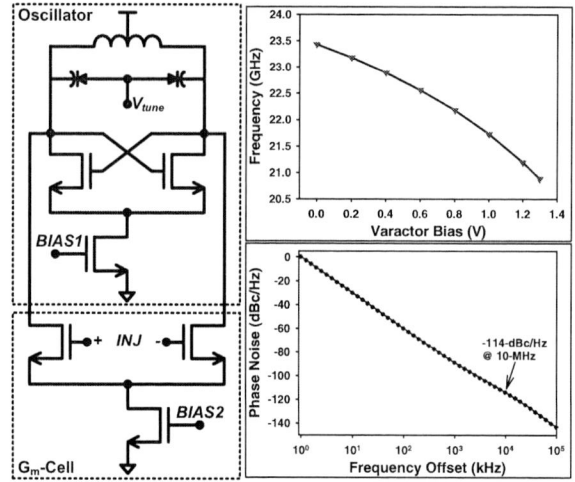

Fig. 4: Schematic of the injection-locked oscillator and simulated results of oscillator under free-running conditions

III. CIRCUIT DESIGN

In this section, we describe the various circuit blocks implemented as parts of the receiver. All the circuits are CMOS-only implementations.

ILO: The ILO oscillator is realized as a negative-g_m LC-oscillator. The natural frequency was tuned by a diode-varactor. The oscillator was designed for a tuning range between 20.9 and 23.4-GHz. The simulated phase-noise at the highest frequency at a 10MHz offset was -114-dBc/Hz.

Even though pMOS-based cross-coupled transistors offer better flicker noise performance, their lower f_T results in higher power requirements. Since the oscillator was designed to operate at mm-wave, the negative-g_m cell was chosen to be nMOS-only. The use of a symmetric octagonal inductor resulted in better phase noise performance and lower area.

An external G_m-cell was used to inject the required frequency into the oscillator. The differential pair was designed to completely swing the tail current from one arm to the other for a 0-dBm differential injection signal. This non-linearity of the G_m cell creates odd harmonics of the injection frequency, out of which the oscillator is locked onto the third harmonic.

Mixer and IF buffer: For the down-conversion mixer, the double-balanced Gilbert cell topology was chosen over the passive-mixer topology, because the passive-mixer requires a larger local oscillator signal for good performance. The output load of the mixer was designed for a bandwidth of

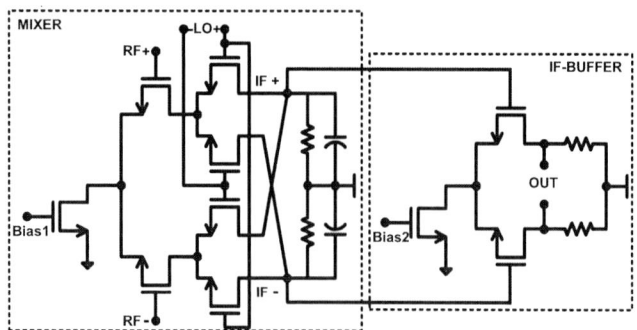

Fig. 5: Mixer and IF buffer schematics

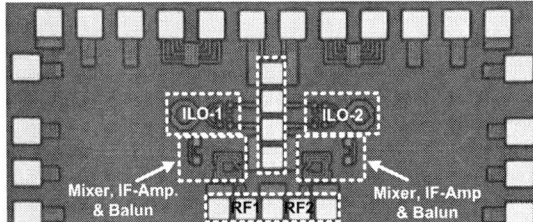

Fig. 6: Die micrograph of the CMOS prototype fabricated in 130-nm SiGe BiCMOS technology

150-MHz. The mixer is preceded by a current-mode logic (CML) buffer at the LO-port. This buffer is used to convert the sinusoidal output from the ILO into an approximate square-wave signal, to improve the mixer gain. It also reduces any LO amplitude mismatch caused at the ILO output due to frequency tuning (for phase generation). Due to the RF-input transistors operating so close to their f_T, the overall mixer conversion gain was 2-dB. An IF buffer drives the output.

IV. MEASUREMENT RESULTS

The circuit was fabricated in the 130-nm SiGe IBM BiCMOS process, but the design was CMOS-only. The die micrograph of the prototype is shown in Fig. 6. The active area occupied (excluding pads and ESD structures) is 0.23-sq. mm. The two-channel receiver draws 16-mA of current from a 1.5-V power supply, excluding biasing circuits, out of which each phase-shifter (ILO and CML buffer) consumes about 6-mA.

Individual channels of a phased-array receiver need to have excellent matching, for optimum performance. In order to evaluate the matching performance, the two oscillators were first characterized under free-running condition. Both oscillators exhibited a tuning range between 22.7 and 24.74-GHz. The frequencies for the oscillators were measured by looking at the carrier frequency leakage at the RF-port of the mixer. Fig. 7 shows the tuning range of the two oscillators and the highest frequency measured on the spectrum analyzer for one oscillator. The two oscillators show excellent matching with the worst case frequency mismatch of about 20-MHz. In order to measure phase-noise, the oscillator signal was downconverted to an IF of 91 MHz. One of the oscillators (Oscillator-I) free-running at 24.372-GHz exhibited a phase noise of -106-dBc/Hz at a 10-MHz offset.

The measurement setup is shown in Fig. 8. The rat-race hybrid converted the single-ended 8.125-GHz signal from the microwave source into a pair of differential signals with approximately 1-dB loss (apart from the usual 3-dB due to

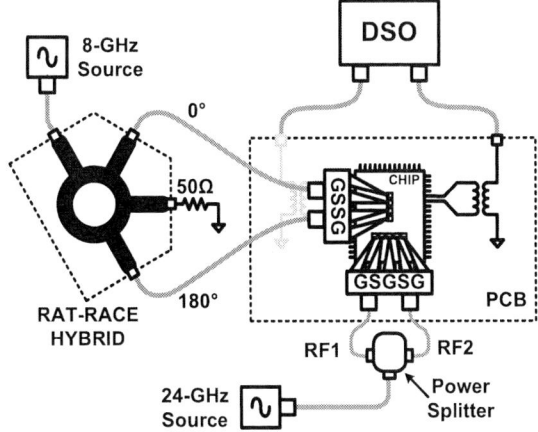

Fig. 8: Measurement setup used to evaluate receiver performance

power splitting). A mm-wave source was used to generate the RF input to the mixer. The signal was power-split into two - equal in phase and magnitude, which were fed to the two channels through RF probes. The IF was measured on a digital storage oscilloscope (DSO) sampling at 5-GSa/s. The 8.125-GHz microwave source exhibited a phase noise of -108.8-dBc/Hz at a 1-MHz offset (Fig. 9(a)). When the one of the oscillators (Oscillator-I) was injection-locked to this signal, the measured phase noise was -97.4-dBc/Hz at a 1-MHz offset (Fig. 9(b)), which matches well with the value predicted from the signal source phase noise (-108.8-dBc/Hz + 9.5-dB = -99.3-dBc/Hz). Also, the in-band phase-noise behavior of the ILO closely follows the signal source (PLL behavior), albeit with a loop-bandwidth increment (Fig. 9).

As discussed earlier, tuning the natural frequency of the VCO in the ILO results in a phase-shift between the injected and output signal. However, this tuning needs to be done within the lock-range. Hence, lock-range estimation is an important step in the measurement and calibration of the system. The injection signal was strong enough to ensure complete switching of the tail current in the G_m-cell (Fig. 4). Hence, the lock-range can be varied by changing the tail-current transistor bias of the G_m-cell. The double-sided lock-range of the ILO was set to 242-MHz by setting the appropriate G_m-cell tail bias ((I_{inj}/I_{osc}) $\approx$ 0.075 at 24.372-GHz). This lock-range was sufficient for the requisite phase-generation in our prototype. For more flexibility, this lock-range should be further enhanced. This can be done by applying a square wave to the G_m-cell, instead of a sinusoid. Simulation results show that the lock range more than doubles using this methodology.

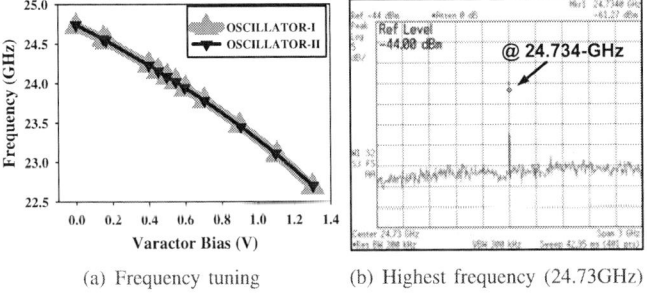

(a) Frequency tuning (b) Highest frequency (24.73GHz)

Fig. 7: Measured oscillator free-running frequency characteristics

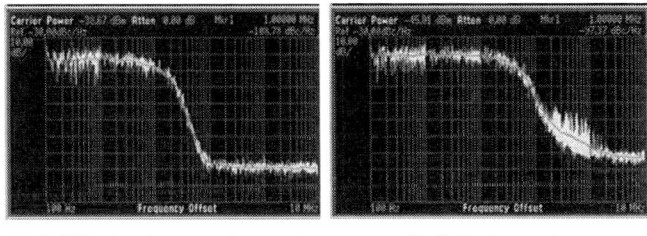

(a) 8-GHz signal source phase noise (b) ILO phase noise

Fig. 9: ILO phase noise characteristics (down-converted to 97-MHz); it is in excellent agreement with signal source phase noise

978-1-4244-6240-7/10 $26.00 © 2010 IEEE

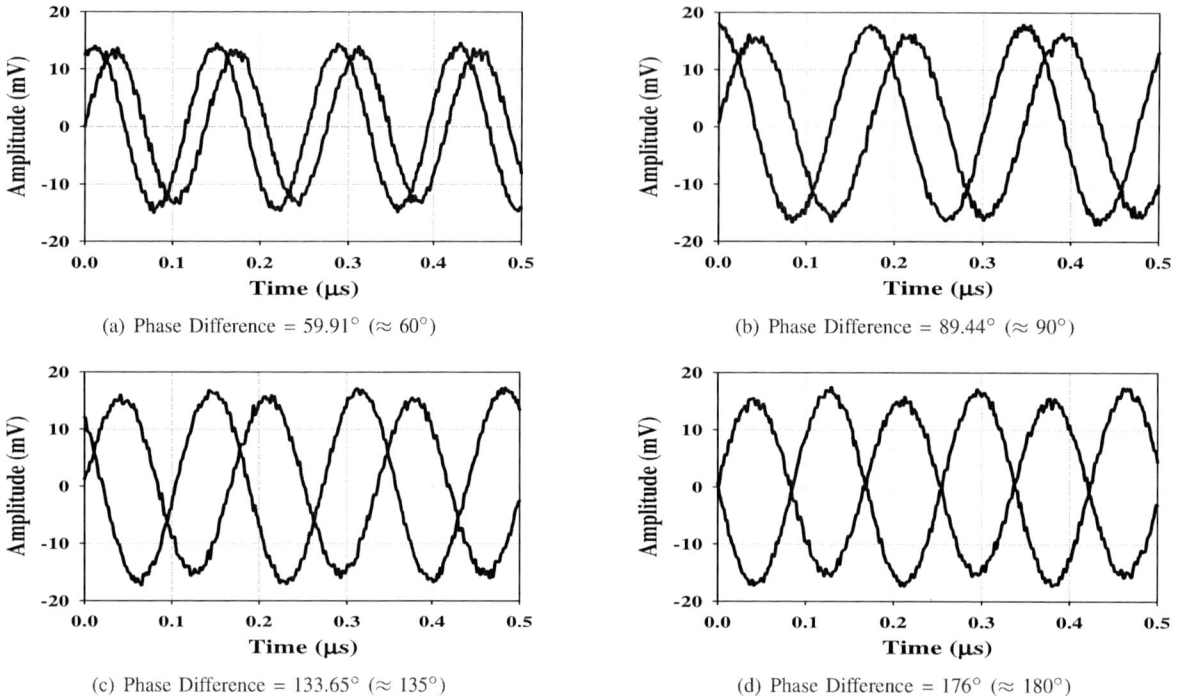

(a) Phase Difference = 59.91° (≈ 60°)

(b) Phase Difference = 89.44° (≈ 90°)

(c) Phase Difference = 133.65° (≈ 135°)

(d) Phase Difference = 176° (≈ 180°)

Fig. 10: Different phase-shifts measured at IF outputs

Traditionally, in a phased-array receiver, the phase-shifter cancels the phase-shift produced by the antenna-array. However, for simplicity of measurement, RF signals were aligned and the LO-phase-shifts were measured at the IF. Fig. 10 illustrates the various phase-shifts that were measured on the oscilloscope. The worst-case amplitude mismatch and phase error were approximately 1.5-dB and 4°, respectively. After de-embedding signal losses from cables and insertion losses from power-divider, balun and RF probes, the mixer-IF-buffer combination showed approximately 0-dB gain.

The accuracy of the phase-shift in a phased-array dictates the quality of beam nulls in the radiation pattern. So for higher rejection from null directions, high phase-accuracy is essential. For each of the phase-shift waveforms, the ILO center frequency (ω_0) was tuned using the varactor bias. An external trim potentiometer ensured fine frequency resolution to achieve accurate phase-shifts. It should be mentioned that in a full-system implementation, a high-resolution digital-to-analog converter (DAC) could be used for varactor bias for accurate control of frequency (and consequently, phase), or a hybrid VCO-DCO [4] could be used, where digital bits are used for coarse control and DAC-controlled varactor bias *fine-tunes* the frequency.

V. CONCLUSIONS

This paper presents a low-power phased-array receiver architecture that is based on sub-harmonic injection-locking. The principal block in this architecture is the injection-locked oscillator, which performs multiple functions, which results in the low power consumption of the receiver. To the author's knowledge, the power consumption per channel (12-mW) for this receiver is one of the lowest reported in recent literature.

Also, the receiver occupies an active area of only 0.23-sq. mm. The reduction in LO-frequency requirement for the receiver allows for simpler LO distribution networks. The two channels of the receiver exhibit a gain error of approximately 1.5-dB with no calibration and a worst-case (best-case) phase error of 4° (0.1°), with minimal calibration. Finally, techniques to improve the flexibility of the phase-shifter are also discussed.

Acknowledgment: The authors thank S. Kudva, N. Lanka and M.R. Ahmadi for their help with design and measurements, and the Trusted Foundry Access Program Office (TAPO) for chip fabrication.

REFERENCES

[1] S. K. Reynolds, *et al.*, "A silicon 60-GHz receiver and transmitter chipset for broadband communications," *IEEE JSSC*, pp. 2820–2831, Dec. 2006.

[2] H. Krishnaswamy and H. Hashemi, "A variable-phase ring oscillator and PLL architecture for integrated phased array transceivers," *IEEE JSSC*, pp. 2446–2463, Nov. 2008.

[3] S. Jeon, *et al.*, "A scalable 6-to-18 GHz concurrent dual-band quad-beam phased-array receiver in CMOS," *IEEE JSSC*, pp. 2660–2673, Dec. 2008.

[4] S. Patnaik, N. Lanka, and R. Harjani, "A dual-mode architecture for a phased-array receiver based on injection locking in 0.13μm CMOS," *IEEE ISSCC*, pp. 490–491,491a, Feb. 2009.

[5] I. Sarkas, *et al.*, "W-band 65-nm CMOS and SiGe BiCMOS transmitter and receiver with lumped I-Q phase shifters," *IEEE RFIC*, pp. 441–444, June 2009.

[6] R. Adler, "A study of locking phenomena in oscillators," *Proc. IRE*, pp. 351–357, June 1946.

[7] N. Lanka, S. Patnaik, and R. Harjani, "A sub-2.5ns frequency-hopped quadrature frequency synthesizer in 0.13-μm technology," *IEEE CICC*, pp. 57–60, Sept. 2009.

[8] J. Kim, B. Jung, P. Cheung, and R. Harjani, "Novel CMOS low-loss transmission-line structure," *IEEE RWS*, pp. 235–238, Sept. 2004.

[9] B. Kleveland, T. H. Lee, and S. S. Wong, "50-GHz interconnect design in standard silicon technology," *IEEE IMS*, pp. 1913–1916, June 1998.

[10] R. Toupe, Y. Deval, F. Badets, and J.-B. Begueret, "A 65-nm CMOS 8-GHz injection locked oscillator for HDR UWB applications," *ESSCIRC*, pp. 106–109, Sept. 2008.

[11] W. L. Chan and J. R. Long, "A 56-65 GHz injection-locked frequency tripler with quadrature outputs in 90-nm CMOS," *IEEE JSSC*, pp. 2739–2746, Dec. 2008.

RTU2C-5

Wafer-Scale W-Band Power Amplifiers Using On-Chip Antennas

Yusuf A. Atesal, Berke Cetinoneri, Ramadan A. Alhalabi, and Gabriel M. Rebeiz

University of California, San Diego, La Jolla, CA, 92093, U.S.A.

Abstract — This paper presents, for the first time, a W-band SiGe power amplifier designed and fabricated together with a high-efficiency on-chip microstrip antenna. The antenna/amplifier results in an effective radiated power (ERP=P_tG_t) > 10 dBm from 88 to 98 GHz, with a peak of 14.6 dBm at 92 GHz. The chip consumes 120 mA from a 1.7 V supply. The antenna/amplifier approach can be extended to a large number of elements (8×8) and allows for efficient wafer-scale power combining and phased-array scanning.

Index Terms — Millimeter-wave integrated circuits, power amplifier (PA), silicon germanium (SiGe) HBT, W-band.

I. INTRODUCTION

Millimeter-wave power amplifiers are still a challenging design aspect for SiGe and CMOS technologies due to the relatively low-powers achieved from a single amplifier. This is due to the low voltage supply available in highly scaled processes (1-2 V) and to the limited Q of the transmission-line and lumped-element inductors (10-20) which precludes the use of on-chip multistage power-combining techniques. A survey of recently published work shows that an output power of 50-100 mW can be achieved at 60-90 GHz using SiGe and CMOS chips using single-ended/balanced designs [1]-[3] and transformers [4]-[5]. However, none of these techniques can be scaled to higher power levels, and achieving 0.1-1 W using on-chip designs has not been proposed or demonstrated.

This paper presents, for the first time, a quasi-optical approach to W-band power combining using on chip antennas (Fig. 1). In this case, each SiGe (or CMOS) amplifier is attached to a planar antenna and the power combining occurs in free-space. A high-efficiency electromagnetically-coupled (EM) microstrip antenna is used, and is defined on a 125 µm quartz substrate placed on top of the silicon wafer. The antenna is designed to be 50 Ω with a -10 dB bandwidth of 91-98 GHz and a gain of 3.0 dB at 94 GHz (radiation efficiency of 50%) [6]. This type of power combining is considered as a phased array with 0° of phase applied to each element (no scanning).

The quasi-optical approach has been demonstrated before using GaAs amplifiers and grids [7], or using antenna-amplifier-antennas [8], but in a hybrid approach and not on SiGe or CMOS RFICs. This power combining technique is scalable to a wafer-scale design, and one can combine 64 power amplifiers using a silicon chip and a

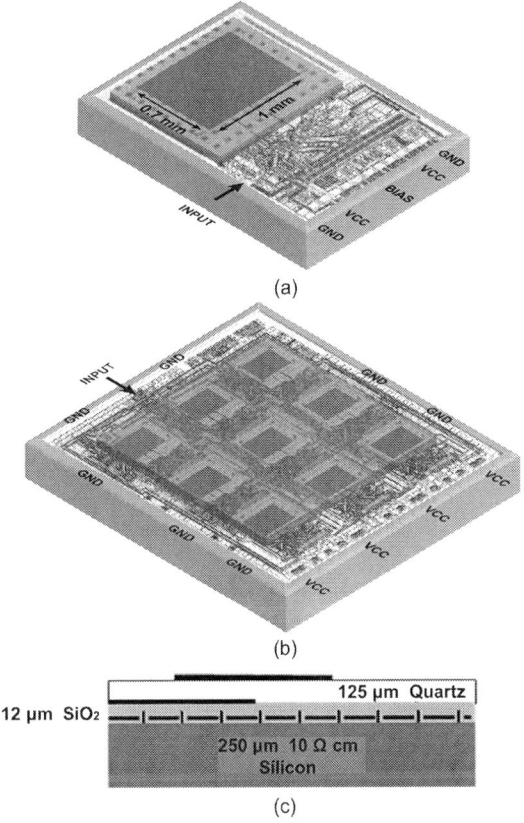

Fig. 1. W-band power amplifiers with on-chip antennas: (a) single element, (b) 3×3 array and (c) cross-section view. The 125 µm quartz wafer is placed on top of the silicon chip. 40×40 µm squares are used under each antenna for metal density rules.

single antenna wafer. Furthermore, the design readily scales to on-chip phased arrays if a phase shifter is included in every cell.

II. DESIGN OF UNIT CELL

A 4-stage PA consisting of common emitter stages is designed using IBM 8HP BiCMOS process with a bipolar f_t of 200 GHz (Fig. 2). The transistors are biased in class A, linear operation, and all biasing circuits, input, output, and interstage matching networks are fully integrated on the chip. Power transistor cells are designed using 5×0.13 µm² cells with a quiescent current of 10-12 mA/µm² for peak f_t. The first two gain stages consist of 2 aggregated power cells (10×0.13 µm²), while the third and

978-1-4244-6240-7/10 $26.00 © 2010 IEEE

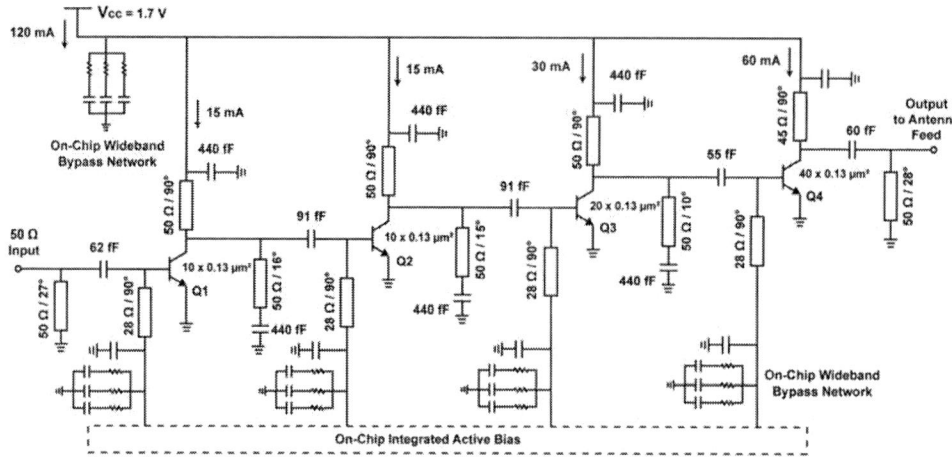

Fig. 2.   Schematic of the 4-stage W-band power amplifier. Transmission-lines are specified at 94 GHz.

fourth stages are made from 4 and 8 power cells, respectively. This ensures that the large signal characteristics are limited by the last stage. Each power cell is surrounded by a deep-trench isolation ring and has a single emitter and dual collector and base fingers. The base impedance seen by each stage is adjusted to be 300 at DC and 15-20 at W-band resulting in an improved emitter breakdown voltage of 4.2 V (BV_{ceo} = 1.7 V) [1].

The collector and base biasing is done using /4 quarter-wave transmission-lines. Each /4 section is followed by a shunt 440 fF MIM capacitor in order to provide a true RF ground at W-band. Interstage matching networks consist of shorted stubs and series capacitors. Shorted stubs are designed using a short t-line in series with a 440 fF MIM capacitor, resulting in a small inductor (10-20 pH) at W-band and open at DC. Metal-oxide-metal (MOM) matching capacitors are designed for < 91 fF. Collector /4 sections and matching stubs are realized using 11-8-11 μm grounded CPW lines with a Q of 11 at 94 GHz, thus minimizing the coupling between the t-lines and the lossy substrate. All transmission-lines, interconnects and MOM capacitors are modeled using full – wave EM

simulations [9].

To facilitate the routing of supply voltage on the wafer, the collector bias (V_{cc}) of all stages are connected to each other and a common V_{cc} is used. Distributed capacitor banks are used all over the V_{cc} plane in order to prevent stage-to-stage coupling through the V_{cc} paths.

The simulated PA has a gain of 14.8 dB at 94 GHz, with a 3-dB bandwidth from 83 to 104 GHz. Input and output return losses are < -10 dB from 83 to 110 GHz. Output P_{1dB} is 9.5 dBm and P_{sat} is 13.3 dBm at 94 GHz.

III. MEASUREMENTS

Two different W-band amplifiers were fabricated using the IBM 8HP process: A test amplifier with input and output CPW probes, and the same amplifier but with an integrated EM probe at its output port for the on-chip microstrip antenna (Fig. 3). Both amplifiers consume 120 mA from a 1.7 V supply. The measured input and output reflection coefficients are shown in Fig. 4 and agree well with simulations. Both amplifiers result in a similar S_{11} even if attached to different loads vs. frequency.

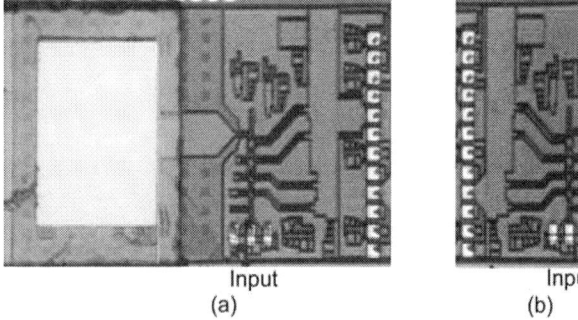

Fig. 3.   Microphotograph of the amplifier with an integrated antenna. The total area, including the pads and the antenna, is 3.0 × 1.3 mm². Also shown is a test amplifier with CPW I/O pads.

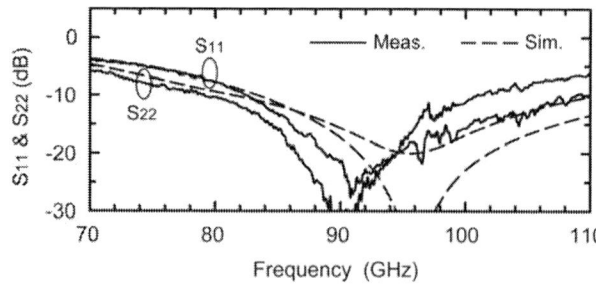

Fig. 4.   Measured input and output return loss of the amplifier. Both the test amplifier and the on-chip antenna/amplifier result in a similar input reflection coefficient.

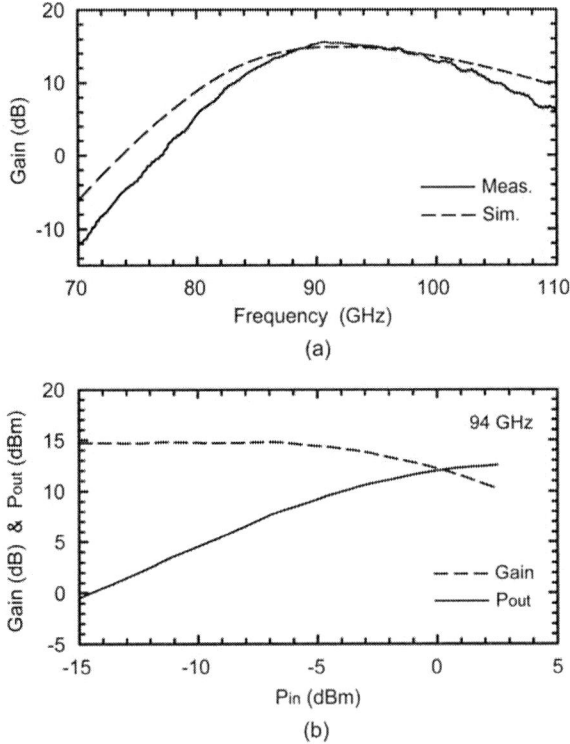

Fig. 5. (a) Measured and simulated gain of the test amplifier and (b) the measured output power at 94 GHz.

The measured small-signal gain vs. frequency of the test amplifier agrees well with simulations (S_{12} is < -40 dB up to 110 GHz and is not shown) (Fig. 5). The test amplifier results in a peak gain of ~15 dB at 89-94 GHz, and > 10 dB gain from 83-103 GHz. The measured P_{1dB} and P_{sat} at 94 GHz are 10 dBm and 13 dBm, respectively, and the measured power-added efficiency at 94 GHz is 8%. Fig. 6 presents the measured P_{sat} vs. collector voltage, and the test amplifier maintains a P_{out} > 12 dBm from 1.6-2.1 V.

The W-band amplifier with the on-chip microstrip antenna is then measured using a far-field test set-up (Fig. 7a). The SiGe chip is located on a standard W-band probe station and is fed using a CPW probe. A W-band horn

antenna with 23 dB gain at 94 GHz is placed at R=25 cm from the microstrip antenna (well into the far-field) and the RF power is measured vs. frequency. RF absorbers are used around the SiGe chip and the W-band horn so as to reduce reflections and standing waves. The output power (P_t) is determined using the Friis equation [10]

$$\frac{P_r}{P_t} = \left(\frac{\lambda}{4\pi R}\right)^2 G_t G_r \quad or \quad P_t G_t = \left(\frac{P_r}{G_r}\right)\left(\frac{4\pi R}{\lambda}\right)^2, \quad (1)$$

where P_r is the received power at the horn, G_r=23 dB, R=25 cm, and G_t is the gain of the on-chip microstrip antenna. The microstrip antenna has a gain of 2.5-3.7 dB and an efficiency of 40-57% at 92-98 GHz (Fig. 7b) [6]. Note that only the effective radiated power (ERP=$P_t G_t$) can be determined with accuracy using (1), and any P_t values include the error in the microstrip antenna gain.

The amplifier with the on-chip microstrip antenna results in a wideband effective radiated power (ERP) which is > 10 dBm from 88 GHz to ~98 GHz, with a peak of 14.6 dBm at 92 GHz (Fig. 8a). The measured power (P_t) from the on-chip antenna-amplifier also agrees with that of the test amplifier over a broad frequency range (Fig. 8b). There is an inherent +/- 1 dB error due to alignment and to the standing waves in the setup. The on-chip antenna/amplifier also shows the same large signal characteristics as the test amplifier (Fig. 9).

IV. EXTENSION TO WAFER-SCALE ARRAYS

A 3×3 antenna/amplifier array was also designed and is currently being tested (Fig. 10 – the chip layout is shown in Fig. 1). The simulated ERP is 25-26 dBm at 94 GHz

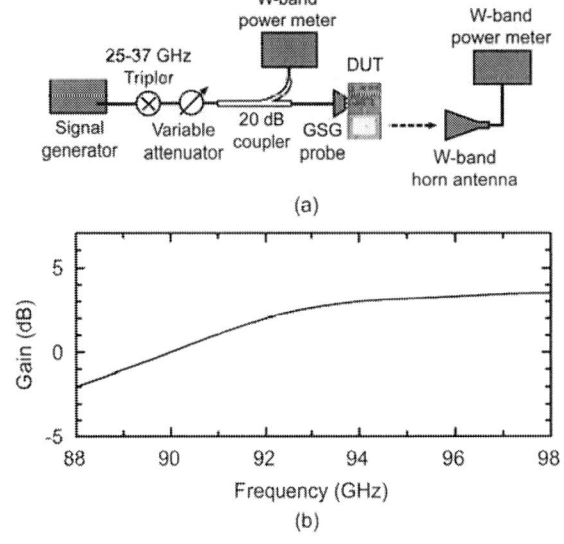

Fig. 7. (a) W-band test setup for the on-chip antenna/power amplifier. (b) Simulated on-chip microstrip antenna gain.

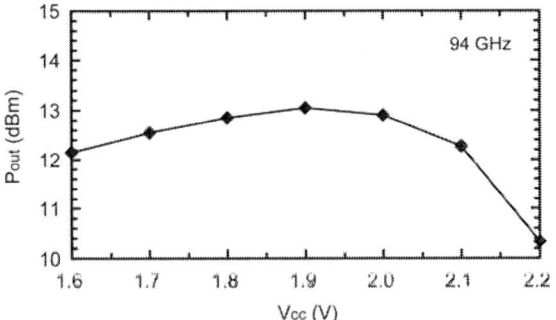

Fig. 6. Measured 94 GHz output power vs. collector voltage.

978-1-4244-6240-7/10 $26.00 © 2010 IEEE

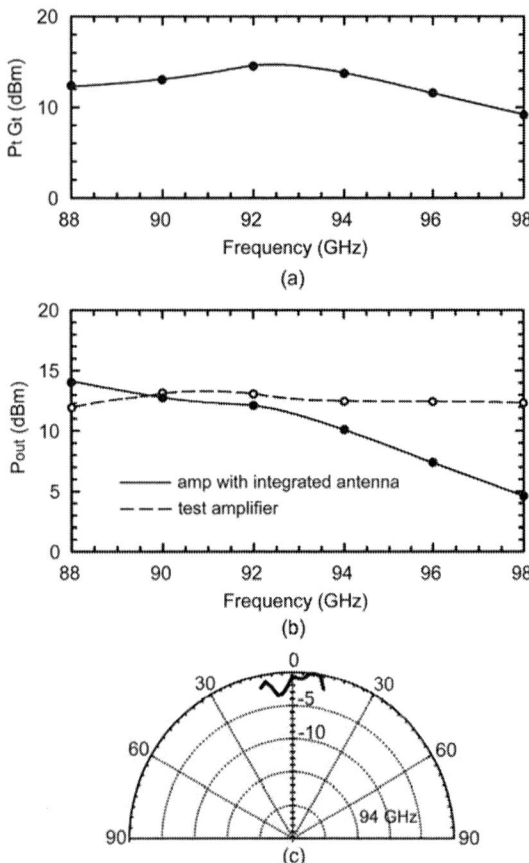

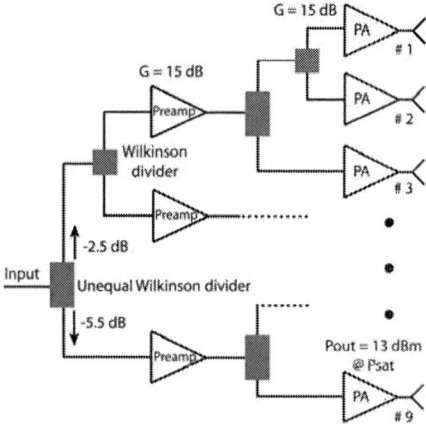

Fig. 10. Block diagram of the 9-element W-band power amplifier.

The same technique can be used for mm-wave phased array transmitters and receivers.

Fig. 8. (a) Measured ERP ($P_t G_t$) for the amplifier with on-chip antenna. (b) Comparison of measured P_{out} from the test amplifier and P_{out} of the amplifier with the on-chip antenna. (c) Measured patterns for +/-12°.

with an antenna gain of 12 dB (Directivity of 15 dB). The results will be presented at the conference.

ACKNOWLEDGEMENT

This work was supported in part by DARPA MTO, Sanjay Raman, Program Monitor. The authors thank Michael Chang for technical discussions.

REFERENCES

[1] M. Chang, and G. M. Rebeiz, "A wideband high-efficiency 79-97 GHz SiGe linear power amplifier with > 90 mW output," *IEEE Bipolar / BiCMOS Circuits and Technology Meeting*, pp. 69-72, October 2008.

[2] J. Lee, C.-C. Chen, J.-H. Tsai, K.-Y. Lin, and H. Wang, "A 68-83 GHz power amplifier in 90 nm CMOS," *IEEE Int. Microwave Symp. Dig.*, pp. 437-440, June 2009.

[3] S. T. Nicolson et al., "A low-voltage SiGe BiCMOS 77-GHz automotive radar chipset," *IEEE Trans. Microwave Theory & Tech.*, vol. 56, no. 5, pp. 1092-1104, May 2008.

[4] T. LaRocca, and M.C.-F. Chang, "60 GHz CMOS differential and transformer-coupled power amplifier for compact design," *IEEE RFIC Symp.*, pp. 65-68, June 2008.

[5] Y.-N. Jen et al., "Design and analysis of a 55-71 GHz broadband distributed active transformer power amplifier in 90-nm CMOS process," *IEEE Trans. Microwave Theory & Tech.*, vol. 57, pp. 1637-1646, July 2009.

[6] J. W. May, R. A. Alhalabi, and G. M. Rebeiz, "A 3 G-bit/s W-band SiGe ASK receiver with a high-efficiency on-chip EM-coupled antenna," *IEEE RFIC Symp. Dig.*, 2010, submitted for publication.

[7] J. Harvey, E. R. Brown, D. B. Rutledge, and R. A. York, "Spatial power combining for high-power transmitters," *IEEE Microwave Magazine*, vol. 1, no. 4, pp. 48-59, December 2000.

[8] Z. Popovic, and A. Mortazawi, "Quasi-optical transmit-receive front ends," *IEEE Trans. Microwave Theory & Tech.*, vol. 46, no. 11, pp. 1964-1975, November 1998.

[9] Sonnet, v. 12, Sonnet Soft. Inc., Syracuse, NY, 1986-2008.

[10] C. A. Balanis, *Antenna Theory*, 2nd ed., Wiley, 2005.

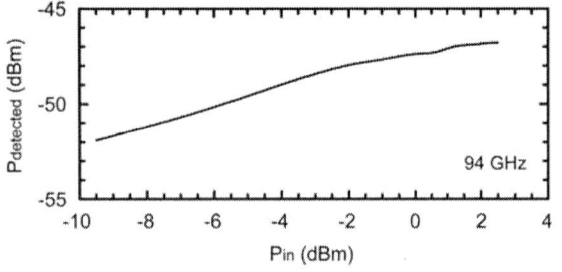

Fig. 9. Measured P_t at the horn (R=25 cm) vs. P_{input} at 94 GHz for the amplifier with the on-chip antenna. The large signal characteristics compare well with Fig. 5 for the test amplifier.

V. CONCLUSION

A W-band antenna/amplifier has been demonstrated at 88-98 GHz and with excellent performance. This approach opens the area to wafer-scale SiGe or CMOS power combining for high-power millimeter - wave transmitters.

978-1-4244-6240-7/10 $26.00 © 2010 IEEE

Application of BSIMSOI MOSFET Model to SOS Technology

James Roach, Lee-Wen Chen, Peter Clarke, and Francis M. Rotella

Peregrine Semiconductor, San Diego, CA, 92121, USA

Abstract — **The BSIMSOI model largely dominates the modeling of silicon-on-insulator (SOI) MOSFET technologies. Silicon-on-sapphire (SOS) technology has many of the advantages of SOI for RF and low-power applications, but with enhanced electrical isolation and heat dissipation, among others. We show that BSIMSOI can reasonably describe state-of-the-art SOS devices as well, including partial and full depletion, as long as differences between SOS and SOI technologies are accounted for in the parameter extraction methodology. For RF switch applications, R_{ON} and C_{OFF} are adequately represented. Also, a spot check at low currents shows that a modeled RF figure of merit, F_T, is not unreasonable.**

Index Terms — **Silicon on insulator technology, silicon on sapphire, BSIMSOI, semiconductor device modeling.**

I. INTRODUCTION

Silicon-on-insulator (SOI) technologies, with a buried oxide layer between the active device region and the silicon substrate, have advantages over bulk silicon for low-power and RF applications due to reduced parasitics, better electrical isolation, and lower substrate losses. The present de facto standard compact model for SOI is the U. C. Berkeley BSIMSOI model [1]-[3]. This model is based on the widely used BSIM MOSFET models, but incorporates additional physics to represent body currents and floating body effects, such as the drain current "kink".

Silicon on sapphire (SOS), with the active silicon layer formed directly on an insulating sapphire substrate, improves on SOI in a number of areas, including further reducing some parasitics, improving the electrical isolation (Fig. 1), and easing thermal management requirements. SOS circuits find widespread adoption in the communications industry due to such advantages.

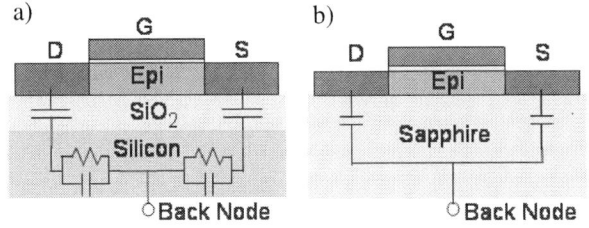

Fig. 1.   Simple cross-section drawings comparing a) SOI to b) SOS. Lumped circuit elements that may be used to model the BOX and substrates are also shown. (Not to scale.)

In this work we show that, with proper consideration of the device physics and careful design of the parameter extraction flow, the BSIMSOI model can also represent the behavior of leading-edge SOS MOSFETs engineered for different electrical operating regions, such as partially depleted (PD) and fully depleted (FD). Fig. 2 shows the effect of the level of channel depletion (doping level) on the measured gate capacitance of MOS structures.

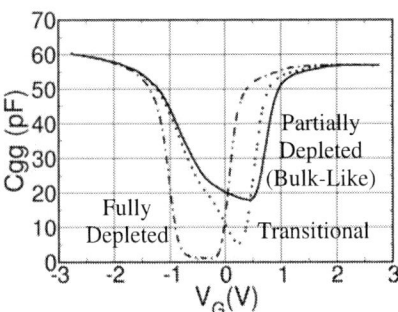

Fig. 2.   Measured NMOS gate capacitance voltage curves for SOS illustrating the effect of the level of channel depletion.

II. PEREGRINE SOS PROCESS

In this section we discuss Peregrine's UltraCMOS™ SOS process, and some differences from SOI, emphasizing the impact on the parameter extraction flow.

A. SOS and SOI

Peregrine's SOS process uses a thin (≤ 0.1 um) epi layer. A benefit of the thin epi is that parasitic junction capacitances are quite small. Also, by controlling the channel doping, we fabricate devices with a range of electrical behavior from partial to full depletion. The fully depleted devices exhibit reduced short channel effects and less severe kink. However, thin epi can, due to higher impedance, make the body contact less efficient at holding a fixed body potential than for thicker epi or bulk silicon. This means that even body-contacted devices may exhibit the kink effect, though less severely than for floating body FETs.

It was mentioned above that some parasitic elements are reduced in SOS compared to SOI. This is due to in part to replacing the buried oxide (BOX) and silicon substrate combination with the single, thick, sapphire insulating substrate. Referring back to Fig. 1, for SOI the parasitic elements from the active device to the back node (and from device to device) consist of capacitance (Cbox) through the BOX, and the (lossy) substrate impedance (often represented as a resistor, or parallel RC network). Only the capacitance though the BOX to the substrate is included in BSIMSOI; the substrate network and coupling to other devices would need to be added externally.

For SOS, no lossy substrate network is required for accurate RF modeling. The sapphire, extending as it does from the active device region to the back node, essentially

978-1-4244-6240-7/10 $26.00 © 2010 IEEE

becomes the BOX: it can be represented by an equivalent BOX thickness and included directly in the BSIMSOI model. The lateral coupling from device to device through the substrate would still be included externally to the BSIMSOI model as it would be for SOI, but is reduced to simple capacitances, with no loss element.

A final point of difference is the effect of self-heating. The SOI BOX consists of a moderately thick SiO_2 layer, typically several tenths of a micron or more. SiO_2 is a poor thermal conductor, greatly reducing the effectiveness of the silicon substrate for removing heat. The sapphire substrate of SOS in contrast, is a decent thermal conductor, allowing significant improvement in heat dissipation. Self-heating is incorporated into BSIMSOI; however, we have not found self-heating to be a significant issue for SOS.

B. General Parameter Extraction Flow

Although there are details specific to BSIMSOI that must be considered, the extraction of the fundamental FET parameters is similar to that for other compact models such as the bulk silicon BSIM family. One generally follows a sequence of steps meant to more or less isolate the effect one or more parameters on various transistor curves to allow easier extraction and more unique parameter sets. Various geometries and bias conditions are employed. Here, we emphasize steps of the importance for BSIMSOI, particularly modeling of the body characteristics. MOSFETs with a body contact are used for extraction.

For BSIMSOI modeling, additional steps are required for extracting parameters controlling the body behavior:
1. Parasitic BJT currents: Extract BJT parameters such as ideality factor and saturation current from the collector current and base current vs. body voltage (V_{BS}) and V_{DS}.
2. Source/drain to body diode parameters: While some parameters are shared with the BJT, there may be other current contributions to the diodes.
3. Impact ionization current: These parameters are obtained from body current (I_B) vs. V_{GS} and/or V_{DS}.

An additional step would typically be needed here for SOI that is not directly applicable to SOS: That is modeling the behavior of the parasitic S and D to substrate MOS capacitor. For SOI, the substrate underneath the S/D regions may be accumulated or depleted (through the BOX), depending on the D and S voltages relative to that of the back node. BSIMSOI implements simple, piecewise equations to model the voltage dependence of this capacitance. A sweep of a few volts on the S and D with the back node grounded is typically sufficient to characterize this element. For SOS, in contrast, due to the thick sapphire, there is little control by the back node of the potential at the back interface (between epi and sapphire), and with no silicon substrate to deplete or accumulate, this capacitance should be voltage independent for SOS. In addition, with roughly two orders of magnitude larger equivalent BOX thickness, this capacitance is very small.

Although fitting using numerical optimization is sometimes unavoidable, where possible we applied analytical least-squared fitting techniques. We found the extraction of the impact ionization parameters a particular challenge. The impact ionization model in BSIMSOI boasts roughly a dozen parameters. In order to make extraction of its parameters more tractable, we used a combination of simplification, optimization, and analytical least-squares fitting. We combined fits using both I_B vs. V_{GS} and I_B vs. V_{DS} data. This allowed us to obtain reasonably unique, consistent parameter sets.

Finally, for RF switch applications, the off capacitance, C_{OFF}, must be accurately represented. This requires an additional drain-to-source capacitance (not captured in BSIMSOI) to be included as a subcircuit element.

III. MEASUREMENTS

N and P MOSFETs of various geometries, with and without a body contact, were fabricated in Peregrine Semiconductor's 2.5 V UltraCMOS™ SOS process. Minimum drawn channel length was 0.25 um. Three levels of channel doping provide devices with various levels of depletion. These devices are designated by their threshold voltages (V_{TH}): the device with highest doping has a V_{TH} of roughly 0.8 V (at V_{BS} = 0 V), and is designated by "H" for "high"; the next lower, with a V_{TH} of 0.45 V is designated "R" for 'regular'; the lowest doped device, with no channel implant, is designated "I" for "intrinsic", has V_{TH} near zero.

Measurements performed include I_D vs. V_{GS} and V_{DS}, including the subthreshold region, as well as the transconductance (g_m) and drain conductance (g_d). Also fully characterized, in more depth for BSIMSOI, are the behavior of the source (S) and drain (D) to body (B) diode currents, parasitic BJT currents, and the impact ionization current, necessary for accurate modeling of the body behavior. These measurements were preformed over a temperature range of -20 C to 125 C. In addition, gate capacitance (C_{gg}) curves were recorded vs. V_{GS}.

Low-frequency C_{OFF} measurements (V_{GS} = -2.5 V, V_{DS} = 0 V) were made on floating body devices. Though not an emphasis of this work, a small number of F_T points were obtained at low drain currents.

IV. RESULTS AND DISCUSSION

For brevity, though PMOS extractions were also performed, we show figures only for NMOS. The NMOS devices will be designated IN, RN, and HN for the different channel doping levels.

A. Threshold Voltage as an Indicator of the Level of Channel Depletion

Though the gate C-V characteristics can give some indication of the level of channel depletion, a more pragmatic indicator for modeling purposes is the trend in V_{TH} with V_{BS}. Fig. 3 illustrates this for long-channel HN, RN, and IN FETs. For the highest doping, the HN V_{TH} vs. V_{BS} curve is nearly flat below V_{BS} = -1 V, then curves downward as V_{BS} increases. The RN V_{TH} is flat to nearly 0 V, then bends down. The IN curve is nearly flat over the

978-1-4244-6240-7/10 $26.00 © 2010 IEEE

entire V_{BS} range. The region of decreasing Vth follows basically the same square-root dependence on V_{BS} that a bulk silicon device does, indicating only partial depletion of the epi under the channel. The flat region indicates that the epi in the channel region is fully depleted. Thus, for the HN and RN, there is a transition from fully depleted behavior to partial depletion, while the IN FET is fully depleted regardless of V_{BS}.

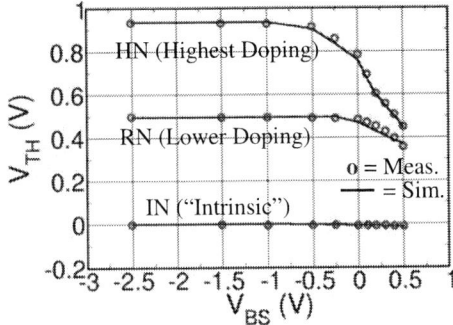

Fig. 3. V_{TH} vs. V_{BS} curves illustrating the effect of doping and the level of channel depletion.

The BSIMSOI model allows three primary selections of model behavior: Always partially depleted (PD), always fully depleted (FD); and a hybrid (sometimes referred to as dynamic depletion, or DD) wherein the transition from full to partial depletion can be captured. Depending on the application, the PD model could be used for the HN FETs and the FD model used for the IN FETs; because the RN Vth transition is near $V_{BS} = 0$ V the hybrid model should be chosen for it. For simplicity and consistency, we employ the hybrid model for all cases.

B. Model Results for the Body Currents

The first body current we will discuss is the collector current (I_C) of the parasitic BJT formed by D, B, and S. Fig. 4 gives the I_C vs. V_{BS} curves at different collector (drain) biases for an IN device with a drawn channel length (L) of 0.25 um. In the active BJT region, ($V_{BS} < 0.8$ V), the current is reasonably well modeled, with some deviation in the high-injection and series resistance dominated regimes.

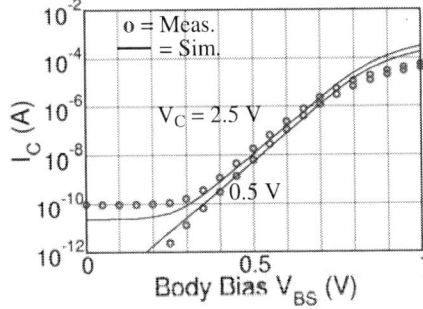

Fig. 4. I_C vs. V_{BS} for the parasitic BJT of an IN FET with W/L = 10/0.25.

For short-channel IN FETs, BJT current gain, beta, can be as high as roughly 100. For the higher doped RN FETs, betas were an order of magnitude lower, and for HN, betas were near 1 or less (probably due to the reduced emitter efficiency at the higher doping). This suggests that the parasitic BJT may have significant influence on the I device behavior, but little on the H device.

We show RN impact ionization characteristics (I_B vs. V_{GS} with V_{DS}) in Fig. 5. We found that we could not obtain fits of equal quality at all channel lengths. Fits at longer channel lengths were typically somewhat better. Impact ionization was stronger for HN and weaker for IN, as expected for the relative doping levels. PMOS devices tended to have more ideal behavior, exhibiting less extension at higher V_{GS}, and typically fit better than the NMOS. Also as expected, the impact ionization currents for PMOS were much lower than for NMOS.

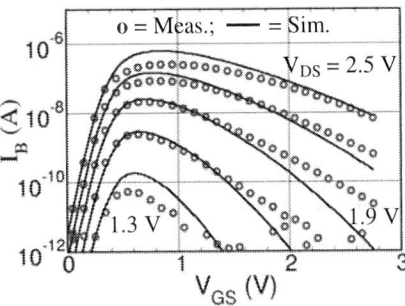

Fig. 5. Impact ionization current vs. V_{GS} for an HN FET with W/L = 10/0.25.

C. Model Results for Floating-Body Devices

Once the BSIMSOI model with its body components from the body contacted devices is in place, the question remains as to how well these models describe the behavior of floating-body devices. Figs. 6-8 show DC curves of common interest. The log(I_D) vs. V_{GS} curves in Fig. 6, for an RN floating-body FET with W/L = 10/0.25, show that the regions near and below threshold are well modeled.

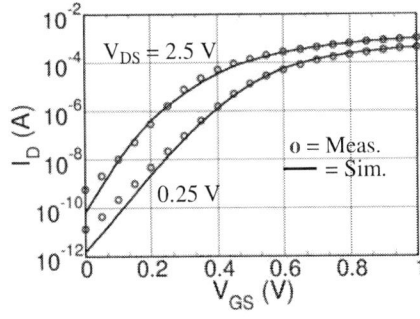

Fig. 6.   Log(I_D) vs. V_{BS} curves for an RN FET with W/L = 10/0.25 showing near- and sub-threshold characteristics.

Fig. 7 shows that the kink in the I_D vs. V_{DS} and g_d vs. V_{DS} characteristics is captured, in this case for an HN

10/0.25 FET (kink is worst for the HN due to the higher impact ionization current). Here, the turn-on position of the kink is quite closely modeled; the magnitude of the kink is somewhat overestimated, with a better match at lower V_{GS} (of interest for analog applications). The only change from the body-contacted device model (aside from a modified width correction for not having a body contact) was to lower an impact ionization turn-on voltage parameter by a few tenths of a volt. Fig. 8 gives the I_D vs. V_{DS} and g_d vs. V_{DS} characteristics for an IN 10/0.25 FET, which, being fully depleted, exhibits very little kink.

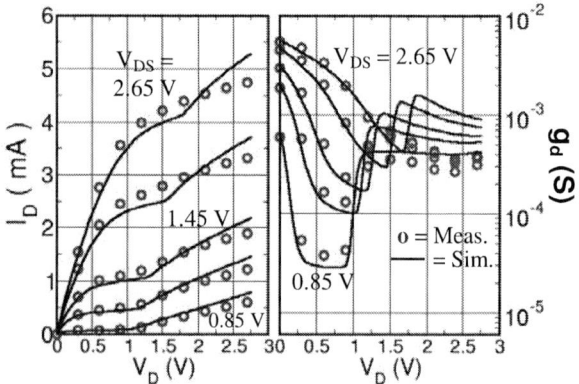

Fig. 7. I_D vs. V_{DS} (left graph) and g_d vs. V_{DS} (right graph) for an HN FET with W/L = 10/0.25 showing strong kink behavior.

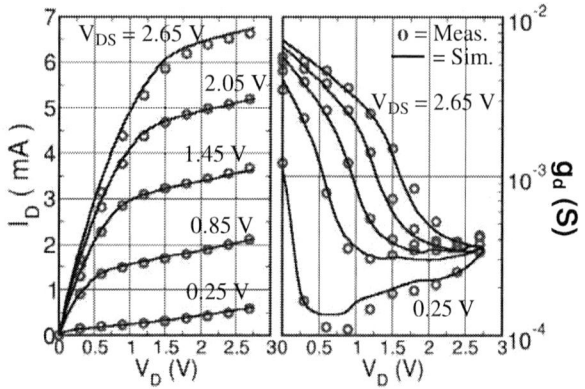

Fig. 8. I_D vs. V_{DS} (left graph) and g_d vs. V_{DS} (right graph) for an IN FET with W/L = 10/0.25 showing very little kink.

D. Applications

For RF switch design, the BSIMSOI model, with the additional drain-to-source capacitance, gives sufficiently accurate results for R_{ON} and C_{OFF}. Their scaling versus channel length is shown in Fig. 9. In addition, though RF gain measurements are not stressed in this work, for low current applications we spot checked F_T for a large IN FET (with 63 gate fingers, each 8 um wide and 0.25 um long). F_T is relatively insensitive to layout parasitics and more indicative of the core model. As may be seen in Fig 10, the simulated values are actually quite reasonable, given that

only estimated layout parasitic contributions were used with the extracted DC and low frequency model. Refinement of the layout parasitic components from S-parameter measurements using standard RF FET modeling techniques would improve the simulation accuracy.

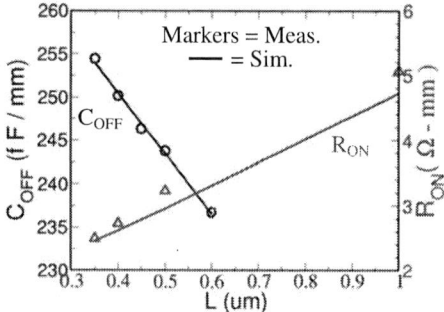

Fig. 9. R_{ON} and C_{OFF} vs. L for an IN FET.

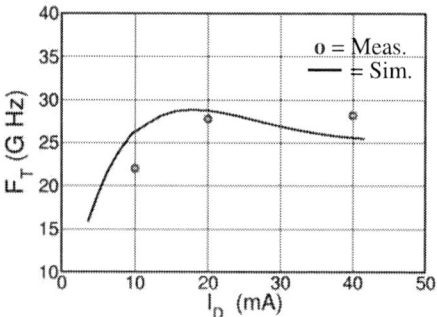

Fig. 10. a) F_T vs. I_D for an IN FET with 63 gate fingers, 8 um finger width, and 0.25 channel length. $V_{DS} = 1.5$ V

V. CONCLUSION

The BSIMSOI model, prevalent in SOI MOSFET modeling, also yields reasonable models for SOS technology, provided channel depletion effects and differences in substrate materials are properly considered. The sapphire substrate of SOS, replacing the BOX and silicon substrate of SOI, can in fact somewhat simplify the overall RF modeling of the FETs.

ACKNOWLEDGEMENT

The authors wish to thank the Peregrine Technology group for useful discussions on the SOS FET construction.

REFERENCES

[1] "BSIMSOI3.1 MOSFET MODEL User's Guide," BSIM Group, University of California, Berkeley, CA, 2003.
[2] "BSIM3.2 Equation List," BSIM Group, University of California, Berkeley, CA, 2004.
[3] Pin Su, K. et al, "A Unified Model for Partial-Depletion and Full-Depletion SOI Circuit Designs: Using BSIMPD as a Foundation," *IEEE CICC Proc.*, pp. 241-244, March 2003.

Modeling of SOI FET for RF Switch Applications

Tzung-Yin Lee and Sunyoung Lee

Skyworks Solutions, Inc.

5221 California Avenue, Irvine, CA 92617

Abstract — **This paper presents the modeling of an SOI FET for RF switch applications. Given that the HF small-signal predictability, i.e. the insertion loss and the isolation, is a common state of the art, the study focuses on the modeling of the non-linearity of the FET. The non-linearity of an SOI FET switch arises from not just the transistor, but also the SOI substrate through various mechanisms. First the non-linearity is caused by the voltage imbalance, a direct result of the substrate loss, in a switch made of many FETs stacked in series. The voltage imbalance is the main non-linearity contributor to a FET switch at high-power levels. Secondly the substrate itself is non-linear and sets the harmonic floor. Besides the substrate, the impact of other important SOI physics, such as the floating-body effect and the parasitic BJT effect, to the switch linearity will also be discussed. Finally a hybrid model that combines PSP as the FET core and a layout-dependent non-linear SOI substrate model is presented, and excellent non-linearity predictability was demonstrated on a real-life RF switch.**

Index Terms — **RF switch modeling, PSP model, SOI model, floating-body effect, parasitic BJT effect**

I. INTRODUCTION

A typical front-end module (FEM), which is frequently employed in a cellular handset as the interface between the external world and the internal RF transceiver and baseband ICs, consists of tens to hundreds of Si, non-Si, and discrete components, including pHEMT switches, GaAs/InGaP PAs, and Si controllers as well as other matching passives. In addition to the demand of smaller footprint, the fast pace of commoditization of the cellular application has driven development of FEMs to use a MOSFET switch to replace the pHEMT switch for its low cost as well as its natural capability to have the controller integrated on the same die. It has been known that, because of its limited high-voltage (HV) isolation from the substrate, the MOSFET would have inferior performance in terms of high-power handling capability, insertion loss, and linearity compared to the pHEMT counterpart. Although the SOIFET was initially developed for high-speed and low-power digital applications [1], it is a natural candidate -- as opposed to the bulk MOSFET -- for high-power switch applications and a few research papers have been published in the past few years, demonstrating its feasibility [2][3].

While Si-based technology is famed for low cost in volume production, it tends to have higher non-recurring engineering (NRE) costs and longer development times.

To mitigate the disadvantage, an accurate model and simulation methodology is required to reduce the number of mask spins to achieve comparable NRE and time-to-market. Also, since a design with higher integration is always more difficult to debug at the hardware level, a predictive model would be a powerful and cost-effective diagnostic tool.

This paper discusses various effects in an SOIFET and their impact to an RF switch application. As the linearity of a switch is very stringent in cellular applications, even very weak non-linear effects in the device need to be accounted for. A typical GSM application requires the switch 2nd and 3rd harmonics to be below -40dBm level at 34dBm input power and below -33dBm under 3:1 VSWR conditions.

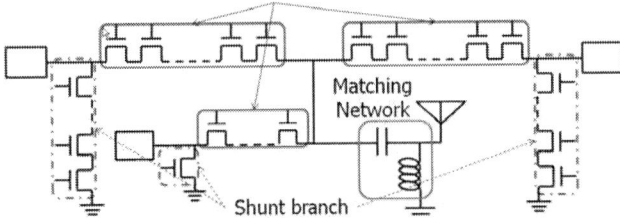

Fig. 1 A simplified switch schematic

The paper is organized as follows. In section II, modeling of a switch will be presented with a focus on the physics that influence the linearity of an RF switch. Section III describes the model implementation for a switch application. Section IV presents the measured data of the switch test structures. Section V concludes the paper with a summary.

II. MODELING OF A SWITCH

In our experience, the small-signal behavior of a switch is not difficult to model, no matter if it is in ON or OFF mode. Most foundry-provided models predict the ON insertion loss right out of the box, even if the model is based on a simple modeling methodology with only DC and large-area AC capacitance measurement. The reverse isolation of such a simple model might be a little bit off. It is because most of modeling methodology is based on measurements taken on transistors in the common-source (CS) configuration, where the OFF capacitance (C_{OFF}) is not a prominent component. The problem can be easily resolved with s-parameter measurements on a few

978-1-4244-6240-7/10 $26.00 © 2010 IEEE

common-gate (CG) structures with high and low gate resistance.

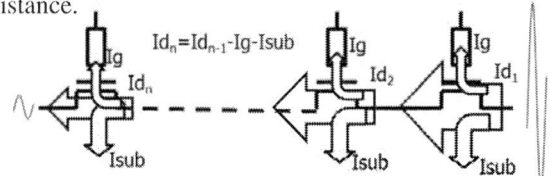

Fig. 2 Different amount of Id flowing through each transistor in a long chain causes voltage imbalance

The challenge of an SOIFET model is its non-linearity prediction. Depending on the mode of operation, the non-linearity of a FET is caused by different mechanisms. A typical switch consists of a number of transistors stacked in series and in shunt to ground as shown in Fig. 1. The number of stacked transistors in series is determined by the maximum voltage for reliable operation of a transistor as well as the maximum voltage switch at the antenna port. When a transistor is fully ON, the non-linearity is mainly contributed by the transistor ON resistance. When a transistor is OFF, the linearity is influenced by not only the C_{OFF}, but also other parasitics. At higher power levels the high voltage swing appearing at the antenna port can turn the first few transistors weakly ON through a significant amount of voltage imbalance, i.e. the voltage drop is not evenly distributed across each transistor in a chain. As shown in Fig. 2, voltage imbalance of a transistor chain is caused by the additional currents required to supply the loss through the substrate and the gate resistor. It can also induce gate-induced drain leakage (GIDL), enhanced by the parasitic bipolar effect in an SOI FET.

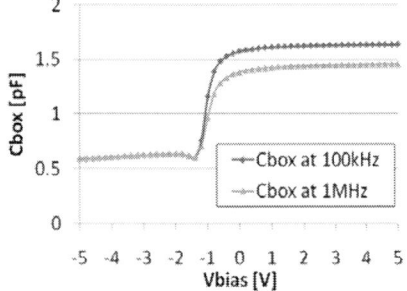

Fig. 3 CV measurement of a large active area over the box oxide

Besides the non-linearity caused by the transistor, the substrate is also an important source of non-linearity. It is commonly perceived that the buried box oxide capacitance to the substrate is a constant, thereby introducing minimal non-linearity to the overall system. However, a simple CV measurement on a large active area over the buried box oxide reveals it behaves just as non-linear as a MOS capacitor. Fig. 2 shows the CV measurement of a large active area over an SOI substrate. The measurement reveals that 1) the capacitance is far

from being a constant over the voltage; 2) the substrate, i.e. the semiconductor side of the MOS cap, behaves more like an N-type substrate rather than a P-type; and 3) the effective capacitance decreases as the measurement frequency increases.

The frequency dependence observed in Fig. 3 is mainly dictated by how fast the accumulation layer underneath the box oxide can be charged and discharged. The good news is that the high substrate resistance limits the bandwidth of the non-linear response of the box oxide capacitance. However, the box capacitance is still fairly non-linear at the GHz range. Fig. 4(a) shows the impedance into the substrate as a function of the bias voltage measured at 1GHz. Although the percentage change of the box capacitance is bandwidth limited, the effective impedance looking into the substrate still appears quite significant at high frequencies. The non-linearity of the box oxide capacitance can also be observed through the HF coupling between two adjacent active structures, as seen in Fig. 4(b).

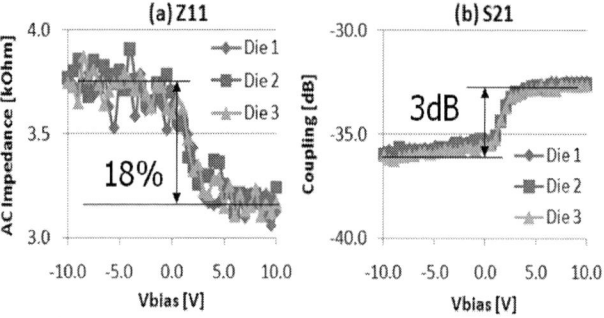

Fig. 4 Two-port s-parameter measurement result on an active-to-active coupling test structure: (a) measured impedance (Z11) looking into the substrate from an active island through the box oxide; (b) measured coupling (S21) between two adjacent active structures at 1GHz.

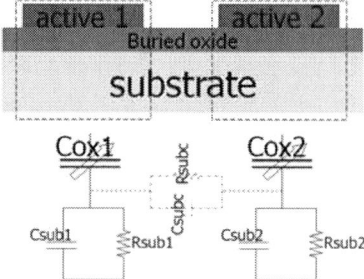

Fig. 5 A simple model for the non-linear substrate

According to an analytical calculation, the amount of AC current going into a substrate of 1000Ohm-cm through a 0.5um thick box oxide represents ~5% of the total AC current going into the drain side of the transistor when it is OFF. The percentage of the substrate current could become higher if the effective substrate resistivity is lower. Assuming a simple non-linear model as shown in

Fig. 5 that produces a small-signal response of Fig. 4, simulation indicates that the non-linear substrate will set the 2nd and 3rd harmonic floors to -50 and -60 dBm levels respectively at an input power of 35dBm, and the result was confirmed with a simple measurement on a plain metal line on the substrate.

III. MODEL IMPLEMENTATION

The PSP model was selected for the SOI FET in the switch simulation for three reasons: 1) being a body-referenced model, PSP is more appropriate than a source-referenced model like BSIM for switch simulation purposes; 2) the bulk version of the PSP model enables simple integration of a user-defined substrate model; and 3) the model is readily available in major simulators such as ADS and Spectre. However, the bulk-version of the PSP model, as compared to PSPSOI [4], is incapable of modeling the floating body effect and the enhanced parasitic BJT effect [5], which are the main physical mechanisms differentiating an SOI FET from bulk CMOS.

The floating-body effect is most severe in a partially depleted (PD) SOI device and is manifested as a threshold voltage (Vth) change and sudden change of the output conductance, i.e. Kink effect. The floating-body effect results mainly from the minority carrier accumulation in the body, which requires certain current flowing in the channel and a relatively high V_{DS} to trigger some impact ionization in the drain depletion region. Fortunately, the bias condition for these effects to occur is usually quite far away from where a switch is operated. Moreover, these effects are usually DC-like effects, i.e. with relatively high time constants. Therefore, as an RF switch is concerned, these effects can be empirically modeled as a slight Vth adjustment, even if it is not necessary.

The parasitic bipolar effect still needs to be carefully considered as it will inevitably enhance the GIDL current, which could influence the linearity of an OFF transistor when the voltage swing is high. The bulk-version of the PSP model has a built-in GIDL current model, however, without the parasitic bipolar effect. The amount of amplification at higher frequencies is determined by how fast the parasitic BJT can track the high speed signal, i.e. by the bandwidth of the transistor. As the amplification of the parasitic BJT is very difficult to be measured at GHz range, the effect is validated indirectly by the large-signal characterization of a single shunt FET. By the harmonic data obtained so far, it is suggested that the GIDL current of an SOI FET at GHz range behaves more like one with little amplification, as the gain of the parasitic bipolar diminishes with the frequency.

SOI tends to have a more severe self-heating effect than the bulk CMOS, as the oxide layer has a much lower thermal conductivity than Si. Since the time constant of the self-heating effect is usually in µS or even mS range, the device temperature rise can be treated simply as an average ambient temperature change for a switch operating at GHz range. The local temperature rise of a particular arm in a switch can be calculated with the insertion loss, the thermal resistance of the switch section, and the power level.

Fig. 6 shows a simplified diagram of the final hybrid model used for SOI switch application. The FET portion of the switch is modeled by the bulk-version of the PSP model. The body node of the device is then connected to the substrate through a MOS varactor. As the total loss into the substrate is the most critical factor to determine the voltage imbalance of a switch made of many (>10) transistors stacked in series in the OFF state, the substrate network is generated with an electro-magnetic (EM) simulation according to the layout. The substrate loss consists of two components: the vertical loss into the substrate and the lateral loss into the adjacent structures. According to the S-parameter characterization on the active-to-active substrate coupling structures of various sizes and spacing, it is discovered that the lateral loss can be more significant than the vertical loss as the spacing between the two un-related active structures decreases. Therefore, a layout-dependent EM approach was adopted here.

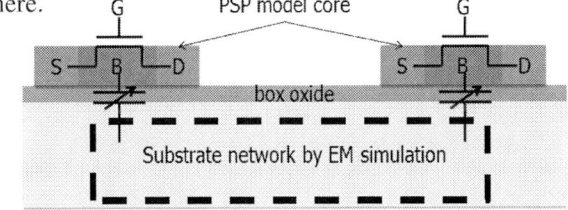

Fig. 6 A simplified diagram of the hybrid model for SOI switch modeling

It is known that EM simulation does not account for any non-linearity. Therefore, the MOS varactor portion has to be excluded from the EM simulation and treated separately. It is done by performing the EM simulation on only the substrate beneath the box oxide with a simple metal contact on top of the substrate, which is equivalent to having the electrical excitation projected from the active area above. The methodology of cascading an external cap on top of the EM simulation of only the substrate is verified to yield the same result of an EM simulation with the box oxide and the active layers. This simple setup makes the EM simulation very efficient for even a very large and complex layout. The EM simulation can be done in the floor planning stage, since it needs only the active area information in the layout.

IV. MEASUREMENT RESULT

The switch test structures were manufactured with a 0.13μm SOI CMOS process. The FET used for the switch design is the 2.5V IO device with thick oxide. The substrate network that accounts for both the vertical loss into the substrate and the lateral loss to the adjacent structures was generated with the 2.5D Momentum simulation. The final switch simulation was performed with ADS Harmonic Balance simulation with RFDE interface, which enables easy co-simulation of schematic and layout-dependent EM simulation result.

Fig. 7 shows the measured harmonic data versus the model of a 12-stacked FET switch of 2.5mm width each. As shown in the figure, the 2nd harmonic is enhanced by the substrate non-linearity by a few dBs. Data of other series switches with different transistor widths indicate that the 2nd harmonic increases with the total FET area, agreeing that the 2nd harmonic correlates with the substrate non-linearity.

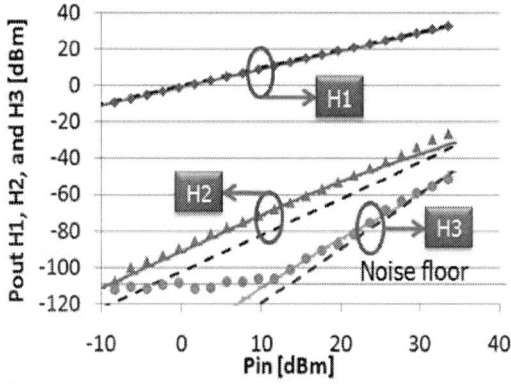

Fig. 7 Harmonic data versus model of a 12-stacked series FET switch of width 2.5mm at VG=2.0V, Freq=900MHz. The solid lines represent the prediction with the non-linear substrate; the dashed lines represent the prediction without the substrate non-linearity.

Fig. 8 shows a comparison of different models for a switch sub-block with a 12-stacked shunt branch and a 12-stacked series branch. As shown in the figure, the simulation of the full-blown model matches the measured data very well. The short-dashed lines show the simulation without any transistor non-linearity. This was done by representing the transistor by just an ON or OFF resistor, according to its state. As shown by the short dashed lines, the 2nd harmonic is pretty much dictated by the substrate non-linearity. As shown by the long dashed lines, the 2nd harmonic prediction could be off by more than 20dB if the substrate non-linearity is not properly accounted for.

V. CONCLUSION

A hybrid modeling methodology that combines the mature bulk PSP model and a layout-dependent nonlinear substrate model is introduced for SOI switch applications. The proposed model is shown to be sufficient for switch harmonic simulation up to a very high power-level of ~35dBm required by cellular applications.

ACKNOWLEDGEMENT

The authors wish to acknowledge their colleagues Ed. Lawrence in Cedar Rapids, IA and Steve Sprinkle in Woburn, MA, in the switch modeling discussion and in providing the system-level characterization data.

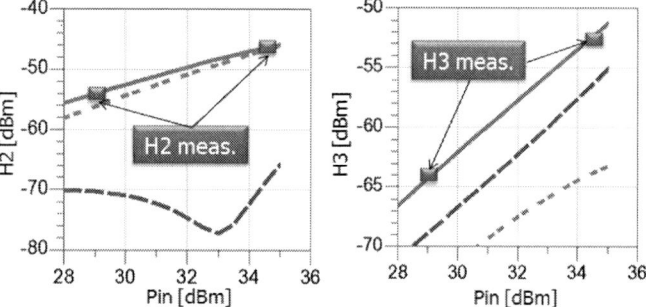

Fig. 8 Harmonic simulation of a 12 stacked shunt switch and 12 series switch together. The two symbols are the measured data at Pin=29 and 34.5 dBm; the solid lines represent the prediction of the full-blown model with the non-linear substrate, while the long-dashed lines with the linear substrate and the short-dashed lines with the transistor modeled as a linear resistor.

REFERENCES

[1] J. Colinge, "Silicon-On-Insulator Technology: Materials to VLSI," Springer, 1997.

[2] J. Costa, M. Carroll, J. Jorgenson, T. Mckay, T. Ivanov, T. Dinh, D. Kozuch, G. Remoundos, D. Kerr, A. Tombak, J. Mcmaken, and M. Zybura, "A Silicon RFCMOS SOI Technology for Integrated Cellular/WLAN RF TX Modules," 2007 IEEE/MTT-S International Microwave Symposium, p. 445-448, June 2007

[3] V. Blaschke, "New Developments in SiGe & CMOS SOI for Wireless Frontend Modules," 2009 Tower/Jazz Technology Conference, Nov. 2009.

[4] W. Wu, X. Li, G. Gildenblat, G. Workman, S. Veeraraghavan, C. McAndrew, R. van Langevelde, G.D.J. Smit, A.J. Scholten, D.B.M. Klaassen and J. Watts, "PSP-SOI: A Surface Potential Based Compact Model of Partially Depleted SOI MOSFETs," IEEE Custom Integrated Circuits Conference, 2007.

[5] J. Fossum, and S Krishnan, "Grasping SOI floating-body effects," IEEE Circuits and Devices Magazine, Volume 14, Issue 4, p. 32 – 37, July 1998.

A High Power CMOS Differential T/R Switch using Multi-section Impedance Transformation Technique

Hyun-Woong Kim[1], Minsik Ahn[2], Ockgoo Lee[1], Chang-Ho Lee[2], and Joy Laskar[1]

[1]Georgia Electronic Design Center, Georgia Institute of Technology, Atlanta, GA 30308, U.S.A
[2]Samsung Design Center, Atlanta, GA 30308, U.S.A

Abstract — A high-power single-pole-double-throw (SPDT) antenna switch using a differential architecture and a multi-section impedance transformation technique is demonstrated in a standard 0.18-µm CMOS process. The differential architecture prevents unwanted channel formation of OFF-state Rx switch transistors by relieving the voltage swing over the Rx switch devices. In addition to this architecture, impedance transformation technique helps to reduce the voltage swing even more, contributing to significant enhancement of power handling capability. A loss of the whole design block including switch and matching networks has been analyzed, considering the integration issue of the front-end circuitries. The measured performance of the differential switch shows input 1-dB compression point (P_{1dB}) of 33.8 dBm with insertion losses of 0.5 dB and 1.1 dB for Tx and Rx modes at 1.9 GHz, respectively.

Index Terms — CMOS switch, differential switch, high-power switch, integration, multi-section impedance transformation.

I. INTRODUCTION

CMOS technology has been one of the strongest candidates to achieve a fully integrated radio frequency (RF) transceiver, keeping step with a recent trend of integration, for its advantages on low cost and high feasibility of integration. However, CMOS technology is not perfectly suitable for implementing high-power-handling components such as power amplifiers (PAs) and antenna switches due to low breakdown voltage and lossy substrate. Also, large parasitic components of the technology make the design more challenging.

There have been numerous efforts to overcome these drawbacks in RF switch design. For example, transistor stacking technique, LC tuned substrate biasing, and resistive body-floating technique were devised to improve the power handling capability of RF switches [1]-[3]. As another attempt to enhance the power performance of switches, the impedance transformation technique and the differential architecture were introduced [4], [5]. However, the insertion loss of the switch is sacrificed in exchange for improvement in power handling capability with the impedance transformation technique.

In this work, impedance transformation technique with a differential architecture has been proposed to implement the switch which handles more than 2-W level of power in a 0.18-µm standard CMOS process. In the proposed design, the degraded loss by impedance transformation technique has been optimized from the perspective of full-integration with PA and switch as shown in Fig. 1.

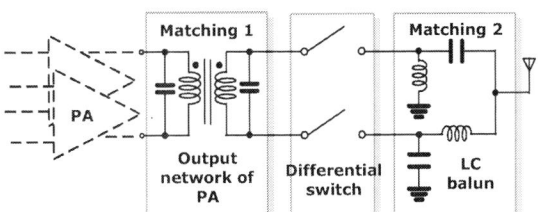

Fig. 1.   Configuration of proposed antenna switch design.

Since the entire loss is affected by various factors such as impedance transformation ratio of the matching networks and an amount of current flowing over the ON-state switches, optimum impedance for the differential switch operation has been selected considering the trade-off between loss and power performance. With these approaches, high power handling capability with the reasonable level of loss has been obtained. Output transformer of PA and the LC balun have been used as two matching networks for this configuration [6].

The stand-alone differential switch shows insertion losses of 0.5 dB and 1.1 dB for Tx and Rx modes, respectively. Power handling capability is characterized by P_{1dB} which is 33.8 dBm for this design.

II. DESIGN OF HIGH POWER CMOS SWITCH

Large voltage swing from a power amplifier is the main cause of the degradation of power performance of RF switches. Channel and junction diodes of OFF-state Rx switches are unintentionally turned on by the large signal, resulting in undesired power losses. Moreover, devices cannot even sustain over a certain level of voltage swing. Therefore, reducing the voltage swing over the OFF-state switch devices is essential to improve overall power handling capability of switches.

A. Differential Switch Design

The differential architecture has superior characteristics in linearity, robustness, and noise immunity to the single-ended configurations [5]. By virtue of these advantages, many state-of-the-art transceiver building blocks such as low noise amplifiers, mixers, and even power amplifiers are designed using differential configurations. Particularly, in the case of switch design, differential architecture helps to divide the voltage stress in half with two identical paths. Therefore, twice of power can be transmitted when it is combined.

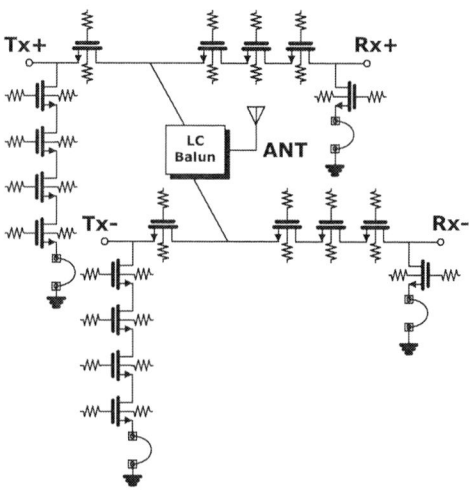

Fig. 2.   Schematic of differential SPDT switch.

Taking these into account, differential architecture is chosen for switch design aiming to the fully-integrated front-end solution with high power handling capability. The proposed differential switch structure is shown in Fig. 2. By implementing the differential switch with two of the identical switches, voltage swing over OFF-state Rx switches is decreased in half as shown in Fig. 3.

In designing the Rx switch, LC resonant circuit is one of the best candidates for high-power applications, but quality factor of on-chip inductor is not high enough to be implemented at 1.9 GHz of frequency range. Instead, a stacked transistor structure has been used to relieve the voltage stress over Rx switch transistors. Three thick-gate-oxide transistors are stacked at Rx series path to improve the power handling capability by protecting switches from device breakdown and unwanted channel formation. Since Tx series and Rx shunt switches are free from the large voltage stress, single thin-gate-oxide devices can be used for these switches to enhance insertion loss in Tx mode. All switch devices employ deep N-well structure. The P-well and deep N-well ports of the devices are biased at negative and positive supply, respectively, to prevent junction diodes from turning-on. Resistors at the gate, body, and deep N-well port are all 10 k Ω to achieve high AC isolations.

B. Impedance Transformation Technique

Given a load impedance, large voltage swing should be applied to transmit a high-power signal. For example, 28-V peak-to-peak voltage (V_{pp}) swing is required to transmit 2-W power with 50-Ω load impedance. Unfortunately, 28-V V_{pp} is too high to be transmitted by CMOS switch since the voltage swing level is strictly limited by device breakdown. Moreover, the voltage swing turns on the OFF-state Rx switches and junction diodes unintentionally. Transistor stacking technique has been used to divide this large voltage swing over stacked

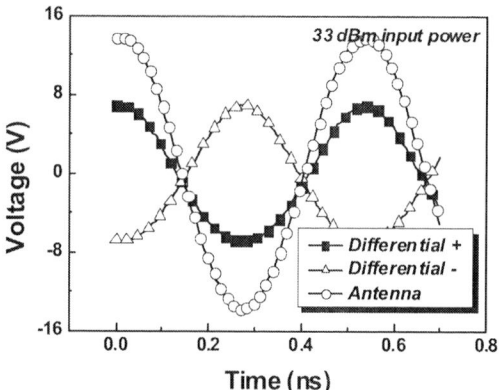

Fig. 3.   Voltage swing reduction by differential architecture.

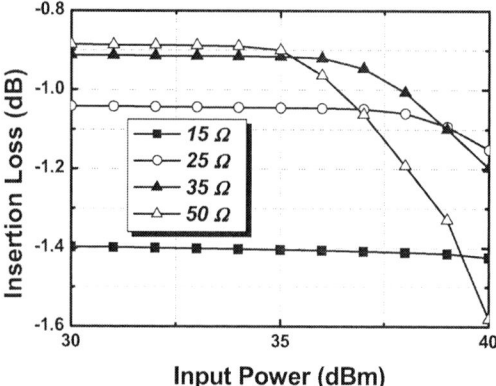

Fig. 4.   Loss and power handling capability of the proposed switch with various operating impedances.

devices, but the number of stacked transistors is also limited because of insertion loss issues at Rx mode and an additional substrate losses through junction diodes [1], [7]. In order to minimize the required number of stacked transistors, voltage swing should be relaxed as much as possible.

Therefore, impedance transformation technique is used with differential architecture. Rather than stacking more switch devices, lowering the operating impedance is better choice to transmit high-power signal since the large voltage swing can be relieved by lowering the operating impedance for a certain level of power. By doing so, the number of stacked transistors is minimized by the diminished voltage swing, and it can help to enhance the insertion loss at Rx mode.

As shown in Fig. 4, power handling capability of the switch is improved with low operating impedance. However, insertion loss of switch is increased along with excessively low operating impedance since the amount of current flowing is also increased as voltage swing reduces, resulting in loss due to on-resistance of switches. Therefore, the impedance should be optimized carefully considering the trade-off between insertion loss and power performance of the switch. In Fig.4, loss and power performance of the proposed design including electromagnetic (EM) simulated transformer and LC balun have been analyzed, and then 35 Ω is selected as the optimum operating impedance.

C. Loss Optimization for Multi-section Impedance Transform

In the proposed design, the multi-section impedance transformation is employed to set the switch operating impedance at 35 Ω. Since the output impedance of a Watt-level CMOS PA is typically low (3~10 Ω) to generate high output power, the operating impedance of the switch can be positioned between the required load impedance of the PA and the antenna impedance, 50 Ω. In other words, the first matching network, which exploits transformer, steps up the low load impedance of the PA to 35 Ω, and the second matching network with LC balun transforms 35 Ω to the antenna impedance.

While improving the power performance of the switch, the loss of the entire signal path, including both switch and matching networks, might be degraded due to excessive transformation steps. That is, the loss of LC balun, the second matching network, is inevitably added to the entire loss. In addition, the reduced operating impedance of switches increases the current flowing through the switches, consequencing the losses from the on-resistances of the switches.

On the other hand, as a merit of the proposed configuration, the efficiency of the matching network can be improved by reducing the impedance transformation ratio [8]. Furthermore, the efficiency of the transformer is enhanced with differential architecture because the quality factor of the inductors used in the transformer is higher when it operates in differential mode than in single-ended mode [9].

Considering all of these factors, insertion loss of the proposed structure has been simulated and compared to the loss of the combination of transformer and single-ended switch without LC balun in Fig. 5. On the basis that the quality factor of inductors in on-chip transformer is around 10, both of on-chip (Q of inductor is 15) and off-chip (Q of inductor is 75) LC balun are acceptable. In the case of off-chip LC balun, loss is obviously reduced.

III. MEASUREMENT RESULTS

Insertion loss, return loss, isolation, and power handling capability were measured to verify the performance of the proposed switch. All of these measurements were performed in chip-on-board (COB) test set-up. Loss and isolation of Tx and Rx switch were measured by S-parameters between each single or differential ports using 4-port network analyzer. Power handling capability was characterized by P_{1dB}, and measured by output power corresponding to input power. Since a PA was not included in this design, an off-chip balun was used to apply the input power to differential switch. Moreover, another off-chip balun was used to combine the differential power at the output of the switch, in order to measure the stand-alone differential switch without LC balun. The losses of PCB board and baluns were de-embedded. In

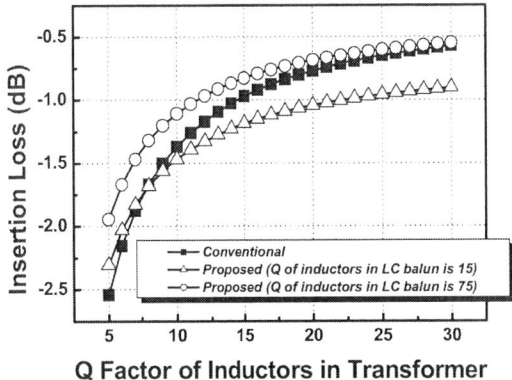

Fig. 5. Comparison of the insertion loss corresponding to the quality factor of inductor in LC balun at 1.9 GHz.

order to prevent forward bias of junction diodes, a negative bias was applied to body, and a positive bias was applied to deep N-well ports. The micro-photograph of differential switch design is presented in Fig. 6. It was fabricated with a standard 0.18-μm CMOS process, and the total size is 0.58 × 0.35 mm².

Fig. 7 and Fig. 8 show S-parameter measurement results for Tx and Rx modes of differential switch. At 1.9 GHz, the measured insertion loss of Tx mode is 0.5 dB, return loss is less than 20 dB, and isolation is 23 dB with four stacked shunt switches. On the other hand, insertion loss of 1.1 dB, return loss of 17 dB, and isolation of 33.5 dB have been measured in Rx mode at 1.9 GHz. In Fig. 9, the power handling capability of the differential switch has been compared to that of the single-ended switch. Input P_{1dB} for the single-ended switch was 32.3 dBm with 3-stacked Rx switch devices and input P_{1dB} has been increased by 1.5 dB with the differential switch.

The performance results of the proposed work are summarized and compared to other switch works in table I. Input P_{1dB} of the proposed switch is more than 2 W, which handles much more power than previous implementations. Since the output impedance of PA is not determined, measurement of whole design with on-chip transformer and LC balun could not be done completely. However, according to the analysis with EM simulated transformer shown earlier, we expect the power handling capability of the differential switch, fully integrated with a PA, to improve when it adopts the multi-section impedance transformation technique.

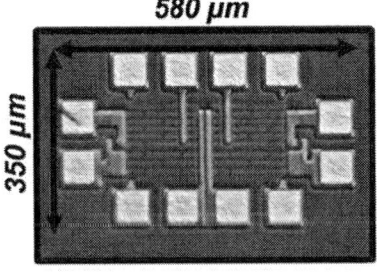

Fig. 6. Micro-photograph of differential switch.

978-1-4244-6240-7/10 $26.00 © 2010 IEEE

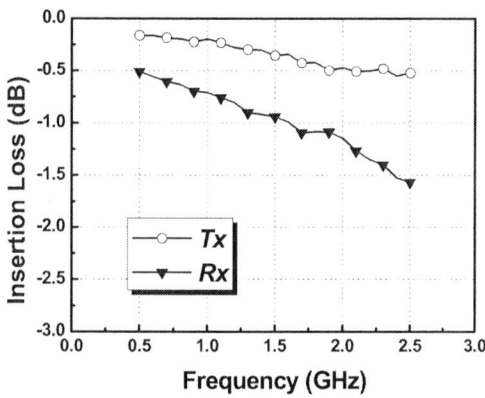

Fig. 7. Measured insertion loss of Tx and Rx mode.

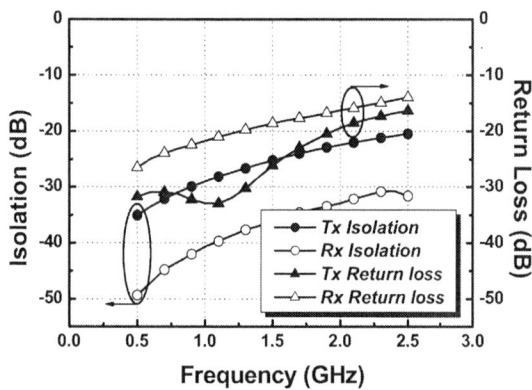

Fig. 8. Measured isolation and return loss of Tx and Rx mode.

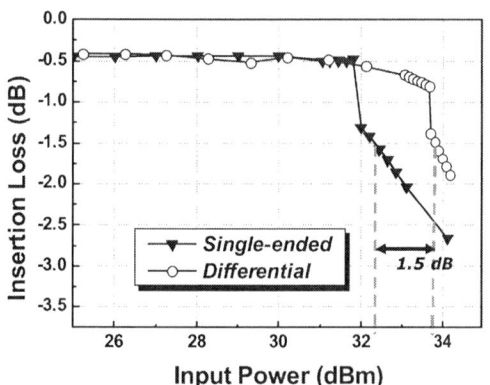

Fig. 9. Measured power handling capability of single-ended and differential switch.

IV. CONCLUSION

A high-power CMOS differential switch using multi-section impedance transformation technique has been demonstrated in a 0.18-μm CMOS process. Large voltage swing from PA is handled by two identical paths adopting differential configuration, and impedance transformation technique contributes the reduction of voltage stress over Rx switches. In order to minimize the entire loss, multi-section impedance transformation has been implemented by splitting up the output matching network of the PA. With the suggested configuration, power handling capability of switch is improved, and total insertion loss is kept at reasonable level. From the measurement results, insertion losses of 0.5 dB and 1.1 dB have been measured for Tx and Rx modes, respectively, and P_{1dB} of 33.8 dBm has been achieved by proposed differential switch structure.

TABLE I

SUMMARY AND COMPARISON OF CMOS RF SWITCHES

Technology	Frequency	IL	Linearity (P_{1dB})	Ref.
0.18-μm triple well	5.2 GHz	1.52 dB(Tx) 1.42 dB (Rx)	28 dBm	[2]
0.13-μm triple well	20 GHz	2.0 dB(Tx) 2.0 dB(Rx)	30 dBm	[5]
90 nm-CMOS	2.4 GHz	0.4 dB(Tx) < 0.2 dB(Rx)	31 dBm	[10]
0.18-μm triple well	1.9 GHz	0.5 dB(Tx) 1.1 dB(Rx)	33.8 dBm	This work

REFERENCES

[1] T. Ohnakado *et al.*, "21.5-dBm power-handling 5-GHz transmit/receive CMOS switch realized by voltage division effect of stacked transistor configuration with depletion-layer-extended transistors (DETs)," *IEEE J. Solid-State Circuits,* vol. 39, no. 4, pp. 577-584, April 2004.

[2] N. A. Talwalkar *et al.*, "Integrated CMOS transmit-receive switch using LC-tuned substrate bias for 2.4-GHz and 5.2-GHz applications," *IEEE J. Solid-State Circuits,* vol. 39, no. 6, pp. 863-870, June 2004.

[3] M.-C. Yeh *et al.*, "Design and analysis for a miniature CMOS SPDT switch using body-floating technique to improve power performance," *IEEE Trans. Microwave Theory Tech.,* vol. 54, no. 1, pp. 31-39, January 2006.

[4] F.-J Huang and K. O, "Single-pole double-throw CMOS switches for 900-MHz and 2.4-GHz applications on p- silicon substrates," *IEEE J. Solid-State Circuits,* vol. 39, no. 1, pp. 35-41, January 2004.

[5] Q. Li and Y. P. Zhang, "CMOS T/R switch design: towards ultra-wideband and higher frequency," *IEEE J. Solid-State Circuits,* vol. 42, no. 3, pp. 563-570, March 2007.

[6] P. Reynaert and M. S. J. Seyaert, "A 2.45-GHz 0.13-μm CMOS PA with parallel amplification," *IEEE J. Solid-State Circuits,* vol. 42, no. 3, pp. 551-562, March 2007.

[7] M. Ahn *et al.*, "A 1.8-GHz 33-dBm P0.1dB CMOS T/R switch using stacked FETs with feed-forward capacitors in a floated well structure," *IEEE Trans. Microwave Theory Tech.,* vol. 57, no. 11, pp. 2661-2670, November 2009.

[8] I. Aoki *et al.*, " Fully integrated CMOS power amplifier design using the distributed active transformer architecture," *IEEE J. Solid-State Circuits,* vol. 37, no. 3, pp. 371-383, March 2002.

[9] J. R. Long, "Monolithic transformers for silicon RF IC design," *IEEE J. Solid-State Circuits,* vol. 35, no. 9, pp. 1368-1382, September 2000.

[10] A. A. Kidwai *et al.*, "A fully integrated ultra-low insertion loss T/R switch for 802.11b/g/n application in 90nm CMOS process," *IEEE J. Solid-State Circuits,* vol. 44, no. 5, pp. 1352-1360, May 2009.

Exploitation of Active Load-pull and DLUT Models in MMIC Design

D. M. FitzPatrick[1], T. Williams[2], J. Lees[1], J. Benedikt[1], S.C. Cripps[1] and P.J. Tasker[1]

[1] Centre for High Frequency Engineering, Cardiff University, Cardiff, Wales, CF24 3AA, UK,

[2] Selex-Galileo SAS, 300 Capability Green, Luton, Bedfordshire, LU1 3PG, UK

Abstract — The use of active load-pull techniques in the design of high efficiency microwave amplifiers has been well documented. This paper describes how it has been applied to the design of a wideband RFIC gain stage. The technique is particularly relevant in new and developing processes where accurate device models are not available and designers otherwise are often forced to use multiple iterations of a design to attempt to encompass the variability in the process. Often in RFIC design, components operate outside of the ideal operating impedance. A look-up table model technique based on measured data which can be used by conventional CAD programs is used to analyse behaviour. This paper shows how the design process and capabilities of the system can be combined to improve the cost effectiveness and performance of RFIC development and with a stable manufacturing process a "first pass" design methodology. The use of the measurement system as an analysis tool is described.

Index Terms — Load-pull, power amplifiers, termination impedance, modelling, on-wafer measurements.

I. INTRODUCTION

When designing RF MMICs the engineer faces many challenges, not least of which is the lack of accurate non-linear models. Most design kits supply scalable equivalent circuit based models, but with little information on their accuracy or applicability (i.e. what are the operating limits of the models). Foundries have to choose how to allocate resources between competing needs; e.g., investment in production equipment and processes to improve reliability and performance or investment in device modelling. Furthermore, the numerous combinations of operating conditions in terms of bias, power levels, impedance environment, and temperature range make the job of modelling to the satisfaction of all customers near impossible. It therefore often falls on the designer to make their own device measurements [1]. Conventional passive load pull systems suffer at high frequencies from system losses and bandwidth limitations; it can also be difficult to integrate the measurements taken into RFIC design packages.

The measurement system at Cardiff University, originally developed for packaged devices targeted at the communications industry [2], has been extended to cover high frequency on-wafer measurements [3]. The time domain waveform based Active Load-Pull System (ALPS) enables the designer to control the impedance environment the test device is situated in at both fundamental harmonic frequencies.

II. OVERVIEW OF THE DESIGN PROCESS

The proposed design process may be summarised in the following steps:

1. Active load-pull of device over output impedance plane across a range of input power levels (sufficient to drive device into 3dB compression).
2. Creation of a Direct Look-Up Table (DLUT) model and matching circuit design.
3. Measure device with matching circuit impedances, (re-iterate matching circuits if necessary).
4. Layout and manufacture MMIC.
5. Test and analysis.

The benefit of the ALPS for MMIC design over passive tuner approaches is that the limitations of impedance imposed by the system losses are overcome. This becomes more significant as frequency increases. A measurement grid is established over the load plane and at each point the input power is swept so that the output power increases to a point where the gain is typically 3 dB into compression. The incident and reflected voltage and currents at both input and output of the device are measured and from this the gain, output power and efficiencies are calculated. A DLUT model is created using the common MDIF (Measurement Data Interchange Format). This contains sets of data for each load impedance point, referenced to the input voltage [4].

The matching circuit design is approached in the conventional fashion. When a satisfactory response has been obtained the measurement system is used to verify the performance of the device with the designed load impedance. During the initial measurements the harmonic impedances are set to 50Ω, (with active harmonic load pull the measurement system can overcome mismatches in the system load). When assessing the designed load the expected harmonic impedances presented to the device by the matching circuit can be used. These impedances can be used to enhance performance [3]. Although manipulating them may also increase circuit complexity, it is important that the impact of harmonic terminations be assessed. If necessary after measurement the matching networks may be adjusted and the process repeated. When satisfactory results are achieved the layout is finalised for production.

In new processes the fabrication method will be under continuing review and device performance may not be consistent from batch to batch. It is important therefore when circuits are analysed to be able to determine the source of any variation between expected and actual results. By including unmatched device cells with the designs variation in device performance can be measured. Thus if there are changes between wafer runs, the measurement system can be used as an analysis tool to determine how circuits will perform within the designed impedance environment.

III. PROCESS DEMONSTRATION

To demonstrate the operation of the methodology a 5-10 GHz 400mW driver stage was designed using a 0.3µm GaAs pHEMT process. No yield or variability information was used so as to replicate the conditions of a new process. A wafer containing 6x100 µm device cells was measured in a class A bias at 5, 7.5 and 10 GHz. The results at the optimum Power Added Efficiency (PAE) loads are summarised in Table 1.

TABLE I
MEASURED PERFORMANCE AT OPTIMUM LOADS

Frequency (GHz)	Optimum PAE at 3dB Compression			
	ΓL (Mag/Ang)	PAE (%)	Pout (dBm)	Gain (dB)
5	0.35/50.3°	50.6	27.1	14.8
7.5	0.41/65.2°	49.8	26.9	12.3
10	0.50/82.7°	46.1	26.5	10.4

A target performance across the band of 26 dBm saturated output power and 35% PAE were set. For the first pass at the output matching circuit design an equivalent circuit to the device output impedance was created. This allows faster optimisation of the circuit than using a nonlinear model. Note, on the wafer the device is placed between two feed lines which require de-embedding to get to the device plane.

Wide bandwidth impedance matching requires some compromise as it is a very difficult task to achieve the desired impedances at each frequency. More matching

elements generally increase loss, which in turn degrade output power and efficiency. Further, there are limits on the achievable transmission line impedances. The circuit solution derived has an impedance trajectory as shown in Figure 1. The compromise is that the performance at the band edges is closer to the optimum than at the centre. Measuring the device with the impedances of the output matching circuit predicted by the circuit simulator, (including harmonic impedances) the performance was found to be within target and is summarised in Table 2. The gain quoted is transducer gain and accounts for power lost due to device input mismatch. Also the PAE at this stage is optimistic as it considers only the load contribution and ignores circuit losses.

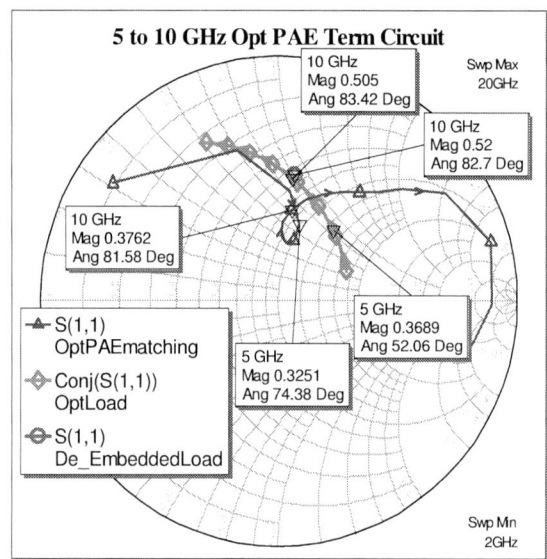

Fig. 1. Designed optimum PAE load impedance, de-conjugate of load equivalent circuit and embedded load.

Input matching and stabilisation circuitry was designed to produce an acceptable compressed gain over the frequency band. The completed design is shown in Figure 2. This comprised a part of the test cell, which included the matching circuits and discrete devices.

TABLE II
PERFORMANCE WITH SIMULATED LOADS

Frequency (GHz)	Driven at 3dB Compression			
	ΓL (Mag/Ang)	PAE (%)	Pout (dBm)	Gain (dB)
5	0.31/85.2°	47.8	27.1	15.4
7.5	0.25/102.3°	44.0	27.2	11.8
10	0.37/99.3°	44.7	26.9	10.2

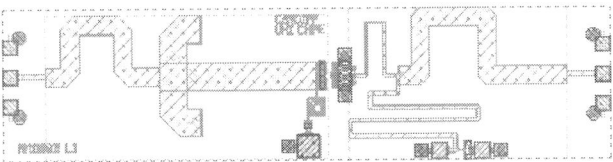

Figure 2. Layout of 5-10 GHz MMIC Stage

IV. MEASUREMENT RESULTS

The manufactured MMIC was measured on the ALPS with a 50Ω termination. The measured results are shown in Figure 3 with the expected performance from the simulation. A small signal measurement of the stage was conducted on the ALPS and a Vector Network Analyser (VNA). The comparison between these measurements and the small signal linear simulation is shown in Figure 4. The simulation was constructed using measured s-parameter data of the device. The VNA operates over the range 10MHz to 26.5 GHz whilst the ALPS over 1 to 40 GHz. The two systems show good correlation in the overlapping range.

V. ANALYSIS OF RESULTS

The actual PAE was lower than in Table 2 as expected. To analyse the actual performance in detail the impedance presented to the device by the matching circuits was measured. This showed that the actual load was different from the designed as shown in Figure 5. Measuring the isolated device on the test cell with the load impedances is shown in Table 3. A direct comparison with the complete MMIC results needs to allow for the input matching. A comparison of the optimum load from the original device and the actual load presented to the device are superimposed on the load pull PAE contours from a device on the same wafer as the MMIC in Figure 6. This

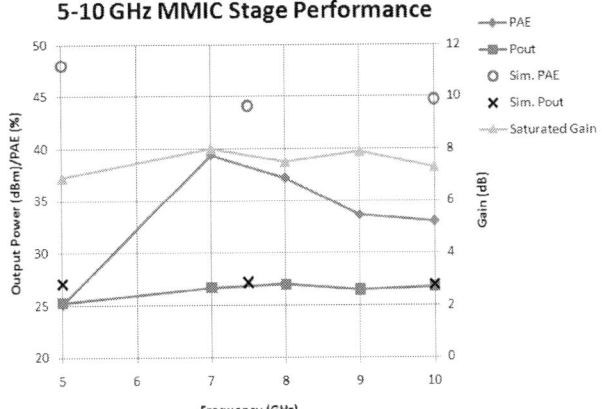

Figure 3, actual measured performance of MMIC vs. predicted.

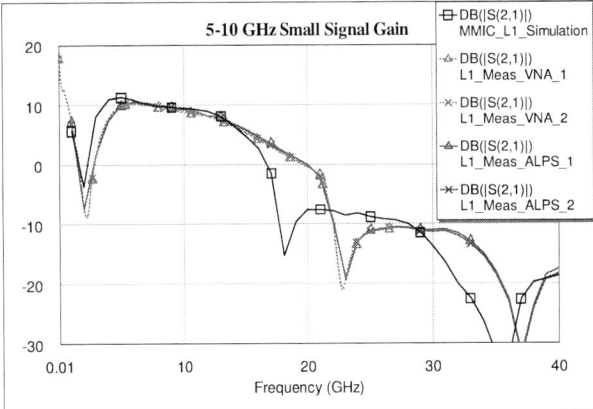

Figure 4. Comparison of small signal gain between linear simulation, VNA and ALPS measurements.

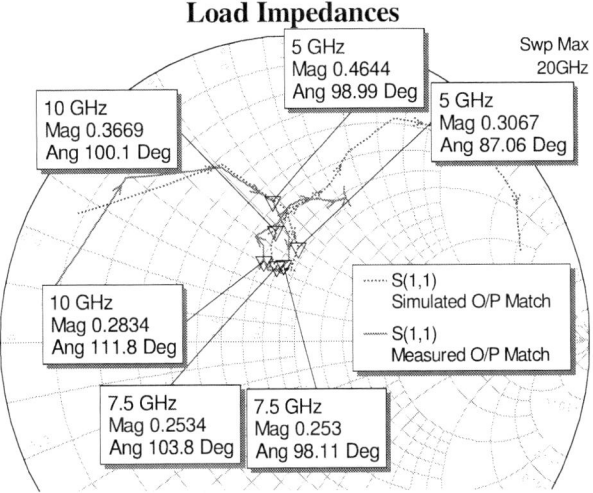

Figure 5. Measured vs. Simulated Load Impedances

shows that not only has the optimum PAE impedance moved, but also the expected performance of the actual circuit. Similar analysis at the other frequencies shows that whilst the performance at 7.5GHz is closer to the design aim that at 5 GHz is further away.

TABLE III
DEVICE PERFORMANCE WITH ACTUAL LOADS

Frequency (GHz)	Device with ALP impedances			MMIC Stage in 50Ω	
	Γ_L (Mag/Ang)	PAE (%)	Pout (dBm)	PAE (%)	Pout (dBm)
5	0.46/99.0°	30.8	25.5	25.1	25.3
7.5	0.25/98.1°	35.7	26.1	38.3	26.9
10	0.28/111.8°	30.7	25.8	33.0	26.8

A further complication is that the input and output matching circuits have losses and these need to be included. To calculate the losses the non 50Ω terminating

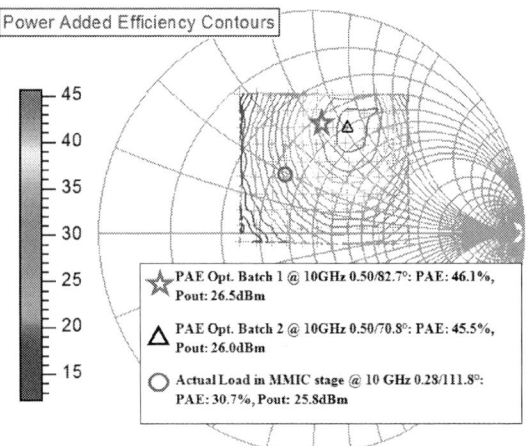

★ PAE Opt. Batch 1 @ 10GHz 0.50/82.7°: PAE: 46.1%, Pout: 26.5dBm

△ PAE Opt. Batch 2 @ 10GHz 0.50/70.8°: PAE: 45.5%, Pout: 26.0dBm

○ Actual Load in MMIC stage @ 10 GHz 0.23/111.8°: PAE: 30.7%, Pout: 25.8dBm

Figure 6. Load Pull Contour plots at 10GHz of device from same wafer as MMIC, with optimum load from original device and load presented by output matching circuit.

impedances of the device must be included, Figure 7. At this point it becomes clear that a CAD model of the device would greatly simplify the calculations necessary. Using the measured data from the device on wafer a DLUT model is generated. This model is loaded into a conventional microwave nonlinear simulator, along with the measured data for the input and output matching circuits (linear s-parameters), Figure 8.

The ability to measure amplifier performance within a varying load impedance environment opens up a number of interesting possibilities for exploring real world scenarios. For example, the harmonic distortion produced under mismatch conditions of varying phase, such as may be encountered in a mixer local oscillator driver.

VI. CONCLUSION

A process has been demonstrated whereby the application of an on wafer ALPS has been used not only in the nonlinear characterisation of a medium power pHEMT device, but also in the assessment of matching

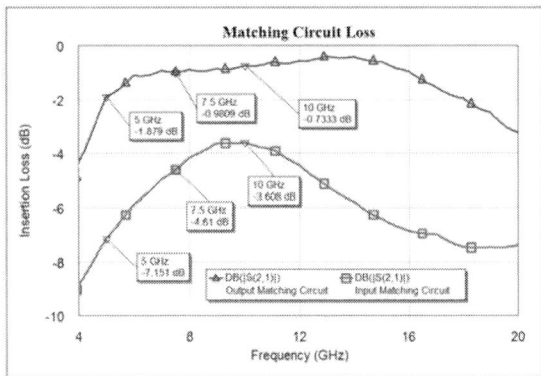

Figure 7. Matching circuit losses, when terminated with device impedances.

solutions and the analysis of implemented amplifier. The utility of a DLUT model in the design of MMIC amplifiers has been shown. Due to matching circuit and device variation from the original devices the target performance was not achieved, however the use of the ALPS enabled the designer to fully explain the variations in performance observed. With a known stable process and proven passive models the designer would be able to accurately predict the large signal performance of their amplifier.

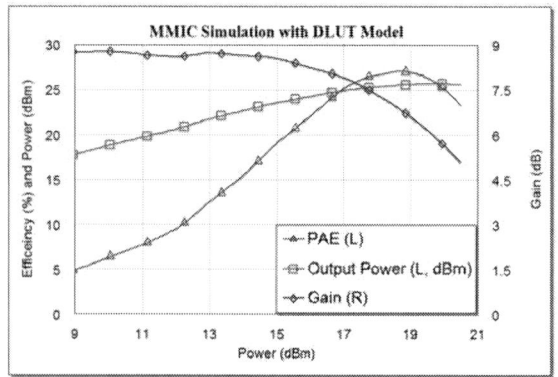

Figure 8. Simulation of DLUT Model with measured linear S-parameters of input and output matching circuits.

ACKNOWLEDGEMENT

This work has been supported by the Electro-Magnetic Remote Sensing group of the Defence Technology Centre in the UK, and assisted by AWR Ltd. in implementing the DLUT within Microwave Office ©.

REFERENCES

[1] C. Schuberth, H. Arthaber, M. L. Mayer, G. Magerl, R. Quay and F. van Raay, "Load Pull Characterisation of GaN/AlGaN HEMTs," International Workshop on Integrated Nonlinear Microwave and Millimeter-Wave Circuits, pages 180-182, January 2006

[2] J. Benedikt, R. Gaddi, P.J. Tasker, et. al. "High power time domain measurement system with active harmonic load-pull for high efficiency base station amplifier design". Microwave Symposium Digest, 2000 IEEE MTT-S International, Vol. 3, pages 1459-1462.

[3] D. M. FitzPatrick, J. Lees, A. Sheikh, J. Benedikt, and P. J. Tasker, "Systematic Investigation of the Impact of Harmonic Termination in the Efficiency Performance of Above Octave Bandwidth Microwave Amplifiers", 39th European Microwave Conference, 2009

[4] Q. Hao, J. Benedikt, and P.J. Tasker, "Nonlinear Data utilization: From Direct Data Lookup to Behavioural Modelling", IEEE Transactions on Microwave Theory and Techniques, vol. 57, no. 6, June 2009

A Mixed-signal Load-Pull System for Base-station Applications

Mauro Marchetti[1], Rob Heeres[2], Michele Squillante[1], Marco Pelk[1], Marco Spirito[1], and Leo C. N. de Vreede[1]

[1]Delft University of Technology, Mekelweg 4, 2628 CD, Delft, The Netherlands
[2]NXP Semiconductors, Gerstweg 2, 6534 AE, Nijmegen, The Netherlands

Abstract — The capabilities of active load-pull are extended to be compatible with the characterization requirements of high-power base-station applications. The proposed measurement setup provides ultra-fast high-power device characterization for both CW, as well as, pulsed, duty-cycle controlled, operation. The realized system has the unique feature that it can handle realistic complex modulated signals like W-CDMA with absolute control of their reflection coefficients vs. frequency.

Index Terms — load-pull, base-station, high power, modulated signals, W-CDMA.

I. INTRODUCTION

Up to date, passive load-pull systems employing mechanical tuners [1][2] have been industry's preferred choice for large-signal characterization due to their simplicity and high power handling capabilities. However, passive tuner systems suffer from loss limitations and electrical delay. Losses in the tuner, interconnects and device test-fixture, limit the maximum magnitude of the reflection coefficient that can be offered to the device under test (DUT), while electrical delay in the tuner and interconnects to the DUT causes large phase variations in the reflection coefficient offered to the DUT vs. frequency, making testing with wideband communication signals (e.g. multi-channel W-CDMA) meaningless [3]. These constrains are even more severe when characterizing high power devices (> 100 W) for base-station applications. Here the low impedance levels of the active device require the use of reflection coefficients with a high magnitude, while the high-Q conditions in the mechanical tuners used to reach these coefficients tend to worsen the phase vs. frequency behavior of these reflection coefficients. In addition, self heating is more pronounced in power devices, demanding duty-cycle controlled pulsed operation, or appropriate testing with signals that have a comparable peak-to-average power ratio as used in the final application.

Active load-pull systems (Fig. 1a) [3]-[7] can, due to the use of injection amplifiers, solve for the losses. However, when aiming for linear DUT operation their practical use in high power applications has always been restricted due to extreme high power and high-linearity requirements of the injection / loop amplifiers. To overcome these limitations, in this work, a novel active load-pull system is presented that is capable of performing high speed load-pull measurements with both CW, as well as, pulsed test signals. The same setup is also capable of handling high power wide-band communications signals, with peak output powers exceeding 150 W, while offering circuit like loading conditions by totally eliminating the impact of electrical delay. These capabilities make the proposed system a perfect candidate for the large-signal characterization of high power devices for base-station applications.

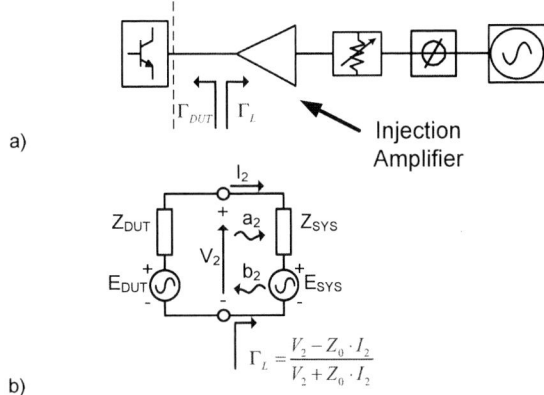

Fig. 1. a) Open-loop active load-pull configuration. b) Thevenin equivalent schematic of an active load-pull configuration. The load impedance offered to the DUT at the reference plane is varied by adjusting the equivalent voltage source E_{SYS} in amplitude and phase. The related power needed to synthesize specific impedances depends strongly on the equivalent system impedance (Z_{SYS}).

II. INJECTION POWER AND LOAD AMPLIFIER LINEARITY

To provide the DUT with a specific Γ_L an injection power is needed, which not only depends on the output power of the DUT and the desired Γ_L, but also on the output impedance of the device [8]. When considering high-power devices, with output impedances in the order of few Ohms, the required injection power to cover the desired Smith chart area can be extremely high (e.g. 2 to 10 times higher than the maximum output power of the DUT). To overcome this issue, typically pre-matching is used, which converts the 50 Ω impedance of the system to a value that is much closer to the output impedance of the DUT. This widely used technique (also applied in passive load-pull) does not only reduce the losses but also lowers the power requirement of the load injection amplifier [8]. E.g. a DUT with an output impedance of 2 Ω and an available output power of 200 W requires, when the system impedance is pre-matched to 10 Ω, an injection power of 360 W to synthesize a load impedance of 1 Ω. Reducing the system pre-matched impedance to 5 Ω, lowers the required injection power for the same load condition to 142.2 W.

When considering multi-tone or modulated signals, the situation becomes more complicated as the linearity of the injection amplifier needs to be taken into account [9]. To study the linearity constrains on the injection amplifier, we consider a two-tone test signal, for which the power injected by the load amplifier at the IM3 frequencies of the two-tone test signal is given by,

$$P_{a_2,IM_3} = 3 \cdot P_{a_2,fund} - 2 \cdot IP_{3,a_2} =$$

$$= 3 \cdot P_{b_2,fund} \cdot \frac{(1-|\Gamma_{LUT}|^2)}{(1-|\Gamma_{SYS}|^2)} \cdot \frac{|Z_{LUT}+Z_0|^2}{|Z_{SYS}+Z_0|^2} \cdot \frac{|Z_L-Z_{SYS}|^2}{|Z_{LUT}+Z_L|^2} - 2 \cdot IP_{3,a_2} \quad (1)$$

where Z_{DUT} and $P_{b2,fund}$ are the output impedance and the available power coming out of the DUT (Fig. 1b), Z_{SYS} is the passive load impedance at the DUT reference plane, $P_{a2,fund}$ and $IP_{3,a2}$ are respectively the power injected by the load amplifier and its output third-order intercept point. Fig. 2 shows the results of a harmonic balance simulation, where the apparent IM_3 of the DUT vs. decreasing output IP_3 of the injection amplifier is shown for different pre-matching conditions of the system impedance. In this experiment the same DUT is used as for the single-tone considerations, (P_{avs}=200 W, output impedance=2 Ω), which is set in the simulation to have an output IP_3 of 63 dBm. For this device the output power is set equal to 50 W per tone, to have the same peak voltage as in the single-tone case. These conditions yield an actual IM_3 of the DUT of -30.35 dBc. From Fig. 2 we can observe that this level is only achieved for sufficiently high IP3 of the injection amplifier. When the injection amplifier is less linear, it will introduce significant IM3 products, which can be approximated by eq. (1), and are also plotted in Fig. 2. Note that IM3 cancelation effects can also occur.

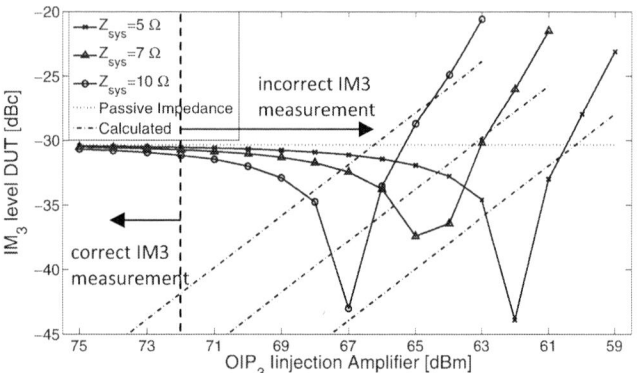

Fig. 2. Harmonic balance simulated IM_3 level of the DUT vs. decreasing output IP_3 of the injection amplifier for different impedance pre-match values. The dotted line is the actual IM3 level as would be achieved with passive matching techniques. The dot-dash line represents the IM3 level due to the $P_{a2,IM3}$ as approximated by Eq. (1). A polynomial model was used for the amplifier linearity.

Consequently, to have reliable linearity measurements in a conventional active load-pull setup, even when pre-matching is used, the injection amplifier linearity (and thus its peak power) needs to be at least 10 times higher than that the of the DUT.

It is obvious that at high power, these amplifiers, if available, will be extremely expensive. For this reason, active load-pull systems that can offer communication standard compliant device testing for e.g. W-CDMA at base-station

power levels (100 W and above) have not been demonstrated up to date.

III. System description

A simplified diagram of the realized open-loop active harmonic load-pull setup is shown in Fig. 3. The system makes use of mixer-based down-conversion after which the data is captured by synchronized (100 MS/s) analogue-to-digital converters. With this hardware configuration it is possible to measure the reflection coefficients of the DUT over a wide frequency band and / or time span.

The reflection coefficients at the device reference plane are synthesized by injecting fully coherent RF signals that are generated by IQ up-converted base-band signals, which are provided by (200 MS/s) arbitrary-waveform-generators (AWGs). Since all data generation and data acquisition of both RF signals and DC parameters are handled through the PXI based DA and AD instrumentation, no mechanical tuners, NWA or DC-parameter analyzer is needed, yielding a cost effective high-end characterization solution.

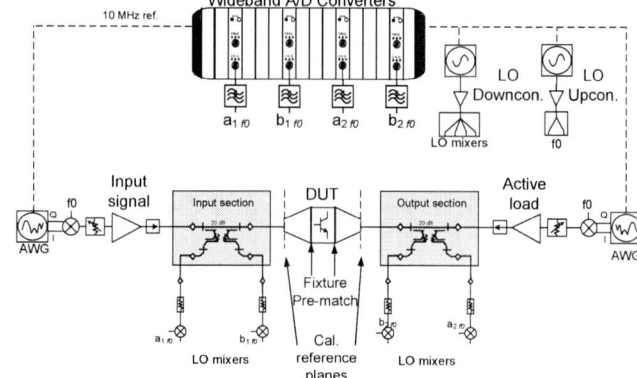

Fig. 3. Simplified schematic of the phase coherent mixed-signal active load-pull setup.

IV. Real Time Pulsed RF Measurements

For single-tone CW signal conditions, the described system configuration is able to generate and measure in a single acquisition thousands of source and load conditions at different power levels. This feature provides ultra-fast load-pull device characterization [10]. For high-power devices, however, the use of CW conditions needs to be omitted to avoid self-heating, which in extreme conditions can even yield device failure. This is especially true for base-station devices which are optimized for operation with complex modulated signals. These signals reach only occasionally their peak values; as result the active device operates most of its time in power back-off. Consequently, in order to create realistic load pull testing conditions, pulsed RF operation is required. For this reason we extend the original concepts of [10] to pulsed operation at much higher peak power levels (above 100 W). These additional features allow the user to perform accurate high power ultra-fast device characterization, while providing full control on the maximum operating conditions of the active device, and thus avoiding voltage and thermal breakdown conditions.

All the waveforms to be injected into the input and output ports of the DUT are defined such that they contain multiple sinusoidal time-segments with different amplitude and phase information (Fig. 4). As a result the device will experience a sequence of time segments with different input powers and loading conditions. Note that for correct operation the system needs to be fully coherent and time aligned. It is also important to remember that both RF and DC bias conditions need to be measured in the proper time segment. Although the tendency is to emphasize the measurement of the RF conditions, the measurement of the DC bias conditions is equally important and quite often difficult for pulsed operation. This can be understood by considering the fact that power devices need sufficient bias decoupling for their correct operation, which increases the time constant on the DC measurement port, troubling the correct measurement of the "DC bias" conditions. To overcome this problem a two step measurement procedure has been followed, which is described below.

A. Phase 1: Calculation of the injection signals

In this phase we only measure the RF properties and perform the necessary iterative calculations in order to find the proper injection signals. To keep the measurement speed at its maximum, the duration of each "time-segment" can be as short as 100 ns. In this phase, the desired duty cycle is achieved by adding a sufficient "idle time" after the stimulus representing the different power and load states. An example is shown in Fig. 4 where the input waveform to the DUT and the load injected waveforms are depicted. In this simplified case the DUT will "experience" two input power levels, one for each pulse, and three different load impedances. The desired injection signals are then optimized to synthesize the desired impedances seen by the DUT. Note that the use of iterations to optimize the injection signals opens the possibility to introduce several features that ensure safe device operation during the measurement, something that is extremely useful especially at high power levels. For example it is possible to exclude from the measurement an area of impedances that might cause device instability [10]. Furthermore, during each iteration, it is possible to check the gain compression of the device for every impedance, thus limiting the input power for those impedances where the DUT gain reduces by more than a user-specified amount. Also the shape of the pulse defined by its rise and fall time can be arbitrarily adjusted which proves to be very important in ensuring safe device operation.

B. Phase 2: Final real-time pulsed measurement

In this last part of the measurement routine, the actual measurement is taken. Now each individual load and power condition is represented by a separate pulse which has the user specified width and duty-cycle. The proper injection signals conditions were found in Phase 1, so no additional iterations

are needed to reach the user specified loading and power conditions. In this final phase also the DC parameters are measured. This one data point at the time scheme is slower than the time segmented approach of Phase 1 (Fig. 4.), but it guarantees the highest accuracy for measurement of the pulsed DC parameters.

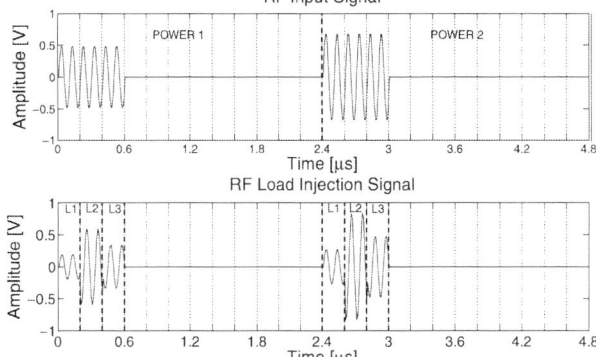

Fig. 4. Time-segmented RF waves for multiple input power (upper plot) and load termination control (lower plot) with pulsed RF. In this simplified example 3 different loads for 2 input power levels are presented to the DUT.

As application example of the proposed active load-pull concept introduced above, a complete load-pull and power sweep for a NXP BLF7G22LS-130 device was carried out at a frequency of 2.14 GHz, with a pulse width of 10 µs, rise and fall times of 100 ns and 10 % duty cycle. During this measurement the power gain compression was limited to a value of 4 dB to avoid device degradation due to undesirable extreme gain compressions, e.g. in higher load line regions. The results are shown in Fig. 5, where the PAE vs. output power at a gain compression level of 3 dB is shown. Note that such a complete device characterization with 25 power levels at each of the 50 load impedances takes less than 3 minutes.

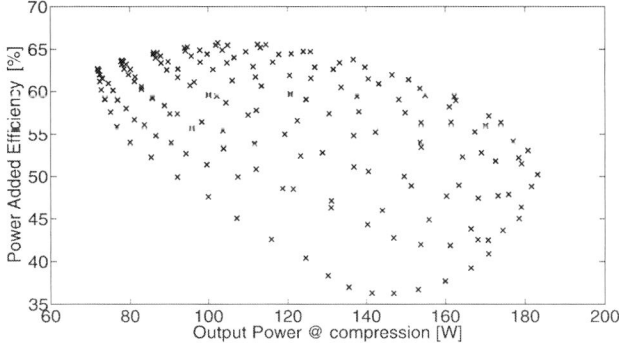

Fig. 5. Measured PAE vs. output power at 3 dB power gain compression level of a NXP Gen7 LDMOS device for different load states and input powers, using pulsed single tone conditions (10 µs pulse width and 10 % duty cycle).

V. MODULATED SIGNALS MEASUREMENTS

The realized system has the unique feature to perform active load-pull device testing with communication standard compliant wide-band modulated signals like W-CDMA. This feature has been described in detail for low-power levels in [3] and has been extended in this work to the power levels that are typically in use for base station applications (e.g. peak envelope power ~ 200 W). As example consider Fig. 6 which shows ACPR and average PAE for a single-channel W-CDMA signal at 2.14 GHz with a peak to average ratio of 9.5 dB.

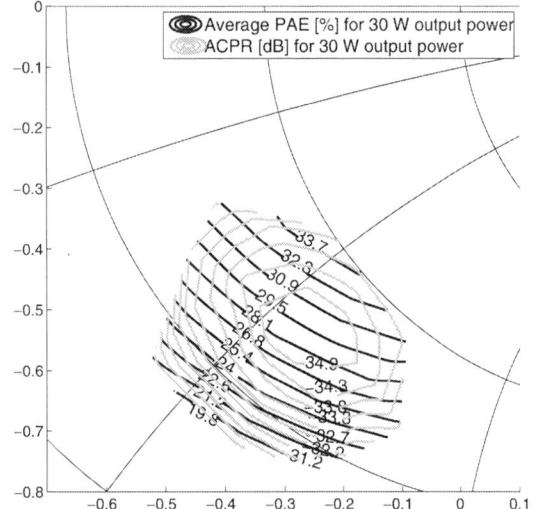

Fig. 6.   Load-pull contours of average power added efficiency and ACPR for an average output power of 30 W. The related peak to average power (PEP) is as high as 150 W.

It should be stressed that in these experiments the maximum saturated power rating of the injection amplifier is only 200 W with an associated 60 dBm output IP$_3$. The reason that the nonlinearity of the injection amplifier does not affect the measurement results showing a non correct IM3 measurement, as was shown in Fig. 2, derives from the iteration process performed to optimize the reflection coefficient of each individual frequency component of the W-CDMA signal (in this experiment 11681 tones with 3 kHz spacing). Due to these iterations the injection amplifier is basically pre-distorted for its own non-linearities, which allows the use of injection amplifiers with a much lower linearity as what is typically required in conventional active load pull systems.

VI. CONCLUSIONS

A cost effective active load-pull system compatible with the requirements of high power, high linearity base-station applications has been presented. It provides ultra-fast large-signal device characterization for both CW and pulsed conditions. For the latter, both the duty-cycle as well as the pulse shape can be independently controlled, limiting at the same time the gain compression of the DUT during the measurement. All these features are crucial in guaranteeing the safe operating conditions of high power DUTs (> 100 W).

In addition, for the first time high power device characterization with realistic W-CDMA signals has been performed. It was shown that the realized system can compensate for the nonlinearities of the injection amplifiers, which normally would obscure the linearity / ACPR measurements. This property allows the use of cheaper injection amplifiers providing a lower Psat than required in conventional active load-pull systems. The ability to eliminate losses and electrical delay, while being completely free in defining the source and load reflection coefficients vs. frequency allows perfect mimicking of in circuit situations, making the system an interesting tool for the RF power amplifier developer.

ACKNOWLEDGEMENT

The authors wish to acknowledge NXP for technical discussions and providing the LDMOS devices. The PANAMA and WING projects are acknowledged for supporting this work.

REFERENCES

[1] Maury Microwave Corporation. [Online]. Available: http://www.maurymw.com/.

[2] Focus Microwaves. [Online]. Available: http://www.focus-microwaves.com/.

[3] M. Marchetti, et al., "Active Harmonic Load-Pull with Realistic Wideband Communications Signals," *IEEE Trans. Microwave Theory and Tech.*, pp. 2979-2988, Dec. 2008.

[4] Y. Takayama, "A new load-pull characterization method for microwave power transistors," *IEEE MTT-S Int. Microwave Symp. Dig.*, Cherry Hill, NJ, Jun. 1976, pp. 218-220.

[5] A. Ferrero, et al., "Novel hardware and software solutions for a complete linear and non linear microwave device characterization," *IEEE Trans. Instrumentation and Measurements*, vol 43, issue. 2, pp. 299-305, Apr. 1994.

[6] T. Williams, et al., "Experimental evaluation of an active envelope load pull architecture for high speed device characterization," in 2005 *IEEE MTT-S Int. Microwave Symp. Dig.*, Long Beach, CA, Jun. 2005, pp. 1509-1512.

[7] H. Arthaber, et al. "A broadband active harmonic load-pull setup with a modulated generator as active load," in Proc. *34th European Microwave Conf.*, Amsterdam, The Netherlands, Oct. 2004, pp. 685-688.

[8] Z. Aboush, et al., "High power active harmonic load-pull system for characterization of high power 100-watt transistors," in *Proc. 35th European Microwave Conf.*, Paris, France, Oct. 2005, pp. 609-612.

[9] M. Spirito, et al., "Active harmonic load-pull for on-wafer out-of-band device linearity optimization," *IEEE Trans. Microwave Theory and Tech.*, vol 54, pp. 4225-4236, Dec. 2006.

[10] M. Squillante, et al., "A Mixed-Signal Approach for High-Speed Fully Controlled Multidimensional Load-Pull Parameters Sweep", in *Proc. 73th ARFTG Conf.*, Boston, MA, Jun. 2009.

RTUIF-01

A 228μW Injection Locked Ring Oscillator based BPSK Demodulator in 65nm CMOS

Qiang Zhu and Yang Xu

Department of Electrical and Computer Engineering

Illinois Institute of Technology, Chicago, IL, 60616, U.S.A.

Abstract — This paper presents an ultra-low power BPSK demodulator based on injection locked oscillators (ILOs). Two second harmonic ILOs are employed to convert BPSK signals to ASK signals which are demodulated by an envelope detector to baseband signals. For sub-GHz applications, the ILOs are implemented using ring oscillators to allow compact chip area and ultra-low power dissipation. The prototype demodulator is fabricated in a 65nm CMOS technology that consumes 228μW of power and occupies 0.014mm² of die area. Measurement results reveal the demodulation of BPSK signal at 750–900MHz carrier with the minimum sensitivity of -33dBm.

Index Terms — CMOS analog integrated circuits, demodulators, injection locked oscillators, phase shift keying, telemetry.

I. INTRODUCTION

Recent advances in RFIC technology enable various innovative and versatile applications through ultra-low power wireless link such as mesh sensor network, remote industrial monitoring and implantable medical device. For the wireless data access, the modulation scheme adapted is critical to the link qualities in terms of bit rate and bit error rate. The binary phase shift keying (BPSK) is one of the most widely used digital modulation scheme in wireless systems such as 802.15.4, GPS and medical telemetry. The demodulation of BPSK signal usually requires coherent detection which is accomplished by carrier recovery circuit such as COSTAS loop. Due to its complexity, however, the COSTAS loop is limited in power consumption reduction.

Alternative approach for non-linear BPSK demodulation has been proposed in [1]. The BPSK signal is firstly converted to ASK signal through injection and locking of two secondary harmonic injection locked oscillators. The ASK signal is then down converted to baseband with an envelope detector. This new method does not require down conversion with mixer and frequency synthesis in the carrier recovery with COSTAS loop. Therefore, it possesses the potential to achieve ultra-low power consumption. The prototype circuits using LC oscillator structure to implement the ILO have been built [2, 3] to demonstrate its effectiveness. In [2], two demodulators working at 500MHz and 2GHz have being

designed using discrete passive components. Both static and dynamic performance of the converter prototypes has been analyzed as a function of the injected power. In [3], an integrated demodulator working at 19GHz has being deployed which consumes just 2.5mW at 0.8V.

The ILO based BPSK demodulator has its limitations when used as wireless data receiver. The sensitivity is limited around -30dBm [3] and the ILO may lock to the interference instead of the targeted signal when strong interference are present [4]. However, it can find its application in biomedical implantable electronic devices that use inductive links for data communications [5]. Due to the close coupling of the inductive link, the interference from external world could be ignored and the received signal power is sufficiently large to drive the demodulator. The inductive link of implantable device works at sub-GHz. The passive components of the LC oscillator may take prohibitive chip area, if not applied in discrete manner. This paper proposes the first ring oscillator ILO based BPSK demodulator. The initial prototype demonstrated demodulation of 750–900MHz carrier BPSK signal with much smaller chip area as well as ultra-low power consumption.

III. PRINCIPLE OF OPERATION

A. Injection locked oscillator

The BPSK demodulator requires two second harmonic ILOs which are implemented using three-stage ring oscillator structure where the block diagram is shown in Fig. 1. The locking of the ring oscillator lies in the mixing of injected signal with the oscillating signal which generates new frequency components that fulfill the oscillating condition. Specifically, when there is no injection, oscillator free runs at ω_r with $\pi/3$ phase shift at each stage. Now, adding the injection signal (single tone at ω_i) through modulating the bias current of the first stage. The first delay cell serves as the mixer between injected signal and oscillating signal. The ω_i is close to $2\omega_r$. In locked state, ILO oscillates at $\omega_i/2$. Ignoring higher order harmonics of the oscillating signal, the mixing in the first stage can be expressed as

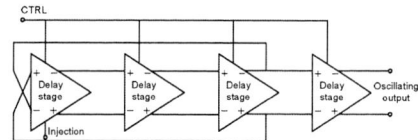

Fig. 1. Ring oscillator ILO block diagram.

$$\text{Acos}\left(\frac{\omega_i}{2}t\right)\left[I_{RF}\cdot\cos(\omega_i t+\alpha)+I_{BIAS}\right]. \quad (1)$$

The frequency component at $\omega_i/2$ has the phase shift of

$$\tan^{-1}\left(\frac{\eta\sin\alpha}{1+\eta\cos\alpha}\right),\ \text{where}\ \eta=\frac{I_{RF}}{2I_{BIAS}}. \quad (2)$$

The additional phase shift at the first stage is compensated by the later stages so that the oscillation condition of total phase change of π is satisfied

$$-\tan^{-1}\left(\frac{\eta\sin\alpha}{1+\eta\cos\alpha}\right)+3\cdot\tan^{-1}(\frac{\omega_i}{2\omega_r}tan\frac{\pi}{3})=\pi. \quad (3)$$

Denote frequency offset $\Delta\omega=\omega_i/2-\omega_r$ and use the following approximation [6],

$$3\cdot\tan^{-1}(\frac{\omega_i}{2\omega_r}tan\frac{\pi}{3})\cong\pi+\frac{3\sin\frac{2\pi}{3}}{2}\frac{\Delta\omega}{\omega_r}, \quad (4)$$

$$\eta\ll 1. \quad (5)$$

One can obtain that

$$\eta\sin\alpha\cong\frac{3\sin\frac{2\pi}{3}}{2}\frac{\Delta\omega}{\omega_r}, \quad (6)$$

and the maximum frequency offset

$$\frac{\Delta\omega_{\max}}{\omega_r}\cong\frac{2\eta}{3\sin\frac{2\pi}{3}}. \quad (7)$$

Therefore, only when ω_i is in the range of $2\omega_r\pm 2\Delta\omega_{\max}$, the ILO is able to lock to the injected signal. And the locking range $4\Delta\omega_{\max}$ is approximately linearly dependent on the injection efficiency η, or the injection power.

B. BPSK demodulation

Consider the case when BPSK signal is injected as illustrated in Fig. 2. Initially, two ILOs lock to the injected signal in frame 1. When there is 180° phase shift at the injected signal in frame 2, ILO1 with $\omega_{r1}<\omega_i/2$ oscillates slower at ω_{r1} to obtain 90^0 phase lagging, while ILO2 with $\omega_{r2}>\omega_i/2$ oscillates faster at ω_{r2} to obtain 90° phase leading, so that both of them get relocked in frame 3. Compared with frame 1, there is 180° phase shift between the oscillating signals of ILO1 and ILO2. Based on this, the demodulator is able to convert the BPSK signal into ASK signal. The block diagram of the

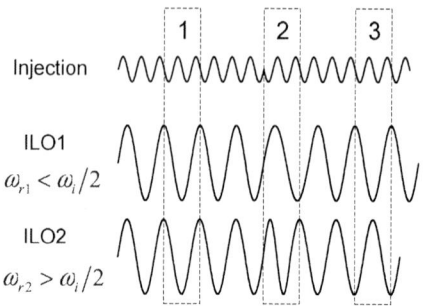

Fig. 2. Relocking of ILOs when there is phase shift of 180° at the injected signal.

demodulator is shown in Fig. 3 [1]. The BPSK modulated signal is injected into two ILOs which are identical except that their free-running frequencies are turned as

$$\Delta\omega=\omega_{r2}-\omega_i/2=\omega_i/2-\omega_{r1} \quad (8)$$

Under sufficient injection power, the locking range is larger than $4\Delta\omega$ so that the two ILOs lock to the injected signal. The oscillating signals of ILO1 and ILO2 are then added in a combiner. According to (6) - (8),

$$\alpha_1=-\alpha_2=\sin^{-1}\frac{\Delta\omega}{\Delta\omega_{\max}}. \quad (9)$$

Therefore, the output of the combiner is

$$\text{Acos}\left(\frac{\omega_i}{2}t-\frac{\alpha_1}{2}\right)+\text{Acos}\left(\frac{\omega_i}{2}t+\frac{\alpha_1}{2}\right)=2A\cos\frac{\alpha_1}{2}\cos\frac{\omega_i}{2}t, \quad (10)$$

where A is the oscillating amplitude. Now, consider the case when the input signal has 180^0 phase shift. Based on the above discussion, the output of combiner then becomes

$$\text{Acos}\left(\frac{\omega_i}{2}t-\frac{\alpha_1}{2}+\frac{\pi}{2}\right)+\text{Acos}\left(\frac{\omega_i}{2}t+\frac{\alpha_1}{2}-\frac{\pi}{2}\right)$$
$$=2A\sin\frac{\alpha_1}{2}\cos\frac{\omega_i}{2}t. \quad (11)$$

By comparing (10) and (11), one can observe that the phase change in the injected BPSK signal is manifested at the combiner output as the amplitude change from $2A\cos\alpha_1/2$ to $2A\sin\alpha_1/2$. Therefore, the BPSK signal is converted to ASK signal by these two ILOs. After that,

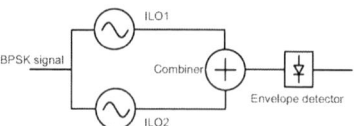

Fig. 3. ILO based BPSK demodulator block diagram.

978-1-4244-6240-7/10 $26.00 © 2010 IEEE

an envelope detector is added to demodulate the ASK signal to baseband so that BPSK demodulation is accomplished.

IV. CIRCUIT IMPLEMENTATION

The ring oscillator ILO is designed to have 3 differential stages connected in a loop, together with another stage serving as an output buffer (Fig. 1). The schematic of a single delay stage is shown in Fig. 4 [7]. For the turning of free-running oscillating frequency, control voltage is applied at the gate of M6 which works in triode region as a variable load. M4 and M5 are cross-coupled loads that clamp the output swing making the stage insensitive to the common mode noise. Bandgap reference is integrated on chip to provide a constant bias current for each stage. The BPSK signal is injected by modulating the bias voltage of the current source at the first stage. The AC coupled injection input is terminated through a 50• off-chip resistive load for impedance matching. To achieve higher injection efficiency, M2 is biased at intermediate inversion region. Each delay stage draws 17.75μA from 1.2V power supply. The ILO has a turning range from 370MHz to 450MHz and an output swing of 280mV.

The differential outputs of the two ILOs are AC coupled to the combiner, which is implemented using balanced differential pair as shown in Fig. 5. Due to the relative large output voltage of the ILO, the combiner is designed to have unit gain in the frequency range of interest. The ASK signal at the combiner's output is then DC coupled to envelope detector for demodulation. An active peak detector shown in Fig. 5 is applied where M7 is used as the nonlinear rectifying element while capacitor C1 serves as the hold capacitor. To be able to drive the 12pF capacitance at the probe of external oscilloscope when testing, an output rail to rail buffer is proceeded which has a bandwidth of 10MHz and power dissipation of 230μW.

The prototype chip is fabricated in TSMC's 65nm CMOS process (Fig. 6). The chip area, including all the

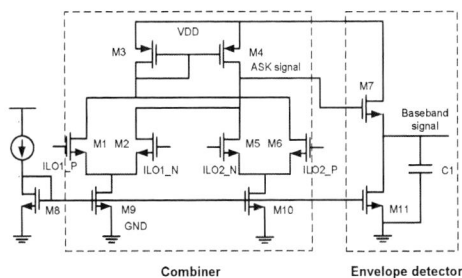

Fig. 5.   Schematic of the combiner and envelope detector.

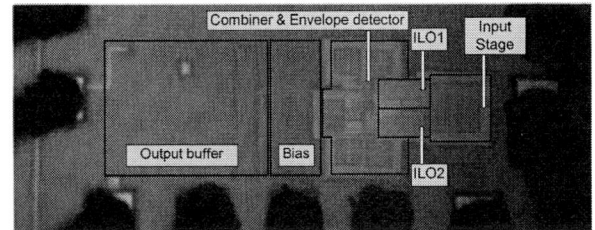

Fig. 6.   Photomicrograph of prototype demodulator.

pads, is 0.20×0.58mm². The core demodulator, including two ILOs, combiner and envelope detector, occupancies an area of 0.014mm². At 1.2V power supply, the total power consumption excluding the output buffer is 228μW, of which the two ILOs dissipate 170.4μW, the combiner and envelope detector take 50.4 μW and the bias circuit consumes 7.2 μW.

V. MEASUREMENT RESULTS

The chip is bonded to SOIC-24 package for testing. The free-running frequencies of the two ILOs are measured as shown in Fig. 8. The frequency turning range from 375MHz to 450MHz is achieved by adjusting the control voltage from 650mV to 520mV. This can cover the BPSK signal with carrier frequency from 750MHz to 900MHz.

The sensitivity of the demodulator relies on the minimum injection power for ILOs to lock. At different injection frequencies, adjust the offset to be ±2.5MHz for 10MHz locking range. The minimum injection power is

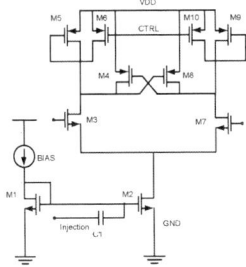

Fig. 5.   Schematic of delay stage of the ring oscillator.

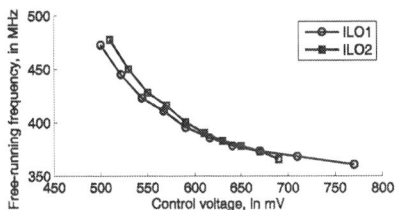

Fig. 7.   Free-running frequencies of two ILOs at different control voltages.

978-1-4244-6240-7/10 $26.00 © 2010 IEEE

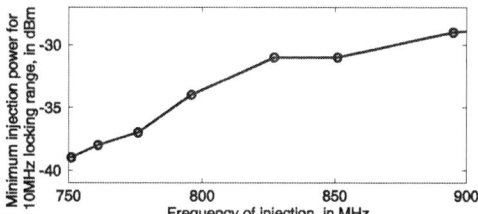

Fig. 8.   Minimum injection power for 10MHz locking range at different injection frequencies.

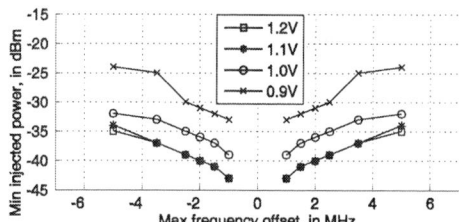

Fig. 9.   Minimum injection power for different frequency offset with different supply voltage.

illustrated in Fig. 8. The lowest one of -39dBm is obtained at 750MHz. It degrades with the increase of injection frequency. At 750MHz, the same measurement is done under different frequency offsets and supply voltages as shown in Fig. 9. The sensitivity deteriorates when supply voltage is below 1.1V.

To verify the demodulation process at 750MHz, the free-running frequencies of the ILO1 and ILO2 are turned to 373.5MHz and 376.5MHz with their control voltage set at 658mV and 650mV. Generate the BPSK signal by mixing a carrier at 750MHz with a bipolar baseband square wave at 1MHz. Under -33dBm injection power and 1.2V power supply, the demodulator output observed from the oscilloscope is shown in Fig. 10. The baseband is successfully demodulated which has amplitude of 416mV. The performance comparison of this demodulator with [2] and [3] are listed in Table 1.

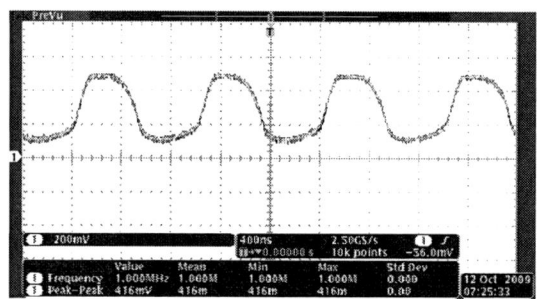

Fig. 10.   Demodulator output of -33dBm 750MHz carrier BPSK injection.

TABLE 1   PERFORMANCE COMPARISON

Ref.	[2]	[2]	[3]	This work
Frequency (MHz)	400-530	1800-2200	19000	750-900
Locking range (MHz)	2.5@ -20dBm	4.5@ -20dBm	28@-40dBm	10@-39dBm
Power (mW)	N/A	N/A	2.5	0.228
Technology	Discrete	Discrete	90nm CMOS	65nm CMOS

VII. CONCLUSION

The ILO based BPSK demodulator using ring oscillator structure for sub-GHz application is proposed. It operates at 750-900MHz carrier BPSK input. It consumes ultra-low power of 228µW from 1.2V power supply. The core demodulator occupancies an area of 0.014mm^2.

ACKNOWLEDGEMENT

The authors like to thank TSMC for chip fabrication, Prof. Dennis Roberson for testing, Keya Kamtikar for layout assistance, Prof. Roc Berenguer for useful discussions.

REFERENCES

[1] Jose Maria lopez-Villegas and Javier Jose Sieiro Cordoba "BPSK to ASK signal conversion using injection-locked oscillators-part I: theory," in *IEEE trans. Microwave theory and techniques,* vol. 53, No. 12, December 2005.

[2] Jose Maria lopez-Villegas and Javier Jose Sieiro Cordoba "BPSK to ASK signal conversion using injection-locked oscillators- part II: experiment," in *IEEE trans. Microwave theory and techniques,* vol. 54, NO. 1, January 2006.

[3] J.G. Macias-Montero, et. al, "A 19GHz, 250pJ/bit non-linear BPSK demodulator in 90nm CMOS," in *Proc. of European solid-state circuits conference,* Athens, Sep. 2009.

[4] B. Razavi, "A study of injection locking and pulling in oscillators," in *IEEE J. Solid state circuits,* vol. 39, no. 9, pp. 1415-1424, Sept. 2004.

[5] Y. Hu and M. Sawan, "A Fully Integrated Low-Power BPSK Demodulator for Implantable Medical Devices," in *IEEE Trans. Circuits Syst. I,* Vol. 52, No. 12, pp.2552-2562, Dec. 2005.

[6] R. J. Betancourt-Zamora, S. Verma, and T. H. Lee, "1-GHz and 2.8-GHz injection-locked ring oscillator prescalers," in *Symp. VLSI Circuits Dig. Tech. Papers,* June 2001, pp. 47–50.

[7] R.J. Betancourt-Zamora, T.H. Lee, "CMOS VCOs for Frequency Synthesis in Wireless Biotelemetry", in *Int'l Symp. Low Power Electronics & Design,* pp. 91-93, August 1998.

RTUIF-02

A 0.13-μm CMOS Wireless Reflector for Phase Sweep Cooperative Diversity

Jean-François Bousquet, Sebastian Magierowski, Geoffrey Messier and Zhixing Zhao
Department of Electrical and Computer Engineering
Schulich School of Engineering, University of Calgary, Calgary, AB, Canada, T2N 1N4
email: jfbousqu@ucalgary.ca

Abstract—A 4-GHz 1.2-V all-analog wireless reflector acting as a cooperative diversity repeater is built in 0.13-μm CMOS technology. Interfaced with a dipole antenna, the circuit achieves 22.3-dB gain for a low power consumption equal to 120 μW. By applying slow phase sweeping at the reflector node, diversity gain is achieved and the coverage area of an indoor wireless network is increased by a factor of 2.5.

Index Terms—active reflector, negative resistance amplifier, cooperative diversity

I. INTRODUCTION

Phase sweep amplify-and-forward (PSAF) is a novel cooperative diversity technique applied to indoor wireless sensor networks. A repeater that acts as an active reflector is placed in proximity of a transmitting sensor to improve coverage. It applies slow phase modulation to the incident data and reflects it towards the destination. Effectively, the channel Doppler frequency is increased. This reduction in the duration of the channel fades results in improved error correction coding performance.

The design procedure for an integrated all-analog active reflector implemented in 0.13-μm CMOS technology has been described in [1]. A differential reflector core presents a negative resistance with variable susceptance and is directly interfaced to an antenna. Precise control circuitry and a dual automatic gain control (DAGC) allow fine tuning to account for process uncertainties and wireless propagation fluctuations.

In this paper we evaluate the performance of a 4-GHz wireless reflector in a controlled wireless environment. Reflection gain of 22.3 dB is realized of a core power consumption of 120 μW. Slow phase modulation is achieved at the repeater and increases the coverage area by a factor of 2.5, a substantial improvement for indoor wireless sensor networks. This novel repeater architecture is an interesting low-power solution to traditional amplify-and-forward repeaters, which require complete transceiver functionality.

The contributions of this paper include 1) the fabrication of a complete wireless reflector 2) the characterization of its gain and phase sweep capability in a wireless environment and 3) the evaluation of the network coverage improvement with the measured reflector performance.

In Sec. II, the all-analog active reflector is described, while in Sec. III the performance of a reflector prototype is demonstrated and in Sec. IV diversity improvement is evaluated.

II. THE INTEGRATED ACTIVE REFLECTOR

The wireless reflector shown in Fig. 1 generates an amplified echo of the incident wave using microwave reflection. The antenna is represented using a narrowband model with a parallel equivalent admittance $Y_{ant} = G_{ant} + jB_{ant}$. It is interfaced to a negative resistance amplifier with admittance $Y_{xc} = G_{xc} + jB_{xc}$.

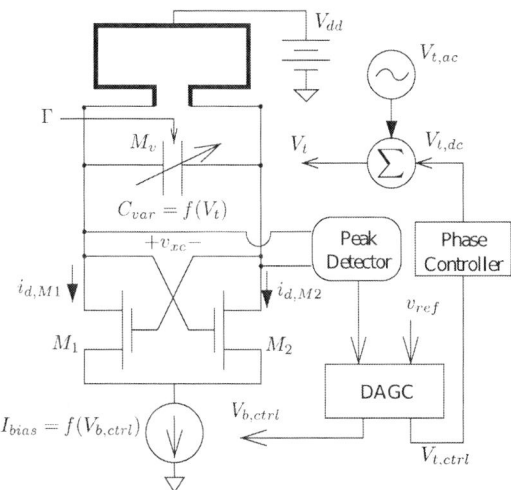

Fig. 1. Reflection amplifier with cross-coupled transistor pair

Using the definition of generalized scattering parameters, the microwave reflection coefficient Γ is [2]:

$$\Gamma = \frac{Y_{ant}}{Y_{ant}^*} \cdot \left(\frac{Y_{ant}^* - Y_{xc}}{Y_{ant} + Y_{xc}} \right) \qquad (1)$$

From Eq. (1), reflection gain is obtained for $G_{xc} < 0$. At resonance (i.e. when $B_{ant} = -B_{xc}$), the gain increases towards infinity as G_{xc} increases to $-G_{ant}$. Stability is maintained at resonance for $|G_{xc}| < G_{ant}$.

The negative conductance G_{xc} is generated by a cross-coupled transistor pair and the differential architecture allows for an easy interface to a balanced antenna structure. Also, a dual varactor provides a variable susceptance B_{xc} that serves to apply phase modulation. Furthermore, the amplifier gain and frequency of operation are controlled using independent signals, I_{bias} and $V_{t,dc}$ respectively. Note that the varactor tuning voltage V_t is equal to $V_{t,dc} + V_{t,ac}$, where $V_{t,dc}$ sets the

978-1-4244-6240-7/10 $26.00 © 2010 IEEE   451

operating frequency, while $V_{t,ac}$ is a 200-Hz periodic signal for phase modulation.

The active reflector core, along with its control circuitry is fabricated in 0.13-μm CMOS technology. The photomicrograph of the die is shown in Fig. 2. The circuit occupies 450×250 μm^2.

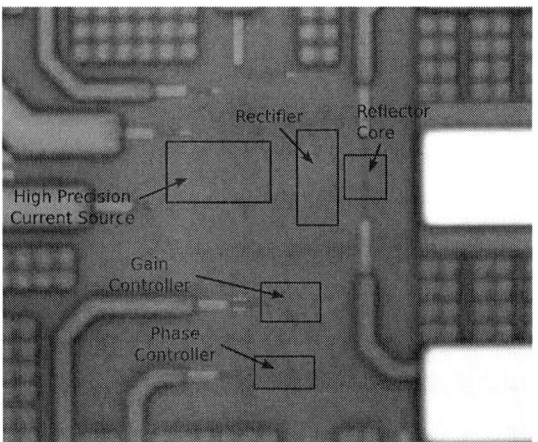

Fig. 2.   Layout of the circuit in 0.13-μm CMOS technology

The reflector gain and operating frequency are sensitive to its control signals I_{bias} and $V_{t,dc}$, respectively. The control voltages $V_{b,ctrl}$ and $V_{t,ctrl}$ shown in Fig. 1 provide 1.5-mV accuracy. A precision current steering circuit converts $V_{b,ctrl}$ to I_{bias} with an accuracy of 0.79 μA. As $V_{b,ctrl}$ is increased up to a maximum value of 1.2 V, the biasing current is decreased from 1.66 mA to zero. Similarly, a tuning voltage V_t with 1-mV precision and spanning a range of [1.0, 1.35] V is generated by $V_{t,ctrl}$ through a current mirror with active load.

The cross-over frequency [3] is a key metric in the design of a cross-coupled pair of transistors and defines the frequency at which the circuit can no longer provide negative resistance due to its parasitics. In practice, the cross-over frequency is very sensitive to measurement imprecision and we define a conductance cutoff frequency $f_{\sqrt{2}}$, which is the frequency at which the conductance has attenuated by a factor of $\sqrt{2}$. Both the transistor and varactor parasitics contribute to the degradation of $f_{\sqrt{2}}$.

The active reflector core consists of a cross-coupled pair of 12-μm wide nFETs in parallel with a pair of 0.24-μm varactors occupying an area of 11.0×5.0 μm^2.

The reflector admittance is measured using a vector network analyzer for different biasing conditions. To maintain linear transistor operation, a low incident power equal to -35 dBm is input to the circuit. This power level is reasonable since the device is expected to reflect a wireless signal for low-power sensor applications.

In Table I the measured low-frequency reflector admittance as well as the conductance cutoff frequency $f_{\sqrt{2}}$ are shown for different bias voltages $V_{b,ctrl}$. The relatively large resistance values $R_{xc} = 1/G_{xc}$ shown allow matching to typical antenna structures. Also, the conductance cutoff frequency $f_{\sqrt{2}}$ is on

the order of 4 GHz and is reduced for lower I_{bias} values. Simulation results are in good agreement with these values and also show a cross-over frequency above 10 GHz for the different biasing conditions.

TABLE I

ACTIVE REFLECTOR CORE MEASUREMENT SUMMARY

$V_{b,ctrl}$	$1/G_{xc}$ (Ω)	C_{xc} (fF)	$f_{\sqrt{2}}$ (GHz)
0.0	-653	201.6	4.51
0.2	-722	200.7	4.41
0.4	-827	199.6	3.79
0.7	-1116	198.1	3.28

The reflector gain and operating frequency is sensitive to process uncertainties and unknown wireless propagation conditions. To adjust the operating frequency and gain, a dual automatic gain control (DAGC) regulation circuit [1] is considered and the procedure, when enabled, consists in sensing a carrier and iteratively tuning I_{bias} and $V_{t,dc}$.

The DAGC relies on a differential rectifier [4] included in the integrated circuit shown in Fig. 2. To improve the rectifier sensitivity a differential cascode preamplifier is connected to the drains of the cross-coupled pair. Also, following the rectifier a post-amplifier provides additional gain.

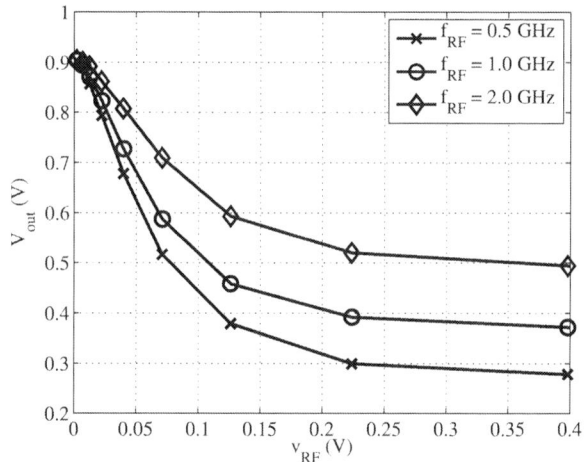

Fig. 3.   Peak rectifier transfer function

The measured transfer function of the rectifier is shown in Fig. 3 for different RF frequencies. We find that the peak detector is sensitive to a cross-coupled voltage on the order of $v_{xc} = 10$ mV. Effectively, for the reflector load impedance found in Sec. III, the rectifier sensitivity is equal to -43.3 dBm and satisfies requirements given in [1] assuming that the repeater is situated approximately 16 cm from the transmitter, outside of the near-field.

In Fig. 3 it can also be observed that the rectifier transfer function is highly dependent on the signal frequency. Actually, measurements indicate that the cascode preamplifier stage cuts off at a lower frequency than expected in simulation. The DAGC baseband algorithm must take into account this frequency dependent behaviour.

978-1-4244-6240-7/10 $26.00 © 2010 IEEE   452

III. THE WIRELESS REPEATER

The wireless repeater is designed to operate at 4 GHz, which is slightly below the conductance cutoff frequency $f_{\sqrt{2}}$ for high I_{bias}. At this frequency, we propose a discrete antenna structure whose dimensions are selected to maximize radiation efficiency. A $\lambda/2$ dipole is selected for a natural differential interface with the cross-coupled pair. The antenna is etched on a RT5870 dielectric and is connected to the integrated active reflector through a wirebond process. The supply voltage is also wirebonded to the chip.

To characterize the reflector wireless performance, it is placed in an anechoic chamber one meter from a horn antenna (Ets-Lindgren's 3160-04). The horn antenna gain is equal to $G_H = 16.7$ dB and its reflection coefficient $S_{11,H}$ is on the order of -20 dB.

As a preliminary test, the reflector stability is confirmed using a spectrum analyzer. In the 4-GHz band, the reflector is unstable for a bias control voltage $V_{b,ctrl}$ as high as 0.79 V, indicating that the antenna presents a high impedance to the reflector circuit. As $V_{b,ctrl}$ is increased to 0.8 V, at which point the bias current I_{bias} is approximately equal to 100 μA, the oscillation is no longer apparent. At such a low power consumption, the negative resistance of the active reflector core is $R_{xc} = -1.89$ kΩ.

Subsequently, a vector network analyzer is connected to the horn antenna and the reflection coefficient $S_{11,meas}$ is measured for $V_{b,ctrl} = 0.8$ V. For a path loss P between the horn antenna and reflector, the measured reflection coefficient $S_{11,meas}$ can be approximated with:

$$S_{11,meas} = 20\log(S_{11,H} + G_H^2 P^2 \Gamma) \text{ (dB)} \quad (2)$$

The first term $S_{11,H}$ is the reflection at the horn antenna and because $S_{11,H} >> G_H^2 P^2 \Gamma$, the reflector gain is not directly apparent. The procedure to extract Γ then consists in measuring $S_{11,H}$ with the reflector deactivated and subtracting this value from $S_{11,meas}$. The path loss P at one meter is 44.5 dB and the horn antenna gain is given above.

To characterize the gain and nonlinearity of the reflector, the device frequency response is measured for variable input power levels P_{rx} incident at the wireless node. The gain and bandwidth of the reflector are shown in Fig. 4 as a function of input power. For an input power equal to -71.9 dBm the gain is 22.3 dB, a relatively high specification for low-power CMOS technology. For an input power above -50 dBm, the reflector respects the system specification bandwidth of 300 kHz for a gain of 13 dB.

In Fig. 4, we find that 1-dB compression occurs for P_{rx} approximately equal to -62 dBm. This is low compared to values reported in [1] and is a consequence of the low biasing conditions that improves power consumption at the cost of linearity. These results indicate that the reflector will be operating in a highly nonlinear region of operation, and we will demonstrate in Sec. IV that the communication performance still shows good improvement.

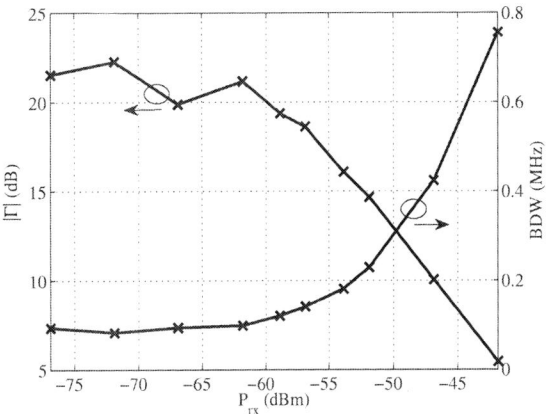

Fig. 4. Measurement of the reflector gain and 3-dB bandwidth as a function of input power

To evaluate the phase modulation performance, the control voltage $V_{t,ctrl}$ is varied and the transfer function is measured. The gain of the reflector at 3.978 GHz is shown in Fig. 5 for a reflector incident power equal to -46.8 dBm. This result confirms that the circuit phase can be varied over a large range and that the gain is attenuated for larger deviations from the resonating frequency.

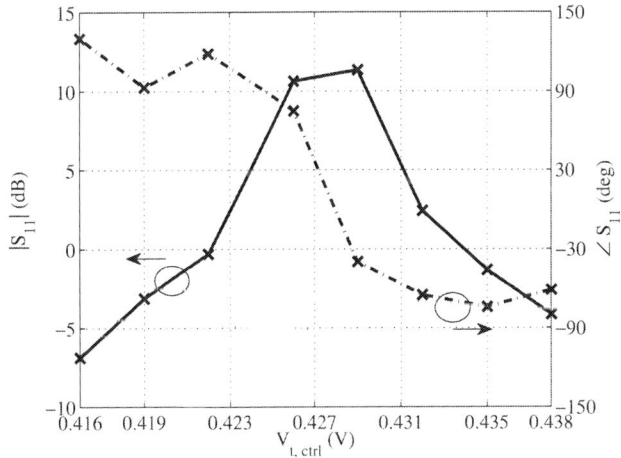

Fig. 5. Evaluation of the reflector gain at 3.978 GHz as a function of $V_{t,ctrl}$ for a reflector incident power equal to $P_{rx} = -46.8$

IV. SYSTEM PERFORMANCE

Phase sweep amplify-and-forward is a cooperative diversity scheme intended for indoor wireless networks. As shown in Fig. 6 the repeater is placed close to the source. The transmit signal x is broadcast to the repeater and destination across channel h_{sr} and h_{sd} respectively. The repeater amplifies the incident signal with gain Γ and slowly phase modulates the reflected signal which is received at the destination through channel h_{rd}. Assuming a noise factor F_r at the repeater,

978-1-4244-6240-7/10 $26.00 © 2010 IEEE

and thermal noise n_r and n_d at the repeater and destination respectively, the received signal at the destination is:

$$y = (h_{sd} + h_{sr}\Gamma h_{rd})x + \sqrt{F_r}\Gamma h_{rd}n_r + n_d \qquad (3)$$

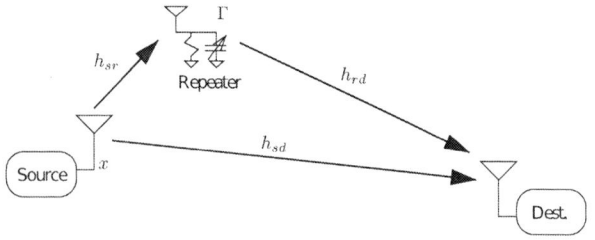

Fig. 6.   Network deployment for cooperative phase sweep amplify-and-forward

As in [1], PSAF is applied to a custom wireless sensor network application. The source sensor transmits 1000-bit encoded frames at 200 kbps. The convolutional code rate is 1/2 and its constraint length is 7. An interleaver spreads adjacent bits along the frame. The signal is QPSK modulated, pulse shaped using a raised cosine filter with a roll off-factor 0.5 and is converted to a 300-kHz analog signal. The transmit power is set to -15 dBm such that the received SNR is 17 dB at $d_{sd} = 10$ m. From [5], at this SNR the probability of error is slightly below 10^{-4} for an optimal, maximal ratio, two-branch diversity scheme. At the destination the received data is decoded using a soft Viterbi algorithm with channel state information.

In order to optimize diversity, the two paths h_{sd} and $h_{sr}\Gamma h_{rd}$ must experience equal average attenuation. In a LOS situation between the source and destination, such as described in [1], the direct path is shorter than the path through the repeater. Using a log-distance large-scale path loss model with exponential factor n equal to 4, 45-dB reflection gain is necessary. This hardware specification is difficult to meet using a discrete antenna prototype. However, non-line of sight scenarios do exist where the h_{sd} path is blocked such that the overall attenuation on the h_{sd} and $(h_{sr} \cdot h_{rd})$ paths becomes more balanced. This would reduce the gain required for the repeater.

The diversity improvement is evaluated using the measured repeater phase sweep transfer function, both for LOS and NLOS scenarios. For the LOS scenario, the paths have unequal average attenuation, while for the NLOS scenario the travelling distances are modeled to be equal.

The node is placed 16 cm from the source. This serves a dual purpose: firstly, to allow a 300-kHz bandwidth, and secondly, to satisfy DAGC sensitivity. The probability of bit error is evaluated as a function of distance between source and destination, d_{sd} and compared to a scenario without diversity. The nominal transmit power without diversity is -11.8 dBm and is adjusted for the scenario with the reflector to maintain an equal SNR level at the receiver.

The BER results are shown in Fig. 7 as a function of distance and corresponding SNR. Firstly, for the LOS scenario,

diversity improvement is only apparent at very high SNR. In this case, the occasional deep fade on the direct path is aided with a varying channel response on the low power path from the repeater. For a NLOS scenario, we find that for SNR values above 24 dB, the bit error rate is improved by 1 to 2 orders of magnitude for separation distances between 5 m and 8 m. This amounts to a coverage area increase by more than a factor of 2.5. Although this is slightly smaller than the four-fold increase cited in [1], this is still an important improvement for indoor wireless sensor networks.

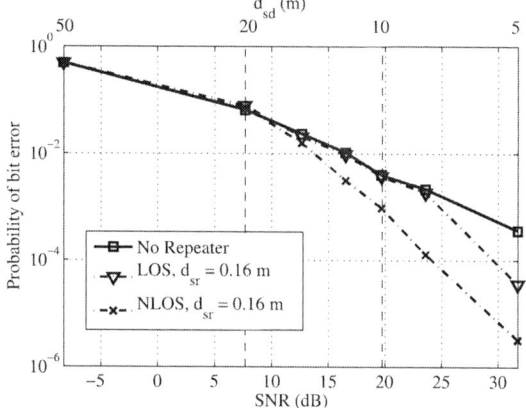

Fig. 7.   Comparison of PSAF probability of error for different channel conditions

V. CONCLUSION

In this work, we have demonstrated the performance of a wireless reflector for phase sweep amplify-and-forward. The active reflector is implemented in 0.13-μm CMOS technology and interfaced to a dipole antenna. A reflection gain of 22.3 dB is observed for a very low power consumption equal to 120 μW. Slow phase modulation at the wireless repeater is shown, and, applied to a PSAF in a non line-of-sight scenario, the wireless reflector provides an increase in the coverage area of an indoor wireless network by a factor of 2.5.

ACKNOWLEDGEMENT

The authors would like to thank CMC Microsystems for chip fabrication support and NSERC and TrLabs for financial support.

REFERENCES

[1] J.-F. Bousquet, S.C. Magierowski, and G.G. Messier, "An integrated active reflector for phase-sweep cooperative diversity," *Circuits and Systems II: Express Briefs, IEEE Transactions on*, vol. 56, no. 8, pp. 624–628, Aug. 2009.

[2] Guillermo Gonzalez, *Microwave Transistor Amplifiers, 2nd Edition*, Prentice-Hall, Upper Saddler River, New Jersey, 1997.

[3] Hugo Veenstra and Edwin van der Heiden, "A 19-23 GHz integrated LC-VCO in a production 70 GHz fT SiGe technology," in *Proceedings of the 29th European Solid-State ESSCIRC*, September 2003, pp. 349 – 352.

[4] R.G. Meyer, "Low-power monolithic RF peak detector analysis," *Solid-State Circuits, IEEE Journal of*, vol. 30, no. 1, pp. 65–67, Jan 1995.

[5] T. S. Rappaport, *Wireless Communications, Principles and Practice*, Prentice Hall, 1996.

978-1-4244-6240-7/10 $26.00 © 2010 IEEE

Design Methodology and Comparison of Rectifiers for UHF-band RFIDs

Francesco Mazzilli*, Prakash E. Thoppay*, Norbert Jöhl†, and Catherine Dehollain*

*Ecole Polytechnique Fédérale de Lausanne, RFIC group, Lausanne, 1015, Switzerland

†Advanced Silicon, Lausanne, 1004, Switzerland

Abstract—Rectifiers are important energy converters and henceforth crucial building blocks for RFID applications. In the first half of the work, we have presented a design methodology for matching the rectifier input impedance with the antenna to maximize the rectifier power conversion efficiency. The proposed design approach uses the fundamental transconductance $(Gm(1))$ analysis to estimate the rectifier input impedance. In the second half, a comparison between various possible single-stage rectifier topologies implemented in a CMOS 0.18 μm technology operating at UHF-band is presented. Using voltage conversion efficiency as the FOM, the optimum rectifier topology for RFID application is determined.

Index Terms—CMOS, impedance matching, radio frequency identification (RFID), rectifiers, ultra-high frequency (UHF).

I. INTRODUCTION

Recently there has been an increase in interest in the study and implementation of energy scavenging transceiver architectures for its use in sensor networks, product tags and access control to name a few [1]. The radio frequency identification (RFID) transponder is one such energy scavenging architecture in which the energy storage element (e.g. battery, super capacitor) in the tag is powered up by the transmitted RF signal using a rectifier, also defined as RF-to-DC converter.

To maximize the available power at the energy storage element tag it is necessary to minimize the power loss across the rectifier which is achieved by matching the rectifier input impedance with the antenna [2]. Moreover, using an impedance matching network, tag read-range increases by boosting the available voltage to the rectifier even in cases wherein the available input power is low [3]. Hence, the rectifier power conversion efficiency (PCE) increases as the available voltage to the rectifier increases. However, to select the nature and the values of the components of the matching network an accurate derivation of the rectifier input impedance has to be done through circuit simulator.

The paper discusses the rectifier design methodology to match its impedance to the antenna thereby maximizing the voltage conversion efficiency. In Section II an analysis and design methodology to match the rectifier input impedance with the antenna is described. In Section III a comparison of various rectifier topologies using voltage conversion efficiency as FOM along with experimental results is shown and conclusions are derived thereof.

II. DESIGN METHODOLOGY FOR THE RECTIFIER

To increase the rectifier power conversion efficiency the rectifier impedance is to be matched with the antenna for maximum power transfer. The rectifier input admittance is modeled as a parallel combination of a capacitor and a resistor mathematically represented as $Y_{rec} = G_{rec} + j * Y_{c_{rec}}$, where G_{rec} and $Y_{c_{rec}}$ represents the nonlinear input conductance and susceptance respectively [4]. To match the rectifier input impedance with the antenna (in this case a 50 Ω is considered) the imaginary part is compensated with a parallel inductor (L_p) which is represented by an equivalent series inductance (L_s) as shown in (Fig. 1). A series inductor topology is used to compensate the rectifier capacitive part as it boosts the voltage across the rectifier by a factor "$Q = \omega C_{rec}/G_{rec}$" (which is the resonant structure quality factor) mathematically represented in (1), where V_{AV} corresponds to the input RF voltage as shown in Fig. 1 thereby increasing the voltage conversion efficiency. Once the capacitive part is compensated by the inductor the rectifier nonlinear resistance is matched to 50 Ω by having proper transistor dimensions (W/L ratio).

$$|V_{IN}| = \frac{|V_{AV}|}{2}\sqrt{1 + Q^2} \qquad (1)$$

In the case of MOSFETs based diode structures to obtain the series inductance value (L_s) it is necessary to compute the rectifier input capacitance which in turn depends on the transistor dimensions for a fixed gate oxide capacitance. The analytical derivation for the rectifier input nonlinear resistance is determined based on the I-V relationship of the MOS device which becomes complicated due to the short channel effects and the quadratic dependency between the current and the voltage. Therefore the aide of large signal analysis in the simulator is used to determine the the rectifier input impedance.

The design procedure for the bridge rectifier shown in (Fig. 2) is described. During the positive cycle of the input RF signal V_{AV} the transistors M4 and M2 are switched on and during the other cycle transistors M3

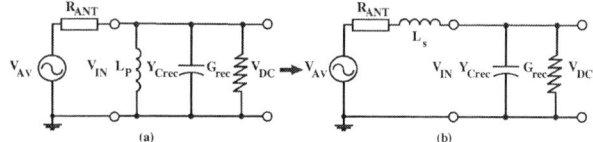

Fig. 1. Simplified schematic of the tag including the antenna, the rectifier equivalent circuit and matching network solutions through an inductor: (a) shunt, (b) series.

and M1 are switched on charging the capacitor C_{OUT} in a single direction thereby rectifying the input RF signal. The input nonlinear resistance determined for a given input

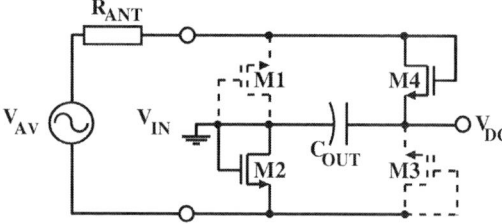

Fig. 2. NMOS bridge rectifier circuit.

power (worst case scenario) using PSS analysis in Cadence simulator for various values of W/L is shown in left hand y-axis of (Fig. 3) and the corresponding value of input capacitance is shown in right hand y-axis of (Fig. 3). From Fig. 3 the input capacitance to achieve 50 Ω is calculated and then the capacitance is compensated by a corresponding series inductance value for the given operating frequency to boost up the voltage across rectifier.

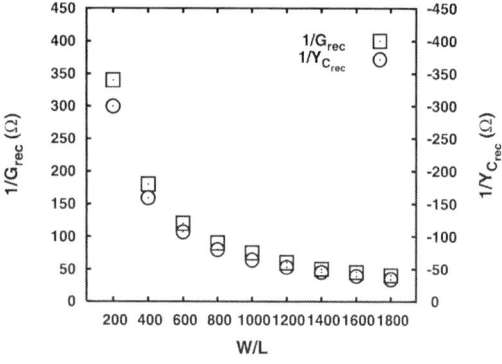

Fig. 3. Simualted nonlinear resistance and input capacitance of the NMOS bridge rectifier.

Fig. 4 compares the rectifier input impedance measured using VNA (Agilent HP8719D) with the results obtained using Cadence Virtuoso simulator. From Fig. 4 it can be seen that the measured value is in good agreement with the

simulated value thereby validating the design procedure for the rectifiers.

Fig. 5 shows the measured voltage efficiency as a function of the input available voltage V_{AV}. An off-chip series inductor $L_s = 12\ nH$ was used to boost the input voltage of the rectifier at input RF signal frequency of 900 MHz. From Fig. 5 it can be seen that the rectifier voltage conversion efficiency matched to the antenna is higher in comparison to without series inductor which is as expected. For the calculations, the rectifier voltage conversion efficiency (VCE) is defined as:

$$\eta = \frac{V_{DC}}{V_{AV,p}} \qquad (2)$$

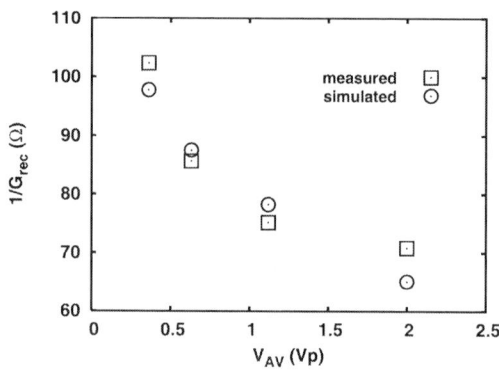

Fig. 4. Simulated and measured real part of the rectifier input impedance with transistor size ratio W/L=500.

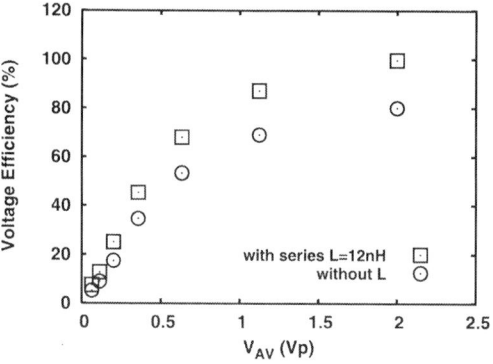

Fig. 5. Comparison between measured DC output voltage of the NMOS bridge rectifier with and without series inductor.

III. COMPARISON OF RECTIFIERS

In the second half, four rectifier configurations shown in (Fig. 6) are compared using voltage conversion efficiency as the FOM. In general VCE can be improved by cascading several stages but in this paper the idea

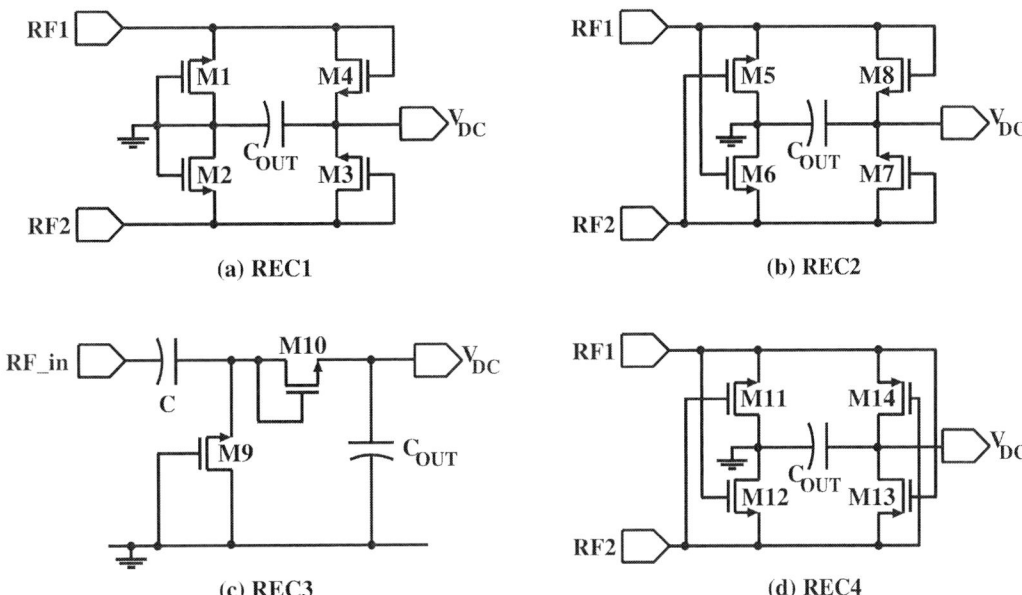

Fig. 6. Rectifier topologies: (a) NMOS differential-drive bridge rectifier, (b) NMOS differential-drive gate cross-connected bridge rectifier, (c) NMOS doubler, (d) NMOS-PMOS differential-drive gate cross-connected bridge rectifier.

is to compare across various rectifier topologies hence a single stage rectifier is used. The rectifier configurations are implemented in CMOS 0.18 μm technology using zero-Vth transistors (Vth=2mV) for NMOS, and low-Vth transistors (Vth=320mV) for PMOS. The use of zero-Vth and low-Vth is to decrease the drop across the transistors and thus increasing the voltage conversion efficiency. A brief description about the operations of various rectifier topologies along with the experimental results is described below.

A. Rec1

Fig. 6.a shows the NMOS differential-drive bridge rectifier normally found in RFID applications where the reader and the tag are in close proximity. For proper rectification using such a topology the input voltage across the rectifiers should be at least twice the threshold voltage, hence using zero-Vth transistors enables such an architecture to operate even at low input power levels.

B. Rec2

Fig. 6.b is a modified version of the common bridge rectifier, where the gate-grounded NMOS transistor are used as switches by cross-coupling each gate thereby decreasing the required input voltage to turn on the transistors and hence an increase in the read range of the RFID tag.

C. Rec3

Fig 6.c is a well known doubler structure where the output V_{DC} ideally is equal to twice the peak input RF

voltage. Many such stages can be used in cascade to increase the output voltage hence increasing the overall voltage conversion efficiency.

D. Rec4

Fig. 6.d is similar to Fig. 6.b where the transistors gate input is cross-coupled and hence act as switches. In Fig. 6.d on contrary to Fig. 6.b PMOS transistors are also used and hence al four transistors acts as switches. A major disadvantage of such a topology is the need for a current controlling circuitry as the current direction reverses when the rectified voltage is higher than the input RF voltage.

E. Experimental Results

Fig. 7 shows the photomicrograph of the four rectifiers, fabricated in 0.18 μm CMOS process. The chip was glued on a PCB used for testing the rectifiers. All transistors width and length are 250 μm/0.5 μm, and output capacitor, C_{OUT}, is 5 pF for the bridge converters and 10 pF for the doubler. The input capacitance, C, as in Fig. 6.c has the same value of its output capacitor. Fig. 8 shows the unloaded voltage efficiency of the fabricated single-stage RF-to-DC converters as a function of the available voltage along with a signal frequency of 900 MHz. The VCE increases more than 100% for Rec2 and Rec4, this is the effect yielded by the input voltage boosting due to the series inductor (L_s), whereas for Rec1 and Rec3, VCE increases toward 90%. Fig. 8 can be distinguished into three regions in which for $V_{AV} < 0.5$ V, Rec4 works efficienctly; then, for 0.5 V$\leq V_{AV} \leq 1$ V is a transistion

978-1-4244-6240-7/10 $26.00 © 2010 IEEE   457

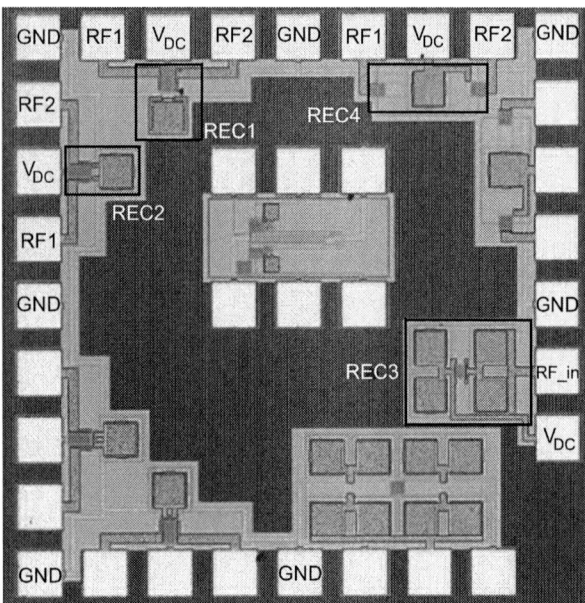

Fig. 7.   Photomicrograph of fabricated chip in 0.18 μm CMOS process.

zone between Rec4 and Rec2 and for $V_{AV} > 1$ V, the output DC voltage of Rec2 increases above V_{AV}.

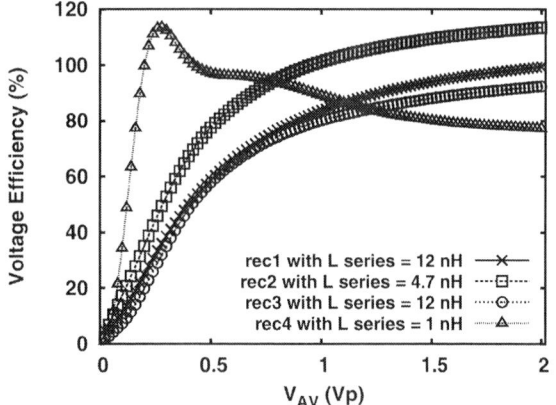

Fig. 8.   Measured unloaded voltage efficiency of the RF-to-DC converters fabricated in 0.18 μm CMOS process.

Fig. 9 shows the voltage efficiency of the RF-to-DC converters under different loads with $V_{AV} =$ 110 mV at 900 MHz. Even if Rec4 for low input available voltage show the best VCE, in loaded condition its efficiency decreases drastically. Whereas, for the others topologies the VCE is constant until 10 $k\Omega$ and Rec2 clearly has the best loaded voltage efficiency.

IV. CONCLUSION

A design and comparison methodology for different RF-to-DC converters in UHF-band for RFIDs has been

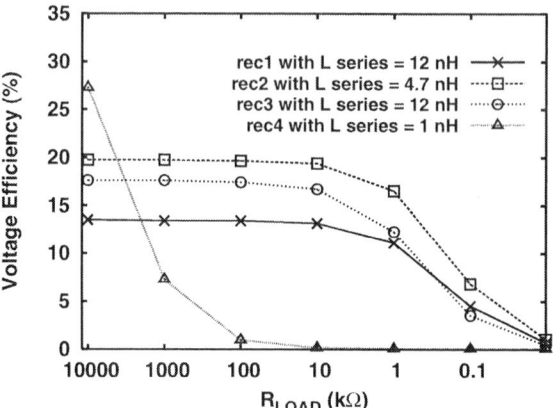

Fig. 9.   Measured loaded voltage efficiency of RF-to-DC converters at input available voltage 110 mV.

proposed. The input impedance obtained using large signal simulation is compared with the results measured. Four different RF-to-DC converters have been designed and fabricated in 0.18 μm CMOS process. FOM as voltage efficiency has been given to compare the perfomances of the rectifiers. This analysis allows the designer to select the proper rectifier according to its operating condition defined by the available input voltage.

ACKNOWLEDGMENT

The research leading to these results has received funding from the European Community's Seventh Framework Programme (FP7/2007-2013) under grant agreement n. 224009.

REFERENCES

[1] Klaus Finkenzeller. *RFID Hanbook: Fundamentals and Applications in Contactless Smart Cards and Identification.* 2nd edition, 2003.
[2] R. Barnett, S. Lazar, and Jin Liu. Design of multistage rectifiers with low-cost impedance matching for passive rfid tags. In *Radio Frequency Integrated Circuits (RFIC) Symposium, 2006 IEEE*, page 4 pp., June 2006.
[3] Nhan Tran, Bomson Lee, and Jong-Wook Lee. Development of long-range uhf-band rfid tag chip using schottky diodes in standard cmos technology. In *Radio Frequency Integrated Circuits (RFIC) Symposium, 2007 IEEE*, pages 281–284, June 2007.
[4] J.-P. Curty, N. Joehl, C. Dehollain, and M.J. Declercq. Remotely powered addressable uhf rfid integrated system. *Solid-State Circuits, IEEE Journal of*, 40(11):2193–2202, Nov. 2005.

A CMOS Ultra-wideband Radar Transmitter
with Pulsed Oscillator

Sungeun Lee, Sanghoon Sim and Songcheol Hong

School of Electrical Engineering and Computer Science, KAIST, Daejeon, 305-701, Korea

Abstract — A design of Ultra-wideband (UWB) radar transmitter is presented. The transmitter which uses a pulsed oscillator consists of pulse generator, switching buffers and control signal generator. The control signal generator includes modulators of binary-phase shift keying (BPSK) and pulse position modulation (PPM) for spreading the spectral lines. It is fabricated using 0.13 μm CMOS technology and the chip size is 910 × 485 μm². The output spectrum is centered at the 22.0 GHz with the 10-dB bandwidth of 2.48 GHz and the pulse width of output pulse is tunable from 630ps to 830ps. Also, the BPSK and PPM modulations are confirmed. In conclusion, the generated pulse complies with FCC's spectral mask.

Index Terms — Ultra-wideband (UWB), UWB radar transmitter, CMOS RFIC, pulse generator, automotive radar, short range radar (SRR), binary-phase shift keying (BPSK), pulse position modulation (PPM)

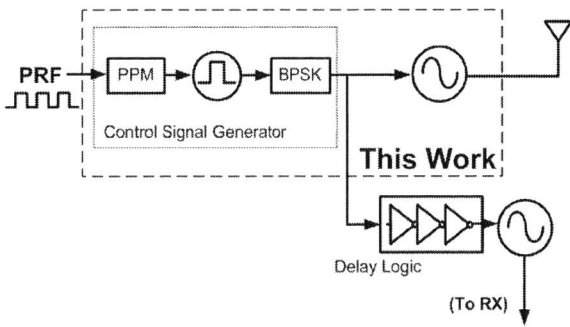

Fig. 1   Block diagram of the proposed UWB Transmitter

I. INTRODUCTION

For last several years, there have been many reports on Ultra-wideband (UWB) systems such as 3 – 10 GHz UWB communication system [1] - [3] and 22 – 29 GHz UWB pulse radar system [4] - [7]. Especially, UWB pulse radar system is approved for short range radar (SRR) for automotive, which is defined by Federal Communications Commission (FCC) in 2002 [8]. Because the 22 – 29 GHz UWB band has a very wide unlicensed bandwidth but the maximum power emission is restricted to -41.3 dBm / MHz, it is suitable for the systems that require high range resolution in a relatively short range (several tens of meters). Definitely, in order to satisfy the requirement of high range resolution, an ultra short pulse (sub-nanosecond) generator is needed.

Until now, the most of UWB pulse generators in the high frequency bands have been designed by using SiGe technology due to its fast switching property [4], [5]. But some papers show the probability to implement the 22-29 GHz UWB radar transmitter or receiver in cost-effective CMOS technology [6], [7], therefore it is no longer doubtful to develop a UWB system on a chip with CMOS technology.

In this paper, the detailed design of the proposed CMOS UWB transmitter consisted of a short pulse generator, switching buffers and its control signal generator, which is operating in the 22 - 29 GHz UWB band, is present. A

simple method to have both binary-phase shift keying (BPSK) and pulse position modulation (PPM) is also introduced. All circuits are implemented in CMOS 0.13 μm mixed signal process.

II. UWB TRANSMITTER DESIGN

Fig. 1 is a block diagram of the proposed UWB transmitter. It mainly consists of a short pulse generator, switching buffers and its control signal generator. The operation principle is as follows. An external clock signal which corresponds to pulse repetition frequency (PRF) inputs to the control signal generator. When the clock is at rising-edge, it produces wanted control signals for the pulse generator as the given modulation code and pulse width signal. This control signals allow short pulse generator (oscillator in above figure) to generate UWB signals and the pulses are transmitted through the wideband antenna.

A. Short Pulse Generator

Generally, there have been three major methods to generate short pulses which can comply with FCC's spectral mask: carrier-based method [2], [4], [6], using pulse shaping filter [5], using ON/OFF pulsed oscillator [3]. First of all, the carrier-based method is not suitable for simple structure and low power operation. Since it needs phase-locked loop (PLL) and wideband single-pole

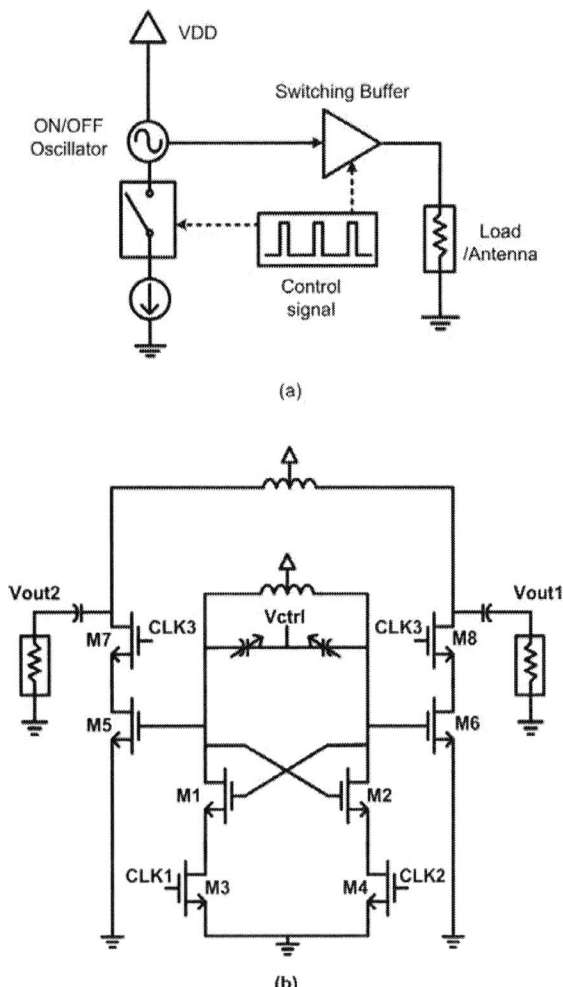

(a)

(b)

Fig. 2   (a) Block diagram (b) Schematic of the modified short pulse generator in [9] and switching buffers

double-through (SPDT) switch, these kinds of transmitters are bound to be relatively complex. And because the voltage-controlled oscillator (VCO) in PLL is always turned-on, the power consumption is much high. Moreover, the finite isolation property of SPDT switch can cause the signal leakage. The second method using pulse shaping filter has difficulty in CMOS process. This method needs ultra short baseband pulse about tens of picoseconds, but it is nearly impossible in available CMOS process.

In this paper, using ON/OFF pulsed oscillator method is accepted. In this case, since conventional UWB systems

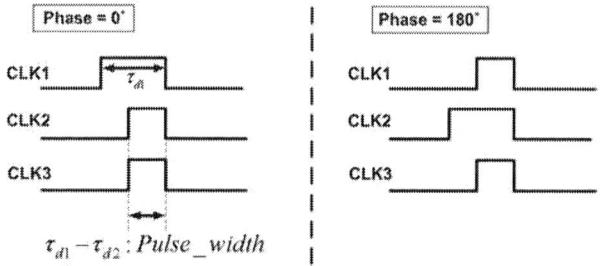

Fig. 3   Timing chart for BPSK modulation. Output phase is determined by just swapping CLK1 and CLK2 [9].

have very low duty-cycle, the pulse generator and switching buffers are usually OFF state, and ON just in a moment of pulse duration. Therefore, it can reduce the power consumption drastically. Fig. 2 shows the block diagram and detailed schematic of the modified short pulse generator in [9] and switching buffers. The pulsed oscillator is basically composed of LC cross-coupled oscillator (M1, M2) and active buffers (M5, M6). The signals from control signal generator can switch on / off the current source (M3, M4) of oscillator and buffers (M7, M8) respectively. To obtain fast switching operation, low Q-factor of LC-tank and high open loop gain is realized [3]. Two varactors are inserted for center frequency tuning and MIM capacitors are used as bypass capacitors.

It is well-known that the spectral lines in an output spectrum modulated by both BPSK and PPM signals can be spread effectively in high-PRF UWB systems [10]. Conventionally, BPSK modulation has been performed in RF region, directly shift the phase of output signals. It has limitations in accuracy and wideband operation. Therefore, a simple BPSK modulation method is proposed in [9]. In Fig. 2, the current sources of oscillator (M3, M4) are separated and controlled asymmetrically for BPSK. It is based on the fact that the phase of output signals can be determined by changing the order of switch-on of two current sources. That is, the phase difference is 180° between when CLK1 is turned on earlier than CLK2 and when CLK1 is late as described in Fig. 3. In this operation, the pulse width difference doesn't need too large, and CLK3 is not dependent on phase. Consequently, for BPSK modulation, the proposed short pulse generator needs just two baseband signal that have different pulse width, and this method is much simpler and more effective than conventional one because it operates in baseband region.

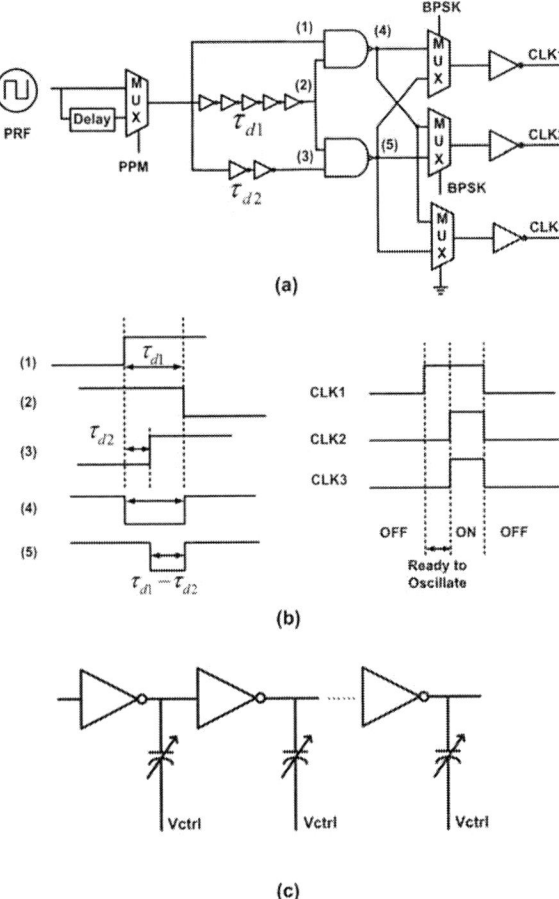

(a)

(b)

(c)

Fig. 4 (a) Block diagram of the control signal generator and BPSK and PPM modulators (b) Each waveforms in control signal generator and modulators (c) Variable delay line for tuning the pulse width

B. Control Signal Generator & Modulators

To obtain the control signals that can burst the proposed pulse generator including switching buffers and modulate the output signal for spreading the spectrum, the control signal generator and modulators (BPSK, PPM) are modified [9]. Fig. 4 (a) is a block diagram of control signal generator and modulators. As described in Fig. 4 (b), there are three passes: no delay pass (node 1), longer delay and inverting pass (node 2), shorter delay and non-inverting pass (node 3). And two NAND gates can generate wanted pulses that have different pulse width. These are CLK1 and CLK2 for switching current source of short pulse generator and these signals can be swapped by multiplex control bit for BPSK modulation as we described in chapter II. And CLK3 for switching buffers always have shorter pulse width regardless BPSK. When only one signal of CLK1 and CLK2 is high, the short pulse generator is just ready to oscillate, and start to

oscillate only when both are ON. The pulse duration which corresponds to τ_{d1} - τ_{d2} is tunable by V_{ctrl} as shown in Fig 4 (c).

And PPM modulation can be realized easily. A delay line and multiplexer play a role of changing the moment of rising edge of external clock signal (Fig. 4 (a)). This delay line can delay the signal to the extent of pulse width and composed of inverter chains. Since the control signals are generated at the moment of this rising edge, the pulse position can be easily modulated.

IV. MEASUREMENT RESULT

Fig. 5 shows the microphotograph of the designed CMOS short pulse generator including switching buffers. It is fabricated using 0.13 µm CMOS mixed signal process. The chip areas are 910µm × 485µm. The control signal generator is separated from RF block, and its outputs are delivered to the short pulse generator through chip-to-chip bond wires. Output spectrums and waveforms are measured using HP8564E Spectrum Analyzer and 86100A Wide-Bandwidth Oscilloscope, respectively.

The supply voltage is 1.2V. It consumes no steady-state current. MP1763C Pulse Pattern Generator is used to make Pseudo-Noise (PN) codes for BPSK and PPM modulations. The measurement results are shown in Fig 6. All cable and connector losses are embedded in evaluation. Fig 6 (a) is an output waveform in time domain. The pulse width is about 830ps which corresponds to 12cm of range resolution and peak-to-peak voltage is about 125 mV. Its pulse width can be tuned from 630ps to 830ps by external control voltage. Fig. 6 (b) is output spectrum without modulations and Fig. 6 (c) is that with both BPSK and PPM modulations at 50 MHz PRF. The output spectrum is centered at the 22.0 GHz with 10-dB bandwidth of 2.48 GHz. And it can be easily confirmed that spectral lines in (b) is spread out by modulation effect. It allows transmitter to use FCC's spectral mask more effectively.

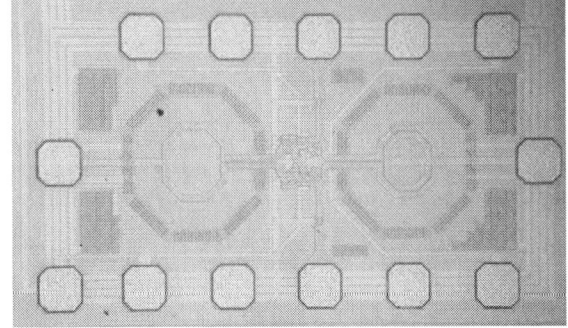

Fig. 5 Chip microphotograph of the proposed UWB radar transmitter which is fabricated using CMOS 0.13 µm mixed signal process (chip size: 910 µm × 485 µm)

978-1-4244-6240-7/10 $26.00 © 2010 IEEE

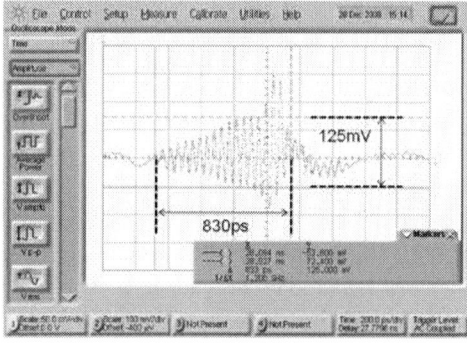

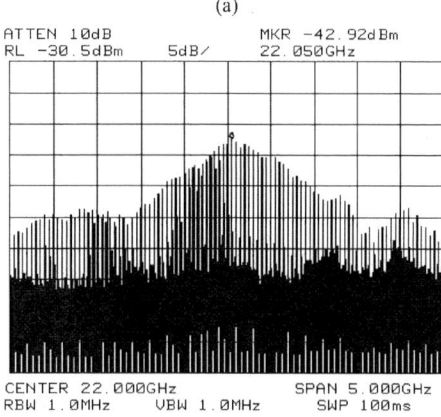

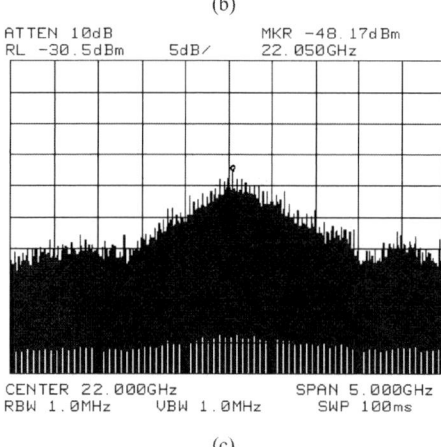

Fig. 6   Measurement Results (a) output waveform (b) output spectrum without modulations (c) output spectrum with both BPSK and PPM modulations.

V. CONCLUSION

A CMOS UWB radar transmitter for short range radar application is proposed. It mainly consists of a short pulse generator including switching buffers and its control

signal generator. To lower the power consumption, we use the ON/OFF pulsed oscillator. A simple BPSK modulation method with the structure is also proposed, which is confirmed by measurements. Although the center frequency of short pulse generator is shifted about 3 GHz due to the parasitic effects, but this UWB transmitter is suitable for simple and low-power UWB radar systems.

REFERENCES

[1] Yunliang Zhu, Jonathan D. Zuegel, John R. Marciante, and Hui Wu, "Distributed Waveform Generator: A New Circuit Technique for Ultra-Wideband Pulse Generation, Shaping and Modulation", *IEEE Journal of Solid-State Circuits*, Vol. 44, No. 3, March 2009.

[2] Rui Xu, Yalin Jin, and Cam Nguyen, "Power-efficient switching-based CMOS UWB transmitters for UWB communications and Radar systems ", *IEEE Trans. Microw. Theory Tech.*, Vol.54, No.8, pp. 3271-3277, August 2006.

[3] Tuan-Anh Phan, Jeongseon Lee, Krizhanobskii, V., Seok-Kyun Han, Sang-Gug Lee, "A 18-pJ/Pulse OOK CMOS Transmitter for Multiband UWB Impulse Radio" *IEEE Microwave and Wireless Components Letters*, Vol 17, Issue 9, Sept. 2007, pp. 688-690.

[4] I. Gresham, A. Jenkins, R. Egri, C. Eswarappa, N. Kinayman, N. Jain, R. Anderson, F. Kolak, R. Wohlert, S. P. Bawell, J. Bennett, and J. P. Lanteri, "Ultra-wideband radar sensors for short-range vehicular applications," *IEEE Trans. Microw. Theory Tech.*, Vol.52, No.9, pp. 2105-2122, September 2004.

[5] Y. Kawano, Y. Nakasha, K. Yokoo, S. Masuda, T. Takahashi, T. Hirose, Y. Oishi, and K. Hamaguchi, "RF Chipset for Impulse UWB Radar Using 0.13-μm InP-HEMT Technology", *IEEE Trans. Microw. Theory Tech.*, Vol. 54, No. 12, pp. 4489-4497, December 2006.

[6] Ahmet Oncu, B. B. M. Wasanthamala Badalawa, Minoru Fujishima, "22-29 GHz UWB Ultra-Wideband CMOS Pulse Generator for Short-Range Radar Applications", *IEEE Journal of Solid-State Circuits*, Vol. 42, No. 7, July 2007.

[7] Vipul Jain, Sriramkumar Sundararaman, and Payam Heydari, "A 22-29-GHz UWB Pulse-Radar Receiver Front-End in 0.18 μm CMOS.", *IEEE Trans. Microw. Theory Tech.*, Vol. 57, No. 8, pp. 1903-1914, August 2009.

[8] "First report and order, revision of part 15 of the commission's rules regarding ultra wideband transmission systems," Federal Communications Commission, Washington, DC, ET Docket 98-153, 2002.

[9] Sanghoon Sim, Dong-Wook Kim, and Songcheol Hong, "A CMOS Pulse Generator with an Embedded BPSK Modulator for UWB Radar Application", *IEEE Journal of Solid-State Circuits, submitted for publication.*

[10] Y. Nakache and A. Molisch, "Spectral shape of UWB signals – influence of modulation format, multiple access scheme and pulse shape." *The 57th IEEE Semiannual Vehicular Technology Conference*, vol. 4, Apr. 2003, pp. 2510-2514.

900MHz/1800MHz GSM Base Station LNA with Sub-1dB Noise Figure and +36dBm OIP3

Domine Leenaerts, Jos Bergervoet, Jan-Willem Lobeek, Marek Schmidt-Szalowski

NXP Semiconductors, Eindhoven, 5656AE, the Netherlands

Abstract — **A sub-1dB NF fully integrated low noise amplifier in a 0.25µm SiGe:C BiCMOS technology targeting GSM base-station applications will be discussed. The two-stage LNA is housed in a HVSON10 package and mounted on a PCB. The LNA measures a NF of 0.75dB in the 900MHz band and 0.9dB in the 1800MHz band. The LNA is matched to 50Ω at the RF I/O pins of the IC and has integrated ESD protection on all IC pins. The LNA achieves an OIP3 of +36dBm, a 1-dB OCP of +19dBm while dissipating 190mW. The LNA performance is in line with the compound technology LNA counterparts.**

Index Terms — **low noise amplifier, base-station, BiCMOS, linearity.**

I. INTRODUCTION

The GSM base station receiver's sensitivity and inter-modulation characteristics require both sub-1dB noise figure (NF) and high linearity for the low noise amplifier (LNA), e.g. output third order intercept point (OIP3) and output 1-dB compression point (1-dB OCP) higher than +30dBm and +15dBm, respectively. The low NF is required for optimal receiver sensitivity, while interference in adjacent channels demand for low third order inter modulation distortion.

These specifications are currently met using compound semiconductor technologies like pHEMT GaAs [1-4]. These technologies inherently reach sub-1dB NF for moderate frequencies. The high supply voltages (e.g. 5V) allow for large signal swings and consequently good 1-dB OCP. There are, however, two main drawbacks of these solutions. The first drawback is the lack of ESD protection. As the solutions are more or less single devices, adding ESD protection is difficult. Besides, the ESD capabilities of compound technologies are less than of the Si-based technologies due to different lattice structure. The second issue is optimum noise and impedance matching. The current solutions [1-4] require additional high quality components to be placed on the PCB to achieve simultaneously noise and impedance matching. The NF values at the transistor input pin are in the 0.6dB range, but for non-50Ω impedance. The NF increases towards 1dB when measured at the 50Ω RF input at the PCB.

There exist only very few examples of Si-based LNAs for these type of applications. In [5] the 0.25µm BiCMOS LNA achieves a 1.35dB NF and +26dBm OIP3 at a moderate gain level of 13dB. For the same technology, the LNA in [6] achieves an NF of 2.6dB and a voltage gain of 32dB. In [7] the 0.5µm BiCMOS LNA achieves a NF of 1.4dB, an OIP3 of +25dBm and 16dB of gain. Sub-1dB NFs have been reported, but they lack the 50Ω impedance match and high linearity [8].

The presented fully integrated LNA overcomes afore-mentioned issues of the compound technology solutions, meanwhile implemented in a Si-based technology. At the IC RF pin, the LNA achieves sub-1dB NF in the GSM band, meanwhile matched to 50Ω without the need of any additional off-chip component. The LNA has integrated ESD protection on all IC pins, including the RF I/O pins.

Section II will discuss the used technology and optimal device conditioning. In section III the complete LNA design will be discussed followed by measurement results in section IV. Comparison with state-of-the-art and conclusions will be given in section V.

II. TECHNOLOGY PERFORMANCE

The two-stage LNA design has been realized in a 0.25µm SiGe:C BiCMOS technology with a cut-off frequency (f_t) of 130GHz [9]. The low-noise NPN device has an emitter width of 0.4µm and has an NF_{min} of 0.5dB at 2GHz for a current density of 0.25mA/µm². This current density will not deliver peak-f_t but the related f_t is 30GHz, still 10 times the operational frequency. Also the available bandwidth for 10 times voltage gain is beyond 10GHz for this current density. In other words the device will certainly have more than 10dB gain at 2GHz. This low noise device can safely be operated at 1.8V supply voltage.

However, higher supply voltages are needed to achieve the high linearity performance requirements. Therefore a second stage is needed which can safely be operated at 3V supply voltage. Clearly, for sub-1dB NF the noise contribution of the second stage remains important. If an overall NF of 0.6dB is the target, the formula of Friis would require a NF of 0.7dB for the second stage, assuming a realistic available gain of 15dB in the first stage. The high voltage device in this technology can still

978-1-4244-6240-7/10 $26.00 © 2010 IEEE

reach such a NF value, requiring a current density of 0.21mA/μm^2. The cut-off frequency is then still 30GHz.

III. LNA DESIGN

The DC voltage gain for a single transistor gain stage with inductive loading is defined as (see Fig. 1):

$$A_v = g_m * Z_{load}(@2GHz) \approx 40I_c * Z_{load}(@2GHz) \quad (1)$$

A DC gain of 60 times or 17dB seems reasonable as the noise contribution of the second stage will then only marginal impact the overall NF. Choosing an emitter length of 10μm will give an optimal noise current of 1mA but would require a large impedance (•1500Ω). Therefore the current has been increased to 1.3mA, sacrificing a little the optimum noise current density as we keep the same device size. The result is a load of 1200Ω or an equivalent inductance of 100nH (at 2GHz). Simulating this initial design reveals an NF_{min} of about 0.5dB with Γ_{opt} = 0.7 + j0.03. The latter means that Re(Zopt) = 300Ω. On the other hand, Zin = 18 – j4, thus there is no simultaneous input impedance match and noise match. Furthermore Gmax is in the order of 15dB and the gain stage has an OIP3 • -12dBm.

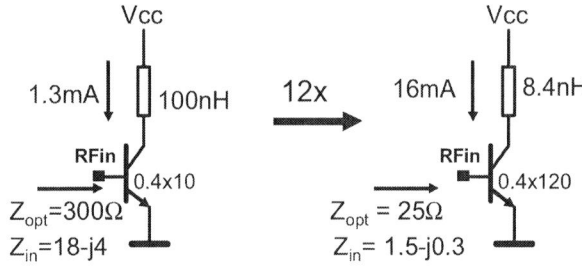

Figure 1. Single stage LNA

The linearity can be improved by setting multiple of these stages in parallel. This will obviously increase the total current, but at the same time Γ_{opt}, Z_{in} and Z_{load} will be decreased by a factor equal to the number of stages in parallel. Doubling the current initially gives 12dB improvement in linearity, but there is a saturation effect. A number of 12 parallel stages seems a reasonable balance between the total current of 16mA and OIP3 of +27dBm. The resulting stage, see Fig. 1, has a NF close to the NF_{min} of 0.5dB.

A similar design strategy can be applied to the second stage. The result is that 30 identical stages need to be placed in parallel, each stage using a high voltage device with an emitter width of 20μm. The resulting OIP3 of this

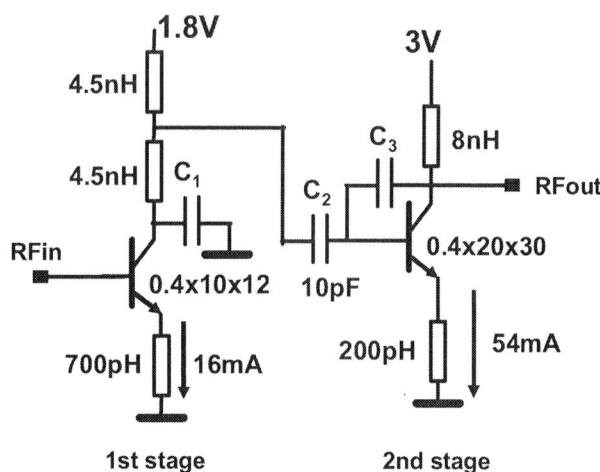

Figure 2. Final LNA design (biasing not shown).

second stage is +40dBm (including emitter degeneration) while the NF is around 0.7dB. The current consumption is 54mA from 3V supply voltage.

Although simulated linearity, noise figure and gain are in the same order of the requested performances, the main issue is now in the input, inter stage and output impedance matching when combining the two stages. The output impedance of the input stage is close to 50Ω, whereas the input impedance of the output stage is in the order of 3Ω. A way to lower the output impedance of the input stage without too much performance degradation is to tap the load of this stage, see Fig. 2. The effect of this action on the input/output impedance of the first stage is rotation of the S_{22}-parameter to the left in the Smith chart, i.e. the output impedance is lowered to values in the order of 5 – 10Ω. As a consequence, also the input impedance is lowered, but this effect can be compensated by adding a degeneration inductance of 700pH in the emitter branch of the first stage and adding a shunt capacitor C1 at the collector node. The tapped inductor can be realized as a 9nH symmetrical inductor with the center tap as output node. Capacitor C2 is added as the two stages need to be biased independently. To flatten the gain a little, a capacitive feedback by means of C3 has been added. The supply voltage for the input stage is 1.8V and for the output stage 3V.

Additional aspects as ESD protection circuitry, I/O bond pad and packaging will influence the final design parameter choices. To minimize layout parasitic effects, routing has been performed in the thick top metal layer only. High quality MiM capacitors have been used. The implemented inductors have Q-factor above 15 at 2GHz.

IV. MEASUREMENTS

The LNA has been realized in a 0.25μm SiGe:C BiCMOS technology. The die size measures 1.1mm x 1.3mm (see Fig. 3). The LNA has been packaged in a HVSON10 and mounted on a PCB (Fig. 4) on which the measurements have been performed.

The input stage consumes 16mA from 1.8V voltage supply. The output stage is biased at 54mA from a 3V voltage supply leading to a total power consumption of 190mW.

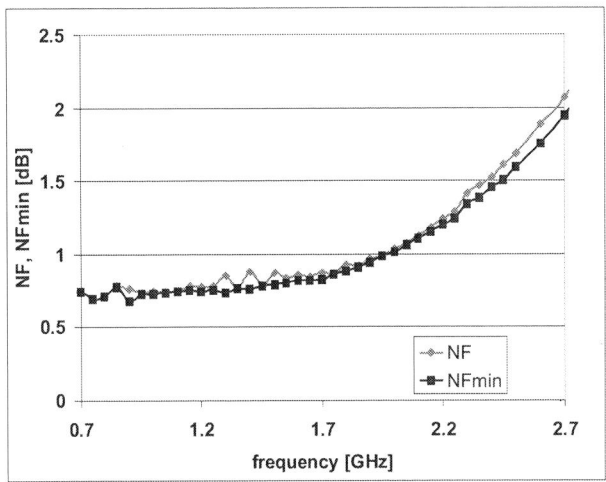

Figure 5. Measured NF and NF$_{min}$

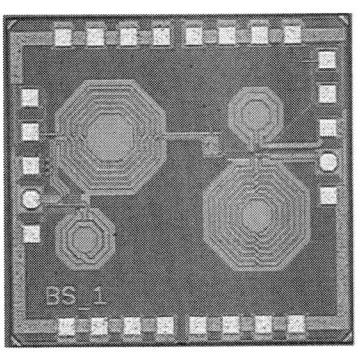

Figure 3. Die micro photo of LNA (left) and demo board (right) with RF input bottom, RF output top.

The measured NF and NF$_{min}$ are shown in Fig. 5 indicating that an almost optimal noise match at the input has been obtained. The NF is 0.75B in the 900MHz GSM band and 0.9dB in the 1800MHz band. The NF remains below 1dB up to 2GHz. Meanwhile also a very good input impedance match is obtained as can be seen from the measured input return loss, i.e. S$_{11}$ (Fig. 6). The input return loss is better than 18dB at the GSM frequencies. The input impedance match is better than -10dB between 750MHz and 2.4GHz indicating wide band impedance matching. The output return loss is better than 7dB in the 900MHz and above 10dB in the 1800MHz band.

The measured transducer gain, i.e. S$_{21}$ is shown in Fig. 7. Measurements indicate a +34dB gain in the 900MHz band and +24dB gain the 1800MHz band. More important than the absolute gain value is the gain flatness in a band of 80MHz. For the 900MHz band the flatness is less than 1dB and in the 1800MHz it is even less than 0.6dB. The measured isolation between RF input and RF output is better than -30dB. The current design is unconditional stable from 600MHz onwards, but this can be improved by adding on-chip traps at the supply nodes.

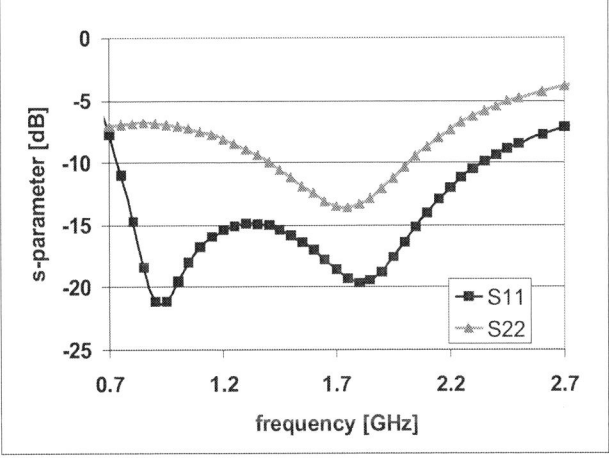

Figure 6. Measured input and output return loss.

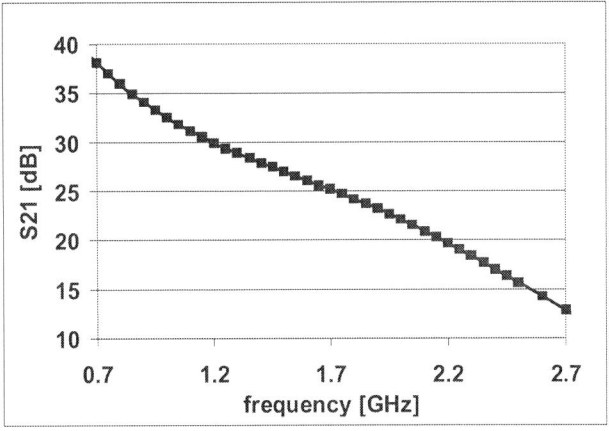

Figure 7. Measured transducer gain (S$_{21}$).

978-1-4244-6240-7/10 $26.00 © 2010 IEEE

Finally linearity measurements have been performed with two tones centered in the 900MHz band and spaced 80MHz (see Fig. 8).The measurements reveal an OIP3 of +36dBm and a corresponding IIP3 of +10dBm derived from the extrapolation of the IM3 and gain curve. One can also calculate the OIP3 for each input power separately (Fig. 9). The measured data reveals an average OIP3 of +35dBm and indicates a 1-dB input compression point of -17dBm or equivalent an 1-dB OCP of +19dBm. Linearity test with tones at 1800MHz show similar performance values.

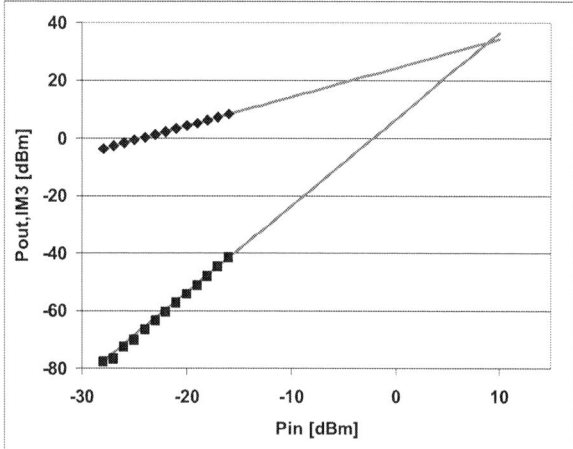

Figure 8. Measured Pout and IM3 power for swept input power.

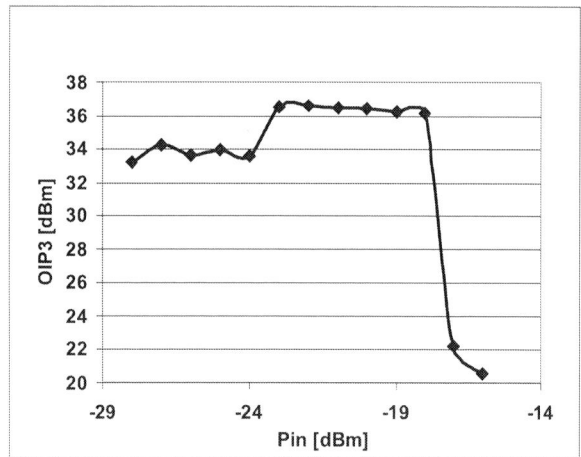

Figure 9. OIP3 for swept input power.

The ESD protection circuitry can handle 1kV ESD Human Body Model discharges.

V. COMPARISON & CONCLUSION

In Table 1 the proposed LNA is compared to state-of-the-art. The compound technologies LNAs have a better NF, albeit for non-50Ω match and without ESD protection. The discussed design out performs other Si-based LNA in terms of noise and linearity at the cost of higher power dissipation. The Si-based two-stage LNA demonstrates that sub 1-dB noise figures combined with high linearity performance is possible. The performance is achieved on a packed sample and where ESD protection circuitry is included. This LNA can trigger higher integration levels in base station receivers.

Table 1: Comparison with state-of-the-art.

	[1]	[2]	[4]	[5]	[7]	This Work**
Frequency [GHz]	0.9	0.9	2	0.9	0.9	0.9
NF [dB]	0.53*	0.6*	0.53*	1.35	1.5	0.75
1-dB OCP [dBm]	+18	+18	+19	+8	0	+19
OIP3 [dBm]	+33	+34	+41	+22	+28	+36
Gain [dB]	18	17	+30	13	16	34
Pdiss [mW]	216	325	465	38	23	190
technology	0.5μm GaAs	0.5μm GaAs	0.5μm GaAs	0.25μm BiCMOS	0.5μm BiCMOS	0.25μm BiCMOS

*non-50Ω match; **includes ESD protection

REFERENCES

[1] MGA-631P8, Low Noise, High Linearity, Active Bias Low Noise Amplifier, Data Sheet, Avago Technologies.
[2] SKY65037, Low Noise Amplifier, Data Sheet, Skyworks
[3] ATF-58143, Low Noise Enhancement Mode Pseudomorphic HEMT in a Surface Mount Plastic Package, Data Sheet, Avago Technologies
[4] T. Chong et al. ,'Design and performance of a 1.6-2.2GHz Low-noise high gain dual amplifier in GaAs E-pHEMT,' proc. APMC pp. 1-4, 2005
[5] O. Boric-Lucbeke et al., 'Si-MMIC BiCMOS Low-noise high linearity amplifiers for base-station applications', proc. Asia-Pacific Microwave Conf. pp.181-184, 2000
[6] T. Tikka, et al., 'Multiband receiver for base-station applications', proc. RWS, pp871-874, 2008
[7] V. Aparin, et al., 'Highly linear SiGe BiCMOS LNA and Mixer for cellular CDMA/AMPS applications', proc. RFIC, pp129-132, 2002.
[8] L. Belostotski, J.W. Haslett, 'Sub-0.2dB Noise Figure Wideband Room-Temperature CMOS LNA With non-50 Ohm Signal Source Impedance', IEEE Solid-State Circuits, vol. 42, pp.2492-2502, 2007
[9] P. Deixler, et al., 'QuBIC4X: an An fT/fmax=130/140GHz SiGe:C-BiCMOS manufacturing technology with elite passives for emerging microwave applications,' proc. BCTM pp.233-236, 2004

A 4.35-mW +22-dBm IIP3 Continuously Tunable Channel Select Filter for WLAN/WiMax Receivers in 90-nm CMOS

Mostafa Savadi Oskooei[1, 2], Nasser Masoumi[2], Mahmud Kamarei[2], and Henrik Sjöland[1]

[1]Department of Electrical and Information Technology, Lund University, SE-221 00 Lund, Sweden

[2]School of Electrical and Computer Engineering, University of Tehran, 14395-515 Tehran, Iran

Abstract — **A low-power high linearity CMOS G_m-C channel select filter for WLAN/WiMax receivers in 90-nm CMOS technology is presented. To reduce power consumption a biquad cell with simple architecture and few devices is used. A simple but efficient technique is also used to improve the linearity of the filter without increasing its power consumption. The cutoff frequency of the sixth order Butterworth low-pass filter can be tuned from 8.1 to 13.5 MHz for WLAN and WiMax applications. The measurement results show an in-band IIP3 of +22 dBm and an input referred noise of 75 nV/√Hz at a power consumption of 4.35 mW from a 1-V supply. The differential filter occupies a chip area of 0.239 mm² excluding pads.**

Index Terms — **Channel select filter, CMOS, G_m-C filter, linear, low-power, WLAN, WiMax.**

I. INTRODUCTION

To support new services and have a high speed and global connection, researchers have in recent years focused on implementing multi-standard transceivers. A highly integrated multi-standard transceiver can be achieved by reconfigurable building blocks. The major challenge is then to design tunable blocks with maximum hardware sharing, while at the same time satisfying the very different specifications of various standards [1].

Many researchers expect the combination of WiMax (IEEE 802.16) and WLAN (IEEE 802.11x) to provide a wireless solution for delivering high speed internet access to businesses, homes, and hotspots [2]. Both WLAN and WiMax radios can be implemented using a multi-standard radio with tunable building blocks. This paper presents a channel select filter satisfying the requirements of both WLAN and WiMax. Since low power consumption is crucial for portable devices, we implement the low-pass filter based on the G_m-C structure rather than the active-RC counterpart [3]. A simple structure is proposed to reduce the power consumption and improve the low linearity of G_m-C filters. Moreover, a simple but efficient technique with no power consumption budget is presented to further improve the linearity of the filter.

This paper organized as follows: In section 2 the structure and circuit of the proposed G_m-C low-pass filter is presented. Section 3 introduces the technique to improve the linearity. Section 4 presents the cutoff frequency tuning method. The measurement results are given in section 5, and finally conclusions are drawn in section 6.

II. FILTER STRUCTURE

A simple block diagram of a second-order G_m-C filter (biquad) is shown in Fig. 1. It provides low-pass and band-pass outputs as indicated, respectively, by LP and BP in the figure. When non-ideal transconductance cells are considered, the effects of non-dominant poles on the stability, frequency response, and quality factor of filter poles must be taken into account. To reduce the effects, the phase margin and unity gain bandwidth of the transconductance cells should be kept high, resulting in increased power consumption [3]-[4]. In order to alleviate the problems caused by non-dominant poles, the transconductors should be designed as simple as possible. The simplest transconductance cell is a single MOS transistor. It offers the maximum achievable bandwidth. Figure 2 shows the proposed complete circuit for implementing a differential version of the block diagram in Fig. 1. The input differential pair (M_1-M_2) and the feedback differential pair (M_3-M_4) implement the input transconductor (G_{m1}) and feedback transconductor (G_{m4}), respectively. Both the input and feedback differential pairs convert their input voltages to currents, which are injected to the BPP and BPN nodes. The negative sign of G_{m4} (i.e. $-G_{m4}$) in the block diagram of Fig. 1 is realized by cross connecting the differential input signals. In the circuit of Fig. 2, the functions of both G_{m2} and G_{m3} cells are simultaneously implemented by M_5-M_6. The transistors M_5 and M_6 are in a common-gate configuration and provide the low resistance of $1/g_{m5}$ and $1/g_{m6}$ to the BPN and BPP nodes, analogous to $1/G_{m2}$ in Fig. 1. M_5 and M_6 also drain the current of $g_{m5}V_{BPN}$ and $g_{m6}V_{BPP}$ from the BPN and BPP nodes and buffer those to the output nodes, performing the function of G_{m3} in Fig. 1. By using this technique one transconductor is eliminated, saving power and circuit complexity. With the same common-gate stage

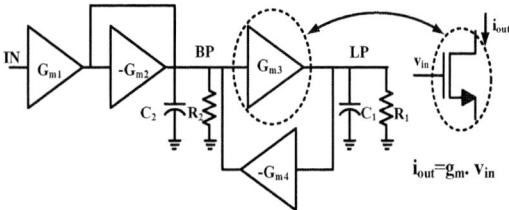

Fig. 1. Block diagram of a G_m-C low-pass biquad.

implementing two transconductors, G_{m2} equals G_{m3}. The proposed circuit reduces four transconductors in Fig. 1 to a simple circuit, and the power consumption and linearity are therefore improved significantly. Letting g_{mi} and g_{mf} represent the transconductance of the input and feedback devices, the pole frequency, quality factor, and DC voltage gain become

$$\omega_0^2 \approx \frac{g_{mf} g_{m6}}{C_1 C_2} \qquad (1)$$

$$Q \approx \sqrt{\frac{g_{mf}}{g_{m6}} \frac{C_2}{C_1}} \qquad (2)$$

$$A_{dc} \approx \frac{g_{mi}}{g_{mf}} \qquad (3)$$

The output common-mode voltage is extracted from the source of the feedback differential pair and transistors M_{21}, M_{22}, M_{15}, and M_{16} form a common-mode feedback circuit.

III. LINEARIZATION TECHNIQUE

As can be seen in Fig. 2, the biquad circuit consists of two differential pairs (input and feedback), a common-gate pair (M_5-M_6), and a source follower pair (M_{17}-M_{18}). Note that the common-gate and the source follower stages are located in the forward path of feedback loop. Therefore, the distortion introduced by these is reduced. The differential pairs on the other hand are not located in the feedback loop, and their nonlinearity is thus not reduced by the feedback. The proposed technique is based on the locating the differential pairs inside the feedback loop, and hence also reducing the input voltage swing of the differential pairs. The circuit of the new biquad is shown in Fig. 3. It is quite similar to Fig. 2, except the connection of the input and feedback differential pairs. The pole frequency and the quality factor are still given by (1) and (2), but the low frequency voltage gain becomes

$$A_{dc} = \frac{2\alpha}{1+2\alpha} \ , \ \alpha = \frac{g_{mi} g_{mf}}{g_{mi} + g_{mf}} R_{out} \qquad (4)$$

where R_{out} represents the resistance seen from OUTP or OUTN to signal ground. The input transconductance is $g_{mi} = g_{m2} = g_{m3}$ and the feedback transconductance is $g_{mf} = g_{m1} = g_{m4}$. The DC voltage gain becomes close to

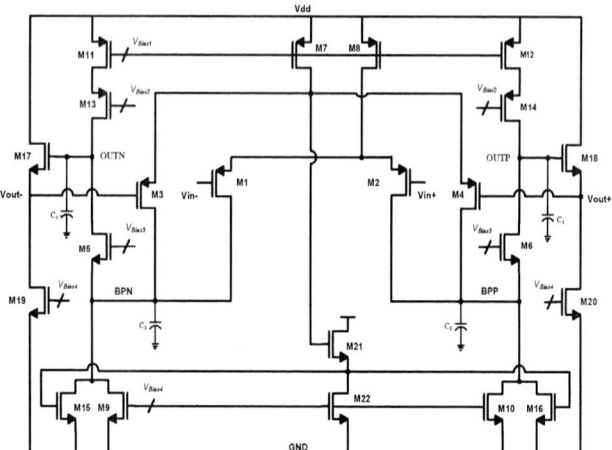

Fig. 2. Biquad circuit.

unity, because α is large. Then, if Vin+ is increased by ΔV and Vin- decreased by ΔV, Vout+ will increase by close to ΔV and Vout- decrease by close to ΔV, so the differential input voltages of the differential pairs M_1-M_2 and M_3-M_4 become very small. Subsequently, the nonlinearity is reduced significantly. Note that in this case the source voltage of differential pair M_3-M_4 increases by ΔV and the source voltage of differential pair M_1-M_2 decreases by ΔV. Therefore, for large signals, the current of M_{19} is increased and the current of M_{20} is reduced. The transconductances of M_3-M_4 are then decreased, while the transconductances of M_1-M_2 are increased. However, according to (4) the gain doesn't change, so the linearity is not significantly degraded. Furthermore, the current reduction of M_3-M_4 is compensated by the current increment of M_1-M_2, preventing output common-mode voltage changes. However, enough head room must be provided for M_{19} and M_{20}. This is accomplished by using source followers (M_{16}-M_{18} and M_{15}-M_{17}) at the outputs. At high frequencies, due to the gain variation and phase shift of the output voltage, the differential input voltage of the differential pairs increase and the linearity improvement is reduced compared to low frequencies. To increase the linearity, differential mode source degeneration resistors were therefore added to the differential pairs. The common-mode feedback circuit (not shown in Fig. 3) controls the output common-mode voltage through the M_{13} and M_{14} current sources.

IV. CUTOFF FREQUENCY TUNING

The tuning of cutoff frequency is accomplished by a combination of coarse and fine tuning. The coarse tuning is performed by switching capacitances to set the bandwidth to roughly 10 or 14 MHz. Then, g_m tuning is used for fine tuning to compensate PVT variations and aging. As equations (1) and (2) indicate, if capacitors C_1

978-1-4244-6240-7/10 $26.00 © 2010 IEEE

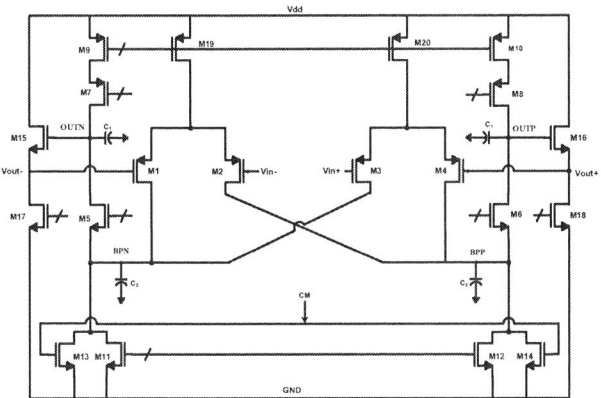

Fig. 3. Schematic of linear biquad.

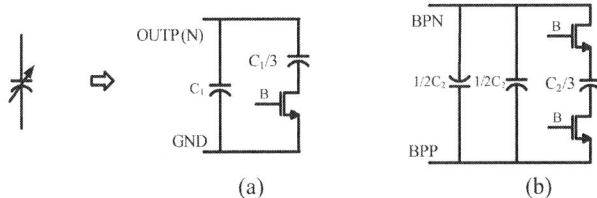

(a)   (b)

Fig. 4. Capacitor array (a) capacitor C_1 and (b) capacitor C_2.

Fig. 5. Chip microphotograph of six-order low-pass filter.

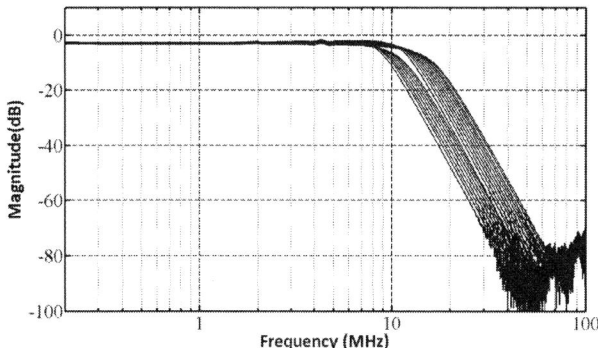

Fig. 6. Frequency tuning for the LPF.

and C_2 are changed to KC_1 and KC_2, the cutoff frequency changes from f_0 to $1/Kf_0$, while the shape of the filter frequency response is maintained. The switching is performed by NMOS transistors located in series with the capacitors; see Fig. 4(a). To reduce the area the capacitors C_2 are connected in differential mode; see Fig. 4(b). To increase the stability of the common-mode feedback loop this technique was not used for C_1. The bandwidth of the filter is 10.5 MHz when all capacitors are connected; therefore 14 MHz bandwidth can be obtained if the switchable part of capacitors C_1 and C_2 (one third of constant part) are disconnected. These values are close to the bandwidths required. The fine tuning is performed by changing the transconductances of the differential pairs and g_{m6}. This can be achieved by changing the reference current in the bias circuit. Then the bias current of the transconductors, which are copies of the reference current (different coefficients), will be altered resulting in a scaling of all transconductances.

V. EXPERIMENTAL RESULTS

The filter was fabricated in a 90-nm CMOS technology. Three stages were cascaded to obtain a sixth-order Butterworth low-pass filter. Figure 5 shows the chip microphotograph. The filter occupies 0.239 mm² (without pads). To achieve the proper shape of filter, pole capacitor and degeneration resistor matching within and among stages is ensured by making a symmetrical and common-

centroid layout. Figure 6 shows the bandwidth tuning capability of the filter. The cutoff frequency can be changed within 8.1-13.5 MHz while the shape of filter is well maintained. Figure 7 shows the result of a two-tone test yielding an in-band IIP3 of +22 dBm (voltage related to 50 Ω) for 1.5 and 2 MHz tones. Figure 8 displays the out-of-band IIP3 and IIP2 of the filter for a bandwidth of 10 MHz, where the out-of-band tones were applied at 20, 32 MHz and 20, 28 MHz for IIP3 and IIP2 measurement, respectively. The fundamental power is set to input tone power multiplied by measured in-band gain. As can be seen out-of-band IIP3 is +19 dBm. The out-of-band IIP2 is high (+54 dBm) thanks to the fully differential architecture. The 1-dB compression point in the presence of an interferer at 50 MHz frequency was also measured. The 1-dB compression point of the filter gain occurred at about +8 dBm interferer amplitude. A single tone test with $f_{in} \approx f_{-3dB}/3$ was also used to measure the total harmonic (THD). The input differential voltage was a sinusoidal at 3 MHz ($\approx 1/3\ f_{-3dB}$) and 0.47 V_P (3.4 dBm over 50 Ω). The third-order harmonic distortion (HD3) was then below -40 dB. The differential input referred noise spectral density is equal to 75 nV/$\sqrt{\text{Hz}}$. Integrated from 100 kHz to 10 MHz this gives 238 μV_{rms}. Therefore, the dynamic range (1% THD) is 63 dB. The total current drawn from a 1-V supply was 4.35 mA. Table I compares the performance of the filter to that of recently works.

VII. CONCLUSIONS

A sixth-order channel select low-pass filter with bandwidth tuning capability for WLAN and WiMax standards was fabricated in 90-nm CMOS technology. A new simple biquad architecture along with a linearization technique is introduced. Furthermore, coarse and fine bandwidth tuning are implemented. The bandwidth can thereby be tuned while maintaining the filter shape. The measurement results indicate that the proposed filter consumes low power, and has sufficient performance for WLAN and WiMax applications.

ACKNOWLEDGEMENTS

This work was supported by the Swedish Agency for Innovation Systems (VINNOVA) Industrial Excellence Center in System Design on Silicon and Iran Telecommunication Research Center (ITRC).

REFERENCES

[1] A. Baschirotto, F. Campi, R. Castello, G. Cesura, R. Guerrieri, L. Lavagno, A. Lodi, P. Malcovati, and M. Toma, "Baseband analog front-end and digital back-end for reconfigurable multi-standard terminals," *IEEE Circuits and Systems Magazine,* vol. 6, no. 1, pp. 8–28, 2006.

[2] L. Lin, N. Wongkomet, D. Yu, C. Lin, M. He, B. Nissim, S. Lyuee, P. l. Yu, T. Sepke, S. Shekarchian, L. Tee, P. Muller, J. Tam, and T. Cho, "A fully integrated 2×2 MIMO dual-band dual-mode direct-conversion CMOS transceiver for WiMAX/WLAN applications," in *Proc. IEEE Solid-State Circuits Conf.*, pp. 416-417, February 2009.

[3] H. Amir-Aslanzadeh, E. J. Pankratz, and E. Sánchez-Sinencio, "A 1-V +31 dBm IIP3, reconfigurable, continuously tunable, power-adjustable active-RC LPF," *IEEE J. Solid-State Circuits,* vol. 44, no. 2, pp. 495–508, February 2009.

[4] K. Hadidi, K. Eguchi, T. Matsumoto, "A 430MHz, -52dB, single transconductor, 3rd-Order low-pass filter and its extension to a 5th-Order, in a 0.5μm CMOS process," in *Proc. IEEE European Solid-State Circuits Conf.*, pp. 390-393, September 1999.

[5] T. Lo, C. Hung, and M. Ismail, "A wide tuning range G_m–C filter for multi-mode CMOS direct-conversion wireless receivers," *IEEE J. Solid-State Circuits,* vol. 44, no. 9, pp. 2515–2524, September 2009.

[6] A. Vasilopoulos, G. Vitzilaios, G. Theodoratos, and Y. Papananos, "A low-power wideband reconfigurable integrated active-RC filter with 73dB SFDR," *IEEE J. Solid-State Circuits*, vol. 41, pp. 1997–2008, September 2006.

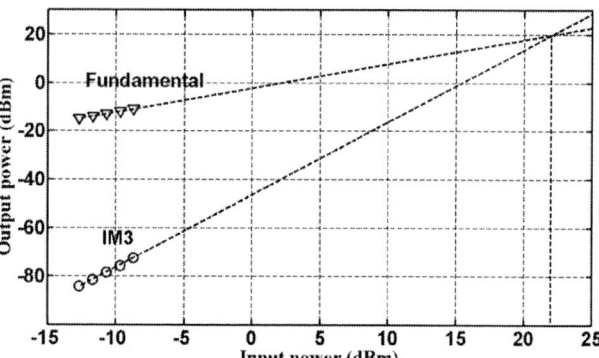
Fig. 7. In-band IIP3 measurement.

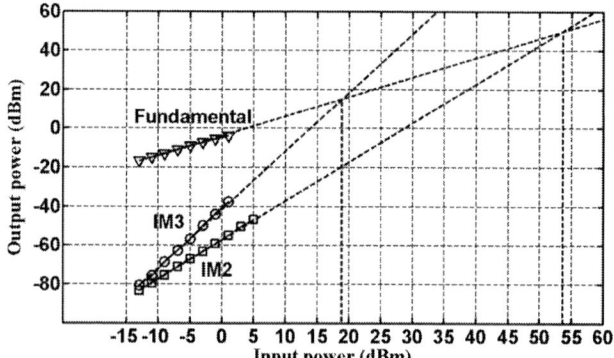
Fig. 8. Out-of-band IIP3 and IIP2 measurements.

TABLE I
CHARACTERISTICS OF FILTER AND COMPARISON

Parameters	This Work	[5]	[6]
Technology (μm)	0.09	0.18	0.12
Topology	G_m-C	G_m-C	Active-RC
Supply voltage (V)	1.0	1.2	1.0
Order	6	3	3,5
Type	Butterworth	Butterworth	Chebyshev Elliptic
f_{-3dB} (MHz)	8.1-13.5	0.5-20	5,10
Continuous Tuning	Yes	No	No
Power (mW)	4.35	4.1-11.1	4.6
IIP3 (dBm)	21.7-22.1	19-22.3	18.8-21.3
Out-of-Band IIP3 (dBm)	17.5-18.9	13-17.5	-
Out-of-Band IIP2 (dBm)	51.9-53.4	30.8-40	-
IRN (nV/√Hz)	75	12-425	85,143
Area (mm²)	0.239	0.23	0.17

978-1-4244-6240-7/10 $26.00 © 2010 IEEE

RTUIF-07

Wideband Trans-Impedance Filter Low Noise Amplifier

Mikko Kaltiokallio*, Aarno Pärssinen** and Jussi Ryynänen*

*SMARAD2/Department of Micro and Nanosciences, Aalto University, Espoo, Finland
**Nokia Research Center, Otaniemi Lablet, Espoo, Finland

Abstract — This paper focuses on the design of wideband low-noise amplifier, which includes transferred-impedance structures to improve interference tolerance. The LNA is implemented as part of simple RF receiver to demonstrate the feasibility of the transferred-impedance circuits in wideband receivers. The LNA itself achieves a gain of 24 and 20 dB, noise figure of 3.4 and 4.9 dB with ICP of -21 and -15 dBm for the interference blocking structure turned off and on, respectively. Added selectivity of 6 dB is achieved by using the structure described in this paper.

Index Terms — Radio receivers, broadband amplifiers, tunable amplifiers, adaptive filters, impedance transformation, passive mixer.

I. INTRODUCTION

With the development of modern cellular systems the radio frequency space allocated for such systems has become more and more congested over the last decade. The problem has arisen partly because the system hardware has been designed for fixed frequencies, i.e. the communication system standards have set strict operating boundaries for the radio systems up to date. The recent research effort has been focused on countering this dilemma. Concepts such as the software defined radio (SDR) and cognitive radio (CR) have been proposed to tackle the frequency congestion and make for more efficient and flexible use of this limited frequency resource. In radio hardware these requirements correspond to the need of wideband interference tolerant RF front-ends, wideband fast synthesizers and spectrum detection methods [1]-[3]

The need for a wideband interference tolerant RF front-end is due to the fact that highly selective pre-select filters cannot be used in large numbers because they are bulky and expensive off-chip components. While in traditional radio systems the receiver is preceded by a pre-select filter to suppress strong interferers and therefore relax the receiver linearity and desensitizing requirements, the wideband front-end in SDR or CR radio has to have other ways to cope with the hostile interferer environment. One way is to make the receiver path hostile to certain frequencies by utilizing a notch filter at the front-end. While this is effective for a limited number of problematic frequencies the problem with this solution is that the frequencies of the large blockers has be to known a priori.

Another solution is to make the receiver path as selective as possible outside reception frequency by utilizing structures that behave as LC-resonators. This work is focused on using such structures in a wideband receiver. The structure called transferred impedance filter (TIF) relies on a passive mixer and capacitors to implement a selective impedance which can be tuned by adjusting the passive mixer local oscillator (LO) frequency. With this technique very wide tuning range can be achieved which makes it very suitable for SDR and CR like receivers. Such concept has been demonstrated for a relatively narrowband solution in [4]-[6]. The interferer suppression and the effect on noise figure (NF) for a wideband receiver has been investigated in this paper. Other issues covered are the LO leakage to the antenna and LNA design with this type of structure.

The paper is divided as follows, in section II the receiver topology is presented. In section III the LNA amplifier and TIF core designs are studied. Measurement results and conclusion are presented in sections IV and V, respectively.

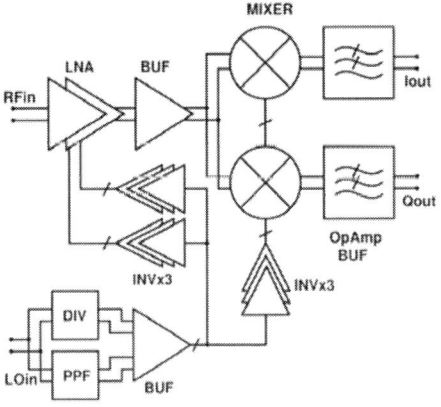

Fig. 1.   RF receiver block diagram.

II. RECEIVER TOPOLOGY

The focus of this work has been in the front-end design and thus the receiver has been designed with that in mind. The goal has been to gain knowledge on the effect of the TIF to the noise figure and selectivity. Therefore the gain partition has been selected such that the noise figure is

978-1-4244-6240-7/10 $26.00 © 2010 IEEE

determined by the LNA and TIF. This means that with the high gain in the front-end the overall receiver linearity is dominated by stages following LNA. Additionally the large-signal noise models are not accurate and thus by using high gain we can remove some uncertainty surrounding the TIF simulations.

In Fig. 1 the system block diagram is presented. The front-end has a two-stage LNA along with the TIF filters in the first stage to provide the additional selectivity. The buffer between the LNA and the passive down-conversion mixer has been designed for high transconductance to ensure good overall gain for the receiver. A simple baseband operational amplifier is used to buffer the signal for the 50-Ohm measurement equipment interface. Moreover, a complete signal path provides a realistic operating environment for the TIF LNA.

The design of LO-chain depicted in Fig. 1 is more complex than for a conventional receiver because the LO-signal has to be provided for three mixers instead of one. The frequency synthesizer design is relaxed by using a divider and a polyphase filter (PPF) for the quadrature generation. The tuning range of the synthesizer is decreased by using a divide-by-two circuit for the lower LO-frequencies (2-3.5 GHz) and a PPF for the higher frequencies (3-5 GHz). A buffer after the divider and PPF is used to combine the quadrature paths and to provide isolation between the two paths. Isolation is necessary since the passive polyphase filter passes the LO-signal while the divider is in use and can therefore corrupt the LO-signal if no isolation is provided.

For best performance the passive mixers need a full rail-to-rail LO-signal. The divider requires little buffering since the output amplitude is large, but the polyphase filter attenuates the LO-signal and thus a cascade of three inverters is used in each of the three inverter chains as shown in Fig. 1. The inverters have been designed such that they can be individually powered down if needed. This enables us to study the performance of the front-end with the TIF-filter turned on or off. Moreover, in realistic environment the filter can be turned on or off depending on the interference scenario. The inverter chains can cover a LO-signal up to 5 GHz without using additional inductors, thus the area penalty of using three separate chains is minimized.

III. TRANS-IMPEDANCE FILTER LNA

The trans-impedance filter LNA is composed of a two-stage LNA and two TIF-filters connected to the input and output of the first stage. In the next two sub-sections the different design aspects of the LNA and the TIF are covered separately.

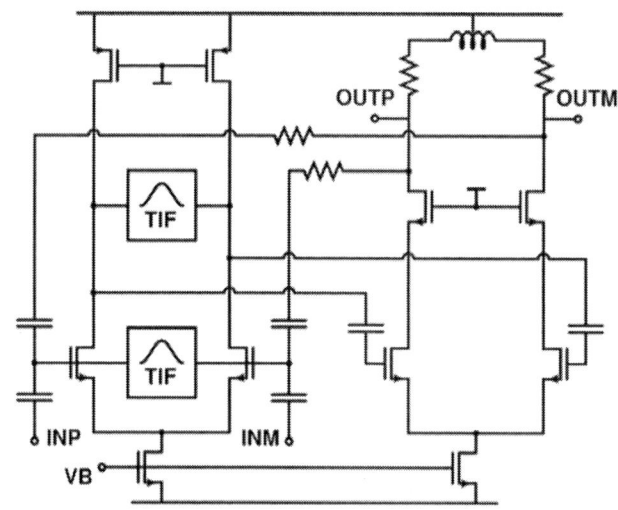

Fig. 2.   Two-stage LNA schematic.

A. LNA Core

The schematic of the two-stage LNA is presented in Fig. 2. As will be discussed in the next sub-section the load impedance has to be designed as high as possible for the TIF to have good selectivity. Thus the first stage uses a common-source (CS) amplifier stage with a high impedance PMOS load. The second stage is a CS amplifier with a cascode stage to increase the overall gain. A shunt-peek load has been used in the second stage to achieve wide bandwidth by compensating the gain response drooping in other stages. A RC-feedback from second stage output to the first stage input has been used to achieve wideband matching. The feedback paths have been cross-connected to maintain correct signal polarity. However for common-mode signals the cross connection is not visible and therefore this type of amplifier tends to oscillate. The current sources have been added to ensure that the amplifier stages have common-mode rejection and keep the TIF LNA from oscillating. As depicted in Fig. 2 the stages utilize DC-blocking capacitors to maintain independent bias levels for the two stages.

The design of a wideband LNA and especially the wideband matching is challenging. The design is made even more challenging with the need for the high impedance nodes for the TIF filters in the LNA which adds another design trade-off.

The placement of the TIF-filters was vigorously studied during the design of the receiver. For the filter to have maximum effect on linearity it should be placed as close to the input of the receiver as possible to decrease the out-of-band gain of the succeeding blocks. However, the increase in noise figure and possible LO leakage can severely degrade the receiver performance if the TIF-filter is placed too close to the input. However, due to the uncertainty of noise simulations it was seen beneficial to place one of the filters at the input so that this performance penalty could be measured. The second filter was placed at the first stage load where it is most effective. The isolation of the first CS-stage is low thus the selective load impedance is also seen at the amplifier input. When turned off, the loading caused by the TIF-filters is minimal and therefore the LNA performance degradation is small.

B. TIF Core

The I-branch of the TIF-Filter core is presented in Fig. 3. The Q-branch depicted in the figure is similar, but has the LO-signals in quadrature to the I-branch LO-signals. The core consists of a passive mixer stage and a filtering capacitor at the bottom of the figure. The input side of the filter has DC-blocking capacitors to ensure that no DC-current flows through the switches. Because LO-signal is full-scale, no DC-blocking capacitors are needed at switch gates. Large DC-feed resistors are used to provide proper switching level for the switches. In this design the DC-level was 0.6V to give a turn-off voltage of -0.6V and turn-on voltage of 0.6V.

The principle of the TIF-filter relies on the lack of reverse isolation in passive mixers. Thus the impedance seen at the baseband side of the mixer is frequency-translated around the LO-frequency and its odd harmonics. Therefore the name transferred-impedance filter. A detailed description of the frequency translation in passive mixers for narrowband system can be found in [4]. Systems utilizing impedance translation can be found in [5]-[6].

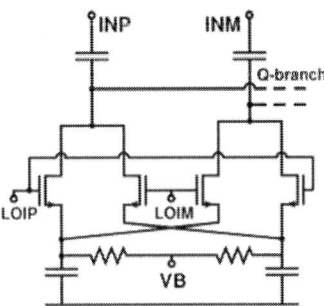

Fig. 3.   TIF core schematic.

The impedance at the input of the TIF-filter can be simply approximated as shown in (1). The overall impedance is determined as a parallel connection of the Z_{RF} and the TIF-filter impedance. Z_{RF} includes the impedances associated with the RF-circuitry i.e. the RF loads and parasitics. The TIF-filter impedance is a series

connection of the effective switch resistance R_{SW} and the frequency-transferred impedance Z_{TIF}. The Z_{TIF} is very large at frequencies around the LO-signal and very low for frequencies far from the LO-signal. Therefore the in-band impedance is set by Z_{RF} and should be as large as possible and the out-of-band impedance is determined by the R_{SW}. which should be as small as possible. The relation of these two impedances sets the limit for the TIF-filter selectivity. Obviously using a LC-resonator for RF-load will result in better selectivity since the resonator has a high Q-value and the TIF-filter performance is enhanced. However such load is narrowband and cannot be used in a wideband design. Thus a PMOS load was used in this design.

$$Z_L = Z_{RF} \| \left(R_{SW} + Z_{TIF} \right) \qquad (1)$$

IV. Measurement Results

The chip was processed on a 65-nm CMOS process and packaged in a quad-flat package together with a printed circuit board for measurements.

The measured in-band gain response of the receiver is presented in Fig. 4 where it can be seen that receiver covers a frequency range of 2-5.5 GHz. The strong gain drooping seen after 5 GHz is due to the LO-chain inverters. In Fig. 5 the input matching is shown with the TIF-filters turned off and on. The influence on matching is strongest around the LO-frequency of 4 GHz.

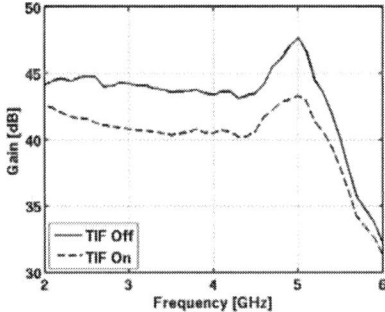

Fig. 4.   Receiver voltage gain response.

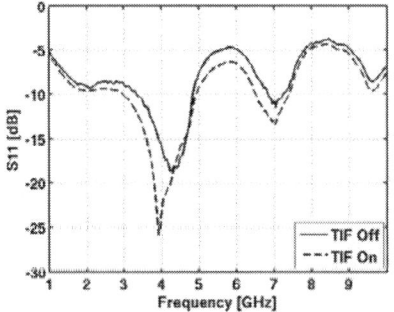

Fig. 5.   Receiver input matching (LO set to 4 GHz).

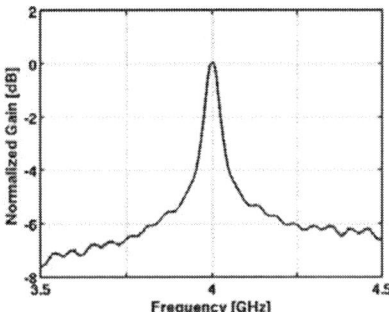

Fig. 6.   TIF relative gain response.

The selectivity of the TIF-filters are shown in Fig. 6 with LO-frequency of 4 GHz. A selectivity improvement of 6 dB is achieved at 250 MHz offset. Slight skewing of the gain response is visible which is due to the IQ cross-talk as described in [4]. The receiver input compression point for a blocker at 500 MHz offset is presented in Fig. 7. The result shows that a linearity improvement of 6 dB is achieved by turning on the TIF-filter. The chip photograph is in Fig. 8. The whole chip area is 890 x 890 um.

The measurement results summary is given in Table I. The LO leakage of the receiver with TIF turned on is -63 dBm. Additionally the LNA gain, linearity and noise have estimated from the measurement results in Table I. The LNA results show that a wideband interference blocking scheme can be implemented with small noise degradation.

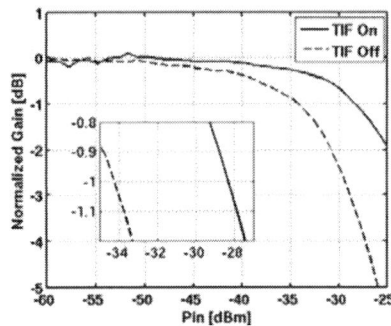

Fig. 7.   Receiver input compression point for a blocker.

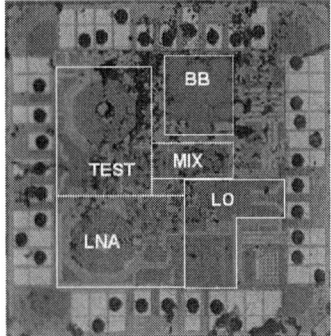

Fig. 8.   Chip photograph of receiver.

V. CONCLUSIONS

This paper described a wideband TIF LNA with a interference blocking scheme. The LNA and TIF design aspects were discussed in detail with consideration about the load impedance and switch design. This paper provides proof that it is possible to design a wideband interference blocking structure by using the impedance frequency translation of the passive mixers. Additionally the TIF connected at the input does not cause severe noise or LO leakage issues.

The measurements show that a 6 dB selectivity improvement can be achieved by using the TIF-filter at 250 MHz offset from LO-frequency. The measured noise figure difference is 1.5 dB between TIF off and on states. The measured LO leakage due to the TIF-filter is -63 dBm.

TABLE I
MEASUREMENT RESULTS

	Receiver		LNA Estimate	
TIF (Off/On)	Off	On	Off	On
Technology	65-nm CMOS		65-nm CMOS	
Area (um)	890 x 890		na	
Supply (V)	1.2		1.2	
Idc (mA)	41	46	14	19
Voltage gain (dB)	44	40	24	20
S11<-6dB (GHz)	1-5	1-6	1-5	1-6
NF (dB)	4.4	5.9	3.4	4.9
ICP (dBm)	-34	-28	-21	-15
IIP3 (dBm)	-24	-20	-12	-8
LO leak(dBm)	na	-63	na	-63

REFERENCES

[1] B. Razavi, "Challenges in the Design of Cognitive Radios," IEEE Custom Integrated Circuits Conference, pp. 391-398, Sept. 2009.

[2] T. Rapinoja, et. al., "A digital frequency synthesizer for cognitive radio spectrum sensing applications," IEEE RFIC Symp. Dig., pp. 423-426, June 2009.

[3] V. Turunen, et. al., "Implementation of Cyclostationary Feature Detector for Cognitive Radios," IEEE CROWNCOM, pp. 1-4, June 2009.

[4] A. Mirzaei, et. al., "Analysis and Optimization of Current-Driven Passive Mixers in Narrowband Direct-Conversion Receivers," IEEE Journal of Solid-State Circuits, vol. 44, no. 10, pp. 2678-2688, Oct. 2009.

[5] T. Werth, C. Schmits, and S. Heinen, "Active Feedback Interference Cancellation in RF Receiver Front-Ends," IEEE RFIC Symp. Dig., pp. 379-382, June 2009.

[6] K. Koli, J. Jussila, P. Sivonen, S. Kallioinen, A. Pärssinen, "A 900MHz Direct ΔΣ Receiver in 65nm CMOS," IEEE Solid-State Circuits Conference, Feb. 2010.

RTUIF-08

A Wideband High-Linearity Mixer in 0.5 μm InP DHBT Technology

Mark Stuenkel and Milton Feng

The University of Illinois at Urbana-Champaign, Urbana, IL, 61801, USA

Abstract — **This paper presents the design, implementation and characterization of a wideband down-conversion mixer in a 0.5 μm InP DHBT process. A modified Gilbert cell topology is implemented to improve circuit linearity and to give a means by which DHBT mixers can operate at lower supply voltages. The measured circuit has a maximum conversion gain of 11.3 dB over a 3 dB bandwidth from 7 to 24 GHz. The input referred 1-dB compression point is greater than -15.6 dB across the entire frequency range.**

Index Terms — **mixers, wideband, linearity, InP, DHBTs**

I. INTRODUCTION

Down-conversion mixers are a key component in the receiver chain of wireless systems and devices, and are increasingly being called upon to down convert signals across ever wider frequency bands. Operating over these wide frequency bands requires that these mixers be very linear so that the mixer does not become desensitized by strong interfering signals at other frequencies in the band and negatively effect the desired IF output. The Gilbert cell, an active mixer topology that is frequently used in receiver chains, has three stages: an RF transconductance stage, an LO current switching stage, and a load. In bipolar technology a typical Gilbert Cell require 3 levels of stacked devices, the mixing quad, transconductance amplifier, and DC current source, which necessitates the use of a large DC voltage supply. In CMOS designs it is commonplace to remove the current source devices, thereby reducing the required voltage supply [1]. This approach is very risky for designs in a submicron InP DHBT process, as the device's collector current is extremely sensitive to the applied base-emitter voltage. Also the process uniformity across wafer and between wafers is not as tight as that found in CMOS processes, leading to slight shifts in the V_{be} turn-on voltages which, given the sensitivity of the device, can lead to significantly different drive currents and biasing than designed.

Heterojunction Bipolar Transistors have two significant sources of nonlinearities, exponential transconductance nonlinearity and base-collector capacitance nonlinearity [2]. In the transconductance stage of a typical Gilbert cell, these nonlinearities cause the injection of a nonlinear current into the mixing stage. This nonlinear current is then modulated by the switching stage and adds to the overall nonlinear behavior of the circuit. This work

proposes a DHBT based mixer in which the transconductance stage typically found in most bipolar Gilbert cell mixer designs is eliminated. This gives a means to reduce the required supply voltage and potentially give a better nonlinearity performance, through the elimination of the G_m stage nonlinearities, while still maintaining precise control over the bias current in this sensitive technology. In this paper the authors propose an updated mixer circuit topology and show that it can be used for high frequency, >24 GHz, broadband frequency downconversion while achieving a high linearity. As this is an InP DHBT design, it is assumed that the power dissipation will be larger than in scaled CMOS.

II. BROADBAND MIXER CIRCUIT TOPOLOGY

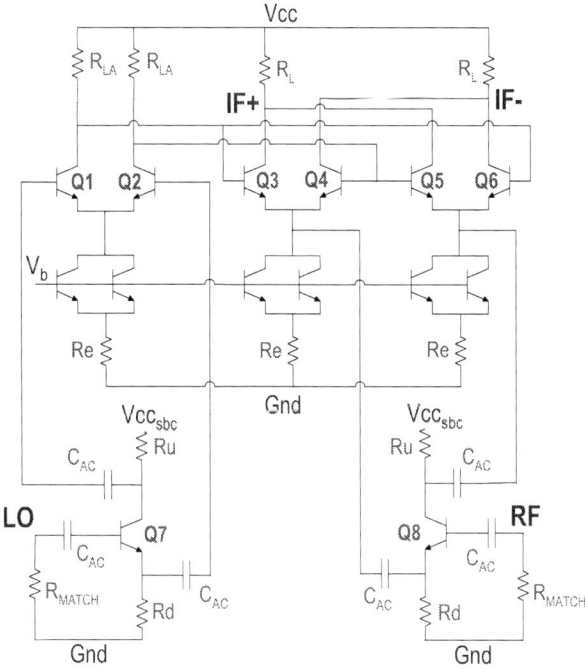

Fig. 1. Simplified schematic of the modified Gilbert cell mixer with the biasing and output buffer networks omitted.

Fig. 1 gives a simplified schematic of the proposed circuit, neglecting biasing networks and the cascode output buffer. The overall circuit consists of five major parts: the intrinsic modified Gilbert cell mixer, a differential LO amplifier, two active baluns or single-ended to differential converters, one for each the RF and

978-1-4244-6240-7/10 $26.00 © 2010 IEEE

LO signal, and a differential output buffer to couple the intrinsic IF output to the 50 Ohm output.

The intrinsic mixer is designed such that each of the devices in the mixing quad, Q3-Q6, is biased at a voltage and current point where each device has its optimum linearity across the band of interest from 5 GHz to 25 GHz. This optimum linearity point was determined from two-tone measurements taken on individual devices, and for 0.5 x 5 μm² devices corresponds to a current of 7 mA at a Vce of 1.1 V. This leads to a significant amount of power being dissipated in each device, but it is necessary to avoid the high current – low voltage effects which greatly degrade the linearity of heterojunction bipolar transistors. Differential RF signals are fed directly into the emitters of the mixing quad transistors; it is therefore important that the current sources have high impedances while being stable for both balanced and unbalance inputs. Careful choice of current source degenerating resistors, Re, is needed to prevent any unbalanced component of the RF signal, due to non-perfect single to balanced conversion, from causing the mixer to become unstable.

On chip active baluns are included, as it was found that they are a reasonably simple and area efficient way of generating relatively phase and magnitude balanced differential signals across a large bandwidth. They also eliminate the need for multiple external baluns for testing, something not readily available. The use of active baluns is detrimental to the noise performance of the overall circuit. As there were no measured device noise data or an accurate noise model for these devices, it was impossible to precisely tune the noise characteristics of the overall circuit. The differential signals from the RF and LO baluns are AC coupled to the mixer core and LO amp respectively. The DC drive current is precisely controlled in the mixer core and LO buffer amplifier through degenerated current sources. Current sources were not used on the active baluns as they led to stability and differential magnitude and phase matching issues.

The lower end of the operational 3-dB bandwidth is designed to be 5 GHz and is set by the series capacitors used to AC couple the differential RF input to the intrinsic mixer core. These series blocking capacitors can be made larger to extend the lower range of the bandwidth, but at a great cost in terms of area. As area is always limited, a lower design goal of 5 GHz was deemed sufficient. Our upper design goal was 25 GHz, for a 1:5 bandwidth.

III. Measurement Results

Fig. 2 shows a photograph of the fabricated mixer as it was laid out for on-wafer testing. The circuit is fabricated in the Teledyne 0.5 um InP Type-I DHBT process. The complete active area of the circuit is 413 μm by 390 μm, for a total intrinsic area of 0.161 mm². While the entire die area, including both the DC and RF pads, totals 0.52 mm².

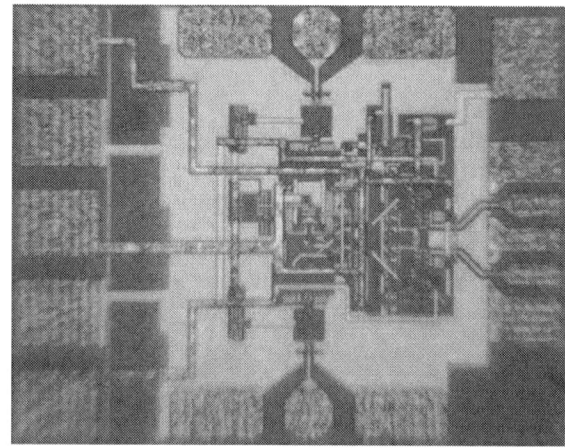

Fig. 2.   Photograph of fabricated modified Gilbert cell mixer.

On wafer measurements were carried out with the use of an Agilent E8364A PNA and an HP 83651 RF sweeper for the LO and RF signals. An Agilent 8565E spectrum analyzer and an Agilent 8975A noise figure analyzer were used for the gain and noise analysis. The fabricated circuit has a differential output, but due to measurement limitations only a single-ended output was measured. One of the differential outputs is externally DC blocked and terminated in the characteristic 50 output match. Extrinsic DC blocking capacitors and the single-ended 50 Ohm termination are the only external components used in the test setup.

As a III-V HBT design, the power supply voltage is required to be significantly larger than a comparable CMOS design. With the switching devices operating at close to optimal linearity, the circuit draws a significant amount of current which leads to a large DC power dissipation. In this instance a dissipation of 146 mW from a 5.2 V supply.

Figure 3 displays the measurements results for the smoothed and raw conversion gain measurements (blue and red respectively), and the 1-dB gain compression linearity measurement (black) across the full measurement range from 1 GHz to 50 GHz. The conversion gain curve given in fig. 3 is for an available LO power of -22 dBm, and an RF power low enough so as the mixer is not compressed. Both the conversion gain and gain compression measurements were measured for a constant IF frequency of 100 MHz with a high side LO ($f_{LO} - f_{RF} = f_{IF}$). A maximum conversion gain of 11.3 dB is measured at a frequency of 12 GHz, with a 3-dB-bandwidth extending from 7 to 24 GHz (designed for 5 to 25 GHz). At an applied LO power of -22 dBm, the circuit under test has a positive conversion gain up to a frequency of 34 GHz. The input referred 1-dB compression point, $P_{1dB,in}$, is greater than -15.6 dBm over the entire measurement range.

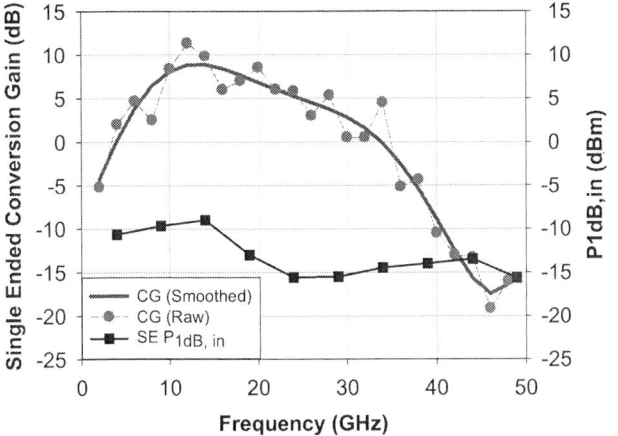

Fig. 3. Measured conversion gain and input referred 1-dB compression point over frequency for an IF of 100 MHz.

The noise figure of the entire circuit under test is shown in fig. 4. It is important to note that this is the total noise figure of all the components comprising the circuit under test and not solely the intrinsic mixer. Fig. 4 shows the single-ended double-sideband noise figure of the circuit under test for a constant IF of 100 MHz, and for an LO frequency that varies from 10 to 26 GHz at an LO power

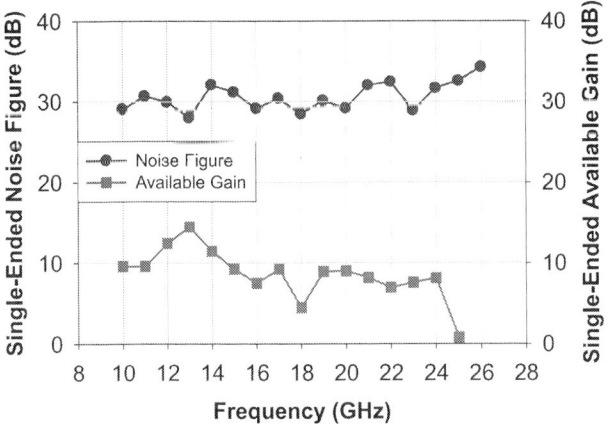

Fig. 4. Measured noise figure and available gain for the entire circuit under test across frequency for an input LO power of -28 dBm.

of -28 dBm. The measured noise figure of the circuit under test varies around 30 dB across the entire range of operations, increasing at higher frequencies as the gain is reduced. These results are higher than other wideband mixers reported in literature. However, as is shown in fig. 5, the active baluns used at both the RF and LO input have a single-ended noise figure of 13 dB, while having a gain of less than 0 dB. These lossy and high noise input stages at both inputs greatly degrade the overall noise figure of the system. As the mixer is a three port system that has a mixture of uncorrelated and correlated noise sources due to the multiple ports and the single-ended to differential conversions, an extracted noise figure of the intrinsic mixer is difficult of obtain, and is not presented in table I

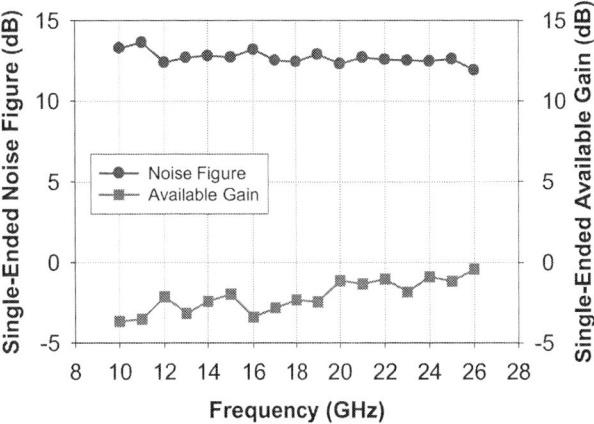

Fig. 5. Measured noise figure and available gain for an individual balun across frequency.

so as to not cause contention. From the balun noise figure and available gain reported in fig. 5, the authors estimate the intrinsic mixer has a noise figure several decibels lower than the data given in fig. 4, especially at high LO power levels, which places this circuit more in line with previously reported circuits [1], [3]-[8]. Further investigation of this circuit with external low loss baluns is needed to precisely determine the noise figure of the intrinsic mixer.

Figure 6 displays the effect that the LO power has on the noise figure, available gain and conversion gain at an LO frequency of 24 GHz and IF frequency of 100 MHz. The results are as expected, that the noise figure decreases and the gains increase as that LO power is increased. The measured available gain is significantly higher than the conversion gain, indicating that a more ideal output match at the IF frequency could yield a larger conversion gain.

978-1-4244-6240-7/10 $26.00 © 2010 IEEE

TABLE I
PERFORMANCE SUMMARY

Work	Date	Technology	Freq (GHz)	CGmax (dB)	Pin,1dB (dBm)	Nfmin (dB)	DC Current (mA)	Vsupply (V)
[3]	2005	GaAs HBT	24	4	-11	x	4	3
[4]	2006	0.35um BiCMOS	3.5~14.5	15	-19	~13	12	5
[5]	2006	0.18um CMOS	5.25	8.9	-16.7	24	7.3	0.9
[1]	2007	0.13um CMOS	3~7	8.2	x	9.6	4.8	1.2
[6]	2007	0.18um CMOS	0.5~7.5	5.7	-16	15	0.63	0.77
[7]	2008	GaAs HBT	1.5~14	20	-17	x	6.8	2.4
[8]	2009	0.18um CMOS	2~11	6.9	-3.5	15.5 (SSB)	14.3	1.8
This Work	2010	InP HBT	7~24	11.3	-13	25.6*	27	5.2

*overall noise figure of active baluns, mixer and output buffer

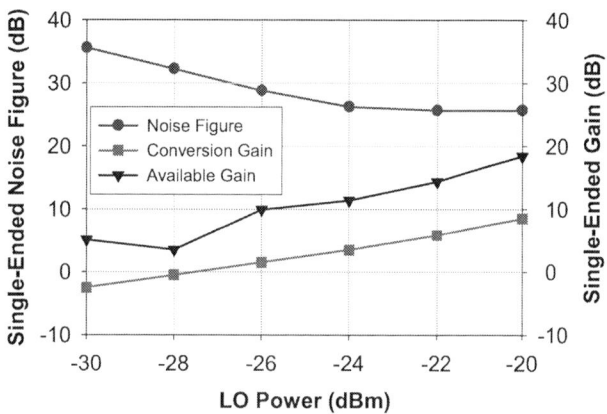

Fig. 6. Measured noise figure, available gain and conversion gain versus LO Power for an LO frequency of 24 GHz and an IF frequency of 100 MHz.

IV. CONCLUSION

In this paper, a high linearity wideband mixer based on a modified Gilbert cell topology is designed and fabricated in the Teledyne InP DHBT process. The traditional Gilbert cell is modified in that it does not have a transconductance stage to inject nonlinear current into the mixing quad. Eliminating the transconductance stage removes the nonlinearities associated with this stage from the overall mixer performance, leading to a higher input-referred 1 dB compression point than is observed in many other state of the art mixers [4]-[7]. Even without a transconductance stage, the designed mixer is able to achieve a maximum conversion gain of 11.3 dB over a 3 dB bandwidth from 7 GHz to 24 GHz.

ACKNOWLEDGEMENT

This work was completed under contracts from the DARPA TFAST program and SPAWAR (N66001-06-1-2004), and the authors wish to thank program manager Dr. Sanjay Raman (DARPA-MTO), and contract managers Dr. Don Mullin (SPAWAR) and Dr. Cindy Hanson

(SPAWAR) for their support. The authors are grateful to Mr. Richard Elder, Dr. Richard Chan and Mr. Frank Stroili of BAE Systems, for their enlightening conversations and output buffer design, and to Dr. Miguel Urteaga from Teledyne Scientific & Imaging for the InP DHBT Foundry Process.

REFERENCES

[1] K. Ckoi, D. H. Shin, and C. P. Yue, "A 1.2-V, 5.8 mW, Ultra-Wideband Folded Mixer," in IEEE RFIC Symposium Digest, pp. 489–492, June 2007.

[2] M. Iwamoto, P. M. Asbeck, T. S. Low, C. P. Hutchinson, J. B. Scott, A. Cognata, X. Qin, L. H. Camnitz, and D. C. D'Avanzo, "Linearity characteristics of GaAs HBT's and the influence of collector design," IEEE Trans. Microwave Theory Tech., vol. 48, pp. 2377–2388, 2000.

[3] M. Huber, S. von der Mark, and G. Boeck, "Ultra low power 24 GHz HBT mixer," in IEEE International Conference on Microwave and Optoelectronics Symposium Digest, pp. 24-27, July 2005.

[4] S.-C. Tseng, C. C. Meng, C.-H. Chang, C.-K. Wu, and G.-W. Huang, "Monolithic broadband Gilbert micromixer with an integrated Marchand balun using standard silicon ic process," IEEE Trans. Microw. Theory Tech., vol. 54, no. 12, pp. 4362–4371, Dec. 2006.

[5] M.-F. Hung, C.-J. Kuo, and S.-Y. Lee, "A 5.25-GHz CMOS folded-cascode even-harmonic mixer for low-voltage applications," IEEE Trans. Microw. Theory Tech., vol. 54, no. 2, pp. 660–669, Feb. 2006.

[6] K.-H. Liang, H.-Y. Chang, and Y.-J. Chang, "A 0.5–7.5 GHz ultra low-voltage low-power mixer using bulk-injection method by 0.18-um CMOS technology," IEEE Microw.Wireless Compon. Lett., vol. 17, no. 7, pp. 531–533, May 2007.

[7] S.-C. Tseng, C. C. Meng, and C.-K.Wu, "GaInP/GaAs HBT wideband transformer Gilbert downconverter with low voltage supply," Electron. Lett., vol. 44, no. 2, pp. 127–128, Jan. 2008.

[8] P-Z Rao, T-Y Chang, C-P Liang, S-J Chung, "An Ultra-Wideband High-Linearity CMOS Mixer With New Wideband Active Baluns," IEEE Trans. Microw. Theory Tech., vol. 57, no. 9, pp. 2184-2192, Sept. 2009.

A High Gain Wideband 77GHz SiGe Power Amplifier

Roee Ben Yishay, Roi Carmon, Oded Katz and Danny Elad

IBM Haifa Research Lab, Mount Carmel 31905 Haifa, Israel

Abstract — **This paper presents a fully integrated 77GHz power amplifier (PA) fabricated in a 0.13 μm SiGe BiCMOS technology. A 4-stages single ended common-emitter topology was utilized to achieve power gain of 19dB at 77GHz with 14.6dBm output power at 1dB compression, saturated power of 16dBm and 12.5% peak PAE. Small signal characteristics show a wideband behavior - Maximal small signal gain of 23dB achieved at 69GHz with 3dB bandwidth of 15GHz (22%) and both input and output matching is better than -10dB from 72GHz to 90GHz. The PA's bias is applied by adjustable bias circuits to provide process and temperature compensation and was measured in room temperature and at 85^0C. It consumes a quiescent current of 100mA from a 2V supply at 1dB compression and occupies area of 1.4mm^2.**

Index Terms — **Millimeter-wave integrated circuits, Power amplifier, Silicon Germanium (SiGe), W-Band.**

I. INTRODUCTION

W-band frequency range offers new applications such as automotive radar at 76-77GHz and point-to-point high bandwidth wireless links at 71-76GHz and 81-86GHz. Much effort focused recently on implementation of key building blocks and complete chipsets for millimeter wave transceivers [1]-[6], especially in advanced SiGe and CMOS technologies, thanks to their low cost and high integration level capabilities. Power amplifier is one of the most challenging components to design due to low power gain at W-band, low breakdown voltages and low quality on-chip passives. All the above impose a strong limit on output power which tends to deteriorate when temperature rises. As a typical transmitted power of 10dBm is required for automotive radar [7], CMOS currently fails to offer a reliable solution with sufficient margins to account for process and temperature variations. However, recent design efforts [4], [8]-[11] proved BiCMOS SiGe based PA may provide adequate performance while taking advantage of the presence of CMOS logic and circuitry for bias circuits, power detectors etc. [12].

The PA presented herein simultaneously delivers high power, high gain and large bandwidth performance with excellent robustness to temperature changes.

II. CIRCUIT DESIGN

The MMIC PA was fabricated in IBM's 0.13μm SiGe BiCMOS8HP process with cutoff frequencies of f_T=200GHz and f_{MAX}=250GHz for the high performance HBT transistors. The 5 metal levels BEOL stack offers two thick 1.25 μm and 4 μm Aluminum top layers for low loss interconnects.

The single-ended PA consists of 2 cascaded identical unit cells. Each unit cell contains two-stage common emitter topology with input, output and interstage matching networks as shown in fig. 1 and seen in the chip's micrograph in fig. 2. The first cell acts as driver and the second one as the power stage. This arrangement is done for measurement purposes, due to the difficulty of driving a single unit cell into compression with the available W-band power source.

Each common emitter transistor in the unit cell consists of three parallel transistors with 8μm/9μm emitter length, placed at the maximal proximity to each other to minimize interconnect losses, while thermal coupling between adjacent NPN's remains negligible thanks to deep trench isolation surrounding each transistor.

Transistor biasing must account for reliability aspects stemming from high current densities that cause junction temperature (Tj) to rise rapidly and defer substantially from ambient temperature. While transistor functionality is guaranteed at room temperature, severe performance degradation may occur when operating at 85^0C or above. Therefore, all transistors are nominally biased at DC current of 0.6-0.8 mA/μm which is slightly below f_{MAX} current density, but serves as a good compromise between power and reliability constrains. Bias circuits are implemented on-chip and provide PTAT current adjusted to compensate for gain drop at high temperature and allow separate linear course tuning for each stage DC current. The power transistors are isolated from the bias circuits by a 4μm wide meandered line, forming a quarter-wave T-line, shorted to ground at RF by 2pF MIM capacitor. To insure reliable operation above BV_{CEO} (1.7V), the bias circuit is designed to introduce moderate 200 impedance to the power transistor base, so that maximum V_{CE} allowed is ~4V [9]. At RF, the transistor's base sees only the preceding stage output impedance transformed by the matching network to low impedance (absolute value of less than 30), typically required for power match of large transistors.

In order to minimize substrate loss and to ensure that the substrate is well strapped to the ground, all interconnects and capacitors are realized over a meshed

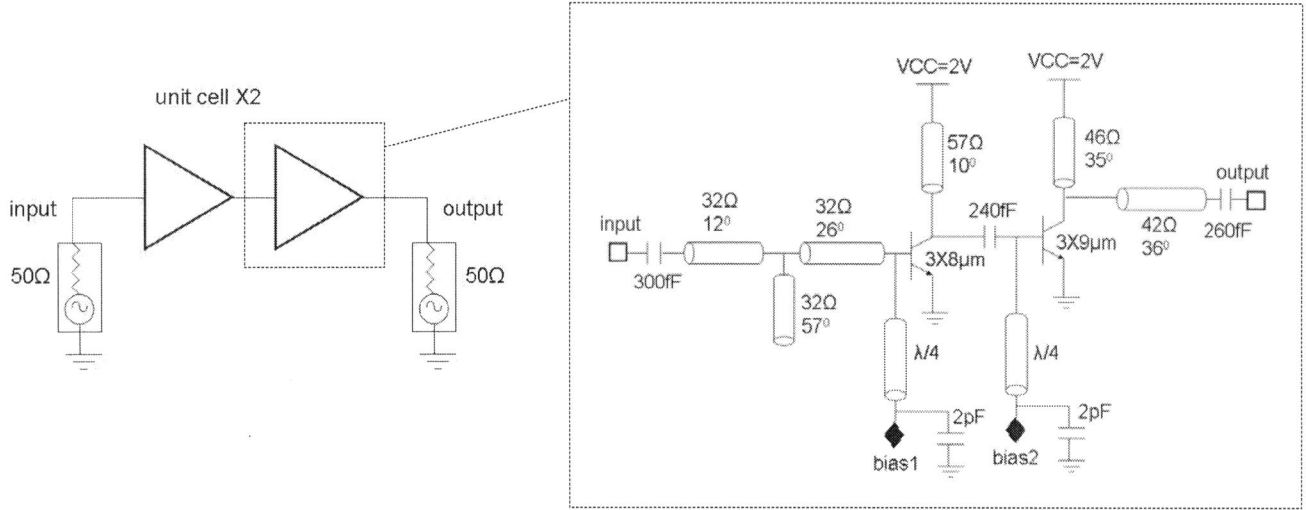

Fig. 1. Schematic diagram of the 77GHz PA

ground-plane, covering the entire circuit area and consisted of the first 3 metals and substrate contacts. All matching networks are realized with MIM capacitors and microstrip T-lines-both shielded and unshielded, (depending on the required characteristics impedance) with loss per unit length of ~0.45dB/mm at 77GHz. In general, the shielding option is preferred when possible, as its model is the most accurate, since the returning current path is well defined by the side-shields which also prevent EM field to interact with nearby excited structures. The use of MIM capacitors as part of the matching networks is restricted to values below 300fF due to their low Q (~7) and to avoid sharp self resonance effects (occurring because of the capacitor's vias inductance), which obstruct a process-robust matching. However, for DC decoupling, where the exact capacitance value is not important, large arrays of 600fF MIM capacitors, operating near their self resonance, where used. Output matching network was designed to provide close to optimal load impedance as determined by load-pull simulation, while maintaining reasonable matching. Input and interstage matching networks are optimized for gain and input return loss respectively. Moreover, matching networks design must also take in account interconnect loss due to low gain and power constrains in W-band. For each matching network a few topologies where optimized, so that the overall loss on all matching networks will not exceed 3dB. RF pads are embedded in the matching networks in order to refrain from inaccurate de-embedding at W-band frequencies. All matching networks and pads were simulated with Agilent Momentum RFDE© full-wave EM simulator up to 3rd harmonic to accurately capture all effects caused by bends, junctions and open/short stubs. All interconnects are sized with respect to electro-migration design rules, to sustain both DC and RMS currents ratings up to 125^{0}C.

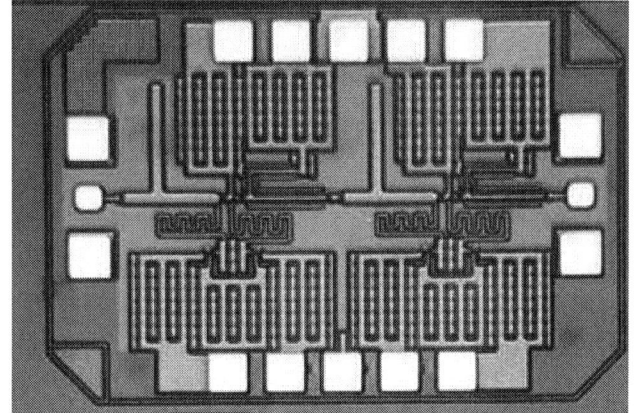

Fig. 2. Die micrograph of the 77GHz PA. Chip size:1.4mm X 0.94mm

III. MEASUREMENT RESULTS

All measurement where taken on-wafer with 110GHz picoprobes in GSG configuration. The power measurements have been performed using Anritsu synthesizer 69347B followed by an external amplifier and WR-10 waveguide while output power was detected by Agilent W8486A 75-110GHz power sensor. For small signal measurements, Agilent E8361A VNA with enabled up conversion up to 110GHz was used.

Fig. 3 shows the measured gain, output power and power added efficiency vs. input power for low, nominal and high collector current density (J_c) modes, while the collectors' voltage remains constant at 2V throughout the measurement. It's evident from measurement that the PA shifts from a linear (class A) operation to a nonlinear (class AB) operation when current density is decreased, and a 1dB gain expansion is observed in the later case. At the high current density mode, small signal gain at 77GHz is 19dB, output 1dB compression point (OCP$_{1dB}$) of 14.6dBm is achieved, where the total quiescent power is 200mW and peak PAE is 12.5%. Nominal and low current density modes yield 18dB and 16dB small signal gain respectively, traded for OCP$_{1dB}$ of 15dBm. The corresponding peak PAE are 13% and 13.5% respectively. Saturated output power is 16dBm. Large signal measurement at 85^{0}C shows less than 0.8dB degradation of P$_{out}$ for a given P$_{in}$, at 77GHz, which means that output power loss can be fully compensated simply by switching to a higher bias mode. This can be done automatically by employing a power detector and a ALC loop, as reported in [14].

Fig. 4 shows the small signal S-parameters measured from 50GHz to 90GHz at 25^{0}C and 85^{0}C in high J$_c$ mode. At 25^{0}C S$_{21}$ peaks at 69GHz to maximum value of 23dB and drops to 19.2dB at 77GHz. In the highest measured frequency (90GHz) the small signal gain is 17.5dB. The PA's 3dB bandwidth extends from 61 to 76GHz which is equivalent to 22% fractional bandwidth. The wideband behavior is even more pronounced in the output matching, which is better than -15dB from 55GHz and up to the maximum measured frequency and the Input matching which is better than -15dB from 76GHz to 86GHz. This may facilitate future designs, aiming at higher frequencies. A comparisons between measured gain at 25^{0}C and 85^{0}C, clearly indicates the PA's robustness to temperature variations-peak gain drops only by 1dB, while at 77GHz only by 0.75dB. Table I summarizes the measured results and compares it with previous work on SiGe based power amplifiers.

IV. CONCLUSION

This paper presented a W-band fully integrated power amplifier consisted of 4 cascaded common-emitter stages, fabricated in a commercially available SiGe BiCMOS technology. It simultaneously achieves high output power, high gain and wideband matching comparable to state of the art published works. Design considerations discussed in this paper proved to yield a reliable and robust high performance operation.

V. ACKNOWLEDGMENT

The authors would like to thank Donald Papae, Charles Rodriguez and Francis Szenher of IBM Fishkill for measurements enablement and assistance.

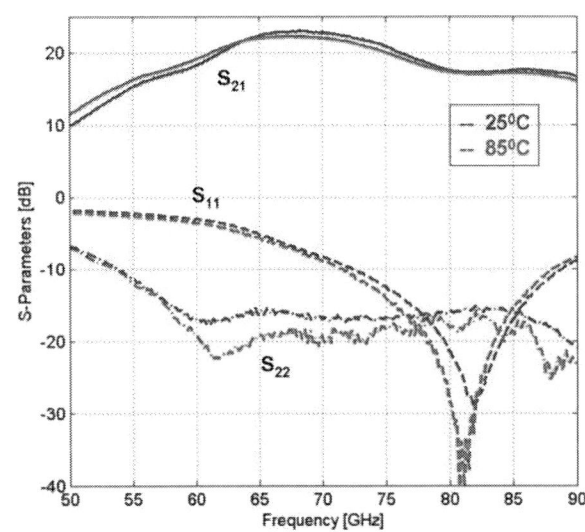

Fig. 4. Measured S-parameters at 25^{0}C and 85^{0}C

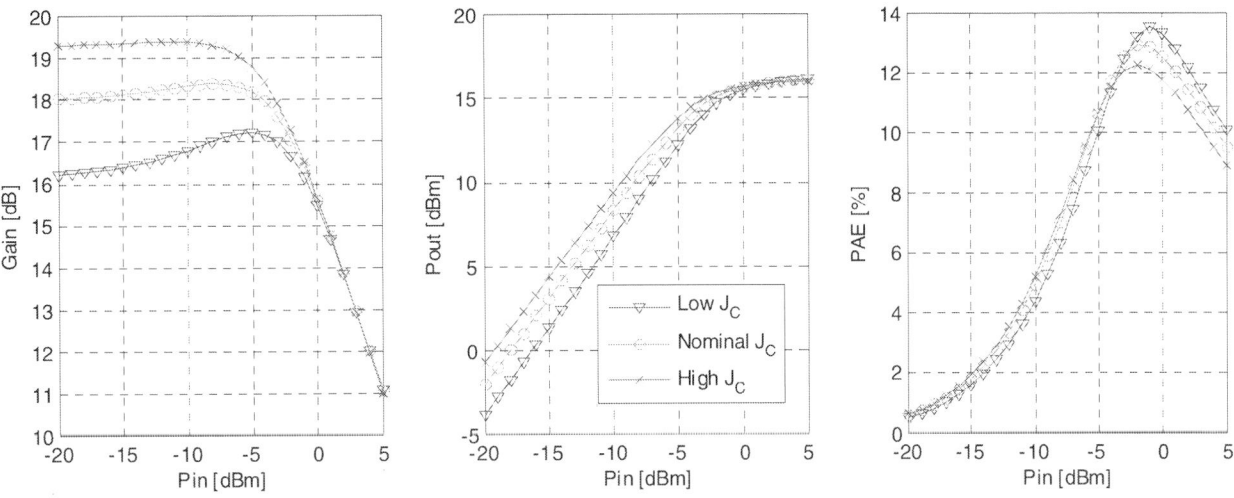

Fig. 3. Measured gain, output power and PAE

TABLE I

SUMMARIZED RESULTS AND COMPARISON TO PREVIOUSLY PUBLISHED SiGe W-BAND PA'S

	This Work	Chang and Rebiez [8]	Nicolson et al. [4]	Komijani and Hajimiri [11]	Afshari et al. [10]	Pfeiffer et al. [9]
Technology	**SiGe 0.13μm** f_T**=200GHz**	SiGe 0.13μm f_T=200GHz	SiGe 0.13μm f_T=230GHz	SiGe 0.13μm f_T=200GHz	SiGe 0.13μm f_T=200GHz	SiGe 0.13μm f_T=200GHz
Topology	**4-stage CE**	Balanced 3-stages CE	1st-stage- CC Stages 2+3 – CE	4-stage CE Power comb.	CascodeX4 Power comb	Balanced 2-stage CE
Frequency	**77GHz**	90GHz	77GHz	77GHz	85GHz	77GHz
OCP_{1dB}	**14.6dBm**	11.8dBm (*)	12dBm	14.5dBm	-	8.6dBm (*)
$S_{21}@f_0$	**19dB**	14.6dB	19dB	17dB	8dB	6.1dB
P_{SAT}	**16dBm**	16dBm (*)	14.5dBm	17.5dBm	21dBm	9.5dBm (*)
PAE_{max}	**12.5%**	9.9%	15.7%	12.8%	3.4%	3.5%
$S_{11}@f_0$	**-15dB**	-15dB	-15dB	-	-	-6.2dB
$S_{22}@f_0$	**-20dB**	-20dB	-9dB	-	-	-9.1dB
DC power	**200mW**	210mW (*)	161mW	270mW	-	162mW (*)
Area	**1.3 mm^2**	2.4 mm^2	0.11 mm^2	0.6 mm^2	2.4 mm^2	1.575 mm^2

(*) Single ended operation

REFERENCES

[1] S. Reynolds, B. Floyd, U. Pfeiffer, T. Beukema, J. Grzyb, C. Haymes, B. Gaucher, and M. Soyuer, "A silicon 60GHz receiver and transmitter chipset for broadband communications," *IEEE J. Solid-State Circuits*, vol. 41, no. 12, pp. 2820-2830, Dec. 2006.

[2] B. Heydari, M. Bohsali, E. Adabi, A. M. Niknejad, "Millimeter-Wave Devices and Circuit Blocks up to 104 GHz in 90 nm CMOS," *IEEE J. Solid State Circuits*, vol. 42, No. 12, pp. 2893 - 2903, Dec. 2007.

[3] M. Khanpour, K. W. Tang, P. Garcia, and S. P. Voinigescu, "A wideband W-band receiver front-end in 65-nm CMOS," *IEEE J. Solid-State Circuits*, vol. 43, no. 8, pp. 1717–1730, Aug. 2008.

[4] S. T. Nicolson, K. H. K. Yau, S. Pruvost, V. Danelon, P. Chevalier, P. Garcia, A. Chantre, B. Sautreuil, and S. P. Voinigescu, "A low-voltage SiGe BiCMOS 77-GHz automotive radar chipset," *IEEE Trans. Microw. Theory and Tech.*, vol. 56, no. 5, pt. 1, pp. 1092–1104, May 2008.

[5] A. Natarajan, A. Komijani, X. Guan, A. Babakhani, and A. Hajimiri, "A 77-GHz phased-array transceiver with on-chip antennas in silicon: Transmitter and local lo-path phase shifting," *IEEE J. Solid-State Circuits*, vol. 41, no. 12, pp. 2807–2819, Dec. 2006.

[6] H. Knapp, B. Dehlink, H.-P. Forstner, E. Kolmhofer, K. Aufinger, J. Bock, and T. F. Meister, "SiGe circuits for automotive radar," *IEEE Silicon Monolithic Integrated. Circuits in RF Systems meeting*, pp. 231–236, Jan. 2007.

[7] L. H. Eriksson and B.. As, "A high performance automotive radar for automatic AICC," *IEEE Aerospace and Electronics System Magazine*, vol. 10, no. 12, pp. 13, Dec. 1995.

[8] M. Chang and G. Rebeiz, " A wideband high-efficiency 79–97 GHz SiGe linear power amplifier with > 90 mW output", *IEEE Bipolar/BiCMOS Circuits and Technology Meeting*, pp. 69-72, Oct. 2008.

[9] U.R. Pfeiffer, et al., "A 77 GHz SiGe Power Amplifier for Potential Applications in Automotive Radar Systems", *IEEE RFIC Symposium Digest*, pp. 91–94, June 2004.

[10] E. Afshari, H. Bhat, X. Li and A. Hajimiri., "Electrical Funnel: A Broadband Signal Combining Method," *IEEE ISSCC Digest*, pp. 751–760, Feb. 2006.

[11] A. Komijani and A. Hajimiri, "A Wideband 77-GHz, 17.5-dBm Fully Integrated Power Amplifier in Silicon," *IEEE J. Solid State Circuit*, pp. 1749–1756, Aug. 2006.

[12] U. R. Pfeiffer and D. Goren, "A 20dBm Fully-Integrated 60GHz SiGe Power Amplifier with Automatic Level Control", *IEEE J. Solid-State Circuits*, vol. 42, issue 7, pp 1455-1463, July 2007.

RTUIF-10

A Broadband Differential Cascode Power Amplifier in 45 nm CMOS for High-Speed 60 GHz System-on-Chip

Morteza Abbasi[1], Torgil Kjellberg[1,2], Anton de Graauw[3], Edwin van der Heijden[3], Raf Roovers[3] and Herbert Zirath[1]

[1]Microwave Electronics Laboratory, Chalmers University of Technology, Göteborg, Sweden
[2]Chalmers Industrial Technologies, Göteborg, Sweden
[3]NXP Semiconductors, Research, Eindhoven, The Netherlands

Abstract—**A compact two-stage differential cascode power amplifier is designed and fabricated in 45 nm standard LP CMOS. The cascode configuration, with the common gate device placed in a separate P-well, provides reliable operating condition for the devices. The amplifier shows 20 dB small-signal gain centered at 60 GHz with a flat frequency response and 1-dB bandwidth of 10 GHz. The broadband large-signal operation is also ensured by providing constant load resistance to both stages over the entire band and coupling them with a dual resonance matching network. The chip delivers 11.2 dBm output power at 1-dB compression and up to 14.5 dBm power in saturation. The power amplifier operates with 2 V supply and draws 90 mA total current which results in 14.4% maximum PAE. The output third order intercept point is measured to be 18 dBm for two-tone measurement at 60 GHz with 0.5 GHz, 1 GHz and 2 GHz frequency separations.**

Index Terms— **60 GHz, power amplifier, 45nm CMOS, cascode, isolated P-well.**

I. INTRODUCTION

The demand for high data-rate communications at 60 GHz band not only puts stringent requirements on millimeter-wave circuit design but also calls for extensive back-end digital processing of signals. For a complete CMOS radio-on-chip solution, it is necessary that the development of millimeter-wave circuits and particularly the power amplifier keep up with the rapid transfer of digital circuitry to the advanced technology nodes.

Due to the voltage limitations of the thin oxide devices in latest CMOS technologies, the power amplifier design still remains as a key challenge in transmitter development. Apart from a few results [1]–[3], most of the previously reported CMOS power amplifiers operating at 60 GHz are implemented in 90 nm and 65 nm technology nodes [4]–[10].

This paper presents a two-stage differential power amplifier, implemented in a 45 nm standard LP CMOS transistor technology. This process has a maximum DC rating of 1.1 V for drain to source voltage which imposes serious limitations for power generation. In order to increase the output voltage swing, in the presented circuit, devices are stacked in cascode configuration. Reliable operation has been demonstrated in [3] by life-time measurements on a single-ended power amplifier with a similar transistor structure. The objectives of the design have been the demonstration of broadband large-signal performance, with high output power and acceptable DC power consumption with a compact design in 45 nm CMOS for integration into future high-speed wireless systems.

II. CIRCUIT DESIGN

Fig. 1 shows the circuit diagram of the two-stage differential cascode power amplifier. As mentioned earlier, stacking the transistors is used to increase the output voltage swing without exceeding the recommended operating conditions for each transistor. By properly biasing the gates of the common source (CS) and common gate (CG) devices, the supply voltage of $V_{dd} = 2$ V divides equally between the two devices. The common gate device is placed in an isolated P-well and the well is biased at half of the supply voltage. This way there will be less than 1 V difference between any two terminals of each device.

In order to avoid the inherent loss of the thin lower metal layers, which are commonly used as the return path in single-ended structures, the amplifier employs a fully differential topology. All interconnects are made in the thicker top metal layers. The matching networks are realized using lumped element inductors and fringe field capacitors and modeled with electromagnetic simulations. The input and output ports are kept differential for interfacing the preceding up-converting mixer and succeeding antenna without the use of balun's.

The layouts of the transistor cells are customized for minimum interconnect parasitics for the differential signals and low DC loss for common-mode bias currents. The amplifier operates in Class A mode for maximum gain and linearity and in a trade-off between output power and DC power consumption the total widths of the devices are chosen 50 μm and 100 μm for the first and second stages respectively. The minimum gate length is used for all the transistors. All supply voltages are provided from a single pad and the gate voltages of each stage as well as the gates of the cascode devices are connected to separate pads for flexibility in bias point adjustments.

The input matching network is designed to compensate for the transistor gain roll-off by reflection matching. Broadband large-signal operation is ensured by optimizing the output matching networks to provide constant load resistance over the

978-1-4244-6240-7/10 $26.00 © 2010 IEEE

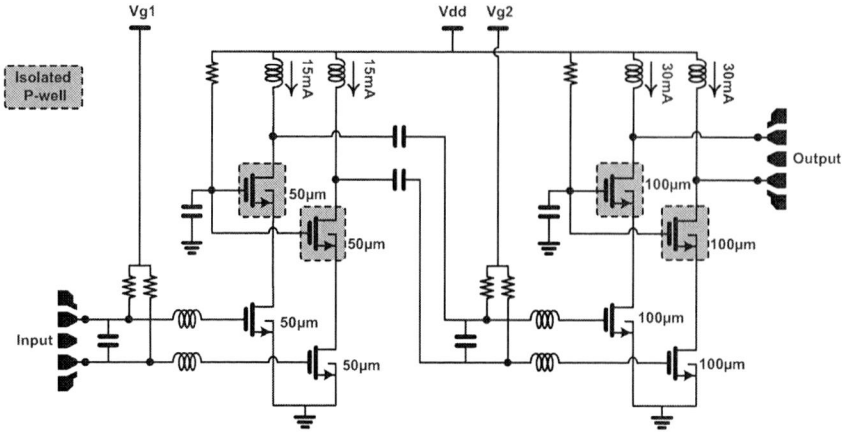

Fig. 1. Circuit schematic of the two-stage differential cascode power amplifier.

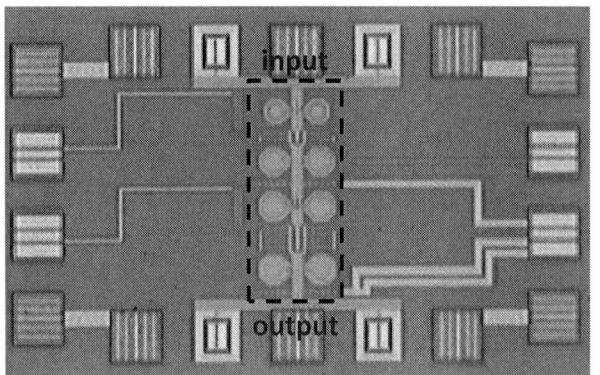

Fig. 2. The chip photograph of the two-stage power amplifier. The active area is $300 \ \mu m \times 130 \ \mu m$.

band of interest for the two stages and using a dual resonance network for the interstage matching network.

The chip photographs of the amplifier is shown in Fig. 2. The careful and symmetric layout for transistors and matching networks resulted in a very compact design suitable for high integration levels. The active chip area is only $300 \ \mu m \times 130 \ \mu m$.

III. MEASUREMENT RESULTS

A. Small-Signal Performance

On-wafer small-signal measurements are performed on the power amplifier by using a four-port network analyzer with mixed mode S-parameters. No sign of instability or oscillation was observed during the measurement at different bias points. The bias point is fixed for further measurements at 2.0 V supply voltage with 30 mA and 60 mA of total bias currents for the first and second stages respectively. The gate of the common gate devices are biased at 1.8 V which makes equal division of supply voltage on drain to source of the transistors.

The amplifier exhibits 20 dB of small-signal gain in differential mode with the 1-dB bandwidth from 55 GHz to 65 GHz as shown in Fig. 3. The flat gain response over such a broad bandwidth is achieved by optimized input port matching and

the interstage matching network. The common mode gain is also shown in Fig. 3. As can be seen, the amplifier provides 15 dB to 20 dB common mode rejection ratio (CMRR) over the entire bandwidth.

The input and output reflection coefficients are plotted in the lower diagram of Fig. 3. The higher impedance mismatch of the input port at the lower edge of the band compensates for the slope of the transconductance to achieve the flat gain response. This mismatch should be taken into account in the design of the preceding circuit. The output port is optimized for maximum power transfer to the antenna and therefore is not perfectly conjugate matched.

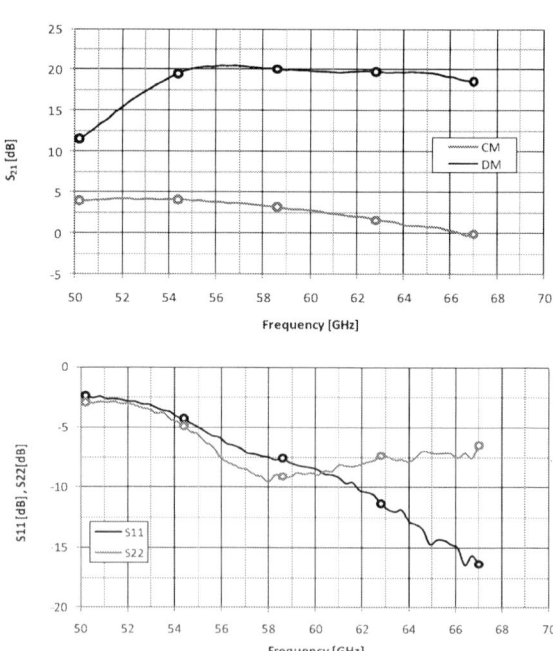

Fig. 3. Measured S-Parameters of the differential power amplifier. The upper diagram shows the gain in differential and common modes and the lower diagram shows the input and output port impedance matchings.

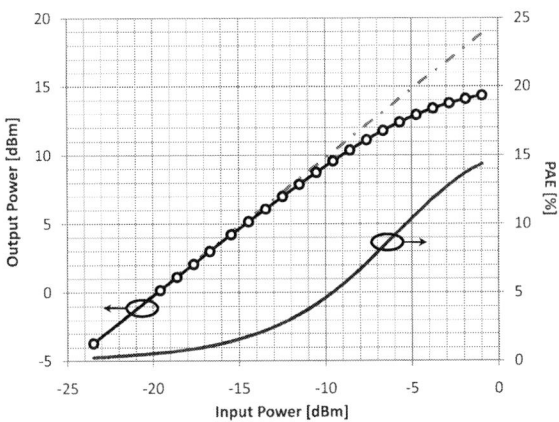

Fig. 4. Measured output power and total PAE versus input power. The measurement is performed at the frequency of 60 GHz.

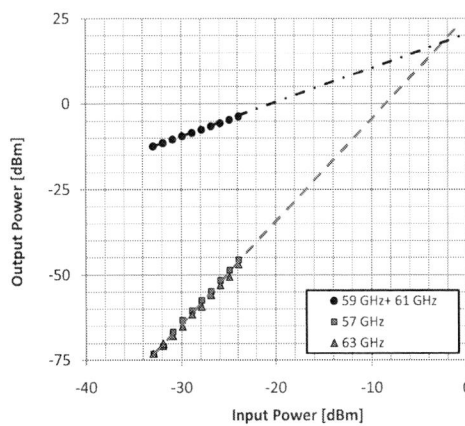

Fig. 6. Measurement results for the two tone excitation of the power amplifier. The output OIP3 is extrapolated to be 18 dBm.

B. Large-Signal Performance

The output power for the power amplifier is measured using differential probes with integrated balun and Agilent V8486A calibrated V-band power detector. The input power is provided from an Agilent E8257D signal generator and the power delivered to the chip is measured for accurate calibration of the results. The measured output power versus input power is shown in Fig 4. As can be seen, the power amplifier delivers 11.2 dBm output power at the 1-dB compression point and 14.5 dBm in saturation. The maximum total power added efficiency (PAE) for this design is 14.4%. When the drain voltage is increased to 2.2 V, the 1-dB compression point and saturated output power can be increased to 11.6 dBm and 15.5 dBm respectively. This operating condition is however not suitable for reliable operation since the devices start to go under stress.

Fig. 5 shows the measured output power at 1-dB compression point over the frequency band. The figure shows that 10 dBm to 11 dBm power can be provided from the amplifier over the broad bandwidth from 54 GHz to 67 GHz.

Since the gate voltages are accessible for tuning the bias

points, different operation modes are also examined. However, as the devices are biased towards class AB operation, the gain drops rapidly and the improvement in power added efficiency becomes marginal.

C. Linearity Measurement

In order to examine the linearity of the power amplifier, two signals around 60 GHz with 2 GHz separation at 59 GHz and 61 GHz are applied to the input. The output spectrum is monitored with the spectrum analyzer after down conversion with a calibrated harmonic mixer. The probable nonlinearities introduced from the compression of the down converting mixer are verified to be negligible. The input signals are set to equal levels and are increased in steps of 1 dB. Three values are recorded; the total power in the desired signals at 59 GHz and 61 GHz, the power at the lower intermodulation product at 57 GHz and the power at the upper intermodulation product at 63 GHz. The measurement results are plotted in Fig 6.

As can be seen in the figure, the linear extrapolations of the power in these frequencies result in output third order intercept point of 18 dBm. The same measurement is repeated when the separation of the two input tones are set to 0.5 GHz and 1 GHz as well and the resulting OIP3 is similar and around 18 dBm.

IV. COMPARISON AND CONCLUSIONS

Table I summarizes the performance of the presented differential cascode power amplifier in comparison with some of the latest reported results for CMOS power amplifiers at 60 GHz. The use of cascode transistors with the separate P-well enables the increase in supply voltage without exceeding the maximum voltage ratings for reliable operation in the C045 process. The larger available voltage swing, use of fully differential topology to reduce losses and optimized coupling of the two stages with dual resonance network, all contributed to increased output power, efficiency, gain and bandwidth of the amplifier. The combination of presented performance parameters compared to other reports at this technology node and even earlier nodes, makes this compact design an outstanding power amplifier result at 60 GHz and an

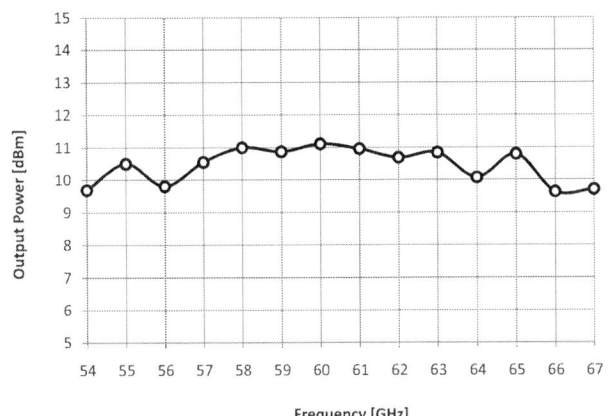

Fig. 5. Measured output power at 1-dB compression point over the frequency band. The input power is around -7.5 dBm.

TABLE I

PERFORMANCE SUMMARY AND COMPARISON TO PREVIOUSLY REPORTED RESULTS

Reference	Description	Gain [dB]	-3 dB Bandwidth	Output Power [dBm] (1dB / Saturated)	PAE max%	Technology
[1]	Three-Stage CS	19	51-61 GHz	NA / 7.9	19.4	45 nm
[2]	Two-Stage Push-Pull	6	50-67 GHz	11 / 13.8	7	45 nm
[3]	Two-Stage Single-ended Cascode	16	50-60 GHz	7.6 / 12	12.3	45 nm
This work	**Two-Stage Differential Cascode**	**20**	**53-67+ GHz**	**11.2 / 14.5**	**14.4**	**45 nm**
[4]	Two-Stage Cascode	16	55-65 GHz	12.7 / 14.5	25	SOI 65 nm
[5]	Three-Stage Differential CS	15.8	57-62 GHz	2.5 / 11.5	11	65 nm
[6]	Four-Stage Power Combining CS	20	57-65 GHz	8.2 / 12	9	90 nm
[7]	DAT Power Combining Cascode	26	55-71 GHz	14.5 / 18	12.2	90 nm
[8]	Power Combining Differential	11.2	54-64 GHz	8.3 / 11.2	3.6	90 nm
[9]	Three-Stage CS	10	55-72 GHz	8.8 / 12.6	6.9	90 nm
[11]	Five-Stage Doherty Cascode	13.5	55-62 GHz	7 / 7.8	3	130 nm

attractive building block for integration with digital processors for high data-rate SOC solutions.

ACKNOWLEDGEMENT

The authors would like to thank their colleagues at NXP Semiconductors, in particular Ralf Pijper and Luuk Tiemeijer for their help and support in device and component measurement and model development.

REFERENCES

[1] E. Cohen, S. Ravid, and D. Ritter, "60 GHz 45 nm PA for linear OFDM signal with predistortion correction achieving 6.1% PAE and -28 dB EVM," in *Proceedings 2009 IEEE Radio Frequency Integrated Circuits Symposium (RFIC 2009)*, 2009, pp. 35 – 8.

[2] K. Raczkowski, S. Thijs, W. De Raedt, B. Nauwelaers, and P. Wambacq, "50-to-67GHz ESD-protected power amplifiers in digital 45nm LP CMOS," in *Solid-State Circuits Conference - Digest of Technical Papers, 2009. ISSCC 2009. IEEE International*, Feb. 2009, pp. 382–383,383a.

[3] T. Kjellberg, M. Abbasi, M. Ferndahl, A. de Graauw, E. van der Heijden, and H. Zirath, "A compact cascode power amplifier in 45-nm CMOS for 60-GHz wireless systems," in *2009 Annual IEEE Compound Semiconductor Integrated Circuit Symposium. Technical Digest 2009*, 2009.

[4] A. Siligaris, Y. Hamada, C. Mounet, C. Raynaud, B. Martineau, N. Deparis, N. Rolland, M. Fukaishi, and P. Vincent, "A 60 GHz power amplifier with 14.5 dBm saturation power and 25% peak PAE in CMOS 65 nm SOI," in *Proceedings of the 35th European Solid-State Circuits Conference. ESSCIRC 2009*, 2009, pp. 168 – 71.

[5] W. Chan, J. Long, M. Spirito, and J. Pekarik, "A 60 GHz-band 1 V 11.5 dBm power amplifier with 11% PAE in 65 nm CMOS," in *2009 IEEE International Solid-State Circuits Conference (ISSCC 2009)*, 2009, pp. 380 – 1.

[6] D. Dawn, S. Sarkar, P. Sen, B. Perumana, M. Leung, N. Mallavarpu, S. Pinel, and J. Laskar, "60 GHz CMOS power amplifier with 20-dB-gain and 12 dBm Psat," in *2009 IEEE MTT-S International Microwave Symposium Digest (MTT)*, 2009, pp. 537 – 40.

[7] Y.-N. Jen, J.-H. Tsai, T.-W. Huang, and H. Wang, "Design and analysis of a 55-71-GHz compact and broadband distributed active transformer power amplifier in 90-nm CMOS process," *IEEE Transactions on Microwave Theory and Techniques*, vol. 57, no. 7, pp. 1637 – 46, Jul. 2009.

[8] Y. Yoshihara, R. Fujimoto, N. Ono, T. Mitomo, H. Hoshino, and M. Hamada, "A 60-GHz CMOS power amplifier with Marchand balun-based parallel power combiner," in *2008 IEEE Asian Solid-State Circuits Conference*, 2008, pp. 121 – 4.

[9] N. Kurita and H. Kondoh, "60 GHz and 80 GHz wide band power amplifier MMICs in 90 nm CMOS technology," in *Proceedings 2009 IEEE Radio Frequency Integrated Circuits Symposium (RFIC 2009)*, 2009, pp. 39 – 42.

[10] D. Chowdhury, P. Reynaert, and A. Niknejad, "A 60 GHz 1 V +12.3 dBm transformer-coupled wideband PA in 90 nm CMOS," in *2008 IEEE International Solid-State Circuits Conference - Digest of Technical Papers*, 2008, pp. 560 – 1.

[11] B. Wicks, E. Skafidas, and R. Evans, "A 60-GHz fully-integrated Doherty power amplifier based on 0.13μm CMOS process," in *2008 IEEE Radio Frequency Integrated Circuits Symposium*, 2008, pp. 69 – 72.

A CMOS LC VCO with Novel Negative Impedance Design for Wide-Band Operation

Chang-Hsi Wu and Guan-Xiu Jian

Department of Electronic Engineering, Lunghwa University of Science and Technology, 333 Taoyuan, 300 Wan-Shou Rd. Sec.1, Kueishan, E-mail: cswu@mail.lhu.edu.tw Phone: 02-82093211 ext: 5623, Fax: 02-82095165, Taiwan, R.O.C.

Abstract — A 5.2 GHz CMOS LC voltage-controlled oscillator (VCO) for UWB receiver, fabricated using CMOS 0.18μm process, is presented in this paper. The tuning range of the proposed VCO is mainly broadened by novel negative resistance and tapping inductance techniques. Measured results of the proposed VCO reveal phase noise of -116.708/Hz at 1 MHz offset and tuning range of 4.567GHz~5.832GHz (24.32%) while consuming only 3.92mW under the supply voltage of 0.8V. The core area is 0.732mm × 0.633mm.

Index Terms — VCO, tuning range, novel negative resistance, tapping inductance, UWB.

I. INTRODUCTION

According to Part 15 of the FCC rules, an ultra-wideband (UWB) signal should have a fractional bandwidth greater than 0.20 of the center frequency or a bandwidth of at least 500 MHz at all times of transmission [1]. Moreover, the FCC ruling allows UWB communication devices to operate at low power (an EIRP of -41.3 dBm/MHz) in an unlicensed spectrum from 3.1 to 10.6 GHz. Recently, according to the Multiband OFDM Alliance Proposal, the UWB spectrum is divided into 14 bands, each with a bandwidth of 528 MHz. As shown in Fig.1, five band groups are defined, consisting of four band groups of three bands each and one band group of two bands [2]. The UWB communication systems bring challenges and opportunities for the RF circuit design.

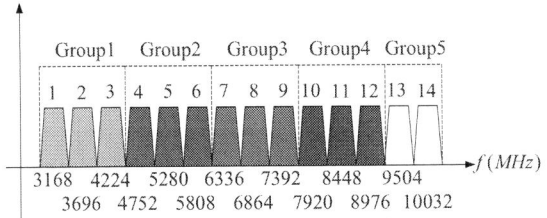

Fig. 1. Allocation of UWB frequency bands and groups.

To improve the tuning range, a novel negative resistance technique was applied to the VCO in this design. Moreover, the tail current source of VCO is replaced by a tapping inductor to reduce supply voltage and power dissipation [3], [4].

In this paper, both design and analysis of the proposed UWB VCO are presented in Section II. Measured results and performance of the designed UWB VCO are described in Section III. Finally, the conclusion is in Section IV.

II. CIRCUIT DESIGN AND ANALYSIS

A. Circuit Design

In the UWB transceiver, a VCO with broad tuning-range design is necessary. To achieve this design, both improving the variable capacitance size and downsizing the parasitical capacitance of the active negative-resistance circuit are considered.

Consider the traditionally active negative-resistance circuit of cross-couple shown in the Fig. 2.

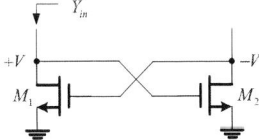

Fig. 2. Traditionally active negative-resistance topology.

The admittance Y_{in} can be written as

$$Y_{in} = SC_{in} - g_{m1} \qquad (1)$$

where

$$C_{in} = C_{db1} + C_{gs2} + C_{gd1}(1 - \frac{1}{K_1}) + C_{gd2}(1 - K_2) \qquad (2)$$

Since the transistors M1 and M2 are matched and the Miller's gains $K_1 = K_2 = -1$, we have $C_{in} = C_{db} + C_{gs} + 4C_{gd}$. In order to enhance the tuning range of VCO, C_{in} should be minimized. Therefore, a novel negative-resistance circuit of reducing the parasitical capacitance as shown in

Fig. 3 is proposed in this paper. The small signal equivalent circuit is shown in the Fig.4.

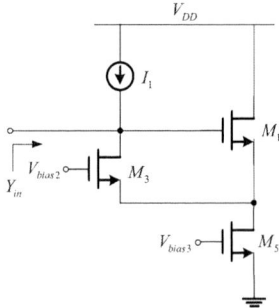

Fig. 3. A novel negative-resistance circuit.

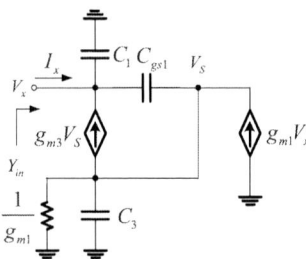

Fig. 4. Small signal model of the novel negative-resistance circuit.

It can be known that

$$C_1 = C_{gd3} + C_{db3} + C_{gd1}$$

$$C_3 = C_{sb1} + C_{gs3} + C_{sb3} + C_{gd5} + C_{db5}$$

From Fig.4, the admittance Y_{in} can be shown as [5~6]

$$Y_{in} = S(C_1 + C_{gs1}) - \frac{(SC_{gs1} + g_{m1})(SC_{gs1} + g_{m3})}{S(C_{gs1} + C_3) + g_{m1} + g_{m3}} \quad (3)$$

Compare equation (1) with equation (3), we can find that the parasitical capacitance of the novel negative-resistance circuit is smaller than that of the traditional negative-resistance circuit. In order to further demonstrate above theoretic result, simulations of susceptances vs. frequency for both circuits in Fig. 2 and Fig. 3 are shown in the Fig.5.

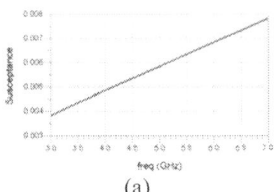
(a)

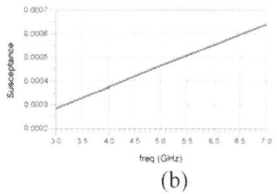
(b)

Fig. 5. Frequency responses of susceptances: (a) the traditional negative-resistance circuit, (b) the novel negative-resistance circuit.

Simulation results of Fig.5 show that the novel negative-resistance circuit has smaller input capacitance than the traditional negative-resistance circuit. It leads to wider tuning-range that the novel negative-resistance circuit can achieve.

From equation (3), the transconductance of the novel negative-resistance circuit can be written as

$$G_m = g_{m1}g_{m3} / (g_{m1} + g_{m3})$$

However, it is too small to produce enough loop gain for oscillation in practice. To enhance the transconductance of the negative-resistance circuit for starting oscillation, the proposed VCO with the modified negative-resistance circuit is shown in the Fig.6 [5].

In Fig.6, a traditional negative-resistance circuit is paralleled with the novel negative-resistance circuit, and then the total transconductance is large enough to raise oscillation. Besides, the parasitical capacitance of this reform negative-resistance circuit is larger than the novel negative-resistance circuit in Fig. 3 but is still smaller than the traditional negative-resistance circuit in Fig. 2, since the sizes of M_7 and M_8 in the proposed circuit are tuned smaller than M_1 and M_2 in the traditional circuit shown in Fig.2 while keeping enough transconductance. So the tuning range is wider than using the traditional negative-resistance circuit. Fig.7 shows frequency response of susceptance of the proposed negative-resistance circuit.

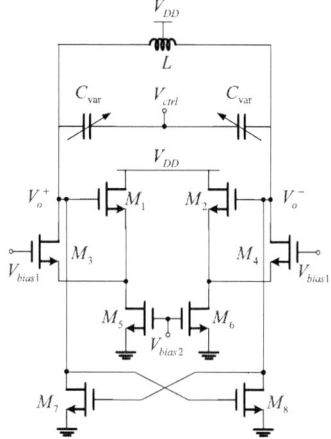

Fig. 6. The proposed VCO schematic.

978-1-4244-6240-7/10 $26.00 © 2010 IEEE   488

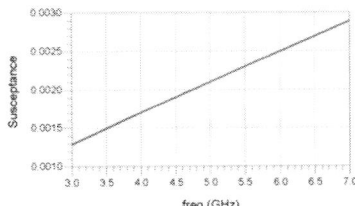

Fig. 7. Frequency response of susceptance of the proposed negative-resistance circuit.

B. Circuit analysis

In order to design the circuit to satisfy the Barkhausen criterion for oscillation, we consider the small-signal equivalent circuit of the proposed VCO shown in Fig.8.

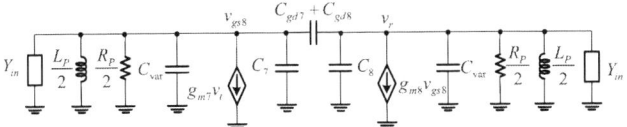

Fig. 8. Small-signal equivalent Circuit of the proposed VCO.

In this equivalent circuit, R_S and L_S represent the equivalent resistance and inductance of the inductor in the proposed VCO circuit shown in Fig. 6.

From the Barkhausen criterion, loop gain of the proposed VCO must satisfy $L = V_r / V_t = 1 \angle 0°$ at oscillation, in other words, $V_r = V_t \neq 0$ when $S = j\omega_0$. By symmetry of the circuit, we have $V_{gs8} = -V_r$. It leads to

$$g_{m7,8} = \frac{2}{R_p} + \frac{2}{j\omega_0 L_p} + j\omega_0 (C_{var} + C_{7,8} + 4C_{gd7,8}) + Y_{in} \quad (4)$$

where Y_{in} is depicts in equation (3). Then the resonant frequency can be found as follows

$$\omega_0 - \sqrt{\frac{2a}{L_p R_p [b - cd - e] - f}} \quad (5)$$

where

$$a = R_p (g_{m1} + g_{m3})$$
$$b = g_{m7,8} (C_{gs1} + C_3)$$
$$c = (C_{var} + C_{7,8} + 4C_{gd7,8})$$
$$d = (g_{m1} + g_{m3})$$
$$e = C_1 (g_{m1} + g_{m3})$$
$$f = L_p (C_{gs1} + C_3)$$

III MEASURED RESULTS

The measured phase noise is -116.7 dBc/Hz at 1MHz offset frequency and is depicted in Fig. 9. The tuning range and the output power are shown in Fig. 10 and Fig.

11, respectively. Fig. 12 is the photograph of the proposed VCO.

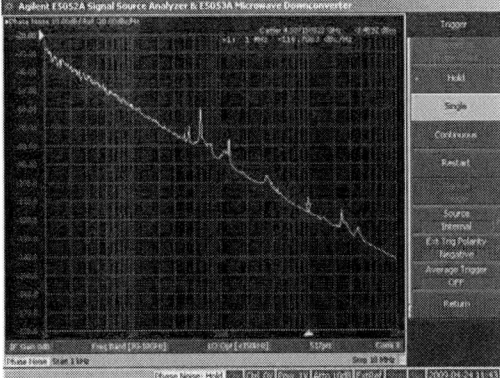

Fig. 9. Measured result of the phase noise.

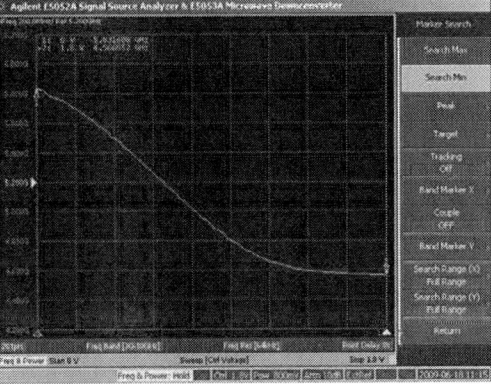

Fig. 10. Measured result of the tuning range.

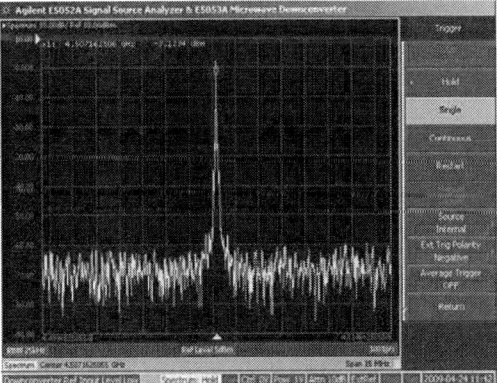

Fig. 11. Measured result of the output power.

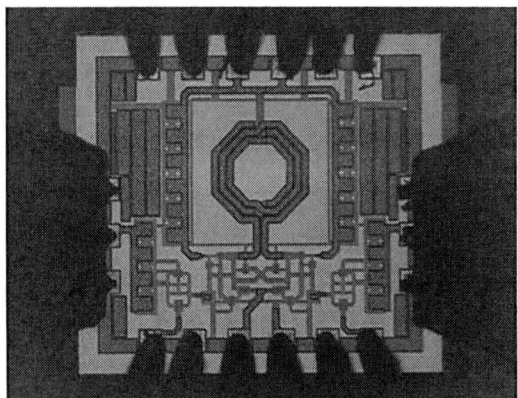

Fig. 12. Chip photograph of the proposed VCO.

TABLE I

Comparison of the measured results with the recently published papers about VCOs.

	This Work	[7]	[8]	[9]
Tuning Range(GHz)	**4.567~5.832**	4.95~5.4	4.39~5.26	5.1~6.1
Phase Noise(dBc/Hz)	**-116.708 @1M**	-115.5 @1M	-113.7 @1M	-110.8 @1M
DC Power(mW)	**3.92**	3	9.7	8.3
FOM(dBc)	**-183.97**	-185.1	-180.0	-182

IV. CONCLUSION

The proposed VCO is implemented in CMOS TSMC 0.18μm process and is designed using a reform negative-resistance topology to increase the tuning range. In addition to reduce the supply voltage and achieve low power consumption, the proposed technique also improves the phase noise. The measured results of the proposed VCO are listed in Table 1 where the phase noise is -116.7 dBc/Hz at 1 MHz offset, the tuning range is 4.567GHz~5.832GHz (24.32%) and the power consumption is 3.92mW from 0.8V supply voltage. The size of the chip is 0.732mm × 0.633mm .

ACKNOWLEDGEMENT

The authors would like to thank the National Chip Implementation Center (CIC) Steering Committee for their valuable comments and suggestions in design and fabrication of this chip. The chip is fabricated by Taiwan Semiconductor Manufacturing Company (TSMC) through Chip Implementation Center (CIC) of Taiwan, R.O.C.

REFERENCES

[1] FCC, First Report and Order 02-48, February 2002.

[2] MultiBand OFDM Alliance SIG, "Multiband.

[3] Jeong-Bae Seo, Kun-Man Park, Jong-Ha Kim, Jin- Hong Park, Young-Sop Lee, Jeong-Hyun Ham, and Tae-Yeoul Yun, " A Low-Noise UWB CMOS Mixer Using Switched Biasing Technique", RFIT2007-IEEE International Workshop on Radio- Frequency Integration Technology, Dec. 9-11 2007.

[4] Pietro Andreani,"Analysis and Design of a 1.8-GHz CMOS LC Quadrature VCO" in IEEE Journal of Solid-State Circuits, vol. 37, no. 12, December 2002.

[5] Behzad Razavi, "Design of Analog CMOS Integrated Circuits",1998.

[6] Thomas H. Lee, " The Design of CMOS Radio- Frequency Integrated Circuits ",2004.

[7] Taeksang Song and Euisik Yoon "A 1-V 5 GHz Low Phase Noise LC-VCO Using Voltage-Dividing and Bias-Level Shifting Technique," 2004 Topical Meeting on Silicon Monolithic Integrated Circuits in RF Systems, pp. 87-90,Sept. 2004.

[8] Young-Jin Moon, Yong-Seong Roh, Chan-Young Jeong and Changsik Yoo, "A 4.39-5.26 GHz LC-Tank CMOS Voltage-Controlled Oscillator With Small VCO-Gain Variation," Microwave and Wireless Component Letters, IEEE, August 2009.

[9] Tuan Thanh Ta; Kameda, S.; Takagi, T.; Tsubouchi, K.; "A 5GHz Band Low Noise and Wide Tuning Range Si-CMOS VCO," Radio Frequency Integrated Circuit Symposium, pp. 571-574, 2009.

RTUIF-12

An 80GHz range Synchronized Push-push Oscillator For Automotive Radar Application

Chama Ameziane[1], Thierry Taris[1], Yann Deval[1], Didier Belot[2], Robert Plana[3]
and Jean-Baptiste Bégueret[1]

[1]IMS-Bordeaux, Talence, France, [2]STMicroelectronics, Grenoble, France, [3]LAAS, Toulouse, France

Abstract — In this paper, we present an injection locked oscillator operating at 81GHz and intended for automotive radar applications. The topology of the voltage controlled oscillator (VCO) is based on a push-push topology suitable for millimeter wave applications. The synthesis technique is based on a synchronization throw an external sub-harmonic signal. The reference signal, around 8GHz, is converted before being injected into the oscillator by inductive coupling. We describe in this work, the details of this synchronization technique. The synthesizer is implemented in a 0.13μm SiGe BiCMOS technology from STMicroelectronics. The synchronized oscillator exhibits a maximum locking range of 3GHz from 81GHz to 84GHz and a phase noise around -108dBc/Hz at 10MHz from the carrier frequency. The 1.1mm² circuit chip consumes 60mA under 1.8V supply.

Index Terms — Frequency synthesizer, millimeter wave oscillator, Tunable oscillator, Voltage controlled oscillator, synchronization, BiCMOS analog integrated circuit, Millimeter wave radar, radar applications.

I. INTRODUCTION

In the sake to improve car traffic safety and driver assistance, automotive systems radar has recently caught much attention. Several frequency bands have been made available for automotive radar applications referred to Short Range Radar (SRR) and Long Range Radar (LRR). An example of Short Range and Long Rang Radar in a virtual seat belt is depicted in the Fig1. A special 4GHz band has been allocated around 79GHz to investigate future solutions. In this aim, transceivers have been rapidly developed. With the growth in frequency, it is increasingly difficult to achieve oscillators that fulfill the standards requirements. New challenges are opened, therefore researches related to millimeter wave synthesizers have been investigated due to their huge impact on the overall chain performances. It is an issue of great interest, especially regarding considerations such as phase noise, tuning range, output power and chip size.

In literature, several oscillators above 80GHz have been published in different technologies. A non-exhaustive list of those works [1]-[6] is reported in Table I.

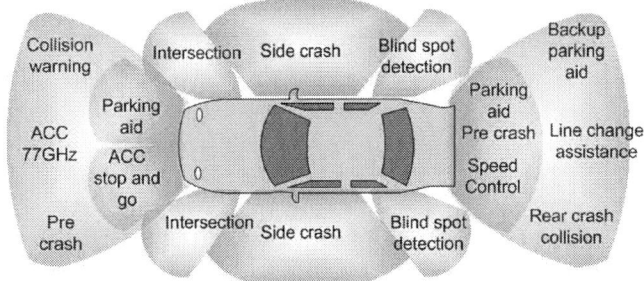

Fig. 1. Virtual seat belt around a vehicle with an association of short range and long range radars

We propose in this work a frequency synthesizer architecture exploiting the principle of injection locked oscillator. This architecture uses a long been known property of oscillators difficult to implement. Therefore, this principle has suffered from classical synthesizer topologies competition.

In this paper, a synchronized push-push oscillator in 0.13μm SiGe BiCMOS process technology is presented. Operating at 82.5GHz, the circuit achieves a -16dBm output power, a phase noise -108dBc/Hz at 10MHz from the carrier frequency and performs a tuning range of 3GHz. The oscillator is synchronized all over the range and consumes 60mA under 1.8V. Due to the hard specifications of the automotive environment, a high phase noise performance must be fulfilled.

II. DESIGN OF MILLIMETER WAVE FREQUENCY SYNTHESIZER

In this section, we first describe the principle of the synchronization

A. Push-push Oscillator

Fig. 2.a) shows the schematic of the VCO. It is based on a Push-push Colpitts topology.

978-1-4244-6240-7/10 $26.00 © 2010 IEEE

TABLE I

COMPARISON OF MMW-VCOs ABOVE 80GHZ

Ref.	Technology	Frequency GHz	Tuning range %	Consumption	Phase noise	FOM[1]
[1]	GaInp/GaAs:HBT	77	3	486 mW	-92 dBc/Hz@1MHz	n.a
[2]	CMOS 90nm	72	2.85	19.2 mW	-112 dBc/Hz@10MHz	176.5
[3]	SiGe BiCMOS	72	8.3	57 mW	-103 dBc/Hz@10MHz	182
[4]	InP-DHBT	74	6	770 mW	-97 dBc/Hz@1MHz	165.5
[5]	SiGe 0.35μm	72	n.a.	190 mW	-102 dBc/Hz@1MHz	175
[6]	0.18 μm CMOS	52.6	n.a.	122.4 mW	-97 dBc/Hz@1MHz	n.a.

$$FOM^1 = 20\log\left(\frac{f_{osc}}{\Delta f}\right) - L(\Delta f) - 10\log(P_{diss})$$

It consists of two symmetrical and individual sub-oscillators, each one operating at 180° out of phase, and working at half output frequency.

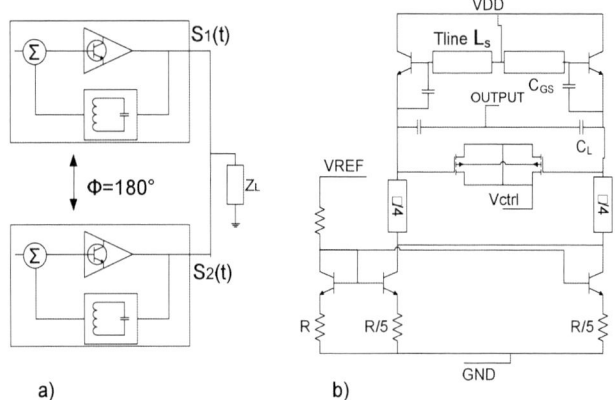

a)   b)

Fig.1 a) Principle of Push-push oscillator. b) 82GHz push-push oscillator schematic.

The signals generated by the two sub-oscillators are expressed in the bellowed equations:

$$s_1 = \sum_0^\infty a_n \sin(\omega_n t + \varphi_n) \qquad (1)$$

$$s_2 = \sum_0^\infty a_n \sin(\omega_n t + \varphi_n + n\pi) \qquad (2)$$

Due to their symmetry, the two oscillators get almost the same spectral components with the same amplitudes a_n but differ in phase by n. The output signal is generated by the sum of the two sub-oscillators. As a result of the sub-oscillators phase differences regarding the fundamental frequency and the odd harmonics, the power distributed to the load is delivered only by the even harmonics. Indeed, frequency contributions of odd harmonics cancel out.

$$s = s_1 + s_2 = \sum_0^\infty 2a_{2n} \sin(\omega_{2n} t + \varphi_{2n}) \qquad (3)$$

Oscillator has been designed in an advanced bipolar technology BiCMOS9mW from STMicroelectronics [7]. The feedback is accomplished by MIM capacitors divider C_L and C_{GS}. In order to achieve the needed frequency range, PMOS varactors are placed in parallel with MIM capacitor C_L to provide a tuning range, according to the following expression:

$$\omega_{sin\,gle\_ended} = \frac{\omega_{push\_push}}{2} = \frac{1}{\sqrt{L_s\left(\frac{(C_L + C_{VAR})C_{GS}}{C_L + C_{VAR} + C_{GS}} + C_{GD}\right)}} \qquad (4)$$

A coplanar line is placed at the transistor base to perform resonance. 50 lines with high quality factor (Q•25@81GHz) are used at the transistor base to reduce 1/f noise. The oscillator is biased by a current mirror degenerated to improve its output impedance and temperature stability. In the aim of power reduction, the first transistor coupled as a diode-connected device is five times smaller than the second transistor that reduces the reference current. High impedance lines of /4 length were added to isolate the biasing stage from the resonant circuit at the operating frequency (Fig.2.b). The VCO has been designed for a center frequency of 82.5GHz with a tuning range of 3GHz.

B. Principle of synchronization

An injection locked oscillator is an oscillator who has an input on which a periodical signal may be applied.

The Fig.3.a) represents the first level of the oscillator synchronization study. Huntoon and Weiss have made a study of the phenomenon without reference to the oscillator inner functioning by replacing the synchronization source with a small variation of the charge impedance. The oscillators synchronization study comes down to have a look to the variation of the oscillations frequency and amplitude versus small impedance variations.

Huntoon and Weiss define elasticity factors as the frequency variation due to the real part and the imaginary part of the previous impedance [8]. Franck Badets provides in its analysis, an expression of these factors for a common emitter Colpitts oscillator in case of parallel injection. The methodology is described in [9].

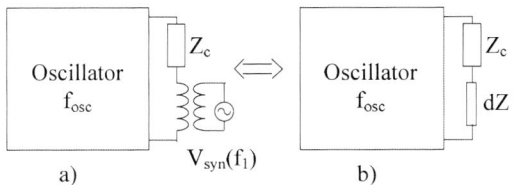

Fig.3. a) Serial injection b) Principle of injection locked oscillator

$$dz = r + jx \quad ; \quad F_r = \left.\frac{\partial F}{\partial r}\right|_{dx=0} ; F_x = \left.\frac{\partial F}{\partial x}\right|_{dr=0} \quad (5)$$

The synchronization voltage can be written as below:

$$V_{sync} = \left|V_{sync}\right| e^{j2\pi f_1 t} \quad (6)$$

With f_1 the synchronization signal frequency. The current I passing through the load is written as below:

$$I = \left|I\right| e^{j\varphi_{osc}(t)} \quad (7)$$

The oscillator's synchronization range is hence given by:

$$\Delta f = 2 \frac{\left|E_F\right|}{\left|I\right|} \left|V_{sync}\right| \quad (8)$$

$$E_F = \left|E_F\right| e^{j\beta} = F_r + jF_x = \sqrt{F_r^2 + F_x^2}\, e^{j\beta} \quad (9)$$

In order to synchronize the VCO on the nth subharmonic of the reference signal, we are going to generate a square signal from the reference signal. To determine the oscillator synchronization range, it is of the most importance to determine the amplitude V_n of the nth order harmonic. The signal Fourier decomposition gives the nth order amplitude :

$$V_n = \frac{\left|V_{sync}\right| \sqrt{2}}{n} \sqrt{1 - \cos(n\pi)} \quad (10)$$

We express bellow, the synchronization range depending on the amplitude V_n and the oscillator components values in case of single-ended Colpitts topology.

$$\Delta f = \frac{\left|V_{sync}\right| \sqrt{2 - 2\cos(n\pi)}}{n \left|I\right|} \frac{\omega^2 C_2 R_L g_m}{2\pi\beta} \frac{g_m^2 + \beta^2 \omega^2 C_1^2}{\left|(\beta+1)g_m^2 - \beta^2 \omega^4 L C_1^2 C_2\right|} \quad (11)$$

Series of inverters transform the reference signal into out of phase square waves in order to generate the subharmonics.

C. Push-push synchronous Oscillator

Fig.4. shows the block diagram of the push-push Colpitts injection locked oscillator as it was implemented. The signal injection is realized by inductive coupling. The square signal generation circuit delivers signals with opposite phases, which will synchronize each suboscillator on the fifth harmonic $f_{single\text{-}ended}/5$ ($f_{push\text{-}push}/10$).

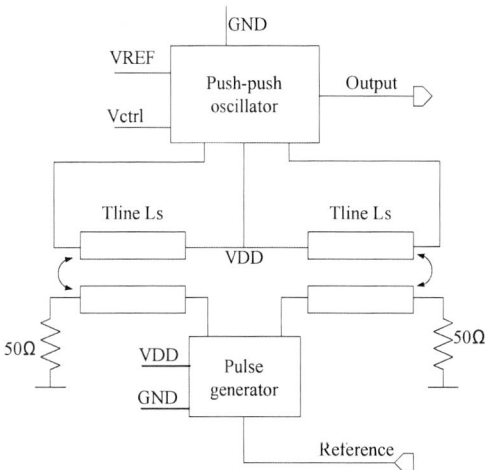

Fig. 4. Synchronized push-push oscillator.

Each oscillator is synchronized separately. Considering the topology choice, the synchronization range is thus twice as important as the one shown in (11).

VI. MEASUREMENT RESULTS

The frequency synthesizer was fabricated in a 0.13μm 250GHz/f_T SiGe BiCMOS process with six metal layers. The top metal was used to realize transmission lines in the circuit. (Fig. 5)

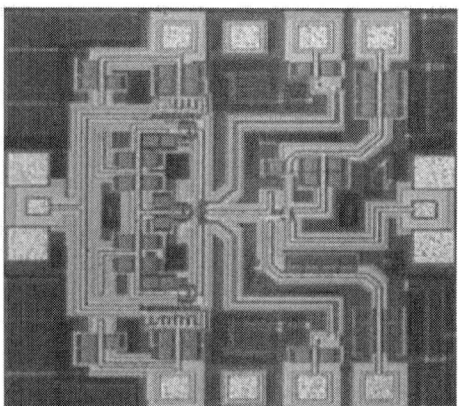

Fig. 5. Chip photograph

The oscillator consumption is 60mA under 1.8V. The oscillator core consumes 20mA. The free running frequency of the VCO is 84GHz. Fig. 6 shows the spectrum with phase noise measurement of synchronized signal by an 8.36GHz reference. The output power of the synthesizer is -16dBm. The phase noise of the synchronous oscillator is -108dBc/Hz @ 10MHz of the carrier frequency. The phase noise of the synthesizer follows the reference phase noise with N=10 multiplier factor.

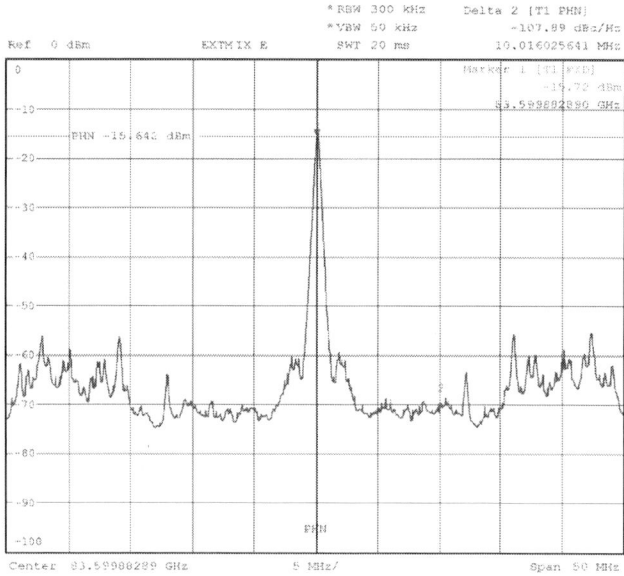

Fig. 6. Spectrum and phase noise of the synchronized oscillator.

The measured performance of the synthesizer is summarized in Table 2.

VII. CONCLUSION

We have presented in this paper the integration of tenth sub harmonic synchronized push-push oscillator. This paper validates the interest of the injection locked oscillator for millimeter wave applications.

TABLE 2
CHARACTERISTICS OF THE SYNTHESIZER

Parameter	Value
Supply voltage	1.8V
Consumption	108mW
FOM[1]	166
Tuning range	3GHz
Output power	-16dBm
VCO free running phase noise	-87dBc/Hz@10MHz
Synchronization range	3GHz (3.75%)
Synthesizer phase noise	-108dBc/Hz@10MHz
Chip size	1.1mm²

ACKNOWLEDGEMENT

The authors thank STMicroelectronics for chip fabrication within the context of the French project VELO.

REFERENCES

[1] Lenk, F. and all, "Low Phase-Noise Monolithic GaInP/GaAs-HBT VCO for 77 GHz," IEEE MTT-S Microwave Symp. Dig., pp. 903-906 Vol.2, June 2003.

[2] De Paola, F.M. and all, "A 71.5-73.5 GHz Voltage-Controlled Standing-Wave Oscillator in 90 nm CMOS Technology," Solide state circuits conference ESSCIRC, pp. 254-257, Sept. 2008.

[3] R Wanner and all, "A SiGe Low Phase Noise Push-push VCO for 72GHz," IEEE MTT-S, pp.1523-1526, June 2005.

[4] R.-E. Makon and all, "Fundamental Low Phase Noise InP-Based DHBT VCOs With High Output Power Operating up to 75 GHz," IEEE Compound Semiconductor Integrated Circuit Symposium, pp159 - 162, October 2004.

[5] H. Li and all, "Wide-band VCOs in SiGe production technology operating up to about 70 GHz," IEEE Microwave and Wireless Components Letters, vol. 13, pp. 425 - 427, Oct. 2003.

[6] Y.-H. Cho and all, "A low phase noise 52-GHz push-push VCO in 0.18 _mbulk CMOS technologies," in IEEE RFIC Symp. Dig. pp. 131–134, June 2005

[7] M. Laurens and all, "A 150GHz fT/fmax 0.13ìm SiGe:C BiCMOS technology," Proc of Bipolar/BiCMOS Circuits and Technology Meeting, pp 199-202, 2003.

[8] R. Huntoon and A. Weiss, "Synchronization of Oscillators," Proceedings of the I.R.E., Vol.35, pp. 1415-1423, December 1947.

[9] F. Badets and all, "A 2.7 V 2.64 GHz Fully Integrated Synchronous Oscillator for WLAN Applications", Proceding of European Solid-State Circuits ESSCIRC, Duisburg, Germany, pp. 508-513, September 1999.

Millimeter Wave CMOS VCO with a High Impedance LC tank

Seung Wan Chai, Jaemo Yang, Bon-Hyun Ku, and Songcheol Hong

Department of EECS, Korea Advanced Institute of Science and Technology (KAIST)
373-1 Guseong-dong, Yuseong-gu, Daejeon, 305-701, Republic of Korea.

Abstract — The proposed VCO achieves low phase noise with high oscillation amplitude by using a high impedance resonator. This resonator has two resonances including the resonance of an additional inductance and a device output capacitance. The proposed VCO is implemented with a 0.13-μm CMOS process and the chip size without the pad is 150um × 250um. The VCO shows phase noise of -94.83dBc/Hz at 1MHz offset, -13.33dBm output power at 55.63GHz oscillation frequency, and the power dissipation of 8.25mW including that of the buffer. In order to provide a comparison with a VCO with a conventional resonator, two VCOs with different resonators are designed. The proposed VCO shows 7.5dB higher output power and 11dB lower phase noise than the VCO with a conventional resonator. The proposed VCO can provide high performance for millimeter wave band applications.

Index Terms — CMOS, 60GHz, millimeter wave, v-band, VCO, RF.

I. INTRODUCTION

Systems utilizing the millimeter wave band, defined as the frequency range from 30 GHz to 300 GHz, have recently been the subject of wide-ranging research. Some frequency bands in this region are currently undergoing standardization and active progress is being made in particular in 60 GHz wireless personal area networks (WPANs), 77 GHz automotive radar systems, and 94 GHz image sensing systems. In particular, the ample unlicensed 60GHz band is a very attractive frequency range for high data-rate communication. Specifically, most countries standardize 7GHz bandwidth in this band. In addition, for low cost and high integration solutions, demand for fabrication with these applications in standard CMOS technology is increasing.

In the millimeter wave band, the performance of CMOS VCOs is strongly affected by the device output capacitance. The device output capacitance increases the phase noise and decreases the signal power and results in high power consumption in the VCO. One potential solution to this problem is a new VCO structure that exploits the device output capacitance.

In this work, a VCO incorporating a resonator with two resonances is proposed. The suggested VCO provides higher oscillation amplitude than the conventional VCO in

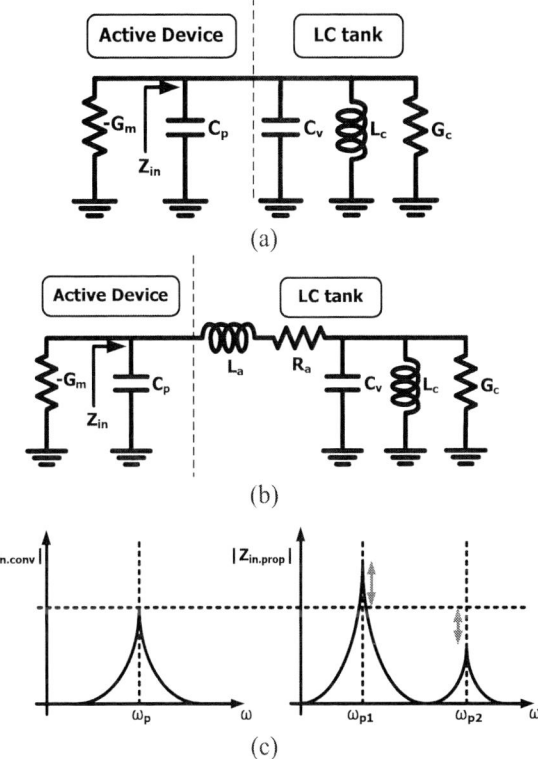

Fig. 1. Characteristic of conventional and proposed LC tank: (a) Equivalent half circuit of the conventional LC tank VCO (b) Equivalent half circuit of the proposed LC tank VCO (c) Comparison of load impedances of two VCOs.

the same bias current, because of the use of a high impedance resonator. According to Lesson's theory [1] and Hajimiri's theory [2], phase noise is in inverse proportion to oscillation amplitude. Therefore, the proposed VCO structure with a high impedance LC tank improves phase noise performance.

The remainder of this paper is organized as follows. In the next section, the concept of the high load impedance LC tank is described and the design of the high voltage swing VCO by the LC tank is presented. In section III, measurement results obtained with the proposed VCO and a conventional VCO are presented. Finally, a summary and conclusion are given in section IV.

II. HIGH VOLTAGE SWING VCO

A. High impedance LC tank resonator

Utilizing a high load impedance LC tank that also uses the additional resonance caused by the output capacitance and an additional inductor, the proposed VCO structure overcomes the problem of degraded phase noise performance by the output capacitance. In the millimeter wave VCO design, the inductor and varactor should be small for high frequency oscillation, and thus the device's output capacitance becomes comparable to that of the varactor. Furthermore, in order to compensate the passive components' loss, which becomes higher at the high frequency region, the transistor should be larger. For these reasons, the capacitance degrades the VCO performance. However, the proposed VCO achieves low phase noise by utilizing the device's output capacitance.

Fig. 1 shows the characteristics of the conventional LC tank and the proposed high impedance LC tank. Fig. 1(a) shows the equivalent half circuit of the conventional LC-tank VCO, where the resonator is composed of an inductor L_c and a varactor C_v. The active devices are employed to provide a negative resistance $-G_m$, and an output parasitic capacitance C_p. The resonance frequency of the conventional VCO is

$$\omega_p = \pm \frac{1}{\sqrt{L(C_p + C_v)}}. \qquad (1)$$

Note that the parasitic capacitance affects the oscillation frequency.

Fig. 1(b) shows the equivalent half circuit of the proposed LC-tank VCO, where the resonator is composed of an additional inductor L_a, and a conventional LC-tank. The additional inductor is located between the varactor and the output capacitance. This inductor makes two resonance points at the LC tank. The resonance points of the proposed tank are given by

$$\omega^2_{p1-p4} = -\frac{1}{2(L_a \| L_c)C_v} - \frac{1}{2L_a C_p}$$
$$\pm \sqrt{\frac{1}{4(L_a \| L_c)^2 C_v^2} + \frac{L_c - L_a}{2L_a^2 L_c C_p C_v} + \frac{1}{4L_a^2 C_p^2}}. \qquad (2)$$

In the millimeter wave region, C_p is almost the same as C_v. Therefore, to simplify equations (1) and (2), let us consider $C_v = C_p = C$ and $L_c = L_a = L$. If equation (1) and (2) are substituted under this assumption, it follows that

$$\omega_p = \pm \frac{1}{\sqrt{2LC}} \qquad (3)$$

$$\omega_{p1,p1'} = \pm j\sqrt{\frac{3-\sqrt{5}}{2LC}} \approx \pm j\frac{0.62}{\sqrt{LC}} \qquad (4)$$

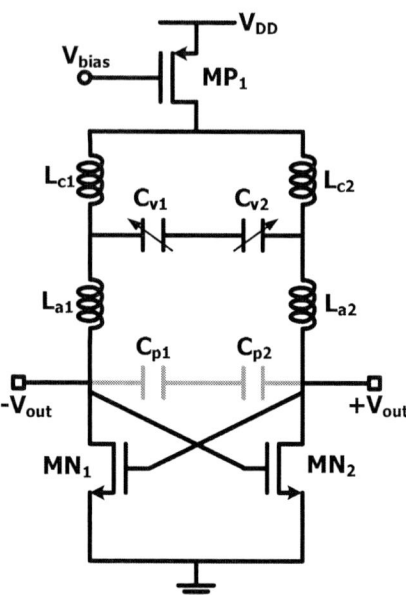

Fig. 2. High voltage swing VCO with high impedance LC tank

$$\omega_{p2,p2'} = \pm j\sqrt{\frac{3+\sqrt{5}}{2LC}} \approx \pm j\frac{1.62}{\sqrt{LC}}. \qquad (5)$$

The results of equation (3)-(5) can be varied by the value of L or C, but the two resonance points are slightly far from each other, i.e. if ω_{p1} is 60GHz, ω_{p2} is 160GHz. In this work, the proposed VCO uses a lower frequency ω_{p1} for high voltage swing operation. If we take ω_{p1} as 60GHz, ω_{p2} will exceed f_{max}. Hence, at the higher frequency, ω_{p2} will not work. As shown in Fig. 1(c), the two resonance points have different impedances at the proposed LC tank's resonance frequency. Comparing the impedance of the two resonance points, ω_{p1} has higher impedance than ω_{p2}. This is the reason why the proposed VCO chooses the proposed high impedance LC tank. The impedance at ω_{p1} is also higher than the impedance of the conventional LC tank at ω_p. Therefore, the proposed VCO achieves higher oscillation amplitude than the conventional VCO at the same bias condition.

B. Design of the proposed VCO and conventional VCO

Fig. 2 shows the schematic of the proposed VCO, where the cross-coupled pair MN_1 and MN_2 is used to provide the required negative conductance and the resonator is a high impedance LC tank resonator, as described in the previous section. The high impedance LC tank resonator is composed of an additional inductor L_a, an output parasitic capacitance, C_p, and a conventional LC tank L_c, C_v. In consideration of the flicker noise contributed by active devices and tuning range contributed by varactor dc bias, a

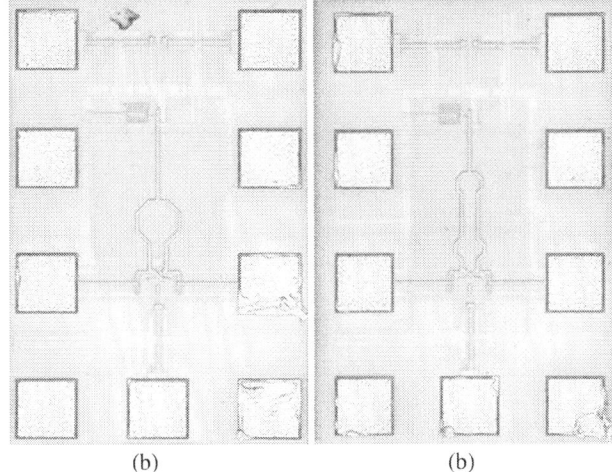

(b)   (b)

Fig. 3. Chip micrographs of (a) the conventional VCO and (b) the proposed high voltage swing VCO (330μm×510μm including pads)

pMOS transistor is employed as the tail current, and consequently lower phase noise can be achieved and lower dc bias at V_{out}.

Two VCOs have been designed to compare the proposed VCO with the conventional VCO. They are composed of same sized components except for their inductors for a fair comparison. The inductors are designed to have the same operating frequency.

Fig. 4 shows chip micrographs of the two VCOs, which are 330μm×510μm with a pad, and 150μm×250μm without a pad. The size of the conventional VCO and the proposed VCO is similar, because the inductor size is very small in the millimeter wave VCO design (conventional inductor: $65×65μm^2$, proposed inductor: $40×110μm^2$).

III. MEASUREMENT RESULTS

The high voltage swing VCO and conventional VCO are measured on wafer probing. Two VCOs are measured

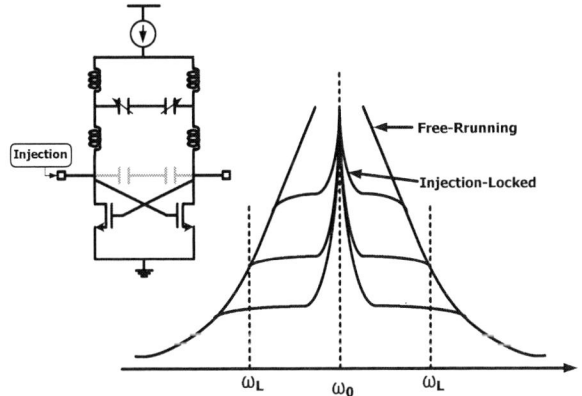

Fig. 4. Measurement phase noise setup by injection-locked.

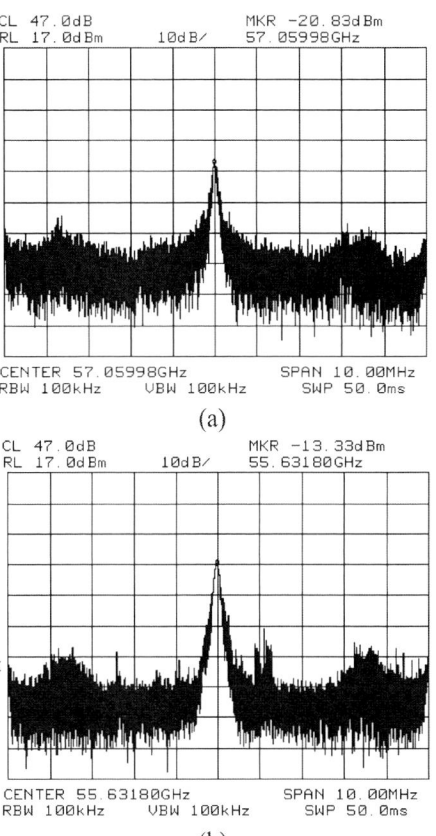

(a)

(b)

Fig. 5. Output spectrum of the two VCOs: (a) the conventional LC tank VCO (b) the proposed LC tank VCO

with 5.4mA bias current from 1.5V supply.

The millimeter wave source is very sensitive to other noise sources, and hence this work uses a low power injection-locked method to measure the VCO's performance. When a low power signal source, located close to the noise floor, is injected at one side of the output node, as shown in Fig. 4, the oscillation frequency is locked by the low power frequency source. The injection signal should be a low power signal, because the injection-locked region is determined by the injection source power levels. At high injection levels, the phase noise of the VCO follows the phase noise of the injected signal source. However, at low injection levels, the injection-locked region is reduced and the phase noise of the VCO follows the phase noise itself in the out band as shown in Fig. 4 [3]-[6].

Fig. 5 shows the output spectrum of the two VCOs, the conventional VCO and the proposed VCO. As anticipated in section II.A, the proposed VCO's output power, -13.33dBm at 55.63GHz, shown in Fig. 5(b), is considerably higher than the conventional VCO's output power, -20.83dBm at 57.06GHz, shown in Fig. 5(a), at the

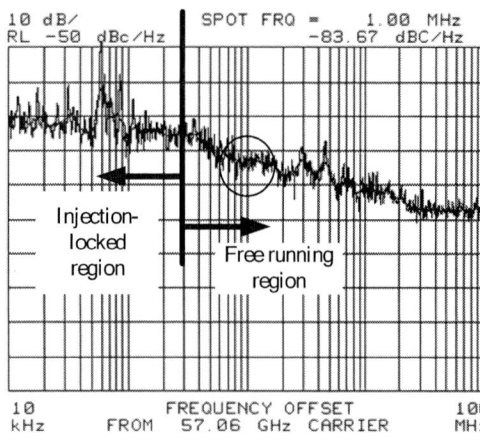

(a)

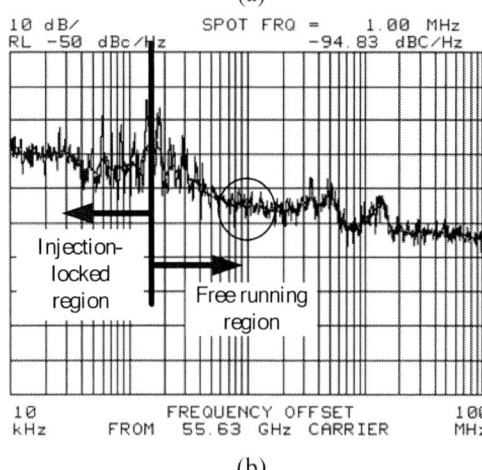

(b)

Fig. 6. Phase noise of two VCOs: (a) the conventional LC tank VCO (b) the proposed LC tank VCO

same bias current. This demonstrates that the proposed VCO has much higher load impedance than the conventional VCO. According to the output power, phase the noise performance of the proposed VCO is also better than that of the conventional VCO, as shown in Fig. 6. Fig. 6(a) shows the phase noise of the conventional VCO, i.e., -83.67dBc/Hz at 1MHz offset, and Fig. 6(b) shows the phase noise of the proposed VCO, i.e., -94.83dBc/Hz at 1MHz offset. The performance of the proposed VCO is summarized in Table I along with that of a previously reported CMOS VCO for comparison.

IV. CONCLUSIONS

The proposed VCO, which has a high voltage swing with a high impedance resonator, is advantageous for application in the millimeter wave region. The device output capacitance of a transistor in a VCO has a

TABLE I
PERFORMANCE SUMMARY OF THE V-BAND CMOS VCOS

	This work	[7]	[8]	[9]
Technology	0.13-μm CMOS	0.13-μm CMOS	0.13-μm CMOS	0.18-μm CMOS
Frequency[GHz]	55.63	59	59.1	63
Tuning Range[GHz]	1.75	5.8	7.09	0.67
Power Cons.[1][mW]	8.25	19.8	7.7	74
PN@1MHz[dBc/Hz]	-94.83	-89	-91	-89
Power Level[dBm]	-13.33	-10	-15	-15
FoM[dBc/Hz]	180.57	174.13	177.56	166.29

1. The power consumption includes buffer power consumption

significant influence on the phase noise performance. However, the present work proposes a new LC tank structure that exploits the output capacitance. The proposed VCO with a high impedance LC tank resonator yields decreased phase noise and increased output power. Since the effect of the output capacitance becomes more serious at the higher frequency region, this structure is effective at the millimeter wave region.

REFERENCES

[1] D. B. Lesson, "A simple model of feedback oscillator noise spectrum," *Proceedings of the IEEE*, vol. 54, pp. 329-330, February. 1966.

[2] T. H. Lee, and A. Hajimiri, "Oscillator phase noise: A tutorial," *IEEE J. Solid-State Circuits*, vol. 35, no. 3, pp. 326-336, March 2000.

[3] R. Adler, "A study of locking phenomena in oscillators," Proceedings *of IRE and Waves and Electronics*, vol. 34, pp. 351-357, June. 1946.

[4] B. Razavi, "A study of injection locking and pulling in Oscillators," *IEEE J. Solid-State Circuits*, vol. 39, no. 9, pp. 1415-1424, September 2004.

[5] X. Zhang, B. J. Rizzi, and J. Kramer, "A new measurement approach for phase noise at close-in offset frequencies of free-running oscillators," *IEEE Trans. Microwave Theory Tech.*, vol. 44, pp 2711-2717, December 1996.

[6] J. Byunghoo, Ramesh Harjani, "High-Frequency LC VCO Design Using Capacitive Degeneration," *IEEE J. Solid-State Circuits*, pp 2359-2370, December 2003.

[7] C. Cao and K. K. O, "Millimeter-wave Voltage-Controlled Oscillators in 0.13μm CMOS Technology," *IEEE J. Solid-State Circuits*, vol.41, no.6, pp.1297-1304, June 2006.

[8] J. Borremans, M. Dehan, K. Scheir, M. Kuijik, and P. Wambacq, "VCO design for 60GHz applications using differential shielded inductors in 0.13μm CMOS," *IEEE Radio Freq. Integrated Circuits Symp.*, pp. 135-138, June 2008.

[9] Hsieh-Hung Hsieh and Liang-Hung Lu, "A 63-GHz Voltage-Controlled Oscillator in 0.18-μm CMOS," *IEEE Symp. on VLSI circuits*, pp 178-179, June 2007.

RTUIF-14

Controlled Dither in 90 nm Digital To Time Conversion Based Direct Digital Synthesizer for Spur Mitigation

S. Talwalkar, T. Gradishar, B. Stengel, G. Cafaro, and G. Nagaraj

Motorola Inc., Plantation, FL, 33322, USA

Abstract — **Dithering is used in many discrete to continuous value conversion functions to provide an effective fractional value. This paper reviews the application of dither to a digital-to-time converter (DTC) based digital synthesizer suitable for many common wireless communication systems. Measurements of a 90 nm CMOS implementation using a 5 bit DTC show extension to effective 8 bits.**

Index Terms — **Delay Line, Digital to Time Conversion, Direct Digital Synthesis, Dither, Fractional Delay, Spur Mitigation.**

I. INTRODUCTION

Digital to time conversion (DTC) based direct digital frequency synthesizers (DDFS) have been reported in [1] and [2]. The DTC architecture offers many advantages of a traditional (digital to sine amplitude based) DDFS architecture such as wide frequency range (e.g., 100 MHz to 2.5 GHz in [1]), fast settling time (~ 5 ns in [1]) and high frequency resolution (15 Hz in [1]). However, the DTC architecture is also prone to the disadvantage of spurious tones (spurs) similar to the traditional DDFS [3]. The DTC architecture involves selectively combining rising and falling edges from a delay locked loop (DLL) to create the output waveform at the desired frequency. Periodic time domain edge location errors translate to the presence of undesirable spurious tones, spurs, in the output spectrum. This paper examines the utility of introducing controlled randomness in the edge placement algorithm for the purpose of correcting the edge errors.

The organization of this paper is as follows. Section II describes the architecture of a digital to time conversion based frequency synthesizer and explains the source of periodic edge location errors. The next section explains the mechanism of introducing randomness for the correction of these edge location errors. Finally measurement results from a DTC based direct digital synthesizer are presented.

II. DTC ARCHITECTURE OVERVIEW

Fig. 1 describes a DTC synthesizer architecture using an example of the reference period split into four nominally uniform delays. Depending on the frequency word (to control the desired output frequency), the tap selection digital logic (running at the reference clock) selects the two multiplexer control signals CS and CR. The multiplexers connect none (shown as GND input in the figure) or one of the buffer outputs to the set or reset inputs of the S-R flip flop.

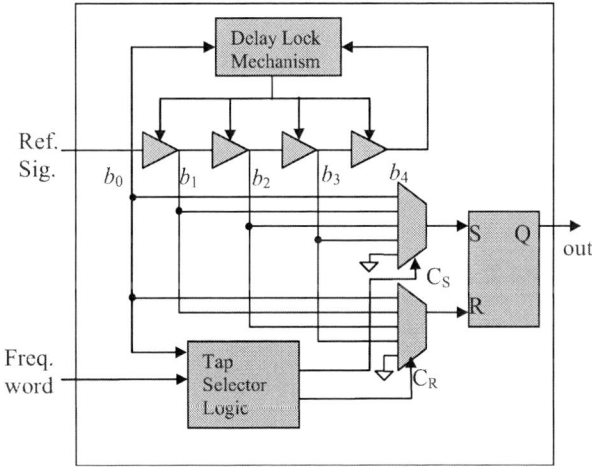

Nominal delay of each buffer = $d = T_{REF}/4$

Fig. 1. The DTC Synthesizer Architecture with four buffers each with a nominal delay of TREF/4

The scheme can generate a square wave output signal with a maximum frequency equal to the reference frequency, whereas the traditional DDFS are theoretically limited to half the reference frequency. Example waveforms for a desired output period 1.5 times the reference period are shown in Fig. 2.

978-1-4244-6240-7/10 $26.00 © 2010 IEEE

GND => no edge is picked for that reference cycle

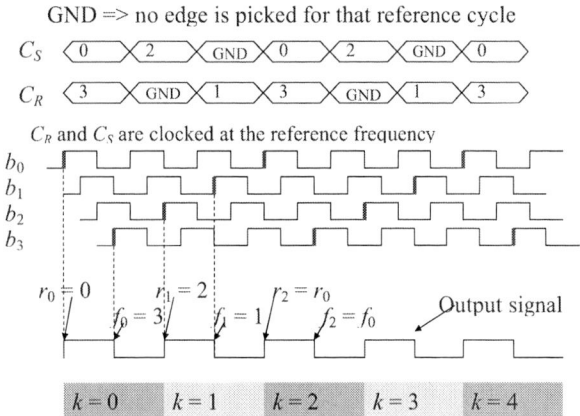

k = index for the desired output period;
r_k (f_k)= buffer number that is used to create the kth rising (falling) edge of the output

Fig. 2. Example control waveforms to cause an output period of 1.5 times the reference period. For each output edge, dotted lines show the corresponding reference edge (highlighted) used to create that output edge

Edge location errors are the root cause of undesirable spurs. There are two distinct types of edge location errors. The first type of error (hereafter referred to as quantization error) is due to the quantization of an edge location to the available number of buffers. For example, consider trying to generate an output period of 1.4 times the reference period in the example of Fig. 1. The other type of edge error (hereafter referred to as buffer errors) is the non-nominal buffer delays due to circuit level non-idealities. Since the buffer sequences used to create rising and falling edges are periodic, so are the edge location errors. This periodicity gives rise to spurious tones in the output spectrum. This work reports the measured reduction of these spurious tones as a result of applying edge location correction by introducing a controlled randomness in the selection of the taps to cause an effective fractional delay. The reduced spur power comes at the cost of an increased noise floor.

III. DITHER INSERTION MECHANISM

The dither is added using a pseudo random generator that is controlled to generate the desired two point probability mass function over neighboring buffers [4]. This theoretically extends the total resolution to 10 bits. For example, to create an effective delay of 2.25 buffers, the probability mass function is designed as Fig. 3.
Probability of selecting buffer 2 = 0.75
Probability of selecting buffer 3 = 0.25

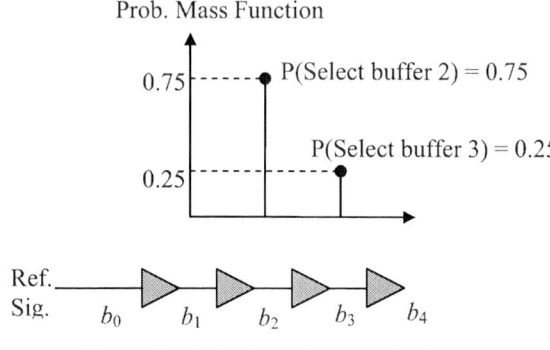

Average delay = (0.75x2+0.25x3) x Nominal Buf. Delay
= 2.25*Nominal Buf. Delay

Fig. 3. Probability mass function imposed to create an effective fractional location at (2.25 * nominal buffer delay)

IV. MEASURED RESULTS

The measured results are reported on a 90 nm CMOS synthesizer that has a frequency range from 2 – 1000 MHz with a 15 Hz or better resolution. The reference frequency is 1 GHz. The DLL has 32 buffers, each with a nominal delay of 31.25 ps (= 1 ns / 32). A 32-buffer delay corresponds to a 5 bit resolution.

Fig. 4 shows the effect of controlled dither to achieve fractional delay in order to mitigate the effects of quantization error. The plot on the top shows the measured spectrum without the introduction of the dither correction for the quantization error while the plot at the bottom shows the spectrum with the dither correction. The output frequency is 996.108949 MHz. Although the total spur spectrum is due to both quantization and buffer errors, the spurs at some frequencies are dominated by quantization errors. Such frequency locations can be determined using analysis similar to [5]. Figures show the desired signal as well as three strongest quantization spurs. In these figures markers are placed on the desired signal and the three strongest quantization spurs. With dithering, the spurs go down by an average of 27 dB. This significant improvement in the spur level comes at the cost of an increase in the noise floor from -141 dBc/Hz to -115 dBc/Hz. Note that the measured images from the instrument were manually edited to selectively magnify the marker table portion of the figure for readability.

978-1-4244-6240-7/10 $26.00 © 2010 IEEE

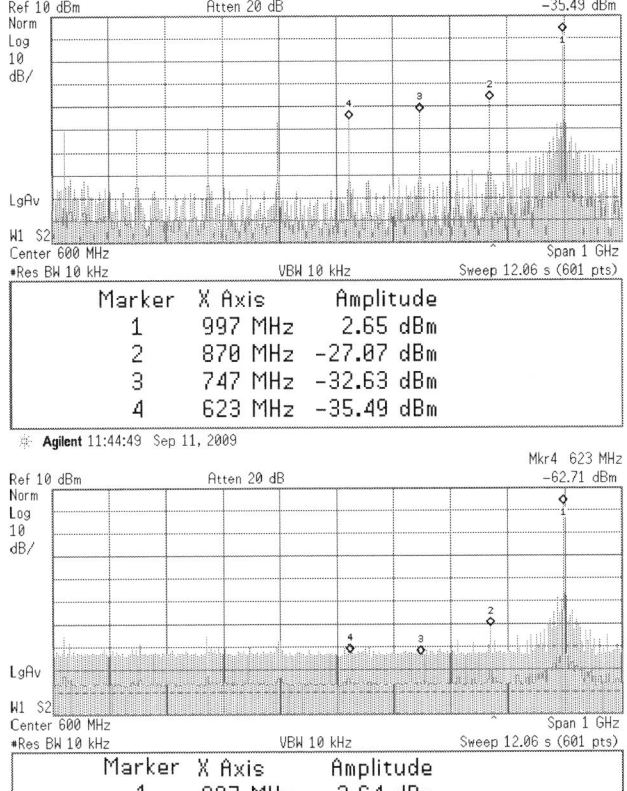

Fig. 4. Measured spectrum showing quantization error spurs before (top) and after (bottom) applying the dither correction

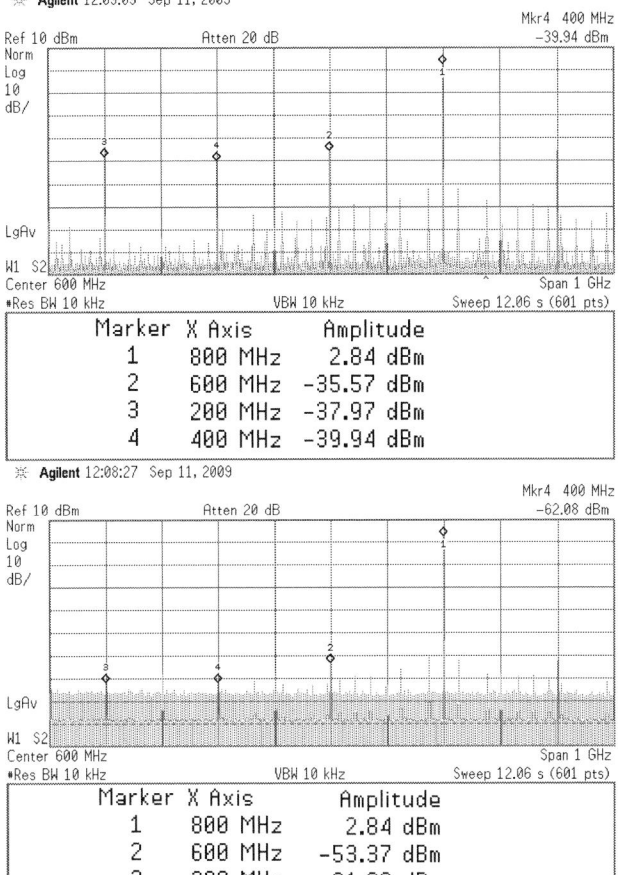

Fig. 5. Measured spectrum showing buffer error spurs before (top) and after (bottom) applying the dither correction

Fig. 5 shows the effect of injecting controlled dither for the purpose of correcting the buffer errors. Here, an output frequency of 800 MHz is chosen deliberately to ensure that there are no quantization errors. All the spurs are due to buffer errors alone. Again, the spectrum at the top is measured without the dither correction while the one at the bottom is with the dither correction. The optimum correction values to introduce with the dither are found by sweeping over possible fractional delay correction values. The spurs are reduced by an average of 21 dB. Again, enabling the dithering raises the noise floor from -143 dBc/Hz to -118 dBc/Hz

It is important to mention that the exact amount of dB reduction of spurs is dependent on the desired output frequency. Measured results demonstrate the utility of controlled dither correction to mitigate the spurs in a digital to time conversion based direct digital frequency synthesizer. The dither reduces average spur power by at least 20 dB. This corresponds to an increase of about 3 bits (of the expected 5 bits) using the conversion factor of 6 dB per bit. We are actively working on determining additional noise mechanisms that may be limiting the effectiveness of dither. The strength of spurs can be mapped to the time domain accuracy of edge placements expressed in terms of the standard deviation of the edge location errors. Dithering reduces the edge error standard deviation from 6 ps to about 0.6 ps. Fig. 6 shows the dependence of spur level on the standard deviation of the edge location error.

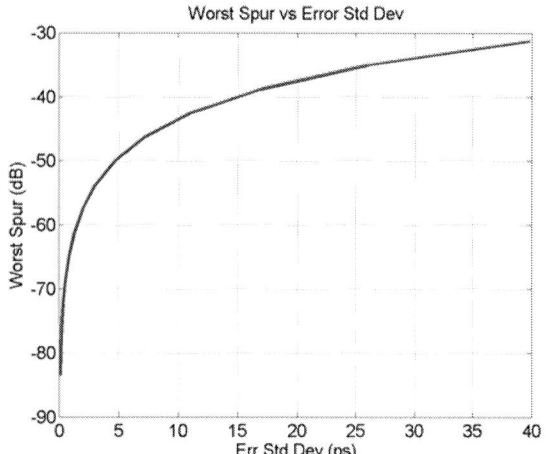

Fig. 6. Relationship between the error standard deviation and the worst case spur found via simulation

The measured phase noise for the synthesizer is shown in Fig. 7. As mentioned before, the insertion of dither raises the noise floor by approximately 20 dBc/Hz. The cycle to cycle jitter of the un-dithered reference has a peak jitter of 20 ps. Dithering degrades this to about 62 ps peak cycle to cycle jitter.

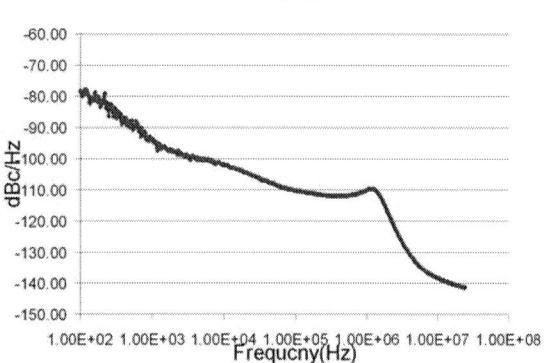

Fig. 7. Measured phase noise

Fig. 8 shows the chip micrograph for the IC that contains the DTC synthesizer capable of injecting controlled dither. The IC is designed in 90 nm CMOS with a 1.2 V supply. The sub-blocks of the IC relevant to the presented work have a die area measuring 0.860 x 1.82 mm2. The corresponding current drain is 65 mA for the digital section (tap selection and dither injection logic) and 57 mA for the analog section (reference generating PLL, 32-buffer delay locked loop and the output network) of the DTC synthesizer. The output power is about +3 dBm into a 50 ohm load.

IV. CONCLUSIONS

Insertion of controlled randomness in the placement of edges in the time domain is shown to be an effective method for spur reduction in the digital to time conversion (DTC) based direct digital frequency synthesis architecture. Measurements on a 90 nm CMOS IC show that the dither helps reduce spurs by 20 to 30 dB. In terms of the time domain error, the introduction of dither correction reduces the edge error standard deviation from about 6 ps to about 0.6 ps. The spur mitigation is achieved at the cost of an increase in the noise floor.

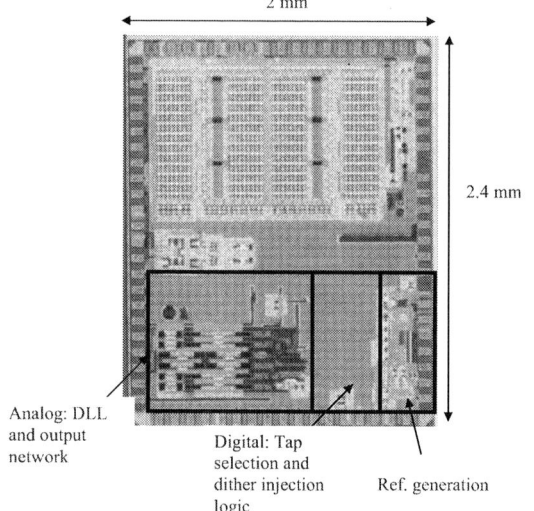

Fig. 8. Chip micrograph showing sub-blocks relevant to the DTC synthesizer

REFERENCES

[1] G. Cafaro, T. Gradishar, J. Heck, S. Machan, G. Nagaraj, S. Olson, R. Salvi, B. Stengel, and B. Ziemer, "100 MHz – 2.5 GHz Direct Conversion CMOS Transceiver for SDR Applications", IEEE Radio Frequency Integrated Circuits Symposium, pp. 189-192, June 2007.

[2] T. Rapinoja, K. Stadius, L. Xu, S. Lindfors, R. Kaunisto, A. Parssinen, and J. Ryynanen, "A digital frequency synthesizer for cognitive radio spectrum sensing applications", IEEE Radio Frequency Integrated Circuits Symposium, pp 423 – 426, June 2009.

[3] H. Nicholas, and H. Samueli, "An analysis of the output spectrum of direct digital frequency synthesizers in the presence of phase accumulator truncation", 41st Annual IEEE Frequency Control Symposium, pp 495 – 502, 1987.

[4] US Patent 7,421,464, Gradishar T., and B. Stengel, "System and method for introducing dither for reducing spurs in digital-to-time converter direct digital synthesis", 2008.

[5] B. Izouggaghen, A. Khouas, and Y. Savaria, "Spurs modeling in direct digital period synthesizers related to phase accumulator truncation", ISCAS 2004, pp 389 – 392.

RTUIF-15

2-4 and 9-12 Gb/s CMOS Fully Integrated ILO-based CDR

O. Mazouffre[1], R. Toupé[1], M. Pignol[2], Y. Deval[1] and J.B. Begueret[1]

[1] IMS Laboratory, University of Bordeaux, Talence, France

[2] CNES (Centre National d'Etudes Spatiales), Toulouse, France

Abstract — **A CDR dedicated to satellite data link is presented. The clock recovery function is made-up of an Injection Locked Oscillator combined with an analog phase alignment circuit. The circuit covers two bit-rate ranges: 2.2 to 4.3 Gb/s and 9.1 to 12.1 Gb/s. It was designed in 130 nm CMOS bulk process from STMicroelectronics. The overall power dissipation is 400 mW in the first bit-rate range and 480 mW in the second including 220 mW for I/O buffers. The eye opening at 10^{-9} of bit error rate is 940 mUI/440 mV at 3.1 Gb/s and 720 mUI/300 mV at 10.3 Gb/s.**

Index Terms — **Clock and data recovery circuit, CDR, Injection Locked Oscillator, ILO, Satellite, CMOS, gigabit.**

I. INTRODUCTION

Usually, fully integrated CDR circuits are designed around a Phase-Locked Loop to recover the clock from the data stream. The PLL is a well-known function that suits terrestrial high-speed data link applications. However, for space applications, PLLs are prone to be sensible to radiation-induced Single Event Transients.

Two PLL building blocks are especially sensitive: the phase detector and the Voltage Controlled Oscillator. The widely used Early/Late phase detector is a digital sequential circuit, which is naturally sensitive to SETs. This phase detector can be hardened, like in [1], but these logic gates are generally slow and not suitable for very high-speed applications. The VCO is another radiation sensitive circuit as shown in [2]. The SETs have the effect of turn-off the oscillation. The sensitivity of both Early/Late and VCO, combined with the relatively low bandwidth of PLLs, could conduct to the lost of a high number of bits during transmission. When the data are stored in memory, the replaying of altered frames can insure their integrity. But for real-time transmission from a detector, replaying data could be impossible and the loss definitive. So, there is a great interest to develop hardened CDRs, especially for scientific probes and satellites.

In our work, we have hardened our CDR circuit thanks to an Injection Locked Oscillator. Indeed, this study [3] has demonstrated that the injection naturally hardens the oscillator. Thanks to the injection process, the VCO recovers rapidly from SETs. However, one drawback of the ILO is that the recovered clock phase depends on the input pattern; consequently, the clock jitter is rather high and some bits can be incorrectly sampled. Another

difficulty with ILOs is that the locking range becomes small for low transitions rate in the data pattern, as shown in [4]-[5]. For instance, the locking range of the 20 Gbit/s ILO based CDR in [5] is only about 0.1% of the nominal bit-rate.

In order to correct these drawbacks, we have added a phase alignment circuit that controls the oscillator free running frequency. With this circuit, the phase of the clock is always aligned on the data, and, the locking range is also widened. This improvement was demonstrated by our previous work [6], where the locking range of the ILO-CDR is multiplied by more than three with the phase alignment enabled.

In addition with the same ILO-CDR, in [7], we have demonstrated that the injection process not only hardens the oscillator, but all the system because of the natural high bandwidth of ILO. The measured equivalent cross-section to heavy-ion radiations of the designed circuit is two to four magnitudes lesser than the one of the commercial CDR HDMP-1034 Glink-Rx used by CNES for Pleiades Satellites.

II. DESIGN

The ILO-CDR presented in this paper is an improved version of our previous circuit in terms of covered bit-rates, jitter, temperature and power-supply sensitivity. More, this new circuit was designed with 75% de-rating on transistors maximum voltages, according to spatial mandatory rules for reliability. We have also duplicated and sized critical biasing functions to mitigate the effect of radiations at the expense of increase of the power dissipation and silicon area. The circuit was designed to covers the -55°C to 125°C temperature range with ± 15% variation on the power supply voltage.

A. CDR Architecture

The circuit is composed of two CDR cores sharing the I/O buffers as shown Fig. 1. The first core was designed to covers 2 to 4 Gb/s while the second covers 9 to 12 Gb/s. The CDR input and outputs are differential with a differential impedance of 100 Ω.

The two cores have the same architecture presented in Fig. 2. The cores are made of six programmable delay lines that generate four data signals that have a relative time shift of 0, 1/4, 2/4 and 3/4 UI (Unit Interval). These

978-1-4244-6240-7/10 $26.00 © 2010 IEEE

signals are connected to two XOR gates inputs. At the XOR gates outputs, the generated signals have the shape of pulses with a width of 1/2 UI. These pulses are generated on rising and falling edges of input data.

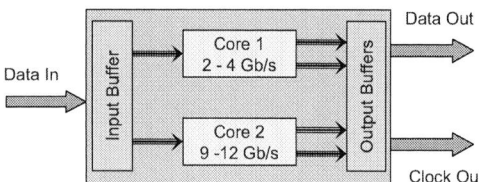

Fig. 1.   CDR architecture.

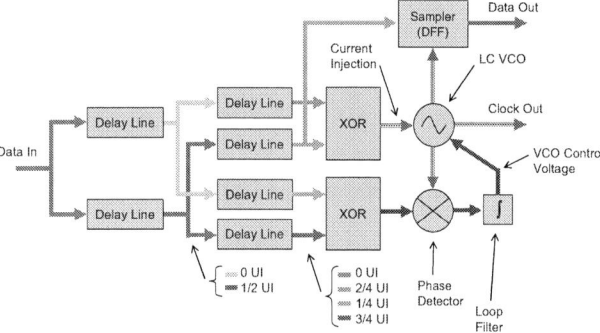

Fig. 2.   Schematic of CDR cores.

The two pulses streams have a time shift of 1/4 UI needed for the operating of the mixer-based phase detector. The first stream achieves the locking by current injection of the LC voltage controlled oscillator. The second stream is applied to the phase detector that realizes the phase alignment function. The loop filter of the phase alignment circuit is a capacitor that, combined with the current output of the phase detector, realizes an integrator. The stability of the loop is achieved thanks to the injection locking that has a first order response.

B. HCMOS9GP Process and CML Gates

The circuit was designed in a HCMOS9GP process from STMicroelectronics. It features 130 nm CMOS transistors with standard and low voltage threshold (V_{TH}), MIM capacitors and inductors. Due to the high operating frequency, we have designed all the logic gates with Current Mode Logic. The CML gates use NMOS differential pair with poly-silicon resistor loads. All transistors are High Speed transistors (HS) with low V_{TH}.

The CML gates are biased by a gm·R constant biasing circuit. So the CML gates amplification is almost independent of power supply voltage, temperature and process variations. For data paths, we used a non-linear active shunt-peaking that reduces the jitter. The typical schematic of theses CML gates is depicted in Fig. 3. The

maximum operating frequency of the CML gates is about 15 GHz.

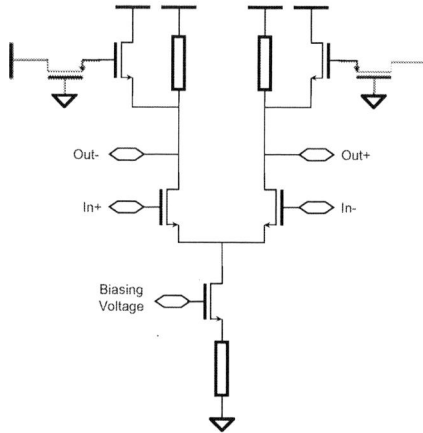

Fig. 3.   Simplified schematic of a CML gate with active shunt-peaking.

C. Voltage Controlled Oscillators

The two VCOs are negative-resistance type with LC loads. The voltage frequency control is achieved by using a couple of accumulation-mode varactors. Integrated inductors are used. The binary frequency control is made of four switched MIM capacitors with size increasing from one to four, so the VCOs present sixteen frequency sub-bands. There is enough overlap of the VCOs sub-bands to achieve continuous covering of the two bit-rate ranges. An OTA-based automatic level control is used to maintain the oscillation level under all operating conditions and limit the stress of transistors due to overvoltage. The simplified schematic of the oscillator used in the CDR cores is depicted in Fig. 4.

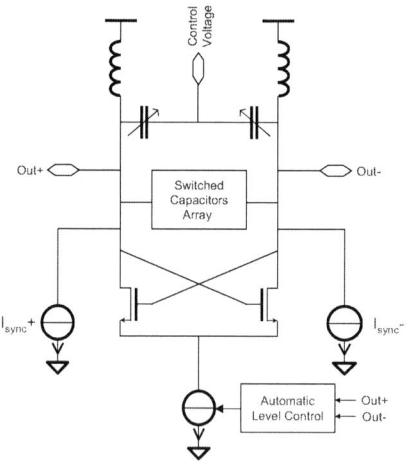

Fig. 4.   Simplified schematic of VCOs.

In post-layout simulation without injection, the first VCO covers 9.2 to 13.7 GHz. At 10 GHz, the VCO phase noise is -79 dBc/Hz @ 100 kHz. The second VCO covers 2.2 to 4.6 GHz.

The synchronization current I_{sync} is directly injected at the VCO outputs. The width of current pulses is 1/2 UI with an amplitude of 3.2 mA. With these values the synchronization range are 370 MHz around 10 GHz and 170 MHz around 3 GHz for a PRBS 2^7-1 input pattern.

D. Programmable Delay Lines

The realization of programmable delays is achieved using variable capacitive loads at the outputs of several CML buffers connected in series. The delay is roughly proportional to the total capacitive load.

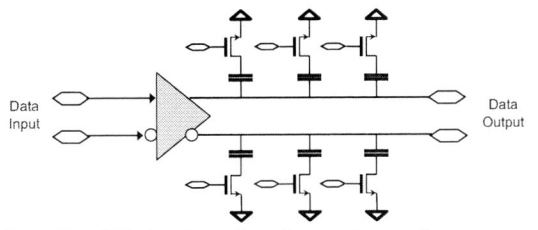

Fig. 5   Simplified schematic of one stage of programmable delays.

Fifteen pairs of switched identical capacitors are used to achieve sixteen different delays (4 bits control word). The switched capacitors are distributed evenly to minimize the capacitive load at each buffer output, thus minimizing the jitter. Combined with CML buffers featuring shunt-peaking, this topology generates nearly zero jitter despite the high number of gates used.

In simulation, for the 9-12 Gb/s CDR core, the delays cover 7.8 to 14 Gb/s bit-rates. For the 2-4 Gb/s core, the delays cover 2 to 4.8 Gb/s. The bit-rate ranges covered by the delays are slightly larger than the VCOs frequency ranges in order to take into account for post-layout extraction incertitude and process dispersions.

III. CDR SIMULATION

The CDR was simulated at room temperature with 1.5 V power supply and a PRBS 2^7-1 input signal. We used extracted capacitors and/or resistors from the layout. The I/O buffers draw 115 mA from the supply. The 9-12 Gb/s core draws 155 mA and the 2-4 Gb/s 120 mA.

After powered on, the CDR needs about 10 to 50 ns to be locked with a fully stabilized biasing. If the VCO is already running, the locking time is reduced to less than few nanoseconds thanks to the current injection.

Without package and printed board circuits, the data output eye opening is about 920 mUI and 830 mV at 10 Gb/s. At 3 Gb/s it is 960 mUI and 850 mV. For the 10 Gb/s delays and VCO configuration, the CDR covers 9.9 to 10.4 Gb/s. For the 3 Gb/s configuration, the CDR covers 2.8 to 3.1 Gb/s. The bit-rate range is mainly defined by the VCO frequency range. The simulated clock

output power is -2.4 dBm at 10 Gb/s and it is almost constant for all bit-rates.

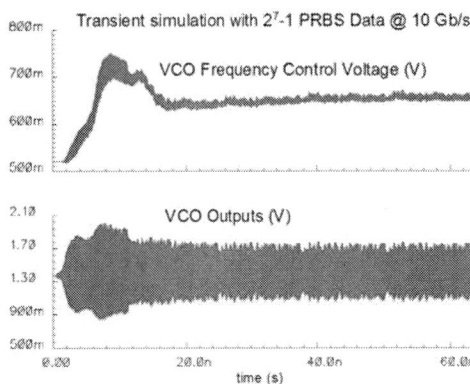

Fig. 6   VCO Outputs and VCO Control voltage waveforms at start-up.

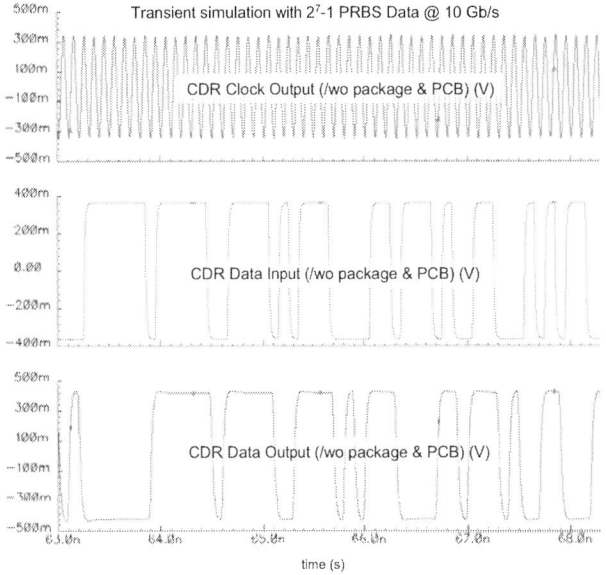

Fig. 7   Data input, clock output and data output waveforms at 10 Gb/s without package and printed board circuit.

IV. MEASUREMENTS

The measurements were realized with a Data Quality Analyser MP1800A from Anritsu using a PRBS 2^{31}-1 data pattern at room temperature. The chip was mounted in QFN64 package and the overall assembled on a glass-epoxy FR4. Because of a weakness in the clock output buffer, the biasing current of output buffers was increased by 20% to achieve 12.1 Gb/s maximum operating bit-rate. We also added a 100 Ω external resistor between the data output to enhanced the eye opening which was degraded by the parasitic effects of the used package. The circuit die is presented in Fig. 8. Its size is 1.6 per 2.4 mm.

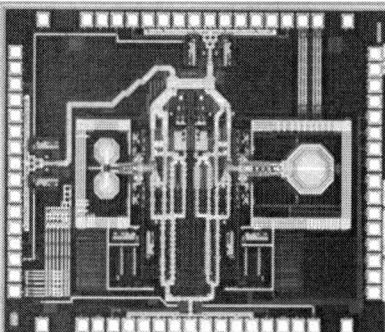

Fig. 8   Die photography.

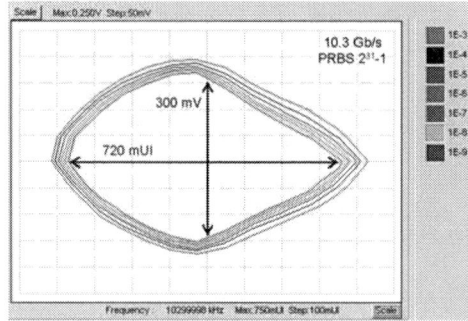

Fig. 9   Measured eye diagram at 10.3 Gb/s.

The CDR is operational from 2.2 to 4.3 Gb/s and 9.1 to 12.1 Gb/s. The circuit current consumption is in agreement with the simulation. At 10.3 Gb/s the eye opening is 720 mUI and 300 mV for 10^{-9} BER, it increases to 940 mUI and 440 mV at 3.1 Gb/s. Measurements and simulation shows that the eye closing at high bit-rate is mainly due to the package (Data Dependant Jitter issue).

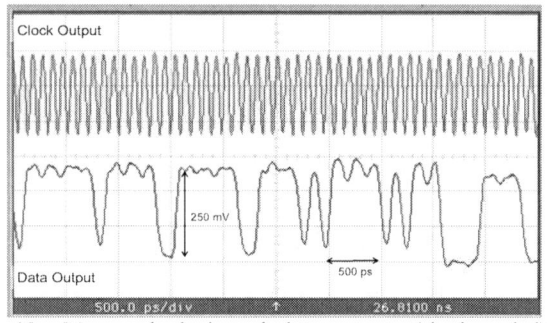

Fig. 10   Measured clock and data outputs (single-ended) at 10.3 Gb/s.

The circuit tolerance to input jitter is high, above the jitter generation limit of MP1800A (220 mUI @ 40 MHz). The summary of the measured performances is presented Table I. The ILO-CDR performances are quite similar to published CMOS CDRs, ie. 350 mW power dissipation and 380 mUI peak to peak output jitter in [8], 290 mW and 540 mUI in [9], considering it was designed for robustness, radiation-hardening and large bitrates ranges.

TABLE I
SUMMARY OF CDR CHARACTERISTICS

Process	130 nm CMOS
Power Dissipation	400-480 mW @ 1.5 V
Bit-rates	2.2-4.3 Gb/s 9.1-12.1 Gb/s
Clock Phase Noise	~ -90 dBc/Hz @ 10 kHz
Data Jitter	70 mUI pp @ 3.1 Gb/s 300 mUI pp @ 10.3 Gb/s
Eye Opening	940 mUI/440 mV @ 3.1 Gb/s 720 mUI/300 mV @ 10.3 Gb/s
BER	< 10^{-12}

V. CONCLUSION

We have proposed a multi Gb/s CDR for spatial applications. The design is based on an innovative radiation hardened architecture using an ILO with a phase alignment circuit. The CDR was implemented in a standard 130 nm CMOS bulk process. It covers 2.2 to 4.3 Gb/s and 9.1 to 12.1 Gb/s bit-rates.

ACKNOWLEDGEMENT

The authors wish to acknowledge Anritsu for the free-of-charge loan of the Data Quality Analyser.

REFERENCES

[1] Ming Zhang et al., "A CMOS design style for logic circuit hardening", Proc. IEEE International Reliability Physics Symposium, 223-229, April, 2005.

[2] Y. Boulghassoul et al., "Effects of technology scaling on the SET sensitivity of RF CMOS voltage-controlled oscillators", IEEE TNS, vol. 52 no. 6, pp. 2426-2432, 2005.

[3] H. Lapuyade et al., "A radiation-hardened injection locked oscillator devoted to radio-frequency applications", IEEE TNS, vol. 53 no. 4, pp. 2040-2046, 2006.

[4] J.B. Bégueret et al., "An innovative open-loop CDR based on injection-locked oscillator for high-speed data link applications", IEEE RFIC 2003, 8-10 June, pp. 313-316.

[5] Jri Lee et al., "A 20Gb/s burst-mode CDR circuit using injection-locking technique", ISSCC 2007, pp. 46-47.

[6] Mazouffre et al., "A 10-Gb/s CMOS fully integrated ILO-based CDR", ESSCIRC 2007, pp. 520-523.

[7] H. Lapuyade et al., "A heavy-ion tolerant Clock and Data Recovery circuit for satellite embedded high-speed data links", IEEE Transactions on Nuclear Science 2007, vol. 54, issue 6, pp. 2080-2085.

[8] Takasoh J. et al., "A 12.5Gbps half-rate CMOS CDR circuit for 10Gbps network applications", VLSI Circuits 2004, pp. 268-271.

[9] Jinghua Li Silva-Martinez J., "A Fully On-Chip 10Gb/s CDR in a Standard 0.18 μm CMOS Technology", RFIC 2007, pp. 237-240.

A 22.5-dB Gain, 20.1-dBm Output Power K-band Power Amplifier in 0.18-μm CMOS

Chi-Cheng Hung, Jing-Lin Kuo, Kun-You Lin, and Huei Wang

Dept. of Electrical Engineering and Graduate Institute of Communication Engineering,
National Taiwan University, Taipei, 10617, Taiwan

Abstract— **A fully integrated power amplifier (PA) at K-band implemented in 0.18-μm CMOS process is presented. With appropriate prematch of power cells and high gain driver stage network design, the power amplifier performs 22.5 dB peak gain and saturation output power of 20.1 dBm. The 3-dB gain bandwidth is from 18–23 GHz, while the output power at 1-dB compression point (OP_{1dB}) from 19–22 GHz is over 15 dBm. To the authors' best knowledge, this is the power amplifier with the highest gain and with good output power in K-band using standard CMOS process.**

Index Terms — **power amplifier (PA), power combining, CMOS, K-band.**

I. INTRODUCTION

THE demand of high data rate wireless communication is blooming in recent years. The K-band (18-26.5 GHz), which includes point to point communications (18-23 GHz), ISM band (24 GHz), and automotive radar applications (24 GHz and 22-29 GHz) [1], is one of the most important frequency bands in modern wireless communication systems.

The CMOS process, which claims for its low cost and high yield, has become the dominant technology in microelectronics industry nowadays. However, due to the low breakdown voltage, low transconductance (g_m), and high substrate loss, it is difficult to obtain sufficient gain and output power in high frequency. Hence, power amplifiers are still critical for the CMOS RF system integration.

A number of K-band power amplifiers in CMOS processes were reported recently. A 24-GHz power amplifier using 2-stages cascode topology was reported using 0.18-μm CMOS process in [2]. The power amplifiers [3]-[8] in similar frequency were reported with good performance. For higher output power, power combining techniques should be developed. This paper demonstrates a high gain power amplifier with high output power in K-band, which is fabricated in standard 0.18-μm CMOS process. The peak gain of the power amplifier is 22.5 dB at 20.5 GHz, while the saturation output power is 20.1 dBm. The OP_{1dB} is also over 15 dBm from 19 to 22 GHz.

II. MMIC PROCESS

This circuit is implemented in TSMC 0.18-μm 1P6M CMOS process [3], which provides deep n-well, and MIM capacitors. The poly is for resistors and the gate of the MOSFETs. The metal layers are used for inter-connection, in which the top metal (metal 6) is thick metal. The device of this process has maximum oscillation frequency (f_{max}) about 58 GHz, and unit current gain frequency (f_T) at 70 GHz.

III. CIRCUIT DESIGN

The schematic of the power amplifier is shown in Fig. 1. It is composed of 3 stages. The first two stages are driver stage, and the third stage is formed by two parallel power cells.

The typical topologies of the CMOS transistors are common source and cascode. For common source amplifier, the turning point of maximum gain (G_{max}) is at lower frequency than that of cascode due to Miller effect of the parasitic drain to gate capacitor. Cascode structure eliminates the parasitic, provides better unilaterality of the amplifier as well as higher turning point of G_{max} [4].

The biasing point of this design is at Class A (V_{GS}= 1.2 V, V_{DS} = 1.8 V) for better linearity. The devices selected in the power cell were determined by the load-pull simulation from the large signal model provided by TSMC, and the G_{max} simulation. The characteristics of device with different total width in each cascode amplifier are plotted in Fig. 2. The larger device provides lower gain and efficiency, but with higher output power. The device with total width of 384 μm was used for this design.

The power stage includes two power cells. The power cells are in-phase combined directly. Two odd-mode suppression resistors of 11 Ω are placed within two power cells for stability consideration [9]. Each power cell was prematched to 100 Ω by an appropriate matching network before binary combining. The matching network includes a

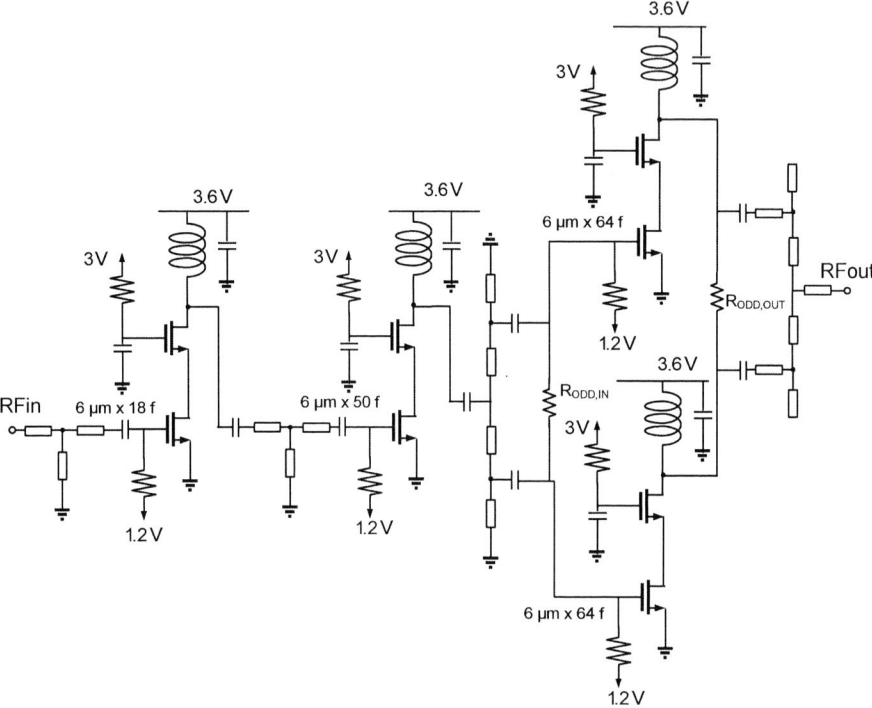

Fig. 1. Schematic of the K-band power amplifier.

peaking inductor, and impedance transform network, which is implemented by thin film micro-strip lines (TFMS) for wide band power match. The complete impedance transform of the optimum load is shown in Fig. 3. The width of the metal is also carefully designed to ensure the high current flow in the PA.

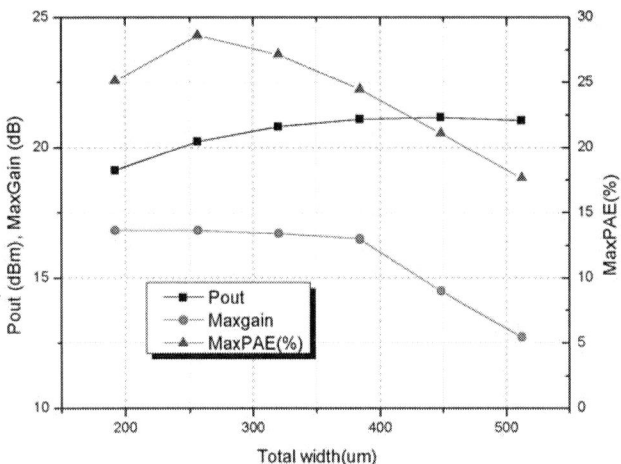

Fig. 2. The selection of the device. The simulation is based on cascode device with each V_{GS}= 1.2 V and V_{DS} = 1.8 V.

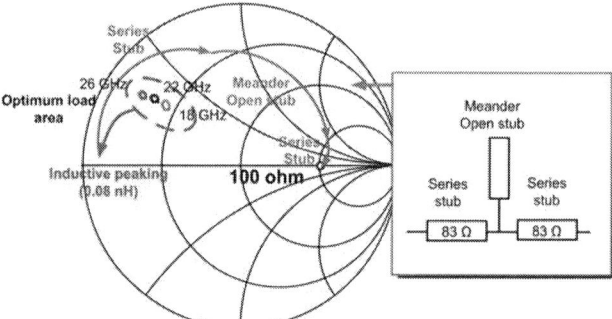

Fig. 3. Prematch network for the impedance transform of the optimum load (18- 26 GHz).

The two stage driver amplifier was designed for sufficient gain. The driver stages should provide sufficient power for power stages. The total width of each MOSFET in the first and second stage is 96 μm and 300 μm, respectively. The matching network is implemented by inductors (used for inductive peaking) and TFMS (used for inter-stage matching for lower loss than lumped elements). Appropriate bypass circuits are placed at each bias point for low frequency stability consideration. The passive structure of the circuit was simulated by full wave electromagnetic software Sonnet [10]. The power amplifier occupied an area of 0.92 mm x 0.86 mm, including all RF and dc pads. The chip photo is shown in Fig. 4.

978-1-4244-6240-7/10 $26.00 © 2010 IEEE

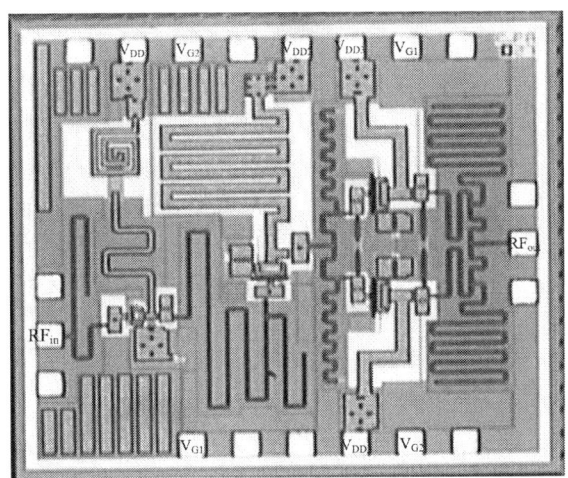

Fig. 4. The chip photo of the power amplifier with total chip area of 0.79 mm².

IV. MEASURED RESULTS

The power amplifier is measured via on-wafer RF probing. The small signal data is measured using Agilent's E8361C. This power amplifier consumes 885.6 mW of quiescent dc power when V_{DD} and the deep n-well are biased at 3.6 V.

The measured small signal gain and return losses are plotted in Fig. 5. The peak gain of this power amplifier is 22.5 dB at 20.5 GHz. The 3-dB bandwidth is from 18–23 GHz. The large signal measurement results of 20 GHz are plotted in Fig. 6. It shows that the OP_{1dB} is over 16 dBm while the P_{sat} is 20.1 dBm. The measured peak power added efficiency (PAE) is 9.3%. The large signal measurement results of the 3-dB gain bandwidth are plotted in Fig. 7. Both of the OP_{1dB} and P_{sat} are within 2-dB of variation.

Table I summaries the performances of previously reported power amplifiers in K-band. This amplifier demonstrates the highest gain with good output power among all prior reported K-band CMOS PAs.

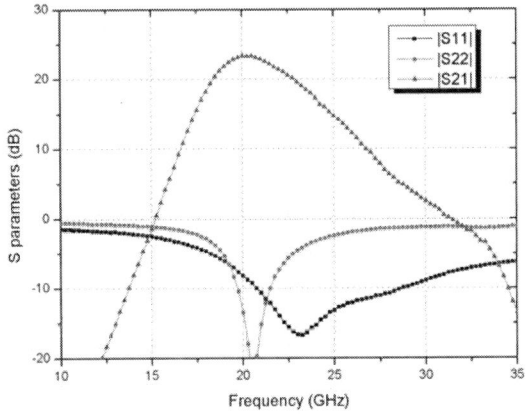

Fig. 5. Measured small signal gain and input/output return loss.

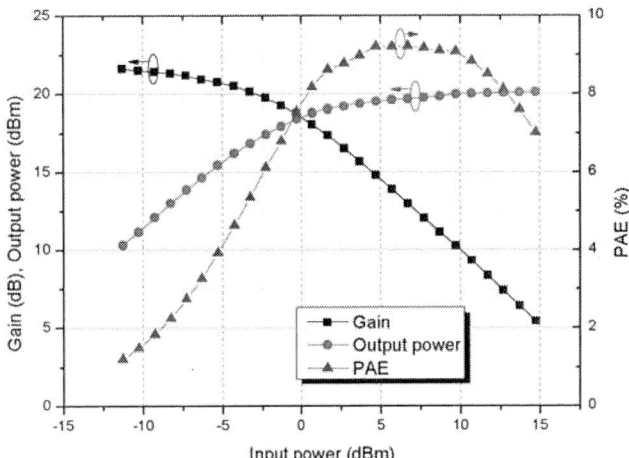

Fig.6. Measured output power, gain, and PAE curves at 20 GHz.

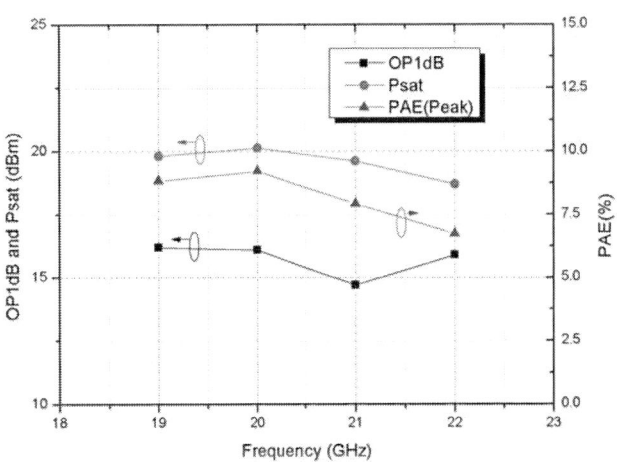

Fig.7. Measured power performance at 3-dB gain bandwidth.

V. CONCLUSION

A K-band power amplifier in 0.18-μm CMOS process is presented in this paper. With appropriate power matching network and high gain driver stage design, this power amplifier achieves 22.5 dB of gain and 20.1 dBm of P_{sat}, 16 dBm of OP_{1dB}, with chip size of 0.79 mm².

ACKNOWLEDGEMENT

The chip is fabricated by TSMC through Chip Implementation Center (CIC), Taiwan, R.O.C. This work was supported in part by the Excellent Research Project of NTU, and the National Science Council of Taiwan R.O.C. (NSC 98-2219-E-002-010, 98R0062-03, NSC 98-2221-E-002-059-MY3). The authors would like to thank Dr. Zuo-Min Tsai, and Pin-Cheng Huang for their helpful suggestions, and Bo-Jr Huang for the help of measurement.

978-1-4244-6240-7/10 $26.00 © 2010 IEEE

REFERENCES

[1] X. Guan and A. Hajimiri, "A 24 GHz CMOS front-end," *IEEE J. Solid State Circuits*, vol. 39, no. 2, pp. 368–373, Feb. 2004.

[2] A. Komijani and A. Hajimiri, "A 24 GHz, +14.5 dBm fully-integrated power amplifier in 0.18-μm CMOS,"*in Proc. IEEE Custom Integr.Circuits Conf.*, Oct. 2004, pp. 561–564.

[3] Jing-Lin Kuo, Zuo-Min Tsai, Huei Wang "A 19.1-dBm fully -integrated 24 GHz power amplifier using 0.18-μm CMOS technology," *in Proceeding of the European Microwave Conference*. Oct. 2008. pp. 558-561.

[4] Yung-Nien Jen, Jheng-Han Tsai, Chung-Te Peng, and Tian-Wei Huang, "A 20 to 24 GHz 16.8 dBm fully integrated power amplifier using 0.18-μm CMOS process," *IEEE Microwave and Wireless Components Lett*, vol. 19, no. 1, pp. 42- 44, Jan. 2009

[5] H. Shigematsu, T. Hirose, F. Brewer, and M. Rodwell, "Millimeter- wave CMOS circuit design," *IEEE Trans. Microwave Theory Tech.* , vol. 53, no. 2, Feb. 2005, pp. 472-477

[6] C. Cao, H. Xu, Y. Su , and K. K. O, "An 18 GHz, 10.9 dBm fully-integrated power amplifier with 23.5% PAE in 130-nm CMOS." *in Proceeding of ESSCIRC*, Grenoble, France, 2005.

[7] Henrique Portela, Viswanathan Subramanian, Georg Boeck "Fully integrated high efficiency K-band PA in 0.18-μm CMOS technology," *in Proceeding of the Microwave and Optical Conference*. Nov. 2009. pp. 393-396.

[8] K. Joshin, Y. Kawano, M. Fujita, T. Suzuki, M, Sato, T. Hirose, "A 24 GHz 90-nm CMOS-based power amplifier module with output power of 20 dBm," *in IEEE International Symposium on Radio-Frequency Integration Technology,* Singapore, Dec. 2009.

[9] R. G. Freitag, "A unified analysis of MMIC power amplifier stability," *IEEE MTT-S International Microwave Symposium Digest*, vol. 1, pp. 297-300, 1992.

[10] *"Sonnet User's Manual, Release 11.52,"* Sonnet Software, Inc., 2007.

TABLE I

PERFORMANCE COMPARISON OF K-BAND POWER AMPLIFIERS IN CMOS.

Reference	Technology	Topology	Frequency [GHz]	Peak Gain [dB]	OP_{1dB} [dBm]	P_{SAT} [dBm]	Peak PAE [%]	Chip Area [mm²]
This work	0.18-μm CMOS	3-stages cascode	18-23	22.5	16.2	20.1	9.3	0.79
[2]	0.18-μm CMOS	2-stages cascode	24	7	11	14.5	6	1.26
[3]	0.18-μm CMOS	2-stages cascode	24	18.8	13.3	19.1	15.6	0.325
[4]	0.18-μm CMOS	2-stages cascode	20-24	16.3	14.3	16.8	10.7	0.35
[5]	0.18-μm CMOS	3-stages cascode	25-28	17	10	14	N/A	2.04
[6]	0.13-μm CMOS	4-stages CS Class E	20	26	N/A	10.2	20.5	0.782
[7]	0.18-μm CMOS	2-stages cascode	24	16.2	13.6	17.5	12.3	0.84
[8]	90 nm CMOS	2-stages cascode	24	22	18.3*	19.8**	N/A	2.4

Note: CS: Common source.

* Output power of 3-dB gain compression point. ** Estimated from the figure.

A 40% PAE Linear CMOS Power Amplifier with Feedback Bias Technique for WCDMA Applications

Hamhee Jeon, Kun-Seok Lee, Ockgoo Lee, Kyu Hwan An, Youngchang Yoon, Hyungwook Kim, Dong Ho Lee, Jongsoo Lee[**], Chang-Ho Lee[***], and Joy Laskar

Georgia Electronic Design Center, Georgia Institute of Technology, Atlanta, GA, 30308, USA

[**]Gwangju Institute of Science and Technology, Gwangjoo, Korea

[***]Samsung Design Center, Atlanta, GA 30308, USA

Abstract — **A highly efficient CMOS linear power amplifier for WCDMA applications with feedback bias technique is presented. The method involves connecting the gates of common-gate devices of the driver stage and the power stage in cascode configurations by a feedback network for enhancing linearity. To achieve high efficiency and linearity simultaneously, large-signal IMD minimum (IMD sweet spot) is properly used at the desired output power level. The proposed PA was fabricated in a 0.18-μm CMOS technology. The experimental results demonstrate a gain of 26 dB, a maximum output power of 26 dBm with 46.4% of peak PAE, and a linear output power of 23.5 dBm with 40% PAE using a 3GPP WCDMA modulated signal. Both simulation and measurement results show an excellent large-signal IMD minimum at the output power using a WCDMA modulated signal.**

Index Terms — **bias, CMOS, efficiency, feedback, linearity, power amplifier, sweet spot, WCDMA**

I. INTRODUCTION

Recent progress of CMOS technology is remarkable, and it makes possible the potentials of a fully integrated RF front-end system. However, the power amplifier (PA) is still a bottleneck for the complete integration because of CMOS technology's inherent characteristics such as low trans-conductance, low reliability, and large parasitic elements.

Because of the low transconductance of CMOS technology, multi-stage cascade methods or larger power cells are required to generate the same output power and gain compared with other III-V compound technologies. However, this creates larger parasitic elements within the PA, and can be the cause of the low efficiency and linearity.

For the reliability of devices, a cascode configuration with a thick-oxide transistor is generally used to prevent oxide breakdown by large voltage swings, or hot carrier degradation which can increase the threshold voltage so that it degrades the performance of the device [1].

However, in the cascode configuration, there might be an additional non-linearity from the common-gate (CG) device compared to the common-source (CS) amplifier topology. The feedback bias technique can improve the PAs' linearity by reducing non-linearity effects that can occur in the CG device of the cascode structure.

The proposed CMOS linear PA is designed with a two-stage and single-ended cascode configuration which incorporates a feedback bias technique to overcome those CMOS technology's drawbacks. Also, to achieve high efficiency and good linearity, large-signal IMD minimum (IMD sweet spot) is properly used. [2]

The CMOS linear PA is designed for WCDMA application and the required specifications are satisfied. The design specifications for WCDMA PAs are listed in Table I [3].

TABLE I. 3GPP WCDMA PA SPECIFICATIONS

Specifications	
Operating Frequency (Band I)	1.92GHz - 1.98GHz
Maximum Output Power (Power class 3)	21 – 25 dBm
ACLR (3.84MHz main channel)	-33dBc @ 5MHz offset -43dBc @ 10MHz offset

II. CIRCUIT DESIGN METHODOLOGY

In linear PA design, high efficiency and good linearity are the most important parameters. As generally known, class A has high linearity with low efficiency, while class B and C have higher efficiency with degraded linearity compared to class A. Since efficiency and linearity are in trade off relationship, class AB can be the best option considering both of them [4].

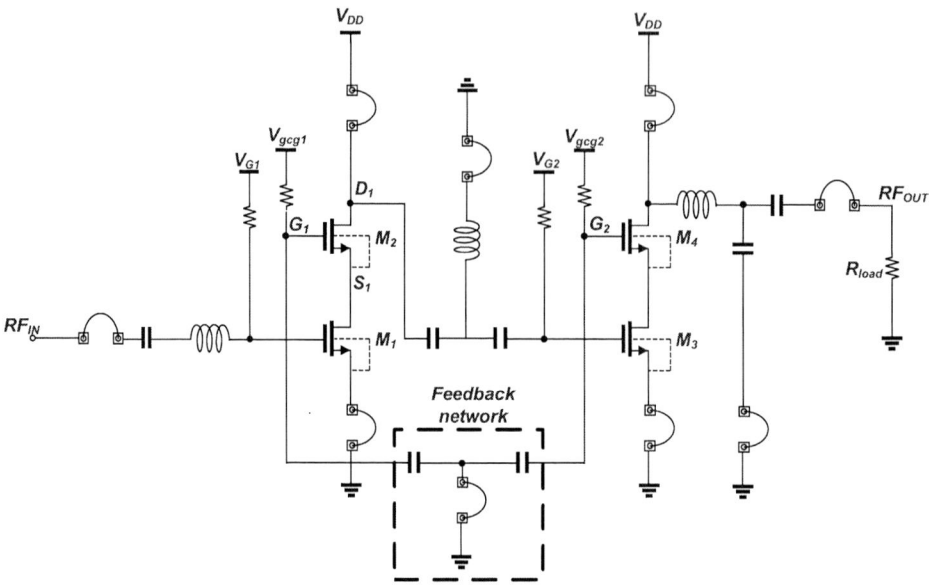

Fig. 1 Simplified schematic of CMOS PA with feedback bias topology

Considering PA operation, the large-signal analysis is more applicable for linearity IMD analysis than small-signal analysis. As the input power increases, higher-order terms of Volterra series expansion become dominant, and the combination with the compressing non-linearity of the saturation-to-linear region transition can result in the IMD sweet spot that occurs close to compression output power level [2]. In this PA design, deep-class AB is employed and the bias voltages are set to have the IMD sweet spot appearing near the compression point. By carefully optimizing a bias point at near threshold, which is a deep-Class AB operation, it is possible to adjust the relative position of the sweet spot and making it possible to achieve good linearity with high efficiency by placing the sweet spot at the desired output power. Based on simulation experiences, under the -25 dBc two-tone IMD3 in harmonic balance simulation was aimed instead of the -33 dBc ACLR for 3GPP WCDMA PA linearity specification in time-consuming envelope simulations.

Fig. 1 shows a simplified schematic of the suggested two stage single-ended CMOS linear PA. A single-ended topology is chosen for easier integration, cost-effectiveness, and avoidance of baluns although a differential configuration has the advantage of the even harmonics control. [5] To reduce the noise coupling with other components on the same silicon substrate, a deep N-well [6] is employed for both the driver and power cells. For the reliability of the devices at 3.4V operation, a 0.4-um thick-oxide NMOS transistor is used in the power stage cascode CG transistor, and 0.18-um NMOS transistors are used for the both of CG and CS in the driver

stage to compensate for the low RF power gain of the thick-oxide CG transistor in the power stage.

III. FEEDBACK BIAS TECHNIQUE

In PA design, parasitic capacitances are unavoidable and exist between the ports of a transistor. In the proposed PA, a feedback bias technique, which connects the gate of the common-gate transistor in the power stage, G_2, and the gate of the common-gate transistor in the driver stage, G_1, by a feedback network, is adopted to enhance linearity using the leakage signals through the parasitic capacitances of the CG-transistor in Fig. 1.

Although the gate of the common-gate device is ideally AC grounded in a cascode topology, there exist signals that are coupled from the both drain and source by gate-source capacitance, C_{gs}, and gate-drain capacitance, C_{dg} respectively. They exist for both the driver and power stage cascode amplifiers. By connecting G_1 and G_2 with the carefully optimized feedback network, the gate leakage signal of G_2 is fed back to G_1 with 180 degree phase shift created by feedback network. Therefore, the signal fed back from G_2 to G_1 is in-phase with the source and drain waveforms, which means that the variation of the CG device's operation is minimized because the voltages between the ports are maintained. It can reduce the additional non-linearity effects that can occur by variations of parasitic capacitances in the CG device within a cascode topology. Fig. 2 shows the waveforms with feedback bias and the waveforms without feedback bias. Because of the large voltage swing, the operation region of the CG transistor continuously changes saturation region, triode region, and cutoff region. With this feedback bias technique, the ratio of the cutoff region of CG transistor is reduced and triode region is also reduced during the turn-on region. By using the technique, the variation of the operation region of CG transistor is minimized, so that it improves the linearity of the proposed PA. However, as the voltage of G_1 follows the source voltage of CG in both positive and negative swings near the DC value of G_1, a non-optimal gain can be achieved compared to the conventional cascode PA which has RF ground at gate node of CG transistor.

The feedback network is a T-network, which consists of an inductor and two capacitors. The inductor is implemented as a bonding wire inductance to ground and the capacitors in the feedback network also act as AC ground for the cascode topology. Thus, there are no additionally required components for the feedback bias technique since the capacitors and bonding wire inductor of the feedback network are essential components to the cascode amplifier topology. Also, the feedback bias technique does not use the signal path of the PA, so that it can improve linearity without affecting operation of the PA.

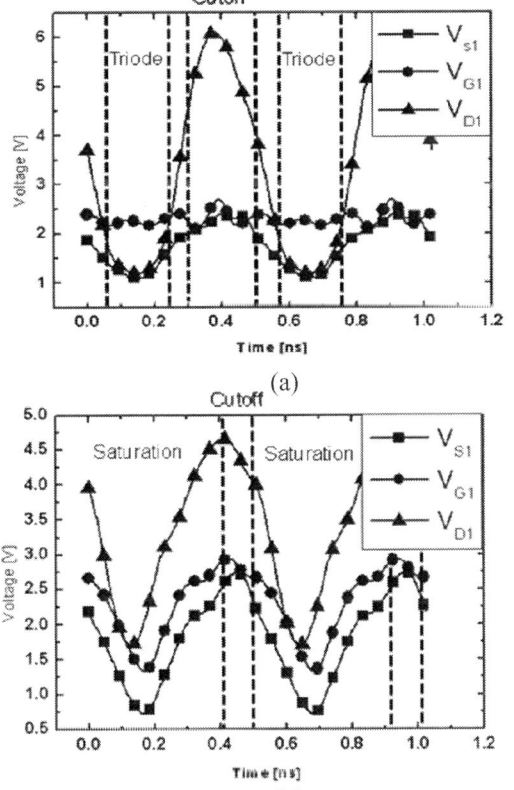

Fig. 2. (a) Voltage waveforms without feedback bias (b) Voltage waveforms with feedback bias

IV. EXPERIMENTAL RESULTS

Fig. 3 shows a photograph of the fabricated CMOS PA, which has 1.6 mm x 0.52 mm chip area. All components are fully integrated to one chip. To verify the chip, an FR-4 PCB evaluation board is used and the chip is mounted on the ground plate of the evaluation board. By connecting multiple bonding wires to the ground, it can minimize source degeneration effects of both driver and power cells which typically can lead to the decrease of gain.

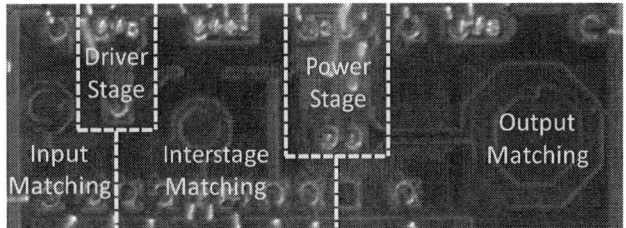

Fig. 3. Photograph of the feedback CMOS power amplifier.

Fig. 4 shows the comparison of simulated and measured Power Added Efficiency (PAE) and gain of the proposed amplifier. From the graph, maximum output power is reduced from 27.5 dBm to 26 dBm and peak efficiency also decreased from 55.4% to 46.4%. The 1-dB gain compression point is 25.4 dBm. The measured output power with a linear 3GPP WCDMA modulated signal is 23.5 dBm with 40% PAE and the DC current consumptions of the driver and power amplifier were 27mA and 143 mA, respectively.

Fig. 5 shows the measured IMD3 and ACLR results. A two-tone test is performed at 1.95 GHz center frequency with 5 MHz tone-spacing. From the measurement results, under the -25dBc two-tone IMD3 guarantees the required linearity ACLR of -33 dBc. Also, the sweet spot appears at the desired output power of 22dBm to achieve lower IMD3.

Comparisons are made between the PA with the feedback bias and the PA without the feedback bias in Fig. 6. The measurement results compare the ACLR performance. As shown in the Fig.6, the feedback bias technique improved ACLR at the desired high power regime having much margin of ACLR. The linear output power is significantly increased from 22.7 dBm to 23.5 dBm using a 3GPP WCDMA modulated signal because of the increase of the ACLR margin.

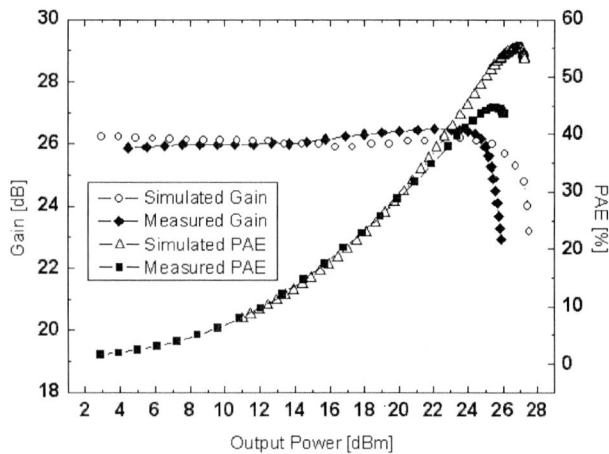

Fig. 4.   Simulated and measured PAE and gain.

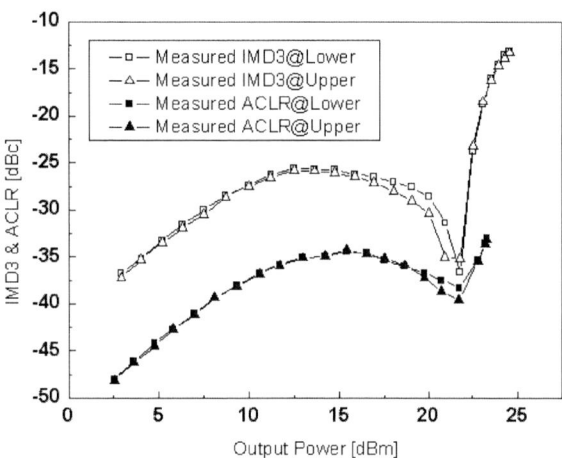

Fig. 5.   Measured IMD3 and ACLR

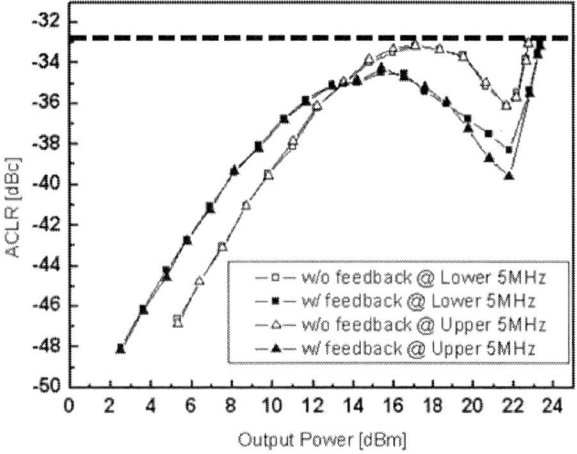

Fig. 6.   Comparison of the measured ACLR with and without feedback bias

V. CONCLUSION

In this paper, the highly efficient CMOS linear PA is demonstrated for WCDMA application. The proposed feedback-biased PA was fabricated in a 0.18-um standard CMOS technology and to achieve high efficiency and good linearity, IMD sweet spot is properly used at the desired output power. The feedback bias technique, which incorporates the gate of common-gate transistors of the driver stage and the power stage in cascode configurations, is used for improving linearity of the common-gate transistor, which is verified in the measurement results. Experimental results demonstrate the power gain of 26 dB, and a max output power of 26 dBm with 46.4% peak PAE and a linear output power of 23.5 dBm with 40% of PAE using a 3GPP WCDMA modulated signal.

ACKNOWLEDGEMENT

The authors would like to thank Samsung Design Center, for their technical assistances and fabrication supports.

REFERENCES

[1] T. Sowlati, and D. M. W. Leenaerts, "A 2.4-GHz 0.18-um CMOS Self-Biased Cascode Power Amplifier," *IEEE J. Solid-State Circuits,* vol. 38, no. 8, pp. 1318-1324, August 2003.

[2] C. Fager, J. C. Pedro, N. B. Carvalho, H. Zirath, F. Fortes and M.J. Rosario "A Comprehensive Analysis of IMD behavior in RF CMOS Power Amplifiers," *IEEE J. Solid-State Circuits,* vol. 39, no. 1, pp. 24-34, January 2004.

[3] UE Radio Transmission and Reception, 3GPP Standard 25.101 (V9.2.0), 2009

[4] S. C. Cripps, RF Power Amplifiers for Wireless Communications. Boston, MA: Artech House, 1999.

[5] J. Kang, J. Yoon, K. Min, D. Yu, J. Nam, Y. Yang and B. Kim, "A Highly Linear and Efficient Differential CMOS Power Amplifier with Harmonic Control," *IEEE J. Solid-State Circuits,* vol. 41, no. 6, pp. 1314-1322, June 2006.

[6] J. Kang, D. Yu, Y. Yang, and B. Kim, "Highly Linear 0.18-um CMOS Power Amplifier with Deep n-Well Structure," *IEEE J. Solid-State Circuits,* vol. 41, no.5, pp. 1073-1080, May 2006.

RTUIF-18

A Switching-Mode Amplifier for Class-S Transmitters for Clock Frequencies up to 7.5 GHz in 0.25µm SiGe-BiCMOS

Stefan Heck[*], Martin Schmidt[*], Alexander Bräckle[*], Frieder Schuller[*],
Markus Grözing[*], Manfred Berroth[*], Hans Gustat[#], Christoph Scheytt[#]

[*]University of Stuttgart, Institute of Electrical and Optical Communications Engineering,
Stuttgart, D-70569, Germany

[#]IHP GmbH, Frankfurt (Oder), D-15236, Germany

Abstract — **This paper presents the first voltage mode H-bridge switching amplifier in a fast complementary SiGe-technology for frequencies in the GHz range. The amplifier is suited as a driver for a high power GaN amplifier in class-S transmitters. It can be operated with pseudo-random digital pulse trains up to 7.5 Gbit/s. The measured broadband output power for a rectangular drive signal with a 50% duty cycle and a frequency of 2 GHz is about 148 mW. The efficiency of the switching stage including its two-stage inverter driver is about 43%. Including the input current-mode-logic (CML) stage, the PAE is about 30%.**

Index Terms — **Bandpass delta-sigma modulation, bipolar integrated circuit, class-S, switching amplifiers.**

I. INTRODUCTION

The coding schemes used in recent mobile communication standards like Universal Mobile Communication System (UMTS) exhibit larger peak to average power ratios (PAPR) than older ones like Global System for Mobile Communications (GSM). This turns out to be a severe problem for the power amplifier (PA) in the transmission chain: conventional linear PAs have to provide a higher back-off when the PAPR is increased and therefore the power efficiency is reduced. Hence, an increase of efficiency is aspired for future PAs while the linearity must not be decreased.

The concept of the class-S architecture offers the desired features [1]. Its block diagram is depicted in Fig. 1. An analogue input signal is converted into a digital pulse train by a delta-sigma modulator (DSM). This pulse train is amplified by a high-efficient switching amplifier and finally demodulated by a bandpass filter.

Efficiency enhancement arises from the high drain efficiency of the switching-mode amplifier. The theoretical limit of the drain efficiency of a class-D amplifier is 100%. It can be used in the class-S architecture. The amplifier can be realized in two different configurations, voltage-mode (VM) or current-mode (CM).

A realization of a class-S transmitter in CMOS is considered as both cost and energy efficient [2]. However,

the required clock frequency for signal frequencies above 2 GHz is not reached yet with current CMOS technologies and the maximum output power is limited due to low breakdown voltages. In contrast to CMOS, SiGe-technologies allow the design of bandpass DSM for wireless application in the lower GHz range [3] - [4] and in addition show higher breakdown voltages.

This paper describes the design and experimental results of an H-bridge switching amplifier [5] in a complementary SiGe-BiCMOS-technology. The paper is organized as follows: section II describes the design of the circuit. Measurement results are presented in section III. Section IV closes with a short conclusion.

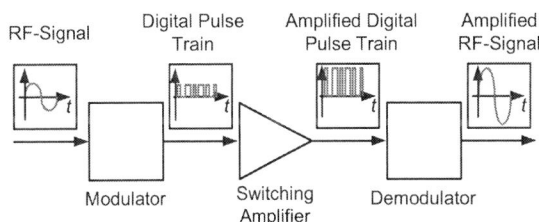

Fig. 1. Block diagram of a class-S power amplifier.

II. CIRCUIT DESIGN

A. Technology

The challenge in the design of the H-bridge switching amplifier is the reduction of switching losses caused by overlap of the direct-path current and the voltage drop across the switching transistor during the transition. This requires high transit frequencies of the transistor in order to reduce the rise and fall times and thus the time interval in which the direct-path current flows. However, this feature is in contrast to the high breakdown voltage necessary to achieve high output power. SiGe bipolar technologies show a good compromise. They have very high transit frequencies while still providing sufficient breakdown voltages. A 0.25µm SiGe-BiCMOS process of

978-1-4244-6240-7/10 $26.00 © 2010 IEEE

the IHP is used in this work. It offers complementary transistors with an f_t/f_{max} of 110 GHz/180 GHz for the npn and 90 GHz/120 GHz for the pnp transistor respectively. Breakdown voltages BV_{CBO}/BV_{CEO} show values about 6 V/2.3 V for the npn and 4 V/2.5 V for the pnp.

B. Circuit Overview

A block diagram of the whole circuit is shown in Fig. 2. The differential input pulse train with a voltage swing of 300 mV is preamplified by a current-mode-logic (CML) driver. A levelshifter provides the voltage levels for the following two-stage inverter drivers (TSID). There are four TSID, one for each transistor of the final switching stage, which has a supply voltage of 4.3 V. The circuit elements are discussed in the following subsections. Fig. 3 shows a chip photo.

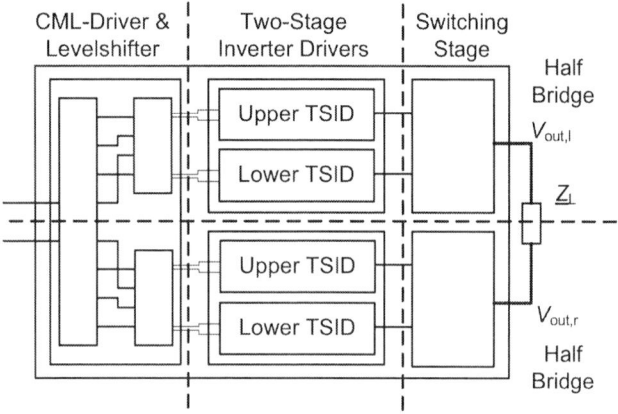

Fig. 2. Block diagram of the H-bridge amplifier including all driver stages.

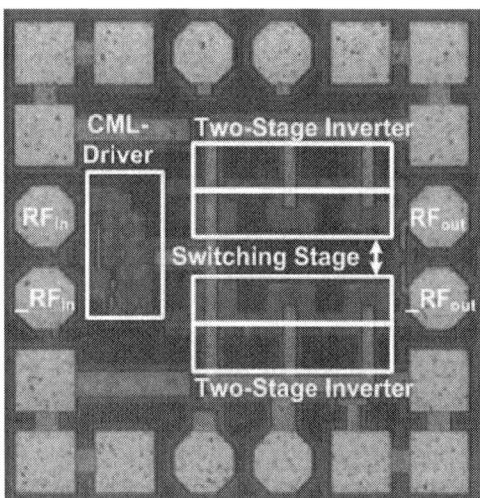

Fig. 3. Chip photo. The chip size is 600 μm x 600 μm.

C. The output switching stage

Fig. 4 shows the circuit of the H-bridge switching amplifier output stage. The H-bridge can be divided into two half bridges which consist of a complementary bipolar circuit. The pnp transistors (T_{11} and T_{12}) are used for pull-up and the npn-transistors (T_{21} and T_{22}) are used for pull-down. Both half bridges are driven in antiphase so either the transistor pair T_{11} and T_{22} or the pair T_{21} and T_{12} is in conducting state. Therefore, only one of the currents i_1 or i_2 is nonzero at the same time. During the off-state of a transistor, the breakdown voltage BV_{CBO} is the maximal voltage across the transistor. The pnp transistor has the smallest BV_{CBO} value and therefore limits the supply voltage of the switching stage.

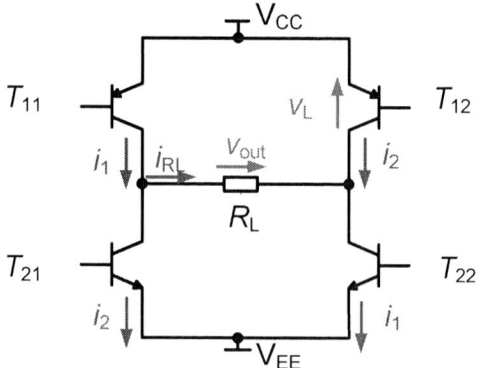

Fig. 4. Circuit of the switching stage.

D. Two-stage inverter driver

Each transistor of the output switching stage is driven by a TSID as shown in Fig. 5. Two inverters are used in the first stage to drive the two transistors of the second stage.

An RC-element is added in the base circuit of each transistor. This allows faster switching between the two states. A resistor is necessary to set the base current. A small resistor causes a higher base current i_B and therefore speeds up the turning-on. However, a large resistance is required to avoid saturation of the transistor. Saturation would cause excessive long time to turn off the transistor. This contradiction can be bypassed with a capacitor in parallel. It behaves like a short circuit while switching and thus provides a high base current only during the transition.

Additionally, a Schottky diode is connected between the base and the collector to avoid saturation. The diode is added only in the driving stages but not in the final switching stage because the inverter with Schottky diodes has high collector-emitter saturation voltages. This would cause high voltage drop losses.

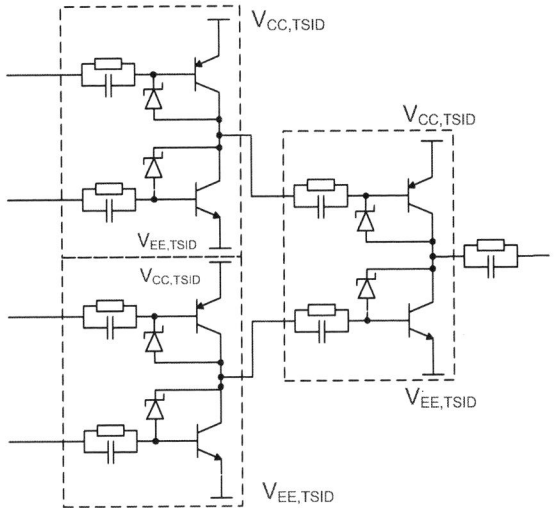

Fig. 5. Circuit of the two-stage inverter driver.

III. EXPERIMENTAL RESULTS

Measurements have been performed in two steps. First, a periodic drive signal has been applied with CW-frequencies from 0.5 to 3.5 GHz which corresponds to a bit rate from 1 to 7 Gbit/s.

In a second step, a pseudo-random binary sequence (PRBS) with bit rates up to 7.5 Gbit/s is used. The PRBS (length 2^7-1) covers the worst case scenario of delta-sigma series: the shortest possible pulse after a long series of logical ones or zeros.

Time domain signals are captured using a sampling oscilloscope. The periodic signals have additionally been investigated using a spectrum analyzer.

Fig. 6 shows the transient curve of the single-ended voltage V_L as drawn in Fig. 4. The peak to peak voltage is about 4 V.

The rise and fall times have been considered over a large frequency range. Both, 10% - 90% and 20% - 80%, are measured. They are shown in Fig. 7. The 90% - 10% transition times increase with lower frequencies whereas the 80% - 20% transition times are lower than 50 ps and only vary about 12 ps.

The output power of the first harmonic has been determined using a spectrum analyzer. Additionally, the "digital" broadband output power P_{po} of the digital pulse train has been calculated according to the following equation:

$$P_{po} = \frac{(\frac{V_L}{2})^2}{50\,\Omega} * \left(1 - 2 * \frac{T_{trans}}{T} + 2 * \frac{T_{trans}}{T} * \frac{1}{3}\right) * 2. \quad (1)$$

The first term in (1) computes the output power for zero rise and fall times. Assuming trapezoid pulses, the output power is reduced by the factor one third during the

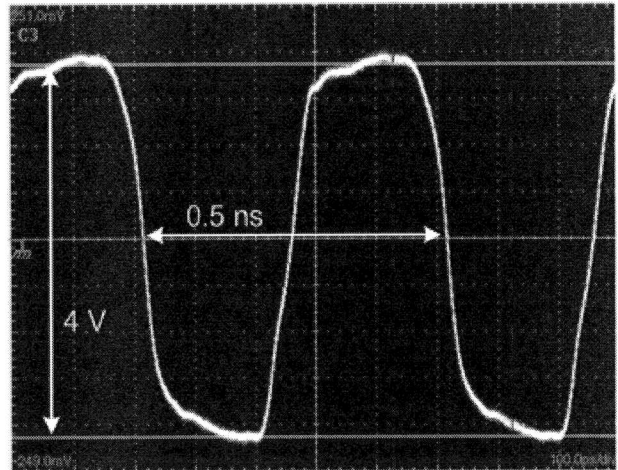

Fig. 6. Output signal of the amplifier with a periodic drive signal. Frequency is 2 GHz which corresponds to a bit rate of 4 Gbit/s.

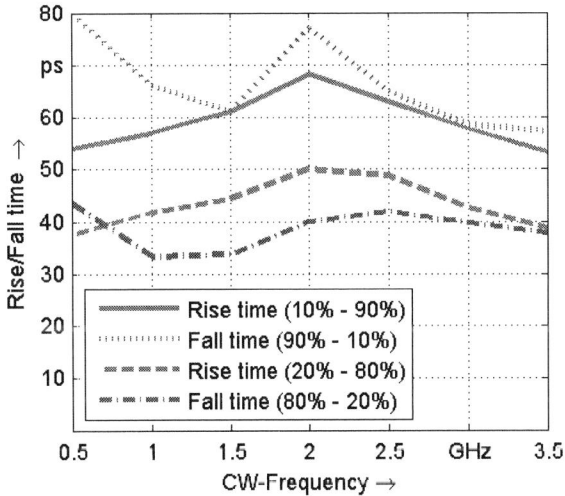

Fig. 7. Rise and fall times versus CW-frequency for a periodic drive signal with a 50% duty cycle.

transition. This is expressed with the term in brackets. The ratio "T_{trans}/T" is the duration of a transition T_{trans} referred to the period T. A transition occurs twice within a period.

The factor 2 at the end of the equation accounts for the two output signals. Fig. 8 shows the filtered and the broadband "digital" powers versus the frequency.

The efficiency of the amplifier is almost equal to the PAE due to the small voltage swing of the input signal. The efficiencies of the amplifier considering the digital output power are plotted in Fig. 9. The upper curve considers the switching stage including the TSID whereas the lower one includes all driver stages.

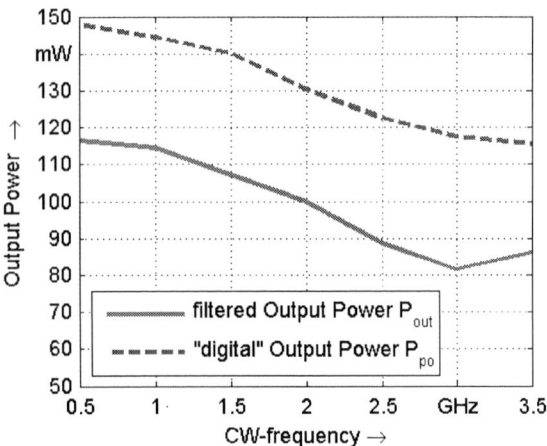

Fig. 8. Output power versus CW-frequency for a periodic drive signal with a 50% duty cycle.

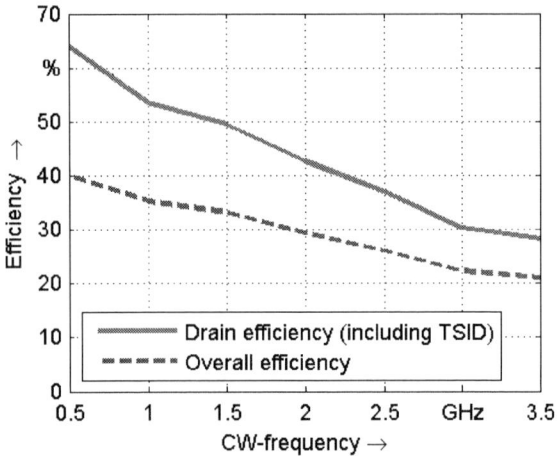

Fig. 9. Efficiency of the switching stage considering the broadband output power.

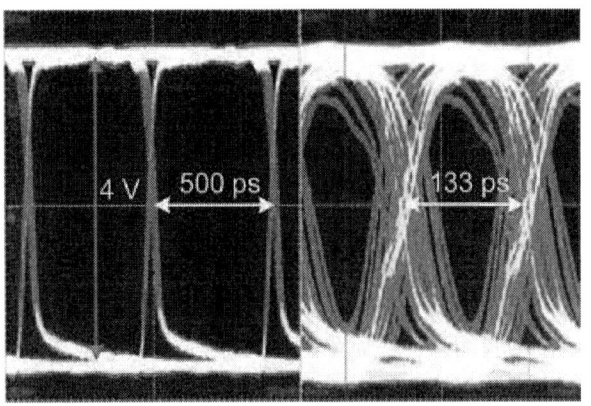

Fig. 10. Eye-diagram of the output signal for a 2 Gbit/s (left) and a 7.5 Gbit/s (right) PRBS drive signal (length 2^7-1).

Delta-sigma signals show an average transition frequency equal to the analogue input signal. For the modulator presented in [3], this frequency is about 2 GHz. The output power as well as the efficiency of the periodic signal at 2 GHz and the values of the delta-sigma modulated signal fit very well in the simulation [6]. For the modulators in [3], the output power of the amplifier would be about 130 mW with a total efficiency of 30%.

The eye diagrams for a 2 Gbit/s and a 7.5 Gbit/s PRBS sequence are shown in Fig. 10. Whereas the switching behavior is nearly perfect at 2 Gbit/s, the circuit operates at its speed limit at 7.5 Gbit/s. The jitter due to inter symbol interference (ISI) increases significantly at this speed.

IV. CONCLUSION

This article presents the first design of an H-Bridge amplifier in a complementary SiGe-BiCMOS technology in the GHz range. It is suitable as a driver in a class-S amplifier. The rise and fall times, the output power as well as the efficiency are demonstrated in the CW-frequency range of 0.5 to 3.5 GHz. The broadband output power of the amplifier driven with a 2 GHz periodic drive signal is 130 mW. The total efficiency is about 30% considering all driver stages.

REFERENCES

[1] M. Iwamoto, A. Jayaraman, G. Hanington, P.F. Chen, A. Bellora, W. Thornton, L.E. Larson and P. Asbeck, "Bandpass delta-sigma class-S amplifier," *IEEE Electronics Letters.*, vol. 36, no. 12, pp. 1010-1012, June 2000.

[2] T. Alpert, M. Schmidt, I. Dettmann, T. Veigel, M. Grözing, and M. Berroth, "Concept for a 12-bit digital bandpass delta-sigma modulator for power amplifier applications," *ESSCIRC 2008*, Sept. 2008.

[3] M. Schmidt, M. Grözing, S. Heck, and M. Berroth, "A 1.55 GHz to 2.45 GHz center frequency continuous-time bandpass delta-sigma modulator for frequency agile transmitters," *IEEE RFIC Symp. Dig.*, pp. 153-156, June 2009.

[4] P. Ostrovskyy, H. Gustat, Ch. Scheytt, and Y. Manoli, "A 9 GS/s 2.1 – 2.2 GHz bandpass delta-sigma modulator for class-S power amplifier," *IEEE IMS Symp. Dig.*, pp. 1129-1132, June 2009.

[5] T. Hung, J. Rode, L. Larson, and P. Asbeck, "Design of H-bridge class-D power amplifiers for digital pulse modulation transmitters," *IEEE Trans. Microwave Theory & Tech.*, vol. 55, no. 12, pp. 2845-2855, June 2007.

[6] S. Heck, A. Bräckle, M. Schmidt, F. Schuller, M. Grözing, M. Berroth, H. Gustat, and Ch. Scheytt, "A SiGe H-bridge switching amplifier for class-S amplifiers with clock frequencies up to 6 GHz," to be published at the German Microwave Conference (GeMiC) 2010, Berlin, Germany.

SiGe Power Amplifier ICs for 4G (WIMAX and LTE) Mobile and Nomadic Applications

V. Krishnamurthy, K. Hershberger, B. Eplett, J. Dekosky, H. Zhao, D. Poulin, R. Rood, and E. Prince

VT Silicon, Inc., Atlanta, GA, 30308, USA

Abstract — SiGe 4G PA development is a key element in enabling integrated 4G front end SiGe ICs. In this paper, we report on a wideband SiGe 4G PA IC which meets WIMAX (802.16e) and LTE specifications. For 802.16e, the SiGe PA produces 25 dBm linear power at Vcc=3.3V for 2.3-2.7 GHz operation with <4% EVM and 18% efficiency while meeting spectral mask and exhibiting -43 dBm/MHz second harmonic levels. For TD-LTE, Band 40 (2.3-2.4 GHz) and FDD-LTE, Band 7 (2.500-2.570 GHz), this SiGe PA produces 28.5 dBm linear power while meeting 3GPP spectral mask and EVM specification for QAM 16.

Index Terms — Power amplifier, PA,, LTE, SiGe, WIMAX, 4G.

I. Introduction

SiGe BiCMOS IC technology provides the potential of integrating all the active RF components for next generation 4G (WIMAX, TD-LTE) RF Front Ends (PA, T/R Switch, LNA) into one IC. SiGe BiCMOS technology enables high performance PAs and LNAs with HBT (Heterojunction Bipolar Transistor) devices while the T/R switch can be developed with a variety of CMOS devices (i.e. triple well NMOS, 5V NMOS). Linear 4G / WLAN SiGe PA performance has been documented extensively in the literature and in products for the past few years [1]-[2]. One key advantage of a SiGe 4G Front End IC is the ability to integrate intelligent controls and digital communications (e.g. SPI Bus) to provide programmability and dynamic optimization for the 4G RF Front End IC.

One key competitive technology to SiGe for 4G RF Front End IC is CMOS. The current drawback with a CMOS solution is the relatively poor 4G PA performance. For WIMAX (802.16e) applications, CMOS PAs have shown efficiencies of 12% at 23 dBm output power for operation in 2.3 GHz to 2.4 GHz band [3]. This is significantly lower than the near 20% efficiencies reported by commercial 4G PAs. The use of Digital Pre-Distortion (DPD) has shown improved CMOS PA linear output power and efficiencies but adds more system complexity and requires close collaboration between the baseband IC and the PA IC [4].

The key component of a 4G RF Front End which has the most impact on the overall performance of a 4G mobile or nomadic device is the power amplifier. The relatively high peak to average power ratio (PAPR) for the WIMAX/LTE uplink, the high output power levels (23 dBm at the antenna), along with the WIMAX/LTE spectral mask requirements impose stringent requirements on the power amplifier. In this paper, a 2.3-2.7 GHz SiGe 802.16e, 2.3-2.4 GHz TD-LTE and 2.500-2.570 GHz FDD-LTE PA is demonstrated. The performance of this SiGe 4G PA is comparable to commercial GaAs PAs and is a key step towards achieving a fully integrated 4G RF Front End IC. For FDD applications, a SiGe 4G PA can enable integration of multi-band/multi-mode PAs (3G and 4G) with intelligent controls and digital communications into one IC.

II. Design Considerations

The 4G SiGe PA is a three stage fully differential design with three power modes (high, medium, low),an integrated input match, an external output match and an external balun. The PA input is a 100 ohm differential input and a single ended 50 Ohm impedance is provided at the balun output. The differential PA input alleviates the need for a balun between the differential transceiver output and the PA. Figure 1 depicts the 4G SiGe PA topology. For future applications, the output match and balun can be combined into a transformer balun in either a Si IPD (Integrated Passive Device), a LTCC (Ceramic) substrate, or fully integrated onto the SiGe IC. The differential topology was chosen to avoid the negative effects of emitter degeneration due to ground wire bonds in single ended designs. The SiGe 4G PA was designed to output 25 dBm for the 802.16e 2.3-2.7 GHz band for a Vcc=3.3V. The RF Front End losses after the PA are estimated to be between 1 to 1.5 dB (low pass filter and T/R switch) for a typical single band Front End Module resulting in 23.5 to 24 dBm delivered to the antenna. The 2.3-2.7 GHz band combines the 2.3-2.4 GHz 802.16e/TD-LTE and 2.5-2.7 GHz FDD-LTE/802.16e bands allowing for global 4G coverage. For a PAPR of approximately 6 dB for the 802.16e uplink signal, the P1dB for this PA was targeted to be approximately 31 dBm.

The PA unit cell is a cascode structure with the common-base and common-emitter device consisting of 3.5V BVCEO SiGe HBTs. To further improve the breakdown characteristics, the impedance presented to the base of HBT devices by the bias circuit is very low (< 300 Ohms) and as a result, the PA voltage swing is dominated by BVCER. For this case, we estimate BVCER to be ~5.5V. For a cascode structure, the stacking of two HBTs further eases the SiGe HBT breakdown requirements since the maximum voltage swing is approximately 6.3V for a Vcc of 3.3V at 31 dBm CW output power. This effect of BVCER on breakdown has been validated in prior work on SiGe HBT PAs [5].

The sizing of the common emitter devices for the three differential stages are as follows: 440um for Stage 1, 880 um for Stage 2, and 3500um for Stage 3. This translates to a 2:1 emitter area ratio between Stage 1 and Stage 2 and a 4:1 emitter area ratio between Stage 2 and Stage 3. Simulation results showed that the sizing of each stage was sufficient to overcome the inter-stage losses and provide enough margin to compensate for the gain variation of each stage.

Linearization for this SiGe PA IC was performed using VT Silicon's patented Linearity Enhancement Technology (LET) [6]. LET consists of optimizing the biasing of the individual branches within each stage along with tailoring the impedances of the bias circuits to obtain reduction in inter-modulation products (i.e. IM3, IM5). Improvements up to 1.5dB in linear power have been experimentally observed in previous linear SiGe PAs with this technique.

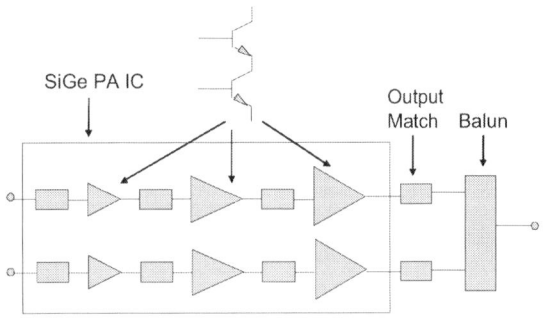

Fig. 1.   Simplified view of the 3-stage fully differential SiGe 4G power amplifier topology

III. MEASUREMENT RESULTS

The SiGe PA IC, measuring approximately 1.7mm x 1.7mm, was epoxy attached and wire bonded onto a four layer BT laminate . The module size is 5mm x 5mm and contains a discrete output match. Figure 2 depicts the SiGe PA module. The PA module is mounted onto a FR-4 evaluation PCB which contains an Anaren balun (part # BD2425N50100A00) with an insertion loss of approximately 0.6 dB. In addition to providing a balun

function, the Anaren balun also serves as a low pass filter and provides reasonable harmonic rejection.

The SiGe PA module was tested using a 16 QAM 10 MHz 802.16e and a TD-LTE uplink signal generated by a Rohde & Schwarz SMJ100A Vector Signal Generator. The EVM and spectral mask measurements were performed with a Rohde & Schwarz FSQ26. The duty cycle for the uplink was set to 100% for the measurements. The reference plane for the power and linearity measurements is immediately after the output match. The measurements were performed with the case temperature held at 25C.

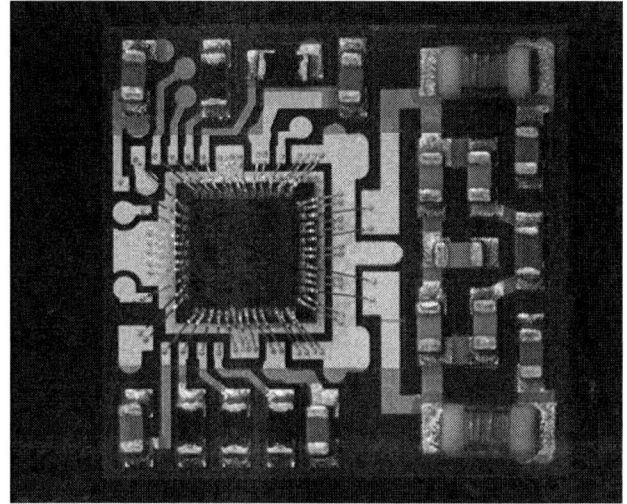

Fig. 2.   Photograph of the SiGe 4G PA laminate module

One of the key metrics for 4G PAs is gain. A sufficient gain is needed from the PA to deliver approximately 25 dBm output power considering that the 4G 802.16e CMOS transceiver can deliver typically -3 dBm to -5 dBm linear power especially at elevated temperatures (i.e. 85C). Fig 3 shows a plot of measured CW gain versus power for the 2.3-2.7 GHz band at 25C.

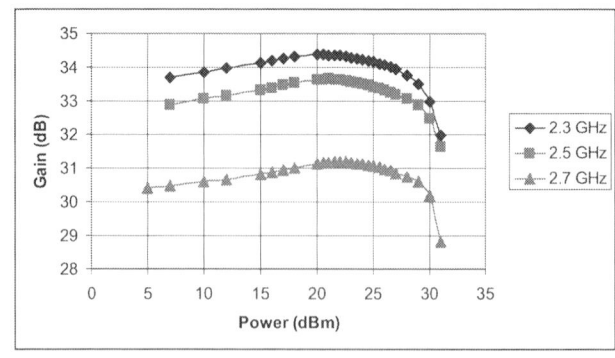

Fig. 3.   4G SiGe PA CW gain for the 2.3-2.7 GHz band

978-1-4244-6240-7/10 $26.00 © 2010 IEEE

We observe that the gain varies from approximately 31 dB to 34 dB at power for the 2.3-2.7 GHz band. This variation is primarily due to detuning of the input and inter-stage matches resulting in a gain roll off at 2.7 GHz. Initial measurements have shown that the PA gain degrades by approximately 2-3dB at 85C indicating that achieving a 25 dBm PA output power with the gain levels of figure 3 is feasible. Also from figure 3, we can infer that the P1dB is approximately 30 dBm for 2.3-2.7 GHz.

The key linearity metrics for 802.16e and TD-LTE are EVM and spectral emission mask (SE. For 802.16e, the specification for a QAM16 WIMAX uplink signal is 6.3% (-24 dB) for EVM. Typically, EVM requirements for the PA itself are set to approximately 4%. One characteristic that is observed in many 4G PAs is the EVM "hump" at low to medium output powers. If the EVM "hump" is above 4%, network performance may suffer. For the SiGe 4G PA, the linearization technique, LET, was used to maintain the IM3 levels below -30 dBc over a wide range of power levels resulting in an EVM below 4%. Figure 4 depicts EVM versus output power for 2.3 GHz to 2.7 GHz.

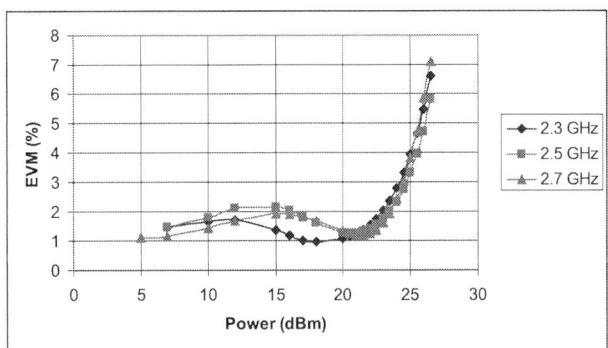

Fig. 4.   EVM (%) versus power for SiGe 4G PA using a 802.16e 10 MHz QAM16 signal

From figure 4, we observe that the EVM is under 4% across 2.3 GHz to 2.7 GHz for PA output powers up to 25 dBm. From the PA gain measurements of figure 3, we estimate that the back off required from P1dB to achieve 4% EVM is 5 dB for 802.16e. The use of the linearization technique has provided at least 1dB improvement in back off since the PAPR for an 802.16e uplink signal is at least 6 dB.

The spectral mask for 802.16e for the 2.5-2.7 GHz band is the most stringent of any wireless standard. For a 10 MHz channel bandwidth, the 6.5 MHz offset from the channel edge is typically the most challenging for mask compliance. At this offset, the spectral mask limit at the antenna is -25 dBm/MHz. For TD-LTE in Band 40 (2.3-2.4 GHz), the spectral mask is significantly more relaxed.

Figure 5 shows a comparison of the spectral masks for the 802.16e, 2.5-2.7 GHz and the TD-LTE 2.3-2.4 GHz band. In figure 5, we observe that the TD-LTE spectral mask is relaxed by 12 dB at the 6.5 MHz offset compared to the 802.16e spectral mask. This should result in a higher compliant output power for the SiGe 4G PA for TD-LTE compared to 802.16e.

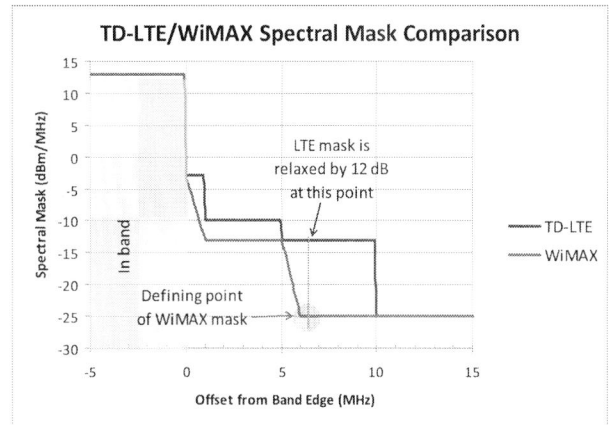

Fig. 5.   Comparison of the TD-LTE and 802.16e spectral mask

The 802.16e spectral mask with a 10 MHz QAM16 uplink signal was measured for the SiGe 4G PA. The PA complied with the 802.16e spectral mask from 2.3-2.7 GHz. Figure 6 shows the spectral mask at an output power of 25 dBm.

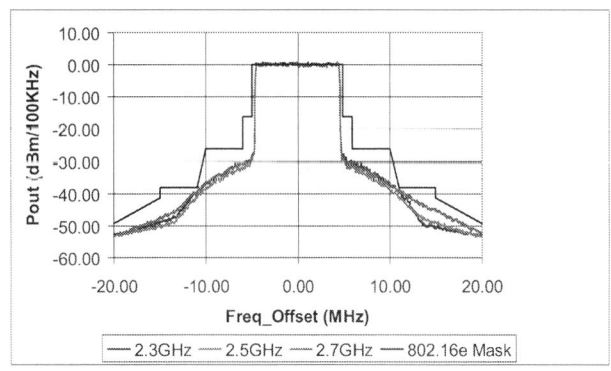

Fig. 6.   802.16e spectral mask for 2.3 GHz to 2.7 GHz at a Vcc =3.3V and a PA output power of 25dBm

In figure 6, we observe that the adjacent channel power (ACP) at the 11.5 MHz offset from the channel center with a 1MHz integration bandwidth at 2.5 GHz is -42 dBc compared to the 802.16e specification of -38 dBc at the antenna. This indicates that there is reasonable margin with spectral mask and therefore further optimization of the PA match and biasing can be performed to improve the PA efficiency at the expense of ACP. EVM and spectral mask were also measured for the TD-LTE standard and the FDD-LTE Band 7. For an EVM limit of 12.5% and 3GPP mask compliance, the SiGe 4G PA was able to produce 28.5 dBm output power for LTE. This is significantly higher than the 25 dBm PA output power needed to deliver 23.5 to 24 dBm at the antenna for TD-LTE assuming 1 to 1.5 dB of post PA losses (balun and T/R switch) for a single band implementation. For FDD-LTE applications, the desired PA output power is approximately 27.5 dBm assuming ~4dB of post PA losses including the duplexer and multi-pole switch.

The current consumption of 4G PAs is critical for mobile and nomadic applications. Figure 7 depicts the DC supply current for a 802.16e 10 MHz QAM 16 uplink waveform. The PA was operated in the high power mode for this measurement. We observe that the current consumption is relatively constant varying from 522mA to 540mA at a Vcc of 3.3V. The power added efficiency (PAE) is approximately 18% across the entire 2.3-2.7 GHz band. For lower power levels, the PA can be placed into a medium or low power mode. Initial measurements at the low power mode reveal approximately 100 mA of current consumption at 0 dBm.

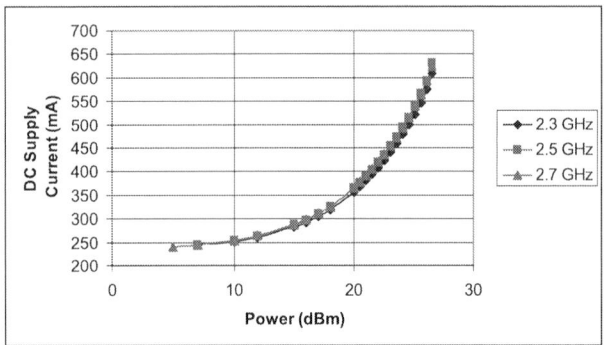

Fig. 7. SiGe 4G PA current versus output power for a QAM16 10 MHz 802.16e uplink signal.

Another important specification is harmonics. For the testing of the harmonics, the measurements were recorded at the output of the balun. The balun provides between 15 dB to 20 dB of common mode rejection at the second harmonics of the 2.3-2.7 GHz band. For a 2.5 GHz 802.16e QAM16 uplink signal with a PA output power of 25 dBm (24.4 dBm following the balun), the second harmonic measured -43 dBm/MHz and the third harmonic measured -23 dBm/MHz. The 802.16e specification for harmonics is -30 dBm/MHz. The second harmonic performance meets the 802.16e specifications but more third harmonic suppression is required. One potential solution is to implement a third harmonic "trap" or a higher order low pass filter characteristic in the output match. For future implementations, the output match and balun can be integrated as a transformer balun with harmonic traps either onto the SiGe PA IC, or onto a separate Si IPD or onto a multi-layer LTCC substrate.

Finally, VSWR ruggedness testing was performed. The PA module with the external balun was subjected to a 10:1 VSWR mismatch for 360 degrees of phase and no instabilities or degradation in performance was observed.

IV. SUMMARY

We have successfully demonstrated a key element in the realization of a fully integrated 4G RF Front End IC; the 4G SiGe PA IC. The 4G SiGe PA produces an output power of 25 dBm for the 2.3-2.7 GHz 802.16e band with a PAE of 18% and second harmonic levels of -34 dBm/MHz after the balun. For TD-LTE Band 40 and FDD-LTE Band 7, this PA produces 28.5 dBm of output power while meeting the 3GPP EVM and spectral mask requirements.

ACKNOWLEDGEMENT

The authors wish to acknowledge Yalin Jin for transformer balun discussions.

REFERENCES

[1] Hsin-Hsing Liao, Hao Jiang, Shanjani, P., King, J., Behzad, A., "A Fully Integrated 2x2 Power Amplifier for Dual Band MIMO 802.11n WLAN Application Using SiGe HBT Technology," *IEEE Journal of Solid State Circuits.*, vol. 44, no.5, pp. 1361-1371, May 2009.

[2] Yan Li, Lopez, J., Lie D.Y.C., Chen K., Wu, S., Tzu-Yi Yang,," SiGe Class-E Power Amplifier With Envelope Tracking For Mobile WIMAX/WiBro Applications,"*IEEE International Symposium on Circuits and Systems*, pp. 2017-2020, May 2009.

[3] Chowdhury, D., Hull, C. D., Degani, O.B., Wang Y., Niknejad, A. M., "A Fully Integrated Dual Mode Highly Linear 2.4 GHz CMOS Power Amplifier For 4G WIMAX Applications," *IEEE Journal of Solid State Circuits*, vol. 44, no. 12, pp. 3393-3402, December 2009

[4] Degani, O., Cossoy, F., Shahaf, S., Chowdhury, D., Hull, C.D., Emanuel, C., Shmuel, R., "A 90nm CMOS Power Amplifier for 802.16e (WIMAX) Applications," *IEEE RFIC Symposium*, pp. 373-376, June 2009

[5] Arvind Keerti, Anh-Vu H. Pham, "RF Characterization of SiGe HBT Power Amplifiers Under Load Mismatch," *IEEE Trans. Microw. Theory and Tech.*, vol. 55, no.2, pp. 207-214, February 2007

[6] V.B. Krishnamurthy, T.K. Khanijoun, K.M. Hershberger, J.T. Reed, P.E, Pace, US Patent #7,573,329, August, 11th, 2009

RTUIF-20

Self-Matched ESD Cell in CMOS Technology for 60-GHz Broadband RF Applications

Chun-Yu Lin[1], Li-Wei Chu[1], Ming-Dou Ker[1,2], Tse-Hua Lu[3], Ping-Fang Hung[3], and Hsiao-Chun Li[3]

[1] National Chiao-Tung University, Hsinchu, Taiwan

[2] I-Shou University, Kaohsiung, Taiwan

[3] Taiwan Semiconductor Manufacturing Company

Abstract — **A self-matched ESD cell library has been implemented in a commercial sub-100nm CMOS process for 60-GHz broadband RF applications. This ESD cell library has reached the 50-Ω input/output matching to reduce the design complexity for RF circuit designer and to provide suitable electrostatic discharge (ESD) protection. Experimental results of this ESD cell library have successfully verified the ESD robustness and the RF characteristics in the 60-GHz frequency band. This self-matched ESD cell library is easily to be used for ESD protection design in the 60-GHz broadband RF applications.**

Index Terms — **Broadband, electrostatic discharge (ESD), ESD cell, V-band, 60 GHz.**

I. INTRODUCTION

With the scaling-down feature size, improving high-frequency characteristics, low power consumption, high integration capability, and low cost for mass production, the nanoscale CMOS technology has been used to implement RF circuits. However, the thinner gate oxide in nanoscale CMOS technology seriously degrades the electrostatic discharge (ESD) robustness of IC products. Therefore, on-chip ESD protection circuits must be added at all input/output ports in RF IC against ESD damages [1]. To minimize the impacts from ESD protection circuit on RF performances, the ESD protection circuit at input/output ports must be carefully designed. Several ESD protection designs have been reported for RF circuits [2], [3]. As the operating frequencies of RF circuits are increasing, on-chip ESD protection designs for RF applications are more challenging.

The frequency band of 57~64 GHz has been allocated for unlicensed usage in the next-generation wireless communications. RF circuits operating at this 60-GHz band have the benefits of excellent interference immunity, high security, multi-gigabit speed, and frequency re-usable. Recently, several CMOS transceivers operating at this frequency band have been reported. Some ESD protection designs for the circuits operating at this frequency band were also presented [4], [5]. In the ESD protection design in [4], the on-chip inductor is used to

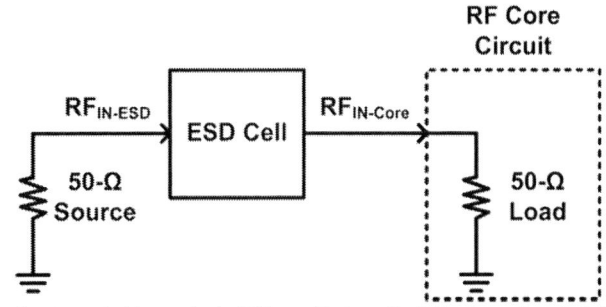

Fig. 1. Self-matched ESD cell for 60-GHz broadband RF applications.

resonate with the capacitive ESD diode to form a series LC resonator. At frequencies above the resonant frequency of the series LC resonator, the impedance becomes large, which means the signal loss from the ESD protection circuit is reduced. Hence, the series LC resonator can be used to protect the 60-GHz circuits. However, the transient voltage across the RF circuits under ESD stress is the total voltage drop across the inductor and the diode. To improve ESD robustness of the RF circuits, the transient voltage across the RF circuits under ESD stress must be reduced, especially for the circuits realized in nanoscale CMOS technology. In another ESD protection design in [5], ESD inductor is co-designed with the 60-GHz power amplifier. The ESD inductor acts part of the RF matching network. Of course, the ESD inductor has been carefully designed to provide ESD current path.

To reduce the design complexity for RF circuit designer, a self-matched ESD-protection cell library in CMOS technology for 60-GHz broadband RF applications is implemented in this work. As shown in Fig. 1, this ESD cell library has reached the 50-Ω input/output matching. Such ESD cell library has been verified in a commercial sub-100nm CMOS process to verify its ESD robustness and broadband RF performances.

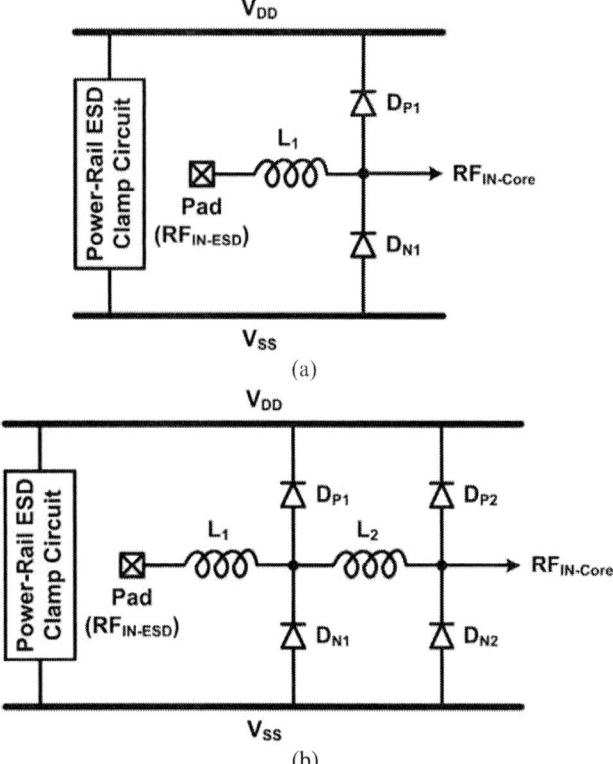

(a)

(b)

Fig. 2. Circuit diagrams of ESD cells with (a) 1-stage ESD protection and (b) 2-stage ESD protection.

II. ESD CELL DESIGN

The distributed ESD protection scheme is used to implement the ESD cells in this work to provide broadband impedance matching around 60 GHz [6], [7]. This ESD cell library provides four ESD cells with different ESD robustness, as listed in Table I. These four ESD cells are designed to sustain 0.5-, 1-, 1.5-, and 2-kV human-body-model (HBM) ESD tests, respectively. Fig. 2(a) shows the circuit diagrams of the 0.5- and 1-kV cells with 1-stage ESD protection. The 1-stage ESD protection is designed with a low-C pad, an on-chip spiral inductor, and a pair of ESD diodes. Similarly, the circuit diagrams of the 1.5- and 2-kV cells with 2-stage ESD protection are shown in Fig. 2(b). The 2-stage ESD protection is designed with a low-C pad, two on-chip spiral inductors, and two pairs of ESD diodes. Besides, the power-rail ESD clamp circuit is added in each ESD cells to provide ESD current paths between V_{DD} and V_{SS}. An RC-inverter-triggered NMOS with ~2000-μm width is used as the power-rail ESD clamp circuit in this work. Since the power-rail ESD clamp circuit is placed between V_{DD} and V_{SS}, it does not contribute any parasitic effects to input/output ports. When the positive-to-V_{DD} (negative-to-

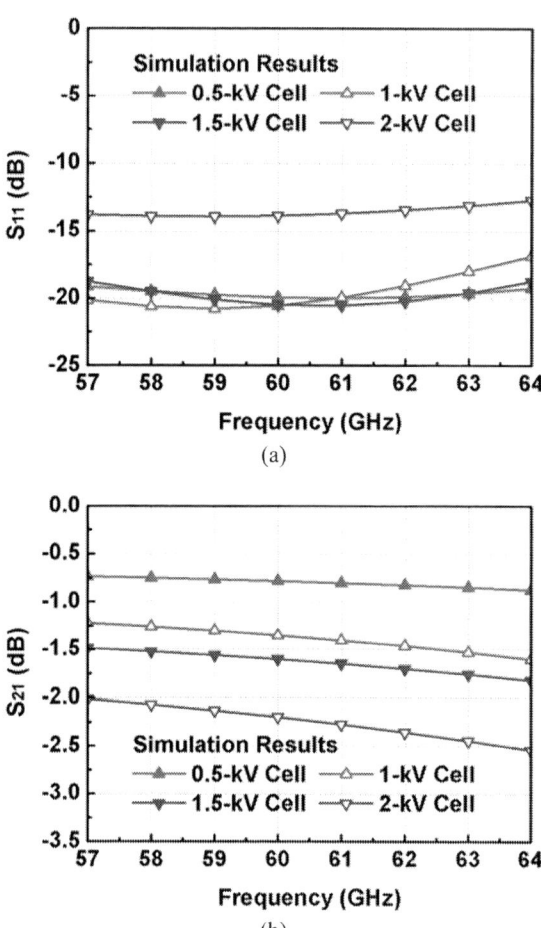

(a)

(b)

Fig. 3. Simulation results of ESD cells on (a) S_{11}, and (b) S_{21}, parameters.

V_{SS}) ESD stress occurs at the pad, the ESD current can be discharged through the diodes D_P (D_N) from the pad to V_{DD} (V_{SS}). Under positive-to-V_{SS} (negative-to-V_{DD}) ESD stress, the ESD current path consists of the diodes D_P (D_N) and the power-rail ESD clamp circuit. The ESD cells can provide the corresponding current discharging paths under all ESD stress modes.

The design parameters of the ESD cells are listed in Table I. The RF characteristics of the ESD cells are simulated by using the microwave circuit simulator ADS with layout parameters. A signal source with 50-Ω impedance drives the RF_{IN-ESD} of the cell, and a 50-Ω load is connected to $RF_{IN-Core}$ to simulate the RF circuit. The reflection (S_{11}) parameters are shown in Fig. 3(a). These ESD cells exhibit good input matching (S_{11} < -12 dB) among 57~64 GHz. The transmission (S_{21}) parameters are compared in Fig. 3(b). At 60-GHz frequency, the 0.5-, 1-, 1.5-, and 2-kV cells have about 0.8-, 1.3-, 1.6-, and 2.2-dB power loss, respectively.

978-1-4244-6240-7/10 $26.00 © 2010 IEEE   524

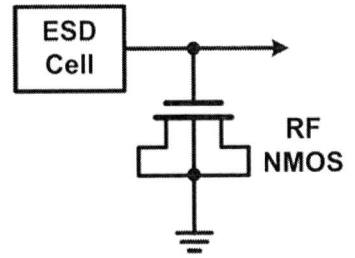

Fig. 4. RF-NMOS emulator to verify ESD protection effectiveness of the proposed ESD cell.

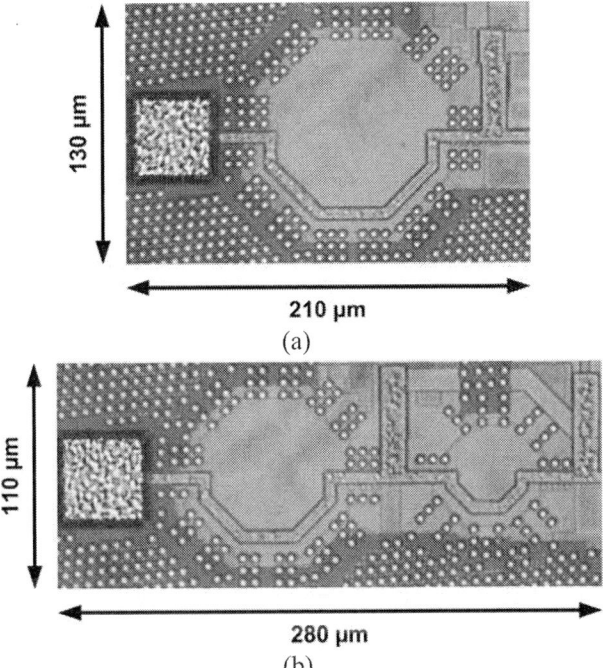

Fig. 5. Die photos of (a) 1-kV cell and (b) 2-kV cell.

One set of these test circuits are arranged with G-S-G style in layout to facilitate the on-wafer RF measurement. Besides, another set of the test circuits are implemented with the RF-NMOS emulator, as shown in Fig. 4. The ESD robustness of the ESD-protected RF circuits can be estimated by the ESD cell with the RF-NMOS emulator. All test circuits have been fabricated in a commercial sub-100nm CMOS process for RF and ESD verifications. Fig. 5 shows the die photos of the 1- and 2-kV cells.

III. EXPERIMENTAL RESULTS

A. RF Performances

The S-parameters of these four ESD cells have been measured around 60 GHz. The voltage supply of V_{DD} (V_{SS}) is 1.2 V (0 V), and the input dc bias is 0.6 V. The source and load resistances to the test circuits are kept at 50 Ω.

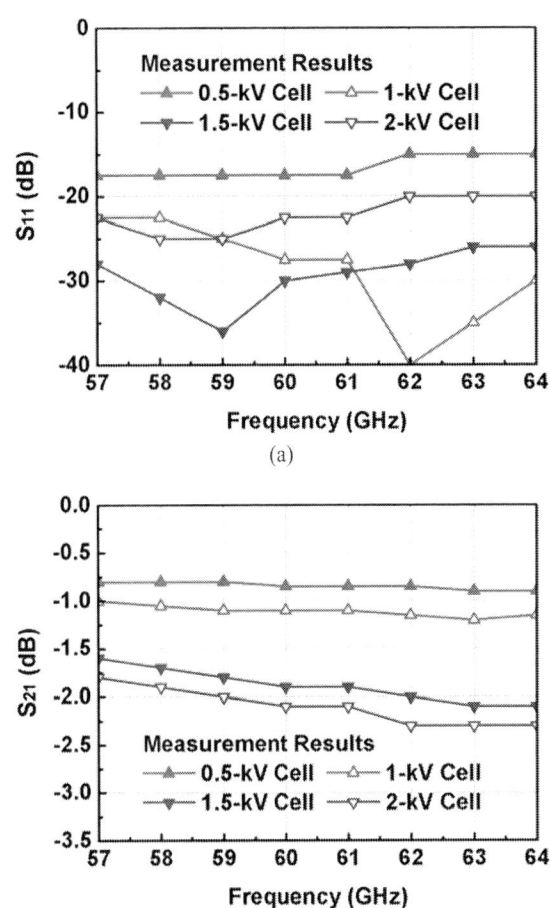

Fig. 6. Measurement results of the fabricated ESD cells on (a) S_{11}, and (b) S_{21}, parameters.

The measured S_{11} and S_{21} parameters versus frequency are shown in Figs. 6(a) and 6(b), respectively. As shown in Fig. 6(a), these ESD cells exhibit good input matching (S_{11} < -20 dB) among 57~64 GHz. At 60-GHz frequency, the 0.5-, 1-, 1.5-, and 2-kV cells have about 0.9-, 1.1-, 1.9-, and 2.1-dB power loss, respectively. The measurement results of the ESD cells are well matching to the simulation results. Hence, applying these ESD cells to RF circuits, the performances can be exactly simulated during the design phase of the RF circuits.

B. ESD Robustness

The HBM ESD robustness of the fabricated ESD cells with the RF-NMOS emulators are evaluated by the ESD tester in human body model (HBM). The positive-to-V_{DD} (PD-mode), positive-to-V_{SS} (PS-mode), negative-to-V_{DD} (ND-mode), and negative-to-V_{SS} (NS-mode) ESD robustness of each ESD cells are listed in Table I. The 0.5-, 1-, 1.5-, and 2-kV cells can sustain 0.75-, 1.5-, 2.25-,

978-1-4244-6240-7/10 $26.00 © 2010 IEEE   525

TABLE I

DESIGN PARAMETERS AND MEASUREMENT RESULTS OF ESD CELLS

		0.5-kV Cell	1-kV Cell	1.5-kV Cell	2-kV Cell
Design Parameters	L_1	0.13 nH	0.13 nH	0.1 nH	0.1 nH
	D_{P1}	8 µm / 0.6 µm	15 µm / 0.6 µm	13 µm / 0.6 µm	17 µm / 0.6 µm
	D_{N1}	8 µm / 0.6 µm	15 µm / 0.6 µm	13 µm / 0.6 µm	17 µm / 0.6 µm
	L_2	N/A	N/A	0.06 nH	0.06 nH
	D_{P2}	N/A	N/A	10 µm / 0.6 µm	13 µm / 0.6 µm
	D_{N2}	N/A	N/A	10 µm / 0.6 µm	13 µm / 0.6 µm
	Cell Area	130 µm x 210 µm	130 µm x 210 µm	110 µm x 280 µm	110 µm x 280 µm
Measurement Results	S_{11} at 60 GHz	< -15 dB	< -25 dB	< -25dB	< -20 dB
	S_{22} at 60 GHz	~ 0.9 dB	~ 1.1 dB	~ 1.9 dB	~ 2.1 dB
	PD-Mode HBM	1 kV	2 kV	2.75 kV	3.5 kV
	PS-Mode HBM	1 kV	1.75 kV	2.25 kV	2.5 kV
	ND-Mode HBM	0.75 kV	1.5 kV	2.25 kV	2.75 kV
	NS-Mode HBM	0.75 kV	1.5 kV	2.25 kV	2.75 kV

and 2.5-kV HBM ESD tests, respectively. These ESD test results are even better than the original specification of these ESD cells. Besides, the power-rail ESD clamp circuit can sustain over 8-kV HBM ESD tests.

IV. CONCLUSION

The ESD cells for 60-GHz broadband RF applications are presented in this work. These ESD cells have reached the 50-Ω input/output matching. This ESD cell library reduces the design complexity for RF circuit designer and provides suitable ESD protection. Verified in a commercial sub-100nm CMOS process, the 0.5-, 1-, 1.5-, and 2-kV cells have about 0.9-, 1.1-, 1.9-, and 2.1-dB power loss, respectively. Besides, they can sustain 0.75-, 1.5-, 2.25-, and 2.5-kV HBM ESD tests, respectively. This ESD cell library is developed to support foundry's customers for them to easily apply ESD protection in the 60-GHz broadband RF circuits.

ACKNOWLEDGEMENT

This work was supported by Taiwan Semiconductor Manufacturing Company (TSMC). The authors would like to thank Mr. C.-P. Jou, Mr. H.-H. Chen, Mr. M.-H. Tsai, Mr. T.-L. Hsu, Mr. M.-H. Song, Mr. J.-C. Tseng, and Mr. T.-H. Chang, of TSMC for their technical discussion to this study. The authors would also like to thank the support from "Aim for the Top University Plan" of the

National Chiao-Tung University and Ministry of Education, Taiwan.

REFERENCES

[1] S. Voldman, *ESD: RF Technology and Circuits*. John Wiley & Sons, 2006.

[2] C.-Y. Lin and M.-D. Ker, "Low-capacitance SCR with waffle layout structure for on-chip ESD protection in RF ICs," in *Proc. IEEE Radio Frequency Integrated Circuit Symp.*, 2007, pp. 749-752.

[3] M.-D. Ker and Y.-W. Hsiao, "On-chip ESD protection strategies for RF circuits in CMOS technology," in *Proc. Int. Conf. Solid-State and Integrated Circuit Technology*, 2006, pp. 1680-1683.

[4] B. Huang, C. Wang, C. Chen, M. Lei, P. Huang, K. Lin, and H. Wang, "Design and analysis for a 60-GHz low-noise amplifier with RF ESD protection," *IEEE Trans. Microwave Theory and Techniques*, vol. 57, no. 2, pp. 298-305, Feb. 2009.

[5] S. Thijs, K. Raczkowski, D. Linten, M. Scholz, A. Griffoni, and G. Groeseneken, "CDM and HBM analysis of ESD protected 60 GHz power amplifier in 45 nm low-power digital CMOS," in *Proc. EOS/ESD Symp.*, 2009, pp. 329-333.

[6] M.-D. Ker and B.-J. Kuo, "Decreasing-size distributed ESD protection scheme for broadband RF circuits," *IEEE Trans. Microwave Theory and Techniques*, vol. 53, no. 2, pp. 582-589, Feb. 2005.

[7] M.-D. Ker and C.-M. Lee, "ESD protection circuits with impedance matching for radio-frequency applications," US Patent 2006/0256489 A1, Nov. 16, 2006.

The Impact of MOSFET Layout Dependent Stress on High Frequency Characteristics and Flicker Noise

Kuo-Liang Yeh, Chih-You Ku, and Jyh-Chyurn Guo

Institute of Electronics Engineering, National Chiao Tung University, Hsinchu, Taiwan
Tel: +886-3-5131368, Fax: +886-3-5724361, E-mail: jcguo@mail.nctu.edu.tw

Abstract —Layout dependent stress in 90 nm MOSFET and its impact on high frequency performance and flicker noise has been investigated. Donut MOSFETs were created to eliminate the transverse stress from shallow trench isolation (STI). Both NMOS and PMOS can benefit from the donut layout in terms of higher effective mobility μ_{eff} and cutoff frequency f_T, as well as lower flicker noise. The measured flicker noise follows number fluctuation model for NMOS and mobility fluctuation model for PMOS, respectively. The reduction of flicker noise suggests the reduction of STI generated traps and the suppression of mobility fluctuation due to eliminated transverse stress using donut structure.

Index Terms — Donut, Shallow-Trench Isolation (STI), Stress, Mobility, Flicker noise

I. INTRODUCTION

With the advancement of CMOS technology to nanoscale regime, the stress introduced from materials and process become more sensitive to the device layout and topography. The shallow trench isolation (STI) process will induce compressive stress and traps, which may have impact on flicker noise (i.e., 1/f noise) in NMOS and PMOS devices. [1] Layout-dependent stress from STI and its impact on high frequency characteristics as well as flicker noise has been investigated but limited to NMOS [2]-[3]. A minor layout modification, namely edge-extended was implemented to reduce the stress and traps introduced by STI [2]. However, the edge-extended layout cannot eliminate the gate-to-STI edge overlap region and leaves STI stress an impact factor. A ring type device was proposed, trying to solve the mentioned problem and identify the influence on flicker noise [3]. However, the study is limited to the stress along the gate width, i.e. transverse to the channel (transverse stress $\sigma_\perp$) and the impact on high frequency performance is unknown. Furthermore, both studies of edge-extended and ring type layouts did not cover PMOS, which is even more important than NMOS for low phase noise design.

In this paper, a new MOSFET layout, namely doughnut (donut) is proposed to create devices free from transverse STI stress, along the gate width.. Meanwhile, an extensive investigation is performed on both NMOS and PMOS devices to explore the STI stress effect on channel current, cutoff frequency (f_T) and flicker noise. For each device structure under a specified bias, the flicker noise is averaged from several different dies to represent statistics of die-to-die variation. This work is aimed to identify the impact from STI stress on high frequency characteristics as well as flicker noise and the results can guide MOSFET layout optimization for RF and analog circuit design.

II. DEVICE FABRICATION AND CHARACTERIZATION

In this work, the devices were fabricated in 90nm CMOS process, with 90nm gate length drawn on the layout L_{drawn} and the total gate width W_{tot} fixed at 64μm. In order to investigate the stress and interface traps generated near STI edge, two types of MOSFET layouts, namely standard and donut are designed and implemented. Standard device means multi-finger structure with finger width W_F=2 μm and finger number N=32. As shown in Fig. 1, donut MOSFETs are constructed as 4-side polygons in which the corners contribute very little to the channel current [4]. In this work, donut devices with two layout dimensions are implemented. In Fig. 1(a), D1S1 represents donut MOSFET in which the space from poly gate to STI edge follows the minimum rule, i.e. 0.3μm, to maximize the compressive stress from STI and along the channel (i.e., longitudinal stress $\sigma_{//}$). Meanwhile, D10S10 shown in Fig. 1(b) denotes donut MOSFET with 10 times larger space between poly gate and STI edge, i.e. 3μm, intentionally to relax $\sigma_{//}$ from STI.

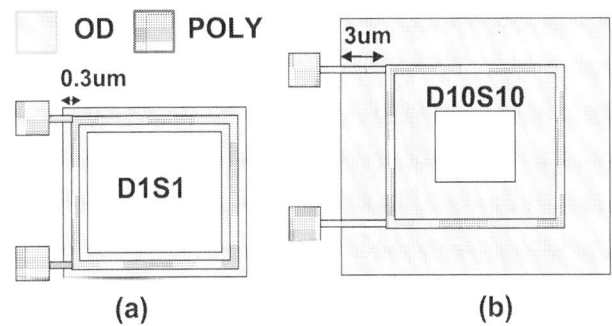

Fig. 1 A brief layout of donut MOSFET (a) D1S1 and (b) D10S10, with two major layers, such as active region (OD) and poly gate (PO)

S-parameters were measured by Agilent E8364B network analyzer for high frequency characterization and AC parameters extraction. Open and short deembedding was performed to remove the parasitic capacitances from the pads as well as interconnection lines and the resistances from all of the metal interconnect. The power spectral density (PSD) of drain current noise, namely S_{ID} was measured by low frequency noise (LFN) measurement system, consisting of Agilent dynamic signal analyzer (DSA 35670) and low noise amplifier (LNA SR570). The LFN measurement generally covers a wide frequency range from 4Hz to 10k Hz. The LFN was measured under various gate-over-drive ($|V_{GT}|$=0.1~0.7V) and $|V_{DS}|$=50mV for both NMOS and PMOS.

III. RESULTS AND DISCUSSION

At first, STI stress introduced in MOSFETs with three different layouts as mentioned (standard, donut D1S1 and D10S10) is illustrated in **Fig.2** to assist an analysis and understanding of layout effect on STI stress and then the electrical characteristics. Note that STI stress is classified as longitudinal stress, denoted as $\sigma_{//}$, which is in parallel with the channel, and transverse stress, namely $\sigma_{\perp}$, which is transverse to the channel. We can see that standard MOSFETs (Fig.2(a)) are subject to $\sigma_{//}$ along the channel length and $\sigma_{\perp}$ along the gate width. On the other, donut MOSFETs are free from $\sigma_{\perp}$. Regarding the stress favorable for mobility enhancement, it has a critical dependence on the device types and orientations, as shown in **Table I** [5]. For NMOS, tensile stress, either $\sigma_{//}$ or $\sigma_{\perp}$ can improve μ_{eff}. As for PMOS, compressive stress in $\sigma_{//}$ or tensile stress in $\sigma_{\perp}$ is the right one for μ_{eff} enhancement.

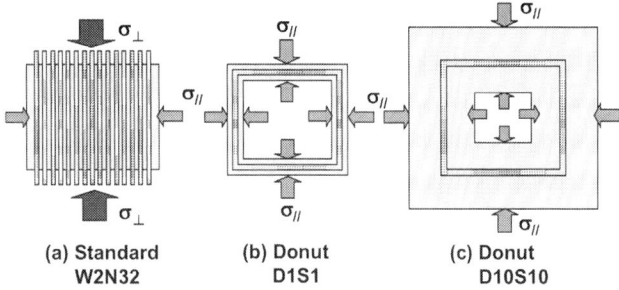

Fig. 2 Schematics of STI stress in MOSFETs with three different layouts (a) standard multi-finger device W2N32 (b) donut device D1S1 (c) donut device D10S10. Longitudinal stress : $\sigma_{//}$ in parallel with the channel, transverse stress : $\sigma_{\perp}$ transverse to the channel.

TABLE I

Stress favorable for mobility enhancement in NMOS and PMOS along longitudinal and transverse directions [5]

Directions	Stress favorable for mobility enhancement	
	NMOS	PMOS
Longitudinal ($\sigma_{//}$)	Tensile	Compressive
Transverse ($\sigma_{\perp}$)	Tensile	Tensile

A. DC Performance of Standard and Donut NMOS

Fig.3(a) presents the maximum transconductance $G_{m,max}$ measured from NMOS. It is found that $G_{m,max}$ of D1S1 is degraded by around 9.7% but that of D10S10 is enhanced by 7.5% as compared with the standard device. The experimental suggests the compressive $\sigma_{//}$ from STI, which is maximized in D1S1 due to the minimum gate to STI space is the primary factor responsible for $G_{m,max}$ degradation. As for D10S10, the much lower $\sigma_{//}$ due to 10 times larger space and eliminated $\sigma_{\perp}$ for donut layout contributes to $G_{m,max}$ improvement. The influence on effective mobility μ_{eff} shown in Fig. 3(b) reveals exactly the same trend. D1S1 suffers 9.2% degradation while D10S10 gain 7.45% enhancement in μ_{eff}. The results justify the mechanism that the layout dependence of $G_{m,max}$ is originated from the effect of STI stress $\sigma_{//}$ and $\sigma_{\perp}$ on electron mobility summarized in Table I.

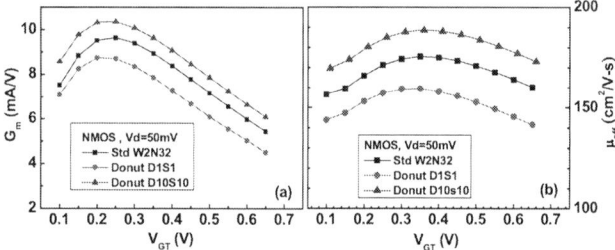

Fig. 3 (a) The transconductance G_m and (b) effective mobility μ_{eff} and extracted from linear I-V for standard and donut NMOS with different poly-gate to STI edge distances, D1S1 and D10S10 defined in Fig.1.

B. DC Performance of Standard and Donut PMOS

As for PMOS, the donut devices D1S1 and D10S10 demonstrate 12.2% and 7.6% higher $G_{m,max}$ than the standard one shown in Fig.4(a). Again, the layout dependence of μ_{eff} illustrated in Fig.4(b) indicates the same trend as that of $G_{m,max}$. The donut PMOS, D1S1 and D10S10 present 12.5% and 6.3% μ_{eff} enhancement compared to the standard device. According to Table I, it is believed that D1S1 with the min. gate to STI edge distance, resulting the highest compressive $\sigma_{//}$ and minimized $\sigma_{\perp}$ can benefit the most in hole mobility. The standard PMOS with relieved $\sigma_{//}$ in multi-finger structure and largest $\sigma_{\perp}$ along narrow width suffers the worst hole mobility.

978-1-4244-6240-7/10 $26.00 © 2010 IEEE

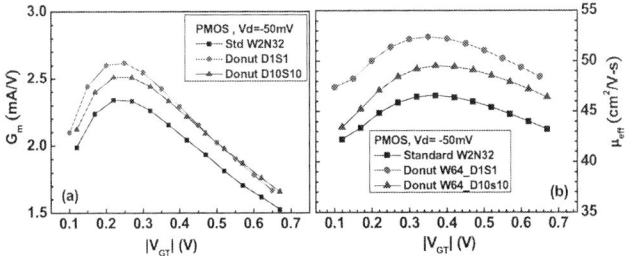

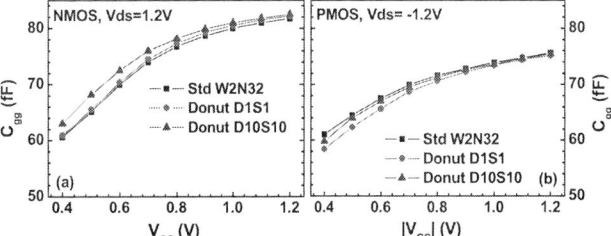

Fig. 4 (a) The transconductance G_m and (b) effective mobility μ_{eff} and extracted from linear I-V for standard and donut PMOS with different poly-gate to STI edge distances, D1S1 and D10S10 defined in Fig.1

C. High Frequency Performance of Donut and Standard MOSFETs

The impact from layout dependent STI stress on high frequency performance is of special concern for RF MOSFETs and circuits design. Fig. 5(a) and (b) illustrate the cutoff frequency f_T measured from NMOS and PMOS with donut and standard layouts. Note that f_T is extracted from the extrapolation of $|H_{21}|$ to unity gain. For NMOS in Fig.5(a), D10S10 gains 5% improvement in the maximum f_T compared to the standard and D1S1. The benefit from donut layout becomes particularly larger for PMOS. As shown in Fig.5(b), D1S1 presents the best performance with the highest f_T and realizes 28% increase in the maximum f_T than the standard device.

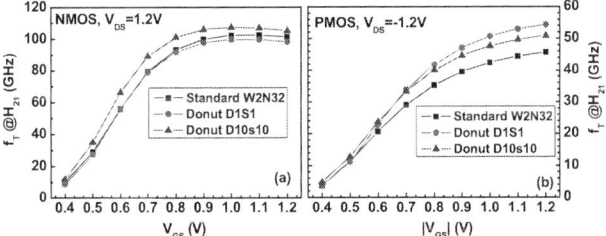

Fig. 5 The cut-off frequency f_t vs. V_{gs} measured for standard and donut devices (a) NMOS (b) PMOS. Standard : multi-finger W2N32. Donut : D1S1 and D10S10.

The resulted improvement on f_T in donut MOSFETs can be consistently explained by the enhancement of μ_{eff} and G_m. Referring to (1), an analytical model for calculating f_T [6], it is predicted that f_T is proportional to G_m and the enhancement of G_m can boost f_T under fixed gate capacitances (C_{gg} and C_{gd}). **Fig.6(a) and (b)** present C_{gg} measured from NMOS and PMOS with three different layouts. The results indicate much smaller difference in C_{gg} between donut and standard layouts, as compared with G_m (Fig.3 and Fig.4). Thus, layout dependence of f_T just follows that of G_m.

$$f_T = \frac{G_m}{2\pi\sqrt{C_{gg}^2 - C_{gd}^2}} \qquad (1)$$

Fig. 6 C_{gg} vs. V_{gs} extracted from Y-parameters for standard and donut devices (a) NMOS (b) PMOS. Standard : multi-finger W2N32. Donut : D1S1 and D10S10.

Regarding other RF performance parameters, such as maximum oscillation frequency, f_{max} and noise figure, NF_{min} (not shown), the donut MOSFETs suffer significant degradation due to inherently larger gate resistances than the standard one with multiple gate fingers. The experimental suggests an innovative donut device layout is required to cover all of the RF and analog performance.

D. Low Frequency Noise of Standard and Donut MOSFETs

Fig. 7(a) and (b) make a comparison of LFN in terms of S_{ID}/I_D^2 between the standard and donut devices for NMOS and PMOS, respectively. The noise spectrum follows 1/f characteristics over a wide frequency domain from 4 to 10K Hz. It means that the measured LFN is a typical flicker noise. The standard device reveals near twice larger S_{ID}/I_D^2 as compared to donut devices for both NMOS and PMOS, under a specified gate overdrive voltage, $|V_{GT}|=0.7V$. In contrast, the donut device D10S10 with the most extended gate to STI-edge distance indicates the lowest S_{ID}/I_D^2. The results can be consistently explained by the fact that D10S10 can keep free from $\sigma_\perp$ as well as interface traps near STI edge, and the smallest $\sigma_{//}$ due to 10 times larger space away from the STI edge compared to D1S1.

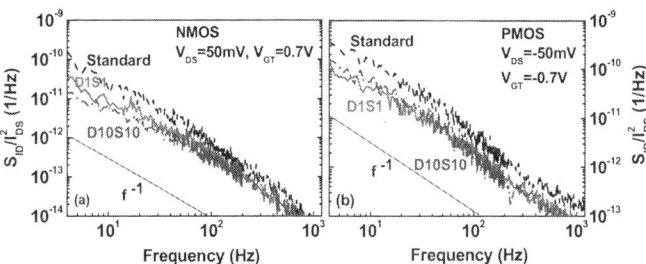

Fig. 7 The low frequency noise S_{ID}/I_{DS}^2 measured for the standard and donut devices (a) NMOS (b) PMOS. Standard : multi-finger W2N32. Donut : D1S1 and D10S10.

To further explore the mechanism responsible for LFN, the measured S_{ID}/I_{DS}^2 at frequency 50Hz are plotted versus

I_{DS} for three different devices, under various $|V_{GT}|$ (0.1~0.7V) shown in Fig.8 (a) and b) for NMOS and PMOS, respectively. For nMOS devices, the measured LFN characteristic is dominated by number fluctuation model given by (2) in which S_{ID}/I_{DS}^2 is proportional to N_t/I_{DS}^2 and that predicts the increase of LFN with increasing the traps density N_t [7]. It is believed that the gate to STI-edge overlap region will suffer the most severe compressive strain as well as interface traps N_t, and the donut devices can eliminate these effects along the gate width, i.e. in the transverse direction. According to previous study, the stress generated traps may aggravate the scattering effect and increase the flicker noise [8]. The mentioned mechanism can explain why the donut devices free from gate to STI-edge overlap region can have the lowest LFN.

$$\frac{S_{ID}}{I_{DS}^2} = \frac{q^2 k_B T \lambda N_t}{f^\gamma} \frac{W C_{ox} \mu_{eff}^2 V_{DS}^2}{L^3} \frac{1}{I_{DS}^2} \qquad (2)$$

N_t : the density of traps at quasi - Fermi level

As for PMOS shown in Fig.8(b), the measured S_{ID}/I_{DS}^2 follows a simple power law of $1/I_{DS}$ and manifests itself governed by mobility fluctuation model, according to Hooge empirical formula expressed in (3) [9]. Note that the Hooge parameter α_H is dimensionless and may vary with biases and process technologies. The reduction of LFN measured from donut PMOS suggests the suppression of mobility fluctuation due to the eliminated compressive $\sigma_\perp$.

$$\frac{S_{ID}}{I_{DS}^2} = \frac{1}{f} \frac{\alpha_H \mu_{eff}}{L^2} \frac{q V_{DS}}{I_{DS}} \qquad (3)$$

α_H : the Hooge parameter

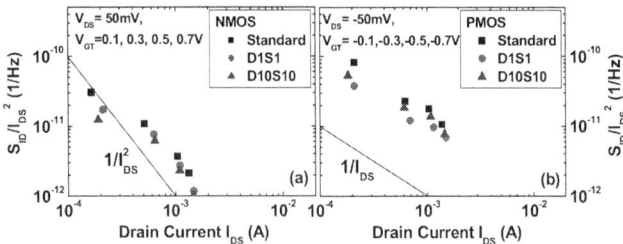

Fig. 8 S_{ID}/I_{DS}^2 vs. I_{DS} under varying $|V_{GT}|$ (0.1~0.7V) for standard and donut devices (a) NMOS (b) PMOS. Standard : multi-finger W2N32. Donut : D1S1 and D10S10.

IV. CONCLUSION

The proposed donut MOSFETs demonstrate the advantages over the standard MOSFETs, such as the lowest S_{ID}/I_{DS}^2 in low frequency domain (1~ 10K Hz) and higher f_T in very high frequency region (100/50 GHz for N/P MOS). The elimination of STI stress and excess traps

along the gate width is validated as the primary mechanism responsible for the enhancement of μ_{eff} as well as f_T, and reduction of LFN. The layout dependent stress mechanism can be applied to both NMOS and PMOS, even though their LFN are governed by different models. An innovative donut device layout for solving the potential degradation of f_{max} and NF_{min} emerges as an interesting and important topic in the future work for RF and analog applications.

ACKNOWLEDGEMENT

This work is supported by NSC98-2221-E009-166-MY3. Besides, the authors acknowledge the support from NDL RF Lab. for noise measurement and Chip Implementation Center (CiC) for device fabrication.

REFERENCES

[1] T. Ohguro, Y. Okayama, K. Matsuzawa, K. Matsunaga, N. Aoki, K. Kojima, H. S. Momose, and K. Ishimaru, "The impact of oxynitride process, deuterium annealing and STI stress to 1/f noise of 0.11μm CMOS", in *Symp. on VLSI Tech.*, 2003, pp. 37–38.

[2] C.-Y. Chan, Y.-S. Lin, Y.-C. Huang, S. S. H. Hsu, and Y.-Z. Juang, "Edge-extended Design for Improved Flicker Noise Characteristics in 0.13-μm RF NMOS", in *IEEE MTT-S Intl. Microwave Symp.* 3-8 June, 2007, pp.441-444.

[3] Y.-L. R. Lu, Y.-C. Liao, W. McMahon, Y-H. Lee, Helen Kung, R. Fastow, and S. Ma, "The Role of Shallow Trench Isolation on Channel Width Noise Scaling for Narrow Width CMOS and Flash Cells", in *Int. Symp. on VLSI TSA*, 21-23 April, 2008, pp. 85–86.

[4] P. L'opez, M. Oberst, H. Neubauer, D. Cabello and J. Hauer, "Performance analysis of high-speed MOS transistors with different layout styles", IEEE Int. Symposium on Circuits and Systems, Kobe, Japan, 23-25 May, 2005.

[5] Y. Luo and D. K. Nayak, "Enhancement of CMOS performance by process-induced stress" IEEE Trans. Semicond. Manuf., vol. 18, no. 1, pp. 63–68, Feb. 2005.

[6] Tajinder Manku, "Microwave CMOS—Device Physics and Design", IEEE Journal of Solid-State Circuits, vol. 34, no. 1, Mar. 1999, pp. 277-285.

[7] G. Reimbold, "Modified 1/f trapping noise theory and experiments in MOS transistors biased from weak to strong inversion-influence of interface states," IEEE Trans. Electron Devices, vol. ED-31, pp. 1190-1198, 1984.

[8] S. Maeda, Y. S. Jin, J. A. Choi, S. Y. Oh, H. W. Lee, J. Y. Woo, M. C. Sun, J. H. Ku, K. Lee, S. G. Bae, S. G. kang, J. H. Yang, Y. W. Kim, K. P. Suh, " Impact of Mechanical Stress Engineering on Flicker Noise Characteristics," in *Symp. on VLSI Tech.*, June 2004, pp.102-103.

[9] F. N. Hooge, *et al.*, "Lattice scattering causes 1/f noise," Phys. Lett. A, vol. 66, pp. 315-316, 1978.

A Novel Low-Profile Low-Parasitic RF Package Using High-Density Build-Up Technology

Chien-Cheng Wei, Ming-Chien Lin, Chin-Ta Fan, Ta-Hsiang Chiang, Ming-Kuen Chiu, Shao-Pin Ru, Nan Ni*, and Albert Cardona*

Tong Hsing Electronic Industries, LTD., 55, Lane 365, Yingtao Road, Yinko, Taipei Hsien, Taiwan 239

Agile RF*, Inc. 93 Castilian Drive, Santa Barbara, CA 93117

Abstract — **This paper presents a low-profile low-parasitic RF package by using the high-density build-up (HD-BU) technology. This package achieves much thinner, fine pitch, and exposed design pattern feature for outstanding electrical and thermal performance. The packaging fabrication is simple and only needs several processes. This HD-BU package provides lower parasitic than other lead-frame types due to the use of very thin bonding pads. Additionally, a capacitor chip is assembled using the proposed technology for packaging demonstration and electrical performance evaluation. Based on the experimental results, the measured capacitances at 1-GHz are quite similar before and after packaging. It indicates that the HD-BU package has low parasitic capacitance even at high-frequency operation, and does not affect the electrical performance for the packaged chip. Therefore, these packages are good candidates for applications requiring low profile, low parasitic and low cost.**

Index Terms — **High-density build-up, lead-frame, BST varactor, RF package.**

I. INTRODUCTION

Wireless communication systems have significantly grown up in different applications during the recent decade. New applications are requiring more demands on high frequency operation, they are putting high pressure on the current RF/microwave components and systems. In the high-frequency packages, too complicated structure may tend to introduce more stray inductive and capacitive losses. Therefore, the electrical performances may be affected easily and degraded by the packaging materials and design [1].

Quad Flat No-Lead (QFN) follows JESD75-5 [2] standard is a very popular, simple, and low-cost packaging type, which has been used widely in RFIC, MMIC, and RF SiP applications. QFN package achieves several desirable attributes including miniaturized size, good electrical and thermal performance [3], but still faces some limitations restricted by the standard rules [4]. Thus, the proposed high-density build-up (HD-BU) is a new packaging technology based on the QFN lead-frame improvement. This HD-BU provides much thinner, fine pitch, and exposed design pattern feature for excellent electrical and thermal performance. The bonding-pad thickness of HD-BU can be down to only several

tens of micrometers, which diminish substantially the parasitic effects in high frequency, and lead to be applied for RF package suitably. Due to the use of very thin bonding pads, the HD-BU packages have much smaller dimension than other conventional packages such as ball grid array (BGA), leadless chip ceramic carrier (LCCC), thin quad flat packages (TQFPs) and thin shrink small outline packages (TSSOPs) [5]. Additionally, HD-BU is also a lead-free, multi-row, array package, ease of fabrication and mass-production solution.

Furthermore, a RF chip, Barium Strontium Titanate (BST) varactor was assembled by using the proposed HD-BU package. The capacitances of the chip were measured and correlated before and after packaging. According to the experimental results, these measurements from bare die and package both exhibit very similar characteristics. It shows that the proposed package has low parasitic, and does not influence the electrical properties seriously. Therefore, this HDBU package will be a good candidate for wireless communication applications. It also provides lower cost and easier fabrication compared to other lead-frame type packages.

Fig. 1   Sample photo of high-density build-up packages

II. HIGH-DENSITY BUILD-UP TECHNOLOGY

The proposed HD-BU technology basically comprises two main processes, consisting of lead-frame manufacture and assembling procedures. The main process flow is systematically depicted in Fig. 1. The copper plate is the main conductor carrier for the HD-BU's substrate, with thickness of

0.4 mm. This lead-frame manufacture is started from using a dry film pasted onto the copper carrier, and forming the circuit diagram with exposure and development processes. In order to provide the bonding pads for die, wire, or any component attachment, nickel (Ni), copper (Cu) and sliver (Ag) are plated with their specific thickness onto the exposed area where the dry film has been removed. The 35μm-thickness Cu layer is the main conductor for electrical signal transmission. The Ni layer (180~380μinch) is a protection layer to avoid the Cu etched during the backside carrier removing process. The Ag layer (7~15μinch) is also another protection layer to prevent the oxidization of Cu layer, and improves the adhesion for the following surface mount technology (SMT), die bonding (D/B), and wire bonding (W/B) processes. However, after stripping the dry film, the bonding pads are protected by sliver films. And then the solder mask can be printed optionally if designer needed.

After accomplishing the stage of lead-frame manufacture, the assembling parts can be started attaching any component onto the bonding pads of the carrier. These attaching processes are the same as any SMT, D/B, and W/B technologies, comprise mounting the surface mount devices with solder paste on the Cu carrier, attaching the dice with epoxy on the Cu carrier, and using the bonding wires to interconnect the microchip and the Cu leads. Then, for protecting the microchip and the required bonding wires, the over-molding process is adopted. It also provides the ability of holding the bonding wires strongly and avoiding the humidity that may damages the circuit.

Finally, backside Cu etching is the most important step in HD-BU processes to finalize the packaging. The backside Cu needs to be etched with the Ammonium Chlorocuprate, to remove the carrier and separate each bonding pad individually. However, after removing the 0.4mm-thickness Cu carrier, the package thickness is determined by the thickness of molding compound. This thickness can be much thinner compared to the packages using the QFN or printed-circuit board (PCB) as their substrate. In general, the lead-frame thickness for standard QFN is around 0.2 ~ 0.5 mm, and it is about the same for FR4 laminate. However, by adopting the HD-BU to manufacture the required lead-frame, the pad thickness can be reduced to only 20 ~ 40 μm, that is 10 % thickness of the QFN lead-frame. Owing to the above reasons, several key attributes of HD-BU technology can be summarized as below:

➢ Low profile package
➢ Multi-row for assembling
➢ Good electrical and thermal performance

Therefore, the packages using the HD-BU technology can be chosen for very thin or very low parasitic applications. And the outstanding thermal performance is also capable to be applied in high-power component that extremely requiring high heat dissipation consideration.

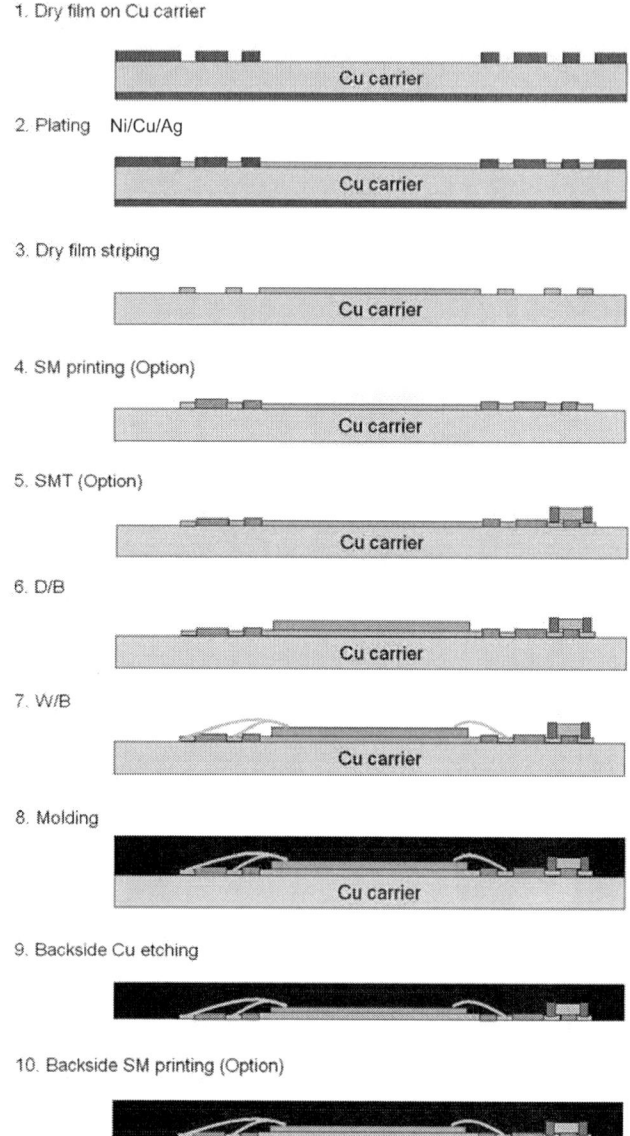

Fig. 2   Process flow for the proposed HD-BU technology

III. HD-BU PACKAGE FOR RF COMPONENT

For demonstrating the proposed HD-BU technology for RF packaging, a capacitor chip, BST (Barium-Strontium-Titanate) varactor was assembled. The designed structure is schematically depicted in Fig. 3, including cross-section and top views. This package only has single die and several wires. The die size is around 1 x 1 mm², and the total package size is 2.1 x 1.7 mm² with one die pad, one dc pad and two RF pads. Three bonding wires are adopted for the varactor's terminal individually, and one wire is used for the voltage bias control. The 35 μm-thickness conductor is fabricated for the bonding pads since the thick Cu carrier is fully removed in step 9 in Fig. 2. The total thickness of entire package is determined by the molding thickness. Therefore, a very thin package can be manufactured easily once a thin mold cap is used during the over molding. All processes are following the HD-BU steps described in section II, include the lead frame manufacture and assembly procedure.

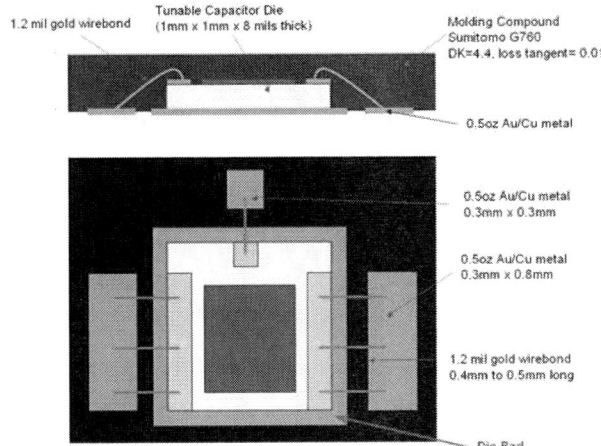

Fig. 3   HD-BU package design for the capacitor chip (BST varactor)

Fig. 4 shows the sample photos of HD-BU package. In Fig. 4(a), it can be seen clearly that the Cu carrier with Ag metallization film (silver color) is still existed under the bonding pads since it is not removed yet at this phase. All the bonding pads are shorted together and do not separate before etching the Cu carrier from the bottom side. Additionally, the finished HD-BU packages are shown in Fig. 4(b). From seeing the left one sample in this photo, it is obvious that all the bonding pads are separated already after over molding and backside carrier removing. Therefore, the thickness of bonding pad can be precisely controlled with plating Ni/Cu/Ag layers, and the total thickness can be very thin to only several tens of micrometers. On the other hand, the molding thickness can be also controlled well, by using a suitable mold cap. In this study, the mold caps with 0.7 mm and 1.0 mm thickness are used, to compare the influence of the electrical performance of the packaged chip.

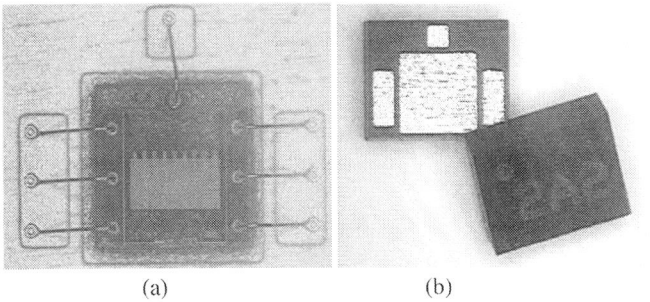

(a)   (b)

Fig. 4 (a) Assembled sample photo before over-molding and backside Cu etching, (b) finished HD-BU samples.

IV. SIMULATION AND MEASUREMENT RESULTS

To characterize the electrical performance of the packaged chip, full-wave EM simulation and practical measurement were both done in this work. Fig. 5 displays the 3D structure of the package in Ansoft HFSS simulator, in order to carry out the capacitance and other parameters. In this figure, the structure is established completely as the package design shown in Fig. 3. This package includes single capacitor chip, several wires, bonding pads, and molding compound except the dc bias parts, since it can be neglected and set directly in the simulator. Regarding the test setup, CASCADE probe station and 1500-µm-pitch G-S probe were adopted to measure the packaged chip's S-parameters. Meanwhile, a power supply is used to provide a bias for varactor's voltage control. Therefore, all measured data can be systematically achieved and compared with the simulated results.

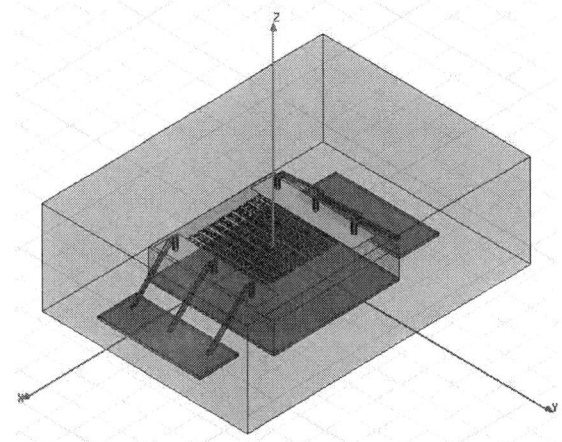

Fig. 5   3D structure of the HD-BU package in Ansoft HFSS

In this paper, two types of BST dice are assembled for comparison, with different capacitance and mold thickness. Fig. 6(a) shows the simulated and measured results for first HD-BU sample with 0.7 mm mold thickness. By adopting the bias voltages of 0 V and 20 V, the simulated capacitances are 4.11 pF and 1.5 pF at 0.5 GHz, and these values are 4.2 pF and 1.6 pF from measurements, respectively. Similar comparison results shown in Fig. 6(b) were also done for another BST die. This one has larger capacitance and 1.0mm mold thickness. For this sample, the simulated capacitances are 5.5 pF and 2.15 pF at the same 0.5 GHz, and the measured values are 5.74 pF and 2.17 pF accordingly. Both figures indicate clearly that the predicted results are well matched to the measured results by using the 3D EM simulator.

Additionally, the capacitances were compared before and after HD-BU packaging. As shown in Table I, the capacitances from bare die and package both have similar values under bias voltages of 0 V, 10 V, and 20 V. Assuming the chips all have the same capacitance, it can be seen clearly that the HD-BU does not affect the performance since the

978-1-4244-6240-7/10 $26.00 © 2010 IEEE   533

assembled part almost have the same capacitance as the chip. These capacitances were all simulated and measured at 1 GHz. However, since the BST capacitors have a very complicated structure and it is hard to model the equivalent circuit, these simulated/measured capacitances are all average values.

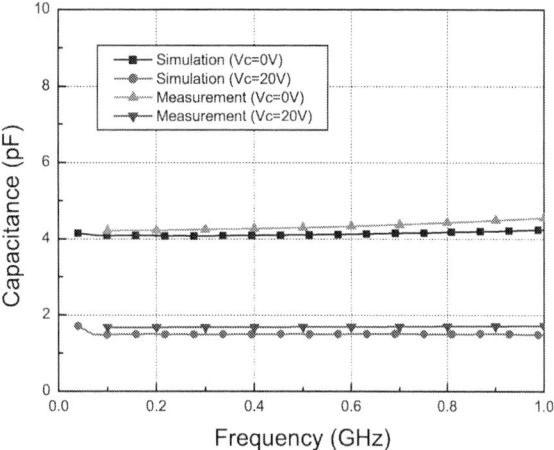

Fig. 6   Simulated and measured capacitances for the packaged varactor under bias voltages of 0 V and 20 V (with 0.7 mm mold thickness)

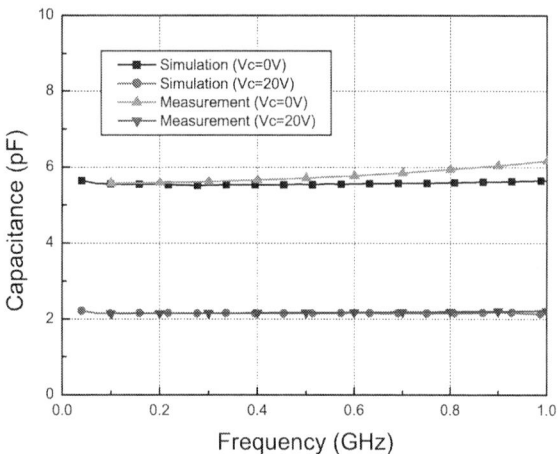

Fig. 7.   Simulated and measured capacitances for the packaged varactor under bias voltages of 0 V and 20 V (with 1.0 mm mold thickness)

Furthermore, in order to compare the HD-BU with other packaging type, another package using flip-chip (FC) process is also simulated for analysis, as listed in Table II. Again, assuming the chips have the same capacitance, it clearly shows that the HD-BU has lower capacitance at 1 GHz than FC, which meaning the parasitic inductance/capacitance is lower. The same conclusion can be seen in the average self-resonant frequency (SRF), the above 3 GHz SRF for HD-BU

is higher than that for FC, saying that this proposed package has a good potential in high-frequency applications.

Table I. Comparisons between bare die and HD-BU package

		0V	10V	20V
Measured Capacitance (pF)	Die	2.5	1.47	1.07
	HDBU	2.59	1.57	1.16
Simulated Capacitance (pF)	Die	2.53	1.45	0.97
	HDBU	2.71	1.64	1.15

Table II. Comparisons for packages using Flip-Chip and HD-BU processes

	Flip Chip	HDBU 1.0mm	HDBU 0.7mm
Measured Capacitance (pF)	6.47	5.64	6.20
SRF(GHz)	2.92	> 3GHz	> 3GHz

V. CONCLUSION

This paper presents a low-profile low-parasitic RF package by using HD-BU technology. This proposed package is based on the improvement of QFN lead-frame, which provides much thinner, fine pitch, and exposed design pattern for outstanding electrical and thermal performance. From the experimental results, the simulated and measured results from bare die and package both exhibit very similar characteristics. It clearly indicates that the proposed package will not influence the electrical characteristics due to the low parasitic of the thin bonding pads. Therefore, these packages are good candidates for the components that requiring thin thickness, low parasitic, high frequency application, and ease of fabrication.

REFERENCES

[1] Tzyy-Sheng Horng, Sung-Mao Wu, Hui-Hsiang Huang, Chi-Tsung Chiu, and Chih-Pin Hung, "Modeling of Lead-Frame Plastic CSPs for Accurate Prediction of Their Low-Pass Filter Effects on RFICs," *IEEE Trans. Microwave Theory & Tech.*, vol. 49, no. 9, pp. 1538-1545, Sept. 2001.

[2] JEDEC Standard, "SON/QFN Package Pinouts Standardized for 1-, 2-, and 3-Bit Logic Functions", July 2004.

[3] Yeong-Lin Lai and Cheng-Yu Ho, "Electrical modeling of quad flat no-lead packages for high-frequency IC applications", in *Proc. IEEE TENCON Dig.*, pp. 344-347, Nov. 2004.

[4] Tan Chee Eng and Geale Fonseka, "QFN Miniaturization Challenges and Solutions", in *Proc. International Electronic Manufacturing Technology Dig.*, 314-319, 2006.

[5] Hirokazu Noma and Tohru Nakanishi, "Etching process analysis based on etchant flow for high-density build-up substrate", in *Proc. Electronics Packaging Technology Conference Dig.*, pp. 289-293, 2004.

A High Quality Factor Varactor Technology Evaluation

Romain Debroucke [1,2], Sebastien Jan [1], Jean-François Larchanché [1] and Christophe Gaquière [2]

[1] STMicroelectronics, 850 rue Jean-Monnet, 38926 Crolles Cedex, France

[2] IEMN, Cité Scientifique Avenue Poincaré, 59652 Villeneuve d'Ascq Cedex, France

Phone: (+33) 4 38 92 28 44 Fax: (+33) 4 38 92 29 54

E-mail: romain.debroucke@st.com

Abstract — **Providing a high quality factor scalable varactor in an integrated technology is a wager. How to insure that your device will give the highest quality factor possible? In order to response this questions, we let the bases of a varactor gauge combining electrical performance and geometrical sizes. Giving a targeted capacitance, it could furnish a qualitive idea of the adequacy with technology performance. It could furnish also a indicator for comparison with other devices. As example of varactor gauge application, we present a comparison between two diode varactor devices in two BiCMOS technologies.**

Index Terms — **Varactor, Quality factor, BiCMOS, Diode, Figure of Merit.**

I. INTRODUCTION

Many works concern quality factor or figure of merit of varactors [1,2] and but generally do not provide sufficient information to orientate the designers on the best choice. Our scope is to set the basis of a tool allowing comparison between integrated CMOS or BiCMOS technologies. Varactors figure of merit depends not only on electrical properties of technologies but also on architecture and so on lithography performance [3]. After having defined a gauge for integrated varactors and a comparison protocol, it will be applied to a new varactor design in a 0.13µm BiCMOS STMicroelectronics dedicated to millimeter wave applications.

II. VARACTOR GAUGE

Let's introduce some definitions before entering the subject. A device represents a parametric structure/architecture. An individual is a device with fixed dimensions. So it is obvious that comparison of two individuals does not provide any trend lines for devices comparison. It is also obvious in the space of interest the two devices performance must recover, i.e. in the case of varactors; devices must share a common space of capacitance. This last condition is necessary but not sufficient for an efficient comparison of two devices and leads us to define a tool called Varactor Family Evaluation.

A. VFE Calculation

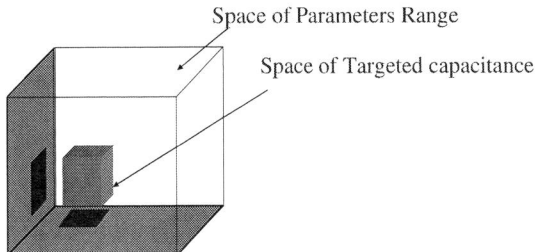

Fig. 1.   Basic definition of VFE

The basic idea of VFE is to evaluate the space of possibilities to synthesize a capacitance value over the whole space of model parameters range (Fig.1). Then:

$$VFE = \frac{VOL_{\text{targetedcapacitance}}}{VOL_{\text{model}}} \qquad (1)$$

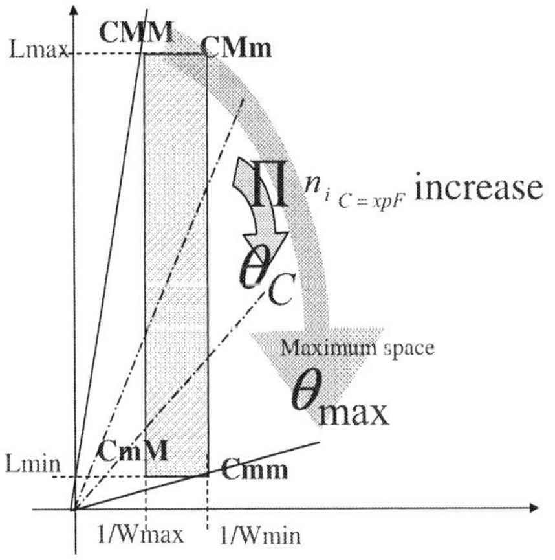

Fig. 2.   VFE applied to planar technologies

This definition could get a more pragmatic formula in a planar technology. Under the assumption of an interdigitated structure, device parameters could be continuous variables and discrete variables. Continuous variables represent the geometrical size of the unit cell; fingers length and finger width for example and discrete variables represent the level of interdigitation (cell number for example). Then a graph (see Fig.2) represents the finger length versus the inverse of finger width. A

978-1-4244-6240-7/10 $26.00 © 2010 IEEE

capacitance value is represented by a straight line crossing the area of geometrical possibilities. As interdigitation level increases, line slope decreases and defines an angle θ_C between its extreme positions. We propose then a simplified definition of VFE with:

$$\beta = \frac{\theta_c}{\theta_{max}} \frac{\prod \Delta n_{i\ C=xpF}}{\prod Max(n_i)} \qquad (2)$$

With $\theta_{max} = atan(L_{max} * W_{max}) - atan(L_{min} * W_{min})$

$$\theta_C = atan\left(\frac{C_{XpF}}{n_{min}}\right) - atan\left(\frac{C_{XpF}}{n_{max}}\right)$$

and Δn_i represents the range of discrete parameters for a targeted capacitance.

B. Application to accumulation mos n-type varactors

We have evaluated VFE for accumulation n-type mos varactor in 65nm CMOS STMicroelectronics technology. On Fig.3, we present VFE varactor with three different oxides: 50Å (GO2 2V5), 31Å (GO2 1V8) and 18Å (GO1). One can see that VFE is maximal for a capacitance of about 1pF whatever oxide thickness is. Those varactors are well-suited for capacitance from 1pF up to 5pF. In this range, the whole set of discrete parameters could be used and it is possible to find the best quality factor achievable in this technology. Outside this range, the space of parameters for a targeted capacitance is limited by the total space of the parameters set and so best quality factor is not relevant of technology performance. On Fig.4, we presented the maximum of Q factor versus finger number for a 50Å oxide varactor. Varactor inside the range 1pF-5pF can have configuration with all finger number and a Q factor optimum appear. Outside this range, the whole set of finger number could not be used, optimum cannot be reached and silicon performances are limited by design rules.

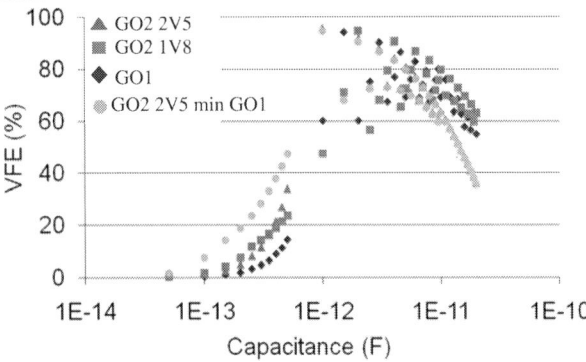

Fig. 3.   VFE evaluation for accumulation MOS n-type varactors in 65 nm STMicroelectronics technology

VFE is also a good indicator of the matching of a varactor with the application. For frequency above 10 GHz, circuit requirements are typically below 100fF of capacitance. Then Fig. 5 decrease dramatically below 1pF meaning that performance of varactor will be limited by parameters range. To overcome this problem, we develop sub-design rule varactor [3]. Thick oxide varactor could have gate length of thin oxide varactor, i.e. in the case of 65nm technology; 50Å could be designed with 65nm gate instead of 0.25µm. This device has been represented by green point on Fig.3. We can see that thick oxide sub-design rule varactors are better suited for low value capacitance.

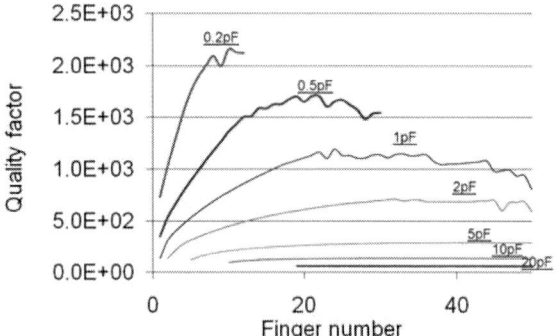

Fig. 4,   Maximum Q factor vs. finger number for a 50Å MOS varactor. Targeted capacitances are indicated on curves.

To conclude, VFE provide a good indicator of a technology. It allows designing the parameters set avoiding the modeling of useless configurations. On the second part, we illustrate VFE usefulness for comparison on diode varactors in BiCMOS technologies.

III. DIODE VARACTOR MODELING

We develop a high Q varactor in a BiCMOS 0.13µm dedicated to millimeter wave technologies. We use the VFE tool to compare performance of high Q varactors with standard diode varactors.

A. High Q varactor structure

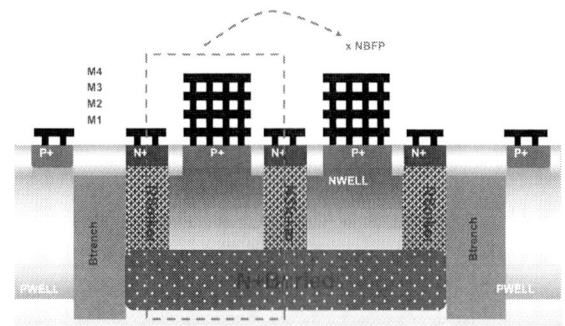

Fig.5.   Cut view of high Q P+/Nwell varactor

The classical structure of a P+/Nwell varactor has been adapted to use the specific layer of bipolar transistor (Fig.5). Among the available layers, we use the N+sinker to reduce the resistive access of anode, the N+Buried layer to have a buried connexion between all anodes and the deep isolation trench to improve the isolation by reducing the parasitic capacitance of the substrate. The layout is an interdigitated structure. The instantiation parameters are: P+finger width and length (Wfp, Lfp), P+ finger number (Nbfp) and cell number (Nbcell).

B. Model

We made a parametric model of the high Q varactor; Low frequency intrinsic capacitance is modeled using a juncap algorithm. Model cover a capacitance from 20fF up to 10pF included all parasitic capacitance with a precision below 5%. Parasitic have been added to low frequency capacitance for a RF model covering frequencies from 50MHz up to 50GHz.

TABLE I

PARAMETERS OF DEVICES UNDER TEST

	Device 1	Device 2
Finger length (µm)	15	15
Finger width (µm)	1	10
Finger number	1	2
Cell number	1	5
Capacitance @0V bias voltage @1GHz	50 fF	5pF
Tuning (C@0V/C@-2.5V)	1.65	1.6
Quality factor @0V bias voltage	500	60

On Fig.6 and Fig.7, we presented comparison of model and measurements for two devices which parameters are presented on Table1. Device 1 measurements are represented by stars (device 2 by circle) and device 1 model is represented by dot line (device 2 by solid lines). The two devices are representative of model accuracy for a design of experiment of 20 devices, asserting scalability of model over instantiation parameter ranges. Once model is finished, performance could be compared to other technology.

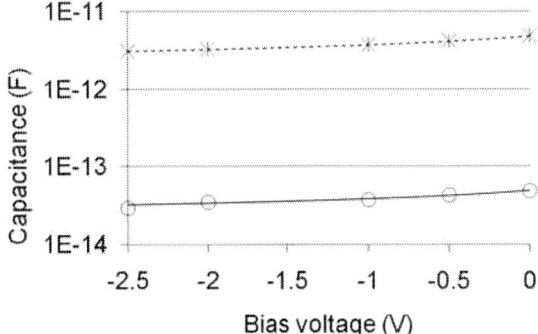

Fig. 6.   Capacitance versus bias voltage at 1GHz

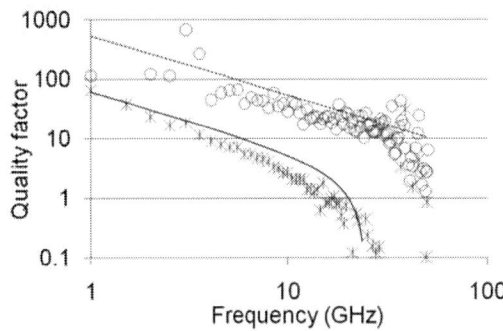

Fig. 7.   Quality factor versus frequency at zero volt of bias voltage

IV. HIGH Q ASSERTION VERIFICATION

Comparison will be done with an early bipolar 0.13 µm not optimized to millimeter wave applications. The structure represented on Fig. 8 shows that the layer N+buried, N+sinker and Trench are not used. N+buried layer is replaced by NISO to improve substrate isolation. This relative simple structure was implemented to have a full compatibility between CMOS and BiCMOS technologies.

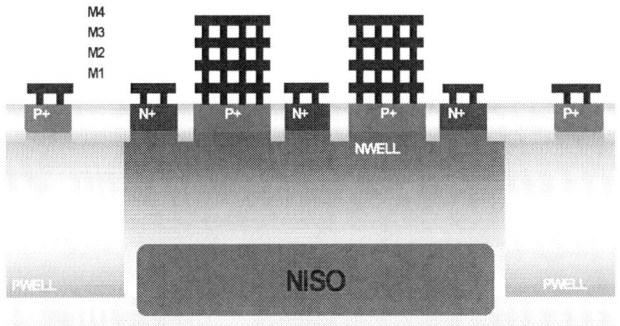

Fig. 8.   Structure of P+/Nwel diode varactor fully compatible CMOS & BiCMOS technologies.

A. VFE Calculation

The instantiation parameters range and surface capacitance are presented on table 2.

TABLE II

PARAMETERS RANGE OF THE TWO TECHNOLOGIES

BiCMOS		0.13 µm	MMW 0.13 µm
Finger length (µm)	Min	15	15
	Max	200	200
Finger width (µm)	Min	1	1
	Max	14	14
Finger number	Min	1	1
	Max	8	10
Cell number	Min	1	1
	Max	1	5
Surface Capacitance @0V (F/m²)		8.98E-4	3.23E-3

On Fig.9, we presented the calculated VFE for the two devices, standard diode varactors are represented by crosses and high Q diode varactors are represented by dots. One can see that standard varactor is well suited for 2pF capacitance and high Q varactor is well suited for 10pF capacitance. This shift is due to surface capacitance increase (due to Nburied layer) and to larger parameters range. Spreading of high Q VFE curve compared to standard varactor one is linked to spreading of geometrical instantiation parameters.

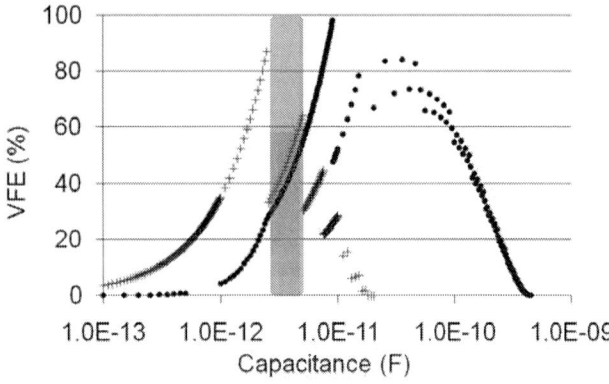

Fig.9. Calculated VFE for standard diode varactor and high Q. Dots (crosses) represent MMW BiCMOS (BiCMOS). Grey zone represents capacitance range of comparison

The two VFE curves share a common range (represented by a grey zone on Fig.9) of capacitance between 2pF and

5pF with a VFE value above 40% which let us enough research space to find an optimum without border effects.

B. Quality factor comparison

To find the best quality factor of a device for a specified capacitance value, we proceed as follow:
- Simulation is made over the whole parameters range. It take about half a day on a computer farm;
- Parameters values are extracted for a given capacitance;
- Best Q is extracted between parameters list previously extracted.

Extracted values show that high Q varactor really exhibit a bigger quality factor than standard varactor in the rnage of 2pF up to 5pF. For example, best Q at 100MHz for a 2pF is about 2.5E3 for High Q varactor and about 1E3 for standard varactor. We find a factor about 2.5 for the whole range 2-5pF. Over 5pF, quality factor of standard varactor is limited by geometrical parameters ranges which reduce the possibilities to synthetize a varactor with a good quality factor. On the other hand, for value below 1pF, standard varactor will provide the best Q factors.

V. CONCLUSION

We have defined a varactor gauge in order to evaluate the performance of a scalable varactor. This tool reveals to be usefull to design the parameters range of varactors. It could also furnish an idea of edge effects and so the possibility to get the best quality factor of a technology. Lastly, it allows comparing technologies by providing recovery area. As example we have presented two varactors developed in 0.13µm BiCMOS technology and we demonstrated the impact of added layers on quality factor.

REFERENCES

[1] S.C.Kelly,J.A.Power and M.O'Neil,"selection and modeling of integrated RF varactors on a 0.35-µM BiCMOS technology." IEEE Trans.on Semi.Manuf.,vol17,no.2,pp 142-149, May 2009.

[2] Y.-J. Chan, C.-F. Huang, C.C. Wu, C.-H. Chen and C.-P. Chao, "Performance consideration of MOS and junction diodes for varactor application", IEEE Trans. On Elec. Dev., vol.54, no.9, pp.2570-2573, September 2007

[3] X. Haifeng and K.O.Kenneth, "High-Q thick-gate MOS varactors with subdesign-rule channel lengths for millimeter-wave applications." IEEE Elect. Device Letters, vol.29,no.4,pp.363-365,   April   2008

Power Improvement for 65nm nMOSFET with High-Tensile CESL and Fast Nonlinear Behavior Modeling

Chia-Sung Chiu[1], Kun-Ming Chen[1], Guo-Wei Huang[1,3], Shu-Yu Lin[1], Bo-Yuan Chen[1],
Cheng-Chou Hung[2], Sheng-Yi Huang[2], Cheng-Wen Fan[2], Chih-Yuh Tzeng[2], and Sam Chou[2]

[1]National Nano Device Laboratories, No. 26, Prosperity Road I, Hsinchu 300, Taiwan
[2]United Microelectronics Corporation, Hsinchu 300, Taiwan
[3]Department of Electronics Engineering, National Chiao Tung University, Hsinchu 300, Taiwan

Abstract — **In this paper, the power gain improvements by stress contact etch stop layer (CESL) in a 65-nm nMOSFET were studied. Compared to the conventional nMOSFET, the device with CESL stress shows an extra 6% power gain enhancement for the increased stress in the channel region. This study also presents the polyharmonic distortion (PHD) model extraction by X-parameters measurement when the power transistor was designed to work far from 50 ohms. By mean of this model, the accurate nonlinear behaviors of nMOSFET were obtained rapidly.**

Index Terms — **Contact etch stop layer (CESL), stress, nMOSFET, X-parameters, PHD.**

I. INTRODUCTION

CMOS power amplifiers potentially are used in wireless products for their low cost and CMOS-compatible process. As a result many CMOS power amplifiers have been designed for integrated radio in place of the more expensive technology for power amplifiers such as GaAs and LDMOS technology [1]-[3].

As CMOS technology shrinks to 65-nm node and beyond, low channel mobility and short channel effect become critical issues and degrade the device performances. Therefore, various strain technologies have been implemented to enhance device mobility. One of the most popular technologies is using high tensile-stress contact etch stop layer (CESL), which can improve mobility and turn-on current in evidence [4][5].

Device characterization is required for power amplifier design, especially in large-signal models. With these models, the performance can be analyzed for varying drive and impedance condition to design the complex circuits. Therefore, many studies have been presented to large signal model development and improvements have been made in recent year [6][7]. These studies using current model or physical equation can predict the large signal operation of active devices accurately. An alternate way to construct a nonlinear model is to use the "black-box" behavioral method proposed by [8], called polyharmonic distortion (PHD) model. This study also presents the nonlinear behavior of nMOSFET using the PHD model.

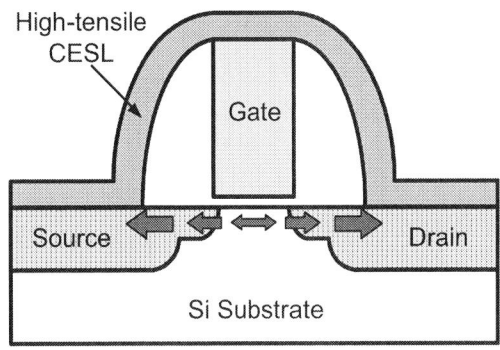

Fig. 1 Tensile strain in channel and source/drain regions of a high-tensile CESL MOSFET

In this paper, for the first time, the high-tensile CESL-stressed 65-nm nMOSFET was analyzed in terms of power performance. Section II describes this device and fabrication. Section III analyzes the DC and small signal performance between the CESL-stress device and control device. In the section IV, we have further examined the power performance and predict the nonlinear characteristics of these devices using the PHD model. It was demonstrated that the presented power improvement method is applicable to CMOS technology.

II. EXPERIMENTS

The devices of nMOSFET with a gate length of 60 nm for this study were prepared using a 65-nm CMOS technology and were fabricated using <100>-channel orientation. In this structure, a high tensile-stress CESL (an 850-Å-thick CESL SiN_x layer) was implemented to induce a higher tensile stress in the channel region. For the control devices, a conventional low-tensile-strength (SiN_x =360 Å) CESL layer was used. The gate width of the test devices was 128 μm. Figure 1 shows the cross-section view of a MOSFET with the high-tensile CESL. In this figure, the advantages of this structure makes CESL induces a higher stress in the channel.

978-1-4244-6240-7/10 $26.00 © 2010 IEEE

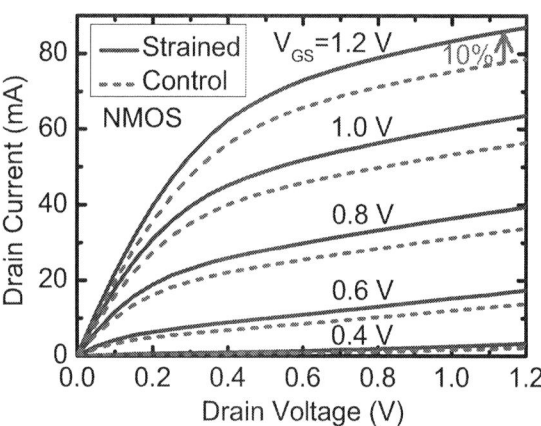

Fig. 2 I_D-V_{DS} characteristic curves of L=60 nm nMOSFET with and without the high-tensile CESL.

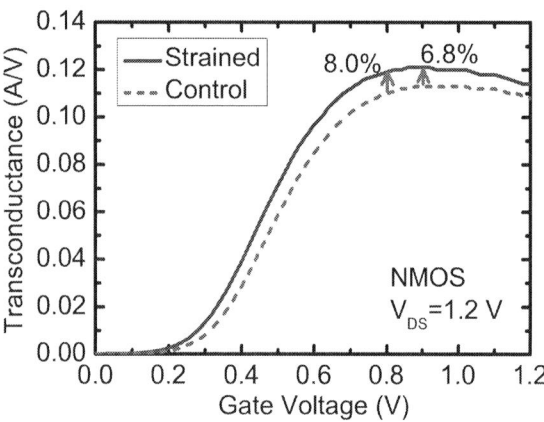

Fig. 3 Transconductance of L=60 nm nMOSFETs with and without the high-tensile CESL.

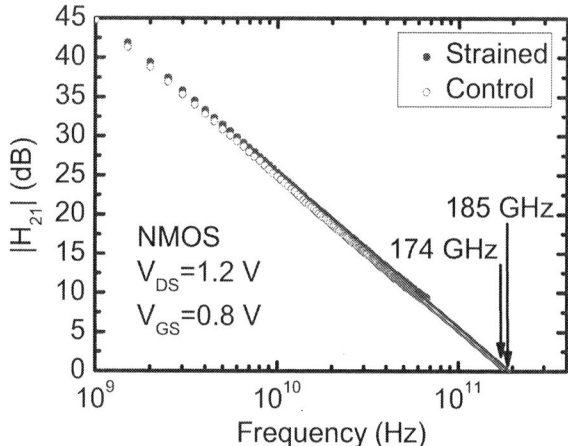

Fig. 4 Magnitude of current gain $|H_{21}|$ of L=60 nm nMOSFETs as a function of frequency for extracting f_T.

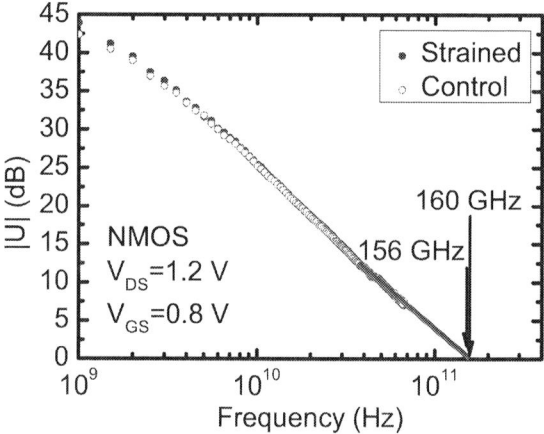

Fig. 5 Magnitude of unilateral power gain $|U|$ of L=60 nm nMOSFETs as a function of frequency for extracting f_{max}.

III. DC AND SMALL SIGNAL PERFORMANCE

Figures 2 and 3 show the DC characteristics of nMOSFETs with gate length L=60 nm. The DC characterizations of these devices were performed using an Agilent semiconductor parameter analyzer (4156C). Based on the Fig. 2, the high-tensile device shows 10% current gain as compared with the control device. The dc transconductance (G_m) has almost 8% enhancement at gate voltage V_{GS}=0.8 V. Since the current of nMOSFET with high-tensile CESL is increased, the RF characteristics of the transistor should be affected as well. To characterize the high-frequency performance, the maximum cutoff frequency (f_T) and the maximum oscillation frequency (f_{max}) of these devices, this study measures S-parameters on-wafer from 1 GHz to 67 GHz using vector network analyzer and then de-embeds them using the OPEN dummy. The cutoff frequency and maximum oscillation frequency are the frequency where the current gain was 0 dB and the frequency where MSG was 0 dB, respectively. As shown in Fig. 4, over 6% improvement in cutoff frequency has been presented at L=60 nm in that the existence of high tensile strain in the channel. Therefore, this improvement attributed to the increase of transconductance. Figure 5 shows the measured maximum oscillation frequency at L=60 nm, where the high-tensile device has only 2.5% gain as compared with the control device. It is because that the less dependence of f_{max} with gm (f_{max} is in proportion to $g_m^{0.5}$)

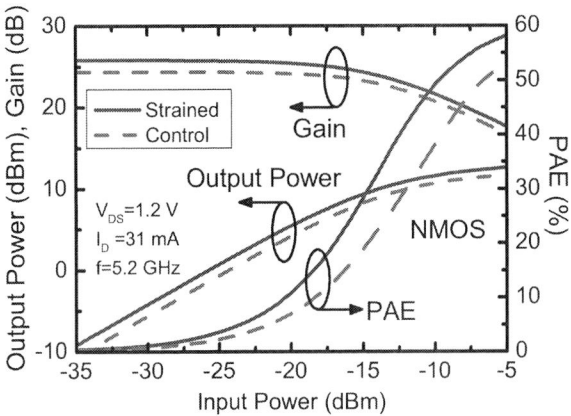

Fig. 6 Power characteristics of L=60 nm nMOSFETs at 5.2 GHz. Measurement results show that the strained device gains an extra output power enhancement as compared to the control device.

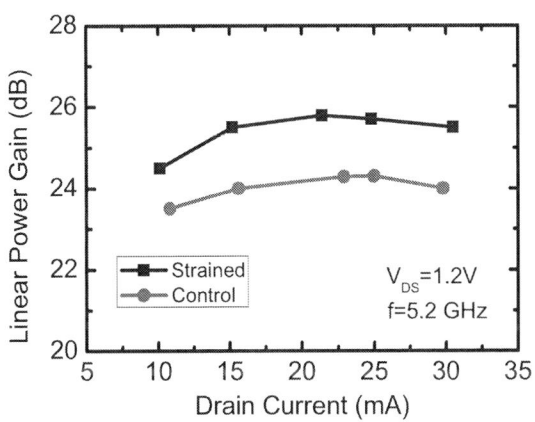

Fig. 7 Linear power gain as a function of drain current for L=60 nm nMOSFETs.

TABLE I
POWER PARAMETERS FOR L=60 NM NMOSFET

NMOS	G_P (dB)	$P_{out,1dB}$ (dBm)	PAE_{1dB} (%)
Control	24.0	7.6	15.8
Strained	25.5	8.9	23.9

IV. POWER PERFORMANCE AND ITS PHD MODEL

Figure 6 shows the power characteristics (gain, output power, and PAE) of nMOSFETs at maximum power gain and maximum output power. At same power consumption, the high-tensile device has better power performance than the control device due to the higher transconductance. As

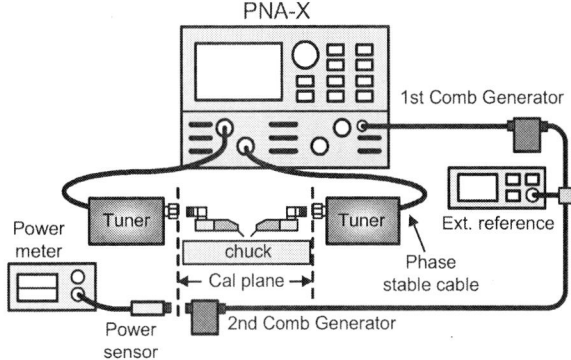

Fig. 8 On-wafer PHD model extraction system (NVNA and High-gamma tuner)

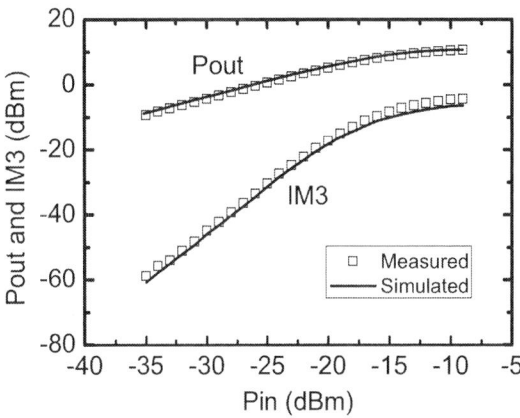

Fig. 9 Measured and simulated results of the IM distortion of L=60 nm nMOSFETs for 5.2 GHz with a tone spacing of 1 MHz and VD = 1.2 V, and ID = 31 mA.

shown in Fig. 7, an almost 1.5 dB enhancement in linear power gain can be achieved. The obtained power parameters for both high-tensile and control devices are summarized in Table I.

This study used an Agilent Nonlinear Vector Network Analyzer (NVNA) capable of nonlinear calibration and measurements to extract the PHD model [9]. Figure 8 illustrates this nonlinear measurement system. This system installed the high-gamma tuners; therefore the PHD model can be extracted from wide impedance range. Using standard nonlinear analysis tool in Agilent Design System (ADS), the measured X-parameters can be immediately used to simulate nonlinear figures of merit such as P_{1dB} and IP_3. The measured and simulated results of the IM distortion of the device for 5.2 GHz were shown in Fig. 9. This figure shows that the PHD model accurately predicts the measured transducer power gain and the third-order intermodulation (IM3) over a wide range of input power.

V. CONCLUSION

RF power performances of CMOS with high-tensile CESL stressors have been studied in this paper. Based on measurement results, a CESL-stressed structure enhances the power characteristics without extra epitaxial or litho process. In addition, the nonlinear behavior of nMOSFETs using PHD model by way of on-wafer NVNA with high-gamma tuner was also presented in this paper. As a result, the high tensile-stress CESL incorporated into nMOSFETs is a potential method to enhance the power performance of device for the RF or mmWave IC design.

ACKNOWLEDGEMENT

The authors would like to acknowledge Agilent Inc., Kenny Liao and Scottie Hsu for their supporting.

REFERENCES

[1] T. Suzuki, Y. Kawano, M. Sato, T. Hirose, and K. Joshin, "60 and 77 GHz power amplifiers in standard 90nm CMOS," *IEEE ISSCC Dig.*, pp. 562-563, Feb 2008.

[2] A. V. Vasylyev, P. Weger, W. Bakalski, and W. Simburger, "17 GHz 50-60mW power amplifiers in 0.13 μm standard CMOS" *IEEE Microwave and Wireless Components Lett.*, vol. 16, no. 1, pp. 37-39, Jan 2006.

[3] N. Kurita and H.. Kondoh, "60 GHz and 80 GHz Wide Band Power Amplifer MMICs in 90nm CMOS Technology," *IEEE RFIC Symp. Dig.*, pp. 39-42, June 2009.

[4] S. Pidin, T. Mori, K. Inoue, S. Fukuta, N. Itoh, E. Mutoh, K. Ohkoshi, R. Nakamura, K. Kobayashi, K. Kawamura, T. Saiki, S. Fukuyama, S. Satoh, M. Kase, and K. Hashimoto, "A novel strain enhanced CMOS architecture using selectively deposited high tensile and high compressive silicon nitride films," in *IEDM Tech. Dig.*, Dec. 2004, pp. 213-216.

[5] Chien-Ting Lin, Yean-Kuen Fang, Wen-Kuan Yeh, Chieh-Ming Lai, Che-Hua Hsu, Li-Wei Cheng, and Guang Hwa Ma, "Impacts of notched-gate structure on contact etch stop layer (CESL) stressed 90-nm nMOSFET," *IEEE Electron Device Lett.*, vol. 28, no. 5, pp. 376-378, May 2007.

[6] D. Bridges, J. Wood, M. Guyonnet, and P.H. Aaen, "A nonlinear electro-thermal model for high power RF LDMOS transistors," in *IEEE MTT-S Int. Microwave Symp. Dig.*, Atlanta, GA, June 2008.

[7] C. Fager, J. Pedro, N. Carvalho, and H. Zirath, "Prediction of IMD in LDMOS transistor amplifiers using a new large-signal model," *IEEE Trans. Microwave Theory Tech.*, vol. 50, no. 12, pp. 2834-2842, Dec. 2002.

[8] J. Verspecht, and D. E. Root, "Polyharmonic Distortion Modeling," *IEEE Microwave Magazine*, vol. 7, no. 3, pp. 44-57, June 2006.

[9] C. S. Chiu, K. M. Chen, G. W, Huang, C. H. Hsiao, K.H. Liao, W. L. Chen, S. C. Wang, M. Y. Chen, Y. C. Yang, K. L. Wang and L. K. Wu, "Characterization of annular-structure RF LDMOS transistors using polyharmonic distortion model," in *IEEE MTT-S Int. Microwave Symp. Dig.*, Boston, MA, Jun. 2009, pp. 977-980.

RTUIF-25

RF benchmark tests for compact MOS models

G.D.J. Smit, A.J. Scholten, and D.B.M. Klaassen

NXP Semiconductors, High Tech Campus 4, 5656 AE Eindhoven, The Netherlands
gert-jan.smit@nxp.com, Phone: +31 40 2729985

Abstract — Next to accurate fits of measurements, smoothness, and robustness, compact MOSFET models should ideally meet a large number of additional requirements. In this paper, we collect and derive a number of such demands that are important for RF-circuit applications. We present, for the first time, a derivation for the required reciprocity of capacitances at zero bias. We also derive from first principles the expected non-quasi-static behavior of a MOSFET at $V_{DS} = 0$ as well as its thermal noise. This leads to a number of benchmark tests that a compact model needs to pass to ensure its suitability for RF-circuit applications. Finally, it is shown that the CMC standard model PSP satisfies all presented requirements.

Index Terms — Benchmark test, compact models, non-quasi-static effects, PSP Model, RF noise

I INTRODUCTION

Any industrial compact MOSFET models has to meet a large set of very diverse requirements. Its ability to accurately fit experimental data is among the most visible and important ones. Smoothness and robustness at all bias conditions are essential properties as well. Less visible—but especially important for analog and RF design—is physical behavior of the model at 'trivial' bias conditions such as $V_{DS} = 0$, where the device is passive. Under those circumstances, several fundamental requirements need to be fulfilled, which can be derived from first principles.

In this paper, we will present some basic analyses of a MOSFET at $V_{DS} = 0$ and derive several fundamental requirements that should be satisfied by compact models. Applications for which passing these tests is particularly important include varactors, transimpedance amplifiers, and passive mixers. Apart from their importance in circuits, the tests discussed in this paper are useful for compact model developers as sanity checks for model equations and verification of their internal consistency.

It is important to note that the presented requirements are nontrivial to fulfill. As compact model equations are usually derived and tested for saturation conditions, their behavior in the linear region typically emerges as a side-effect. If care is not taken, the resulting behavior may be grossly unphysical. Moreover, in compact models the dc behavior, capacitances, and NQS-effects are often treated almost independent of each other, leading to a high risk of inconsistencies among them. Such flaws are even observed in well-established industrial standard models.

Previously [1,2], a number of tests for compact MOSFET models have been described to benchmark dc aspects and capacitances of the model. They were successfully applied to the new industry standard PSP MOSFET model [3]. In this work, we instead focus on the modeling of non-quasi-static (NQS) effects and RF noise. Some of the test results were published by us before [1, 4–6]. In this paper, for the first time, the complete set is published, including their physical origin and derivation.

It is important to note that all tests presented in this paper are in fact tests of the internal *consistency* of the model (for example, between the NQS model on one hand and the dc and capacitance model on the other hand). Hence, compact models should pass these tests for *any* choice of the model parameters.

II TESTS

A Reciprocity of capacitances

At $V_{DS} = 0$, a MOSFET is a linear, passive device and therefore its capacitances must be reciprocal (that is, $C_{ij} = C_{ji}$ for all i, j). The Compact Model Council [7] requires its standard models to satisfy this demand, but the physical origin is not traceable in literature. A proof of capacitance reciprocity has been published for a set of isolated metallic conductors [8], but that does not apply to a MOSFET, which has a conductive path between source and drain. Here we present for the first time a derivation of the reciprocity requirement for MOSFETs.

Generally speaking, the small-signal power dissipation p of a linear n-port is given by

$$p = \frac{1}{2} \sum_{k=1}^{n} \mathrm{Re}(v_k i_k^*), \tag{1}$$

where the v_k and i_k are (complex) phasors characterizing the amplitude and phase of the small-signal perturbation at port k of the voltage and current, respectively. Using the definition of Y-parameters, p can be written as

$$p = \frac{1}{4} \sum_{k,l=1}^{n} (Y_{kl} + Y_{lk}^*) v_k v_l^*, \tag{2}$$

where '*' denotes complex conjugation. Because it is passive, the n-port network is not allowed to generate power, hence p must be nonnegative for any choice of the v_k. In mathematical terms, the self-adjoint matrix $(Y + Y^\dagger)$ is required to be positive definite (here, $Y^\dagger$ is the conjugate transpose of Y), which is equivalent to all of its eigenvalues being positive. Consequently, the same must hold for any of its 2×2 submatrices:

$$\begin{pmatrix} Y_{kk} + Y_{kk}^* & Y_{kl} + Y_{lk}^* \\ Y_{lk} + Y_{kl}^* & Y_{ll} + Y_{ll}^* \end{pmatrix} \text{ with } k \neq l. \quad (3)$$

Choosing certain k, l and splitting the Y-parameters into their real and imaginary parts as $Y_{ij} = g_{ij} + i\omega C_{ij}$, such a submatrix can be written as

$$\begin{pmatrix} 2g_{kk} & 2g_{kl}^s + i\omega(C_{kl} - C_{lk}) \\ 2g_{kl}^s - i\omega(C_{kl} - C_{lk}) & 2g_{ll} \end{pmatrix}, \quad (4)$$

with $g_{kl}^s = (g_{kl} + g_{lk})/2$. For both eigenvalues to be nonnegative, it is required that

$$4(g_{kl}^s)^2 + \omega^2(C_{kl} - C_{lk})^2 \leq 4g_{kk} \cdot g_{ll} \quad (5)$$

for any ω.

Applying this to a MOSFET, if either k or l refers to the gate or bulk, we have that g_{kk} and/or g_{ll} vanishes (we disregard leakage currents). Therefore, we must have $C_{kl} = C_{lk}$. For the only remaining case (when k refers to source, l to drain), we observe that only between source and drain the real part of the small-signal currents is nonzero. Therefore, $\text{Re}(i_S) = -\text{Re}(i_D)$ and hence $g_{SS} = -g_{DS}$ and $g_{SD} = -g_{DD}$. Substitution in Eq. (5) gives that $(g_{SS} - g_{DD})^2 + \omega^2(C_{SD} - C_{DS})^2 \leq 0$, from which it follows that $C_{DS} = C_{SD}$. Note that we did *not* assume the device to be symmetric under S-D-exchange, so this proof is also valid for, e.g., LDMOS devices.

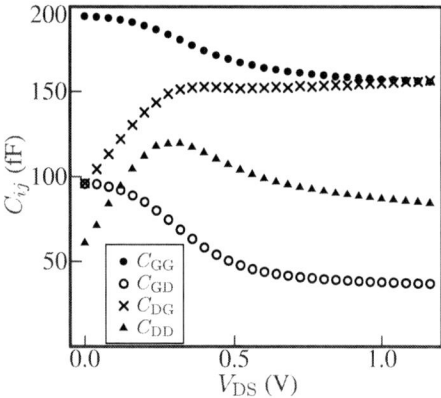

Fig. 1: Measured capacitances of a $L = 100$ nm device versus drain bias (taken from S-parameter measurements at $f = 10$ GHz). As expected, at $V_{DS} = 0$ V we have that $C_{DG} = C_{GD}$.

In summary, we have proven that at $V_{DS} = 0$ the capacitances must be reciprocal, which is our first test. It is illustrated by measurements in Fig. 1. As has been shown before [9], the PSP model passes this test.

B NQS behavior at $V_{DS} = 0$

From this moment onwards, we assume the MOSFET to be laterally homogeneous. Because there is no dc bias drop over the channel, a homogeneous MOSFET of length L at $V_{DS} = 0$ and biased above threshold can be represented by a distributed network of resistances and capacitances (see Fig. 2). This description allows for a first-principles analysis of the transmission line properties of the device that can serve a benchmark for non-quasi-static MOSFET models.

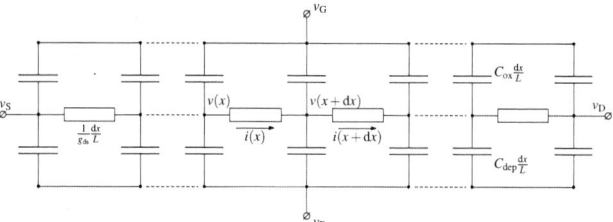

Fig. 2: Network representing a MOSFET channel at $V_{DS} = 0$ and its distributed capacitance to gate and bulk.

By applying Kirchoff's law to this network, the following second-order linear differential equation can be derived:

$$\frac{d^2v(x)}{dx^2} = a^2 \left(v(x) - \frac{C_{ox}v_G + C_{dep}v_B}{C_{ox} + C_{dep}} \right). \quad (6)$$

Here, $a^2 = i\omega(C_{ox} + C_{dep})/(g_{ds} \cdot L)$ and $v(x)$ is the local small-signal potential inside the channel. Moreover, g_{ds} is the total channel conductance, while C_{ox} and C_{dep} are the total gate-to-channel and bulk-to-channel capacitances, respectively. Note that g_{ds}, C_{ox}, and C_{dep} are functions of gate and bulk bias. Using the boundary conditions at source and drain, this differential equation can be solved explicitly and from there, analytic results can be derived for all elements of the Y-parameter matrix. The result, developed to third order in ω, is given by

$$Y = \begin{pmatrix} 1 & 0 & -1 \\ 0 & 0 & 0 \\ -1 & 0 & 1 \end{pmatrix} g_{ds} +$$

$$\begin{pmatrix} 2\alpha & -3 & \alpha \\ -3 & 6 & -3 \\ \alpha & -3 & 2\alpha \end{pmatrix} \frac{i\omega C_{ox}}{6} +$$

$$\begin{pmatrix} 8\alpha^2 & -15\alpha & 7\alpha^2 \\ -15\alpha & 30 & -15\alpha \\ 7\alpha^2 & -15\alpha & 8\alpha^2 \end{pmatrix} \frac{\omega^2 C_{ox}^2}{360 g_{ds}} +$$

$$\begin{pmatrix} -32\alpha^3 & 63\alpha^2 & -31\alpha^3 \\ 63\alpha^2 & -126\alpha & 63\alpha^2 \\ -31\alpha^3 & 63\alpha^2 & -32\alpha^3 \end{pmatrix} \frac{i\omega^3 C_{ox}^3}{15120 g_{ds}^2}$$

$$+ O(\omega^4) \quad (7)$$

where $\alpha = 1 + C_{\text{dep}}/C_{\text{ox}}$. Here (and in the following) the indices 1, 2, and 3 refer to drain, gate, and source, respectively. The bulk-related values are left out due to the limited available space; they can be easily reconstructed from the fact that row and column sums are zero. Note that in strong inversion we generally have $C_{\text{ox}} \gg C_{\text{dep}}$, so $\alpha \approx 1$.

This leads to our second test, namely that the lowest order (in ω) NQS corrections at $V_{\text{DS}} = 0$ should satisfy (in very good approximation) Eq. (7).

To demonstrate that these first-order corrections are not automatically captured correctly by all NQS models, we briefly discuss a popular method to implement NQS effects in compact MOSFET models, namely the so-called delay-time method [10]. It amounts to introducing a phenomenological delay time τ that characterizes how the NQS terminal charges relax to their QS counterparts, as in

$$\frac{dQ_{\text{NQS}}}{dt} = -\frac{Q_{\text{NQS}} - Q_{\text{QS}}}{\tau}. \tag{8}$$

In a small-signal representation, the NQS charges can be directly expressed as function of the QS charges:

$$Q_{\text{NQS}} = \frac{1}{1 + i\omega\tau} Q_{\text{QS}} \approx (1 - i\omega\tau - \omega^2\tau^2)Q_{\text{QS}}. \tag{9}$$

Although this delay time can have, in principle, a different value for each of the four terminal charges, demands on symmetry and reciprocity require all τ to be identical when $V_{\text{DS}} = 0$. The most natural choice for the value of the delay time turns out to be $\tau = \alpha C_{\text{ox}}/(12g_{\text{ds}})$. Now, the lowest order NQS correction to the conductance (ω^2 term) for the delay-time model becomes

$$\begin{pmatrix} 10\alpha^2 & -15\alpha & 5\alpha^2 \\ -15\alpha & 30\alpha & -15\alpha \\ 5\alpha^2 & -15\alpha & 10\alpha^2 \end{pmatrix} \frac{\omega^2 C_{\text{ox}}^2}{360 g_{\text{ds}}} \tag{10}$$

and the lowest-order correction to the capacitances (ω^3 term)

$$\begin{pmatrix} -70\alpha^3 & 105\alpha^2 & -35\alpha^3 \\ 105\alpha^2 & -210\alpha^2 & 105\alpha^2 \\ -35\alpha^3 & 105\alpha^2 & -70\alpha^3 \end{pmatrix} \frac{i\omega^3 C_{\text{ox}}^3}{30240 g_{\text{ds}}^2}. \tag{11}$$

Comparison with the exact results in Eq. (7) shows that only for few of the elements the exact results are reproduced, while several are off by up to a factor of 2. Moreover, all bulk-related elements are off by a large factor. For example the ω^2 term of Y_{BB} is off by a factor $\alpha/(\alpha-1) \gg 1$ and Y_{GB} vanishes completely, while it has a finite value in the exact result. This shows that delay-time models, though certainly having their merits, suffer from their phenomenological nature and do not give consistent results.

Because models based on channel segmentation [11] mimic the actual physical mechanism behind NQS effects,

such models *do* naturally (though approximately) produce the exact results. The same holds for the spline-collocation method employed in PSP; see, e.g., Ref. [5].

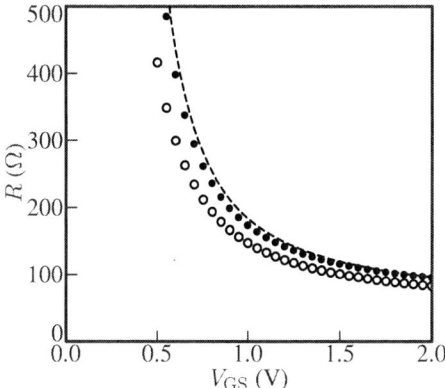

Fig. 3: R_{GG} (solid symbols) and R_{BB} (open symbols) as a function of V_{GS}. The dashed line is the prediction according to Eq. (12), the symbols represent the simulated PSP value (at **SWNQS** = 5).

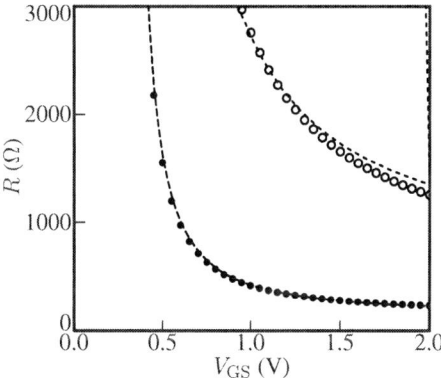

Fig. 4: R_{GD} (solid symbols) and R_{BD} (open symbols) as a function of V_{GS}. The dashed lines are corresponding predictions according to Eq. (12), the symbols represent the simulated PSP value (at **SWNQS** = 5).

C Effective series resistance

Y-parameters are a natural way to represent impedances as a *parallel* connection of a capacitance and a conductance. However, if this conductance is zero (most elements of the 0th-order Y-parameter matrix in Eq. (7) are zero), there can still be an effective resistance *in series* with the capacitance. These resistances can be calculated as $R_{ij} = \text{Re}(1/Y_{ij})$ and the full matrix R is given by

$$\frac{1}{g_{\text{ds}}} \begin{pmatrix} 1 & -\alpha/6 & -1 & -\beta/6 \\ -\alpha/6 & 1/12 & -\alpha/6 & \star \\ -1 & -\alpha/6 & 1 & -\beta/6 \\ -\beta/6 & \star & -\beta/6 & 1/12 \end{pmatrix} + O(\omega), \tag{12}$$

where $\beta = \alpha/(\alpha-1)$. (The elements indicated by $\star$ have a leading term $\propto \omega^{-2}$ and are left out). Note that—contrary to the first term in Eq. (7)—this matrix has no vanishing elements. Also note that a QS model cannot reproduce these results, even though they are valid in the limit $\omega \to 0$, because the ω^2 term from Eq. (7) is needed to compute R. Analogous to distributed gate resistance, one can recognize (for example) $Y_{GG} = 1/(12g_{ds})$ as the effective resistance of a two-sided connected channel.

From this follows as our third test that the effective dc series resistances predicted by the NQS model at $V_{DS} = 0$ should satisfy Eq. (12). In Figs. 3 and 4, the nontrivial elements of the matrix in Eq. (12) are plotted as a function of V_{GS}, showing that the PSP model passes this benchmark test.

D Thermal noise

The thermal noise of an n-port network can be characterized by its so-called CY-matrix. This matrix contains all noise current spectral densities; its elements are given by $CY_{kl} = \langle i_k i_l^* \rangle$. In the case of a passive linear n-port network, the CY-matrix is directly related to its admittance matrix [12]: it is given by

$$2kT(Y + Y^\dagger). \tag{13}$$

This relationship follows from the fact that if both the n-port and the elements connected to its outputs are passive and it is in thermal equilibrium with its environment, there can be no net power transfer from the network to its environment or vice versa [12].

From Eq. (13), it follows directly that in this case the elements of the CY-matrix (lowest order in ω) are given by $S_{id} = \langle i_d i_d^* \rangle = 4kT g_{ds}$,

$$S_{ig} = \langle i_g i_g^* \rangle = 4kT \frac{C_{ox}^2 \omega^2}{12g_{ds}}, \tag{14}$$

and the correlation coefficient is given by

$$c = \frac{\langle i_g i_d^* \rangle}{\sqrt{S_{id} \cdot S_{ig}}} = \frac{\alpha C_{ox} \omega}{4\sqrt{3}g_{ds}}. \tag{15}$$

Here, S_{id} is simply the Nyquist noise generated by the channel conductance g_{ds}. In the expression for S_{ig}, one can recognize the Nyquist noise of the 'effective channel resistance' $1/(12g_{ds})$ (see Eq. (12)) coupling through the gate capacitance. Contrary to statements in the literature [13], the preceding derivation demonstrates that the induced gate noise S_{ig} is a pure equilibrium effect. Eq. (15) shows that the well-known ω-independent imaginary part of c should vanish at $V_{DS} = 0$. The ω-dependent real part in Eq. (15) can be neglected in practice. This concludes the fourth test.

In Ref. [6], it is illustrated that the PSP model fulfills this requirement. Moreover, in that paper, a number of additional tests are discussed for $V_{DS} \neq 0$.

III Conclusion

We have presented a number of benchmark tests for compact MOSFET models biased at $V_{DS} = 0$: (1) the capacitance matrix must be symmetric, (2) first-order NQS corrections to conductance and capacitance should be given by Eq. (7), (3) the effective series resistances should satisfy Eq. (12), and (4) the CY-matrix of a MOSFET should (to lowest order in ω) be given by Eq. (13). Passing these tests is a prerequisite for models to be reliably used in RF applications like varactors and passive mixers. It is also demonstrated that the PSP model satisfies all requirements discussed in this paper.

References

[1] X. Li et al., "Benchmarking the PSP Compact Model for MOS Transistors," in Proc. of the 2007 IEEE Int. Conf. on Microelectronic Test Structures, pp. 259–264, March 2007.

[2] X. Li et al., "Benchmark Tests for MOSFET Compact Models With Application to the PSP Model," IEEE Trans. Electron Devices, vol. 56, no. 2, pp. 243–251, February 2008.

[3] see: http://pspmodel.asu.edu

[4] G.G. Gildenblat et al., "PSP: An Advanced Surface-Potential-Based MOSFET Model for Circuit Simulation," IEEE Trans. Electron Devices, vol. 53, no. 9, pp. 1979–1993, September 2006.

[5] H.-L. Wang, "A Unified Non-Quasi-Static MOSFET Model for Large and Small Signal Simulations," IEEE Trans. Electron Devices, vol. 53, no. 9, pp. 2035–2043, September 2006.

[6] A.J. Scholten, R. van Langevelde, L.F. Tiemeijer, and D.B.M. Klaassen, "Compact modeling of noise in CMOS," in Proc. of the IEEE 2006 Custom Integrated Circuits Conference, pp. 711–716, 2006.

[7] see: http://www.geia.org/index.asp?bid=597

[8] V. Lorenzo and B. Carrascal, "Green's functions and symmetry of the coefficients of a capacitance matrix," Am. J. Phys., vol. 56, no. 6, p. 565, June 1988.

[9] A.J. Scholten et al., "The New CMC Standard Compact MOS Model PSP: Advantages for RF Applications," IEEE J. Solid-State Circuits, vol. 44, no. 5, pp. 1415–1424, May 2009.

[10] M. Chan, K.Y. Hui, C. Hu, and P.K. Ko, "A Robust and Physical BSIM3 Non-Quasi-Static Transient and AC Small-Signal Model for Circuit Simulation," IEEE Trans. Electron Devices, vol. 45, no. 4, pp. 834–841, April 1998.

[11] A.J. Scholten, L.F. Tiemeijer, P.W.H. de Vreede, and D.B.M. Klaassen, "A large signal non-quasi-static MOS model for RF circuit simulation," in IEDM Tech. Dig., pp. 163–166, 1999.

[12] R.Q. Twiss, "Nyquist's and Thevenin's Theorems Generalized for Nonreciprocal Linear Networks," J. Appl. Phys., vol. 26, no. 5, pp. 599–602, May 1955.

[13] R.P. Jindal, "Effect of Induced Gate Noise at Zero Drain Bias in Field-Effect Transistors," IEEE Trans. Electron Devices, vol. 52, no. 3, pp. 432–434, March 2005.

A 1.8V 74mW UHF RFID Reader Receiver with 18.5dBm IIP3 and -77dBm Sensitivity in 0.18μm CMOS

Xuguang Sun, Baoyong Chi, Chun Zhang, Ziqiang Wang and Zhihua Wang

Institute of Microelectronics of Tsinghua University, Beijing, 100084, P. R. China

Abstract — A UHF RFID reader receiver is implemented in 0.18μm CMOS. The direct-conversion receiver consists of an LNA, passive mixers, baseband PGAs and LPFs. As high as 18.5dBm measured IIP3 of the RF front-end is achieved by using passive mixers driven by 25% duty cycle square wave LO. The receiver has a sensitivity of -77dBm in the normal mode and -87dBm in the LBT mode. The total power dissipation in the normal mode is 74mW from 1.8V power supply.

Index Terms — RFID, UHF, reader, receiver, CMOS.

I. INTRODUCTION

Although the global radio frequency identification (RFID) market keeps growing fast in recent years, the major contribution to the market boosting is provided by the high frequency (HF) RFID. The market growth in the ultra-high frequency (UHF) counterpart is still far behind most people's expectations. One of the most important reasons is that the high cost of UHF RFID, both tags and readers, is still the bottleneck for massive applications. Therefore, more and more attentions have been drawn to the low cost UHF RFID tag and reader development.

In a passive UHF RFID system, the design challenge of the reader is even harder than that of the tag, since most tags have been designed to be as "simple" as possible due to the constraints of limited power and chip area, leaving most processing work to readers. The reader is required to receive the returned signals which are backscattered by tags while transmitting a continuous wave (CW) to power up the passive tags simultaneously. Due to the limited isolation between transmitter (Tx) and receiver (Rx), there is always some leakage from Tx to Rx. Typically, the power of this leakage carrier is higher than 0dBm [1]. Such a strong leakage could easily saturate RF front-end of the receiver, and induce severe noise performance degradation. Therefore, the RF front-end of a reader needs to have an extremely high linearity to handle the carrier leakage problem and the tradeoff between linearity and noise figure becomes the most challenge issue for RFID reader receiver.

In this paper, an 860~960MHz CMOS RFID reader receiver chip with 18.5dBm RF front-end IIP3 and -77dBm sensitivity is presented. Compared with the previous works in CMOS [1]-[3], this receiver shows higher linearity with similar sensitivity and power consumption.

II. RECEIVER ARCHITECTURE

The architecture of the presented reader receiver is shown in Fig. 1. The receiver mainly consists of a low noise amplifier (LNA), passive down-conversion mixers, baseband programmable gain amplifiers (PGA) and low pass filters (LPF).

The reader is supposed to have two modes: the normal mode and the listen-before-talk (LBT) mode. In the normal mode, the reader communicates with tags and the existence of the carrier leakage calls for an extremely high linearity of the RF front-end. As a result, an LNA is difficult to be used in this mode. In the LBT mode, the reader needs to check if the frequency channel is occupied by other users before it talks to prevent reader collisions. In this condition, the linearity is not a problem since the transmitter is turned off and there is no carrier leakage. Therefore, an alternative RF receiving path with an LNA is used to improve sensitivity, as shown in the dashed box in Fig. 1.

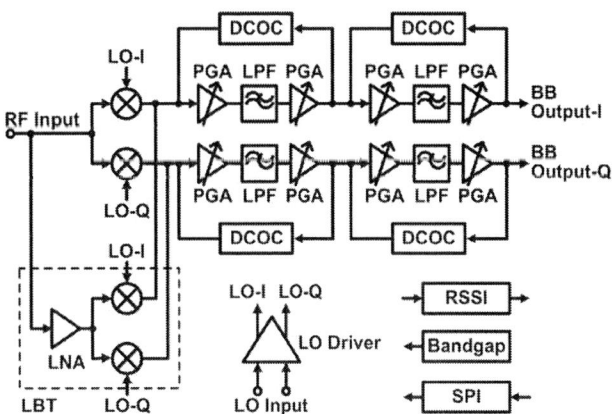

Fig. 1.   Receiver architecture.

To cope with the carrier leakage problem, the direct-conversion architecture is the best choice. Since the LO frequency equals RF input carrier frequency, the received signals are mixed down to the baseband and the carrier leakage component is converted to DC which can be removed by AC-coupling. Since the power of the carrier leakage could achieve higher than 0dBm, the 1-dB compression point (P1dB) of the RF front-end is supposed to be at least 5dBm to handle such a strong in-band blocker. As a result, passive mixers are chosen to meet this stringent linearity requirement.

In the baseband, there are two branches in quadrature, which are I-path and Q-path. Each branch includes four PGAs to provide a wide gain control range. Each PGA has a gain control range from -10dB to 20dB in a step of 2dB. The baseband low pass filter is an active-RC filter with a programmable 3-dB bandwidth from 100kHz to 1.6MHz according to different Rx data rate. Like other direct-conversion receivers, the baseband analog modules are susceptible to the DC offset, especially in CMOS technology. To prevent the DC offset from saturating the receiver, two off-chip passive DC offset cancellation (DCOC) loops are added to each baseband receiving path.

III. KEY BUILDING BLOCKS

A. Passive Down-conversion Mixer

Due to its extremely high linearity and low 1/f noise, the passive mixer is suitable for the RF front-end of the receiver. The structure of the passive down-conversion mixer is shown in Fig. 2.

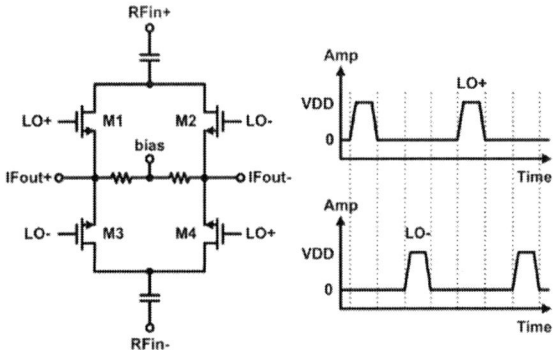

Fig. 2.   Passive mixer schematic and 25% duty cycle square wave LO.

The gate of the switch transistor (M1~M4) in the mixer is driven by square wave LO signals. To reduce the conversion loss of the passive mixer, the LO signals are shaped to have a 25% duty cycle [4] by an LO driver. The input impedance of the passive mixer roughly equals the on-resistance of the switch transistor because the RF signal is shorten to ground by load capacitors at mixer output. As a result, the on-resistance of the switch is set to 50Ω approximately by adjusting W/L of the transistor to obtain a better input matching performance.

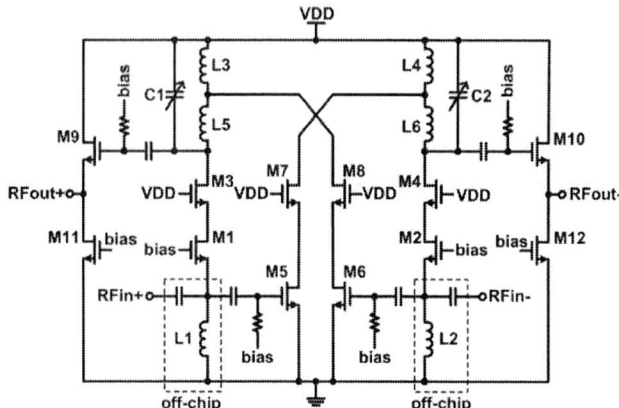

Fig. 3.   LNA and output buffer schematic.

B. Low Noise Amplifier

When working in the LBT mode, the receiver uses an LNA to improve detection sensitivity. Fig. 3 is the schematic of the LNA with an output buffer. The LNA has a common-gate (CG) stage (M1~M2) to achieve wideband input matching and high linearity. To reduce noise figure of CG amplifier, a common-source (CS) stage (M5~M6) with current-to-voltage combiner load (L3~L6, C1~C2) is added in parallel with CG stage to form a noise-cancelling structure [5]. The load capacitor C1 and C2 is tunable to adjust resonant frequency and the source follower output buffer (M9~M10) is used to drive the subsequent passive mixers.

C. Programmable Gain Amplifier

Fig. 4 shows the topology of the PGA, which is a source degeneration amplifier. The voltage gain of PGA is determined by resistance ratio of load resistor RL and source degeneration resistor RS, which can be controlled by 4-bit digital signals. The gain control range is from -10dB to 20dB in a step of 2dB. Transistors M1~M8 compose a gm-boosting structure, which could increase the effective transconductance of M1 and M2, in order to reduce the gain error caused by process variation.

D. Programmable Baseband Filter

The baseband low pass filter in the receiver is used as a channel select filter to reject out-of-band interferers and noise. The structure of the filter is illustrated in Fig. 5. The LPF is a fourth-order Chebyshev-II active-RC filter which is implemented in biquad. The bandwidth of the filter is programmable from 100kHz to 1.6MHz by means of switch resistor and capacitor arrays, in order to optimize noise performance according to various Rx data rate.

978-1-4244-6240-7/10 $26.00 © 2010 IEEE   548

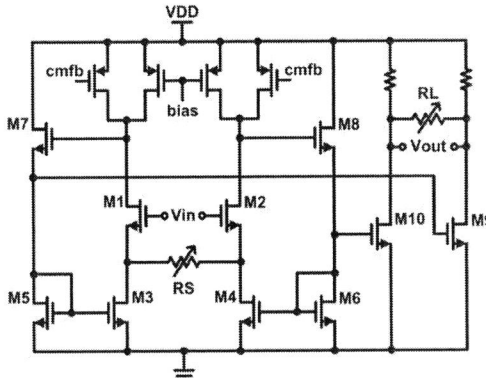

Fig. 4.   PGA schematic.

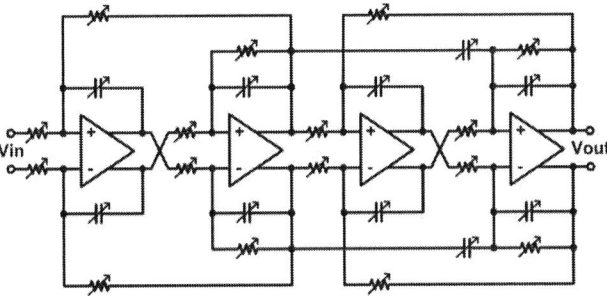

Fig. 5.   Programmable baseband filter schematic.

E. DCOC Loop

To remove the annoying DC offset in the baseband, an off-chip passive DCOC loop is presented in Fig. 6. A first order low pass filter (R1~R2, C1) extracts DC and low frequency components from the second PGA output and then feedback to the input of IF modules. Since the cut-off frequency of the low pass filter is supposed to be much lower than the useful signal frequency, the resistors and the capacitor are too large to be integrated on chip. Compared with the active DCOC circuits, the passive loop does not induce extra noise to the main amplification stage, but at a cost of several off-chip components.

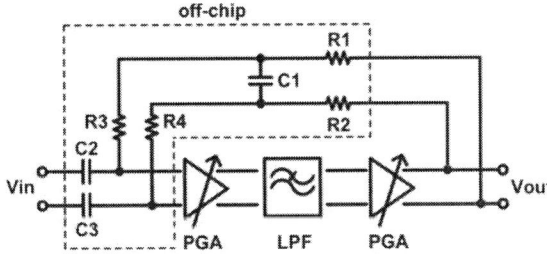

Fig. 6.   Off-chip passive DCOC loop.

IV. MEASUREMENT RESULTS

The presented receiver is fabricated in 0.18μm CMOS. Fig. 7 shows the microphotograph of the chip. Total die area including ESD I/O pads is 2.8mm × 2.1mm.

In the normal mode, the measured P1dB and IIP3 of the proposed RF front-end are shown in Fig. 8. It could be seen from this figure that the P1dB is as high as 7dBm and the IIP3 achieves 18.5dBm.

Fig. 9 depicts the measured gain control characteristic and gain error of the PGA. The gain control range is from -10dB to 20dB in a step of 2dB and the maximum gain error is less than 0.4dB. As shown in Fig. 10, the measured frequency response of the analog baseband module fits the simulation results well. The out-of-band rejection achieves 45dB and the pass band ripple is less than 1dB.

Fig. 7.   Microphotograph of the presented receiver.

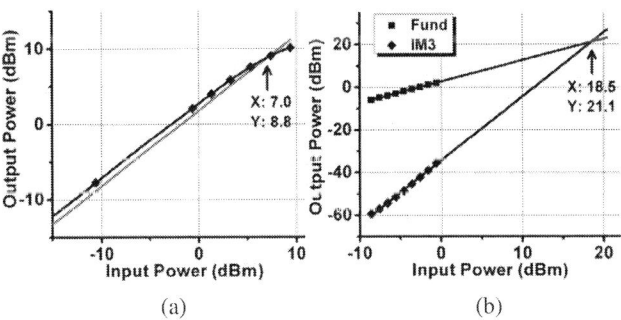

Fig. 8.   RF front-end (a) P1dB and (b) IIP3 measurement results in the normal mode.

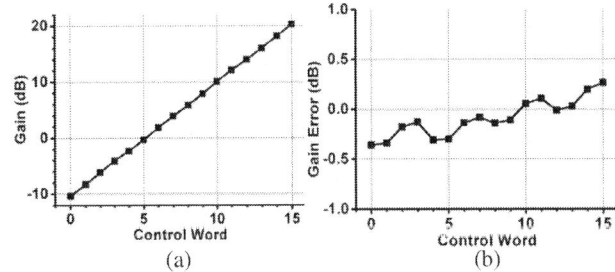

Fig. 9.   PGA (a) gain control and (b) gain error measurements.

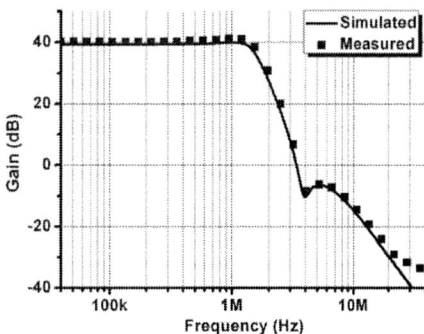

Fig. 10. Simulated and measured frequency response of the analog baseband module in 1.6MHz bandwidth.

TABLE I

SUMMARY OF RECEIVER PERFORMANCE

Parameter	Measured Results	
	Normal	LBT
Operation Frequency	860~960 MHz	
P1dB (RF front-end)	7dBm	-7dBm
IIP3 (RF front-end)	18.5dBm	8.5dBm
Gain Range	3~83dB	16~96dB
Sensitivity (SNR:10dB) 100kHz IF BW 200kHz IF BW 400kHz IF BW 800kHz IF BW 1.6MHz IF BW	-77dBm -75dBm -74dBm -72dBm -71dBm	-87dBm -85dBm -84dBm -82dBm -79dBm
Power Consumption	74mW	122mW
Supply Voltage	1.8V	
Die Area	2.8mm×2.1mm	

The major chip performance is summarized in Table I. In the condition of 40kHz Rx data rate and 10dB required output signal-to-noise ratio (SNR), the sensitivity of the receiver could achieve -77dBm in the normal mode. When working in the LBT mode, the sensitivity of the receiver improves about 10dB at a cost of another 27mA current consumption from LNA. The comparison between state-of-the-arts and this work is given in Table II. It could be seen that our receiver has higher linearity with similar sensitivity and power consumption.

V. CONCLUSION

This paper presents an 860~960MHz UHF RFID reader receiver. The extremely high linearity of RF front-end is achieved by using passive mixers driven by 25% duty cycle square wave LO. To optimize noise performance, the bandwidth of LPFs is programmable. The DC offset is cancelled by off-chip passive DCOC loops. Measured results show that the proposed receiver has an IIP3 of 18.5dBm and a sensitivity of -77dBm in the normal mode.

ACKNOWLEDGEMENT

This research was partly supported by National Natural Science Foundation of China (No. 60806008) and partly supported by National High Technology Research and Development Program of China (863 Program) (No. 2008AA04A102).

REFERENCES

[1] I. Kwon, Y. Eo, H. Bang, K. Choi, S. Jeon, S. Jung, D. Lee and H. Lee, "A single-chip CMOS transceiver for UHF mobile RFID reader," *IEEE J. Solid-State Circuits*, vol. 43, no. 3, pp. 729-738, March 2008.

[2] P.B. Khannur, X. Chen, D. L. Yan, D. Shen, B. Zhao, M. Kumarasamy Raja, Y. Wu, A.B. Ajjikuttira, W. G. Yeoh, R. Singh, "An 860 to 960MHz RFID Reader IC in CMOS," *IEEE RFIC Symp. Dig.*, pp. 269-272, June 2007.

[3] W. Wang, S. Lou, K. W. C. Chui, S. Rong, C. F. Lok, H. Zheng, H. T. Chan, S. W. Man, H. C. Luong, V. K. Lau and C. Y. Tsui, "A single-chip UHF RFID reader in 0.18μm CMOS process," *IEEE J. Solid-State Circuits*, vol. 43, no. 8, pp. 1741-1754, August 2008.

[4] M. Camus, B. Butaye, L. Garcia, M. Sie, B. Pellat, T. Parra, "A 5.4mW 0.07mm² 2.4GHz front-end receiver in 90nm CMOS for IEEE 802.15.4 WPAN," *IEEE ISSCC Dig. Tech. Papers*, pp. 368-620, February 2008.

[5] J. Jussila and P. Sivonen, "A 1.2-V highly linear balanced noise-cancelling LNA in 0.13-μm CMOS," *IEEE J. Solid-State Circuits*, vol. 43, no. 3, pp. 579-587, March 2008.

[6] I. Kipnis, S. Chiu, M. Loyer, J. Carrigan, J. Rapp, P. Johansson, D. Westberg and J. Johansson, "A 900MHz UHF RFID reader transceiver IC," *IEEE ISSCC Dig. Tech. Papers*, pp. 214-598, February 2007.

TABLE II

COMPARISON WITH STATE-OF-THE-ARTS

Parameter		[1]	[2]	[3]	[6]	This work
IIP3 (RF front-end)	Normal	18.5dBm	0dBm	18dBm	21dBm	18.5dBm
	LBT	1.5dBm	-10dBm	0dBm		8.5dBm
Sensitivity	Normal	-70dBm	-85dBm	-70dBm (-5dBm jammer)	-96dBm	-77dBm
	LBT	-80dBm	-96dBm	-90dBm		-87dBm
Power Consumption	Normal	Rx 7.2mW	Total 540mW	Rx 124mW	Total 1.2W	Rx 74mW
	LBT	Rx 11.5mW				Rx 122mW
Supply Voltage		1.8V	1.8V	1.8V	5V(RF)/3.3V(Analog)	1.8V
Technology		0.18μm CMOS	0.18μm CMOS	0.18μm CMOS	0.18μm SiGe BiCMOS	0.18μm CMOS

978-1-4244-6240-7/10 $26.00 © 2010 IEEE

AUTHOR INDEX

Abbasi, Morteza 483
Acar, M. 359
Afsahi, Ali 351
Ahn, Minsik 435
Akhnoukh, Atef 247
Akhtar, Siraj 3
Akima, Hiroshi 299
Alhalabi, Ramadan A. 75, 423
Allstot, David J. 1
Aloui, Sofiane 187
Amaya, Rony E. 219
Ameziane, Chama 491
An, Kyu Hwan 347, 511
Arnaud, C. 223
Asbeck, Peter M. 91
Ashckenazi, O. 387
Atesal, Yusuf A. 423
Ayazi, Farrokh 295
Bae, Jong-Dae 163, 167
Baeyens, Yves 59, 327
Baghaei-Nejad, Majid 99
Bakkaloglu, Bertan 135, 259, 367
Balsara, Poras T. 7
Banerjee, Bhaskar 7
Banin, R. 387
Bar, P. 223
Barajas, Enrique 175
Bashir, Imran 7
Bassat, Assaf Ben 131, 387
Beattie, William 279
Beer, Z. 387
Bégueret, Jean-Baptiste 491, 503
Behzad, Arya 351
Belot, Didier....................187, 215, 491
Ben-bassat, I. 403
Benech, P. 315
Benedikt, J. 439
Bergervoet, Jos 463
Berroth, Manfred 515
Bhan, Vivek 19
Bien, Franklin 83
Boers, Michael 307
Boglione, Luciano 191
Boret, S. 203
Borremans, J. 375
Bos, Lynn 399
Bousquet, Jean-François 339, 451
Bräckle, Alexander 515
Burghartz, Joachim N. 247
Cafaro, G. 499
Callaghan, Lori 291
Candra, Panglijen 211
Cardona, Albert 531
Carmon, Roi 479
Carpentier, J.-F. 223

Carrara, Francesco 159
Carroll, Mike 243
Carusone, Anthony Chan 43
Cetinoneri, Berke 423
Chai, Seung Wan 495
Chaki, Shin 151
Chakraborty, Sudipto 259, 367
Chan, Ka-Un 391
Chang, Chia-Jun 391
Chang, Chiao-Ling 391
Chang, Jae-Hong 163, 167
Chang, Jin-Fa 335
Chang, Mau-Chung Frank 47, 59, 123
Chen, Austin Ying-Kuang 327
Chen, Bo-Yuan 539
Chen, Chi-Chen 335
Chen, Chu-Yu 195
Chen, Gaopeng 379, 383
Chen, Hongyi 207
Chen, Jianwu 379
Chen, Jiashu 311
Chen, Kun-Ming 539
Chen, Lee-Wen 427
Chen, Yihao 87
Chen, Young-Kai 59, 327
Chevalier, Pascal 43
Chi, Baoyong 279, 547
Chiang, Patrick Y. 179, 279
Chiang, Ta-Hsiang 531
Chiu, Chia-Sung 539
Chiu, Ming-Kuen 531
Cho, Chang-Hyuk 35
Cho, Joshua 303
Cho, SeongHwan 263
Choi, Jaehyouk 83
Choi, Jeong-Hyun 163, 167
Choi, Jeong-Ki 363
Choi, Jinsung 227
Choi, Kihwa 111
Choi, Wooyeol 343
Choo, Wooseung 51, 163, 167
Chou, Sam 539
Chow, T. P. 231
Chu, C. 403
Chu, Li-Wei 523
Chuang, Huey-Ru 195
Chung, Hangun 167
Clarke, Peter 427
Cohen, Emanuel 411
Copani, Tino 135, 259, 367
Corrao, N. 315
Cossoy, F. 387
Craninckx, Jan 107, 375, 399
Cripps, S. C. 439
David, J. B. 215

AUTHOR INDEX

De Flaviis, Franco 139
De Graauw, Anton J. M. 359, 483
De Groot, Harmke 235
De Vreede, Leo C. N. 247, 359, 443
Debroucke, Romain 535
Dec, Aleksander 299
Deferm, Noël 323
Degani, Ofir 387, 403
Dehollain, Catherine 455
Dekosky, J. 519
Deng, Zhiming 183, 319
Deval, Yann 215, 491, 503
Doherty, Mark 395
Dolmans, Guido 235
Dülger, Fikret 287
Dunn, Jim 211
Durand, C. 203
Edmondson, Dan 3
Eilat, Y. 387
El Hassan, M. 215
Elad, Danny 479
Elahi, Imtinan 3
Eliezer, Oren 7
El-Nozahi, Mohamed 103
Endo, Kunihiro 151
Entesari, Kamran 103
Eo, Yun Seong 119
Eplett, B. 519
Fan, Cheng-Wen 539
Fan, Chin-Ta 531
Fan, Siqiang 207
Fang, Qiang 207
Feng, Milton 475
Feygin, Gennady 287
FitzPatrick, D. M. 439
Floyd, Brian A. 415
Fournier, J.-M. 315
Fridman, A. 387
Fridman, Amir 403
Ganger, Jeff 19
Gaquière, Christophe 535
Garcia, Patrice 43, 223
Garnier, C. 223
Geis, Arnd 107, 399
Ghaffari, Amir 267
Gharpurey, Ranjit 27
Ghosh, Diptendu 27
Ghouchani, Shadi Saberi 55
Giammello, Vittorio 63
Gianesello, F. 71, 203
Gielen, Georges 99
Gillenwater, Todd 243
Gilreath, Leland 67
Gloria, D. 71, 203
Gómez, Didac 175

Gonzalez, José Luis 175
Gradishar, T. 499
Griffith, Danielle 3, 287
Gross, S. 387
Grözing, Markus 515
Gu, Qun Jane 47, 59
Gudem, Prasad S. 407
Guo, Jyh-Chyurn 527
Gustat, Hans 515
Hajimiri, Ali 39
Hara, G. 387
Harame, David 211
Harjani, Ramesh 419
Harpe, Pieter 235
Hart, Adam 43
Hau, Gary 143
Hausmann, Kurt 19
Hayashi, Yoshihiro 283
He, Wei 87
Heck, Stefan 515
Heeres, Rob 443
Heinemann, Bernd 79
Hella, Mona 231
Helmy, Ahmed A. 103
Heo, Jungwook 167
Hershberger, Douglas B. 211
Hershberger, K. 519
Herzinger, Stefan 15
Heydari, Payam 67
Hijioka, Ken'ichiro 283
Hirano, Yoshihito 151
Hodel, Uwe 355
Hong, Songcheol 459, 495
Hsu, Cheng-Ying 195
Hu, Changhui 179
Hu, Julie 291
Hu, Kangmin 179
Huang, Chun-Wen Paul 395
Huang, Guo-Wei 539
Huang, Ming-Feng 23
Huang, Oscar 415
Huang, Sheng-Yi 539
Huang, Xiongchuan 235
Huang, Yan-Yu 271
Huard, V. 315
Hung, Cheng-Chou 539
Hung, Chi-Cheng 507
Hung, Ping-Fang 523
Inoue, Akira 151
Itoh, Tatsuo 123
Iversen, Christian 243
Jain, Vipul 67
Jakobson, Claudio 411
Jan, Sebastien 535
Jann, B. 403

AUTHOR INDEX

Jayaraman, Karthik 279
Jensen, J.C. 403
Jeon, Hamhee 511
Jian, Guan-Xiu 487
Jian, Heng-Yu 47, 59
Jin, Zhi 379, 383
Joblot, S. 223
Jöhl, Norbert 455
Jung, Seung Hwan 119
Kaltiokallio, Mikko 471
Kam, Dong G. 415
Kamarei, Mahmud 467
Kang, Daehyun 227
Kang, Sanghoon 163, 167
Katz, Oded 479
Keehr, Edward A. 39
Kelly, Dylan 91
Ker, Ming-Dou 523
Kerherve, Eric 187, 215
Kerr, Dan 243
Khalil, W. 367
Khan, Qadeer 279
Khanna, Rahul 179
Kiaei, Sayfe 135, 259, 367
Kidwai, Adil A. 131, 403
Kim, Bumman 227
Kim, Dongsu 227
Kim, Eung Jung 35
Kim, Eun-Hee 363
Kim, Huijung 163, 167
Kim, Hyungseok 135
Kim, Hyungwook 347, 511
Kim, Hyun-Woong 435
Kim, Jaewook 263
Kim, Jihoon 343
Kim, Jihwan 347
Kim, Junghyun 155
Kim, Kyutae 155
Kim, Stephen T. 83
Kim, Sunghwan 407
Kim, Unha 155
Kim, Woonyun 35, 347
Kiper, Halil 3
Kirschenmann, Mark 19
Kjellberg, Torgil 483
Klaassen, D. B. M. 543
Klar, Heinrich 355
Klumperink, Eric A. M. 267
Ko, Jinho 363
Ko, Sangsoo 51
Kornegay, Kevin T. 347
Krishnamurthy, V. 519
Ku, Bon-Hyun 495
Ku, Chih-You 527
Kundu, Sandipan 275

Kuo, Jing-Lin 507
Kwon, Yong-Il 251
Kwon, Youngwoo 155, 343
Lai, Zongsheng 87
Lam, Lui 395
Larchanché, Jean-François 535
Larson, Lawrence E. 239, 407
Laskar, Joy 35, 83, 271, 347, 435, 511
Lavasani, Hossein Miri 295
Le Pennec, F. 71
Lee, Chang-Ho 35, 271, 347, 435, 511
Lee, Dong Ho 511
Lee, Hai-Young 251
Lee, Jen-How 335
Lee, Jongsoo 511
Lee, Joonhee 263
Lee, Kang Hyuk 119
Lee, Kun-Seok 511
Lee, Kwyro 363
Lee, Meng-Chang 7
Lee, Ockgoo 435, 511
Lee, Sungeun 459
Lee, Sunyoung 431
Lee, Tzung-Yin 431
Lee, Young Jae 119
Leenaerts, Domine 463
Lees, J. 439
Leung, Lincoln L.K. 127
Leuschner, Stephan 355
Li, Chin-Lung 391
Li, Hsiao-Chun 523
Li, Mingyuan 351
Lim, Kyutae 83
Lin, Chun-Yu 523
Lin, Jenshan 327
Lin, Kun-You 507
Lin, Lin 207
Lin, Ming-Chien 531
Lin, Shu-Yu 539
Lin, Ying-Hsi 391
Lin, Yo-Sheng 335
Lin, Yu-Sheng 195
Liu, Duixian 415
Liu, Huaping 179
Liu, Jian 207
Liu, Xinyu 379, 383
Lobeek, Jan-Willem 463
Long, John R. 171, 199, 247
Lu, Tse-Hua 523
Luong, Howard C. 127
Ly, A. 403
Machiels, Brecht 115
Macias-Montero, Jose Gabriel 247
Magierowski, Sebastian 339, 451
Maimon, T. 387

AUTHOR INDEX

Marchetti, Mauro 443
Martineau, B. 203
Mason, Phil 243
Masoumi, Nasser 467
Masuda, Toru 331
Mateo, Diego 175
May, Jason W. 75
Mazouffre, O. 503
Mazzilli, Francesco 455
Mehta, Jaimin 7
Messier, Geoffrey 451
Min, Seungkee 259
Mohieldin, Ahmed Nader 287
Moquillon, L. 315
Mor, E. 387
Morelle, J. 223
Mueller, Jan-Erik 355
Mukherjee, Tamal 111, 147
Murphy, David 123
Nagaraj, G. 499
Nagase, Hirokazu 283
Nakahara, Kazuhiko 151
Nakamura, Takahiro 331
Nariman, Mohammad 139
Natarajan, Arun S. 415
Nauta, Bram 267
Nazimov, A. 387
Nejedlo, Jay 179
Ng, Alan W.L. 127
Ni, Nan 531
Nicolson, Sean 415
Niknejad, Ali M. 183, 311, 319
Normatov, G. 387
Norris, George B. 19
Öjefors, Erik 79
Oskooei, Mostafa Savadi 467
Oster, A. 403
Otis, Brian 255, 291
Otsuka, Hiroshi 151
Paff, Aaron 303
Pala, V. 231
Palmisano, Giuseppe 63, 159
Pamarti, Sudhakar 371
Pan, Hsuan-yu Marcus 239
Pan, Wanling 295
Pandey, Jagdish 255
Paramesh, Jeyanandh 55, 111, 275
Parat, G. 223
Park, Byeong-ha 51, 163, 167
Park, Min 31
Park, Sang-Ku 251
Park, T. J. 251
Park, Youngrak 343
Pärssinen, Aarno 471
Patel, Bijit 3

Patnaik, Satwik 419
Pekarik, John J. 171, 247
Pelk, Marco 443
Peng, Han 231
Pennisi, Salvatore 3
Perrott, Michael H. 31
Person, C. 71
Petit, D. 223
Pfeiffer, Ullrich R. 79
Phelps, Richard A. 211
Pierres, Jean-Blaise 243
Pignol, M. 503
Pilard, R. 71, 203
Pinarello, Sandro 355
Plana, Robert 187, 491
Porat, S. 403
Poulin, D. 519
Presti, Calogero D. 91
Pretl, Harald 15
Prince, E. 519
Quémerais, T. 315
Radiom, Soheil 99
Ragonese, Egidio 63
Rainey, BethAnn 211
Rascoe, Jay 211
Rassel, Robert M. 211
Rauber, B. 203
Ravid, Shmuel 411
Raynaud, C. 203
Razafimandimby, S. 223
Rebeiz, Gabriel M. 75, 423
Rentala, Vijay 3
Reynaert, Patrick 115, 323
Reynolds, Scott K. 415
Rhee, Woogeun 51
Ritter, Dan 411
Rivel, Shahar 131
Rivkind, A. 387
Roach, James 427
Rofougaran, Reza 139
Rolain, Yves 107
Rood, R. 519
Roovers, Raf 483
Rotella, Francis M. 427
Ru, Shao-Pin 531
Ruberto, M. 387
Ruby, Richard 291
Ryckaert, Julien 399
Ryynänen, Jussi 471
Sadhwani, Ram 131, 403
Saeidi, Behzad 303
Samala, Sreekiran 3
Sánchez-Sinencio, Edgar 103
Sankaran, Swaminathan 3
Saputra, Nitz 171

AUTHOR INDEX

Schelmbauer, Werner 15
Scheytt, Christoph 515
Schmidt, Martin 515
Schmidt-Szalowski, Marek 463
Scholten, A. J. 543
Schuller, Frieder 515
Schwartz, Daniel B. 19
Shahaf, S. 387
Shahar, N. 387
Shahramian, Shahriar 43
Shang, H. 403
Sharkansky, M. 403
Shashidharan, Sridhar 259, 367
Shepherd, Wayne 19
Shi, Chunqi 87
Shi, Yun 211
Shih, Han-Jung 391
Shiramizu, Nobuhiro 331
Sim, Sanghoon 459
Singh, Mahendra 143
Sjöland, Henrik 467
Smit, G. D. J. 543
Smith, M. 403
Song, Taejoong 83
Spears, Eddie 243
Spirito, Marco 199, 443
Squillante, Michele 443
Srinivasan, Venkatesh 3
Staszewski, Robert Bogdan 7, 11
Stengel, B. 499
Stevens, Mark 259
Steyaert, Michiel 115
Stuenkel, Mark 475
Su, Pin-En 371
Su, Yu 3
Sun, Xuguang 547
Sun, Yuanfeng 51
Suyama, Ken 299
Sweeney, Susan L. 211
Tal, Nir 7
Talwalkar, S. 499
Tam, Sai-Wang 123
Tanabe, Akira 283
Tang, He 207
Taris, Thierry 491
Tasker, P. J. 439
Taskov, Georgi 303
Telzhensky, N. 387
Theilmann, Paul T. 91
Thoppay, Prakash E. 455
Tian, Xiaowei 211
Tombak, Ali 243
Toupe, R. 503
Tsai, Chih-Yu 391
Tsai, Meng-Hsun 391

Tsai, Ming-Da 415
Tzeng, Chih-Yuh 539
Vaillancourt, William 395
Valdes-Garcia, Alberto 415
Vallur, Prasanth 287
Van Der Heijden, Edwin 483
Van Der Heijden, M. P. 359
Van Der Plas, Geert 399
Van Schie, M. C. A. 359
Vandenbosch, Guy 99
Vandersteen, Gerd 107
Vazny, Rastislav 15
Vemulapalli, Sudheer 11
Vengattarmane, K. 375
Voinigescu, Sorin P. 43
Waheed, Khurram 7, 11
Wang, Albert 207
Wang, Frank 47
Wang, Huei 507
Wang, Jingchao 95
Wang, Leon 147
Wang, Po-Chih 391
Wang, Wen-Shan 391
Wang, Xiaoyan 235
Wang, Xin 207
Wang, Zhihua 51, 95, 279, 547
Wang, Ziqiang 547
Washio, Katsuyoshi 331
Waters, Gregory L. 2
Webster, Richard T. 191
Wei, Chien-Cheng 531
Weigel, Robert 15
Williams, T. 439
Woo, Wangmyong 271
Wu, Chang-Hsi 487
Wu, Danyu 379, 383
Wu, Jin 383
Wu, Liang 127
Wu, Yi-Cheng 47, 59
Xie, Haolu 207
Xu, Ping 87
Xu, Shuai 87
Xu, Yang 447
Xu, Zhiwei 47, 59
Yamanaka, Koji 151
Yamauchi, Kazuhisa 151
Yan, Han 247
Yang, Jaemo 495
Yao, Hsin-Cheng 67
Yeh, Kuo-Liang 527
Yishay, Roee Ben 479
Yoon, Youngchang 347, 511
Yousef, H. 387
Yu, Alvin Hsing-Ting 123
Yu, Hyun Kyu 119

AUTHOR INDEX

Yu, Xueyi 51
Yun, Seok-Oh 363
Zaguri, S. 387
Zajac, M. 387
Zhan, Jing-Hong Conan 415
Zhang, Chun 95, 547
Zhang, Gary 207
Zhang, Runxi 87
Zhao, Bin 207
Zhao, H. 519
Zhao, Hui 207
Zhao, Lixin 379
Zhao, Yi 199
Zhao, Zhixing 339, 451
Zheng, Le 67
Zheng, Li-Rong 99
Zhu, Qiang 447
Zirath, Herbert 483
Zoicas, Vasile 7